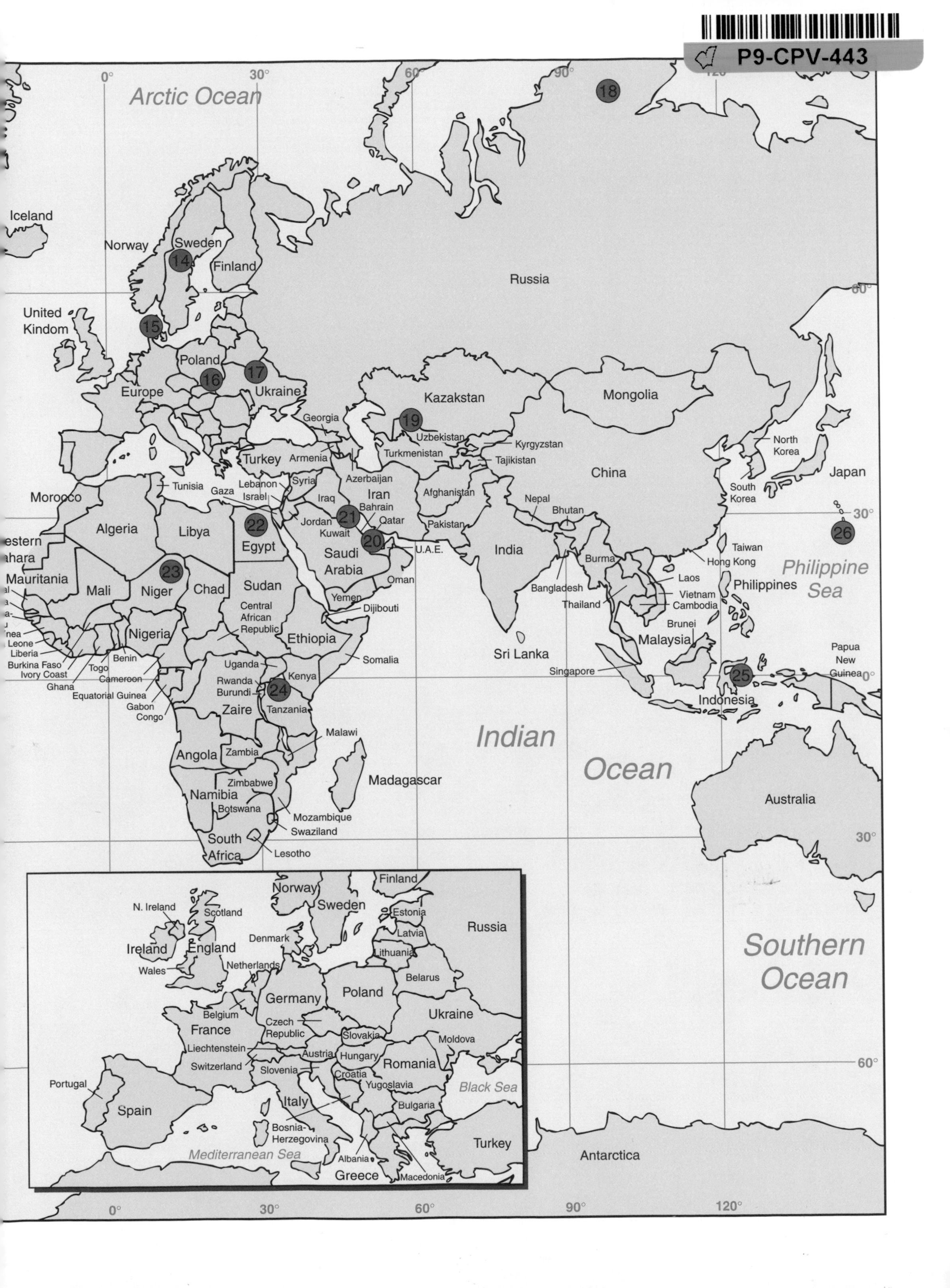

0°
30°
60°
90°
120°
Arctic Ocean
18
Iceland
Norway
Sweden
14
Finland
Russia
60°
United
Kindom
15
Poland
16
17
Europe
Ukraine
Kazakstan
Mongolia
19
Georgia
Uzbekistan
Kyrgyzstan
Turkmenistan
Tajikistan
Turkey
Armenia
North
Korea
Azerbaijan
China
Japan
Tunisia
Lebanon
Syria
Gaza
Israel
Iraq
Iran
Afghanistan
South
Korea
Morocco
Bahrain
Nepal
Bhutan
30°
Algeria
Libya
22
Jordan
21
Qatar
Pakistan
26
Egypt
Kuwait
20
U.A.E.
India
Taiwan
Saudi
Arabia
Burma
Hong Kong
Philippine
Sea
Mauritania
23
Oman
Laos
Mali
Niger
Chad
Sudan
Bangladesh
Philippines
Vietnam
Yemen
Thailand
Cambodia
Central
African
Republic
Dijibouti
Nigeria
Ethiopia
Brunei
Leone
Liberia
Benin
Malaysia
Papua
New
Guinea
Somalia
Sri Lanka
Burkina Faso
Togo
Ivory Coast
Uganda
Kenya
Singapore
25
0°
Cameroon
Rwanda
Burundi
24
Ghana
Equatorial Guinea
Indonesia
Gabon
Zaire
Tanzania
Congo
Malawi
Indian
Ocean
Angola
Zambia
Madagascar
Zimbabwe
Namibia
Australia
Botswana
Mozambique
Swaziland
30°
South
Africa
Lesotho
Norway
Finland
Sweden
N. Ireland
Scotland
Estonia
Russia
Latvia
Denmark
Ireland
England
Lithuania
Netherlands
Wales
Belarus
Poland
Germany
Ukraine
Belgium
France
Czech
Republic
Slovakia
Moldova
Liechtenstein
Austria
Hungary
Romania
Switzerland
Slovenia
Croatia
Portugal
Yugoslavia
Black Sea
Spain
Italy
Bulgaria
Bosnia-
Herzegovina
Turkey
Mediterranean Sea
Albania
Greece
Macedonia
Southern
Ocean
60°
Antarctica
0°
30°
60°
90°
120°

ENVIRONMENT

Second Edition

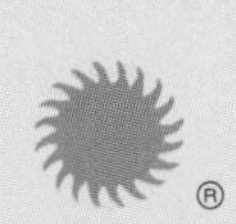

Saunders College Publishing
Harcourt Brace College Publishers

Fort Worth Philadelphia San Diego New York
Orlando Austin San Antonio Toronto
Montreal London Sydney Tokyo

ENVIRONMENT

Second Edition

Peter H. Raven
Missouri Botanical Garden

Linda R. Berg
St. Petersburg Junior College

George B. Johnson
Washington University

Publisher: Emily Barrosse
Executive Editor: Edith Beard-Brady
Product Manager: Erik Fahlgren
Associate Editor: Marc Sherman
Project Editor: Margaret Mary Anderson
Production Manager: Joanne Cassetti
Art Director: Carol Bleistine
Text and Cover Designer: Chazz Bjanes

Cover credit: Four-color illustration by Pat Morrison

1997 World Population Data Sheet reproduced with permission by Population Reference Bureau

Frontmatter photo credits: Half-Title: Frans Lanting/Minden Pictures; facing Title Page: Bruce H. Morrison, Phototake, NYC; Foreword: Mike Goldwater–Network/Matrix; Preface: American Farmland Trust; Contents Opener: © 1997 Marilyn Kazmers/Dembinsky Photo Associates.

Printed in the United States of America

ENVIRONMENT, Second Edition
0-03-018679-X

Library of Congress Catalog Card Number: 97-67174

7890123456 032 10 987654321

FOREWORD

► The year 2001: the stuff of predictions, fantasy, and new beginnings. This is only the second time since humans began tracking the years in our modern system that we will experience the dawning of a new millennium. A watershed event like this offers a rare opportunity for collective reflection about our current situation—the enormous changes and accomplishments that humans have engendered and the unprecedented challenges that face us.

This text offers a detailed and comprehensive situation analysis of planet Earth. It describes the physical and living systems of the planet and examines the myriad ways in which humans depend on, interact with, and affect those systems. Before you begin to consider these matters, however, we think that it is essential to consider the habits, patterns, and belief systems that got us to this point. We would like to lay them out here for you before you begin your study of this book.

As we near the beginning of the new millennium, we are also experiencing the waning years of the Industrial Revolution, a very distinctive period of human history. It is our hope that this text will help you play a role in ushering in the next revolution, which has variously been called the Environmental Revolution or the Knowledge Revolution.

The period of history that we called the Industrial Revolution flourished over the last 250 years as humans applied their creativity and intelligence to transforming nature's wealth into products by using large amounts of energy, thus serving human desires and needs. The rich deposits of ancient fossilized organisms fueled this revolution. As we pumped and dug them out of the Earth's crust, extracted their energy value, and dispersed the waste products resulting from this conversion into the biosphere and the atmosphere, we began with increasing speed to change the characteristics of our whole planet, and, in turn, its ability to support us over the years to come.

These realities have forced us, especially during the past 50 years, to focus our attention on the fact that the Earth is virtually a closed system. The only thing that comes in from the outside in large quantities is energy in the form of sunlight; the only thing that leaves in large quantities is radiated energy. A small fraction of the Sun's energy is transformed by the process of photosynthesis carried out in green cells into the energy of chemical bonds that, in turn, make possible all life. All of the products we make and everything else that makes up the physical and biological systems on Earth, including all of the waste we generate—the enormous mass of substances left after we have used what we wanted, the accumulation of matter that we have mined from the Earth's crust and released into the biosphere and the atmosphere—all is part of the closed system on Earth.

Our industrial economy was developed on the implicit premise that you could actually throw something away. In fact, there is no "away" on planet Earth. Whatever accumulates here ultimately damages the life-support systems on which we depend and limits their sustainability. Only by preserving the productivity and diversity of biological systems can life on Earth ultimately be maintained.

Another major fallacy is the idea that our economic activity does not depend on the functioning of the planet's ecosystems. In truth, the only enterprises on the planet that keep running are these systems. As U.S. Under Secretary, Global Affairs, Department of State, Tim Wirth said, "The economy is a wholly-known subsidiary of the environment." As a result of our failure to appreciate and act upon that relationship, we now find ourselves facing an unprecedented challenge—one greater than any that has confronted humans during their roughly two-million-year history on Earth. The process of the systematic erosion of our biological support systems is accelerating year by year, and economics includes virtually no feedback mechanisms to help signal us that this is a very serious problem.

Once we understand the challenge, however, and see that a different set of premises is needed to drive our industrial activity, we can begin to turn our creativity and intelligence toward redesigning everything we do—from industry, transportation, and agriculture, to building cities—so that these processes do not destroy but rather maintain or even restore the life-support systems on which we all depend. By doing this, we will be in a position to usher in the next great revolution of human history.

When nations are at war, their people instantly recognize the severity of the challenge facing them and swiftly rally to the cause, anxious to protect the values that they hold dear. The challenge of making the transition from the unsustainable, materials-based economy of the Industrial Revolution to an information-based economy, centered on expanding fields of knowledge such as informatics and genetics in medicine and biology, is daunting. We will meet this challenge, however, and the transition will be accomplished. The way in which we meet it will determine the quality of life for our children, our grandchildren, and all who come after us on this incredibly beautiful planet that nurtures us and forms the setting for all of our activities.

In the global transition that coincides with the birth of a new millennium, the opportunities for individuals, communities, and nations are enormous. With knowledge based on a realistic world view comes the opportunity for countless new inventions, new jobs, spiritual progress, and the further refinement of human civilization. In the spirit of helping to make it possible for us to realize these opportunities, we offer this book. We hope that it will be useful to you as you develop your own pathways, and we count on your determination and willingness to make a difference in your own way. As never before, we need one another's abilities, expressed fully, to secure our common future in the best way possible.

Peter H. Raven
St. Louis, Missouri
July 1997

PREFACE

► The challenge of creating and maintaining a sustainable environment is probably the single most pressing issue that will confront students throughout their lives. Today, environmental science is not only relevant to students' personal experience but vital to the future of the entire planet. As humans increasingly alter Earth's land, water, and atmosphere on local, regional, and global levels, the resulting environmental problems can seem insurmountable. Armed with the proper tools, however, students need not find these issues overwhelming. *Environment,* 2nd edition, equips students with the most essential of these tools: an understanding of the concepts that underlie the problems.

One of our principal goals in preparing this book is to convey to students an appreciation of the remarkable complexity and precise functioning of natural ecosystems. *Environment,* 2/e, begins with an exploration of the basic ecological principles that govern the natural world and considers the many ways in which humans affect the environment. From the opening pages, we acquaint students with current environmental issues—issues that have many dimensions and that defy easy solutions. Later chapters examine in detail the effects of human activities, including overpopulation, energy production and consumption, depletion of natural resources, and pollution.

Although we do not sugarcoat these problems—many are very serious indeed—we try to avoid the gloomy predictions of disaster so common in environmental science textbooks today. Instead, students are encouraged to take active, positive roles, using the practical and conceptual tools presented in this book, to meet the environmental challenges of today and tomorrow.

Environment, 2/e, integrates important information from a number of different fields, such as biology, geography, chemistry, geology, physics, economics, sociology, natural resources management, law, and politics. Because environmental science is an interdisciplinary field, this book is appropriate for use in environmental science courses offered by a variety of departments, including (but not limited to) biology, geology, geography, and agriculture.

This book is intended as an introductory text for undergraduate students, both science and nonscience majors. Although relevant to all students, *Environment,* 2/e, is particularly appropriate for those majoring in education, journalism, political science/government, and business, as well as the traditional sciences. We assume our students have very little prior knowledge of how ecosystems work, the dynamics of how matter and energy move through ecosystems, and how populations affect and are affected by ecosystems. These important ecological concepts and processes are presented in a straightforward, unambiguous manner.

INSTRUCTIONAL FEATURES OF THE SECOND EDITION

Environment, 2/e, is written in an interesting, conversational style that will help students remember important concepts. The up-to-date coverage of environmental topics includes many unique applications and interesting case studies throughout. Numerous learning aids are used.

1. **Chapter outlines** reflect the main headings within each chapter and provide students with an overview of the material covered.
2. **Tables and graphs,** many new to the second edition, summarize and organize information.
3. Carefully rendered **illustrations,** including data that are graphically presented, and stunning **color photographs** support concepts covered in the text, elaborate on relevant issues, and add visual detail. Many illustrations and photographs are new to the second edition.
4. **New terms** are in boldface, permitting easy identification and providing emphasis. Occasional, boxed **Mini-Glossaries** define closely related terms, or potentially confusing new terms, and are located within the chapter for handy reference.
5. A new feature of the second edition is **Follow-Up** sections that provide the very latest information about unfolding environmental stories.
6. **Focus On** boxes spark student interest, present applications of concepts discussed, and familiarize students with current issues.
7. **You Can Make a Difference** boxes suggest specific courses of action or lifestyle changes students can make to improve the environment.
8. **Envirobriefs,** most of which are new to this edition, provide additional topical material about relevant environmental issues. Envirobriefs can be used as a starting point for lively classroom discussions.
9. **Meeting the Challenge** boxes discuss pressing environmental dilemmas that defy easy answers. Examples of how others have responded to challenges allow the student to consider effective problem-solving.
10. **Chapter summaries** are presented as numbered statements at the end of each chapter to provide a quick review of the material presented.
11. **Thinking About the Environment** questions encourage critical thinking and highlight important concepts and applications. A new feature in the second edition is **quantitative questions,** which are marked by an asterisk. Solutions to these problems appear in Appendix VII.
12. **Research Projects** at the end of each chapter direct students to conduct more in-depth investigations related to the chapter. A web site address that provides the student with a wealth of current information is included in this section.
13. **Suggested Reading** provides current references for further learning.
14. **Interviews** with prominent individuals who are making a positive impact on the environment explore the issues and trends they consider to be most vital, and emphasize the importance of active involvement in environmental affairs. Their web sites are included.
15. **Appendices** provide useful supplemental material. Appendix I examines the fundamentals of chemistry. Appendix II is a guide to environmental organizations and government agencies. Appendix III serves as a reference for students interested in pursuing careers in the field. Appendix IV provides useful conversions of units of measure used in the text. Appendix V, which is new, examines the fundamentals of presenting data graphically. Appendix VI, also new, is a useful list of abbreviations, formulas, and acronyms used in the text. Appendix VII, also new, provides the solutions and answers to the quantitative "Thinking About the Environment" questions. Appendix VIII, also new, adds questions about population based on the data sheet described below in item 18.
16. A separate **glossary** is provided, allowing rapid location of definitions.
17. A **world map** highlighting some of the environmentally sensitive areas discussed in the text appears inside the front cover and is conveniently chapter-referenced.
18. **World Population Data Sheet,** provided by the Population Reference Bureau, is folded into the text and intended to be pulled out for classroom use.

ORGANIZATION

Educators present the major topics of an introductory environmental science course in a variety of orders. We make no pretense that we have found the best way to organize the subject of environmental science. However, we have put our best efforts into writing the seven parts and their chapters so that they can be successfully presented in any number of sequences.

Part 1 Humans in the Environment

Chapter 1, **Our Changing Environment,** introduces environmental science and highlights several current environmental issues. Chapter 2, **Solving Environmental Problems,** expands on the nature of science and presents the steps that should be followed to resolve environmental problems. Part 1 includes an interview with Adam Werbach of the Sierra Club.

Part 2 The World We Live In

Part 2 provides a detailed introduction to the principles of ecology. It is organized around the ecosystem, which is the fundamental unit of ecology. Chapter 3, **Ecosystems and Energy,** discusses the linear flow of energy through ecosystems, and Chapter 4, **Ecosystems and Living Organisms,** examines the living things that comprise ecosystems. In Chapter 5, **Ecosystems and the Physical Environment,** the cycling of materials in ecosystems and the influence of climate on ecosystems are discussed. Chapter 6, **Major Ecosystems of the World,** examines the major biomes of the terrestrial environment as well as major aquatic ecosystems. Human activities relating to economics and government policies are covered in Chapter 7, **Economics, Government, and the Environment.** Part 2 includes an interview with Nobel Prize–winning chemist F. Sherwood Rowland of the University of California–Irvine.

Part 3 A Crowded World

The principles of population ecology are discussed in Chapter 8, **Understanding Population Growth.** Although the human population is the focus of this chapter, the fact that human populations comply with the same principles of population ecology as other organisms is emphasized. Chapter 9, **Facing the Problems of Overpopulation,** examines sociological and cultural factors that affect human population growth. Part 3 includes an interview with Nafis Sadik of the U.N. Population Fund.

Part 4 The Search For Energy

The environmental impact of the human quest for energy is considered in this section. Chapter 10, **Fossil Fuels,** discusses the problems associated with the use of oil, coal, and natural gas. In Chapter 11, **Nuclear Energy,** the use of nuclear power as a viable energy source in the future is considered. Chapter 12, **Renewable Energy and Conservation,** surveys energy alternatives to fossil fuels and nuclear energy. Part 4 includes an interview with L. Hunter Lovins of the Rocky Mountain Institute.

Part 5 Our Precious Resources

Overusing and abusing our natural resources is considered in detail in Part 5. Chapter 13, **Water: A Fragile Resource,** describes the problems that can arise from an overabundance or a lack of water resources. In Chapter 14, **Soils and Their Preservation,** the significance of soil, the least appreciated natural resource, is explained. Chapter 15, **Minerals: A Nonrenewable Resource,** discusses some of the problems associated with our increasing use of minerals. In Chapter 16, **Preserving Earth's Biological Diversity,** the importance of organisms is developed. Chapter 17, **Land Resources and Conservation,** examines how we use land and the important ecological contributions made by natural areas. Chapter 18, **Food Resources: A Challenge for Agriculture,** explains the challenge of providing enough food for the ever-expanding human population. Part 5 includes an interview with Jon Piper of The Land Institute.

Part 6 Environmental Concerns

The effects of pollution are examined in this part. Chapter 19, **Air Pollution,** looks at the local effects of air pollution, including indoor air pollution, whereas Chapter 20, **Global Atmospheric Changes,** considers regional and global effects of air pollution: acid deposition, global climate change, and stratospheric ozone destruction. Chapter 21, **Water and Soil Pollution,** discusses the closely related issues of water and soil pollution. Pesticides pollute air, water, soil, and food and contaminate so much of the biosphere that they are considered in a separate chapter, Chapter 22, **The Pesticide Dilemma.** Chapter 23, **Solid and Hazardous Wastes,** examines the problem of disposing materials we no longer need or want. Part 6 contains an interview with entrepreneur-environmentalist Paul Hawken.

Part 7 Tomorrow's World

Part 7 concludes the book with Chapter 24, **Tomorrow's World,** which presents the opinions of the authors on social responsibilities, identifying some of the most critical issues that must be grappled with today in order to assure a better tomorrow. Part 7 contains an interview with Irwin Price of George Washington University about green campuses.

SUPPLEMENTS

The package accompanying *Environment,* 2/e includes several items developed specifically to augment students' understanding of environmental issues and concerns. Together, these ancillaries provide instructors and students with interesting and helpful teaching and learning tools and take full advantage of both electronic and print media. New additions to the ancillary package are noted.

For the Instructor and Student

NEW! Environment Case Studies by Larry Underwood, Northern Virginia Community College.

New case studies on key environmental issues provide integrative, interactive, and collaborative assignments for students. The issues addressed include wildlife management, land use, water resources, air and water pollution, alternative energy strategies, solid and toxic waste disposal, and soil conservation. Eight distinct regions in North America—seven in the United States and one in Canada—are covered by this set of topics. Monthly updates for each case study specific to these regions are available on the accompanying web site.

NEW! Saunders Web Site co-managed by Chris Migliaccio, Miami-Dade Community College, Wolfson Campus.

Issues covered in the text and case studies change rapidly. Our web site will track these issues and keep you up to date as they occur. See

http://www.saunderscollege.com/environment2/

The site will include components for student self-testing, based on each text chapter's contents, as well as links and activities for environmental activism. The site offers the following features:

- *Case Studies.* To support the new printed ancillary *Environment Case Studies,* we provide updated links and information regarding issues covered in each case.
- *Student Self-Test.* Chapter reviews, questions, and critical thinking exercises test student knowledge of important concepts explained in *Environment*, 2/e, and direct students back to the text, the Study Guide, or to other resources on the web when they need more guidance.
- *Careers in the Environment.* Highlights organizations, schools, and associations offering jobs in the environmental sector. E-mail and web site links are provided.
- *Instructor Resource Center.* Instructors can submit and exchange their ideas about teaching and learning using *Environment* 2/e. Read online or download chapters from the Instructor's Manual. The *Instructor's Curriculum Integration* is a guide on how to integrate Saunders technology into all aspects of environmental science instruction. *PowerPoint™* slides are also available to download.
- *Hot Topics.* Tracks the most dynamic topics in each chapter and provides monthly updates on these developments. A *News flash* component highlights the biggest environmental news event every month, and explores what it means for students.
- *Environmental Resources on the Web.* An extensive listing of environmental resources available on the World Wide Web with access to each resource.
- *The Green Campus.* Some schools and corporations are implementing programs to keep their campuses green. We'll help you stay abreast of these projects, and we will provide guidelines for developing similar programs on your campus.

NEW! World Population Data Sheet by the Population Reference Bureau, Inc.

Current and detailed population data are presented in a convenient, poster-size wall chart for students. Classroom or homework assignment in Appendix VIII of the text is based on use of this chart. The data sheet comes folded inside each new text.

NEW! Student Study Guide by Charlene Waggoner of Bowling Green State University.

This innovative guide not only helps students learn the details of environmental science but fosters critical thinking skills and scientific reasoning. Sections include: Learning the Language, Checking What You Know, Making Connections, Making Decisions, and Doing Science.

Environment Laboratory Manual by Robert Wolff, Trinity Christian College. This laboratory manual assists in the teaching of laboratory, problem-solving, and thinking skills in environmental science.

Videos. Two videos feature Peter Raven discussing two critical topics: destruction of the world's rain forests and species extinction.

For the Instructor

Instructor's Manual/Test Bank by Dennis Woodland of Andrews University. This resource contains chapter outlines, objectives, lecture outlines, and teaching tips. The Test Bank consists of short-answer essays, multiple-choice, true/false, and fill-in questions.

NEW! MediaActive™ CD-ROM. The Saunders MediaActive™ will contain art from the text for presentation use by instructors.

NEW! PowerPoint™ Slides can be edited by instructors, who can use all images from MediaActive™ for presentation.

ExaMaster™+ Computerized Test Banks. Instructors can create or modify tests derived from the Printed Test Bank. Included is the capability to print out tests with answer keys and student answer sheets. (Available in Macintosh, IBM, and Windows formats.)

Overhead Transparency Acetates. The full-color set includes 150 transparencies with figures from the text. Overhead transparencies have been reformatted with large-print labeling for easy viewing in any classroom.

Saunders College Publishing may provide complimentary instructional aids and supplements or supplement packages to those adopters qualified under our adoption policy. Please contact your sales representative for more information. If as an adopter or potential user you receive supplements you do not need, please return them to your sales representative or send them to:

Attn: Returns Department
Troy Warehouse
465 South Lincoln Drive
Troy, MO 63379

MAJOR CHANGES IN THE SECOND EDITION

Issues in environmental science change rapidly, and, as a result, the second edition of *Environment* is a major revision. A complete list of all changes and updates to the second edition is too long to fit in the Preface, but a list of some of the more important changes follows:

Chapter 1 Our Changing Environment

Expanded discussion of the nature of environmental science. New topics include green architecture, endocrine disrupters (with new Table 1–1 of supporting data on the population of juvenile alligators in Lake Apopka), the closing of the Georges Bank fishery (with supporting data graphed in Figure 1–5), declining bird populations, reintroducing wolves to Yellowstone, and oil lakes in Kuwait.

Chapter 2 Solving Environmental Problems

Expanded sections on the scientific analysis of environmental problems, the scientific method, the role of government in solving environmental problems today, and risk management and cost-benefit analysis. New photographs in this chapter include the dramatic chapter introduction photo of Red Cross workers dealing with the Ebola virus in Zaire.

Chapter 3 Ecosystems and Energy

New material on energy efficiency as it relates to the second law of thermodynamics and on ecosystem productivity, including new Table 3–1 (net primary productivities of selected ecosystems) and Figure 3–15 (satellite data on Earth's primary productivity). New mini-glossary (different forms of energy). New art includes Figures 3–8 (food web) and 3–9 (energy flow).

Chapter 4 Ecosystems and Living Organisms

New sections on keystone species; predation; species diversity, ecotones, and the edge effect (with new graph, Figure 4–10); and evolution by natural selection (with new Figure 4–11). New envirobrief on Biosphere II. New art includes Figures 4–1 (rotting log) and 4–13 (secondary succession). New photos of primary succession (Figure 4–12).

Chapter 5 Ecosystems and the Physical Environment

New sections on tornadoes, tropical cyclones, plate tectonics, volcanoes, and earthquakes (with Table 5–1 of major earthquakes). New Focus On wildfires. New mini-glossary and Figure 5–17 on Köppen's climate zones. There are many new figures in this chapter, such as Figures 5–3 (improved nitrogen cycle), 5–10 (layers of the atmosphere), 5–12 (Coriolis force), and 5–16 (coastal upwelling).

Chapter 6 Major Ecosystems of the World

New material on freshwater wetlands, mangrove forests, marine ecosystems (including new mini-glossary of marine environment terms), and coral reefs. New Focus On the Everglades. Many new pieces of art, such as Figures 6–1 (improved figure of world's biomes), 6–12 (typical river), 6–13 (lake zonation), and 6–21 (development of coral reefs).

Chapter 7 Economics, Government, and the Environment

New chapter introduction on unfunded mandates and their effect on environmental legislation. New You Can Make a Difference box on buying a waste-discharge permit. New Table 7–3 of major environmental incidents since 1970. New Figure 7–1 (how economics is related to the environment) is very significant.

Chapter 8 Understanding Population Growth

Expanded section on population ecology to include new sections on population density, density-dependent and density-independent factors, *r*- and *K*-selection (with new Table 8–1 on *r*- and *K*-selection), and survivorship curves. New Focus On predator/prey dynamics of wolves and moose on Isle Royale (with supporting data graphed). New envirobrief on Russia's death rate.

Chapter 9 Facing the Problems of Overpopulation

New introduction on declining fertility rates in Egypt. Discussion of U. N. Population Summit and Fourth World Conference on Women's Rights. New Table 9–3 relates contraceptive use to lower fertility rates in selected countries. Remarkable photographs, such as Figure 9–2, which compares material possessions of typical American and Indian families. New graph of external debt in developing countries (Figure 9–3).

Chapter 10 Fossil Fuels

New Meeting the Challenge box on reclamation of surface-mined land. New Tables 10–3 (energy output of various kinds of transportation) and 10–4 (U.S. National Energy Policy). New graphs of distribution of world's coal deposits (Figure 10–4), oil resources (Figure 10–12), and natural gas deposits (Figure 10–13). New diagrams of fluidized bed combustion (Figure 10–8) and coal gasification (Figure 10–16).

Chapter 11 Nuclear Energy

New section on the future of nuclear power. New Table 11–2 compares the environmental impacts of coal versus nuclear energy. New follow-up section on Chernobyl. New envirobrief on Russia's disposal of radioactive wastes. New graphs: Figures 11–2 (distribution of world's uranium deposits) and 11–11 (worldwide plutonium buildup). Improved Figures 11–3 (nuclear fuel cycle) and 11–7 (nuclear power plant).

Chapter 12 Renewable Energy and Conservation

New chapter introduction on variable speed wind turbines. Material is updated throughout (for example, sections on solar thermal energy, photovoltaics, geothermal energy, OTEC). New tables on types of renewable energy (Table 12–1) and energy-efficiency upgrades in commercial buildings (Table 12–3). New quantitative Thinking About the Environment question involves data on world increase in wind power.

Chapter 13 Water: A Fragile Resource

New introduction on San Francisco Bay water accord. Follow-ups on Aral Sea, Mono Lake, 1993 floods, and Columbia River. Expanded section on international water tensions. New envirobrief on xeriscaping. New graphs show composition of seawater (Figure 13–2), water on Earth (Figure 13–3), and water budget for United States (Figure 13–10). Dramatic historical photograph of subsidence (Figure 13–7).

Chapter 14 Soils and Their Preservation

New chapter opener on China's Great Green Wall. New envirobriefs on soil and climate change and on Everglades soil. Updates on conservation tillage, soil erosion in United States, and Conservation Reserve Program. New line art includes Figures 14–8 (major soil types, redrawn with guidance from a soil scientist) and 14–13 (worldwide soil degradation).

Chapter 15 Minerals: A Nonrenewable Resource

New chapter introduction on reforming the 1872 mining law. New Meeting the Challenge box on industrial ecosystems. New materials on processing mineral resources and remediating acid mine drainage by constructing wetlands. New Table 15–2 of mineral reserves and life indices of world reserves. New graph (Figure 15–7) of United States consumption, as percent of world total consumption, of selected metals.

Chapter 16 Preserving Earth's Biological Diversity

New topics include Guam rail and brown tree snakes, solving crimes against wildlife, attempts to reintroduce the thick-billed parrot, and the work of the Biological Resources Division. New Focus On Lake Victoria. New Tables 16–2 (number of organisms listed as endangered or threatened in the United States) and 16–3 (selected animal species that are no longer classified as endangered). New historical photograph of bison skulls (Figure 16–9) is very dramatic.

Chapter 17 Land Resources and Conservation

New Meeting the Challenge box on Trout Creek Mountain Working Group. New sections on coastal demographics, deforestation of boreal forests, salvage logging, the Tongass National Forest, ecologically sustainable forest management, and the wise-use movement. New introduction on recently added national parks. New Table 17–3 of farm areas threatened by urban and suburban sprawl. Dramatic photograph of deforestation (Figure 17–12).

Chapter 18 Food Resources: A Challenge for Agriculture

New material on organic foods; grain stockpiles; preservation of plant and animal germplasm; diet, overnutrition, and cancer; polyculture; livestock mega-farms; and the U.N. Conference on Protecting the Marine Environment. New Table 18–4 shows effects of overfishing. New graphs of total and per-capita world grain production (Figure 18–2), world grain carryover stocks (Figure 18–3), and world fertilizer use (Figure 18–9).

Chapter 19 Air Pollution

New topics include Chattanooga's air pollution success story, effects of air pollution on children's health, relationship between carbon monoxide and congestive heart failure, global distillation effect, and hydrocarbon emissions by green plants (with new graph, Figure 19–3). Follow-ups on air pollution in Mexico City and Los Angeles. New Meeting the Challenge box on clean cars, clean fuel.

Chapter 20 Global Atmospheric Changes

New material on sulfur dioxide emissions and acidified lakes and soil; forest decline; CFC smuggling; 1996 U.N. Climate Change Convention; iron hypothesis; and links among acid deposition, global warming, and ozone depletion. New Table 20–2 of world CFC production. New graphs of mean global temperatures (Figure 20–6) and per-capita CO_2 emissions of selected countries (Figure 20–15). New map (Figure 20–18) shows relationship between exposure to UV-B radiation and latitude.

Chapter 21 Water and Soil Pollution

New information on purification of drinking water, key provisions of 1996 Safe Drinking Water Act, hypoxia and the dead zone in the Gulf of Mexico, the chlorine dilemma, soil remediation, and combined sewer overflow. New Focus On, map, and table (Table A) on the Great Lakes. Two new envirobriefs on pollution in the wake of motorboats and planting life back into a dead marsh.

Chapter 22 The Pesticide Dilemma

New sections on IPM in Asian rice fields (includes new graph of supporting data, Figure 22–13), resistance management, health effects of pesticides on children, pesticides as endocrine disrupters, mobility of pesticides in the atmosphere, and Food Quality Protection Act of 1996. New Focus On Bhopal. Remarkable historical photograph of DDT spraying on a beach (Figure 22–2).

Chapter 23 Solid and Hazardous Wastes

New chapter opener on reusing and recycling computers. New material on chemical accidents, green chemistry, the military and the environment, radioactive waste from nuclear weapons facilities, environmental justice, and charging for trash by the bag. New Table 23–2 on phytoremediation. Super photograph of a year's worth of a typical American family's trash and recycling (Figure 23–1). New graphs of sources of solid waste in the United States (Figure 23–3) and disposal of MSW (Figure 23–4).

Appendix II: How to Make a Difference

Many environmental organizations were added, along with phone numbers, e-mail addresses, web sites, and mailing addresses.

Appendix V: Graphing

New to the second edition, this appendix gives some basic information about graphing that will be needed by students when answering certain quantitative questions at the ends of chapters.

Appendix VI: Abbreviations, Formulas, and Acronyms Used in this Text

New to the second edition, this appendix provides a handy reference when an unfamiliar abbreviation or formula appears in the text.

Appendix VII: Solutions to Quantitative "Thinking About the Environment" Questions

New to the second edition, this appendix not only gives the answers to quantitative questions (those with an asterisk in the Thinking About the Environment sections), but also shows the student how to go about solving the problems.

Appendix VIII: World Population Data Sheet Assignment

The Population Reference Bureau's World Population Data Sheet, found inside the text, is used to answer a series of questions. The student then places numbers corresponding to the answers on the map provided at the end of this exercise. The world map located inside the front cover helps locate countries and political entities.

Glossary

Many new terms were added to the glossary, including aerosol, aerosol effect, agroforestry, antagonism, artificial insemination, bellwether species, benthic environment, biotic pollution, bycatch, calendar spraying, combined sewer overflow, combined sewer system, conservation biology, cost-benefit analysis, datum, demand-side management, density-dependent factor, density-independent factor, detritus, dilution, dioxins, ecologically sustainable forest management, ecological risk assessment, ecosystem management, ecotone, edge effect, embryo transfer, endocrine disrupters, environmental chemistry, environmental justice, environmental sustainability, epicenter, even-age harvest, fault, fee-per-bag approach, flowing-water ecosystem, focus, forest edge, freshwater wetland, Gaia hypothesis, germplasm, global distillation effect, green architecture, hot spot, hydrology, hypoxia, illuviation, industrial ecosystem, keystone species, krill, *K*-selected species, *K* selection, *K* strategist, lava, life index of world reserves, magma, mangrove forest, marine snow, marsh, Montreal Protocol, National Environmental Policy Act, natural gas, natural selection, neotropical birds, nest parasitism, nuclear reactor, Ogallala aquifer, oil, oncogenes, open management, pesticide treadmill, phosphorus cycle, phytoremediation, plate boundary, polyculture, population density, population ecology, principle, principle of inherent safety, reactor vessel, reclaimed water, reservoir, resistance management, resource recovery, riparian buffer, risk management, *r*-selected species, *r* selection, *r* strategist, rural land, salvage logging, scout-and-spray, seismic wave, standing-water ecosystem, subduction, survivorship, swamp, synergism, synthetic botanical, tropical cyclone, ultraviolet radiation, upwelling, vapor extraction, variable, vitrification, wildlife corridor, wildlife ranching, wise-use movement, and world grain carryover stocks.

Index

The index was expanded to include more geographical areas and general ideas. For example, Canada has 38 listings.

ACKNOWLEDGMENTS

The development and production of *Environment*, 2/e, was a process involving interaction and cooperation among the author team and between the authors and many individuals in our home and professional environments. We appreciate the valuable input and support from editors, colleagues, and students. We thank our families and friends for their understanding, support, and encouragement as we struggled through many revisions and deadlines.

The Editorial Environment

Preparing this book has been hard work, but working with the outstanding editorial and production staff at Saunders College Publishing has been enjoyable. We thank our Editor-in-Chief Emily Barrosse and Acquisitions Editor Edith Beard-Brady for their support, enthusiasm, and ideas.

Senior Associate Editor Marc Sherman expertly guided us through the revision process, coordinated the final stages of development, and provided us with valuable suggestions before the project went into production. We are indebted to Christine Rickoff and Alison Levine, who conducted and edited the interviews.

We thank Ed Dolan, co-author of *Economics*, 7/e (Dryden Press, Hinsdale, Illinois, 1994). Ed provided the inspiration and much of the content for Chapter 7, Economics, Government, and the Environment. We thank Mary Catherine Hager, who researched and wrote many of the text's Envirobriefs. We appreciate the efforts of Greg Young, who updated the guide to environmental organizations and government agencies in Appendix II. Donna Rodgers of the Missouri Botanical Garden was instrumental in researching some of the data used in the second edition.

We thank Photo Researcher George Semple for helping us find the wonderful photographs that enhance the text. Carlyn Iverson helped conceptualize and draw much of the new art. Art Director Carol Bleistine coordinated the art development and ensured that a consistent standard of quality was maintained throughout the illustration program. We greatly appreciate our Project Editor, Margaret Mary Anderson, for guiding us through the many deadlines of production. Project Editors Nancy Lubars and Robin Bonner also contributed to the text in its early stages. We appreciate the excellent work of copyeditor Merry Post and proofreader Betty Gittens.

We are grateful to Travis Moses-Westphal for overseeing and coordinating the development of the New Media supplemental package. Photo and Permissions Editor George Semple and Permissions Editor Eleanor Garner obtained permissions for the vast amount of timely information so important to this book.

Our colleagues and students have provided us with valuable input and have played an important role in shaping *Environment*, 2/e. We thank them and ask for additional comments and suggestions from instructors and students who use this text. You can reach us through our editors at Saunders College Publishing; they will see that we get your comments. Any errors can be corrected in subsequent printings of the book, and more general ideas can be incorporated into future editions.

The Professional Environment

The success of *Environment*, 2/e, is due largely to the quality of the many professors and specialists who have read the manuscript during various stages of its preparation and provided us with valuable suggestions for improving it. In addition, the reviewers of the first edition made important contributions that are still part of this book. They are as follows:

Reviewers of the Second Edition
Mark Belk, *Brigham Young University*
Andy Friedland, *Dartmouth College*
Bob Galbraith, *Crafton Hills College*
James Horwitz, *Palm Beach Community College*
Patricia Johnson, *Palm Beach Community College*
Karen Kakiba-Russell, *Mount San Antonio College*
Andrew Neill, *Joliet Junior College*
Maralyn Renner, *College of the Redwoods*
Jeffrey Schneider, *State University of New York at Oswego*
Judith Schultz, *University of Cincinnati*
Jerry Towle, *California State University–Fresno*
Alicia Whatley, *Troy State University*
Ray Williams, *Rio Hondo College*

Specialist Reviewers of the Second Edition
Clarence R. Allen, *California Institute of Technology*
Robert Anderson, *Walnut Acres*
Tom Boden, *Carbon Dioxide Information Analysis Center, Oak Ridge National Laboratory*
Wayne Bowers, *Oregon Department Fish & Wildlife*
Lois Chalmers, *Institute for Energy and Environmental Research*
Robert Colby, *Chattanooga-Hamilton County Air Pollution Control Bureau*
Mark Coscarelli, *Office of the Great Lakes, Michigan Department of Environmental Quality*
Steve Davis, *South Florida Water Management District*
Jenny Day, *Can Manufacturers Institute*
Dan Diamond, *Energy Information Administration, Department of Energy*
Janine Dinan, *Environmental Protection Agency*
Dan Dudek, *Environmental Defense Fund*
Jerry F. Franklin, *University of Washington*
Barry Gilbert, *Ozone Policy & Strategies Group, EPA*
Louis J. Guillette, Jr., *University of Florida*
Keith Haberern, architect/engineer
Charles Hall, retired, *Lockheed Martin*
Gary Hartshorn, *World Wildlife Fund*
William K. Jordan III, *University of Wisconsin–Madison Arboretum*
Kay Livingston, *R. Frazier, Ltd.*
Tom Lovejoy, *Smithsonian Institution*
Beth Marks, *The Antarctica Project*
Tracy Alaimo Mattson, *Automotive Recyclers Association*
Ralph Mayo, *National Marine Fisheries Service*
Peter M. Molton, *Battelle Pacific Northwest Laboratories*
Winnie Park, *South Florida Water Management District*
Scott Peters, *Nuclear Energy Institute*
David Pimentel, *Cornell University*
Paul R. Portney, *Resources for the Future*
Donna Rodgers, *Missouri Botanical Garden*
Stuart Sipkin, *Central Region Geologic Hazards Team, Geological Survey, U.S. Department of the Interior*
Jo Stevens, *Office of Water Education, California Department of Water Resources*
Stephen Viederman, *Jessie Smith Noyes Foundation*
Les Vilcheck, *South Florida Water Management District*
Ray Weil, *University of Maryland*
Dan Weintraub, *Family Planning International Assistance*
Peter Weisser, *California Department of Water Resources*
Shannon Weller, *American Farmland Trust*
Alicia Whatley, *Troy State University*
Alan Woodward, *Florida Game and Fresh Water Fish Commission*

Reviewers of the First Edition and 1995 Update
Lynne Bankert, *Erie Community College*
Richard Bates, *Rancho Santiago Community College*
George Bean, *University of Maryland*
Bruce Bennett, *Community College of Rhode Island–Knight Campus*
Kelly Cain, *University of Arizona*
Ann Causey, *Prescott College*
Gary Clambey, *North Dakota State University*
Harold Cones, *Christopher Newport College*
Bruce Congdon, *Seattle Pacific University*
Donald Emmeluth, *Fulton–Montgomery Community College*
Kate Fish, *EarthWays, St. Louis, Missouri*
Neil Harriman, *University of Wisconsin–Oshkosh*
Denny Harris, *University of Kentucky*
John Jahoda, *Bridgewater State College*
Jan Jenner, *Talladega College*
David Johnson, *Michigan State University*
Norm Leeling, *Grand Valley State University*
Joe Lefevre, *SUNY at Oswego*
Gary Miller, *University of North Carolina at Asheville*
James T. Oris, *Miami University, Ohio*
Robert Paoletti, *Kings College*
Richard Pemble, *Moorhead State University*
Ervin Poduska, *Kirkwood Community College*
C. Howard Richardson, *Central Michigan University*
W. Guy Rivers, *Lynchburg College*
Barbra Roller, *Miami Dade Community College–South Campus*
Richard Rosenberg, *The Pennsylvania State University*
Judith Schultz, *University of Cincinnati*
Lynda Swander, *Johnson County Community College*
Jack Turner, *University of South Carolina at Spartanburg*
Jeffrey White, *Indiana University*
James Willard, *Cleveland State University*

P.H.R.
L.R.B.
G.B.J.

CONTENTS OVERVIEW

CONTENTS

► **Two Massai girls stand under a tree in Kenya.** *(Carlos Navajas/Image Bank)*

PART 1

Humans in the Environment

Environmental Action

Adam Werbach is on a mission. As the 46th president of the Sierra Club, and the youngest person to ever hold that position, he wants to engage the "Generation X'ers"—a term he dislikes—in efforts to protect the environment they will inherit. Mr. Werbach is well-suited to this task, having begun his career in environmental advocacy by getting his second grade classmates to sign a petition to oust then Secretary of the Interior James Watt. While in high school he founded the Sierra Student Coalition, which grew to 30,000 members under his leadership and played an integral role in the passage of the California Desert Protection Act. The mission he decided to accept is to lead the 104-year-old organization in its ongoing quest to protect America's environmental heritage. ◄

ADAM WERBACH

Tells us about the Sierra Club. When was it founded? What has it accomplished?

The Sierra Club was founded by naturalist John Muir over 100 years ago. The Club has always been active on a range of environmental issues, including opposition to dam construction in the Grand Canyon, passage of the Endangered Species Act, and efforts to protect large areas of the California desert. We have a successful track record of supporting the establishment, enlargement, and protection of wilderness areas and national parks. Yosemite National Park, Glacier National Park, and Kings Canyon National Park were established with Sierra Club support. Efforts of the Sierra Club and others led to passage of the National Environmental Policy Act and creation of the Environmental Protection Agency. One hundred years ago we had about 350 members. Today there are approximately 600,000 Sierra Club members throughout the United States and Canada.

You began your career as an environmental activist at an early age. What prompted your involvement?

Well, it happened accidentally. My parents received a direct-mail piece for a campaign to oust Secretary of the Interior James Watt. I didn't know what it was about, in fact, I thought it had something to do with electricity, but I circulated the petition among my classmates and got them to sign their names. This was my first experience in seeing that my participation on an issue could have an impact. I was asked to make a difference. I was given a chance to help. I was hooked. My classmates' signatures were part of 1.1 million signatures the Sierra Club obtained in that drive.

You followed your early experience with environmental activism by forming the Sierra Club Student Coalition. Why did you want to form a student group?

Student activism can make a big difference on environmental issues, as the student coalition has proved many times. When the California Desert Protection Act was in trouble, the student coalition began knocking on dorm rooms, "dorm-storming," to get students to call Congress to make their views known. Student activists made 600 calls to a senator's office to get him to postpone a vacation so he could vote in favor of the act. The act passed by one vote. Students made a difference there. Through the student coalition a unique brand of effective activism can be harnessed on environmental issues that concern us all.

When you took office as president of the Sierra Club, the 1996 Presidential and Congressional elections were still ahead. You said, "...Goal number 1 is November." What did you hope to accomplish?

We hoped to make the environment the number one issue people voted on. The 104th Congress was the worst Congress ever. *Period.* We saw a full-scale assault on environmental protections that were enacted over the last 20 years with the universal support of the American people. The Sierra Club is committed to educating and turning out the millions of Americans who will make a candidate's record on the environment the deciding factor in their vote.

Those elections are over. How did the Sierra Club efforts fare? In particular, how did students help?

The Sierra Club endorsed over 220 environment-friendly candidates for the House and Senate, focusing special attention on 66 high-priority races. In many of those priority races, college and high-school activists from the Sierra Student Coalition volunteered to distribute flyers, stuff envelopes, phonebank, rally other students, get out the vote—whatever it took. And they got results. Pro-environment candidates won in two out of every three high-priority races, whereas seven of ten Club-endorsed candidates won. In Massachusetts, students at Harvard and Clark helped underdog Senator John Kerry (D-MA), one of our best friends in Congress, survive a tough challenge from Governor William Weld (R-MA). With help from students, the environment won the day for many other candidates as well. The elections proved that voters, and particularly younger voters, will support environmentally friendly candidates. The environment claimed its rightful place at the table of American politics.

What is on the current agenda for the Sierra Club?

We have three national campaigns for the current year that focus on protecting and restoring America's forests, water and wetlands, and endangered or threatened species and their habitats by reforming national and state forest laws, policies, and practices. In addition, we will continue our efforts in public education, grassroots organizing, and public policy lobbying dedicated to protecting America's environment. The Sierra Club will also continue its international campaigns on issues of green trade, global warming, population, development funding, and human rights.

We want President Clinton to focus on environmental issues. We want him to reform the Forest Service so that our national forests are protected, and to appoint a National Parks Service Director who is dedicated to reforming the ailing national park system. The whole body of U.S. environmental protections remains a target for certain Congressional leaders. It's vital that President Clinton continue to oppose and veto all efforts to weaken environmental laws.

Another of our priority efforts concerns the tough new clean-air standards recently issued by the Environmental Protection Agency aimed at curbing ozone and fine particulates. In truth, the proposed regulations don't go far enough in protecting American's health. Thanks to the Clean Air Act, kids today are breathing healthier air than the stuff we inhaled when we were growing up, watching *Three's Company*. When I grew up in Los Angeles, we used to have T-ball practice cancelled once a month because the air was so bad. Kids should be able to play T-ball. But some 60 million Americans are still waiting—often while coughing and wheezing—for healthy air. And each new year brings thousands of premature deaths from heart and lung disease exacerbated by air pollution.

The National Association of Manufacturers, together with other industry lobbying groups, is hoping to pressure Congress into torpedoing the new clean-air regulations. That's a fight we plan to win. Many of us are already taking steps as individuals to clean the air—taking public transit or carpooling, composting leaves instead of burning them, and so forth. But to reach the goal of clean air for all, we need to make sure Congress and the Clinton Administration stand up to industrial polluters who put their profits before our health. It's essential that the President stand firm behind the scientists and environmental policy experts in his own EPA.

Whether it's cleaning up a stream, recycling, or lobbying Congress to keep our air and water healthy and safe, every American has the power and responsibility to protect our environment.

–Adam Werbach

The Sierra Club is now in its second century as an environmental organization. What do you see as the future goals of the club?

We're placing a lot of emphasis on outreach. We want to work with hunters and anglers, farmers, the faith community, and young people. All of us want a sustainable environment for ourselves and for future generations. Whether it's cleaning up a stream, recycling, or lobbying Congress to keep our air and water healthy and safe, every American has the power and responsibility to protect our environment. We need everyone in this battle.

You said it is time for your generation to become more involved in environmental activism. Do you think they will heed the call?

We've been told we're Generation X, we've been told we're apathetic, we've been told we're disengaged. Call us Generation D because we've inherited a degenerated world of pollution and debt. The world is degenerated, the generation is not. This is the generation that will turn it around.

▶ Web site: **www.sierraclub.org**
E-mail: **adam.werbach@sierraclub.org**

Environmental Science
Our Impact on the Environment
The Goals of Environmental Science
Meeting the Challenge: *The Earth Summit*

An energy-efficient house that is constructed of recycled materials.
(Courtesy of National Home Builders Association)

Our Changing Environment

The attractive colonial house shown in the opening photo looks very traditional but is far from conventional. The National Association of Home Builders planned and built this home with three ideas in mind: increasing energy efficiency, conserving wood, and recycling discarded materials. The house is designed for a time in the not-so-distant future when the cost of energy will be higher than it is today, wood will be less plentiful and therefore very expensive, and sanitary landfills will be too full to accept much trash.

Let us examine what makes this house, located in an upscale neighborhood in Bowie, Maryland, so unusual. The frame of the house, which is normally made of wood, is constructed of steel, two-thirds of which was recycled from old auto parts. Sawmill wastes—wood chips and wood fibers—were used to make the hardboard siding. The roofing panels, which look like cedar, contain 52 percent resin from old computer housings. The wall and attic insulation came from recovered newspaper, and the ceiling tiles from wood fiber that contains 22 percent recycled newspaper. Water pipes, fittings, gutters, and downspouts are made of 70 percent recycled copper. The deck was constructed of a wood substitute composed of 50 percent sawdust and 50 percent post-consumer polyethylene, recycled from plastic grocery bags.

In the back yard, solar panels collect the sun's energy to meet most of the family's hot water requirement. An array of photovoltaic cells produces enough electricity from solar energy to power the security system and outside lights. Energy is saved by the energy-efficient double-paned windows and advanced-design, earth-coupled heat pump, which extracts heat from the ground through copper pipes buried in the yard, thereby operating at a lower energy cost.

The house in Bowie is an example of a growing trend known as **green architecture**. Construction firms across the country are increasingly converting trash such as used tires, aluminum cans, plastic milk jugs, and old copper pipes into building materials. Some of the techniques being tried today will eventually become economical for large-scale manufacturing, helping us to extend our resource base.

The more efficient use of materials and energy in home construction is just one way that people have responded in recent years to pressing environmental concerns. In this chapter we examine several current environmental issues. These problems are as controversial as the reintroduction of wolves to Yellowstone, as threatening as the ozone hole in our atmosphere, and as avoidable as the closing of a productive fishery due to overfishing. ◀

ENVIRONMENTAL SCIENCE

How humankind can best live within Earth's environment is the theme of what is loosely called **environmental science**, the interdisciplinary study of how humanity affects other organisms and the nonliving physical environment. Environmental science encompasses many complex and interconnected problems involving human numbers, Earth's natural resources, and environmental pollution. Environmental science is interdisciplinary because it uses and combines information from many disciplines: biology (particularly ecology), geography, chemistry, geology, physics, economics, sociology (particularly demographics, the study of population dynamics), natural resources management, law, and politics.

One of the most important and most frequently used terms in environmental science is **environmental sustainability**. Broadly speaking, environmental sustainability is the ability of the environment to function indefinitely without going into a decline from the stresses imposed by human society on natural systems (such as soil, water, air, and biological diversity) that maintain life. When the environment is used sustainably, humanity's present needs are met without endangering the needs of future generations. Environmental sustainability applies at many levels, including community, regional, national, and global levels.

Many experts in environmental science think that human society is not operating sustainably because of the following problems:

1. We are using nonrenewable resources such as fossil fuels as if they were present in unlimited supplies.
2. We are using renewable resources such as fresh water faster than they can be replenished naturally.
3. We are polluting the environment with toxins as if the capacity of the environment to absorb them were limitless.
4. Our numbers continue to grow despite Earth's finite ability to support us.

We examine environmental sustainability throughout the text, in terms of the natural environment, human population, natural resources, and pollution.

OUR IMPACT ON THE ENVIRONMENT

One of the best ways to gain a sense of the scope of environmental science is to examine some of the problems that today's environmental scientists identify and attempt to solve. Some of the problems are truly global in scope, such as destruction of Earth's ozone shield; others are more regional, such as the oil spill in Kuwait; and still others are local, such as the loss of wolves in Yellowstone

THE SCOOP ON NEWSPAPER

Most people do not realize that adding to landfills is not the only environmental concern associated with newspapers. Newspaper manufacturing contributes to deforestation and soil erosion, requires substantial energy consumption for processing and transporting raw materials, and uses abundant nonrenewable resources such as petroleum and copper for inks. Known environmental hazards abound in newspaper production: metals in inks are associated with many human health problems, heavy metals and dioxins are emitted when sludge is incinerated, greenhouse gases are released at landfills, and volatile organic compounds (VOCs) are breathed by printing plant workers. Any single manufacturing process can affect the environment in a variety of ways; in fact, the newspaper is more environmentally benign than many other manufactured goods. Newspaper manufacturers could reduce their environmental impact, however, by using chlorine dioxide instead of chlorine (to reduce dioxin releases), replacing petroleum-based inks with those based on soybean oil and heavy metals in inks with vegetable-based pigments, implementing new printing processes that eliminate VOCs, and generally reducing their bulk by making advertisements available to consumers on demand by computer rather than in print.

Park. Some problems are cases of upsetting the balance among native organisms by introducing foreign species; the invasion of zebra mussels in the Great Lakes is a problem of this sort. Other problems involve the complete destruction of entire ecosystems, such as filling in the coastal wetlands of the Atlantic coast or dynamiting coral reefs to catch fishes.

Humans do not live alone, nor are we above the laws of nature. On the contrary, we have many partners who share Earth with us, and we would not live long without them. Think of the number of organisms that had to live in order for you to get through this day: much of the oxygen you breathe was produced by plants, as were the fibers in the paper of this page and in the cloth of your cotton or linen clothing. If you eat meat, it was an animal once, and so was the leather of your shoes; animals produced the wool of your sweaters and socks.

Every human lives within a complex community of organisms, and our acts do have consequences for those organisms. One of the principal goals of environmental science is to identify ways to avoid upsetting the balance of the biological systems that support us. When imbalances do arise, one of the tasks of environmental science is to suggest how to deal with them in the most constructive way possible. For example, destruction of Earth's ozone shield by industrial chemicals would have a disastrous effect if allowed to continue; pollution of the environment by industrial wastes and overuse of natural resources such as fresh water and forest trees are other problems of this kind. Environmental scientists are called on to find solutions to all these problems.

To gain some idea of the diversity of problems dealt with by today's environmental scientists, look at a newspaper. You will see many stories about the environment and our impact on it. Following (in no particular order) are just a few of the stories that have made the news in the last few years. Every one of them affects the quality of our lives.

Human Numbers

The view in Figure 1–1 is a portrait of about 300 million people. It is a satellite photograph of most of the United States, Canada, and Mexico at night. The tiny specks of light represent cities, whereas the great metropolitan areas, such as New York along the northeastern seacoast, are ablaze with light.

The central problem of environmental science, the one that links all others together, is that there are many people in this picture—and soon there will be many more. In 1950 only eight cities in the world had populations larger than 5 million; the largest was New York, with 7.9 million. By 1994 the largest city, Tokyo—Yokohama, had 26.5 million inhabitants, and the combined population of the world's ten largest cities was 150.8 million (see Table 9–1). There are currently 20 megacities with populations greater than 8 million. By the year 2010, there will be 28 megacities, 22 of them in developing countries such as Brazil, India, and Indonesia. (See Chapter 8 for a discussion of developed and developing countries.)

In 1987 the human population as a whole passed a significant milestone: 5 billion individuals (Figure 1–2). Since then, the population has grown to 5.77 billion in 1996 (latest data available). From 1995 to 1996, the global population increased by 68 million people, a num-

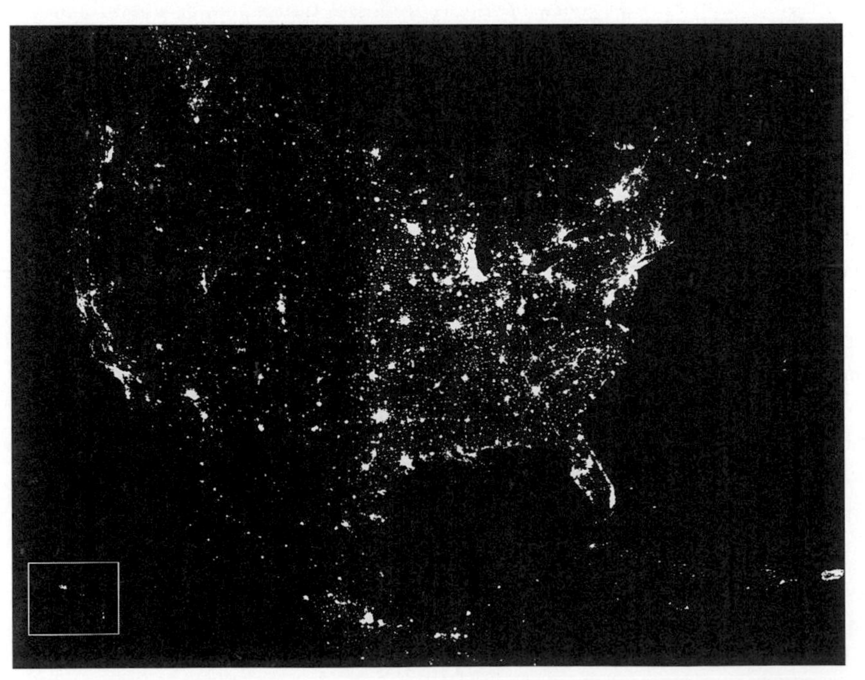

Figure 1–1

A satellite view of most of North America at night shows cities and major metropolitan areas. (Hansen Planetarium, SLC, UT, USA, and W. T. Sullivan III, Univ. of Washington)

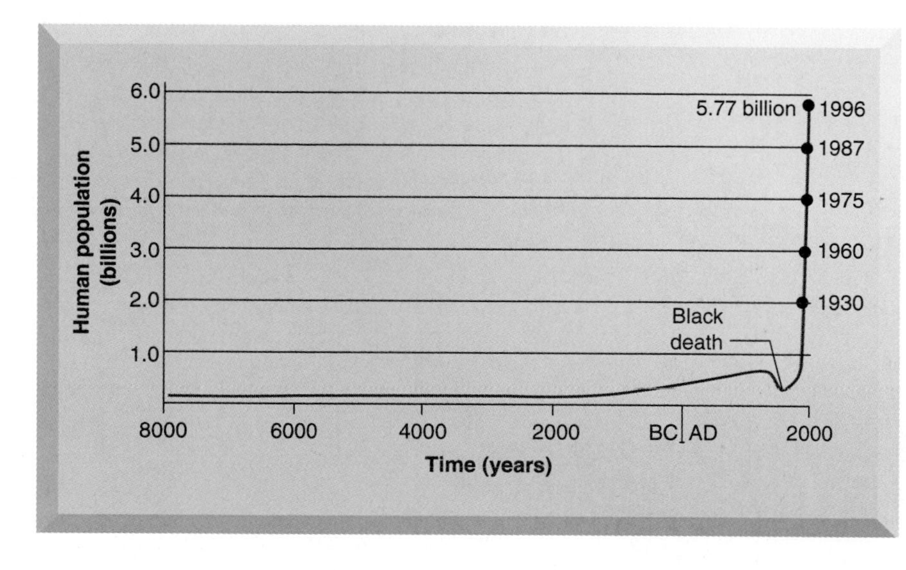

Figure 1–2

Human population numbers for the last ten thousand years since the advent of agriculture. It took thousands of years for the human population to reach 1 billion, 130 years to reach 2 billion, 30 years to reach 3 billion, 15 years to reach 4 billion, and 12 years to reach 5 billion. At our present growth rate the sixth billion will be added by 1999.

ber greater than the combined populations of Italy and Greece! At this growth rate, the world's population will surpass 6 billion people in 1999—and then increase by almost 2 billion more (the entire world's population in 1930) during the following 20 years. All these people consume a lot of food and water, use a great deal of energy and raw materials, and produce much waste.

Despite the vigorous involvement of most countries with family planning, population growth rates cannot be expected to change overnight. Three billion people will be added to the world in the next three to four decades, so even if we continue to be concerned about the overpopulation problem and even if our solutions are very effective, the coming decades may very well be clouded with tragedy. The World Bank estimates that 1.3 billion people are now living in extreme poverty (Figure 1–3), and according to the United Nations, an estimated 800 million people consume less than 80 percent of the recommended daily levels of food (calories).

Given the family planning efforts that are under way and considering the direction of growth trends, it is estimated that the world population may stabilize by the end of the 21st century. Population experts have made various projections for the population at that time, from almost twice what it is now (about 10.4 billion people) to almost four times our current population (greater than 20 billion people).

No one knows whether Earth can support so many people indefinitely. Finding ways for it to do so represents one of the greatest challenges of our times. Among the tasks that we must accomplish is feeding a world population almost twice as large as the present one without destroying the biological communities that support us. The quality of life available to our children and grandchildren will depend to a large extent on our ability to achieve this goal.

A factor as important as population *size* is a population's level of *consumption*. Inhabitants of the United States and other developed countries consume many more resources per person than do citizens of developing countries such as Nigeria, India, and Peru. Our high rate of resource consumption affects the environment as much as, or more than, the explosion in population that is occurring in other parts of the world. *Thus, as human numbers and consumption increase worldwide, so does humanity's impact on Earth, posing new challenges to us all.* Success in achieving sustainability in population size and consumption will require the cooperation of all the world's peoples. (Human population issues are examined further in Chapters 8 and 9.)

Figure 1–3

A slum in Kenya, East Africa. Many of the world's people live in extreme poverty. (K. Doyle/Visuals Unlimited)

Figure 1–4
A young alligator hatches from eggs taken by University of Florida researchers from Lake Apopka, Florida. Few Lake Apopka eggs actually hatch, and many of the young alligators that do hatch have abnormalities in their reproductive system. This young alligator may not leave any offspring. (St. Petersburg Times)

Endocrine Disrupters

The activities of Earth's growing human population are known to have a profound, sometimes global, effect on the environment. Is it possible that some of the thousands of chemicals that humans produce and use could threaten the health of animals, including the human species? Unfortunately, the answer may be yes. Mounting evidence suggests that hundreds of widely used chemicals are **endocrine disrupters** that interfere with the actions of the endocrine system (the body's hormones). These chemicals include certain plastics such as polycarbonate found in many water jugs; chlorine compounds such as PCBs and dioxins; the heavy metals lead and mercury; and some pesticides such as DDT, kepone, dieldrin, chlordane, and endosulfan. Although some of these chemicals have been banned in the United States, they are still used by many countries, particularly in the tropics.

Hormones are chemical messengers that regulate growth, reproduction, and other activities. Some endocrine disrupters mimic the estrogens, a class of female sex hormones, by fitting into their cellular receptors and sending false signals to the body that interfere with the normal functioning of the reproductive system. Both males and females of many species, including humans, produce estrogen and have estrogen receptors in their bodies. Additional endocrine disrupters interfere with the endocrine system by changing how the body synthesizes and uses hormones other than estrogen. Like hormones, endocrine disrupters are active at very low concentrations and, as a result, may cause significant health effects at relatively low doses.

Many endocrine disrupters appear to alter reproductive development in males and females of various animal species. Accumulating evidence indicates that fishes, birds, reptiles such as turtles and alligators, and other animals exposed to these environmental pollutants exhibit reproductive disorders and are often left sterile. For example, male alligators in Lake Apopka, Florida, which was contaminated with DDT and other agricultural chemicals, have low levels of testosterone and elevated levels of estrogen. Their reproductive organs are feminized, and they have abnormally small penises. The mortality rate for eggs in this lake is extremely high—only 18 percent hatch, and half of these die within ten days (Figure 1–4 and Table 1–1).

Humans may be equally at risk from endocrine disrupters, as the number of reproductive disorders, infertility cases, and hormonally related cancers appears to be increasing. A controversial Danish study reported that sperm counts in men from 21 nations, including the United States, dropped an average of 50 percent between 1940 and 1990. Scientists do not know if there is a link between this apparent decline and environmental factors, however. The incidence of birth defects in the male reproductive system appears to be increasing rapidly in several countries, and testicular cancer in men in their 20s and 30s appears to be rising almost everywhere that data are available. Deaths from breast cancer have been increasing in the United States by about one percent per

Table 1–1
The Population of Juvenile Alligators in Lake Apopka After a Chemical Spill in 1980

Year	Number of Alligators, Size 30 to 120 Centimeters, per Kilometer of Shoreline*
1980	23.5**
1981	12.0
1982	6.0
1983	3.0
1984	2.0
1985	2.0
1986	2.5
1987	2.5
1988	1.5
1989	0.5
1990	2.0
1991	4.0
1992	3.0

*Counts are adjusted for changes in water level.
**Chemical spill occurred just prior to the 1980 surveys.

year since the 1940s. Currently, one out of every eight women can expect to develop breast cancer by age 85. Although some of this increase can be attributed to earlier diagnosis, medical researchers are focusing their attention on endocrine disrupters as a possible cause.

Definite links between environmental endocrine disrupters and human health problems cannot be made at this time because of the limited number of human studies. Human exposure to endocrine-disrupting chemicals needs to be quantified so we know exactly how much of these chemicals we are dealing with in various communities. Complicating such assessments is the fact that humans are also exposed to *natural,* hormone-mimicking substances in the plants we eat. It is not known how the hundreds of different endocrine-disrupting chemicals interact. Sometimes, for example, two or more pollutants interact in such a way that their combined effects are more severe than the sum of their individual effects, a phenomenon known as **synergism**. The combined effects of two or more pollutants may also be less severe than the sum of their individual effects, a phenomenon known as **antagonism**. Our ignorance of possible synergisms and antagonisms among the many different endocrine-disrupting chemicals makes it very difficult to conclusively understand their individual and collective effects. (The effects of toxic chemicals, including several endocrine disrupters, are discussed in Chapters 19, 21, 22, and 23.)

The hypothesis that endocrine disrupters could affect the reproductive health of animals, including humans, was not widely recognized until the 1990s. *One lesson we can learn from this is that serious environmental problems sometimes take us by surprise.*

Collapse of the Georges Bank Fishery

In late 1994, the National Marine Fisheries Service of the U.S. Commerce Department closed Georges Bank, once one of the world's richest fishing grounds. The moratorium on fishing the Georges Bank, a 6,600 square mile area off the coast of New England in the North Atlantic Ocean, was not a surprise to any observer of the fishing industry. Catches of cod, haddock, and yellowtail flounder had been steadily declining there for the past ten years (Figure 1–5). During that period, for example, the had-

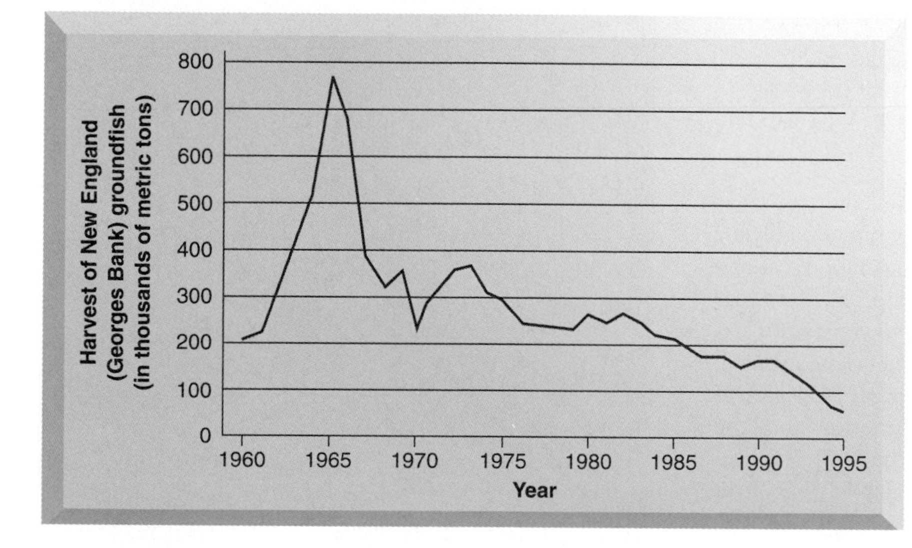

Figure 1–5
The harvest of groundfishes (cod, haddock, and flounder) from the Georges Bank area has declined steadily in recent years. A moratorium on fishing was declared in December 1994. (National Marine Fisheries Service)

dock population dropped by 75 percent, primarily due to overfishing. (Overfishing means that fishes are harvested faster than they can replace themselves.) Haddock and similar economically important fishes have reached **commercial extinction** in the Georges Bank area because their numbers are so low that they are unprofitable to harvest. As a result, thousands of people have lost their jobs. No one knows for sure how long it will take for fish populations in the Georges Bank area to recover, but estimates range from 4 to 20 years. (As we go to press, there is no sign of a recovery.)

The collapse of the Georges Bank fishery is not unique. The Alaskan salmon fishery declined during the 1960s and 1970s, the Peruvian anchovy fishery collapsed in the 1970s, and the Newfoundland cod fishery along the Grand Banks closed in the 1990s. Fish shortages are occurring all over the globe as a result of overfishing. According to the National Marine Fisheries Service, 40 percent of commercially important species have been overfished in U.S. waters. The problem is particularly acute in the Pacific Northwest, Nova Scotia, and the Gulf of Mexico.

Overfishing occurs because the worldwide demand for fish has grown, and, to meet the demand, there are more fishing boats than ever before. In addition, more and more fishermen are using high-tech methods to find and harvest fishes. Advanced technologies include the use of sonar and Comsat satellite images to locate fishes. Tracking with sonar allows the nets to remain in the water until the very last fish is caught.

International fishing fleets roam the ocean, competing with one another, and as fish populations have dwindled, the competition has become fierce. In 1994, more than 30 fights occurred at sea over fishing rights. (The topic of declining fisheries is revisited in Chapter 18.)

One lesson that the closing of the Georges Bank fishery teaches us is that as our technology advances, such as more effective fishing techniques, so does the impact we have on the environment. It is a theme you will encounter repeatedly as you proceed through the text.

Declining Bird Populations

Evidence is mounting that the populations of many birds that were once abundant have been dropping steadily for more than a decade. Across the North American continent, population declines have been noted in every major group of birds. In the eastern United States, for example, 70 percent of the 300 to 400 species of neotropical birds have declining numbers. **Neotropical birds** spend the winter in Central and South America and the Caribbean and then migrate north to the United States and Canada to breed during the summer. Cerulean warblers, olive-sided flycatchers, yellow-billed cuckoos, black-capped vireos, and rose-breasted grosbeaks are examples of neotropical birds with declining numbers.

Neotropical birds are faced with changing environments in both their winter and summer homes, and loss of habitat appears to be the main reason for their decline. These migratory birds are under stress from the burning of tropical rain forests in Central and South America as well as the fragmenting of forested areas in North America to accommodate suburban development, agriculture, and logging. (The effects of habitat disruption on organisms are discussed further in Chapter 16.)

Several studies have shown that fragmentation of forests increases the likelihood of reproductive failure for neotropical birds, many of which are insect-eaters that nest only in forests. As a result of forest fragmentation, the nests are more likely to be located near a **forest edge**, the often sharp boundary between the forest and surrounding farmlands or residential neighborhoods, rather than deep in the forest. Nests that are within 100 meters (328 feet) of a forest edge are more vulnerable to predation by small mammalian predators such as raccoons, opossums, snakes, and feral cats (domestic cats that have gone wild). Blue jays, American crows, and other egg-eating birds do most of their hunting along the forest edge.

Nest parasitism by brown-headed cowbirds is also more significant along the forest edge. Female cowbirds

Figure 1–6
Cowbird eggs in the nest of a wood thrush. When the wood thrush nestlings hatch, they will have to compete for food with their larger, more aggressive foster nestlings. More of the speckled cowbird eggs are laid in songbird nests along the forest edge than deep within a forest. (Courtesy Illinois Natural History Survey)

lay their eggs in the nests of other bird species and leave all parenting jobs to the hosts (Figure 1–6). In a 1989 to 1993 study that monitored 5000 nests of neotropical birds, the majority of nests were parasitized in areas with less than 55 percent forest cover and in many cases, there were more cowbird eggs in the nest than eggs that belonged there. Nest parasitism causes reproductive failure because the host parents provide food for the larger, more aggressive cowbird babies, while their own young starve.

Cowbirds do not nest or feed in intact forests. Before Europeans settled in North America, cowbirds followed the migratory herds of bison across the Great Plains, foraging on seeds and insects stirred up by the bison. With the settlement of the continent by Europeans, many forests were cut to provide agricultural lands, and the cowbird expanded its range into fields and cattle pastures. The cowbirds began parasitizing new bird species that had not yet evolved ways to resist them. (Some birds, such as robins, gray catbirds, and Baltimore orioles, reject cowbird eggs.)

An important conclusion we can make regarding declining bird populations is that human activities such as forest fragmentation often unwittingly benefit some species at the expense of others.

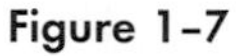

Figure 1–7

The first gray wolves were reintroduced into Yellowstone National Park on January 12, 1995. Carrying the first wolf are U.S. Fish and Wildlife Director Mollie Beattie (left, now deceased), Interior Secretary Bruce Babbit (center), and Yellowstone's Superintendent Mike Finley (right). (Lacy Atkins/Reuter)

Reintroducing Wolves to Yellowstone

In 1995, one of the most controversial reintroductions of an endangered species took place in Yellowstone Park and the central Idaho wilderness. Fourteen gray wolves (*Canis lupus*) were captured in Canada and released into Yellowstone Park in Wyoming and the Frank Church–River of No Return Wilderness in Idaho by the U.S. Fish and Wildlife Service (FWS) (Figure 1–7). More wolves were released in 1996, and additional releases are planned in the next five years. It is hoped that survivors will form packs and reproduce to form two populations of about 100 wolves each by 2002. The ultimate goal of the reintroduction is to remove the gray wolf from the endangered species list.

Wolves originally ranged from northern Mexico to Greenland, but they have been trapped, poisoned, snared, and hunted to extinction in most places. They were wiped out of Yellowstone during the 1930s by a federal program. By 1960, the only wolves remaining in the lower 48 states were small populations in Minnesota. Under the protection offered by the 1973 Endangered Species Act (ESA), however, small packs of wolves have spread from Canada into Montana and Washington and from Minnesota into Michigan and Wisconsin.

The wolf reintroduction is supported by biologists and environmentalists because it will restore the natural balance between predators and prey. The wolves' main prey are elk, deer, and bighorn sheep. The packs of wolves that roam these wilderness areas will reduce the burgeoning elk populations, which, in the absence of natural predators, are at an all-time high. When elk numbers are too large, they overgraze their habitat and thousands starve during hard winters. Biologists are also excited by the reintroduction of wolves because it will give them a unique opportunity to study the ecological impact of a top predator.

The reintroduction of wolves did not occur without a fight. The FWS was given the green light only after holding more than 150 public hearings, conducting numerous scientific studies, and considering 160,000 comments and public testimonies. Two lobbying groups that represent farmers and ranchers filed suit challenging the program. Farmers and ranchers who live in the area were against the reintroduction because they feared that their livestock and pets would be attacked.

Under a compromise developed to deal with this concern, ranchers are allowed to kill wolves that attack their cattle and sheep, and federal officers can remove any wolf that threatens humans or livestock. To get around the ESA, which forbids killing an endangered species, the wolf was declared an "experimental nonessential" in the area. In addition, the Defenders of Wildlife, an environmental organization that concentrates on protecting wild plant and animal species in their natural environments, has a "wolf compensation fund" that reimburses ranchers for full market value of livestock lost to wolves. (Chapter 16 expands on endangered species and the ESA.)

The reintroduction of wolves has had a positive economic impact on tourism in the Yellowstone area. There has been a surge in tourism, and surveys indicate that vis-

WELCOME HOME

The restoration of wolves to Yellowstone National Park in early 1995 replaced the only missing link in the park's original ecosystem. However natural the wolf might be in Yellowstone, its return promises substantial changes for other residents. The top predator's effects on its surroundings will likely range from altering relationships among predator and prey species to transforming vegetation profiles. Coyotes are potential prey for wolves, and decreases in their populations could reduce the threat they pose to local sheep flocks. Mountain lions, the wolf's chief rival, could be restricted to the rockier, higher locations that represent that animal's more natural terrain. Broader impacts on the Yellowstone ecosystem should arise from the pruning effect wolves will have on prey populations such as elk, deer, and bison. Yellowstone's booming elk population could be reduced as much as 20 percent, which would relieve heavy grazing pressure on plants and encourage a more lush and varied plant composition. Richer vegetation should support more herbivores like snow hares, which in turn will support small predators such as foxes.

itors want to see wolves more than any other animal in the park. During the first year after wolves were released, thousands of people reported catching glimpses of wolves through binoculars or hearing their song.

The reintroduction of wolves to Yellowstone is a vivid lesson that environmental issues are often complicated by differing opinions about what should be done. Compromise is essential to solving environmental problems.

Oil Spills in Kuwait

During February 1991, it was impossible to read a news magazine or watch a news show on television without seeing pictures of thick clouds of black smoke pouring out of hundreds of Kuwait's oil fields, which had been ignited by Iraqi soldiers during the Persian Gulf War. The 650 fires burned for months before being completely extinguished on November 6, 1991, and the media filled the news with dire predictions of the long-term effects of the fires. Doctors feared serious health effects for people in the immediate region, and some scientists predicted global and regional problems such as severe acid rain, absence of a monsoon season in Asia, and global climate change.

Although the total effects of the fires in Kuwait will take years to study and evaluate, an understanding of their short-term impact has begun to emerge. Some people who live in the immediate vicinity of the fires and who inhaled the smoke suffer health problems, including respiratory illnesses, and some people have even died from ailments associated with the fires.

The fires apparently did not have a large, long-term impact on global climate. Fortunately, the smoke plumes stayed in the lower atmosphere, where they were subjected to wind and precipitation. Had much of the smoke risen into the stratosphere, where weather does not occur, the smoke particles could have remained for many years, exerting a cooling influence on global climate. Local weather changes did occur, however. During the spring of 1991, when the sky was still black with smoke, temperatures in Kuwait were unseasonably low—sometimes as much as 11°C (20°F) below normal. Acid deposition, including acid rain, caused by sulfur oxides in the smoke fell in areas downwind from the fires.

When the Iraqis detonated the Kuwaiti oil wells, many sabotaged wells did not burn but instead poured their oil across the desert, forming hundreds of shallow lakes (Figure 1–8). In 1995, four years after the conflict, at least 100 lakes, containing 3 to 4 million barrels of oil, had not been drained. These oil lakes, some of which are one meter (3.3 feet) deep and one kilometer (0.6 mile) long, have infiltrated the soil and killed plants, birds, and insects. Even where the lakes have been drained, a layer of sandy tar remains behind, preventing plants from growing in the oil-logged soil. Because the tar is waterproof, the depressions fill with oily water when rainfall occurs.

Figure 1–8

Oil is pumped from an oil lake near Kuwait City. Many lakes of oil remain in Kuwait from oil wells sabotaged during the Persian Gulf War. (Thomas Hartwell/SYGMA)

The desert ecosystem is fragile, and the war may have caused irreparable damage to the desert soil. Before the war, a thin layer of pebbles covered most of the Kuwaiti soil, reducing erosion and preventing the formation of sand dunes. Because 214,000 bunkers, trenches, and pits were dug during the conflict, the gravel pavement was disrupted, and now the underlying sand forms dunes when the winds blow. It is clear that the Kuwaiti desert, particularly its soil, will take many years to recover from the environmental damage of the Persian Gulf War. (Chapter 10 considers the environmental impacts of oil, including oil spills.)

One lesson to be learned from the oil spills in Kuwaiti deserts is that certain ecosystems—those that are already stressed by natural factors such as lack of precipitation—are more sensitive to disruption and degradation by human activities.

Figure 1–9

Zebra mussels clog a pipe. The zebra mussel has caused hundreds of millions of dollars in damage in addition to displacing native clams and mussels. (Illinois Department of Conservation)

The Introduction of Exotic Species

Cargo-carrying ocean vessels carry water from their ports of origin in order to increase their ships' stability in the open ocean. When they reach their destinations, this water, known as ballast water, is discharged into local bays, rivers, or lakes. Ballast water may contain clams, mussels, worms, small fishes, and crabs, in addition to millions of microscopic aquatic organisms. These organisms, if they become successfully established in their new environment, can threaten the area's ecology and contribute to the possible extinction of native organisms.

In 1982, an unwanted American visitor—a jellyfish-like organism called a ctenophore—arrived somewhere in the Black Sea. Carried there in the ballast water of a ship, the ctenophore quickly became established because it had an ample food supply (microscopic plankton) and no natural predators in its new environment. As its numbers increased, it consumed so much plankton that larger fishes were deprived of their food supply. The ctenophore's success has all but eliminated commercially important fishes from the Black Sea. Fish catches, once an important part of the local Russian economy, have declined dramatically since the late 1980s. Marine biologists say the establishment of ctenophores in the Black Sea is one of the most devastating biological invasions to occur in the past 50 years.

One of North America's greatest biological threats is the zebra mussel, a native of the Caspian Sea that was probably introduced through ballast water flushed into the Great Lakes by a foreign ship in 1985 or 1986. Since then, the tiny freshwater mussel, which clusters in extraordinary densities, has massed on hulls of boats, piers, buoys, and, most damaging of all, on water intake systems. The zebra mussel's strong appetite for algae, phytoplankton, and zooplankton is also cutting into the food supply of native fishes, mussels, and clams, threatening their survival. By 1994, the zebra mussel had progressed from the Great Lakes into the Mississippi River. It is found as far south as New Orleans, as far north as Duluth, Minnesota, and as far east as the Hudson River in New York. Some scientists hypothesize that by the year 2000 it will have colonized all freshwater bodies in North America that meet its wide-ranging ecological requirements. Since the zebra mussel invaded North America, an estimated $5 billion dollars have been spent to repair damage such as clogged pipes (Figure 1–9).

Unfortunately, such invasions are not unusual. The Great Lakes contain at least 139 introduced species, and worldwide, most regions are thought to contain 10 to 30 percent exotic species. *Each introduction seems a unique accident when it happens, but the overall problem is general and very serious because our highly mobile society has facilitated the movement of foreign organisms.* Although not all exotic species are harmful, all too often introduced species produce havoc in ecosystems and are costly in economic terms. (Chapter 16 expands on the problems of introduced species.)

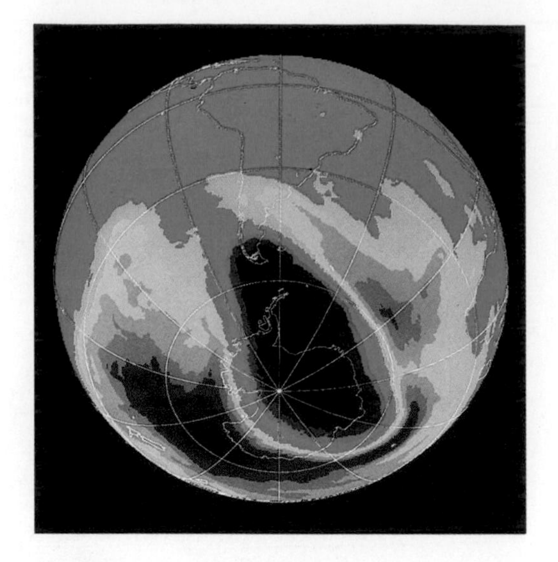

Figure 1–10

A computer-generated image of part of the Southern Hemisphere on October 17, 1994, reveals the ozone "hole" (black and purple areas) over Antarctica and the tip of South America. Relatively low ozone levels (blue and green areas) extend into much of South America as well as Central America. Normal ozone levels are shown in yellow, orange, and red. The ozone hole is not stationary but moves about as a result of air currents. (Courtesy NASA)

Damage to the Atmosphere: Stratospheric Ozone Depletion

The swirling colors of Figure 1–10 are a view of the South Pole from a satellite. This is not the picture your eye would see, but rather a computer reconstruction of ozone levels in the **stratosphere**, the layer of the atmosphere that extends from 10 to 45 kilometers (6.2 to 28 miles) above Earth's surface. As you can easily see, there is a large ozone "hole" over Antarctica, within which the ozone concentration is much lower than elsewhere; it covers an area about the size of the United States. Ozone thinning was first hypothesized by two chemists at the University of California based on calculations they made in 1974 of the atmospheric effects of chlorofluorocarbons (CFCs). CFCs, a group of chemicals familiar as aerosol propellants in spray cans, which are now banned in most countries, and as the Freon cooling agent in refrigerators and air conditioners, are the main cause of ozone thinning.

The first direct observations of ozone thinning were reported in 1985 by British scientists. The thinning is not a permanent feature but a seasonal phenomenon, evident only for a few months at the onset of the Antarctic spring in September. In 1990 the minimum ozone concentration in the hole was 50 percent lower than the minimum ten years earlier. And by 1992 there was clear evidence that stratospheric ozone was also being depleted over the Arctic.

Environmental scientists are worried about the Antarctic ozone loss because it is related to a gradual thinning of stratospheric ozone worldwide. In the heavily populated mid-latitudes of the planet, ozone levels have been decreasing for several decades. Why is this worrisome? Stratospheric ozone absorbs harmful ultraviolet radiation from sunlight. Without its protection, human skin cancers caused by ultraviolet radiation would become far more common. The increased ultraviolet radiation would also damage other species.

As a direct result of public awareness of the problem, strong laws were eventually passed restricting the use of CFCs in the United States and Canada. In 1990, some 90 other nations agreed to a total ban on CFC production by the year 2000, and in 1992, the phase-out date for CFCs and similar compounds was moved up to 1996. Because CFCs are very stable, they can survive in the atmosphere 120 years or more, so the changes initiated by the release of CFCs may not be quickly reversible. (Stratospheric ozone depletion is examined further in Chapter 20.)

An important lesson to learn from this story is that we often must act on serious environmental threats based on what we already know, even when there are obvious gaps in our knowledge. Had we waited until all scientific evidence on the ozone hole had been gathered before we took any action, the problem would have been much graver than it is.

Global Climate Change and Increasing Carbon Dioxide Levels

During the past two centuries, as the world's population has grown to ten times its former size, the level of carbon dioxide (CO_2) in the atmosphere has increased dramatically (Figure 1–11). The causes of the increase in atmospheric CO_2 are no mystery: the burning of fossil fuels such as coal, oil, and natural gas, and the clearing and burning of forests by farmers. Environmental scientists are growing increasingly concerned that the rising levels of CO_2 may change Earth's climate. Carbon dioxide levels rose from 315 parts per million (ppm) in 1958 to 360.9 ppm in 1995 (latest data available). Just as the panes of glass in a greenhouse let light in but do not allow heat out, so CO_2 in the atmosphere allows solar radiation to pass through but does not allow heat to radiate back into space. Instead, the heat is radiated back to Earth's surface. As CO_2 accumulates in the atmosphere, enough heat may be trapped to gradually warm the planet.

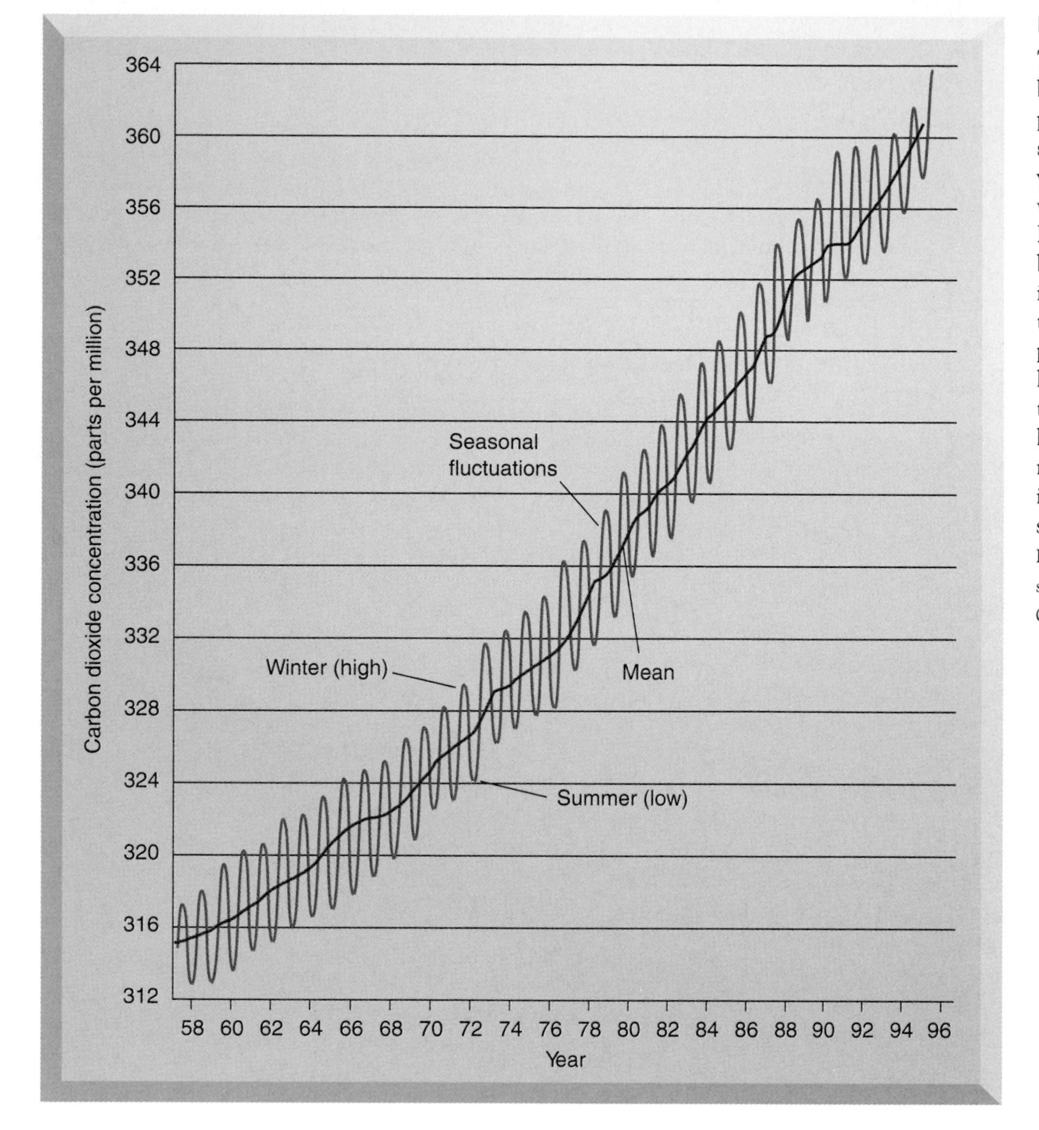

Figure 1-11

The concentration of carbon dioxide in the atmosphere has shown a slow but steady increase for many years. These measurements were taken at the Mauna Loa Observatory, far from urban areas where carbon dioxide levels are high because of the large number of factories, power plants, and motor vehicles. The seasonal fluctuation corresponds to winter (a high level of CO_2) and summer (a low level of CO_2) and is caused by greater photosynthesis in summer. (Dave Keeling and Tim Whorf, Scripps Institute of Oceanography, La Jolla, California)

The 1980s and early 1990s saw some of the warmest years in U.S. weather records. Environmental scientists estimate that if trends are not changed, Earth's mean temperature could rise 1.5 to 4.5°C (2.5 to 7.5°F) by the middle of the next century, making the atmosphere warmer than it has been at any time in the last 100,000 years. This might produce major shifts in patterns of rainfall and initiate melting of the West Antarctic ice sheet, as did the last warm period 120,000 years ago. This melting would cause the ocean level to rise. Such a rise is alarming, as it might put many major cities at least partly under water. With each new warm year in the 1990s, the possibility seems more likely that a significant warming trend has begun. Hard conclusions about long-term trends are difficult to reach, but an increasing number of environmental scientists are concerned. (Global warming is examined further in Chapter 20.)

One important conclusion that we can make about climate change is that the effects of human activities locally and regionally can have global repercussions, even changing the composition of the atmosphere.

THE GOALS OF ENVIRONMENTAL SCIENCE

Unlike biology, geology, chemistry, and physics—sciences that seek to establish general principles about how the natural world functions—environmental science is an *applied* science that has deep roots in problem solving. Environmental scientists search for viable solutions to environmental problems—solutions that are based as much as possible on scientific knowledge.

(text continues on page 17)

MEETING THE CHALLENGE

The Earth Summit

In June 1992, representatives from around the world met in Rio de Janeiro, Brazil, for a summit conference officially called the UN Conference on Environment and Development. Countries attending the conference examined environmental problems that cross international borders and are truly global in nature: pollution and deterioration of the planet's atmosphere and ocean, a decline in the number and kinds of organisms, and destruction of forests.

It is easy for the representatives of a country to say their country supports a cleaner environment, but the actual specifics—what is going to be done, how soon, how much it is going to cost, and who is going to pay—are very difficult to agree on. The issues that were discussed at the 1992 Earth Summit included:

1. *Climate change:* A treaty to curb carbon dioxide emissions, thereby reducing the greenhouse effect.
2. *Biological diversity:* A treaty to decrease the rate of extinction of the world's endangered species.
3. *Deforestation:* A statement of principles regarding **deforestation**, the destruction of the world's forests.
4. *Agenda 21:* A complex action plan for the 21st century in which developed nations would provide money to help developing countries become industrialized without harming the environment.
5. *Earth Charter (The Rio Declaration):* A philosophical statement about environment and development.

The pre-summit negotiations for all of these topics were complex. In general, the poorer developing nations felt that their top priority had to be economic survival rather than saving the environment. They expressed a willingness to follow the environmental mandates of industrialized countries, but only if those countries contributed money to help them protect the environment. Industrialized nations, for their part, acknowledged that they had some responsibility toward developing nations but they wanted developing countries to focus on slowing their rates of population growth and industrialization in order to preserve the environment.

Although the Earth Summit did not accomplish all that environmentalists had hoped it would, it was a resounding success in many ways. It was the largest international gathering ever to concentrate on serious environmental issues, and, because it received so much international attention, it increased worldwide awareness of global issues. Also, the Earth Summit demonstrated just how far apart developed and developing nations stand on many issues. This awareness will be needed in future negotiations of international environmental issues.

► *FOLLOW-UP*

1. *Climate change treaty.* More than 165 nations have signed the climate change treaty, which is now considered legally binding. The United States ratified the treaty in 1992. However, the United States and most other large producers of CO_2 are not expected to meet their targets mandated by the treaty for the year 2000. Worldwide emissions continue to increase by about 20 percent each decade.

2. *Biological diversity treaty.* More than 165 nations have signed the biological diversity treaty, which is now considered legally binding. President Clinton signed the treaty in June 1993. It was then approved by the Senate Foreign Relations Committee, but Congress failed to approve it. As of early 1997, it has not been ratified by the United States.

3. *Agenda 21.* Agenda 21, an approach to sustainable development, recommends more than 2500 actions to deal with our most urgent environmental, health, and social problems. Although it represents collective action and shared responsibility between developed and developing nations, Agenda 21 does not offer a firm commitment for financial assistance by richer nations to implement the agreement. A high-level Commission on Sustainable Development, which consists of 53 members, including the United States, meets annually to monitor implementation of Agenda 21.

4. *Earth Charter.* The Earth Charter was established as a fragile compromise on issues involving the Northern and Southern Hemispheres. UN leaders want participating nations to adopt a new Earth Charter that is more progressive and inspiring in its goals.

5. *Treaty on desertification.* During the Earth Summit, many delegates spoke about the need to deal with **desertification**, the degradation of once-fertile arid and semiarid land into nonproductive desert. Caused by natural factors (drought) as well as human factors (such as overgrazing, improper irrigation, deforestation), desertification appears to be accelerating. The desertification treaty, officially called the International Convention to Combat Desertification, should go into effect in 1997 or thereafter.

Environmental problems are uniformly complex, however, and so our scientific understanding of them is often less complete than we would like it to be. Environmental scientists are often called on to reach a scientific consensus before the data are complete. As a result, they often make recommendations to elected officials based on probabilities rather than precise answers.

The science of **ecology**, a discipline of biology that studies the interrelationships between organisms and their environment, is a basic tool of environmental science, and so we begin our study with a detailed treatment of ecology (Chapters 3 through 6). Using what we have learned about ecology, we then directly address human population growth (Chapters 8 and 9) and three of the major results of that growth: the increasing need for energy (Chapters 10 to 12), depletion of resources (Chapters 13 to 18), and rising pollution (Chapters 19 to 23).

Many of the environmental problems we consider in this book are serious ones that must be addressed—if not by us, then unavoidably by our children. Environmental science is not, however, simply a "doom and gloom" listing of problems coupled with predictions of a bleak future. To the contrary, its focus, and our focus as individuals and as world citizens, is on identifying, understanding, and solving problems that we and our ancestors have generated. A great deal is being done, and more must be done—at individual, community, country, and worldwide levels—to address the problems of today's world.

Environmental issues are complex, however, and have multiple causes and repercussions. As a result, environmental problems interact with numerous competing interests and cannot be solved by science alone (see Meeting the Challenge: The Earth Summit). We fill a considerable amount of space in this text examining some successful approaches to environmental problems and exploring other problems that defy easy solutions.

SUMMARY

1. Environmental science is the interdisciplinary study of how humanity affects the relationships among other organisms and the nonliving physical environment. Environmental science encompasses many complex and interconnected problems involving human numbers, Earth's natural resources, and environmental pollution. Environmental science is interdisciplinary because it uses and combines information from many disciplines: biology, geography, chemistry, geology, physics, economics, sociology, natural resources management, law, and politics.

2. The world's human population is expected to surpass 6 billion in 1999. The increasing population is placing a nonsustainable stress on the environment, as humans consume ever-increasing quantities of food and water, use more and more energy and raw materials, and produce enormous amounts of waste and pollution.

3. Among the many challenges to the environment posed by human activities is the release of materials that may harm the environment: the use of chemicals that are endocrine disrupters; the spillage of pollutants such as oil in Kuwait; the release of chemicals such as CFCs that attack the ozone shield in the stratosphere; and the production of carbon dioxide that may alter global climate.

4. Other environmental challenges are caused by our attempts to manage organisms with which we share the planet: the gray wolf, an endangered species, is being reintroduced in Yellowstone Park and the central Idaho wilderness because humans killed all of them there earlier in the century; we are responsible for foreign species such as zebra mussels and ctenophores that invade our ecosystems, sometimes in dramatic fashion.

5. Our "progress" in developing natural resources has led to other challenges: the Georges Bank area has been fished to commercial extinction, and neotropical bird populations are declining, in part because their breeding habitat—the forested areas in North America—is increasingly fragmented by development.

6. All of these challenges to the environment can and must be addressed. Environmental science is the study of such challenges. Environmental scientists attempt to identify how human activities affect the environment and search for constructive solutions to the problems that arise.

THINKING ABOUT THE ENVIRONMENT

1. How does the "house of trash" discussed in the chapter introduction reduce the amount of waste deposited in local landfills? How does it save energy? How does it conserve wood?

2. Discuss several ways in which humans are making the environment unsuitable for other organisms.

3. How has recent history demonstrated the root causes of environmental problems?

4. From the standpoint of human well-being, what are some advantages of our industrialized society? What are some disadvantages?

5. Do you think it is possible for the world to sustain its present population of 5.7 billion people indefinitely? Why or why not?

6. Do you think that all or most environmental problems can be solved by technology? Why or why not?

* 7. Use the data from Table 1–1 to graph the decline in the population of juvenile alligators in Lake Apopka after the pesticide spill in 1980. Refer to Appendix V for information and help in graphing.

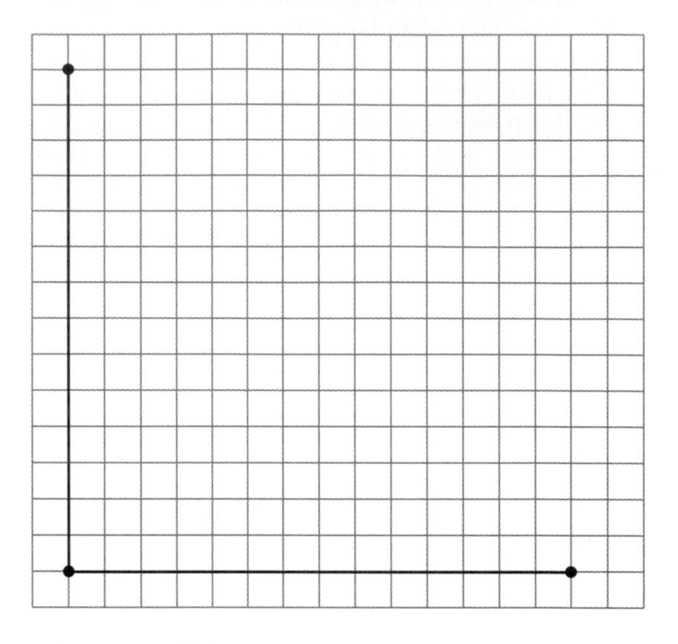

*Solutions to questions preceded by asterisks appear in Appendix VII.

RESEARCH PROJECTS

Select one of the environmental issues introduced in this chapter and research it further. Interview representatives from industry, government, and environmental organizations that are involved in this issue.

Obtain a copy of "Environmental Unknowns" by Norman Myers (*Science*, Vol. 269, July 21, 1995), in which the author contends that some of the greatest environmental problems facing us in the near future may still be unknown to us. Analyze this paper for your class.

Write a brief essay at the beginning of the semester about if/how you think this course may change the way you live your life. At the end of the semester, read your original statement and write a somewhat longer essay describing your views about living in the environment. Explain why your views have changed (or have not changed) as a result of this course.

Investigate the relationship between environmental science and environmentalism, the movement of people who are concerned about the environment. Write a short paper contrasting the overall goals of each and distinguishing among the four basic environmental ideologies: deep environmentalists, soft technologists, accommodators, and cornucopians.

On-line materials relating to this chapter are on the ► World Wide Web at: **http://www.saunderscollege.com/lifesci/environment2** (select Chapter 1 from the Table of Contents).

SUGGESTED READING

Bongaarts, J. "Can the Growing Human Population Feed Itself?" *Scientific American,* March 1994. We may be able to feed the 10 billion people that will populate the planet by the middle of the next century, but the environmental costs will be extremely high.

Burns, C. "Wildlife Service Dances with Wolves," *The Christian Science Monitor,* March 15, 1995. The controversial plan to reestablish wolves in the Rocky Mountains is explored.

Carlton, J. T. "Marine Bioinvasions: The Alteration of Marine Ecosystems by Nonindigenous Species," *Oceanography* Vol. 9, No. 1, 1996. The introduction of exotic marine species has compromised the integrity of many marine ecosystems.

Colburn, T., D. Dumanoski, and J. P. Myers. *Our Stolen Future,* New York: Dutton, 1996. Drawing on work from many fields, this controversial book records the rapidly unfolding story of commonly used chemicals that disrupt the activities of hormones.

Diamond, J. "Easter's End," *Discover,* August 1995. The story of the environmental collapse of Easter Island and its significance to planet Earth.

Hansen, J., A. Lacis, R. Ruedy, M. Sato, and H. Wilson. "How Sensitive Is the World's Climate?" *Research & Exploration* (A scholarly publication of the National Geographic Society), Spring 1993. Although interpreting climate data is difficult, current data indicate that the climate is warming as a result of human activities.

McLachlan, J. A., and S. F. Arnold. "Environmental Estrogens," *American Scientist* Vol. 84, September–October 1996. Environmental estrogens appear to be affecting the reproduction, immune systems, and behavior of many animal species.

Partfit, M. "Diminishing Returns: Exploiting the Ocean's Bounty," *National Geographic,* November 1995. As fish stocks decline, tensions increase over fishing grounds.

Pearce, F. "Devastation in the Desert," *New Scientist* Vol. 146, No. 1971, April 1, 1995. Four years after the Persian Gulf War, much of the desert ecosystem of Kuwait shows environmental damage.

Robinson, S. K. "Nest Gains, Nest Losses," *Natural History,* July 1996. Includes a fascinating account of how humans enabled the cowbird to significantly expand its range at the expense of other bird species.

Zimmer, C. "Son of Ozone Hole," *Discover,* October 1993. A new computer model of atmospheric circulation presents some disquieting news about the ozone hole over Antarctica.

Red Cross workers carry the body of a woman killed by the Ebola virus.
(Corinne Dufka/Bettmann/Reuters)

Solving Environmental Problems

In 1995 an outbreak of the deadly Ebola virus occurred in Zaire, a country in Central Africa. Ebola causes an often-fatal disease in which the victim hemorrhages internally and bleeds from the mouth, eyes, ears, and other body orifices. Ebola outbreaks previously had occurred in 1976 in Zaire and Sudan and in 1979 in Sudan.

Blood samples of people thought to be infected were flown to the U.S. Centers for Disease Control and Prevention in Atlanta for positive identification of the virus. As soon as Ebola was identified in the 1995 outbreak, global health care teams were sent into Central Africa to prevent its spread. Health officials confirmed that Ebola is spread by direct contact with infected body fluids. Because victims hemorrhage, infected blood appears to be the principal method of infection. Infection-control methods, such as using gloves, gowns, and masks, were implemented. Patients were quarantined, and travel out of the infected area was monitored. Global cooperation and rapid response successfully contained the 1995 Ebola outbreak, but not before almost 250 people died.

Ebola serves as a grim reminder that disease-causing agents can strike quickly and fatally. According to the Centers for Disease Control and Prevention, more than 200 new, continual, or re-emerging diseases have the potential to strike globally.

Many human endeavors may unintentionally contribute to outbreaks of infectious disease. The disruption of natural environments, for example, may give disease-causing agents an opportunity to break out of their isolation. Activities such as cutting down forests, building dams, and intensive agriculture may bring more humans into contact with new animal viruses and other rare disease-causing agents that could potentially infect them. Social factors may also contribute to disease epidemics. For example, human populations increasingly concentrate in large cities, permitting the rapid spread of infectious diseases. Global travel also has the potential to contribute to the rapid spread of disease. These and other factors interact in complex ways to cause the outbreak or increase the severity of infectious diseases.

In studying environmental problems that face the world today, from the emergence of once-obscure diseases to pollution of freshwater lakes, it is important to remember that much can be done to improve our situation. The role of environmental science is not only to identify problems but also to suggest and evaluate potential solutions. Although the choice to implement a proposed solution is almost always a matter of public policy, environmental scientists play key roles in educating both government officials and the general public. ◀

SOLVING ENVIRONMENTAL PROBLEMS: AN OVERVIEW

As discussed in Chapter 1, environmental science is the interdisciplinary study of how humanity affects other living organisms and the nonliving physical environment. One goal of environmental science is to improve our understanding of how humans alter the natural processes that make up Earth's life support system. This basic information is needed by policy makers and others to produce wise environmental decisions, and in that sense environmental science is fundamentally a problem-solving science. Before we begin a detailed examination of the environmental problems that face our society today, it is useful to consider the many elements that go into solving environmental problems. How is information gathered, and at what point can conclusions be regarded as certain? Who makes the decisions, and what are the tradeoffs?

The problems described in Chapter 1 are not insurmountable. A combination of scientific investigation and public action can solve them. Pollution of Earth's atmosphere, land, and water can be halted, and resources can be protected for the future. How can this be achieved? Viewed simply, there are five components in the solving of an environmental problem:

1. *Scientific assessment.* The first stage of addressing any environmental problem is scientific assessment, the gathering of information. Data must be collected and experiments or simulations performed to construct a model that describes the situation. Such a model can be used to predict the future course of events.
2. *Risk analysis.* Using the results of the scientific investigation, it is possible to analyze the potential effects of intervention—what could be expected to happen if a particular course of action were followed, including any adverse effects the action might generate.
3. *Public education.* When a clear choice can be made among alternative courses of action, the public must be informed. This involves explaining the problem, presenting all the available alternatives for action, and revealing the probable cost and results of each choice.
4. *Political action.* The public, through its elected officials, selects a course of action and implements it.
5. *Follow-through.* The results of any action taken should be carefully monitored, both to see if the environmental problem is being solved and, more basically, to judge and improve the initial evaluation and modeling of the problem.

THE SCIENTIFIC ANALYSIS OF ENVIRONMENTAL PROBLEMS

The key to the successful solution of any environmental problem is rigorous scientific evaluation. It is important to understand clearly just what science is, as well as what it is not. Most people think of science as a body of knowledge—a collection of facts about the natural world. However, science is also a dynamic *process,* a particular way to investigate the natural world. Science seeks to reduce the apparent complexity of our world to general principles, which can then be used to solve problems or provide new insights.

Scientists collect objective **data** (singular, *datum*), the information with which science works. Data are collected by observation and experimentation and then analyzed or interpreted. Scientific conclusions are inferred from the available data and are not based on faith, emotion, or intuition.

Science is an ongoing enterprise, and scientific concepts must be re-evaluated in light of newly discovered data. Thus, scientists can never claim to know the "final answer" about anything because scientific knowledge changes.

Several areas of human endeavor are not scientific. Ethical principles often have a religious foundation, and political principles reflect social systems. Some general principles, however, derive not from religion or politics, but from the physical world around us. If you drop an apple, it will fall whether or not you wish it to, despite any laws you may pass forbidding it to do so. Science is devoted to discovering the general principles that govern the operation of the natural world.

The Scientific Method

The process that scientists use to answer questions or solve problems is called the **scientific method**. Although there are many versions of the scientific method, it basically involves five steps:

1. Recognize a question or unexplained occurrence in the natural world. Because science is based on knowledge accumulated previously, after a problem is recognized, one determines what is already known about it by investigating the relevant scientific literature.
2. Develop a **hypothesis**, or educated guess, to explain the problem. A good hypothesis makes a prediction that can be tested. The same factual evidence can be used to formulate several alternative hypotheses, each of which must be tested.
3. Design and perform an experiment to test the hypothesis. An experiment involves the collection of data by making observations and measurements.
4. Analyze and interpret the data to reach a conclusion. Does the evidence match the prediction stated in the hypothesis? In other words, do the data support or refute the hypothesis?
5. Share new knowledge with the scientific community. This is done by publishing articles in scientific journals or books and by presenting the information at scientific meetings. Sharing new knowledge with the scientific community permits other scientists to repeat the experiment or design new experiments that either verify or refute the work.

Inductive and Deductive Reasoning

Scientists use inductive and deductive reasoning. Discovering general principles by the careful examination of specific cases is called **inductive reasoning**. The scientist begins by organizing data into manageable categories and asking the question "What does this information have in common?" He or she then seeks a unifying explanation for the data. Inductive reasoning is the basis of modern experimental science.

As an example of inductive reasoning, consider the following:

Fact: Gold is a metal that is heavier than water.

Fact: Iron is a metal that is heavier than water.

Fact: Silver is a metal that is heavier than water.

Conclusion based on inductive reasoning: All metals are heavier than water.

Even if inductive reasoning makes use of correct data, the conclusion may be either true or false. As new data come to light, they may show that the generalization arrived at through inductive reasoning is false. Science has shown, for example, that the density of lithium, the lightest of all metals, is about half that of water. When one adds this information to the preceding list, a different conclusion must be formulated, in this case: *Most* metals are heavier than water. Inductive reasoning, then, produces new knowledge but is error-prone.

Science also makes use of **deductive reasoning**, which proceeds from generalities to specifics. Deductive reasoning adds nothing new to knowledge, but it can make relationships among data more apparent. For example:

General rule: All birds have wings.

A specific example: Robins are birds.

Conclusion based on deductive reasoning: All robins have wings.

This is a valid argument. The conclusion that robins have wings follows inevitably from the information given. Deductive reasoning is used by scientists to determine the type of experiment or observations necessary to test a hypothesis.

The Importance of Prediction

A successful scientific hypothesis needs to be not only valid but useful—it needs to tell you something you want to know. A hypothesis is most useful when it makes predictions because the predictions provide a way to test the hypothesis' validity. A hypothesis that makes a prediction that your experiment refutes must be rejected. The more verifiable predictions a hypothesis makes, the more valid that hypothesis is. There is something very satisfying about a successful prediction, because the prediction being tested to verify the hypothesis is generated by the hypothesis itself, and the result is not known ahead of time.

Experimental Controls

Most often, the processes we want to learn about are influenced by many factors. We call each factor that influences a process a **variable**. In order to evaluate alternative hypotheses about one variable, it is necessary to hold all the other variables constant so that we do not get misled or confused by them.

To test a hypothesis about a variable, we carry out two forms of the experiment in parallel (Figure 2–1). In the experimental test we alter the chosen variable in a known way. In the **control** test we do not alter that variable. We make sure that in all other respects the two tests are the same. We then ask, "What is the difference, if any, between the outcomes of the two tests?" Any difference that we see must be due to the influence of the vari-

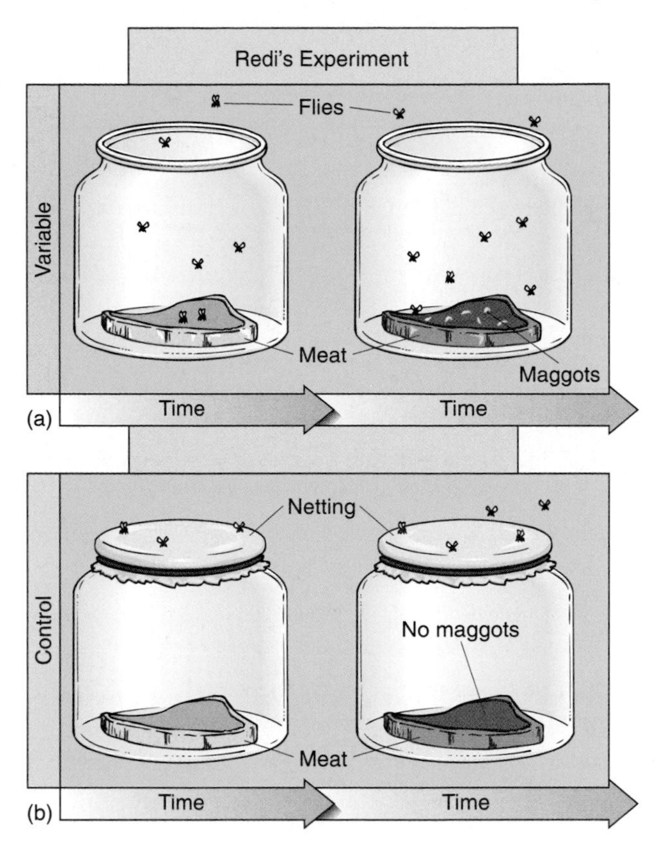

Figure 2–1

The variable and control of a classical experiment conducted by the Italian physician Francesco Redi (1621–1697). Redi's experiment supported the hypothesis that living things are produced only by other living things. (a) In the variable group, flies could enter the open jars and lay their eggs. Maggots eventually were seen on the meat and then flies were seen. (b) In the control group, the jars were screened. No maggots or flies appeared.

able that we changed, because all other variables remained the same. Much of the challenge of experimental science lies in designing control tests and in successfully isolating a single variable from all other variables.

Theories and Principles

A hypothesis supported by a large body of observations and experiments becomes a **theory**. A theory relates many data that previously appeared to be unrelated. A good theory grows as additional information becomes known. It predicts new data and suggests new relationships among phenomena.

By demonstrating the relationships among classes of data, a theory simplifies and clarifies our understanding of the natural world. Theories are the solid ground of science, the explanations of which we are most sure. This definition contrasts sharply with the general public's use of the word "theory," implying *lack* of knowledge, or a guess—as in, "I have a theory about the assassination of John Kennedy." In this book, the word "theory" is always used in its scientific sense, to refer to a broadly conceived, logically coherent, and very well supported explanation.

Some theories—for example, Newton's theory of gravity, Darwin's theory of evolution, and Einstein's theory of relativity—are so well established that the likelihood of their being rejected in the future is very small. Because they have withstood repeated testing and are the strongest statements we can make about the natural world, these theories are promoted to our highest level of confidence, that of scientific **principles**.

Yet there is no absolute truth in science—only varying degrees of uncertainty. The possibility always remains that future evidence will require a hypothesis, a theory, or even a principle to be revised. A scientist's acceptance of a hypothesis, theory, or principle is always provisional.

SCIENTIFIC DECISION-MAKING AND UNCERTAINTY: AN ASSESSMENT OF RISKS

Each of us takes risks every day of our lives. Walking on stairs involves a small risk, but a risk nonetheless, because some people die from falls on stairs. Using household appliances is slightly risky, because some people die from electrocution when they operate appliances with faulty wiring or use appliances in an unsafe manner. Driving in an automobile or flying in a jet offers risks that are easier for most of us to recognize. Yet few of us hesitate to fly in a plane, and even fewer hesitate to drive in a car because of the associated risk.

Estimating the risks involved in a particular action so that they can be compared and contrasted with other risks is known as **risk assessment**. The four steps involved in risk assessment for adverse health effects are summarized in Table 2–1. Once a risk assessment has been performed, its results are evaluated with relevant political, social, and economic considerations to determine whether there is a need to reduce or eliminate a particular risk and if so, what should be done. This evaluation, which includes the development and implementation of laws to regulate hazardous substances, is known as **risk management**.

Risk assessment helps us estimate the probability that an event will occur and enables us to set priorities and manage risks in an appropriate way (Table 2–2). As an example, consider a person who smokes a pack of cigarettes a day and drinks well water containing traces of the cancer-causing chemical trichloroethylene (in acceptable amounts as established by the Environmental Protection Agency, or EPA). Without knowledge of risk assessment, this person might buy bottled water in an attempt to reduce his or her chances of getting cancer. Based on risk assessment, the annual risk from smoking a pack of cigarettes per day is 3.6×10^{-3}, whereas the

Table 2-1
The Four Steps of Risk Assessment for Adverse Health Effects

Step	What it Answers
1. Hazard identification	Does exposure to a substance cause an increased likelihood of an adverse health effect such as cancer or birth defects?
2. Dose-response assessment	What is the relationship between amount of exposure (the dose) and the seriousness of the adverse heath effect?
3. Exposure assessment	How much, how often, and how long are humans exposed to the substance in question? For hazardous air pollutants, emissions are measured and analyzed to determine the relationship between emissions and concentrations in the environment. Where humans live relative to the emissions is also considered.
4. Risk characterization	What is the probability of an individual or population having an adverse health effect? Risk characterization combines data from dose-response assessment and exposure assessment (Steps 2 and 3).

Table 2-2
*Some Risks of Daily Living**

Relative Risk**	Type of Risk
0.2	Disease from PCBs in diet
0.3	Disease from DDT and DDE in diet
1	Disease from drinking 1 quart of municipal water per day (contains traces of chloroform)
18	Dying by electrocution in any given year
60	Disease from drinking 12 ounces of diet cola per day (contains saccharin)
367	Falls, fires, poisonings in the home
667	Respiratory illness caused by air pollution (for those living in eastern U.S.)
800	Dying in auto accident in any given year
2800	Disease from drinking 12 ounces of beer per day (contains ethyl alcohol)
12,000	Disease from smoking one pack of cigarettes per day

*These are hypothetical worst case risks, not "real" risks.
**Lower numbers indicate less risk. As an example of how to interpret the information in this chart, consider drinking water versus beer. The risk of disease from drinking 12 ounces of beer per day is 2800 times greater than the risk from drinking 1 quart of city water per day.

annual risk from drinking water with EPA-accepted levels of trichloroethylene is 2×10^{-9}. This means that this person is *180 million times* more likely to get cancer from smoking than to get it from drinking such low levels of trichloroethylene. Knowing this, the person in our example would, we hope, be persuaded to stop smoking.

One of the most perplexing dilemmas of risk assessment is that people often ignore substantial risks but get extremely upset about minor risks. The average life expectancy of smokers is more than eight years shorter than that of nonsmokers; almost one-third of all smokers die from diseases caused or exacerbated by their habit. Yet many people get much more upset over a one-in-a-million chance of getting cancer from pesticide residues on food than they do over the relationship between smoking and cancer. Perhaps part of the reason for this attitude is that behaviors such as diet, smoking, and exercise are parts of our lives that we can control *if we choose to.* Risks over which most of us have no control, such as pesticide residues, tend to evoke more fearful responses.

A Balanced Perspective on Risks

Threats to our health, particularly from toxic chemicals in the environment, make big news. Many of these stories are more sensational than factual. If they were completely accurate, people would be dying left and right, whereas, in fact, human health is better today than at any time in our history, and our life expectancy continues to increase rather than decline.

This does not mean that we should ignore chemicals that humans introduce into the environment. Nor does it mean we should discount the stories that are sometimes sensationalized by the news media. These stories serve an important role in getting the regulatory wheels of the government moving to protect us as much as possible from the dangers of our technological and industrialized world.

People should not expect no-risk foods, no-risk water, or no-risk anything else. Risk is inherent in all our actions and in everything in our environment. We do, however, have the right to expect the risks to be minimized. We should not ignore small risks just because larger ones exist. However, it is extremely important that we have an adequate understanding of the nature and size of risks before deciding what actions are appropriate to avoid them.

Cost-Benefit Analysis of Risks

Before the benefits of scientific risk assessment were understood or widely appreciated, politicians and government agencies tended to respond to the environmental issues that received the most publicity. As data on actual risks became available, however, it was discovered that some highly publicized environmental problems are astronomically expensive to correct and at the same time do not pose as much of a threat as many of the less-publicized problems. As a result, decision makers have increasingly adopted an approach known as **cost-benefit analysis** to address environmental problems, particularly those that involve human health and safety. In cost-benefit analysis, estimated cost is compared with potential benefits to determine how much of a particular toxic chemical or pollutant society can tolerate.

Cost-benefit analysis is an important mechanism to help decision makers formulate and evaluate environmental legislation, but cost-benefit analysis is only as good as the data and assumptions on which it is based. Corporate estimates of the cost to control pollution are often many times higher than the actual cost turns out to be. For example, during the debate over phasing out leaded gasoline in 1971, the oil industry predicted that the cost during the transition would be $7 billion per year, but the actual cost was less than $500 million per year.

Despite the wide range that often occurs between projected and actual costs, the cost portion of cost-benefit analysis is often easier to determine than are the health and environmental benefits. The cost of installing air pollution control devices at factories is relatively easy to estimate, for example, but how does one put a price tag on the benefits of a reduction in air pollution? What is the value of reducing respiratory problems in children and the elderly, two groups that are very susceptible to air pollution? How much is clean air worth?

Another problem with cost-benefit analysis is that the risk assessments on which they are based are far from perfect. Scientists admit that even the best risk assessments are based on assumptions that, if changed, could substantially alter the estimated risk. Risk assessment by its very nature is an uncertain science (see Focus On: Determining Health Effects of Environmental Pollutants).

To summarize, cost-benefit analyses and risk assessments are useful in evaluating and solving environmental problems, but decision makers must recognize their limitations when developing new government regulations.

ECOLOGICAL RISK ASSESSMENT

Doing a risk assessment as it relates to human health is relatively easy compared to doing one for the environment. Yet the Environmental Protection Agency (EPA) and other regulatory and environmental monitoring groups are increasingly trying to evaluate *ecosystem* health. While there is no formal method for **ecological risk assessments,** the EPA is in the process of establishing guidelines for estimating the probable effects of a wide range of human activities on ecosystems. Given the hazards and exposure levels of human-induced chemical and nonchemical stressors, ecological consequences can range from good to bad, or from acceptable to unacceptable.

An ecological risk assessment may become a required framework for organizing research and methods prior to introducing a new product, for determining cleanup options for a Superfund site, or for predicting the effects of a variety of other human activities. Since ecological risk assessment is a new and evolving field, the EPA guidelines will provide some flexibility to accommodate new developments.

The proposed EPA guidelines include three phases: problem formulation, analysis, and risk characterization. Problem formulation defines the questions to be answered, identifies goals of the ecological risk assessment, and establishes the analysis plan. During the analysis phase, exposures to stressors (human-induced changes that tax the environment) are evaluated. At the risk characterization phase, the risk is estimated by carefully examining the "stressor-response profile," that is, by relating the environmental effects of the stressor to its concentration in the environment. A report is prepared that summarizes the ecological risk, along with its significance. After discussion between the risk assessor and the risk manager, the ecological risk assessment is used to manage ecological risks.

FOCUS ON

Determining Health Effects of Environmental Pollutants

The current role of the federal government in assessing human health risks caused by hazardous substances in the environment is spelled out in such legislation as the Clean Air Act and the Toxic Substances Control Act. This box discusses how the scientific community assesses the risk of human exposure to hazardous substances. The federal government uses these results to make difficult decisions: if, when, and how human exposures to potentially hazardous substances should be regulated, for example, in the workplace.

Hazardous Substances The science of poisons is known as **toxicology**. It encompasses the effects of chemicals on living organisms as well as ways to counteract their toxicity. Most chemicals classified as toxins are not restricted to the air or the soil or water, but are found in many places in the environment.

Often the only disease evaluated in chemical risk assessment is cancer, but environmental contaminants are linked to several other serious diseases. Certain chemicals cause birth defects, damage the immune response, cause reproductive problems (recall endocrine disrupters, discussed in Chapter 1), or attack the nervous system or other body systems. Cancer, then, is not the only disease that is caused or aggravated by toxins.

Identifying Carcinogenic Substances The most common method of determining whether a chemical causes cancer is to expose laboratory animals such as rats to extremely large doses of that chemical and see whether they develop cancer. This method is indirect and uncertain, however. For one thing, although humans and rats are both mammals, they are different organisms and may respond differently to exposure to the same chemical. Another problem is that the rats are exposed to massive doses of the suspected carcinogen relative to their body size, whereas humans are usually exposed to much lower amounts. (Large doses are needed to elicit a positive response in a small group of test animals.) It is assumed that one can work backwards from the huge doses of chemicals and the high rates of cancer they cause in rats to determine the rates of cancer that might be expected in humans exposed to lower amounts of the same chemicals (see figure). However, there is little evidence to indicate that extrapolating backwards is scientifically sound. Even if you are reasonably sure that exposure to high doses of a chemical causes the same effects for the same reasons in both rats and humans, you cannot assume that these same mechanisms work at low doses in humans. In short, extrapolating from one species to another and from high doses to low doses is fraught with uncertainty.

Although scientists do not currently have a reliable way to determine if exposure to small amounts of a substance causes cancer in humans, the EPA is planning to change how toxic chemicals are evaluated and regulated. Methods are being developed that will give direct evidence of the risk involved in exposure to low doses of chemicals that cause cancer. Once implemented, these methods should be more accurate in assessing risk.

Epidemiological evidence, including studies of human groups who have been accidentally exposed to high levels of suspected carcinogens, is also used to determine whether chemicals are carcinogenic. For example, in 1989 epidemiologists in Germany established a direct link between cancer and a group of chemicals called dioxins. They observed the incidence of cancer in workers exposed to high concentrations of dioxins during an accident at a chemical plant in 1953, and found unexpectedly high levels of cancers of both the digestive and respiratory tracts.

Risk Assessment of Chemical Mixtures Humans are frequently exposed to various combinations of chemical compounds. Such chemical mixtures are present in the air we breathe, the food we eat, and the water we drink. Cigarette smoke contains a mixture of chemicals, for example, as does automobile exhaust. However, the vast majority of toxicology studies have been performed on single chemicals rather than chemical mixtures, and for

A CASE HISTORY: THE RESCUE OF LAKE WASHINGTON

Just as generals study old battles in order to learn how battlefield decisions are made, we can analyze how environmental problems are solved by studying an environmental battle that was successfully waged in the 1950s. The battle was fought over the pollution of Lake Washington, a large (86-square-kilometer or 33.2-square-mile), deep freshwater lake that forms the eastern boundary of the city of Seattle (Figure 2–2). During the first part of the 20th century, the Seattle metropolitan area expanded eastward toward the lake from the shores of Puget Sound, an inlet of the Pacific Ocean. As this expansion occurred, Lake Washington came under increasingly intense environmental pressures. Recreational use of the lake expanded greatly, and so did its use for waste disposal. Sewage arrangements in particular had a major impact on the lake.

good reason. Mixtures of chemicals can interact in a variety of ways, increasing the level of complexity to risk assessment, a field already complicated by many uncertainties. Moreover, toxicologists point out that there are simply too many chemical mixtures to evaluate all of them.

Chemical mixtures can interact by additivity, synergy, or antagonism. When a chemical mixture is *additive,* the effect is exactly what one would expect given the individual effects of each component of the mixture; for example, if a chemical with a toxicity level of one is mixed with a different chemical, also with a toxicity level of one, the combined effect of exposure to the mixture is two. Recall from Chapter 1 that a chemical mixture that is *synergistic* has a greater combined effect than would be expected; for example, two chemicals, each with a toxicity level of one, might have a combined toxicity of three. An *antagonistic* interaction in a chemical mixture results in a smaller combined effect than would be expected; for example, the combined effect of two chemicals, each with toxicity levels of one, might be 1.3.

If toxicological studies of chemical mixtures are lacking, how is risk assessment for chemical mixtures assigned? Risks are assigned to mixtures by additivity—that is, by adding the known effects of each compound in the mixture. Such an approach sometimes overestimates or underestimates the actual risk involved, but it represents the best approach currently available. The alternative—that of waiting for years or decades until numerous studies have been designed, funded, and completed—is unacceptable.

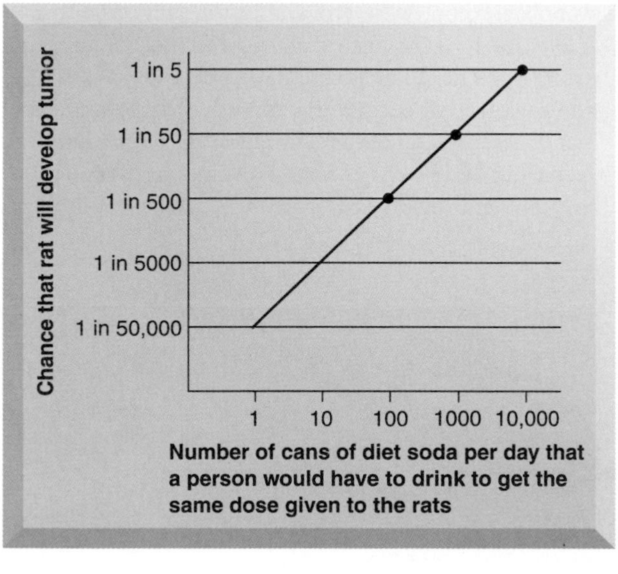

Chance that rat will develop tumor
1 in 5
1 in 50
1 in 500
1 in 5000
1 in 50,000
1
10
100
1000
10,000
Number of cans of diet soda per day that a person would have to drink to get the same dose given to the rats

Attempting to extrapolate backward from data on the incidence of cancer in rats exposed to high levels of a substance to predict the incidence of cancer in humans exposed to trace amounts of that substance may not be scientifically sound.

Birth of an Environmental Problem

Seattle began discharging raw sewage into the waters of Lake Washington at the beginning of the 20th century, as the city first began to expand eastward from Puget Sound (Figure 2–3). By 1926 the lake shore was sufficiently unpleasant that the city of Seattle passed a bond issue for a system of sewer lines to divert the city's sewage from Lake Washington to a treatment plant that discharged directly into Puget Sound. By 1936, the sewage discharge into Lake Washington from Seattle had stopped, and the lake again became a pleasant place where people came to boat, swim, and fish.

Like many cities in the United States, Seattle is ringed by suburbs with individual municipal governments. These suburbs expanded rapidly in the 1940s, generating an enormous waste disposal problem. Between 1941 and 1954, ten suburban sewage treatment plants began operating at points around the lake, with a combined daily discharge of 75.7 million liters (20 million gallons) into

Figure 2–2
Lake Washington is a large freshwater lake near Seattle, Washington. (Mark E. Gibson)

Figure 2–3
Sewage discharge. (Doug Wechsler)

Lake Washington. Each plant treated the raw sewage to break down the organic material within it and release the effluent—that is, treated sewage—into the lake.

By the mid-1950s, although raw sewage dumping had ended, a great deal of treated sewage had been poured into the lake. Try multiplying 75.7 million liters per day by 365 days per year by 5 to 10 years: enough effluent was dumped to give about 25 to 50 liters (6.5 to 13 gallons) of it to every man, woman, and child living today.

The effects of this discharge on the lake were first noted by G. Comita and G. Anderson, doctoral students at the University of Washington in Seattle. Their studies of the lake's microscopic single-celled organisms in 1950 indicated that large masses of *Oscillatoria*, a filamentous cyanobacterium, were growing in the lake (Figure 2–4). These are long strings of photosynthetic bacterial cells strung together. Their abundance in Lake Washington was unexpected, because the growth of large numbers of cyanobacteria requires a plentiful supply of nutrients, and deepwater lakes such as Lake Washington do not usually have enough dissolved nutrients to support cyanobacte-

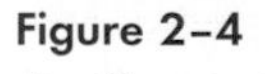

Figure 2–4
***Oscillatoria* is a photosynthetic cyanobacterium whose filaments are composed of chains of cells.** (J. R. Waaland/Biological Photo Service)

rial growth.[1] Deepwater lakes are particularly poor in the essential nutrient phosphorus. The amount of filamentous cyanobacteria in Lake Washington's waters hinted that the lake was somehow changing, becoming richer in dissolved nutrients such as phosphorus.

Sounding the Alarm

One of the first public alarms was sounded on July 11, 1955, in a technical report by the Washington Pollution Control Commission. Its author, citing the work of Comita and Anderson, concluded that the treated sewage effluent that was being released into the lake's waters was raising the lake's levels of dissolved nutrients to the point of serious pollution. Whereas primary treatment (see Chapter 21), followed by chlorination of the sewage, was ridding it of bacteria, it was not eliminating many chemicals, particularly phosphorus, a major component of detergents. In essence, the treated sewage was fertilizing the lake by enriching it with dissolved nutrients.

The process of nutrient enrichment of freshwater lakes is well understood by ecologists, who call it **eutrophication** (see Chapter 21). Eutrophication is undesirable because, as Comita and Anderson had already begun to observe, high nutrient levels contribute to the growth of filamentous cyanobacteria. These photosynthetic organisms need only three things in order to grow: light for photosynthesis, which they get from the sun; carbon atoms, which they get from carbon dioxide dissolved in water; and nutrients such as nitrogen and phosphorus, which were provided by the treated sewage. Without the nutrients, cyanobacteria cannot grow; supply them, and soon mats of cyanobacteria form a green scum over the surface of the water, and the water begins to stink from the odor of rotting organic matter (Figure 2–5).

Then the serious problem begins: the deep-water bacteria that decompose the masses of dead cyanobacteria multiply explosively, consuming vast quantities of oxygen in the process, until the lake's deeper waters become so depleted that they can no longer support other organisms that require oxygen to live. Fishes can no longer extract enough oxygen through their gills, and neither can the myriad of tiny invertebrates that populate freshwater lakes.

The local newspaper, the *Seattle Times,* mentioned the Pollution Control Commission's technical report in a July 11, 1955, article, "Lake's Play Use Periled by Pollution." The article did not grab the public's attention, but a month later, something else did: the annual Gold Cup

Figure 2–5
A eutrophic pond in the Santa Monica Mountains National Recreation Area is covered with the blue-green scum of filamentous cyanobacteria. (John D. Cunningham/Visuals Unlimited)

yacht races, when spectators saw magnificent sailboats slicing cleanly through green scum and smelled the odor of rotting cyanobacteria. These detractions to what had been a popular summer holiday raised protest among spectators and lakeshore residents.

Local authorities discounted the possibility that the cyanobacteria were the result of treated sewage discharged into the lake, blaming them instead on the unusually sunny weather. But on the very day of the yacht race, G. Anderson collected a water sample from the lake that was to forever banish such sunny explanations. The sample contained a filamentous cyanobacterium that neither Anderson nor earlier investigators had ever encountered in large numbers in the lake: *Oscillatoria rubescens.* The presence of this cyanobacterium proved to be a vital clue. When Anderson's professor at the University of Washington, W. T. Edmondson, reviewed the literature on eutrophication, he came across the name *Oscillatoria rubescens* again and again in the lists of organisms found in polluted lakes.

To Edmondson, the abundance of *Oscillatoria* in Lake Washington was a clear warning. On October 13, 1955, the University of Washington *Daily* ran a story, "Edmondson Announces Pollution May Ruin Lake," in which Edmondson announced the appearance of large masses of *Oscillatoria* and its likely meaning. From this point on, the scientific case was clear: the eutrophication of Lake Washington was demonstrably at an advanced stage, and unless it was reversed, it would soon destroy the water quality of the lake.

Scientific Assessment

The purpose of scientific assessment of an environmental problem is, first, to identify that a problem exists and,

[1] *Low* levels of nutrients are desirable in freshwater lakes because they permit the controlled growth of photosynthetic organisms that are the base of the food web. When a body of water contains a *high* level of nutrients such as phosphorus, the photosynthetic organisms are present in vast numbers, upsetting the natural balance in the lake.

second, to build a sound set of observations on which to base a solution. Lake Washington's microscopic life had been the subject of a detailed study in 1933. Thus, when the telltale signs of pollution first appeared in 1950, they were quickly detected by Edmondson's students as changes from the previous study. Without the earlier study's careful analyses of the many microorganisms living in the lake, understanding of the changes that were occurring would have been delayed.

Edmondson examined and compared the earlier study of the lake and confirmed that there had indeed been a great increase in dissolved nutrients in the lake's water. Surmising that the added nutrients were the result of sewage treatment waste discharge into the lake by suburban communities, Edmondson formed the hypothesis that treated sewage was introducing so many nutrients into the lake that its waters were beginning to support the growth of photosynthetic cyanobacteria.

Edmondson's hypothesis made a clear prediction: the continued addition of phosphates and other nutrients to the lake would change its surface into a stinking mat of rotting cyanobacteria, unfit for swimming or drinking, and the beauty of the lake would be only a memory. Bolstering his prediction was the fact that lakes near other cities, such as Madison, Wisconsin, had deteriorated after receiving discharges of treated sewage.

The appearance of large numbers of *Oscillatoria* in 1955 confirmed Edmondson's prediction: pollution was progressing in a classic pattern, its seriousness signaled by this almost-universal indicator of future trouble.

Making a model

Edmondson constructed a graphical model of the lake, which predicted that the decline could be reversed: if the pollution was stopped, the lake would clean itself at a predictable rate, reverting to its previous, unpolluted state within five years.[2] Could anything be done to reverse the process? In April 1956, Edmondson outlined three steps that would be necessary in any serious attempt to save the lake: (1) comprehensive regional planning by the many suburbs that ringed the lake, (2) complete elimination of sewage discharge into the lake, and (3) research to identify the key nutrients that were causing the cyanobacteria to grow. His proposals received widespread publicity in the Seattle area, and the stage was set to bring scientists and civic leaders together.

Risk Analysis

It is one thing to suggest that the addition of treated sewage to Lake Washington stop, and quite another to devise an acceptable alternative. Further treatment of sewage can remove some nutrients, but it is not practical to remove all of them. The alternative is to dump the sewage somewhere else—but where? In this case, officials decided to discharge the treated sewage into Puget Sound. In their plan, a ring of sewers to be built around the lake would collect sewage treatment discharges, treat them further, and then transport them to be discharged at great depth into Puget Sound.

Why go to all the trouble and expense of treating the discharges further, if you are just going to dump them? And why bother discharging them deep under water? Because it is important that the solution to one problem not produce another. The plans to further treat the discharge and release it at great depth were formulated in an attempt to minimize the environmental impact of diverting Lake Washington's discharge into Puget Sound. It was assumed that sewage effluent would have less of an impact on the great quantity of water in Puget Sound than on the much smaller amount of water in Lake Washington. Also, nutrient chemistry in marine water is different from that in fresh water. Puget Sound is naturally rich in nutrients, and phosphate does not control cyanobacterial growth there as it does in Lake Washington. The growth of photosynthetic bacteria and algae in Puget Sound is largely limited by tides, which mix the water and transport the tiny organisms into deeper water where they cannot get enough light to grow rapidly.

Practically any course of action that can be taken to reverse an environmental problem has its own impacts on the environment, which must be assessed when evaluating potential solutions (see Focus On: Environmental Impact Statements). Environmental impact analyses often involve studies by geographers, chemists, and engineers as well as ecologists and other biologists. Furthermore, the decision whether to implement a plan to restore or protect the environment is almost always affected by nonscientific factors and concerns. Any proposal is inevitably and rightly constrained by existing laws and by the citizens who will be affected by the decision.

Public Education

Despite the technical bulletin published by the Washington Pollution Control Commission in 1955, local sanitation authorities were not convinced that urgent action was necessary. Public action required further education, and it was at this stage that scientists played a key role. Edmondson and other scientists wrote articles for the general public that contained concise explanations of what nutrient enrichment is and what problems it causes. As these articles were picked up by the local newspapers, the general public's awareness of the problem increased.

In December 1956, Edmondson wrote a letter in an effort to alert a committee established by the mayor of Seattle to examine regional problems affecting Seattle

[2]In freshwater lakes, iron reacts with phosphorus to form an insoluble complex that sinks to the bottom of the lake and is buried in the sediments. Thus, if additional phosphorus were not introduced into the lake from sewage effluent, the lake would slowly recover.

FOCUS ON

Environmental Impact Statements

The **National Environmental Policy Act** (NEPA), passed by Congress in 1969, is the cornerstone of U.S. environmental policy. It requires that the federal government consider the environmental impact of a proposed federal action, such as financing highway or dam construction, when making decisions about that action. The NEPA provides the basis for developing detailed **environmental impact statements** (EISs) to accompany every federal recommendation or proposal for legislation. These EISs are supposed to help federal officials and the public make informed decisions. Each EIS must include the following:

1. The nature of the proposal and why it is needed.
2. The environmental impact of the proposal, including short-term and long-term effects and any adverse environmental effects.
3. Alternatives to the proposed course of action, including a no-action alternative, that will lessen the adverse effects.

The draft of an EIS must be available for public scrutiny and review by other federal agencies for at least 45 days before the final EIS is prepared. Individuals, citizen groups, or environmental groups can send letters commenting on the proposal, and these letters must be evaluated when preparing the final EIS.

The final decision on the proposal cannot be made until at least 30 days after the final EIS is available or 90 days after the draft EIS is available, whichever is later. After the final decision is made, any challenges to the decision are made in court.

As a result of the NEPA, individuals, citizen groups, and environmental groups have filed hundreds of lawsuits against agencies of the federal government. These suits have forced the federal government and individuals doing business with the federal government to focus on the environmental impacts of their projects.

The NEPA has influenced environmental legislation in many states and other countries. Thirty-six states have passed similar legislation requiring EISs for state-funded projects. Australia, Canada, France, New Zealand, and Sweden are some of the countries that now require EISs for government-sponsored projects.

Although almost everyone agrees that the NEPA has been successful in helping federal agencies reduce adverse environmental impacts of their activities and projects, it is not without its critics. Environmentalists complain that EIS reports are sometimes incomplete or that reports are ignored when decisions are made. Other critics think the EIS reports delay important projects because the EISs are too involved, take too long to prepare, and are often the targets of lawsuits.

and its suburbs. Edmondson explained that even well treated sewage would soon destroy the lake, and that Lake Washington was already showing signs of deterioration. He received an encouraging response and prepared for the committee a nine-page report, in nontechnical language, of his scientific findings. After presenting his data showing that the mass of cyanobacteria varied in strict proportion to the amounts of nutrients being added to the lake, Edmondson posed a series of questions: "How has Lake Washington changed?" "What will happen if nothing is done to halt nutrient accumulation?" "Why not poison the cyanobacteria and then continue to discharge the effluent?" He then answered the questions, outlined two alternative courses of public action—do nothing, or stop adding nutrients to the lake—and made a clear prediction about the consequences of each.

Political Action

Edmondson's report was widely circulated among local governments, but implementing its proposals presented serious political problems because there was no governmental mechanism that would permit the many local suburbs to act together on regional matters such as sewage disposal. In late 1957 the state legislature passed a bill permitting a public referendum in the Seattle area on the formation of a regional government with six functions: water supply, sewage disposal, garbage disposal, transportation, parks, and planning. The referendum was defeated in March 1958, apparently because suburban voters felt that the plan was an attempt to tax them for the city's expenses.

Understanding the urgency of Edmondson's proposals, an advisory committee immediately submitted to the voters a revised bill limited to sewage disposal. Over the summer there was widespread discussion of the lake's future, and when the votes were counted on September 9, 1958, the revised bill had passed by a wide margin.

At the time it was passed, the Lake Washington plan was the most ambitious and most expensive pollution control project in the United States. Every household in the area had to pay $2 a month in additional taxes for construction of a massive trunk sewer to ring the lake, collect all the effluent, treat it, and discharge it into Puget Sound.

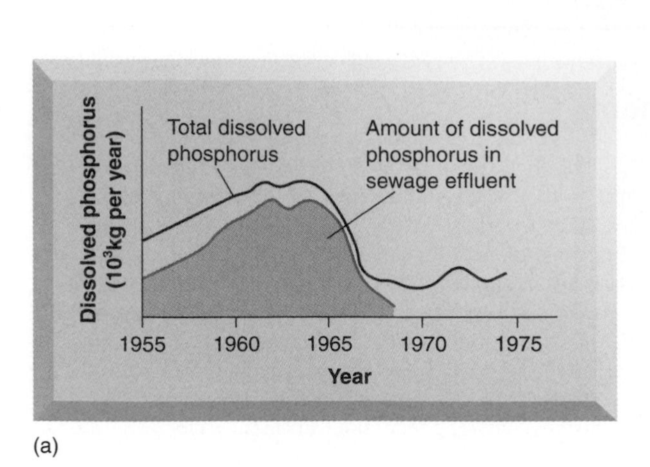

(a)

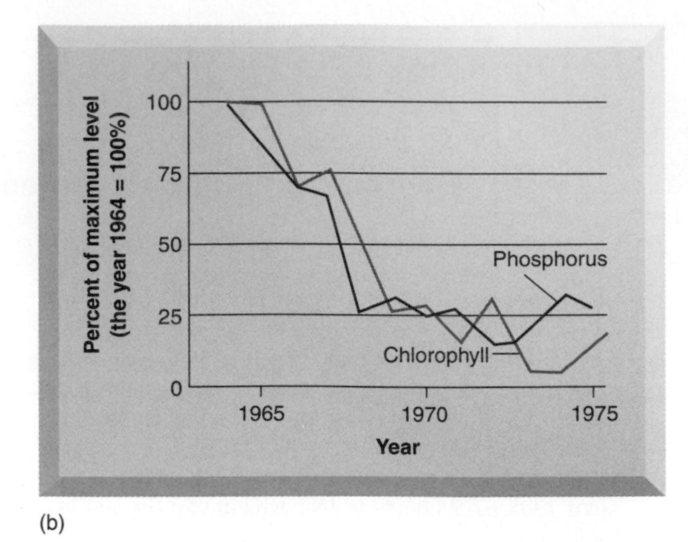

(b)

Figure 2–6

Relationship between nutrient additions to Lake Washington from sewage treatment plants and the growth of cyanobacteria. (a) Dissolved phosphorus in Lake Washington from 1955 to 1975. The amount of phosphorus contributed by sewage effluent is indicated in the shaded area. (b) Cyanobacterial growth during Lake Washington's recovery, 1965 to 1975. As the level of phosphorus dropped in the lake, the number of cyanobacteria also declined, as measured indirectly by the amount of chlorophyll.

Implementing the plan of action

Ground-breaking ceremonies for the new project were held in July 1961. As Edmondson had predicted, the lake had deteriorated further. Visibility in lake water declined from 4 meters (12.3 feet) in 1950 to less than 1 meter (3.1 feet) in 1962, the water being clouded with cyanobacteria. On October 5, 1963, a suburban newspaper dubbed Lake Washington "Lake Stinko." In 1963 the first of the waste treatment plants around the lake began to divert its effluent into the new trunk sewer. One by one, the others diverted theirs, until the last effluent was diverted in 1968. The lake's deterioration stopped by 1964, and then its condition began to improve (Figure 2–6).

Follow-Through

By carefully analyzing what was happening in the lake, Edmondson could predict that the lake would recover fully. Not all environmental scientists agreed with him, many arguing that dissolved phosphorus, the key nutrient regulating cyanobacterial growth, would not dissipate for decades, if ever. A lot depended on assumptions about the chemical makeup of the sediment at the bottom of the lake. Edmondson's hypothesis was correct. Water transparency returned to normal within a few years (Figure 2–7). *Oscillatoria* persisted until 1970, but eventually it disappeared. By 1975 the lake was back to normal.

Every environmental intervention is an experiment, and continued monitoring is necessary because environ-

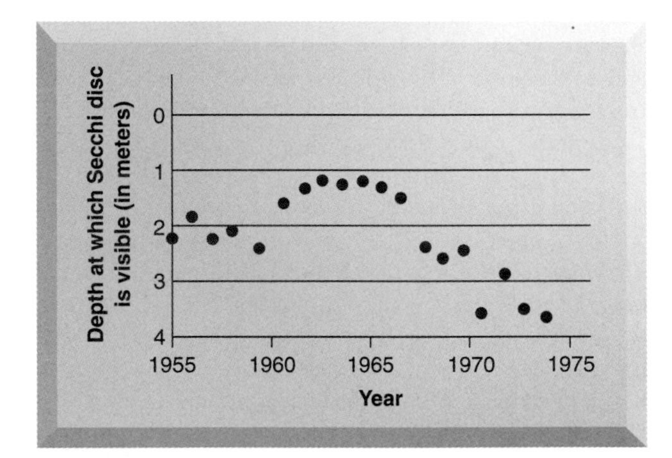

Figure 2–7

Effect of pollution and recovery of water transparency in Lake Washington. Measurements were taken in July or August from 1955 to 1975 using a round disc called a Secchi disc attached to a rope. The greater the water transparency, the deeper the Secchi disc can be lowered and still be visible. If the Secchi disc disappears at one meter, it means the water is very cloudy from the growth of cyanobacteria. If the Secchi disc can be lowered over three meters before disappearing from sight, it means there are few cyanobacteria present.

mental scientists work with imperfect tools. There is a great deal we do not know, and every added bit of information increases our ability to deal with future problems. The unanticipated always lurks just beneath the surface of any experiment carried out in nature.

It was not anticipated, for example, that water transparency would continue to increase. By 1980, the lake was clearer than at any time in recent memory, with visibility exceeding 12 meters (39.4 feet) at times. Today, more than 15 years later, the lake remains clear. Before the recovery, the presence of filamentous cyanobacteria such as *Oscillatoria* had limited the population size of a microscopic organism called *Daphnia* because cyanobacterial filaments clog *Daphnia*'s feeding apparatus. The disappearance of *Oscillatoria* and other filamentous cyanobacteria allowed the lake's *Daphnia* population to flourish and become dominant among the many kinds of invertebrates that live there. Because *Daphnia* are very efficient eaters of nonfilamentous algae, levels of these algae fell, too, so that the water became even clearer.

Another unanticipated change observed since 1988 is that the lake water has become increasingly alkaline. The cause of this chemical change in the lake's water is unknown at present, but it has been suggested that the development of land around the east side of the lake is involved. The kinds of chemicals draining into a lake from a natural drainage basin are quite different from those draining from storm drains, and it has been hypothesized that this change is responsible for the lake's altered chemistry. Additional scientific studies should help clarify the situation in Lake Washington.

WORKING TOGETHER

The reversal of the pollution of Lake Washington is a particularly clear example of how environmental science can work to identify, address, and help solve environmental problems. Many environmental problems facing us today are far more complex than Lake Washington, however, and public attitudes are different. Lake Washington's pollution problem was solved only because the many small towns involved in the problem cooperated in seeking a solution.

In the 1990s, confrontation over an environmental problem frequently makes it difficult to reach an agreement. Even scientists disagree among themselves and call for additional research to help them arrive at a consensus. In such an atmosphere, politicians often compromise by adopting a "wait-and-see" approach.

Such delays are really a form of negative action because the consequences of many environmental problems are so serious that they must be acted on before a scientific consensus is reached. The need for additional scientific studies should not prevent us from taking action on such serious regional and global issues as stratospheric ozone depletion and global climate change (both discussed in Chapter 1), pollution in the Mediterranean Sea, and acid rain. We need to recognize the uncertainty inherent in environmental problems; consider a variety of possible approaches; weigh the cost, benefits, and probable outcomes of each; and set in motion a policy that is flexible enough to allow us to modify it as additional information becomes available.

In the final analysis, then, environmental scientists identify a problem and often suggest a solution, but implementation depends on a political decision that is influenced by social and economic agendas as well as scientific evidence. Unfortunately, humans have not yet learned to cooperate well at any level. We desperately need to develop mechanisms, including laws, to induce us to do so (see Focus On: The Tragedy of the Commons).

ENVIRONMENTAL LITERACY

Because responses to environmental problems depend on the public's awareness and understanding of the issues and the underlying scientific concepts involved, environmental education is critical to appropriate decision-making. The emphasis on environmental education has grown dramatically over the years:

- As of 1994, 99 percent of primary and secondary schools in the United States had environmental education programs.
- Three international treaties supporting environmental education came into force between 1975 and 1990.
- Project GLOBE (Global Learning and Observations to Benefit the Environment), established in 1994, encourages U.S. students to conduct ecosystem studies.
- In 1990, 22 university presidents from 13 nations issued a declaration of their commitment to environmental education and research at their institutions; 157 university presidents from 38 countries have since followed suit.

Unfortunately, most strides in environmental education have taken place in developed nations. Developing nations lag behind because of economic constraints and the stresses caused by rapid population growth.

FOCUS ON

The Tragedy of the Commons

Garrett Hardin is a professor of human ecology who writes about human environmental dilemmas. In 1968 he published a short essay, "The Tragedy of the Commons," in which he contended that our inability to solve complex environmental problems is the result of a struggle between short-term individual welfare and long-term societal welfare.

Hardin used the commons to illustrate this struggle. In medieval Europe, the inhabitants of a village shared pasture land, called the commons, and each herder could bring animals onto the commons to graze. The more animals a herder brought onto the commons, the greater the advantage to that individual. When every herder in the village brought as many animals onto the commons as possible, however, the plants were killed from overgrazing, and the entire village suffered (see figure). One of the outcomes of the eventual destruction of the commons was private ownership of land, because when each individual owned a parcel of land, it was in that individual's best interest to protect the land from overgrazing.

Hardin's parable has relevance today. The "commons" are those parts of our environment that are available to everyone but for which no single individual has responsibility—the atmosphere and the ocean, for example. These modern-day commons, sometimes collectively called the **global commons**, are experiencing increasing environmental stress. Because they are owned by no individual, jurisdiction, or country, they are susceptible to overuse. Although their exploitation may benefit only a few, everyone on Earth must pay for the environmental cost of exploitation.

Clearly, the world needs legal and economic policies to prevent the short-term degradation of our global commons and ensure the long-term well-being of our natural resources. There are no quick fixes, because solutions to global environmental problems are not as simple or short-term as solutions to some local problems are. Most environmental ills are inextricably linked to other persistent problems, such as poverty, overpopulation, and social injustice—problems that are beyond the ability of a single nation to resolve. Clearly, cooperation and international commitment are essential if we are to alleviate poverty, stabilize the human population, and preserve our environment for future generations.

The effects of overgrazing in Kenya. When too many animals graze on a given piece of land, the grasses die. Notice the piles of leaf-stripped brush. The farmers have cut branches off trees to feed their starving cattle. (C. Houston/Visuals Unlimited)

Environmental education will play an important role in promoting this cooperation. The more environmental education people have, the more they understand risks and the more capable they are of making good decisions. Environmental education teaches students how to think and encourages active participation in solving environmental problems. This sort of education will be particularly important in developing countries, which have relatively few of the world's scientists and engineers and therefore are less equipped to work out clear, scientifically based alternatives for themselves. With assistance from countries such as the United States, they may be able to avoid repeating some of the mistakes developed countries have made. International cooperation will be increasingly needed as environmental problems become more global in scope.

THE ROLE OF GOVERNMENT IN SOLVING TODAY'S ENVIRONMENTAL PROBLEMS

Lake Washington's environmental drama occurred several decades ago, before our extensive network of federal, state, and local environmental management agencies was in place (Figure 2–8). Today, for example, thousands of government officials interact with citizens, private consultants, environmental advocacy groups, and others in solving environmental problems. There is a wealth of federal, state, and local environmental legislation (see Table 7–1). Also, many environmental problems are discovered as part of established pollution control programs that conduct routine monitoring programs or investigate complaints from citizens.

ENVIROBRIEF

PAINTING THE WHITE HOUSE GREEN

With the help of a team of 100 energy experts, President Clinton has established the White House as a model for achieving energy efficiency. Electricity and water consumption at the White House complex could be reduced by as much as 30 percent by 2001 through such actions as the following:

- Installing chlorofluorocarbon-free refrigerators in the Clinton family quarters.
- Modifying the fountain on the White House lawn so that it reuses water.
- Installing reduced-flow faucets, showers, and toilets.
- Replacing incandescent bulbs in hallways with more efficient compact fluorescent bulbs.
- Installing double-paned windows.

Although such modifications can be costly, efforts are being made to control expenditures; the only measures taken at the White House are those that will eventually save both money and energy.

Figure 2–8

Carol Browner, Administrator of the Environmental Protection Agency (EPA), appointed by President Clinton. Established in 1970, the EPA implements and enforces U.S. environmental laws. (U.S. Environmental Protection Agency)

SUMMARY

1. Science is a systematic process to investigate the natural world. Science seeks to reduce the apparent complexity of our world to general principles that can be used to solve problems or provide new insights.

2. The scientific method encompasses five steps. (1) After making careful observations, a scientist recognizes and states the problem or unanswered question, then (2) develops a hypothesis to explain the problem. (3) An experiment is designed and performed to test the hypothesis. (4) The results obtained from the experiment are analyzed and interpreted to reach a conclusion, which is (5) shared with the scientific community.

3. A well designed experiment has two parts, a control and a variable (that part of the experiment that is identical to the control in all ways except one). Any difference in the outcome between the control and the variable must be the result of the variable.

4. Both inductive and deductive reasoning are used in the scientific method. Inductive reasoning begins with specific examples and seeks to draw a conclusion or discover a unifying rule or general principle on the basis of those examples. Inductive reasoning provides new knowledge but is error-prone. Deductive reasoning, which operates from generalities to specifics, adds nothing new to knowledge, but it can make relationships among data more apparent.

5. A hypothesis is an educated guess. Hypotheses are most useful when they make predictions that can be tested. A theory is a higher level of scientific interpretation—one that is well supported by scientific evidence. A principle is a theory that is almost universally accepted.

6. Damage to the environment can often be reversed, as demonstrated by the rescue of Seattle's polluted Lake Washington. The pouring of treated sewage into the lake had raised its level of nutrients to the point where the lake supported the growth of filamentous cyanobacteria. Disposal of the sewage in another way solved the lake's pollution problem.

7. A successful approach to solving an environmental problem usually involves five components: scientific assessment, risk analysis, public education, political action, and follow-through. Scientific assessment of environmental issues involves identifying potential environmental problems and suggesting possible solutions.

8. Risk analysis is the evaluation of the potential consequences of solutions to environmental problems. Some of these consequences may be new environmental problems; others may raise political or social issues.

9. An element of risk is inherent in everything we do. Risk assessment, the estimation of risks for comparative purposes, helps us set priorities and manage risks. Cost-benefit analysis, in which estimated cost is compared with potential benefits, is increasingly used to determine how much of a particular toxic chemical or pollutant society can tolerate.

10. Public education involves placing into the public arena the results of scientific assessment and risk analysis. The public is thus made aware of the problem and of the consequences of alternative actions.

11. Political action is the implementation of a particular plan of action by elected or appointed officials; ultimately, the decisions are made by the voting public. Follow-through is the assessment of the effects of an action that was taken in an attempt to correct a perceived environmental problem.

12. The National Environmental Policy Act (NEPA) requires that environmental impact statements (EISs) be prepared to aid decision-making about federally funded projects. EISs ensure that environmental values are given appropriate consideration in planning and decision-making.

THINKING ABOUT THE ENVIRONMENT

1. How are human-induced changes in the environment thought to be related to the emergence of formerly rare disease-causing agents?

2. Would the five steps used to solve environmental problems (scientific assessment, risk analysis, public education, political action, and follow-through) be effective

in containing an outbreak of a dangerous disease like Ebola? Why or why not?

3. When Sherlock Holmes amazed his friend Watson by determining the general habits of a stranger based on isolated observations, what kind of reasoning was he using? Explain.

4. The Seattle area was presented with two alternatives for dealing with the pollution of Lake Washington: (1) do nothing or (2) divert the sewage. What other alternatives might have been proposed? Why do you think they were not proposed?

5. The Lake Washington case is presented as a model because it clearly demonstrates the five components of solving any environmental problem. The final outcome in the Lake Washington case—dumping treated sewage into Puget Sound—is not an ideal solution, however. Explain why.

6. Name an environmental problem that appears to have received adequate scientific assessment and to have been extensively reported by the press, and yet remains unsolved. What do you think has made the difference between the fate of this issue and the fate of the pollution of Lake Washington?

7. The National Environmental Policy Act (NEPA) is sometimes called the "Magna Carta of environmental law." What is meant by such a comparison?

8. How is the collapse of the Georges Bank fishery, discussed in Chapter 1, an example of the tragedy of the commons? Name several additional examples of the global commons other than those mentioned in the chapter.

* 9. The annual death rate from sitting in a classroom with an asbestos ceiling is estimated at 0.05 people per million who are exposed to the risk. This risk is equivalent to 1 person per ___?___ million people.

RESEARCH PROJECTS

Read the classic paper by Garrett Hardin (listed in the Suggested Readings) and write a short essay that makes a connection between Hardin's commons and the Lake Washington case study.

Some environmentalists charge that requirements for risk assessment studies and cost-benefits analyses are being used to delay and weaken environmental legislation. Investigate a piece of environmental legislation up for consideration by Congress now and try to determine if "paralysis by analysis" is indeed occurring.

Choose a federal agency that has international projects, such as the U.S. Air Force, Navy, or State Department, and investigate whether the NEPA applies to their activities in foreign countries. Cite specific examples to support your findings.

On-line materials relating to this chapter are on the ► World Wide Web at **http://www.saunderscollege.com/lifesci/environment2** (select Chapter 2 from the Table of Contents).

SUGGESTED READING

Edmondson, W. T. *The Uses of Ecology: Lake Washington and Beyond.* Seattle: University of Washington Press, 1991. The most definitive piece on Lake Washington to date, including details of the public action.

Hardin, G. "The Tragedy of the Commons," *Science* Vol. 162, 1968. This classic paper has as much relevance for the natural and social sciences today as it did 30 years ago.

LeGuenno, B. "Emerging Viruses," *Scientific American,* October 1995. Examines why Ebola and other hemorrhagic fever viruses appear to be spreading.

Monbiot, G. "The Tragedy of the Enclosure," *Scientific American,* January 1994. An essay that argues that commonly owned land protects the environment more effectively than private ownership.

Morgan, M. G. "Risk Analysis and Management," *Scientific American*, July 1993. Techniques to analyze risks to human health, safety, and the environment can be used effectively to develop sound public policy.

Platt, A. E. *Infecting Ourselves: How Environmental and Social Disruptions Trigger Disease.* Worldwatch Paper 129, Worldwatch Institute, April 1996. The author builds a case for understanding environmental causes of infectious diseases as an important way to prevent their outbreak.

Ross, J. F. "Risk: Where Do Real Dangers Lie?" *Smithsonian*, November 1995. Risk analysis is critical in making public policies, but misunderstanding of risk by government officials, the media, and the public complicates the picture.

Wynne, B., and S. Mayer. "How Science Fails the Environment," *New Scientist* Vol. 138, No. 1876, 5 June 1993. Some of the limitations of science in evaluating environmental issues.

► A bull moose wades in a lake in Denali National Park, Alaska. *(Bruce H. Morrison/Phototake)*

PART

2 The World We Live In

Protecting the Ozone

F. Sherwood Rowland knows how to ask the right questions. When he and his colleague Mario Molina sounded the alarm about the effect of chlorofluorocarbons on the ozone layer, they had come to a conclusion that began with the question, "What does happen to all those CFCs?" Since they were invented in the late 1920s, chlorofluorocarbons (CFCs) were thought to be chemically stable, "miracle compounds" that could replace flammable and toxic chemicals then used throughout industry as refrigerants. Drs. Rowland and Molina hypothesized that these compounds were not quite so benign and did break down in the stratosphere, resulting in depletion of the ozone layer. After much subsequent experimentation and many more questions—by Dr. Rowland and other researchers—the scientific community was persuaded that CFCs do destroy atmospheric ozone. Since 1987 more than 150 countries have ratified the Montreal Protocol, a treaty to ban the use of CFCs. Dr. Rowland, along with Dr. Molina, and Paul Crutzen of the Max Planck Institute, won the Nobel Prize for Chemistry in 1995. Dr. Rowland is the Donald Bren Research Professor of Chemistry and Earth System Science at the University of California, Irvine. ◀

F. SHERWOOD ROWLAND

Chlorofluorocarbons were considered to be chemically stable compounds, and this perception, combined with their great utility, led to widespread CFC use throughout industry. You thought these compounds might react with something?

These chemicals *are* very stable, and this stability is the prime reason for their industrial utility. Nevertheless, we decided to explore their stability in the Earth's atmosphere. We started by asking whether any chemical reactions would occur in the lower atmosphere, and eliminated the possibilities one by one. We always knew that very short wavelength ultraviolet radiation can cause their decomposition, and that such UV radiation reached the atmosphere from the sun. But this UV radiation does not reach the lower atmosphere because it is absorbed by the ozone layer in the stratosphere. However, after an average time of many decades, the drifting CFC molecules get above most of the ozone, are exposed to some of this very energetic UV radiation, and are destroyed with the release of atoms of chlorine. We then found that these chlorine atoms would enter into a catalytic chain reaction that would seriously deplete the ozone layer itself.

What is the ozone layer? Why it is important to preserve it?

The two most abundant gases in the Earth's atmosphere are nitrogen and oxygen. Almost all of the oxygen atoms are found as the two-atom molecular oxygen, O_2. However, highly energetic solar UV radiation can break up O_2 into O atoms, each of which can react with another O_2 molecule to form the three-atom form of oxygen known as ozone, O_3. About 90 percent of the Earth's ozone is found far above us in the stratosphere, and serves as a shield protecting the surface from the most energetic solar UV. As far as we know, life did not evolve on land until the atmosphere had ozone in the stratosphere to shield out the harmful UV. Some of the ozone is also found near the surface where it displays a destructive side when it comes into direct contact with biological species—including humans. Surface-level ozone is an important toxic ingredient in urban smog.

This stratospheric ozone layer is vital to us and to all other species living on the sunlit Earth because it both establishes the temperature structure of the atmosphere and protects us against damage from solar ultraviolet radiation. Substantial reductions in its concentration could have a very negative effect on humans and the rest of the biosphere. This high energy radiation is biologically damaging because it is readily absorbed by DNA in our skin, inducing coding errors into the replication of DNA. The accumulation of such errors can cause skin cancer, eye cataracts, and weaken the immune system. Comparable effects can occur with other biological species, upsetting the fragile balance of an entire ecosystem.

Wasn't ultraviolet radiation already affecting the composition of the ozone layer?

Yes, ozone is a very reactive molecule. Two simultaneous processes are going on in the atmosphere. The energetic UV in sunlight is constantly ripping up O_2 and making more O_3, and natural processes for the O_3 quickly return it to the O_2 form. There are roughly one million O_2 molecules for every O_3 molecule in the atmosphere, with the amount of ozone remaining fairly constant over the years. However, the atomic chlorine from the CFCs provides an additional process for returning O_3 to O_2, with the result that there is less O_3 now than there was 20 or 30 years ago. This new process will continue until the CFCs eventually disappear from the atmosphere, but this will take more than 100 years.

You said that no processes destroyed CFCs in the lower atmosphere, so you reasoned they could be destroyed after rising to the stratosphere?

We found that the CFCs remained undisturbed in the lower atmosphere for decades. They are invulnerable to visible sunlight, nearly insoluble in water, and resistant to oxidation. CFCs display an impressive durability in the atmosphere's lower levels. But at altitudes above 18 miles, with 99 percent of all air molecules lying beneath them, CFCs show their vulnerability. At this height, high-energy ultraviolet radiation from the sun impinges directly on the CFC molecules, breaking them apart into chlorine atoms and residual fragments.

It was then that you asked the next question, "What happens to those constituent atoms released in the stratosphere?"

Yes. There was already much research on chemical kinetics—the study of how quickly molecules react with one another and how such reactions take place. From this basic research we determined that most of the chlorine atoms combine with ozone, the form of oxygen that protects Earth from ultraviolet radiation. When chlorine and ozone react, they form the free radical chlorine oxide, which in turn becomes part of a chain reaction. As a result of that chain reaction, a single chlorine atom can remove as many as 100,000 molecules of ozone.

This stratospheric ozone layer is vital to us and to all other species living on the sunlit Earth because it both establishes the temperature structure of the atmosphere and protects us against damage from solar ultraviolet radiation.

–F. Sherwood Rowland

This was controversial news and it took a long time to be accepted.

In 1974, when our work was first published, very little direct information existed about the CFCs or any other chlorine-containing compounds in the stratosphere. Natural sources of chlorine such as ocean salt and volcanoes also needed careful measurement. Hundreds of scientists began working on these problems—formulating and testing hypotheses, carrying out chemical experiments in the laboratory, looking at the stratosphere with instruments on airplanes, balloons, satellites. Then, when the Antarctic ozone hole was discovered in 1985, the scientific community had very little information about the chemical and meteorological conditions in the stratosphere above that continent. Several expeditions went to Antarctica in 1986 and 1987 to carry out direct measurements in the stratosphere. By the end of the 1980s, the major scientific questions had all been answered, and it was clear to the atmospheric scientific community that stratospheric ozone depletion was occurring, and that the cause of this loss could be traced to the CFCs and other human-made chlorine compounds.

Your experiences with disseminating information to the public about CFCs have increased your concern about scientific literacy.

From my own experience, I see that the most serious problems are related to faulty communication about science among the various segments of society—including the scientific segment itself. Each of us is bombarded daily by messages from television, radio, magazines, newspapers, and so on. We live in the midst of massive information flow, but those items connected with science itself are often badly garbled, sometimes with potentially serious negative consequences. The remedy must lie in greater emphasis by all of us in increasing both the base level of knowledge of science and communication about science with all levels of society. This is much easier to say than to do, because one does not sense an overwhelming general demand for more information about scientific matters. But more effort is required from all of us.

▶ Web site: **www.chem.uci.edu/research/faculty/rowland.html**

Tube worms living in hydrothermal vents on the deep-ocean floor.
(D. Foster, Science VU–WHOI/Visuals Unlimited)

Ecosystems and Energy

In 1977 an oceanographic expedition aboard the submersible research craft *Alvin* studied the Galapagos Rift, a deep cleft in the ocean floor off the coast of Ecuador. The expedition revealed, on the floor of the deep ocean, a series of hydrothermal vents where seawater apparently had penetrated and been heated by the hot rocks below. During its time within the Earth, the water had also been charged with mineral compounds, including hydrogen sulfide, H_2S.

At the tremendous depths (greater than 2500 meters) of the Galapagos Rift, there is no light for photosynthesis. But the hot springs support a rich and bizarre variety of organisms that contrasts with the surrounding "desert" of the deep-ocean floor. Most of the species found in these oases of life were new to science. For example, giant, blood-red tube worms almost 3 meters (10 feet) in length cluster in great numbers around the vents. Other animals around the hydrothermal vents include clams, crabs, barnacles, and mussels.

The mystery is, what is the ultimate source of energy for the species in this dark environment? Most deep-sea communities depend on the organic (carbon-containing) material that drifts down from surface waters; that is, they depend on energy derived from photosynthesis. But the Galapagos Rift community is too densely clustered and too productive to be dependent on chance encounters with organic material from surface waters.

The base of the food web in these aquatic oases consists of certain bacteria that can survive and multiply in water so hot (exceeding 200°C) that it would not even remain in liquid form were it not under such extreme pressure. These bacteria function as producers, but they do not photosynthesize. Instead, they possess enzymes—organic catalysts—that cause hydrogen sulfide to react with oxygen, producing water and sulfur or sulfate. Such chemical reactions provide the energy required to support these bacteria and other organisms in deep-ocean hydrothermal vents. Many of the Galapagos Rift animals consume the bacteria directly by filter feeding. Others, such as the giant tube worms, get their energy from bacteria that live inside their bodies.

In this chapter, we examine the importance of energy to Earth's ecosystems. An *ecosystem* encompasses all the interactions among the plants, animals, and microorganisms living together in a particular area and between those organisms and their physical environment. The interactions can be self-sustaining—that is, the ecosystem can continue to operate—only as long as energy is provided to the system. Virtually all of Earth's organisms depend on the Sun for energy; the organisms in deep-sea hydrothermal vents are an interesting and unique exception. ◀

THE HOUSE WE LIVE IN

The concept of ecology was first developed in the 19th century by Ernst Haeckel, who also devised its name—*eco* from the Greek word for "house" and *logy* from the Greek word for "study." Thus, ecology literally means the study of one's house. The environment—one's house—consists of two parts, the **biotic** (living) environment, which includes all organisms, and the **abiotic** (nonliving, or physical) environment, which includes such physical factors as temperature, sunlight, and precipitation. **Ecology**, then, is the study of the interactions among organisms, and between organisms and their abiotic environment.

What Ecologists Study

The focus of ecology can be local and very specific or global and quite generalized, depending on the scientist's view. One ecologist might determine the temperature or light requirements of a single species of oak, another might study all the organisms that live in a forest where the oak is found, and another might examine how matter flows between the forest and surrounding communities.

How does the field of ecology fit into the organization of the biological world? As you may know, one of the characteristics of life is its high degree of organization (Figure 3–1). Atoms are organized into molecules, which are organized into cells. In multicellular organisms, cells are organized into tissues, tissues into organs such as the brain and stomach, organs into body systems such as the nervous system and digestive system, and body systems into individual organisms such as dogs, ferns, and so on.

The levels of biological organization that interest ecologists are those above the level of the individual organism. Organisms occur in **populations**, members of the same species[1] that live together in the same area at the same time. A population ecologist might study a population of polar bears or a population of marsh grass. (Populations are discussed in Chapters 8 and 9.)

[1]A species is a group of similar organisms whose members freely interbreed with one another in the wild and do not interbreed with other sorts of organisms.

Populations are organized into communities. A **community** consists of all the populations of different species that live and interact together within an area. A community ecologist might study how organisms interact with one another—including who eats whom—in an alpine meadow community or in a coral reef community (Figure 3–2). *Ecosystem* is a more inclusive term than *community* because an **ecosystem** is a community together with its physical environment. Thus, an ecosystem includes not only all the biotic interactions of a community but also the interactions between organisms and their abiotic environment. An ecosystem ecologist might examine how temperature, light, precipitation, and soil characteristics affect the organisms living in a desert community or a coastal bay ecosystem.

All of Earth's communities of organisms are organized into the **biosphere**. The organisms of the biosphere depend on one another and on the other divisions of the Earth's physical environment: the atmosphere, hydrosphere, and lithosphere. The **atmosphere** is the gaseous envelope surrounding the Earth; the **hydrosphere** is the Earth's supply of water—liquid and frozen, fresh and salty; and the **lithosphere** is the soil and rock of Earth's crust. The **ecosphere** encompasses the biosphere and its interactions with the atmosphere, hydrosphere, and lithosphere. Ecologists who study the biosphere or ecosphere examine the complex global interrelationships among the Earth's atmosphere, land, water, and organisms.

Ecology is the broadest field within the biological sciences, and it is linked to every other biological discipline. The universality of ecology also brings subjects into view that are not traditionally part of biology. Geology and earth science are extremely important to ecology, especially when scientists examine the physical environment of planet Earth. Because humans are biological organisms, all of our activities have a bearing on ecology. Even economics and politics have profound ecological implications (see Chapter 7).

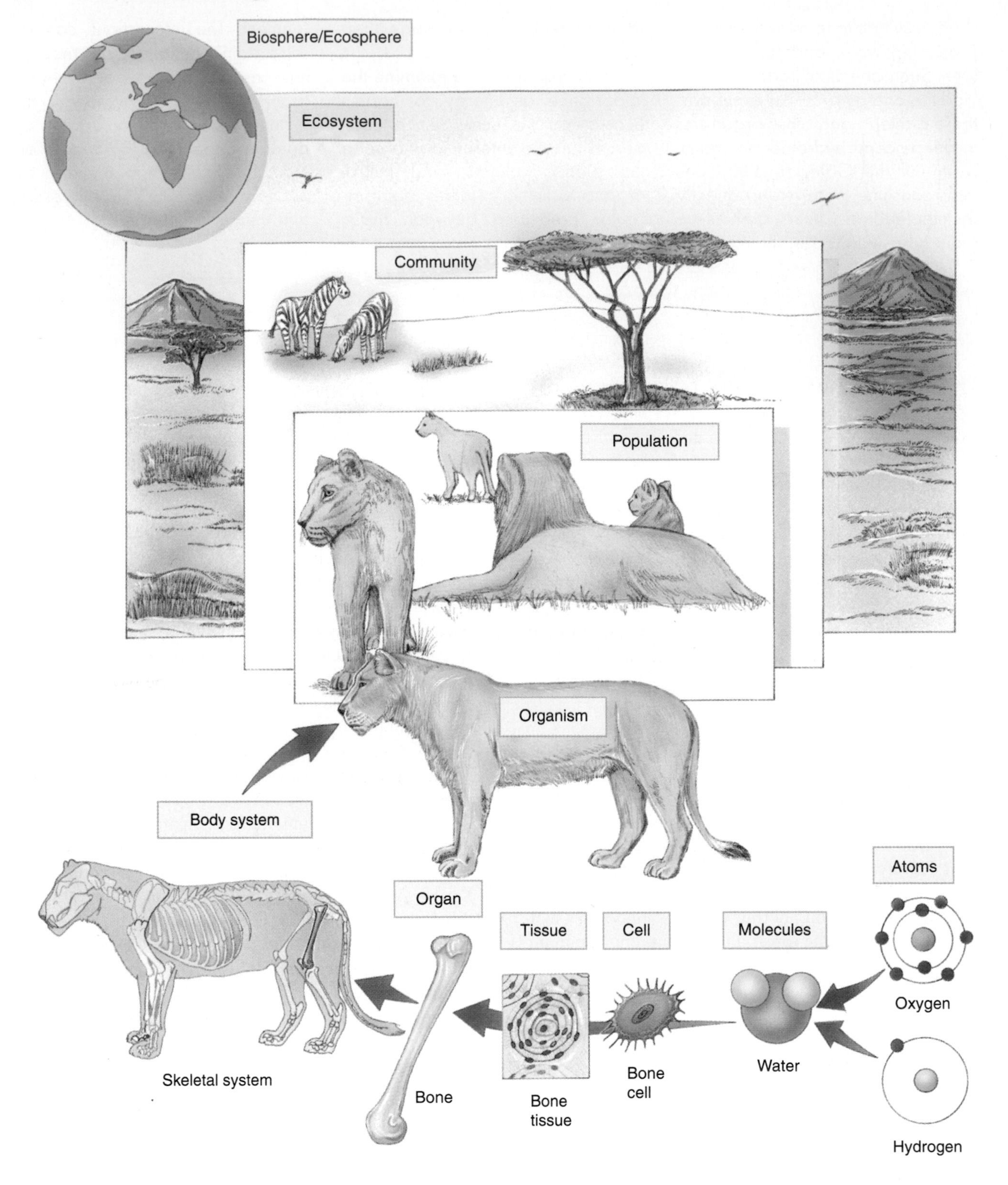

Figure 3–1

Levels of biological organization. Starting at the simplest level, atoms are organized into molecules, which are organized into cells. Cells are organized into tissues, tissues into organs, organs into body systems, and body systems into individual multicellular organisms. A group of individuals of the same species is a population. Populations of different species interact to form communities. A community and its abiotic environment is an ecosystem, while all communities of organisms on Earth comprise the biosphere. Ecologists study the highest levels of biological organization: populations, communities/ecosystems, and the biosphere/ecosphere.

Figure 3–2

A coral reef community in the Red Sea. Coral reef communities have the greatest species diversity and are the most complex kind of aquatic community. (©1997 Marilyn Kazmers/Dembinsky Photo Associates)

Mini-Glossary of Ecology Terms

population: A group of organisms of the same species that live together.

community: All the organisms found in a particular environment. Includes all the populations of different species that are living together.

ecosystem: A community and its abiotic environment. Includes all the interactions among organisms and between organisms and their abiotic environment.

biosphere: All of the Earth's organisms. Includes all the communities on Earth.

ecosphere: The largest, worldwide ecosystem. It encompasses all the organisms on Earth and their interactions with each other, the land, the water, and the atmosphere.

THE INHABITANTS OF ECOSYSTEMS

Ecosystems often contain an astonishing assortment of organisms that interact with each other and are interdependent in a variety of ways. Consider for a moment a salt marsh in the Chesapeake Bay on the east coast of the United States. This bay is one of the world's richest estuaries, which are semi-enclosed bodies of water found where fresh water drains into the ocean. Biological diversity and productivity abound wherever fresh water and salt water form a **salinity gradient**—a gradual change from unsalty fresh water to salty ocean water—as they do in the Chesapeake Bay. The salinity gradient in the bay results in three distinct marsh communities: freshwater marshes at the head of the bay, brackish (moderately salty) marshes in the middle bay region, and salt marshes on the ocean side of the bay. Each community has its own characteristic organisms.

Sail to one of the salt marsh islands in the Chesapeake Bay, such as South Marsh Island, and you can explore a salt marsh community that is relatively unaffected by humans. A salt marsh presents a monotonous view—miles and miles of flooded meadows of cordgrass (*Spartina*). High salinity—although not as high as that of ocean water—and twice-daily tidal inundations establish a challenging environment to which only a few plants have adapted.

Nutrients such as nitrates and phosphates, which drain into the marsh from the land, promote rapid growth of both cordgrass and microscopic algae. These organisms are eaten directly by some animals, and when they die, their remains, called detritus, provide food for many inhabitants of both the salt marsh and the bay.

A casual visitor to a salt marsh would observe two different types of animal life: insects and birds. Insects such as salt marsh mosquitoes and horseflies number in the millions. Birds nesting in the salt marsh include seaside sparrows, laughing gulls, and clapper rails. Migratory birds spend time in the salt marsh as well.

Study the salt marsh carefully and you will find it has numerous other species (Figure 3–3). Large numbers of invertebrates seek refuge in the water surrounding the cordgrass. Here they eat, hide from predators to avoid being eaten, and reproduce. Many of them gather in the intertidal zone (the shore between the high tide mark and the low tide mark) because food—detritus, algae, protozoa, and worms—is abundant there. A variety of shrimps, lobsters, crabs, barnacles, and other crustaceans live in the salt marsh. The marsh crab, for example, is a common crustacean that eats cordgrass and small animals, such as grass shrimp, as well as detritus. Mollusks include the marsh periwinkle, a snail that moves along the cordgrass, skimming off attached algae for its food. Marsh periwinkles climb up the cordgrass to avoid becoming prey for larger marsh animals, such as terrapins, which are semiaquatic turtles.

Almost no amphibians inhabit salt marshes, because the salty water dries out their skin, but a few reptiles have adapted—the northern diamondback terrapin, for example. It spends its time basking in the sun or swimming in the water searching for food—snails, crabs, worms, in-

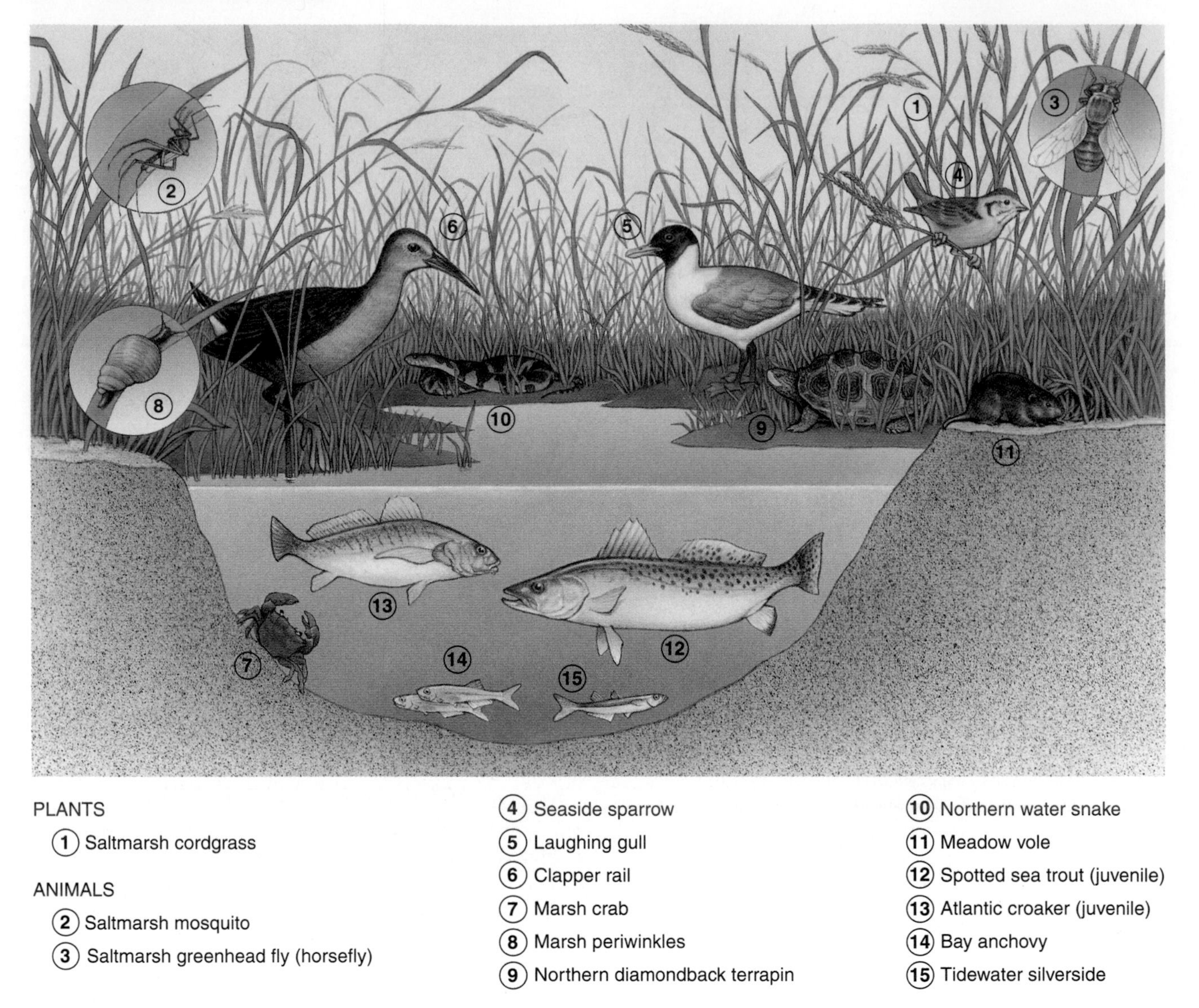

Figure 3–3
Some of the organisms in a Chesapeake Bay salt marsh. Salt marshes, which teem with life, are found in the transitional areas between ocean and land.

sects, and fishes. Although a variety of snakes abound in the dry areas adjacent to salt marshes, only the northern water snake, which preys on fishes, is adapted to brackish water.

Mammals are represented in the salt marsh by the meadow vole, a small rodent that constructs its nest of cordgrass on the ground above the high-tide zone. Meadow voles are excellent swimmers and scamper about the salt marsh day and night. Their diet consists mainly of insects and cordgrass.

The Chesapeake Bay marshes are an important nursery for numerous marine fishes—spotted sea trout, Atlantic croaker, striped bass, and bluefish, to name just a few. These fishes typically spawn (reproduce) in the open ocean, and the young then enter the estuary, where they eat smaller fishes and invertebrates and grow into juveniles. Both juveniles and adults feed in the estuary as well as the ocean. Other fishes, such as bay anchovies, bull minnows, and tidewater silversides, never leave the estuary, spending their summers in the salt marsh shallows and their winters burrowed in the mud or swimming in the deeper waters of the bay.

Add to all these visible plant and animal organisms the unseen microscopic world of the salt marsh, which includes countless numbers of protozoa, fungi, and bacteria, and you can begin to appreciate the complexity of a salt marsh community.

Ecosystems such as the Chesapeake Bay salt marsh teem with life. Where do these organisms get the energy to live? And how do they harness this energy?

THE ENERGY OF LIFE

Energy is the capacity or ability to do work. In organisms, the biological work that requires energy includes processes such as growing, moving, reproducing, and maintaining and repairing damaged tissues.

Energy exists in several different forms: chemical, radiant, heat, mechanical, nuclear, and electrical. Energy can exist as stored energy—called **potential energy**—or as **kinetic energy**, the energy of motion (Figure 3–4). You can think of potential energy as an arrow on a drawn bow. When the string is released, this potential energy is converted to kinetic energy as the motion of the bow propels the arrow. Thus, energy can change from one form to another.

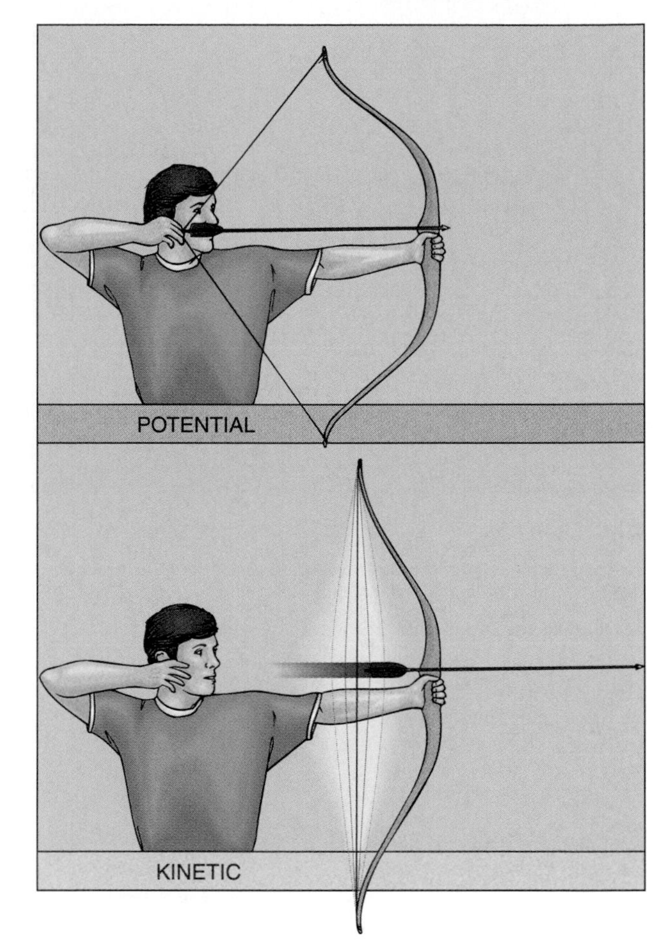

Figure 3–4
Potential and kinetic energy can be represented by a bow and arrow. Potential energy is stored in the drawn bow and is converted to kinetic energy as the arrow speeds toward its target.

Mini-Glossary of Forms of Energy

chemical energy: Energy stored in the chemical bonds of molecules. For example, food contains chemical energy.

nuclear energy: Energy found within atomic nuclei (discussed in Chapter 11).

radiant, or solar, energy: Energy transported from the sun as electromagnetic waves.

heat energy: Thermal energy that flows from a hotter object to a colder one.

mechanical energy: Energy in the movement of matter.

electrical energy: Energy that flows as charged particles.

The study of energy and its transformations is called **thermodynamics**. There are two laws about energy that apply to all things in the universe: the first and second laws of thermodynamics.

The First Law of Thermodynamics

According to the **first law of thermodynamics**, energy cannot be created or destroyed, although it can be transformed from one form to another. As far as we know, the energy present in the universe at its formation, approximately 15 to 20 billion years ago, equals the amount of energy present in the universe today. This is all the energy that can ever be present in the universe. Similarly, the energy of any object and its surroundings is constant. An object may absorb energy from its surroundings, or it may give up some energy into its surroundings, but the total energy content of that object and its surroundings is always the same.

As specified by the first law of thermodynamics, then, organisms cannot create the energy that they require to live. Instead, they must capture energy from the environment to use for biological work, a process involving the transformation of energy from one form to another. In photosynthesis, for example, plants absorb the radiant energy of the sun and convert it into the chemical energy contained in the bonds of carbohydrate molecules. Similarly, some of that chemical energy may later be transformed by some animal that eats the plant into the mechanical energy of muscle contraction, enabling it to walk, run, slither, fly, or swim.

The Second Law of Thermodynamics

As each energy transformation occurs, some of the energy is changed to heat energy that is then given off into the cooler surroundings. This energy can never again be used by any organism for biological work; it is "lost" from the biological point of view. However, it is not really gone from a thermodynamic point of view because it still exists in the surrounding physical environment. For example, the use of food to enable us to walk or run does not destroy the chemical energy that was once present in the food molecules. After we have performed the task of walk-

Figure 3–5

Entropy can be represented by beakers filled with marbles of two different colors. The beaker on the left, in which all marbles of the same color are located together, represents a highly organized system with low entropy. The beaker on the right, in which the marbles are randomly arranged, regardless of color, represents a system with greater entropy. (Dennis Drenner)

ing or running, the energy still exists in the surroundings as heat energy.

The **second law of thermodynamics** can be stated most simply as follows: when energy is converted from one form to another, some usable energy—that is, energy available to do work—is degraded into a less usable form, heat that disperses into the environment. As a result, the amount of usable energy available to do work in the universe decreases over time.

It is important to understand that the second law of thermodynamics is consistent with the first law; that is, the total amount of energy in the universe is *not* decreasing with time. However, the energy available to do work is being degraded to less-usable energy with time.

Less-usable energy is more diffuse, or disorganized. **Entropy** is a measure of this disorder or randomness; organized, usable energy has a low entropy, whereas disorganized energy such as heat has a high entropy (Figure 3–5). Entropy is continuously increasing in the universe in all natural processes. It may be that at some time, billions of years from now, all energy will exist as heat uniformly distributed throughout the universe. If that happens, the universe will cease to operate because no work will be possible. Everything will be at the same temperature, so there will be no way to convert the thermal energy of the universe into usable mechanical energy.

Another way to explain the second law of thermodynamics, then, is that entropy, or disorder, in an isolated system spontaneously tends to increase over time. (The word *spontaneously* in this context means that entropy occurs naturally rather than being caused by some external influence.)

As a result of the second law of thermodynamics, no process requiring an energy conversion is ever 100 percent efficient, because much of the energy is dispersed as heat, resulting in an increase in entropy. For example, an automobile engine, which converts the chemical energy of gasoline → heat energy → mechanical energy, is between 20 and 30 percent efficient. That is, only 20 to 30 percent of the original energy stored in the chemical bonds of the gasoline molecules is actually transformed into mechanical energy. In our cells, energy utilization is about 50 percent efficient, with the remaining energy given to the surroundings as heat.

Organisms have a high degree of organization, and at first glance, they appear to refute the second law of thermodynamics. As organisms grow and develop, they maintain a high level of order and do not appear to become more disorganized. However, organisms are able to maintain their degree of order over time only with the constant input of energy. That is why plants must photosynthesize and animals must eat.

Photosynthesis and Cell Respiration

Photosynthesis is the biological process in which light energy from the sun is captured and transformed into the chemical energy of carbohydrate molecules. Photosynthetic pigments such as **chlorophyll**, which is green and gives plants their green color, absorb radiant energy. This energy is used to manufacture a carbohydrate called glucose ($C_6H_{12}O_6$) from carbon dioxide (CO_2) and water (H_2O), with the liberation of oxygen (O_2):

$$6CO_2 + 12H_2O + \text{radiant energy} \longrightarrow C_6H_{12}O_6 + 6H_2O + 6O_2$$

The chemical equation for photosynthesis is read as follows: 6 molecules of carbon dioxide plus 12 molecules of water plus light energy are used to produce 1 molecule of glucose plus 6 molecules of water plus 6 molecules of oxygen. (See Appendix I for a review of basic chemistry.)

Photosynthesis, which is essential for almost all life, is performed by plants, algae,[2] and a few bacteria. Photosynthesis provides these organisms with a ready supply of energy in glucose molecules that they can use as the need arises. The energy can also be transferred from one

[2]Algae, photosynthetic aquatic organisms that range from single cells to seaweeds well over 50 meters in length, were originally classified as plants. Most biologists classify algae as protists (simple eukaryotic organisms) rather than as plants, because algae lack many of the structural features of plants.

Figure 3–6
The chemical energy produced by photosynthesis and stored in seeds and leaves is transferred to the black-tailed prairie dog as he eats. (Barbara Gerlach/Visuals Unlimited)

organism to another—for instance, from plants to the organisms that eat plants (Figure 3–6). Photosynthesis also produces oxygen, which is required by organisms when they break down food.

The chemical energy that plants store in carbohydrates and other molecules is released within cells of plants, animals, or other organisms through **cell respiration**. In this process, molecules such as glucose are broken down in the presence of oxygen and water into carbon dioxide and water, with the release of energy:

$$C_6H_{12}O_6 + 6O_2 + 6H_2O \longrightarrow 6CO_2 + 12H_2O + \text{energy}$$

Cell respiration makes the chemical energy stored in glucose and other food molecules available to the cell for biological work. All organisms respire to obtain energy.

THE FLOW OF ENERGY THROUGH ECOSYSTEMS

The passage of energy in a one-way direction through an ecosystem is known as **energy flow**. Energy enters an ecosystem as radiant energy (sunlight), some of which is trapped by plants during photosynthesis. The energy, now in chemical form, is stored in the bonds of organic molecules such as glucose. When these molecules are broken apart by cell respiration, the energy becomes available to do work such as repairing tissues, producing body heat, or reproducing. As the work is accomplished, the energy escapes the organism and dissipates into the environment as heat (recall the second law of thermodynamics). Ultimately, this heat energy radiates into space. Thus, once energy has been used by an organism, it becomes unavailable for reuse.

Producers, Consumers, and Decomposers

The organisms of a community can be divided into three categories based on how they get nourishment: producers, consumers, and decomposers (Figure 3–7). Most communities contain representatives of all three groups, which interact extensively with one another.

Sunlight is the source of energy that powers almost all life processes. **Producers**, also called **autotrophs** (Greek *auto,* "self," and *tropho,* "nourishment"), manufacture complex organic molecules from simple inorganic substances, generally carbon dioxide and water, usually using the energy of sunlight to do so. In other words, producers perform the process of photosynthesis. By incorporating the chemicals they manufacture into their own bodies, producers become potential food resources for other organisms. Whereas plants are the most significant producers on land, algae and certain types of bacteria are important producers in aquatic environments. In the salt marsh community, cordgrass, algae, and photosynthetic bacteria are all important producers. In the deep-sea hydrothermal vents discussed in the chapter introduction, nonphotosynthetic bacteria are the producers.

Animals are **consumers**; that is, they use the bodies of other organisms as a source of food energy and body-building materials. Consumers are also called **heterotrophs** (Greek *heter,* "different," and *tropho,* "nourishment"). Consumers that eat producers are called **primary consumers**, which usually means that they are exclusively **herbivores** (plant eaters). Cattle and deer are examples of primary consumers, as is the marsh periwinkle in the salt marsh community. **Secondary consumers** eat primary consumers, whereas **tertiary consumers** eat secondary consumers. Both secondary and tertiary consumers are flesh-eating **carnivores** that eat other animals. Lions and spiders are examples of carnivores, as are the northern diamondback terrapin and the northern water snake in the salt marsh community. Other consumers, called **omnivores**, eat a variety of organisms, both plant and animal. Bears, pigs, and humans are examples of omnivores; the meadow vole, which eats both

(a)

Figure 3–7

Producers, consumers, and decomposers. (a) The trees and other vegetation of the tropical rain forest are producers. (b) The three-toed sloth is a consumer in the tropical rainforest ecosystem. It eats mostly leaves; other consumers eat meat or detritus. (c) Decomposers such as this rainforest cup fungus, which is growing on a rotting log, reduce dead organisms to their mineral constituents, plus carbon dioxide and water. (a, Philip Sze/Visuals Unlimited; b, A. Kerstitch/Visuals Unlimited; c, James L. Castner)

(b)

(c)

insects and cordgrass in the salt marsh community, is also an omnivore.

Some consumers, called **detritus feeders** or **detritivores**, consume **detritus**, which is organic matter that includes animal carcasses, leaf litter, and feces. Detritus feeders, such as snails, crabs, clams, and worms, are especially abundant in aquatic environments, where they burrow in the bottom muck and consume the organic matter that collects there. For example, marsh crabs are detritus feeders in the salt marsh community. Earthworms are terrestrial (land-dwelling) detritus feeders, as are termites, beetles, snails, and millipedes. Detritus feeders work together with microbial decomposers to destroy dead organisms and waste products. An earthworm, for example, actually eats its way through the soil, digesting much of the organic matter contained there.

Many consumers do not fit readily into a single category such as herbivore, carnivore, omnivore, or detritivore. These organisms modify their food preferences to some degree as the need arises.

Decomposers, also called **saprotrophs**, are microbial heterotrophs that break down organic material and use the decomposition products to supply themselves with energy. They typically release simple inorganic molecules, such as carbon dioxide and mineral salts, that can then be reused by producers. Bacteria and fungi are important examples of decomposers. Dead wood, for example, is invaded first by sugar-metabolizing fungi that consume the wood's simple carbohydrates, such as glucose and maltose. When these carbohydrates are exhausted, other fungi, often aided by termites with symbiotic bacteria in their guts, complete the digestion of the wood by breaking down cellulose, a complex carbohydrate that is the main component of wood.

Communities such as the Chesapeake Bay salt marsh contain a balanced representation of all three ecological categories of organisms—producers, consumers, and decomposers—and all of these have indispensable roles in ecosystems. Producers provide both food and oxygen for the rest of the community. Detritus feeders and decomposers are necessary for the long-term survival of any community because, without them, dead organisms and waste products would accumulate indefinitely. Without microbial decomposers, important elements such as potassium, nitrogen, and phosphorus would permanently remain in dead organisms and therefore be unavailable for use by new generations of organisms. Consumers also play an important role in communities by maintaining a balance between producers and decomposers.

A CHAIN OF POISON

Toxic substances that resist breakdown can accumulate in the tissues of organisms. The poison can become more concentrated as it moves through the food web—a phenomenon called biological magnification—harming organisms and causing other unforeseen and often disastrous effects. Just such a catastrophe occurred in Borneo. It began when the World Health Organization sprayed large quantities of the insecticide DDT to control mosquitoes carrying deadly malaria. The mosquitoes died, but so did a predatory wasp that naturally controlled caterpillars. Without the wasp around, the caterpillars bred rapidly and proceeded to devour the thatched roofs of local homes, causing many to collapse. More spraying was done indoors to kill house flies, and the DDT entered the food web. Gecko lizards, which ate and had controlled the fly population, began ingesting high concentrations of the poison. The dying lizards were caught and eaten by house cats, who received massive doses of DDT and also died. The lack of cats led to an abundance of rats, which ate the people's food and threatened to cause the worst plague of all—bubonic plague. Eventually, the government of Borneo grew so concerned that it dropped healthy cats into villages by parachute to hold back the growing rat population. Biological magnification in food webs is discussed further in Chapter 22.

The Path of Energy Flow: Who Eats Whom in Ecosystems

In an ecosystem, energy flow occurs in **food chains**, in which energy from food passes from one organism to the next in a sequence. Producers form the beginning of the food chain by capturing the sun's energy through photosynthesis. Herbivores and omnivores eat plants, obtaining both the chemical energy of the producers' molecules as well as building materials from which they construct their own tissues. Herbivores are in turn consumed by carnivores and omnivores, who reap the energy stored in the herbivores' molecules. At every step in a food chain are decomposers, which respire organic molecules in the carcasses and body wastes of all members of the food chain.

Simple food chains rarely occur in nature, because few organisms eat just one kind of organism. More typically, the flow of energy and materials through an ecosystem takes place in accordance with a range of choices of food for each organism involved. In an ecosystem of average complexity, numerous alternative pathways are possible. Thus, a **food web**, which is a complex of interconnected food chains in an ecosystem, is a more realistic model of the flow of energy and materials through ecosystems (Figure 3–8). (See Focus On: Changes in Antarctic Food Webs for an examination of how humans have affected the complex food web in Antarctic waters.)

The most important thing to remember about energy flow in ecosystems is that it is linear, or one-way. That is, energy can move along a food web from one organism to the next as long as it has not been used to do biological work. Once energy has been used by an organism, however, it is lost as heat and is unavailable for use by any other organism in the ecosystem (Figure 3–9 on page 54).

Each level, or "link," in a food chain or food web is called a **trophic level** (recall that the Greek *tropho* means nourishment). The first trophic level is formed by pro-

(text continues on page 54)

Figure 3–8

A food web for a deciduous forest. This food web is greatly simplified compared to what actually happens in nature. Groups of species are lumped into single categories such as "spiders," many other species are not included, and numerous links in the web are not shown.

FOCUS ON

Changes in Antarctic Food Webs

Although the icy waters around Antarctica may seem to be a very inhospitable environment, a rich variety of life is found there. The base of the food web is microscopic algae, which are present in vast numbers. These algae are eaten by a huge population of tiny shrimplike animals called **krill** (see figure), which in turn support a variety of larger animals. One of the main consumers of krill is baleen whales, which filter krill out of the frigid water. Baleen whales include blue whales, humpback whales, and right whales. Krill are also consumed in great quantities by squid and fishes. These in turn are eaten by other carnivores: toothed whales such as the sperm whale; elephant seals and leopard seals; king penguins and emperor penguins; and birds such as the albatross and petrel.

Humans have had an impact on the complex Antarctic food web as they have on most other ecosystems. Before the advent of whaling, baleen whales consumed huge quantities of krill. During the past 150 years—until a 1986 global ban on hunting all large whales—whaling steadily reduced the numbers of large baleen whales in Antarctic waters. Some whale populations are so decimated that they are on the brink of extinction. As a result of fewer whales eating krill, more krill has been available for other krill-eating animals, whose populations have increased. Seals, penguins, and smaller baleen whales have replaced the large baleen whales as the main eaters of krill.

Now that commercial whaling is regulated, it is hoped that the number of large baleen whales will slowly increase, and that appears to be the case for many species. As of 1994, the only whale population that did not appear to be growing in response to the moratorium on whaling was the southern blue whale. It is not known whether baleen whales will return to or be excluded from their former position of dominance in terms of krill consumption in the food web. Biologists will monitor changes in the Antarctic food web as the whale populations recover.

Recently, a human-related change has developed in the atmosphere over Antarctica that has the potential to cause far greater effects on the entire Antarctic food web—depletion of the ozone layer in the stratospheric region of the atmosphere. This ozone "hole" allows more of the Sun's ultraviolet radiation to penetrate to the Earth's surface. Ultraviolet radiation contains more energy than visible light. It is so energetic that it can break the chemical bonds of some biologically important molecules, such as DNA.

Scientists are concerned that the ozone hole over Antarctica could possibly damage the algae that form the base of the food web in the southern ocean. A 1992 study confirmed that increased ultraviolet radiation is penetrating the surface waters around Antarctica and that algal productivity has declined by at least 6 to 12 percent, probably as a result of increased exposure to ultraviolet radiation. The problem of stratospheric ozone depletion is discussed in detail in Chapter 20.

Krill are tiny, shrimplike animals that eat photosynthetic algae in the Antarctic food web. Vast numbers of krill are consumed by whales, squids, and fishes. (George Holton/Photo Researchers, Inc.)

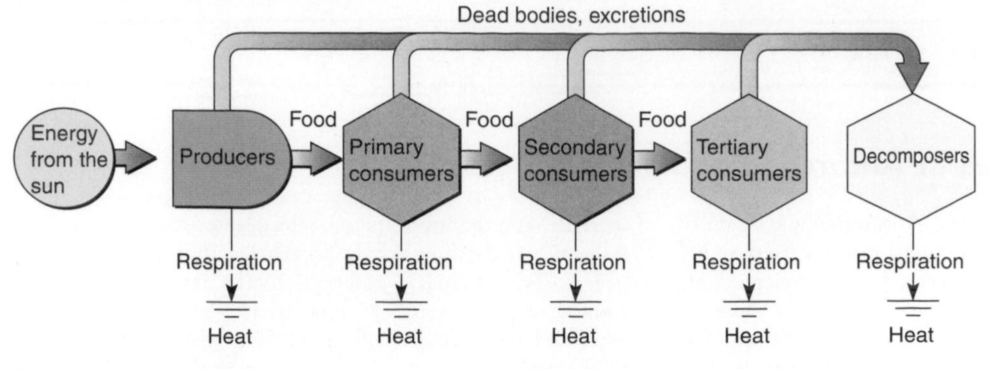

Figure 3–9

Energy flow occurs linearly—that is, in a one-way direction—through ecosystems. Energy enters ecosystems from an external source (the sun) and exits as heat loss. Much of the energy acquired by a given level of the food chain is used for respiration at that level and escapes into the surrounding environment as heat; this energy is therefore unavailable to the next level of the food chain.

ducers (organisms that photosynthesize), the second by primary consumers (herbivores), the third by secondary consumers (carnivores), and so on (Figure 3–10).

Ecological Pyramids

An important feature of energy flow is that, as a result of the second law of thermodynamics, most of the energy going from one trophic level to the next in a food chain or food web dissipates into the environment. The relative energy values of trophic levels are often graphically represented by **ecological pyramids**. There are three main types of pyramids—a pyramid of numbers, a pyramid of biomass, and a pyramid of energy.

A **pyramid of numbers** shows the number of organisms at each trophic level in a given ecosystem, with greater numbers illustrated by a larger area for that sec-

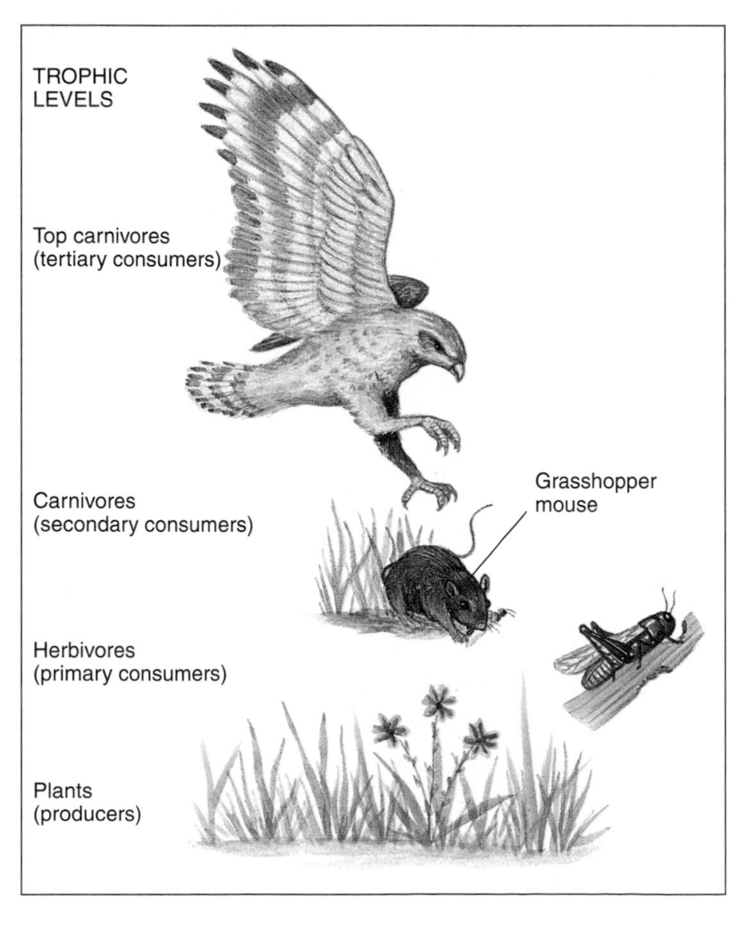

Figure 3–10

Trophic levels show how chemical energy in food flows through an ecosystem. As stipulated by the second law of thermodynamics, most energy at each level is released into the environment as heat.

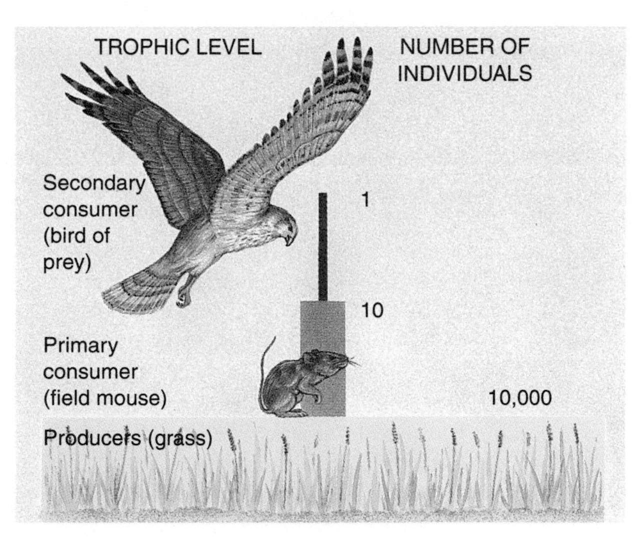

Figure 3–11

A pyramid of numbers is based on the number of organisms found at each trophic level. It is not as useful as other ecological pyramids because it provides no information about biomass or energy relationships between one trophic level and the next.

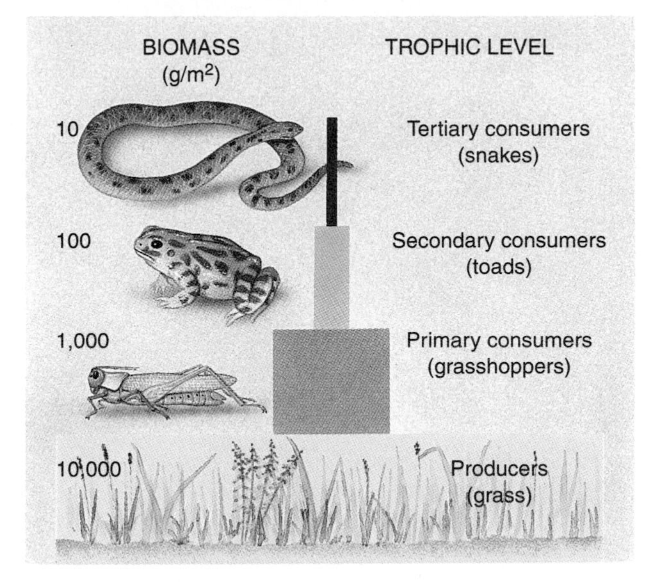

Figure 3–12

A pyramid of biomass for a hypothetical area of a temperate grassland. Pyramids of biomass are based on the biomass at each trophic level and generally have a pyramid shape with a large base and progressively smaller areas for each succeeding trophic level.

tion of the pyramid (Figure 3–11). In most pyramids of numbers, the organisms at the lower end of the food chain are the most abundant, and each successive trophic level is occupied by fewer organisms. Thus, in an African grassland the number of herbivores, such as zebras and wildebeests, is far greater than the number of carnivores, such as lions. Inverted pyramids of numbers, in which higher trophic levels have *more* organisms than lower trophic levels, are often observed among decomposers, parasites, tree-dwelling herbivorous insects, and similar organisms. One tree can provide food for thousands of leaf-eating insects, for example. Pyramids of numbers are of limited usefulness because they do not indicate the biomass of the organisms at each level, and they do not indicate the amount of energy transferred from one level to another.

A **pyramid of biomass** illustrates the total biomass at each successive trophic level. **Biomass** is a quantitative estimate of the total mass, or amount, of living material; it indicates the amount of fixed energy at a particular time. Biomass units of measure vary: biomass may be represented as total volume, as dry weight, or as live weight. Typically, pyramids of biomass illustrate a progressive reduction of biomass in succeeding trophic levels (Figure 3–12). On the assumption that there is, on the average, about a 90 percent reduction of biomass for each trophic level,[3] 10,000 kilograms of grass should be able to support 1000 kg of grasshoppers, which in turn support 100 kg of toads. By this logic, the biomass of toad eaters such as snakes could be, at the most, only about 10 kg. From this brief exercise you can see that, although carnivores may eat no vegetation, a great deal of vegetation is still required to support them.

[3]The 90 percent reduction in biomass is an approximation; actual field numbers for biomass reduction in nature vary widely.

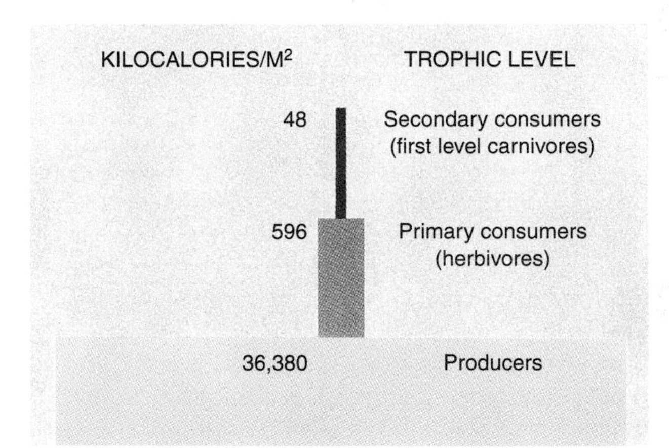

Figure 3–13

The functional basis of ecosystem structure—energy flow—is represented by a pyramid of energy. Note the substantial loss of usable energy from one trophic level to the next.

A **pyramid of energy** illustrates the energy content, often expressed as kilocalories per square meter, of the biomass of each trophic level (Figure 3–13). These pyramids, which always have large energy bases and get pro-

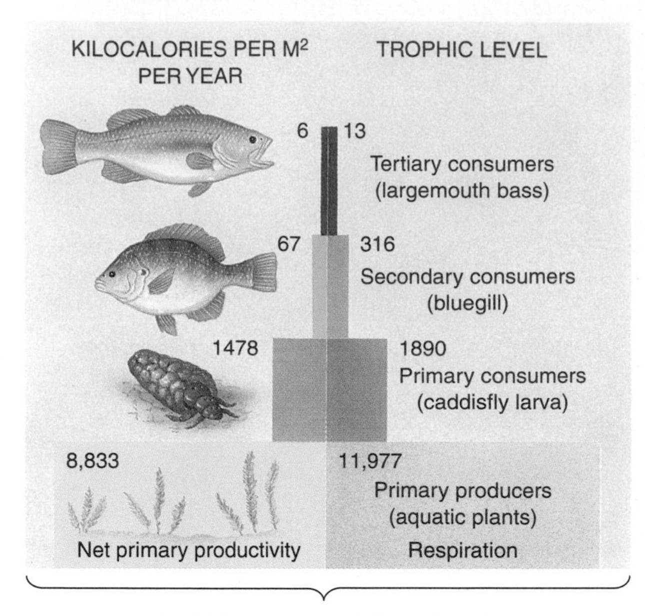

Figure 3–14

A pyramid of energy for a river ecosystem, illustrating gross primary productivity and net primary productivity. Measurements are in kilocalories per square meter per year.

MORE THAN OUR FAIR SHARE

It should come as no surprise that humans consume far more of the Earth's resources than any of the other millions of animal species. When put in terms of the portion of the biosphere's production that is consumed, the impact of humans becomes staggering. When both direct and indirect human impacts are accounted for, humans are estimated to use 25 percent of the global annual net primary production (NPP) and an astounding 40 percent of the annual NPP of terrestrial ecosystems. Essentially, humans' disproportionate use of global productivity is crowding out other species and may contribute to the loss of many species, some potentially useful to humans, through extinction or genetic impoverishment. Clearly, at these levels of consumption and exploitation of the Earth's resources, explosive human population growth becomes a serious threat to the planet's ability to support its occupants.

gressively smaller through succeeding trophic levels, show that most energy dissipates into the environment when going from one trophic level to the next. Less energy reaches each successive trophic level from the level beneath it because some of the energy at the lower level is used by those organisms to perform work and some of it is lost (remember, no biological process is ever 100 percent efficient). Energy pyramids explain why there are few trophic levels: *food webs are short because of the dramatic reduction in energy content that occurs at each trophic level.* (In Chapter 18, the eating habits of humans as they relate to food chains and trophic levels are discussed; see You Can Make a Difference: Vegetarian Diets.)

Productivity of Producers

The **gross primary productivity (GPP)**[4] of an ecosystem is the *rate* at which energy is captured during photosynthesis. Thus, GPP is the total amount of photosynthetic energy captured in a given period of time. Of course, plants must respire to provide energy for their own life processes, and cell respiration acts as a drain on photosynthesis. Energy that remains in plant tissues after cell respiration has occurred is called **net primary productivity (NPP)**. That is, NPP is the amount of biomass found in excess of that broken down by a plant's cell respiration. NPP represents the *rate* at which this organic matter is actually incorporated into plant tissues so as to produce growth (Figure 3–14).

net primary productivity = gross primary productivity − plant respiration
(plant growth) (total photosynthesis)

Only the energy represented by NPP is available for consumers, and of this energy only a portion is actually used by them. Both GPP and NPP can be expressed as energy per unit area per unit time—for example, kilocalories of energy fixed by photosynthesis per square meter per year—or in terms of dry weight—that is, grams of carbon incorporated into tissue—per square meter per year. In Figure 3–14, for example, the GPP of the plants in a river ecosystem is 20,810 kilocalories per m^2 per year. However, the plants respire 11,977 kilocalories per m^2 per year, leaving an NPP of 8833 kilocalories per m^2 per year.

[4]Gross and net primary productivities are referred to as *primary* because plants occupy the first position in food webs.

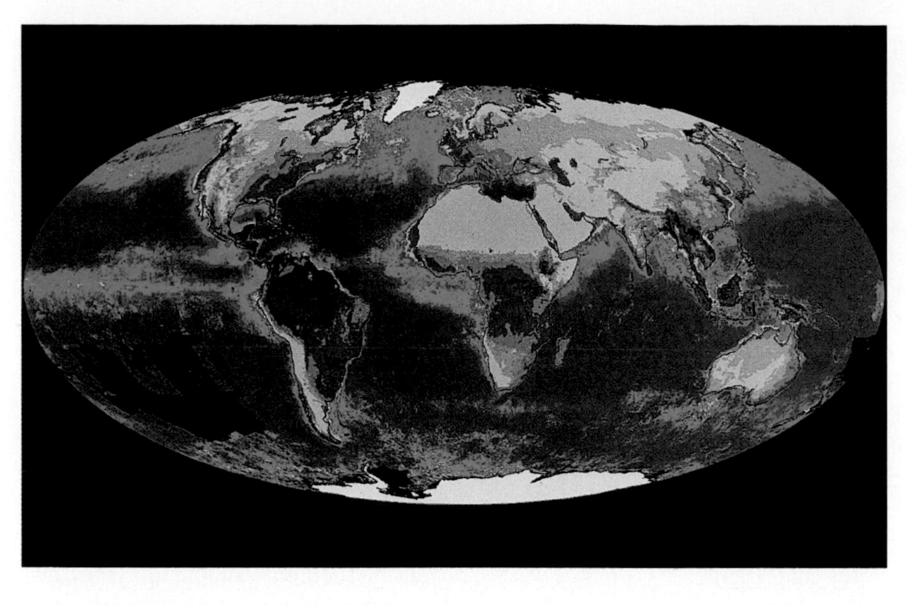

Figure 3–15
Earth's primary productivity as measured by satellite. On land, the most productive areas, such as tropical rain forests, are dark green, whereas the least productive ecosystems (deserts) are yellow. In the ocean and other aquatic ecosystems, the most productive regions are red, followed by orange, yellow, green, and blue (the least productive). Data are not available for the black areas. (NASA/Goddard Space Flight Center)

Ecosystems differ strikingly in their productivity (Figure 3–15 and Table 3–1). On land, tropical rain forests have the highest productivity, probably due to their abundant rainfall, warm temperatures, and intense sunlight. As you might expect, tundra with its harsh, cold winters and deserts with their lack of precipitation are the least productive terrestrial ecosystems. Wetlands, swamps and marshes that connect terrestrial and aquatic environments, are extremely productive. The most productive aquatic ecosystems are algal beds, coral reefs, and estuaries. The lack of available mineral nutrients in the sunlit region of the open ocean makes it extremely unproductive, equivalent to an aquatic desert. (Earth's major aquatic and terrestrial ecosystems are discussed in Chapter 6.)

Table 3–1
Average Net Primary Productivities for Selected Terrestrial Ecosystems

Type of Ecosystem	Average NPP (grams of tissue/ m^2 of land/year)
Tropical rain forest	1800
Temperate deciduous forest	1250
Northern coniferous forest	800
Savanna	700
Temperate grassland	500
Tundra	140
Desert	70

MORE IS SOMETIMES LESS

Conventional ecological theory held that the more productive the habitat, the greater the biodiversity it supported. Now ecologists are seeing a recurring pattern worldwide: habitats do become more productive and diverse as resources increase, but at some point diversity actually declines with increasing productivity. For instance, the resource-poor depths of the Atlantic Ocean's abyssal plain have a higher species diversity than the productive shallow waters near the coasts; intermediate depths exhibit the greatest diversity. Scientists have little solid data to help explain the pattern, which holds true with rodents in Israel, birds in South America, and large mammals in Africa. Mathematical ecosystem models, however, suggest that a patchy distribution of resources reduces competition and allows the coexistence of a greater variety of organisms. The bad news for global biodiversity is that we are constantly enriching our environment, for example, with nitrogen inputs from fossil fuels and cattle. This continual fertilization may make the Earth's ecosystems more and more productive, a shift that could cost the world a substantial loss of biodiversity. (Factors that affect species diversity, the number of species within a community, are discussed in Chapter 4.)

SUMMARY

1. The study of the relationships between organisms and their abiotic environment is called ecology. Ecologists study populations, communities/ecosystems, and the biosphere/ecosphere. A population is all the members of the same species that live together. A community is all the populations of different species living in the same area; an ecosystem is a community and its environment. The biosphere is all the communities on Earth—in other words, all of the Earth's organisms. The ecosphere, the largest ecosystem on Earth, comprises the interactions among the biosphere, atmosphere, lithosphere, and hydrosphere.

2. The first law of thermodynamics states that energy can be converted from one form to another, but it can neither be created nor destroyed. The second law of thermodynamics states that entropy continually increases in the universe as usable energy is converted to a lower quality, less-usable form—that is, heat.

3. Organisms, like everything else in the universe, obey the two laws of thermodynamics. Because the first law stipulates that organisms cannot create energy, producers must photosynthesize, and consumers and decomposers must eat, in order to obtain the energy that is required for biological work.

4. Almost all ecosystems obtain their energy from the sun. Photosynthesis, performed by producers, converts radiant energy from the sun into the chemical energy stored in the bonds that hold food molecules together. The chemical energy of food is released and made available for biological work by the process of cell respiration. All organisms respire.

5. Energy flows through an ecosystem linearly, from the sun to producer, to consumer, to decomposer. Much of this energy is converted to less-usable heat as the energy moves from one organism to another, as stipulated in the second law of thermodynamics.

6. Organisms in ecosystems assume the roles of producer, consumer, and decomposer. Producers, or autotrophs, are the photosynthetic organisms that are at the base of almost all food chains. Producers include plants, algae, and some bacteria. Consumers, which feed on other organisms, are almost exclusively animals. Microbial decomposers feed on the components of dead organisms and organic wastes, degrading them into simple inorganic materials that can then be used by producers to manufacture more organic material. Both consumers and decomposers are heterotrophs.

7. Trophic relationships may be expressed as food chains or, more realistically, food webs, which show the multitude of alternative pathways that energy can take among the producers, consumers, and decomposers of an ecosystem. Ecological pyramids express the progressive reduction in numbers of organisms, biomass, and energy found in successively higher trophic levels.

8. Gross primary productivity (GPP) of an ecosystem is the rate at which organic matter is produced by photosynthesis. Net primary productivity (NPP) expresses the rate at which some of this matter is incorporated into plant bodies; NPP is less than GPP because of the losses resulting from cell respiration by plants.

THINKING ABOUT THE ENVIRONMENT

1. Why do deep-sea organisms cluster around hydrothermal vents? What is their energy source?

2. Why do we begin our study of environmental science with basic ecological principles that govern the natural world? Why, for example, is it important to know about energy flow through ecosystems?

3. What is the simplest stable ecosystem that you can imagine?

4. Could a balanced ecosystem be constructed that contained only producers and consumers? Only consumers and decomposers? Only producers and decomposers? Explain the reasons for your answers.

5. How are the following forms of energy significant to organisms in ecosystems? (a) radiant energy, (b) mechanical energy, (c) chemical energy, (d) heat.

6. Why is the concept of a food web generally preferred over that of a food chain?

7. Draw a food web containing organisms found in a Chesapeake Bay salt marsh.

8. Suggest a food chain that might have an inverted pyramid of numbers—that is, greater numbers of organisms at higher trophic levels than at lower trophic levels.

9. Is it possible to have an inverted pyramid of energy? Why or why not?

10. Relate the pyramid of energy to the second law of thermodynamics.

*11. Graph the information given in Table 3–1. Would it be more appropriate to construct a line graph or bar graph? Why?

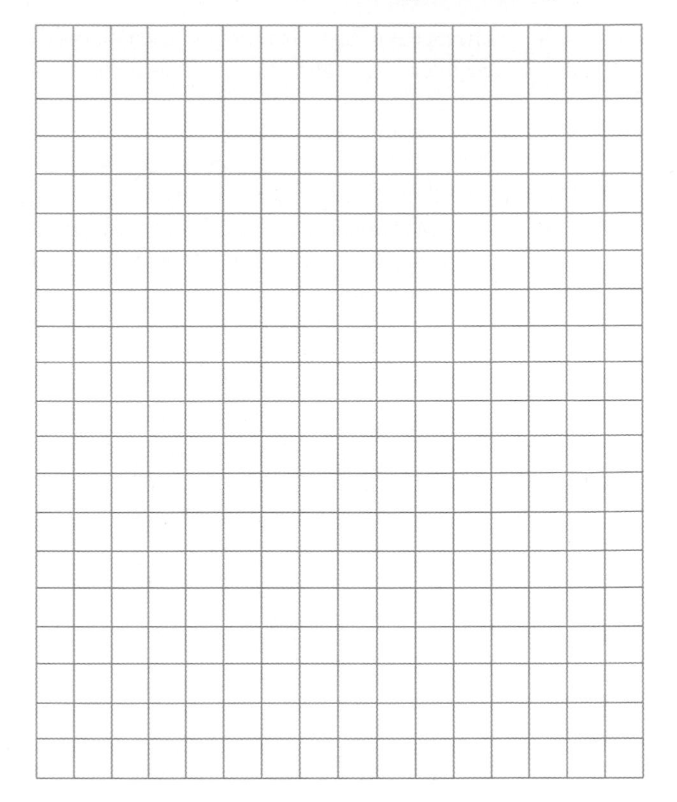

RESEARCH PROJECTS

Study a local ecosystem and make a list of all the organisms at each trophic level. Construct a pyramid of energy based on the most recent data. Describe any changes to the ecosystem in the past 50 years.

Construct your own ecosystem in a terrarium or aquarium. Make sure that you include representative organisms for several trophic levels.

On-line materials relating to this chapter are on the ► World Wide Web at **http://www.saunderscollege.com/lifesci/environment2** (select Chapter 3 from the Table of Contents).

SUGGESTED READING

Brewer, R. *The Science of Ecology.* 2d ed. Philadelphia: Saunders College Publishing, 1994. A good general textbook on the principles of ecology, including ecosystem ecology.

Grall, G. "Pillar of Life," *National Geographic* Vol. 182, No. 1, July 1992. A complex community of organisms exists on wharf pilings in the Chesapeake Bay.

Horton, T. "Chesapeake Bay—Hanging in the Balance," *National Geographic* Vol. 183, No. 6, June 1993. Pollution and overharvesting of its crabs, oysters, and rockfish are threatening the life of the Chesapeake Bay.

Lutz, R. A., and R. M. Haymon. Rebirth of a Deep-Sea Vent. *National Geographic* Vol. 186, No. 5, November 1994. Undersea lava flows recently buried some hydrothermal vent communities off the west coast of Mexico. The author follows their quick rebirth as nearby vent species migrated into the area.

Pimm, S. L., J. H. Lawton, and J. E. Cohen. "Food Web Patterns and Their Consequences," *Nature* Vol. 350, 25 April 1991. A well written review article on current ecological knowledge of food webs.

Ross, J. F. "Hardly a Mouse or a Molecule Moves Here without Being Noticed," *Smithsonian,* July 1996. The Smithsonian Environmental Research Center studies how the Chesapeake Bay functions as an ecological unit.

An aquarium-like microcosm is the simplest stable ecosystem to be devised. (Courtesy of Ecosphere Associates, Ltd.)

Ecosystems and Living Organisms

A balanced aquarium—that is, an aquarium containing fishes and plants along with decomposer bacteria—has always been popular as an illustration of ecosystems. If the aquarium is properly set up, it should be possible for its inhabitants to survive indefinitely even if it is totally sealed off from the outside world. Unfortunately, when this experiment has actually been tried, the organisms inside the aquarium have usually died in a very short period of time.

This outcome is of more than academic interest, in part because of the development of space flight. Spacecraft sent on journeys lasting years cannot be expected to carry all the food and oxygen the astronauts will need. It is obvious that space vehicles must be balanced ecosystems, growing their own food, recycling their own wastes, and producing their own oxygen by photosynthesis.

In 1977 a NASA scientist developed the first stable, sealed ecosystems; these simple ecosystems contained shrimp, algae, and bacteria. By extensive experimentation, researchers developed a controlled mixture of as few as 100 species of organisms (mostly bacteria) that worked well together. It was necessary to include bacteria in addition to shrimp and algae to convert the toxic ammonia excreted by the shrimp into nitrite. Nitrite, also toxic, is in turn converted to nitrate by a different species of bacteria. The nitrate is then used as a nitrogen source by the algae.

The entire system, called a **microcosm**, is set up inside a sealed glass sphere resembling a paperweight. It remains balanced—that is, the organisms within survive and reproduce—as long as it receives adequate light and warmth. These simple ecosystems may provide us with clues about the management of spacecraft, including our own spaceship, planet Earth.

Regardless of whether an ecosystem is a microcosm or a forest, its organisms are continually interacting with one another and adapting to changes in the environment. Each species confronts the challenge of survival in its own unique fashion. This chapter is concerned with making sense of community structure and diversity by finding common patterns and processes in a wide variety of communities. ◄

BIOLOGICAL COMMUNITIES

As you may recall from Chapter 3, the term "community" has a far broader sense in ecology than in everyday speech. For the biologist, a **community** is an association of different populations of organisms that live and interact together in a given environment (see Focus On: The Kingdoms of Life for an overview of the categories of organisms found in communities). The organisms that comprise a community interact with one another and are interdependent in a variety of ways. Species compete with one another for food, water, living space, and other resources. Organisms kill and eat other organisms. Species form intimate associations with one another. Each organism plays one of three main roles in community life: producer, consumer, or decomposer. The unraveling of the many interactions and interdependencies of organisms living together as a community is one of the goals of community ecologists.

Communities vary greatly in size, lack precise boundaries, and are rarely completely isolated. They interact with and influence one another in countless ways, even when the interaction is not readily apparent. Furthermore, communities are nested within one another like Chinese boxes; that is, there are communities within communities. A forest is a community, but so is a rotting log in that forest. The log contains bacteria, fungi, slime molds, bark beetles, carpenter ants, termites, and perhaps even voles or mice (Figure 4–1). The microorganisms living within the gut of a termite in the rotting log also form a community. On the other end of the scale, the entire living world can be considered a community.

Organisms exist in a nonliving environment that is as essential to their lives as are their interactions with one another. Minerals, air, water, and sunlight are just as much a part of a honeybee's environment, for example, as the flowers that it pollinates and from which it takes nectar. A biological community and its abiotic environment together comprise an **ecosystem**. (The abiotic environment of ecosystems is considered in Chapter 5.)

Keystone Species

Within an ecosystem, certain species are more important than others. Such species, called **keystone species**, are

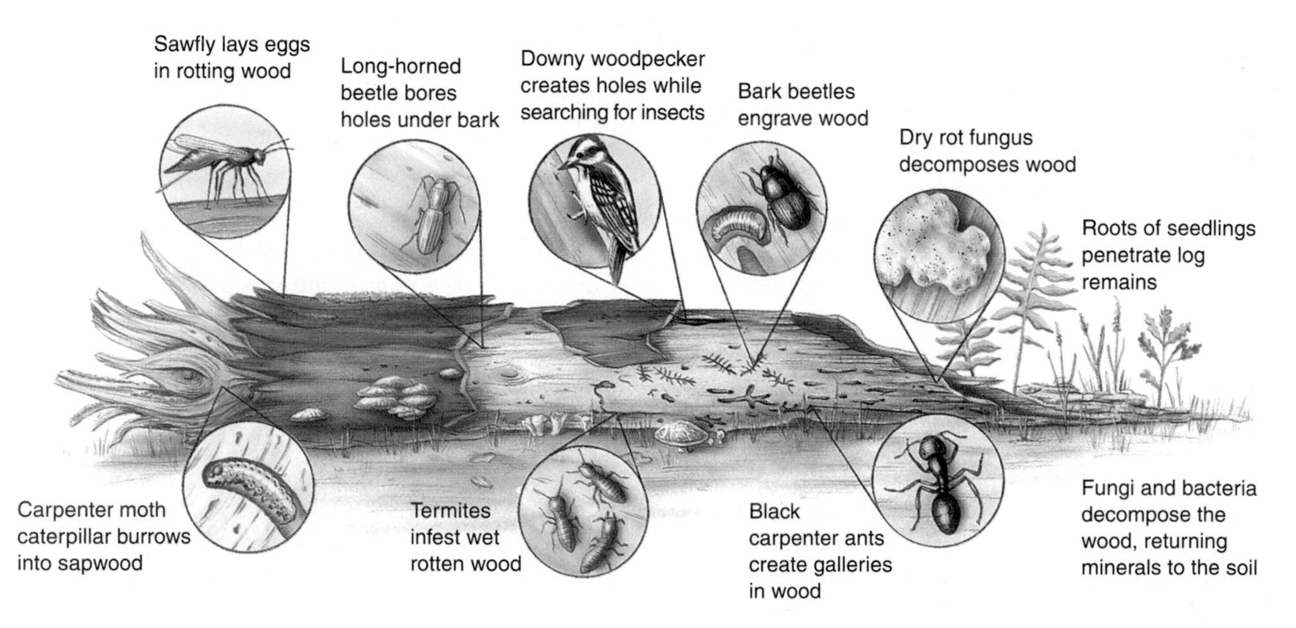

Figure 4–1

Detritivores and decomposers consume a rotting log, which teems with a variety of fungi, animals, and plants.

A ROTTING LOG: BETTER OFF DEAD

Traditional forest management in the United States and Europe includes the removal of dead trees, a practice that forest ecologists now recognize as short-changing a forest's productivity and magnifying its vulnerability to environmental stress. When trees die and remain in forests, they end up supporting a vast array of organisms—including more than 1300 species of birds, mammals, reptiles, and amphibians—and are often more valuable dead than alive. A fallen tree is invaded by insects, plants, and fungi as it undergoes a series of decay steps. First the tree is opened up by wood-boring insects and termites, who forge paths through bark that are later followed by other insects, plant roots, and fungi. Mosses and lichens that establish on the tree's surface trap rainwater and extract nutrients, and fungi and bacteria speed decay, thus providing nutrients for other inhabitants. As decay progresses, small mammals burrow into the now soft wood and eat the fungi, insects, and plants. Ultimately, nutrients from the tree are returned to the soil in the form of animal carcasses, decaying leaves, and such. Because these decay processes can last hundreds of years, dead trees provide enormous long-term benefits, including the storage of moisture, nutrients, and carbon dioxide (which helps mitigate global warming); protection from erosion; and the steady release of nutrients.

crucial in determining the nature and structure of the entire ecosystem—that is, its species composition and its functioning. Other species of a community depend on or are greatly affected by the keystone species. Keystone species are usually not the most abundant species in the ecosystem. Although present in relatively small numbers, they exert a profound influence on the entire ecosystem because keystone species often affect the available amount of food, water, or some other resource.

Identifying and protecting keystone species is a crucial goal of conservation biologists because if a keystone species disappears from an ecosystem, many other organisms in that ecosystem may also disappear. One example of a keystone species is a top predator such as the gray wolf. Where wolves were hunted to extinction, the populations of deer and other herbivores increased explosively (recall the reintroduction of wolves to Yellowstone, discussed in Chapter 1). As these herbivores overgrazed the vegetation, many plant species that could not tolerate such grazing pressure disappeared. Many smaller animals such as insects also were lost from the ecosystem because the plants that they depended on for food were now less abundant. Thus, the disappearance of the wolf resulted in an ecosystem with considerably less biological diversity.

Because fig trees produce a continuous crop of fruits, they appear to be a keystone species in tropical rain forests of Peru. Monkeys, fruit-eating birds, and other vertebrates of the forest do not normally consume large quantities of figs in their diets. During the time of year when other fruits are less plentiful, however, fig trees become very important in sustaining fruit-eating vertebrates. Should the fig trees disappear, most of the fruit-eating vertebrates would also be eliminated. Thus, protecting fig trees in such tropical rain forest ecosystems increases the likelihood that monkeys, birds, and other vertebrates will survive.

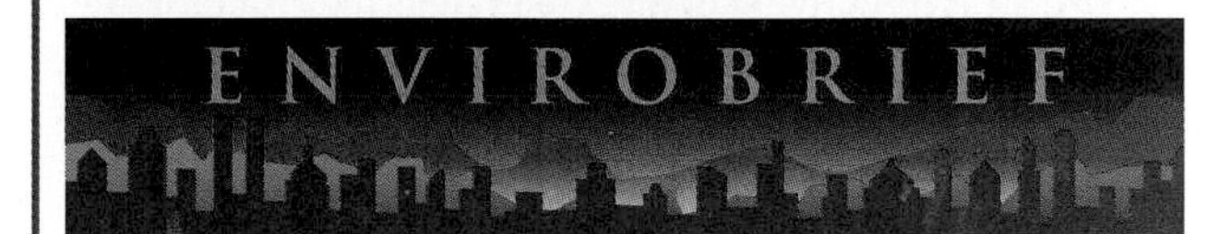

CREDIBLE SCIENCE IN THE BIOSPHERE

Biosphere 2, the artificial self-contained ecosystem that began operating in 1991 in the Arizona desert, is a 1.27-hectare (3-acre) greenhouse that contains a miniature world, including a rain forest, savanna, desert, ocean, mangrove estuary, and agricultural regions. After supporting nonscientist inhabitants in a two-year test of survival in a sealed environment, the facility's credibility suffered from serious operating difficulties and a barrage of criticism from the scientific community. Critics challenged the original idea of Biosphere 2—that of a holistic, self-contained prototype of a Martian space colony—as being too theatrical and lacking in application of the scientific process. Many scientists recognized the tremendous potential of Biosphere 2 as a unique site for conducting controlled environmental studies with potentially global applications but were frustrated that such experiments were not allowed.

In August 1994 Biosphere 2 started a new direction when owner Ed Bass announced the appointment of a new scientific director and the start of a nonprofit joint venture with Columbia University's Lamont-Doherty Earth Observatory. The new arrangement opens the door for outside scientists to propose and carry out ecosystem experiments within the facility. Participating scientists will have to contend with the structure's limitations, such as the inability to conduct control or duplicate experiments, but most are excited by the challenge. The scientists plan to examine the responses of the many different plants in the structure to increased levels of carbon dioxide. This information could be used to help understand the future of planet Earth as carbon dioxide levels continue to rise in the 21st century.

FOCUS ON

The Kingdoms of Life

For hundreds of years, biologists regarded organisms as falling into two broad categories—plants and animals. With the development of microscopes, however, it became increasingly obvious that many organisms did not fit very well into either the plant kingdom or animal kingdom. For example, bacteria have a prokaryotic cell structure: they lack a nuclear envelope and other internal cell membranes. This feature, which separates bacteria from all other organisms, is far more fundamental than the differences between plants and animals, which have similar cell structures. Hence, it became clear that bacteria were neither plants nor animals. Furthermore, certain microorganisms such as *Euglena,* which is both motile and photosynthetic, seem to possess characteristics of both plants and animals.

These and other considerations have led to the six-kingdom system of classification used by many biologists today: Archaebacteria, Eubacteria, Protista, Fungi, Plantae, and Animalia (see figure). Bacteria, with a prokaryotic cell structure, are now divided into two groups on the basis of significant biochemical differences. The archaebacteria frequently live in oxygen-deficient environments and are often adapted to harsh conditions; these include hot springs, salt ponds, and hydrothermal vents (see Chapter 3 introduction). The thousands of kinds of remaining bacteria are collectively known as the eubacteria.

The remaining four kingdoms are composed of organisms with a eukaryotic cell structure. Eukaryotic cells have a high degree of internal organization, containing such organelles as nuclei, chloroplasts (in photosynthetic cells), and mitochondria. Eukaryotes that are single-celled or relatively simple multicellular organisms, such as algae, protozoa, slime molds, and water molds, are classified as members of the kingdom Protista.

In addition to the kingdom Protista, there are three specialized groups of multicellular organisms—fungi, plants, and animals—that evolved independently from different groups of Protista. The kingdoms Fungi, Plantae, and Animalia differ from one another in—among other features—their types of nutrition: fungi secrete digestive enzymes into their food and then absorb the predigested nutrients; plants use radiant energy to manufacture food molecules by photosynthesis; and animals ingest their food, then digest it inside their bodies.

Although the six-kingdom system is a definite improvement over the two-kingdom system, it is not perfect. Most of its problems concern the kingdom Protista, which includes some organisms that may be more closely related to members of other kingdoms than to certain other protists. For example, green algae are protists that are clearly similar to plants but do not appear to be closely related to other protists, such as slime molds and red algae.

INTERACTIONS AMONG ORGANISMS

No organism exists independent of other organisms. The producers, consumers, and decomposers of a community interact with one another in a variety of complex ways, and each forms associations with other organisms. Three main types of interactions occur among species in a community: predation, symbiosis, and competition. We now examine predation and symbiosis. Competition will be discussed later in the chapter (and also in Chapter 8).

Predation

Predation is the consumption of one species, the *prey,* by another, the *predator.* It includes both animals eating other animals and animals eating plants. Predation has resulted in an "arms race," with the evolution of predator strategies—more efficient ways to catch prey—and prey strategies—better ways to escape the predator. (Evolution is discussed later in the chapter.)

Pursuit and ambush

The brown pelican sights its prey—a fish—while in flight. Less than two seconds after diving into the water at a speed as great as 72 kilometers (45 miles) per hour, it has its catch (Figure 4–2). Killer whales, which hunt in packs, often herd salmon or tuna into a cove so that they are easier to catch. Any trait that increases hunting efficiency, such as the speed of a brown pelican or the cunning of killer whales, favors predators that pursue their prey. Because these carnivores must be able to process information quickly during the pursuit of prey, their brains are generally larger—relative to body size—than those of the prey they pursue.

Ambush is another effective way to catch prey. The goldenrod spider, for example, is the same color as the white or yellow flowers in which it hides. This camouflage keeps unwary insects that visit the flower for nectar from noticing the spider until it is too late. Predators that are able to *attract* prey are particularly effective at ambushing. For example, a diverse group of deep sea fishes known as angler fish possess rodlike luminescent lures close to their mouths to attract prey.

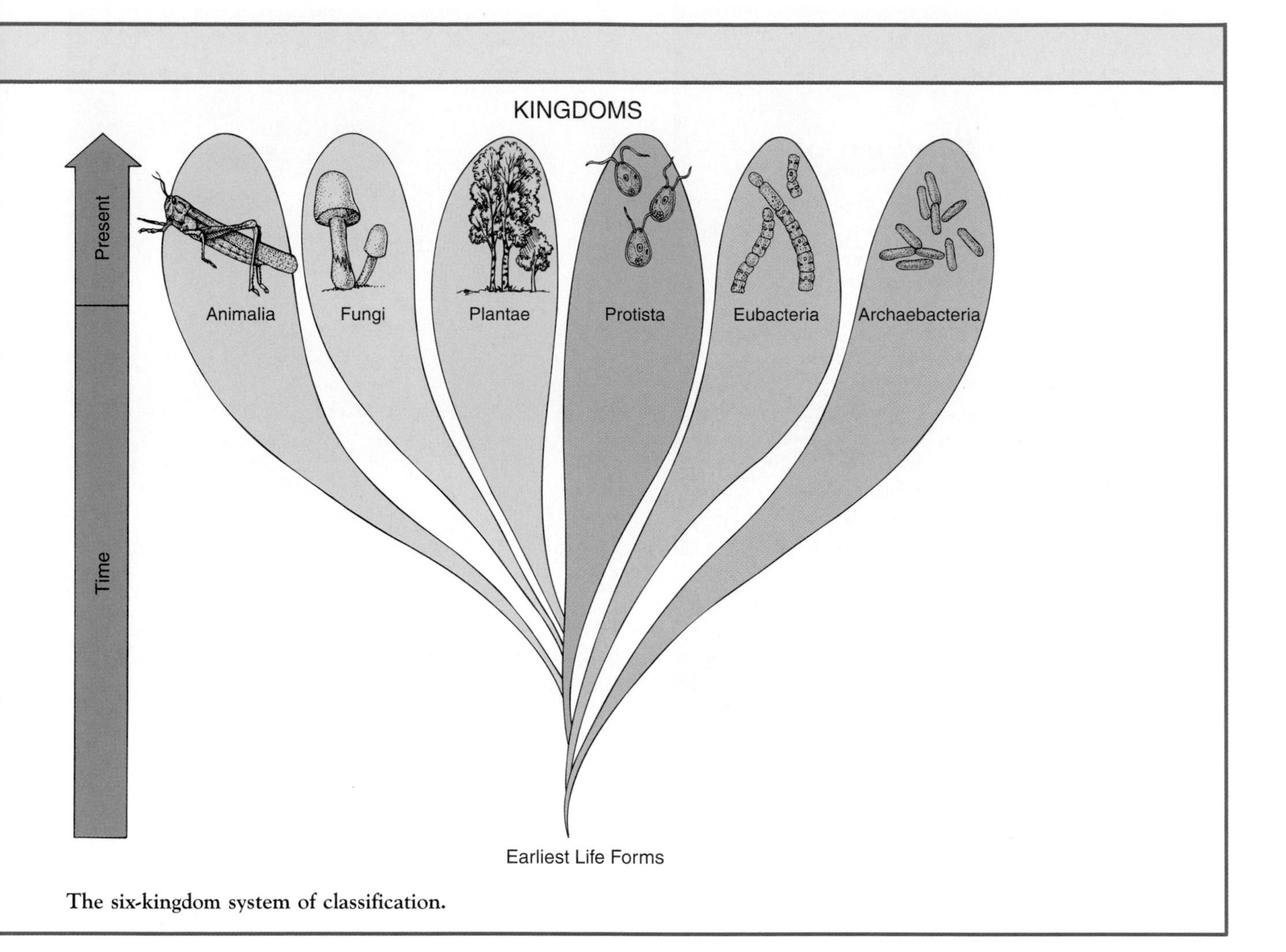

The six-kingdom system of classification.

Figure 4–2

A brown pelican alights in the water, having just caught a fish. To be an effective predator strategy, pursuit of prey requires speed. (©1997 SharkSong/M. Kazmers/Dembinsky Photo Associates)

Plant defense against herbivores

Plants cannot escape predators by fleeing, but they possess adaptations that protect them from being eaten. The presence of spines, thorns, tough leathery leaves, or even thick wax on leaves discourages foraging herbivores from grazing. Other plants produce an array of protective chemicals that are unpalatable or even toxic to herbivores. The active ingredients in such plants as marijuana, opium poppy, tobacco, and peyote cactus may discourage the foraging of herbivores. The nicotine found in tobacco, for example, is so effective at killing insects that it is a common ingredient in many commercial insecticides.

Milkweeds are an excellent example of the evolutionary arms race between plants and herbivores. Milkweeds produce alkaloids and cardiac glycosides, chemicals that are poisonous to all animals except for a small group of insects. During the course of evolution, these insects acquired the ability to either tolerate or metabolize the milkweed toxins. As a result, they can eat milkweeds without being poisoned. These insects avoid competition from other herbivorous insects since few other insects are able to tolerate milkweed toxins. Predators also learn to

Figure 4–3

The common milkweed is protected by its toxic chemicals. Its leaves are poisonous to most herbivores except monarch caterpillars (*shown*) and a few other insects. (Patti Murray)

Figure 4–4

The poison arrow frog (*Dendrobates tinctorius*) advertises its poisonous nature with its conspicuous coloring, warning away would-be predators. (Animals Animals ©1995 Michael Fogden)

avoid these insects, which accumulate the toxins in their tissues and are usually brightly colored to announce that fact. The black, white, and yellow banded caterpillar of the monarch butterfly is an example of a milkweed feeder (Figure 4–3).

Defensive adaptations of animals

Many animals, such as woodchucks, flee from predators by running to their underground burrows. Others have mechanical defenses, such as the barbed quills of a porcupine and the shell of a pond turtle. Some animals live in groups—for example, a herd of antelope, colony of honeybees, school of anchovies, or flock of pigeons. Because the group has so many eyes, ears, and noses watching, listening, and smelling for predators, this social behavior decreases the likelihood of a predator catching one of them unaware.

Chemical defenses are also common among animal prey. The South American poison arrow frog has poison glands in its skin. Its bright yellow **warning coloration** prompts avoidance by experienced predators (Figure 4–4). Snakes or other animals that have tried to eat a poisonous frog do not repeat their mistake! Other examples of warning coloration occur in the striped skunk, which sprays acrid chemicals from its anal glands, and the bombardier beetle, which sprays harsh chemicals at potential predators.

Some animals hide from predators by blending into their surroundings. Such camouflage is often enhanced by the animal's behavior. There are many examples of camouflage. Certain caterpillars resemble twigs so closely that you would never guess they are animals until they move (Figure 4–5). Some katydids resemble leaves not only in color but in the pattern of veins in their wings. Pipefish are almost perfectly camouflaged in green eel grass. Such camouflage has been preserved and accentuated by evolution.

Symbiosis

Symbiosis is any intimate relationship or association between members of two or more different species. Usually, one species lives in or on another species. The partners of a symbiotic relationship, called **symbionts**, may benefit from, be unaffected by, or be harmed by the relationship. Symbiosis is the result of **coevolution**, the interdependent evolution between two interacting species (see Focus On: Coevolution). The thousands, or even millions, of symbiotic associations in nature fall into three categories: mutualism, commensalism, and parasitism.

Figure 4–5
Camouflage helps animals hide from predators. Shown is a geometrid larva, a caterpillar that resembles a twig. Can you find it? (James L. Castner)

Mutualism: sharing benefits

Mutualism is a symbiotic relationship in which both partners benefit. The association between nitrogen-fixing bacteria of the genus *Rhizobium* and legumes, which are plants such as peas, beans, and clover, is an example of mutualism. Nitrogen-fixing bacteria live in nodules in the roots of legumes and supply the plants with all of the nitrogen they need. The legumes supply sugar to their bacterial symbionts.

Another example of mutualism is the association between reef-building coral animals and microscopic algae. These symbiotic algae, which are called **zooxanthellae** (pronounced *zoh-zan-thel'ee*), live inside cells of the coral, where they photosynthesize and provide the animal with carbon and nitrogen compounds as well as oxygen. Zooxanthellae have a stimulatory effect on the growth of corals, causing calcium carbonate skeletons to form around their bodies much faster when the algae are present. The corals, in turn, supply their zooxanthellae with waste products such as ammonia, which the algae use to make nitrogen compounds for both partners.

Mycorrhizae are mutualistic associations between fungi and the roots of about 80 percent of all plants. The fungus absorbs essential minerals, especially phosphorus, from the soil and provides them to the plant. In return, the plant provides the fungus with food produced by photosynthesis. Plants grow more vigorously in the presence of mycorrhizae (Figure 4–6), and they are better able to tolerate environmental stresses such as drought and high soil temperatures. Indeed, some plants cannot maintain themselves under natural conditions if the fungi with which they normally form mycorrhizae are not present.

Commensalism: taking without harming

Commensalism is a type of symbiosis in which one organism benefits and the other one is neither harmed nor helped. One example of commensalism is the relationship between two kinds of insects, silverfish and army ants. Certain kinds of silverfish move along in permanent association with the marching columns of army ants and share the food caught in their raids. The army ants derive no apparent benefit or harm from the silverfish. Another example of commensalism is the relationship between a tropical tree and its **epiphytes**, smaller plants, such as mosses, orchids, and ferns, that live attached to

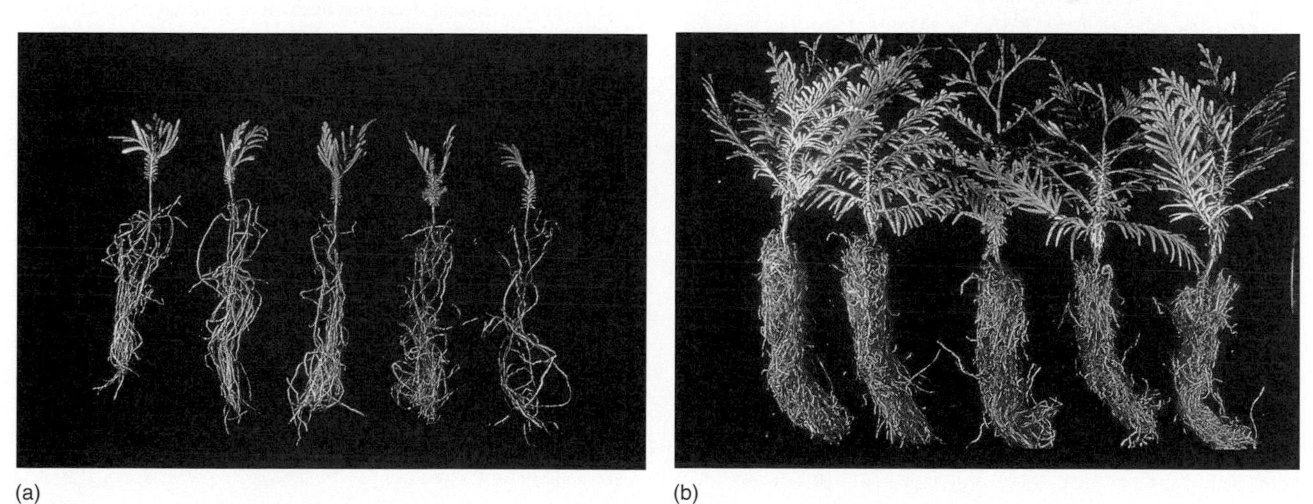

Figure 4–6
An example of mutualism. Western red cedar seedlings respond to mycorrhizal fungi.
(a) Seedlings grown in low phosphorus in the absence of the fungus. (b) Comparable seedlings grown in low phosphorus with the fungus. (a,b, Randy Molina/U.S. Forest Service)

FOCUS ON

Coevolution

Flowering plants and their animal pollinators provide an excellent example of coevolution. Because plants are rooted in the ground, they lack the mobility that animals have when mating. Many flowering plants rely on animals to help them mate. Bees, beetles, hummingbirds, bats, and other animals transport the male reproductive structures, called pollen, from one plant to another, in effect giving plants mobility. How has this come about?

During the millions of years over which these associations developed, flowering plants evolved several ways to attract animal pollinators. One of the rewards for the pollinator is food—nectar (a sugary solution) and pollen. Plants often produce food that is precisely correct for one type of pollinator. The nectar of flowers that are pollinated by bees, for example, usually contains between 30 percent and 35 percent sugar, the concentration that bees need in order to make honey. Bees will not visit flowers with lower sugar concentrations in their nectar. Bees also use pollen to make bee bread, a nutritious mixture of nectar and pollen that is eaten by their larvae.

Plants also possess a variety of ways to get the pollinator's attention, most of which involve colors and scents. Showy petals visually attract the pollinator much as a neon sign or golden arches attract a hungry person to a restaurant. Scents are also an effective way to attract pollinators. Insects have a well developed sense of smell, and many insect-pollinated flowers have strong scents, which may also be very pleasant perfumes to humans. A few specialized kinds of flowers may have unpleasant odors. The carrion plant, for example, is pollinated by carrion flies and smells like rotting flesh (see figure). Flies move from one flower to another, looking for a place to deposit their eggs, and in the process pollen is transferred from one flower to another.

During the time plants were acquiring specialized features to attract pollinators, the animal pollinators coevolved specialized body parts and behaviors that enabled them to both aid pollination and obtain nectar and pollen as a reward. Many insects have mouthparts that fit into certain flowers much as a lock fits into a key; for example, the long, coiled mouthparts of butterflies and moths fit well in the long, tubular flowers often visited by these insects. Even though other insect species may be attracted to the flowers, they may not be able to obtain the rewards because they lack these specialized mouthparts.

Animal behavior has also coevolved, sometimes in bizarre ways. The flowers of certain orchids, for example, resemble female wasps in coloring and shape. The resemblance between one orchid species and female wasps is so strong that male wasps mount the flowers and attempt to copulate with them. During this misdirected activity, a pollen sac usually attaches to the back of the wasp. When the frustrated wasp departs and attempts to copulate with another orchid flower, pollen grains are transferred to the flower.

The carrion plant (*Stapelia variegata*). This East African desert plant gets its name because its coloration and bad smell resemble that of decaying flesh. It is pollinated by carrion flies, several of which have alighted on the flower. (Gary J. James, Biological Photo Service)

the bark of the tree's branches (Figure 4–7). The epiphyte anchors itself to the tree but does not obtain nutrients or water directly from the tree. Its location on the tree enables it to obtain adequate light, water (as rainfall dripping down the branches), and required minerals (washed out of the tree's leaves by rainfall). Thus, the epiphyte benefits from the association, whereas the tree is apparently unaffected.

Mini-Glossary of Symbiosis

symbiosis: Any intimate relationship between two or more different species. Includes mutualism, commensalism, and parasitism.

mutualism: A symbiotic relationship in which both partners benefit.

commensalism: A symbiotic relationship in which one partner benefits and the other partner is unaffected.

parasitism: A symbiotic relationship in which one partner—the parasite—obtains nutrients at the expense of the other—the host.

Parasitism: taking at another's expense

Parasitism is a symbiotic relationship in which one member, the **parasite**, benefits and the other, the **host**, is adversely affected. The parasite obtains nourishment from its host, but although a parasite may weaken its host, it rarely kills it. (A parasite would have a difficult life if it kept killing off its hosts!) Some parasites, such as ticks, live outside the host's body; other parasites, such as tapeworms, live within the host. Parasitism is a very successful lifestyle; more than 100 parasites live in or on the human species alone!

When a parasite causes disease and sometimes the death of a host, it is called a **pathogen**. Crown gall disease, which is caused by a bacterium, occurs in many different kinds of plants and results in millions of dollars of damage each year to ornamental and agricultural plants. Crown gall bacteria, which also live on detritus (organic debris) in the soil, enter plants through small wounds such as those caused by insects. They cause galls (tumor-like growths), often at a plant's crown—that is, between the stem and the roots, at or near the soil surface. Although plants seldom die from crown gall disease, they are weakened, grow more slowly, and often succumb to other pathogens.

Many parasites do not cause disease. For example, humans can become infected with the pork tapeworm by eating poorly cooked pork that is infested with immature tapeworms. Once the tapeworm is inside the human digestive system, it attaches itself to the wall of the small intestine and grows rapidly by absorbing nutrients that pass through the small intestine. A single pork tapeworm that lives in a human digestive tract does not cause any

Figure 4–7
An example of commensalism. Epiphytes are small plants that grow attached to the bark of a tree. (Carlyn Iverson)

noticeable symptoms, although some weight loss may be associated with a multiple infestation.

We have examined some of the ways that species interact to form interdependent relationships within the community. Now we examine the way of life of a species in its ecosystem.

THE ECOLOGICAL NICHE

A biological description of a species includes: (1) whether it is a producer, consumer, or decomposer; (2) whether it is a predator and/or prey; and (3) the kinds of symbiotic associations it forms. Other details are needed, however, to provide a complete picture.

Every organism is thought to have its own role within the structure and function of an ecosystem; we call this role its **ecological niche**. An ecological niche, which is difficult to define precisely, takes into account *all* aspects of the organism's existence—that is, all physical, chemical, and biological factors that the organism needs to survive, remain healthy, and reproduce. Among other things, the niche includes the local environment in which an organism lives—that is, its **habitat**. An organism's niche also encompasses what it eats, what eats it, what organisms it competes with, and how it interacts with and is influenced by the nonliving components of its environment, such as light, temperature, and moisture. The niche, then, represents the totality of an organism's adaptations, its use of resources, and the lifestyle to which it is suited. Thus, a complete description of an organism's ecological niche involves numerous dimensions.

The ecological niche of an organism may be much broader potentially than it is in actuality. Put differently, an organism is usually capable of using much more of its environment's resources or of living in a wider assortment

(a)

(b)

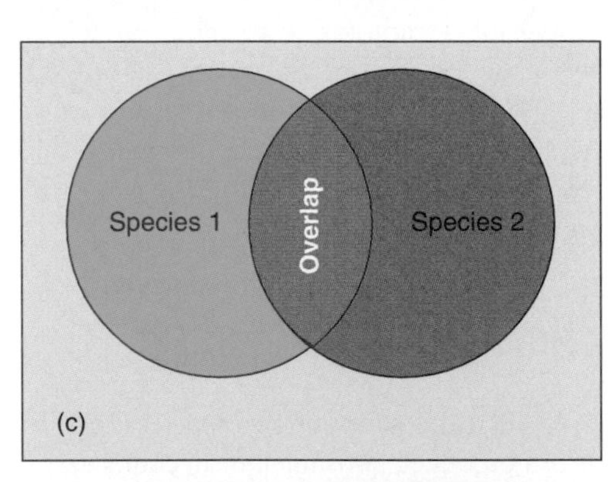

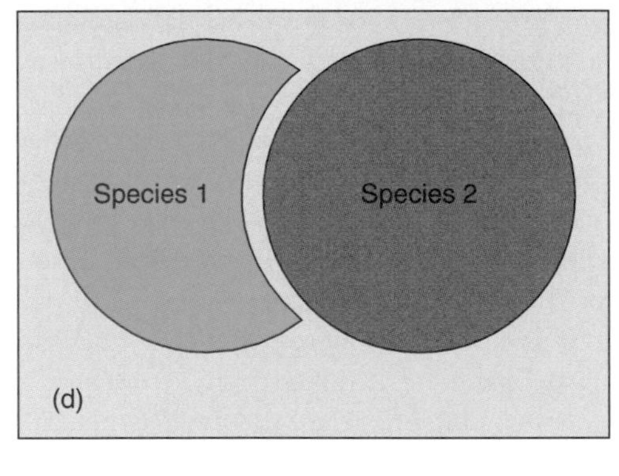

Figure 4–8

Competition can restrict an organism's realized niche. (a) The green anole is native to Florida. (b) The brown anole was introduced in Florida. (c) The fundamental niches of the two lizards initially overlapped. Species 1 is the green anole, and species 2 is the brown anole. (d) The brown anole was able to outcompete the green anole, restricting its niche. (a, ©1997 Ed Kanze/Dembinsky Photo Associates; b, Connie Toops)

of habitats than it actually does. The potential ecological niche of an organism is its **fundamental niche**, but various factors such as competition with other species may exclude it from part of its fundamental niche. Thus, the lifestyle that an organism actually pursues and the resources that it actually uses make up its **realized niche**.

An example may help clarify the distinction between fundamental and realized niches. The green anole, a lizard native to Florida and other southeastern states, perches on trees, shrubs, walls, or fences during the day and waits for insect and spider prey (Figure 4–8a). In past years these little lizards were widespread in Florida. Several years ago, however, a related species, the brown anole, was introduced from Cuba into southern Florida and quickly became common (Figure 4–8b; also see Chapters 1 and 16 for discussions of the introductions of non-native species into new habitats). Suddenly the green anoles became rare—apparently driven out of their habitat by competition from the slightly larger brown lizard. Careful investigation disclosed, however, that green anoles were still around. They were now confined largely to the vegetation in wetlands and to the foliated crowns of trees, where they were less easily seen.

The habitat portion of the green anole's fundamental niche includes the trunks and crowns of trees, exterior house walls, and many other locations. Where they became established, brown anoles were able to drive green anoles out from all but wetlands and tree crowns, so that the green anole's realized niche became much smaller as a result of competition (Figure 4–8c,d). Because natural communities consist of numerous species, many of which compete to some extent, the complex interactions among species produce the realized niche of each.

Competitive Exclusion

When two species are very similar, as are the green and brown anoles, their fundamental niches may overlap. However, many ecologists think that no two species can indefinitely occupy the same niche in the same community, because **competitive exclusion** eventually occurs. In competitive exclusion, one species is excluded from a niche by another as a result of competition between species—that is, *interspecific competition.* Although it is possible for different species to compete for some necessary resource without being total competitors, two species with absolutely identical ecological niches cannot coexist. Coexistence *can* occur, however, if the overlap in the two species' niches is reduced. In the lizard example, direct competition between the two species was reduced as the brown anole excluded the green anole from most of its former habitat until the only place that remained open to it was wetland vegetation and tree crowns.

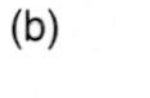

The initial evidence that competition between species determines an organism's realized niche came from a series of experiments conducted by the Russian biologist G. F. Gause in 1934. In one study Gause grew populations of two species of *Paramecium* (a type of protozoa), *P. aurelia* and the larger *P. caudatum* (Figure 4–9). When grown in separate test tubes, the population of each species quickly increased to a high level, which was maintained for some time thereafter. When grown together, however, only *P. aurelia* thrived. *P. caudatum* dwindled and eventually died out. Under different sets of culture conditions, *P. caudatum* prevailed over *P. aurelia.*

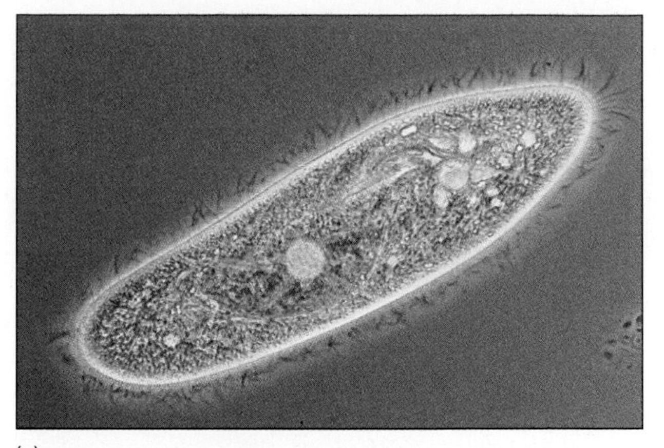

(a)

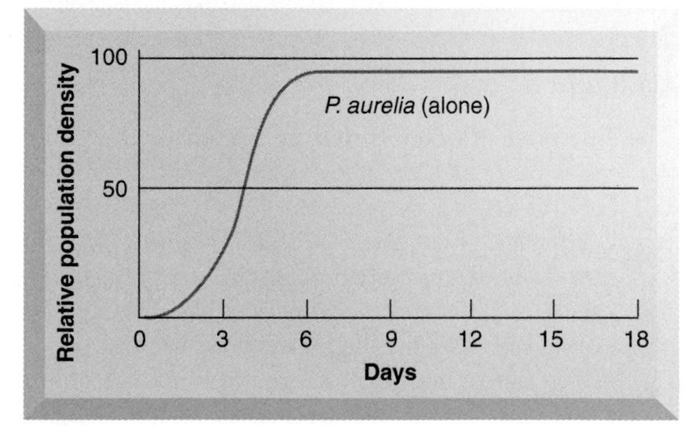

(b)

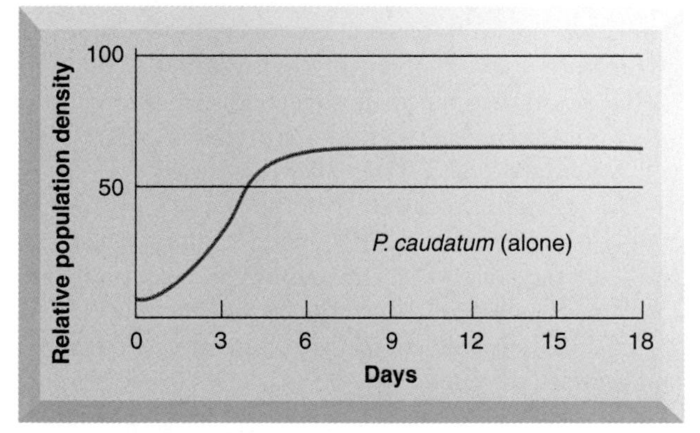

(c)

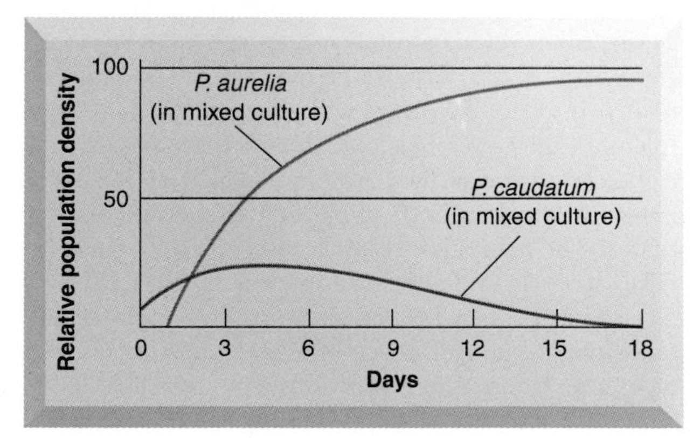

(d)

Figure 4–9

Competition among paramecia. (a) *Paramecium* is a complex single-celled protist. The top graph (b) shows how a population of *Paramecium aurelia* grows in separate culture (that is, in a single-species environment), whereas (c) shows a separate culture of *Paramecium caudatum.* The bottom graph (d) shows how these two species grow together in a single culture (that is, in competition with each other). *P. aurelia* outcompetes *P. caudatum* and drives it to extinction. (a, M. Abbey/Photo Researchers, Inc.; b,c,d, After Gause 1934)

Gause interpreted this to mean that, although one set of conditions favored one species, a different set favored the other. Nonetheless, because both species were similar, given time, one or the other would eventually triumph at the other's expense.

Apparent contradictions to the competitive exclusion principle sometimes occur. In Florida, for instance, native fishes and introduced (non-native) fishes seem to coexist in identical niches. Similarly, botanists have observed closely competitive plant species in the same location. Although such situations seem to contradict the concept of competitive exclusion, the realized niches of these organisms may differ in some way that biologists do not yet understand.

Limiting Factors

The environmental factors that actually determine an organism's realized niche can be extremely difficult to identify. For this reason, the concept of ecological niche is largely abstract, although some of its dimensions can be experimentally determined. The environmental variable that, because it is scarce or unfavorable, restricts the niche of an organism is called a **limiting factor**.

What factors actually determine the niche of a species? A niche is basically determined by the total of a species' structural, physiological, and behavioral adaptations. Such adaptations determine, for example, the tolerance an organism has for environmental extremes. If any feature of its environment lies outside the bounds of its tolerance, then the organism cannot live there. Just as you would not expect to find a cactus living in a pond, you would not expect water lilies in a desert.

Most limiting factors that have been investigated are simple variables such as the mineral content of soil, extremes of temperature, and amount of precipitation. Such investigations have disclosed that any factor that exceeds an organism's tolerance, or that is present in quantities smaller than the minimum required, limits the occurrence of that organism in an ecosystem. By their interaction, such factors help to define an organism's ecological niche.

Limiting factors often affect only one part of an organism's life cycle. For instance, although adult blue crabs can live in almost fresh water, they cannot become permanently established there because their larvae cannot tolerate fresh water. Similarly, the ring-necked pheasant, a popular game bird, has been introduced widely in North America but does not survive in the southern United States. The adult birds do well, but the eggs cannot develop properly in the high southern temperatures.

We have seen that an organism's ecological niche takes into account all aspects of that organism's existence. Now we examine the number of niches available in different communities, which in turn affects the number of species those communities contain.

SPECIES DIVERSITY

Species diversity, the number of species present, varies greatly from one community to another. Tropical rain forests and coral reefs are examples of communities with extremely high species diversity. In contrast, geographically isolated islands and mountaintops exhibit low species diversity.

What determines the number of species in a community? There seems to be no single answer, but several factors appear to be significant, such as the abundance of potential ecological niches, location at the margins of adjacent communities, geographical isolation, habitat stress, dominance of one species over others, and geological history.

Species diversity is related to the *abundance of potential ecological niches*. An already complex community offers a greater variety of potential ecological niches than does a simple community (Figure 4–10). It may become even more complex if organisms potentially capable of filling those niches evolve or migrate into the community. Thus, it appears that species diversity is self-reinforcing.

Species diversity is usually greater *at the margins of adjacent communities* than in their centers. This is because

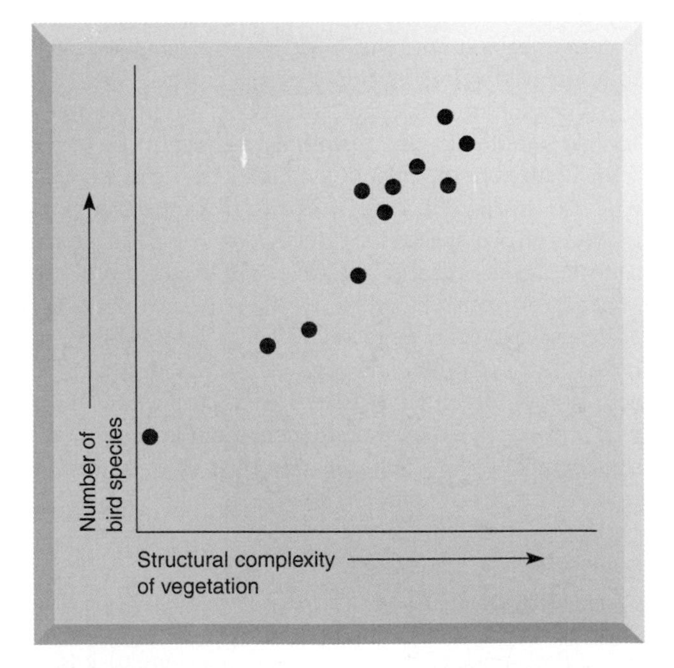

Figure 4–10

As community complexity increases, more species are able to live in the community. A community in which the vegetation is structurally complex—a forest, for example—provides birds with more kinds of food and hiding places than a grassland. (After MacArthur and MacArthur 1961)

the **ecotones**—transitional zones where two or more communities meet—contain all or most of the ecological niches of the adjacent communities as well as some unique to the ecotone. This change in species composition produced at ecotones is known as the **edge effect**.

Species diversity is inversely related to the *geographical isolation* of a community. Isolated island communities tend to be much less diverse than are communities in similar environments found on continents. This is due partly to the difficulty encountered by many species in reaching and successfully colonizing the island. Also, sometimes species become locally extinct as a result of random events. In isolated environments such as islands or mountaintops, extinct species cannot be readily replaced. Isolated areas are likely to be small and to possess fewer potential ecological niches.

Generally, species diversity is inversely related to the *environmental stress of a habitat*. Only those species capable of tolerating extreme environmental conditions can live in an environmentally stressed community. Thus, the species diversity of a polluted stream is low compared to that of a nearby pristine stream. Similarly, the species diversity of high-latitude (further from the equator) communities exposed to harsh climates is less than that of lower-latitude (closer to the equator) communities with milder climates. Although the equatorial countries of Colombia, Ecuador, and Peru occupy only 2 percent of the Earth's land, they contain an astonishing 45,000 native plant species. The continental United States and Canada, with a significantly larger land area, possess a total of 19,000 native species of plants. Ecuador alone contains more than 1300 native species of birds—twice as many as the United States and Canada combined.

Species diversity is reduced when any *one species enjoys a decided position of dominance within a community* so that it is able to appropriate a disproportionate share of available resources, thus crowding out, or outcompeting, other species. In an experiment conducted in the Chihuahuan desert of southeastern Arizona, the removal of a single dominant species, the kangaroo rat, from several plots resulted in an increased diversity of rodent species. This was attributed to less competition for food and to an altered habitat, because the number of plants increased dramatically after the removal of the kangaroo rats.

Species diversity is greatly affected by *geological history*. Tropical rain forests are very old, stable communities that have undergone few climatic changes in the entire history of the Earth. During this time, myriad species evolved in tropical rain forests, having experienced few or no abrupt climatic changes that might have led to their extinction. In contrast, glaciers have repeatedly altered temperate and arctic regions during Earth's history. An area recently vacated by glaciers will have a low species diversity because few species will as yet have had a chance to enter it and become established.

BAD NEWS FOR RAIN FORESTS

The Pacific island of Krakatoa has provided scientists with a perfect long-term study of succession in a tropical rain forest. In 1883 a volcanic eruption destroyed all life on the island. Biologists have surveyed the ecosystem in the more than 100 years since the devastation to document the return of life forms. As of 1991, scientists had found that the progress of succession was extremely slow. For example, a portion of Krakatoa's forest might have only one-tenth the tree species diversity of an undisturbed tropical rain forest. The lack of plant diversity has in turn limited the number of colonizing animal species—in a forest area where zoologists would expect more than 100 butterfly species, there were only 2 species. Krakatoa's slow recovery has surprised researchers, who expected a tropical ecosystem to mature in decades rather than the apparent millennia it will take Krakatoa. A sobering lesson can be applied to other tropical forests that are being destroyed by humans, such as in Zaire or the Amazon. Like Krakatoa, these ecosystems may recover only very slowly from extreme devastation, certainly not in our lifetimes.

EVOLUTION: HOW POPULATIONS CHANGE OVER TIME

We have seen that communities vary in species diversity, but where did Earth's remarkable diversity of species come from in the first place? **Evolution**, which can be defined as a genetic change in a population of organisms that occurs over time, explains the diversity of life present on Earth today. The concept of evolution dates back to the time of Aristotle (384–322 B.C.), but Charles Darwin (1809–1882), a 19th century naturalist, discovered the mechanism of evolution that is still accepted by the scientific establishment today. Darwin proposed the theory of evolution by natural selection in his monumental book, *Origin of Species by Means of Natural Selection,* which was published in 1859. Since that time, scientists have accumulated an enormous body of observations and experiments that support Darwin's theory. Although biologists still do not agree completely on some aspects by which evolutionary changes occur, the concept that evolution by natural selection has taken place is now well documented.

Natural Selection

Darwin's mechanism of evolution by **natural selection** consists of four observations about the natural world: overproduction, variation, limits on population growth, and survival to reproduce.

1. *Overproduction.* Each species produces more offspring than will survive to maturity. Natural populations have the reproductive potential to continuously increase their numbers over time. For example, if each breeding pair of elephants produces 6 offspring during a 90-year lifespan, in 750 years a single pair of elephants will have given rise to a population of 19 million! Yet elephants have not overrun the planet.
2. *Variation.* The individuals in a population exhibit variation in their traits. Some of these traits improve the chances of an individual's survival and reproductive success, whereas other traits do not. It is important to remember that the variation necessary for evolution by natural selection must be inheritable so that it can be passed on to offspring.
3. *Limits on population growth.* There is only so much food, water, light, growing space, and so on available to a population, and organisms compete with one another for the limited resources available to them. Because there are more individuals than the environment can support, not all of the offspring will survive to reproductive age. Other limits on population growth include predators and diseases.
4. *Survival to reproduce, or "survival of the fittest."* Offspring that possess the most favorable combination of characteristics will be most likely to survive, reproduce, and pass their traits on to the next generation. Natural selection thus causes an increase of favorable traits and a decrease of unfavorable traits within a population, which results in the individual members of a population becoming better adapted to local conditions (Figure 4–11). Over time these changes accumulate in geographically separated populations and may be significant enough to cause new species to evolve.

SUCCESSION: HOW COMMUNITIES CHANGE OVER TIME

A community of organisms does not spring into existence full-blown but develops gradually, through a series of stages. The process of community development over time, which involves species in one stage being replaced by different species, is called **succession**. An area is initially colonized by certain organisms that are replaced over time by others, which themselves may be replaced much later by still others.

Ecologists initially thought that succession inevitably led to a stable and persistent community, such as a for-

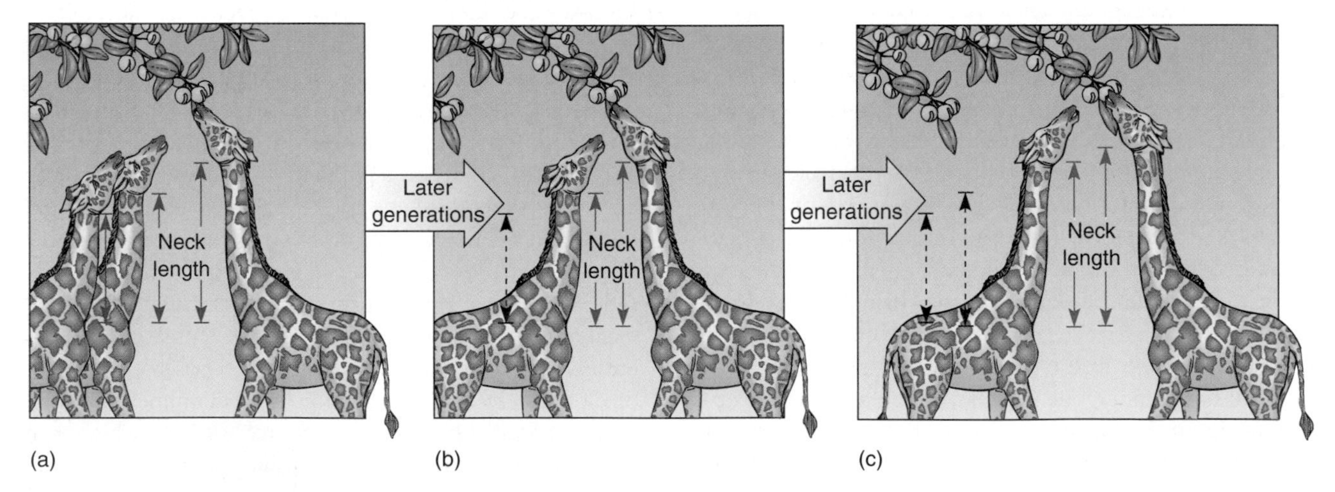

Figure 4–11

How the giraffe got its long neck, as explained by Darwin's theory of evolution by natural selection. (a) Ancestral giraffes had slightly varying neck lengths, which were determined by heredity. (b) The animals with longer necks were more likely to obtain food and therefore to survive and reproduce. As a result, the animals with longer necks became more common in the population. (c) After many generations, the average neck length increased to that of the modern giraffe.

est, known as a *climax community.* But more recently, the traditional view has fallen out of favor. The apparent stability of a "climax" forest, for example, is probably the result of how long trees live relative to the human lifespan. It is now recognized that mature "climax" communities are not in a state of permanent equilibrium, but rather in a state of continual flux. A mature community changes in species composition and in the relative abundance of each species, despite the fact that it retains an overall uniform appearance.

Succession is usually described in terms of the changes in the species composition of the vegetation of an area, although each successional stage also has its own characteristic kinds of animals and other organisms. The time involved in ecological succession is on the scale of tens, hundreds, or thousands of years, not the millions of years involved in the evolutionary time scale.

Primary Succession

Primary succession is the change in species composition over time in an environment that has not previously been inhabited by organisms. No soil exists when primary succession begins. Bare rock surfaces, such as recently formed volcanic lava and rock scraped clean by glaciers, are examples of sites where primary succession might occur (Figure 4–12).

Although the details vary from one site to another, lichens are often the most important element in the **pioneer community**, which is the initial community that develops during primary succession on bare rock. Lichens secrete acids that help to break the rock apart, beginning the process of soil formation. Over time, the lichen community may be replaced by mosses and drought-resistant ferns, followed in turn by tough grasses and herbs. Once enough soil has accumulated, grasses and herbs may be replaced by low shrubs, which in turn would be replaced by forest trees in several distinct stages. Primary succession on bare rock, which may take hundreds or thousands of years, might proceed from a pioneer community to a forest community as follows: lichens → mosses → grasses → shrubs → trees.

(a) (b) (c)

Figure 4–12

Primary succession has occurred in Glacier Bay, Alaska, as glaciers retreated there during the past 200 years. (a) The barren landscape exposed after the retreat of the glacier is initially colonized by lichens, then mosses. (b) At a later time, dwarf trees and shrubs colonize the area. (c) Still later, spruces dominate the community. (a,c, Wolfgang Kaehler; b, Glenn N. Oliver/Visuals Unlimited)

Mini-Glossary of Succession

succession: A process of community development that involves a changing sequence of species.

pioneer community: The first organisms to colonize (or recolonize) an area.

primary succession: Ecological succession in an environment that has not previously been inhabited.

secondary succession: Ecological succession in an environment that has previously been inhabited.

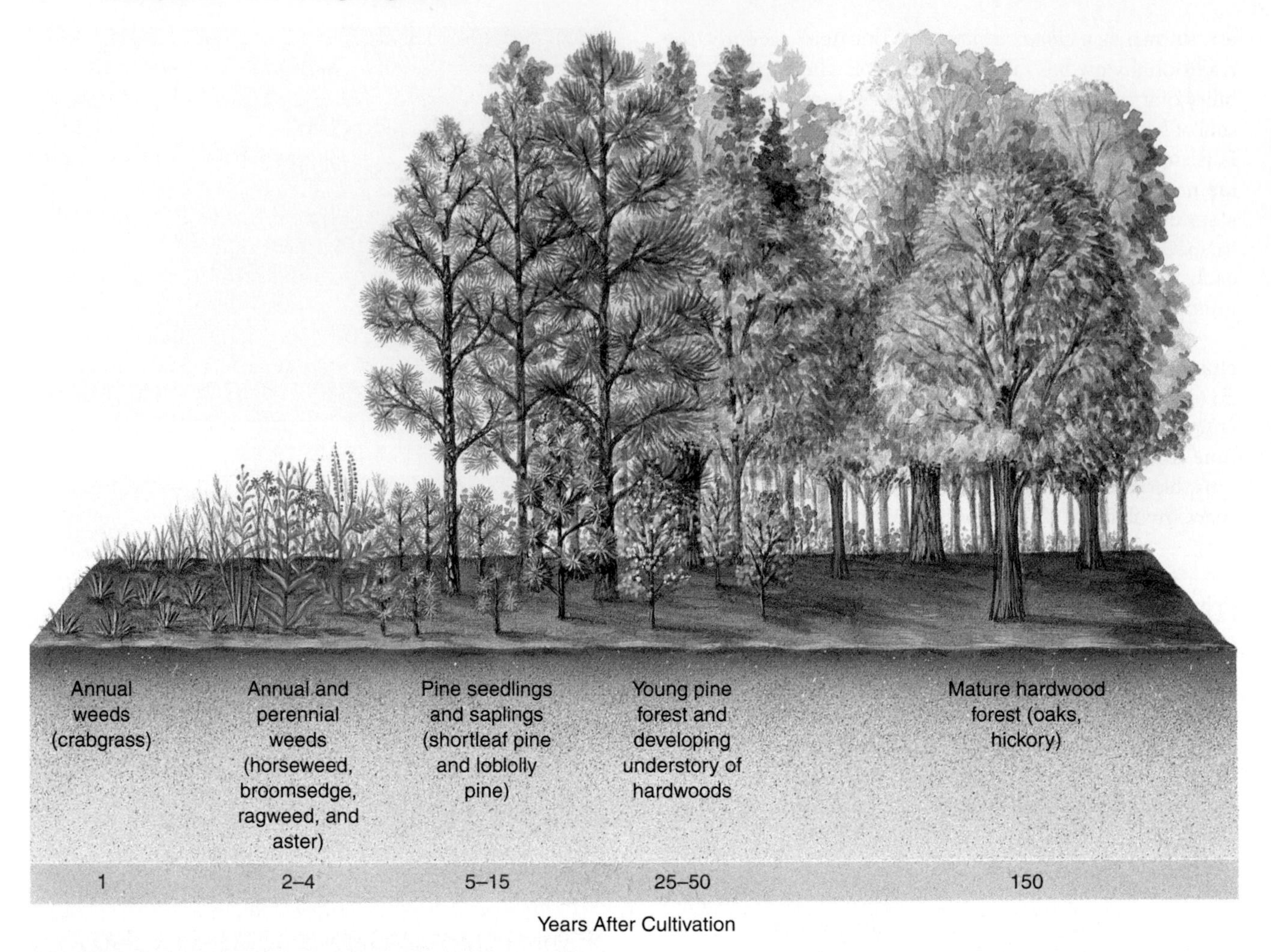

Figure 4–13
Secondary succession on an abandoned field in North Carolina.

Secondary Succession

Secondary succession is the change in species composition over time in an environment already substantially modified by a preexisting community. Soil is already present at these sites. An open area caused by a forest fire and abandoned farmland are common examples of sites where secondary succession occurs.

Secondary succession on abandoned farmland has been studied extensively. Although it takes more than 100 years for secondary succession to occur at a single site, it is possible for a single researcher to study old field succession in its entirety by observing different sites in the same area that have been undergoing succession for different amounts of time. The biologist may examine court records to determine when each field was abandoned.

Abandoned farmland in North Carolina is colonized by a predictable succession of communities (Figure 4–13). The first year after cultivation ceases, the field is dominated by crabgrass. During the second year, horseweed, a larger plant that outgrows crabgrass, is the dominant species. Horseweed does not dominate more than one year, however, because decaying horseweed roots inhibit the growth of young horseweed seedlings. In addition, horseweed does not compete well with other plants that become established in the third year. During the third year after the last cultivation, other weeds—broomsedge, ragweed, and aster—become established. Typically, broomsedge outcompetes aster, because broomsedge is drought-tolerant, whereas aster is not.

In years 5 to 15, the dominant plants in an abandoned field are pines such as shortleaf pine and loblolly pine. Through the buildup of litter, such as pine needles and branches, on the soil, pines produce conditions that cause the earlier dominant plants to decline in importance. Over time, pines give up their dominance to hardwoods such as oaks. The replacement of pines by oaks depends primarily on the environmental changes produced by the pines. The pine litter causes soil changes, such as an increase in water-holding capacity, that are necessary in order for young oak seedlings to become established. In addition, hardwood seedlings are more tolerant of shade than young pine seedlings. Secondary succession on abandoned farmland might proceed as follows: crabgrass → horseweed → broomsedge and other weeds → pine trees → hardwood trees.

Animal life during secondary succession

As secondary succession in North Carolina proceeds, a progression of animal life follows the changes in vegetation. Although a few animals—the short-tailed shrew, for example—are found in all stages of abandoned farmland succession, most animals appear with certain stages and disappear with others. During the crabgrass and weed stages of secondary succession, the environment is characterized by open fields that support grasshoppers, meadow voles, cottontail rabbits, and birds such as grasshopper sparrows and meadowlarks. As young pine seedlings become established, animals of open fields give way to animals common in mixed herbaceous and shrubby habitats. Now white-tailed deer, white-footed mice, ruffed grouse, robins, and song sparrows are common, whereas grasshoppers, meadow mice, grasshopper sparrows, and meadowlarks disappear. As the pine seedlings grow into trees, animals of the forest replace those common in mixed herbaceous and shrubby habitats. Cottontail rabbits give way to red squirrels, and ruffed grouse, robins, and song sparrows are replaced by warblers and veeries. Thus, each stage of succession supports its own characteristic animal life.

SUMMARY

1. A biological community consists of a group of organisms that interact and live together. A living community and its abiotic environment constitute an ecosystem. Within an ecosystem, keystone species help determine the species composition and functioning of the entire ecosystem. Although present in relatively small numbers, keystone species often affect the available amount of food, water, or some other resource for the entire community.

2. Predation is the consumption of one species (the prey) by another (the predator). Two effective predator strategies are pursuit and ambush. Adaptations that protect plants from being eaten include spines; thorns; tough, leathery leaves; and protective chemicals. Strategies that help animals avoid being killed and eaten include flight, association in groups, and camouflage, as well as a variety of mechanical and chemical defenses.

3. Symbiosis is any intimate association between two or more different species. Both partners benefit from a mutualistic association. In commensalism, one organism benefits and the other is unaffected. In parasitism, the parasite benefits and the host is harmed.

4. The distinctive lifestyle and role of an organism in a community are its ecological niche. The niche takes into account all aspects of the organism's existence—that is, all of the physical, chemical, and biological factors that the organism needs to survive, to remain healthy, and to reproduce.

5. Organisms are potentially able to exploit more resources and play a broader role in the life of their community than they actually do. The potential ecological niche of an organism is its fundamental niche, whereas the niche an organism actually occupies is its realized niche.

6. Many ecologists think that no two species can occupy the same niche in the same community for an indefinite period of time because of competitive exclusion. In this process, one species is excluded by another as a result of competition for limited resources.

7. Community complexity is expressed in terms of species diversity. Species diversity is often great when there are many potential ecological niches, at the margins of adjacent communities, when a community is not isolated or stressed, when one species does not dominate others, and when communities have a long history.

8. Evolution can be defined as a genetic change in a population of organisms that occurs over time. Charles Darwin proposed the theory of evolution by natural selection, which is based on four premises. (1) Each species produces more offspring than will survive to maturity. (2) The individuals in a population exhibit inheritable variation in their traits. (3) Organisms compete with one another for the resources needed to survive. (4) Those individuals with the most favorable combination of traits are most likely to survive and reproduce, passing those genetic characters on to the next generation.

9. Succession is the orderly replacement of one community by another. Primary succession begins in an environment that has not previously been inhabited. Secondary succession begins in an area where there was a preexisting community and a well-formed soil.

THINKING ABOUT THE ENVIRONMENT

1. What might the study of microcosms show us about conditions for stability on our own spaceship, planet Earth?

2. Some biologists think that protecting keystone species would help preserve biological diversity in an ecosystem. Explain.

3. Describe how evolution has affected predator-prey relationships.

4. Why is a realized niche usually narrower, or more restricted, than a fundamental niche?

5. What portion of the human's fundamental niche are we occupying today? Do you think our realized niche is changing? Why or why not?

6. Describe two determinants of species diversity and give an example of each.

7. Give an example of a coevolutionary relationship in which humans are involved.

8. Is human survival dependent on Earth's web of life? Explain.

*9. Examine and interpret the figure shown below, to answer the following questions.
   a. When (approximately what day) does the population of *P. aurelia* exceed that of *P. caudatum?*
   b. When does the population of *P. caudatum* peak?
   c. When does the population of *P. aurelia* peak?

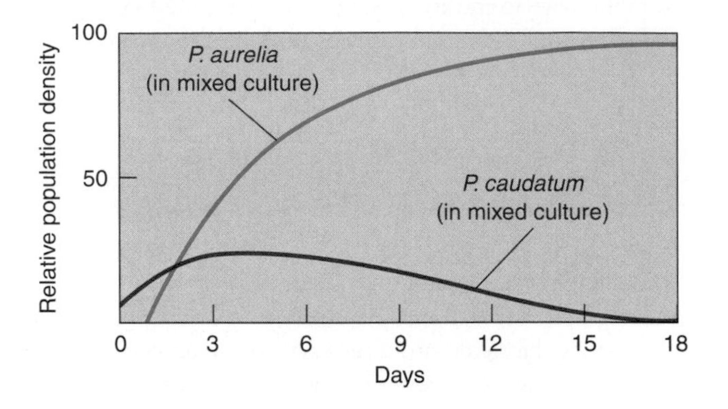

RESEARCH PROJECTS

Choose an organism found in a local natural area such as a pond or forest. Describe its ecological niche, including other species that it depends on. Predict what would happen to your organism if those species disappeared from the local environment.

Choose a local town and identify changes in the kinds of plant and animal life as the town progressed from a natural area to a human settlement.

On-line materials relating to this chapter are on the ► World Wide Web at **http://www.saunderscollege.com/lifesci/environment2** (select Chapter 4 from the Table of Contents).

SUGGESTED READING

Beardsley, T. "Recovery drill," *Scientific American,* November 1990. Some conventional ideas about how communities should respond to environmental catastrophes are being challenged by the recovery of Mount St. Helens.

Boucher, D. H. "Growing Back after Hurricanes," *BioScience* Vol. 40, No. 3, March 1990. The significance for communities of periodic environmental catastrophes such as hurricanes is causing ecologists to reconsider the idea of a climax community.

Conniff, R. "Yellowstone's Rebirth amid the Ashes Is Not Neat or Simple, but It's Real," *Smithsonian,* September 1989. Secondary succession of the forests that were burned during the September 1988 fire in Yellowstone.

DeVries, P. J. "Singing Caterpillars, Ants, and Symbiosis," *Scientific American,* October 1992. Fascinating symbiotic partnership between ants and caterpillars protects the caterpillars from predators.

Gillis, A. M. "Sea Dwellers and Their Sidekicks," *BioScience* Vol. 43, No. 9, October 1993. Unusual mutualistic relationships between bioluminescent (light-producing) bacteria and squid or fishes.

Power, M. E., et al. "Challenges in the Quest for Keystones," *BioScience* Vol. 46, No. 8, September 1996. Examines the significance of keystone species and some of the difficulties associated with identifying them.

Tumlinson, J. H., W. J. Lewis, and L. E. M. Vet. "How Parasitic Wasps Find Their Hosts," *Scientific American,* March 1993. One way that wasps identify their caterpillar hosts is by recognizing chemicals produced by the plants that caterpillars feed on.

Walker, G. "Secrets from Another Earth," *New Scientist,* May 18, 1996. Examines recent developments at Biosphere 2.

According to the Gaia hypothesis, the Earth is a huge "organism" capable of self-maintenance. (NASA)

Ecosystems and the Physical Environment

Almost completely isolated from everything in the Universe but sunlight, Earth has often been compared to a vast spaceship whose life-support system consists of the organisms that inhabit it. These living things modify the composition of gases in the atmosphere, transfer energy, and recycle waste products with great efficiency.

Broadly speaking, Earth is alive in the sense that it is capable of self-maintenance. Organisms interact with the abiotic environment to produce and maintain the chemical composition of the atmosphere, the global temperature, the ocean's salinity, and other characteristics. Thus, Earth's environment and organisms depend on one another and work together as a homeostatic mechanism. (Biological systems have homeostatic mechanisms to help maintain a steady state or constant environment.) Because of these relationships, the entire planet is comparable to a living organism. The series of hypotheses that Earth's organisms adjust the environment to improve conditions for life has been called, collectively, the **Gaia hypothesis.** (*Gaia* is derived from the Greek *Gaea*, which is the Earth personified as a goddess.) The Gaia hypothesis was first proposed in the early 1970s by James Lovelock, a British chemist, and Lynn Margulis, an American biologist.

As an example of the Gaia hypothesis, consider the Earth's temperature. It is generally accepted that the temperature has remained relatively constant at a level suitable for life over the past 3.5 to 4 billion years that life has existed. Yet there is evidence that the sun has been heating up during that time. Why has the Earth's temperature not increased? Gaia proponents hypothesize that Earth has remained the same temperature because the level of atmosphere-warming CO_2 has dropped during that time. This happened because the living Earth compensated for increased sunlight by "fixing" CO_2 into calcium carbonate shells of countless billions of marine phytoplankton (photosynthetic plankton). As the phytoplankton died, their shells sank to the ocean floor, thus removing CO_2 from the system. This planetary temperature mechanism is an example of a feedback loop between the abiotic environment and organisms, which mutually interact to regulate Earth's temperature.

Many scientists are reluctant to accept all aspects of the Gaia hypothesis. Virtually everyone agrees that the environment modifies organisms and that organisms modify the environment; these parts of the Gaia hypothesis are easily testable, and much evidence supports them. Moreover, the dependence of organisms on the physical environment is indisputable. However, it is impossible to test the premise that organisms *deliberately* adjust the global environment to their advantage. ◄

THE CYCLING OF MATERIALS WITHIN ECOSYSTEMS

In Chapter 3 we learned that energy flows in one direction through an ecosystem. In contrast, matter, the material of which organisms are composed, moves in numerous cycles from one part of an ecosystem to another—that is, from one organism to another and from living organisms to the abiotic environment and back again (Figure 5–1). We call these cycles of matter **biogeochemical cycles**.

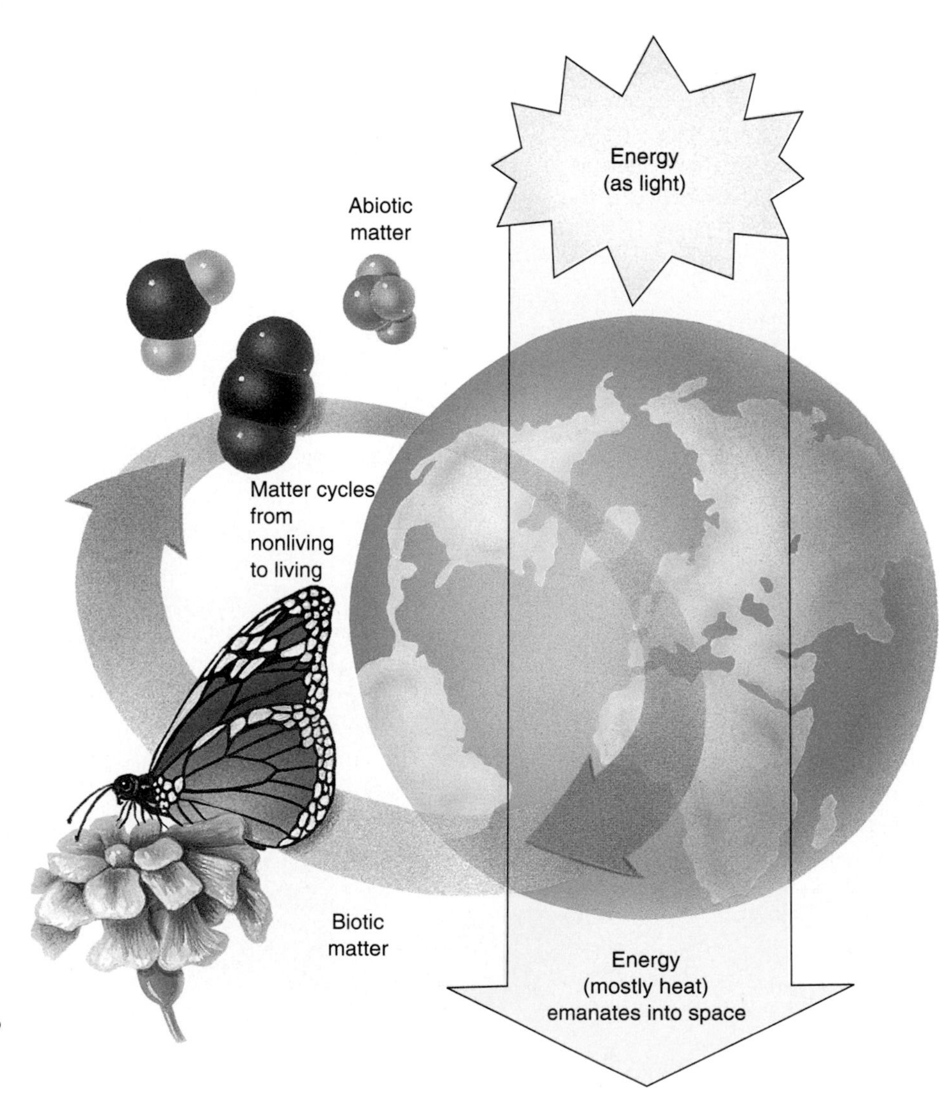

Figure 5–1

Although energy flows one way through ecosystems, matter continually cycles from the abiotic to the biotic components of ecosystems and back again.

With respect to matter, the Earth is essentially a *closed system*; for all practical purposes, matter cannot escape from Earth's boundaries. The materials used by organisms cannot be "lost," although they can end up in locations that are outside the reach of organisms. Usually, however, materials are reused and are often recycled both within and among ecosystems.

Four different biogeochemical cycles of matter—carbon, nitrogen, phosphorus, and water—are representative of all biogeochemical cycles. These four cycles are particularly important to organisms, as these materials are used to make the chemical compounds of cells. Carbon, nitrogen, and water have gaseous components and so cycle over large distances of the atmosphere with relative ease. Phosphorus, however, is an element that is completely nongaseous, and as a result, only local cycling occurs easily.

The Carbon Cycle

Proteins, carbohydrates, and other molecules essential to life contain carbon, so organisms must have carbon available to them. Carbon makes up approximately 0.03 percent of the atmosphere as a gas, carbon dioxide (CO_2). It is also present in the ocean as dissolved carbon dioxide—that is, carbonate ($CO_3{}^{2-}$) and bicarbonate ($HCO_3{}^{-}$)—and in rocks such as limestone. The global movement of carbon between the abiotic environment, including the atmosphere, and organisms is known as the **carbon cycle** (Figure 5–2).

During photosynthesis (see Chapter 3), plants, algae, and certain bacteria remove carbon dioxide from the air and *fix*, or incorporate, it into complex chemical compounds such as sugar. The overall equation for photosynthesis is

$$6CO_2 + 12H_2O \xrightarrow{\text{light}} \underset{\text{sugar (glucose)}}{C_6H_{12}O_6} + 6O_2 + 6H_2O$$

Plants use sugar to make other compounds. Thus, photosynthesis incorporates carbon from the abiotic environment into the biological compounds of producers. Those compounds are usually used as fuel for cell respiration (see Chapter 3) by the producer that made them, by a consumer that eats the producer, or by a decomposer that breaks down the remains of the producer or consumer. The overall equation for cell respiration is

$$C_6H_{12}O_6 + 6O_2 + 6H_2O \longrightarrow 6CO_2 + 12H_2O + \text{energy for biological work}$$

Thus, carbon dioxide is returned to the atmosphere by the process of cell respiration.

Sometimes the carbon in biological molecules is not recycled back to the abiotic environment for some time.

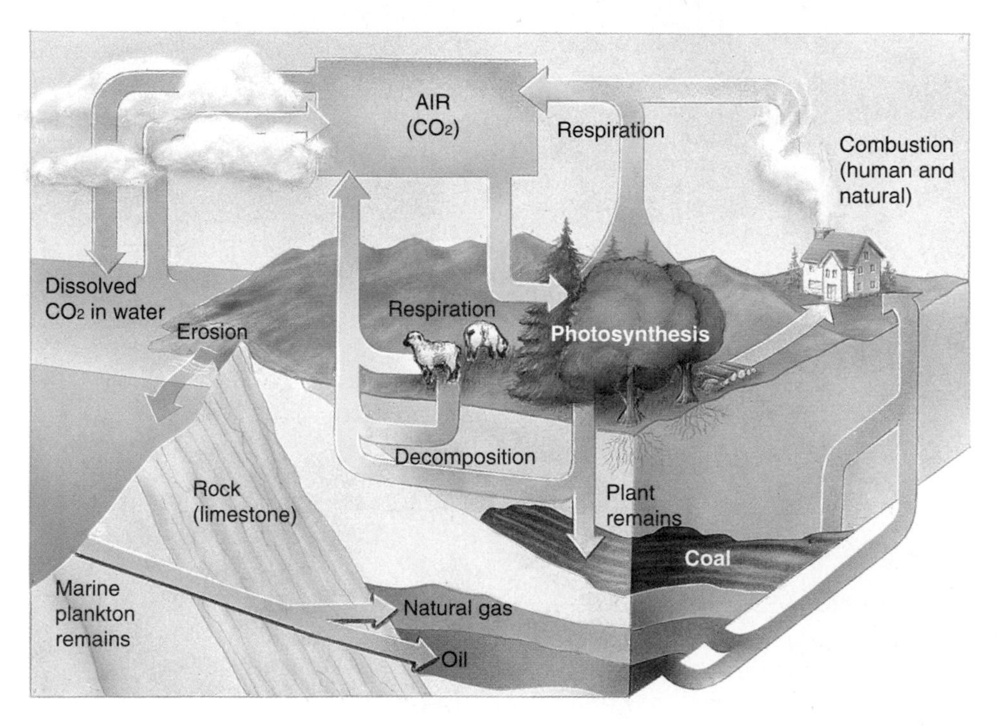

Figure 5–2
A simplified diagram of the carbon cycle. Carbon, in the form of carbon dioxide, enters organisms from the abiotic environment when plants and other producers photosynthesize. Carbon returns to the environment when organisms respire, are decomposed, or are burned (combustion). Fossil fuels, which are carbon-containing compounds formed from the remains of ancient organisms, and from the carbon of limestone rock and shells of marine organisms, may take millions of years to cycle back to the biotic world.

For example, a large amount of carbon is stored in the wood of trees, where it may stay for several hundred years or even longer. In addition, millions of years ago, vast coal beds formed from the bodies of ancient trees that did not decay fully before they were buried. Similarly, the oils of unicellular marine organisms probably gave rise to the underground deposits of oil and natural gas that accumulated in the geological past. Coal, oil, and natural gas, called **fossil fuels** because they formed from the remains of ancient organisms, are vast deposits of carbon compounds—the end products of photosynthesis that occurred millions of years ago (see Chapter 10).

The carbon in coal, oil, natural gas, and wood can be returned to the atmosphere by the process of burning, or **combustion**. In combustion, organic molecules are rapidly oxidized—that is, combined with oxygen—and thus converted into carbon dioxide and water, with an accompanying release of heat and light.

An even greater amount of carbon that leaves the carbon cycle for millions of years is incorporated into the shells of marine organisms. When these organisms die, their shells sink to the ocean floor and are covered by sediments, forming seabed deposits thousands of feet thick. The deposits are eventually cemented together to form a sedimentary rock called limestone. The crust is dynamically active, and over millions of years, sedimentary rock on the bottom of the sea floor may lift to form land surfaces. The summit of Mount Everest, for example, is composed of sedimentary rock. After limestone is exposed by the process of geological uplift, it slowly erodes away by chemical and physical weathering processes. This returns the carbon to the water and atmosphere, where it is available to participate in the carbon cycle once again.

Thus, photosynthesis removes carbon dioxide from the abiotic environment and incorporates it into biological molecules, and cell respiration, combustion, and erosion return the carbon in biological molecules to the water and atmosphere of the abiotic environment.

The carbon cycle and global warming

Human activities have disturbed the balance of the carbon cycle. From the advent of the Industrial Revolution to the present, humans have burned increasing amounts of fossil fuels—coal, oil, and natural gas. This trend, along with a greater combustion of wood as a fuel and the burning of large sections of tropical forest, has released carbon dioxide into the atmosphere at a rate greater than the natural carbon cycle can handle.

The slow and steady rise of CO_2 in the atmosphere may cause changes in climate called global warming. Global warming could result in a rise in sea level, changes in precipitation patterns, death of forests, extinction of organisms, and problems for agriculture. It could force the displacement of thousands or even millions of people, particularly from coastal areas. A more thorough discussion of increasing atmospheric CO_2 and global warming is found in Chapter 20.

The Nitrogen Cycle

Nitrogen is crucial for all organisms because it is an essential part of biological molecules such as proteins and nucleic acids. At first glance it would appear that a shortage of nitrogen for organisms is impossible: the atmosphere is 78 percent nitrogen gas (N_2), a 2-atom molecule. But molecular nitrogen is so stable that it does not readily combine with other elements. Therefore, molecular nitrogen must first be broken apart before the nitrogen can combine with other elements to form proteins and nucleic acids.

There are five steps in the **nitrogen cycle**, in which nitrogen cycles between the abiotic environment and organisms: nitrogen fixation, nitrification, assimilation, ammonification, and denitrification (Figure 5–3). Bacteria are involved in all of these steps except assimilation.

Nitrogen fixation

The first step in the nitrogen cycle, biological **nitrogen fixation**, is the conversion of gaseous nitrogen (N_2) to ammonia (NH_3). The process gets its name from the fact that nitrogen is *fixed* into a form that organisms can use. Although considerable nitrogen is also fixed by combustion, volcanic action, lightning discharges, and by industrial processes, all of which supply enough energy to break up molecular nitrogen, most nitrogen fixation is thought to be biological. It is carried out by nitrogen-fixing bacteria, including cyanobacteria, in soil and aquatic environments. Nitrogen-fixing bacteria employ an enzyme called **nitrogenase** to break up molecular nitrogen and combine it with hydrogen.

Because nitrogenase functions only in the absence of oxygen, the bacteria that use nitrogenase must insulate the enzyme from oxygen by some means. Some nitrogen-fixing bacteria live beneath layers of oxygen-excluding slime on the roots of certain plants. Other important nitrogen-fixing bacteria, *Rhizobium*, live inside special swellings, or **nodules**, on the roots of legumes such as beans or peas and some woody plants (Figure 5–4a). The relationship between *Rhizobium* and its host plants is mutualistic: the bacteria receive carbohydrates from the plant, and the plant receives nitrogen in a form that it can use.

In aquatic environments most of the nitrogen fixation is done by cyanobacteria. Filamentous cyanobacteria have special oxygen-excluding cells called **heterocysts** that function as the sites of nitrogen fixation (Figure 5–4b).

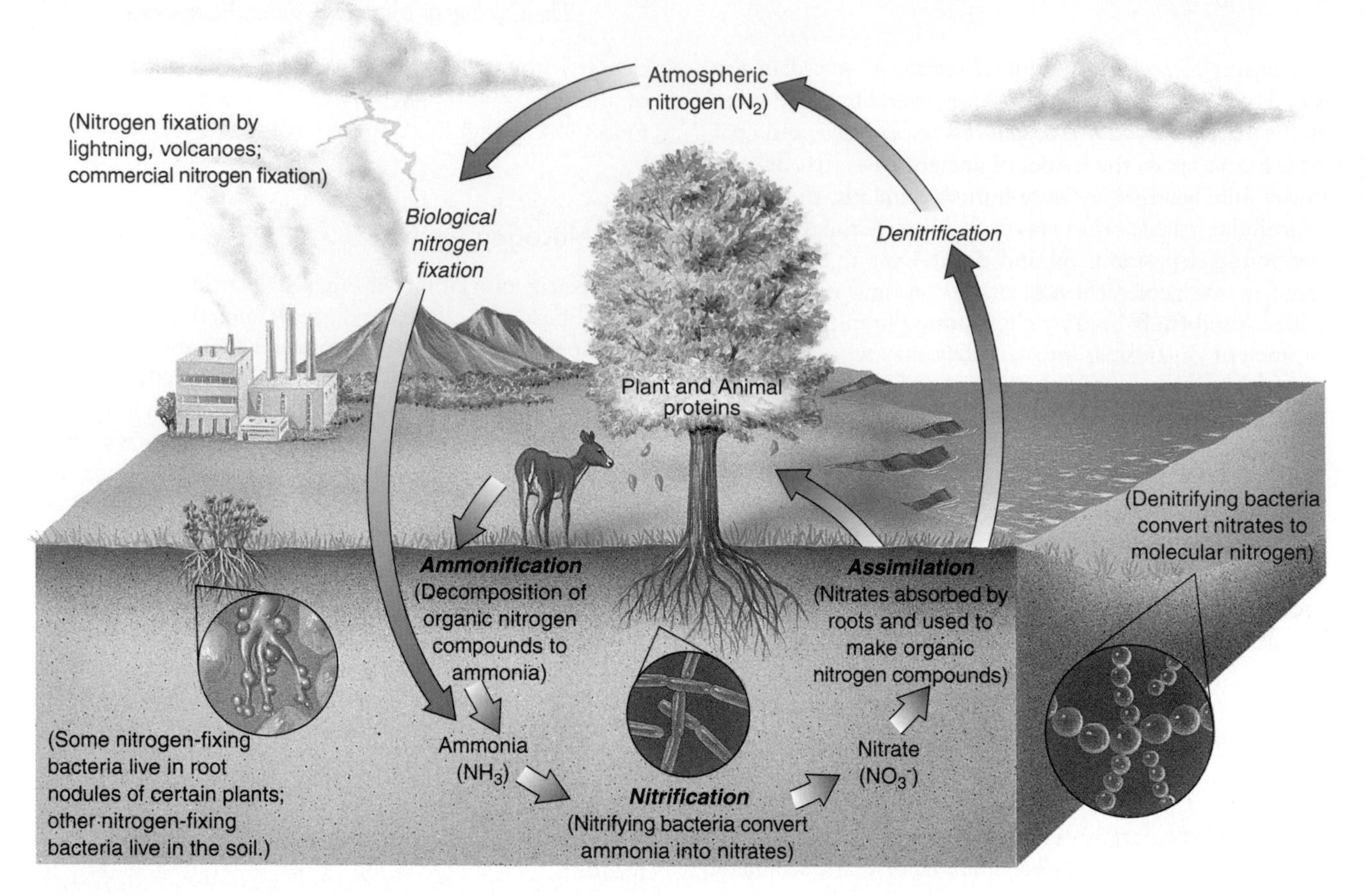

Figure 5–3

A simplified diagram of the nitrogen cycle. Nitrogen-fixing bacteria convert atmospheric nitrogen (N_2) into ammonia (NH_3). Ammonia is converted to nitrate (NO_3^-) by nitrifying bacteria in the soil. Plants assimilate nitrate, producing proteins and nucleic acids in the process; then animals eat plant proteins and produce animal proteins. When organisms die, their nitrogen compounds are broken down by ammonifying bacteria, and ammonia is released for reuse. Nitrogen is returned to the atmosphere by denitrifying bacteria, which convert nitrate to molecular nitrogen.

Figure 5–4

Nitrogenase, the enzyme involved in nitrogen fixation, only functions in the absence of oxygen. (a) Root nodules of a pea plant provide an oxygen-free environment for nitrogen-fixing *Rhizobium* bacteria that live in them. (b) *Anabaena*, a cyanobacterium, has oxygen-excluding heterocysts in which nitrogen fixation occurs. (a, Hugh Spencer/Photo Researchers, Inc.; b, Dennis Drenner)

Some water ferns have cavities in which cyanobacteria live, in a manner comparable to how *Rhizobium* lives in the root nodules of legumes. Other cyanobacteria fix nitrogen in symbiotic association with certain plants or as the photosynthetic partners of many kinds of lichens.

The reduction of nitrogen gas to ammonia by nitrogenase is a remarkable accomplishment of organisms that is achieved without the tremendous heat, pressure, and energy required to do the same thing during the manufacture of commercial fertilizers. Even so, nitrogen-fixing bacteria must consume the energy in 12 grams of glucose or the equivalent in order to fix a single gram of nitrogen biologically.

Nitrification

The conversion of ammonia (NH_3) to nitrate (NO_3^-) is called **nitrification**. Nitrification, a two-step process, is accomplished by soil bacteria. First the soil bacteria *Nitrosomonas* and *Nitrococcus* convert ammonia to nitrite (NO_2^-). Then the soil bacterium *Nitrobacter* oxidizes nitrite to nitrate. The process of nitrification furnishes these bacteria, called nitrifying bacteria, with energy.

Assimilation

In **assimilation**, plant roots absorb either nitrate (NO_3^-) or ammonia (NH_3) that was formed by nitrogen fixation and nitrification, and incorporate the nitrogen of these molecules into plant proteins and nucleic acids. When animals consume plant tissues, they also assimilate nitrogen by taking in plant nitrogen compounds and converting them to animal compounds.

Ammonification

The conversion of biological nitrogen compounds into ammonia is known as **ammonification**. Ammonification begins when organisms produce nitrogen-containing waste products such as urea (in urine) and uric acid (in the wastes of birds). These substances, plus the nitrogen compounds that occur in dead organisms, are decomposed, releasing the nitrogen into the abiotic environment as ammonia (NH_3). The bacteria that perform this process both in the soil and in aquatic environments are called ammonifying bacteria. The ammonia produced by ammonification enters the nitrogen cycle and is once again available for the processes of nitrification and assimilation.

Denitrification

The reduction of nitrate (NO_3^-) to gaseous nitrogen (N_2) is called **denitrification**. Denitrifying bacteria reverse the action of nitrogen-fixing and nitrifying bacteria by returning nitrogen to the atmosphere as nitrogen gas. Denitrifying bacteria are anaerobic, meaning they prefer to live and grow where there is little or no free oxygen. For example, they are found deep in the soil near the water table, an environment that is nearly oxygen-free.

Mini-Glossary of the Nitrogen Cycle

nitrogen fixation: The conversion of atmospheric nitrogen to ammonia, performed by nitrogen-fixing bacteria, including cyanobacteria.

nitrification: The conversion of ammonia to nitrate, performed by nitrifying bacteria.

assimilation: The conversion of inorganic nitrogen (nitrate or ammonia) to the organic molecules of organisms.

ammonification: The conversion of organic nitrogen (biological molecules containing nitrogen) to ammonia, performed by ammonifying bacteria.

denitrification: The conversion of nitrate to nitrogen gas, performed by denitrifying bacteria.

The nitrogen cycle and water pollution

Humans affect the nitrogen cycle by producing large quantities of nitrogen fertilizer, both ammonia and nitrate, from nitrogen gas. Although this process in itself is not harmful, the overuse of commercial fertilizers on the land can cause water quality problems. Rain washes nitrate fertilizer into rivers and lakes, where it stimulates the growth of algae. As these algae die, their decomposition robs the water of dissolved oxygen, which in turn causes other aquatic organisms, including many fishes, to die of suffocation. Nitrates from fertilizers can also leach (dissolve and wash down) through the soil and contaminate groundwater. Many people who live in rural areas drink groundwater, and groundwater contaminated by nitrates is dangerous, particularly for infants and small children. The effects of nitrate contamination on the environment and on human health are discussed in Chapters 18 and 21.

The Phosphorus Cycle

In the **phosphorus cycle**, phosphorus, which does not exist in a gaseous state and therefore does not enter the atmosphere, cycles from the land to sediments in the ocean and back to the land (Figure 5–5). As water runs over rocks containing phosphorus, it gradually wears away the surface and carries off inorganic phosphate (PO_4^{3-}) molecules.

The erosion of phosphorus rocks releases phosphorus into the soil, where it is taken up by plant roots in the form of inorganic phosphates. Once in cells, phosphates are used in a variety of biological molecules, including nucleic acids. Animals obtain most of their required phosphate from the food they eat, although in some localities drinking water may contain a substantial amount of inorganic phosphate. Thus, like carbon and nitrogen, phosphorus moves through the food web as one organism consumes another. Phosphorus released by decomposers becomes part of the soil's pool of inorganic phosphate that can be reused by plants.

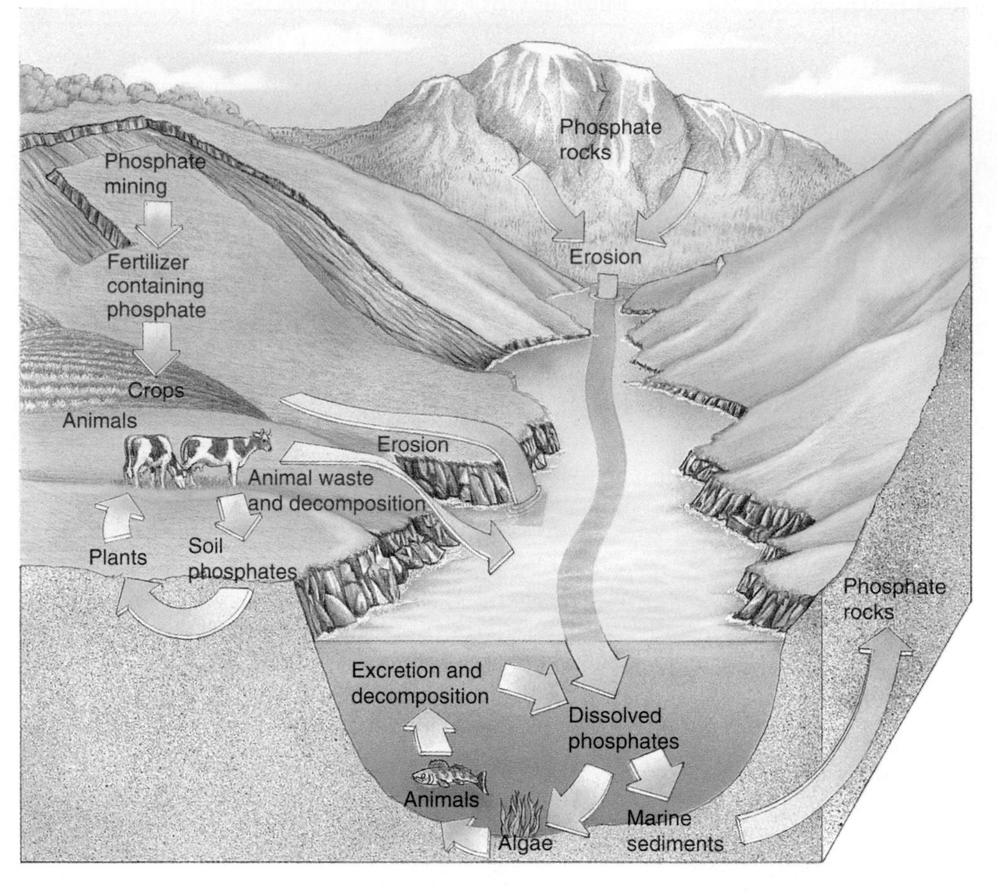

Figure 5-5

A simplified diagram of the phosphorus cycle. Recycling of phosphorus (as phosphate, PO_4^{3-}) is slow because no biologically important form of phosphorus is gaseous. Phosphates that become part of marine sediments may take millions of years to solidify into rock, uplift as mountains, and erode to again become available to organisms.

Phosphorus cycles through aquatic communities in much the same way it does through terrestrial communities. Dissolved phosphorus enters aquatic communities through absorption by algae and plants, which are then consumed by plankton and larger organisms. These, in turn, are eaten by a variety of fin fishes and shellfishes. Ultimately, decomposers that break down wastes and dead organisms release inorganic phosphorus into the water, where it is available to be used by aquatic producers again.

Phosphate can be lost from biological cycles. Some phosphate is carried from the land by streams and rivers to the ocean, where it can be deposited on the sea floor and remain for millions of years. The geological process of uplift may someday expose these sea floor sediments as new land surfaces, from which phosphate will once again be eroded.

A very small portion of the phosphate in the aquatic food web finds its way back to the land. A few fishes and aquatic invertebrates are eaten by sea birds, which may defecate on land where they roost. Guano, the manure of sea birds, contains large amounts of phosphate and nitrate. Once on land, these minerals may be absorbed by the roots of plants. The phosphate contained in guano may enter terrestrial food webs in this way, although the amounts involved are quite small.

Humans and the phosphorus cycle

Humans affect the natural cycling of phosphorus by accelerating its long-term loss from the land. Corn grown in Iowa, which contains phosphate absorbed from the soil, may be used to fatten cattle in an Illinois feedlot. Part of the phosphate absorbed by the roots of the corn plants thus ends up in the feedlot wastes, which probably wash into the Mississippi River eventually. Beef from the Illinois cattle may be consumed by people living far away—in New York City, for instance. Hence, more of the phosphate ends up in human wastes and is flushed down toilets into the New York City sewer system. Sewage treatment rarely removes phosphates, and so they cause water quality problems in rivers and lakes (see Chapters 2 and 21). To compensate for the steady loss of phosphate from their land, farmers must add phosphate fertilizer to their fields. More than likely, that fertilizer is produced in Florida from the large deposits of phosphate rock that are mined there.

In natural communities, very little phosphorus is lost from the cycle, but few communities today are in a natural state. Phosphorus loss from the soil is accelerated by land-denuding practices, such as the clearcutting of timber, and by erosion of agricultural and residential lands. For practical purposes, phosphorus that washes from the land into the sea is permanently lost from the terrestrial phosphorus cycle, for it remains in the sea for millions of years.

The Hydrologic Cycle

Water continuously circulates from the ocean to the atmosphere to the land and back to the ocean, providing us with a renewable supply of purified water on land. This complex cycle, known as the **hydrologic cycle**, results in a balance between water in the ocean, on the land, and in the atmosphere (Figure 5–6).

Water moves from the atmosphere to the land and ocean in the form of precipitation—rain, snow, sleet, or hail. When water evaporates from the ocean surface and from soil, streams, rivers, and lakes on land, it forms clouds in the atmosphere. In addition, **transpiration**, the loss of water vapor from land plants, adds water to the atmosphere. Roughly 97 percent of the water absorbed from the soil by a plant is transported to the leaves, where it is lost by transpiration.

Water may evaporate from land and reenter the atmosphere directly. Alternatively, it may flow in rivers and streams to coastal **estuaries** where fresh water meets the ocean. The movement of water from land to ocean is called **runoff**, and the area of land being drained by runoff is called a **watershed**. Water also percolates, or seeps, downward through the soil and rock to become **groundwater**. Groundwater may reside in the ground for hun-

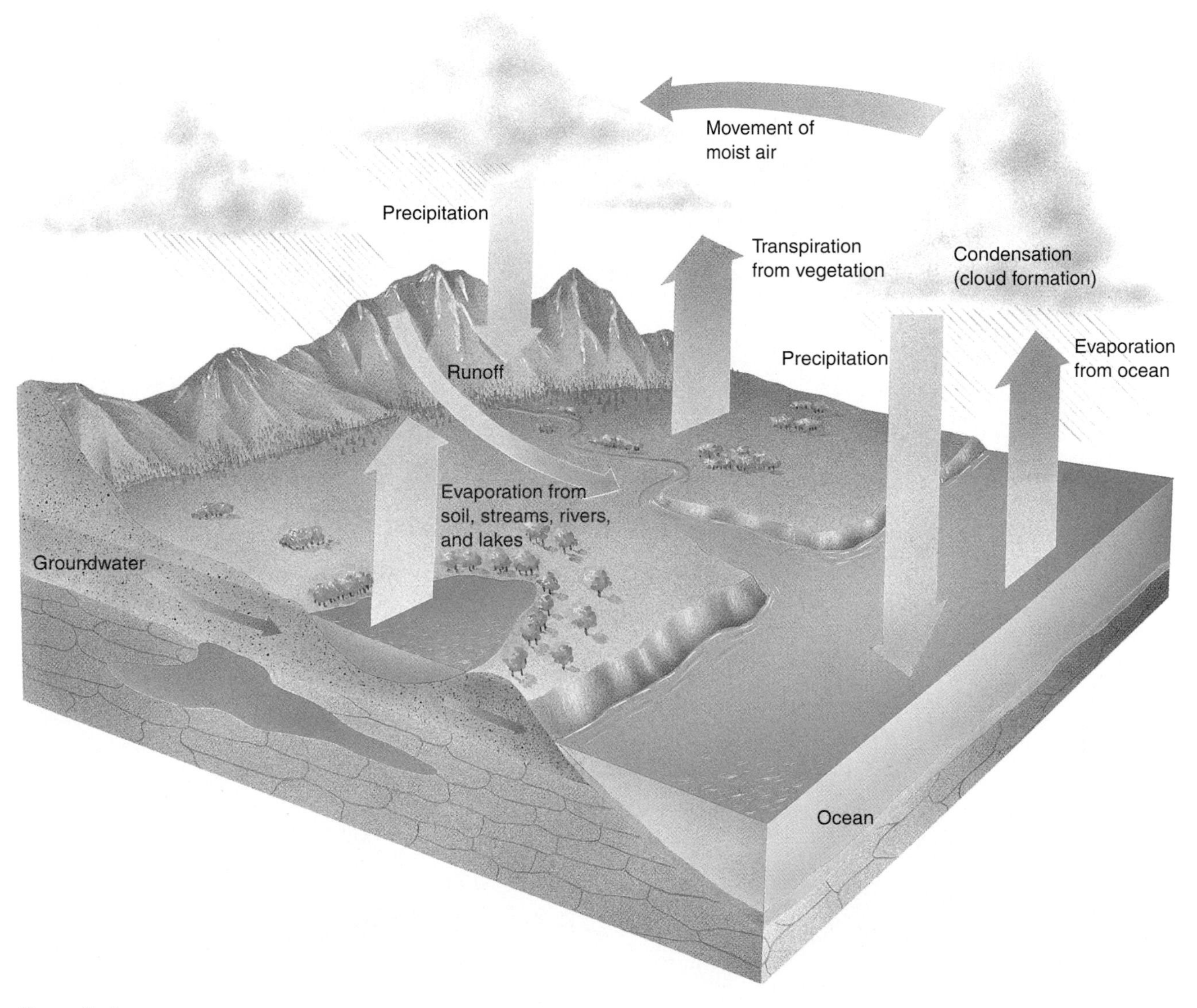

Figure 5–6

A simplified diagram of the hydrologic cycle. Water cycles from the ocean to the atmosphere to the land and back to the ocean. Although some water molecules are unavailable for thousands of years—locked up in polar ice, for example—all water molecules eventually travel through the hydrologic cycle.

dreds to many thousands of years, but eventually it supplies water to the soil, to vegetation, to streams and rivers, and to the ocean.

Regardless of its physical form—solid, liquid, or vapor—or location, every molecule of water eventually moves through the hydrologic cycle. Tremendous quantities of water are cycled annually between the Earth and its atmosphere. The volume of water entering the atmosphere each year is estimated at about 389,500 cubic kilometers (95,000 cubic miles). Approximately three-fourths of this water reenters the ocean directly as precipitation over water; the remainder falls on land. (Chapter 13 discusses the local and global supply of fresh water, and Chapter 21 considers water pollution issues.)

We have seen how living things depend on the abiotic environment to supply energy (see Chapter 3) and essential materials (in biogeochemical cycles). We now consider five aspects of the physical environment that also affect organisms: solar radiation that warms the Earth; the atmosphere; the ocean; weather and climate; and internal planetary processes.

SOLAR RADIATION

The sun makes life on Earth possible. It warms the planet, including the atmosphere, to habitable temperatures. Without the sun's energy, the temperature would approach absolute zero (−273°C) and all water would be frozen, even in the ocean. The hydrologic cycle, carbon cycle, and other biogeochemical cycles are powered by

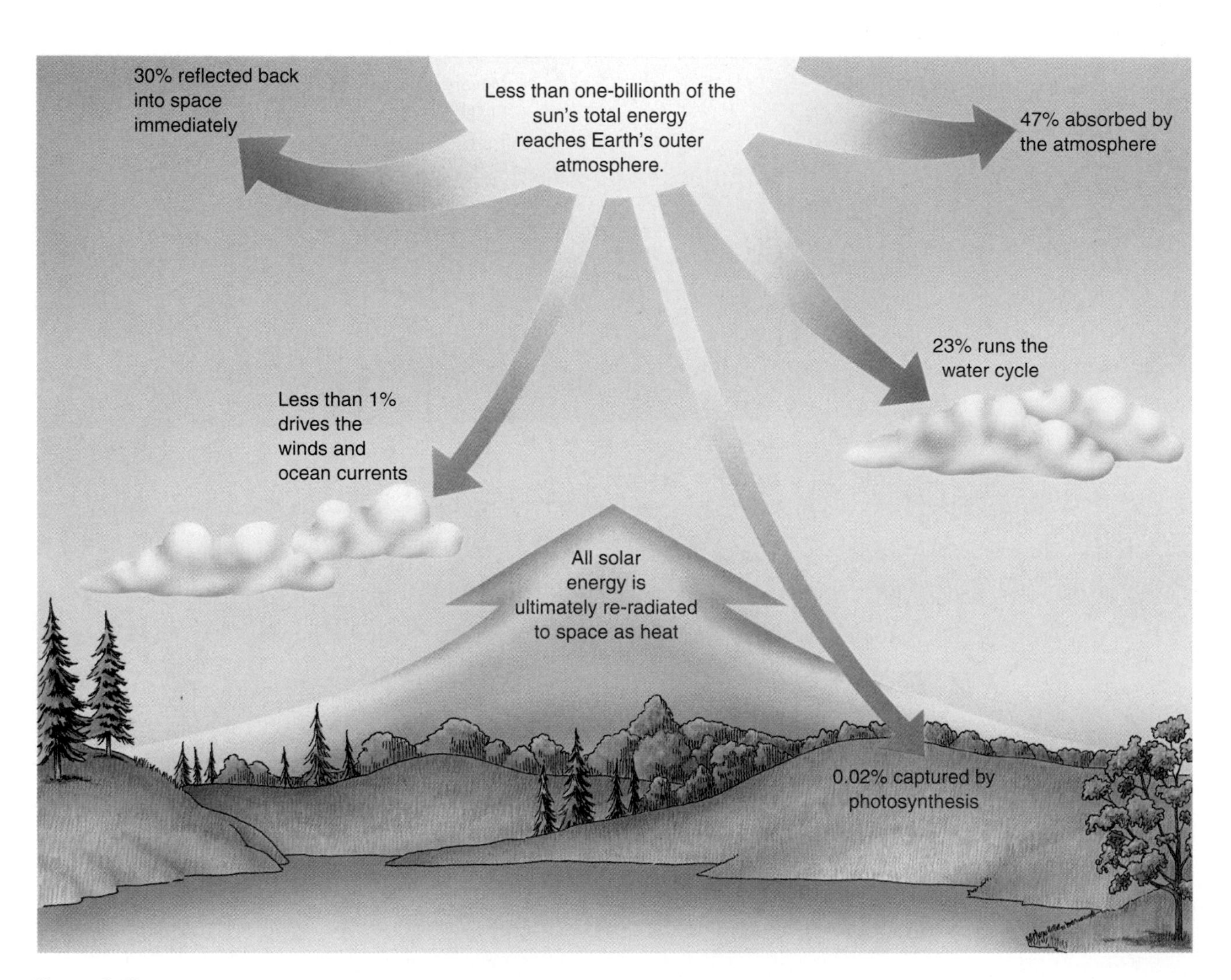

Figure 5–7

The fate of solar radiation that reaches the Earth. Most of the energy produced by the sun never reaches the Earth. The solar energy that does reach the Earth warms the planet's surface, drives the water cycle and other biogeochemical cycles, produces our climate, and powers almost all life through the process of photosynthesis.

the sun, and it is the primary determinant of climate. The sun's energy is captured by photosynthetic organisms, which use it to make the food molecules required by almost all forms of life. Most of our fuels—wood, oil, coal, and natural gas, for example—represent solar energy captured by photosynthetic organisms. Without the sun, almost all life would cease.

The sun's energy is the product of a massive nuclear fusion reaction (see Chapter 11) and is emitted into space in the form of electromagnetic radiation—especially visible light and infrared and ultraviolet radiation, which are not visible to the human eye. Approximately one-billionth of the total energy released by the sun strikes our atmosphere, and of this tiny trickle of energy a minute part operates the ecosphere.

On average, 30 percent of the solar radiation that falls on Earth is immediately reflected away by clouds and, to a lesser extent, by surfaces, especially snow, ice, and the ocean (Figure 5–7). The remaining 70 percent is absorbed by the Earth, where it runs the water cycle, drives winds and ocean currents, powers photosynthesis, and warms the planet. Ultimately, however, all of this energy is lost by the continual radiation of long-wave infrared (heat) energy into space.

Temperature Changes with Latitude

The most significant local variation in Earth's temperature is produced because the sun's energy does not reach all places uniformly. A combination of the Earth's roughly spherical shape and the tilt of its axis produces a great deal of variation in the exposure of the surface to the energy delivered by sunlight.

The principal effect of the tilt is on the angles at which the sun's rays strike different areas of the planet at any one time (Figure 5–8). On the average, the sun's rays hit vertically near the equator, making the energy more concentrated and producing higher temperatures. Near the poles the sun's rays hit more obliquely, and, as a result, their energy is spread over a larger surface area. Also, rays of light entering the atmosphere obliquely near the poles must pass through a deeper envelope of air than those entering near the equator. This causes more of the sun's energy to be scattered and reflected back to space, which in turn further lowers temperatures near the poles. Thus, because the solar energy that reaches polar regions is less concentrated, temperatures are lower.

Temperature Changes with Season

Seasons are primarily determined by Earth's inclination on its axis. Since the Earth's inclination on its axis is always the same (23.5°), during half of the year (March 21 to September 22) the Northern Hemisphere tilts *toward* the sun, and during the other half (September 22 to March 21) it tilts *away* from the sun (Figure 5–9). The orientation of the Southern Hemisphere is just the opposite at these times.

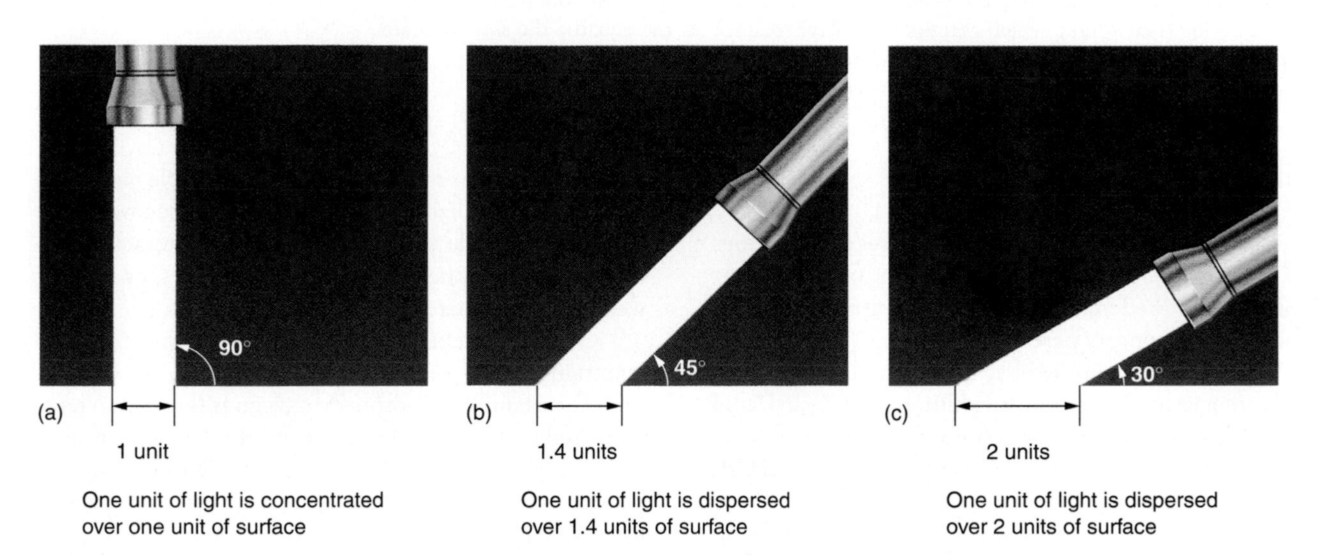

Figure 5–8

The angle at which the sun's rays strike the Earth varies from one geographical location to another due to the Earth's spherical shape and its inclination on its axis. (a) Sunlight (represented by the flashlight) that shines vertically (near the Equator) is concentrated on the Earth's surface. (b,c) As one moves toward the poles, the light hits the surface more and more obliquely, spreading the radiation over larger and larger areas.

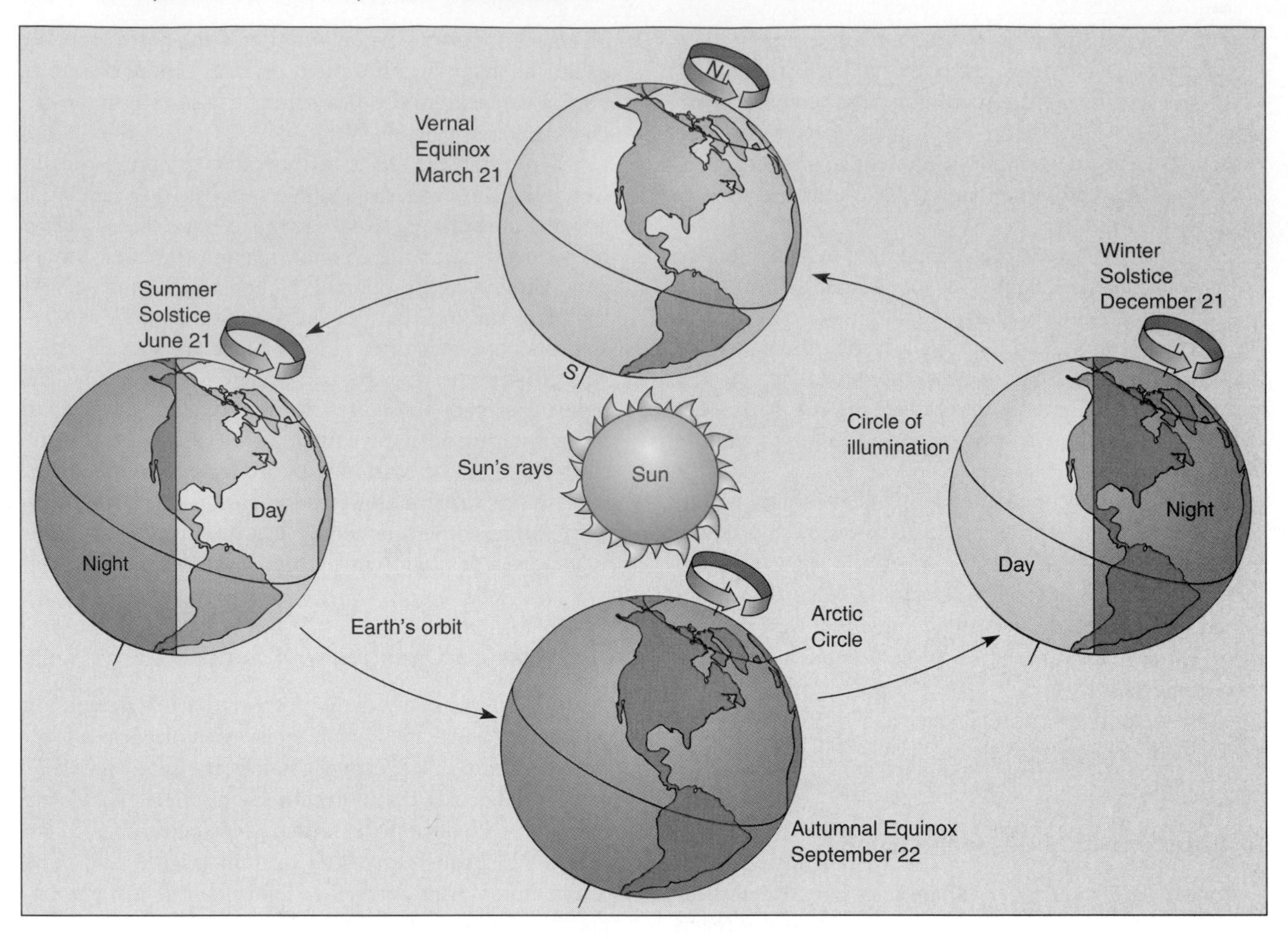

Figure 5–9

Seasonal changes in temperature. The Earth's inclination on its axis remains the same as it travels around the sun. Thus, the sun's rays hit the Northern Hemisphere obliquely during the winter months and more directly during the summer. In the Southern Hemisphere, the sun's rays are oblique during their winter, which corresponds to our summer. At the equator, the sun's rays are approximately vertical on March 21 and September 22.

THE ATMOSPHERE

The atmosphere is an invisible layer of gases that envelops the Earth. Oxygen (21 percent) and nitrogen (78 percent) are the predominant gases in the atmosphere, accounting for about 99 percent of dry air; other gases, including argon, carbon dioxide, neon, and helium, make up the remaining 1 percent. In addition, water vapor and trace amounts of various air pollutants, such as methane, ozone, dust particles, and chlorofluorocarbons (CFCs), are present in the air. The atmosphere becomes less dense as it extends outward into space; most of its mass is found near the Earth's surface.

The atmosphere performs several ecologically important functions. It protects the Earth's surface from most of the sun's ultraviolet radiation and x rays and from lethal amounts of cosmic rays from space. Without this shielding by the atmosphere, life as we know it would cease to exist. While the atmosphere protects the Earth from high-energy radiations, it allows visible light and some infrared radiation to penetrate, and they warm the surface and the lower atmosphere. This interaction between the atmosphere and solar energy is responsible for weather and climate.

Organisms depend on the atmosphere, but they also maintain and, in certain instances, modify its composition. For example, atmospheric oxygen is thought to have increased to its present level as a result of millions of years of photosynthesis. The level is maintained by a balance between oxygen-producing photosynthesis and oxygen-using respiration.

Layers of the Atmosphere

The atmosphere is composed of a series of five concentric layers—the troposphere, stratosphere, mesosphere, thermosphere, and exosphere (Figure 5–10). These layers vary in altitude and temperature with latitude and season. The **troposphere**, the layer of atmosphere closest to the Earth's surface, extends up to a height of approximately 10 kilometers (6.2 miles). The temperature of the troposphere decreases with increasing altitude by about −6°C for every kilometer. Weather, including turbulent wind, storms, and most clouds, occurs in the troposphere.

In the next layer of atmosphere, the **stratosphere**, there is a steady wind but no turbulence. The temperature is more or less uniform (−45°C to −75°C) in the lower stratosphere; commercial jets fly here. The stratosphere extends from 10 to 45 kilometers (6.2 to 28 miles) above the Earth's surface and contains a layer of ozone that is critical to life because it absorbs much of the sun's damaging ultraviolet radiation. The absorption of ultraviolet radiation by the ozone layer heats the air, and so temperature increases with increasing altitude in the stratosphere.

The **mesosphere**, the layer of atmosphere directly above the stratosphere, extends from 45 to 80 kilometers (28 to 50 miles) above the Earth's surface. Temperatures drop steadily in the mesosphere to the lowest in the atmosphere—as low as −138°C.

The **thermosphere**, which extends from 80 to 500 kilometers (50 to 310 miles), is characterized by steadily rising temperatures. Gases in the extremely thin air of the thermosphere absorb x rays and short-wave ultraviolet radiation. This absorption drives the few molecules present in the thermosphere to great speeds, raising their temperature in the process to 1000°C or more. The aurora, a colorful display of lights in dark polar skies, is produced when charged particles from the sun hit oxygen or nitrogen molecules in the thermosphere. The thermosphere is important in long-distance communication because it reflects outgoing radio waves back toward the Earth without the aid of satellites.

The outermost layer of the atmosphere, the **exosphere**, begins about 500 kilometers (310 miles) above Earth's surface. The exosphere continues to thin until it converges with interplanetary space.

Atmospheric Circulation

In large measure, differences in temperature caused by variations in the amount of solar energy reaching different locations on Earth drive the circulation of the atmosphere. The very warm surface near the equator heats the air that is in contact with it, causing this air to expand and rise. As the warm air rises, it cools, and then it sinks again. Much of it recirculates almost immediately

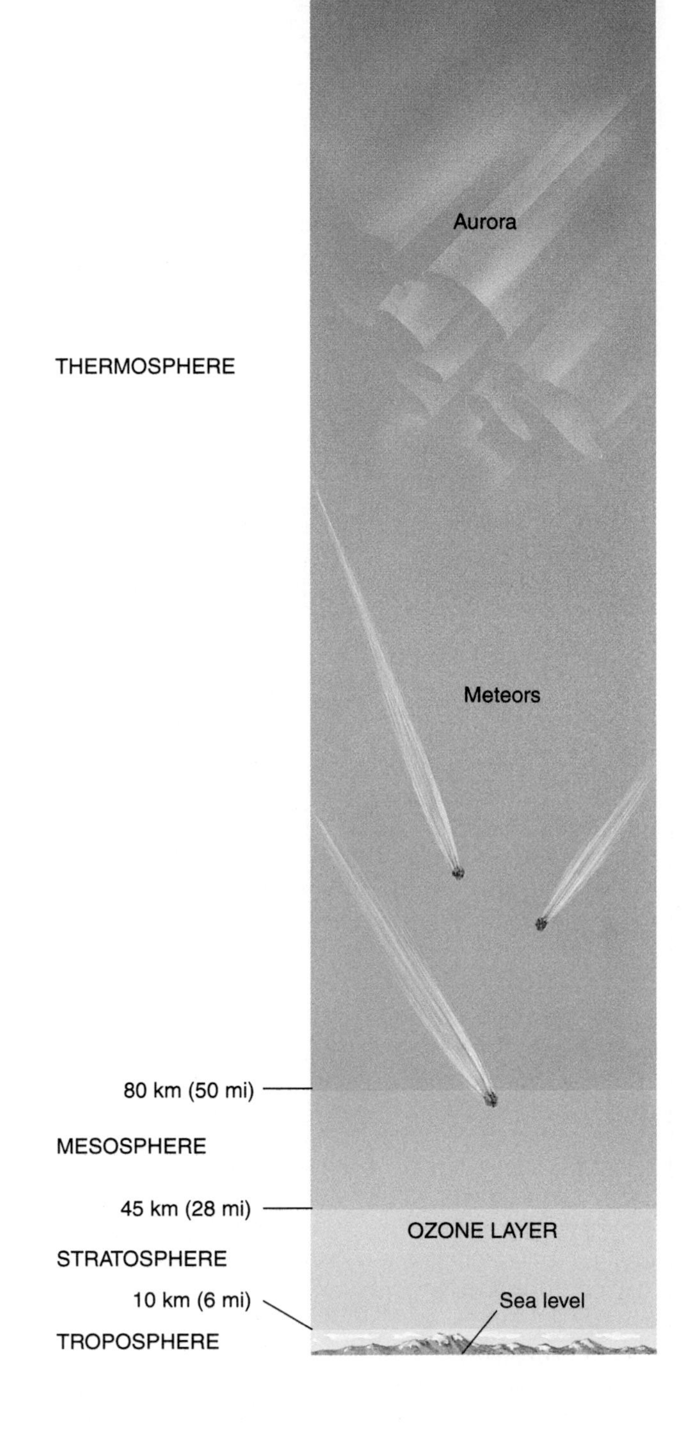

Figure 5-10

Layers of the atmosphere. The outermost layer of the atmosphere, the exosphere, has no distinct boundary separating it from interplanetary space.

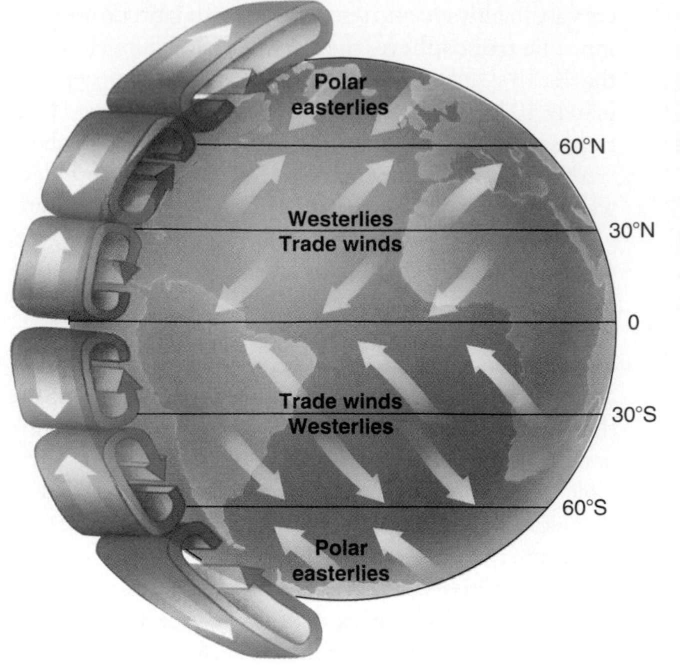

Figure 5–11

Atmospheric circulation transports heat from the equator to the poles. The greatest solar energy input occurs at the equator, heating air most strongly in that area. The air rises and travels to the poles, but is cooled in the process so that much of it descends again at around 30 degrees latitude in both hemispheres. At higher latitudes the patterns of movement are more complex.

to the same areas it has left, but the remainder of the heated air flows toward the poles, where eventually it is chilled. Similar upward movements of warm air and its subsequent flow toward the poles occur at higher latitudes—farther from the equator—as well (Figure 5–11). As air cools by contact with the polar ground and ocean, it sinks and flows toward the equator, generally beneath the sheets of warm air that simultaneously flow toward the poles. The constant motion of air transfers heat from the equator toward the poles, and as the air returns, it cools the land over which it passes. This continuous turnover does not equalize temperatures over the surface of the Earth, but it does moderate them.

Surface winds

In addition to global circulation patterns, the atmosphere exhibits complex horizontal movements that are commonly referred to as **winds**. The nature of wind, with its turbulent gusts, eddies, and lulls, is difficult to understand or predict. It results in part from differences in atmospheric pressure and from the rotation of the Earth.

The gases that constitute the atmosphere have weight and exert a pressure that is, at sea level, about 1013 millibars (14.7 pounds per square inch). Air pressure is variable, however, changing with altitude, temperature, and humidity. Winds tend to blow from areas of high atmospheric pressure to areas of low pressure, and the greater the difference between the high- and low-pressure areas, the stronger the wind.

The Earth's rotation also influences the direction of wind. Earth rotates from west to east, which causes moving air to be deflected from its path and swerve to the right of the direction in which it is traveling in the Northern Hemisphere and to the left of the direction in which it is traveling in the Southern Hemisphere. This tendency is known as the **Coriolis effect**.

The Coriolis effect can be visualized by imagining that you and a friend are standing about 10 feet apart on a merry-go-round that is turning clockwise (Figure 5–12). Suppose you throw a ball directly at your friend. By the time the ball reaches the place where your friend was, he or she is no longer in that spot. The ball will have swerved far to the left of your friend. This is how the Coriolis effect works in the Southern Hemisphere.

To visualize how the Coriolis effect works in the Northern Hemisphere, imagine you and your friend are standing on the same merry-go-round, only this time it is moving counterclockwise. Now when you throw the ball, it will swerve far to the right of your friend.

HUMANS CHANGE EARTH'S ROTATION

A geophysicist at the Goddard Space Flight Center announced in 1996 that humans may have unwittingly altered the rotation rate of planet Earth. Because we have constructed so many large reservoirs to hold water, and because they are located primarily in the Northern Hemisphere, rather than being randomly distributed around the globe, enough of Earth's mass has been shifted to speed up its rotation. Currently, 88 mega-reservoirs impound some 10 trillion tons of water. Before the reservoirs were built, this water was located in the ocean, which has most of its mass in the Southern Hemisphere. The effect has been compared to that of a whirling skater who pulls her arms in to turn faster. Because natural factors in the environment, such as the pull of the tides, are gradually slowing the Earth's rotation, the human influence is actually curbing the natural rate of deceleration. The shift in Earth's mass has also changed the location of the axis on which Earth rotates. The North Pole has moved about 2 feet since the early 1950s.

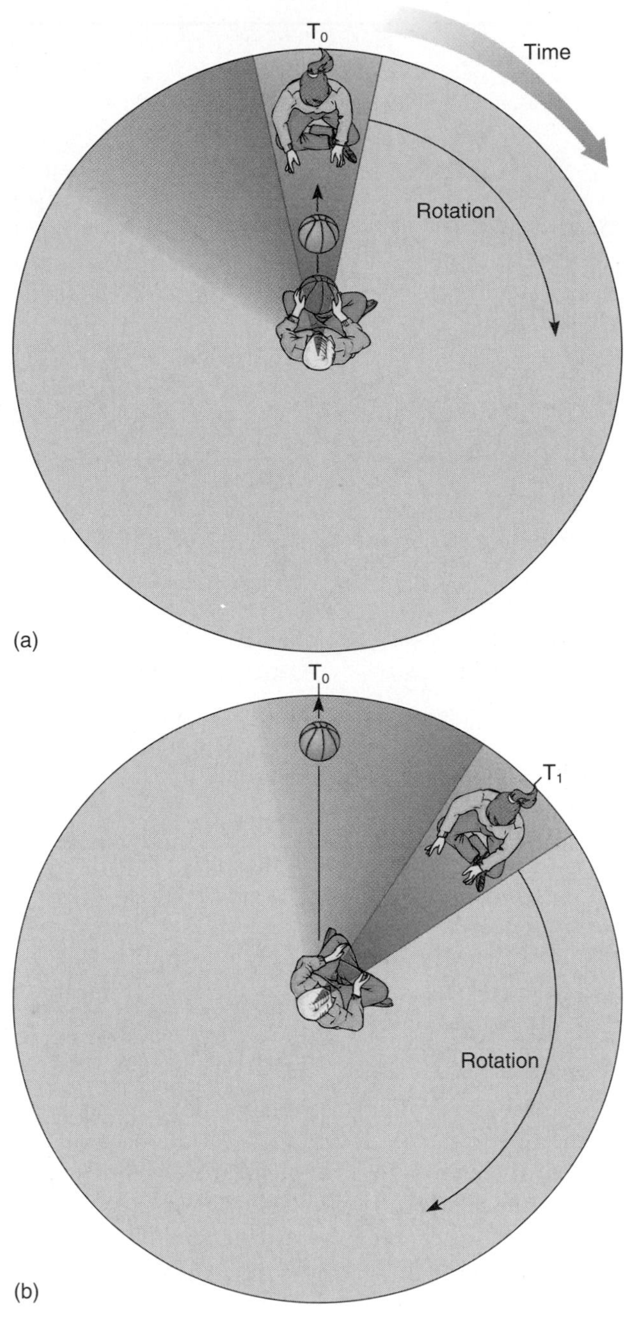

Figure 5–12

A merry-go-round (shown from above) can be used to demonstrate the Coriolis effect. The center of the merry-go-round corresponds to the South Pole, and the outer edge to the equator. (a) If you throw a ball to a friend at time zero T_0, when the merry-go-round is rotating clockwise, the ball at time T_1 (b) appears to curve to the left instead of going straight. Because of Earth's rotation, winds appear to curve to the left in the Southern Hemisphere and to the right in the Northern Hemisphere.

The atmosphere has three **prevailing winds**—major surface winds that blow more or less continually (see Figure 5–11). Prevailing winds that blow from the northeast near the North Pole or from the southeast near the South Pole are called **polar easterlies**. Winds that blow in the mid-latitudes from the southwest in the Northern Hemisphere or from the northwest in the Southern Hemisphere are called **westerlies**. Tropical winds that blow from the northeast in the Northern Hemisphere or the southeast in the Southern Hemisphere are called **trade winds**.

THE GLOBAL OCEAN

The global ocean is a huge body of salt water that surrounds the continents and covers almost three-fourths of the Earth's surface. It is a single, continuous body of water, but geographers divide it into four sections that are separated by the continents: the Pacific, Atlantic, Indian, and Arctic Oceans. The Pacific Ocean is the largest by far: it covers one-third of Earth's surface and contains more than half of Earth's water.

Patterns of Circulation in the Ocean

The persistent prevailing winds blowing over the ocean produce mass movements of surface ocean water known as **currents** (Figure 5–13). The prevailing winds generate *circular* ocean currents called **gyres**. For example, in the North Atlantic, the tropical trade winds tend to blow toward the west, whereas the westerlies in the mid-latitudes blow toward the east. This helps establish a clockwise gyre in the North Atlantic. Thus, surface ocean currents and winds tend to move in the same direction, although there are many variations on this general rule.

The paths traveled by surface ocean currents are influenced by the Coriolis effect. The Earth's rotation from west to east causes surface ocean currents to swerve to the right in the Northern Hemisphere, helping establish a circular, clockwise pattern of water currents. In the Southern Hemisphere, ocean currents swerve to the left, thereby moving in a circular, counterclockwise pattern.

The varying **density** (mass per unit volume) of seawater affects deep ocean currents. Water that is colder is denser than warmer water.[1] Thus, colder ocean water sinks and flows under warmer water, generating currents far below the surface. Deep ocean currents often travel in different directions and at different speeds than do surface currents, in part because the Coriolis effect is more pronounced at greater depths.

[1]The density of water increases with decreasing temperature down to 4°C.

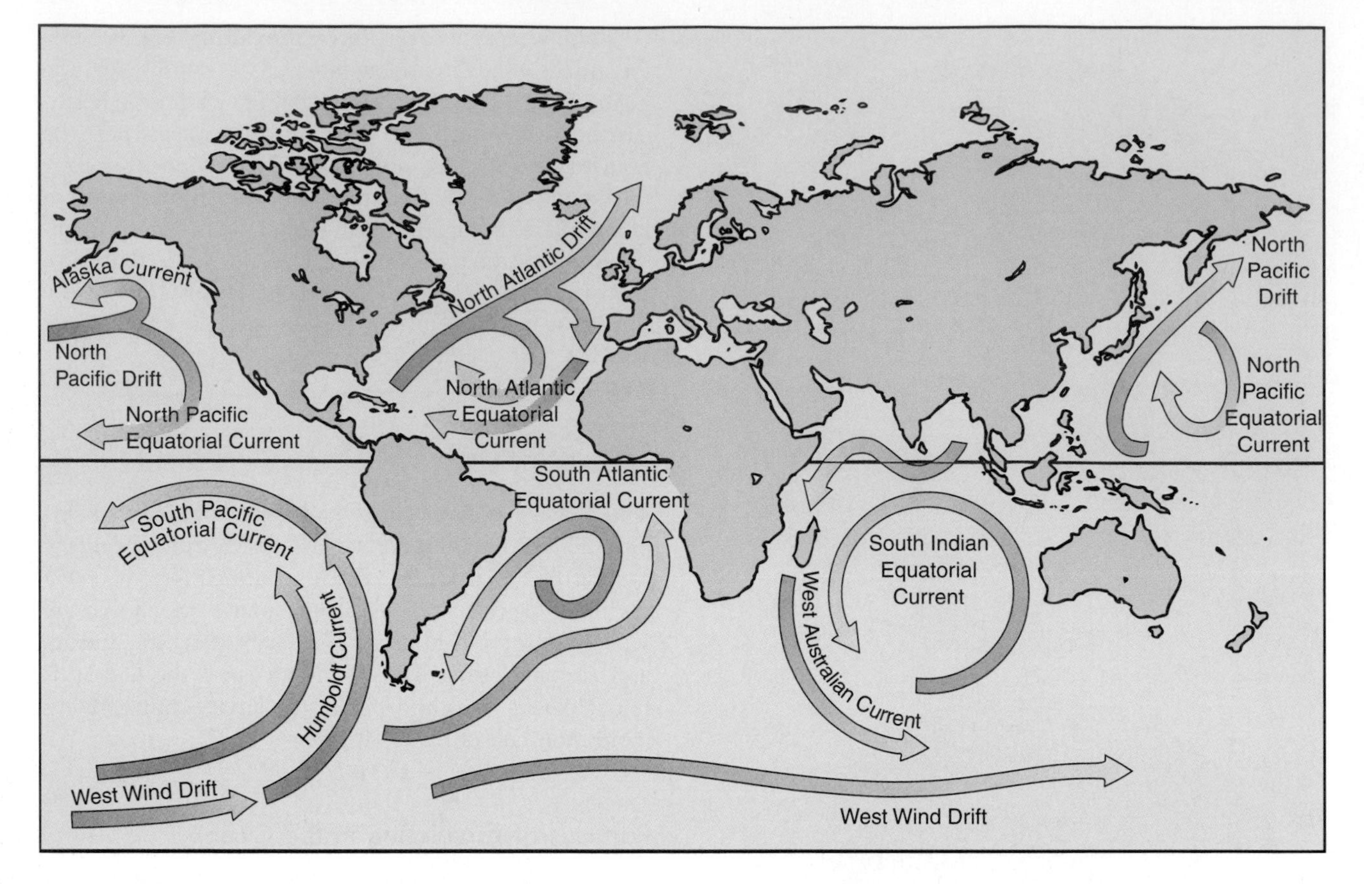

Figure 5–13

Earth's major surface ocean currents. The basic pattern of ocean currents is caused largely by the action of winds. The main ocean current flow—clockwise in the Northern Hemisphere and counterclockwise in the Southern Hemisphere—results partly from the Coriolis effect.

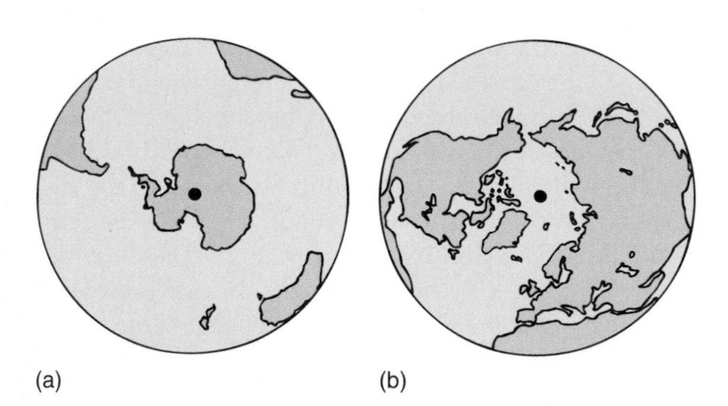

Figure 5–14

The Northern and Southern Hemispheres have greatly differing proportions of land and water, with far more water occurring in the Southern Hemisphere. (a) The Southern Hemisphere as viewed from the South Pole. (b) The Northern Hemisphere as viewed from the North Pole. Ocean currents are freer to flow in a circumpolar manner in the Southern Hemisphere.

The position of land masses also affects ocean circulation. As you can see in Figure 5–14, the ocean is not distributed uniformly over the globe: there is clearly more water in the Southern Hemisphere than in the Northern Hemisphere. Therefore, the circumpolar (around the pole) flow of water in the Southern Hemisphere is almost unimpeded by land masses.

The Ocean Interacts with the Atmosphere

The ocean and the atmosphere are strongly linked, with wind from the atmosphere affecting the ocean currents and heat from the ocean affecting atmospheric circulation. One of the best examples of the interaction between ocean and atmosphere is the **El Niño–Southern Oscillation (ENSO)** event. ENSO is a periodic warming of surface waters of the tropical East Pacific that alters both ocean and atmospheric circulation patterns and results in unusual weather in areas far from the tropical Pacific (Figure 5–15). Normally, westward-blowing trade winds restrict the warmest waters to the western Pacific near Australia. Every three to seven years, however, the trade

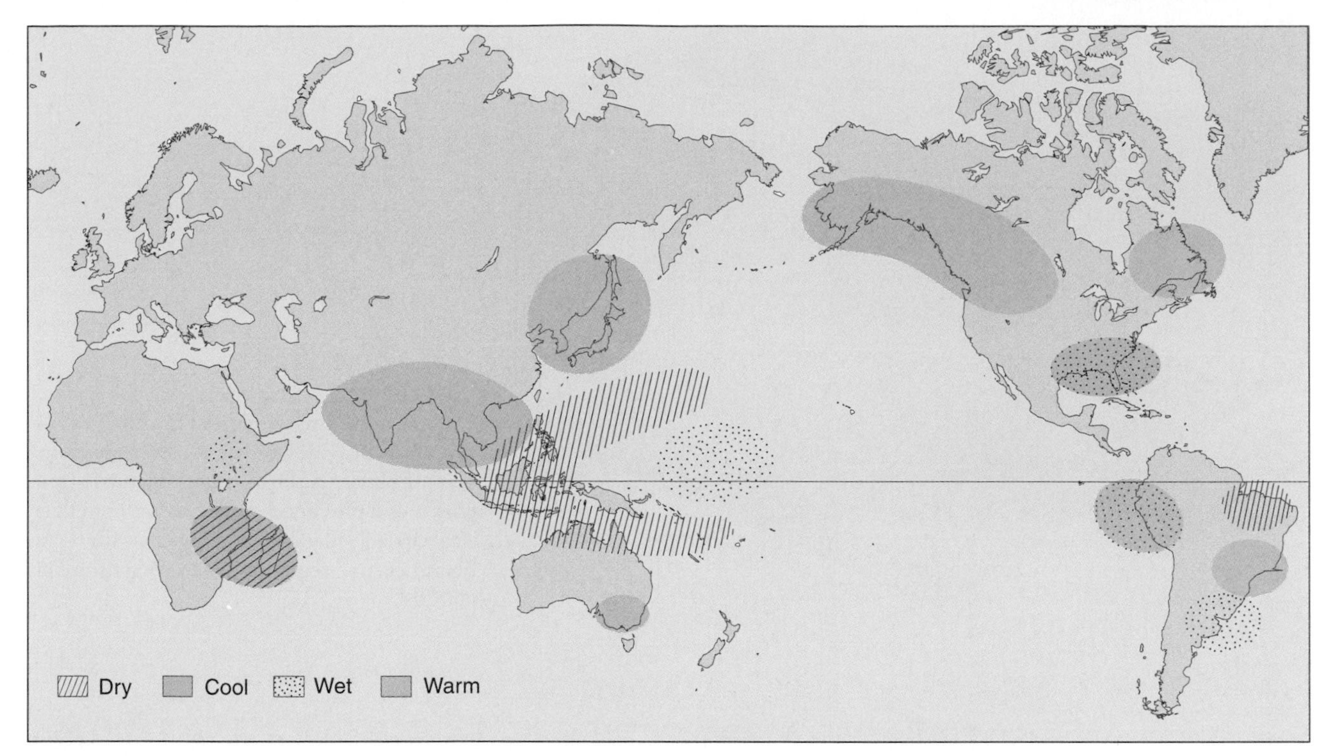

Figure 5–15
Climate patterns associated with ENSO. ENSO events can drastically alter the climate in many areas remote from the Pacific Ocean. As a result of ENSO, some areas are drier, some wetter, some cooler, and some warmer than usual.

FORECASTING WITH CONFIDENCE

In January 1995 the National Weather Service (NWS) began releasing 15-month forecasts, the longest advance time ever incorporated into its predictions. The long-range forecasts are made possible by a new ability to predict El Niño, a large pool of warm water that appears in the tropical Pacific every three to seven years, affecting air circulation and disrupting global weather patterns. NWS meteorologists employ three approaches in their forecasts, two of which are based on predicting El Niño and its influences. A statistical method analyzes past data to predict an El Niño occurrence months in advance; a computer model simulates future El Niño conditions by coupling Pacific conditions with atmospheric responses; and the third method, optimal climate models, is independent of El Niño but instead assumes existing climate patterns will continue and adds a high degree of quantitative analysis to long-term trends.

winds weaken and the warm mass of water expands eastward to South America, increasing surface temperatures in the East Pacific to 3 to 4 degrees above normal. Ocean currents, which normally flow westward in this area, slow down, stop altogether, or even reverse and go eastward. The phenomenon is called *El Niño* (Spanish, *the child*) because the warming usually reaches the fishing grounds off Peru just before Christmas. Most ENSOs last from one to two years.

ENSO has a devastating effect on the fisheries off South America. The higher temperatures and accompanying changes in ocean circulation patterns prevent nutrient-laden deeper waters from **upwelling** (coming to the surface) off the west coast of South America (Figure 5–16). This results in a severe decrease in the populations of anchovies and many other marine fishes. For example, during the 1982–83 El Niño, the worst ever recorded, the anchovy population decreased by 99 percent. Other species such as shrimp and scallops, however, thrive during an ENSO event.

ENSO also alters global air currents, directing unusual weather to areas far from the tropical Pacific. The 1994–95 ENSO, for example, resulted in heavy snows in parts of the western United States. This ENSO was also responsible for the torrential rains that flooded California, Arizona, and western Europe. In addition, the effects

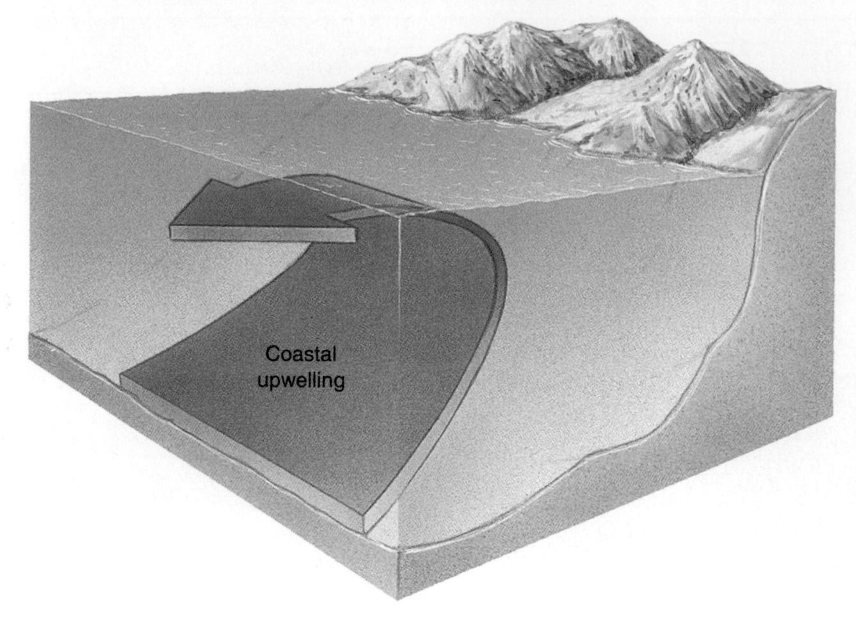

Figure 5–16

Coastal upwelling along the South American coast provides nutrients for large numbers of phytoplankton, which in turn support a complex food web. Coastal upwelling weakens considerably during years with ENSO events, temporarily reducing fish populations.

of this ENSO have been linked to droughts in Australia and Indonesia.

WEATHER AND CLIMATE

Weather refers to the conditions in the atmosphere at a given place and time; it includes temperature, atmospheric pressure, precipitation, cloudiness, humidity, and wind. Weather changes from one hour to the next and from one day to the next.

Climate comprises the average weather conditions that occur in a place over a period of years. The two most important factors that help determine an area's climate are temperature—both average temperature and temperature extremes—and precipitation—both average precipitation and seasonal distribution. Other climate factors include wind, humidity, fog, and cloud cover. Lightning is an important aspect of climate in some areas because it starts fires (see Focus On: Wildfires).

Day-to-day variations, day-to-night variations, and seasonal variations in climate factors are also important dimensions of climate that affect organisms. Temperature, precipitation, and other aspects of climate are influenced by latitude, elevation, topography, vegetation, distance from the ocean, and location on a continent or other land mass. Unlike weather, which changes rapidly, climate changes slowly—over hundreds or thousands of years.

Earth has many different climates, and because each is relatively constant for many years, organisms have adapted to them. The many different kinds of organisms on Earth are here in part because of the large number of different climates—from cold, snow-covered polar climates, to tropical climates where it is hot and rains almost every day. A Russian meteorologist, Wladimir Köppen, developed the most widely used system for classifying climates in the early part of the 20th century. He based his scheme on the observation that various types of vegetation (see Chapter 6) are associated with different climates, particularly temperature and precipitation. Figure 5–17 shows a world climate map modified from Köppen. Note that there are six climate zones—tropical, dry, mild, continental, polar, and high elevation—and that each is subdivided into climate types. For example, the three types of continental climates are warm summer, cool summer, and subarctic. Continental climates are found only in the Northern Hemisphere; there is no land at the corresponding latitudes in the Southern Hemisphere. All three types of continental climate are characterized by cold winters with snow.

Mini-Glossary of Köppen's Climate Zones

tropical climate zone: Every month is warm, mean temperature is over 18°C (64°F).

dry climate zone: Evaporation exceeds precipitation in most months.

mild climate zone: Distinct winter and summer seasons; winters are mild, mean temperature in coldest month is above −3°C (27°F).

continental climate zone: Distinct winter and summer seasons; winters are cold, mean temperature in coldest month is below −3°C (27°F).

polar climate zone: Distinct winter and summer seasons; winters are long and extremely cold, mean temperature in the warmest month is below 10°C (50°F).

high elevation climate zone: Climate changes as elevation increases and from one side of the mountain to the other; characteristic of high plateaus and mountains.

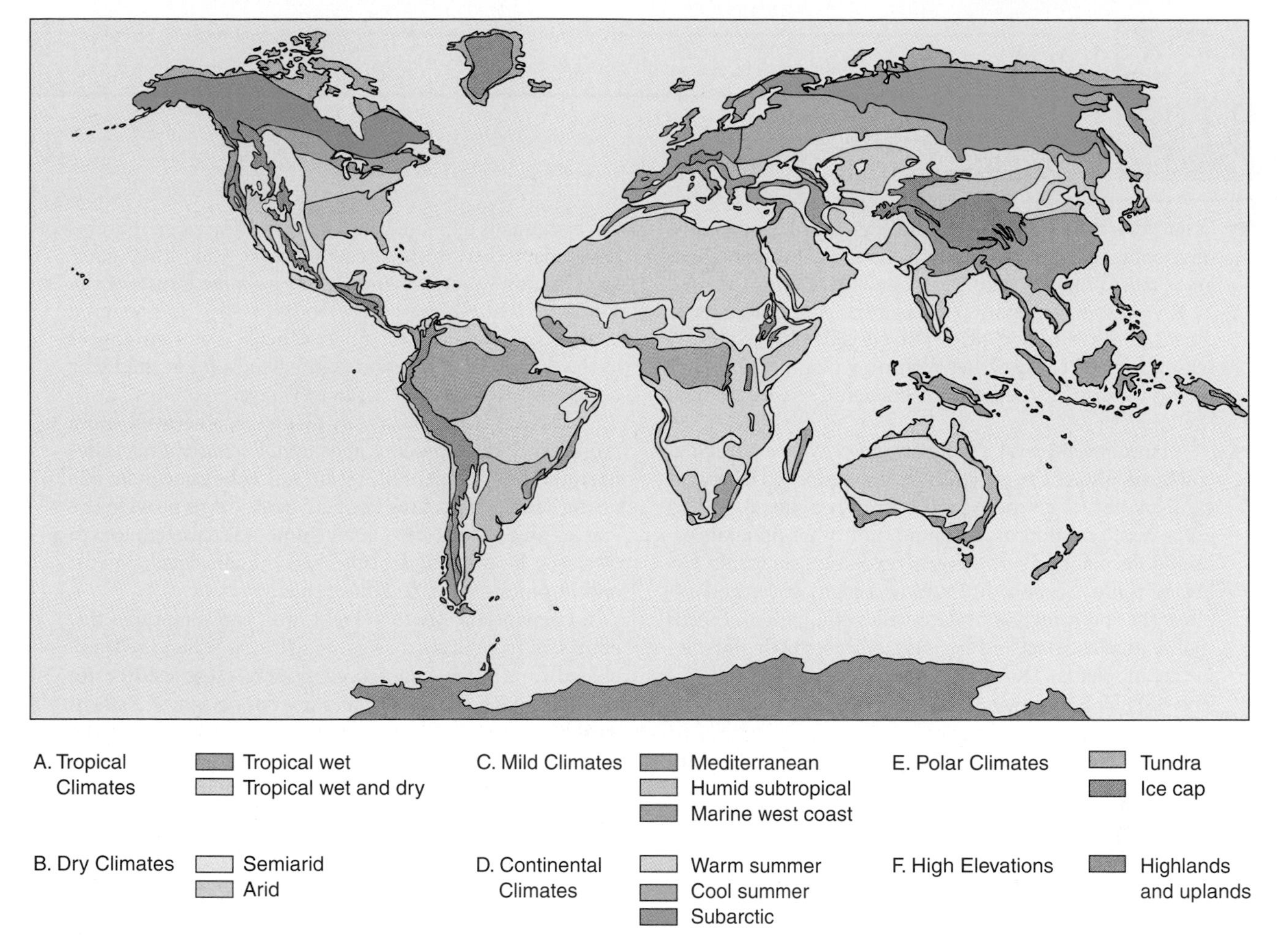

Figure 5–17

A map of Earth's climate regions, as determined mainly by temperature and precipitation. Each climate region supports characteristic vegetation.

Precipitation

Precipitation refers to any form of water, such as rain, snow, sleet, and hail, that falls to the Earth from the atmosphere. Precipitation varies from one location to another and has a profound effect on the distribution and kinds of organisms. One of the driest places on Earth is in the Atacama Desert in Chile, where the average annual rainfall is 0.05 centimeter (0.02 inch). In contrast, Mount Waialeale in Hawaii, Earth's wettest spot, receives an average annual precipitation of 1200 centimeters (472 inches).

Differences in precipitation depend on several factors. The heavy rainfall of some areas of the tropics results mainly from the equatorial uplift of moisture-laden air. High surface-water temperatures cause the evaporation of vast quantities of water from tropical parts of the ocean, and prevailing winds blow the resulting moist air over land masses. Heating of the air by land surface that has been warmed by the sun causes moist air to rise. As it rises, the air cools, and its moisture-holding ability decreases (cool air holds less water vapor than warm air). When the air reaches its saturation point—that is, when it cannot hold any additional water vapor—clouds form and water is released as precipitation. The air eventually returns to Earth on both sides of the equator near the Tropics of Cancer and Capricorn (latitudes 23.5° north and 23.5° south). By then most of its moisture has precipitated, and the dry air returns to the equator. This dry air makes little biological difference over the ocean, but its lack of moisture produces some of the great tropical deserts, such as the Sahara Desert.

Air is also dried by long journeys over land masses. Near the windward (the side from which the wind blows) coasts of continents, rainfall may be heavy. However, in the temperate zones—the areas between the tropics and the polar zones—continental interiors are usually dry, because they are far from the ocean that replenishes water in the air passing over it.

FOCUS ON

Wildfires

Wildfires—fires started by lightning—are an important environmental force in many geographical areas. Those areas most prone to wildfires have wet seasons followed by dry seasons: vegetation that grows and accumulates during the wet season dries out enough during the dry season to burn easily. When lightning hits the ground, it ignites the dry organic material, and a fire spreads through the area.

Fires have several effects on the environment. First, combustion frees the minerals that were locked in dry organic matter. The ashes remaining after a fire are rich in potassium, phosphorus, calcium, and other minerals essential for plant growth. Thus, vegetation flourishes following a fire. Second, fire removes plant cover and exposes the soil, which stimulates the germination of seeds that require bare soil and encourages the growth of shade-intolerant plants. Third, fire can cause increased soil erosion because it removes plant cover, leaving the soil more vulnerable to wind and water.

Grasses adapted to wildfire have underground stems and buds, which are unaffected by a fire sweeping over them. After the aerial parts have been killed by fire, the underground parts send up new sprouts. Fire-adapted trees such as bur oak and ponderosa pine have a thick bark that is resistant to fire. In contrast, fire-sensitive trees such as many hardwoods have a thin bark. Certain pines such as jack and lodgepole pine depend on fire for successful reproduction, because the heat of the fire opens the cones so that the seeds can be released.

Fires were a part of the natural environment long before humans appeared, and many terrestrial ecosystems have adapted to fire. African savanna, California chaparral, North American grasslands, and pine forests of the southern United States are some of the fire-adapted ecosystems (see Chapter 6). Fire helps maintain grasses as the dominant vegetation in grasslands, for example, by removing fire-sensitive hardwood trees.

The influence of fire on plants became even more pronounced once humans appeared. Because humans deliberately and accidentally set fires, fires became more frequent. Humans set fires for many reasons: to provide the grasses and shrubs that many game animals require; to clear the land for agriculture and human development; and in times of war, to reduce enemy cover.

Humans also try to prevent fires, and sometimes this effort can have disastrous consequences. When fire is excluded from a fire-adapted ecosystem, organic litter accumulates. As a result, when a fire does occur, it is much more destructive. The deadly fire in Colorado during the summer of 1994, which claimed the lives of 14 fire fighters, was blamed in part on decades of suppressing fires in the region. Prevention of fire also converts grassland to woody vegetation and facilitates the invasion of fire-sensitive trees into fire-adapted forests.

Humans sometimes conduct *controlled burns*, a tool of ecological management in which the organic litter is deliberately burned before it accumulates to dangerous levels (see figure). Controlled burns are also used to suppress fire-sensitive trees, thereby maintaining the natural fire-adapted ecosystem.

Fire is a necessary tool of ecological management. Here, a prescribed burn will help control potentially worse future fires. (U.S. Department of Agriculture)

MILKING MOUNTAINS FOR MOISTURE

In many parts of the world, moisture-laden clouds (fog) pass tantalizingly close to extremely arid regions without releasing rainfall. In the Middle Eastern Sultanate of Oman, people have "harvested" water for centuries from such clouds using the leaves of olive trees that grow near mountain summits. Small tanks built at the foot of the trees collect droplets that form on the leaves. The idea has been taken one step further in Chile's Atacama Desert. Fifty low-cost "captors" that resemble volleyball nets have been built along a ridge of the Andes Mountains. Moist clouds from the Pacific Ocean pass through the captors, releasing 8500 liters of fresh water each day. The water is channeled by an aqueduct to the coastal village of Caleta Chungungo, which formerly depended on weekly truck shipments for its drinking water. Gardens are now being grown with this new source of water. When more captor nets are built, the village plans to develop a fish-processing plant.

Rain shadows

Moisture is removed from humid air by mountains, which force air to rise. The air cools as it gains altitude, clouds form, and precipitation occurs—primarily on the windward slopes of the mountains. As the air mass moves down on the other side of the mountain, it is warmed, thereby lessening the chance of precipitation of any remaining moisture. This situation exists on the west coast of North America, where precipitation falls on the western slopes of mountains that are close to the coast. The dry land on the side of the mountains away from the prevailing wind—in this case, east of the mountain range—is called a **rain shadow** (Figure 5–18).

Tornadoes

A tornado, or twister, is a powerful, rotating funnel of air associated with severe thunderstorms. Tornadoes form when a mass of cool, dry air collides with warm, humid air, producing a strong updraft of spinning air on the underside of a cloud. The spinning funnel is called a tornado when it descends from the cloud and touches the ground. Wind velocity in a strong tornado may reach 480 kilometers per hour (300 miles per hour). Tornadoes range from 1 meter to 3.2 kilometers (2 miles) in width. They last from several seconds to as long as 7 hours and travel along the ground for several meters to more than 320 kilometers (200 miles).

Tornadoes are more destructive than any other kind of storm. They can destroy buildings, bridges, freight trains, and even blow the water out of a river or small lake, leaving it empty. Tornadoes also kill people; more than 10,000 Americans have died in tornadoes during the 20th century. Although tornadoes occur in other countries, the United States, which is known as the severe-storm capital of the world, has more tornadoes—typically about 1000 per year—than anywhere else. They are most common in the spring months throughout the Great Plains and Midwestern states (especially Texas, Oklahoma, and Kansas), as well as states along the Gulf of Mexico coast (especially Florida).

Tropical Cyclones

Tropical cyclones are giant, rotating tropical storms with winds of at least 119 kilometers per hour (74 miles per hour); the most powerful have wind velocities greater

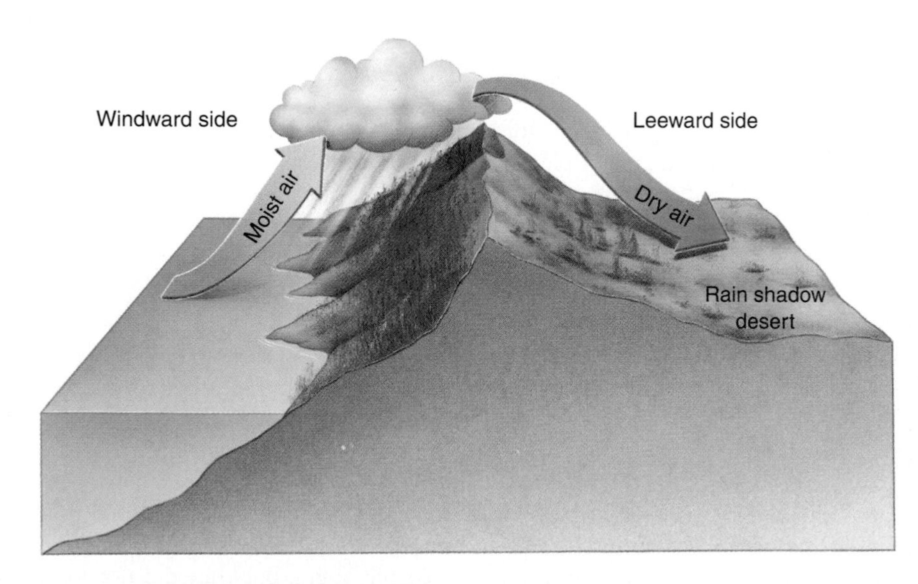

Figure 5–18
A rain shadow refers to arid or semiarid land that occurs on the far side (leeward side) of a mountain. Prevailing winds blow warm, moist air from the windward side. Air cools as it rises, releasing precipitation so that dry air descends on the leeward side. Such a rain shadow exists east of the Cascade Range in Washington state.

Figure 5–19

A satellite image of Hurricane Andrew as it approaches Florida in August 1992. Andrew devastated South Florida, causing 40 deaths and $30 billion in damage. It was the most expensive hurricane in U.S. history. (NOAA)

than 250 kilometers per hour (155 miles per hour). They form as strong winds pick up moisture over warm seas and start to spin as a result of the rotation of the Earth. The spinning causes an upward spiral of massive clouds as air is pulled upward.

Tropical cyclones are very destructive when they hit land, not so much from their strong winds as from resultant *storm surges*, waves that rise as much as 7.5 meters (25 feet) above the ocean surface. Known as *hurricanes* in the Atlantic, *typhoons* in the Pacific, and *cyclones* in the Indian Ocean, tropical cyclones are most common during summer and autumn months when ocean temperatures are warmest. With their spiral of clouds measuring about 800 kilometers (500 miles) in diameter, tropical cyclones are easy to recognize in satellite photographs (Figure 5–19).

Some years produce more hurricanes than others. The 1995 hurricane season in the Atlantic Ocean, for example, was one of the busiest in the 20th century, with 19 named tropical storms, 11 of which became hurricanes. Some of the factors that influence hurricane formation in the North Atlantic include precipitation in western Africa and water temperatures in the eastern Pacific Ocean. A wetter than usual rainy season in the western Sahel region of Africa appears to translate into more hurricanes, as does a dissipation of ENSO, which results in cooler water temperatures in the Pacific.

INTERNAL PLANETARY PROCESSES

The Earth's crust is composed of seven large plates, plus a few smaller ones, that float on the mantle (the layer of hot, soft rock lying beneath the crust and above the core) (Figure 5–20). The land masses are situated on some of these plates. As the plates move horizontally across the Earth's surface, the continents change their relative positions. The movement of the crustal plates is called **plate tectonics**.

Any area where two plates meet, called a **plate boundary**, is a site of intense geological activity (Figures 5–21 and 5–22). Earthquakes and volcanoes are common in such a region. Both San Francisco, noted for its earthquakes, and the volcano Mount Saint Helens in Washington are situated where two plates meet. Where land masses are on the boundary between two plates, mountains may form. The Himalayas formed when the plate carrying India rammed into the plate carrying Asia. When two plates grind together, one of them sometimes descends under the other, in a process known as **subduction**. When two plates move apart, a ridge of molten rock from the mantle wells up between them; the ridge continually expands as the plates move farther apart. The Atlantic Ocean is getting larger because of the buildup of lava along the mid-Atlantic ridge, where two plates are separating.

Volcanoes

Most volcanic activity is caused by the movement of crustal plates on the hot, soft rock of the outer mantle. In places, the rock reaches the melting point, forming pockets of molten rock called **magma**. When one plate slides under or away from an adjacent plate, magma may rise to the surface, often forming volcanoes. Magma that reaches the surface is called **lava**.

Volcanoes occur at three locations: subduction zones, spreading centers, and above hot spots. Subduction zones around the Pacific Basin have given rise to hundreds of volcanoes around Asia and the Americas known as the "ring of fire." Plates that are spreading apart also form volcanoes. For example, Iceland is a volcanic island that formed along the mid-Atlantic ridge. The Hawaiian Islands also have a volcanic origin, but they did not form

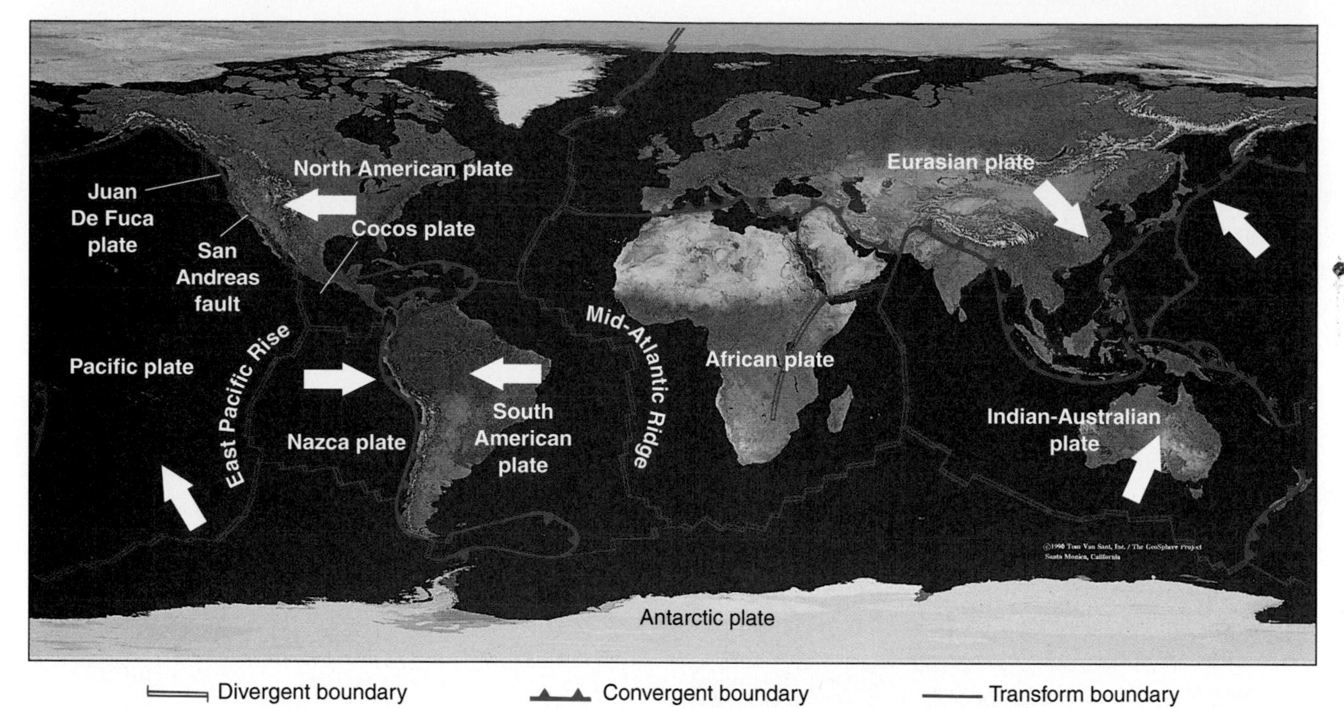

Figure 5–20

Earth's crust consists of seven main plates: African, Eurasian, Indian-Australian, Antarctic, Pacific, North American, and South American. Plate boundaries are shown in red, and arrows show the directions of plate movements. The three types of boundaries are explained in Figure 5–22. (© 1990 Tom Van Sant/The Geosphere Project, Santa Monica, California)

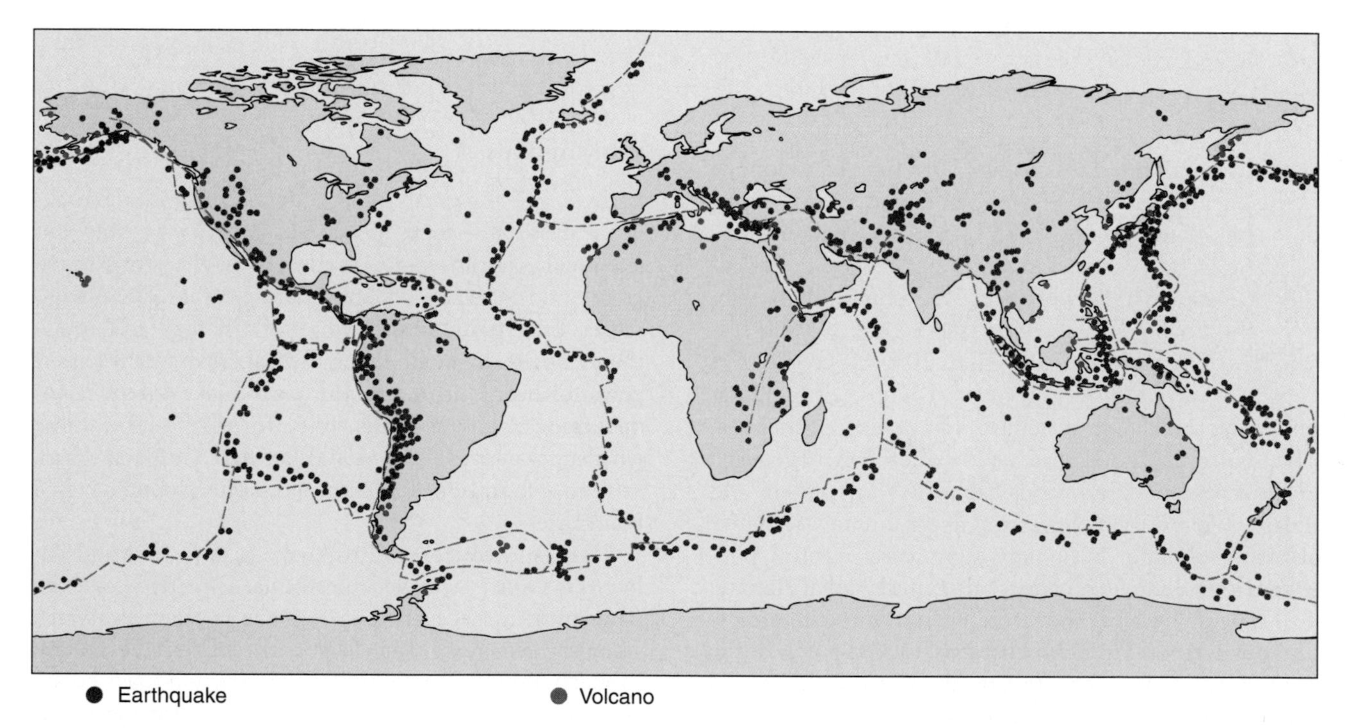

Figure 5–21

Volcanoes and earthquakes commonly occur at plate boundaries. Volcanoes are shown in (teal), earthquakes in (red).

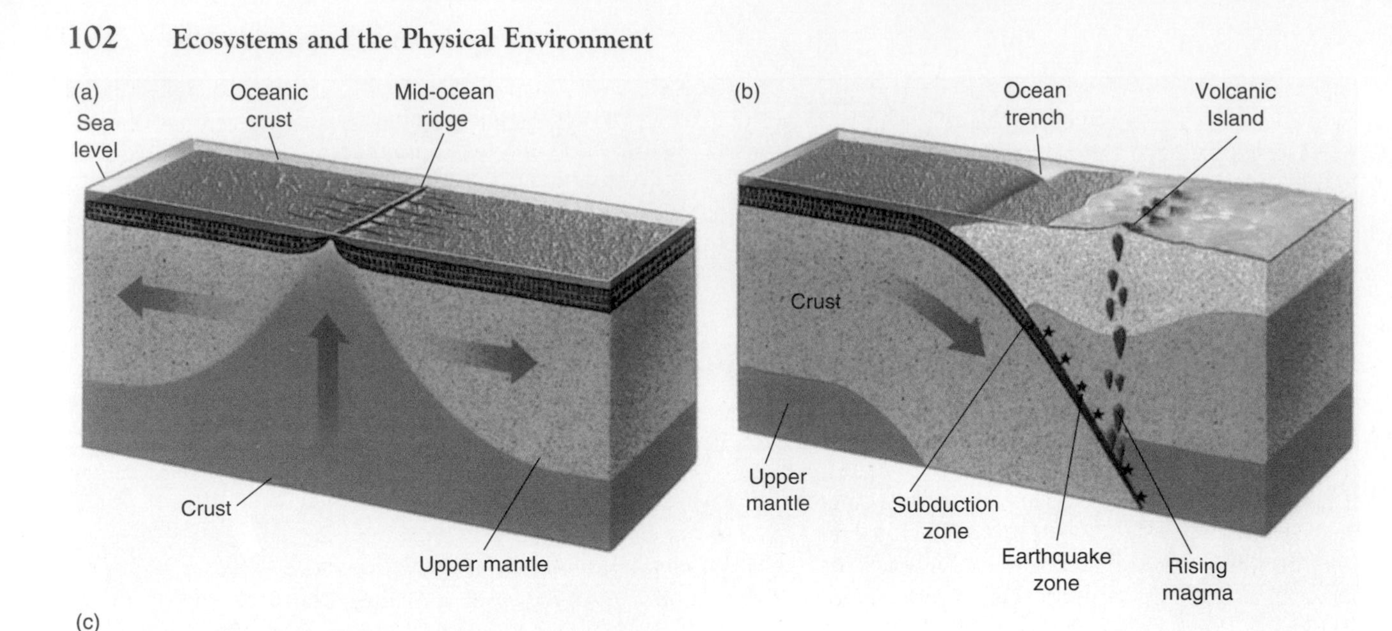

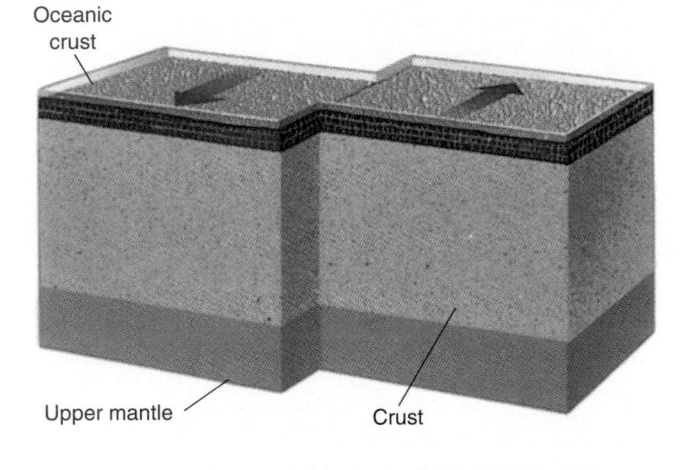

Figure 5–22

The three types of plate boundaries. All three occur both in the ocean and on land. (a) Two plates move apart at a divergent plate boundary. (b) When two plates collide at a convergent plate boundary, subduction may occur. (c) At a transform plate boundary, plates move horizontally in opposite but parallel directions.

at plate boundaries. This chain of volcanic islands formed as the Pacific plate moved over a **hot spot**, a rising plume of magma that flowed from an opening in the crust.

The largest volcanic eruption in the twentieth century occurred in 1991 when Mt. Pinatubo in the Republic of the Philippines exploded (see Figure 20–10). Despite the evacuation of more than 200,000 people, 338 deaths occurred, mostly from the collapse of buildings under the thick layer of wet ash that blanketed the area. The volcanic cloud produced when Mt. Pinatubo erupted extended upwards some 30 kilometers (48 miles). We are used to hearing about human activities affecting climate, but many significant natural phenomena, including volcanoes, also affect global climate. The magma and ash that were ejected into the atmosphere when Mt. Pinatubo erupted blocked much of the sun's warmth and brought an end to several years of warmer-than-usual global temperatures that had been observed in the late 1980s and early 1990s (see Chapter 20).

Earthquakes

Forces inside the Earth sometimes push and stretch rocks in the crust. The rocks absorb this energy for a time, but eventually, as the energy accumulates, the stress is too great and the rocks suddenly shift or break. The energy, released as **seismic waves**, vibrations that spread through the rocks rapidly in all directions, causes one of the most powerful events in nature, an earthquake. Most earthquakes occur along **faults**, fractures in the crust along which rock moves forward and backward, up and down, or from side to side. Fault zones are often found at plate boundaries.

The site where an earthquake begins, often far below the surface, is called the **focus**. Directly above the focus, at the surface, is the earthquake's **epicenter**. When seismic waves reach the surface, they cause the ground to shake. Buildings and bridges collapse, and roads break. One of the instruments used to measure seismic waves is a seismograph, which helps seismologists (scientists who study earthquakes) determine where an earthquake started, how strong it was, and how long it lasted. One way the magnitude of energy released by an earthquake can be measured is by the *Richter scale*, invented by Charles Richter, a California seismologist, in 1935. Each unit on the Richter scale represents about 30 times more released energy than the unit immediately below it. As an example, a magnitude 8 earthquake is 30 times more powerful than a magnitude 7 earthquake and 900 times more powerful than a magnitude 6 earthquake. The Richter scale makes it easy to compare earthquakes, but it tends to underestimate the energy of very large quakes.

Although the public is used to hearing about the Richter scale, seismologists typically do not use it. There are several ways to measure the magnitude of an earthquake, just as there are several ways to measure the size of a person (for example, height, weight, and amount of body fat). Most seismologists use a more precise scale called the *moment magnitude scale* to measure earthquakes, especially those larger than magnitude 6.5. The moment magnitude scale calculates the total energy released by a quake.

More than one million earthquakes are recorded by seismologists each year (Table 5–1), but most of these are too small to be felt (readings of about 2 on the Richter scale). A magnitude 5 earthquake usually causes property damage. Every five years or more, a great earthquake occurs with a reading of 8 or higher. Such quakes usually cause massive property destruction and kill large numbers of people. Few people die directly from the seismic waves, but rather from collapsed buildings or fires started by ruptured gas lines.

Landslides and tsunamis are some of the side effects of earthquakes. A landslide is an avalanche of rock, soil, and other debris that slides swiftly down a mountainside. A 1970 earthquake in Peru resulted in a landslide that buried the town of Yungay and killed 17,000 people. A tsunami, a giant sea wave caused by an underwater earthquake or volcanic eruption, can sweep through the water at more than 750 kilometers (450 miles) per hour. Although a tsunami may be only about 1 meter high in deep ocean water, it can build to a wall of water 30.5 meters (100 feet)—as high as a 10-story building—when it comes ashore, often far from where the original earthquake triggered it. Tsunamis have caused thousands of deaths, particularly along the Pacific coast. Today the Pacific Tsunami Warning System monitors submarine earthquakes and warns people of approaching tsunamis.

One of the most geologically active places in North America is California's San Andreas fault, which runs parallel to the California coast from the Mexican border to northern California—a length of more than 1100 kilometers (700 miles). The Pacific plate (the west side of the San Andreas fault) is sliding northward relative to the North American plate (the east side) at a rate of about 3.5 centimeters (1.4 inches) per year. In 1906, much of San Francisco, which is located near the San Andreas fault, was destroyed by a magnitude 8.3 earthquake and the fire, caused by ruptured gas lines, that followed it. In 1989 a magnitude 6.9 earthquake along the San Andreas fault in Loma Prieta, about 90 kilometers south of San Francisco, killed 67 people and caused $6 billion in damage to the San Francisco Bay area.

Not all earthquakes occur at plate boundaries. Some occur on smaller faults that crisscross the large plates—the major earthquakes that damaged Northridge, California in January 1994 and Kobe, Japan in January 1995, for example. Such earthquakes pose a seismic hazard that is difficult to evaluate because major quakes occur along a given small fault only every 1000 to 5000 years, in contrast to a larger fault line, which may have a major quake every century or so.

Table 5–1
Major Earthquakes 1985–1995

Year	Location	Moment Magnitude	Damage
1995	Kobe, Japan	6.9	5502 deaths; at least $30 billion damage
1994	Northridge, California	6.7	61 deaths; $20 billion
1993	Marharashtra, India	6.2	9748 deaths; $280 million
1990	Luzon Island, The Philippines	7.7	1621 deaths; $2 billion
1990	Northwestern Iran	7.4	50,000 deaths; $7 billion
1989	San Francisco Bay Area (Loma Prieta)	6.9	62 deaths; $6 billion
1988	Armenia	6.8	55,000 deaths; $14 billion
1985	Mexico	8.0	9500 deaths; $4 billion

SUMMARY

1. In contrast to energy, which moves in one direction through ecosystems, matter is cyclic. All materials vital to life are continually recycled from the environment to organisms and back to the environment. Biogeochemical cycles are cycles of matter such as carbon, nitrogen, phosphorus, and water.

2. Carbon enters plants, algae, and cyanobacteria as carbon dioxide (CO_2), which is incorporated into organic molecules by photosynthesis. Cell respiration by plants, by animals that eat plants, and by decomposers returns CO_2 to the atmosphere, making it available for producers again.

3. There are five steps in the nitrogen cycle. Nitrogen fixation is the conversion of nitrogen gas to ammonia. Nitrification is the conversion of ammonia to nitrate, one of the main forms of nitrogen used by plants. Assimilation is the biological conversion of nitrates or ammonia into proteins and other nitrogen-containing compounds by plants; the conversion of plant proteins into animal proteins is also part of assimilation. Ammonification is the conversion of organic nitrogen to ammonia. Denitrification converts nitrate to nitrogen gas.

4. The phosphorus cycle has no biologically important gaseous compounds. Phosphorus erodes from rock in the form of inorganic phosphates, which are absorbed from the soil by plant roots. Phosphorus enters other organisms through the food web and is released back into the environment as inorganic phosphate by decomposers. When phosphorus washes into the ocean and is deposited in seabeds, it can be lost from biological cycles for millions of years.

5. The hydrologic cycle, which continuously renews the supply of water that is essential to life, involves an exchange of water among the land, the atmosphere, and organisms. Water enters the atmosphere by evaporation and transpiration, and leaves the atmosphere as precipitation. On land, water filters through the ground or runs off to lakes, rivers, and the ocean.

6. Sunlight is the primary source of energy available to the biosphere. Of the solar energy that reaches the Earth, 30 percent is immediately reflected away and the remaining 70 percent is absorbed, including 0.02 percent that is absorbed by plants. Ultimately, all absorbed solar energy is radiated into space as infrared (heat) radiation.

7. A combination of the Earth's roughly spherical shape and the tilt of its axis concentrates solar energy at the equator and dilutes solar energy at the poles. The tropics are therefore hotter and less variable in climate than are temperate and polar areas. Seasons are determined by the inclination of the Earth's axis.

8. The atmosphere protects the Earth's surface from most of the sun's ultraviolet radiation and x rays as well as from lethal amounts of cosmic rays from space. Visible light and some infrared radiation penetrate the atmosphere and warm the surface and the lower atmosphere.

9. Atmospheric heat transfer from the equator to the poles produces a movement of warm air toward the poles and a movement of cool air toward the equator, thus moderating the climate. In addition to these global circulation patterns, the atmosphere exhibits complex horizontal movements called winds that result in part from differences in atmospheric pressure and from the rotation of the Earth (the Coriolis effect).

10. The global ocean is a single, continuous body of water that surrounds the continents and covers almost three-fourths of the Earth's surface. Surface ocean currents result largely from prevailing winds. Other factors that contribute to ocean currents include the Coriolis effect, the varying density of water, and the position of land masses.

11. The El Niño–Southern Oscillation (ENSO) event is a periodic warming of surface waters of the tropical East Pacific that alters both ocean and atmospheric circulation patterns and results in unusual weather in areas far from the tropical Pacific.

12. An area's climate comprises the average weather conditions that occur there over a period of years. Temperature (both average temperature and temperature extremes) and precipitation (both average precipitation and seasonal distribution) are the two most important climatic factors. Precipitation is greatest where warm air passes over the ocean, absorbing moisture, and is then cooled, such as when humid air is forced upward by mountains. Deserts develop in the rain shadows of mountain ranges or in continental interiors.

13. A tornado is a powerful, rotating funnel of air associated with severe thunderstorms. A tropical cyclone is a giant, rotating tropical storm with high winds. Tropical cyclones are called hurricanes in the Atlantic, typhoons in the Pacific, and cyclones in the Indian Ocean.

14. Earth's crust consists of seven large plates, plus a few smaller ones, that float on the mantle. As the plates move horizontally, the continents change their relative positions. The movement of the crustal plates is called plate tectonics. Plate boundaries are sites of intense geological activity, such as mountain building, volcanoes, and earthquakes.

THINKING ABOUT THE ENVIRONMENT

1. Why might industrial polluters suppose that the Gaia hypothesis gives them permission to pollute the air, water, and soil indefinitely?

2. As discussed in the chapter, the Earth is a closed system regarding matter. Is the Earth a closed or open system regarding energy? Explain.

3. Why is the cycling of matter essential to the continuance of life?

4. How might humans disturb the global temperature balance?

5. What basic forces determine the circulation of the atmosphere? Describe the general directions of atmospheric circulation.

6. What is the energy source that powers the wind?

7. Explain why mountain climbers are easily sunburned even when the temperature is below freezing.

8. How do ocean currents affect climate on land?

9. What are some of the environmental factors that produce areas of precipitation extremes, such as rain forests and deserts?

10. Evaluate the area where you live with respect to natural dangers. Is there a threat of possible earthquakes, volcanic eruptions, hurricanes, tornadoes, or tsunamis?

* 11. Convert the temperature range in the lower stratosphere (−45°C to −75°C) to °F.

* 12. In 1994, 79,107 wildfires were reported to the National Interagency Fire Center in Boise, Idaho, and 4.1 million acres burned. What was the average number of acres that burned per fire event?

RESEARCH PROJECTS

Choose a recent environmental catastrophe (flood, earthquake, wildfire, mud slide, etc.) and describe how human activities precipitated or exacerbated the catastrophe. Identify any measures that might minimize such catastrophes in the future. If possible, document a case of an area that has learned from past mistakes, and how other areas can benefit from the experience.

On-line materials relating to this chapter are on the ► World Wide Web at **http://www.saunderscollege.com/lifesci/environment2** (select Chapter 5 from the Table of Contents).

SUGGESTED READING

Anderson, R. "Cloud Harvest," *Natural History*, November 1995. How a coastal fishing village along the Atacama Coast of northern Chile harvests water from the clouds.

Baskin, Y. "Under the Influence of Clouds," *Discover*, September 1995. Researchers are only beginning to understand how clouds affect global temperature by absorbing and reflecting light and heat.

Broecker, W. S. "Chaotic Climate," *Scientific American,* November 1995. Earth's climatic system has occasionally exhibited abrupt warming and cooling shifts.

Davies-Jones, R. "Tornadoes," *Scientific American*, August 1995. We are beginning to understand the thunderstorms that spawn tornadoes, but much remains to be learned about the twisters themselves.

Gore, R. "Living with California's Faults," *National Geographic*, April 1995. Examines the causes of earthquakes in California.

Krakauer, J. "Geologists Worry about Dangers of Living Under the Volcano," *Smithsonian*, July 1996. Mount Rainier, which looms over Tacoma, Washington, and Puget Sound, is an active volcano that scientists consider to be extremely dangerous.

Parfit, M. "The Essential Elements of Fire," *National Geographic* Vol. 190, No. 3, September 1996. Controlled fires keep some ecosystems healthy, and their absence can lead to disastrous consequences when a fire does occur.

Pimm, S. L., et al. "Hurricane Andrew," *BioScience* Vol. 44, No. 4, April 1994. Scientists are assessing the damage caused by 1992's Hurricane Andrew to South Florida's diverse ecosystems.

Reid, T. R. "Kobe Wakes to a Nightmare," *National Geographic* Vol. 188, No. 1, July 1995. All about the earthquake that struck Japan in January 1995, killing 5500 people and destroying much of Kobe.

Tibbetts, J. "Farming and Fishing in the Wake of El Niño," *BioScience* Vol. 46, No. 8, September 1996. How ENSO events affect global climate and food production.

Yulsman, T. "Dynamic Earth," *Earth*, November 1994. Spectacular satellite photographs provide visual evidence of plate tectonics, volcanism, glaciation, and erosion.

Thick, epiphyte-covered lianas grow into the canopy using a tree trunk for support.
(Mark Moffett/Minden Pictures)

Major Ecosystems of the World

Climate, particularly temperature and precipitation, influences the distribution of organisms (see Chapter 5). In each major kind of climate a distinctive type of vegetation develops. For example, tropical rainforest plants are associated with warm, humid climates.

A fully developed tropical rain forest has at least three distinct stories, or layers, of vegetation. The topmost story consists of the crowns of occasional, very tall trees, some 50 meters (164 feet) or more in height, which are exposed to direct sunlight. The middle story, which reaches a height of 30 to 40 meters (100 to 130 feet), forms a continuous canopy of leaves that lets in very little sunlight to support the sparse understory. Smaller plants that are specialized for life in the shade, as well as the seedlings of taller trees, comprise the understory.

The vegetation of tropical rain forests is not dense at ground level except near stream banks or where a fallen tree has opened the canopy. The continuous canopy of leaves overhead produces a dark, extremely moist environment.

Tropical rainforest trees support extensive epiphytic communities of plants such as ferns, mosses, orchids, and bromeliads. (Epiphytes are smaller plants that grow nonparasitically on the large trees.) Although epiphytes grow in crotches of branches, on bark, or even on the leaves of their hosts, they use their host trees primarily for physical support.

Because little light penetrates to the understory, many of the plants living there are adapted to climb already established host trees rather than to invest their meager photosynthetic resources in building the cellulose tissues of their own trunks. Lianas, which are woody tropical vines, some as thick as a human thigh, twist up through the branches of the huge rainforest trees. Once in the canopy, lianas grow from the upper branches of one forest tree to another, connecting the tops of the trees together and providing a walkway for many of the canopy's residents. They and herbaceous vines provide nectar and fruit for many tree-dwelling animals.

Tropical rainforest plants are restricted to warm, humid climates and are not uniformly distributed throughout the Earth. In like manner, cordgrasses are adapted to salt marshes, where fresh water and saltwater mix, and do not grow in other ecosystems. In this chapter we examine the Earth's major biomes and aquatic ecosystems, including the physical factors that largely determine which organisms characterize each. ◄

EARTH'S MAJOR BIOMES

A **biome** is a large, relatively distinct terrestrial region characterized by similar climate, soil, plants, and animals, regardless of where it occurs in the world (Figure 6–1). Because it is so large in area, a biome encompasses many interacting ecosystems. In terrestrial ecology, a biome is considered the next level of ecological organization above that of community and ecosystem. A biome's boundaries are determined by invisible climatic barriers, with tem-

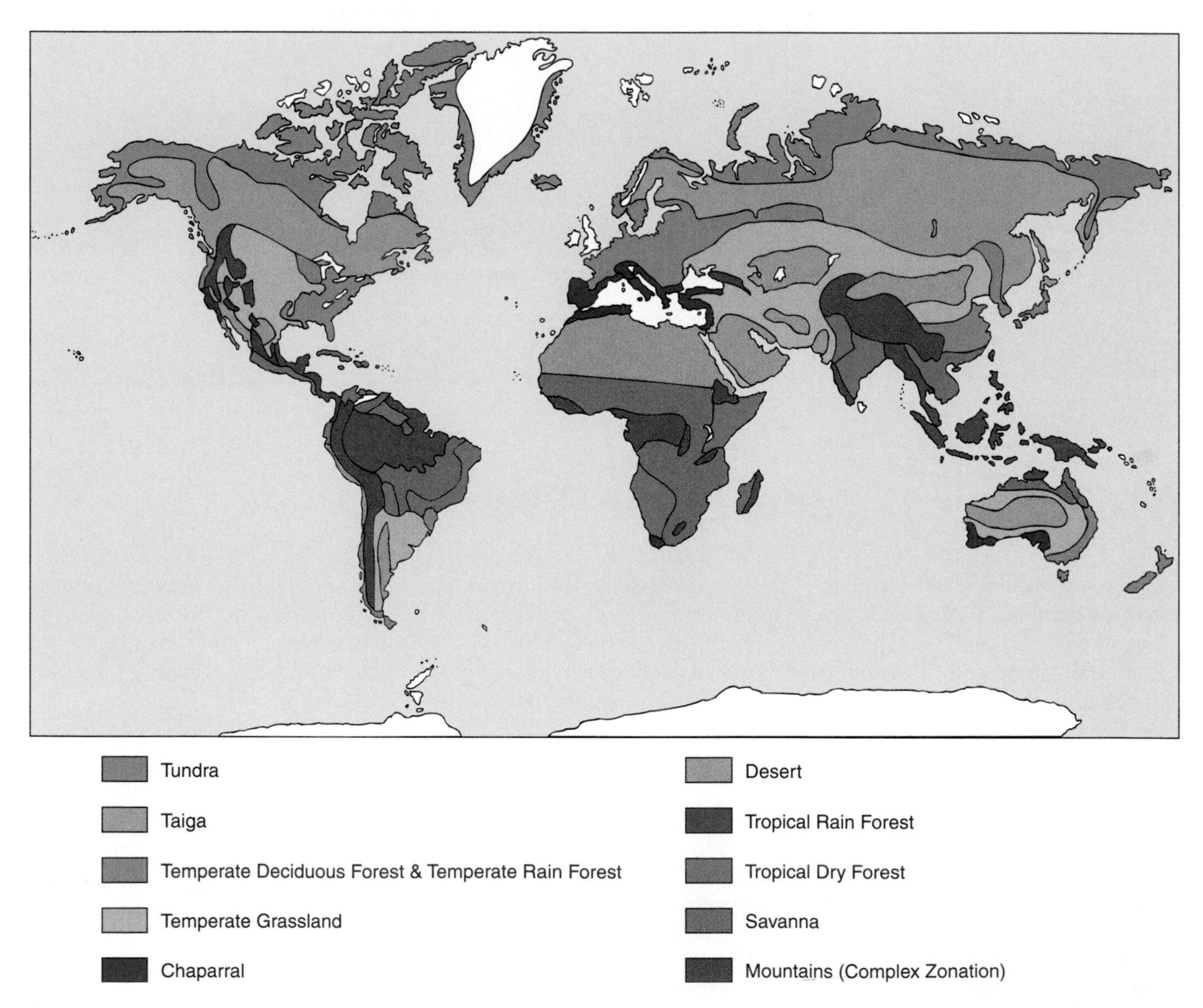

Figure 6–1

A highly simplified map showing the distribution of the world's biomes. Although sharp boundaries are shown, biomes actually grade together at their boundaries. (Adapted from Odum 1971)

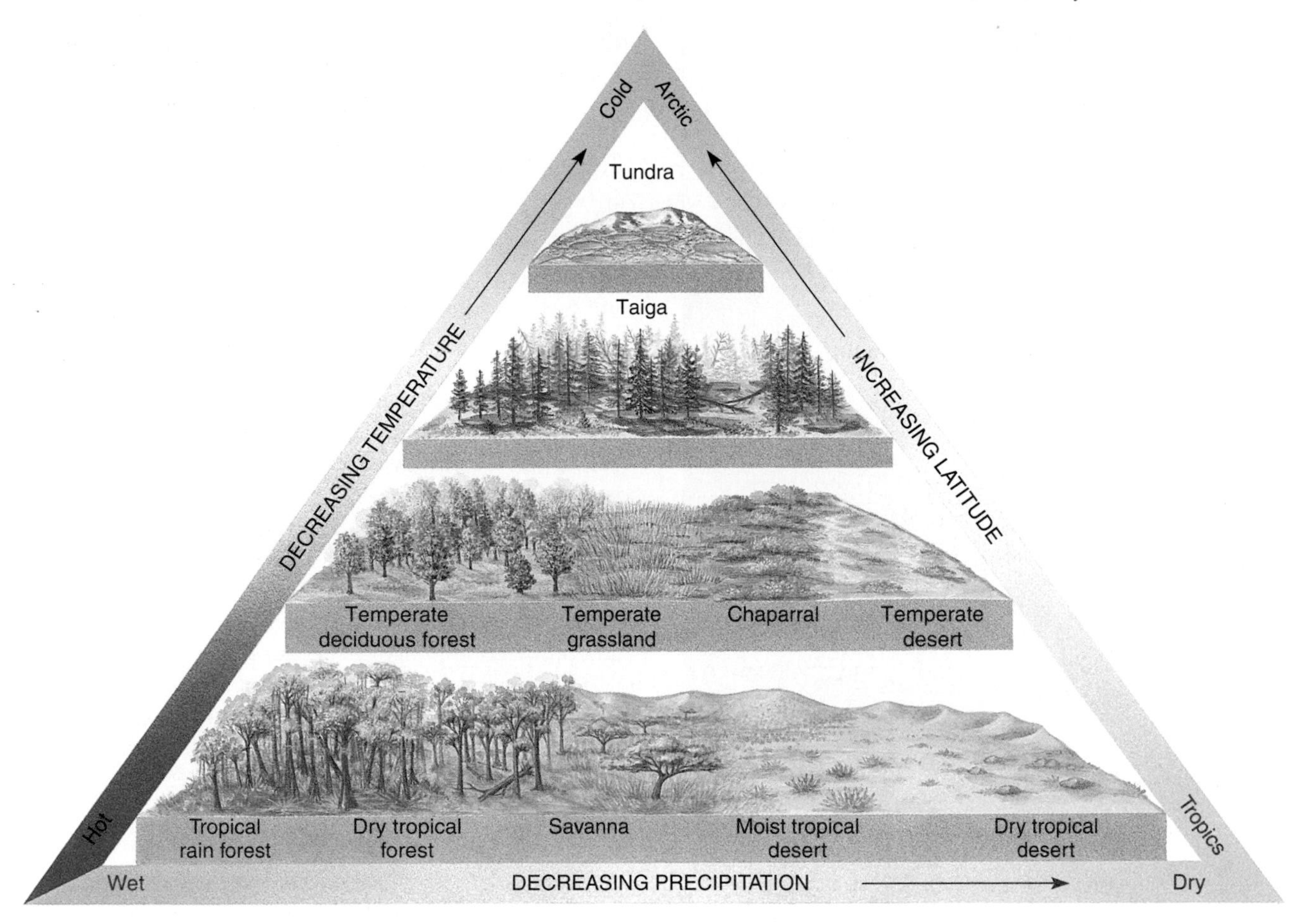

Figure 6–2

The biomes are distributed primarily in accordance with two climatic factors, temperature and precipitation. In the higher latitudes, temperature is the more important of the two. In temperate and tropical zones, precipitation is a significant determinant of community composition.

perature and precipitation being most important (Figure 6–2). Altitude also affects ecosystems: changes in vegetation with increasing altitude resemble the changes in vegetation observed with movement from warmer to colder climates (see Focus On Vertical Zonation: The Distribution of Vegetation on Mountains, page 112).

Nine biomes are discussed: tundra, taiga, temperate rain forest, temperate deciduous forest, temperate grassland, chaparral, desert, savanna, and tropical rain forest.

Tundra: Cold Boggy Plains of the Far North

Tundra occurs in the extreme northern latitudes wherever the snow melts seasonally (Figure 6–3). The Southern Hemisphere has no equivalent of the arctic tundra because it has no land in the corresponding latitudes. Tundra has long, harsh winters and very short summers. Although the growing season, with its warmer temperatures, is short (from 50 to 160 days depending on location), the days are long. In many places the sun does not set at all for many days in midsummer, although the amount of light at midnight is one-tenth that at noon. There is little precipitation (10 to 25 centimeters, or 4 to 10 inches, per year) over much of the tundra, with most of it falling during the summer months.

Tundra soils tend to be geologically young, since most of them were formed when glaciers retreated after the last Ice Age.[1] These soils are usually nutrient-poor and have little organic litter. Although the soil melts at the surface during the summer, tundra has a layer of permanently frozen ground, called **permafrost**, that varies in depth and thickness. Permafrost interferes with drainage—soil is usually waterlogged during the summer—and prevents the roots of larger plants from becoming established. The limited precipitation in combination with low temperatures, flat topography (surface features), and the per-

[1]Glacier ice, which occupied about 29 percent of the Earth's land during the last Ice Age, began retreating about 17,000 years ago. Today, glacier ice occupies about 10 percent of the land surface.

Figure 6–3
Arctic tundra. During its short growing season, small, hardy plants grow in tundra, the northernmost biome that encircles the Arctic Ocean. (Michio Hoshino/Minden Pictures)

mafrost layer produces a landscape of broad, shallow lakes and ponds, sluggish streams, and bogs.

Few species are found in the tundra, but individual species often exist in great numbers. Tundra is dominated by mosses; lichens, such as reindeer moss; grasses; and sedges, which are grasslike plants. Shrubs and trees are small and confined to sheltered locations. As a rule, tundra plants seldom grow taller than 30 centimeters (12 inches).

The year-round animal life of the tundra includes lemmings, voles, weasels, arctic foxes, snowshoe hares, ptarmigan, snowy owls, and musk-oxen. In the summer, caribou migrate north to the tundra to graze on sedges, grasses, and dwarf willow. Dozens of birds also migrate north in summer to nest and feed on abundant insects. Mosquitoes, blackflies, and deerflies survive the winter as eggs or pupae and occur in great numbers during summer weeks. There are no reptiles or amphibians.

Tundra regenerates very slowly after it has been disturbed. Even casual use by hikers can cause damage. Large portions of the arctic tundra have suffered long-lasting injury, likely to persist for hundreds of years, as a result of oil and natural gas exploration and military use (see the discussion on the Arctic National Wildlife Refuge in Chapter 10).

Taiga: Evergreen Forests of the North

Just south of the tundra is the **taiga**, or **boreal forest**, which stretches across North America and Eurasia, covering approximately 11 percent of the Earth's land (Figure 6–4). A biome comparable to the taiga is not found in the Southern Hemisphere. Winters are extremely cold and severe, although not as harsh as in the tundra. The growing season of the boreal forest is somewhat longer than that of the tundra. Taiga receives little precipitation, perhaps 50 centimeters (20 inches) per year, and its soil is typically acidic, mineral-poor, and characterized by a deep layer of partly decomposed pine and spruce needles at the surface. Permafrost is patchy and, where found, is often deep under the soil. Taiga has numerous ponds and lakes in water-filled depressions that were dug in the ground by grinding ice sheets during the last Ice Age.

Deciduous trees such as aspen and birch, which shed their leaves in autumn, may form striking stands in the taiga, but overall, spruce, fir, and other conifers (cone-

Figure 6–4
The taiga, or boreal forest. These coniferous forests occur in cold regions of the Northern Hemisphere adjacent to the tundra. (Charlie Ott/Photo Researchers, Inc.)

bearing evergreens) clearly dominate. Conifers have many drought-resistant adaptations, such as needle-like leaves with a minimal surface area for water loss. Such an adaptation enables conifers to withstand the "drought" of the northern winter months (roots cannot absorb water when the ground is frozen).

The animal life of the boreal forest consists of some larger species such as caribou, which migrate from the tundra to the taiga for winter; wolves; bears; and moose. However, most mammals are medium-sized to small, including rodents, rabbits, and fur-bearing predators such as lynx, sable, and mink. Most species of birds are abundant in the summer but migrate to warmer climates for winter. Insects are abundant, but there are few amphibians and reptiles except in the southern taiga.

Most of the taiga is not well suited to agriculture because of its short growing season and mineral-poor soil. However, the boreal forest yields vast quantities of lumber, pulpwood for paper products, animal furs, and other forest products. See Chapter 17 for a discussion of deforestation of this biome.

Temperate Rain Forest: Lush Temperate Forests

A coniferous **temperate rain forest** occurs on the northwest coast of North America. Similar vegetation exists in southeastern Australia and in southern South America. Annual precipitation in this biome is high, from 200 to 380 centimeters (80 to 152 inches), and is augmented by condensation of water from dense coastal fogs. The proximity of temperate rain forest to the coastline moderates the temperature so that the seasonal fluctuation is narrow; winters are mild and summers are cool. Temperate rain forest has relatively nutrient-poor soil, although its organic content may be high. Needles and large fallen branches and trunks accumulate on the ground as litter that takes many years to decay and release nutrients back to the soil.

The dominant vegetation in the North American temperate rain forest is large evergreen trees such as western hemlock, Douglas fir, western red cedar, Sitka spruce, and western arborvitae (Figure 6–5). Temperate rain forest is rich in epiphytic vegetation such as mosses, club mosses, lichens, and ferns. Squirrels, deer, and numerous bird species are among the animals found in the temperate rain forest.

Temperate rain forest is one of the world's richest wood producers. It is also one of the world's most complex ecosystems. Care must be taken to avoid overharvesting the original (never-logged) old-growth forest, however, because such an ecosystem takes hundreds of years to develop (see Chapters 7 and 17). The logging industry typically harvests old-growth forest and replants the area with a monoculture of trees (of a single species) that it can harvest in 40- to 100-year cycles. Thus, the old-growth forest ecosystem, once harvested, never has a chance to redevelop.

Figure 6–5

Temperate rain forest. Trees of the temperate rain forest in Washington are conifers and include Douglas fir, western hemlock, and western red cedar. Moisture-loving ferns, mosses, and lichens grow on the trees as well as on the ground. This temperate biome is characterized by high amounts of precipitation. (© 1997 Terry Donnelly/Dembinsky Photo Associates)

Temperate Deciduous Forest: Broad-leaved Trees That Shed Their Leaves

Seasonality—hot summers and cold winters—is characteristic of **temperate deciduous forest**, which occurs in temperate areas where precipitation ranges from about 75 to 126 centimeters (30 to 50 inches) annually. Typically, the soil of a temperate deciduous forest consists of a topsoil rich in organic material and a deep, clay-rich lower layer. As organic materials decay, mineral ions are released. If they are not immediately absorbed by the roots of living trees, these ions leach into the clay, where they may be retained.

The temperate deciduous forests of the northeastern and mideastern United States are dominated by broad-leaved hardwood trees, such as oak, hickory, and beech,

(text continues on page 113)

FOCUS ON

Vertical Zonation: The Distribution of Vegetation on Mountains

Hiking up a mountain is similar to traveling toward the North Pole with respect to the major life zones encountered (see figure). This altitude-latitude similarity occurs because as one climbs a mountain, the temperature drops just as it does when one travels north. The types of organisms living on the mountain change as the temperature changes.

The base of a mountain in Colorado, for example, might be covered by deciduous trees, which shed their leaves every autumn. Above that altitude, where the climate is colder and more severe, one might find a coniferous forest called subalpine forest, which resembles the northern taiga. Higher still, where the climate is very cold, a kind of tundra occurs, with vegetation composed of grasses, sedges, and small tufted plants; it is called alpine tundra to distinguish it from arctic tundra. At the very top of the mountain, a permanent ice or snow cap might be found, similar to the nearly lifeless polar land areas.

There are important environmental differences between high altitudes and high latitudes, however, that affect the types of organisms found in each place. Alpine tundra typically lacks permafrost and receives more precipitation than arctic tundra. Also, high elevations of temperate mountains do not have the great extremes in daylength that are associated with the changing seasons in biomes at high latitudes. Furthermore, the intensity of solar radiation is greater at high elevations than at high latitudes. For example, at high elevations the sun's rays pass through less atmosphere, which results in a greater exposure to ultraviolet (UV) radiation—that is, less UV is filtered out by the atmosphere—than occurs at high latitudes.

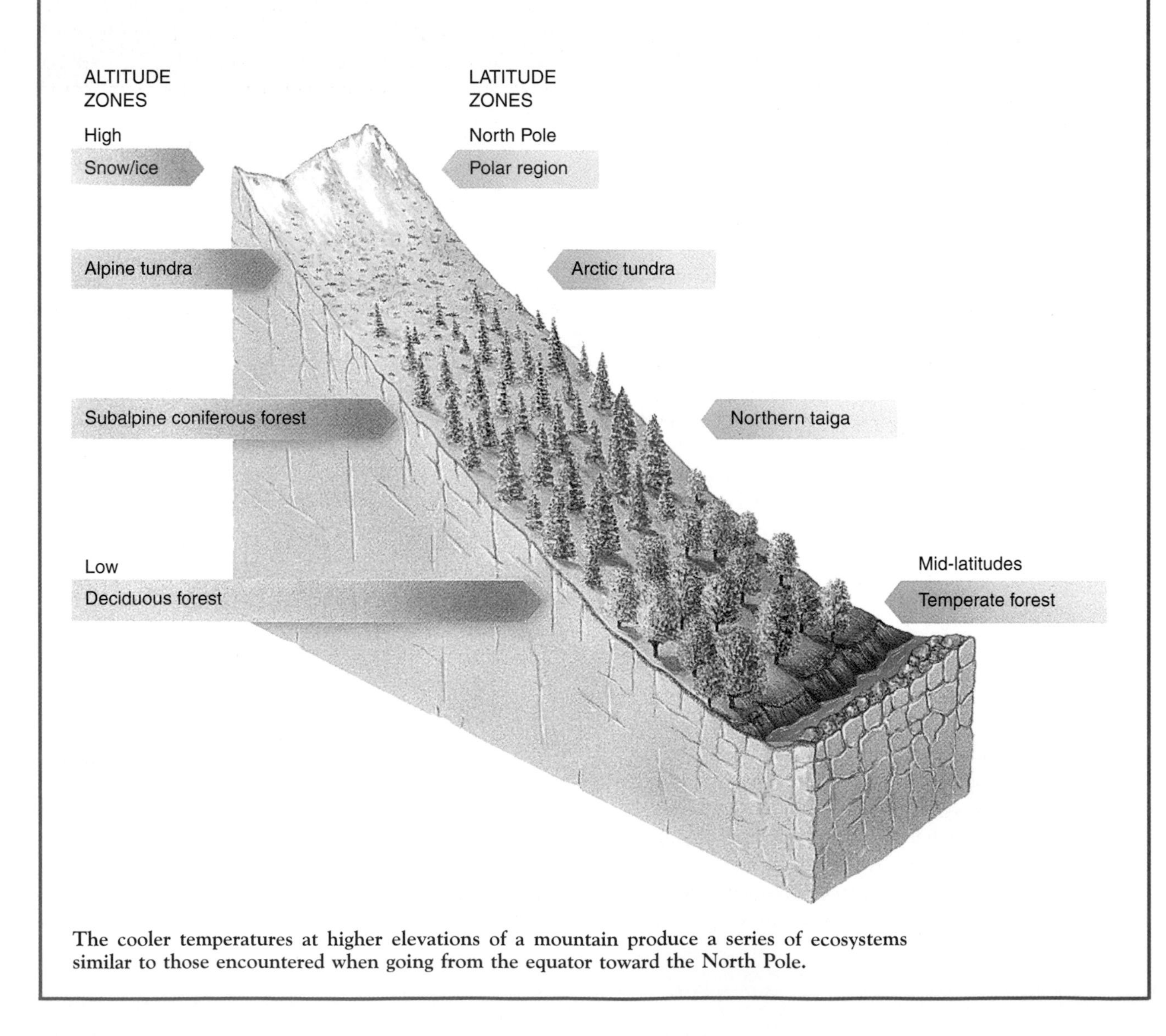

The cooler temperatures at higher elevations of a mountain produce a series of ecosystems similar to those encountered when going from the equator toward the North Pole.

Figure 6–6

Temperate deciduous forest. The broad-leaf trees that dominate this biome are deciduous and will shed their leaves before winter. (John C. Weaver/Vermont Dept. of Travel and Tourism)

that lose their foliage annually (Figure 6–6). In the southern reaches of the temperate deciduous forest, the number of broad-leaved evergreen trees, such as magnolia, increases. The trees of the temperate deciduous forest form a dense canopy that overlies saplings and shrubs.

Temperate deciduous forest originally contained a variety of large mammals, such as puma, wolves, and bison, which are now absent, plus deer, bears, and many small mammals and birds. Both reptiles and amphibians abounded, together with a denser and more varied insect life than exists today.

Much of the original temperate deciduous forest was removed by logging and land clearing. Where it has been allowed to regenerate, temperate deciduous forest is often in a seminatural state—that is, highly modified by humans.

Worldwide, deciduous forests were among the first biomes to be converted to agricultural use. In Europe and Asia, for example, many soils that originally supported deciduous forests have been cultivated by traditional agricultural methods for thousands of years without a substantial loss in fertility. During the 20th century, however, intensive agricultural practices have been widely adopted that have resulted in the degrading of perhaps 15 to 20 percent of the Earth's total agricultural land. Most damage to farmland has been done since the end of World War II. We say more about soil degradation in Chapter 14.

Grasslands: Temperate Seas of Grass

Summers are hot, winters are cold, and rainfall is often uncertain in **temperate grasslands**. Annual precipitation averages 25 to 75 centimeters (10 to 30 inches). In grasslands with less precipitation, minerals tend to accumulate in a well defined layer just below the topsoil. These minerals tend to wash out of the soil in areas with more precipitation. Grassland soil has considerable organic material because the above-ground portions of many grasses die off each winter and contribute to the organic content of the soil, while the roots and rhizomes (underground stems) survive underground. Also, many grasses are sod formers; that is, their roots and rhizomes form a thick, continuous underground mat.

Moist temperate grasslands, also known as tallgrass prairies, occur in the United States in Iowa, western Minnesota, eastern Nebraska, and parts of other Midwestern states. Here there are few trees, except for oaks that grow near rivers and streams, and grasses grow in great profusion in the thick, rich soil. Periodic wildfires help maintain grasses as the dominant vegetation in grasslands (see Focus On: Wildfires in Chapter 5). Tallgrass prairies were formerly dominated by several species of grasses that, under favorable conditions, grew as tall as a person on horseback, and the land was covered with herds of grazing animals, such as bison and pronghorn elk. The principal predators were wolves and coyotes. Smaller animals included prairie dogs and their predators (foxes, black-footed ferrets, and various birds of prey), grouse, reptiles such as snakes and lizards, and great numbers of insects.

Shortgrass prairies (Figure 6–7) are temperate grasslands that receive less precipitation than the moister grasslands just described but more precipitation than deserts. In the United States, shortgrass prairies occur in the eastern half of Montana, the western half of South Dakota, and parts of other Midwestern states. Shortgrass prairies are dominated by grasses that grow knee high or lower. The plants grow in less abundance than in the moister grasslands, and occasionally some bare soil is exposed. Native grasses of shortgrass prairies are drought-resistant.

The North American grassland, particularly the tallgrass prairie, was so well suited to agriculture that little of it now remains. More than 90 percent has vanished under the plow, and the remaining is so fragmented that almost nowhere can we see even an approximation of what European settlers saw when they settled the Midwest. Today, the tallgrass prairie is considered North

Figure 6–7

Temperate grassland. This biome is mostly treeless but contains a profusion of grasses and other herbaceous flowering plants. A shortgrass prairie of the North American plains is shown. Other temperate grasslands have more luxurious vegetation because they receive more annual precipitation than shortgrass prairies. (David Muench 1992)

Figure 6–8

Chaparral. Chaparral vegetation consists mainly of drought-resistant evergreen shrubs and small trees. Hot, dry summers and mild, rainy winters characterize the chaparral. This photo shows the Santa Monica Mountains in California. (John Cunningham/Visuals Unlimited)

America's rarest biome. It is not surprising that the American Midwest, the Ukraine, and other moist temperate grasslands became the breadbaskets of the world because they provide ideal growing conditions for crops such as corn and wheat, which are also grasses.

Chaparral: Thickets of Evergreen Shrubs and Small Trees

Some temperate environments have mild winters with abundant rainfall combined with very dry summers. Such **mediterranean climates**, as they are called, occur not only in the area around the Mediterranean Sea but also in the North American Southwest, southwestern and southern Australia, central Chile, and southwestern South Africa. In the North American Southwest this mediterranean-type community is known as **chaparral**. Chaparral soil is thin and often not very fertile. Fires occur naturally and frequently in this environment, particularly in late summer and in autumn.

Vegetation in mediterranean climates looks strikingly similar in different areas of the world, even though the individual species are quite distinct. The area is usually dominated by a dense growth of evergreen shrubs but may contain drought-resistant pine or scrub oak trees (Figure 6–8). During the rainy winter season the environment may be lush and green, but the plants lie dormant during the hot, dry summer. Trees and shrubs often have **sclerophyllous leaves**—hard, small, leathery leaves that resist water loss. Many plants are also specifically fire-adapted and actually grow best in the months following a fire. Such growth is possible because fire releases minerals that were present in above-ground parts of plants that burned. The underground parts are not destroyed by the fire, however, and with the new availability of essential minerals, the plants sprout vigorously during winter rains. Mule deer, wood rats, chipmunks, lizards, and many species of birds are some of the common animals of the chaparral.

The fires that occur at irregular intervals in California chaparral may be quite costly to humans because they sometimes consume expensive homes built on the hilly chaparral landscape. Unfortunately, efforts to prevent the naturally occurring fires sometimes backfire. Denser, thicker vegetation tends to accumulate over several years; then, when a fire does occur, it is much more severe. Removing the chaparral vegetation, whose roots hold the soil in place, can also cause problems—witness the mud slides that sometimes occur during winter rains in these areas.

Deserts: Arid Life Zones

Deserts are dry areas found in both temperate and tropical regions. The low water content of the desert atmosphere results in daily temperature extremes of heat and cold, so that a major change in temperature occurs in each single 24-hour period. Deserts vary greatly, depending on the amount of precipitation they receive, which is generally less than 25 centimeters (10 inches) per year. A few deserts are so dry that virtually no plant life occurs in them, as for example, the African Namib Desert and the Atacama Desert of northern Chile and Peru. As a result of sparse vegetation, desert soil is low in organic material but is often high in mineral content. In some regions, the concentration of certain soil minerals reaches toxic levels.

Plant cover is sparse in deserts, so much of the soil is exposed. Both perennials—plants that live for more than two years—and annuals—plants that complete their life cycles in one growing season— occur in deserts. However, annuals are common only after rainfall. Plants in North American deserts include cacti, yuccas, Joshua trees, and widely scattered bunchgrass (Figure 6–9). Desert plants tend to have reduced leaves or no leaves, an adaptation that conserves water. Other desert plants shed their leaves for most of the year, growing only during the brief moist season. Desert plants are noted for **allelopathy**, an adaptation in which toxic substances secreted by roots or shed leaves inhibit the establishment of competing plants nearby. Many desert plants are provided with spines, thorns, or toxins to resist heavy grazing pressure often experienced in this food- and water-deficient environment.

Desert animals tend to be small. During the heat of the day, they remain under cover or return to shelter periodically. At night they come out to forage or hunt. In addition to desert-adapted insects, there are many specialized desert reptiles, such as the desert tortoise, desert iguana, Gila monster, and Mohave rattlesnake. Desert mammals include rodents such as gerbils and jerboas in African and Asian deserts and kangaroo rats in North American deserts. There are also mule deer and jack-rabbits in American deserts (Figure 6–9), oryxes in African deserts, and kangaroos in Australian deserts. Carnivores such as the African fennec fox and some birds of prey, especially owls, live on the rodents and rabbits.

American deserts have been altered by humans in several ways. Off-road vehicles damage desert vegetation, which sometimes takes years to recover, and certain cacti and desert tortoises are rare as a result of poaching. Also, houses, factories, and farms built in desert areas require vast quantities of water, which must be imported from distant areas.

Figure 6–9

Desert. Inhabitants of deserts are strikingly adapted to the demands of their environment. The moister deserts of North America contain large cacti such as the giant saguaro, which grows 15 to 18 meters (50 to 60 feet) tall. The smaller cacti, called chollas, have a dense covering of barbed spines. (*Inset*) An antelope jack-rabbit. This animal is so-named because its huge ears resemble those of a jackass. Jack-rabbits never drink water but instead obtain the liquid they need from the plants on which they feed. (© 1997 Willard Clay/Dembinsky Photo Associates; *Inset*, © 1997 Stan Osolinski/Dembinsky Photo Associates)

Figure 6–10
Savanna. Grasslands such as this one, with widely scattered *Acacia* trees, are found in eastern Africa. Parts of South America, Australia, and India also have savannas. Savannas formerly supported large herds of grazing animals and their predators, both of which are swiftly vanishing under pressure from pastoral and agricultural land use. (Frans Lanting/Minden Pictures)

Savanna: Tropical Grasslands

The **savanna** biome is a tropical grassland with widely scattered clumps of low trees (Figure 6–10). Savanna is found in areas of low rainfall or seasonal rainfall with prolonged dry periods. The temperatures in tropical savannas vary little throughout the year, and seasons are regulated by precipitation, rather than by temperature as in temperate grasslands. Annual precipitation is 85 to 150 centimeters (34 to 60 inches). Savanna soil is somewhat low in essential mineral nutrients, in part because it is strongly leached. Savanna soil is often rich in aluminum, which resists leaching, and in places the aluminum reaches levels that are toxic to many plants. Although the African savanna is best known, savanna also occurs in South America, western India, and northern Australia.

Savanna is characterized by wide expanses of grasses interrupted by occasional trees such as *Acacia*, which bristles with thorns that provide protection against herbivores. Both trees and grasses have fire-adapted features, such as extensive underground root systems, that enable them to survive seasonal droughts as well as periodic fires.

Spectacular herds of hoofed mammals—such as wildebeest, antelope, giraffe, zebra, and elephants—occur in the African savanna. Large predators, such as lions and hyenas, kill and scavenge the herds. In areas of seasonally varying rainfall, the herds and their predators may migrate annually.

Savannas are rapidly being converted to rangeland for cattle and other domesticated animals, which are replacing the big herds of wild animals. In places, severe overgrazing has converted marginal savanna to desert (see Chapter 17).

Tropical Rain Forests: Lush Equatorial Forests

Tropical rain forests occur where temperatures are warm throughout the year and precipitation occurs almost daily. The annual precipitation of tropical rain forest is typically from 200 to 450 centimeters (80 to 180 inches). Much of this precipitation comes from locally recycled water that enters the atmosphere by transpiration (loss of water vapor from plants) of the forest's own trees.

Tropical rain forests commonly occur in areas with ancient, highly weathered, mineral-poor soil. Little organic matter accumulates in such soils: Because temperatures are high year round, decay organisms and detritus-feeding ants and termites decompose organic litter quite rapidly. Nutrients from the decomposing material are quickly absorbed by roots. Thus, the minerals of tropical rain forests are tied up in the vegetation rather than the soil. Tropical rain forests are found in Central and South America, Africa, and Southeast Asia.

Tropical rain forest is very productive—that is, its plants capture a lot of energy by photosynthesis. High productivity is stimulated by the abundant solar energy and precipitation. Productivity is high here despite the scarcity of mineral nutrients in the soil.

Of all the biomes, the tropical rain forest is unexcelled in species diversity and variety. No single species dominates this biome. Often one can travel for hundreds of meters without encountering two members of the same species of tree.

As discussed in the chapter introduction, the trees of tropical rain forests form a dense, multi-layered canopy overhead (Figure 6–11a). The trees of tropical rain forests are typically evergreen flowering plants. Their roots are often shallow and concentrated near the surface in a mat only an inch or so (a few centimeters) thick; the root mat catches and absorbs almost all mineral nutrients released from leaves and litter by decay processes. Swollen bases or braces called buttresses hold the trees upright and aid in the extensive distribution of the shallow roots (Figure 6–11b).

Rainforest animals include the most abundant and varied insects, reptiles, and amphibians on Earth. The birds, often brilliantly colored, are also varied. Most rainforest mammals, such as sloths and monkeys, live only in the trees, although some large, ground-dwelling mammals, including Asian elephants, are also found in rain forests.

Unless strong conservation measures are initiated soon, human population growth and industrialization in tropical countries may spell the end of tropical rain forests in your lifetime. At current rates of deforestation, scientists estimate that almost all tropical rain forests will be destroyed by 2030. It is likely that many rainforest organisms will become extinct before they have even been scientifically described. The ecological impacts of tropical rainforest destruction are discussed extensively throughout this text.

AQUATIC ECOSYSTEMS

Not surprisingly, aquatic life zones are different in almost all respects from terrestrial life zones. For example, recall that in biomes, temperature and precipitation are the major determinants of plant and animal inhabitants, and light is relatively plentiful, except in certain habitats such as the rainforest floor. Significant environmental factors in aquatic ecosystems are very different. Temperature is less important in watery environments because the water

(a) (b)

Figure 6–11

Tropical rain forest. Except at riverbanks, tropical rain forest has a closed canopy that admits little light to the rainforest floor. (a) A broad view of tropical rainforest vegetation along a riverbank. (b) Tropical rainforest trees typically possess elaborate buttresses that support them in the shallow, often wet soil. Shown are the buttress roots on Australian banyan. (a, Frans Lanting/Minden Pictures; b, John Arnaldi)

itself tends to moderate temperature, and water is obviously not an important limiting factor in aquatic ecosystems.

The most fundamental division in aquatic ecology is probably between freshwater and saltwater environments. **Salinity**, which is the concentration of dissolved salts such as sodium chloride in a body of water, affects the kinds of organisms present in aquatic ecosystems, as does the amount of dissolved oxygen. Water greatly interferes with the penetration of light, so floating aquatic organisms that photosynthesize must remain near the water's surface, and vegetation attached to the bottom can grow

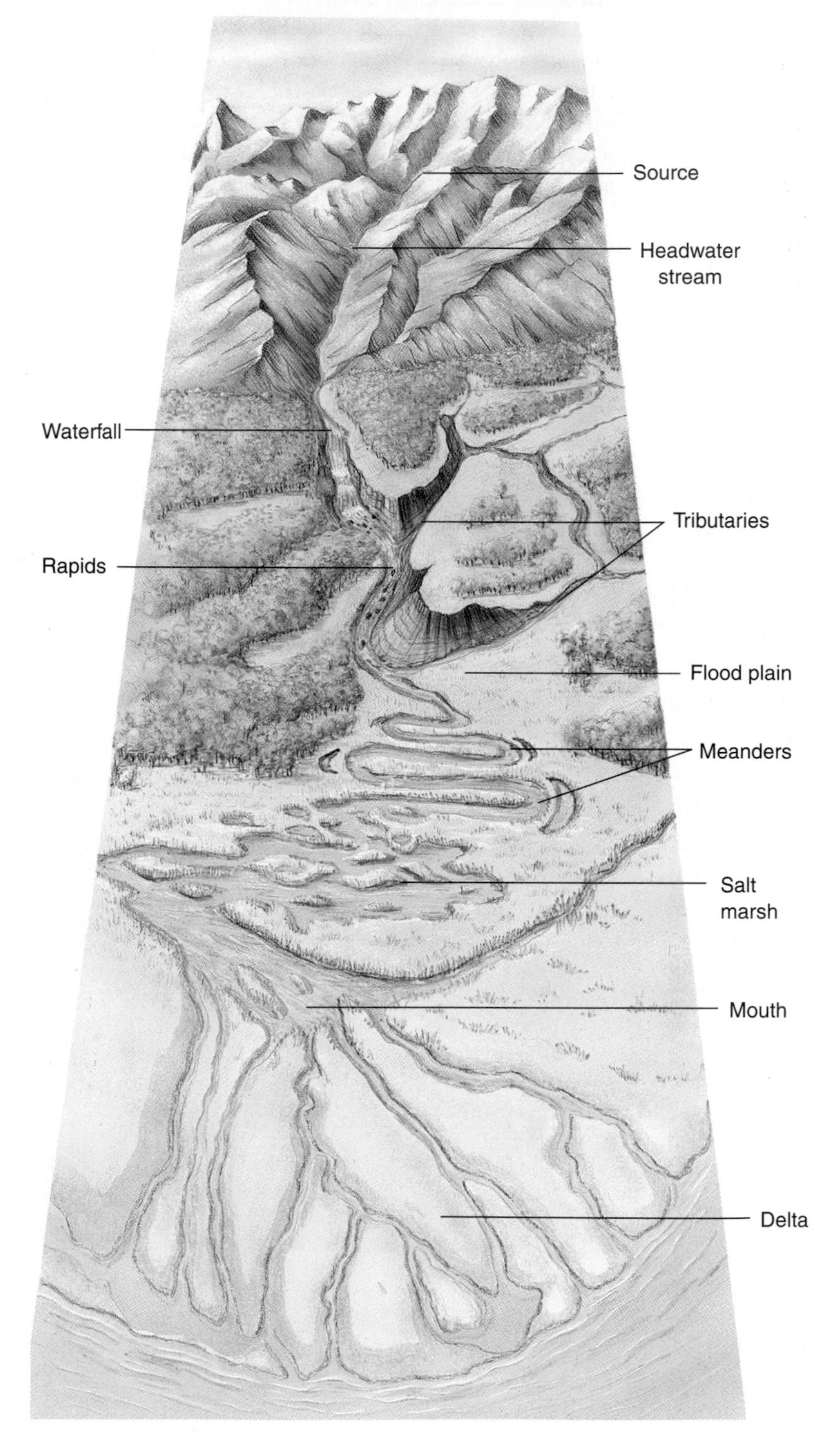

Figure 6–12

Features of a typical river. The river begins at a source, often high in mountains and fed by melting snows or glaciers. Headwater streams flow downstream rapidly, often over rocks (as rapids) or bluffs (as waterfalls). Along the way, smaller streams called tributaries feed into the river, adding to its flow. As the river's course levels out, the river flows more slowly and winds from side to side, forming bends called meanders. The flood plain is the relatively flat area on either side of the river that is subject to flooding. Some flood plains are quite large: the Mississippi River's flood plain is up to 130 kilometers (80 miles) wide. Near the ocean, the river may form a salt marsh where fresh water from the river and saltwater from the ocean mix. The delta is a fertile, low-lying plain at the river's mouth that forms from sediments deposited by the slow-moving river as it empties into the ocean.

only in shallow water. In addition, low levels of essential mineral nutrients limit the number and distribution of organisms in certain aquatic environments.

Aquatic ecosystems contain three main ecological categories of organisms: free-floating plankton, strongly swimming nekton, and bottom-dwelling benthos. **Plankton** are usually small or microscopic organisms that are relatively feeble swimmers. For the most part, they are carried about at the mercy of currents and waves. They are unable to swim far horizontally, but some species are capable of large daily vertical migrations and are found at different depths of water at different times of the day or at different seasons. Plankton are generally subdivided into two major categories, phytoplankton and zooplankton. **Phytoplankton**, which are free-floating photosynthetic cyanobacteria and algae, are producers that form the base of most aquatic food webs. **Zooplankton** are nonphotosynthetic organisms that include protozoa (animal-like protists) and the larval (immature) stages of many animals.

Nekton are larger, more strongly swimming organisms such as fishes, whales, and turtles. **Benthos** are bottom-dwelling organisms that fix themselves to one spot (oysters and barnacles), burrow into the sand (many worms and echinoderms), or simply walk about on the bottom (lobsters and brittle stars).

FRESHWATER ECOSYSTEMS

Freshwater ecosystems include rivers and streams (flowing-water ecosystems), lakes and ponds (standing-water ecosystems), and marshes and swamps (freshwater wetlands).

Although freshwater ecosystems occupy a relatively small portion of Earth's surface, they have an important role in the hydrologic cycle: They assist in recycling precipitation that flows as surface runoff to the ocean. Large bodies of fresh water also help moderate daily and seasonal temperature fluctuations on nearby land.

Rivers and Streams: Flowing-water Ecosystems

Many different conditions exist along the length of a river or stream (Figure 6–12). The nature of a **flowing-water ecosystem** changes greatly between its source—where it begins—and its mouth—where it empties into another body of water. For example, headwater streams—the small streams that are the sources of a river—are usually shallow, cold, swiftly flowing, and therefore highly oxygenated. In contrast, rivers downstream from the headwaters are wider and deeper, not as cold, slower-flowing, and therefore less oxygenated.

The kinds of organisms found in flowing-water ecosystems vary greatly from one stream to another, depending primarily on the strength of the current. In streams with fast currents, the inhabitants may have adaptations such as suckers to attach themselves to rocks so that they are not swept away, or they may have flattened bodies to enable them to slip under or between rocks. Organisms in large, slow-moving streams and rivers do not need such adaptations, although they are typically streamlined like most aquatic organisms to lessen resistance when moving through water. Where the current is slow, organisms of the headwaters are replaced by those characteristic of ponds and lakes.

In addition to having currents, flowing-water ecosystems differ from other freshwater ecosystems in their dependence on the land for much of their energy. In headwater streams, for example, up to 99 percent of the energy input comes from detritus such as leaves carried from the land into streams and rivers by wind or surface drainage. Downstream, rivers contain producers and therefore have a slightly lower dependence on detritus as a source of energy than in the headwaters.

Human activities have several adverse impacts on rivers and streams, including water pollution (see Chapter 21) and the effects of dams (see Chapters 12 and 13), which are built to contain the water of rivers or streams. Both pollution and dams change the nature of flowing-water ecosystems downstream.

Lakes and Ponds: Standing-water Ecosystems

Standing-water ecosystems are characterized by zonation. A large lake has three basic layers or zones: the littoral, limnetic, and profundal zones (Figure 6–13). Smaller lakes and ponds typically lack a profundal zone. The **littoral zone** is a shallow-water area along the shore of a lake or pond. Emergent vegetation, such as cattails and burreeds, plus several deeper-dwelling aquatic plants and algae, live in the littoral zone. The littoral zone is the most productive section of the lake–that is, photosynthesis is greatest here—in part because it receives nutrient inputs from surrounding land that stimulate the growth of plants and algae. Animals of the littoral zone include frogs and their tadpoles; turtles; worms; crayfish and other crustaceans; insect larvae; and many fishes, such as perch, carp, and bass. Surface dwellers such as water striders and whirligig beetles are found in the quieter areas.

The **limnetic zone** is the open-water area away from the shore; it extends down as far as sunlight penetrates. The main organisms of the limnetic zone are microscopic phytoplankton and zooplankton. Larger fishes also spend most of their time in the limnetic zone, although they

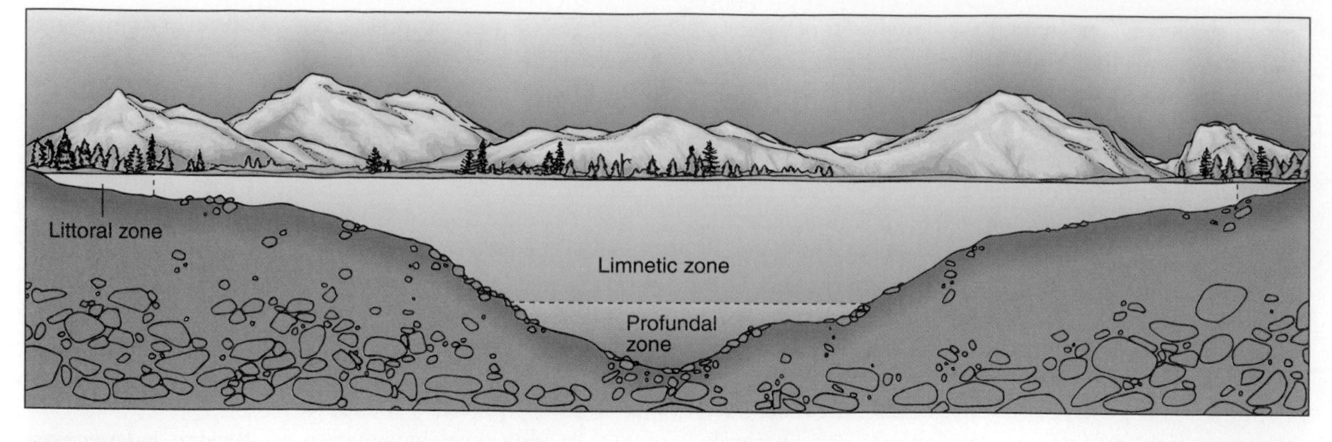

Figure 6–13
Zonation in a large lake. The littoral zone is the shallow-water area around the lake's edge. The limnetic zone is the open, sunlit water away from the shore. The profundal zone, under the limnetic zone, is below where light penetrates.

may visit the littoral zone to feed and reproduce. Owing to its depth, less vegetation grows here than in the littoral zone.

The deepest zone, the **profundal zone**, is beneath the limnetic zone of a large lake. Because no light penetrates to this depth, plants and algae do not live in the profundal zone. Food drifts into the profundal zone from the littoral and limnetic zones. Bacteria decompose dead organisms that reach the profundal zone, liberating the minerals contained in their bodies. These minerals are not effectively recycled, because there are no producers to absorb them and incorporate them into the food web. As a result, the profundal zone tends to be both mineral-rich and anaerobic (without oxygen), with few organisms occupying it other than anaerobic bacteria.

Thermal stratification and turnover in temperate lakes

The marked layering of large temperate lakes caused by how far light penetrates is accentuated by **thermal stratification**, in which the temperature changes sharply with depth. Thermal stratification occurs because the summer sunlight penetrates and warms surface waters, making them less dense.[2] In the summer, cool and therefore denser water remains at the lake bottom, separated from the warm and therefore less dense water above by an abrupt temperature transition called the **thermocline** (Figure 6–14a).

In temperate lakes, falling temperatures in autumn cause a mixing of the lake waters called the **fall turnover** (Figure 6–14b). (Because there is little seasonal temperature variation in the tropics, such turnovers are not common there.) As the surface water cools, its density increases and eventually it displaces the less dense, warmer, mineral-rich water beneath. The warmer water then rises to the surface where it, in turn, cools and sinks. This process of cooling and sinking continues until the lake reaches a uniform temperature throughout.

When winter comes, the surface water cools below 4°C, its temperature of greatest density, and if it is cold enough, ice forms. Ice, which forms at 0°C, is less dense than cold water; thus, ice forms on the surface, and the water on the lake bottom is warmer than the ice on the surface.

SUBMERGED LAKE DISCOVERED

During the 1990s, satellite measurements and radio-echo surveys provided evidence for the existence of a "new" freshwater lake. Similar in size to Lake Ontario, the lake had not been discovered earlier because it lies in a remote location–under about 4 kilometers (2.5 miles) of ice in East Antarctica! The lake, which has not yet been named, appears to have a mean depth of 125 meters (410 feet). Biologists are anxious to sample and examine the lake's water for signs of microorganisms that, if present, have been isolated from the rest of the living world since the lake first formed, about 1 million years ago.

[2]Recall that the density of water is greatest at 4°C; both above and below this temperature, water is less dense.

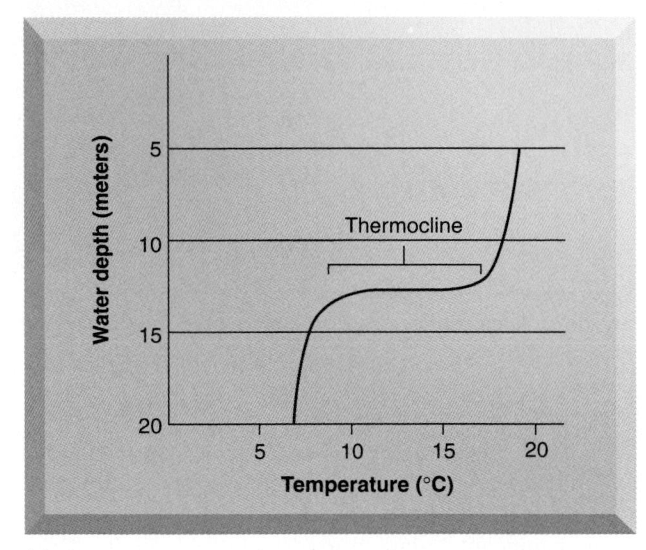

(a)

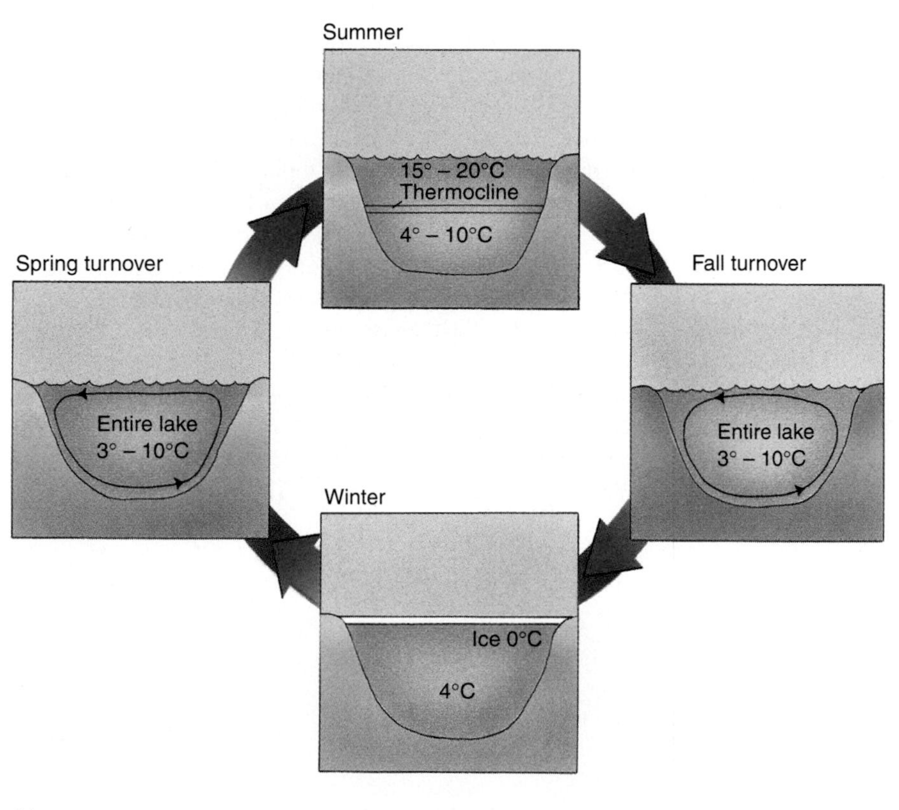

(b)

Figure 6-14

Thermal stratification in a temperate lake. (a) Temperature differences by depth during the summer. There is an abrupt temperature transition, called the thermocline, between the upper warm layer and the bottom cold layer. (b) During fall and spring turnovers, a mixing of upper and lower layers of water brings oxygen to the oxygen-depleted depths of the lake and minerals to the mineral-deficient surface waters.

In the spring, a **spring turnover** occurs as ice melts and the surface water reaches 4°C. Surface water again sinks to the bottom, and bottom water returns to the surface. As summer arrives, thermal stratification occurs once again.

The mixing of deeper, nutrient-rich water with surface, nutrient-poor water during the fall and spring turnovers brings essential minerals to the surface. The sudden presence of large amounts of essential minerals in surface waters encourages the development of large algal populations, which form temporary **blooms** (population explosions) in the fall and spring.

Marshes and Swamps: Freshwater Wetlands

Freshwater wetlands are usually covered by shallow water for at least part of the year and have characteristic

Figure 6–15

Freshwater swamps are inland areas permanently saturated or even covered by water and dominated by trees. Shown is a cypress swamp in Water Spring, South Carolina. (Milton H. Tierney, Jr./Visuals Unlimited)

soils and water-tolerant vegetation. Wetland soils are waterlogged for variable periods and therefore anaerobic. Most wetland soils are rich in accumulated organic materials, in part because anaerobic conditions discourage decomposition. Freshwater wetlands include marshes, in which grasslike plants dominate, and swamps, in which woody trees or shrubs dominate (Figure 6–15). Freshwater wetlands also include hardwood bottomland forests (lowlands along streams and rivers that are periodically flooded), prairie potholes (small, shallow ponds that formed when glacial ice melted at the end of the last Ice Age), and peat moss bogs (peat-accumulating wetlands where mosses dominate).

Wetland plants, which are highly productive, provide enough food to support a wide variety of organisms. Wetlands are valued as habitats for migratory waterfowl and many other bird species, beaver, otters, muskrats, and game fishes. Wetlands help to control flooding by acting as holding areas for excess water when rivers flood their banks. The floodwater stored in wetlands then drains slowly back into the rivers, providing a steady flow of water throughout the year. Wetlands also serve as groundwater recharging areas. One of their most important roles is to help cleanse and purify water by trapping and holding pollutants in the flooded soil where they are gradually detoxified by chemical interactions.

At one time wetlands were considered wastelands, areas that needed to be filled in or drained so that farms, housing developments, and industrial plants could be built on them. Wetlands are also breeding places for mosquitoes and therefore were viewed as a menace to public health. Today, however, the crucial environmental services that wetlands provide are widely recognized, and wetlands have some legal protections (see Focus On The Everglades: Paradise Almost Lost, page 124).

ESTUARIES: WHERE FRESH WATER AND SALTWATER MEET

Where the ocean meets the land, there may be one of several kinds of ecosystems: a rocky shore, a sandy beach, an intertidal mud flat, or a tidal estuary. An **estuary** is a coastal body of water, partly surrounded by land, with access to the open ocean and a large supply of fresh water from rivers. Estuaries usually contain **salt marshes**, shallow, swampy areas dominated by grasses, in which the salinity fluctuates between that of seawater and fresh water (Figure 6–16; also, see Chapter 3). Many estuaries undergo significant variations in temperature, salinity, depth of light penetration, and other physical properties in the course of a year. To survive there, estuarine organisms must be able to tolerate this wide range of conditions.

Estuaries are among the most fertile ecosystems in the world, often having a much greater productivity than either the adjacent ocean or the fresh water upriver. This high productivity is brought about by (1) the transport of nutrients from the land into rivers and creeks that empty into the estuary; (2) the action of tides, which promotes a rapid circulation of nutrients and helps remove waste products; and (3) the presence of many plants, which provide an extensive photosynthetic carpet and also mechanically trap much potential food material in their roots and stems. As plants die, they decay, forming the base of a detritus food web. Most commercially important fin fishes and shellfish spend their larval stages in estuaries among the protective tangle of decaying plants.

Figure 6–16

A salt marsh at the Brigantine Wildlife Reserve in New Jersey. Saltwater and fresh water mix in salt marshes. (Karen L. Milstein/Wings & Things Photography)

Salt marshes have often appeared worthless, empty stretches of land to uninformed people. As a result, they have been used as dumps and become severely polluted or have been filled with dredged bottom material to form artificial land for residential and industrial development. A large part of the estuarine environment has been lost in this way, along with many of its benefits, which are the same as those of freshwater wetlands: biological habitats, sediment and pollution trapping, flood control, and groundwater supply.

Mangrove forests, the tropical equivalent of salt marshes, cover perhaps 70 percent of tropical coastlines (Figure 6–17). Like salt marshes, mangrove forests provide valuable environmental services. Their interlacing roots are breeding grounds and nurseries for several commercially important fishes and shellfish, and their branches are nesting sites for many species of birds. Mangrove roots stabilize the submerged soil, thereby preventing coastal erosion and providing a barrier against the ocean during storms. Scientists at the Mangrove Ecosystem Research Centre in Hanoi, North Vietnam, for example, have evidence that mangroves are more effective than concrete sea walls in controlling floodwater from tropical storms. Mangroves are under assault from coastal development and unsustainable logging. Some countries, such as the Philippines, Bangladesh, and Guinea-Bissau, have lost 70 percent or more of their mangrove swamps.

MARINE ECOSYSTEMS

Although lakes and the ocean are comparable in many ways, there are also many differences. The depths of even the deepest lakes do not approach those of the ocean abysses, which are extremely deep areas that extend more than 6 kilometers (3.6 miles) below the sunlit surface. The ocean is profoundly influenced by tides and currents. Gravitational pulls of both sun and moon produce two

(text continues on page 126)

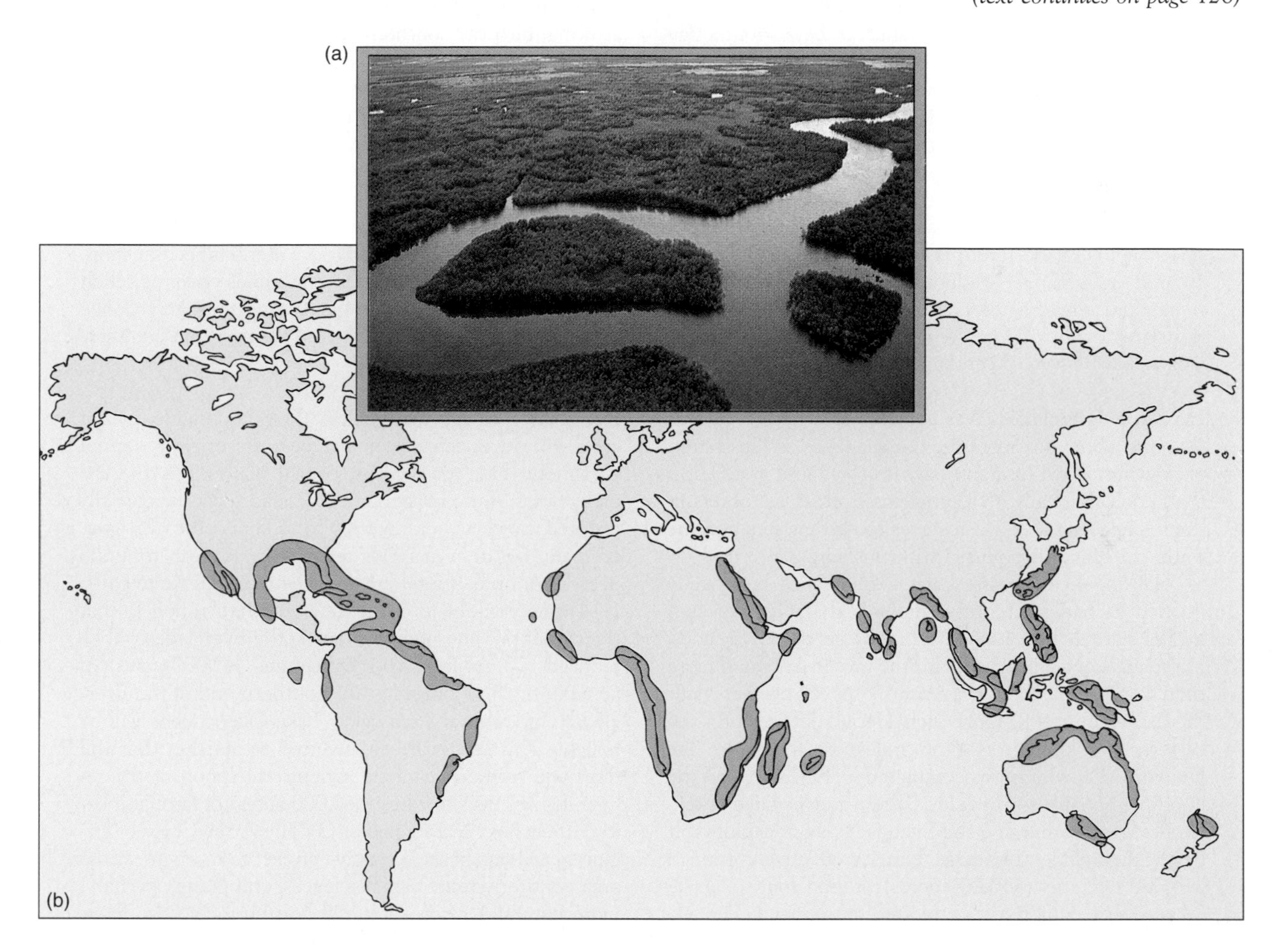

Figure 6–17

Mangrove forests often grow in tropical and subtropical coastal mud flats where tidal waters fluctuate. (a) A mangrove swamp along the Pacific coast of Costa Rica. (b) Distribution of mangrove forests worldwide. (a, Walt Anderson/Visuals Unlimited)

FOCUS ON

The Everglades: Paradise Almost Lost

The Everglades in the southernmost part of Florida is a vast expanse of predominantly sawgrass wetlands dotted with small islands of trees. At one time 80 kilometers (50 miles) wide, 160 kilometers (100 miles) long, and 15 centimeters to 0.9 meter (6 inches to 3 feet) deep, the "river of grass" drifted south in a slow-moving sheet of water from Lake Okeechobee to Florida Bay (see figure). The Everglades is a haven for wildlife, including alligators, snakes, otters, raccoons, and thousands of wading birds and birds of prey—great blue herons, snowy egrets, great white herons, roseate spoonbills, and osprey, to name just a few. The region's natural wonders were popularized in an environmental classic, *The Everglades: River of Grass*, written by Marjory Stoneman Douglas in 1947.

South of the Everglades is Florida Bay, a shallow estuary dotted with many tiny islands, or keys. Florida Bay and the Florida Keys are greatly affected by the water leaving the Everglades. Both Florida Bay and the Everglades are important nurseries for commercially important fishes, shrimp, lobster, and stone crab. With all the wildlife and recreational opportunities in the Everglades and Florida Bay, it is not surprising that the local economy relies heavily on tourism and commercial fishing.

The Everglades today is about half its original size of 1.6 million hectares (4 million acres), and it has many serious environmental problems. Most water bird populations are down by 90 percent in recent decades, and the area is now home to 50 endangered or threatened species.

How the Everglades Was Almost Destroyed Let us examine a brief history of the Everglades as an illustration of how misguided human activities can cause more harm than good. Basically, two problems override all others in the Everglades today—it receives too little water, and the water it receives is polluted with nutrients.

During heavy rains Lake Okeechobee historically flooded its banks, creating wetlands that provided biological habitat and helped to recharge the Everglades. However, when a hurricane hit the lake in 1928 and more than 1800 people died, the Army Corps of Engineers built the 12.2-meter- (40-foot) high Hoover Dike along the eastern 240 kilometers (150 miles) of the lake. The Hoover Dike, which was completed in 1932, stopped the flooding, but it also prevented the water in Lake Okeechobee from recharging the Everglades. Four canals built by the Everglades Drainage District effectively drained 214,000 hectares (530,000 acres) of land south of Lake Okeechobee, which was converted to farmland. The fertilizers and pesticides used in this area eventually make their way to the Everglades, where they alter native plant communities. Phosphorus in fertilizer is particularly harmful because it encourages the growth of non-native cattails that overrun the native saw grasses.

After several tropical storms caused flooding damage in South Florida in 1947, Congress authorized the Army Corps of Engineers to construct an extensive system of canals, levees, and pump stations to prevent flooding, provide drainage, and supply water to South Florida. These structures, which divert excess water to the Atlantic Ocean rather than the Everglades, stopped the periodic floods in South Florida. However, the drier lands produced by the levees, canals, and pumps encouraged accelerated urban growth, particularly along the East coast, and the expansion of agriculture, primarily sugar growing, into the southernmost parts of the Everglades.

Thus, more than 60 years of engineering projects have reduced the quantity of water flowing into the Everglades, and the water that does enter is polluted from agricultural runoff. Urbanization has also contributed to the Everglades' problems by fragmenting the ecosystem. Like the Everglades, Florida Bay is showing signs of environmental decline. Because it receives so little water from the Everglades, Florida Bay's water has become more salty. Blooms of cyanobacteria sometimes cover as much as 40 percent of the bay. Its sea grasses, fishes, and sponge populations have all declined, and tourism and commercial fishing have suffered.

Restoration of the Everglades Florida's unique river of grass will never return completely to its original natural condition; there are too many sugar plantations and too many cities in the region. The Everglades can be partially restored, however, and in 1996 Florida and the U.S. government began a massive restoration project to undo decades of human interference. The plan has three parts. (1) Farmers will be forced to clean up their runoff so that the amount of phosphorus entering the Everglades will be reduced. (2) At least 16,180 hectares (40,000 acres) of agricultural land located at the southern end of the Everglades Agricultural Area below Lake Okeechobee will be bought. This land will be converted to marshes that will filter and further clean the agricultural runoff of the remaining 267,000 hectares (660,000 acres) of farmland before it reaches the Everglades. (3) The Army Corps of Engineers will undertake a massive project to re-engineer the area's entire system of canals, levees, and pumps so that a more natural flow of water will be restored to the Everglades. A great deal of scientific research must be done be-

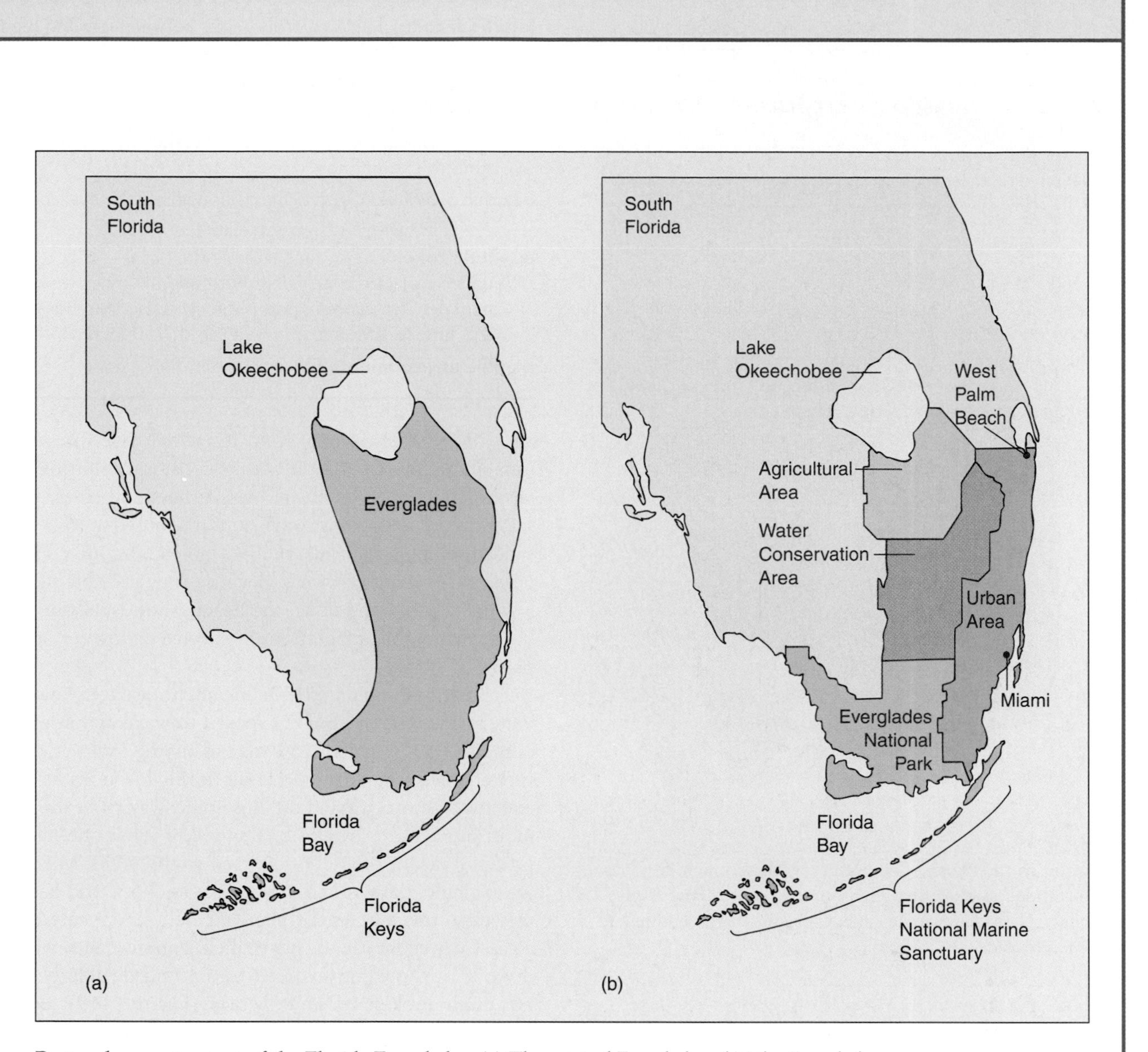

Past and present extent of the Florida Everglades. (a) The original Everglades. (b) The Everglades today.

fore the restoration can be completed. Many questions remain about how best to restore a more natural water flow, repel invasions of foreign species, and reestablish the native biological diversity.

Restoration plans have not come without bitter debates that have pitted the state's tourism, fishing, and environmental interests against the region's urban developers and sugar farmers, who are prosperous and politically powerful. (One reason the sugar industry is so prosperous is that the federal government provides it with price supports that prevent U.S. consumers from purchasing less expensive imported sugar.) After an expensive and often acrimonius battle, a Florida referendum that would have taxed sugar growers a penny for every pound of sugar they produce was defeated in 1996. The tax, along with additional money already provided by both state and federal taxpayers, would have helped finance the restoration of the Everglades.

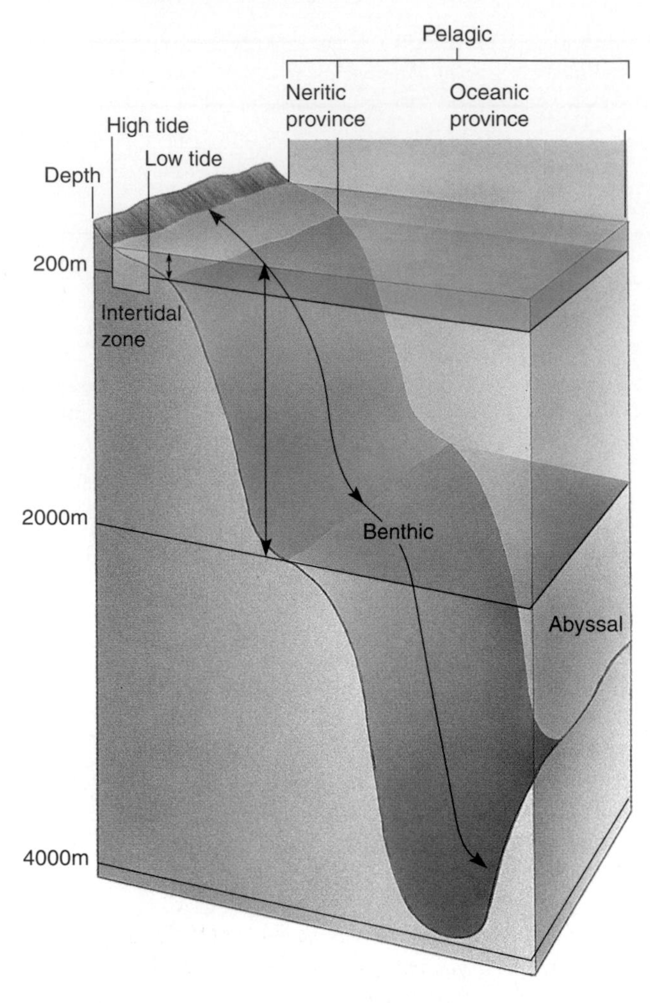

Figure 6–18

Zonation in the ocean. The ocean has three main life zones: the intertidal zone, benthic environment, and pelagic environment. The pelagic environment consists of the neritic and oceanic provinces.

high tides a day throughout the ocean, but the height of those tides varies with the phases of the moon (a full moon causes the highest tides), the season, and the local topography.

The immense marine environment is subdivided into several zones: the intertidal zone, benthic (ocean floor) environment, and pelagic (ocean water) environment (Figure 6–18). The pelagic environment is in turn divided into two provinces, the neritic province and the oceanic province.

The Intertidal Zone: Transition between Land and Ocean

The area of shoreline between low and high tides is called the **intertidal zone**. Although the high levels of light and nutrients, together with an abundance of oxygen, make the intertidal zone a biologically productive habitat, it is also a very stressful one. If an intertidal beach is sandy, inhabitants must contend with a constantly shifting environment that threatens to engulf them and gives them scant protection against wave action. Consequently, most sand-dwelling organisms, such as mole crabs, are continuous and active burrowers. Because they are able to follow the tides up and down the beach, they usually do not have any notable adaptations to survive drying out or exposure.

A rocky shore provides a fine anchorage for seaweeds and marine animals, but is exposed to wave action when immersed during high tides and to drying and temperature changes when exposed to air during low tides. A typical rocky-shore inhabitant has some way of sealing in moisture, perhaps by closing its shell, if it has one, plus a powerful means of anchoring itself to the rocks. Mussels, for example, have tough, threadlike anchors, and barnacles have special cement glands. Rocky-shore intertidal algae (seaweeds) usually have thick, gummy coats, which dry out slowly when exposed to air, and flexible bodies not easily broken by wave action (Figure 6–19). Some rocky-shore community inhabitants hide in burrows or crevices at low tide, and some small crabs run about the splash line, following it up and down the beach.

Mini-Glossary of the Marine Environment

intertidal zone: The area of shoreline between low and high tides.

pelagic environment: The open ocean environment; divided into neritic and oceanic provinces.

neritic province: Open ocean that overlies the ocean floor from the shoreline to a depth of 200 meters.

oceanic province: Open ocean that overlies the ocean floor at depths greater than 200 meters.

euphotic zone: The surface layer of the ocean—that is, the upper part of the pelagic environment—where enough light penetrates to support photosynthesis; extends from the surface to a maximum depth of 150 meters.

benthic environment: The ocean bottom or floor.

Coral Reefs: An Important Part of the Benthic Environment

The **benthic environment** is the ocean floor. Most of the ocean floor consists of sand and mud in which many marine animals, such as worms and clams, burrow. Bacteria are common in marine sediments, but until recently it was assumed that bacteria did not extend very far into the sediments. In 1994 in the journal *Nature*, however, bacteria were reported in ocean sediments more than 500 meters (1640 feet) below the ocean floor at several different sites in the Pacific Ocean.

Benthic communities in shallow ocean waters that are particularly productive include kelp forests, seaweed

Figure 6–19

Sea palms (*Postelsia*), sturdy seaweeds adapted to life in a rocky intertidal zone, are exposed at low tide. The sea palm is common on the rocky Pacific coast from Vancouver Island to California. Its base is firmly attached to the rocky substrate, enabling it to withstand pounding wave action. (William E. Ferguson)

beds, and coral reefs. We restrict our discussion here to coral reefs, which are ecologically important because they both provide habitat for a wide variety of marine organisms and protect coastlines from shoreline erosion.

Coral reefs are found in warm (usually greater than 21°C), shallow seawater (Figure 6–20). The living portions of coral reefs must grow in shallow waters where light penetrates. Many coral reefs are composed principally of red coralline algae that require light for photosynthesis. Other coral reefs consist of colonies of millions of tiny coral animals, which also require light for the large number of symbiotic algae, known as **zooxanthellae**, that live and photosynthesize in their tissues (see Chapter 4). Although species of coral exist without zooxanthellae, only those with zooxanthellae build reefs. In addition to obtaining food from the zooxanthellae that live inside them, coral animals capture food at night with stinging tentacles that paralyze small animals that drift nearby.

Coral reefs grow slowly, as coral organisms build on the calcareous remains of countless organisms before them. The basic pattern of coral reef development was first outlined by the famous naturalist Charles Darwin, best known for his theory of evolution (see Chapter 4). Coral reefs start in warm, shallow water—for example, directly attached to the shore of a volcanic island—to form a **fringing reef** (Figure 6–21). After the volcano becomes inactive, it slowly erodes and sinks back into the ocean. As this occurs, the channel between the island and the reef widens, and the reef becomes a **barrier reef**. A barrier reef is therefore separated from the nearby land by open water. If the volcano erodes or sinks completely below the water level, the outcome is a circular coral reef, called an **atoll**, that surrounds a central lagoon of quiet water. An atoll is the final stage in the development of coral reefs associated with ocean volcanoes.

The waters in which coral reefs are found are often poor in nutrients. Yet other factors are favorable for a high productivity, including the presence of zooxanthellae, a favorable temperature, and year-round sunlight.

Coral reef ecosystems contain hundreds of species of fishes and invertebrates, such as giant clams, sea urchins,

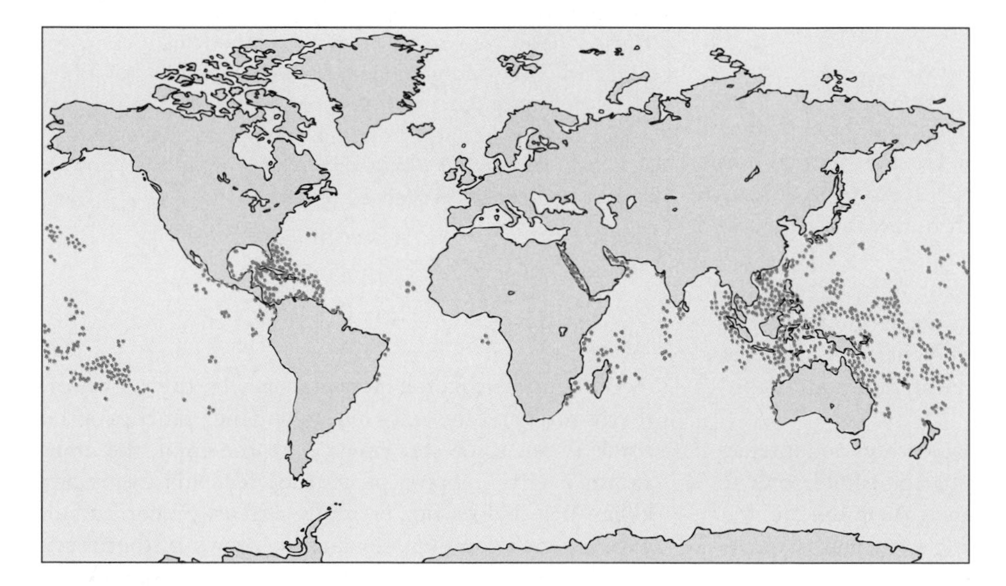

Figure 6–20

Distribution of the more than 6000 coral reefs worldwide.

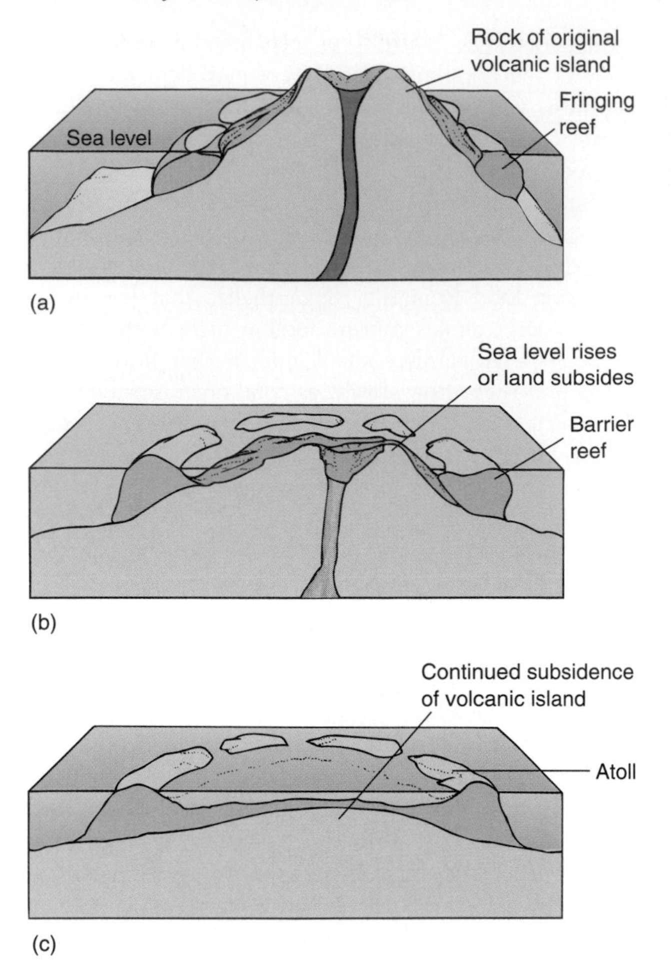

Figure 6–21

Coral colonies form reefs that change over time. (a) A fringing coral reef grows in the shallow water around the shores of a volcanic island. (b) Over time, as the coral continues to grow, the island sinks or the sea level rises, forming a barrier reef. (c) More time passes, and the island sinks below the surface, leaving an atoll.

IMPORTANCE OF CORAL

Coral formations are important ecosystems, as rich in species as tropical rain forests, yet the world's coral reefs are being degraded and destroyed. Of 109 countries with large reef formations, 90 are damaging them. Throughout the tropical western Atlantic, 30 to 50 percent of coral species are either rare or endangered. In some areas, silt washing downstream from clearcut inland forests has smothered reefs under a layer of sediment. High salinity resulting from the diversion of fresh water to supply the growing human population is thought by some scientists to be killing Florida reefs. Overfishing, pollution from sewage discharge and agricultural runoff, hurricane damage, disease, land reclamation, tourism, and the mining of corals for building material are also taking a heavy toll.

Since 1987, corals in the tropical Atlantic and Pacific have suffered extensive bleaching, a mysterious phenomenon in which corals lose their zooxanthellae and starve to death. Scientists are not sure what causes coral bleaching, which was first observed in the Caribbean in 1964. They suspect several environmental stresses. For example, coral bleaching is associated with warmer seawater temperatures (about 1°C above average). Other contributing factors may include low wind velocities and higher levels of ultraviolet radiation that occur during clear-water conditions such as those associated with climatic events like El Niño. Coral regeneration cannot keep pace with all these assaults: A new coral colony requires 20 years to grow to the size of a human head. Biologists have established the Global Coral Reef Monitoring Network, in which all of Earth's coral reefs will be surveyed for the first time. Knowledge gained from this survey may help persuade the United States and other countries to protect and conserve their coral reefs.

sea stars, sponges, brittle stars, sea fans, and shrimp. Many are brightly colored to advertise the fact that they are poisonous. The conspicuous lion fish, for example, defends itself from potential predators with numerous spiny projections that contain a powerful poison.

Many unusual relationships occur at coral reefs. Certain tiny fishes, for example, swim inside the mouths of larger fishes and remove potentially harmful parasites. Fishes sometimes line up at these cleaning stations, waiting their turn to be serviced.

The complex multitude of relationships and interactions that occur at coral reefs seems comparable only to the tropical rain forest among biomes. As in the rain forest, however, competition is intense, particularly competition for light and space to grow. *Podillopora* corals, for example, have a remarkable defense mechanism that protects them from coral browsers such as the crown-of-thorns sea star. Tiny crabs live among the branches of the coral. When a sea star crawls onto the coral, the crabs swarm over it, nipping off its tube feet and eventually killing it if it does not promptly retreat! Since the late 1960s, population explosions of the crown-of-thorns sea stars have occurred on many coral reefs, causing extensive devastation of the coral. It is not known what causes these outbreaks, nor is it known if human factors such as pollution play a part.

The Neritic Province: Shallow Waters Close to Shore

The **neritic province** is open ocean that overlies the ocean floor from the shoreline to a depth of 200 meters (650 feet). Organisms that live in the neritic province are all floaters or swimmers. The upper reaches of the pelagic environment comprise the **euphotic** region, which extends from the surface to a maximum depth of 150 meters (488 feet) in the clearest open ocean water. Sufficient light penetrates the euphotic zone to support photosynthesis. Large numbers of phytoplankton, particularly diatoms in cooler waters and dinoflagellates in warmer waters, produce food by photosynthesis and are thus the base of food webs. Small zooplankton, including tiny crustaceans, jellyfish, comb jellies, protists such as foraminiferans, and larvae of barnacles, sea urchins, worms, and crabs, feed on phytoplankton.

Zooplankton are consumed by plankton-eating nekton such as herring, sardines, squid, baleen whales, and manta rays (Figure 6–22), which in turn become prey for carnivorous nekton such as sharks, tuna, porpoises, and toothed whales. Nekton are mostly confined to the shallower neritic waters—less than 60 meters, or 195 feet, deep—because that is where their food is.

Figure 6–22

The huge manta ray, which makes its home in the neritic province, looks more fearsome than it is. It swims in the open ocean, peacefully grazing on plankton and small fishes. A full-grown manta ray can have a "wingspan" of 5.5 meters (more than 18 feet) and weigh 1.4 metric tons (3000 pounds). This manta ray was photographed in the Pacific Ocean off the coast of Mexico. Note the two remoras clinging to its body. Remoras feed on scraps of food left by their host. (Alex Kerstitch/Visuals Unlimited)

The Oceanic Province: Most of the Ocean

The average depth of the ocean is 4000 meters (more than 2 miles). The **oceanic province** is the part of the open ocean that overlies the ocean floor at depths greater than 200 meters (650 feet). It is the largest marine environment, comprising about 75 percent of the ocean's waters. The oceanic province is characterized by cold temperatures, high hydrostatic pressure, and an absence of sunlight; these environmental conditions are uniform throughout the year.

Most organisms of the oceanic province depend on **marine snow**, organic debris that drifts down into their habitat from the upper, lighted regions. Organisms of this little-known realm are filter-feeders, scavengers, or predators. Many are gelatinous invertebrates, some of which attain great sizes. Fishes of the oceanic province are strikingly adapted to darkness and scarcity of food. For example, the rare gulper eel has huge jaws that enable it to swallow large prey. An organism that encounters food infrequently needs to eat as much as possible when food is available. Many animals of the oceanic province have illuminated organs that enable them to see one another for mating or food capture.

INTERACTION OF LIFE ZONES

Although we have discussed terrestrial and aquatic life zones as discrete entities, none of them exists in isolation. When parts of the Amazon rain forest flood annually, for example, fishes leave the stream beds and swim all over the forest floor, where they play a role in dispersing the seeds of many species of plants. And in the Antarctic, whose waters are much more productive than its land areas, the many seabirds and seals form a link between marine and terrestrial environments. Although these animals are supported exclusively by the ocean, their waste products, cast-off feathers, and the like, when deposited on land, support whatever lichens and insects live there.

Some inhabitants of terrestrial and aquatic life zones cover great distances—in the case of migratory fishes and birds, even global distances. For example, many young albacore tuna migrate from the California coast across the Pacific Ocean to Japan! Some species of flycatchers spend their summers in Canada and the United States and their winters in Central and South America. Like the flycatchers, many other migratory birds commonly spend critical parts of their life cycles in entirely different countries, which can make their conservation difficult (see Chapter 1). It does little good, for instance, to protect a songbird in one country if the inhabitants of the next put it in the cooking pot as soon as it arrives. Such large-scale interactions among various ecosystems make ecological concepts difficult for many people to grasp and apply.

SUMMARY

1. A biome is a large, relatively distinct terrestrial region with characteristic climate, soil, plants, and animals, regardless of where it occurs. Because it is so large, a biome encompasses many interacting ecosystems. A biome's boundaries are determined by climate more than any other factor.

2. Tundra is the northernmost biome. It is characterized by a permanently frozen layer of subsoil called the permafrost, by low-growing vegetation adapted to extreme cold, and by a very short growing season. The taiga, or boreal forest, lies south of the tundra and is dominated by coniferous trees.

3. Temperate rain forest, such as occurs on the northwest coast of North America, receives very high precipitation and is dominated by large conifers. Temperate deciduous forest, which occurs where precipitation is relatively high, is dominated by broad-leaved trees that lose their leaves seasonally.

4. Temperate grasslands typically possess deep, mineral-rich soil, have moderate but uncertain precipitation, and are well suited to growing grain crops. Tropical grasslands, called savannas, have widely scattered trees interspersed with grassy areas. Chaparral and other mediterranean climates are characterized by thickets of small-leaved evergreen shrubs and trees and a climate of wet, mild winters and very dry summers.

5. Deserts, found where there is little precipitation, are communities whose organisms have specialized water-conserving adaptations. Deserts occur in both temperate and tropical areas.

6. Tropical rain forests are characterized by mineral-poor soil and very high rainfall that is evenly distributed through the year. Rain forests have high species diversity.

7. In aquatic ecosystems, important environmental factors include salinity, amount of dissolved oxygen, and availability of light for photosynthesis. Aquatic life is ecologically divided into plankton (free-floating), nekton (strongly swimming), and benthos (bottom-dwelling). The microscopic phytoplankton are photosynthetic and are the base of the food web in most aquatic communities.

8. Freshwater ecosystems include flowing-water (rivers and smaller streams), standing-water (lakes and ponds), and freshwater wetlands. In flowing-water ecosystems, the water flows in a current. Flowing-water ecosystems have few phytoplankton and depend on detritus from the land for much of their energy. The organisms in flowing-water ecosystems vary greatly depending on the current, which is swifter in headwaters than downstream.

9. Freshwater lakes are divided into zones on the basis of water depth. The marginal littoral zone contains emergent vegetation and heavy growths of algae. The limnetic zone is open water away from the shore. The deep, dark profundal zone holds little life other than decomposers.

10. Freshwater wetlands, lands that are transitional between freshwater and terrestrial ecosystems, are usually covered by shallow water and have characteristic soils and vegetation.

11. An estuary is a coastal body of water, partly surrounded by land, with access to the ocean and a large supply of fresh water from rivers. Estuaries are very productive, in part because they receive a high input of nutrients from the adjacent land. One of the important roles of estuaries is to provide a nursery for the young of many aquatic organisms.

12. Four important marine environments are the intertidal zone, benthic environment, neritic province, and oceanic province. The intertidal zone is the shoreline area between low and high tides. Organisms of the intertidal zone possess adaptations to resist wave action and the extremes of being covered by water (high tide) and exposed to air (low tide).

13. The benthic environment is the ocean floor. Coral reefs are particularly important benthic communities in shallow ocean waters. Coral reefs, which have high species diversity and high productivity, are classified as fringing reefs, barrier reefs, and atolls.

14. The neritic province is open ocean that overlies the ocean floor from the shoreline to a depth of 200 meters. Organisms that live in the neritic province are all floaters or swimmers. Phytoplankton in the euphotic zone are the base of the food web.

15. The oceanic province is that part of the open ocean that overlies the ocean floor at depths greater than 200 meters. The uniform environment is one of darkness, cold temperature, and high pressure. Animal inhabitants of the oceanic province are either predators or scavengers that subsist on detritus that drifts in from other areas of the ocean.

THINKING ABOUT THE ENVIRONMENT

1. Why do most of the animals of the tropical rain forest live in trees?

2. What climate and soil factors produce each of the major terrestrial biomes?

3. Describe representative organisms of the forest biomes described in the text: taiga, temperate deciduous forest, temperate rain forest, tropical rain forest.

4. In which biome do you live? If your biome does not match the description given in this book, how do you explain the discrepancy?

5. Which biomes are best suited for agriculture? Explain why each of the biomes you did not specify is unsuitable for agriculture.

6. What environmental factors are most important in determining the kinds of organisms found in aquatic environments?

7. Distinguish between freshwater wetlands and estuaries. Between flowing-water and standing-water ecosystems.

8. What would happen to the organisms in a river with a fast current if a dam were built? Explain your answer.

9. List and briefly describe the four main marine environments.

10. Which aquatic ecosystem is often compared to tropical rain forests? Why?

*11. Convert the three ocean depths in Figure 6–18 from metric (meters) to the English equivalent (feet). Appendix IV can help you with this conversion.

RESEARCH PROJECTS

Choose a biome or major aquatic ecosystem and investigate the problems it faces. Identify any human activities—for example, major construction projects or heavy development—that affect the ecosystem's inherent qualities. Describe current efforts to rectify past abuses and assess their chances of success.

On-line materials relating to this chapter are on the World Wide Web at **http://www.saunderscollege.com/lifesci/environment2** (select Chapter 6 from the Table of Contents).

SUGGESTED READING

Chadwick, D. H. "Roots of the Sky," *National Geographic* Vol. 184, No. 4, October 1993. Biological diversity flourishes in the patchy remnants of the North American prairie.

Frederickson, J. J., and T. C. Onstott. "Microbes Deep inside the Earth," *Scientific American*, October 1996. More than 9000 strains of bacteria and fungi that live in sedimentary rocks and other environments within the Earth's crust have been identified.

Goulding, M. "Flooded Forests of the Amazon," *Scientific American*, March 1993. During the rainy season, vast sections of the Amazonian rain forest are inundated, making them a unique aquatic ecosystem as well as terrestrial biome.

Levine, J. S. "Dusk and Dawn Are Rush Hours on the Coral Reef," *Smithsonian*, October 1993. Examines the night life of a coral reef.

"Life in the Deep," *Discover*, November 1993. A photo essay of several unusual organisms adapted to various parts of the marine environment.

Mairson, A. "The Everglades: Dying for Help," *National Geographic*, April 1994. Florida's river of grass in Everglades National Park is suffering from decades of development upstream.

McClanahan, L. L., R. Ruibal, and V. H. Shoemaker. "Frogs and Toads in the Desert," *Scientific American*, March 1994. Some amphibians have special adaptations for dry climates.

Moffett, M. "These Plants Claw and Strangle Their Way to the Top," *Smithsonian*, September 1993. A striking assemblage of vines is found in tropical forests.

Robison, B. H. "Light in the Ocean's Midwaters," *Scientific American*, July 1995. Biologists are only beginning to explore the little-known realm of animal life in the ocean's dark middle depths.

Rützler, K., and I. C. Feller. "Caribbean Mangrove Swamps," *Scientific American*, March 1996. The authors provide a fascinating glimpse of the complexities of a mangrove swamp in Belize.

The Safe Drinking Water Act is an unfunded mandate that requires local water systems to monitor their drinking water for contaminants. (© Charlie D. Winters 1993)

Economics, Government, and the Environment

In 1995 a bill requiring the federal government to pay for future programs it imposes on state and local governments was passed by Congress and signed into law by the president. This law, which targets "unfunded mandates," is likely to have far-reaching impacts on future environmental legislation, as well as future federal mandates in education, transportation and safety, and health and welfare. Prior to passage of the unfunded mandates bill, Congress frequently passed directives to states and local governments without providing a way to pay for the cost of compliance. Congress must now determine the cost of proposed federal orders that it passes along to states and local governments, and any federal mandates that exceed $50 million must be paid by the federal government. The law does not apply to *existing* environmental laws, such as those covering clean air or clean water, only future laws.

One of the reasons that state and local governments rebelled against unfunded mandates is that these regulations are expensive and sometimes have questionable benefits. In one well publicized example, cities in Ohio tested for 52 pesticides and other contaminants in drinking water, as required by the Safe Drinking Water Act. One of these pesticides, used mostly in Hawaii before 1977, had *never* been found in Ohio drinking water. However, because the law was inflexible, Ohio cities were required to continue testing for it. (The rules have since been modified so that the discontinued pesticide no longer has to be monitored.)

Despite this example of regulatory inflexibility, most state and local governments do not appear to be against federal environmental mandates as much as the fact that they are *unfunded*. The Safe Drinking Water Act, for example, sets national standards for public drinking water, thereby protecting human health, and nobody wants to drink pesticide-laced water. State and local governments say they cannot afford the Safe Drinking Water Act, despite its commendable goals, because it is only one of *many* unfunded mandates that they must pay for.

The 1995 unfunded mandate law will probably limit national environmental legislation in the future. In certain cases, Congress may decide that a piece of environmental legislation is important enough to spend federal money to pay for it. Given our budget deficit (forecast to be $121 billion in fiscal year 1998), however, the federal government will probably be reluctant to issue most environmental mandates. ◄

AN ECONOMIST'S VIEW OF POLLUTION

As the human population has grown, its appetite for natural resources—plants, animals, water, minerals, air, land, and so on—has seriously stressed the Earth's environment. Pressures for continued economic and industrial development have proven difficult to resist, and such development has almost always entailed the disruption of natural ecosystems.

In this chapter we examine one form of such disruption, industrial pollution. Lessons about the economics of industrial pollution also apply to other environmental issues where disruption of ecosystems is a consequence of economic activity. For now we focus on the two factors that most critically influence decisions in which development and the environment come into conflict: economic pressures and government policy.

Economics is the study of how people use their limited resources to try to satisfy their unlimited wants. Developing hypotheses, testing models, and analyzing observations and data, economists try to understand the consequences of the ways in which people, businesses,

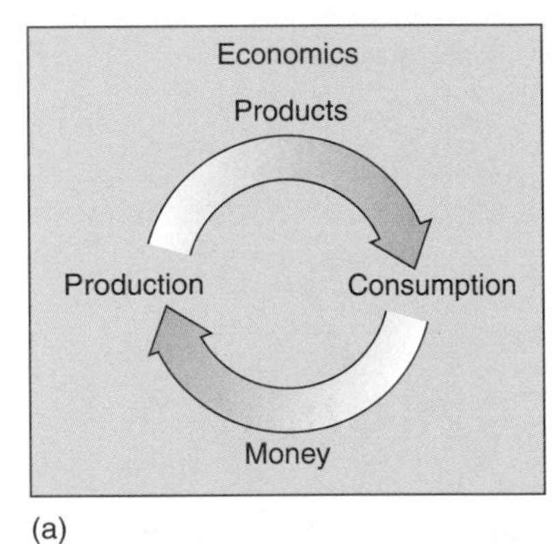

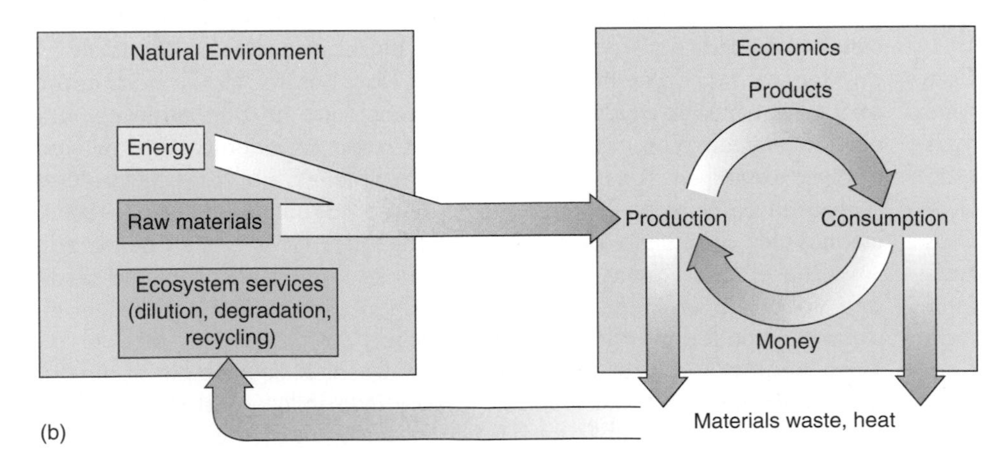

Figure 7-1

How economics is related to the environment. (a) Goods and services (products) and money (to pay for the products) flow between businesses (production) and consumers (consumption). (b) Economics depends on the natural environment to provide resources (energy and raw materials) and ecosystem services.

and governments allocate their limited resources. In a free-market system such as that of the United States, economists study the prices of goods and services and how those prices influence the amount of a given good or service that is produced and consumed. Like any scientists conducting experiments, economists try to predict the consequences of particular economic actions. When the actions involve economic development, their predictions may contribute to policy decisions that have significant environmental consequences. If, as citizens, we wish to affect these policy decisions, we need to understand how economists view the world.

Seen through an economist's eyes, the world is one large marketplace where resources are allocated to a variety of uses and where goods—a car, a pair of shoes, a hog—and services—a haircut, a tour of a museum, an education—are consumed and paid for (Figure 7–1). In a free market, the price of a good is determined by its supply and by the demand for it. If something in great demand is in short supply, its price will be high. High prices encourage suppliers to produce more of a good or service, as long as the selling price is higher than the cost of producing the good or service. This interaction of consumer demand, producer's supply, prices, and costs underlies much of what happens in our country's economy, from the price of a hamburger to the cycles of economic expansion (increase in economic activity) and recession (slowdown in economic activity).

An important aspect of the operation of a free-market system is that the person consuming a product should be the one to pay for all the cost of producing that product. When consumption or production of a product has a harmful side effect that is borne by people not directly involved in the market exchange for that product, the side effect is called an **external cost**. Because external cost is usually not reflected in a product's price, the market system does not operate in the most efficient way. For example, if an industry makes a product and, in so doing, also releases a pollutant into the environment, the product is bought at a price that reflects the cost of making it, but *not* the cost of the damage to the ecosystem by the pollutant. Because this damage is not included in the product's price and because the consumer may not be aware that the pollution exists or that it harms the environment, the cost of the pollution has no impact on the consumer's decision to buy the product. As a result, consumers of the product buy more of it than they would if its true cost, including the cost of pollution, were known. The failure to add the price of environmental damage to the cost of products generates a market force that increases pollution. From the perspective of economics, then, one of the root causes of the world's pollution problem is the failure to consider external costs in the pricing of goods.

ENVIROBRIEF

A CORPORATE APPROACH TO ENVIRONMENTALISM

Many U.S. multinational corporations are finding that their environmental policies can do more than boost public relations; cleaning up wastes can often cut their own cost substantially. 3M provides an impressive corporate example. Its "3P" program, "Pollution Prevention Pays," has made tremendous strides since it went into effect in 1975:

- By 1991, the company had saved $537 million through its cleanup efforts.
- As of 1993, the company's air pollution, solid and hazardous waste, and water pollution around the world had been cut in half, eliminating more than 134,000 tons of air pollution, 16,900 tons of water pollution, and 426,000 tons of sludge and solid waste.
- The company's goals include the reduction of 1987 air, land, and water discharges by 90 percent by the year 2000 and achieving zero pollution and sustainable development thereafter.

Most pollution prevention ideas are suggested and implemented by 3M employees.

HOW MUCH POLLUTION IS ACCEPTABLE?

In order to come to grips with the problem of assigning a proper price to pollution, economists first attempt to answer the basic question "How much pollution should be allowed?" One can imagine two extremes: an uninhabited wilderness in which no pollution is produced and an uninhabitable sewer that is completely polluted from excess production of goods. In an uninhabited wilderness, the environmental quality would be the highest possible, but many goods that are desirable to humans would be scarce or nonexistent. In an uninhabitable sewer, millions and millions of goods could be produced, but environmental quality would be extremely low. In our world, a move toward a better environment almost always entails a cost in terms of goods.

How do we, as a country and as part of the larger international community, decide where we want to be on the environment–material prosperity continuum? Economists answer such questions by analyzing the marginal costs of environmental quality and of other goods. A **marginal cost** is the additional cost associated with one more unit of something. The tradeoff between environmental quality versus more goods involves balancing two kinds of marginal costs: (1) the cost, in terms of environmental quality, of enduring more pollution (the marginal cost of pollution); and (2) the cost, in terms of giving up goods, of eliminating pollution (the marginal cost of pollution abatement).

The Marginal Cost of Pollution

The **marginal cost of pollution** is the added cost for all present and future members of society of an additional unit of pollution. For each type of pollution, such as sulfur dioxide, which causes acid rain, economists add up the harm done by each additional unit of pollution, such as another ton of sulfur dioxide added to the atmosphere. As the total amount of pollution increases, the harm done by each additional unit usually also increases, so that the curve showing the marginal cost of pollution slopes upward (Figure 7–2). At low pollution levels, the environment may be able to absorb the damage so that the marginal cost of one added unit of pollution is near zero. As the quantity of pollution increases, the marginal cost rises, and at very high levels of pollution the cost soars.

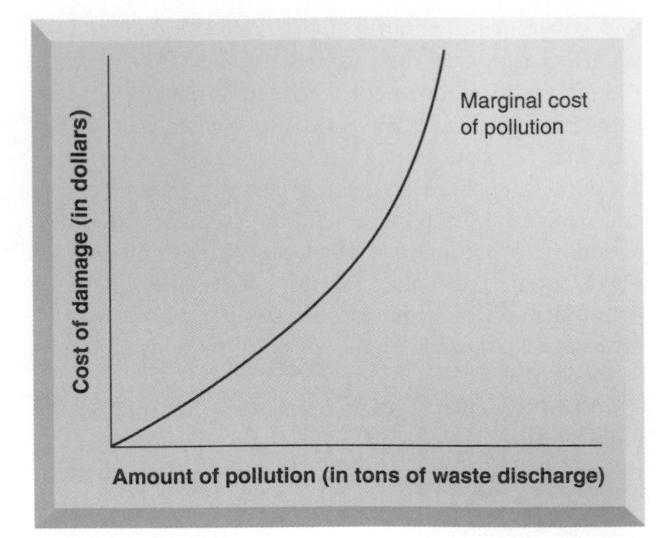

Figure 7–2
The marginal cost of pollution may be represented by an upward sloping curve, which shows that as the level of pollution rises, the social cost (in terms of human health and a damaged environment) increases sharply.

Mini-Glossary of the Economics of Pollution

marginal cost of pollution: The cost, in environmental quality, of a unit of pollution that is emitted into the environment.

marginal cost of pollution abatement: The cost to dispose of a unit of pollution in a nonpolluting way.

optimum amount of pollution: The amount of pollution that is economically most desirable. It is determined by plotting two curves, the marginal cost of pollution and the marginal cost of pollution abatement. The point where the two curves meet is the optimum amount of pollution from an economic standpoint.

emission charge policy: A government policy that controls pollution by charging the polluter for each unit of emissions, that is, by establishing a tax on pollution.

waste-discharge permit policy: A government policy that controls pollution by issuing permits allowing the holder to pollute a given amount. Holders are not allowed to produce more emissions than are sanctioned by their permits.

emission reduction credit (ERC): A waste-discharge permit that can be bought and sold by companies producing emissions. Companies have a financial incentive to reduce emissions because they can recover some or all of their cost of pollution abatement by the sale of the ERCs that they no longer need.

command and control: Pollution control laws that work by setting pollution ceilings. Examples include the Clean Water and Clean Air Acts.

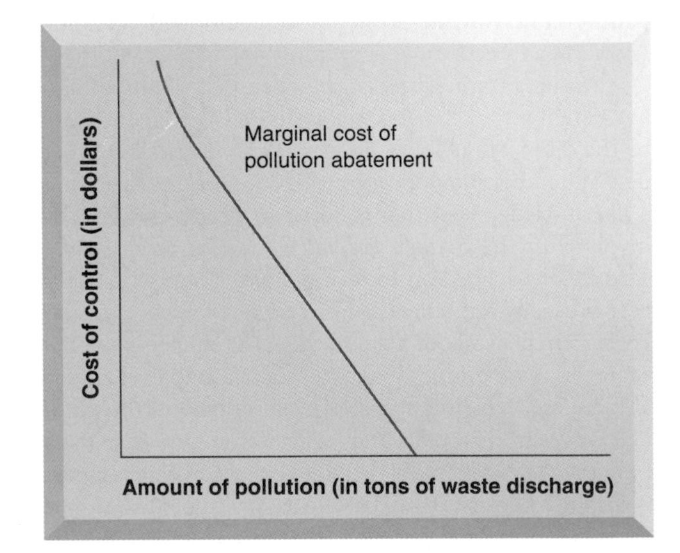

Figure 7–3
The marginal cost of pollution abatement is represented by a downward sloping curve, which shows that as more and more pollution is eliminated from the environment, the cost of removing each additional (marginal) unit of pollution increases.

The Marginal Cost of Pollution Abatement

The **marginal cost of pollution abatement** is the added cost for all present and future members of society of reducing a given type of pollution by one unit. This cost tends to rise as the level of pollution falls (Figure 7–3).

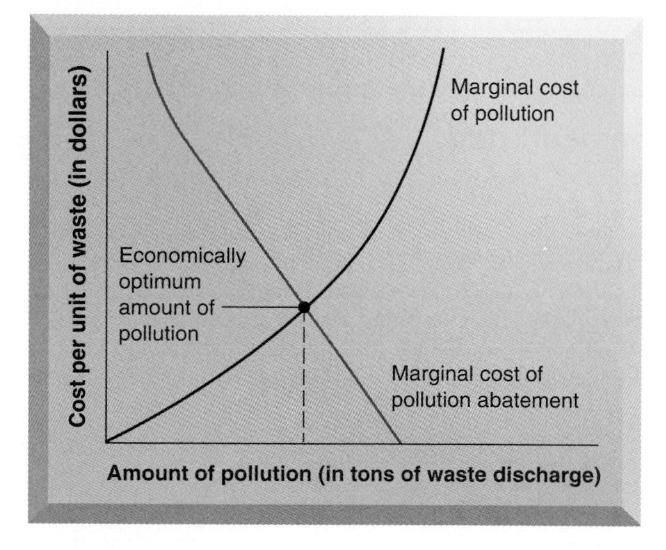

Figure 7–4

Economists identify the optimum amount of pollution as the amount whose marginal cost of pollution equals the marginal cost of pollution abatement (the point at which the two curves intersect). If more pollution than the optimum is allowed, the social cost will be unacceptably high. If less than the optimum amount of pollution is allowed, the pollution abatement cost will be unacceptably high.

PRICING THE RAIN FOREST

The economic value of a tropical rain forest is normally calculated only by the value of its tree-related products: lumber and wood that can be converted into pulp for paper. Botanical researchers have recently demonstrated that in some areas the non-woody resources of rain forests, such as edible fruits, nuts, latex for rubber, and medicinal extractions might be nearly twice as valuable per acre as wood. The challenge is one of changing public policy to reflect these economic opportunities. While timber brings much-needed foreign exchange to poor tropical nations, nonwood products can be harvested manually and traded at local markets by thousands of individual collectors, farmers, and middlemen, with much less damage to the forest. Cooperative international efforts, such as those of Conservation International, Shaman Pharmaceuticals, and various ecotourism firms, aim to develop viable markets for nonwood resources while contributing to local economies and preserving rainforest diversity.

It is relatively inexpensive, for example, to reduce automobile exhaust emissions by half, but costly devices are required to reduce the remaining emissions by half again. For this reason, the curve showing the marginal cost of pollution abatement slopes downward. At high pollution levels the marginal cost of eliminating one unit of pollution is low, but as more and more pollution is eliminated, the cost rises.

How Economists Estimate the Optimum Amount of Pollution

In Figure 7–4, the two marginal-cost curves from Figures 7–2 and 7–3 are plotted together on one graph, called a cost-benefit diagram. Economists use this diagram to identify the point at which the marginal cost of pollution equals the marginal cost of abatement—that is, the point where the two curves intersect. As far as economics is concerned, this point represents an **optimum amount of pollution**. If pollution exceeds this amount (is right of the point of intersection), the harm done (measured on the marginal-cost-of-pollution curve) exceeds the cost of reducing it (the marginal-cost-of-pollution-abatement curve), and it is economically efficient to reduce pollution. On the other hand, if pollution is less than this amount (is left of the intersection point), then the marginal cost incurred to reduce it exceeds the marginal cost of the pollution. In the latter situation the gain in environmental quality is more than offset monetarily by the decrease in resources available for other uses. In such a situation, an economist would argue, pollution should be increased.

Flaws in the Optimum Pollution Concept

There are two major flaws in the economist's concept of optimum pollution. First, it is difficult to actually measure the monetary cost of pollution. Second, the risk of ecosystem disruption is rarely taken into account.

The difficulty of measuring pollution cost

Economists usually measure the cost of pollution in terms of damage to property, damage to health in the form of medical expenses or time lost from work, and the monetary value of animals and plants killed. It is difficult, however, to place a value on damage to natural beauty—how much is a scenic river worth, or the sound of a bird singing? And how does one assign a value to the extinction of a species? Also, when pollution covers a large area and involves millions of people—as, for example, acid rain does—assessing pollution cost is extremely complex (see Focus On: Natural Resources, the Environment, and the National Income Accounts).

FOCUS ON

Natural Resources, The Environment, and The National Income Accounts

Much of our economic well-being flows from natural, rather than human-made, assets—our land, rivers, the ocean, natural resources such as oil and timber, and indeed the air that we breathe. Ideally, for the purposes of economic and environmental planning, the use and misuse of natural resources and the environment should be appropriately measured in the national income accounts.* Unfortunately, they are not. There are at least two important conceptual problems with the way the national income accounts currently handle the economic use of natural resources and the environment.

1. **Natural resource depletion.** If a firm produces some output but in the process wears out a portion of its plant and equipment, the firm's output is counted as part of gross domestic product (GDP), but the depreciation of capital is subtracted in the calculation of net domestic product (NDP). Thus NDP is a measure of the net production of the economy, after a deduction for used-up capital. In contrast, when an oil driller drains oil from an underground field, the value of the oil produced is counted as part of the nation's GDP; but no offsetting deduction to NDP is made to account for the fact that nonrenewable resources have been used up. In principle, the draining of the oil field should be considered a type of depreciation, and the net product of the oil company should be accordingly reduced. The same point applies to any other natural resource that is depleted in the process of production.
2. **The cost and benefits of pollution control.** Imagine that a company has the following choices: It can produce $100 million worth of output and in the process pollute the local river by dumping its wastes. Alternatively, by using 10 percent of its workers to dispose properly of its wastes, it can avoid polluting but will only get $90 million of output. Under current national income accounting rules, if the firm chooses to pollute rather than not to pollute, its contribution to GDP will be larger ($100 million rather than $90 million), because the national income accounts attach no explicit value to a clean river. In an ideal accounting system, the economic cost of environmental degradation would be subtracted in the calculation of a firm's contribution to GDP, and activities that improve the environment—because they provide real economic benefits—would be added to GDP.

Discussing the national income accounting implications of resource depletion and pollution may seem to trivialize these important problems. But in fact, since GDP and related statistics are used continually in policy analyses, abstract questions of measurement may often turn out to have significant real effects. For example, economic development experts have expressed concern that some poor countries, in attempting to raise measured GDP as quickly as possible, have done so in part by over-exploiting their natural resources and impairing the environment. Conceivably, if "hidden" resource and environmental costs were explicitly incorporated into official measures of economic growth, these policies might be modified. Similarly, in industrialized countries political debates about the environment have sometimes emphasized the impact on conventionally measured GDP of proposed pollution control measures, rather than the impact on overall economic welfare. Better accounting for environmental quality might serve to refocus these debates to the more relevant question of whether, for any given environmental proposal, the benefits (economic and noneconomic) exceed the cost.

*See any introductory macroeconomics text for a discussion of the national income accounts.

Sources: Jonathan Levin, "The Economy and the Environment: Revising the National Accounts," *IMF Survey*, 4 June 1990; Washington: International Monetary Fund, 1990.

The risk of disrupting the ecosystem

In adding up pollution costs, economists do not take into account the possible disruption or destruction of an ecosystem. As you have seen in the last few chapters, the web of relationships within an ecosystem is extremely complex and may be vulnerable to pollution damage, often with disastrous results. For an economist to simply add up the cost of lost elements in a polluted ecosystem is like sitting in an airplane adding up the cost of damaged items as someone repeatedly shoots a gun in the cockpit. The sum of costs does not reflect the very real danger that the shooting will cause the plane to crash.

ECONOMIC STRATEGIES FOR POLLUTION CONTROL

Economists have traditionally approached the problem of pollution control in terms of supply and demand. Because their suggested solutions have a significant impact on government policy, it is important for any student of environmental science to understand the nature of their arguments and proposed solutions.

The Demand Curve

To understand the economist's analysis of the problem of pollution control, examine Figure 7–5. You will recognize it as the diagram presented in Figure 7–4, but we now look at the curve we called "marginal cost of pollution abatement" from a different point of view—that of the industrial polluter. Labeled "Demand for pollution opportunities," the curve now expresses how much a firm would be willing to pay for the opportunity to dump an additional unit of pollution into the environment. From the point of view of cost effectiveness, the firm would pay any amount smaller than the marginal cost of pollution abatement. That is, it would pay for the opportunity to pollute as long as polluting costs less than not polluting. In this sense, an industry's demand to pollute is determined by its desire to avoid abatement costs.

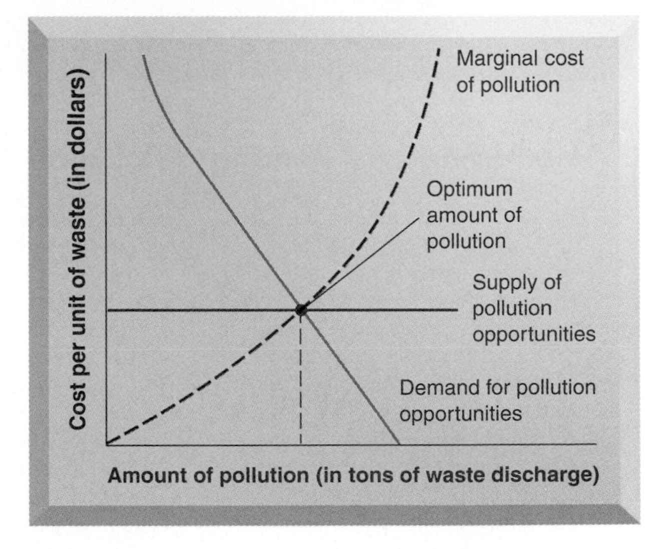

Figure 7–5

Supply and demand of pollution opportunities. The supply curve of pollution opportunities is a horizontal line, whereas the demand curve of pollution opportunities is equivalent to the marginal cost of pollution abatement. It is possible to adjust the supply curve so that supply and demand curves intersect at the optimum amount of pollution.

The Supply Curve

In Figure 7–5, the horizontal axis is, in effect, a supply curve of pollution opportunities. When the cost of pollution (measured on the vertical axis) is zero, the opportunities for pollution are unlimited, and events are determined simply by the demand curve. As the cost of pollution rises, the horizontal line rises, and its intersection with the demand curve moves to the left—that is, the demand for pollution falls. For any cost of pollution, supply and demand are equal at the point at which their curves intersect.

The Intersection of Supply and Demand

From an economist's point of view, the problem of excessive pollution can be looked at as an error in cost estimation: because the polluter is not charged for dumping wastes into the environment and therefore the cost of pollution is not included in the product's price, the supply curve is too low, and so the intersection of supply and demand curves is too low. As a result, it is less expensive to pollute than to dispose of wastes properly. Given this view, economists propose a simple and straightforward solution: raise the supply curve by raising the cost of polluting. This approach is often referred to by economists as "adopting a market-oriented strategy" because it seeks to use the economic forces of a free market to alleviate the pollution problem.

A very popular market-oriented strategy for controlling pollution involves raising the supply curve of Figure 7–5 up to the optimum pollution value by imposing an **emission charge** on polluters. In effect, this charge is a tax on pollution. Several European countries, for example, have encouraged drivers to switch to unleaded gasoline by imposing extra taxes on leaded gasoline, which releases more pollutants into the air when it burns. Sweden taxes the active ingredients in pesticides. Finland and Norway impose charges on nonreturnable containers, as do some communities in the United States. Such charges increase the cost of polluting and shift the supply curve of pollution opportunities upward. It is hoped that polluters react to the increase in cost caused by the emission charge by decreasing pollution, because the cost of abatement is now less than the cost of polluting. In actual practice, this approach has not been notably successful because taxes are almost always set too low to have much effect on the behavior of people or companies.

Many economists argue for a different market-oriented strategy: raising the supply curve to the optimum pollution level by issuing a fixed number of **waste-discharge permits**, each of which allows the holder to emit a certain amount of a given pollutant. The permits, called **emission reduction credits (ERCs)**, can then be freely bought and sold. Economists believe that this approach is an efficient way to move the market toward the point of optimum pollution: potential polluters desiring to add more pollution to the environment will increase their cost by buying additional ERCs, whereas those polluting less than the optimum level will sell their excess ERCs and so lower their cost. Because using marketable permits ensures that pollution does not fall below the op-

YOU CAN MAKE A DIFFERENCE

Purchasing and Retiring a Pollution Permit

The 1990 amendments to the Clean Air Act established a pollution permit system based on a free-market approach. Each permit that a coal-fired power plant possesses allows it to discharge one ton of sulfur dioxide yearly; most utilities emit many thousands of tons of sulfur dioxide. Companies have to pay a fine of $2000 for every excess ton of sulfur dioxide emitted beyond what its permits allow.

The federal government is distributing a maximum of 9.5 million waste-discharge permits, which can be bought and sold each year at the Chicago Board of Trade. Thus, when a utility installs pollution control equipment and thereby reduces its emissions, it can sell its waste-discharge permits to utilities that are still heavy polluters.

Anybody can purchase the emission allowances. For example, students at the University of Maryland School of Law purchased a waste-discharge permit for one ton of sulfur dioxide emissions. They retired the emission allowance, which means that another utility cannot purchase it. As a result, one less ton of sulfur dioxide can be released into the air each year.

The price of each waste-discharge permit has fluctuated between $125 and $450, but is expected to rise as more allowances are permanently retired. This increase should stimulate companies that have lagged in installing pollution control equipment because they will be able to sell their waste-discharge permits at a higher price to help pay for the equipment.

How would *you* like to make a difference by owning a ton of pollution? Contact the Clean Air Conservancy (1-800-2-BUY-AIR) and investigate how to bid for a waste-discharge permit at the Chicago Board of Trade. Then work with your environmental science class or campus environmental group to raise the money for the permit. Be creative: organize activities or sponsor contests and charge a small fee for participation. This way you will get many people involved and show that any problem—even one as global as air pollution—can be successfully tackled if individuals work together.

timum amount and allows firms with higher abatement costs to control their pollution less, it prevents pollution abatement from impeding economic development and achieves pollution reduction at low cost. (See You Can Make a Difference: Purchasing and Retiring a Pollution Permit.)

HOW MUCH DOES A CLEAN ENVIRONMENT COST?

Pollution prevention, cleanup, and compliance with environmental regulations is big business in the United States. In a 1994 report to Congress, the EPA estimated that the private sector—that is, U.S. industries—spent $53.7 billion on pollution abatement (capital expenses, operation, and maintenance) in 1987. That expenditure is increasing annually and is projected to be $88.7 billion by the year 2000.

The federal government's pollution control expenses are also on the rise. The EPA's budget for fiscal year 1995 was set at $7.2 billion, an increase of $500 million over the 1994 budget. EPA presidential initiatives for 1995 invested $1.2 billion in businesses developing advanced pollution remediation systems.

This approach was used by the Environmental Protection Agency (EPA) in mandating lower levels of lead in gasoline. Refineries were assigned quotas of lead, which they could trade with each other. Since 1974 the EPA has also issued air pollution permits as marketable ERCs. When a company wishes to move into a city that fails to meet the standards of the 1970 Clean Air Act, it buys ERC polluting rights from established firms that have cut their own emissions. The Clean Air Act of 1990 includes a similar plan to cut sulfur dioxide emissions with tradable permits for coal-burning electricity utilities.

GOVERNMENT AND ENVIRONMENTAL POLICY

Given that it is difficult to assess how much, if any, pollution should be allowed, how should we govern levels of pollution in our society? Historically, acceptable levels of pollution have been determined by rough guess, using the guiding principle that cleaner is safer. While this might seem imprecise, setting such guidelines has generally proven effective. To enforce the determined limit on pollution, legislation is usually enacted (Table 7–1).

Historically, most pollution control efforts have involved what economists called **command and control**—the passage of laws that impose rules and regulations and set limits on levels of pollution. Sometimes such laws state that a specific pollution control method must be used—such as catalytic converters in cars to decrease polluting emissions in exhaust. In other cases, a quantitative goal

Table 7–1
Some Important Environmental Legislation

General
National Environmental Policy Act of 1970 (NEPA)
International Environmental Protection Act of 1983
National Environmental Education Act of 1990

Conservation of Energy
Energy Policy and Conservation Acts of 1978, 1980
Northwest Power Act of 1980
National Appliance Energy Conservation Act of 1987
Energy Policy Act of 1992

Conservation of Wildlife
Anadromous Fish Conservation Act of 1965
Fur Seal Act of 1966
National Wildlife Refuge System Acts of 1966, 1976, 1978
Species Conservation Acts of 1966, 1969
Marine Mammal Protection Act of 1972
Marine Protection, Research, and Sanctuaries Act of 1972
Endangered Species Acts of 1973, 1982, 1985, 1988
Fishery Conservation Acts of 1976, 1978, 1982
Whale Conservation and Protection Study Act of 1976
Fish and Wildlife Improvement Act of 1978
Fish and Wildlife Conservation Act of 1980

Conservation of Land
General Revision Act of 1891
Taylor Grazing Act of 1934
Soil Conservation Act of 1935
Wilderness Act of 1964
Land and Water Conservation Fund Act of 1965
Multiple Use Sustained Yield Act of 1968
Wild and Scenic Rivers Act of 1968
National Trails System Act of 1968
Coastal Zone Management Acts of 1972, 1980
National Reserves Management Acts of 1974, 1976
Forest and Rangeland Renewable Resources Acts of 1974, 1978
Federal Land Policy and Management Act of 1976
National Forest Management Act of 1976
Public Rangelands Improvement Act of 1978
Soil and Water Conservation Act of 1977
Surface Mining Control and Reclamation Act of 1977
Antarctic Conservation Act of 1978
Endangered American Wilderness Act of 1978
Alaska National Interest Lands Act of 1980
Coastal Barrier Resources Act of 1982
Emergency Wetlands Resources Act of 1986
California Desert Protection Act of 1994
Federal Agriculture Improvement and Reform Act of 1996 (the latest version of the Farm Bill, which has been amended and renamed every 5 years or so since the 1930s)

Air Quality and Noise Control
Clean Air Acts of 1970, 1977, 1990
Noise Control Act of 1965
Quiet Communities Act of 1978
Asbestos Hazard and Emergency Response Act of 1986

Water Quality and Management
Refuse Act of 1899
Water Resources Research Act of 1964
Water Quality Act of 1965
Water Resources Planning Act of 1965
Water Pollution Control Act of 1972 (amendment to Water Quality Act of 1965)
Ocean Dumping Act of 1972
Ocean Dumping Ban Act of 1988
Safe Drinking Waters Acts of 1974, 1986, 1996
Water Resources Development Act of 1986
Great Lakes Toxic Substance Control Agreement of 1986
Clean Water Acts of 1977, 1981, 1987 (up for renewal)
Water Quality Act of 1987 (amendment of Clean Water Acts)

Control of Pesticides
Food, Drug, and Cosmetics Acts of 1938, 1954, 1958
Insecticide, Fungicide, and Rodenticide Acts of 1947, 1972, 1988
Food Quality Protection Act of 1996

Management of Solid and Hazardous Wastes
Solid Waste Disposal Act of 1965
Resource Recovery Act of 1970
Hazardous Materials Transportation Act of 1975
Toxic Substances Control Act of 1976
Resource Conservation and Recovery Acts of 1976, 1984
Low-Level Radioactive Policy Act of 1980
Comprehensive Environmental Response, Compensation, and Liability (Superfund) Acts of 1980, 1986
Nuclear Waste Policy Act of 1982
Marine Plastic Pollution Control Act of 1987
Oil Pollution Act of 1990

is set. For example, the 1990 Clean Air Act set a goal of a 60 percent reduction in nitrogen oxide emissions in passenger cars by the year 2003. Usually, all polluters must comply with the same rules and regulations regardless of their particular circumstances.

Overall, legislative approaches to pollution control have been successful (discussed later in the chapter). For example, the air in many of the world's cities, although still polluted, is far cleaner than it was 20 years ago. Levels of sulfur-dioxide emissions from coal-fired power plants have been cut in several large industrial countries, including the United States, United Kingdom, Canada, and Japan (Table 7–2, page 144).

In countries where laws have not been passed to control pollution, the situation is far worse. Cities in developing countries, which typically do not regulate air emissions, have far filthier air than those in developed countries. Water and air pollution in Central and Eastern Europe, which never attempted to control pollution while under communist rule,[1] are particularly bad (see Focus On: Economics and the Environment in Central and Eastern Europe).

[1]Many laws and regulations *existed* but were not enforced.

(text continues on page 144)

FOCUS ON

Economics and the Environment in Central and Eastern Europe

The fall of the Soviet Union and communist governments in Central and Eastern Europe during the late 1980s revealed a grim legacy of environmental destruction (see figures). Water is so poisoned from raw sewage and chemicals that it cannot be used for industrial purposes, let alone for drinking. Unidentified chemicals leak out of dump sites into the surrounding soil and water, while nearby, fruits and vegetables are grown in chemically laced soil. Power plants pour soot and sulfur dioxide into the air, producing a persistent chemical haze. Buildings and statues are eroded, and entire forests are dead because of air pollution and acid rain. Crop yields are falling despite intensive use of chemical pesticides and fertilizers.

How does this massive pollution affect human health? Many Central and Eastern Europeans suffer from asthma, emphysema, chronic bronchitis, and other respiratory diseases as a result of breathing the filthy, acrid air. By the time most Polish children are ten years old, for example, they suffer from chronic respiratory diseases or heart problems. The levels of cancer, miscarriages, and birth defects are also extremely high. Life expectancies are lower than in other industrialized nations; in 1996 the average Eastern European lives to age 68, which is nine years less than the average Western European.

The economic assumption behind communism was one of high production and economic self-sufficiency—regardless of cost or damages—and so pollution in communist-controlled Europe went unchecked. Meeting industrial production quotas always took precedence over environmental concerns, even though production was not carried out for profit. The communist regimes supported heavy industry—power plants, chemicals, metallurgy, large machinery—at the expense of the more environmentally benign service industries. As a result, Central and Eastern Europe is overindustrialized, and because most of its plants were built during the post-World War II period, it lacks the pollution abatement equipment now required in factories in most industrialized countries.

In addition, communism did little to encourage resource conservation, which is a very effective way to curb pollution. Because high energy subsidies and lack of competition allowed power plants to provide energy at prices far below its actual cost, neither industries nor individuals had a strong incentive to conserve energy.

Communism also took a toll on the environment because a repressive government run by a single political party cannot be held accountable. Political opposition to any aspect of government operations, including environmental damage, was unthinkable, and people who wanted to scrutinize pollution information found that it was unavailable, having been classified as top secret. Only the collapse of communism allowed the citizens of Central and Eastern Europe to begin to assess the full extent of damage to their environments.

Clearly the new Central and Eastern European governments face an intimidating task. While switching from communism to democracy with a free-market economy, they also face the overwhelming responsibility of improving the environment. They are trying to formulate environmental policies based on the experiences of the United States and Western Europe over the past several decades.

The United States, Japan, and Western Europe are supplying Central and Eastern European countries with some scientific and technical assistance and have promised money to help in environmental reconstruction. Water and air pollution move readily across political borders, so Western European countries in particular have a real incentive to help Central and Eastern European nations reduce pollution. In doing so, they help to protect their own environments.

How long will it take? Optimistically, experts predict that years, even decades, will be required to clean up the pollution legacy of communism. How much will it cost? The figures are staggering. It is estimated that improving the environment in what was formerly East Germany alone will cost up to $300 billion.

FOLLOW-UP

Following the collapse of communism during the late 1980s and early 1990s, the environment of countries in the former Soviet bloc is worse now than ever. The legacies of communism—antiquated chemical factories, smelters, and power plants—continue to degrade the environment. Pollution-induced health problems persist, including high rates of lung diseases, birth defects, low birth weights, and water-borne diseases such as hepatitis.

To compound existing problems, capitalism has introduced new environmental concerns. More people are driving automobiles, which generate air pollution, and fewer are using public transportation. New, fancy packaging for products has generated a solid waste problem, and recycling, which was practiced widely under the communists, is no longer common. Western companies are locating in this region to take advantage of the lax environmental standards. (It was initially hoped that Western companies would introduce modern, pollution-cutting production methods, but this has generally not occurred.)

In many cases, new environmental laws are in place, but they are simply not enforced. The main reason that Central and Eastern Europe has made little progress in cleaning up the environment is that economic recovery has been slow. The transition from a state-run economy to capitalism has been difficult, and severe budgetary problems have forced the environment to take a back seat to political and economic reform.

Pollution in Central Europe. (a) Smoke billows from a chemical plant in Copsa Mica, Romania. (b) A pipe from a chemical plant dumps pollution into a lake near the city of Cottbus in what was formerly East Germany. (a, V. Leloup/Figaro/Gamma Liaison; b, Peter Hirth/Sovfoto/Eastfoto)

Table 7–2
Sulfur Dioxide Emissions (Thousand Metric Tons) in Selected Countries

Country	1975	1993 (or latest year available)
Canada	5,319	3,030 (in 1992)
France	3,328	1,221 (in 1992)
Japan	2,586	876 (in 1990)
United Kingdom	5,368	3,188
United States	25,510	20,622 (in 1992)
Former West Germany	3,334	875 (in 1992)
Former East Germany	4,114	3,021 (in 1992)
Former Czechoslovakia	2,900	2,564 (in 1990)
Hungary	—	827 (in 1992)
Bulgaria	—	1,030 (in 1990)

Figure 7–6
Loading up logs in a white pine forest near Lake Huron. By 1897, most of the forests in Michigan had been cut. (The Bettmann Archive)

Seeking to limit pollution by legislation is only one of many ways in which governments act to protect the environment. In order to better understand the key role governmental policy plays in environmental protection, let us look briefly at the history of environmental legislation in this country.

National Forests

From the establishment of the first permanent English colony at Jamestown, Virginia, in 1607, the first two centuries of our country's history were a time of widespread environmental destruction. The great forests of the Northeast were leveled within a few generations, and soon after the Civil War in the 1860s, loggers began deforesting the Midwest at an appalling rate. Within 40 years they deforested an area the size of Europe, stripping Minnesota, Michigan, and Wisconsin of virgin forest. By 1897 the sawmills of Michigan had processed 160 billion board feet of white pine, leaving less than 6 billion board feet standing in the whole state (Figure 7–6). There has been nothing like this unbridled environmental destruction since.

In 1875 a group of public-minded citizens formed the American Forestry Association, with the intent of influencing public opinion against the wholesale destruction of America's forests. Sixteen years later, in 1891, the General Revision Act gave the president the authority to establish forest reserves on public (federally owned) land. Benjamin Harrison, Grover Cleveland, and Theodore Roosevelt used this law to put 43 million acres of forest, primarily in the West, out of the reach of loggers. In 1907 angry Northwest congressmen pushed through a bill rescinding the president's powers to establish forest reserves. Theodore Roosevelt responded by designating 21 new national forests that totaled 16 million acres, *then* signing the bill that would have prevented him into law. Today, national forests have multiple uses, from biological habitats to recreation to timber harvest (see Meeting the Challenge: Old-Growth Forests of the Pacific Northwest: Jobs Versus the Environment; also see Chapter 17).

National Parks and Monuments

The world's first national park was established in 1872 after a party of Montana explorers reported on the natural beauty of the canyon and falls of the Yellowstone River; Yellowstone National Park now includes parts of Idaho, Montana, and Wyoming. In 1890 the Yosemite and Sequoia parks in California were established by the Yosemite National Park Bill, largely in response to the efforts of a single man, John Muir, and the organization he founded, the Sierra Club (Figure 7–7).

In 1906 Congress passed the Antiquities Act, which authorized the president to set aside as national monuments sites, such as the Badlands in South Dakota, that had scientific, historic, or prehistoric importance. By 1916 there were 13 national parks and 20 national monuments, under the loose management of the U.S. Army. Today there are 369 national parks and monuments.

Some environmental battles were lost. John Muir's Sierra Club fought such a battle with the city of San Francisco over its efforts to dam a river and form a reservoir in the Hetch Hetchy Valley (Figure 7–8), which lay within Yosemite National Park and was as beautiful as Yosemite Valley. In 1913 Congress voted to approve the dam. But the controversy generated a strong sentiment that the nation's national parks should be better protected, and in

Figure 7–7

President Theodore Roosevelt and John Muir on Glacier Point above Yosemite Valley, California. (The Bettmann Archive)

1916 Congress created the National Park Service to manage the national parks and monuments for the enjoyment of the public, "without impairment." It was this clause that gave a different outcome to another battle, fought in the 1950s between environmentalists and dam builders over the construction of a dam within Dinosaur National Monument. Nobody could deny that to drown the canyon under 400 feet of water would "impair" it. This victory for conservation established the "use without impairment" clause as the firm backbone of legal protection afforded our national parks and monuments.

The Power of Public Awareness

Until 1970 the voice of environmentalists in the United States was heard primarily through societies such as the Sierra Club and the National Wildlife Federation. There was no generally perceived mass "environmental movement" until the spring of 1970, when the first Earth Day transformed the specialized interests of a few societies into a pervasive popular movement. On Earth Day 1970, an estimated 20 million people in the United States demonstrated their support of improvements in resource conservation and environmental quality. By Earth Day 1990, the movement had spread worldwide (Figure 7–9, page 148). Approximately 200 million people in 141 nations demonstrated to increase public awareness of the importance of individual efforts in sustaining the Earth ("Think globally, act locally").

Galvanized by ecological disasters such as the 1969 Santa Barbara oil spill and by overwhelming public support for the Earth Day movement, in 1970 the National Environmental Policy Act (NEPA) was signed into law. A key provision stated that before beginning any development project involving federal lands or funds, public agencies and private individuals and corporations would have to examine the likely environmental consequences. The required environmental impact statements (EISs) would be monitored by a newly established federal board, the Council on Environmental Quality, which reports directly to the president.

Because this council had no enforcement powers, the NEPA was thought to be innocuous, generally more a statement of good intentions than a regulatory policy. During the next few years, however, environmental activists took people, corporations, and the government to court to challenge their EISs or to use the statements to

(text continues on page 148)

(a)

(b)

Figure 7–8

Hetch Hetchy Valley in Yosemite. Before (a) and after (b) Congress approved a dam to supply water to San Francisco. (a, National Archives; b, Wallace Kleck/ Terraphotographics)

MEETING THE CHALLENGE

Old-Growth Forests of the Pacific Northwest: Jobs Versus the Environment

In western Oregon, site of the most commercially valuable public forest land in the United States, a drama involving people's jobs versus the environment unfolded during the late 1980s and early 1990s. At stake were thousands of jobs and the future of 3 million acres of old-growth (virgin) coniferous forest, along with the existence of organisms that depend on the forest. One of these forest animals, the northern spotted owl, came to symbolize the confrontation (see figure).

Biologists regard the few remaining old-growth forests of Douglas fir and spruce in the Pacific Northwest as living laboratories that demonstrate the complexity of a natural ecosystem that has not been extensively altered by humans. To environmentalists, the old-growth forests, with their 2000-year-old trees, are a national treasure to be protected and cherished. These stable forest ecosystems provide biological habitats for many species, including the northern spotted owl and 40 other endangered or threatened species. Like other, less well known species, the northern spotted owl is found primarily in old-growth forests. Provisions of the Endangered Species Act require the government to protect the habitat of endangered species so that their numbers can increase. Enforcement of this law required the court-ordered suspension in 1991 of logging where the owl lives—that is, in about 3 million acres of federal forest.

The timber industry bitterly opposed such an action, stating that thousands of jobs would be lost if the northern spotted owl habitat were to be set aside. Many rural communities in the Pacific Northwest do not have diversified economies; timber is their main source of revenue. Thus, a major confrontation over the future of the old-growth forest ensued between the timber industry and environmentalists. Strong feelings were expressed on both sides—witness the bumper sticker reading, "Save a logger, kill an owl."

The situation was more complex than simply jobs versus environment, however. The timber industry was already declining in terms of its ability to support people in the Pacific Northwest. During the decade between 1977 and 1987, logging in Oregon's national forests increased by over 15 percent, whereas employment dropped by 15 percent—an estimated 12,000 jobs—during the same period. The main cause of this decline was automation of the timber industry.

In addition, the timber industry in that region had not been operating *sustainably*—that is, they removed trees faster than the forest could regenerate. If the industry continued to log at their 1980s rates, most of the remaining old-growth forests would have disappeared within 20 years.

Fortunately, timber is not as important to the economy of the Pacific Northwest as it used to be, a change that started long before the northern spotted owl controversy. States in the Pacific Northwest have rapidly become diversified, and by the late 1980s, the timber industry's share of the economy in Oregon and Washington was less than 4 percent.

The northern spotted owl, an endangered species found only in old-growth forests in the Pacific Northwest. (Jack Wilburn © 1993 Animals Animals)

Although some had predicted economic disaster in 1991 when logging was blocked in the old-growth forests, Oregon's unemployment rate in 1994 was the lowest in 25 years. Oregon attracted several high-technology companies, which produced more jobs than had been lost from logging. (Interestingly, a 1994 analysis by the Institute for Southern Studies concluded that states that protect their natural resources and have good environmental records have the best long-term prospects for economic development. An important factor in Oregon's economic growth is its "quality of life," which in turn is a result of its forests and other natural resources.)

► *FOLLOW-UP*

President Clinton recognized that a broad new agenda was needed in the dwindling forests of the Pacific Northwest—an agenda that took into account the multiple issues and interests involved. In 1993, early in his first term of office, Clinton convened a Timber Summit in Portland, Oregon, with members of all sides of the issue (see figure). The Clinton Plan that arose from this summit represented a compromise that Clinton hoped would satisfy both environmental and timber interests. It was approved by a federal judge on December 22, 1994.

Thanks to a healthy infusion of federal aid to the area, some timber workers are being retrained for other kinds of careers. State programs, such as the "Jobs for the Environment" project funded by the state of Washington, have also helped reduce unemployment. Hundreds of former loggers are employed by "Jobs for the Environment" and similar programs to restore watersheds and salmon habitat in the forests they used to harvest.

As a result of the Clinton Plan, logging has resumed on federal forests of Washington, Oregon, and northern California. However, only about one-fifth of the logging that occurred during the 1980s will be permitted. The plan, along with previous congressional and administrative actions, reserves about 75 percent of federal timberlands to provide protection for the northern spotted owl and other species, especially salmon and other fishes. Fishes are protected by **riparian buffers**, no-logging areas that are being established along streams to reduce the sedimentation caused by soil erosion.

As happens with many compromises, neither environmentalists nor timber-cutting interests are happy with the Clinton Plan. Some environmentalists do not think the plan is scientifically sound and worry that it takes too many risks with conservation values. Timber-cutting interests are challenging the legality of the plan and trying to revoke or revise the laws that require protection of biodiversity. They have also asked Congress to pass legislation allowing a greater harvest of timber because of economic hardship. Still other proposals involve turning over a significant acreage of federal timberlands to local counties.

President Clinton and other delegates at the Timber Summit in Portland, Oregon, on April 2, 1993. (© Cynthia Johnson/Gamma Liaison)

Figure 7–9
School children join with environmentalists to celebrate Earth Day 1990 in Hong Kong. (Carl Ho/Reuter)

Table 7–3
Selected Environmental Incidents Since 1970

Year	Incident
1973	Arab oil embargo precipitated energy crisis.
	Construction of Tellico dam in Tennessee blocked by Endangered Species Act.
1978	Love Canal, New York, evacuated because of chemical waste contamination.
1979	Accident at Three Mile Island nuclear power plant in Pennsylvania almost caused a meltdown.
1984	Toxic chemical release at Union Carbide plant in Bhopal, India, killed several thousand people and injured as many as 200,000.
1985	Scientists first reported ozone hole over Antarctica.
1986	Times Beach, Missouri, evacuated because of chemical waste contamination.
	Meltdown at Chernobyl nuclear power plant in the Ukraine contaminated large area of Europe and exposed thousands of people to radiation.
1989	Exxon Valdez ran aground in Prince William Sound, spilling 11 million gallons of crude oil.
1991	Logging controversy ignited when court order blocked logging in 3 million acres of federal forests in Pacific Northwest to protect northern spotted owl.
	Iraqi Army set fire to Kuwaiti oil fields during Persian Gulf War.

block proposed development. The courts decreed that EISs had to be substantial documents that thoroughly analyzed the environmental consequences of anticipated projects on soil, water, and organisms and that EISs must be made available to the public for its scrutiny. These rulings put very sharp teeth into the law—particularly the provision for public scrutiny, which placed intense pressure on federal agencies to respect EIS findings (see Chapter 2).

The NEPA revolutionized environmental protection in this country. In addition to overseeing federal highway construction, flood and erosion controls, military projects, and many other public works, federal agencies own nearly one-third of the land in the United States. Their holdings include extensive fossil fuel and mineral reserves as well as millions of acres of public grazing land and public forests. Since 1970 very little has been done to any of them without some sort of environmental review.

During the period since Earth Day 1970, Congress has passed almost 40 major environmental laws that address a wide range of issues, such as endangered species, clean water, clean air, energy conservation, hazardous wastes, and pesticides. Some of these laws were motivated by environmental crises (Table 7–3). These laws greatly increased federal regulation of pollution, creating a tough interlocking mesh of laws to improve environmental quality.

However, the laws are not perfect (recall the chapter introduction). Economists and industries have argued that they make pollution abatement unduly complex and expensive. Nor have the laws always worked as intended. The Clean Air Act of 1977, for example, required coal-burning power plants to outfit their smokestacks with expensive "scrubbers" to remove sulfur dioxide from their emissions, but made an exception for tall smokestacks (Figure 7–10). This loophole led directly to the proliferation of tall stacks that have since produced acid rain throughout the Northeast. The Clean Air Act of 1990, described in Chapter 19, goes a long way toward closing such loopholes.

Figure 7–10
Tall smokestacks, which emit sulfur dioxide from coal-burning power plants, were permitted by the Clean Air Act of 1977. (Albert Copley/Visuals Unlimited)

GREEN HOTELS

Boston's Saunders Hotel Group, owners of the Park Plaza, Copley, and Lenox hotels, initiated an environmental program in 1989 that has taken surprising initiatives and benefitted the corporation as well as the environment.

- Thermopane windows reduce annual heating costs by $84,000.
- Toilets that used 7 gallons a minute have been replaced with ones that use 1.5 gallons; shower heads have been similarly replaced with low-flow models, and a laundry-water recycling system further reduces water use.
- Porcelain from old toilets is recycled as roadfill; old bathroom fixtures are sold as scrap metal.
- 2,400 phone books are recycled annually; enough paper is recycled to save 300 trees each year.
- Old bedspreads and linen tablecloths are turned into aprons and stuffed animals.
- Used furniture is donated to homeless shelters or sold to hotel employees.
- Only environmentally safe cleaning chemicals are used; aerosols and polystyrene are banned.
- Guests can make voluntary donations, which are matched by the hotels, to a fund benefitting environmental groups.
- Each year, the hotel group saves more than $142,000 in water, $96,000 in energy, and $41,000 in solid waste.

Legislative Approaches to Environmental Problems Have Been Effective Overall

On balance, despite imperfections, environmental command and control legislation has had positive and substantial effects. Since 1970:

- Eight national parks have been established, and the national wilderness system now totals more than 100 million acres.
- Millions of acres of farmland that are particularly vulnerable to erosion have been withdrawn from production, reducing soil erosion by more than 60 percent.
- Many previously endangered species are better off than they were in 1970, and the American alligator, California gray whale, and bald eagle have recovered enough to be removed from the endangered species list. (However, dozens of other species have suffered further decline or extinction since 1970.)

Although we still have a long way to go, pollution control efforts have been particularly successful. Since 1970:

- The level of lead in the air has dropped by 98 percent with the phaseout of leaded gasoline.
- The amount of toxic chemicals released into water and air from industrial sources has declined by 43 percent.
- Hydrocarbon emissions from motor vehicles have declined from 10.3 million tons to 5.5 million tons.
- Emissions of sulfur dioxide, carbon monoxide, and soot have been reduced by more than 30 percent.
- The number of secondary sewage treatment facilities (in which bacteria break down organic wastes before the water is discharged into rivers and streams) has increased by 72 percent in the last ten years.
- Certain toxic chemicals, such as DDT, asbestos, and dioxins, have been banned.

SUMMARY

1. Economists view pollution in a market economy as a failure in pricing. In addition, the impact of pollution on third parties is not taken into account in the marketplace.

2. From an economic point of view, the appropriate amount of pollution is a tradeoff between harm to the environment and inhibition of development. The cost in environmental quality of a unit of pollution that is emitted into the environment is known as the marginal cost of pollution. The cost to dispose of a unit of pollution in a nonpolluting way is called the marginal cost of pollution abatement.

3. According to economists, the use of resources for pollution abatement should be increased only until the cost of abatement equals the cost of the damage done by the pollution. This results in the optimum amount of pollution—the amount that is economically most desirable.

4. There are two main flaws in the economic approach to pollution: the true cost of environmental damage by pollution is difficult to determine, and the risks of unanticipated environmental catastrophe are not taken into account in assessing the potential environmental damage of pollution.

5. Market-oriented strategies to control pollution include emission charges and waste-discharge permits. An emission charge policy controls pollution by charging the polluter for each given unit of emissions—that is, by establishing a tax on pollution. As the imposed cost—that is, the emission charge—on the supply of permissible pollution rises, the "demand" for pollution falls. Pollution can also be controlled by waste-discharge permits, which allow the holder to pollute a given amount; holders are not allowed to produce more emissions than are sanctioned by their permits, although they can buy and sell permits as needed.

6. Government controls pollution by imposing legal limits on amounts of permissible pollution—that is, by command and control. Such laws began in this country in 1970, with the passage of the National Environmental Policy Act. By requiring environmental impact statements that are open to public scrutiny, this law initiated serious environmental protection in the United States. Later laws, such as the Endangered Species Act, the Clean Air Act, and the Clean Water Act, have added to the environmental safety net, with some success.

THINKING ABOUT THE ENVIRONMENT

1. Discuss the pros and cons of the 1995 unfunded mandate law as it relates to future environmental legislation.

2. Does your life style reflect an environmental ethic or an economic ethic? Elaborate.

3. How would your life be different if society were run under an environmental ethic rather than under an economic ethic?

4. If you were a member of Congress, what legislation would you introduce to deal with each of the following problems?
   a. Poisons from a major sanitary landfill are polluting your state's groundwater.
   b. Acid rain from a coal-burning power plant in a nearby state is harming the trees in your state. Loggers and foresters are upset.
   c. There is a high incidence of cancer in the area of your state where heavy industry is concentrated.

5. How would an economist approach each of the problems listed in question 4? How would an environmentalist?

6. How might pollution abatement legislation actually cause a net increase in the number of jobs in an area?

* 7. Interpret the following graphs. In each case, examine the amount of pollution indicated by the red dashed line. Is it more or less than the economically optimum amount of pollution? Is the marginal cost of pollution higher or lower than if the pollution level was at its optimum? Is the cost of pollution abatement higher or lower than if the pollution level were at its optimum?

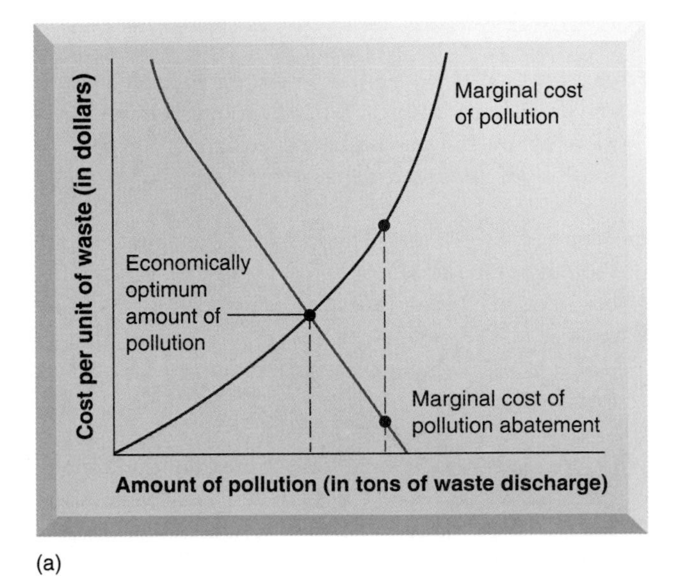

(a)

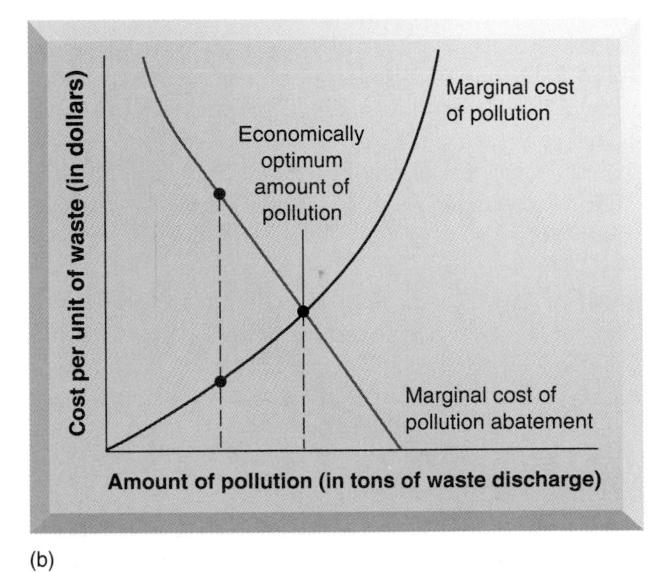

(b)

RESEARCH PROJECTS

Contact your Congressperson's office and ask for his or her record on environmental issues. Then compare this statement to an environmental organization's rating of the Congressperson. Ask the Congressperson to explain discrepancies, if any.

As a library assignment, study a major environmental impact statement that is at least five years old and assess its accuracy of projected cost and benefits. Then update the currency of the project and explain how the EIS affected the project's status. Identify any new factors that might be relevant if a new EIS regarding the same project were to be undertaken now.

Some scholars contend that government institutions are ineffective in dealing with environmental issues that could affect future generations because political time frames are so short. After all, ecosystem issues, such as evolution of biological diversity, biogeochemical cycles, and succession are measured in centuries or even longer, whereas political time frames are measured in years or months, especially as an election approaches. Investigate the U.S. government's response to the threat of global climate change and write a brief essay supporting or refuting this scholarly observation.

Most biologists think that old-growth forests cannot be logged sustainably because it would take too long (at least 800 to 1000 years) for logged trees to replace themselves. They view the northern spotted owl controversy as an issue over how much of these ancient forests we want to maintain intact on public land. Investigate the extent of logging on public land in the mid-1990s, both in the Pacific Northwest and nationwide. Determine what percentage of the public forest land is protected from logging, and the nature of this "protection."

On-line materials relating to this chapter are on the ► World Wide Web at **http://www.saunderscollege.com/lifesci/environment2** (select Chapter 7 from the Table of Contents).

SUGGESTED READING

Arrow, K., et al. "Economic Growth, Carrying Capacity, and the Environment," *Science* Vol. 268, April 28, 1995. Eleven distinguished scholars warn that continued economic growth does not take into account the limited carrying capacity of the planet's ecosystems.

Dolan, E. G., and D. E. Lindsey. *Economics*. 7th ed. Hinsdale, Ill.: Dryden Press, 1994. Chapter 32, "Externalities and Environmental Policy," presents the economists' view of pollution in an understandable fashion. Numerous examples help clarify technical economic concepts.

Edwards, M. "Lethal Legacy: Pollution in the Former U.S.S.R.," *National Geographic* Vol. 186, No. 2, August 1994. This article contains vivid photographs that document the extent of pollution and its accompanying health problems in the former Soviet Union.

"Environment and the Economy," *Science* Vol. 260, 25 June 1993. This issue contains eleven special reports on maintaining a healthy economy and a sustainable environment.

Graham, F., Jr. "Earth Day: 25 Years," *National Geographic*, April 1995. Highlights several environmental activists who have made substantial contributions.

O'Riordan, T., et al. "The Legacy of Earth Day: Reflections at a Turning Point," *Environment* Vol. 37, No. 3, April 1995. Provides an overview of environmental accomplishments during the past 25 years as well as what remains to be done.

Porter, M. E., and C. van der Linde. "Green and Competitive: Ending the Stalemate," *Harvard Business Review*, September–October 1995. This article is based on the premise that environmental standards can trigger innovations that lower the cost of a product or improve its value.

Power, T. M. "The Wealth of Nature," *Issues in Science and Technology*, Spring 1996. Environmental quality is more significant to economic development in western states than mining, logging, or ranching.

de Stiguer, J. E. "Three Theories from Economics about the Environment," *BioScience* Vol. 45, No. 8, September 1995. Examines three economic theories (Malthus, neoclassical economics, and Mill's steady-state theory) to determine which is most appropriate for today's world.

Woodard, C. "Cheap Solution to Rife Pollution Eludes Eastern Europe," *The Christian Science Monitor*, August 9, 1996. Many of Eastern Europe's environmental problems, which first came to light with the end of communism, are unresolved.

Zachary, G. P. "A 'Green Economist' Warns Growth May Be Overrated," *The Wall Street Journal*, June 25, 1996. Examines the views of Herman Daly, an economist at the University of Maryland who believes that "sustainable growth" is an oxymoron.

► **Pedestrians and traffic pack a street in Shanghai, China.**
(Jeff Greenberg/Photo Researchers, Inc.)

PART 3

A Crowded World

A Population Policy for the World

Dr. (Mrs.) Nafis Sadik is the Executive Director of the United Nations Population Fund (UNFPA) and holds the rank of Under-Secretary-General. On her appointment in 1987, she became the first woman to hold such a rank in the United Nations system. The UNFPA is the world's largest source of multilateral assistance to population programs; and as Executive Director, Dr. Sadik manages a worldwide staff of 900 and a budget of more than $300 million.

Dr. Sadik—a national of Pakistan—was educated in India where her father was a government finance officer. Although the roles for women were restrictive when she was a child, her father encouraged her interest in medicine. After obtaining a medical degree, Dr. Sadik served an internship in gynecology and obstetrics at City Hospital in Baltimore, Maryland. She completed further studies in public health and family planning at The Johns Hopkins University and held the post of Research Fellow in physiology at Queens University, Kingston, Ontario. ◄

NAFIS SADIK

What prompted your interest in family planning? How did you get started in your family-planning career?

After completing my residency I returned to Pakistan where I became a civilian doctor in army hospitals. While there, I worked with women in rural villages as part of the army's community service program. I would tell new mothers they shouldn't get pregnant again for two years, but the women would say that their husbands or mothers-in-law had other ideas—especially if the baby was a girl. It made me realize that these poor women really had no control over their lives. In Pakistan in the 1950s, there were no family planning services. I decided to start a program and asked my commanding officer for money to buy contraceptives. He nearly fell off his chair, but finally gave me some money with a warning that if there were any complaints he would say it was all my own doing.

What does the UNFPA do? What are the objectives of the organization?

The United Nations Population Fund has several objectives and three program areas. The objectives include providing reproductive health care, including family planning and sexual health; formulating population strategies in support of sustainable development; and advocacy for issues related to population, reproductive health, and the autonomy and equality of women. Another goal is to advance the strategy endorsed by the 1994 International Conference on Population and Development (ICPD), which emphasized the inseparability of population and development and focused on meeting individuals' needs rather than demographic targets. The key to this new approach is empowering women and expanding access to education, health services, and employment opportunities. Additionally, we want to promote cooperation and coordination among U.N. organizations, governments, non-governmental organizations (NGOs), and the private sector in addressing issues of population and development, reproductive health, gender equality, and women's empowerment.

The UNFPA is concerned with sustainable development. How would you define sustainable development?

The term "sustainable development" is used to communicate the idea that the processes by which people satisfy their needs and improve their quality of life in the present should not compromise the ability of future generations to meet their own needs. At any level of development, human impact on the environment is a function of population size, per-capita consumption, and the environmental damage caused by the technology used to produce what is consumed. Currently, people living in developed countries have the greatest impact on the global environment. But as standards of living rise in developing countries, the environmental consequences of population growth in these countries will be amplified.

Is this more of an issue for developed countries that use a disproportionate amount of world resources?

The debate over environmental challenges cannot be reduced to assigning blame. Patterns of consumption and resource use in the industrialized countries of the North are certainly responsible for much environmental degradation in both the North and South. But rapidly growing populations, whatever their levels of consumption, also place a greater burden on resources and the environment. Ensuring sustainability will require people to make changes, in both the way they think about their environment and how they live in it. The high-consumption, high-waste lifestyle of the top fifth of the world's population, most of whom live in the North, cannot continue without imperiling the right of the bottom fifth to satisfy their basic needs.

What can be done? What kinds of programs does the UNFPA support to stabilize world population growth?

The gradual slowing of population growth already under way is part of the answer to this environmental dilemma. With slower growth rates, countries will have more time to prepare for the still-inevitable, if smaller, population increases to come—time to build schools, dig sewers, and lay water pipes.

A second part of the answer lies in the kind of North-South partnership envisioned in both Agenda 21, the sustainable development plan produced at the 1992 Earth Summit, and the Programme of Action emerging from the ICPD in 1994. Developed countries need to push their industries toward efficiency and developing technologies that minimize damage to natural systems. For both North and South, the ultimate goal should be sustainability in all areas of economic activity, including agriculture, industry, forestry, fisheries, transportation, and tourism.

> **Ensuring that women and their partners have the right to choose will support a global trend toward smaller families and help countries find a balance between population and resources.**
>
> —*Nafis Sadik*

North-South cooperation is also vital to success in ending absolute poverty, a critical element of the answer to our ongoing environmental dilemma. For those eking out a living, whether on a steeply sloped hillside in the Philippines or in an illegal shanty town in Istanbul, environmentally sound practices are a luxury and no choice at all. As one Egyptian environmental activist said, "You can't ask people to dispose of garbage properly if there's nowhere to put it; you can't really talk about water conservation without the technology to make it happen.... The fact is that there are few alternatives to the way most people currently live their lives."

You have said that empowering women is critical to the success of slower population growth and sustainable development. How so?

In every part of the world, women are primarily responsible for the fetching of water, gathering of fuel, and the preparation of food, in both rural and urban areas. Yet women only rarely own the resources with which they labor. In addition, those working to change environmentally unsound practices or improve agricultural productivity typically direct their efforts at men rather than women. Women make up two-thirds of the world's poorest people and are nearly twice as likely as men to be illiterate. They receive less education and less food, and have few legal rights. Removing the constraints on women's effectiveness as resources managers—through education, access to credit and land, and the enforcement of legal rights—would, therefore, not only benefit women as individuals, but also contribute to the environmental and economic well-being of their families and communities.

There is also a pressing need for women to be able to exercise their reproductive rights—reproductive self-determination—and to have access to reproductive health care. Reproductive health care includes, among other measures, prenatal care; assisted deliveries; postnatal care, including referrals for complications of pregnancy; and screening for sexually-transmitted diseases, including HIV/AIDS. Improved access to reproductive health, including family planning, would prevent many of the 70,000 annual deaths due to unsafe abortions, and would slow the spread of sexually transmitted diseases and HIV/AIDS. Most of the 585,000 maternal deaths each year from pregnancy related causes could be averted with relatively low-cost improvements in health care systems, by expanding emergency obstetric care, and by making modern contraceptive services more widely available.

In addition to adequate health and the ability to determine the number and spacing of one's children, reproductive self-determination involves the right to marry voluntarily and to form a family, and freedom from sexual violence and coercion. Reproductive rights and reproductive health care are critical both to the empowerment of individual women, and to the economic and social life of communities, nations, and the world. Ensuring that women and their partners have the right to choose will support a global trend toward smaller families and help countries find a balance between population and resources.

▶ Web site: **www.unfpa.org**

The Population Reference Bureau's World Population Data Sheet, a poster located within this text, provides population data for all countries and geopolitical entities with populations of 150,000 or more.

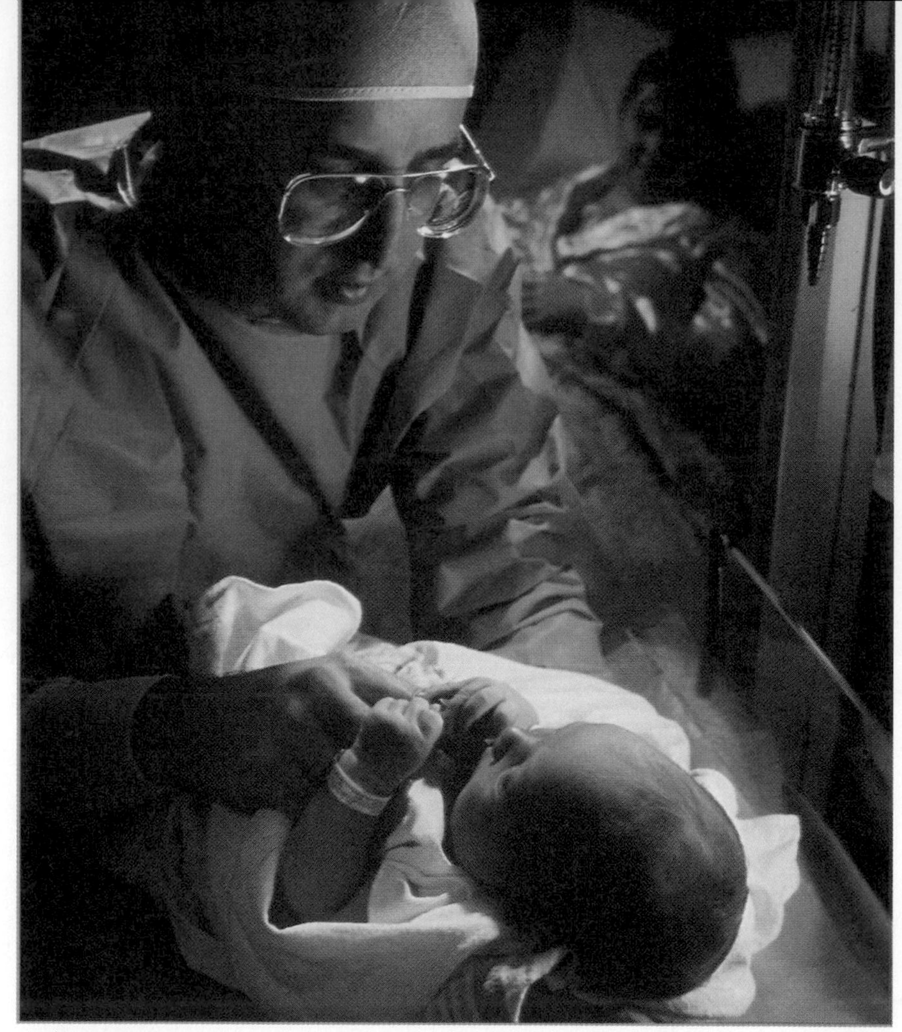

A doctor examines a newborn infant. Medical advances have contributed to a decline in infant deaths in many countries. (Charles Gupton/Uniphoto)

Understanding Population Growth

If you are a typical college student, you were probably born in the mid-1970s. At that time the human population was slightly less than 4 billion. Today there are about 5.8 billion humans, and it is likely that our numbers will increase to 8 billion or more during your lifetime. This tremendous increase in population is a measure of our biological success as a species. Humans have been able to provide more food and better nutrition for themselves and their offspring by increasing the productivity of agriculturally important crops and animals through selective breeding. We have made great strides in the fight against diseases with life-saving sanitation practices, ever-advancing medical techniques, and newly developed medicines. All of these factors have not only increased our numbers, but have also increased the likelihood that we will live longer.

Unfortunately, our biological success has created innumerable problems for us and the other organisms on our planet—we are in danger of overwhelming the Earth with too many people. The Earth has limited resources, and the human population is using up, encroaching upon, fouling, and wasting them. Pollution, extinction of species, degradation and loss of natural resources, and depletion of energy reserves in today's world are all related to human population growth.

All living things tend to reproduce in greater numbers than can survive, but an organism's environment—which includes food, water, and living space—limits the size of its population. If the human population continues to increase at its current rate, we may reach the limits imposed by our world environment. (Many scientists feel there is substantial evidence we have already passed the Earth's limits.) If that occurs, the human species could be significantly reduced in numbers or even destroyed. In this chapter, we focus on the dynamics of population growth characteristic of organisms and then describe the current state of the human population. Chapter 9 examines the consequences of continued population growth and explores ways to limit the expanding world population.

PRINCIPLES OF POPULATION ECOLOGY

Because the human population is central to so many environmental problems and their solutions, it is important that we understand how populations increase or decrease. The biological principles that affect the populations of other organisms also apply to human populations.

Population ecology deals with the number of individuals of a particular species that are found in an area and how and why those numbers change or remain fixed over time. Population ecologists try to determine the population processes that are common to all populations. They study how a population responds to its environment, that is, to competition for resources, predation, disease, and other environmental pressures. Population growth cannot increase indefinitely because of such environmental controls.

Population Density

By itself, the size of a population tells us relatively little. Population size is more meaningful when the boundaries of the population are defined. Consider, for example, the difference between 1000 mice in 100 hectares (250 acres) and 1000 mice in one hectare (2.5 acres).

Sometimes a population is too large to study in its entirety. Such a population is examined by sampling a part of it and then expressing the population in terms of density—as, for example, the number of grass plants per square meter or the number of cabbage aphids per square inch of cabbage leaf. **Population density**, then, is the number of individuals of a species per unit of area or volume at a given time.

What Causes Populations to Change in Size?

Populations of organisms, whether they are sunflowers, eagles, or humans, change over time. On a *global* scale, this change is due to two factors: the number of births and the number of deaths in the population (Figure 8–1a). For humans, the **birth rate** is usually expressed as the number of births per 1000 people per year, and the **death rate** is the number of deaths per 1000 people per year.

The rate of change, or **growth rate (*r*)**, of a population is the birth rate (*b*) minus the death rate (*d*). Growth rate is also referred to as **natural increase** in human populations.

$$r = b - d$$

As an example, consider a hypothetical human population of 10,000 in which there are 200 births per year, or 20 births per 1000 people, and 100 deaths per year, or 10 deaths per 1000 people.

$$r = \underbrace{\frac{20}{1000}}_{b} - \underbrace{\frac{10}{1000}}_{d}$$

$$r = 0.02 - 0.01 = 0.01, \text{ or 1 percent per year}$$

Another way to express the growth rate of a population is to determine the **doubling time**, which is the amount of time it would take for the population to double in size, assuming that its growth rate does not change. Doubling time (t_d) is approximated by dividing the number 70 by the growth rate expressed as percent per year.[1]

$$t_d = \frac{70}{r}$$

In our example, in which $r = 1$ percent per year, the doubling time would be 70/1 = 70 years.

In addition to birth and death rates, migration, which is movement from one region or country to another, must be considered when changes in populations on a *local* scale are examined. There are two types of migration: **immigration**, by which individuals enter a population and thus increase the size of the population, and **emigration**, by which individuals leave a population and thus decrease

[1]This is a simplified formula for doubling time. The actual formula involves mathematics that is beyond the scope of this text.

its size. The growth rate of a local population of organisms must take into account birth rate (b), death rate (d), immigration (i), and emigration (e) (Figure 8–1b). The growth rate is equal to the value of birth rate minus death rate, plus the value of immigration minus emigration:

$$r = (b - d) + (i - e)$$

For example, the growth rate of a population of 10,000 that has 100 births (10 per 1000), 50 deaths (5 per 1000), 10 immigrants (1 per 1000), and 100 emigrants (10 per 1000) in a given year would be calculated as follows:

$$r = \left(\underbrace{\frac{10}{1000}}_{b} - \underbrace{\frac{5}{1000}}_{d}\right) + \left(\underbrace{\frac{1}{1000}}_{i} - \underbrace{\frac{10}{1000}}_{e}\right)$$

$$r = (0.01 - 0.005) + (0.001 - 0.01)$$

$$r = 0.005 - 0.009 = -0.004, \text{ or } -0.4 \text{ percent per year}$$

Can you calculate the doubling time for this population?[2]

Mini-Glossary of Population Terms

birth rate: The number of births per 1000 individuals.

death rate: The number of deaths per 1000 individuals.

growth rate: The natural increase of a population, expressed as percent per year.

doubling time: The number of years it will take a population to double in size, given its current growth rate.

immigration: The migration of individuals into a population from another area or country.

emigration: The migration of individuals from a population, bound for another area or country to live.

zero population growth: The condition when a population is no longer increasing because the birth rate equals the death rate.

infant mortality rate: The number of infant deaths per 1000 live births.

replacement-level fertility: The number of children a couple must have to "replace" themselves.

total fertility rate: The average number of children born to each woman during her lifetime.

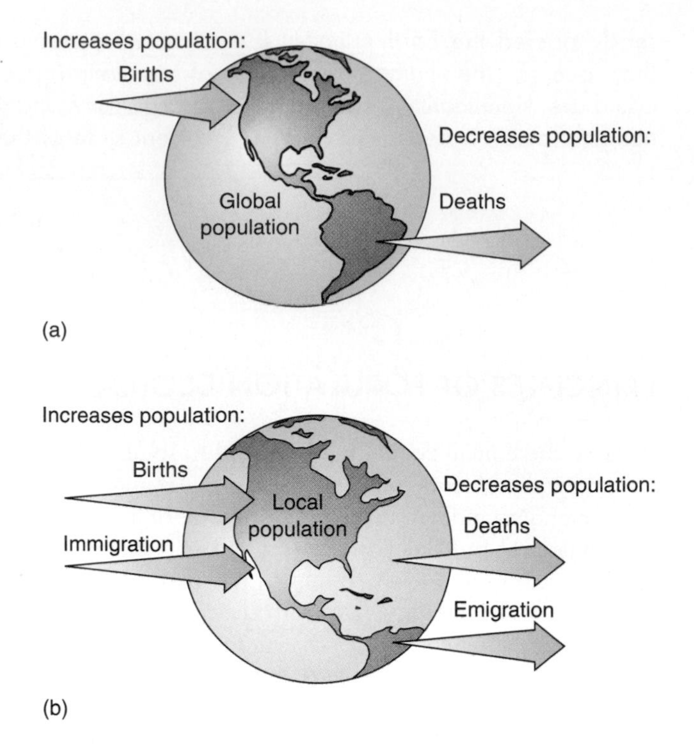

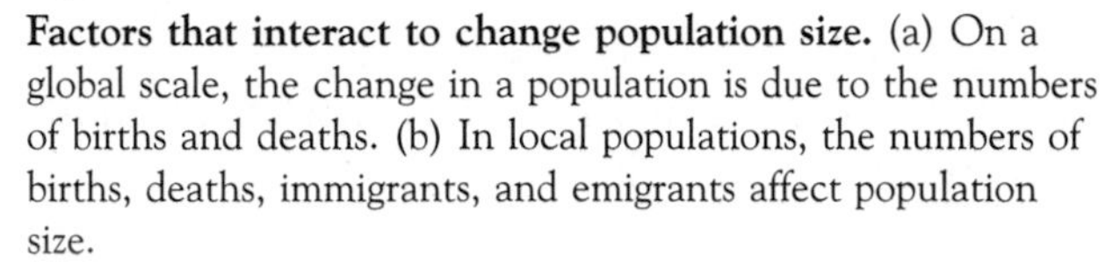

Figure 8–1
Factors that interact to change population size. (a) On a global scale, the change in a population is due to the numbers of births and deaths. (b) In local populations, the numbers of births, deaths, immigrants, and emigrants affect population size.

Maximum Population Growth

The maximum rate at which a population could increase under ideal conditions is known as its **biotic potential**. Different species have different biotic potentials. A particular species' biotic potential is influenced by several factors, including the age at which reproduction begins, the percentage of the life span during which the organism is capable of reproducing, and the number of offspring produced during each period of reproduction.

Generally, larger organisms, such as humans and elephants, have lower biotic potentials, whereas microorganisms have the greatest biotic potentials. Under ideal conditions, certain bacteria can reproduce by splitting in half every 20 to 30 minutes. At this rate of growth, a single bacterium would increase to a population of more than 1 million in just 10 hours (Figure 8–2a), and the population from a single individual would exceed 1 billion in 15 hours! If one were to plot the population number versus time, the graph would have the "J" shape that is characteristic of **exponential growth**, the constant reproductive rate that occurs under optimal conditions (Figure 8–2b). When a population grows exponentially, the larger that population gets, the faster it grows. It doubles, then doubles again, then again, and so on.

[2]Answer: It is not possible to calculate the doubling time for this population because its natural increase is a negative number. This population is declining, not increasing.

Time (hours)	Number of bacteria
0	1
0.5	2
1.0	4
1.5	8
2.0	16
2.5	32
3.0	64
3.5	128
4.0	256
4.5	512
5.0	1,024
5.5	2,048
6.0	4,096
6.5	8,192
7.0	16,384
7.5	32,768
8.0	65,536
8.5	131,072
9.0	262,144
9.5	524,288
10.0	1,048,576

(a)

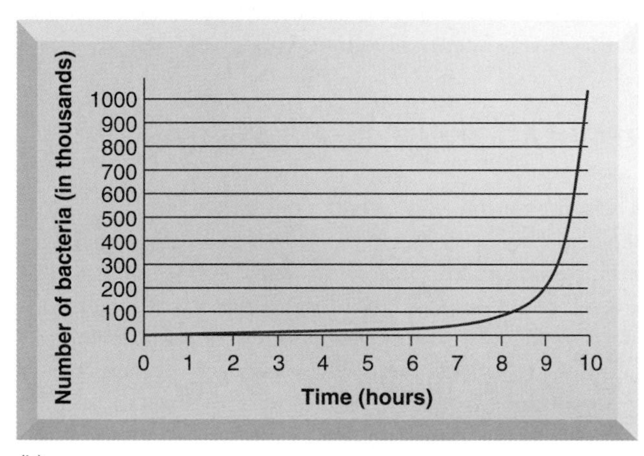

(b)

Figure 8–2

Exponential growth. (a) When bacteria divide every 30 minutes, their numbers increase exponentially. This set of figures assumes a zero death rate, but even if a certain percentage of each generation of bacteria died, exponential growth would still occur; it would just take longer to reach the very high numbers. (b) When these data are graphed, the curve of exponential growth has a characteristic "J" shape.

Regardless of which organism one is considering, whenever the population is growing at its biotic potential, population number plotted versus time gives a curve of the same shape. The only variable is time. That is, it may take longer for an elephant population than for a bacterial population to reach a certain size, but both populations will always increase exponentially under ideal conditions.

ENVIRONMENTAL RESISTANCE AND CARRYING CAPACITY

Certain populations may exhibit exponential growth for a short period of time. However, organisms cannot reproduce indefinitely at their biotic potentials, because the environment sets limits, which are collectively called **environmental resistance**. Environmental resistance includes such unfavorable environmental conditions as the limited availability of food, water, shelter, and other essential resources as well as limits imposed by disease and predation. Using the earlier example, bacteria would never be able to reproduce unchecked for an extended period of time, because they would run out of food and living space, and poisonous body wastes would accumulate in their vicinity. With crowding, they would also become more susceptible to parasites and predators. As their environment changed, their birth rate (*b*) would decline and their death rate (*d*) would increase due to shortages of food, increased predation, increased competition, and stress. The environmental conditions might worsen to a point where *d* would exceed *b* and the population would decrease. The number of organisms in a population, then, is controlled by the ability of the environment to support it. As the population increases, so does environmental resistance.

Over longer periods of time, the growth rate for most organisms decreases to around zero. This leveling out occurs at or near the limit of the environment's ability to support a population. The **carrying capacity (K)** represents the largest population that can be maintained for an indefinite period of time by a particular environment.

When a population regulated by environmental resistance is graphed over longer periods of time (Figure 8–3), the curve has a characteristic "S" shape. The curve shows the population's initial exponential increase (note the "J" shape at the start, when environmental resistance is low), followed by a leveling out as the carrying capacity of the environment is approached. Although the S-curve is a simplification of actual population changes over time, it does appear to fit the population growth observed in many populations that have been studied in the laboratory and a few that have been studied in nature. For example, G. F. Gause grew a single species, *Paramecium caudatum*, in a test tube (see Figure 4–9). He supplied a constant but limited amount of food daily and replenished the media occasionally to eliminate the buildup of metabolic wastes. Under these conditions, as shown in Figure 8–3, the population of *P. caudatum* increased exponentially at first. The paramecia became so numerous

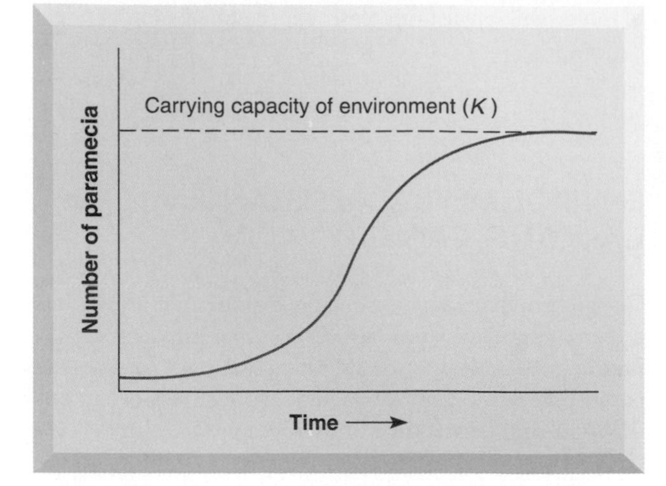

Figure 8–3

In many laboratory studies, including Gause's investigation with *Paramecium caudatum*, exponential population growth slows as the carrying capacity of the environment is approached. This produces a curve with a characteristic "S" shape.

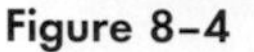

Figure 8–4

A moose browses on vegetation along the shoreline of Isle Royale in Lake Superior. Individual moose compete with one another for food during the long, cold winter months. (Rolf O. Peterson, Michigan Technological University)

that the water was cloudy with them. But then their growth rate declined and their population leveled off.

Sometimes, when a population exceeds the carrying capacity and environmental degradation results, a population crash occurs. The moose population on Isle Royale, the largest island in Lake Superior, provides a vivid example (Figure 8–4). Isle Royale differs from most islands in that large mammals can walk to it when the lake freezes over in winter. The minimum distance to be walked is 24 kilometers (15 miles), however, so this has happened infrequently. Around 1900, some moose reached the island for the first time. By 1935 the moose population on the island had increased to about 3000 and consumed almost all the edible vegetation. In the absence of this food resource, nearly 90 percent of the moose starved. By 1948 the population had recovered to its peak number once again, and again most of the moose starved.

Carrying Capacity Is Not Fixed

Different environments vary in their carrying capacity, and the carrying capacity of a single habitat may vary from season to season or year to year. As an example, consider red grouse populations in northwest Scotland at two locations only 2.5 kilometers (1.6 miles) apart. At one location the population density remained stable during a three-year period, but at the other it almost doubled in 2 years and then declined to its initial density once again. The reason was likely a difference in habitat. The area where the population density increased had been burned, and young plant growth produced after the burn was beneficial to the red grouse. So carrying capacity for a given organism is determined in large part by external factors in the environment, which are variable.

REGULATION OF POPULATION SIZE

The observation that small populations tend to grow rapidly, whereas larger populations grow less rapidly or even decline in size suggests that certain mechanisms operate to regulate population size. Factors that affect population size fall into two categories, density-dependent and density-independent. These two sets of factors vary in importance from one species to another, and in most cases probably interact in complex ways to determine the size of a population.

Density-Dependent Factors

If a change in population density alters how an environmental factor affects a population, then the environmental factor is said to be a **density-dependent factor**. As population density increases, density-dependent factors tend to slow population growth by causing an increase in death rate and/or a decrease in birth rate. The effect of these density-dependent factors on population growth increases as the population density increases; that is, density-dependent factors affect a larger proportion, not just a larger number, of the population. Density-dependent factors can also exert an effect on population growth when population density declines. A decrease in population density results in density-dependent factors enhancing population growth by decreasing death rate

FOCUS ON

Predator-Prey Dynamics on Isle Royale

As discussed earlier in the chapter, during the early 1900s, a small herd of moose wandered across the ice of frozen Lake Superior to an island, Isle Royale. In the ensuing years, until 1949, moose became successfully established there, although the population experienced oscillations. The story became more complex beginning in 1949, when a few Canadian wolves wandered across the frozen lake and discovered abundant moose prey on the island. The wolves also remained and became established.

Moose have been the wolf's primary winter prey on Isle Royale. Wolves hunt in packs by encircling a moose and trying to get it to run so it can be attacked from behind. (The moose is the wolf's largest, most dangerous prey. A standing moose is more dangerous than one that is running because when standing, it can kick and slash its attackers with its hooves.)

Since 1958, wildlife biologists have studied the populations of moose and wolves on Isle Royale (see figure). Not unexpectedly, they found that the two populations fluctuated over the years. Generally, as the population of wolves increased, the population of moose declined. Wolves tended to reduce the population of moose, but they did not eliminate them from the island. Studies indicate that wolves primarily feed on the very old and very young in the moose population. Healthy moose in their peak reproductive years are not eaten.

Despite the fact that both populations appeared to be in a state of dynamic equilibrium, a new episode recently began in the Isle Royale story. The wolf population has plunged, from 50 animals in 1980 to 16 animals in 1995. As expected, the moose population is increasing; in 1995 there were 2424 moose on Isle Royale. Biologists think that the extreme genetic uniformity of the wolf population is one of the reasons for its decline. Genetic studies of the Isle Royale wolves indicate that they are all descended from the same female.

Populations that lack genetic variability cannot adapt well to changing environmental conditions. Such populations often have a low reproductive success. In 1994, for example, only two wolf pups were born on the island. A genetically uniform population is also more susceptible to disease. Analysis of their blood revealed the presence of antibodies to canine parvovirus, indicating that the wolves have been exposed to this deadly disease.

It is not known if the wolf population on Isle Royale will recover or disappear altogether. Given their genetic uniformity, however, the outlook for their long-term survival is poor.

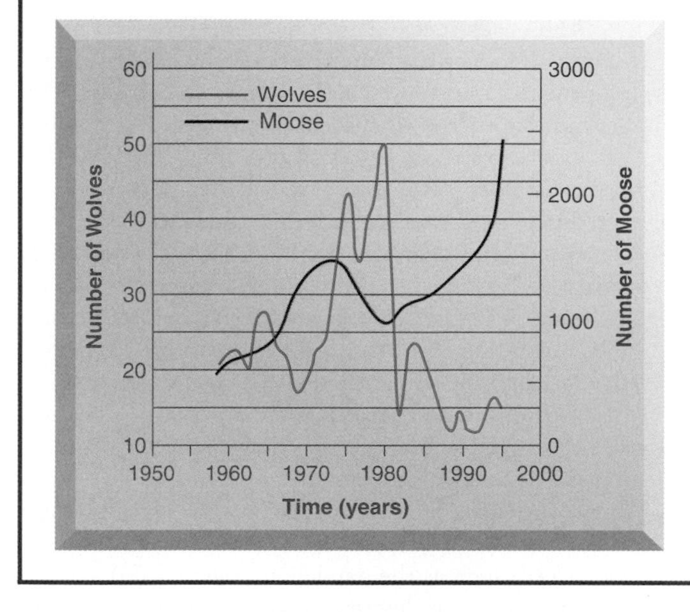

Fluctuations in the populations of wolves and moose on Isle Royale, 1959 to 1995. (Rolf O. Peterson, Michigan Technological University)

and/or increasing birth rate. Density-dependent factors, then, tend to help maintain a population at a relatively constant size that is near the carrying capacity of the environment.

Predation, disease, and competition are examples of density-dependent factors. As the density of a population increases, for example, predators are more likely to find an individual of a given prey species (see Focus On: Predator-Prey Dynamics on Isle Royale). Also, when population density is high, the members of a population encounter one another more frequently, and the chance of their transmitting infectious disease organisms increases.

Most studies of density dependence have been conducted in laboratory settings where all density-dependent factors except one are controlled experimentally. But populations in natural settings are exposed to a complex set of variables that continually change. As a result, in natural communities it can be difficult to evaluate the relative effects of different density-dependent factors. Ecologists from the University of California at Davis noted that on tropical islands inhabited by lizards, few spiders occur, whereas more spiders and more species of spiders are found on lizard-free islands. Deciding to study these observations experimentally, the researchers staked out plots of vegetation (mainly sea grape shrubs) on Bahamian islands and enclosed some of them with lizard-proof screens. Some of the plots were emptied of all lizards, and web-building spiders were introduced into all the plots. At the end of a two-year study period, spider population densities averaged 2.5 times higher in the lizard-free enclosures than in enclosures with lizards. Moreover, the species diversity of spiders was higher in the lizard-free areas. Even in this relatively simple experiment, the results may be explained by a combination of two density-dependent factors, predation (lizards eat spiders) and competition (lizards compete with spiders for insect prey). In this experiment, the effects of the two density-dependent factors in determining spider population size cannot be evaluated separately.

Density-Independent Factors

Any environmental factor that affects the size of a population but is not influenced by changes in population density is called a **density-independent factor**. Random weather events that reduce population size serve as density-independent factors. These often affect population density in unpredictable ways. A severe blizzard, hurricane, or fire, for example, may cause extreme and irregular reductions in a vulnerable population and thus might be considered largely density independent.

Consider a density-independent factor that influences mosquito populations in arctic environments. These insects produce several generations per summer and achieve high population densities by the end of the season. A shortage of food does not seem to be a limiting factor for mosquitoes, nor is there any shortage of ponds in which to breed. Instead, winter puts a stop to the skyrocketing mosquito population. Not a single adult mosquito survives winter, and the entire population must grow afresh the next summer from the few eggs and hibernating larvae that survive. Thus, severe winter weather is a density-independent factor that affects arctic mosquito populations.

It is difficult to envision many density-independent limiting factors that bear absolutely *no* relationship to population density. Social animals, for example, are often able to resist dangerous weather conditions by means of their collective behavior, as in the case of sheep huddling together in a snowstorm. In this case it appears that the greater the population density, the better their ability to resist the environmental pressure of a snowstorm.

REPRODUCTIVE STRATEGIES

Imagine an organism that is a perfect "reproductive machine." In other words, this hypothetical organism produces the maximum number of offspring, and the majority of these offspring survive to reproductive maturity. Such an organism would have to mature soon after it was born so that it could begin reproducing at an early age. It would reproduce frequently throughout its life and produce large numbers of offspring each time. Further, it would have to be able to care for its young in order to assure their survival.

In nature, such an organism does not exist because if an organism puts all its energy into reproduction, it would not expend any energy toward ensuring its own survival. Animals must use energy to hunt for food, and plants need energy to grow taller than surrounding plants to obtain adequate sunlight. Nature, then, requires organisms to make tradeoffs in the expenditure of energy. Organisms, if they are to be successful, must do what is required to survive as individuals as well as to survive as a species.

Thus, each species has its own life history strategy—its own reproductive characteristics, body size, habitat requirements, migration patterns, and so on—designed around this energy compromise. Although many different life history strategies exist, organisms often fall into two main groups with respect to their reproductive characteristics, *r*-selected species and *K*-selected species.

Populations described by the concept of ***r* selection** have a strategy in which evolution has selected traits that contribute to a high growth rate. Recall that *r* designates the growth rate. Because such organisms have a high *r*, they are known as ***r* strategists** or ***r*-selected species**. Small body size and large broods are typical of many *r* strategists (Table 8–1), which are usually opportunists found in variable, temporary, or unpredictable environments where the probability of long-term survival is minimal. Some of the best examples of *r* strategists are common weeds such as the dandelion (Figure 8–5a).

In populations described by the concept of ***K* selection**, evolution has selected traits that maximize the chance of surviving in an environment where population size is near the carrying capacity (*K*) of the environment. Because the population of a species with *K* selection is maintained at or near the carrying capacity, its individuals have little need for a high reproductive rate. These organisms, called ***K* strategists** or ***K*-selected species**,

Table 8–1
A Comparison of Selected Features of r-Selected and K-Selected Species

Feature	r-Selected Species	K-Selected Species
Development time to maturity	Short	Long
Body size	Small	Large
Lifespan	Short	Long
Number of offspring per reproductive event	Many	Few
Age at first reproductive event	Early	Late
Number of reproductive events per lifetime	Often only one	Often more than one
Parental care invested in offspring	No	Yes

do not produce large numbers of offspring. They characteristically have long life spans with slow development, late reproduction, large body size, and a low reproductive rate (Table 8–1). Animals that are *K* strategists also invest in parental care of their young. *K* strategists are found in relatively constant or stable environments, where they have a high competitive ability.

Tawny owls are *K* strategists that pair-bond for life, with both members of a pair living and hunting in adjacent, well defined territories (Figure 8–5b). They reproduce in accordance with the resources, especially food, present in their territories. In an average year, 30 percent of the birds do not breed. If food supplies are more limited than initial conditions had indicated, many of those that breed fail to incubate their eggs. Rarely do the owls lay the maximum number of eggs, and often they delay breeding until late in the season, when the rodent populations on which they depend have become large. Thus,

(a)

(b)

Figure 8–5

Most organisms have one of two reproductive strategies, *r* selection or K selection. (a) Dandelions (*Taraxacum officinale*) are *r* strategists—annuals that mature early and produce many small seeds. A dandelion population fluctuates from year to year but rarely approaches the carrying capacity of its environment. (b) Tawny owls (*Strix aluco*) are *K* strategists: They maintain a fairly constant population size at or near the carrying capacity. They mature slowly, delay reproduction, and have a relatively large body size. (a, Marion Lobstein; b, Stephan Dalton/Photo Researchers, Inc.)

the behavior of tawny owls contributes to the regulation of their population so that its number stays at or near the carrying capacity of the environment. Starvation, an indication that the tawny owl population has exceeded the carrying capacity, rarely occurs.

Survivorship

The various reproductive strategies that organisms possess, from *r* selection to *K* selection, are associated with different patterns of **survivorship**, the proportion of individuals in a population that survive to a particular age. Figure 8–6 is a graph of the three main survivorship curves that ecologists recognize. In Type I survivorship, as exemplified by humans, the probability of survival decreases with age; that is, deaths are concentrated later in life. *K*-selected species are characterized by Type I survivorship. Humans and other *K*-selected species are long-lived organisms whose young have a high likelihood of survival.

In Type III survivorship, the probability of survival increases with age; that is, deaths are concentrated early in life, and those that survive early death subsequently have a high rate of survival. Type III survivorship is characteristic of oysters and other *r*-selected species. Young oysters have three free-swimming larval stages before settling down and secreting a shell. These larvae are vulnerable to predation, and few survive to adulthood.

In Type II survivorship, which is intermediate between Types I and III, the probability of survival does not change with age. Death is spread evenly across all age groups, resulting in a linear decline in survivorship. This constancy probably results from essentially random events that cause death with little age bias. Certain annual plants and some lizards have a Type II survivorship.

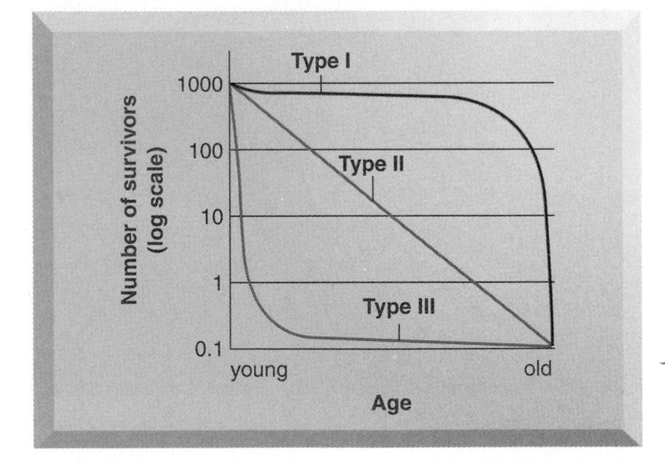

Figure 8–6

These generalized survivorship curves represent the ideal survivorships of organisms in which death is greatest in old age (Type I), spread evenly across all age groups (Type II), and greatest among the young (Type III). The survivorship of most organisms can be compared to these curves.

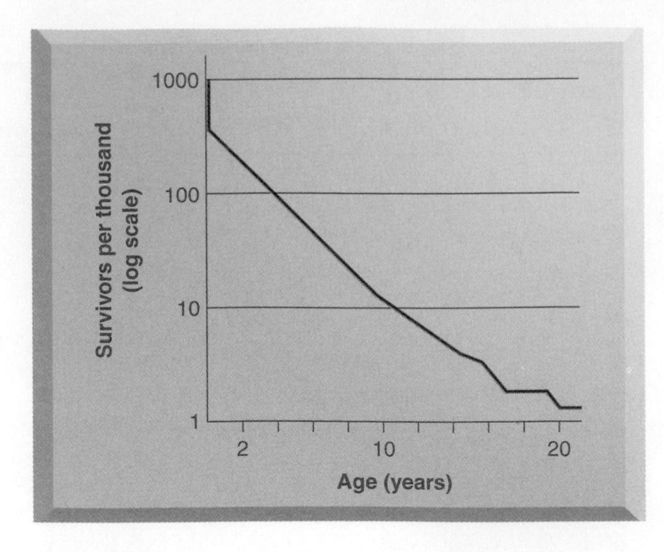

Figure 8–7

The survivorship curve for a herring gull population reveals Type III survivorship as chicks and Type II survivorship as adults. Data collected from Kent Island, Maine, 1934 to 1939. Baby gulls were banded to establish identity.

The three survivorship curves are generalizations, and not all organisms fit one of the three. Many organisms have one type of survivorship curve early in life and another type as adults. Herring gulls, for example, have a Type III survivorship curve early in life and a Type II curve as adults (Figure 8–7). Note that most death occurs almost immediately despite the protection and care given to the chicks by the parent bird. Herring gull chicks die from predation or attack by other herring gulls, inclement weather, infectious disease, or starvation following the death of the parent. Once the chicks become independent, their survivorship increases dramatically, and death occurs at about the same rate throughout their remaining lives. As a result few or no herring gulls die from degenerative diseases of "old age" that cause death in most humans.

THE HUMAN POPULATION

Now that we have examined some of the basic concepts of population biology, we can apply those concepts to the human population. Examine Figure 8–8, which shows the worldwide increase in population since the development of agriculture, about 10,000 years ago. Now look back at Figure 8–2b and observe how the human population is increasing exponentially. The characteristic J-curve of exponential growth reflects the decreasing amount of time it has taken to add each additional billion people to our

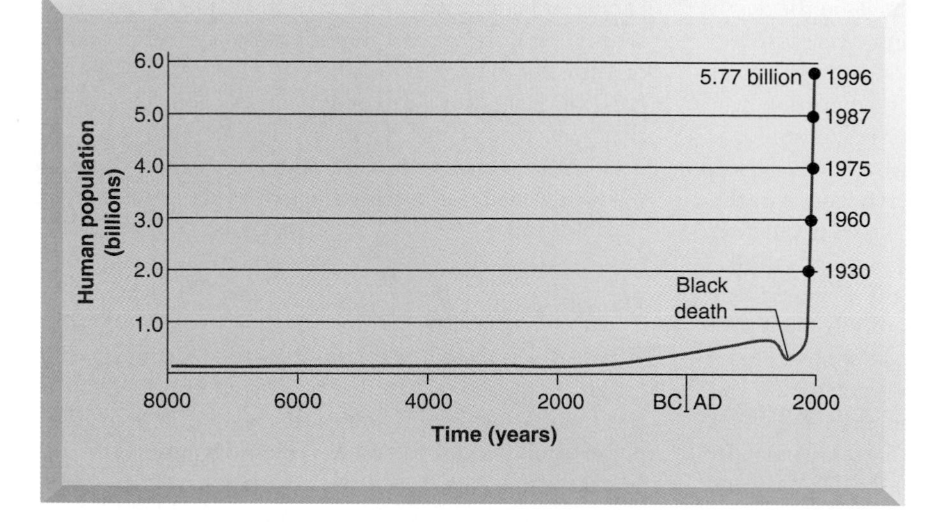

Figure 8–8

The human population has been increasing exponentially, as shown by the J-shaped curve of human population growth.

numbers. It took thousands of years for the human population to reach 1 billion, a milestone that took place in 1800. It took 130 years to reach 2 billion (in 1930), 30 years to reach 3 billion (in 1960), 15 years to reach 4 billion (in 1975), and 12 years to reach 5 billion (in 1987). The population is projected to reach 6 billion in 1999.

One of the first people to recognize that the human population cannot continue to increase indefinitely was Thomas Malthus, a British economist (1766–1834). He pointed out that human population growth was not always desirable—a view contrary to the beliefs of his day—and that the human population was capable of increasing faster than the food supply. He maintained that the inevitable consequences of population growth were famine, disease, and war.

Our world population was 5.77 billion in 1996 and increased by approximately 69 million from 1995 to 1996. This increase is not due to an increase in the birth rate (*b*). In fact, the worldwide birth rate has actually declined slightly during the past 200 years. The population increase is due instead to a large decrease in the death rate (*d*), which has occurred primarily because of greater food production, better medical care, and improved sanitation practices. For example, from about 1920 to 1996, the death rate in Mexico fell from approximately 40 to 5 per 1000 individuals, whereas the birth rate dropped from approximately 40 to 27 per 1000 individuals (Figure 8–9). Because Mexico's birth rate is currently much greater than its death rate, it has a high growth rate.

Projecting Future Population Numbers

The human population has reached a turning point. Although our numbers continue to increase, the worldwide growth rate (*r*) has declined over the past several years, from a peak in 1965 of about 2 percent per year to a 1996 growth rate of 1.5 percent per year. Population experts at the United Nations and the World Bank have projected that the growth rate will continue to slowly decrease until zero population growth is attained. It is projected that **zero population growth**—when the birth rate equals the death rate—will occur toward the end of the 21st century.

The United Nations periodically publishes population projections for the 21st century. The latest U.N. figures available forecast that the human population will be between 7.8 billion (their "low" projection) and 12.5 billion (their "high" projection) in the year 2050.

Another group that examines population trends is the International Institute for Applied Systems Analysis (IIASA) in Vienna, Austria. The latest (1996) population projections of the IIASA were obtained by combining statistically the predictions of 12 population experts.

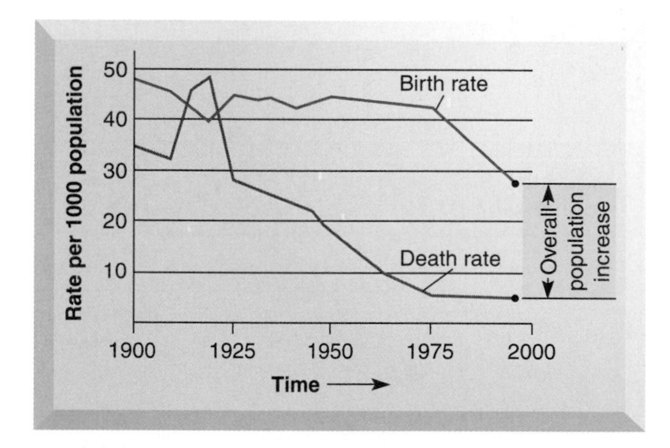

Figure 8–9

In Mexico, the birth and death rates have generally declined in this century. Because the death rate has declined much more than the birth rate, however, Mexico has experienced a high growth rate. (The high death rate prior to 1925 was caused by the Mexican Revolution.)

IIASA also took into account future changes in the death rate due to such factors as starvation and disease. (Most population projections just examine changes in the birth rate.) IIASA projects that the world's population will reach 10 billion in 2050 and peak at 11 billion in 2075.

Such population projections are "what if" exercises: given certain assumptions about future tendencies in the birth rate, death rate, and migration, an area's population can be calculated for a given number of years into the future. Population projections indicate the changes that may be upcoming, but they must be interpreted with care because they vary depending on the assumptions made. For example, in projecting that the world population will be 7.8 billion in the year 2050, demographers assume that the average number of children born to each woman in all countries will have declined to 1.7 in the 21st century. In 1996, the average number of children born to each woman on Earth was 3.0. If that decline does not occur, our population could be significantly higher. For example, if the population were to continue to increase at its 1996 growth rate, there would be almost 13 billion humans in the year 2050.

The main unknown factor in this population growth scenario is Earth's carrying capacity. No one knows how many humans can be supported by Earth, and projections and estimates vary widely, depending on what assumptions about resource consumption, waste generation, and the like are made. For example, if we want a high level of material well-being for all people, then the Earth will clearly be able to support far fewer humans than if everyone lives just above the subsistence level.

It is also not clear what will happen to the human population when the carrying capacity is approached. Optimists suggest that the human population will stabilize because of a decrease in the birth rate and an increase in the death rate (more people will die because the Earth cannot support them). Some experts take a more pessimistic view and predict that the widespread degradation of our environment caused by our ever-expanding numbers will make the Earth uninhabitable for humans and that a massive wave of human suffering and death will occur. Some experts think that human population has already exceeded the carrying capacity of the environment.

Demographics of Countries

Whereas worldwide population figures illustrate overall trends, they do not describe other important aspects of the human population story, such as population differences from country to country. **Demographics**, the applied science that deals with population statistics, provides interesting information on the populations of countries. As you probably know, not all countries have the same rates of population increase. Countries can be classified into two groups—developed and developing—depending on growth rates and other factors, such as degree of industrialization and relative prosperity (Table 8–2).

Developed countries (also called **highly developed countries**), such as the United States, Canada, France, Germany, Sweden, Australia, and Japan, have low rates of population growth, and are highly industrialized relative to the rest of the world. Developed countries have the lowest birth rates in the world. Indeed, some developed countries such as Germany have birth rates just below those needed to sustain their populations and are thus declining slightly in numbers. Highly developed countries also have very low **infant mortality rates** (the number of infant deaths per 1000 live births). The infant mortality rate of the United States was 7.5 in 1996, for example, compared with a worldwide rate of 62. Highly de-

Table 8–2
Comparison of 1996 Population Data in Developed and Developing Countries

	Developed	Developing	
	(Highly Developed) United States	(Moderately Developed) Brazil	(Less Developed) Ethiopia
Fertility rate	2.0	2.8	6.8
Doubling time at current rate	114 years	41 years	23 years
Infant mortality rate	7.5 per 1000	58 per 1000	120 per 1000
Life expectancy at birth	76 years	66 years	50 years
Per capita GNP (U.S. $; 1994)	$25,860	$3,370	$130
Women using modern contraception	65%	56%	3%

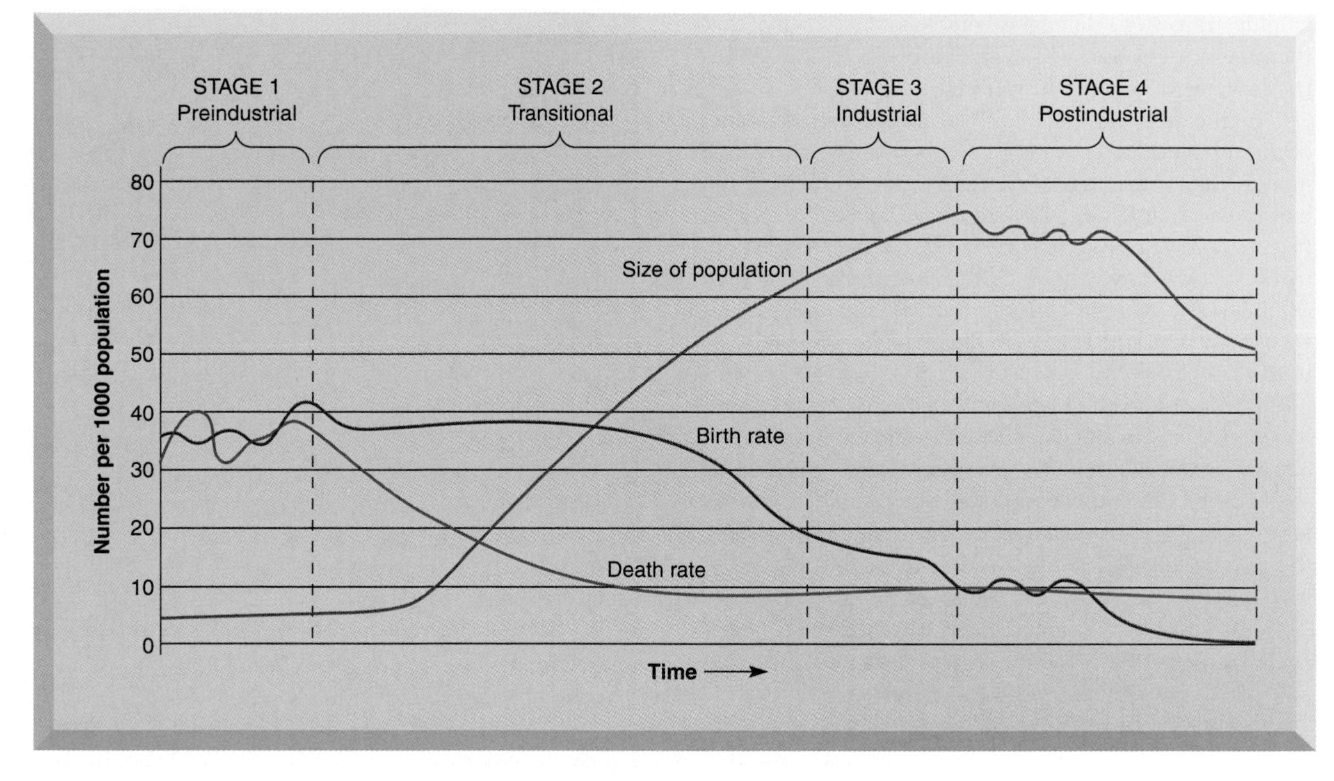

Figure 8–10

The demographic transition consists of four demographic stages through which a population progresses as its society becomes industrialized. Note that the death rate declines first, followed by a decline in the birth rate.

veloped countries also have longer life expectancies (76 years in the United States versus 66 years worldwide) and high average per-capita GNPs ($25,860 in the United States versus $4740 worldwide).[3]

Developing countries fall into two subcategories, moderately developed and less developed. Mexico, Turkey, Thailand, and most South American nations are examples of **moderately developed countries**. Their birth rates and infant mortality rates are higher than those of highly developed countries. Moderately developed countries have a medium level of industrialization, and their average per-capita GNPs are lower than those of highly developed countries. **Less developed countries** include Bangladesh, Niger, Ethiopia, Laos, and Cambodia. These countries have the highest birth rates, the highest infant mortality rates, the shortest life expectancies, and the lowest average per-capita GNPs in the world.

A country's doubling time can identify it as a highly, moderately, or less developed country: the shorter the doubling time, the less developed the country. At 1996 rates of growth, for example, the doubling time is 19 years for Togo, 24 years for Laos, 32 years for Mexico, 37 years for India, 114 years for the United States, and 630 years for Belgium.

It is also instructive to examine **replacement-level fertility**, that is, the number of children a couple must produce in order to "replace" themselves. Replacement-level fertility is usually given as 2.1 children in developed countries and 2.7 children in developing countries. The number is always greater than 2.0 because some children die before they reach reproductive age. Higher infant mortality rates are the main reason that replacement levels are greater in developing countries than in developed countries. Worldwide, the **total fertility rate**—the average number of children born to each woman—is currently 3.0, which is well above replacement levels.

Demographic stages

Demographers recognize four demographic stages, based on their observations of Europe as it became industrialized and urbanized (Figure 8–10). These stages converted Europe from relatively high birth and death rates to relatively low birth and death rates. Because all highly developed and moderately developed countries with more advanced economies have gone through this demo-

[3]GNP stands for gross national product, the total value of a nation's annual output in goods and services.

graphic transition, demographers generally assume that the same progression will occur in less developed countries as they become industrialized.

In the first stage, called the **preindustrial stage**, birth and death rates are high and population grows at a modest rate. Although women have many children, the infant mortality rate is high. Intermittent famines, plagues, and wars also increase the death rate, so the population grows slowly. If we use a European country such as Finland to demonstrate the four demographic stages, we can say that Finland was in the first demographic stage in the late 1700s.

As a result of the improved health care and more reliable food and water supplies that accompany the beginning of an industrial society, the second demographic stage, called the **transitional stage**, is characterized by a lowered death rate. Because the birth rate is still high, the population grows rapidly. Finland in the mid-1800s was in the second demographic stage.

The third demographic stage, the **industrial stage**, is characterized by a decline in the birth rate and takes place at some point during the industrialization process. The decline in the birth rate slows population growth despite a relatively low death rate. For Finland, this occurred in the early 1900s.

The fourth demographic stage, sometimes called the **postindustrial stage**, is characterized by low birth and death rates. In countries that are heavily industrialized, people are better educated and more affluent; they tend to desire smaller families and take steps to limit family size. The population grows very slowly or not at all in the fourth demographic stage. This is the situation in such developed countries as the United States, Canada, Australia, Japan, and Europe, including Finland.

Once a country reaches the fourth demographic stage, is it correct to assume it will continue to have a low birth rate indefinitely? The answer is that we do not know. Low birth rates may be a permanent response to the socioeconomic factors that are a part of an industrialized, urbanized society. On the other hand, low birth rates may be a temporary response to socioeconomic factors such as the changing roles of women in developed countries. No one knows for sure.

Why has the population stabilized in many developed countries? The reasons are complex. The decline in birth rate has been associated with an improvement in living standards. However, it is difficult to say whether improved socioeconomic conditions have resulted in a decrease in birth rate, or a decrease in birth rate has resulted in improved socioeconomic conditions. Perhaps both are true. Another reason for the decline in birth rate in developed countries is the increased availability of family planning services. Still other factors influence birth rate, including education, particularly of women, and urbanization of our society. These factors are considered in greater detail in Chapter 9.

Table 8–3
Fertility Changes in Selected Developing Countries

Country	Total Fertility Rate* 1960–65	Total Fertility Rate* 1996
Bangladesh	6.7	3.7
Brazil	6.2	2.8
China	5.9	1.8
Egypt	7.1	3.6
Guatemala	6.9	5.1
India	5.8	3.4
Kenya	8.1	5.4
Mexico	6.8	3.1
Nepal	5.9	5.2
Nigeria	6.9	6.0
Thailand	6.4	2.2

*Total fertility rate = average number of children born to each woman during her lifetime.

The population in many developing countries is beginning to approach stabilization. The fertility rate must decline in order for a population to stabilize; see Table 8–3 and note the general decline in total fertility rates in selected developing countries from the 1960s to 1996. The total fertility rate in developing countries has decreased from an average of 6.1 children per woman in 1970 to 3.4 in 1996. In the past decade, fertility rates have declined by at least 25 percent in countries such as Brazil, Indonesia, and Mexico.[4]

Age Structure of Countries

In order to predict the future growth of a population, it is important to know its **age structure**, the number and proportion of people at each age in a population. The number of males and number of females at each age, from birth to death, can be represented in an **age structure diagram** (Figure 8–11).

[4]Although fertility rates in these countries have declined, it should be remembered that they are still greater than replacement-level fertility. Consequently, the populations of these countries are still increasing.

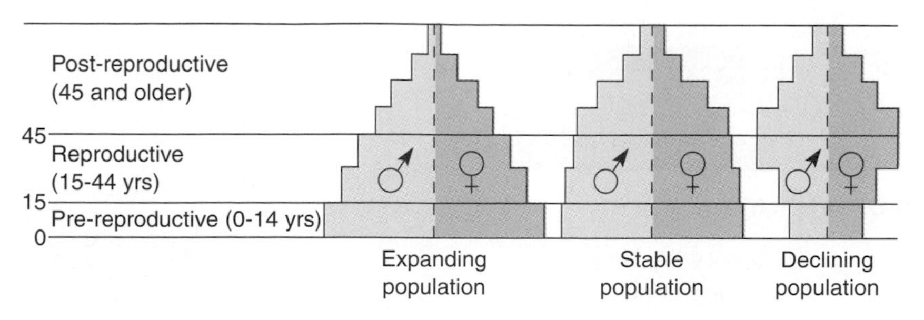

Figure 8–11

Generalized age structure diagrams for an expanding population, a stable population, and a population that is decreasing in size. Each diagram is divided vertically in half, one side representing the males in a population and the other side the females. The bottom third of each diagram represents pre-reproductive humans (from 0 to 14 years of age); the middle third, reproductive humans (15 to 44 years); and the top third, post-reproductive humans (45 years and older). The widths of these segments are proportional to the population sizes—a broader width implies a larger population.

Figure 8–12

Age structure diagrams for countries with (a) fast, (b) slow, and (c) declining population growth. These age structure diagrams indicate that less developed countries such as Nigeria have a much higher percentage of young people than highly developed countries. As a result, less developed countries are projected to have greater population growth than highly developed countries.

Predicting population changes using age structure diagrams

The overall shape of an age structure diagram indicates whether the population is increasing, stable, or shrinking. The age structure diagram of a country with a very high growth rate—for example, Nigeria or Bolivia—is shaped like a pyramid (Figure 8–12a). Because the largest percentage of the population is in the pre-reproductive age group, the probability of future population growth is very great. A strong *population growth momentum* exists because

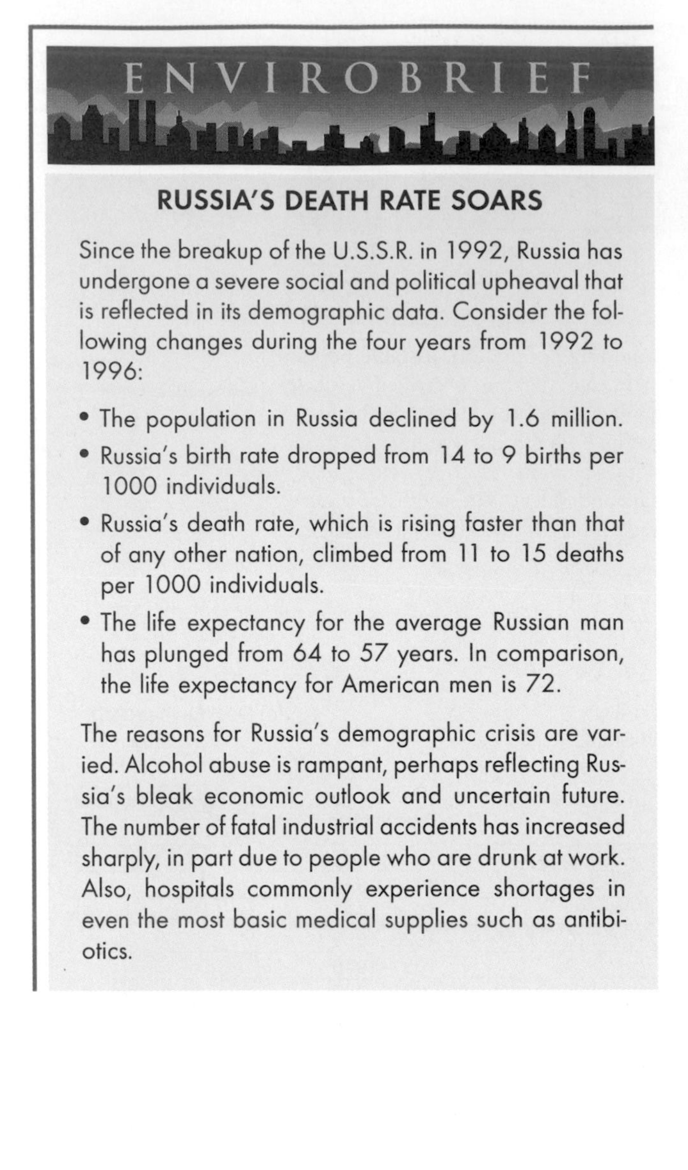

ENVIROBRIEF

RUSSIA'S DEATH RATE SOARS

Since the breakup of the U.S.S.R. in 1992, Russia has undergone a severe social and political upheaval that is reflected in its demographic data. Consider the following changes during the four years from 1992 to 1996:

- The population in Russia declined by 1.6 million.
- Russia's birth rate dropped from 14 to 9 births per 1000 individuals.
- Russia's death rate, which is rising faster than that of any other nation, climbed from 11 to 15 deaths per 1000 individuals.
- The life expectancy for the average Russian man has plunged from 64 to 57 years. In comparison, the life expectancy for American men is 72.

The reasons for Russia's demographic crisis are varied. Alcohol abuse is rampant, perhaps reflecting Russia's bleak economic outlook and uncertain future. The number of fatal industrial accidents has increased sharply, in part due to people who are drunk at work. Also, hospitals commonly experience shortages in even the most basic medical supplies such as antibiotics.

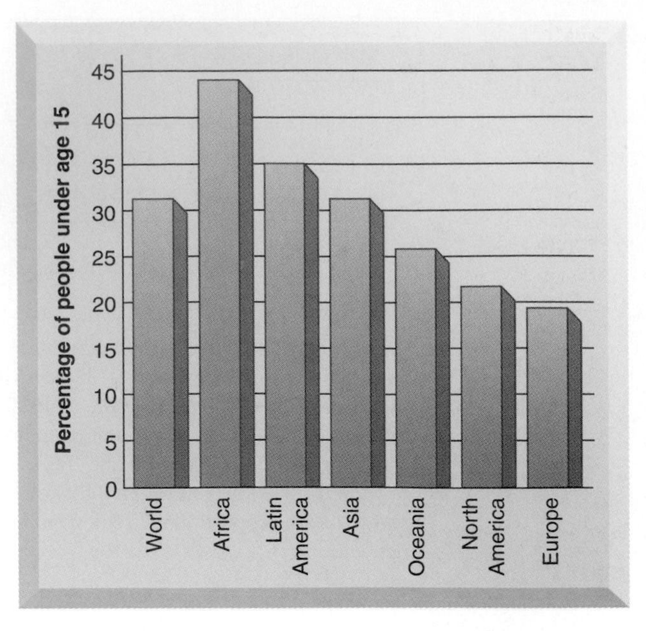

Figure 8–13
The percentage of the population under age 15 for various regions of the world in 1996. The higher this percentage, the greater the potential for population growth when those people reach their reproductive years.

when all these children mature they will become the parents of the next generation, and this group of parents will be larger than the previous group. Thus, even if the fertility rate of such a country is at replacement level, the population will continue to grow.

In contrast, the more tapered bases of the age structure diagrams of countries with slowly growing, stable, or declining populations indicate a smaller proportion of children to become the parents of the next generation (Figure 8–12b and c). The age structure diagram of a stable population, one that is neither growing nor shrinking, demonstrates that the numbers of people at pre-reproductive and reproductive ages are approximately the same. Also, a larger percentage of the population is older—that is, post-reproductive—than in a rapidly increasing population. Many countries in Europe have stable populations.

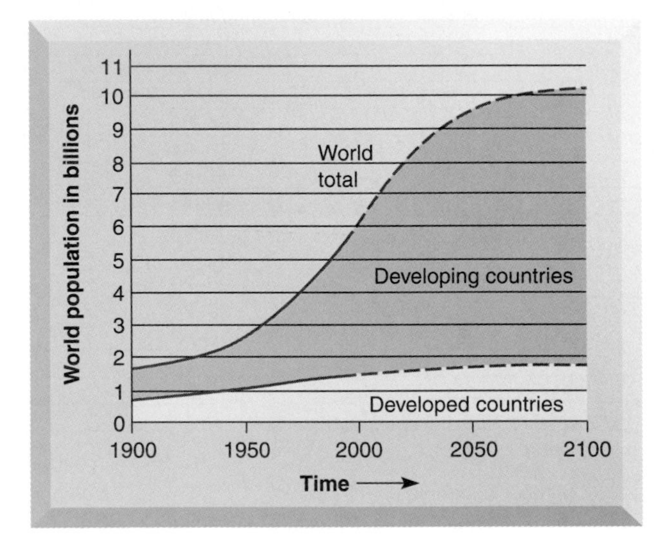

Figure 8–14
Most of the worldwide increase in population since 1950 has occurred in developing countries.

In a population that is shrinking in size, the pre-reproductive age group is *smaller* than either the reproductive or post-reproductive group. Russia, Bulgaria, and Hungary are examples of countries with slowly shrinking populations.

Worldwide, 32 percent of the human population is under age 15 (Figure 8–13). When these people enter their reproductive years, they have the potential to cause a large increase in the growth rate. Even if the birth rate does not increase, the growth rate will increase simply because there are more people reproducing.

Most of the worldwide population increase that has occurred since 1950 has taken place in developing countries as a result of the younger age structure and the higher-than-replacement-level fertility rates of their populations (Figure 8–14). In 1950, 66.8 percent of the world's population was in developing countries in Africa, Asia (minus Japan), and Latin America. Between 1950 and 1996, the world's population more than doubled in size, but most of that growth occurred in developing countries. As a reflection of this, in 1996 the number of people in developing countries had increased to 80 percent of the world's population. Because different age structures in populations imply different potential changes when we project to the future, most of the population increase that will occur during the next century will also take place in developing countries. As you know, these countries are least able to support such growth. By 2020, it is estimated that about 85 percent of the people in the world will live in developing countries.

Demographics of the United States

The United States has one of the highest rates of population increase of all the developed countries. For example, the U.S. population increased by 2 million from 1995 to 1996. This translates to a 1996 percent annual increase of 0.6, which is high compared to many developed countries—for example, Western Europe's 1996 percent annual increase was 0.1, and Japan's was 0.2. These figures take only birth and death rates into account; migration is not considered.

Although the birth rate has decreased slightly in the United States during the past few years, we are still experiencing an increase in population growth. There are two reasons for this. First, our population has a built-in population growth momentum because of the Baby Boom, the large wave of births that followed World War II (Figure 8–15). The babies born then are now in their reproductive years. Thus, although the number of children born per female has declined (in 1996 the total fertility rate was 2.0), there is an increase in births because of the greater number of females who are bearing children.

A second reason for the large growth rate in U.S. population is immigration, which has a greater effect on population size in the United States than in most other nations (see Focus On: A History of U.S. Immigration).

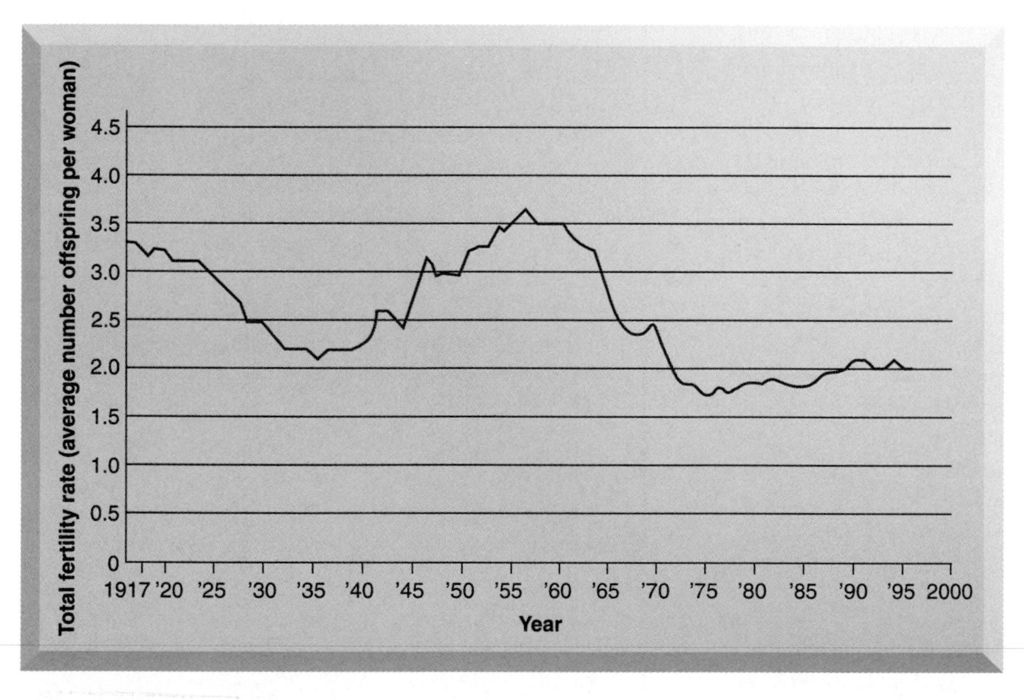

Figure 8–15

The total fertility rate in the United States from 1917 to 1996. Note the high average number of children born per woman during the Baby Boom years (1945 to 1962).

From 1990 to 1993 the United States accepted 924,000 international immigrants annually, and the number has accelerated rapidly since then. The number of *illegal* immigrants who gained access to the United States and were not deported is not known with any certainty, but was estimated by the Census Bureau as about 4 million in 1995.

The United States does not have a formal population policy, and the closest it has come to such a policy occurred under President Nixon's leadership, when a commission was established to examine U.S. population growth. The commission spent three years gathering data and listening to testimonies from experts and the public before concluding in 1972 that the United States would gain no substantial benefits from continued population growth. The commission therefore recommended that the United States should try to stabilize its population. Despite this recommendation, U.S. population has continued to increase.

A LOOK AT U.S. POPULATION

Here are some facts about the changing and complex population within the United States, based on 1993 data (latest available).

The three states with the greatest number of residents are California (31.2 million), New York (18.2 million), and Texas (18 million).

The two states with the highest population densities are New Jersey (1062 people per square mile of land area) and Rhode Island (957 people per square mile). The two states with the lowest population densities are Alaska (1 person per square mile) and Wyoming (5 people per square mile).

Between 1990 and 1993, Nevada had the highest growth rate (4.8 percent), with Idaho second (2.7 percent). The District of Columbia had the largest population decline (−1.5 percent), and populations in Massachusetts, Rhode Island, Connecticut, and North Dakota declined slightly.

Florida and Texas had the highest total immigration (1990–1993 data), with a whopping 164,000 and 135,000 people added per year, respectively, as compared to New York, which lost 57,000 people due to emigration. These figures include both international immigrants and U.S. citizens migrating from other states.

In 1993, Utah had the highest percentage of its population below the age of 15 (29.7 percent), whereas Florida had the highest percentage of its population over age 65 (18.6 percent).

In 2020 the 3 states with the greatest number of residents are projected to be California (48 million), Texas (25.6 million), and Florida (19.4 million).

ENVIROBRIEF

CLOSING THE GATES ON A FLOOD OF REFUGEES

At a time when movements of people worldwide are reaching unprecedented rates, the United States and other developed countries are becoming more reluctant hosts. In 1995 the United Nations reported that 27.4 million people qualified for refugee assistance under international law. These refugees, who left their homes because of war, political strife, ethnic conflict, poverty, or environmental degradation, cause serious ecological, social, and political problems. Because of their desperate plight, most refugees were once welcome in the United States and a few other asylum nations, but this hospitable attitude is disappearing:

- The end of the cold war eliminated the incentive for Western nations to accept refugees from communist countries.
- U.S. policy now leans against assisting nations that have repressive regimes and shows less tolerance for refugees from those nations.
- The U.S. public often fails to distinguish between refugees and illegal immigrants, who are the object of much national hostility.
- Some population control organizations advocate restricting immigration to curb population growth.

The United Nations and the international community are providing some assistance to refugees, but their resources are stretched to the limit because the number of refugees requiring assistance has increased by 12.5 million between 1990 and 1995.

FOCUS ON

A History of U.S. Immigration

Prior to 1875, there were no immigration laws in the United States, and thus no such thing as illegal immigration. In 1875, however, Congress passed a law denying convicts and prostitutes entrance to the United States. In 1882 the Chinese Exclusion Act was passed, and in 1891 the Bureau of Immigration was established. Thus, by the late 1800s a policy of selective exclusion was officially established in the United States and began to shape the population of this country.

During the early 20th century, Congress set numerical restrictions on immigration, including quotas allowing only a certain number of people from each foreign country to immigrate. This severely restricted entrance to the United States by people from Asia, southern and Eastern Europe, and Mexico. With these stronger laws, illegal immigration began to increase. U.S. immigration policy relaxed during World War II, when labor shortages made it possible for workers from Mexico, Barbados, Jamaica, and British Honduras to gain temporary residence in the United States.

In 1952 the Immigration and Nationality Act was passed, and although it has been revised since then (it is now called the Immigration Reform and Control Act, or IRCA), it is still the basic immigration law in effect. The act applies penalties to U.S. citizens who harbor illegal aliens; in 1986 it was expanded to include employers who knowingly hire undocumented immigrants. Its intent is to control and eventually eliminate the flow of illegal aliens into the United States.

Most experts feel that the IRCA has failed to curb the number of illegal aliens entering the United States, although they admit that the situation could be worse without the act. The size of the illegal immigrant population dropped dramatically after passage of the IRCA, but it began to increase again in 1989. By January 1992, it had reached an estimated size of 2.5 to 4.0 million, as compared with an estimated peak population of 5.0 million in June 1986. The number of illegal immigrants caught at the U.S.–Mexican border exhibited a similar trend. The employer sanctions of the IRCA have not appeared to work, primarily because the market for production of illegal identification documents has mushroomed and enforcement of penalties has been weak. In general, the U.S. Immigration and Naturalization Service has been overwhelmed, and expanded budgets to stem border crossings have not been funded by Congress as originally planned.

Contrary to popular opinion, the vast numbers of illegal immigrants do not strain the federal government but actually contribute to its wealth in terms of cheap labor and income taxes. It is state and local governments that suffer from the strain of increased services and growing crime problems. During the 1990s six states—California, Texas, Florida, Arizona, New Jersey, and New York—filed suit against the federal government to be reimbursed for the costs associated with illegal immigration. All suits were dismissed, however.

Working conditions for illegal immigrants—and often for legal ones—appear to have declined since the IRCA was revised. In the 1990s, job discrimination was on the rise in response to employers' fears of the legislation, and many "sweat shop" enterprises that prey on illegal immigrants entered the underground economy.

SUMMARY

1. Populations have certain properties that individual organisms lack, such as birth rates and death rates. Population density is the number of individuals of a species per unit of area or volume at a given time.

2. On a global scale, the rate of change in a population is due to two factors, the number of births and the number of deaths: populations increase in size as long as the birth rate is greater than the death rate. In addition to birth rate and death rate, migration must be considered when examining changes in local populations. The number of individuals emigrating from an area and the number immigrating into an area affect its population size and growth rate.

3. Biotic potential is the maximum rate at which an organism or population could increase under ideal conditions. Although certain populations exhibit exponential growth for limited periods of time, eventually the growth rate decreases to around zero. This occurs when the carrying capacity of the environment is reached. Thus, population size is modified by environmental resistance, which includes the limited availability of food, water, shelter, and other resources, as well as limits imposed by disease and predation.

4. Density-dependent factors are most effective at limiting population growth when the population density is high; predation, disease, and competition are examples. Density-independent factors limit population growth but are not influenced by changes in population density; hurricanes and fires are examples.

5. Each organism has its own life history strategy. An *r* strategy emphasizes a high growth rate; these organisms often have small body sizes, high reproductive rates, short life spans, and inhabit variable environments. A *K* strategy emphasizes maintenance of a population near the carrying capacity of the environment; these organisms often have large body sizes, low reproductive rates, long life spans, and inhabit stable environments.

6. There are three general types of survivorship curves. Type I survivorship, in which death is greatest in old age, is typical of *K*-selected species. Type III survivorship, in which death is greatest among the young, is typical of *r*-selected species. In Type II survivorship, death is spread evenly across all age groups.

7. The principles of population biology that are used to understand populations of other organisms also apply to humans. Currently, human population is increasing exponentially, although the growth rate has declined slightly over the past several years. Demographers project that the world population will stabilize by the end of the 21st century.

8. Highly developed countries have the lowest birth rates, the lowest infant mortality rates, the longest life expectancies, and the highest per-capita GNPs. Developing countries have the highest birth rates, the highest infant mortality rates, the shortest life expectancies, and the lowest per-capita GNPs.

9. The age structure of a population greatly influences its population changes. It is possible for a country to have replacement-level fertility and still experience population growth if the largest percentage of the population is in the pre-reproductive years.

THINKING ABOUT THE ENVIRONMENT

1. Describe the effect of each of the following on population size: birth rate, death rate, immigration, and emigration.

2. Explain the J-shaped and S-shaped population growth curves in terms of biotic potential and carrying capacity.

3. Draw a graph to represent the *long-term* growth of a population of bacteria cultured in a test tube containing a nutrient medium.

4. Give three examples each of density-dependent and density-independent factors that affect population growth.

5. Draw the three main survivorship curves and relate them to *r* selection and *K* selection.

6. A female elephant bears a single offspring every two to four years. Based on this information, do you think elephants are *r*-selected or *K*-selected species? Which survivorship curve do you think is representative of elephants? Explain your answers.

7. Should the rapid increase in world population be of concern to the average citizen in the United States? Why or why not?

8. Why is a declining birth rate not necessarily a reliable indicator of future population growth trends?

9. If all the women in the world suddenly started bearing children at replacement-level fertility rates, would the population stop increasing immediately? Why or why not?

10. What is replacement-level fertility? Why is it higher in developing countries?

11. In Bolivia, 41 percent of the population are younger than 15 and 4 percent are older than 65. In Austria, 18 percent of the population are younger than 15 and 15 percent are older than 65. Which country will have the highest population growth momentum over the next two decades? Why?

12. Should the United States increase or decrease the number of legal immigrants? Present arguments in favor of both sides.

* 13. The 1996 population of the Netherlands was 15.5 million, and its land area is 990 square miles. The 1996 population of the United States was 265.2 million, and its land area is 3,615,200 square miles. Which country has the greatest population density?

* **14.** If a population of 100,000 has 2700 births per year and 900 deaths per year, what is its growth rate, in percent per year? What is its doubling time?

* **15.** The population of India in 1996 was 949.6 million, and its growth rate was 1.9 percent per year. Calculate the 1997 population of India. Assuming that the 1996 growth rate remains constant, what will its population be in 2000? In what year will India's population be double that of its 1996 population?

* **16.** The world population in 1996 is 5.771 billion, and the annual growth rate is 1.5 percent. If the 1996 birth rate is 24 per 1000 people, what is the 1996 death rate?

RESEARCH PROJECTS

Choose a country and assess its population trends for the next 25 years. Do steps need to be taken to stabilize the population? Why or why not, and, if so, what steps do you recommend? Does the country have modern contraceptive methods available to the population? What factors affect the country's choices for population control?

Investigate population trends in the United States. Does the U.S. government have a national population policy? If not, do you think it should? Why or why not?

Some environmentalists see a direct link between liberal immigration laws and environmental degradation in the United States. Contact Population-Environment Balance (see Appendix II), an organization that supports population stabilization in the United States, and investigate whether there is a connection between immigration and the environment.

On-line materials relating to this chapter are on the ► World Wide Web at **http://www.saunderscollege.com/lifesci/environment2** (select Chapter 8 from the Table of Contents).

SUGGESTED READING

Bongaarts, J. "Population Policy Options in the Developing World," *Science* Vol. 263, 11 February 1994. Discusses various policy options that could be implemented during the next century to lessen population growth in developing countries.

Cohen, J. E. "Population Growth and Earth's Human Carrying Capacity," *Science* Vol. 269, 21 July 1995. Explores the question of how many people the Earth can support.

Raloff, J. "The Human Numbers Crunch," *Science News* Vol. 149, June 22, 1996. An overview of the challenges that population experts think we will face during the next 50 years.

Sachs, A. "Population Increase Slightly Down." In *Vital Signs 1996*, by L. R. Brown, C. Flavin, and H. Kane, New York: W. W. Norton & Company, 1996. This short article summarizes world population trends in 1996.

Tickell, C. "The Human Hazard," *Our Planet* Vol. 7, No. 3, 1995. The author, former United Kingdom Ambassador to the United Nations, discusses the rapid increase in environmental refugees.

United States Population Data Sheet. 11th ed. Population Reference Bureau, Washington, D.C., 1994. Provides a current demographic picture of the United States. Includes population changes by region and state.

1996 World Population Data Sheet. Population Reference Bureau, Washington, D.C., 1996. A chart published annually that provides current population data for all countries: birth rates, death rates, infant mortality rates, total fertility rates, and life expectancies, as well as other pertinent information.

Women at the Dakalaia Health Clinic in Egypt learn about family planning and birth control. (Donna DeCesare/ Impact Visuals)

Facing the Problems of Overpopulation

The location of the 1994 U.N. International Conference on Population and Development—in Cairo, Egypt—was particularly appropriate. Egypt's population of almost 62 million has tripled since 1950, and because the majority of the land in Egypt is uninhabitable desert, most Egyptians live in crowded strips along the Nile River. Yet despite its population growth, Egypt's 1996 fertility rate, which is the average number of children born to each woman, is 3.6, down from 5.3 just one decade ago and down from 7.0 during the 1960s.

Egypt's successful, national family planning program, established in 1965, became particularly effective in the late 1980s when its administrators decided to communicate through religion and the media. Beginning in the late 1980s, teams of religious leaders and health care workers started holding neighborhood meetings to educate people about birth control.

Traditionally, Islamic leaders have opposed any form of birth control as an impediment against God's wish to create life. However, the Prophet Mohammed's words have recently been reinterpreted by Islamic scholars in Egypt and several other Islamic nations as sanctioning birth control. Moreover, the Grand Mufti, the Chief Cleric and highest interpreter of Islamic law in Egypt, approves of most forms of birth control.

Television entered the campaign, with a popular soap opera called "And the Nile Flows On" that presents the desirability of marrying at a later age and the benefits of small families. The hero in "And the Nile Flows On" is a handsome, well educated Muslim cleric who debunks the idea that birth control is sinful. The villain is the local mullah, a religious leader whose wife is a midwife and who therefore profits from every birth in the village. In addition to the soap opera, contests, such as "Wedding of the Month," ask prospective brides questions about birth control. The winner's wedding is broadcast on national television and is attended by television celebrities. Similar programs on the radio award cash prizes to those who correctly answer the most questions about Egypt's population problem.

Like Egypt, many other countries are experiencing fertility decline, but we have far to go to achieve population stabilization. In this chapter we examine world hunger, environmental problems, underdevelopment, poverty, and urban problems, all of which are aggravated by rapid population growth. We then discuss factors that influence the fertility rate, including cultural traditions, the social and economic status of women, and the availability of family planning services. If population growth is slowed and resource consumption per person is decreased, the world will be in a better position to tackle many of its serious environmental problems. ◀

THE HUMAN POPULATION CRISIS

Most people would agree that all people in all countries should have access to the basic requirements of life: a balanced diet, clean water, decent shelter, and adequate clothing. Over the next century, however, it will become increasingly difficult to meet these basic needs, especially in countries that have not achieved population stabilization. Moreover, it is probable that the social, political, and economic problems resulting from continued population growth in these countries will affect other countries that have already achieved stabilized populations and high standards of living. For these reasons, population growth should be of concern to the entire world community, regardless of where it is occurring.

As our numbers increase during the next 100 years, environmental deterioration, hunger, persistent poverty, and health issues will continue to challenge us. Already the need for food for the increasing numbers of people living in environmentally fragile arid lands, such as parts of sub-Saharan[1] Africa, has led to overuse of the land for grazing and crop production. When overuse occurs in combination with an extended period of drought, these formerly productive lands have been degraded into unproductive deserts (see Chapter 17). Although it is possible to reclaim such arid lands, efforts to do so are made difficult by the large numbers of people and their animal herds trying to live off the land.

You may recall from Chapter 8 that when the carrying capacity of the environment is reached, the population will stabilize or crash due to a decrease in the birth rate, an increase in the death rate, or a combination of both. No one knows whether the Earth can support 10 billion or even 6 billion humans. It is extremely difficult to estimate the carrying capacity for humans, in part because we must make certain assumptions about our *quality* of life as well. For example, do we assume that everyone in the world should have the same standard of living as Americans currently do? If so, then the Earth would be able to support a fraction of the humans it could support if everyone in the world had only the barest minimum of food, clothing, and shelter needed to survive. It is therefore very difficult to determine Earth's carrying capacity for humans. It may be that we have already reached our carrying capacity and that the numerous environmental problems we are experiencing will cause the worldwide population increase to come to a halt.

On a national level, developing countries have the largest rates of population increase and often have the fewest resources to support their growing numbers. If a country is to support its human population, it must have either the agricultural land to raise enough food for those people or enough of other natural resources, such as minerals or oil, to provide buying power to purchase food.

Population and World Hunger

Many of the world's people—an estimated 840 million in 1996—do not get enough food to thrive, and in certain areas of the world people, especially children, still starve to death. South Asia and sub-Saharan Africa are the two regions of the world with the greatest food insecurity (Figure 9–1). The cause of world hunger, however, is anything but clear.[2] Experts agree that population, world hunger, poverty, and environmental problems are interrelated, but they do not agree about the most effective way to stop world hunger.

Those who assert that population growth is the root cause of the world's food problems point out that countries with some of the highest fertility rates are also the ones with the greatest food shortages. They argue that it is imperative to reduce population growth, even through drastic measures such as the establishment of world pop-

[1]Sub-Saharan Africa refers to all African countries located south of the Sahara Desert.

[2]Famines are a notable exception. A famine can usually be attributed to bad weather, insects, war, or some other disaster.

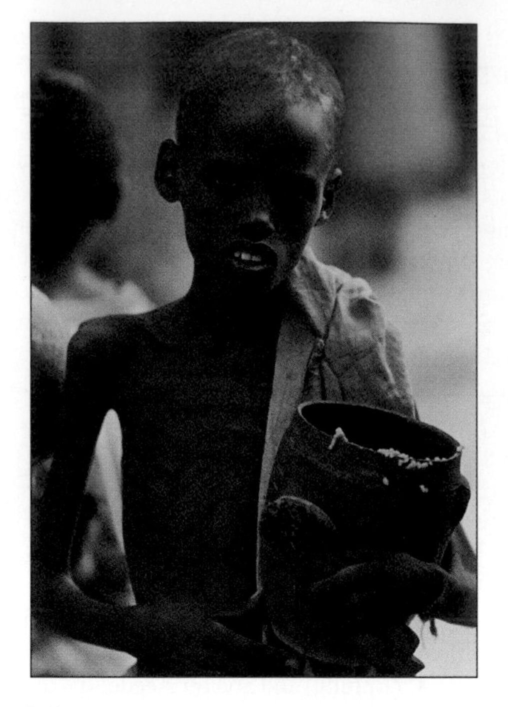

Figure 9–1

A young Somali boy receiving rice and beans from a Red Cross food distribution center. Overpopulation is not the only reason that there is not enough food for all the world's peoples. In Somalia, underlying causes include drought and continuing civil strife. (Reuters/Bettmann)

ulation quotas. Under such a system, a country that exceeded its assigned population size would not be eligible for relief from the international community during times of food shortages.

Some people think the way to tackle world food problems is not by controlling population but by promoting the economic development of countries that are unable to produce adequate food for their people. They presume that development would provide the appropriate technology for the people living in those countries to increase their food production. Also, once a country becomes more developed, its fertility rate should decline, helping to lessen the population problem.

A third group of people maintains that neither controlling population growth nor enhancing economic development alone will solve world food problems. They argue that the inequitable distribution of resources is the primary cause of world hunger. According to this view, there are enough resources, land, and technologies to produce food for all humans, but people on the lower end of the socioeconomic scale in many developing countries do not have access to the resources they need to support themselves. In other words, the principal cause of hunger is that people are unable to afford food. (Poverty and hunger are revisited in Chapter 18.)

These differing viewpoints indicate that the relationship between world hunger and population may be affected by economic development as well as by poverty and the uneven distribution of resources. Regardless of whether population growth is the main cause of world hunger, however, it is clear that world food problems are exacerbated by population pressures.

Population, Resources, and the Environment

The relationships among population growth, use of natural resources, and environmental degradation are also complex. We address the details of resource management and environmental problems in later chapters, but for now, we can make two useful generalizations. (1) The resources that are essential to an individual's survival are small, but a rapidly increasing number of people (as we see in developing countries) tends to overwhelm and deplete a country's soils, forests, and other natural resources (Figure 9–2a). (2) In developed nations, individual resource demands are large, far above requirements for survival. In order to satisfy their desires rather than their basic needs, people in more affluent nations exhaust resources and degrade the global environment through extravagant consumption and "throwaway" lifestyles (Figure 9–2b).

Types of resources

When examining the effects of population on the environment, it is important to distinguish between the two types of natural resources: nonrenewable and renewable. **Nonrenewable resources,** which include minerals (such as aluminum, tin, and copper) and fossil fuels (coal, oil, and natural gas), are present in limited supplies and are depleted by use. They are not replenished by natural processes within a reasonable period of time. Fossil fuels, for example, take millions of years to form.

In addition to a nation's population, several other factors affect how nonrenewable resources are used—including how efficiently the resource is extracted and processed, and how much of it is required or consumed by different groups of people. Nonetheless, the inescapable fact is that Earth has a finite supply of nonrenewable resources that sooner or later will be exhausted. In time, technological advances may enable us to find or develop substitutes for nonrenewable resources. And slowing the rate of population growth will help us buy time to develop such alternatives.

Renewable resources include trees in forests; fishes in lakes, rivers, and the ocean; fertile agricultural soil; and fresh water in lakes and rivers. Nature replaces these resources, and they can be used forever as long as they are not overexploited in the short term. In developing countries, forests, fisheries, and agricultural land are particularly important renewable resources because they provide food. Indeed, many people in developing countries are subsistence farmers, able to harvest just enough food so that they and their families can survive.

(a)

(b)

Figure 9–2

People and natural resources. (a) The rapidly increasing number of people in developing countries overwhelms their natural resources, even though individual resource requirements may be low. Shown is a typical Indian family, from Ahraura Village, India, with all their possessions. (b) People in developed countries consume a disproportionate share of natural resources. Shown is a typical American family, from Pearland, Texas, with all their possessions. (a, b, Peter Ginter/*Material World*)

Rapid population growth can cause renewable resources to be overexploited. For example, when fisheries are overharvested, there will be too few fishes to function as a food source (see Chapters 1 and 18). A similar problem arises when land that is inappropriate for farming—such as mountain slopes or tropical rain forests—is used to grow crops. Although this practice may provide a short-term solution to the need for food, it does not work in the long run, because when these lands are cleared for farming, their agricultural productivity declines rapidly and severe environmental deterioration occurs. Renewable resources, then, are *potentially* renewable. They must be used in a sustainable way—that is, in a manner that gives them time to replace or replenish themselves.

The effects of population growth on natural resources are particularly critical in developing countries. The economic growth of developing countries is often tied to the exploitation of natural resources. These countries are faced with the difficult choice of exploiting natural resources to provide for their expanding populations in the short term or conserving those resources for future generations. (It is instructive to note that the economic growth and development of the United States and of other highly developed nations came about through the exploitation—and in some cases the destruction—of their resources.)

Population: numbers versus resource consumption Whereas it is true that resource issues are clearly related to population size (more people use more resources), an equally if not more important factor is a population's *resource consumption*. People in developed countries are conspicuous consumers; their use of resources is greatly out of proportion to their numbers. A single child born in a developed country such as the United States, for example, causes a greater impact on the environment and on resource depletion than do a dozen or more children born in a developing country. Many natural resources are needed to provide the air conditioners, disposable diapers, cars, video cassette recorders, and other comforts of life in developed nations. Thus, the disproportionately large consumption of resources by developed countries affects natural resources and the environment as much as does the population explosion in the developing world.

Population and environmental impact: a simple model

Human impact on the environment is difficult to assess, but it can be estimated by using the three factors that are most important in determining this impact: (1) the number of people, (2) the environmental effects of the technologies used to obtain and consume the resources used, and (3) the amount of resources used per person.

$$\text{environmental impact} = \text{population} \times \text{technological impact} \times \text{consumption per person}$$

In order to determine environmental impact of CO_2 emissions from cars, for example, multiply the number of people driving cars (population) times the CO_2 emissions per mile (technological impact) times the number of miles driven per person (consumption per person). This model demonstrates that developing cleaner technologies will not by itself reduce pollution and environmental degradation; to accomplish this, population and per-capita consumption must also be controlled.

A country is *overpopulated* if the level of demand on its resource base results in damage to the environment. If we combine our three factors in order to compare human impact on the environment in developing and developed countries, we see that a country can be overpopulated in two ways. **People overpopulation** occurs when the environment is worsening from too many people, even if those people consume few resources per person. People overpopulation is the current problem in many developing nations. In contrast, **consumption overpopulation** occurs when each individual in a population consumes too large a share of resources. The effect of consumption overpopulation on the environment is the same as that of people overpopulation—pollution and degradation of the environment. Many affluent developed nations suffer from consumption overpopulation: *developed nations represent only 20 percent of the world's population, yet they consume significantly more than half of its resources.*

BORN CONSUMERS

From birth, people from highly developed nations are rampant consumers, using far more resources per person than residents of other nations. Most highly developed nations consume far more than their fair share of resources. For example, highly developed nations, which make up approximately 20 percent of the world's population, account for the lion's share of total resources consumed:

- 86 percent of aluminum used.
- 76 percent of timber harvested.
- 68 percent of energy produced.
- 61 percent of meat eaten.
- 42 percent of the fresh water consumed.

Rich nations also generate 75 percent of the world's pollution and waste.

Economic Effects of Continued Population Growth

The relationship between economic development and population growth is difficult to evaluate. Some economists have argued that population growth stimulates economic development and technological innovation. Others hold that developmental efforts are hampered by a rapidly expanding population. Most major technological advances are now occurring in countries where population growth is low to moderate, an observation that seems to support the latter point of view.

The National Research Council of the U.S. National Academy of Sciences published a report in 1986 that examined whether large increases in population were a deterrent to economic development. Their panel of experts took into account the complex interactions among global problems such as underdevelopment, hunger, poverty, environmental problems, and population growth. While concluding that population stabilization alone would not eliminate other world problems, the panel determined that for most of the developing world, economic development would profit from slower population growth. Thus, population stabilization would not guarantee higher living standards but would probably promote economic development, which in turn would raise the standard of living.

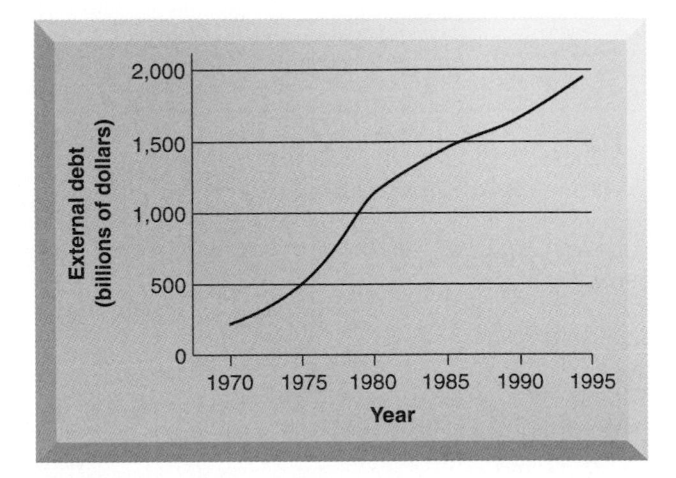

Figure 9–3
The external debt in developing countries, from 1970 to 1994.

Debt in developing countries

If a country's standard of living is to be raised, its economic growth must be greater than its population growth; if a population doubles every 40 years, then its economic goods and services must more than double during that time. Until recently, many developing nations were able to realize economic growth despite increases in population, largely because of financial assistance, usually in the form of loans from banks and governments of developed nations, or multilateral institutions such as the World Bank and International Monetary Fund.

However, it has become increasingly difficult for many developing countries to continue raising their standards of living, because the tremendous debts they have accumulated while funding past development preclude future loans. For example, sub-Saharan African countries, excluding South Africa, are overwhelmed by a massive foreign debt of $180 billion in 1995. This amount, estimated to be ten percent higher than the entire region's output in goods and services, is severely impeding economic development.

As of 1994, developing nations owe more than $1.9 trillion to developed nations, foreign banks, and multilateral institutions (Figure 9–3). Because many of their loans were in default during the 1980s and early 1990s, much of this debt has been restructured—that is, reduced and renegotiated so that the debtor nations will be able to repay it. However, the very poorest nations still have significant unrestructured debts.

Population and Urbanization

The social, environmental, and economic aspects of population growth are influenced not only by an excess of people but also by the geographical distribution of people in rural areas, cities, and towns. Throughout recent history, people have increasingly migrated to cities. When Europeans first settled in North America, the majority of the population was farmers in rural areas. Today approximately 5 percent of the people in the United States are involved in farming, and three-fourths of the U.S. population lives in cities. The increasing convergence of a population in cities is known as **urbanization.**

How many people does it take to make an urban area or a city? The answer varies from country to country; it can be anything from 100 homes clustered in one place to a population of 50,000 residents. The important distinction between rural and urban areas is not how many people live there but how people make a living. Most people residing in a rural area have occupations that involve harvesting natural resources—such as fishing, logging, and farming. In urban areas, most people have jobs that are not directly connected with natural resources.

Cities have grown at the expense of rural populations for several reasons. With advances in agriculture, including the increased mechanization of farms, more and more people can be supported by fewer and fewer farmers. Consequently, there are fewer employment opportunities for people in rural settings. Cities have traditionally provided more jobs because cities are the sites of industry, economic development, and educational and cultural opportunities.

The advantages of urban life notwithstanding, today many problems are faced by cities in developed and developing countries. Consider homelessness. Every country, even a highly developed country such as the United States, has people who lack shelter living in cities. Urban problems such as homelessness are usually more pronounced in the cities of developing nations, however. In many cities in India, for example, thousands of homeless people sleep in the streets each night (Figure 9–4).

Urbanization is a worldwide phenomenon, but the percentage of people living in cities compared to rural settings is higher in developed countries than in developing countries. In 1996, urban inhabitants made up 75 percent of the total population of developed countries, but only 35 percent of the total population of developing countries.

Urban growth

Although proportionately more people still live in rural settings in developing countries, urbanization has been increasing there. Illustrating the greater urban growth of developing nations is the fact that most of the largest cities today are in developing countries. In 1950, three of the ten largest cities were in developing countries: Shanghai, Buenos Aires, and Calcutta. In 1994, seven of the ten largest cities were in developing countries: São Paulo,

Figure 9–4
Homeless people sleep on a street in Bombay, India. Homelessness, a serious problem in the cities of developed countries, is an even greater problem in cities of developing nations. (Louis Psihoyos/Matrix)

Table 9–1
The World's Ten Largest Cities

1950	1994
1. New York, USA, 12.3*	1. Tokyo, Japan, 26.5
2. London, United Kingdom, 8.7	2. New York, USA, 16.3
3. Tokyo, Japan, 6.7	3. São Paulo, Brazil, 16.1
4. Paris, France, 5.4	4. Mexico City, Mexico, 15.5
5. Shanghai, China, 5.3	5. Shanghai, China, 14.7
6. Buenos Aires, Argentina, 5.0	6. Bombay, India, 14.5
7. Chicago, USA, 4.9	7. Los Angeles, USA, 12.2
8. Moscow, USSR, 4.8	8. Beijing, China, 12.0
9. Calcutta, India, 4.4	9. Calcutta, India 11.5
10. Los Angeles, USA, 4.0	10. Seoul, South Korea, 11.5

*Population in millions.

Mexico City, Shanghai, Bombay, Beijing, Calcutta, and Seoul (Table 9–1).

Urbanization is increasing in developed nations, too, but at a much slower rate. Consider the United States as representative of developed nations. Here, most of the migration to cities occurred during the past 150 years, when an increased need for industrial labor coincided with a decreased need for agricultural labor. The growth of U.S. cities over such a long period of time was slow enough to allow important city services such as water, sewage, education, and adequate housing to keep pace.

In contrast, the faster urban growth in developing nations has outstripped the capacity of many cities to provide basic services. It has also outstripped their economic growth. Consequently, cities in developing nations are faced with graver challenges than are cities in developed countries. These challenges include substandard housing for most residents (Figure 9–5), exceptionally high unemployment, and inadequate or nonexistent water, sewage, and waste disposal. Rapid urban growth also strains school, medical, and transportation systems. The Second U.N. Conference on Human Settlement, which was held in Istanbul, Turkey, in 1996, considered these urban issues and others, such as poverty, crime, and the potential of epidemics in a densely populated city.

At first glance, it appears that the concentration of people into cities has an overall harmful effect on the environment. Urban areas rely on the surrounding nonurban environment to provide natural resources—the food, water, energy, building supplies, and other materials that they require. Moreover, the concentration of human activities in urban areas produces pollution that contaminates surrounding rural areas as well as the cities them-

Figure 9–5
The substandard housing seen in this section of Port au Prince, Haiti, is common in cities of developing nations. Rapid urban growth, together with overpopulation, poverty, and lack of economic development, has caused severe problems in many cities of developing nations. (Les Stone/SYGMA)

FOCUS ON

How Compact Development Helps The Environment

From all accounts, it appears that people in the future will live primarily in urban areas. Today's cities are associated with the loss to suburban sprawl of surrounding nonurban areas and with an increase in pollution produced by such high concentrations of people. Most workers have to commute dozens of miles through traffic-snarled streets, from suburbs where they live to downtown areas where they work. Because development is so spread out in the suburbs, automobiles are a necessity to accomplish everyday chores. This heavy dependence on motor vehicles as our primary means of transportation increases air pollution and causes other environmental problems (see Chapter 10). Also, most cities have blocks and blocks of abandoned, vacant areas, while the suburbs continue to expand outward, swallowing natural areas and farmland.

Urban areas do not have to be this way, and a well planned city actually benefits the environment by reducing pollution and preserving nonurban areas. The solution to urban growth is **compact development,** in which cities are designed so that housing is close to shopping and jobs, all of which are connected by public transportation. Dependence on motor vehicles and their associated pollution is reduced as people walk, cycle, or take public transit such as buses or light rails to work and shop. With compact development, fewer parking lots and highways are needed, so there is more room for parks, open space, housing, and businesses.

Curitiba, a city of more than 2 million people in Brazil, provides a good example of compact development. Curitiba's city officials and planners have had notable successes in public transportation, traffic management, land use planning, waste reduction and recycling, and community livability. For example, the city possesses an inexpensive, efficient mass transit system that uses clean, modern buses. Because Curitiba relies less on automobiles than do comparable Brazilian cities, it has less traffic congestion and significantly cleaner air, both of which are major goals of compact development. Over several decades, Curitiba purchased and converted flood-prone properties along rivers in the city to a series of interconnected parks that are crisscrossed by bicycle paths. This move reduced flood damage and increased the per-capita amount of "green space" from 0.5 square meter in 1950 to 50 square meters today, a significant accomplishment considering Curitiba's rapid population growth during the same period. Another example of Curitiba's creativity is its labor-intensive Garbage Purchase program, in which poor people exchange filled garbage bags for bus tokens, surplus food, or school notebooks.

Like Curitiba, most cities can be carefully reshaped over a period of several decades to make better use of space and to reduce dependence on motor vehicles. City planners and local and regional governments are increasingly adopting measures that will provide the benefits of compact development in the future.

selves. However, urbanization also has the potential to provide tangible environmental benefits (see Focus On: How Compact Development Helps the Environment).

Does urbanization affect the rate at which the population grows? Urbanization appears to be a factor in *decreasing* fertility rates, perhaps because family planning services, including access to contraceptives, are more readily available in urban settings.

REDUCING THE FERTILITY RATE

Migration (moving from one place to another) used to be a solution for overpopulation, but not today. As a species, we have expanded our range throughout Earth; there is no habitable location to which we can migrate that does not already support humans. Nor is increasing the death rate an acceptable means of regulating population size. Clearly, the way to control our expanding population is by reducing the number of births. Because the total fertility rate is influenced by cultural traditions, women's social and economic status, and family planning, we examine each of these factors.

Culture and Fertility

The values and norms of a society—what is considered right and important, and what is expected of a person—constitute a part of that society's culture. With respect to fertility and culture, a couple is expected to have the number of children that are determined by the traditions of their society.

High fertility rates are traditional in many cultures. The motivations for having lots of babies vary from culture to culture, but overall *a major reason for high fertility rates is that infant and child mortality rates are high.* In order for a society to endure, it must continue to produce enough children who survive to reproductive age. Thus, if infant and child mortality rates are high, fertility rates must also be high to compensate. Although the worldwide infant and child mortality rates have been decreasing, it will take longer for fertility levels to decline. Part of the reason for this slowness is cultural: changing anything that has been traditional, including large family size, usually takes a long time.

Higher fertility rates in some developing countries are also due to the important economic and societal roles of children. In some societies, children usually work in family enterprises such as farming or commerce, contributing to the family's livelihood (Figure 9–6). The International Labor Organization estimates that, worldwide, 100 to 200 million children under the age of 15 work; more than 95 percent of these children live in developing countries. When these children become adults, they provide support for their aging parents. In contrast, children in developed countries have less value as a source of labor, because they attend school and because human labor is less important in a mechanized society. Further, developed countries provide many social services for the elderly, so the burden of their care does not fall entirely on their offspring.

Many cultures place a higher value on male children than on female children. In these societies, a woman who bears many sons achieves a high status; thus, there is a social pressure that keeps the fertility rate high.

ENVIROBRIEF

SUCCESSFUL SHOTS

Since the 1980s, worldwide immunization programs have enjoyed tremendous success:

- The U.N. Expanded Programme for Immunizations administers vaccines against tuberculosis, diphtheria, pertussis, tetanus, polio, and measles.
- In 1990, 80 percent of infants from developing nations were immunized, up from 25 percent in the 1980s.
- Vaccination rates continue to rise during the 1990s.
- An estimated 3 million lives were saved in 1992, at an average cost of $15 per child in poorer nations.
- Researchers strive for further immunization success at lower cost by developing vaccines that require fewer doses and maintain their potency at tropical temperatures.

Figure 9–6

Children gathering firewood near the Ethiopian village of Meshal. In developing countries, fertility rates are high partly because children contribute to the family by working. More children mean a greater family income. (Mike Goldwater-Network/Matrix)

Religious values are another aspect of culture that affects fertility rates. For example, several studies done in the United States point to differences in fertility rates among Catholics, Protestants, and Jews. In general, Catholic women have a higher fertility rate than either Protestant or Jewish women, and women who do not follow any religion have the lowest fertility rate of all. However, it is difficult to conclude that the observed differences in fertility rates are the result of religious differences alone. Other variables, such as race (certain religions are associated with particular races) and residence (certain religions are associated with urban or with rural living), complicate any generalizations that might be made.

The Social and Economic Status of Women

In many societies, women do not have the same rights as men. Sons are more highly valued than daughters, so girls are often kept home to work rather than being sent to school. In some African countries, for example, only 2 to 5 percent of girls are enrolled in secondary school (high school). Worldwide, some 90 million girls are not given the opportunity to receive a primary (elementary school) education. Laws, customs, and lack of education often limit women to low-skilled, low-paying jobs. In such societies, marriage is usually the only way for a woman to achieve social influence and economic security.

Evidence is accumulating that *the single most important factor affecting high fertility rates may be the low status of women in many societies*. A significant way to tackle population growth, then, is to improve the social and economic status of women. We say more about this later, but for now we examine how marriage age and educational opportunities, especially for women, affect fertility.

Marriage age and fertility

The fertility rate is affected by the average age at which women marry, which in turn is determined by the laws and customs of the society in which they live. Women who marry are more apt to bear children than women who do not marry, and the earlier a woman marries, the more children she is likely to have.

The percentage of women who marry and the average age at marriage vary widely among different societies. There is always a correlation between marriage age and fertility rate, however. Consider Sri Lanka and Bangladesh, two developing countries in South Central Asia. In Sri Lanka, the average age at marriage is 25, and the average number of children born per woman is 2.3. In contrast, in Bangladesh, the average age at marriage is 17, and the average number of children born per woman is 3.7.

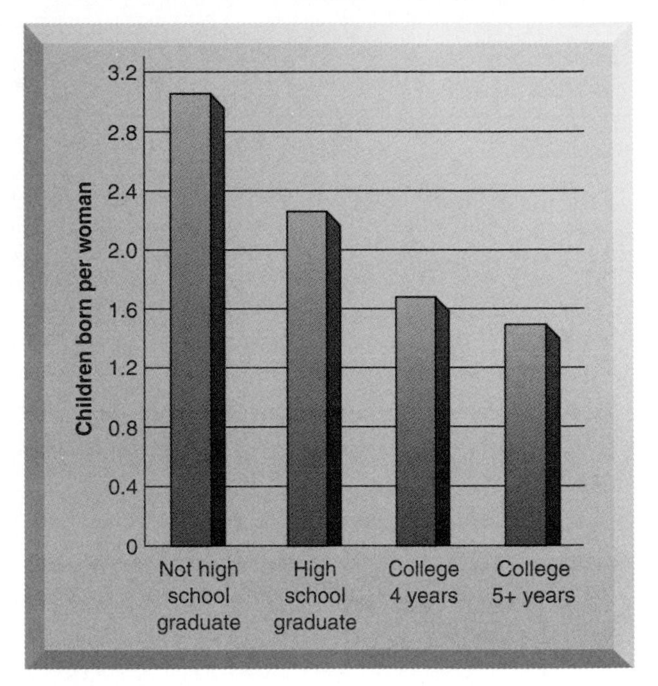

Figure 9–7
The number of children a woman has varies with the amount of education she has received. Shown are the 1987 fertility rates for 35- to 44-year-old women in the United States with differing levels of education.

Educational opportunities and fertility

Women with more education tend to marry later and have fewer children. Figure 9–7 shows the fertility rates of women in the United States with different education levels. Providing women with educational opportunities delays their first childbirth, thereby reducing the number of childbearing years and increasing the amount of time between generations. Education also opens the door to greater career opportunities and may change women's lifetime aspirations. In the United States, it is not uncommon for a woman in her thirties or forties to give birth to her first child, after establishing a career.

In developing countries there is also a strong correlation between the average amount of education a woman receives and fertility rate. For example, women in Botswana with a secondary education have an average of 3.1 children each, women with a primary education have 5.1 children each, and women with no formal education have 5.9 children each.

Education increases the probability that women will know how to control their fertility, and it provides them with knowledge to improve the health of their families, which results in a decrease in infant and child mortality. Education also increases women's options, opening doors to other careers and ways of achieving status besides having babies.

Education may have an indirect effect on fertility rate, as well. Children who are educated have a greater chance of improving their living standards, partly because they have more employment opportunities. Parents who recognize this may be more willing to invest in the education of a few children than in the birth of many children whom they cannot afford to educate. The ability of better educated people to earn more money may be one of the reasons why smaller family size is associated with increased family income, although another obvious reason is that fewer children are fewer mouths to feed and, thus, are less of a drain on the family income.

Family Planning Services

Socioeconomic factors may encourage people to want smaller families, but *reduction in fertility will not become a reality without the availability of health and family planning services*. The governments of most countries recognize the importance of educating people about basic maternal and child health care. Family planning services provide information on reproductive physiology and contraceptives, as well as the actual contraceptive devices, to those who wish to control the number of children they produce or to space their children's births. Family planning does not try to force people to limit their family sizes, but rather

Table 9–2
Birth Control Methods

Method	Failure Rate*	Mode of Action	Advantages	Disadvantages
Oral contraceptives	0.3; 5	Inhibit ovulation; may also affect endometrium and cervical mucus and prevent implantation	Highly effective; regulate menstrual cycle	Minor discomfort in some women; possible thromboembolism; hypertension, heart disease in some users; possible increased risk of infertility; should not be used by women who smoke
Depo-Provera	About 1	Inhibits ovulation	Effective; long-lasting	Fertility may not return for 6 to 12 months after use discontinued
Progesterone implantation (Norplant)	About 1	Inhibits ovulation	Effective; long-lasting	Irregular menstrual bleeding in some women
Intrauterine device (IUD)	1; 5	Not known; probably stimulates inflammatory response	Provides continuous protection; highly effective	Cramps; increased menstrual flow; spontaneous expulsion; increased risk of pelvic inflammatory disease and infertility; not recommended for women who have not completed childbearing
Spermicides (foams, jellies, creams)	3; 20	Chemically kill sperm	No side effects (?)	Some evidence linking spermicides to birth defects
Condom, female	25	Mechanically prevents sperm from entering vagina	No side effects; some protection against sexually transmitted diseases, including AIDS	Could break

Modified from: Solomon, Berg, Martin, Villee, *Biology*, 4th ed. (Philadelphia: Saunders College Publishing, 1996).
*The lower figure is the failure rate of the method; the higher figure is the rate of method failure plus failure of the user to use the method correctly. Based on number of failures per 100 women who use the method per year in the United States.
†The failure rate is lower when the diaphragm is used together with spermicides.

attempts to convince people that small families (and the contraceptives that promote small families) are acceptable and desirable. The major birth control methods in use today are shown in Table 9–2.

Contraceptive use is strongly linked to lower fertility rates (Table 9–3). Research has shown that 90 percent of the decrease in fertility in 31 countries was a direct result of the increased use of contraceptives. In developed countries, where fertility rates are at replacement levels or lower, the percentage of married women of reproductive age who use contraceptives is often greater than 65 percent. Fertility declines have been noted in developing countries where contraceptives are readily available. During the 1970s and early 1980s, use of contraceptives in East Asia and many areas of Latin America increased significantly, and these regions experienced a corresponding decline in birth rate. In areas where contraceptive use remained low, such as parts of Africa, there was little or no decline in birth rate.

Family planning centers provide information and services primarily to women. As a result, in the male-dominated societies of many developing countries, such

Table 9-2 (continued)
Birth Control Methods

Method	Failure Rate*	Mode of Action	Advantages	Disadvantages
Condom, male	2.6; 10	Mechanically prevents sperm from entering vagina	No side effects; some protection against sexually transmitted diseases, including AIDS	Interruption of foreplay to put it on; slightly decreased sensation for male; could break
Contraceptive diaphragm (with jelly)†	3; 14	Diaphragm mechanically blocks entrance to cervix; jelly is spermicidal	No side effects	Must be prescribed (and fitted) by physician; must be inserted prior to coitus and left in place for several hours after intercourse
RU 486 (abortion pill)	5	Induces miscarriage in early pregnancy	Safer alternative to a surgical abortion	Some bleeding and cramps; not yet available in many countries
Rhythm‡	13; 21	Abstinence during fertile period	No side effects (?)	Not very reliable
Douche	40	Flush semen from vagina	No side effects	Not reliable; sperm are beyond reach of douche in seconds
Withdrawal (coitus interruptus)	9; 22	Male withdraws penis from vagina prior to ejaculation	No side effects	Not reliable; sperm in the fluid secreted before ejaculation may be sufficient for conception
Sterilization				
Tubal ligation	0.04	Prevents ovum from leaving uterine tube	Most reliable method	Often not reversible
Vasectomy	0.15	Prevents sperm from leaving scrotum	Most reliable method	Often not reversible
Chance (no contraception)	About 90			

‡There are several variations of the rhythm method. For those who use the calendar method alone, the failure rate is about 35. However, if the body temperature is taken daily and careful records are kept (temperature rises after ovulation), the failure rate can be reduced. When women use some method to help determine the time of ovulation, and have intercourse *only* more than 48 hours *after* ovulation, the failure rate can be reduced to about 7.

services may not be as effective as they could otherwise be. Polls of women in developing countries reveal that many who say they do not want additional children still do not practice any form of birth control. When asked why they do not use birth control, these women frequently respond that their husbands want additional children.

Reduction in fertility will not result from family planning alone, especially where cultural traditions and religious beliefs prohibit birth control. Only when people are convinced that having smaller families will somehow benefit them in such ways as a better standard of living or better health will they embrace family planning.

GOVERNMENT POLICIES AND FERTILITY

The involvement of governments in childbearing is well established. Laws determine the minimum age at which people may marry and the amount of education that is compulsory. Governments may allot portions of their budgets to family planning services, education, health care,

Table 9–3
1996 Contraceptive Use and Fertility Rates in Selected Countries

Country	Percent Married Women Using Any Type of Contraception	Total Fertility Rate
China	83	1.8
Germany	75	1.3
United States	73	2.0
Thailand	66	2.2
Mexico	65	3.1
South Africa	53	4.1
Indonesia	55	2.9
India	41	3.4
Pakistan	12	5.6
Nigeria	6	6.0

old-age security, or incentives for smaller or larger family size. The tax structure, including additional charges or allowances based on family size, also affects fertility.

In recent years, the governments of many developing countries have recognized the need to limit population growth and have formulated policies such as economic rewards and penalties designed to achieve this goal. Most countries sponsor family planning projects, many of which are integrated with health care, education, economic development, and efforts to improve women's status. Many of these projects are supported by the U.N. Fund for Population Activities.

Population control measures have been instituted in many developing countries (recall the chapter introduction). Here we examine those in China, India (the two most populous nations), Nigeria, and Mexico.

China

China, with a mid-1996 estimated population of 1.22 billion people, has the largest population in the world. Recognizing that its rate of population growth had to decrease or the quality of life for everyone in China would be compromised, the Chinese government in 1979 instigated an aggressive plan to push China into the third demographic stage. (Recall from Chapter 8 that the third demographic stage is characterized by a decline in the birth rate along with a relatively low death rate.) Announcements were made of incentives to promote later marriages and one-child families. Local jurisdictions were assigned the task of reaching this goal. A couple who signed a pledge to limit themselves to a single child might be eligible for such incentives as medical care and schooling for that child, cash bonuses, preferential housing, and retirement funds. Penalties were also instituted, including fines and the surrender of all privileges if a second child was born.

China's aggressive plan brought about an immediate and drastic reduction in fertility, from 5.8 births per woman in 1970 to 2.1 births per woman in 1981. However, it compromised individual freedom of choice. In some instances, social pressures from the community induced women who were pregnant with a second child to get an abortion. Moreover, based on the disproportionate number of male versus female babies reported born in recent years, it is suspected that thousands of newborn baby girls were killed or abandoned by their parents who, if required to conform to the one-baby policy, wanted a boy. In China, sons traditionally provide old-age security for their parents and—for this and other cultural reasons—are valued more highly than daughters. As a result of the increased ratio of male to female births, demographers project that by the middle of the next century, marriageable males will outnumber marriageable females by one million.

Recently, China's population control program has used less coercive education and publicity campaigns, more than penalties, to achieve its goals (Figure 9–8). China trains population specialists at institutions such as the Nanjing College for Family Planning Administrators. In addition, thousands of secondary school teachers have been taught how to integrate population education into the curriculum. China's 1996 total fertility rate was 1.8, and the Chinese government continues its policy of strict population control.

India

India is the second most populous nation, with an estimated mid-1996 population of 950 million. It was the first country to establish government-sponsored family planning, in the 1950s. Unlike China, India did not experience immediate results from its efforts to control population growth, in part because of the diverse cultures, religions, and customs in different regions of the country. For example, Indians speak many different languages, which makes a broad program of family planning education difficult.

In 1976 the Indian government became more aggressive. It introduced incentives to control population growth and controversial programs of compulsory sterilization in several states. If a man had three or more living children, he was compelled to obtain a vasectomy. Compulsory sterilization was a failure; it had little effect on the birth rate and was exceedingly unpopular. It may have been partly responsible for Indira Gandhi being voted out of office in 1977.

More recently, India has integrated development and family planning projects. For example, adult literacy and

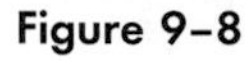

Figure 9–8

A billboard campaign in China promotes the one-child family. Note that the child is a male rather than a female. (E. F. Anderson/Visuals Unlimited)

population education programs have been combined. Multi-media advertisements and education have been used to promote voluntary birth control. India has also emphasized improving health services to lower infant and child mortality rates, improving women's status, and increasing birth spacing. These changes have had an effect: India's total fertility rate has declined from 5.3 in 1980 to 3.4 in 1996.

Despite these gains, India is facing severe problems. Population pressure has caused the deterioration of India's environment in the past few decades, and 40 percent of Indians live below the official poverty level.

Mexico

Mexico, with a mid-1996 estimated population of 94.8 million, is the second most populous nation in Latin America. (Brazil, with a population of 160.5 million, is the most populous.) Mexico has a tremendous potential for growth because 36 percent of its population is under 15 years of age. Even with a low birth rate, the population would still increase in the future because of the large number of women having babies.

In 1973 the Mexican government instigated several measures to reduce population growth, such as educational reform, family planning, and health care. Mexico has had great success in reducing its fertility level, from 6.7 births per woman in 1970 to 3.1 births per woman in 1996.

Mexico's goal includes not only population stabilization, but balanced regional development. Its urban population is 71 percent of its total population, and most of these people live in Mexico City. Although Mexico is largely urbanized compared to other developing countries, its urban-based industrial economy has not been able to absorb the great number of people in the work force, and unemployment is very high. Consequently, many Mexicans have migrated, both legally and illegally, to the United States.

Mexico's recent efforts at population control include multi-media campaigns. As in Egypt, popular television and radio soap operas carry family planning messages. Booklets on family planning are distributed, and population education is being integrated into the public school curriculum. Social workers receive training in family planning as part of their education.

Nigeria

Nigeria is part of the sub-Saharan region that currently has the world's most rapid population growth. Nigeria has the largest population of any African country: in mid-1996 its population was estimated at 103.9 million people. Its total fertility rate is 6.0 births per woman, and it has great reproductive potential because 45 percent of the population is under 15 years of age. The average life expectancy in Nigeria is 56 years, which is low in part because of high infant and child mortality rates.

The Nigerian government has recognized that its economic goals are more likely to be attained if its rate of population growth decreases. In 1986 Nigeria developed a national population policy that is an integration of population and development projects. Part of the plan involves improving health care, including training nurses and other health care professionals. Population education is being used to encourage later marriages and birth spacing.

THE 1994 GLOBAL SUMMIT ON POPULATION AND DEVELOPMENT

As mentioned in the chapter introduction, the third U.N. International Conference on Population and Development was held in Cairo in 1994 to address population issues and the role of government in controlling population growth. The two previous U.N. population conferences, in 1974 and 1984, focused on setting national and global population targets and meeting those targets through family planning. Departing from this tradition, the 1994 conference drafted a 20-year World Programme of Action that provides specific measures to be addressed at the *local* level. It focuses on reproductive rights, empowerment of women, and reproductive health. *Reproductive rights* refers to the rights of individuals to make informed decisions about their fertility and reproductive health. *Empowerment of women* is concerned with improving women's status through education and economic opportunities; it is thought that such empowerment will give women more choices and therefore greater control over their reproductive rights. *Reproductive health* includes access to affordable contraception, modern obstetrical practices, and improved maternal and child heath care, as well as control of sexually transmitted diseases. For example, most of the more than 500,000 women who die each year as a result of pregnancy and childbirth live in developing countries and do not have access to medical care during pregnancy and childbirth.

The conference also recognized a wide range of issues that had been largely ignored during previous conferences: the responsibilities of men in family planning and raising children (Figure 9–9), violence against women, female genital mutilation, unsafe abortions, and prenatal sex selection.

Despite the bitter debates over contentious issues such as abortion, the 1994 plan was approved by 180 nations. Although the plan is not a treaty and is therefore not legally binding, it is expected to set the agenda for national and international programs to slow population growth. Local laws, cultural values, and religious beliefs will be respected in adopting the provisions of the plan.

The World Programme of Action was further supported by the Fourth World Conference on Women's Rights, which met in 1995 in Beijing. The conference em-

Figure 9–9
The importance of male involvement in family planning and child rearing has been overlooked in the past, when family planning agencies focused almost exclusively on women. Shown is a man assuming family responsibilities in Manila, Philippines. (Peter Barker/Panos Pictures)

REPRODUCTION AND WOMEN—DIFFERENT COUNTRIES, DIFFERENT RIGHTS

One of the many issues raised at the U.N. population summit in Cairo in 1994 was the reproductive rights of women, and particularly the vastly different government policies that pressure or force women to adhere to specific reproductive plans:

- Urban women in China are limited to one child through allegedly enforced contraception, sterilization, or abortion.
- In Japan, oral contraceptives (birth control pills) are illegal, in part because it is feared that their widespread use will further lower the fertility rate. The only legal birth control options available to Japanese couples are condoms and abortion.
- In certain European countries (Germany, Hungary, Poland, Belgium, Luxembourg, and Portugal) and Japan, women are strongly encouraged, with cash payments or promises of benefits such as payment of obstetric expenses and child care, to bear several children.
- Women in Bangladesh and India are encouraged to limit reproduction to no more than two children, and many undergo substandard tubal ligations.
- Saudi Arabian and Libyan couples do not have legal access to contraceptives.
- Abortions are illegal in most Latin American nations, where many maternal deaths are caused by botched illegal abortions.
- Russian women are forced to rely primarily on abortions as birth control because quality contraceptives are often unavailable.

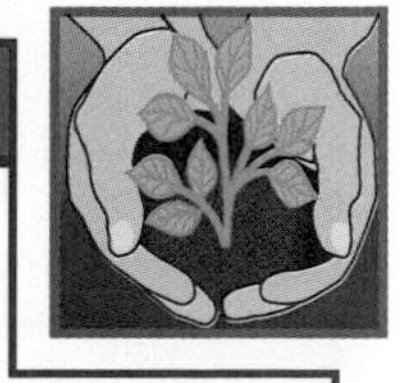

YOU CAN MAKE A DIFFERENCE

Sponsor a Child in a Developing Country

It is easy to conclude that there is nothing you can do to help improve the lives of people who live in developing countries. But you can make a difference in one child's life and, in doing so, help the child's family and community. Several nonprofit organizations offer the opportunity for an individual or group to sponsor a child by donating a fixed amount each month, quarter, or year to assist in the child's education, nutrition, and health care. In addition, your donations help these organizations carry out all-important family and community projects, including the improvement of health facilities, clean water projects, teacher training, community gardens, and family planning. The average sponsorship fee is less than $25 per month, which is roughly equivalent to the monthly cost of cable TV.

When you sponsor a child through one of these organizations, you typically correspond with the child and receive reports from case workers and community workers about the child's progress in school, the family's economic status, and any community projects being undertaken by the field staff in conjunction with community and family members.

If you decide to sponsor a child, contact one of the nonprofit organizations that run these programs. Two of them are PLAN International USA, 155 Plan Way, Warwick, RI 02886, 1-800-556-7918; and Save the Children, 54 Wilton Road, Westport, CT 06880, 1-800-243-5075. (Save the Children also has sponsorship programs for needy children in the United States.) You can sponsor a child as an individual, or your instructor might want to organize a sponsorship by the entire class that is carried on by each subsequent class taking this course. Remember that once you begin sponsorship of a child, that child and his or her family will probably count on your assistance for several years, although you are not legally bound to continue. Thus, you should be ready to maintain your commitment long after you finish this course.

phasized that expanding women's access to health care, education, and employment are important not only in their own right, but because this access will slow population growth.

ACHIEVING POPULATION STABILIZATION

In this chapter we have considered global problems—including hunger, poverty, underdevelopment, and environmental issues—that are exacerbated by an increase in population. Population stabilization is critically important if we are to effectively tackle these serious problems. All countries must develop policies to bring about an immediate reduction in the rate of population growth. (See Focus On: Population Concerns in Europe for viewpoints that oppose population stabilization.) But what kinds of policies will help achieve population stabilization?

Developing countries should increase the amount of money that they allot to public health (to reduce infant and child mortality rates) and family planning services, particularly the dissemination of information on birth control. Governments should take steps to increase the average level of education, especially of women, and women must be given more employment opportunities. National population policies will not work without local community involvement and acceptance, and community leaders can provide feedback on the effectiveness of programs.

Developed countries can help developing nations achieve a decrease in population growth by providing financial support for the U.N. Fund for Population Activities and by supporting the research and development of new birth control methods. People living in developed countries can also help developing nations by sponsoring children (see You Can Make a Difference: Sponsor a Child in a Developing Country).

Most important, developed nations need to face their own population problems, particularly the environmental costs of consumption overpopulation. Policies to support recycling of resources, eliminate needless production, and discourage overconsumption and the throwaway mentality should be formulated. These policies will also show developing nations that developed countries are serious about facing the issues of overpopulation at home as well as abroad.

On a personal level, individuals in developed countries should examine their own consumptive practices and take steps to reduce them. Individual efforts to reduce material consumption—sometimes called *voluntary simplicity*—are also effective, not only for their collective effects but because they often influence additional people to reduce unnecessary consumption. Before purchasing a product, ask yourself if you *really* need it. A material lifestyle—two cars in the garage, steaks on the grill, a television in every room, and similar luxury items—may provide short-term satisfaction. Such a lifestyle is not as fulfilling in the long term, however, as learning about the world, interacting with others in meaningful ways, and contributing to the betterment of your family and community. In the end, a rich life is measured not by what you own but by what you have done.

FOCUS ON

Population Concerns in Europe

Population has stabilized in Europe, where most countries have lower-than-replacement-level fertility rates, and several countries have experienced a slight decline in population. As a result of fewer births, the proportion of elderly people in the European population is increasing. Population control has thus become a controversial issue in Europe, with two opposing viewpoints emerging.

Those who favor population growth are called pronatalists. Pronatalists think that the vitality of their region is at risk because of declining birth rates and that the decrease in population might result in a loss of economic growth. They presume their countries' positions in the world will weaken and that their cultural identity will be diluted by immigrants from non-European countries. (The next wave of immigrants to Europe is expected to be primarily from North Africa and the Middle East.) Pronatalists are also concerned about the possibility that pension and old-age security systems will be overwhelmed by the large numbers of elderly in Europe unless a larger work force is available to contribute to those systems. The most outspoken pronatalists assert that women should marry young and have many children for the good of society. Pronatalists favor government policies that provide incentives for larger families and penalties for smaller families.

The opponents of pronatalists do not view European society as declining; rather, they contend European influence is increasing, because they judge "power" in economic terms rather than in terms of population. Generally, they are not as opposed to immigration, and they maintain that Europe should not be making a concerted effort to increase its fertility rates when overpopulation is such a serious problem in much of the world. Further, they point out that technological innovations have eliminated many jobs in Europe, and the consequential unemployment would only be made worse by an increase in the labor force caused by a rise in birth rate. Opponents of pronatalists also question whether the elderly are a burden to society. Moreover, they argue that the elderly are not the only ones the work force supports. Their view is that the cost of providing for the young, especially for many years of education, cancels out any perceived benefit from an increased birth rate.

Advertisement showing the desirability of population growth in Ireland. (Courtesy of Industrial Development Authority of Ireland [IDA Ireland])

SUMMARY

1. Many human problems, such as hunger, resource depletion, environmental problems, underdevelopment, poverty, and urban problems, are exacerbated by the rapid increase in population. Those countries with the greatest food shortages are also the ones with the highest fertility rates. The relationship between world hunger and population growth may be influenced by the degrees of economic development, poverty, and inequitable distribution of resources.

2. Nonrenewable resources are present in a limited supply and cannot be replenished in a reasonable period on the human time scale. Slowing the rate of population growth will give us more time to find substitutes for nonrenewable resources as they are depleted. Renewable resources are replaced by natural processes and can be used forever, provided they are used in a sustainable way. Overpopulation causes renewable resources to be overexploited. When this happens, renewable resources become nonrenewable.

3. One model of the effect of population on the environment has three factors: the number of people, the amount of resources used per person, and the effect on the environment of using those resources. This model shows that there are two kinds of overpopulation: people overpopulation and consumption overpopulation.

4. Developing countries have people overpopulation, in which the population increase degrades the environment even though each individual uses few resources. In contrast, developed countries have consumption overpopulation, in which each individual in a stable population consumes a large share of resources and the result is environmental degradation.

5. Underdevelopment and poverty are associated with high fertility rates. Most economists think that slowing population growth helps promote economic development.

6. As a nation develops economically, the proportion of the population living in cities increases. In developing nations, most people live in rural settings, but their rates of urbanization are rapidly increasing. This makes it difficult to provide city dwellers with basic services such as housing, water, sewage, and transportation systems. Urbanization appears to be a factor in decreasing population growth, however.

7. The relationships between fertility rate and cultural, social, governmental, and economic factors are intricate. It appears that a combination of four factors is primarily responsible for high fertility rates. These include high infant and child mortality rates, the important economic and societal roles of children in some cultures, the low status of women in many societies, and a lack of health and family planning services.

8. The 1994 U.N. International Conference on Population and Development drafted a 20-year World Programme of Action that focuses on reproductive rights, empowerment of women, and reproductive health.

THINKING ABOUT THE ENVIRONMENT

1. Keep a list of the natural resources you use in a single day. How would this list compare to a similar list made by a poor person in a developing country?

2. How does a rapidly expanding population affect nonrenewable resources? Renewable resources?

3. Discuss this statement: The current human population crisis causes or exacerbates all environmental problems.

4. Explain how a single child born in the United States can have a greater effect on the environment and natural resources than a dozen or more children born in Kenya.

5. Discuss some of the ways in which population and economic growth are related.

6. What are some of the problems brought on by rapid urban growth in developing countries?

7. What is the relationship between fertility and marriage age? Between fertility and educational opportunities for women?

8. Explain the rationale behind this statement: It is better for developed countries to spend millions of dollars on family planning in developing countries now than to have to spend billions of dollars on relief efforts later.

* 9. The United States produces 30 times more of the pollutant carbon dioxide *per capita* than India, but India's population (949.6 million) is much larger than that of the United States (265.2 million). Which country adds more carbon dioxide to the atmosphere? How many times more?

RESEARCH PROJECTS

Contact the Census Bureau and study U.S. population growth during the past 20 years. Evaluate the impact of population growth in at least two different domains (for example, the environment, city services, freshwater shortages, job availability). Develop recommendations for a national domestic population policy based on your evaluation.

Investigate the role of the United States in establishing and maintaining population programs in developing nations, particularly through U.N. family planning programs. Determine how the U.S. role changed as a result of politics by comparing the Reagan/Bush and Clinton administrations' support of such programs.

On-line materials relating to this chapter are on the World Wide Web at **http://www.saunderscollege.com/lifesci/environment2** (select Chapter 9 from the Table of Contents).

SUGGESTED READING

Ashford, L. S. and J. A. Noble. "Population Policy: Consensus and Challenges," *Consequences* Vol. 2, No. 2, 1996. This article discusses how a consensus emerged at the 1994 International Conference on Population and Development (the population summit). It then examines what it will take to implement the summit's World Programme of Action.

Dasgupta, P. S. "Population, Poverty, and the Local Environment," *Scientific American*, February 1995. The three issues mentioned in the title of this article interact with one another in complex ways.

Ezcurra, E., and M. Mazari-Hiriart. "Are Mega-Cities Viable?" *Environment* Vol. 38, No. 1, January–February 1996. This article highlights many of the problems that plague Mexico City and other rapidly expanding urban areas.

Feeney, G. "Fertility Decline in East Asia," *Science* Vol. 266, 2 December 1994. Discusses recent fertility declines in Japan, Taiwan, South Korea, and China.

Harrison, P. "Sex and the Single Planet," *The Amicus Journal*, Winter 1994. Examines the link between population growth and environmental degradation.

Rabinovitch, J. and J. Leitman. "Urban Planning in Curitiba," *Scientific American*, March 1996. The Brazilian city of Curitiba is a model of compact development.

Riley, N. E. "China's 'Missing Girls': Prospects and Policy," *Population Today* Vol. 24, No. 2, February 1996. The low status of women in China has resulted in an imbalance in the number of reported births of girls versus boys.

Robey, B., S. O. Rutstein, and L. Morris. "The Fertility Decline in Developing Countries," *Scientific American*, December 1993. Some developing countries are experiencing a rapid decline in fertility rates, largely due to a greater availability of contraceptives and increased coverage by the media.

Sachs, A. "Men, Sex, and Parenthood," *World Watch* Vol. 7, No. 2, March–April 1994. An argument for shifting the focus of family planning to include men as well as women.

U.S. Consumption and the Environment. A briefing paper by the Union of Concerned Scientists, February 1994. Discusses the environmental costs of the U.S. lifestyle. A copy of this briefing paper may be obtained free by calling the Publications Department of the Union of Concerned Scientists, (617) 547-5552.

► A diver checks an underwater oil pipe leading from an offshore platform in the North Sea to Norway. (Terje Rakke/Image Bank).

PART 4

The Search for Energy

Efficiency: Less Energy, More Power

L. Hunter Lovins is President of Rocky Mountain Institute in Old Snowmass, Colorado, which she co-founded with her husband and colleague, Amory Lovins, in 1982. RMI's interrelated programs in Energy, Water, Agriculture, Economic Renewal, The Hypercar Center, Green Development, Corporate Sustainability, and Security are designed to foster the efficient, sustainable use of resources as a path to global security. Ms. Lovins holds B.A.s in Political Science and Sociology from Pitzer College, California, and a J.D. from Loyola University. Internationally recognized for their expertise in energy policy, the Lovinses have been consulted by heads of state, government agencies, and multinational corporations around the world. ◄

L. HUNTER LOVINS

What are the primary objectives of Rocky Mountain Institute's Energy Program? Explain what you call the "end-use/least-cost" approach, and how it relates to energy issues.

Our Energy Program grew from Amory's and my longstanding efforts to find solutions to the energy problem that make sense economically and environmentally. Amory developed the end-use/least-cost approach to analyze first, how much and what kinds of energy we need for each task, and second, how to get that energy at the lowest cost.

What has this analysis shown?

Many people believe that we are running out of energy, thus the logical solution is to obtain more of it. In response to the 1973 Arab oil embargo, President Nixon proposed to build, by the year 2000, 450–800 new nuclear reactors and 500–800 new coal fire plants, primarily to produce more electricity. During 1976–1985 alone, this would have cost over $1 trillion, three quarters of all available domestic investment money. The plan was discarded for obvious reasons.

Of course we are not running out of energy. There is still a lot of gas, coal, and oil. But we don't need all of it and can't afford it. About 58% of America's delivered energy needs are for *heat,* over half of that at low temperatures, and 34% of our energy needs are for liquid fuels, primarily for *transportation.* Only 8% of our energy needs are for *electricity.*

What are the alternatives?

Most of our heating needs can be met through efficiency (using the energy we already have more productively) and through various forms of renewable energy sources. The efficiency technologies generally cost between half a cent and 2 cents per kilowatt-hour. Compare that to running an existing coal or nuclear plant, which costs about 2–5 cents per kilowatt-hour. Many renewables now compete with the costs of just running coal or nuclear plants.

Do we need to combine the new technologies with energy conservation?

Conservation does not mean being colder in the winter and hotter in the summer. Our focus is on technological efficiency to do the same job as before, using less energy and less money. If people have incentive and opportunity, they will make wise energy decisions for themselves, and in the process, preserve resources for the future.

RMI created and owns E Source, the premier source of information on electric efficiency. Tell us about that.

E Source has more than 230 members from over 35 countries, both developed and developing. Most members are utilities, industries, and governments. Saving energy can mean the difference between profit and loss. Consider the Southwire Corporation, the largest independent U.S. producer of cable, rod, and wire. Its ten plants in six states are quite energy-intensive. In the early 1980s, faced with falling prices for its products and rising costs, Southwire cut its energy use per pound of product in half over eight years. That savings *equaled* the company's profit for that period. Without an energy saving program, they would have operated at a loss and perhaps followed other such firms into bankruptcy. Their efficiency program may have saved over 4000 jobs. That's why so many companies subscribe to E Source. It's not just a public relations scheme to give themselves a "clean, green" image. We provide hard numbers to improve their financial bottom lines.

Your prescription for improving the utility industry's bottom line centers around the idea of the "negawatt" or saved watt of electricity. How can a utility make money by selling less energy?

Utility companies have many costs associated with providing energy to their customers: building new power plants, buying fuel, running the plant, dispatching the energy, etc. It was once true that the more electricity consumers used and the more utilities sold, the cheaper it was for everyone. For complex reasons, in the early 1970s this began to change. Today, it is more financially advantageous for utilities to produce *less* electricity, avoiding the costs of production, and saving money and energy. That is the negawatt, a saved watt. It is identical in every way to a generated watt, except that it is cheaper to create. Consumers are not interested in buying raw energy, but in getting access to energy services—heat, light, hot water, etc. If those services can be provided more cheaply through efficiency than by generating more energy, then it is just good business for utilities to supply those services, and to enable their consumers to use energy more efficiently.

> **If people have incentive and opportunity, they will make wise energy decisions for themselves, and in the process, preserve resources for the future.**
>
> –*L. Hunter Lovins*

What does the future look like for utilities?

Five West Coast utilities now project that they will be able to meet 75–100% of their new power needs in the 1990s from efficiency, and the rest from such renewables as wind power, solar photovoltaics, or small hydroelectric plants. They are not planning to build any large thermal power plants. Many of the renewable energy facilities will be built by private entrepreneurs. Utilities will still exist, but ultimately may come to resemble phone companies distributing energy services to customers without necessarily owning or operating power plants.

Could the negawatt idea be used by other industries? For example, when the ultralight, super fuel-efficient hypercars become available, could oil companies make money by selling less gas rather than by charging us more per gallon?

Yes. The big oil companies today are essentially banks, pools of capital in search of something to finance. They believe the business they're in is extracting and selling oil. They really should provide energy services to their customers by selling "negabarrels." They should be financing the conversion to hypercars, taking their return as a bank does. Few oil companies have yet begun to redefine their business in this way, but they will eventually. As new efficiency technologies are brought onto the market, oil prices are likely to be durably low.

To remedy our most dire energy problems, what are some of the newest technologies developing now? What technologies are in need of more research?

At the moment, almost a dozen manufacturers have made prototype cars that get 63–138 mpg, and that isn't even close to state-of-the-art. Using recent advances in aerodynamics, ultralight-weight materials, new engine and hybrid electric drive technology, and computer-aided design, you could get 150 mpg in a zippy, comfortable, affordable station wagon. Oil efficiency has not received nearly enough attention, so Rocky Mountain Institute will be working in that area.

There have been dramatic developments in renewable energy sources. Among the most important to pursue aggressively are: further refinements in new solar absorbing surfaces that can yield high industrial temperatures even in cloudy weather; sustainably producing biofuels from farm and forestry wastes; making photovoltaics even cheaper; and using sunlight to break down water directly into hydrogen. It's already so much cheaper to save fuel than to burn it that we could now abate virtually all of the greenhouse gases associated with global warming, not at a cost but at a profit. Unless government and industry foster the transition to an efficient, sustainable energy path, we will forego these opportunities, and commit ourselves to a future of economic and environmental degradation.

▶ Web site: **www.rmi.org**

10

The 1990 war in the Persian Gulf demonstrated America's dependency on foreign oil. (A. Tannenbaum/SYGMA)

Fossil Fuels

U.S. dependence on energy resources from other countries has been dramatically demonstrated several times in the past few decades. In 1973, the Organization of Petroleum Exporting Countries (OPEC) restricted oil shipments to the United States, precipitating an energy crisis. Total oil consumption was cut by only about 5 percent in response to the restrictions, but the United States was thrown into a panic. The embargo resulted in escalating prices for gasoline and home heating oil. Long lines at filling stations were commonplace, and in some states motorists were restricted to buying gasoline every other day. Car sales dropped, and people who did buy cars generally purchased the more fuel-efficient foreign makes. Because cars get better mileage at moderate speeds, the freeway speed limit was reduced by federal law to 55 miles per hour as an energy conservation measure.

The OPEC oil embargo of 1973 was not the only oil crisis Americans have faced in recent years. In 1979 crude oil prices skyrocketed from $13 to $34 a barrel due to an oil shortage touched off by the Iranian revolution, although the effects of this oil crisis were not as crippling as those of 1973. By the 1980s, the oil scares of the seventies were largely forgotten as oil prices declined. Americans purchased more cars, both domestic and foreign, than ever before, and gasoline was so cheap and abundant that its consumption increased. Between 1985 and 1989, oil imports increased from 3.2 million barrels (134 million gallons) a day to 5.8 million barrels (244 million gallons). At the same time, domestic oil production in the United States declined.

In the 1990s, a greater proportion of our total oil supply is being used to provide gasoline for automobiles than was used in 1973.

In 1994, 46 percent of all refined crude oil in the United States was converted to gasoline. However, the United States is still vulnerable when it comes to energy supplies and is even willing to go to war to ensure a dependable energy supply (witness the 1990 war in the Persian Gulf). In this chapter we examine the various fossil fuels that supply us with most of our energy, and discuss the environmental problems associated with their production and use. ◀

ENERGY CONSUMPTION IN DEVELOPED AND DEVELOPING COUNTRIES

Human society depends on energy. We use it to warm our homes in winter and cool them in summer; to grow and cook our food; to extract and process natural resources for manufacturing items we use daily; and to power various forms of transportation. Many of the conveniences of modern living depend on a ready supply of energy.

A conspicuous difference in per-capita energy consumption exists between developed and developing nations (Figure 10–1). As you might expect, developed nations consume much more energy per person than developing nations. Although only 22.3 percent of the

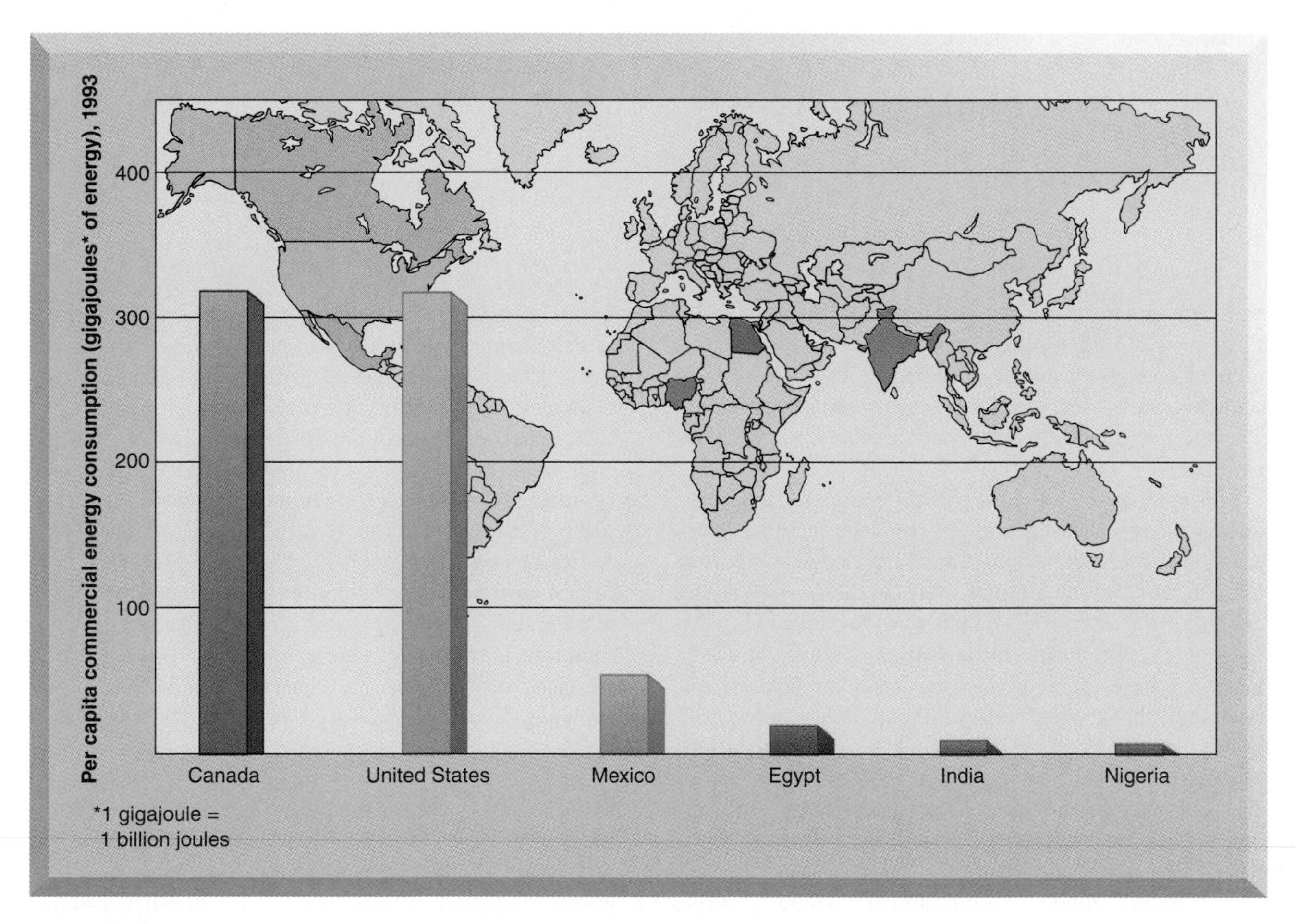

Figure 10–1

Annual per-capita commercial energy consumption in selected countries, 1993. Note that energy consumption per person in developed nations is much higher than it is in developing countries.

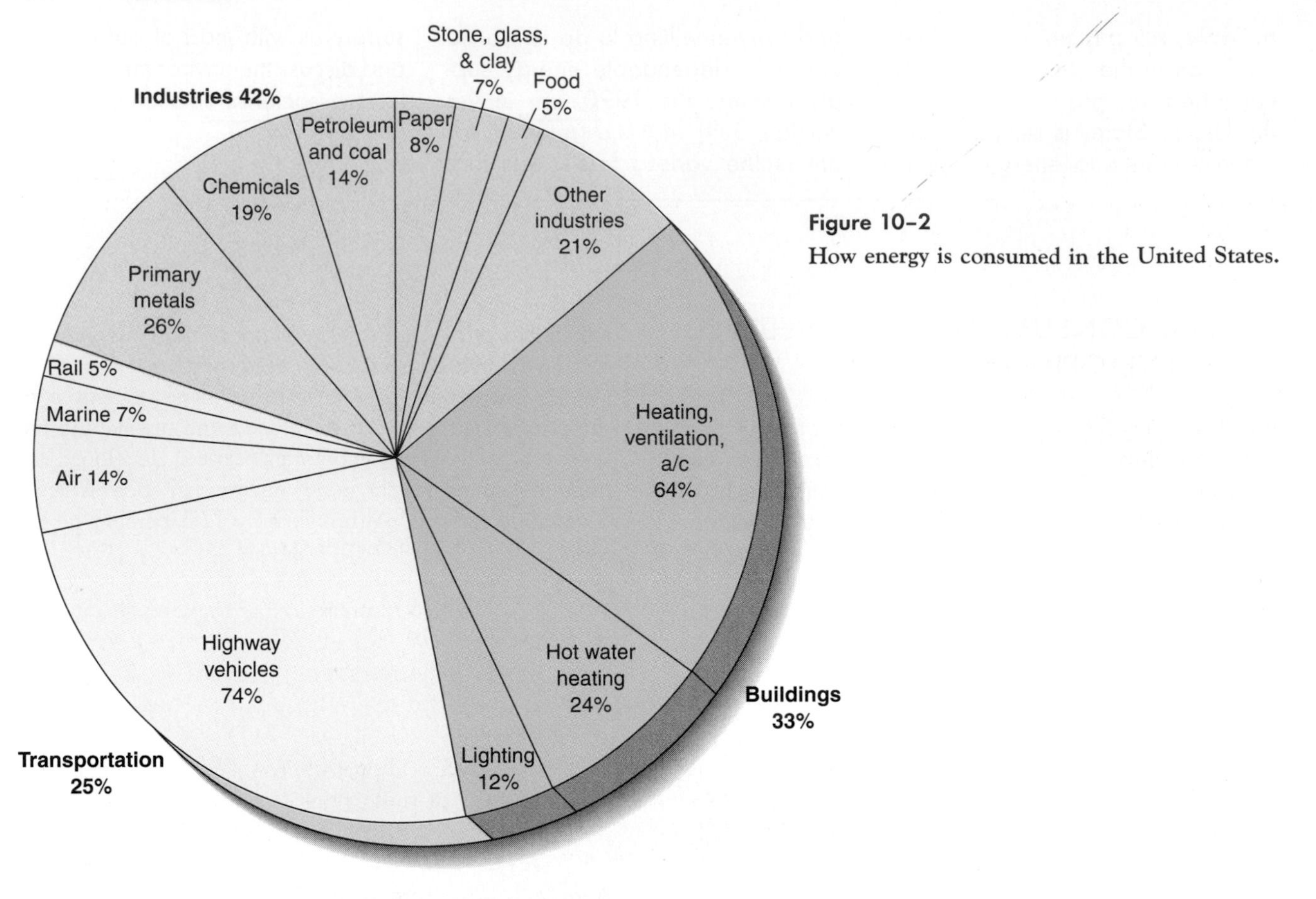

Figure 10–2
How energy is consumed in the United States.

world's population lived in developed countries in 1993, these people used approximately 68 percent of the commercial energy consumed worldwide.[1] That means that each person in developed countries uses approximately 7.4 times as much energy as each person in developing countries.

A comparison of energy requirements for food production clearly illustrates the energy consumption differences between developing and developed countries. Farmers in developing nations rely on their own physical energy or the energy of animals to plow and tend fields. In contrast, agriculture in developed countries involves many machines, such as tractors, automatic loaders, and combines, that require fuel. Additional energy is required to produce the fertilizers and pesticides widely used in industrialized agriculture. A higher energy input is one of the reasons that the agricultural productivity of developed countries is greater than that of developing countries.

Currently, energy consumption is increasing worldwide at about 1.6 percent annually, with most of the increase occurring in developing countries. One of the goals of developing countries is to improve their standard of living. One way to achieve this is through economic development, a process usually accompanied by a rise in per-capita energy consumption. Furthermore, the world's energy requirements will also increase during the 21st century, as the human population continues to increase. Most of this population growth will be in developing countries (see Chapter 8).

In contrast, the population in developed nations is more stable, and those nations' per-capita energy consumption may be at or near saturation. Also, it is possible that additional energy demands may be more than compensated for by increased energy efficiency of such items as appliances, automobiles, and home insulation.

Figure 10–2 illustrates how energy is used in the United States. Approximately 42 percent of the energy we consume is used by industry, which encompasses the production of chemicals, minerals, food, and additional energy resources. Another 33 percent of our consumed energy makes buildings comfortable through heating, air conditioning, lighting, and hot water. The remainder of the energy we consume provides for transportation, primarily for motor vehicles.

[1]Worldwide commercial energy use does not take into account the use of firewood, which meets the energy needs of a substantial portion of the population in many developing countries.

FOSSIL FUELS

Energy is obtained from a variety of sources, including fossil fuels, nuclear reactors (see Chapter 11), and solar and alternative energy sources (see Chapter 12). Today, most of the energy required in North America is supplied by fossil fuels: oil, natural gas, and coal. A **fossil fuel** is composed of the partially decayed remnants of organisms. Most of the fossil fuels that we use today were formed millions of years ago.

Fossil fuels are nonrenewable resources; that is, the Earth has a finite, or limited, supply of them. Although coal and other fossil fuels are still being formed by natural processes today, they are forming too slowly to replace the fossil fuel reserves we are using. Because fossil fuel formation does not keep pace with use, when our present supply of fossil fuels has been used up, we will have to make a transition to other, sustainable forms of energy (see Chapter 12).

How Fossil Fuels Were Formed

Three hundred million years ago, the climate of much of the Earth was mild and warm, and plants grew year round. Vast swamps were filled with plant species that have long since become extinct. Many of these plants—horsetails, ferns, and club mosses—were large trees (Figure 10–3).

Plants in most environments decay rapidly after death, due to the activities of decomposers such as bacteria and fungi. As the ancient swamp plants died, either from old age or from storm damage, they fell into the swamp, where they were covered by water. Their watery grave prevented the plants from decomposing much; wood-rotting fungi cannot act on plant material where oxygen is absent, and anaerobic bacteria, which thrive in oxygen-deficient environments, do not decompose wood very rapidly. Over time, more and more dead plants piled up. As a result of periodic changes in sea level, layers of sediment (materials deposited by gravity) accumulated, covering the plant material. Aeons passed, and the heat and pressure that accompanied burial converted the plant material into a carbon-rich rock called coal, and the layers of sediment into sedimentary rock. Much later, geological upheavals raised these layers so that they were nearer the Earth's surface.

Oil was formed when large numbers of microscopic aquatic organisms died and settled in the sediments. As these organisms accumulated, their decomposition depleted the small amount of oxygen that was present in the sediments. The resultant oxygen-deficient environment prevented further decomposition; over time, the dead remains were covered and buried deeper in the sediments. The heat and pressure caused by burial aided in the conversion of these remains to the mixture of **hydrocarbons** (molecules containing carbon and hydrogen) known as oil.

Natural gas, composed primarily of the simplest hydrocarbon, **methane**, was formed in essentially the same

Figure 10–3

Reconstruction of a Carboniferous swamp. The plants of the Carboniferous period included giant ferns, horsetails, and club mosses. (No. GEO85638c, Field Museum of Natural History, Chicago)

way as oil, only at higher temperatures. Over millions of years, as the organisms were converted to oil or natural gas, the sediments covering them were transformed to sedimentary rock.

Deposits of oil and natural gas are often found together. Because oil and natural gas are less dense than rock, they tend to move upward through porous rock layers and accumulate in pools beneath nonporous or impermeable rock layers. Most oil and natural gas deposits are found in rocks that are less than 200 million years old.

COAL

Although coal had been used as a fuel for centuries, it was not until the 18th century that it began to replace wood as the dominant fuel in the Western world, and since then it has had a significant impact on human history. It was coal, for example, that powered the steam engine and supplied the energy needed for the Industrial Revolution. Today coal is used primarily by utility companies to produce electricity and, to a lesser extent, by heavy industries such as steelmaking.

Coal occurs in different grades, largely as a result of the varying amounts of heat and pressure to which it was exposed during formation. Coal that was exposed to high heat and pressure during its formation is drier, is more compact (and therefore harder), and has a higher heating value. Lignite, bituminous coal, and anthracite are the three most common grades of coal (Table 10–1).

Lignite is a soft coal, brown or brown-black in color, with a soft, woody texture. It is moist and produces little heat compared to other coal types. Lignite is often used to fuel power plants. Sizable deposits of it are found in the western states, and the largest producer of lignite in the United States is North Dakota.

Bituminous coal, the most common type, is also called **soft coal** even though it is harder than lignite. It is dull to bright black with dull bands. Many bituminous coals contain sulfur, a chemical element that causes severe environmental problems when the coal is burned in the absence of pollution-control equipment. Bituminous coal is nevertheless used extensively by electric power plants because it produces a lot of heat. In the United States, bituminous coal deposits are found in the Appalachian region, near the Great Lakes, in the Mississippi valley, and in central Texas.

The highest grade of coal, **anthracite** or **hard coal**, was exposed to extremely high temperatures during its formation. It is a dark, brilliant black and burns most cleanly—that is, produces the fewest pollutants per unit of heat released—of all the types of coal because it is not contaminated by large amounts of sulfur. Anthracite also has the highest heat-producing capacity of any grade of coal. Most of the anthracite in the United States is located east of the Mississippi River, particularly in Pennsylvania.

Coal is usually found in underground layers, called seams, that vary from 2.5 centimeters (1 inch) to more than 30 meters (100 feet) in thickness. Because they are easily located, geologists think that most or all major coal deposits have probably been identified. Scientists working with coal are therefore concerned less about finding new deposits than about the safety and environmental problems associated with coal (discussed shortly).

Coal Mining

The two basic types of coal mines are surface and subsurface (underground) mines. The type of mine chosen depends on the location of the coal bed relative to the surface as well as on surface contours. If the coal bed is within 30 meters (100 feet) or so of the surface, **surface mining** (also called *open pit mining* and *strip mining*) is usually done. This process involves using bulldozers, giant power shovels, and wheel excavators to remove the ground covering the coal seam. The coal is then scraped out of the ground and loaded into railroad cars or trucks. Approximately 60 percent of the coal mined in the United States is obtained by surface mining.

Table 10–1
*A Comparison of Different Kinds of Coal**

Type of Coal	Color	Water Content (%)	Noncombustible Compounds (%)	Carbon Content (%)	Heat Value (BTU/pound)
Lignite	Dark brown	45	20	35	7000
Bituminous coal	Black	5–15	20–30	55–75	12,000
Anthracite	Black	4	1	95	14,000

*Courtesy of Thompson and Turk, *Modern Physical Geology*, 2nd ed. (Philadelphia: Saunders College Publishing, © 1997).

When the coal is deeper in the ground or runs deep into the Earth from an outcrop on a hillside, it is mined underground. **Subsurface mining** accounts for approximately 40 percent of the coal mined in the United States.

Surface mining has several advantages over subsurface mining: it is usually less expensive, safer for miners, and generally allows a more complete removal of coal from the ground. However, surface mining disrupts the land much more extensively than subsurface mining and has the potential to cause several serious environmental problems (discussed shortly).

Coal Reserves

Coal, the most abundant fossil fuel in the world, is found primarily in the Northern Hemisphere. The largest coal deposits are in the United States, Russia, China, and India (Figure 10–4), but deposits are also found in Canada, Germany, Poland, South Africa, Australia, Mongolia, and Brazil. The United States has 23.6 percent of the world's coal supply in its massive deposits.

According to the World Resources Institute, known world coal reserves could last for more than 200 years at the present rate of consumption. Additional coal resources that are currently too expensive to develop[2] have the potential to provide enough coal to last for a thousand or more years (at current consumption rates).

Safety and Environmental Problems Associated with Coal

Although we usually focus on the environmental problems caused by mining and burning coal, there are also significant human safety and health risks in the mining process itself. Underground mining is a hazardous occupation (Figure 10–5). According to the Department of Energy, during the 20th century more than 90,000 American coal miners have died in mining accidents, although the number of deaths per year has declined significantly since the earlier part of the century. Those not killed or maimed in accidents have an increased risk of cancer and **black lung disease**, a condition in which the lungs are

[2]For example, some coal deposits are buried more than 5000 feet inside the Earth's crust. Drilling a shaft that deep would cost considerably more than the current price of coal would justify.

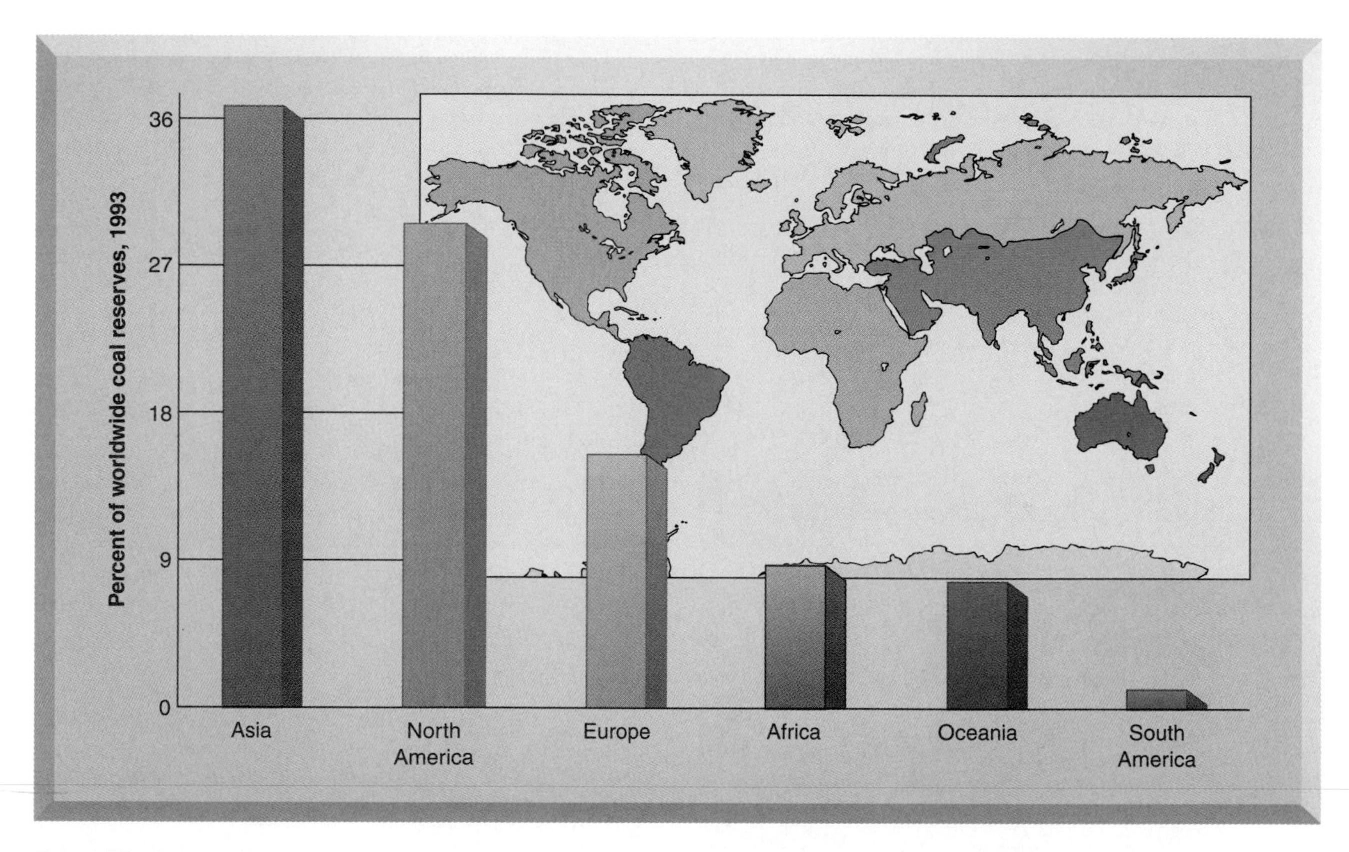

Figure 10–4

Distribution of coal deposits by continent, as percentages of the 1993 proved reserves (that is, of coal known to exist). The majority of the world's coal deposits are located in the Northern Hemisphere. (Please note: throughout the text we have included Russia as part of Europe even though it spans both Europe and Asia geographically.)

Reclamation of Coal-Mined Land

In the deregulatory political climate of the mid-1990s, some members of Congress think that environmental monitoring should be scaled back and environmental laws should be weakened to promote economic growth. However, most of these laws have been extremely effective in protecting the environment. Consider the 1977 Surface Mining Control and Reclamation Act (SMCRA). In the 20 years since this law's passage, more than 34,000 permits regulating the reclamation of more than 1.6 million hectares (4 million acres), an area the size of Connecticut and Rhode Island combined, have been issued. In addition, some of the most dangerous (from a health and safety viewpoint) abandoned coal mine lands have been reclaimed (see figure). The Office of Surface Mining of the Department of the Interior gives annual awards to those coal companies that have developed outstanding restoration sites.

Sometimes this restored land is used for biological habitat, such as rangelands for wild grazing animals in western states or wetlands in midwestern and eastern states. The Eastern Kentucky Regional Airport is located on a reclaimed mining site, as is a tree farm in Garrett County, Maryland. Other post-mining land uses include camping and recreational areas such as golf courses, farmlands, sanitary landfills, cemeteries, and housing developments.

The SMCRA stipulates that mined land must be reclaimed as soon as possible. Old highwalls are *backfilled* (covered with excavated rock from current mining operations) and graded to resemble the original pre-mining terrain. Then topsoil is spread, and the area is prepared for planting.

Until the area has vegetation established, soil erosion will continue to be a problem, so sediment and erosion-control structures are constructed. For example, many sites contain *sedimentation ponds*, which confine sediment-laden runoff and allow the sediment to settle out before the water leaves the area. Water from nearby streams is monitored during mining and reclamation to ensure that it remains clean. In some places, wells are dug to monitor groundwater quality as well.

Soil and water contamination by acid and toxic materials such as selenium is prevented by identifying these materials and excavating and safely disposing of them. In arid regions such as Wyoming, the toxic *spoils* (excavated rock) may be placed at the bottom of a nearby mine pit, which is then covered with additional rock and topsoil, then planted. In moister regions, acid-producing rock is kept "high and dry," since water on the open pit floor would cause acid mine drainage.

Because excavated rock often occupies more space than unexcavated rock, excess spoil usually remains after the mine is regraded to its original contours. This material is often placed in a nearby valley, which is eventually covered with topsoil and replanted.

The permanent establishment of plant cover on all land affected by coal mining is required by the SMCRA. The plants selected for revegetation are usually native to the area and thus adapted to the climate. Generally, seeds are planted, but greater success is often observed when seedlings are hand-planted (a more expensive technique). The coal mining company must maintain the vegetation for a minimum of five years in the East and ten years in the arid West, after which time the vegetation is considered to be successfully reestablished.

Reclaimed coal-mined land. Prior to reclamation, this 45-acre site in West Virginia had huge piles of coal refuse, dangerous highwalls, and 11 abandoned underground mine openings. All health and safety hazards have been removed, and the land is similar to what it was before coal mining occurred. (Chuck Meyers, Office of Surface Mining)

Figure 10–5

An underground coal mine in Illinois. The dangers to workers from underground coal mining, which produces about 40 percent of the coal in the United States, include deaths and injuries due to accidents and respiratory diseases. (U.S. Department of Energy)

coated with inhaled coal dust so that the exchange of oxygen between the lungs and the blood is severely restricted. It is estimated that these diseases are responsible for the deaths of at least 2000 miners in the United States each year.

Environmental impacts of the mining process

Coal mining, especially surface mining, has substantial effects on the environment. In surface mining, vegetation and topsoil are completely removed, causing a loss of habitat for organisms and increasing soil erosion and water pollution (Figure 10–6).

Figure 10–6

Surface mining in Gillette, Wyoming. Before coal can be mined, the area's vegetation, soil, and bedrock overlying the coal deposit must be removed. (Chuck Meyers, Office of Surface Mining)

Prior to passage of the 1977 Surface Mining Control and Reclamation Act (SMCRA), abandoned surface mines were usually left as large open pits. Cliffs of excavated rock called *highwalls*, some more than 30 meters (100 feet) high, were left exposed. Acid and toxic mineral drainage from such mines, along with the removal of topsoil, which was buried or washed away by erosion, prevented most plants from naturally recolonizing the land. Streams were polluted with sediment and acid mine drainage. Dangerous landslides occurred on land that was unstable due to the lack of vegetation.

Surface-mined land can be restored to prevent such degradation and to make the land productive for other purposes, but restoration is extremely expensive. Without laws requiring coal companies to reclaim the land, few restorations of surface-mined land took place. The SMCRA requires coal companies to restore areas that have been surface mined, beginning in 1977 (see Meeting the Challenge: Reclamation of Coal-Mined Land). The SMCRA protects the environment by requiring permits and inspections of active coal mining operations and reclamation sites. It also prohibits coal mining in sensitive areas such as national parks, wildlife refuges, wild and scenic rivers, and sites listed on the National Register of Historic Places. In addition, the SMCRA stipulates that surface-mined land abandoned prior to 1977 (more than 0.4 million hectares, or 1 million acres) gradually be restored, using money from a tax that coal companies pay on currently mined coal. However, so many abandoned coal mines exist that it is doubtful that they can all be restored.

Environmental impacts of burning coal

Burning any fossil fuel releases carbon dioxide, CO_2, into the atmosphere. You may recall from the discussion of the carbon cycle in Chapter 5 that a natural equilibrium exists between the CO_2 in the atmosphere and the CO_2 dissolved in the ocean. Currently we are releasing so much CO_2 into the atmosphere through our consumption of fossil fuels that the Earth's CO_2 equilibrium has been disrupted. Because the concentration of CO_2 in the atmosphere is increasing and CO_2 prevents heat from escaping from the planet (it acts like glass in a greenhouse), global temperature may be affected. An increase of a few degrees in global temperature caused by higher levels of CO_2 and other greenhouse gases may not seem very serious at first glance, but a closer look reveals that such an increase is potentially harmful. For example, a global rise in temperature may cause polar ice to melt, raising sea levels and flooding coastal areas. This would increase coastal erosion and put many coastal buildings (and the people that live in them) at higher risk from violent storms. Other serious environmental consequences of global climate change are considered in Chapter 20. The CO_2 problem is made more severe by the burning of coal than by the burning of other fossil fuels, because coal burning releases more CO_2 per unit of heat produced.

Figure 10–7

Dead fir trees enveloped in acid fog on Mt. Mitchell, North Carolina. Forest decline was first documented in Germany and Eastern Europe. More recently, it has been observed in the eastern United States, particularly at higher elevations. Acid deposition is an important factor contributing to forest decline. (John Shaw)

Coal burning also contributes more of other air pollutants than does the combustion of either oil or natural gas. Bituminous coal contains sulfur and nitrogen that, when burned, are released into the atmosphere as sulfur oxides (SO_2 and SO_3) and nitrogen oxides (NO, NO_2, and N_2O). Both sulfur oxides and the nitrogen oxides NO and NO_2 form acids when they react with water. These reactions result in a type of air pollution known as **acid deposition**, in which acid falls from the atmosphere to the surface as precipitation **(acid precipitation)** or as dry acid particles. The combustion of coal is partly responsible for acid deposition, which is particularly prevalent downwind from coal-burning power plants. Normal rain is slightly acidic (pH 5.6), but in some areas the pH of acid precipitation has been measured at 2.1, equivalent to that of lemon juice. (See Appendix I for a review of pH.) Acid deposition has acidified lakes and streams, where it has caused aquatic animal populations to decline. It has also been linked to some of the forest decline that has been documented worldwide (Figure 10–7).

Although it is relatively easy to identify and measure pollutants in the atmosphere, it is difficult to trace their exact origins. They are dispersed by air currents and are often altered as they react chemically with other pollutants in the air. Even so, it is clear that some nations suffer the damage of acid deposition caused by pollutants produced in other countries, and as a result acid deposition has become an international issue. The environmental repercussions of acid deposition are considered in detail in Chapter 20.

It is possible to reduce sulfur emissions associated with the combustion of coal by installing desulfurization systems, or **scrubbers**, in smokestacks. As the polluted air passes through a scrubber, chemicals in the scrubber react with the pollution and cause it to precipitate out. In lime scrubbers, a chemical spray of water and lime neutralizes acidic gases such as sulfur dioxide, which remain behind as a sulfur-containing sludge that itself becomes a disposal problem. A large power plant may produce enough sludge annually to cover 2.6 square kilometers (1 square mile) of land 0.3 meter (1 foot) deep. Currently, most power plants place the sludge in holding ponds and landfills. More recently, scrubbers that use magnesium and other chemicals have been developed; these scrubbers concentrate on **resource recovery** by removing sulfur from polluted emissions as a marketable product rather than as a polluted slurry. For example, the magnesium sulfate produced in scrubbers may be used in the dyeing and tanning industries.

The 1990 Clean Air Act (see Chapter 19) stated that the nation's 111 dirtiest coal-burning power plants were required to cut sulfur dioxide emissions by 1995 (or by 1997 if they committed to buying scrubbers). Compliance resulted in a total annual decrease of 5 million metric tons nationwide. This represents a significant portion of the total amount of sulfur dioxide emitted in the United States each year (20.6 million metric tons in 1993). In the second phase of the Clean Air Act, more than 200 additional power plants must make SO_2 cuts by the year 2000, resulting in a total annual decrease of 10 million metric tons nationwide. A nationwide cap on SO_2 emissions from coal-burning power plants will be imposed after the year 2000. Utilities must also cut nitrogen oxide emissions by 2 million tons per year, beginning in 1995.

While CO_2 emissions remain a significant problem, new methods for burning coal (called **clean coal technologies**) are being developed that will not contaminate the atmosphere with sulfur oxides and will significantly reduce nitrogen oxide production. Clean coal technologies include coal gasification (considered shortly, in the discussion of synfuels) and fluidized-bed combustion.

Fluidized-bed combustion mixes crushed coal with particles of limestone in a strong air current during combustion (Figure 10–8). This coal-burning process has greater efficiency and several additional advantages. Because fluidized-bed combustion takes place at a lower temperature than regular coal burning, fewer nitrogen oxides are produced. (Higher temperatures cause atmospheric nitrogen and oxygen to combine, forming nitrogen oxides.) Also, because the sulfur in coal reacts with the

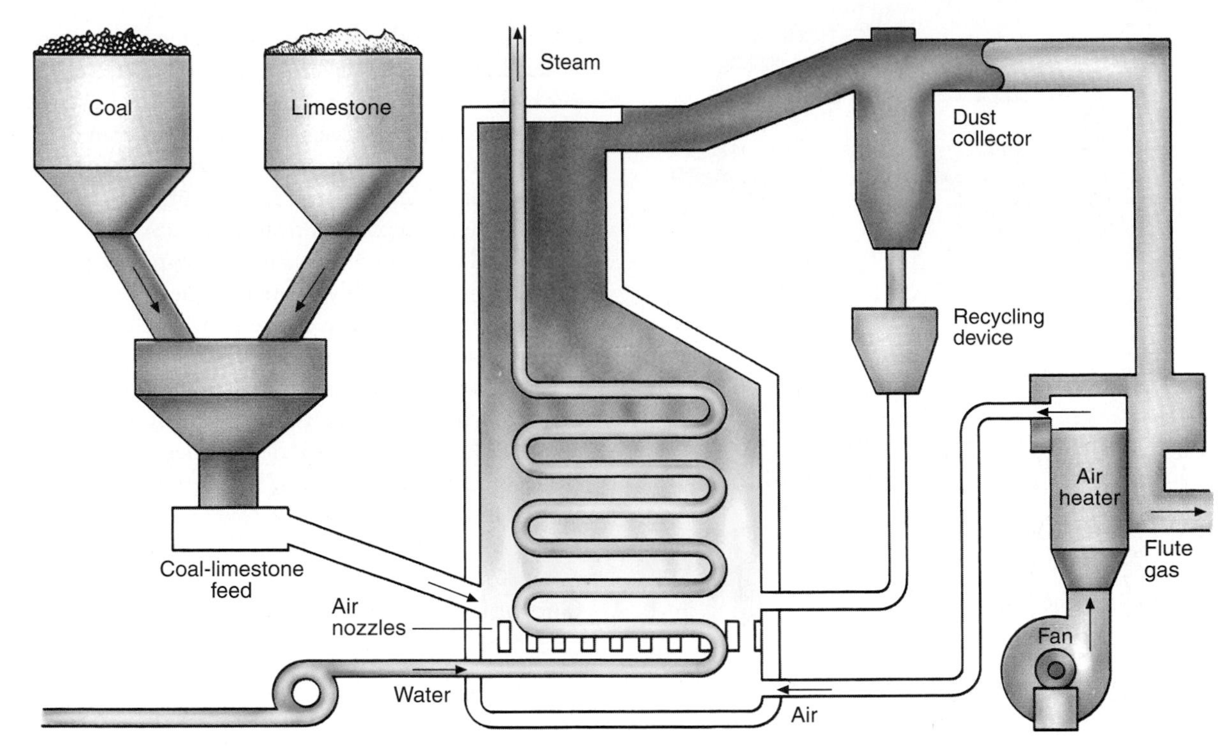

Figure 10–8

Fluidized-bed combustion of coal. Crushed coal and limestone are suspended in air. As the coal burns, most of the sulfur dioxide in the coal is neutralized by the limestone. The heat generated during combustion is used to convert water to steam, which can be used to power various industrial processes.

calcium in limestone to form calcium sulfate, which then precipitates out, sulfur is removed from the coal *during* the burning process, so scrubbers are not needed to remove it *after* combustion.

In the United States several large power plants are testing fluidized-bed combustion, and a few small power plants that use this technology are already in commercial operation. The 1990 Clean Air Act provides incentives for utility companies to convert to clean coal technologies.

OIL AND NATURAL GAS

From the 1600s through the 1800s, wood was the predominant fuel in the United States. By 1900, coal had taken its place as the most important energy source. Beginning in the 1940s, oil and natural gas became increasingly important, largely because they are easier to transport and they burn more cleanly than coal. In 1993, oil and natural gas supplied approximately 56 percent of the energy used in the United States. Globally in 1993, 40 percent of the world's energy was provided by oil and 23 percent by natural gas. In comparison, other major energy sources used worldwide include coal (27 percent), hydroelectric power (2.5 percent), and nuclear power (7 percent) (Figure 10–9).

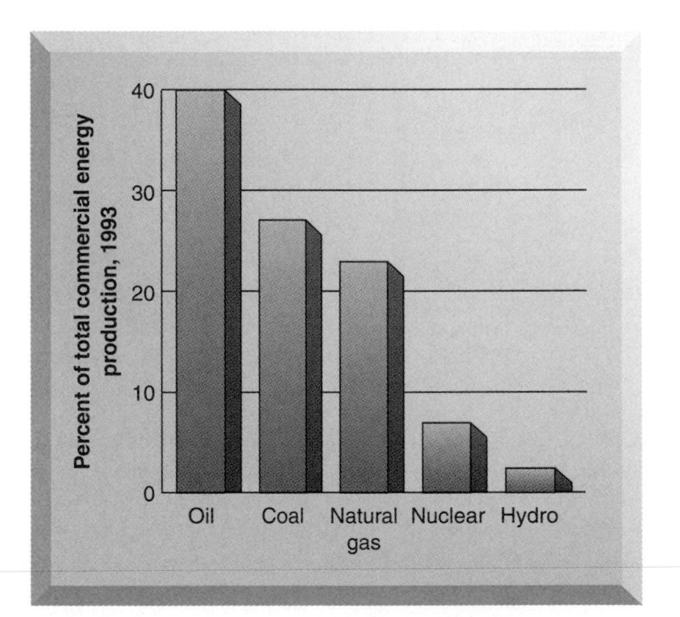

Figure 10–9

The world's main commercial energy sources—oil, coal, natural gas, nuclear, and hydroelectric power—in 1993. Note the overwhelming importance of oil, coal, and natural gas.

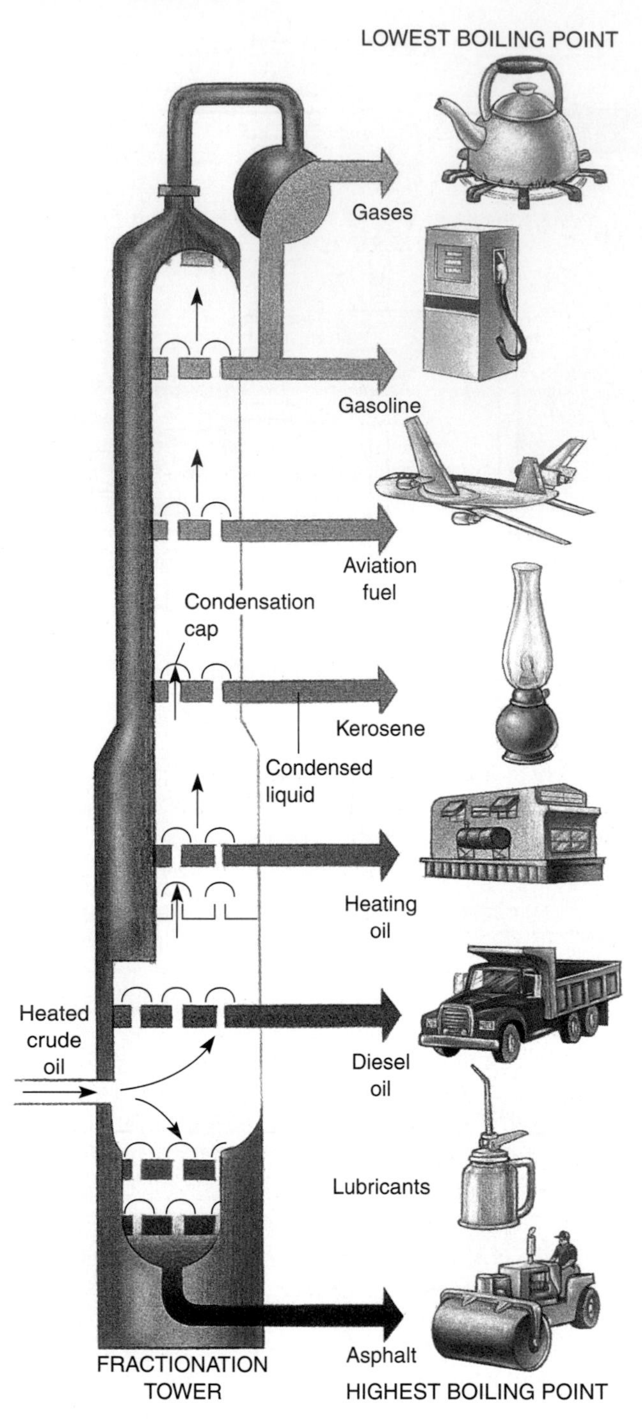

Figure 10–10

Crude oil is separated into a variety of products during refining. Different components of crude oil have different boiling points. After being heated, they are separated from one another in a fractionation tower, which may be 100 feet tall. The lower the boiling point, the higher the compounds rise in the tower.

Petroleum, or **crude oil**, is a liquid composed of hundreds of hydrocarbon compounds. During petroleum refining, the compounds are separated into different products—such as gases, gasoline, heating oil, diesel oil, and asphalt—based on their different boiling points (Figure 10–10). Oil also contains **petrochemicals**, compounds that are used in the production of such diverse products as fertilizers, plastics, paints, pesticides, medicines, and synthetic fibers.

In contrast to petroleum, natural gas contains only a few different hydrocarbons: methane and smaller amounts of ethane, propane, and butane. The propane and butane are separated from the natural gas, stored in pressurized tanks as a liquid called **liquefied petroleum gas**, and used primarily as fuel for heating and cooking in rural areas. Methane is used to heat residential and commercial buildings, to generate electricity in power plants, and for a variety of purposes in the organic chemistry industry. Methane is usually distributed by being pumped through pressurized pipelines or shipped (as a solid) in refrigerated tankers. (At very low temperatures, natural gas becomes a solid.)

Geological Exploration for Oil and Natural Gas

Exploration is continually under way in search of new oil and natural gas deposits, which are usually found together under one or more layers of rock. Usually oil and natural gas deposits are discovered indirectly by the detection of **structural traps**, geological structures that tend to trap any oil or natural gas that is present (recall that oil and natural gas tend to migrate upward until they reach an impermeable rock layer). Two examples of structural traps are anticlines and salt domes (Figure 10–11).

An **anticline** is an upward folding of rock **strata** (layers). Sometimes the strata that arch upward include both porous and impermeable rock. If impermeable layers overlie porous layers, it is possible that any oil or natural gas present will work its way up through the porous rock to accumulate under the impermeable layer.

Many important oil and natural gas deposits (for example, oil deposits known to exist in the Gulf of Mexico) have been found in association with **salt domes**, underground columns of salt. Salt domes develop when extensive salt deposits form at the Earth's surface due to the evaporation of water. All surface water contains dissolved salts. The salts dissolved in ocean water are so concentrated that they can be tasted, but even fresh water contains some dissolved material.

If a body of water lacks a passage to the ocean, as an inland lake often does, the salt concentration in the water gradually increases.[3] If such a lake were to dry up, a massive salt deposit called an **evaporite deposit** would remain. Evaporite deposits may eventually be covered by layers of sediment, which convert to sedimentary rock after millions of years. Because salt is less dense than rock, the rock layers settle, and the salt deposit tends to rise in

[3]The Great Salt Lake in North America is an example of a salty inland body of water that formed in this way. Although three rivers empty into the Great Salt Lake, water escapes from the lake only by evaporation, accounting for its high salinity—four times as high as that of ocean water.

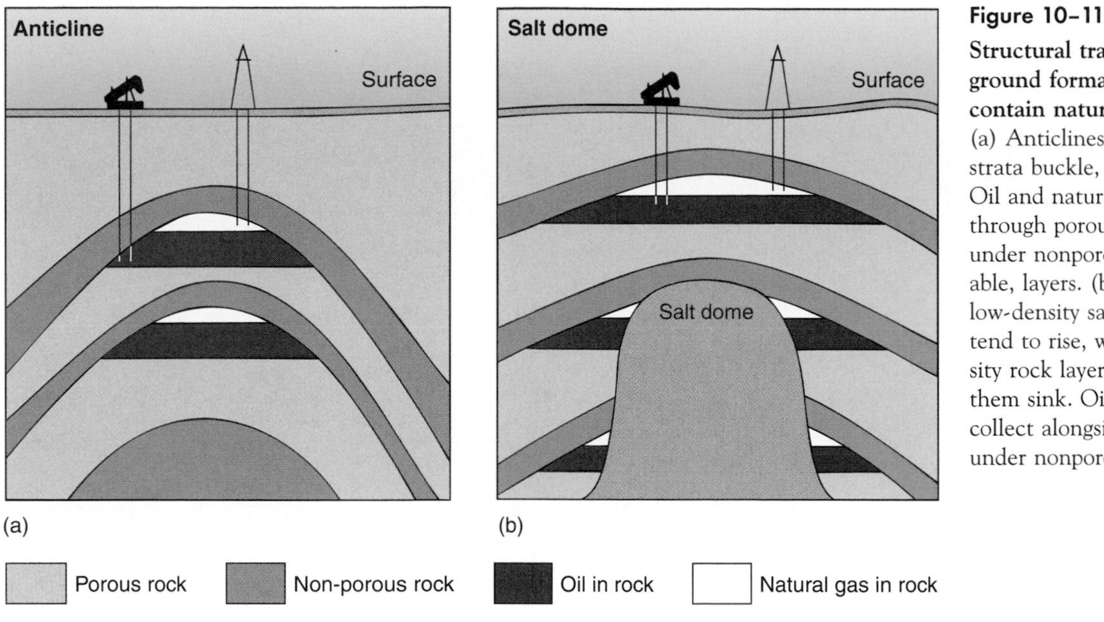

Figure 10–11
Structural traps are underground formations that may contain natural gas or oil. (a) Anticlines form when rock strata buckle, or fold upward. Oil and natural gas seep through porous rock and collect under nonporous, or impermeable, layers. (b) Salt domes are low-density salt formations that tend to rise, whereas high-density rock layers surrounding them sink. Oil and natural gas collect alongside the salt dome under nonporous rock strata.

a column—a salt dome. The ascending salt dome, together with the rock layers that buckle over it, provides a trap for oil or natural gas.

Geologists use a variety of techniques to identify structural traps that might contain oil or natural gas. One method is to drill test holes in the surface and obtain rock samples. Another method is to produce an explosion at the surface and measure the echoes of sound waves that bounce off rock layers under the surface. These data can be interpreted to determine whether or not structural traps are present. It should be emphasized, however, that many structural traps do not contain oil or natural gas.

Searching for oil and natural gas is very expensive. It costs millions of dollars just to do the basic geological analyses to find structural traps. And once oil or natural gas has been located, drilling and operating the wells cost additional millions.

Reserves of Oil and Natural Gas

Although oil and natural gas deposits exist on every continent, their distribution is uneven, and a disproportionate share of total oil deposits are clustered relatively close to each other (Figure 10–12). Enormous oil fields containing more than half of the world's total estimated reserves are situated in the politically unstable Persian Gulf region.[4] In addition, major oil fields are known to exist in Venezuela, Mexico, Russia, Libya, and the United States (in Alaska and the Gulf of Mexico). Almost half of the world's proved recoverable reserves of natural gas are located in two countries, Russia and Iran (Figure 10–13). Because the United States has more deposits of natural gas than Western Europe, use of natural gas is more common in North America (see Focus On: New Roles for Natural Gas, page 212).

[4]The Persian Gulf region includes Iran, Iraq, Kuwait, Oman, Saudi Arabia, Syrian Arab Republic, United Arab Emirates, and Yemen.

It is unlikely that major new oil fields will be discovered in the continental United States, which has been explored for oil more extensively than any other country. In the last two decades, the success rate of searches for new oil fields has declined, as has the amount of exploration.

There is reason to believe that large oil deposits exist on the **continental shelves**, the relatively flat underwater areas that surround continents, and areas adjacent to the continental shelves. Despite problems, such as storms at sea and the potential for major oil spills, many countries engage in offshore drilling for this oil. New technologies, such as platforms the size of football fields, enable oil companies to drill several thousand feet for oil, making oil fields that were once considered inaccessible open for tapping. For example, as much as 18 billion barrels (756 billion gallons) of oil and natural gas may exist in the deep water of the gulf of Mexico just outside the continental shelf, from Texas to Alabama.

Environmentalists generally oppose opening the outer continental shelves for oil and natural gas exploration because of the threat a major oil spill would pose to marine and coastal environments. Coastal industries, including fishing and tourism, also oppose oil and natural gas exploration in these areas.

How long will oil and natural gas supplies last?

A major problem associated with our dependence on oil and natural gas is their limited supplies. It is difficult to say with certainty how long it will be before the world runs out of oil and natural gas, because there are so many unknowns. We do not know how many additional oil and

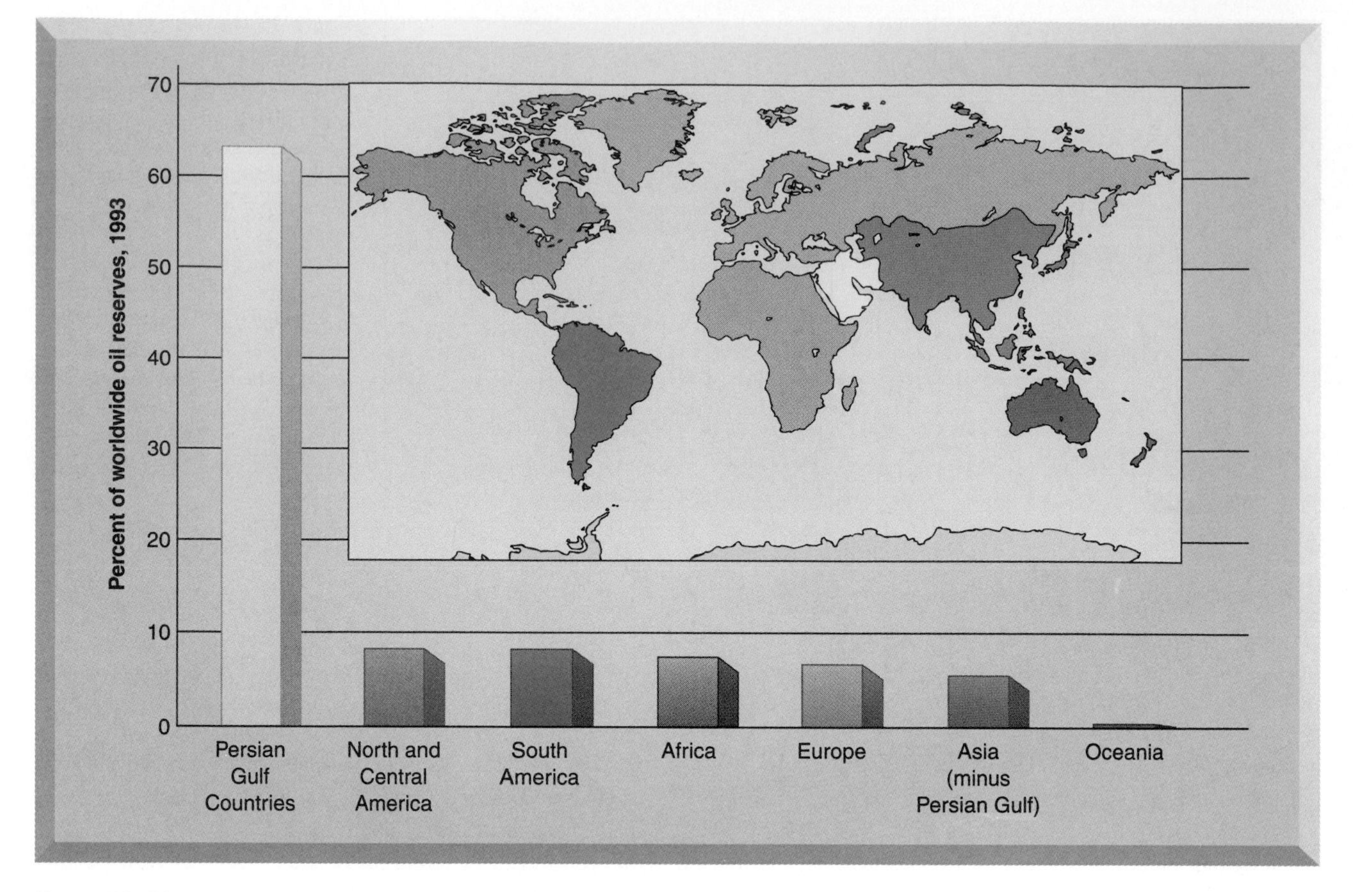

Figure 10–12

The world's oil deposits are not evenly distributed. The Persian Gulf region contains huge oil deposits in a relatively small area, while other regions have very few. Shown are regional percentages of the 1993 worldwide estimate of proved recoverable oil reserves.

natural gas reserves will be discovered, nor do we know if or when technological breakthroughs will allow us to extract more fuel from each deposit. The answer to how long these fuels will last also depends on whether worldwide consumption of oil and natural gas increases, remains the same, or decreases. Economic factors influence oil and natural gas consumption; as reserves are exhausted, prices will increase, which can drive down consumption and stimulate greater energy efficiency and the search for additional deposits and alternative energy sources.

Despite adequate oil and natural gas supplies for the near future, at some point their production will peak and then decline over a period of several decades. Many experts think we will begin to have serious problems with oil and natural gas supplies sometime during the 21st century. Estimates on when we will run out of oil and natural gas vary from several decades to 100 years, but they are only guesses. The only thing we can say with certainty is that, at projected rates of consumption, oil and natural gas reserves will be depleted before those of coal.

Global Oil Demand and Supply

One difficult aspect of oil consumption is that the world's major oil producers are not its major oil consumers. In 1993, 39 percent of the world's total commercial energy was consumed by North America and Western Europe,[5] yet these same countries produced only 21 percent of the world's oil. In contrast, the Persian Gulf region consumed 2.9 percent of the world's energy and produced 29 percent of the world's oil.

In the United States, a severe economic burden has resulted from the large amount of oil that is imported; beginning in 1994, the United States has imported more than one-half of its oil at an annual cost of more than $50 billion. The U.S. Department of Energy and other experts project that the United States will be importing almost 100 percent of its oil by 2015.

[5]The United States, Canada, Mexico, Austria, Belgium, France, Germany, Netherlands, and Switzerland are included in these calculations of North America and Western Europe.

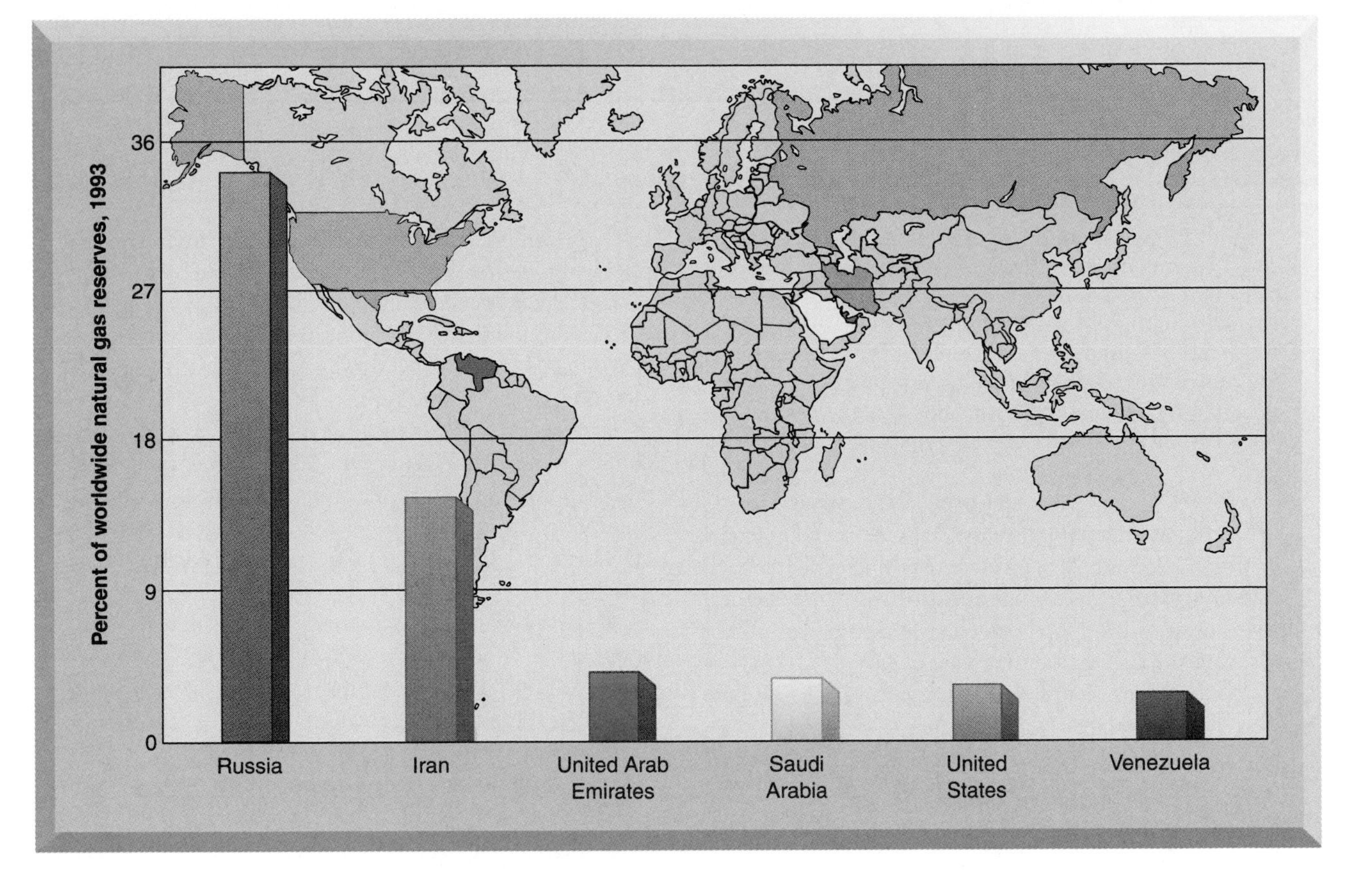

Figure 10–13

The six countries with the greatest natural gas deposits. Data shown as percentages of the 1993 worldwide estimate of proved recoverable natural gas reserves. Note that Russia and Iran together possess 49 percent of the world's natural gas deposits.

The imbalance between oil consumers and oil producers will probably worsen in the years to come, because the Persian Gulf region has much higher proven reserves than other countries. For example, at current rates of production, North America's oil reserves will run out decades before those of the Persian Gulf nations, which have 63.2 percent of the known world oil reserves and can produce oil at current rates for more than a century.

Environmental Problems Associated with Oil and Natural Gas

Two sets of environmental problems are associated with the use of oil and natural gas: the problems that result from burning the fuels (combustion) and the problems involved in obtaining them in order to burn them (production and transport). We have already mentioned the CO_2 emissions that are a direct result of the combustion of fossil fuels. As with coal, the burning of oil and natural gas produces CO_2. Every gallon of gasoline you burn in your automobile, for example, releases an estimated 9 kilograms (20 pounds) of CO_2 into the atmosphere. As CO_2 accumulates in the atmosphere, it prevents planetary heat from radiating into space. Global climate may be warming more rapidly now than it did during any of the warming periods following the ice ages. The environmental impact of rapid global climate change could be catastrophic.

Another negative environmental impact of burning oil is acid deposition. Although oil does not produce appreciable amounts of sulfur oxides, it does produce nitrogen oxides, mainly through gasoline combustion in automobiles, which contributes approximately half the nitrogen oxides released into the atmosphere. (Coal combustion is responsible for the other half.) Nitrogen oxides contribute to acid deposition. The burning of natural gas, on the other hand, does not pollute the atmosphere as much as the burning of oil; natural gas is the cleanest of the fossil fuels.

One of the concerns in oil and natural gas production is the environmental damage that may occur during their transport, which is often over long distances by pipelines or by ocean tankers. A serious spill along the route creates an environmental crisis, particularly in aquatic ecosystems, where the oil slick can travel.

FOCUS ON

New Roles for Natural Gas

Natural gas is gaining in popularity because it is a relatively clean, efficient source of energy. Natural gas contains almost no sulfur, a contributor to acid deposition. In addition, natural gas produces far less CO_2 and fewer hydrocarbons than do gasoline and coal. (Carbon dioxide contributes to global warming, and hydrocarbons produce photochemical smog.) Finally, natural gas does not produce harmful particulate matter the way coal and oil fuels do.

Use of natural gas is increasing in three main areas—generation of electricity, transportation, and commercial cooling. One example of an increasingly popular approach that uses natural gas is **cogeneration**, in which natural gas is used to produce both electricity and useful heat for water and space heating (see Chapter 12). Cogeneration systems that use natural gas are able to provide electricity cleanly and efficiently. Natural gas can also be used to help control emissions and costs at already-existing, coal-fueled electric power plants. The natural gas is piped to the power plant and burned instead of coal.

The use of compressed natural gas as a fuel for trucks, buses, and automobiles is increasing. Natural gas as a vehicular fuel offers impressive environmental advantages over gasoline: natural gas vehicles emit 80 to 93 percent fewer hydrocarbons, 90 percent less carbon monoxide, 90 percent fewer toxic emissions, and virtually no soot. Natural gas as a fuel choice also fares well in economic terms: it costs about the same as gasoline, is abundant in North America, and is becoming more readily available. As of 1996, more than 80,000 vehicles that use natural gas for fuel are on the road. The city of Los Angeles is the largest operator of natural gas-powered transit buses in North America.

Natural gas can efficiently fuel residential and commercial cooling systems. One example is the use of a desiccant-based (air-drying) cooling system, which is ideal for supermarkets, where humidity control is more important than temperature control.

Given the emphasis on reducing dependence on oil and coal, new technologies for the practical use of natural gas should continue to develop in the coming decades.

THE BACKYARD OIL SPILL

Correlation between the amount of oil dumped by home auto mechanics in the United States every 2.5 weeks and the amount spilled by the Exxon Valdez: 1:1

Estimated annual volume, in gallons, of used oil from American vehicles: 600,000,000.

Estimated annual volume, in gallons, of oil changed by American drivers themselves: 350,000,000.

Estimated annual volume, in gallons, improperly discarded by American drivers: 240,000,000.

Estimated annual volume, in gallons, that goes into storm sewers: 176,000,000.

Number of quarts of used motor oil needed to contaminate 250,000 gallons of drinking water: 1.

The 1989 Alaskan oil spill

On March 24, 1989, the supertanker *Exxon Valdez* slammed into Bligh Reef and spewed 260,000 barrels (10.9 million gallons) of crude oil into Prince William Sound off the coast of Alaska, creating the largest oil spill in U.S. history. As it spread, the black, tarry gunk eventually covered thousands of square kilometers of water (Figure 10–14) and contaminated hundreds of kilometers of shoreline. According to the U.S. Fish and Wildlife Service and the Alaska Department of Environmental Conservation, more than 30,000 birds (sea ducks, loons, cormorants, bald eagles, and other species) and between 3500 and 5500 sea otters died as a result of the spill. The area's killer whale population declined, and salmon migration was disrupted. Throughout the area, there was no fishing season that year.

Within hours of the spill, scientists began to arrive on the scene to advise both Exxon Corporation and the government on the best way to try to contain and clean up the spill. But it took much longer for any real action to be taken. Eventually, nearly 12,000 workers took part in the cleanup; their activities included mechanized steam cleaning and rinsing, which actually killed shore-

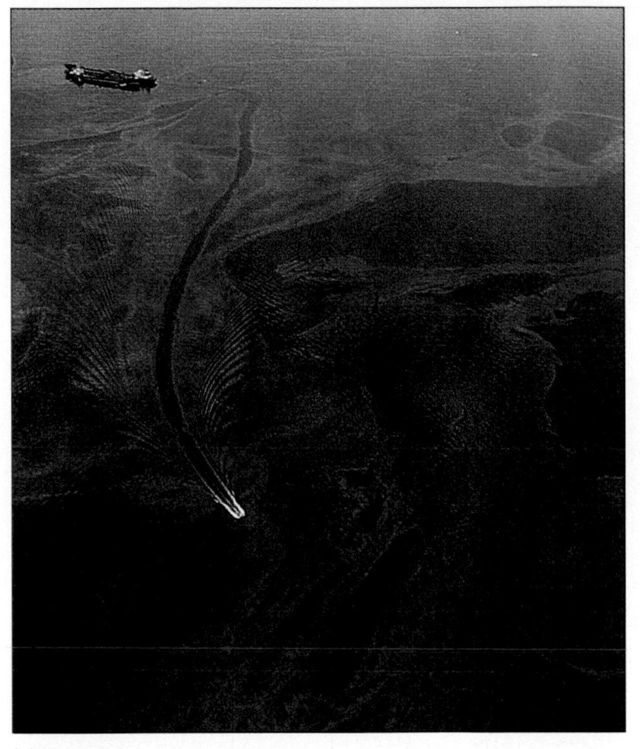

(a)

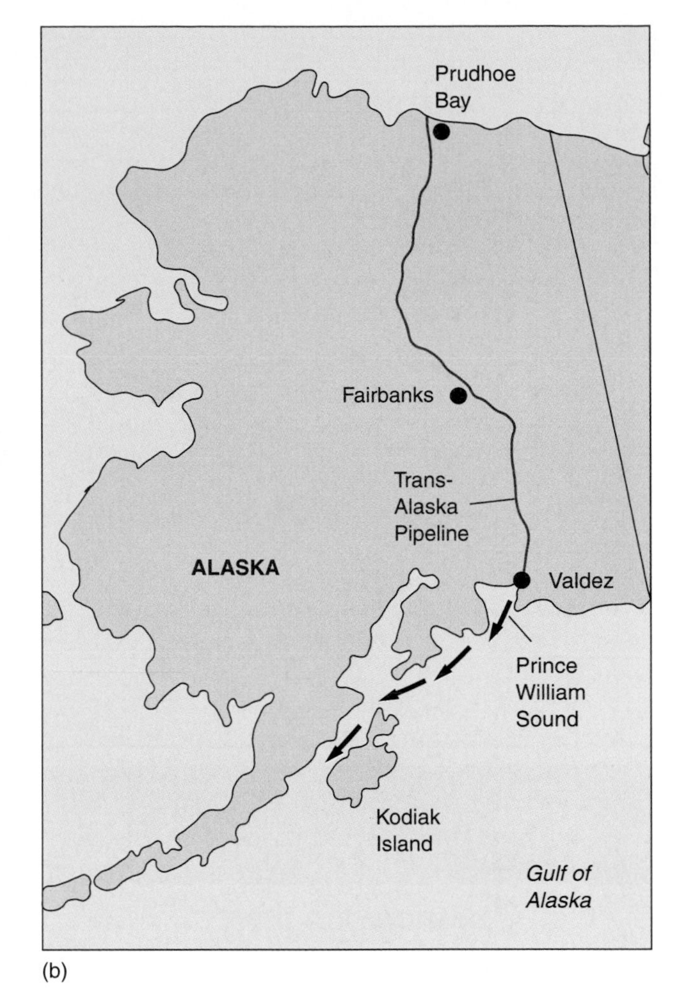

(b)

Figure 10–14

The 1989 Alaskan oil spill. (a) An aerial view of the massive oil slick. (b) The extent of the spill (black arrows). Water currents caused it to spread rapidly for hundreds of miles. Countless animals, such as sea otters and ocean birds, died. (a, Natalie Fobes/Tony Stone Images)

line organisms such as barnacles, clams, mussels, eelgrass, and rockweed.

In 1989, Exxon declared the cleanup "complete." But it left behind contaminated shoreline, continued damage to some species of birds, fishes, and mammals, and a reduced commercial salmon catch, among other problems. In 1991, Exxon agreed to pay Alaska a settlement of $1.1 billion, and in 1994, a jury awarded 34,000 fishermen and other Alaskans $5 billion; Exxon is appealing the award. Additional lawsuits have been filed and will take years to settle. The final cost to Exxon is expected to exceed $10 billion.

One positive outcome of the disaster was passage of the Oil Pollution Act of 1990. This legislation requires double hulls on all oil tankers that enter U.S. waters by 2015. (Had the *Exxon Valdez* possessed a double hull, the disaster probably would not have occurred because only the outer hull would have broken.) Because tankers are extremely expensive to build, however, most oil tankers will have single hulls until the 2015 deadline.

The 1991 Persian Gulf oil spill

The world's most massive oil spill occurred in 1991, during the Persian Gulf War, when about 6 million barrels (250 million gallons) of crude oil—more than 20 times the amount of the *Exxon Valdez* spill—were deliberately dumped into the Persian Gulf. Many oil wells were also set on fire, and lakes of oil spilled into the desert around the burning oil wells (see Chapter 1). Cleanup efforts along the coastline and in the desert were initially hampered by the war, and environmentalists fear that it may take a century or more for the area to completely recover.

To Drill or Not to Drill: A Case Study of the Arctic National Wildlife Refuge

In order to understand the complexities of energy issues, let us examine the proposed opening of the Arctic National Wildlife Refuge to oil exploration. This controversy has been a major environment-versus-economy conflict off-and-on since 1980. On one side are those who seek to protect rare and fragile natural environments; on the other side are those whose higher priority is the development of the last major U.S. oil supplies. Pressure to open the refuge subsided for about five years following the Alaskan oil spill, when public sentiments were strongly against oil companies. However, in 1995, pro-

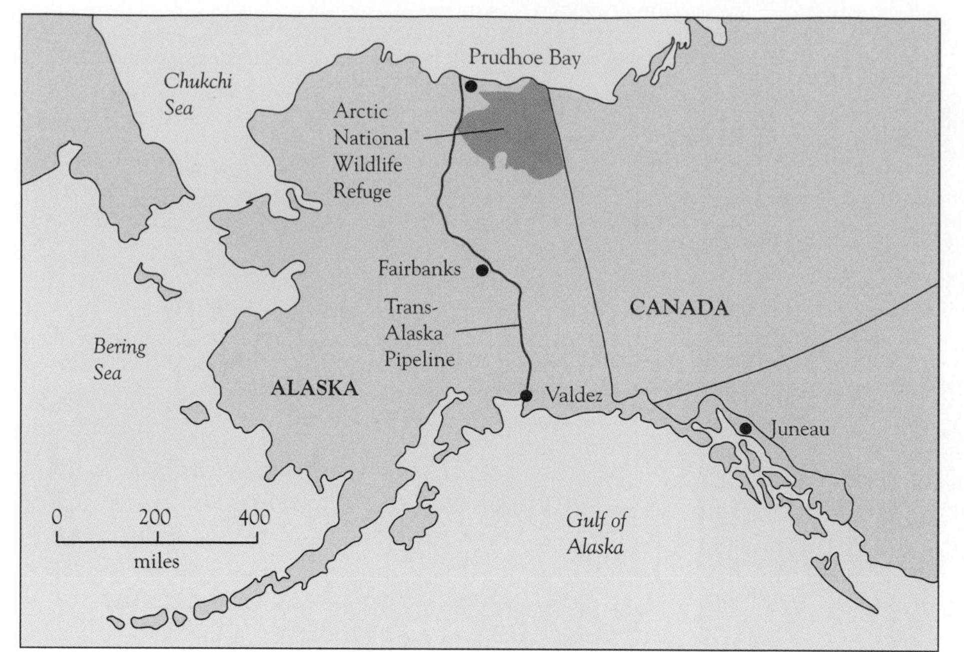

Figure 10–15
The Arctic National Wildlife Refuge, which is located in the northeastern part of Alaska, is situated close to the Trans-Alaska pipeline, which begins at Prudhoe Bay and extends south to Valdez.

OFFSHORE RIG DEBATE

When aging offshore oil platforms have reached the end of their usefulness, what should be done with them? Should they be completely dismantled, a complex task that costs millions of dollars per platform, or should underwater parts be left behind to serve as artificial reefs? For years, oil companies have left the bases of obsolete oil platforms anchored in the floor of the Gulf of Mexico. (They cut the legs off at a depth safe for ships.) Now many of the rigs off the coast of Southern California are nearing retirement age, and environmentally conscious Californians are beginning to debate their future. Oil companies want to leave the underwater legs behind because they save substantial money in removal costs. Moreover, the vertical underwater reefs support a large number of marine organisms.

Researchers at the University of California at Santa Barbara who studied a single underwater rig in the Santa Barbara Channel catalogued more than 25,000 fishes in the vicinity of the reef, as well as thousands of mussels, crabs, and starfish. Many environmentalists and fisherman, however, think the rigs should be completely removed because their presence could provide oil companies with a justification for not cleaning up oil seepage in the area as extensively as they should. Environmentalists are concerned that oil companies will say they cannot clean the area because the cleanup could jeopardize the artificial reefs.

development interests became more vocal, partly because in 1994, for the first time, the United States imported *more* than half the oil it used. Although a 1995 study by the Department of the Interior concluded that oil drilling in the refuge would harm the area's ecology, both the Senate and the House of Representatives passed measures to allow it. (The bill was vetoed by President Clinton in the fall of 1995.)

History of the Arctic National Wildlife Refuge

In 1960 Congress declared a section of northeastern Alaska protected because of its distinctive wildlife. In 1980 Congress expanded this wilderness area to form the Arctic National Wildlife Refuge—7.3 million hectares (18 million acres) of untouched northern forests, tundra wetlands, and glaciers (Figure 10–15). The Department of the Interior was given permission to conduct a study of the potential for oil discoveries in the area, but exploration and development could proceed only with congressional approval.

The refuge, which has been referred to as "America's Serengeti," is home to an extremely diverse community of organisms, including polar bears, arctic foxes, peregrine falcons, musk-oxen, Dall sheep, wolverines, and snow geese. It also contains the calving area for a large migrating herd of caribou. The Porcupine caribou herd,

SIBERIA'S DISAPPEARING WILDERNESS

Although most people picture Siberia as a vast, unspoiled wilderness, Russian oil and natural gas exploration is changing that image. Western Siberia produces 60 percent of Russia's oil and most of the natural gas piped to Europe. Exploitation of these hydrocarbon resources has come rapidly, since the 1970s, and at a tremendous ecological cost to northern ecosystems–tundra and boreal forests. Because Russian drilling technology is primitive, spills from faulty pipelines are commonplace, far more drilling sites are constructed than would be necessary with more modern methods, and oil is extracted wastefully. Abandoned drilling equipment litters boreal forests, frequent fires destroy large areas, and an astounding number of roads, power lines, and pipelines fragment natural habitat. Oil wastes and spills are often burned or dumped into creeks, where they present perhaps their greatest threat, the contamination of lakes, swamps, and rivers. A massive 1994 spill in the Russian Republic of Komi that was estimated to be 730,000 barrels (30.7 million gallons)–almost three times the size of the *Exxon Valdez* spill–sent oil into the Pechora River and its tributaries. Tundra regions, like other northern ecosystems, experience very slow growth and a sluggish ability to rid pollutants, making them poor prospects for long-term recovery from such environmental damage.

named after the Porcupine River in Canada where the herd winters, contains more than 150,000 head. Dominant plants in this coastal plain of tundra include mosses, lichens, sedges, grasses, dwarf shrubs, and small herbs. Under a thin upper layer of soil is the permafrost layer, which contains permanently frozen water.

Although it is biologically rich, the tundra is an extremely fragile ecosystem, in part because of its harsh climate. The organisms living here are adapted to their environment, but they live "on the edge." Any additional stress has the potential to harm or even kill them. Thus, arctic organisms are particularly vulnerable to human activities.

Support for oil exploration in the refuge

Supporters cite economic considerations as the main reason for searching for oil in the refuge. They point out that the United States is spending a large proportion of its energy budget to purchase foreign oil. Development of domestic oil would help to improve the balance of trade and make us less dependent on foreign countries for our oil.

The oil companies are eager to develop this particular site because it is near Prudhoe Bay, where large oil deposits are already being tapped. Prudhoe Bay has a sprawling industrial complex to support oil production, including roads, pipelines, gravel pads, and storage tanks. The Prudhoe Bay oil deposits peaked in production in 1985 and have declined in productivity since then. As a result, the oil industry is looking for sites that can make use of the infrastructure already in place.

The study conducted by the Department of the Interior on the possibility of oil in the wildlife refuge was made public in 1987. It concluded that there is a 19 percent chance of finding oil there, which the oil industry considers enough to justify exploration.

Supporters of oil exploration argue that it will have little lasting impact on the environment or on organisms. They say that there has been little environmental disruption or contamination of Prudhoe Bay by the oil industry; further, they point out that the number of caribou in that area has actually increased.

Opposition to oil exploration in the refuge

Conservationists think that oil exploration poses permanent threats to the delicate balance of nature in the Alaskan wilderness, in exchange for a very temporary oil supply. They reason that the money that would be spent searching for oil would be better used for research into alternative, renewable energy sources and energy conservation—a more permanent solution to the energy problem. They further argue that "drain America first" policies will only increase our future dependence on foreign oil supplies.

Opponents of oil exploration also refute supporters' claims that Prudhoe Bay has been developed with little environmental damage. Studies such as one conducted by the U.S. Fish and Wildlife Service document considerable habitat damage and declining numbers of wolves and bears in the Prudhoe Bay area;[6] biologists directly attribute the increase in the caribou population to the decline in the number of predators. The oil industry and conservationists *do* agree on one point: it is not financially practical to restore developed areas in the Arctic to their natural states. Thus, development in the Arctic causes permanent changes in the natural environment.

SYNFUELS AND OTHER POTENTIAL FOSSIL FUEL RESOURCES

Synthetic fuels, or **synfuels**, are used in place of oil or natural gas. They are synthesized from coal and other sources and may be liquid or gaseous. Synfuels include

[6]The top predators are usually more susceptible to environmental disruption than are the organisms occupying lower positions in a food web.

tar sands, oil shales, gas hydrates, liquefied coal, and coal gas. All synfuels emit CO_2 when burned, and many have other negative environmental effects, such as land damage from surface mining. Although synfuels are more expensive to produce than fossil fuels, they may become more important as fossil fuel reserves decline.

Tar sands are underground sand deposits permeated with tar or oil so thick and heavy that it does not move. The oil in tar sands deep in the ground cannot be pumped out unless it is heated underground with steam to make it more fluid. If tar sands are close to the Earth's surface, however, they can be surface mined. Once oil is obtained from tar sands, it must be refined like crude oil. Major tar sands are found in Alberta, Canada, and in Venezuela. World tar sand reserves are estimated to contain half again as much fuel as world oil reserves.

"Oily rocks" were discovered by western American pioneers whose rock hearths caught fire and burned. In order to yield their oil, these sedimentary rocks, called **oil shales**, must be crushed, heated, and refined after they are mined. Because the mining and refinement of oil shales require the expenditure of a great deal of energy, it is not yet cost efficient to process oil shales. Large oil shale deposits are located in the United States, Russia, China, and Canada. Wyoming, Utah, and Colorado have the largest deposits in the United States. Like tar sands, oil shale reserves may contain half again as much fuel as world oil reserves.

Gas hydrates are reserves of ice-encrusted natural gas deep underground in porous rock. Massive deposits have been identified in the arctic tundra, deep under the permafrost, and in the deep ocean sediments of the continental slope. The U.S. oil industry is not particularly interested in extracting natural gas from gas hydrates at present, because of the expense involved. In a pilot program in Siberia, Russian scientists successfully removed natural gas from hydrate deposits by pumping methanol (an alcohol) into the hydrate region. Natural gas that is associated with ice readily dissolves in methanol, which can then be pumped out of the ground. The Japan National Oil Corporation is currently involved in a pilot project to extract gas hydrates from the Sea of Japan by 1999. Natural gas is the fastest growing source of energy in Japan, which currently gets its supplies from politically unstable countries in the Middle East and the former Soviet Union.

Coal has also been used to produce a nonalcohol liquid fuel. This process, called **coal liquefaction**, was first developed before World War II, but its expense prevented it from replacing gasoline production. Research in the 1980s resulted in a series of technological improvements that have lowered the cost of coal liquefaction, but it is still not cost-competitive. It may become commercially attractive when the cost of gasoline rises or when new innovations reduce the cost of producing liquid coal even further.

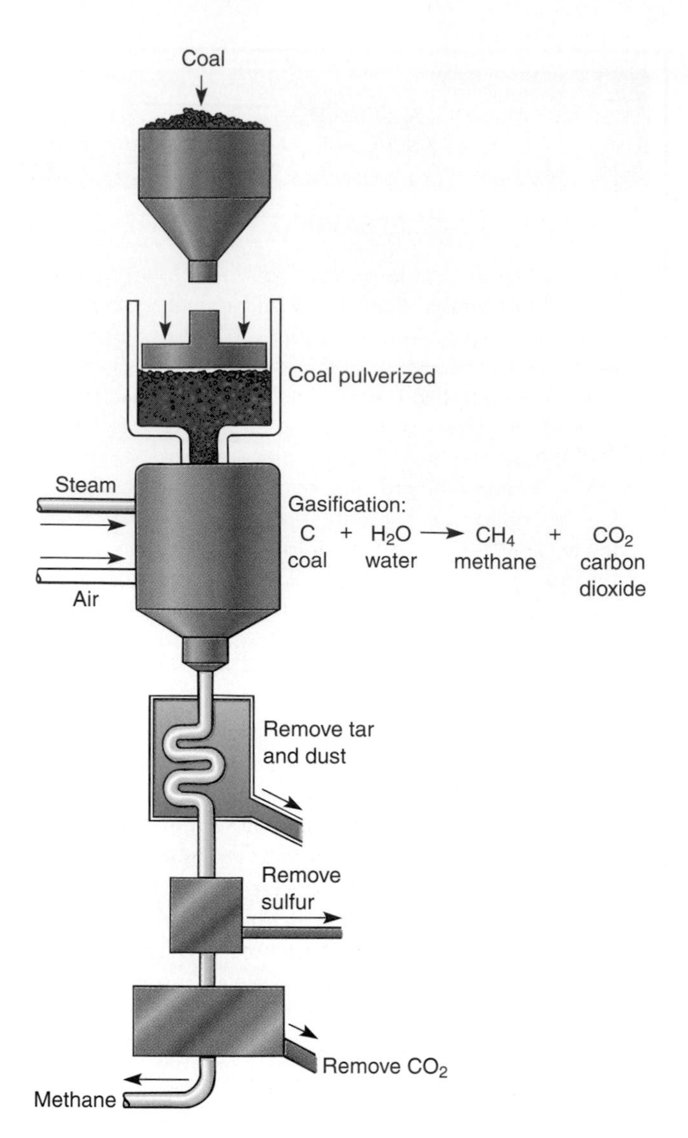

Figure 10–16
One method of coal gasification, in which the combustible gas methane is generated from coal.

Another synfuel is a gaseous product of coal. Coal gas has been produced since the 19th century. As a matter of fact, it was the major fuel used for lighting and heating in American homes until oil and natural gas replaced it in this century. Production of the combustible gas methane from coal is called **coal gasification** (Figure 10–16). A promising coal gasification technique was developed at Stanford University, and several demonstration power plants that convert coal into gas have been constructed in the United States. One advantage of coal gas over solid coal is that it burns cleanly. Because sulfur is removed during coal gasification, scrubbers are not needed when coal gas is burned. Like other synfuels, coal gas is more expensive to produce than fossil fuels.

Environmental Impacts of Synfuels

Although synfuels are promising energy sources, they have many of the same undesirable effects as fossil fuels. Their combustion releases enormous quantities of CO_2 into the atmosphere, thereby contributing to global warming. Some synfuels, such as coal gas, require large amounts of water during production and would therefore be of limited usefulness in arid areas, where water shortages are already commonplace. Also, mining the fossil fuels that are needed to produce synfuels damages the land. Enormously large areas of land would have to be surface mined in order to recover the fuel in tar sands and oil shales.

AN ENERGY STRATEGY FOR THE UNITED STATES

The United States needs a comprehensive energy strategy for several reasons: (1) the supply of fossil fuels is limited; (2) the production, transport, and use of fossil fuels pollute the environment; (3) our heavy dependence on foreign oil makes us economically vulnerable. Because of the complex nature of energy issues, any policy that is adopted by our political leaders has to have many approaches. Although there is no way to completely eliminate our vulnerability to disruptions in foreign oil supplies and to oil price increases, we can lessen the effects of such events through a comprehensive energy strategy. Such a strategy must provide us with a secure supply of energy, encourage us to use less energy, and protect the environment. The following elements should be included in a comprehensive national energy policy.

Increase Energy Efficiency and Conservation

Between 1975 and 1994, energy efficiency in the United States improved by about 27 percent,[7] but despite these gains, the United States uses more energy than any other country. There is room for great improvement on all fronts, from individuals conserving heating oil by weatherproofing their homes, to groups of commuters conserving gasoline by carpooling, to corporations developing more energy-efficient products. The automobile industry could be required to increase the average new-car gasoline mileage, for example.

One way to encourage energy conservation is to eliminate government subsidies that keep energy prices artificially low. *When prices reflect the true costs of energy, including the environmental costs incurred by its production, transport, and use, energy will be used more efficiently* (see Focus On: Energy Subsidies and the Real Price of Fuel). Gasoline prices in the United States do not reflect the true cost of gasoline and are unrealistically low. During the early 1990s, for example, Europeans paid two to three times more for gasoline than Americans did (Table 10–2). It has been demonstrated many times that the price of gasoline affects the level of gasoline consumption: lower prices encourage greater consumption. Over the next few years, a more realistic price for gasoline should be introduced to encourage people to buy fuel-efficient automobiles, carpool, and use public transportation (Table 10–3).

Other gasoline-conserving measures could be adopted, such as reinstating the 55-mile-per-hour speed limit in areas where the limit is now as high as 75 mph. Fuel consumption increases by approximately 50 percent

[7]Based on total energy consumption (in Btu's) per dollar of gross domestic product.

Table 10–2
Comparison of Gasoline Prices in Selected Countries

Country	Gasoline Price (Dollars per Gallon, 1993)*
United States	$1.31
Great Britain	$2.77
Germany	$3.25
France	$3.41
Italy	$3.77
Sweden	$4.20
Japan	$4.55

*Retail price, including taxes, for premium gasoline on January 1, 1993.
Source: U.S. Department of Energy

Table 10–3
Comparison of Energy Output for Different Kinds of Transportation

Method of Transportation	Energy Output (In BTUs)* per Person, per Mile
Automobile (Driver Only)	6530
Rail	3534
Carpool	2230
Vanpool	1094
Bus	939

*BTU stands for British thermal unit, an energy unit equivalent to 252 calories.

FOCUS ON

Energy Subsidies and the Real Price of Fuel

The price you pay for gasoline at the fuel pump is not its actual price. Its real cost is subject to a national energy pricing policy, which attempts to stabilize energy prices and reduce our dependence on sources of foreign energy supplies. Most governments practice some type of energy pricing policy, often in the form of energy subsidies or energy taxes.

An energy subsidy is equal to the difference between the world market price and the domestic market price for a particular fuel. For example, if gasoline trades for $1 per gallon on the world market, but costs only $.75 within a particular country, then that country's government is subsidizing energy at a cost of $.25 per gallon. Energy subsidies are often aimed at reducing the price of fuel for consumers, with the intent of stimulating economic growth. Low costs for consumers encourage a high use of energy, which in turn accelerates the depletion of a nonrenewable energy resource. When a government removes its energy subsidies, consumers have to pay more for fuel, which encourages them to use energy more efficiently so they can save money. The resulting decrease in energy use reduces a country's dependence on costly fuel imports and lessens the harmful environmental impacts of fuel production and consumption.

Whereas subsidies reduce the cost of fuel for consumers, energy taxes *increase* the cost. Energy taxes serve to raise a government's revenues. In addition, they encourage consumers to conserve energy, thereby helping to reduce a country's dependence on foreign energy supplies. Energy taxes reflect some of the hidden costs of energy consumption, such as pollution control and cleanup.

Energy pricing policies apply not only to gasoline (that is, oil) but also to electricity, natural gas, and coal. So the next time you fill up at the gas station, turn on a light, cook dinner, or adjust the thermostat, keep in mind that the price you pay for energy is determined by more than just the current day's price for a barrel of foreign oil.

if a car is driven at 75 mph rather than 55 mph; fuel consumption increases by 30 percent at 65 mph rather than 55 mph. In addition, federal financial support for transportation could be shifted from highway construction to public transportation. We say more about energy conservation in Chapter 12. (Also see You Can Make a Difference: Getting Around Town.)

Secure Future Energy Supplies

A comprehensive national energy strategy will probably include the environmentally sound and responsible development of domestically produced fossil fuels, especially natural gas.

There are two types of opposition to this element of a national energy strategy; one is economic and the other is environmental. Some think it is better to deplete foreign oil reserves while prices are reasonable and save domestic supplies for the future. Most economists argue against this view, however, because of the U.S. trade deficit; we do not currently finance our oil imports by exporting goods and services of equal value. Many environmentalists oppose the development and increased use of domestic fossil fuels, largely due to the environmental problems already discussed.

Everyone, environmentalists included, recognizes the need for a dependable energy supply. Securing a future energy supply is a *temporary* solution, however, because fossil fuels are nonrenewable resources that will eventually be depleted, regardless of how efficient our use or how much we conserve. Having a secure energy supply for the short term will, however, allow us time to develop alternative energy sources for the long term.

Improve Energy Technology

Research and development must be expanded for all possible alternatives to fossil fuels, especially renewable energy sources such as solar and wind energy (see Chapter 12). Our long-term energy policy goal should be to shift to energy sources that are less harmful to the environment.

Who should pay for the research costs of improving energy conservation and developing alternative forms of energy? The answer is that we all should share in these costs because we will all share in the benefits. The proceeds of a gasoline tax have been suggested as a means of financing programs to achieve a sustainable energy future. Some policy makers have suggested a tax of as much as 50 cents per gallon. This may sound excessive until

YOU CAN MAKE A DIFFERENCE

Getting Around Town

Can you imagine getting around town without a car? How would you get to class, the grocery store, the laundromat? Hopping in your car for every errand seems like the natural thing to do. According to the American Automobile Association, American motorists drive an average of 10,100 miles annually and burn 507 gallons of gasoline in the process.

However, consider this: from the production of gasoline to the disposal of old automobiles, the car has a significant negative impact on the environment. Acid deposition and global warming are just two of the problems caused by the combustion of gasoline. Vehicle exhaust, photochemical smog, and chronic low-level exposure to toxins are all health threats to car owners and to those who live in areas with a high density of cars (see Chapter 19). Dumping of engine oil, fumes from the burning of tires and batteries, and automobile junkyards threaten both our health and our environment.

There is something you can do about it. Granted, you may not be able to give up your car entirely, but you can cut down on its use wherever possible. For example, try the following:

1. Consider whether you need to drive to accomplish a task. Sometimes a phone call can be substituted for a trip in the automobile.
2. Carpool to class, to work, to the grocery store, to social events. One car on the road is better than three or four.
3. Buy a good bicycle; it is less expensive than a car to buy and maintain, and it is great for local transportation. It is also good exercise.
4. Ride the bus or the train whenever possible. Think about how jammed the road would be if all 50 people on a bus were driving individual cars!
5. Walk to class or work if you live within a mile or so. You will need to allow yourself a little extra time, but once you get into the habit, it is easy. Walking is good exercise, too.
6. Modify your driving habits to save gasoline. Minimize braking, and do not let your engine idle for more than 1 minute. Keep your car well tuned, replace air and oil filters often, and keep your tires inflated at the recommended pressure. Remove any unnecessary weight from your car. All of these measures help boost gasoline mileage.
7. When you purchase a motor vehicle, use the EPA's official mileage ratings as one of the criteria to help you select a model. Avoid sport utility vans, which have extremely low mileages per gallon. The Chevrolet Suburban gets only 13 miles to the gallon in city traffic, for example, and the Jeep Grand Cherokee only 14 miles to the gallon. Sport utility vehicles also emit more air pollution (unburned fuel, soot particles, and other pollutants) than most other motor vehicles.

U.S. gasoline prices are compared with those in Japan and Europe, which are much higher (see Table 10–2). More expensive gasoline does not seem to have hampered Japan's or Europe's economic competitiveness.

Accomplish the First Three Objectives Without Further Damaging the Environment

The environmental costs of using a particular energy source must be weighed against its benefits when it is considered as a practical component of an energy policy. If domestic supplies of fossil fuels are developed with as much attention to the environment as possible, they will not only help reduce our dependence on foreign oil, but also give us time to develop alternative forms of energy. One suggestion has been to add a 5-cent tax on each barrel of domestically produced oil to establish a reclamation fund for undoing some of the environmental damage caused by mining and production of fossil fuels.

How Politics Influences the National Energy Strategy

Each presidential administration develops an energy policy that is implemented by the Department of Energy and other federal agencies. Thus, the nation's energy policy reflects the political views of the president, Congress, and the American public. It is also subject to major changes with every change in the presidential office.

The Clinton energy strategy has five components: (1) increase the efficiency of energy use; (2) develop a balanced domestic energy resource portfolio; (3) invest in science and technology advances; (4) reinvent environmental protection; and (5) engage the international market (Table 10–4).

Table 10–4
An Overview of the U.S. National Energy Policy

Increase the efficiency of energy use

Improve efficiencies (fuel economy) in transportation

Develop markets for alternative fuels for transportation

Invest in energy-efficient improvements for buildings

Use energy-efficient technologies in industry

Develop a balanced domestic energy resource portfolio

Enhance domestic oil production

Increase utilization of natural gas

Increase long-term investments in renewable energy

Support clean coal technology development

Increase safety in the nuclear power industry

Invest in science and technology advances

Support research in materials science, such as solar voltaic cells and lightweight materials for batteries and fuel cells

Support research in geoscience, such as underground imaging, computer modeling, and remote sensing, all of which are necessary to discover new energy resources

Support research in chemistry, such as combustion processes and increasing combustion efficiency

Support research in biological and environmental science, for example, to anticipate health and environmental effects of energy use

Support research in new, advanced energy sources, such as hydrogen technologies and fusion

Reinvent environmental protection

Encourage cost-effective pollution prevention

Focus research on technologies that increase productivity, reduce pollution, and cut energy use

Try new approaches to environmental regulations to achieve more cost-effective environmental results

Engage the international market

Strengthen energy policies here and abroad

Prepare for global energy supply disruptions

Respond to global environmental challenges

Promote the use of U.S. energy technology abroad

Source: *Sustainable Energy Strategy: Clean and Secure Energy for a Competitive Economy*, July 1995, National Energy Policy Plan, ISBN 0-16-048183-X

SUMMARY

1. Oil, natural gas, and coal are fossil fuels that formed several hundred million years ago from plant and animal remains. They are nonrenewable resources. None of them is a perfect energy source; their combustion produces several pollutants—in particular, large amounts of CO_2, a greenhouse gas that prevents heat from escaping from the Earth.

2. Coal is formed when partially decomposed plant material is exposed to heat and pressure for aeons. There are several grades of coal. Lignite is soft coal that is low in sulfur and produces less heat than other grades; bituminous coal is usually high in sulfur and produces a lot of heat; anthracite is low in sulfur and produces more heat than the other coal types.

3. Coal is present in greater quantities than oil or natural gas, but its use has a greater potential to harm the environment. Underground coal mining is dangerous and unhealthful. Surface mining disturbs large areas of land, which are difficult to restore.

4. Coal produces more CO_2 emissions per unit of heat than do other fossil fuels, and there is currently no way to reduce or eliminate CO_2 emissions where coal is burned. Burning soft coals that contain sulfur contributes to acid deposition. There are several ways to control sulfur emissions, including fluidized-bed combustion and the installation of scrubbers in smokestacks.

5. Oil and natural gas deposits occur in association with structural traps such as anticlines and salt domes. Although world oil and natural gas reserves are large, they will probably be depleted during the 21st century.

6. The environmental problems associated with the use of oil and natural gas include damage to the environment where the wells are located, accidental oil spills during transport and storage, and CO_2 emissions when the oil and natural gas are burned.

7. Synfuels (tar sands, oil shales, gas hydrates, liquid coal, and coal gas) are liquid or gaseous fuels that substitute for oil or natural gas. Synfuels are currently more expensive to produce than oil or natural gas, but they may be used more in the future, as fossil fuel reserves decline. Synfuels have many of the environmental drawbacks associated with traditional fossil fuels.

8. The goal of a national energy strategy should be to ensure adequate energy supplies without harming the environment. Such a policy can be accomplished by increasing energy efficiency, securing future energy supplies, improving energy technology, and avoiding harm to the environment.

THINKING ABOUT THE ENVIRONMENT

1. It has been suggested that the Industrial Revolution was concentrated in the Northern Hemisphere largely because coal is located there. What is the relationship between coal and the Industrial Revolution?

2. Several politicians have commented that the United States is "the OPEC of coal." What does this mean?

3. Based on what you have learned about coal, oil, and natural gas, which fossil fuel do you think the United States should exploit in the short term (during the next 20 years)? Explain your rationale.

4. In your estimation, which fossil fuel has the greatest potential for the next century? Why?

5. Which of the negative environmental impacts associated with fossil fuels is most serious? Why?

6. Which major consumer of oil is most vulnerable to disruption in the event of another energy crisis: electric power generation, motor vehicles, heating and air conditioning, or industry? Why?

7. Why are synfuels promising, but not a panacea for our energy needs?

* 8. Draw a graph showing daily per-capita gasoline consumption (in gallons) from 1950 to 1993. Use the information provided in Table A. You will need to convert barrels of gasoline to gallons (1 barrel = 42 gallons) and then

determine the per-capita gasoline consumption. (Divide gasoline consumption in gallons per day by the population; round to the nearest tenth of a gallon.)

* 9. If the military cost of securing a steady supply of oil from the Persian Gulf, conservatively estimated at $40 billion per year, is added to America's fuel bill, imported oil actually costs much more than Americans think it does. Calculate the increase per gallon of gasoline if the military cost were spread equally across all gasoline consumed, both domestic and foreign, in the United States. Use the 1993 data (gasoline, million barrels per day) provided in Table A to get started.

Table A

Per-Capita Gasoline Consumption in the United States, 1950 to 1993

Year	Gasoline (Million Barrels/Day)	Gasoline (Million Gallons/Day)	Population (Millions)	Gasoline Consumption (Gallons/Day/Person)
1950	2.72	114.24	151.3	0.8
1955	3.66		165.1	
1960	4.13		179.3	
1965	4.59		193.5	
1970	5.78		203.2	
1975	6.67		215.5	
1980	6.58		226.5	
1985	6.83		238.7	
1990	7.23		248.7	
1993*	7.48		257.9	

*Latest data available.
Source: U.S. Department of Energy

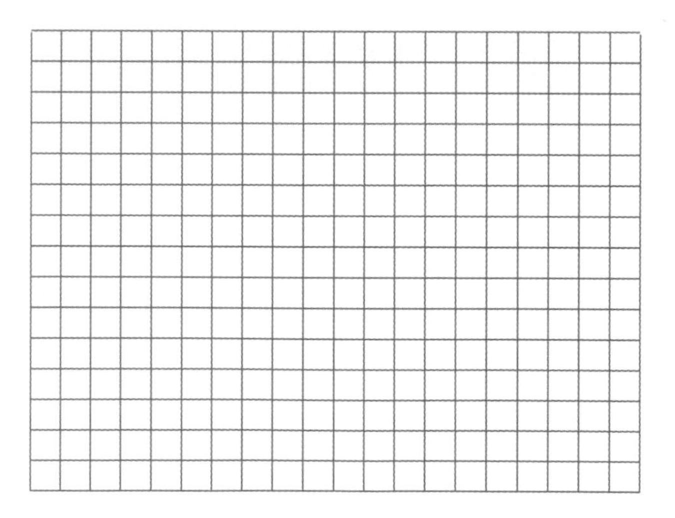

RESEARCH PROJECTS

Some analysts contend that the energy crises of the 1970s (discussed in the introduction) were caused in part by policies of the U.S. government that encouraged domestic oil consumption but discouraged domestic oil production. Investigate oil price controls and entitlements during the 1970s to determine what role, if any, these policies had on oil imports and the energy crises.

Contact a major oil, coal, or natural gas company and ask how its environmental protection program has changed in the past ten years. Ask for documentation of expenditures, and determine what impact such costs have on the final cost of delivered energy.

On-line materials relating to this chapter are on the World Wide Web at **http://www.saunderscollege.com/lifesci/environment2** (select Chapter 10 from the Table of Contents).

SUGGESTED READING

Corcoran, E. "Cleaning Up Coal," *Scientific American*, May 1992. A review of new technologies that could clean up coal's dirty environmental image.

Doherty, J. "The Arctic National Wildlife Refuge: The Best of the Last Wild Places," *Smithsonian*, March 1996. Discusses the riches of the Arctic National Wildlife Refuge.

Earle, S. A. "Persian Gulf Pollution: Assessing the Damage One Year Later," *National Geographic* Vol. 181, No. 2, February 1992. How researchers started evaluating the environmental destruction in Kuwait and the Persian Gulf.

Feldman, D. L. "Revisiting the Energy Crisis: How Far Have We Come?" *Environment* Vol. 37, No. 4, May 1995. Examines the legacy of the 1973 Arab oil embargo.

Flavin, C. "The Bridge to Clean Energy," *World Watch* Vol. 5, No. 4, July–August 1992. Natural gas is increasingly viewed as a better energy alternative than either coal or oil.

Hubbard, H. M. "The Real Cost of Energy," *Scientific American*, April 1991. Analyzes how the U.S. government has subsidized energy and what should be done to bring market prices in line with true energy costs.

Mackenzie, J. J. "Heading Off the Permanent Oil Crisis," *Issues in Science and Technology*, Summer 1996. The author contends that the short-term spike in gasoline prices during the spring of 1996 forecasts future high prices for oil.

Monastersky, R. "The Mother Lode of Natural Gas," *Science News* Vol. 150, November 9, 1996. Gas hydrates may be more widespread than previously thought.

Pain, S. "The Two Faces of the Exxon Disaster," *New Scientist* Vol. 138, 22 May 1993. Scientists disagree over the effects of the *Exxon Valdez* oil spill, with Exxon researchers downplaying the damage.

Tunali, O. "A Billion Cars: The Road Ahead," *World Watch* Vol. 9, No. 1, January–February 1996. What the skyrocketing demand for automobiles in Asia, Latin America, and Eastern Europe means to the environment and to the quality of human life.

Wiens, J. A. "Oil, Seabirds, and Science: The Effects of the *Exxon Valdez* Oil Spill," *BioScience* Vol. 46, No. 8, September 1996. How the Alaskan oil spill compromised the scientific process.

The nuclear power plant at Salem, New Jersey. (Courtesy Public Service Electric & Gas Co.)

Nuclear Energy

In this chapter we examine the facts and controversies of nuclear power. In order to arrive at conclusions that are intelligent and informed, we must first understand some of the basic science behind nuclear power and how nuclear technology is used to produce energy.

As a way to obtain energy, nuclear power is fundamentally different from the combustion that produces energy from fossil fuels. Combustion is a chemical reaction. In ordinary chemical reactions, atoms of one element do not change into atoms of another element, nor does any of their mass (matter) change into energy. The energy released in combustion and other chemical reactions comes from changes in the chemical bonds that hold the atoms together. Chemical bonds are associations between electrons, so ordinary chemical reactions involve the rearrangement of electrons (see Appendix I).

In contrast, nuclear energy involves changes within the *nuclei* of atoms; small amounts of matter from the nucleus are converted into very large amounts of energy. There are two different reactions that release nuclear energy: fission and fusion. In **fission**, larger atoms of certain elements are split into two smaller atoms of different elements, whereas in **fusion**, two smaller atoms are combined to make one larger atom of a different element. In each case, the mass of the end product(s) is less than the mass of the starting material(s) because a small quantity of the starting material is converted to energy.

Nuclear reactions produce 100,000 times more energy per atom than do chemical reactions such as combustion. In nuclear bombs this energy is released all at once, producing a tremendous surge of heat and power that destroys everything in its vicinity. When nuclear energy is used to generate electricity, the nuclear reaction is *controlled* to produce smaller amounts of energy in the form of heat, which can then be converted to electricity. ◀

ATOMS AND RADIOACTIVITY

All atoms are composed of positively charged protons, negatively charged electrons, and electrically neutral neutrons (Figure 11–1). Protons and neutrons, which have approximately the same mass, are clustered in the center of the atom, making up its nucleus. Electrons, which possess little mass in comparison to protons and neutrons, orbit around the nucleus in distinct regions. Atoms that are electrically neutral possess identical numbers of positively charged protons and negatively charged electrons.

The **atomic mass** of an element is equal to the sum of protons and neutrons in the nucleus. Each element contains a characteristic number of protons per atom, called its **atomic number**. In contrast, the number of neutrons in each atom of a given element may vary, resulting in atoms of one element with different atomic masses. Forms of a single element that differ in atomic mass are known as **isotopes**. For example, normal hydrogen, the lightest element, contains one proton and no neutrons in the nucleus of each atom. The other two isotopes of hydrogen are **deuterium**, which contains one proton and one neutron per nucleus, and **tritium**, which contains one proton and two neutrons per nucleus. Many isotopes are stable, and some are unstable; the unstable ones, called **radioisotopes**, are said to be **radioactive** because they spontaneously emit **radiation**, a form of energy. The only radioisotope of hydrogen is tritium.

As a radioactive element emits radiation, its nucleus changes into the nucleus of a different element, one that is more stable; this process is known as **radioactive decay**. For example, the radioactive nucleus of one isotope of uranium, U-235,[1] decays over time into lead (Pb-207). Each radioisotope has its own characteristic rate of decay. The period of time required for one-half of a radioactive substance to change into a different material is known as its **radioactive half-life**. There is enormous variation in the half-lives of different radioisotopes (Table 11–1). For example, the half-life of iodine (I-132) is only 2.4 hours, whereas the half-life of an isotope of uranium (U-234) is 250,000 years.

[1]Uranium-235 (U-235 or ^{235}U) is an isotope of uranium with an atomic mass of 235.

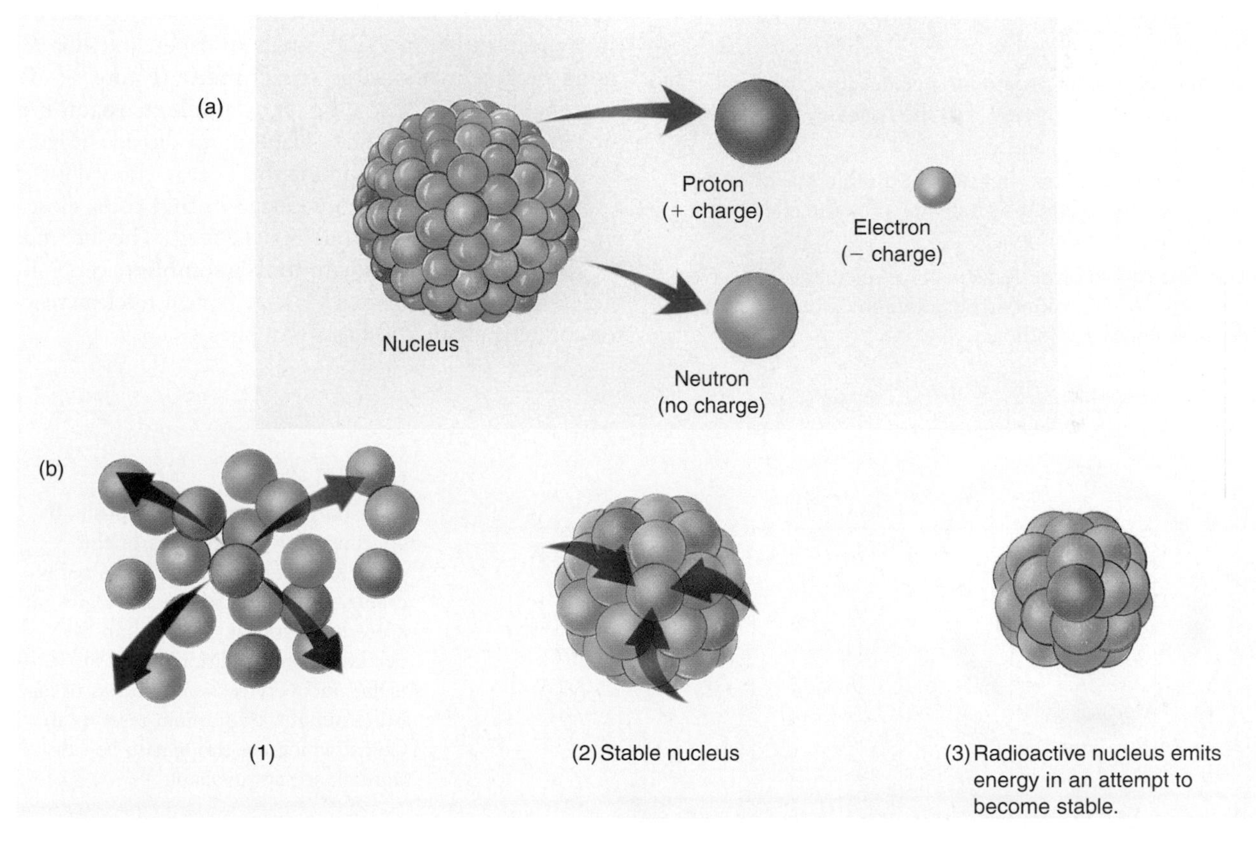

Figure 11–1

Atomic structure. Matter is composed of atoms. (a) Atoms contain a nucleus made of positively charged particles (protons) and particles with no charge (neutrons). Circling the nucleus is a cloud of small, negatively charged particles called electrons. (b) Particles with like charges tend to repel one another, which would cause the nucleus to fall apart (1). Atomic forces keep the positively charged protons together (2). Neutrons tend to stabilize the nucleus. Some atoms have too few or too many neutrons to maintain nuclear stability (3). These atoms are radioactive.

Table 11–1
Some Common Radioactive Isotopes Associated with the Fission of Uranium

Radioisotope	Half-Life (years)
Cerium-144	0.8
Cesium-137	30
Iodine-131	0.02 (8.1 days)
Krypton-85	10.4
Neptunium-237	2,130,000
Plutonium-239	24,400
Plutonium-240	6600
Radon-226	1600
Ruthenium-106	1.0
Strontium-90	28
Tritium	13
Xenon-133	0.04 (15.3 days)

Mini-Glossary of Nuclear Energy Terms

nuclear energy: The energy released by nuclear fission or fusion.

fission: The splitting of an atomic nucleus into two smaller fragments, accompanied by the release of a large amount of energy.

fusion: The joining of two lightweight atomic nuclei into a single heavier nucleus, accompanied by the release of a large amount of energy.

isotope: One of two or more forms of an element that have the same atomic number but different atomic masses. Some isotopes are radioactive.

radioactive decay: The emission of energetic particles or rays from unstable atomic nuclei.

radioisotope: A radioactive isotope of an element.

half-life: The amount of time it takes for half of the atoms in a radioactive substance to decay.

CONVENTIONAL NUCLEAR FISSION

Uranium ore, the mineral fuel used in conventional nuclear power plants, is a nonrenewable resource present in limited amounts in the Earth's crust. Uranium deposits are usually located in sedimentary rocks, but how they got there is not well understood. One possibility is that groundwater containing dissolved uranium seeped through the sediments, which gradually became infiltrated with uranium.

Substantial deposits of uranium are found in Australia (25.7 percent), Africa (24 percent), and North America (21.9 percent) (Figure 11–2). Uranium ore contains three isotopes, U-238 (99.28 percent), U-235 (0.71 percent), and U-234 (less than 0.01 percent). Because U-235, the isotope that is used in conventional fission reactions, is such a minor part (0.71 percent) of uranium ore, uranium ore must be refined after mining to increase the concentration of U-235 to about 3 percent; this refining process is known as **enrichment** (Figure 11–3).

The uranium fuel used in a **nuclear reactor** is processed into small pellets of uranium dioxide (Figure 11–4), each of which contains the energy equivalent of a ton of coal. The pellets are placed in **fuel rods**, closed pipes, often as long as 3.7 meters (12 feet). The fuel rods are then grouped into square **fuel assemblies**, generally of 200 rods each (Figure 11–5). A typical nuclear reactor contains about 250 fuel assemblies.

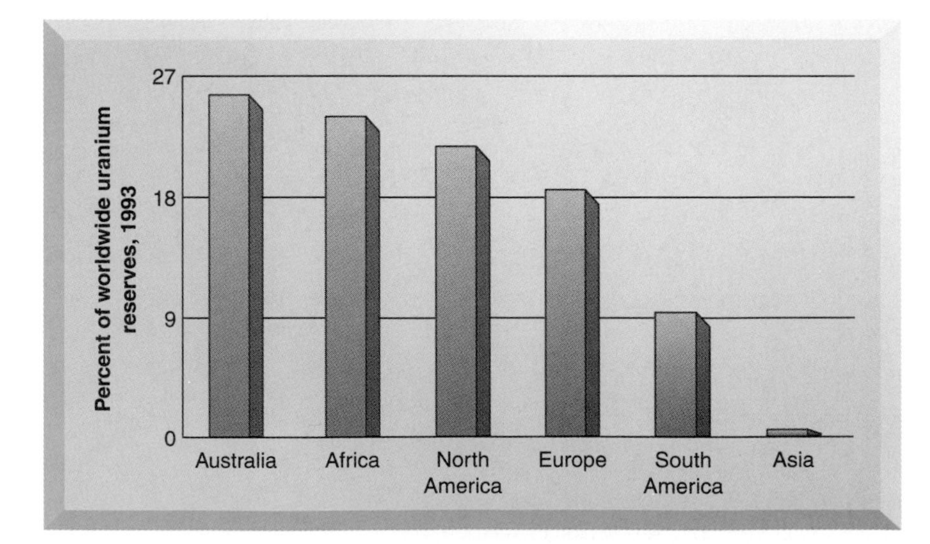

Figure 11–2
Distribution of uranium deposits by continent, as percentages of the 1993 known reserves that are recoverable, using existing technologies, at a production cost of less than $80 per kilogram. Additional deposits with higher recovery costs are known to exist. (Estimates of uranium reserves in China, which are thought to be substantial, are not available.)

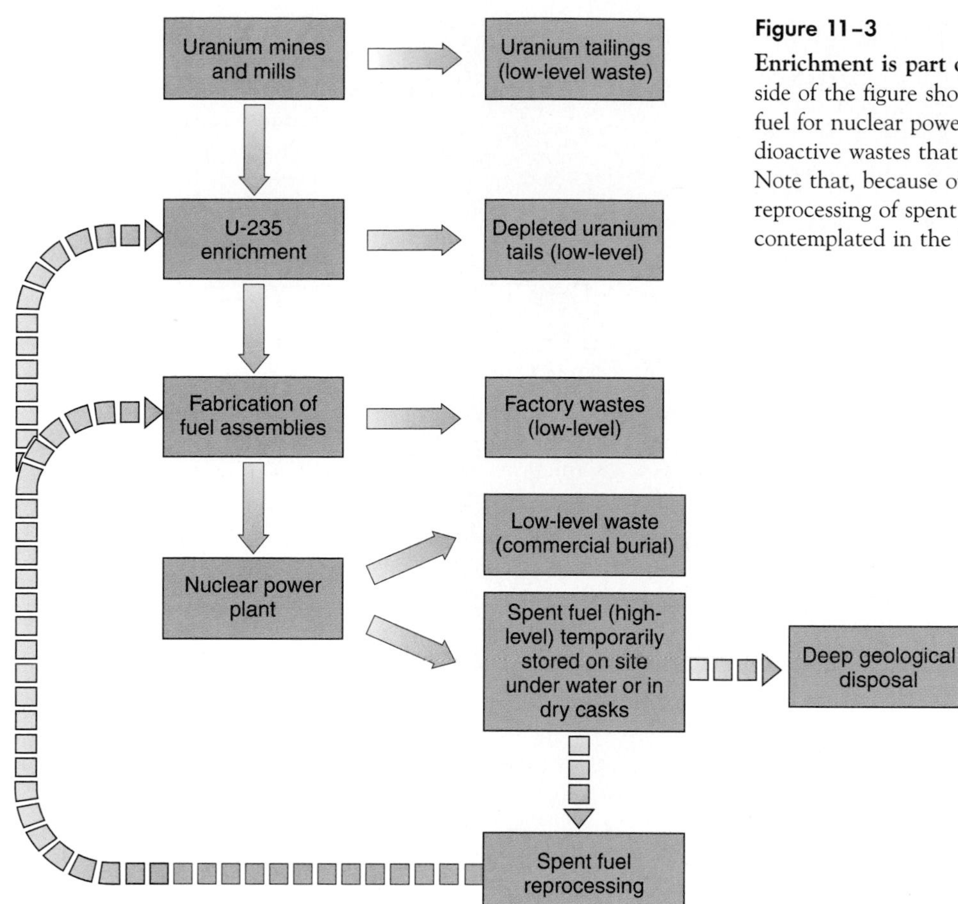

Figure 11–3

Enrichment is part of the nuclear fuel cycle. The left side of the figure shows how mined uranium becomes fuel for nuclear power plants. The right side shows radioactive wastes that must be handled during the cycle. Note that, because of economic and political reasons, reprocessing of spent fuel is not under way or currently contemplated in the United States.

Figure 11–4

Uranium dioxide pellets, held in a gloved hand, contain about 3 percent uranium-235, which is the fission fuel in a nuclear reactor. Each pellet contains the energy equivalent of one ton of coal. (Courtesy of Westinghouse Electric Corp., Commercial Nuclear Fuel Division)

In nuclear fission U-235 is bombarded with neutrons (Figure 11–6). When the nucleus of an atom of U-235 is struck by and absorbs a neutron, it becomes unstable and splits into two smaller atoms, each approximately half the size of the original uranium atom. In the fission process, two or three neutrons are also ejected from the uranium atom. They collide with other U-235 atoms, generating a chain reaction as those atoms are split and more neutrons are released to collide with additional U-235 atoms.

The fission of U-235 releases an enormous amount of heat, which is used in a nuclear power plant to transform water into steam, which is, in turn, used to generate electricity. Production of electricity is possible because the fission reaction is controlled. (Recall that an uncontrolled fission reaction results in a nuclear explosion. Even if the control mechanism failed, a military-type nuclear explosion cannot take place in a nuclear power reactor.) Fission reactions in the reactor of a nuclear power plant can be started or stopped, increased or decreased, thus allowing the desired amount of heat energy to be produced.

Figure 11–5

The uranium pellets are loaded into long fuel rods, which are grouped into square fuel assemblies (shown). (Courtesy of Westinghouse Electric Corp., Commercial Nuclear Fuel Division)

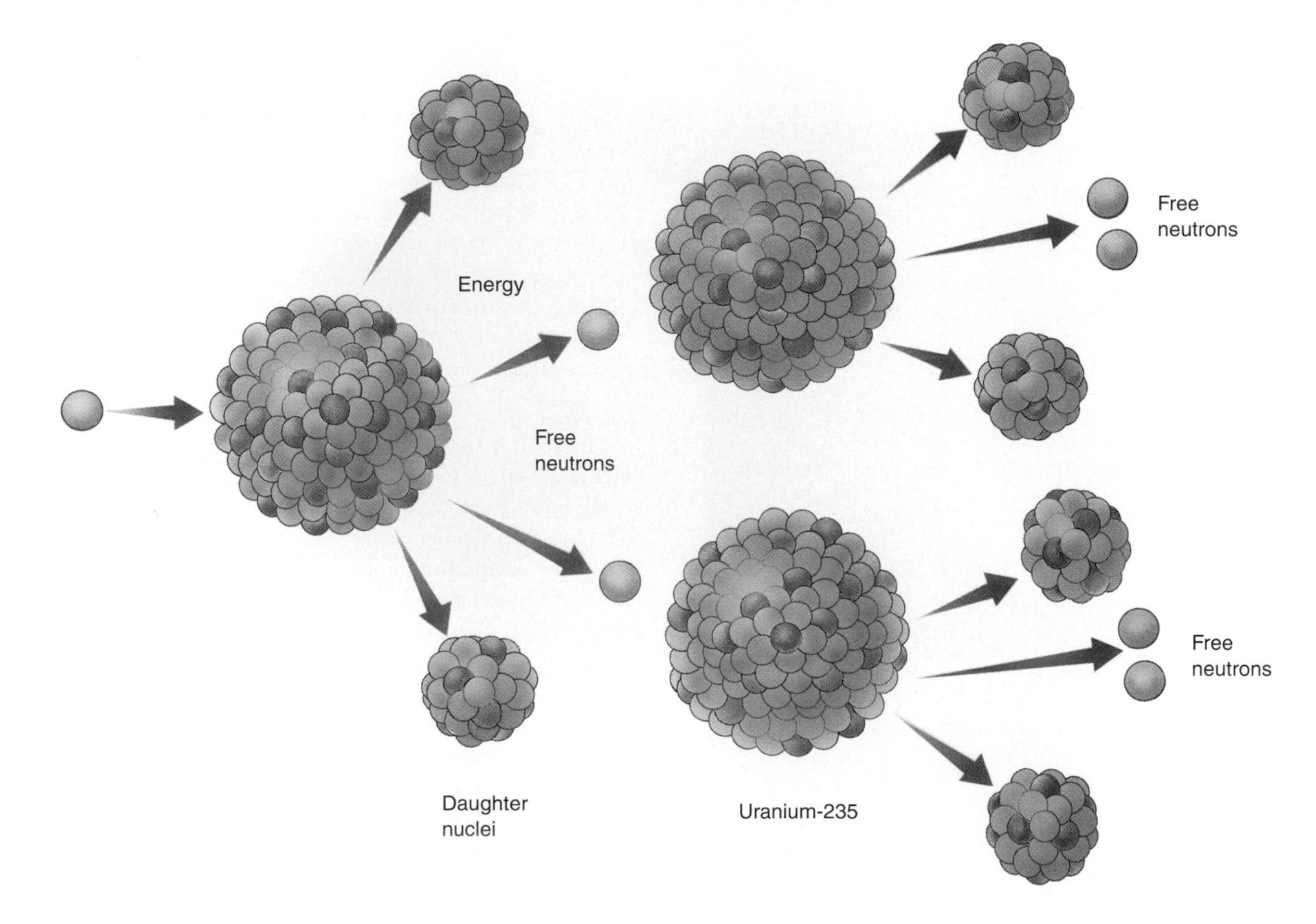

Figure 11–6

Nuclear fission. Neutron bombardment of a U-235 nucleus causes it to split into two smaller atomic fragments and several free neutrons. The free neutrons bombard nearby U-235 nuclei, causing them to split and release still more free neutrons in a process called a chain reaction.

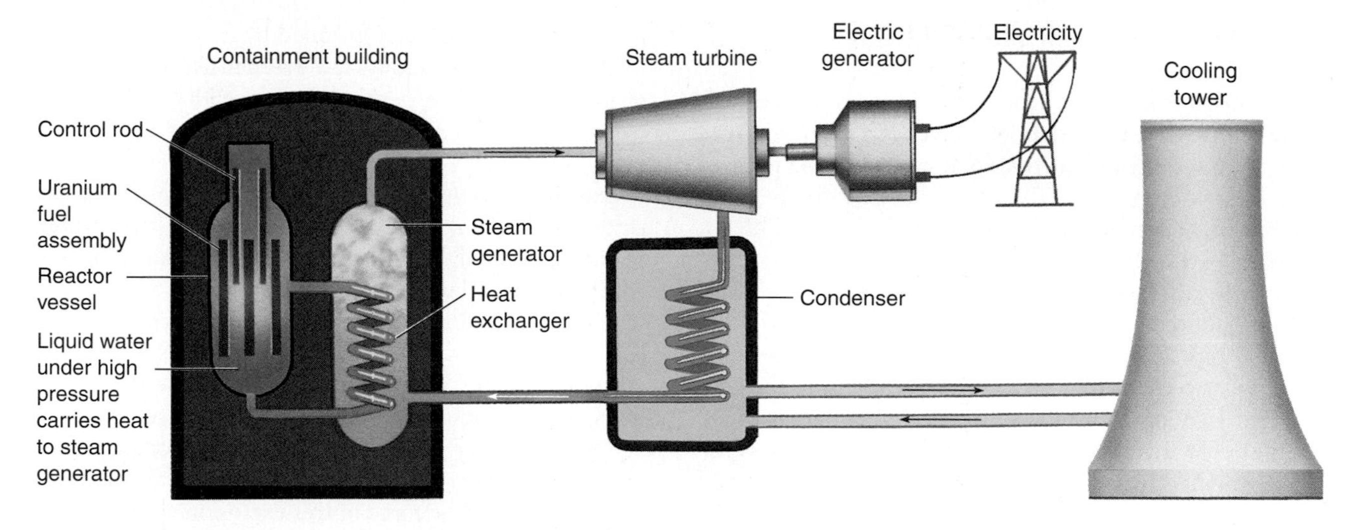

Figure 11–7

How a nuclear power plant generates electricity. Fission of U-235 that occurs in the reactor vessel produces heat, which is used to produce steam in the steam generator. The steam drives a turbine to generate electricity. The steam then leaves the turbine and is pumped through a condenser before returning to the steam generator. Excess heat is controlled by pumping hot water from the condenser to a massive cooling tower. After it is cooled, the water is pumped back to the condenser. The power plant shown is a pressurized water reactor. Approximately two thirds of all nuclear power plants in the United States are of this type.

How Electricity Is Produced from Nuclear Power

A typical nuclear power plant has four main parts: (1) the **reactor core**, where fission occurs; (2) the **steam generator**, where the heat produced by nuclear fission is used to produce steam from liquid water; (3) the **turbine**, which uses the steam to generate electricity; and (4) the **condenser**, which cools the steam, converting it back to a liquid (Figure 11–7).

Fission takes place in the reactor core, which contains the fuel assemblies. Above each fuel assembly is a **control rod** made of a special metal alloy that is capable of absorbing neutrons. The plant operator signals the control rod to move either up out of or down into the fuel assembly. If the control rod is out of the fuel assembly, free neutrons collide with the fuel rods and fission of uranium takes place. If the control rod is completely lowered into the fuel assembly, it absorbs the free neutrons, and fission of uranium no longer occurs. By exactly controlling the placement of the control rod, the plant operator can produce the exact amount of fission required.

A typical nuclear power plant has three water circuits. The **primary water circuit** (orange circuit in Figure 11–7) heats water, using the energy produced by the fission reaction. This circuit is a closed system that circulates water under high pressure through the reactor core, where it is heated to about 293°C (560°F). Because it is under such high pressure, this superheated water cannot expand to become steam and so remains in a liquid state.

From the reactor core, the very hot water circulates to the steam generator, where it boils water held in a **secondary water circuit** (blue circuit in Figure 11–7), converting the water to steam. The pressurized steam goes to and turns the turbine, which in turn spins a generator to produce electricity. After it has turned the turbine, the steam in the secondary water circuit goes to a condenser, where it is converted to a liquid again. The cooling is necessary to obtain the pressure differential that helps turn the turbine blades.

A **tertiary water circuit** (green circuit in Figure 11–7) provides cool water to the condenser, which cools the spent steam in the secondary water circuit. As the water in the tertiary water circuit is heated, it moves from the condenser to a **cooling tower**, where it is cooled before circulating back to the condenser.

Safety Features of Nuclear Power Plants

The reactor core where fission occurs is surrounded by a huge steel potlike structure called a **reactor vessel**—a safety feature designed to prevent the accidental release of radiation into the environment. The reactor vessel and steam generator are placed in a **containment building**, an additional line of defense against accidental radiation leaks. Containment buildings have steel-reinforced concrete walls 0.9 to 1.5 meters (3 to 5 feet) thick and are built to withstand severe earthquakes, the high winds of hurricanes and tornadoes, and even planes crashing into them.

BREEDER NUCLEAR FISSION

Uranium ore is mostly U-238, which is not fissionable and is therefore a waste product of conventional nuclear fission. In **breeder nuclear fission**, however, U-238 is converted to plutonium, Pu-239, a human-made isotope that is fissionable. Breeder reactors can use U-235, Pu-239, or thorium (Th-232) as fuel. Some of the neutrons that are emitted in breeder nuclear fission are used to produce additional plutonium from U-238 (Figure 11–8). A breeder reactor thus makes more fissionable fuel than it uses.

Because breeder fission can use U-238, it has the potential to generate much larger quantities of energy from uranium ore than traditional nuclear fission can. For example, if the U-238 stored in nuclear waste sites across the United States could be taken out and used in breeder reactors, it would supply the entire country with electricity for at least 100 years! When one adds the uranium reserves in the ground to these nuclear waste stockpiles, breeder fission has the potential to supply the entire country with electrical energy for several centuries.

Although the first breeder reactor experiments were performed in the United States, leadership in developing this technology has been assumed by Europe and Japan. In the whole world, only a few breeder fission plants are operational, and the development of additional breeder reactors will be a slow process, as many technical and safety problems have yet to be resolved. For example, for reasons too complex to consider here, breeder fission reactors use liquid sodium as a coolant, rather than water. Sodium is a highly reactive metal that reacts explosively with water and burns spontaneously in air at the high temperatures maintained in the breeder reactor. Thus, a leak in a breeder reactor's plumbing system is dangerous. Also, should a leak be large enough to cause the loss of much of the liquid sodium coolant, the temperature within the reactor might get high enough to cause an uncontrolled nuclear fission reaction—that is, a small nuclear explosion. The force of this explosion would almost certainly rip open the containment building, releasing radioactive materials into the atmosphere.

Public and governmental distrust of breeder reactors is greater than their misgivings about conventional fission, because plutonium is used not only in breeder nuclear fission, but also in nuclear weapons (discussed later in this chapter).

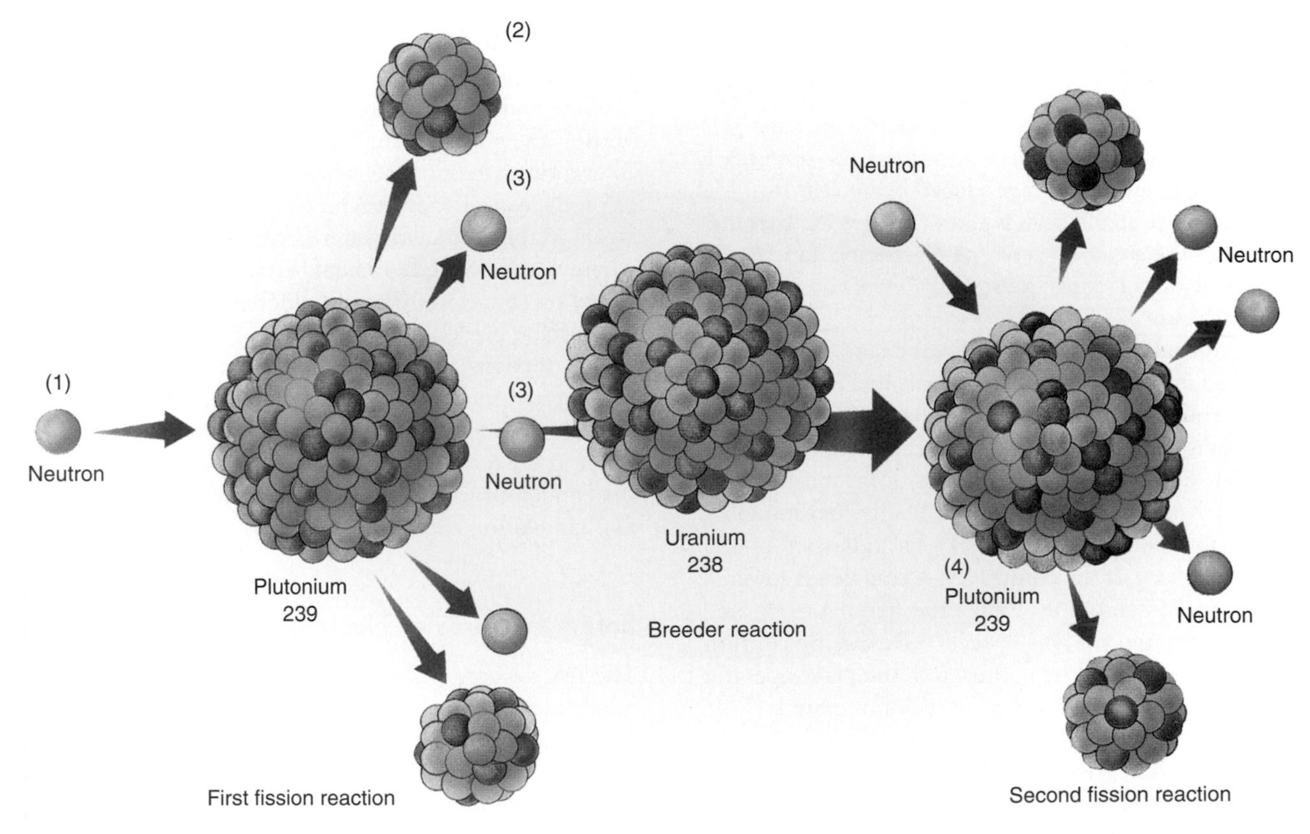

Figure 11–8

Breeder nuclear fission. A neutron (1) from a previous fission reaction bombards plutonium-239, causing it to split into smaller radioactive fission fragments (2). Additional neutrons (3) are ejected in the process, some of which may collide with uranium-238 to form plutonium-239 (4). Neutron bombardment of this plutonium-239 molecule causes it to split, and the process continues.

IS NUCLEAR ENERGY A CLEANER ALTERNATIVE THAN COAL?

One of the reasons proponents of nuclear energy argue for the widespread adoption of nuclear power is that nuclear energy has less of an environmental impact than fossil fuels, particularly coal (Table 11–2). They point out that the combustion of coal to generate electricity is responsible for more than one-third of the air pollution in this country. As discussed in Chapter 10, coal is an extremely dirty fuel, especially since we have used up most of our reserves of cleaner-burning coal. Today most coal-burning power plants burn soft coal that produces sulfur-containing emissions that interact with moisture in the atmosphere to form acid precipitation. In addition, the combustion of coal releases carbon dioxide, a greenhouse gas that traps solar heat in our atmosphere and may be causing global warming.

In comparison, nuclear energy emits very few pollutants into the atmosphere. It does, however, generate nuclear waste, such as spent fuel, that is highly radioactive and therefore very dangerous (to be discussed shortly). The extreme health and environmental hazards caused by this waste require that special measures be taken in its storage and disposal. Nuclear power plants also produce other nuclear wastes, such as radioactive coolant fluids and gases in the reactor.

Opponents of nuclear energy contend that the fact that coal is a dirty fuel is not so much an argument in favor of nuclear energy as it is an argument in favor of a cleaner alternative to coal. They point out that pollution control devices can significantly lessen the air pollution produced by coal-burning power plants. As provisions of the 1990 Clean Air Act go into effect, more and more coal-fired power plants are installing pollution control equipment.

The replacement of coal-burning power plants with nuclear power does not significantly lessen the threat of global warming, because only 15 percent of the greenhouse gases come from power plants in the first place. Most greenhouse gases are produced by automobile emissions and industrial processes, both of which are currently unaffected by nuclear power. Also, the uranium-mining through uranium-enrichment steps in the nuclear fuel cycle require the combustion of fossil fuels, meaning that nuclear energy indirectly contributes to the greenhouse effect.

IS ELECTRICITY PRODUCED BY NUCLEAR ENERGY CHEAP?

In the United States, 109 nuclear power plants supply about 21 percent of the nation's electricity. The cost of producing electricity from a particular energy source varies over time, but in some areas of the country electricity can be produced more cheaply by nuclear energy than by coal-fired plants.

Proponents of nuclear energy usually point to France, which in 1995 obtained 76 percent of its electricity from nuclear energy. In France, electricity generated by nuclear plants is 27 percent less expensive than that generated from coal. Nuclear energy proponents believe there is no reason why the United States could not achieve the same economic efficiency.

Opponents of nuclear power dismiss the example of France because the French government heavily subsidizes the nuclear industry. Among other things, the French government funds research and development, waste disposal, and insurance coverage for its nuclear industry. The government-owned utility, Electricité de France, currently (in 1994) has a debt of $30 billion, yet government-approved rate increases are hard to obtain.

The true costs of nuclear energy, like those of coal and other fossil fuels, are not always obvious in utility bills, whether in France or in the United States. The gen-

Table 11–2
*A Comparison of the Environmental Impacts of Coal-Fired and Nuclear Power Plants**

Annual Impact	Coal (Surface-Mined)	Nuclear
Land used (acres)	17,000	1900
Water discharges (tons)	40,000	21,000
Air emissions (tons)	380,000	6200
Radioactive emissions (curies)	1	28,000
Occupational deaths	0.5 to 5	0.1 to 1
Other environmental issues	Global climate change due to CO_2 emissions	Possible link to nuclear weapons

*Environmental impacts include extraction, processing, transportation, and conversion.

eration of electricity using nuclear energy is very expensive when all costs are taken into account, including the cost of building and maintaining the nuclear power plant, the cost of storing and disposing of spent fuel and other radioactive wastes, and the cost of dismantling the nuclear power plant after it is too old to safely and economically produce energy (discussed shortly).

The High Costs of Building a Nuclear Power Plant

In the United States, no nuclear power plants have been ordered since 1976 and only one remains under construction in 1997. One of the reasons electrical utilities will not commit to building new nuclear power plants is their high costs. These costs must be paid for by the utility customers in the form of higher prices for their electricity. Traditionally designed nuclear power plants are large and take years to plan and build. In comparison, electric plants powered by fossil fuels can be much smaller, and therefore easier and less expensive to build. Also, the regulatory process by which permits are obtained to build a nuclear power plant is cumbersome and adds to the expense.

The initial cost estimates for building a nuclear power plant are high, and the actual costs are usually much higher than the forecasts. Cost overruns, which are borne by the utility and its customers, occur partly because the slow permitting process makes construction fall far behind schedule. Consider the Seabrook nuclear reactor in Seabrook, New Hampshire, which is one of the most controversial plants ever licensed.[2] Seabrook obtained its operating license from the Nuclear Regulatory Commission in 1990, after numerous delays had put its construction 11 years behind schedule. The plant cost $6.45 billion, *12 times* the original estimate.

CAN NUCLEAR ENERGY DECREASE OUR RELIANCE ON FOREIGN OIL?

International crises, such as the oil embargo of the early 1970s and the Persian Gulf crisis in the early 1990s, occasionally threaten the supply of oil to the United States. Supporters of nuclear energy point out that our dependence on foreign oil would be lessened if all oil-burning power plants were converted to nuclear plants. This claim is not as convincing as it seems, however, because oil is responsible for generating only 6 percent of the electricity in the United States; we rely on oil primarily for automobiles and for home heating. Thus, the replacement of electricity generated by oil with electricity generated by nuclear energy would do little in the short term to lessen our dependence on foreign oil, because we would still need oil for heating buildings and driving automotive vehicles. As electric heat pumps and electric motor vehicles become more common in the future, however, nuclear power plants have the potential to heat buildings and power automobiles.

[2]One of the main controversies surrounding Seabrook was concern over the evacuation plan that would be used should an accident occur at the plant. Seabrook is located in a small coastal town without large, multilane highways. The evacuation area includes several tourist spots that are crammed with thousands of visitors during summer months. In reviewing the evacuation plan for Seabrook, the Nuclear Regulatory Commission chose not to count the tourists.

SAFETY IN NUCLEAR POWER PLANTS

Approximately 390 conventional nuclear power plants are in operation worldwide, and accidents have occurred in some of them. Although conventional nuclear power plants cannot explode like atomic bombs, accidents can happen in which dangerous levels of radiation might be released into the environment and result in human casualties. At high temperatures the metal encasing the uranium fuel melts, releasing radiation; this is called a **meltdown**. Also, the water that is used in a nuclear reactor to transfer heat can boil away during an accident, contaminating the atmosphere with radioactivity.

The probability of a major accident occurring is considered low by the nuclear industry, but public perception of the risk is high because it is an involuntary risk, unlike smoking or driving in an automobile. Also, the consequences of such an accident are drastic and life-threatening, both immediately and long after the accident has occurred. We now consider two accidents, one in the United States (Three Mile Island) and the other at Chernobyl in Ukraine.

Three Mile Island

The most serious nuclear reactor accident in the United States occurred in 1979 at the Three Mile Island power plant in Pennsylvania, the result of human error after a valve failure. A partial meltdown of the reactor core took place. Had there been a complete meltdown of the fuel assembly, dangerous radioactivity might have been emitted into the surrounding countryside. Fortunately, the containment building kept virtually all the radioactivity released by the core material from escaping. Although a small amount of radiation entered the environment, there were no substantial environmental damages and no immediate human casualties. A study conducted within a 10-mile radius around the plant ten years after the accident concluded that cancer rates were in the normal range and that there was no association between cancer rates and radiation emissions from the accident.

(a)

(b)

Figure 11–9

Chernobyl. (a) The site of the explosion is indicated by the arrow. The upper part of the reactor was completely destroyed. (b) Since the accident, the damaged area has been completely entombed in a concrete and metal sarcophagus. (a, Novosti/Gamma Liaison; b, Patrick Landmann/Gamma Liaison)

The seriousness of the situation at Three Mile Island elevated public apprehension about nuclear power. It took six years for Three Mile Island to be repaired and reopened. The reactor that was involved in the accident at Three Mile Island was destroyed during the partial meltdown. However, a second reactor that was undamaged during the accident is currently in operation. In the aftermath of the accident, public wariness prompted construction delays and cancellations of several new nuclear power plants across the United States. On the positive side, the accident at Three Mile Island prompted new safety regulations—including evacuation plans for the areas surrounding nuclear power plants—and reduced the complacency that had been commonplace in the nuclear industry.

Chernobyl

The worst accident ever to occur at a nuclear power plant took place at the Chernobyl plant, located in what is now Ukraine, on April 26, 1986, when one or possibly two explosions ripped a nuclear reactor apart and expelled large quantities of radioactive material into the atmosphere (Figure 11–9). The effects of this accident were not confined to the area immediately surrounding the power plant: in the atmosphere radiation quickly spread across large portions of the Northern Hemisphere. The Chernobyl accident affected and will continue to affect many nations, especially Ukraine and other European countries.

The first task faced after the accident was to contain the fire that had broken out after the explosion and prevent it from spreading to other reactors at the power plant. Local fire fighters, many of whom later died from exposure to the high levels of radiation, battled courageously to contain the fire. In addition, 116,000 people who lived within a 30-kilometer (18.5-mile) radius around the plant had to be quickly evacuated and resettled. Ultimately, more than 170,000 people had to permanently abandon their homes.

Once the danger from the explosion and fire had passed, the radioactivity at the power station had to be cleaned up and contained so that it would not spread.

Dressed in protective clothing, workers were transported to the site in radiation-proof vehicles, and initially the radioactivity was so high that they could stay in the area for only a few minutes at a time. There are few photographs of the cleanup because camera film was quickly ruined by the radiation. After the initial cleanup, the damaged reactor building was encased in 300,000 tons of concrete. Then the surrounding countryside had to be decontaminated: highly radioactive soil was removed, and buildings and roads were scrubbed down.

Although as much cleanup as can be done in the immediate vicinity of Chernobyl is largely accomplished, the people in Ukraine face many long-term problems. Much of the farmland and forests are so contaminated that they cannot be used for more than a century. Loss of agricultural production is one of the largest costs of the Chernobyl accident for the local economy.

Inhabitants in many areas of Ukraine still cannot drink the water or consume locally produced milk, meat, fruits, or vegetables. Mothers do not nurse their babies because their milk is contaminated by radioactivity. The health of approximately 350,000 Ukrainians is being constantly monitored.

In the investigation that ensued after the accident, it became apparent that there had been two fundamental causes. First, the design of the nuclear reactor was flawed—the reactor was not housed in a containment building and was extremely unstable at low power. This type of reactor, called an RBMK reactor, is not used commercially in either North America or Western Europe because nuclear engineers consider it too unsafe. Russia, Ukraine, and several adjacent countries still have a number of RBMK reactors in operation.

Second, human error contributed greatly to the disaster. Many of the Chernobyl plant operators lacked scientific or technical understanding of the plant they were operating, and they made several major mistakes in response to the initial problem. As a result of the disaster at Chernobyl, the government developed a retraining program for operators at all the nuclear power plants in the country. In addition, safety features were added to existing reactors.

One of the disquieting consequences of Chernobyl was the lack of predictability of the course taken by spreading radiation (Figure 11–10). Chernobyl's radiation cloud dumped radioactive fallout over some areas of Europe and Asia, leaving other areas relatively untouched. This unevenness makes it difficult to plan emergency responses for a possible future nuclear accident.

As the health effects of the accident at Chernobyl are monitored over the years, the death toll is expected to rise. In 1995, the Ukrainian health minister said the death toll to date was 125,000; most of these were workers who participated in the cleanup, children, and pregnant women.[3] In addition, many other people in nearby countries received dangerous doses of radiation (see Focus On: The Effects of Radiation on Organisms). An increase in the frequency of birth defects and mental retardation in newborns has been documented in Belarus, Ukraine, and other parts of Europe, along with an alarming incidence of thyroid cancer in children. Breast cancer, stomach cancer, and other organ cancers are not expected to appear in large numbers until 20 years or more after the accident. Psychological injuries to those living under the cloud of Chernobyl are also being assessed because many of these individuals are suffering from debilitating stress.

The world has learned much from this nuclear disaster. Most countries are taking nuclear power more seriously, hoping to prevent accidents. Safety features that are commonplace in North American and European reactors are being incorporated into new nuclear power plants around the world. In addition, nuclear engineers learned a great deal from the cleanup and entombment of Chernobyl; this knowledge will be useful when old nuclear power plants are dismantled. Doctors learned more about effective treatment of people who have been exposed to massive doses of radiation. In the years to come, health researchers will learn more about the relationship between cancer and radiation as they follow the health of the thousands who were exposed to radiation from Chernobyl.

► *FOLLOW-UP*

Although the 1986 explosion at the Chernobyl nuclear plant remains the world's worst commercial nuclear accident, other reactors at the site continue to operate. Ukraine and other countries in the former Soviet Union desperately need inexpensive sources of electricity, thus 57 potentially unsafe reactors remain operational in the region, including the two at Chernobyl. Nuclear experts regard ten of these reactors as posing significant safety risks. Inspections of Chernobyl by the International Atomic Energy Agency (IAEA) in 1994 revealed numerous potential safety hazards. The most serious threat is posed by cracks in the concrete encasing the radioactive remains of the destroyed reactor. Radioactive dust could spread from the site, or worse, rain seeping into the cracks could contaminate groundwater or contribute to another massive explosion.

[3]Western scientists regard such assessments with caution, in part because an increase in intensive screening for diseases that occurred after the accident has also uncovered previously undiagnosed diseases that were unrelated to the accident. A European Union health study released in 1996, for example, said that Chernobyl's death toll had been exaggerated by disease screening.

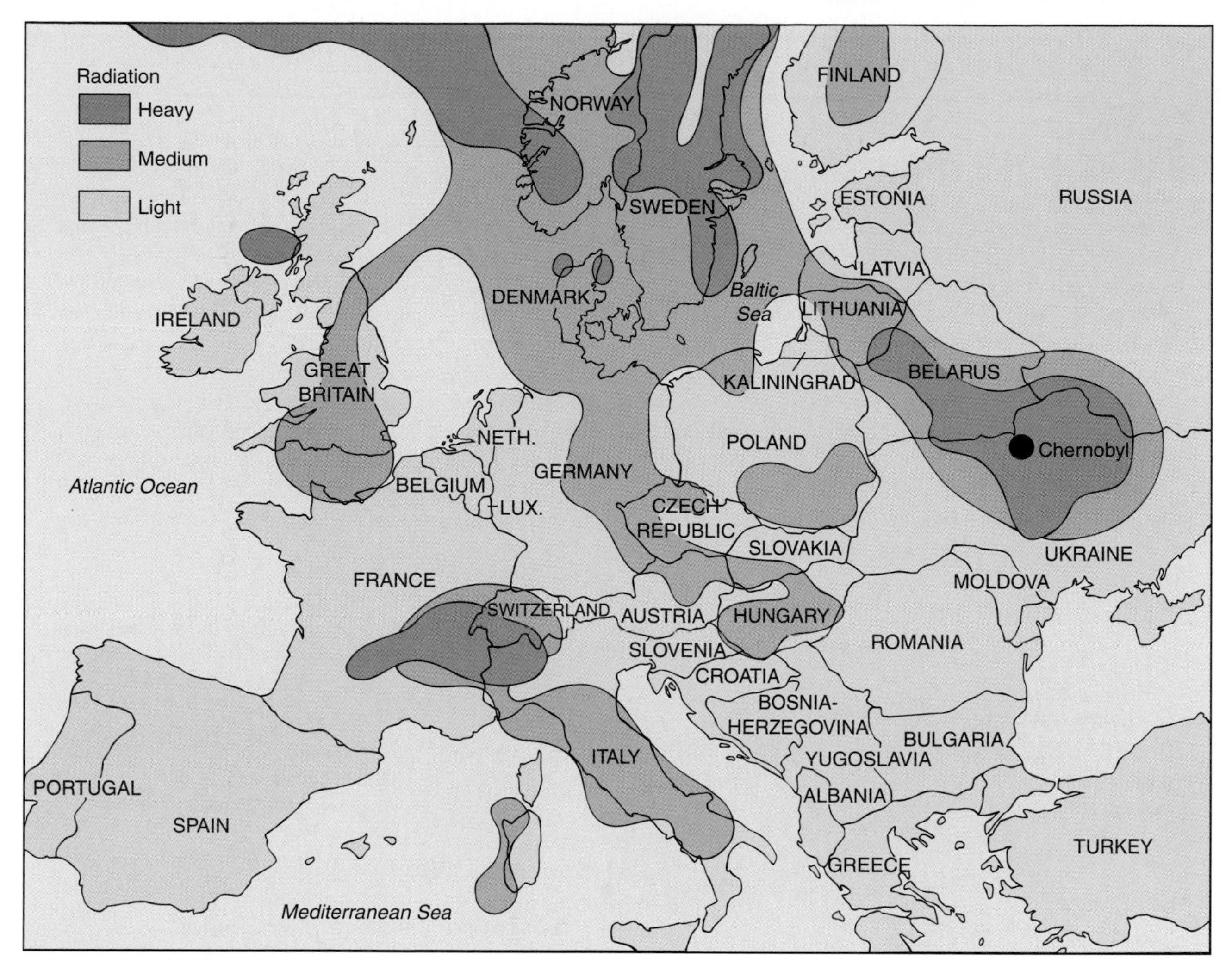

Figure 11–10
Locations that were hardest hit by radioactive fallout from the accident at Chernobyl.

The Ukrainian government does not think it can close Chernobyl without developing alternative energy sources, and it wants assistance from Western nations to finish building five Russian-designed reactors whose safety standards are questioned by the IAEA. The $1.5 billion in aid offered by Western nations and the World Bank in 1994 does not approach the estimated $14 billion sought by Ukraine. Apparently Ukraine will take no safety measures nor develop new energy without the aid, putting the West in the position of either providing the needed funds somehow or continuing to face the prospect of another possible nuclear accident.

The Link Between Nuclear Energy and Nuclear Weapons

Fission is involved in both the production of electricity by nuclear energy and the destructive power of nuclear weapons. Uranium-235 and plutonium-239 are the two fuels commonly used in atomic fission weapons. As you know, plutonium is produced in breeder reactors. It is also possible to reprocess spent fuel from conventional fission reactors to make weapons-grade plutonium.

In 1994 there were about 270 metric tons (298 tons) of weapons-grade plutonium worldwide, much of it sur-

FOCUS ON

The Effects of Radiation on Organisms

Humans and other organisms are continually exposed to low levels of background radiation from several natural sources, including cosmic rays from outer space and radioactive elements in the Earth's crust (see figure; also see the discussion of radon in Chapter 19). This radiation, often referred to as **ionizing radiation**, contains enough energy to eject electrons from atoms, which results in the formation of positively charged atoms called ions.

One of the most dangerous effects of ionizing radiation is the damage it does to DNA, the genetic material of organisms. Because DNA provides a blueprint for all characteristics of an organism and directs the activities of cells, damage to DNA is almost always harmful to the organism. Changes in DNA are known as **mutations**. When mutations occur in reproductive cells (that is, eggs or sperm), the changes can be passed on to the next generation, where they might result in birth defects, mental retardation, or genetic disease. Pronounced effects of high doses of radiation on subsequent generations have been documented in experimental animals (mice, for example), but no such ev-

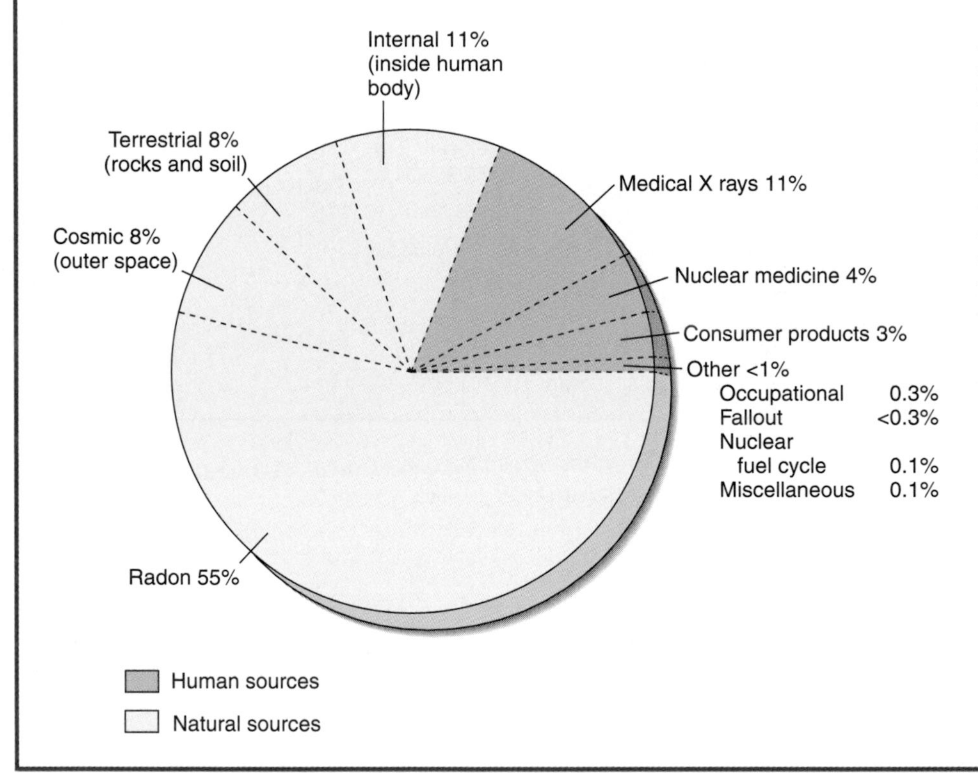

Average sources of ionizing radiation in the United States. Most of the radiation (82 percent) to which people are exposed comes from natural (and unavoidable) sources. The nuclear fuel cycle is responsible for an average of only 0.1 percent of the radiation in the environment.

plus from scrapped nuclear weapons. An additional 930 metric tons (1026 tons) of plutonium wastes exist from commercial reactors, and this amount is expected to double in the next ten years (Figure 11–11). Storing plutonium is both an environmental and a security nightmare because it only takes several kilograms to make a nuclear bomb as powerful as the ones that destroyed Nagasaki and Hiroshima in World War II. Of special concern to international security is the political instability of Russia, because it increases the chance that terrorist groups could steal the wastes and use them to make nuclear weapons.

Many countries are using or contemplating using nuclear power to generate electricity. The possession of nuclear power plants gives these countries access to the fuel needed for nuclear weapons (by reprocessing spent fuel). Responsible world leaders are concerned about the consequences of terrorist groups and rogue nations building nuclear weapons. These concerns have caused many peo-

idence exists for humans. Extensive studies of offspring of the survivors of the World War II atomic bombs at Hiroshima and Nagasaki, for example, suggest that there has been some genetic damage, but the results are not statistically significant.

If mutations occur in nonreproductive cells of the body, they may alter the functioning of those cells. This can contribute to health problems during an individual's lifetime, including an increased likelihood of cancer. Mutations in normal genes involved in the control of growth and development may convert them to **oncogenes**, which are cancer-causing genes. Certain oncogenes appear to be particularly common and are found in a variety of cancers.

Exposure to very high levels of radiation may cause such severe physiological damage that death occurs. Radiation exposure that is extensive but not great enough to cause death may cause numerous medical problems, including burns on the skin, an increased chance of developing cataracts, temporary male sterility, and several types of cancer such as leukemia and cancers of the bone, thyroid, skin, lung, and breast (see figure). For example, a higher incidence of leukemia was correlated with greater exposures to radiation in the survivors of Hiroshima and Nagasaki.

A question that remains to be answered definitively is whether low-level radiation such as that around nuclear power plants causes a higher incidence of cancer in people who live and work nearby. Some recent studies, such as one on the incidence of leukemia in adults living near a nuclear power plant in Massachusetts, show a direct correlation. Other U.S. and French studies of cancer rates near nuclear power plants do not confirm a cancer risk from exposure to low levels of radiation.

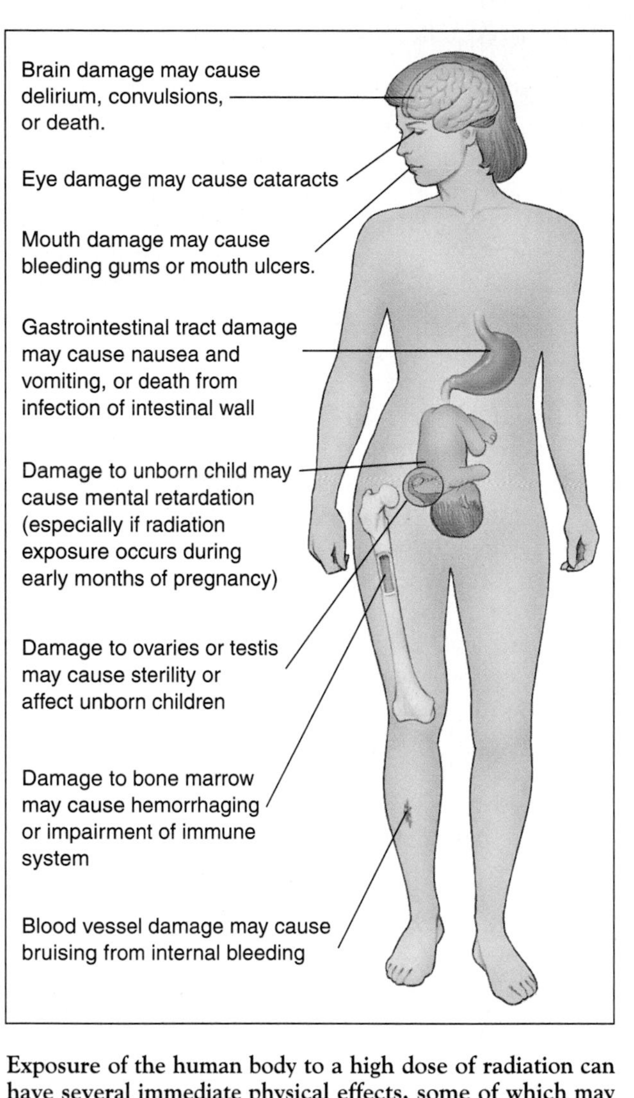

Exposure of the human body to a high dose of radiation can have several immediate physical effects, some of which may contribute to death.

ple to shun nuclear energy, particularly breeder fission, and to seek alternatives that are not so intimately connected with nuclear weapons.

► *FOLLOW-UP*

The United States needs to get rid of more than 50 tons of surplus plutonium from the dismantling of nuclear warheads and ensure that Russia, which also has a large plutonium stockpile, does the same. After reviewing dozens of possible solutions, in late 1996 the Clinton Administration decided to pursue two options: (1) vitrify some of it into glass logs (discussed shortly) and (2) convert some of it to a mixed oxide form and then burn it in commercial nuclear power reactors. Many nuclear scientists and arms-control experts agree with the Clinton plan, although all agree that there is no perfect solution to such a complex problem.

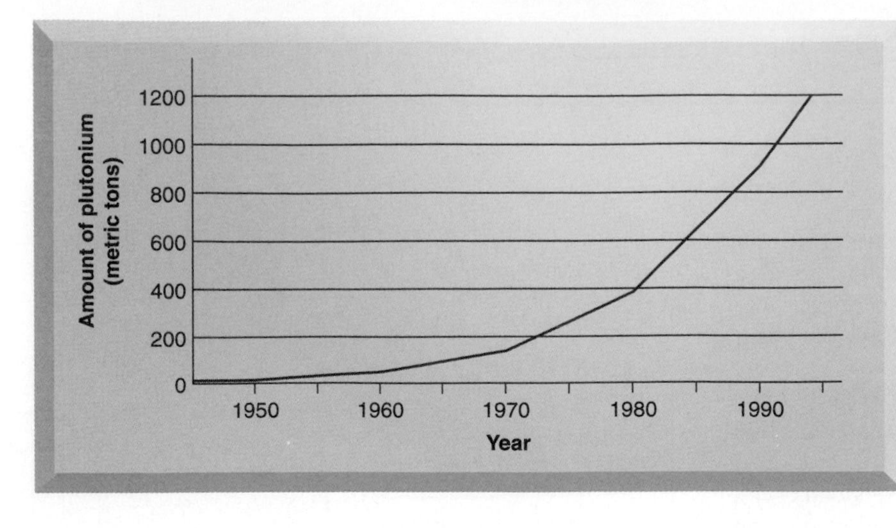

Figure 11–11
Worldwide plutonium buildup from 1945 to 1995. Estimates include both separated and unseparated plutonium from military and commercial sources. The U.S. Department of Energy released the federal inventory of its plutonium stocks in 1996—a total of 99.5 metric tons (110 tons) at ten sites in nine states.

RADIOACTIVE WASTES

Radioactive wastes are classified as either low-level or high-level. **Low-level radioactive wastes** are radioactive solids, liquids, or gases that give off small amounts of ionizing radiation. Produced by nuclear power plants, university research labs, hospitals (nuclear medicine), and industries, low-level radioactive wastes include glassware, tools, paper, clothing, and other items that have been contaminated by radioactivity. The Low-Level Radioactive Policy Act, passed in 1980, specified that all states must have facilities to handle low-level radioactive wastes by 1996. However, few states—only South Carolina and Washington—have complied with this requirement, and as a result, low-level radioactive waste is currently stored at some 800 locations, most of them urban, across the country.

High-level radioactive wastes are radioactive solids, liquids, or gases that initially give off large amounts of ionizing radiation. Examples of high-level radioactive wastes produced during nuclear fission include the reactor metals (fuel rods and assemblies), coolant fluids, and air or other gases found in the reactor. Produced by nuclear power plants and nuclear weapons facilities, high-level radioactive wastes are among the most dangerous hazardous wastes produced by humans.

Fuel rods, which absorb neutrons, thereby forming radioisotopes, can be used for only about three years, after which they become the world's most highly radioactive waste (Table 11–3). As the radioisotopes decay, they

Table 11–3
Nuclear Power Plants and Waste in Selected Countries

Country (In Alphabetical Order)	Number of Operable Commercial Reactors, 1994	Spent Fuel Inventories (Tons of Radioactive Metal)*
Argentina	2	X**
Canada	22	17,700
France	56	7300***
Japan	47	7500
South Korea	9	700
South Africa	2	100
Sweden	12	2360
Taiwan	6	900
United Kingdom	34	30,900
United States	109	21,800

*As of 1990.
**Data not available.
***Does not include 16,500 tons produced by dual-use military and civilian reactors.

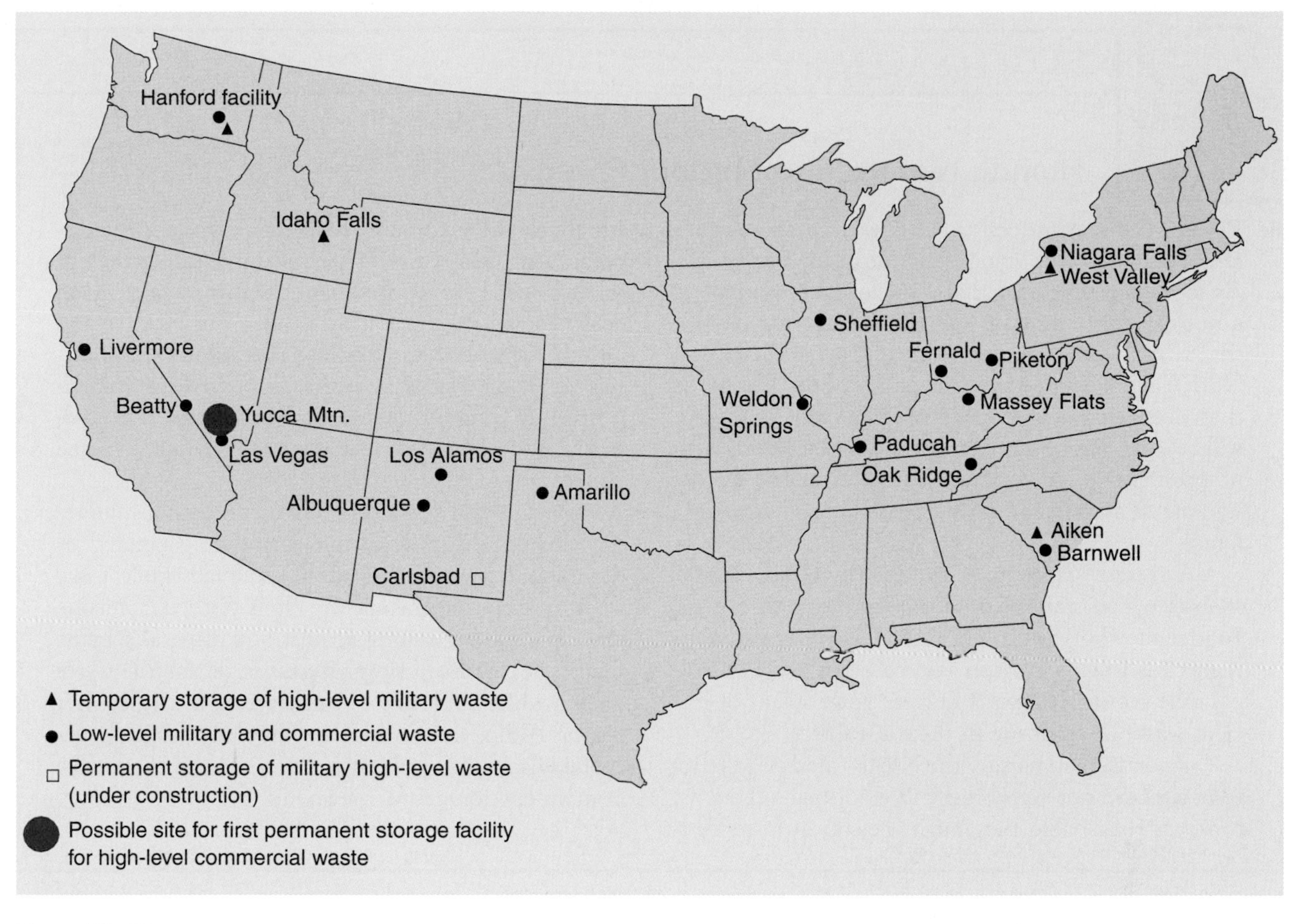

Figure 11–12

Some of the major nuclear waste storage sites. Nuclear wastes from power plants and military activities are now stored in about 80 sites in 41 states. Many of these sites are almost full.

produce considerable heat, are extremely toxic to organisms, and remain radioactive for thousands of years. Their dangerous level of radioactivity requires that they be handled in special ways. Secure storage of these materials must be guaranteed for thousands of years, until they can decay sufficiently to be safe.

Clearly, the safe disposal of radioactive wastes is one of the main difficulties that must be overcome if nuclear power is to realize its potential. Many people question whether we can safely guarantee the storage of wastes that must be isolated from organisms for millennia. High-level radioactive wastes must be stored in an isolated area where there is minimal possibility they can contaminate the environment. The storage site must also have geological stability and little or no water flowing nearby, which might transport the waste away from its original site.

What are the best sites for the long-term storage of high-level radioactive wastes? Many scientists recommend storing the wastes in stable rock formations deep in the ground. Another suggestion for the long-term storage of radioactive wastes is aboveground mausoleums, which would be built in remote locations. If we built mausoleums, however, we would not be able to simply store the wastes and forget about them. Mausoleums would have to have adequate security to guarantee their safety.

Other long-term possibilities that have been considered include storage in Antarctic ice sheets and burial beneath the ocean floor. Because ocean disposal has the potential to harm the marine environment, international agreements currently prohibit ocean disposal of high-level wastes. More recently, President Clinton initiated an international agreement that prohibits ocean disposal of low-level radioactive wastes as well.

Most experts today support underground geological disposal in rock formations. The selection of these sites is complicated by people's reluctance to have radioactive wastes stored near their homes (see Focus On: Human Nature and Nuclear Energy).

The problem of long-term storage of high-level radioactive wastes is demonstrated by the fact that only one country, Germany, has a permanent storage facility for radioactive wastes. In the United States, there are about 80 sites at which domestic radioactive wastes have been "temporarily" stored for decades (Figure 11–12). Most commercially operated nuclear power plants store their

FOCUS ON

Human Nature and Nuclear Energy

The NIMBY response casts a pall over the promise of nuclear energy. NIMBY stands for "**n**ot **i**n **m**y **b**ack**y**ard." As soon as people hear that a nuclear power plant or a nuclear waste disposal site may be situated nearby, the NIMBY response rears its head. Part of the reason NIMBYism is so prevalent in nuclear power issues is that, despite the assurances given by experts that a site will be safe, no one can guarantee *complete* safety, with no possibility of an accident. (Recall from the discussion of risk assessment in Chapter 2 that *nothing* is risk-free.)

A sister response to NIMBY is the NIMTOO response, which stands for "**n**ot **i**n **m**y **t**erm **o**f **o**ffice." Politicians who wish to get reelected are sensitive to their constituents' concerns and are not likely to support the construction of a nuclear power plant or nuclear waste disposal site in their districts.

Given human nature, the NIMBY and NIMTOO responses are not surprising. But emotional reactions, however reasonable they might be, do little to constructively solve complex problems. Consider the disposal of nuclear waste. There is universal agreement that we need to safely isolate nuclear waste until it can decay enough to cause little danger. But NIMBY and NIMTOO, with their associated demonstrations, lawsuits, and administrative hearings, prevent us from effectively dealing with nuclear waste disposal. Every potential disposal site is near someone's home, in some politician's state.

Most people agree that our generation has the responsibility to dispose of nuclear waste generated by the nuclear power plants we have already built. Only we want to put it in someone else's state, in someone else's backyard. Arguing against any disposal scheme that is proposed will simply result in letting the waste remain where it is now . . . at existing nuclear power plants. Although this may be the only *politically* acceptable solution, at least for now, it is unacceptable from an environmental viewpoint.

spent fuel in huge pools of water on-site. None of these plants was designed for long-term storage of spent fuel, and by 1998 about 25 nuclear power plants will have filled their on-site storage sites. Because they have nowhere to send their spent fuel, these plants must either expand their on-site storage—an expensive proposition because of legal battles that usually occur before expansion is granted—or shut down. To date, six commercially operated plants have expanded on-site storage by building aboveground, air-cooled concrete and steel casks.

The U.S. government has a contract with the nation's nuclear utilities in which it promises to take their spent fuel beginning in 1998. At least 27 states have sued the Department of Energy to get it to meet this obligation, and the courts have ruled that the federal government must accept waste in 1998. Despite this ruling more than half of the nuclear utilities are negotiating with the Mescalero Apache tribe in New Mexico. The Mescaleros want to build a private, temporary storage site for up to 30,000 metric tons of high-level radioactive wastes until the permanent storage facility opens in Yucca Mountain in 2015, the earliest date it will be open. They calculate they could earn up to $25 million a year for the 40 years that the facility, which could open as early as 2002, operates. Many of New Mexico's non-Indian residents fiercely oppose such a site, however.

Yucca Mountain

In 1982 the passage of the Nuclear Waste Policy Act put the burden of developing permanent waste sites on the federal government and required the first site to be operational by 1985.[4] In a 1987 amendment to this bill, the federal government identified Yucca Mountain in Nevada as the only candidate for a permanent storage site for high-level nuclear wastes from commercially operated power plants (Figure 11–13). (Low-level radioactive wastes from the manufacture of nuclear weapons are to be stored in deep underground caves near Carlsbad, New Mexico by 1998.)

The Yucca Mountain site, some 113 kilometers (70 miles) northwest of Las Vegas, is controversial because it is near a volcano (its last eruption may have been as recent as 20,000 years ago) and active earthquake fault lines. There are concerns that earthquakes might possibly disturb the site and raise the water table, which could result in radioactive contamination of air and groundwater.

Since 1987, scientists and engineers have been conducting feasibility studies at Yucca Mountain. Although

[4]The deadline for completion of an operational high-level radioactive waste repository has been postponed from 1985 to 1989 to 1998 to 2010 to 2015.

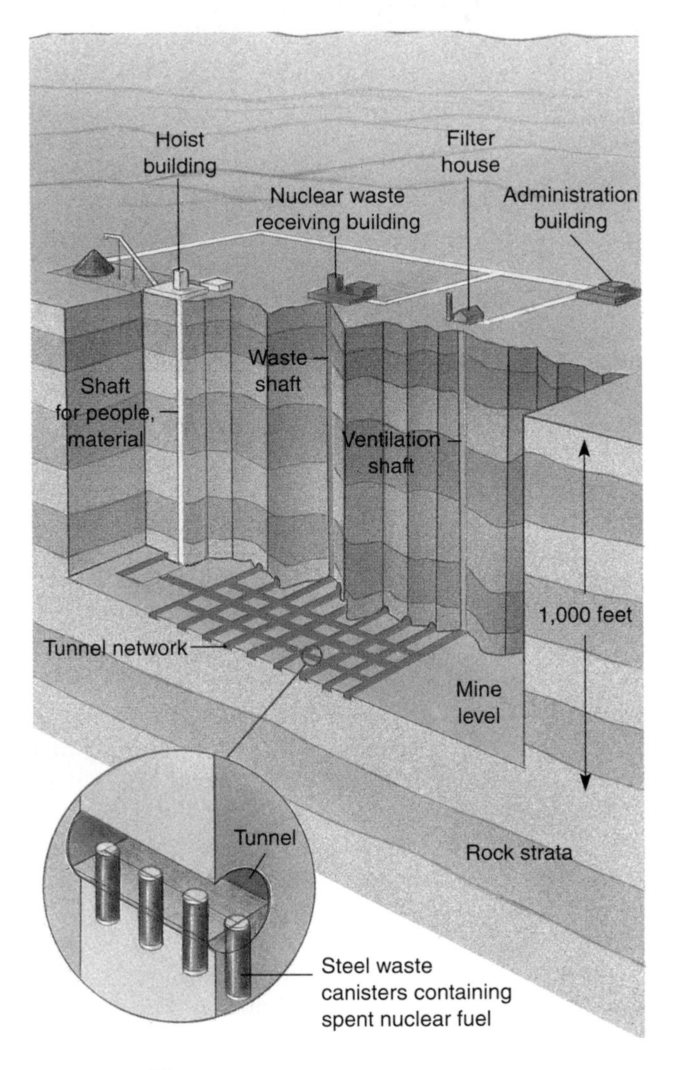

Figure 11–13

In the United States, permanent storage sites for high-level radioactive wastes will probably be deep underground in rock formations. Shown is the kind of nuclear waste facility designed for Yucca Mountain, which would be a three-square-mile complex of interconnected tunnels located in dense volcanic rock 305 meters (1000) feet beneath the mountain.

A NUCLEAR WASTE NIGHTMARE

Over the past three decades, Soviet (and now Russian) practices for nuclear waste disposal have violated international standards:

- Billions of gallons of liquid nuclear wastes have been pumped directly underground, without being stored in protective containers.
- Although Russian officials claim that layers of clay and shale at the sites, which are all near major rivers, prevent leakage, they also admit that some wastes have already leaked more than expected.
- Highly radioactive wastes were dumped into the ocean in amounts double the combined total of dumped wastes from 12 other nuclear nations.
- The underwater wastes include 18 nuclear reactors from submarines and an icebreaker, 6 of which contained live fuel and are considered the most dangerous of the wastes.
- Both underground injection and underwater dumping of nuclear wastes continue because Russia lacks alternatives for nuclear waste processing. Potential health and environmental hazards associated with these wastes are unknown because so little data exist for these types of long-term storage.

the studies will probably not be finished until 2001, preliminary (as of 1996) results indicate that the site is safe. The possibility of a volcanic eruption at Yucca Mountain is considered very remote (1 chance in 10,000 during the next 10,000 years), although scientists still need to study the indirect effects of possible volcanic activity in nearby areas.

A magnitude 5.6 earthquake that occurred in 1992 about 20 kilometers (about 12 miles) from Yucca Mountain allowed scientists to collect data on earthquake effects. Scientists were already monitoring water table elevation, and they measured a one-meter change caused by the earthquake. Because the water table is some 800 meters beneath the mountain crest, however, water elevation changes are not considered a serious problem.

Yucca Mountain has several precariously balanced boulders, which indicate that the mountain has not experienced strong ground shaking in recent memory; had an earthquake caused the ground to shake, the boulders would have fallen. Geologists are currently dating the exposed surfaces of these precariously balanced rocks to estimate how long they have been in place.

The suitability of the Yucca Mountain site suffered a setback in 1995, however, when nearby groundwater was found to contain radioactive tritium from aboveground nuclear tests conducted in the area since the late 1940s. The presence of tritium at a depth of 305 meters (1000 feet) indicates that groundwater travels more freely near the mountain than was formerly thought. This hydrologic issue will dominate the ongoing scientific assessment. Nevada, which opposes the selection of their state for a nuclear waste site, plans to use this hydrologic evidence to challenge the choice of Yucca Mountain.

A very real concern with long-term storage of radioactive wastes is that humans living thousands of years from now may not recognize that Yucca Mountain is dangerous. What kind of warning sign or monument do you place at the site that will last 12,000 or more years that will be intelligible to future generations who may not speak or read English?

High-Level Liquid Waste

High-level liquid wastes are dangerously unstable and very hard to monitor. For that reason, they must be converted to solid form before they can be stored at Yucca Mountain. The U.S. government is planning to store high-level liquid wastes in enormous glass logs 10 feet high and 2 feet in diameter that are themselves contained in stainless steel canisters. The glass contains boron, which absorbs neutrons and so will prevent the logs from exploding. Solidifying liquid waste into solid glass logs, known as **vitrification**, is an established practice in Germany, where the logs are stored underground in a salt mine, and France, where they are stored in a surface facility. The United States began producing glass logs in South Carolina (the Savannah River site) and New York (the West Valley site) in 1996.

Radioactive Wastes with Relatively Short Half-Lives

Some radioactive wastes are produced directly from the fission reaction. U-235, the reactor fuel, may split in several different ways, forming smaller atoms, many of which are radioactive. Most of these, including strontium-90 (half-life 28 years), cesium-137 (half-life 30 years), and krypton-85 (half-life 10.4 years), have relatively short-term radioactivities. In 300 to 600 years they will have decayed to the point where they are safe.

The safe storage of fission products with relatively short half-lives is of concern because fission produces larger amounts of these materials than of the materials with extremely long half-lives. Also, health concerns exist because many of the shorter lived fission products mimic essential nutrients and tend to concentrate in the body, where they continue to decay, with harmful effects (see discussion of bioaccumulation in Chapter 22). For example, one of the common fission products, strontium-90, is chemically similar to calcium. If strontium-90 were to be accidentally released into the environment from radioactive waste that had not been stored properly, it could be incorporated into human and animal bones and teeth in place of calcium. In like manner, cesium-137 replaces potassium in the body and accumulates in muscle tissue, and iodine-131 concentrates in the thyroid gland.

Decommissioning Nuclear Power Plants

Nuclear power plants can operate for only 25 to 40 years before certain critical sections, such as the reactor vessel, become brittle or corroded. At the end of their operational usefulness, however, nuclear power plants cannot simply be abandoned or demolished, because many parts have become contaminated with radioactivity.

Three options exist when a nuclear power plant is closed: storage, entombment, and decommissioning. If an old plant is put into storage, it is simply guarded by the utility company for 50 to 100 years, during which time some of the radioactive materials decay. This decrease in radioactivity makes it safer to dismantle the plant later, although accidental leaks during the storage period are still a concern.

Entombment, in which the entire power plant is permanently encased in concrete, is not considered a viable option by most experts because the tomb would have to remain intact for at least one thousand years. (Recall how the concrete sarcophagus surrounding the damaged Chernobyl reactor developed cracks in just a few years.) It is likely that accidental leaks would occur during that time. Also, we cannot guarantee that future generations would inspect and maintain the "tomb."

The third option for the retirement of a nuclear power plant is to **decommission**, or dismantle, the plant immediately after it closes. The workers who dismantle the plant must wear protective clothing and masks. Some portions of the plant are too "hot" (radioactive) to be safely dismantled by workers, although advances in robotics may make it feasible to tear down these sections. As the plant is torn down, small sections of it are transported to a permanent storage site.

Several small nuclear power plants have been decommissioned. Shippingport, the nation's first commercial nuclear power plant, was dismantled in 1989 and transported by barge more than 8000 miles from its working site in Pennsylvania to Hanford Nuclear Reservation in Washington state (see Chapter 23). The decommissioning of a large nuclear power plant will not be possible, however, until there are permanent storage sites for all the radioactive pieces.

Decommissioning nuclear power plants is the responsible thing to do once a plant is no longer operable. There are risks, however, including dangers to workers during the decommissioning process and accidental discharges of radiation into the environment either during dismantling or during transport of radioactive debris to a permanent site.

Worldwide, 84 nuclear power plants were permanently retired as of 1995, and many nuclear power plants are nearing retirement age. In 1995 approximately 32 operational plants were 25 years old or older; by 2000 there will be 98. As we enter the 21st century, we may find that we are paying more in our utility bills to close old plants than to have new plants constructed. For example, decommissioning the Yankee Rowe nuclear reactor in Massachusetts, shut down in 1991, will cost some $370 million; the plant's construction costs in 1960 were $186 million (in today's dollars).

FUSION: NUCLEAR ENERGY FOR THE FUTURE?

The atomic reaction that powers the stars, including our sun, is fusion. In fusion, two lighter atomic nuclei are brought together under conditions of high heat and pressure in such a way that they combine, producing a larger nucleus. The energy produced by fusion is considerable; it makes the energy produced by the burning of fossil fuels seem trifling by comparison: 30 milliliters (1 ounce) of fusion fuel has the energy equivalent of 266,000 liters (70,000 gallons) of gasoline.

Isotopes of hydrogen are the fuel of fusion (Figure 11–14). In the fusion reaction, the nuclei of deuterium and tritium combine to form helium, releasing huge amounts of energy in the process. Deuterium, also called heavy hydrogen, is present in water and is relatively easy to separate from normal hydrogen. Tritium, which is weakly radioactive, is not found in nature; this human-made hydrogen isotope can be formed during the fusion reaction by bombarding another element, lithium, with neutrons. The isotope of lithium that is required, Li-6, is found in seawater and in certain types of surface rocks.

Supporters of nuclear energy view fusion as the best possible form of energy, not only because its fuel, hydrogen, is available in virtually limitless supply, but also because fusion will produce little in the way of radioactive pollution. Unfortunately, many technological difficulties have been encountered in efforts to stage a controlled fusion reaction. It takes a phenomenally high temperature (100,000,000°C) to make atoms fuse, and once the reaction starts, no one knows whether it can be regulated.

Fusion research is currently being conducted in several countries, particularly Germany, Japan, and Russia. Fusion research is very expensive. Over the past four decades, the United States has invested more than $10 billion in its fusion energy program. (Federal funding for the U.S. fusion energy program at Princeton University was canceled in 1997.) While progress has been made, there is no guarantee that fusion will ever be successful at producing commercial electric power.

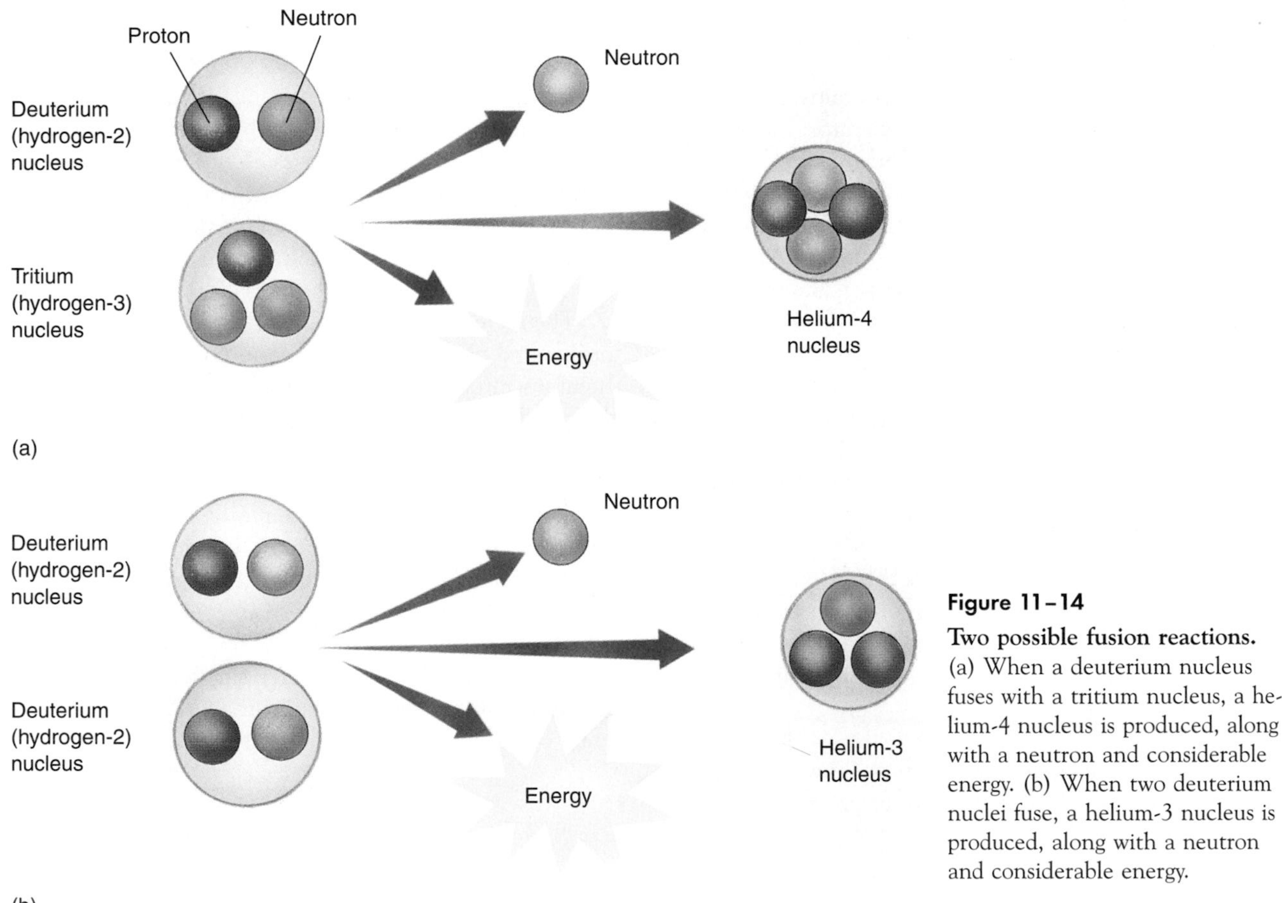

Figure 11–14

Two possible fusion reactions. (a) When a deuterium nucleus fuses with a tritium nucleus, a helium-4 nucleus is produced, along with a neutron and considerable energy. (b) When two deuterium nuclei fuse, a helium-3 nucleus is produced, along with a neutron and considerable energy.

THE FUTURE OF NUCLEAR POWER

In an effort to promote nuclear energy, nuclear and utility executives in the early 1990s developed a plan addressing the safety and economic issues associated with nuclear power. They envisioned building a series of "new generation" nuclear reactors, designed to be ten times safer than current reactors. They also gave serious consideration to the financial risks involved in building a nuclear power plant. According to their plan, costs could be held in line by standardizing nuclear power plants rather than custom-building each one. When France standardized its plant designs, the costs were lowered considerably. The plan also calls for building schedules to be improved and the regulatory process to be streamlined.

New Designs for Nuclear Power Plants

New reactors have been designed that are considered by nuclear experts to be much safer than the reactors currently operating in this country. Some of the new reactors would not have a meltdown even in a worst-case scenario. Because the fuel cannot melt, radiation cannot be released into the environment during the plant's operation. Some of the designs use helium rather than water to turn the turbines and cool the system. Helium, a gas that is much less corrosive than steam, is less likely to cause leaks by corroding pipes. However, the new generation of nuclear power plants, although smaller, simpler in design, less expensive to build, and safer to operate, will still have some of the unresolved problems of traditional designs, such as what to do with high-level radioactive wastes.

SUMMARY

1. Nuclear power involves the nucleus of an atom. In fission, atomic nuclei are split apart, whereas in fusion atomic nuclei are combined. Both fission and fusion result in a significant release of energy, in comparison to the chemical combustion of fossil fuels.

2. A typical nuclear power plant contains a reactor core, where fission occurs; a steam generator; a turbine; and a condenser. Safety features include a steel reactor vessel and a containment building. In a nuclear power reaction, U-235 is bombarded with neutrons, which split the nucleus into two smaller atoms plus additional neutrons. These neutrons, in turn, collide with additional U-235 atoms. The fission of U-235 releases heat that converts water to steam, which is used to generate electricity.

3. In breeder nuclear reactors, both U-235 and Pu-239 are split to release energy. The large quantity of U-238 in uranium fuel is converted into Pu-239; thus, breeder reactors make more fuel than they use.

4. Supporters say nuclear energy is better than alternatives because it is less polluting and more economical, and its fuel, uranium, is plentiful. They also suggest that increased use of nuclear power will decrease our dependence on foreign oil.

5. Problems associated with nuclear power include questions about safety in nuclear power plants. Radioactive fallout from the 1986 accident at the Chernobyl power plant resulted in widespread environmental pollution as well as serious local contamination. Chernobyl stimulated a worldwide reassessment of nuclear power.

6. In all countries that use nuclear power, permanent waste disposal sites are urgently needed to house spent fuel, which is highly radioactive, as well as radioactive parts of dismantled power plants. The United States has tentatively selected Yucca Mountain in Nevada for a permanent storage site for high-level nuclear wastes, pending the results of scientific studies.

7. The increase in global supplies of weapons-grade plutonium and plutonium wastes from commercial reactors threatens international security because it increases the chance that certain nations and terrorist groups could use them to make nuclear weapons.

8. Commercial fusion as a source of energy has yet to become a reality, but fusion may have the potential to produce unlimited energy with very little radioactive waste in the future.

9. New reactor designs are safer, smaller, simpler in design, and less expensive to build than the reactors currently operating in the United States.

THINKING ABOUT THE ENVIRONMENT

1. Compare the environmental effects of coal combustion and conventional fission for the generation of electricity.

2. Breeder reactors produce more fuel than they consume. Does this mean that if we use breeder reactors, we will have a perpetual supply of plutonium for breeder fission? Why or why not?

3. Can accidents at nuclear power plants be prevented in the future? Why or why not?

4. How does the disposal of nuclear wastes pose technical problems? Political problems?

5. Why is decommissioning nuclear power plants such a major task?

6. Are you in favor of the United States developing additional nuclear power plants to provide us with electricity in the 21st century? Why or why not?

* **7.** Uranium-235 has an atomic mass of 235 and an atomic number of 92. Calculate the number of protons and neutrons in an atom of U-235.

* **8.** Assume that you start with 1 kilogram of uranium-234, which has a half-life of 250,000 years.
  a. How many grams of U-234 will remain after 250,000 years?
  b. How many years will it take for 750 grams of U-234 to decay?
  c. How many grams of U-234 will remain after 1 million years?

RESEARCH PROJECTS

Contact the nuclear power plant closest to you and ask about its size (generating capacity), construction costs, and how long it has been operating. What is the cost of electricity generated (kilowatt per hour charge), and how does the cost compare to nearby fossil-fueled power plants? Find out how the spent fuel rods are stored and what measures are being taken to dispose of radioactive wastes.

Obtain a copy of the Executive Summary of the National Academy of Science's "Management and Disposition of Excess Weapons Plutonium," published by the National Academy Press in 1994. Present to your class the three options presented in the summary for long-term plutonium disposition.

On-line materials relating to this chapter are on the ► World Wide Web at **http://www.saunderscollege.com/lifesci/environment2** (select Chapter 11 from the Table of Contents).

SUGGESTED READING

Ahearne, J. F. "The Future of Nuclear Power," *American Scientist* Vol. 81, January/February 1993. Examines what would make nuclear power an appealing option for generating electricity.

Furth, H. P. "Fusion," *Scientific American*, September 1995. Nuclear fusion still holds promise to be an important way to generate electric power in the 21st century.

League of Women Voters. *The Nuclear Waste Primer.* New York: Lyons and Burford, 1993. A well balanced, comprehensive treatment of nuclear waste.

Morgan, M. G. "What Would It Take to Revitalize Nuclear Power in the United States?" *Environment* Vol. 35, No. 2, March 1993. Although some people advocate the use of nuclear power because it does not produce greenhouse gases, the author thinks six key problems should be resolved before nuclear power is expanded.

Ryan, M. "Power Move: The Nuclear Salesmen Target the Third World," *World Watch* Vol. 7, No. 2, March–April 1994. The IAEA of the United Nations encourages developing nations to use nuclear energy to provide electricity.

Shcherbak, Y. M. "Ten Years of the Chernobyl Era," *Scientific American*, April 1996. A follow-up ten years after the nuclear disaster at Chernobyl.

Wheelwright, J. "For Our Nuclear Wastes, There's a Gridlock on the Road to the Dump," *Smithsonian*, May 1995. The amount of high-level radioactive wastes in the United States continues to increase, but nobody can agree on where to put it.

Whipple, C. G. "Can Nuclear Waste Be Stored Safely at Yucca Mountain?" *Scientific American*, June 1996. Scientific tests being conducted at Yucca Mountain to determine if it is a suitable site for long-term storage of nuclear wastes will yield useful information but not absolute conclusions.

Williams, O., and P. N. Woessner. "The Real Threat of Nuclear Smuggling," *Scientific American*, January 1996. The authors argue that nuclear smuggling poses a real danger and that security at nuclear stockpile sites around the world is lax.

Wolfe, B. "Why Environmentalists Should Promote Nuclear Energy," *Issues in Science and Technology*, Summer 1996. The author contends that, as developing nations increase their per-capita energy use during the next century, serious damage will be done to the global environment by increased fossil fuel combustion. An increased reliance on nuclear power should help prevent global warming associated with fossil fuel use.

Variable-speed wind turbines have lowered the price of wind-generated electricity.
(Courtesy of Kenetech Windpower, Inc.)

Renewable Energy and Conservation

In recent years **wind farms**, arrays of nonpolluting wind turbines, have sprung up in many open landscapes, the rotors of each turbine spinning furiously to generate electricity when the wind blows. However, almost all wind turbines are designed to operate at a single optimal speed; when the wind gusts, a heavy brake is applied to the rotors so they do not exceed the optimal speed. Such wind turbines cannot take advantage of the extra power, and because gusty winds strain the machines, some of the costly components wear out more rapidly.

A new prototype wind turbine was developed in the early 1990s by a joint venture among several companies and utilities called the Variable-Speed Wind Turbine Development Alliance. This group, which includes the research and development group of electric utilities, designed gearless turbines to operate in winds ranging from 15 to 97 kilometers per hour (9 to 60 miles per hour). Early tests have been promising, and electric utilities are cautiously optimistic that this new technology can provide substantial amounts of electricity economically. Indeed, costs to generate electricity using wind power have declined from about $.30 per kilowatt hour (kWh) in 1980 to roughly $.04 per kWh with the improved turbines. This cost makes wind power competitive with conventional power plants. Calculations by the California Energy Commission indicate that wind energy is less costly than fossil fuel or nuclear energy when *all* expenses—such as costs associated with the restoration of coal-mined land or the decommissioning of nuclear power plants—are taken into account over the life of a power plant.

Wind energy and other alternative energy sources are becoming increasingly important, both because they are renewable and because their use generally has less environmental impact than the use of fossil fuels or nuclear power. The most promising alternative energy technologies use the sun's energy. *Direct* solar energy can be used to heat water, heat buildings, and generate electricity. Biomass (such as wood, agricultural wastes, and fast-growing plants), wind, and hydropower (the energy of flowing water) are all examples of *indirect* solar energy. Currently, the non-solar renewable energy source that is most widely used is geothermal energy—the heat of the Earth. In this chapter we examine each of these renewable energy sources. ◄

ALTERNATIVES TO FOSSIL FUELS AND NUCLEAR POWER

In the Philippines, 100-hectare (247-acre) wood plantations are being managed for present and future wood-fired power plants. Denmark has built large clusters of windmills to generate electricity, and Brazil has significantly reduced its reliance on imported oil by converting sugarcane crops into alcohol fuels. Photovoltaic solar cells are being manufactured in India and Algeria. Hungary and Mexico use increasingly large amounts of geothermal energy. In the United States, at least six power plants generate electricity from the combustion of old tires. As these examples show, most countries are trying new approaches in the endless quest for energy.

Alternatives to fossil fuels and nuclear power are receiving a great deal of attention these days, and for good reason: alternative energy sources are not only renewable, but they cause fewer environmental problems than fossil fuels or nuclear power. We have seen that there are several concerns about using fossil fuels for energy. Reserves of oil, gas, and even coal are limited and will eventually be depleted, and burning these fossil fuels for energy has negative environmental consequences such as global warming, air pollution, acid rain, and oil spills. Some people have suggested that nuclear energy can be used when we run out of fossil fuels, but as we saw in Chapter 11, several serious problems are associated with the use of nuclear fission. In addition, uranium, the fuel for nuclear fission, is a nonrenewable resource.

Given the rapidly expanding world population, future energy needs will probably demand the exploitation of most energy sources. The recognized need for a long-term solution has prompted worldwide interest in **renewable energy sources**, those sources that are replenished by natural processes so that they can be used indefinitely (Table 12–1). Among the most attractive renewable energy sources is **solar energy**, in part because its use has little negative impact on the environment. Re-

Table 12–1
Selected Types and Uses of Renewable Energy Sources

Type of Renewable Energy	How Energy is Used
Solar Energy—Direct	
Active solar energy	Space heating and cooling; hot water
Passive solar energy	Space heating
Solar thermal	Electricity
Photovoltaics	Electricity
Solar Energy—Indirect	
Biomass	Heat (direct combustion); electricity; liquid & gaseous fuels
Wind	Electricity
Hydropower	Electricity
Other Renewable Energy Sources	
Geothermal	Space heating; hot water; electricity
Tidal	Electricity

Table 12–2
1995 Generating Costs of Electric Power Plants

Energy Source	Generating Costs (Cents per Kilowatt Hour)
Hydropower	4–7¢
Biomass	2–8¢
Geothermal	3–8¢
Wind	4¢
Solar thermal	9¢
Photovoltaics	25–50¢
Natural gas	4¢
Coal	4–5¢
Nuclear power	5–25¢

newable energy is currently more expensive than energy produced by fossil fuels and nuclear power (Table 12–2), but as technological advances are made, costs are expected to decrease.

Renewable forms of energy are not the complete and final solution to our energy problem, however. All energy resources must be used with conservation in mind, and we must continue to find ways to enhance energy efficiency and decrease waste. *Energy conservation is our single most important long-term energy solution.*

DIRECT SOLAR ENERGY

The sun produces a tremendous amount of energy, most of which dissipates into outer space; a very small portion is radiated to the Earth (see Figure 5–7). Solar energy is different from fossil and nuclear fuels because it is perpetually available; we will run out of solar energy only when the sun's nuclear fire burns out. Also solar energy is dispersed over the Earth's entire surface rather than concentrated in highly localized areas, as are coal, oil, and uranium deposits. In order to make solar energy useful to us, we must concentrate it.

Recall from Chapter 5 that solar radiation varies in intensity depending on the latitude, season of the year, time of day, and degree of cloud cover.

1. Areas at lower latitudes–that is, closer to the equator—receive more solar radiation annually than do latitudes closer to the North and South poles.
2. More solar radiation is received during summer than during winter, because the sun is directly overhead in the summer and lower on the horizon in winter (see Figure 5–9).
3. Solar radiation is most intense when the sun is high in the sky (noon) than when it is low in the sky (dawn or dusk).
4. Clouds absorb some of the sun's energy, thereby reducing its intensity.

Because of its lack of cloud cover and lower latitude, the southwestern United States receives the greatest amount of solar radiation annually, whereas the Northeast receives the least (Figure 12–1).

Although the technology exists to use solar energy directly, it has not been adopted widely, largely because the initial costs associated with converting to solar power are high. However, the long-term energy savings of solar power offset the high start-up costs. Trapping the sun's energy using current technology is also inefficient, meaning that relatively little of the sun's energy that hits the solar panels (collecting devices) is actually used. With new technological developments, the efficiency of solar energy collection is increasing, making it a more cost-effective alternative source of energy. Solar energy is projected to become increasingly important in the future (see Focus On: Cooking with Sunlight, page 253).

Heating Buildings and Water

You have probably noticed that, in winter or summer, the air inside a car that is sitting in the sun with its windows rolled up becomes much hotter than the surrounding air. Similarly, the air inside a greenhouse remains warmer than the outside air during cold months. (Greenhouses usually require additional heating in cold climates, but far less than might be expected.) This kind of warming occurs partly because the material—such as glass—that envelops the air inside the enclosure is transparent to visible light but impenetrable to infrared radiation (heat). Thus, visible light from the sun penetrates the glass and warms the surfaces of objects inside, which in turn give off **infrared radiation**—invisible waves of heat energy. Because infrared radiation cannot penetrate glass, heat does not escape, and the area surrounded by glass grows continuously warmer.

In **passive solar heating**, solar energy is used to heat buildings without the need for pumps or fans to distribute the collected heat. Certain design features can be put to use in a passive solar heating system to warm buildings in winter and help them remain cool in summer (Figure

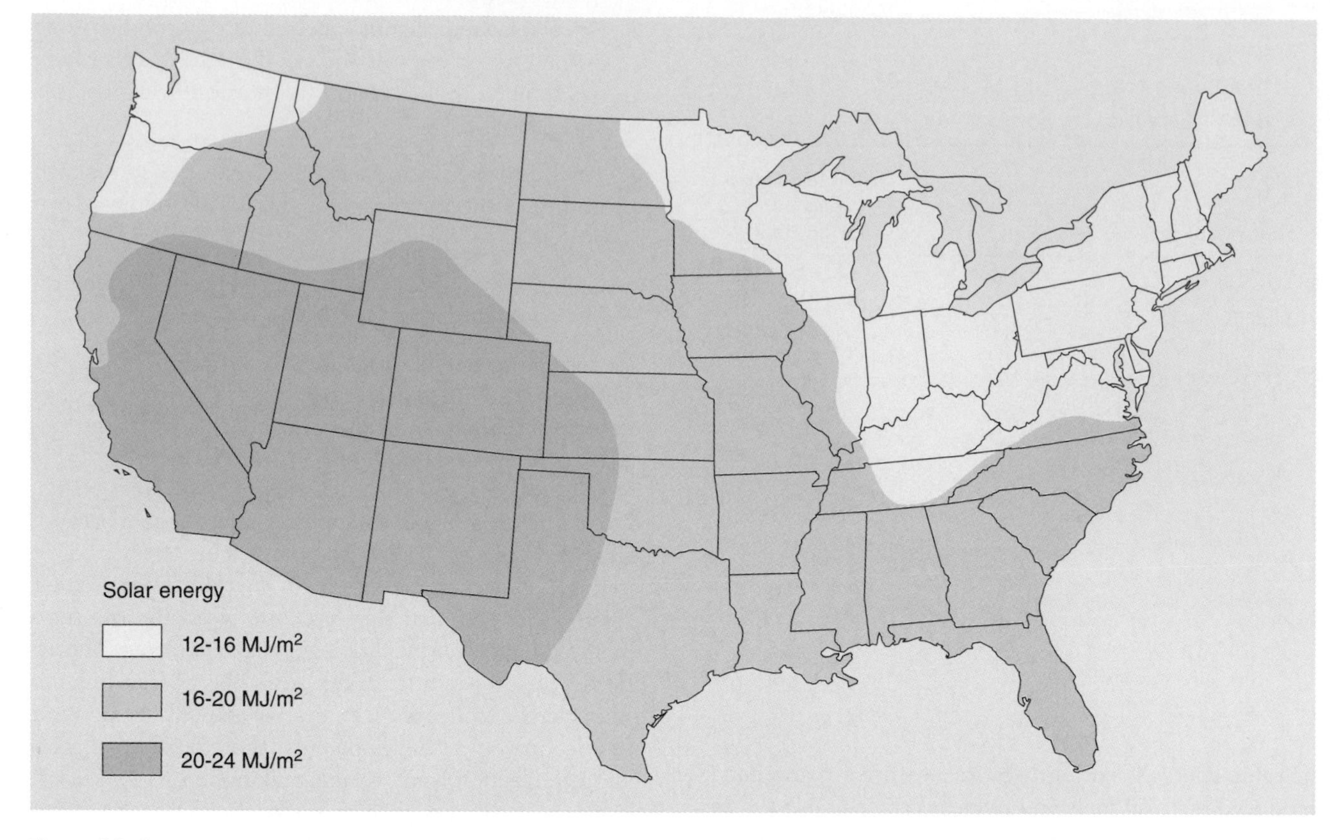

Figure 12–1

Solar energy distribution over the United States. This map shows the average daily total of solar energy that would be received on a solar collector that tilts to compensate for latitude. The units are in megajoules per square meter. The Southwest is the best area in the United States for year-round solar energy collection.

12–2). In the Northern Hemisphere, for example, large south-facing windows receive more total sunlight during the day than windows facing other directions. The sunlight entering through the windows provides heat that is then stored in floors, walls, or containers of water. This stored heat can be transmitted throughout the building naturally by convection, which is the circulation that occurs because warm air rises and cool air sinks. Buildings with passive solar heating systems must be well insulated so that accumulated heat does not escape. Passive solar heating is most effective in sunny climates; in cloudy regions it must be augmented by some other form of energy.

In **active solar heating**, a series of collection devices mounted on a roof or in a field is used to gather solar energy. The most common collection device is a flat solar panel or plate of black metal (which absorbs the sun's energy) enclosed in an insulated box (Figure 12–3a). The absorbed heat is transferred to liquid or air inside the panel, which is then pumped to a storage tank that is usually located in the basement. Although an active solar heating system can be used for both space heating (Figure 12–3b) and hot water heating (Figure 12–4), it is especially cost-effective for water. Because approximately 8 percent of the energy consumed in the United States goes toward heating water, active solar heating has the potential to supply a significant amount of the nation's energy demand.

The use of solar energy for space heating will undoubtedly become more important as other energy supplies dwindle. Furthermore, when diminishing supplies of fossil fuels force their prices higher, solar heating, now costlier than more conventional forms of space heating, will become more competitive. Solar air conditioning, although technically feasible, is expensive and so is not yet commercially available for homes.

Solar Thermal Electric Generation

Electricity can be produced by several different systems that employ **solar thermal electric generation**, in which concentrated sunlight produces high temperatures. In one such system, trough-shaped mirrors, guided by comput-

(text continues on page 254)

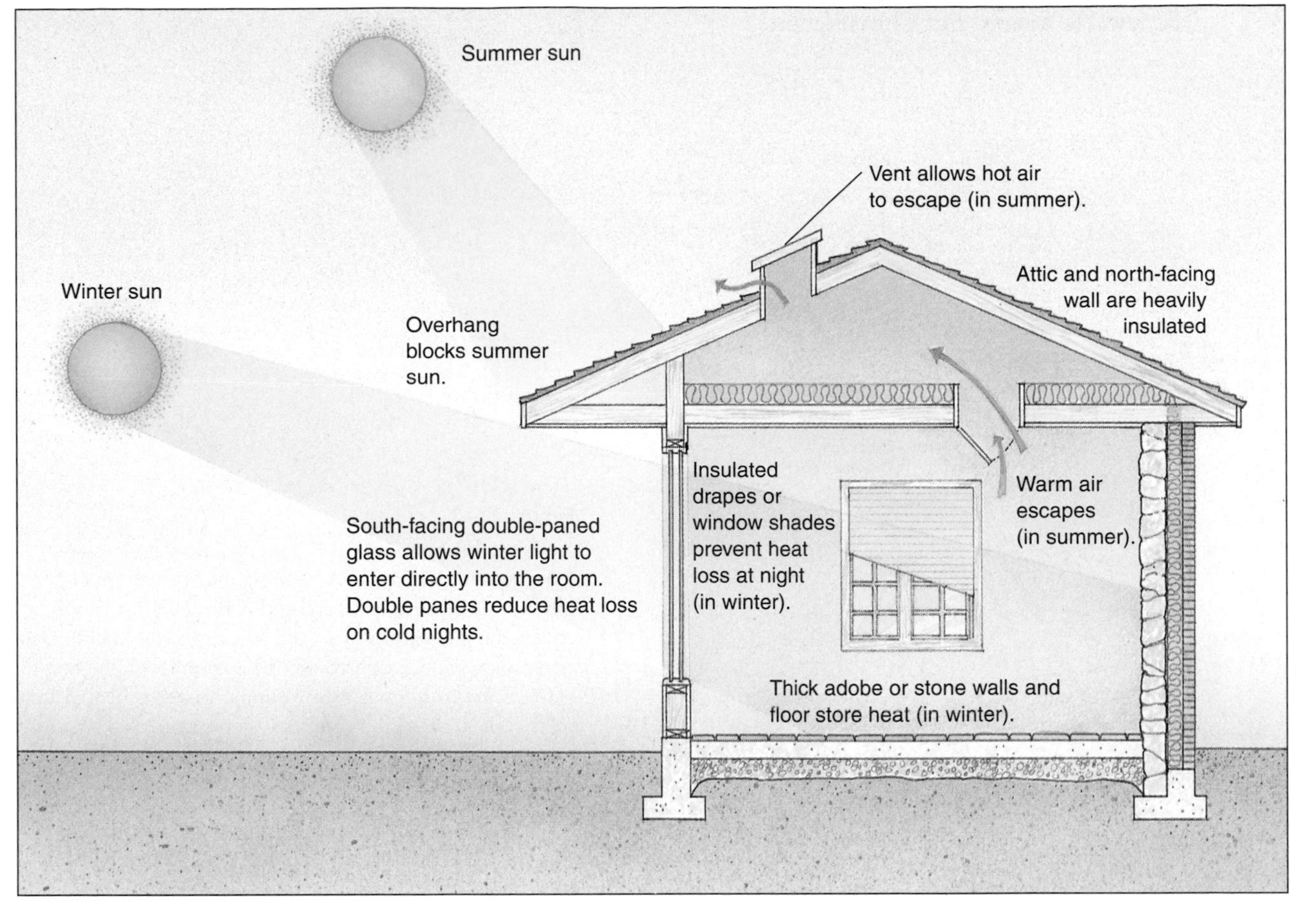

(a)

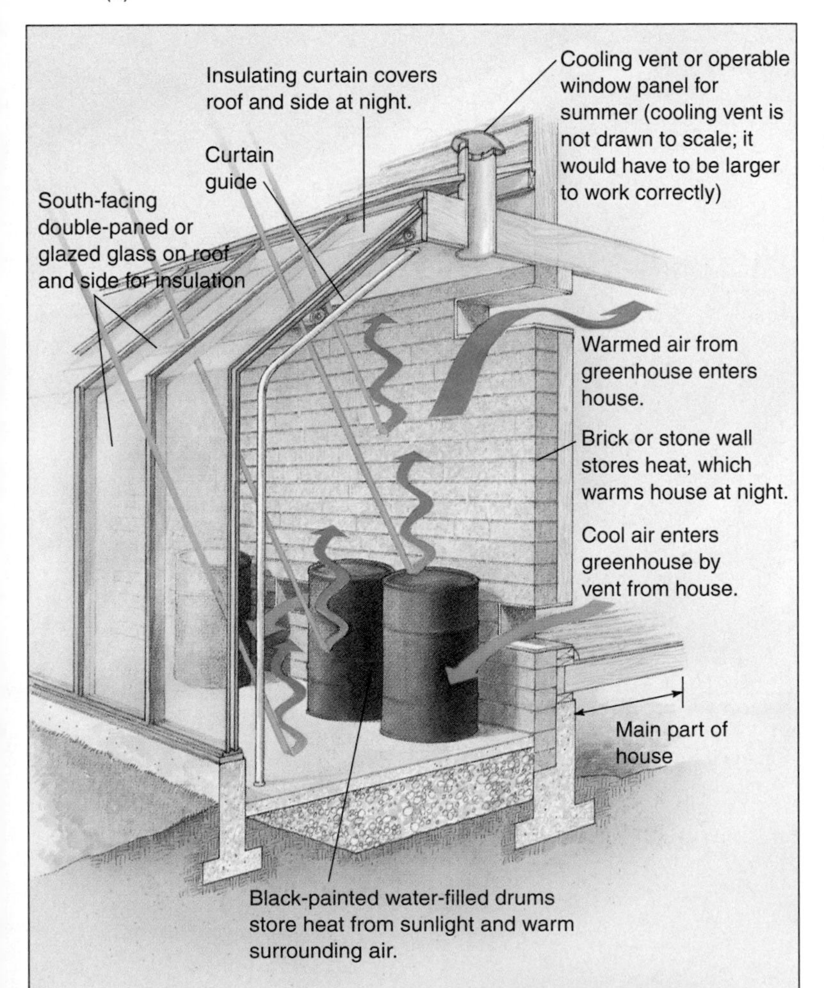

(b)

(c)

Figure 12–2

Passive solar heating designs. (a) Several passive designs are incorporated into this home. The south-facing window on the left admits light, which warms the room. The roof overhang allows sunlight to enter the room in winter when the sun is lower in the sky, but blocks the sun's rays in summer. Insulation is a must, particularly for the attic and north-facing walls. Insulating curtains or blinds are drawn over windows at night to help reduce heat loss. (b) A solar greenhouse can be added to existing homes. Light entering the greenhouse is generally used to warm water-filled containers. The heat from these containers provides warmth to the rest of the house at night. The glass in the greenhouse walls and roof is double-paned or glazed for additional insulation at night. (c) An earth-sheltered home is another passive solar design. Much of an earth-sheltered home is built underground. Such homes use less energy for heating and cooling because of the natural insulation properties of the earth. (c, Courtesy of Solar Design, Inc.)

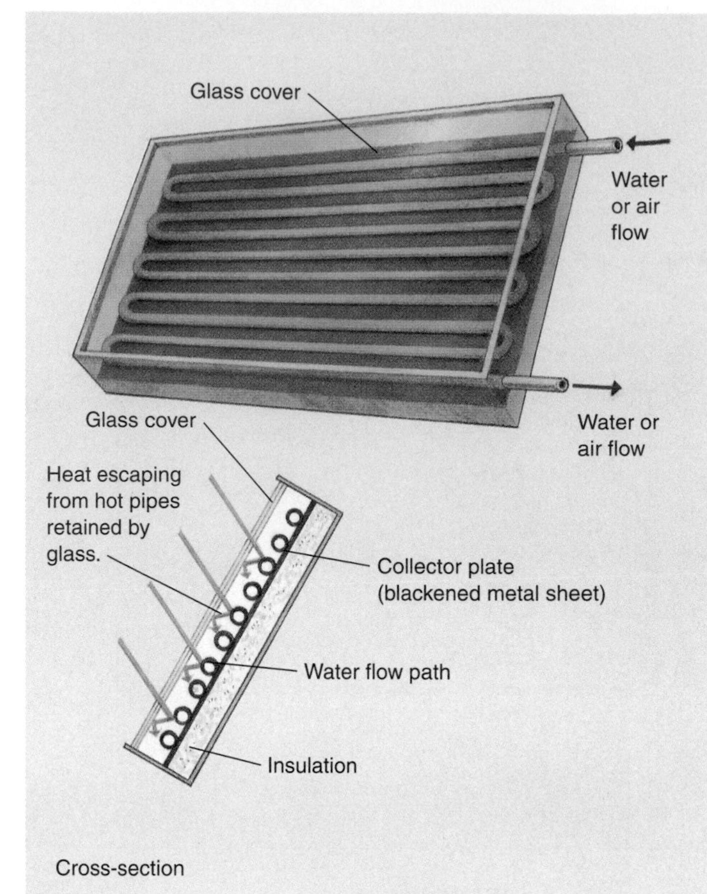

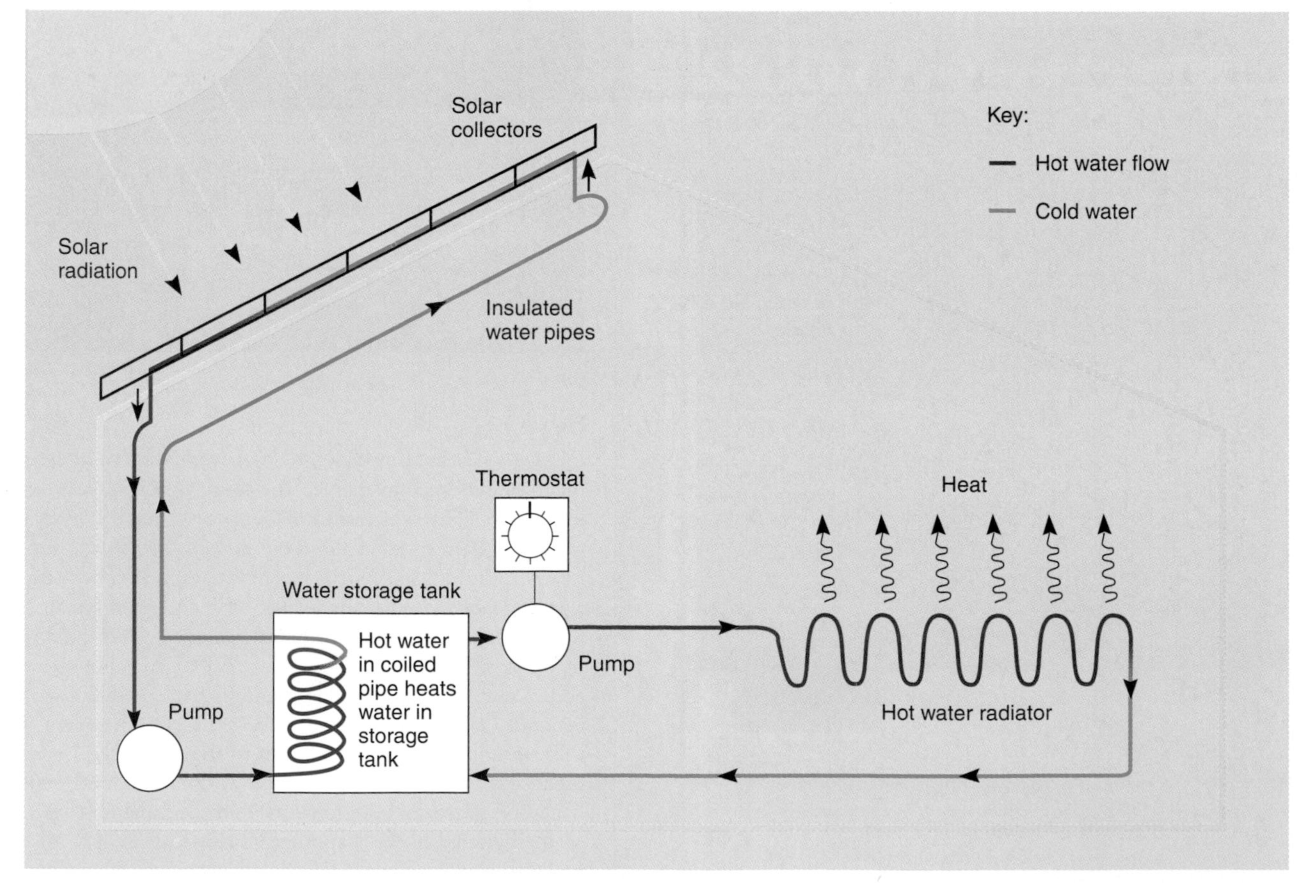

(b)

Figure 12–3

Active solar heating. (a) Solar collection devices are mounted on the roof of a building. Each solar panel is a box with a black metal base and glass covering. Sunlight enters the glass and warms the pipes and the liquid or air that is flowing through them. (b) A simplified diagram of active solar heating. In this example, water is heated in solar collectors on the roof and then pumped to a water storage tank where it heats water for a hot water radiator. The cooled water is then pumped back to the roof to be warmed again.

FOCUS ON

Cooking with Sunlight

Maybe you tried to fry an egg on a hot, summer sidewalk when you were a child. Or maybe you used a magnifying glass to burn a hole in a leaf. Probably neither technique worked very well, but you had the right idea: solar energy can be harnessed for cooking.

Recent designs for solar ovens literally capture solar energy in a box. The new solar cooker, which has a glass top, transmits solar light into the box, but does not transmit out the infrared wavelengths (heat) that would normally escape. Pots containing the food to be cooked are placed inside the box on a black metal plate. The solar oven can reach a temperature of 177°C (350°F) and can be used to boil, bake, simmer, and sauté foods. In average sunlight, a person can cook a full meal in 2 to 4 hours. The solar oven works like a Crock Pot in that the foods retain moisture and more vitamins and minerals than they would if cooked by conventional methods.

The solar cooker can be built from inexpensive, easily available materials at a cost as low as $3. This is a crucial factor for the people who stand to gain the most from the use of the new technology. Approximately one-half million people in areas such as Africa, Central America, India, and China, where fragile ecosystems are continually being destroyed by deforestation, are already using solar box cookers. People can build the structure in a few hours and use it for the majority of their cooking, preparing steaming hot meals without burning a single stick of firewood (see figure).

Given that more than 25 percent of the world's people use wood fuel for cooking, the solar cooker may very well prove to be one of the most important developments in the use of solar energy.

A woman in Kenya prepares her first meal using a solar oven that she just built. In average sunlight, food will cook in two to four hours. Solar ovens, which can be built from inexpensive, easily available materials, reduce the number of trees destroyed for fuel. (© 1993 Jon Reader, Courtesy of Earthwatch Expeditions)

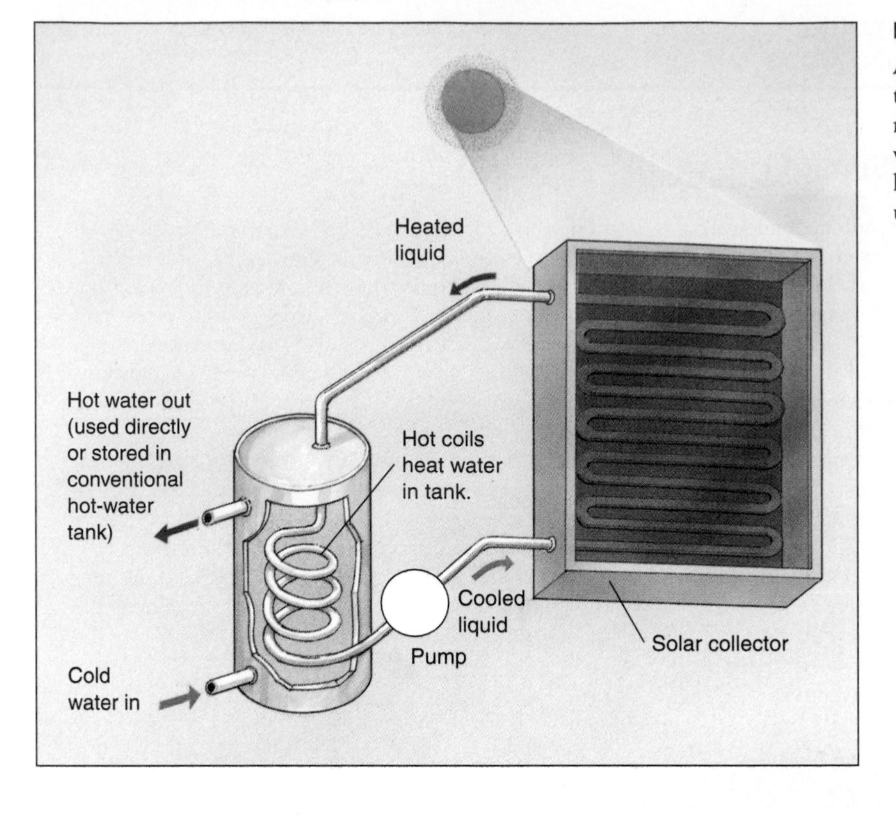

Figure 12–4

Active solar water heating. Although active solar heating to heat buildings is only needed during cooler months, active solar water heating can be used year-round. The liquid that is heated in solar collectors is used to heat water.

ers, track the sun for optimum efficiency, center sunlight on oil-filled pipes, and heat the oil to 390°C (735°F) (Figure 12–5a,b). The hot oil is circulated to a water storage system and used to change water into superheated steam, which turns a turbine to generate electricity. Alternatively, the heat can be used in industrial processes or for desalinization—that is, removal of salt from water—and water purification.

One problem with this form of energy is that a backup system—typically natural gas—must be available to generate electricity at night and during cloudy days when solar power is not operating. A large solar thermal system of this type is currently operating in the Mojave Desert in southern California.

A solar thermal system at White Cliffs, New South Wales, Australia, uses solar-collecting dishes that resemble radio telescopes (Figure 12–5c). This advanced design is more efficient in winter and at higher latitudes. It generates temperatures as high as 1500°C, as compared to 390°C in the trough design.

The solar power tower is a solar thermal system with a tall building or tower surrounded by numerous mirrors. The computer-controlled mirrors move to follow the sun, focusing solar radiation on a central receiver at the top of the tower. There a circulating liquid—molten salt—is heated by the concentrated sunlight to produce steam, which is used to generate electricity. Because molten salt retains heat, some of the heat may be stored to be used for electricity generation during the night, when solar energy is unavailable. Solar power towers are being tested in the United States, several European countries, and Japan.

Solar thermal energy systems are currently more efficient at trapping direct sunlight than other solar technologies. With improved engineering, manufacturing, and construction methods, solar thermal energy is becoming cost-competitive with fossil fuels. For example, electricity currently costs \$.09 per kWh at the Solar Electric Generating System in southern California. In addition, the environmental benefits of solar thermal plants are significant because they do not produce air pollution or contribute to acid rain or global warming.

Photovoltaic Solar Cells

It is possible to convert sunlight directly into electricity by using **photovoltaic (PV) solar cells**. Photovoltaic cells are wafers or thin films of crystalline silicon that are treated with certain metals in such a way that they generate electricity—that is, a flow of electrons—when solar energy is absorbed. Photovoltaic solar cells are

(a)

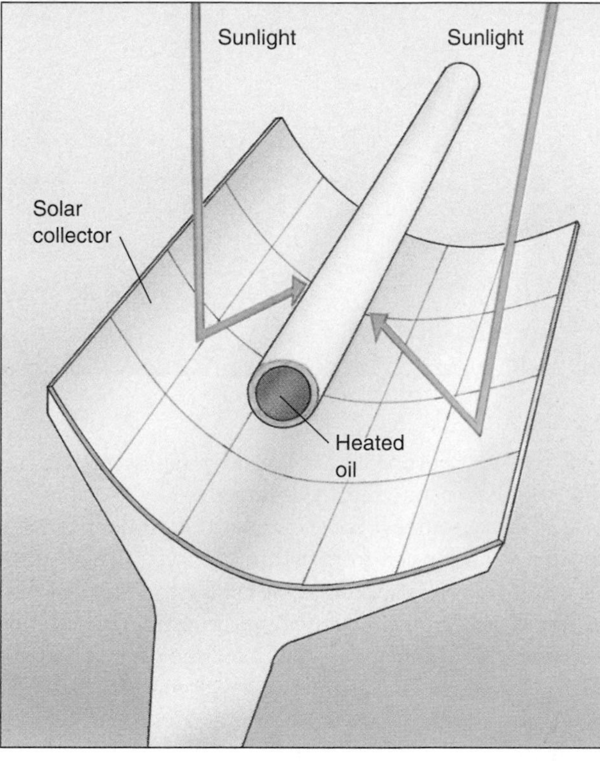

(b)

(c)

Figure 12–5

Solar thermal electric generation. (a) A solar thermal plant located in New Mexico uses troughs to focus the sun's energy on a fluid-filled tube. (b) How electricity is generated using the solar thermal technology shown in part (a). Sunlight is concentrated onto oil-filled pipes. The heated oil is circulated to a water tank where it provides the heat to generate steam, which is then used to produce electricity. (c) A solar thermal plant in White Cliffs, Australia, concentrates the sun's energy from each dish-shaped reflector to a small, central point, enabling water passing through that point to be heated to steam that can generate electricity in a steam turbine. A computer steers the dishes so they face the sun throughout the day. (a, Sandia National Laboratories; c, Science Photo Library)

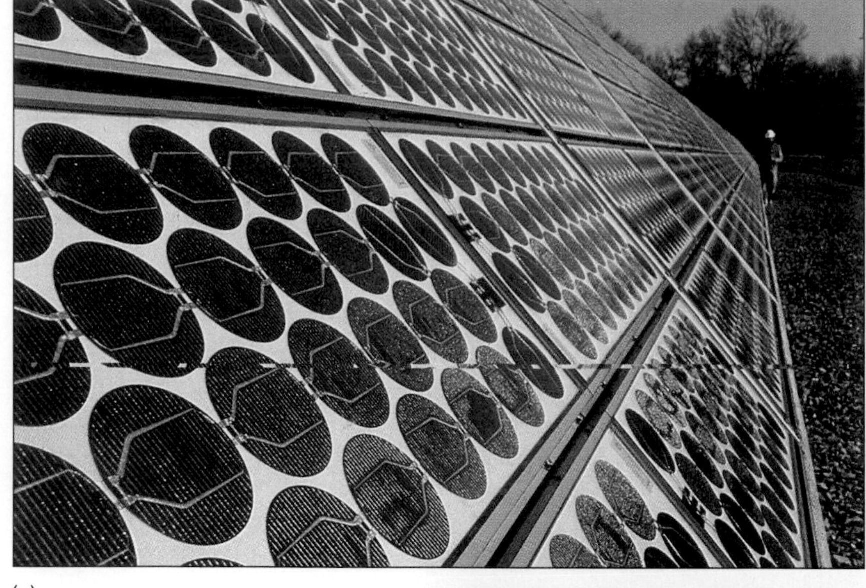

(a)

(b)

Figure 12–6

Photovoltaic cells convert solar energy into electricity. (a) Arrays of photovoltaic cells are used to generate electricity at Beverly High School in Beverly, Massachusetts. About 10 percent of the electricity needed by the school is supplied by its photovoltaic system. (b) A small solar panel on the roof of this rural home in Sri Lanka powers indoor lighting as well as a radio and television. (a, U.S. Department of Energy; b, Solar Electric Light Fund)

arranged on large panels that are set up to absorb sunlight, even on cloudy or rainy days (Figure 12–6a).

Our current photovoltaic solar cell technology, which is also used to power satellites, watches, and calculators, has several limitations that prevent the cells' widespread use to generate electricity. Photovoltaic solar cells are not very efficient at converting solar energy to electricity, although efficiencies have been steadily improving in the past decade. Other disadvantages are that PV cells are expensive to produce, and the number of solar panels needed for large-scale use requires a great deal of land.

On the positive side, PVs generate electricity with no pollution and minimal maintenance. They can be used on any scale, from small, portable modules to large, multimegawatt power plants. Also, the cost of producing electricity from photovoltaics has steadily declined from 1970 to the present; in the mid-1990s, the cost is about $.25 to $.50/kWh. Future technological progress may make photovoltaics economically competitive with conventional energy sources. For example, the production of a new type of solar cell from a thin film of a semiconductor material promises to decrease costs; thin-film cells are also much more efficient in converting solar energy into electricity. As another example of technological progress, a research group in Australia is developing efficient PV cells using a lower grade of silicon, making them considerably more economical to manufacture.

One of the main benefits of photovoltaic devices for utility companies is that they can be purchased in small modular units that become operational in a short amount of time. A utility company can purchase photovoltaic elements to increase its generating capacity in small increments, rather than committing a billion dollars or more and a decade or more of construction for a massive conventional power plant. Used in this supplementary way, the photovoltaic units can provide the additional energy needed, for example, to power air conditioners and irrigation pumps on hot, sunny days.

Although there has been steady improvement in the design and materials of photovoltaic solar cells, until their efficiency is improved it is unlikely that they can be used to generate electricity on a large scale, because we simply do not have enough space. At current efficiencies, for example, several thousand acres of solar panels would be required to absorb enough solar energy to produce the electricity generated by a single, large conventional power plant.

The use of photovoltaic solar cells for smaller generating requirements is well established, however. In remote areas that are not served by any electrical power plant (such as rural areas of developing countries), it is more economical to use photovoltaic solar cells for electricity than to extend power lines. Photovoltaics is the energy choice to pump water, refrigerate vaccines, grind grain, charge batteries, and supply rural homes with lighting. By 1995, an estimated 250,000 households in developing countries of Asia, Latin America, and Africa had installed photovoltaic solar cells on the roofs of their homes. A PV panel the size of two pizza boxes can supply a rural household with enough electricity for five lights, a radio, and a television (Figure 12–6b).

Solar Hydrogen

Solar electricity generated by photovoltaics can be used to split water into the gases oxygen and hydrogen. Hydrogen can also be produced using conventional energy sources such as fossil fuels, but then, of course, the environmental and energy security problems of fossil fuels are not avoided. For that reason, we limit this discussion to hydrogen fuel production using solar electricity.

Hydrogen is a clean fuel; it produces water and heat when it is burned, and produces no sulfur oxides, carbon monoxide, hydrocarbon particulates, or CO_2 emissions. It does produce some nitrogen oxides, but in amounts that are fairly easy to control. Hydrogen has the potential to provide energy for transportation in the form of hydrogen-powered electric automobiles (see Chapter 19) as well as for heating buildings and producing electricity.

It may seem wasteful to use electricity generated from solar energy to make hydrogen, which can then be used to generate more electricity. However, electricity that is generated by existing photovoltaic cells cannot be stored long-term; it must be used immediately. Hydrogen offers a convenient way to store solar energy as chemical energy. It can be transported by pipeline, possibly less expensively than electricity can be transported by wire (Figure 12–7).

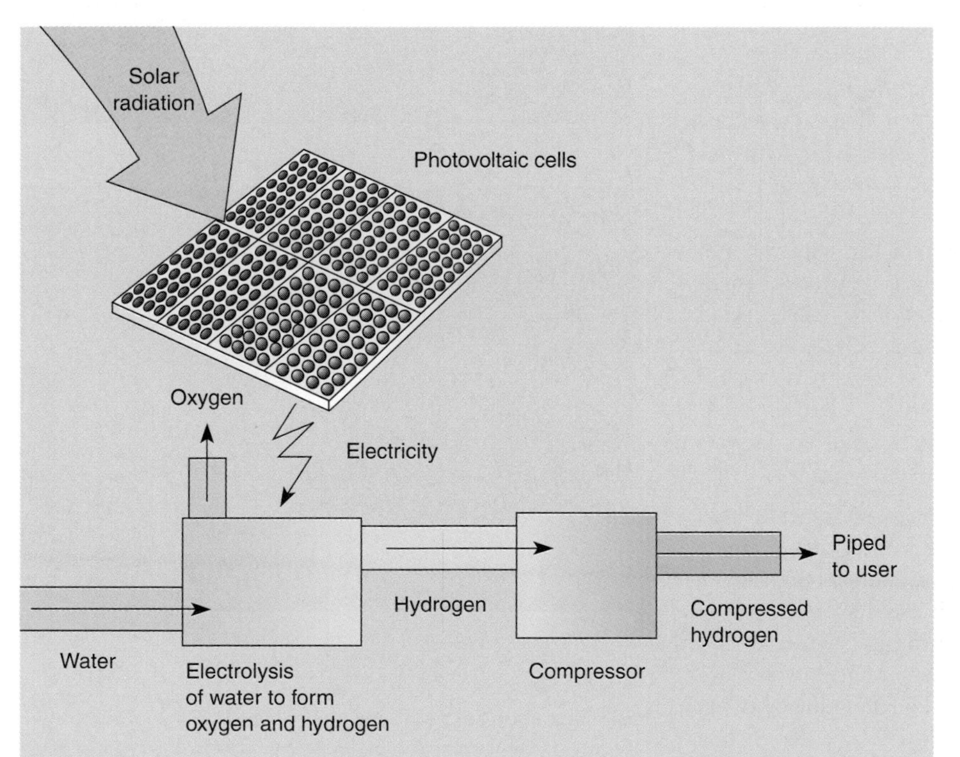

Figure 12–7

Solar hydrogen. Photovoltaic electricity can be used to split water in a process called electrolysis. This produces hydrogen gas, which represents a chemical form of solar energy. Hydrogen can be transported by pipelines to users. When burned in the presence of oxygen, it produces usable energy and water.

Production of hydrogen from solar electricity currently has an efficiency of 8 percent, which means that only 8 percent of the solar energy absorbed by the photovoltaic cells is actually converted into the chemical energy of hydrogen fuel. Scientists are working to improve the efficiency, which will decrease costs and make solar hydrogen fuel more attractive. A pilot plant that makes solar hydrogen fuel is being tested in Saudi Arabia under the joint sponsorship of Germany and Saudi Arabia. In addition, Mercedes-Benz and Ballard Power Systems of Vancouver, Canada, jointly developed a minivan powered by a hydrogen fuel cell that is virtually emissions-free. They are now working to lower the fuel cell's cost so it can be mass produced economically.

Solar Ponds

Because water absorbs solar energy, it is possible to build a pond of water specifically to collect solar energy. **Solar ponds** are generally dug one to several meters deep and are frequently lined with black plastic. Because a large percentage of solar radiation penetrates to the bottom of the pond, the temperature near the bottom may be as high as 100°C; the water near the surface remains at air temperature.

Under normal conditions, warm water rises to the top of a body of water because it is less dense than cool water. However, the water at the bottom of a solar pond is made denser than the surface water by the addition of salt; brackish water—water in which salt is dissolved—is denser than fresh water. Therefore, minimal mixing occurs between the warm, dense bottom layer and the cooler, less dense surface layer.

One of the problems associated with solar ponds is the large amount of land needed. Also, there is a potential for environmental contamination: if the brackish water leaked into the surroundings, it would harm plants and other organisms.

The use of solar ponds is still in the experimental stage. One experiment is taking place in Israel; a small power plant there generates electricity from a solar pond that contains salt water from the Dead Sea.

INDIRECT SOLAR ENERGY

There are several ways to use the sun's energy indirectly. Combustion of **biomass**—wood and other organic matter—is an example of indirect solar energy, because the energy contained in biomass is produced by green plants that use solar energy for photosynthesis. Windmills can harness the energy of **wind**—surface air currents that are caused by the solar warming of air. The damming of rivers and streams to generate electricity is a type of **hydropower**—the energy of flowing water—which exists because of the hydrologic cycle that is driven by solar energy (see Chapter 5).

Biomass Energy

Biomass consists of such materials as wood, fast-growing plant and algal crops, crop wastes, and animal wastes (Figure 12–8). Biomass contains chemical energy, the source of which can be traced back to radiant energy from the sun, which was used by photosynthetic organisms to form the organic molecules of biomass. Biomass is a renewable form of energy as long as it is managed properly.

Biomass fuel, which can be a solid, liquid, or gas, is burned to release its energy. Solid biomass such as wood is burned directly to obtain energy. Biomass—particularly firewood, charcoal (wood that has been turned into coal by partial burning), animal dung, and peat (partly decayed plant matter found in bogs and swamps)—supplies a substantial portion of worldwide energy. At least half of the world's population relies on biomass as their main source of energy. In developing countries, for example, wood is the primary fuel for cooking.

(a)

(b)

Figure 12–8

Biomass is organic material that is used as a source of energy. (a) Firewood is the major energy source for most of the developing world. (b) Animal dung is an important energy source in Ethiopia and many parts of the world. (a, Robert E. Ford/Terraphotographics; b, Mike Andrews © 1993 Earth Scenes)

Biomass can also be converted to liquid fuels, especially **methanol** (methyl alcohol) and **ethanol** (ethyl alcohol), which can then be used in internal combustion engines (Figure 12–9a). However, a major disadvantage of alcohol fuels, whether they are produced from biomass, natural gas, or coal, is that 30 to 40 percent of the energy in the starting material is lost in the conversion to alcohol. Some experts have calculated that alcohol production from a crop such as corn consumes more energy than it yields, when all energy costs associated with growing and harvesting a crop for alcohol production are assessed. Therefore, alcohol fuels produced from such crops should not be considered a renewable energy source.

It is possible to convert biomass, particularly animal wastes, into **biogas**. Biogas, usually composed of a mixture of gases, can be stored and transported easily like natural gas. It is a clean fuel whose combustion produces fewer pollutants than either coal or biomass. In India and China, several million family-sized **biogas digesters** use microbial decomposition of household and agricultural wastes to produce biogas that is used for heating and cooking (Figure 12–9b). When biogas conversion is complete, the solid remains are removed from the digester and used as fertilizer.

(a)

(b)

Figure 12–9

Biomass can be converted to liquid and gaseous fuels. (a) A methanol-powered bus on a street in Denver, Colorado. (b) A concrete biogas digester in Chu Zhang Village, near Beijing. The 135 digesters in this village of 600 people offer the benefit of improved sanitation as well as fuel for cooking. Human and animal wastes and crop residues are converted by bacteria to clean methane gas. Note the tubing that leads directly to individual homes in the village. (a, U.S. Department of Energy; b Marc Sherman)

Advantages of biomass use

Biomass is attractive as a source of energy because it reduces dependence on fossil fuels and because it can make use of wastes, thereby reducing our waste disposal problem (see Chapter 23). For example, the Mesquite Lake Resource Recovery Project in southern California burns cow manure in special furnaces to generate electricity for thousands of homes. The manure is too salty and contaminated with weed seeds to be used as fertilizer, so its use as an energy source helps solve the problem of its disposal.

Biomass is usually burned to produce energy, so the pollution problems caused by fossil fuel combustion, particularly carbon dioxide emissions, are not completely absent in biomass combustion. However, the low levels of sulfur and ash produced by biomass combustion compare favorably with the high levels produced when bituminous coal is burned. It is possible to offset the CO_2 that is released into the atmosphere from biomass combustion by increasing tree planting. As trees photosynthesize, they absorb atmospheric CO_2 and lock it up in organic molecules that make up the body of the tree, thereby providing a carbon "sink" (see Chapter 20). Thus, if biomass is regenerated to replace the biomass used, there is no net contribution of CO_2 to the atmosphere and to global warming.

Disadvantages of biomass use

Some problems are associated with use of biomass, especially from plants. For one thing, biomass production requires land and water. The use of agricultural land for energy crops competes with the growing of food crops, so shifting the balance toward energy production might decrease food production, contributing to higher food prices. For this reason, some scientists are interested in the commercial development of certain desert shrubs, which produce oils that could be used for fuel. The shrubs do not require prime agricultural land, although care would have to be taken to ensure that the desert soils were not degraded or eroded by overuse.

As mentioned earlier, at least half of the world's population relies on biomass as its main source of energy. Unfortunately, in many areas people burn wood faster than they replant it. Intensive use of wood for energy has resulted in severe damage to the environment, including soil erosion (see Chapter 14), deforestation and desertification (see Chapter 17), air pollution (see Chapters 19 and 20), and degradation of water supplies (see Chapter 21).

Crop residues, another category of biomass that includes cornstalks, wheat stalks, and wood wastes at paper mills and sawmills, are increasingly being used for energy. At first glance, it may seem that crop residues, which normally remain in the soil after harvest, would be a good source of energy if they were collected and burned. After all, they are just waste materials that will eventually decompose. As it turns out, however, crop residues left in the ground prevent erosion by helping to hold the soil in place. Also, their decomposition serves to enrich the soil by making the minerals that were originally locked up in the plant residues available for new plant growth. If all crop residues were removed from the ground, the soil would eventually be depleted of minerals and its future productivity would decline. Forest residues, which remain in the soil after trees are harvested, fill similar ecological roles.

The use of crop and forest residues for biomass would have to be carefully managed: some of the residues could be removed for energy, and the rest would have to be left in the soil to maintain soil fertility.

ENVIROBRIEF

UNLIKELY PROPONENTS OF RENEWABLE ENERGY

In a surprising assessment of the world's energy future, Shell International has predicted that world energy production will be dominated by renewable sources by the year 2050. The oil company's perspective differs sharply from the more conservative outlook of most other energy corporations, which believe that fossil fuels will continue to provide the bulk of the world's energy. In one of its scenarios of future energy development, Shell projects that the growing energy demands from heavily populated developing nations such as China and India, along with increased competition in the energy industry, will stimulate development of energy-efficient renewable technologies. Some energy companies agree with Shell's assessment of the future. Bechtel Enterprises, once an important manufacturer of nuclear power plants, and PacifiCorp, a utility company that operates several coal-fired power plants in the northwestern United States, have formed a partnership called Energy Works to develop solar and other renewable energy technologies.

Wind Energy

Wind, which results from the warming of the atmosphere by the sun, is an indirect form of solar energy in which the radiant energy of the sun is transformed into mechanical energy—the movement of air molecules. Wind is sporadic over much of the Earth's surface, varying in direction and magnitude; and, like direct solar energy, wind power is a highly dispersed form of energy. Harnessing wind energy to generate electricity has great potential, however, and wind is becoming increasingly important in supplying our energy needs. It is currently the most cost-competitive of all forms of solar energy, and new technological advances (recall the chapter introduction) suggest that wind energy could become an important source of electricity within the next decade.

Harnessing wind energy is most profitable in rural areas that receive fairly continual winds, such as islands, coastal areas, mountain passes, and grasslands. The world's largest cluster of wind turbines is located in Tehachapi, a mountainous area at the southern end of the Sierra Nevada in California. Although the United States was the world leader in producing wind-generated electricity in the 1980s, European and Asian countries have dominated new wind power generation in the 1990s. In 1994, for example, almost half of the world's new turbines were installed in Germany. India is currently leading Asia in the installation of wind farms.

In the continental United States, some of the best locations for large-scale electricity generation from wind energy are off the coasts of New England and the Pacific Northwest and on the western Great Plains. Current U.S. wind power projects are underway in Maine, Minnesota, Montana, and Texas.

The use of wind power does not cause major environmental problems, although one concern is reported bird kills. The California Energy Commission estimated that as many as 567 birds, many of them raptors (birds of prey), turned up dead in the vicinity of the 7000 turbines at Altamont Pass in California during a two-year study; most had collided with the turbines. Kenetech Windpower, the largest operator at Altamont Pass, has embarked on a study of how to create a compatible coexistence between birds and wind turbines. Among other actions they have implemented, Kenetech Windpower is testing the effectiveness of various warning devices such as painted blades and noise. They have also installed antiperching devices to discourage raptors from roosting on the towers, and they are testing the feasibility of tracking migratory birds by radar so that wind turbines can be shut down as the birds approach.

Because it produces no waste, wind is a clean source of energy. It produces no emissions of sulfur dioxide, carbon dioxide, or nitrogen oxides. Every kWh of electricity generated by wind power rather than fossil fuels prevents 1 to 2 pounds of the greenhouse gas carbon dioxide from entering the atmosphere. A major problem with increasing our use of wind power is aesthetics: wind machines detract from the beauty of the landscape. Fortunately, most of the locations that are appropriate for large-scale wind power are not densely populated. Combining wind farms with cattle grazing, as is done in Altamont, California, is a very productive and profitable use of land.

Hydropower

The sun's energy drives the hydrologic cycle, which encompasses evaporation from land and water, transpiration from plants, precipitation, and drainage and runoff. As water flows from higher elevations back to sea level, we can harness its energy. Unlike the sun's energy, which is highly dispersed, hydropower is a more concentrated energy. The potential energy of water held back by a dam is converted to kinetic energy as the water falls over a spillway (Figure 12–10), where it turns turbines to generate electricity.

Currently, hydropower produces approximately 20 percent of the world's electricity, making it the form of solar energy in greatest use. Developed countries have already built dams at most of their potential sites, but this is not the case in many developing nations. There—particularly in undeveloped, unexploited parts of Africa and South America—hydropower is still a great potential source of electricity. However, even if all potential sites worldwide were used, the energy generated would be less than 15 percent of the total energy needed.

One of the problems associated with hydropower is that building a dam changes the natural flow of a river. A dam causes water to back up, flooding large areas of

(a)

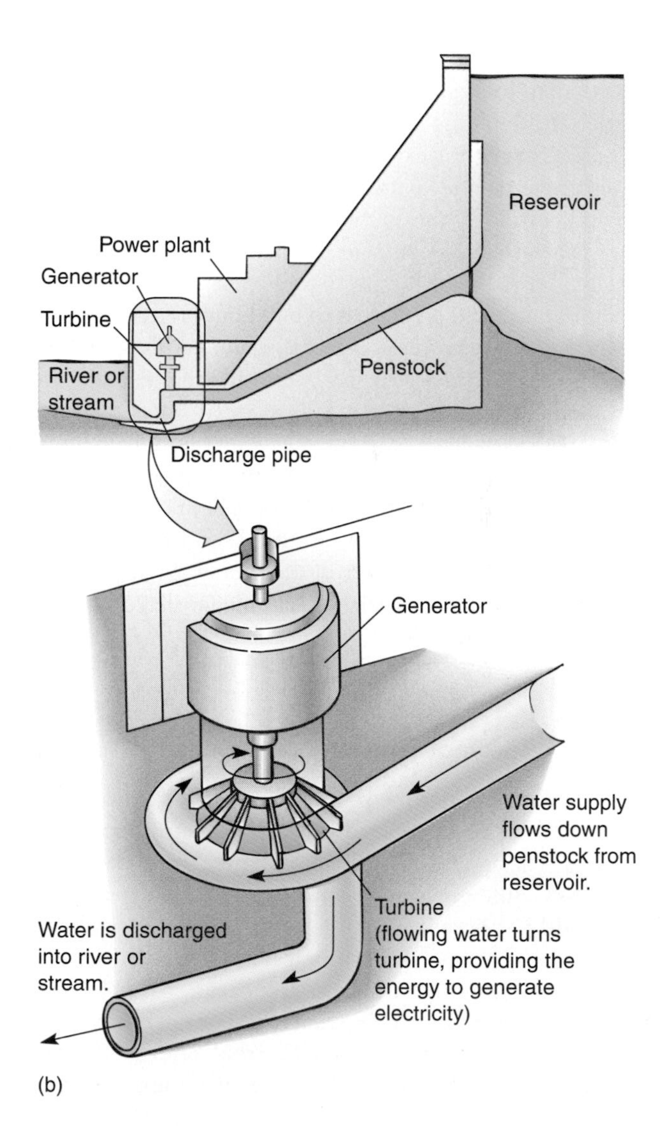

(b)

Figure 12–10

Hydroelectric power may be obtained from water held back by a dam. (a) The McNary Dam supplies some of the electricity used in the Pacific Northwest. (b) How a hydroelectric power plant works. A controlled flow of water released down the penstock turns a turbine, which generates electricity. (a, U.S. Department of Energy)

land and forming a reservoir, which destroys plant and animal habitats. Below the dam, the once-powerful river is reduced to a relative trickle. The natural beauty of the countryside is affected, and certain forms of wilderness recreation are made impossible or less enjoyable.

In arid regions, the creation of a reservoir behind a dam results in greater evaporation of water, because the reservoir has a larger surface area in contact with the air than the stream or river did. As a result, serious water loss and increased salinity of the remaining water may occur (see Chapter 13).

Dams destroy farmlands and displace people. When a dam breaks, people and property downstream may be endangered. In addition, waterborne diseases such as schistosomiasis may spread throughout the local population. **Schistosomiasis** is a tropical disease, caused by a parasitic worm, that can damage the liver, urinary tract, nervous system, and lungs. It is estimated that as much as half the population of Egypt suffers from this disease, largely as a direct result of the Aswan Dam, built on the Nile River in 1902 to control flooding but used since 1960 to provide electrical power.

The ecological, environmental, and personal impacts of a dam may not be acceptable to the people living in a particular area. Laws have been passed to prevent or restrict the building of dams in certain locations. In the United States, for example, the Wild and Scenic Rivers Act prevents the hydroelectric development of certain rivers, although the number of rivers protected by this law is less than one percent of the nation's total river systems.

Dams cost a great deal to build but are relatively inexpensive to operate. A dam has a limited life span, usually 50 to 200 years, because over time the reservoir fills in with silt until it cannot hold enough water to generate electricity. This trapped silt, which is rich in nutrients, is prevented from enriching agricultural lands downstream. For example, the gradual depletion of agricultural productivity downstream from the Aswan Dam in Egypt is well documented. Egypt now relies on heavy applications of chemical fertilizers to maintain fertility of the Nile valley and its delta.

Ocean Thermal Energy Conversion

In the future it may be possible to generate power using **ocean temperature gradients**, the differences in temperature at various ocean depths. There may be as much as a 24°C difference between warm surface water and very cold, deeper ocean water. Ocean temperature gradients, which are greatest in the tropics, are the result of solar energy warming the surface of the ocean.

The generation of electricity from ocean temperature gradients is known as **ocean thermal energy conversion (OTEC)**. Warm surface water is pumped into the power plant, where it heats a liquid, such as ammonia, to the boiling point. (Liquid ammonia has a very low boiling point, −33°C.) The ammonia steam drives a turbine to generate electricity as ammonia is cooled by the very cold water from the ocean depths. Alternatively, warm seawater can be made to boil in a vacuum chamber, forming steam to drive the turbine as the vapor is cooled and condensed back into a liquid by cold seawater.

The latest pilot OTEC project, sponsored by the state of Hawaii, the federal government, and a nonprofit consortium with funds from the Japanese government, opened in 1993 in Hawaii. In addition to generating a small amount of electricity, it supports several spinoff projects. After the cold water from the ocean depths is used in the power plant, it is still colder than surface ocean water, so it is used to air condition buildings. Then it is piped to nearby aquaculture facilities to provide clean, nutrient-laden seawater for growing algae, fishes, and crustaceans.

Although OTEC is technologically possible, the potential impact of bringing massive quantities of cold water to the surface in a tropical area needs to be considered carefully before it is adopted on a large scale. Such water properties as dissolved gases, turbidity (cloudiness), nutrient levels, and salinity gradients (differences in salt concentrations) are bound to be altered along with the temperature, and these changes would probably have a profound effect on marine organisms.

Ocean Waves

Ocean waves are produced by winds, which are caused by the sun, so wave energy is considered an indirect form of solar energy. Like other types of flowing water, wave power has the potential to turn a turbine, thereby generating electricity. Norway, Great Britain, Japan, and several other countries are investigating the production of electricity from ocean waves.

The wave power station uses a simple technology. Essentially, the concrete, hollow power-plant box is sunk into a gully off the coast to catch waves. As each new wave enters the chamber (about every 10 seconds), the rising water in the chamber pushes air into a vent that contains a turbine, causing the turbine to spin. In turn, the spinning turbine drives a generator. When the wave recedes, it draws the air back into the chamber and the moving air continues to drive the turbine.

More testing and improvements in design need to be made, and engineers cautiously warn the public about past failures of wave power stations. For example, the world's first commercial power station, located off the coast of northern Scotland, sank in 1995 during a storm, less than one month after it opened. Although a great deal of power can be obtained from the ocean, the power plants themselves need to be able to harness that energy at a rate that is efficient and productive. Currently, harnessing wave power is very inefficient, and more research will have to be done to make generation of electricity from wave motion practical.

OTHER RENEWABLE ENERGY SOURCES

Geothermal energy and tidal energy are renewable energy sources that are not direct or indirect results of solar energy. **Geothermal energy**, which is actually a form of nuclear energy, is the heat produced in the Earth by the natural decay of radioactive elements. In locations where radioactive elements are close to the surface, this heat can be tapped. **Tidal energy**, which is caused by the change in water level between high and low tides, has also been exploited to generate electricity on a limited scale.

Geothermal Energy

Geothermal energy, the heat produced in the Earth's interior by natural nuclear processes, is carried to the surface through volcanoes and groundwater. Heated groundwater flows upward as hot water or steam; thus, natural hot springs are frequently found in areas where radioactive elements, whose decay produces the heat, are close to the surface. Because it takes many years for natural processes to replace groundwater, geothermal energy is sometimes considered nonrenewable or very slowly renewable.

Geothermal energy from hot springs has been exploited for thousands of years for bathing, cooking, and heating buildings. More recently, geothermal energy based on moderate temperature differences between the surface and a depth of perhaps 61 meters (200 feet) is being employed to heat and cool buildings; this energy source can be tapped almost anywhere, not just in areas that are known to be geothermally hot. In addition, several small geothermal power plants are in use today that use the steam from underground reservoirs to turn turbines and generate electricity (Figure 12–11).

Iceland, situated on the mid-Atlantic ridge, a boundary between two continental plates, is an island of intense volcanic activity and numerous hot springs and therefore can make optimum use of geothermal energy. Recall from Chapter 5 that volcanoes are commonly located at the boundary between two geological plates. Like hot springs, volcanic activity is a product of geothermal energy and is associated with radioactive elements close to the surface. Because of its location, Iceland uses geothermal energy to meet a substantial portion of its energy requirements. Two-thirds of Icelandic homes are heated with geothermal energy. In addition, most of the fruits and vegetables

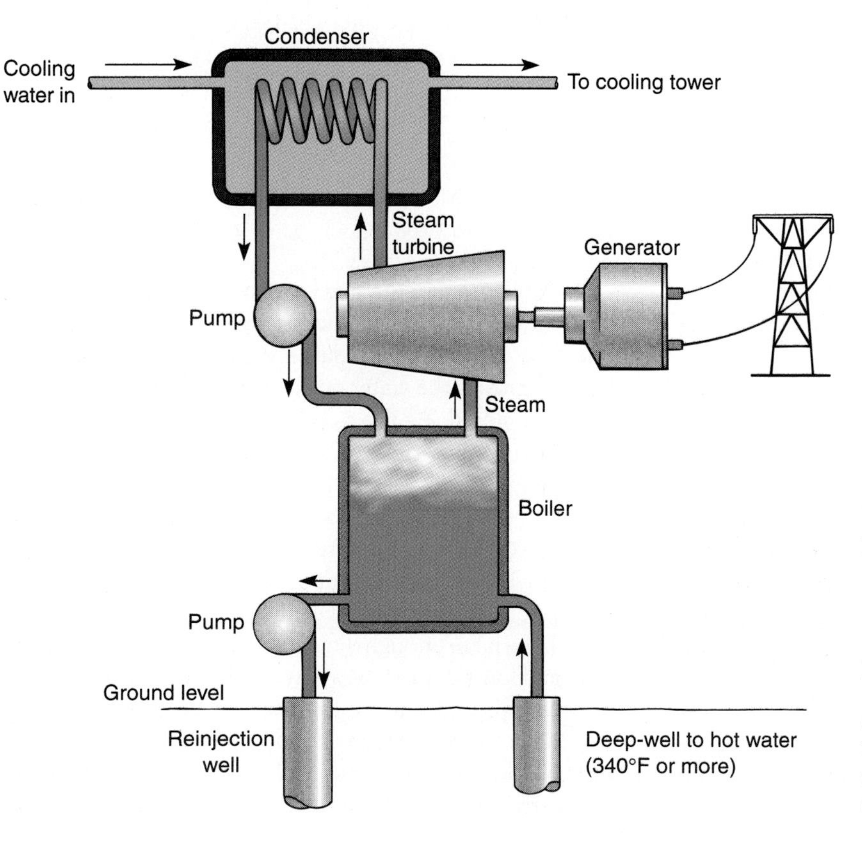

Figure 12–11

One design for a geothermal power plant. Steam separated from hot water that was pumped from underground is used to turn a turbine and generate electricity. After its use, the steam is condensed and pumped back into the ground. By reinjecting spent water into the ground, geothermal energy remains a renewable energy source because the cooler, reinjected water can be reheated and used again.

required by the people of Iceland are grown in geothermally heated greenhouses.

Countries that are increasing their use of geothermal energy include the Philippines, Japan, Italy, Nicaragua, Mexico, and the United States. Currently, the world's total geothermal energy production is fairly small, although certain countries such as Nicaragua and the Philippines use geothermal energy to generate at least 25 percent of their electricity.

One of the environmental concerns associated with geothermal energy is that the surrounding land may subside, or sink, as the water from hot springs and their connecting underground reservoirs is removed. Conventional geothermal energy also produces several pollutants, including hydrogen sulfide and carbon dioxide. Also, geothermal energy becomes nonrenewable if the hot groundwater is tapped faster than it can be replaced by natural processes.

Geothermal energy from hot, dry rocks

Conventional use of geothermal energy relies on hot springs—that is, on groundwater bringing geothermal energy to the surface from subsurface, hot rocks. Scientists in both Australia and the United States are studying the feasibility of extracting subsurface geothermal energy in dry areas. For example, in the mid-1980s the Los Alamos National Laboratory in New Mexico produced enough hot dry-rock geothermal energy to provide electricity for a town of 2000 people. They drilled a shaft to the subsurface hot rock, used hydraulic pressure to fracture the rock, and then pumped water into the fractured area under high pressure to make an artificial underground reservoir. When the pressurized water returned to the surface by way of a second well, it turned to steam, which drove an electricity-generating turbine.

The hot dry-rock system has an additional benefit over conventional geothermal energy because it produces less pollution. Although this method requires the development of sophisticated technology if it is to become a practical reality, scientists are optimistic that it could be used in many locations that cannot employ conventional geothermal energy.

ENVIROBRIEF

COOL TECHNOLOGY FROM THE REFRIGERATOR WARS

Because refrigerators can guzzle 20 percent of a home's consumed electricity, a group of 24 electric utility companies sponsored the Super Efficient Refrigerator Program, a contest challenging appliance manufacturers to design a more energy-efficient and environmentally friendly refrigerator. Whirlpool won the contest—and the $30 million prize—in 1993 with these design elements:

- No ozone-depleting chlorofluorocarbons (CFCs), a contest requirement.
- A defrost-control unit that employs a "fuzzy logic" microchip to initiate defrosting only as needed, rather than at set intervals.
- A new drainpipe that is bent to keep hot air from reaching the refrigerator's interior.
- A redesigned compressor to handle CFC-free coolants while functioning more efficiently.
- Overall operating costs equivalent to those of burning a 75-watt light bulb.

As a result of the competition, utilities will need to build fewer power plants, Whirlpool and other manufacturers will market new technologies more rapidly, and consumers will gain efficient appliances that operate at lower financial and environmental costs.

Tidal Energy

Tides, the alternate rising and falling of the surface waters of the ocean and seas that occur twice each day, are the result of the gravitational pull of the moon and the sun. Normally, the difference in water level between high and low tides is about 0.5 meter (1 or 2 feet). However, certain coastal regions with narrow bays have extremely large differences in water level between high and low tides. The Bay of Fundy in Nova Scotia, for example, has the largest tides in the world, with up to 16 meters (53 feet) difference between high and low tides.

By building a dam across a bay, it is possible to harness the energy of large tides to generate electricity. In one type of system, the dam's floodgates are opened as high tide raises the water on the bay side. Then the floodgates are closed. As the tide falls, water flowing back out to the ocean over the dam's spillway is used to turn a turbine and generate electricity.

Currently, existing power plants that make use of tidal power are in operation in France, Russia, China, and Canada. However, tidal energy cannot become a significant resource worldwide, because few geographical locations have large enough differences in water level between high and low tides to make power generation feasible. The most promising locations for tidal power in North America include the Bay of Fundy in Nova Scotia, Passamaquoddy Bay in Maine, Puget Sound in Washington, and Cook Inlet in Alaska.

Other problems associated with tidal energy include the high cost of building a tidal power station and the potential environmental problems. The greatest tidal energy is found in estuaries, coastal areas where river currents meet ocean tides. Because the mixing of fresh and salt waters creates a nutrient-rich environment, estuaries are among the most productive aquatic environments in the world. Fishes and countless invertebrates migrate there to spawn. Building a dam across the mouth of an estuary would prevent these animals from reaching their breeding habitats. Estuaries are also popular sites for recreation, which would be severely curtailed by a tidal dam.

ENERGY SOLUTIONS: CONSERVATION AND EFFICIENCY

Human energy needs will continue to increase, if only because the human population is growing. We must therefore place a high priority not only on developing alternative energy sources but also on energy conservation and energy efficiency. **Energy conservation** is moderating or eliminating wasteful or unnecessary energy-consuming activities, whereas **energy efficiency** is using technology to accomplish a particular task with less energy. As an example of the difference between energy conservation and energy efficiency, consider gasoline consumption by automobiles. Energy conservation measures to reduce gasoline consumption would include carpooling and reducing driving speed, whereas energy efficiency measures would include making more fuel-efficient automobiles. Both conservation and efficiency accomplish the same goal, that of saving energy.

Many energy experts consider energy conservation and efficiency to be the most promising energy "sources" available to humans, because they not only save energy for future use but also buy us time to explore new energy alternatives. Energy conservation and efficiency also cost less than development of new sources or supplies of energy. A system that supports energy conservation and efficiency makes good economic sense, as well. The adoption of energy-efficient technologies generates new business opportunities, including the development, manufacture, and marketing of those technologies.

In addition to economic benefits and energy resource savings, there are important environmental benefits from greater energy efficiency and conservation. For example, using more energy-efficient appliances could cut our carbon dioxide emissions by millions of tons each year, thereby slowing global climate change. Energy conservation and efficiency also reduce air pollution, acid precipitation, and other environmental damage related to energy production and consumption.

Energy Consumption Trends and Economics

Energy consumption in the United States increased by 13 percent during the period from 1973 to 1993, whereas the economy (as represented by the gross domestic product) increased by 25 percent. This means that the United States used almost 10 percent less energy to generate each dollar of gross domestic product in the mid-1990s than in 1973.

During the past few decades, many industries have improved their energy efficiencies. New aircraft, for example, are much more fuel-efficient than older models. Similarly, technological improvements in the papermaking industry make it possible to use less energy to manufacture paper today than was used just a few years ago. The energy savings from such improvements in efficiency translate into greater profits for the companies employing them.

A country's or region's total commercial energy consumption divided by its gross domestic product gives its **energy intensity** (Figure 12–12). Despite gains in effi-

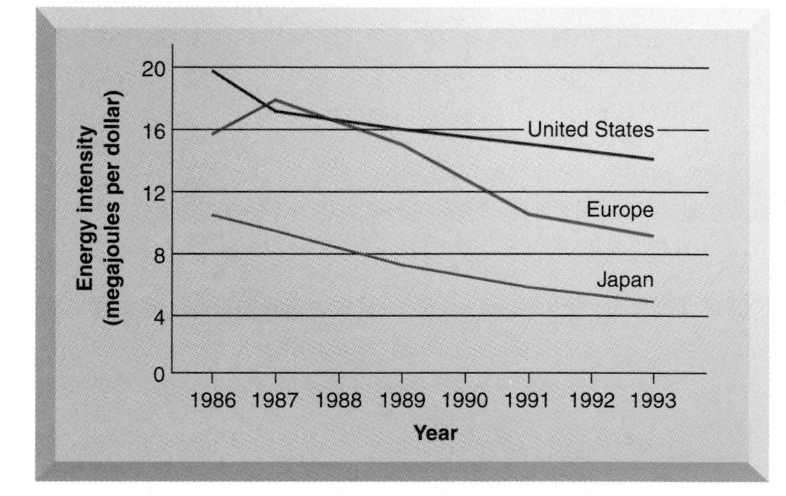

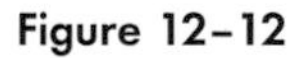

Figure 12–12

Energy intensity of the United States, Europe, and Japan. The three areas indicated here have been equalized for comparison purposes by conversion of all gross domestic products to U.S. dollars.

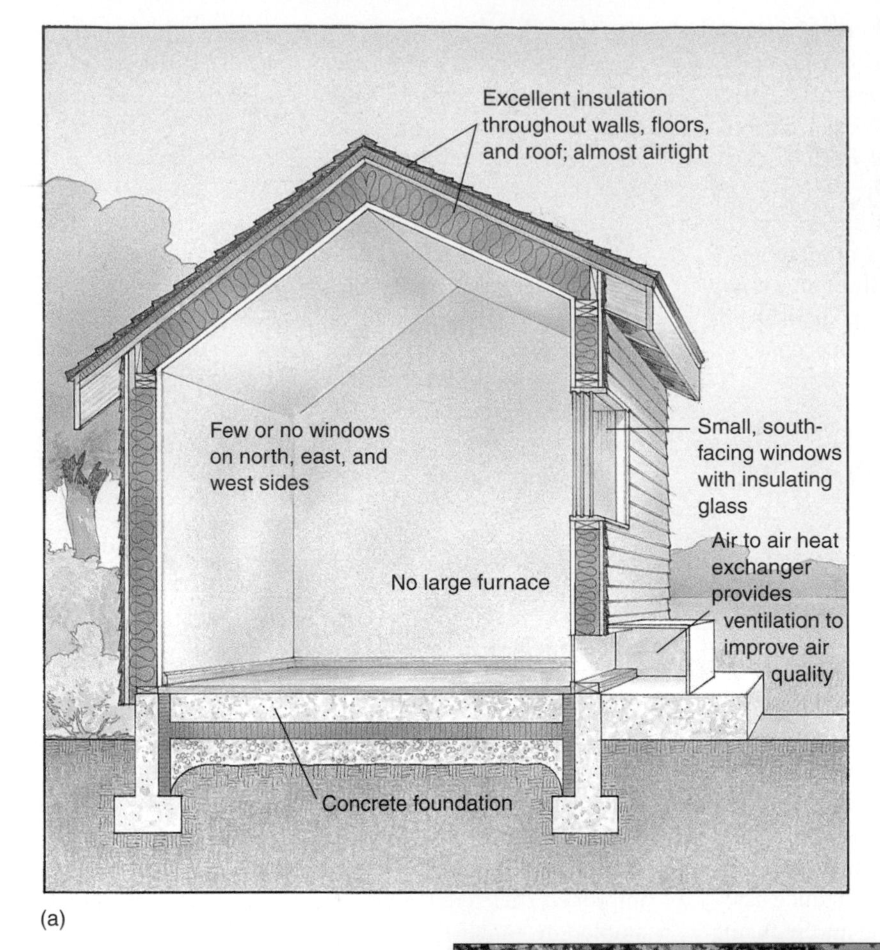

Figure 12–13

Superinsulated buildings. (a) Some of the characteristics of a superinsulated home, which is so well insulated and airtight that it does not require a furnace in winter. Heat from the inhabitants, light bulbs, the stove, and other appliances provides almost all the necessary heat. (b) A superinsulated office building in Toronto, Canada, has south-facing windows with insulating glass. The building is so well insulated that it uses no furnace. (b, Ontario Hydro)

ciency, the energy intensity of the United States is still considerably higher than the energy intensity of Japan or Europe.

Energy trends in developing countries

Per-capita consumption of energy in developing nations is substantially less than it is in industrialized countries (see Figure 10–1), although the fastest increase in energy consumption today is occurring in the developing nations. As developing nations boost their economic development, their energy demands will continue to increase. This is partly because the "new" industrial and agricultural processes being adopted in developing countries often represent older technologies that are less energy-efficient. Also, the burgeoning populations in developing countries will raise energy demands.

Developing countries are faced with the need for economic development and the need to control environmental degradation. At first glance, these two goals appear to be mutually exclusive. However, both goals can be realized by the use of new technologies now being developed in industrialized nations to achieve greater energy efficiency. For example, it would cost Brazil $44 billion to build power plants to meet its projected electricity needs for the near future; this cost could be avoided by investing $10 billion in more efficient refrigerators, lighting, and electric motors. The energy efficiency approach in Brazil would not only cause fewer environmental problems but also foster and expand the growth of manufacturing industries devoted to energy-efficient products.

Energy-Efficient Technologies

The development of more efficient appliances, automobiles, and buildings has been a major factor in the recent reduction of energy consumption in developed countries. Compact fluorescent light bulbs, introduced in the late-1980s, require 25 percent of the energy used by regular incandescent bulbs and last nine times longer; the energy-efficient bulbs are expensive, but they more than pay for themselves in energy savings. New condensing furnaces require approximately 30 percent less fuel than conventional gas furnaces. "Superinsulated" homes in Sweden and the United States use 70 to 90 percent less heat than do homes insulated by standard methods (Figure 12–13).

The National Appliance Energy Conservation Act sets national appliance efficiency standards for refrigerators, freezers, washing machines, clothes dryers, dishwashers, air conditioners, and water heaters. Refrigerators built in the mid-1990s, for example, consume 80 to 90 percent less energy than those built in the early 1980s. The energy cost that consumers will save as a direct result of energy savings required by this law is estimated at $28 billion by the year 2000.

Automobile efficiency has improved dramatically as a result of the use of lighter materials and designs that reduce air drag. The average fuel efficiency of passenger cars increased from 15.9 miles per gallon (mpg) in 1981 to 21.4 mpg in 1994. Using current technology, automobiles with fuel efficiencies of 60 to 65 mpg could be routinely manufactured within the next decade or so.

Cogeneration

One nontraditional energy technology with a bright future is **cogeneration**, which is the production of two useful forms of energy from the same fuel. Typically, cogeneration involves the generation of electricity, and then the steam produced during this process is used rather than wasted. Cogeneration is mostly done on a small scale, and its use is increasing. Modular cogeneration systems enable hospitals, hotels, restaurants, and other businesses to harness steam that would otherwise be wasted. In a typical cogeneration system, electricity is produced in a traditional manner—that is, some type of fuel provides heat to form steam from water. Normally, the steam used to turn the electricity-generating turbine would be cooled before being pumped back to the boiler to be reheated. In cogeneration, after the steam is used to turn the turbine, it supplies energy to heat buildings, cook food, or operate machinery before it is cooled and pumped back to the boiler as water (Figure 12–14).

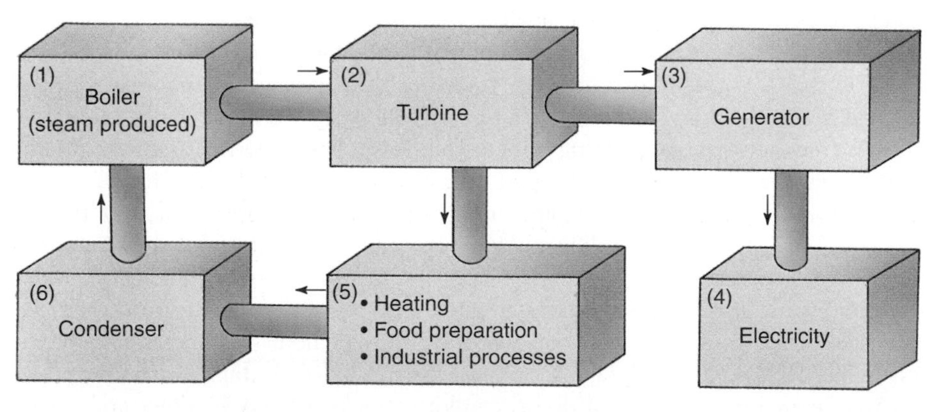

Figure 12–14

In a cogeneration system, electricity is produced in the usual manner. In this example, fuel combustion occurs in a boiler (1). The heat produced is used to make steam, which turns a turbine (2) that generates electricity in a generator (3). The electricity that is produced (4) is used in-house or sold to a local utility. Cogeneration involves using the waste heat (leftover steam) from electricity generation to do useful work, such as cooking, space heating, or operating machinery (5). Any residual steam that is not used is piped into a condenser (6) and recycled back to the boiler.

Table 12–3
Energy-Efficiency Upgrades in Selected Commercial Buildings

Project	Energy Payback Time*	Unexpected Benefits Attributed to Project**
Energy-efficient lighting (post office in Nevada)	6 years	6% increase in mail sorting productivity
Energy-efficient (metal-halide) lighting (aircraft assembly plant in Washington)	2 years	Up to 20% better quality control
Energy-efficient lighting (drafting area of utility company in Pennsylvania)	About 4 years	25% lower absenteeism; 12% increase in drawing productivity
Energy-efficient lighting and air conditioning (office building in Wisconsin)	0 years (paid for by utility rebates); energy savings estimated at 40%	16% increase in worker productivity
Energy-saving daylighting, passive solar heating, heat recovery system (bank in Amsterdam)	3 months	15% lower absenteeism

*How long it takes for energy savings to cover the cost of the project.
**Lighting quality as well as lighting efficiency is improved, resulting in greater worker comfort.

Cogeneration can also be accomplished on a large scale. One of the largest cogeneration systems in the United States is a natural gas-fired plant completed in 1995 in Oswego, New York, that produces electricity for the local utility. The waste steam produced in the generation of electricity provides thermal energy to a nearby industry. The overall efficiency of the Oswego cogeneration system is 54 percent, as compared to 33 percent efficiency in a typical fossil-fueled power plant.

Energy savings in commercial buildings

Energy costs often account for 30 percent of a company's operating budget. Unlike cars, which are traded in every few years, buildings are usually used for 50 or 100 years, so a company housed in an older building normally does not have the benefits of new energy-saving technologies. It makes good economic sense for these businesses to invest in energy improvements, which often pay for themselves in a few years (Table 12–3).

To get businesses to install new energy-efficient technologies, many energy-services companies, which specialize in improving energy efficiency, offer their assistance in such a way that the business makes little or no financial outlay. Here is how it works. An energy-services company makes a detailed assessment of how a business can improve its energy efficiency. In developing its proposal, the energy-services company guarantees a certain amount of energy savings. It also provides the funding to accomplish the improvements, which may be as simple as fine-tuning existing heating, ventilation, and air conditioning systems or as major as replacing all existing windows and lights. The reduction in utility costs is used to pay the energy-services company, but once the bill is paid, the business benefits from substantial energy savings.

Energy savings in homes

When buying a new home, a smart consumer should demand energy efficiency. Although a more energy-efficient house might cost a little more, depending on the technologies employed, the improvements usually pay for themselves in two or three years. Any time spent in the home after the payback period means substantial energy savings. Energy efficiency will almost certainly be an important part of the design of homes of the future.

Some energy-saving improvements, such as thicker wall insulation, are easier to install while the home is being built. Other improvements can be made in older homes to enhance energy efficiency and, as a result, reduce the cost of heating the homes. Examples include installing thicker attic insulation, installing storm windows and doors, caulking cracks around windows and doors, replacing inefficient furnaces, and adding heat pumps.

Many of the same improvements also provide energy savings when a home is air conditioned. Additional cool-

WHERE ENERGY CONSERVATION IS ACADEMIC

The State University of New York (SUNY) at Buffalo invested $18 million in energy conservation, beginning in 1994. It is one of the largest energy conservation projects ever attempted by an American university.

- SUNY-Buffalo should save $3.2 million each year; annual energy savings will be 8 megawatts of electricity, the equivalent used by 4000 households.
- The project combines quick-fix measures such as installing more efficient light bulbs with major modifications like converting some electric heating and cooling systems to natural gas. Also, heat-recovery systems are being installed to recapture air ventilated through science laboratories.
- In a unique partnership, the project is partially funded by Niagara Mohawk Power, the local utility.

ing efficiency is achieved by insulating the air conditioner ducts, especially in the attic; buying an energy-efficient air conditioner; and shading the south and west sides of a house with trees. Window shades and awnings can also help reduce the heat a building gains from its environment. Ceiling fans can supplement air conditioners by making a room feel comfortable at a higher thermostat setting.

Other home improvements that result in substantial energy savings include replacing incandescent bulbs with energy-efficient compact fluorescent light bulbs, wrapping insulation around water heaters and water pipes, and installing low-flow shower heads to reduce the amount of hot water used.

How does a homeowner learn which improvements will result in the most substantial energy savings? In addition to reading the many articles on energy efficiency that appear in newspapers and magazines, a good way to learn about your home is to have an energy audit done. Most local utility companies can send an energy expert to your home to perform an audit for little or no charge. The audit will determine the total energy consumed during the heating season and where most heat is lost in your home (through the ceiling, floors, walls, or windows). Based on this assessment, the energy expert will then make recommendations about how you can reduce your heating bills.

Electric Power Companies and Energy Efficiency

Recent changes in the regulations governing electric utilities have enabled utilities to make more money by generating *less* electricity. Such programs provide incentives to save energy and thereby reduce power plant emissions that contribute to environmental problems.

Traditionally, to meet future power needs, electric utilities planned to build new power plants or purchase additional power from alternative sources. Now they can avoid these massive expenses by **demand-side management**, in which they help electricity consumers save energy. Some utilities support energy conservation and efficiency by offering cash awards to consumers who install energy-efficient technologies. The reward might be a $100 rebate to a homeowner who purchases an energy-efficient refrigerator, or thousands of dollars to a large industry that overhauls its entire production to increase energy efficiency. At least one power company offers a 10 percent reduction on monthly bills for 10 years to those whose homes qualify as being energy-efficient.

Some utilities also give customers energy-efficient compact fluorescent light bulbs, air conditioners, or other appliances. They then charge slightly higher rates or a small leasing fee, but the greater efficiency results in savings for both the utility company and the consumer. The utility company makes more money from selling less electricity because rates are slightly higher, and the consumer saves because less energy is used by the efficient light bulbs or appliances.

What About Energy Conservation?

Simple measures such as lowering your thermostat during the winter and raising it during the summer, turning off lights when you leave a room, and driving more slowly result in small energy (and cost!) savings. You also contribute to energy conservation by making use of carpools or public transportation. The cumulative effect of many people taking similar measures is substantial.

Energy conservation and efficiency are sound ideas for all of us. Energy saved today will be available for our grandchildren. The energy we save now will help to slow down climate change and environmental degradation so that our consumption will not become an overwhelming burden on future generations. The energy we save now will give us additional time to develop and improve alternative energy sources (see You Can Make a Difference: Using Your Government).

YOU CAN MAKE A DIFFERENCE

Using Your Government

One of the most important ways you can help decrease energy consumption is to become an involved citizen. Let your elected officials know that you support legislative measures that result in energy conservation. Governments generally respond to energy shortages by making it easier to find and exploit additional sources of energy, rather than by conserving energy. Do not let this happen.

1. Write to your senators and representatives. Ask them to pass laws requiring all energy-consuming technologies to be rated for energy efficiency. The ratings should be visibly displayed, whether the equipment or appliance is for domestic, commercial, or industrial purposes.
2. Encourage elected officials to support tax credits for homeowners and businesses that make energy-conserving improvements in their residences or workplaces.
3. Tell officials that you want them to support an increase in available funding for research into renewable energy technologies such as solar and wind power.
4. Encourage state legislators to pressure public utilities to spend more of their research time and budget on renewable energy technology.
5. Contact the two major legislative branches of the U.S. Congress and the President on the World Wide Web. The E-mail address for the U.S. House of Representatives is www.house.gov; for the U.S. Senate, it is www.senate.gov; the President's E-mail address is www.whitehouse.gov.

Go ahead and make yourself heard. Be a squeaky wheel. After all, you have the power of your vote when election day rolls around. You have nothing to lose, but an elected official who turns a deaf ear to energy concerns could easily lose at the polls.

SUMMARY

1. Alternative energy sources are increasingly being examined with our future energy needs in mind, because they are renewable and generally cause less environmental impact than fossil fuels or nuclear energy. However, using alternative energy sources is generally feasible only in restricted locations.

2. Many forms of renewable energy are dispersed and therefore tend to be inefficiently used. Their inefficiency is partly a result of the need to collect or concentrate the energy so that it is useful.

3. Solar energy, the radiant energy derived from the sun, can be used directly, as heat, or indirectly, as for example, wind, which results from the sun heating the atmosphere.

4. Direct solar energy can be used either actively or passively to heat buildings and water. Solar thermal electric generation produces electricity by concentrating sunlight onto a fluid. Photovoltaic cells convert solar energy directly into electricity.

5. Forms of indirect solar energy include biomass (wood, agricultural wastes, and fast-growing plants), wind, and hydropower (the energy of flowing water). Biomass can be burned directly to produce heat or electricity or converted to gas or liquid fuels. Biomass is already being used for energy on a large scale, particularly in developing nations.

6. Wind and hydropower have potential as indirect sources of solar energy in certain areas. Ocean thermal energy conversion makes use of the temperature gradient of the ocean.

7. Geothermal energy can be obtained from hot rocks near the Earth's surface. The established technology for extracting geothermal energy from hot rocks involves steam or hot water.

8. The most promising alternative energy "sources" are energy conservation and improved energy efficiency. Using technology to increase energy efficiency results in energy savings that conserve our conventional fuel supplies. Cogeneration, the production of two forms of energy (usually, electricity and useful heat) from the same fuel, is an example of increased energy efficiency.

THINKING ABOUT THE ENVIRONMENT

1. Solar energy in all its forms could be considered an indirect form of nuclear energy. Why?

2. Explain the following statement: Unlike fossil fuels, solar energy is not resource-limited but is technology-limited.

3. Japan wishes to make use of solar power, but it does not have extensive tracts of land for building large solar power plants. Which solar technology do you think is best suited to Japan's needs? Why?

4. Give an example of how one or more of the alternative energy sources discussed in this chapter could have a negative effect on each of the following aspects of ecosystems: soil preservation, natural water flow, production of foods used by natural plant and animal populations, and maintenance of the diversity of organisms found in an area.

5. Biomass is considered an example of indirect solar energy because it is the result of photosynthesis. Given that plants are the organisms that photosynthesize, why are animal wastes also considered biomass?

6. Why might some environmentalists support the development of wind power while others oppose it?

7. Why is it easier to obtain energy from a small river with a steep grade than from the vast ocean currents?

8. One advantage of the various forms of renewable energy, such as solar thermal and wind energy, is that they cause no net increase in carbon dioxide. Is this true for biomass? Why or why not?

9. Explain how energy conservation and efficiency are major "sources" of energy.

10. Evaluate which forms of energy other than fossil fuels and nuclear power have the greatest potential where you live.

11. List energy conservation measures that you could adopt for each of the following aspects of your life: washing laundry, lighting, bathing, buying a car, cooking, driving a car.

* **12.** Examine the following data on global wind power generating capacity from 1990 to 1995. Calculate the percent increase over the preceding year and place your answers in Table A.

Table A
Global Wind Power Generating Capacity, 1990 to 1995

Year	Capacity (Megawatts)	Percent Increase over Preceding Year
1990	1930	11.6
1991	2170	
1992	2510	
1993	2990	
1994	3680	
1995 (preliminary)	4880	

RESEARCH PROJECTS

Contact Shell International and get a record of its investment in alternative energy. What percentage of its funding for future energy supplies does this investment represent? In your opinion, is it sufficient? Explain.

Obtain a copy of the article by Eric Hirst (see Suggested Reading). After reading about utility investment in demand-side management programs, contact your local electric utility and investigate their participation in such programs.

On-line materials relating to this chapter are on the World Wide Web at **http://www.saunderscollege.com/lifesci/environment2** (select Chapter 12 from the Table of Contents).

SUGGESTED READING

Anderson, I. "Sunny Days for Solar Power," *New Scientist* Vol. 143, No. 1932, 2 July 1994. Two solar technologies, solar thermal and photovoltaics, promise to be commercially viable in the near future.

Branch, M. A. "Smart Buildings," *Earthwatch*, July–August 1993. Sustainable architecture encompasses greater energy efficiency, the use of alternative energy forms, the use of recycled building materials, good indoor air quality, and a host of other ideas.

Hirst, E. "A Bright Future: Energy Efficiency Programs at Electric Utilities," *Environment* Vol. 36, No. 9, November 1994. An increasingly important activity of electric utilities is providing services that help customers reduce their electric bills.

Hoagland, W. "Solar Energy," *Scientific American*, September 1995. Using solar power to produce electricity and nonpolluting fuels shows great promise.

Mestel, R. "White Paint on a Hot Tin Roof," *New Scientist*, March 25, 1995. Discusses simple, relatively inexpensive measures that would result in billions in energy savings in the United States.

Pimentel, D., et al. "Renewable Energy: Economic and Environmental Issues," *BioScience* Vol. 44, No. 8, September 1994. Examines the pros and cons of different types of alternative energy.

Reader, J. "A Boxful of Miracles," *Earthwatch*, March–April 1994. How solar ovens are changing people's lives in Kenya and other African nations.

Roodman, D. M. "The Obsolescent Incandescent," *World Watch* Vol. 6, No. 3, May–June 1993. Discusses the advantages of compact fluorescent light bulbs.

Wilbanks, T. J. "Improving Energy Efficiency," *Environment*, Vol. 36, No. 9, November 1994. A comprehensive analysis of energy efficiency improvements that benefit the environment without requiring sacrifices in the availability or cost of energy services.

Sockeye salmon in a freshwater stream in Alaska. This photograph shows some of our most important resources: Earth's atmosphere, forests, water, and organisms. *(Kennan Ward/The Stock Market)*

Our Precious Resources

Natural Systems Agriculture

Jon K. Piper has been the ecologist at The Land Institute since 1985, where he teaches classes and directs the program in Natural Systems Agriculture. He holds a degree in biology from Bates College and a Ph.D. in botany from Washington State University. Piper has received grants from the National Science Foundation, the Eppley Foundation for Research, and the Charles A. Lindbergh Foundation. He has published over 25 scientific articles and co-authored Farming in Nature's Image with Judith Soule. ◄

JON K. PIPER

What is Natural Systems Agriculture? What elements does it draw from ecology and what does it take from agriculture?

Research in Natural Systems Agriculture represents a new synthesis between ecology and agriculture, with a goal to develop grain agricultural systems that function like natural ecosystems. Because the practice of agriculture has changed the face of the Earth more than any other human activity, the time is ripe for such a synthesis. This research is attracting increasing attention from both ecologists and agronomists, who have previously worked in isolation, or even at cross purposes.

Since 1976, The Land Institute has been investigating the feasibility of a grain agriculture that protects topsoil and does not rely on environmentally hazardous fertilizers, herbicides, and insecticides. The long-term goal is to find combinations of perennial plants that, when planted together, will provide the restorative properties of a year-round cover as well as yielding significant amounts of edible grain. Grain-producing mixtures of perennial grasses, legumes, and sunflowers would mimic the vegetation structure and ecological function of native grassland ecosystems in some fundamental ways.

In what ways are the systems created by Natural Systems Researchers like naturally occurring ecosystems?

Natural systems, such as grasslands, have two important features that can contribute to sustainability. First, such systems are composed primarily of perennial plants, which protect and improve soil quality over time, maintain soil biodiversity, promote nutrient cycling, reduce weed growth, and support populations of beneficial insects. Second, natural systems comprise diverse arrays of plant species. Benefits of biodiversity include nitrogen supplied by legumes, management of insects that prey on crops, reduction of some plant diseases, and ecosystem resilience. The research in Natural Systems Agriculture aims to provide the ecosystem-level benefits of a diverse, prairie-like vegetative structure using mixtures of high-yielding edible grains.

How will Natural Systems Agriculture help to protect the environment?

During the last few decades, about one third of the world's arable land area has been lost through soil erosion. Although modern agricultural methods are highly productive, they are sustainable only as long as topsoil is intact and fossil fuel supplies are affordable. With topsoil on roughly 90 percent of U.S. cropland being lost faster than it is being formed, we need to rethink agriculture in terms of preserving this limited "ecological capital." Natural Systems Agriculture will help prevent soil erosion by providing a year round cover for cropland, protecting topsoil from short periods of heavy rain or high wind, which is when unprotected land loses most of its topsoil.

There are two types of problems associated with soil erosion, on-site and off-site effects. The on-site effects of erosion include reduced productivity and increased input costs to farmers to compensate for lowered soil fertility and water-holding capacity. It has been estimated that 10 percent of all energy used in U.S. agriculture is expended simply to offset the losses of soil nutrients and water caused by erosion.

Soil loss from cropland has severe consequences off-site as well. The major effect is surface water pollution due to sediment deposition, pesticide residues, and nutrient loading from chemical fertilizers and animal manure. Sediment load both reduces surface water quality and increases the likelihood of flooding, while nitrates from fertilizer and pesticides in drinking water pose a human health risk. If estimated off-site and on-site costs of soil erosion are combined, the total cost to the U.S. may be in the billions.

Several promising plant species have been identified as good candidates for Natural Systems Agriculture. These species include wild perennials that may undergo selection for improved seed yield, as well as annual grain crops into which the perennial habit could be introduced.

How can we "introduce" the perennial habit into annual crops or screen wild perennials for improved seed yields?

Two recent insights are particularly relevant. The first is the finding that a relatively small number of genes accounts for a large proportion of the observed variation between wild and domesticated plants. John Doebley has shown that genes located in five particular stretches of DNA account for most of the considerable differences in shape between corn (maize) and the Central American grass teosinte. This suggests that a relatively few mutations, with large effects, may be responsible for most of the differences between domesticated and wild-type plants.

> **Research in Natural Systems Agriculture represents a new synthesis between ecology and agriculture, with a goal to develop grain agricultural systems that function like natural ecosystems.**
>
> *–Jon K. Piper*

The second insight concerns recent discoveries of the high level of similarity in the genetic constitutions of various grasses as well as the fact that the genes coding for such traits as seed size, shattering, and daylength sensitivity are positioned very closely to one another. Because these genes are so close to each other on the chromosome, chances are good that as plant species evolved these genes moved together from species to evolving species. This raises the possibility that genes in different species, such as some cereals and wild grasses, may be identical.

If some of these genes are identical, the daunting time frame for developing perennial grains may be shortened considerably because these genes could then be moved easily and quickly through breeding or genetic engineering. This would allow us to bring together the most agriculturally desirable traits from wild perennials and domesticated crops in a shorter time frame than would otherwise be necessary.

What is the greatest challenge that faces the implementation of Natural Systems Agriculture?

Although interest in the potential and feasibility of perennial grains has increased substantially in recent years, on-farm use of perennial grain systems is still several years down the road. A common objection raised regarding the development of perennial grains is the assumption that a plant cannot simultaneously invest in both large/numerous seeds and the vegetative organs required for the perennial growth habit. This assumption has been challenged by several studies that have shown that, within the ranges investigated, there are no strict "trade-offs" between increased seed yield, vegetative growth, and the likelihood of future reproduction.

Another objection concerns the expected long-term time frame needed to develop high-yielding perennial grains. As mentioned above, if the preliminary research in genetics pans out, this time frame could be significantly shortened.

How do the yields of crops grown with Natural Systems Agriculture compare to yields of crops grown more traditionally?

Under favorable growing conditions, high yields of some perennial grasses can range from 1500 to 2000 kilograms per hectare (kg/ha). Maximum seed yields ranging from 1720 to 2090 kg/ha have been recorded for some perennial legumes and sunflowers in non-irrigated, unfertilized plots at The Land Institute. These yields are comparable to the benchmark yield of Kansas wheat (approximately 2000 kg/ha).

What other projects is The Land Institute working on?

There are two other major programs underway at The Land Institute. The Sunshine Farm Project is exploring the possibilities of farming without fossil fuels, and calculating a less damaging ecological cost for food production. We are raising livestock and growing conventional crops without pesticides, nitrogen fertilizer, or irrigation. We rely instead on innovative management techniques, biological process, and renewable energy sources.

Our other major research efforts are happening at Matfield Green, a small community nestled in the scenic Flint Hills of Kansas. Here we are exploring such issues as why small rural communities are losing their young people to cities, how this depopulation is related to land use, and how rural schools can be catalysts for change. Our work is geared toward developing conceptual tools that will help communities minimize dependence on non-renewable resources and maximize possibilities for cultural innovation and adaptation.

▶ E-mail: **theland@igc.apc.org**

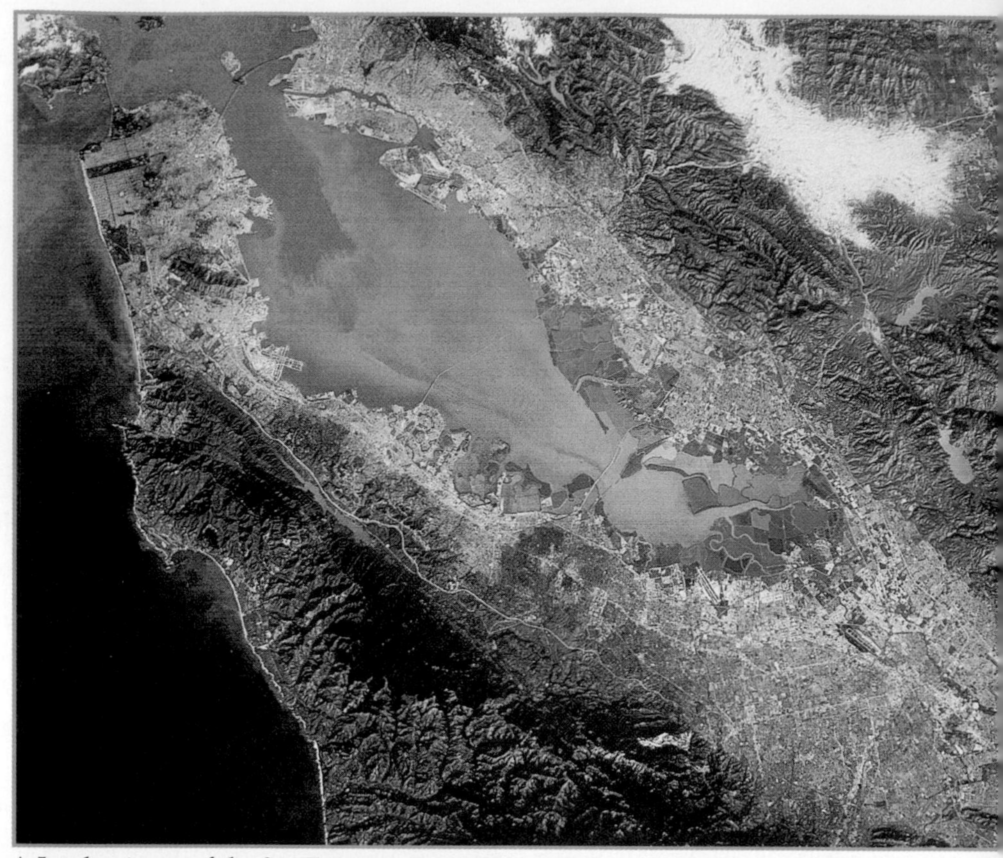

A Landsat image of the San Francisco Bay area. (Earth Satellite Corporation/Science Photo Library)

Water: A Fragile Resource

The San Francisco Bay and its delta, a 4144-square-kilometer (1600-square-mile) estuary that drains 45 percent of California's land area, is both environmentally and economically important. Home to 120 species of fishes, it is the largest, most productive U.S. estuary on the Pacific coast. It is also highly developed. The cities of San Francisco (the peninsula on the upper left), San Jose (bottom right), and Oakland (upper right, opposite San Francisco) are evident in this Landsat image. The bay and its delta, which extends eastward almost to Sacramento, provide drinking water for two-thirds of the state—some 20 million people—and irrigation water for farmland that supplies 45 percent of the nation's vegetables and fruits.

Years of water diversion for municipal and agricultural uses have harmed the delta, because when less fresh water flows through the delta toward the Pacific Ocean, more salt water intrudes from the ocean. As a result, some municipalities have experienced salty tapwater—a violation of the Clean Water Act—and populations of certain fish species have plummeted. When several species, notably the delta smelt and chinook salmon, were declared endangered, the diversion of water became a violation of the Endangered Species Act.

Various federal and state agencies and interest groups such as water users and environmentalists have argued for years over the extent of the bay's environmental problems and who is to blame. Developing a viable solution was impossible in such a confrontational atmosphere.

In 1994, however, compromise and cooperation among the various groups during a year of intense negotiations initiated by the Clinton Administration resulted in a historic agreement, called the Bay-Delta Program, to protect the San Francisco Bay and its delta. Since 1994, the Bay-Delta Program has successfully specified the environmental problems in San Francisco Bay and its delta and devised three alternate solutions to these problems. The environmental impacts of these three solutions are under review by the public, as well as by federal and state agencies. A final environmental impact statement that recommends a single comprehensive solution should be released in 1998. The project is expected to begin implementation in 1999 and take several decades to complete. ◀

THE IMPORTANCE OF WATER

The view of planet Earth from outer space reveals that it is different from other planets in the Solar System. Earth is a predominantly blue planet because of the water that covers three-fourths of its surface. Water has a tremendous effect on our planet: it helps shape the continents, it moderates our climate, and it allows organisms to survive.

Life on planet Earth would be impossible without water. All life forms, from simple bacteria to complex multicellular plants and animals, contain water. Humans are composed of approximately 70 percent water by weight, and we depend on water for our survival as well as for our convenience: we drink it, cook with it, wash with it, travel on it, and use an enormous amount of it for agriculture, manufacturing, mining, energy production, and waste disposal.

Although Earth has plenty of water, it is distributed unevenly, and serious water supply problems exist. In regions where fresh water is in short supply, such as deserts, obtaining it is critically important. Because the use of water for one purpose decreases the amount available for other purposes, serious conflicts often arise over how water should be used. Even regions with readily available fresh water are not without problems, however, and maintaining the quality and quantity of water is a top priority.

Worldwide, freshwater use is increasing, in part because the human population is expanding and in part because, on the average, each person is using more water. A 1996 report published in the journal *Science* estimates that humans are now using 54 percent of all fresh water flowing in rivers and streams. If the human population continues to increase as projected, and if the per-capita water use continues to increase as it has during the past several decades, worldwide freshwater supplies may run out during the next century.

To meet the growing need for water, we try to augment our supply by building dams to create **reservoirs** (artificial lakes in which water is stored for later use) and by diverting river water. In many areas, the quantity of water is not as critical as its quality, and steps must be taken to ensure a supply of clean water. All of these efforts to obtain and maintain a steady supply of clean water involve considerable expense.

This chapter examines some of the ecological processes and human activities that affect the availability of water. Water quality, including water pollution, is such a significant issue that it is covered separately, in Chapter 21.

Properties of Water

Water is a molecule, H_2O, consisting of two atoms of hydrogen and one atom of oxygen, that can exist in any of three forms: solid (ice), liquid, and vapor (water vapor or steam). Water is a polar molecule; that is, one end of the molecule has a positive electrical charge, and the other end has a negative charge (Figure 13–1). The negative (oxygen) end of one water molecule is attracted to the positive (hydrogen) end of another water molecule, forming a **hydrogen bond** between the two molecules. Hydrogen bonds are the basis for many of water's physical properties including its high melting/freezing point (0°C, 32°F) and high boiling point (100°C, 212°F). Because most of the Earth has a temperature between 0°C and 100°C, most water exists as the liquid on which organisms depend.

Water absorbs a great deal of solar heat without its temperature rising substantially. It is this high heat capacity that allows the ocean to have a moderating influence on climate, particularly along coastal areas. Another consequence of water's high heat capacity is that the ocean does not experience the wide temperature fluctuations that are common on land.

Water must absorb a lot of heat before it **vaporizes**, or changes from liquid to vapor.[1] When it does evaporate, it carries the heat (called its heat of vaporization) with it into the atmosphere. Thus, evaporating water has a cooling effect. That is why your body is cooled when perspiration evaporates from your skin.

[1]2250 joules (540 calories) of energy are required to convert one gram of water from a liquid to a vapor.

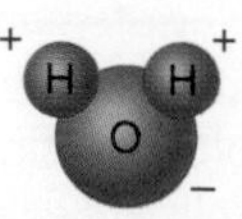

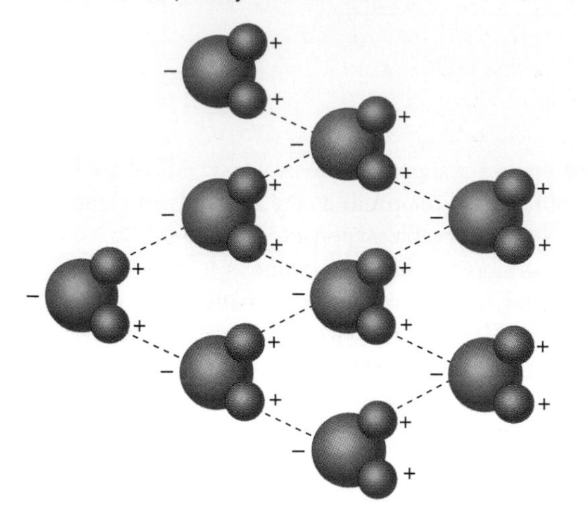

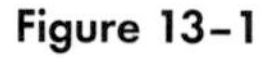
Figure 13–1

Many of the unusual properties of water are a result of its chemistry. (a) Water, which consists of two hydrogen atoms and one oxygen atom, is a polar molecule with positively and negatively charged areas. (b) The polarity of water molecules causes hydrogen bonds (represented by dashed lines) to form between the positive area of one water molecule and the negative areas of others. Each water molecule can form up to four hydrogen bonds with other water molecules.

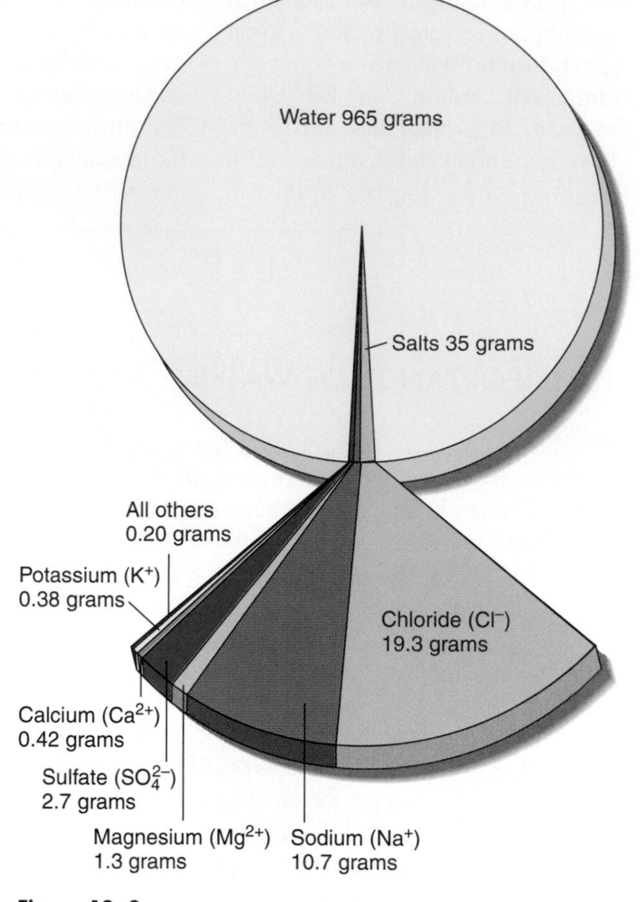

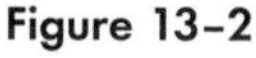
Figure 13–2

The chemical composition of 1 kilogram (2.2 pounds) of seawater. Seawater contains a variety of dissolved salts, which are present as ions.

Water is sometimes called the "universal solvent," and although this is an exaggeration, many materials do dissolve in water. In nature, water is never completely pure, because it contains dissolved gases from the atmosphere and dissolved mineral salts from the land. Seawater, for example, contains a variety of dissolved salts including sodium chloride, magnesium chloride, magnesium sulfate, calcium sulfate, and potassium chloride (Figure 13–2). Water's dissolving ability has a major drawback, however: many of the substances that dissolve in water also pollute it.

Water partially obeys the general physical rule that heat expands and cold contracts. As water cools, it contracts and becomes denser until it reaches 4°C (39°F), the temperature at which it is densest. When the temperature of water falls *below* 4°C, however, it becomes less dense. Thus, ice (at 0°C) floats on denser, slightly warmer liquid water. Because of this, water freezes from the top down rather than from the bottom up, so that aquatic organisms can survive beneath a frozen surface.

The Hydrologic Cycle and Our Supply of Fresh Water

Water continuously circulates through the abiotic environment, from the ocean to the atmosphere to the land and back to the ocean, in a complex cycle known as the **hydrologic cycle** (see Chapter 5). The result is a balance among water in the ocean, water on the land, and water in the atmosphere. The cycle thus continually renews the supply of fresh water on land, which is essential to terrestrial organisms.

Approximately 97 percent of the Earth's water is in the ocean and contains a high amount of dissolved salts (Figure 13–3). Seawater is too salty for human con-

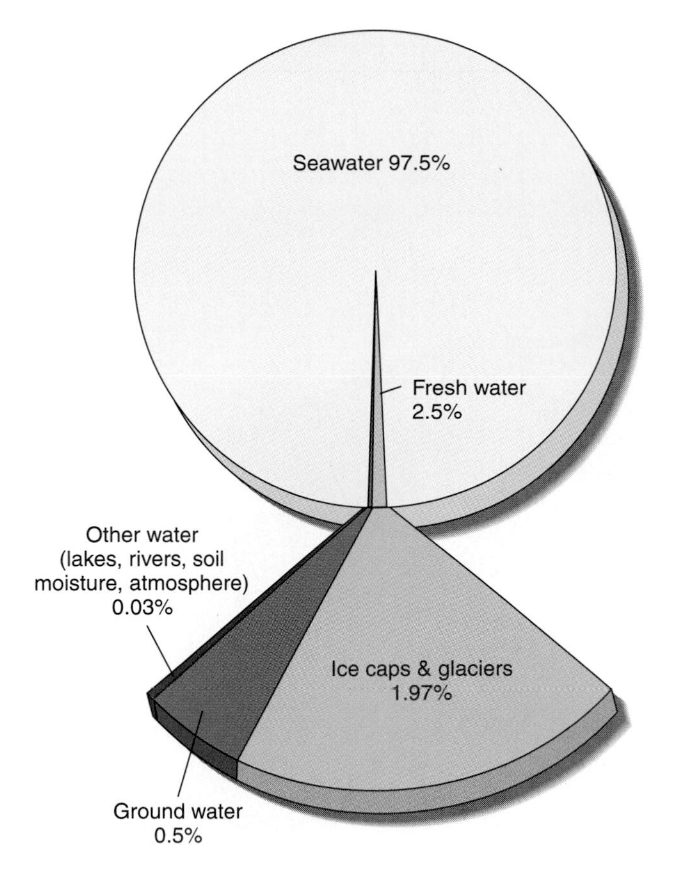

Figure 13–3

Although three-fourths of the Earth's surface is covered with water, less than 1 percent is available for humans. Most water is salty, frozen, or inaccessible in the soil and atmosphere.

sumption and for most other uses. For example, if you watered your garden with seawater, your plants would die. Most fresh water is unavailable for easy human consumption because it is frozen as polar or glacial ice or is in the atmosphere or soil. Lakes, creeks, streams, rivers, and groundwater account for only a small portion of the Earth's fresh water.

Surface water is fresh water found on the Earth's surface in streams and rivers, lakes, ponds, reservoirs, and **wetlands**—areas of land that are covered with water for at least part of the year. Surface waters are replenished by the **runoff** of precipitation from the land and are therefore considered a renewable, although finite, resource. A **drainage basin**, or **watershed**, is the area of land that is drained by a river.

The Earth contains underground formations that collect and store water in the ground. This water originates as rain or melting snow that seeps into the soil and finds its way down through cracks and spaces in rock until it is stopped by an impenetrable layer; there it accumulates as **groundwater**. Groundwater flows slowly—typically covering distances of several millimeters to a few meters per day. Eventually it is discharged into rivers, wetlands, springs, or the ocean.

The underground caverns and porous layers of rock in which groundwater is stored are called **aquifers** (Figure 13–4), and they can be either unconfined or confined. Those whose contents are replaced by surface water directly above them are **unconfined aquifers**; in other words, the layers of rock above an unconfined aquifer are porous. The upper limit of an unconfined aquifer, below which the ground is saturated with water, is the **water table**.

A **confined aquifer**, also called an **artesian aquifer**, is a groundwater storage area between impermeable layers of rock. The water in a confined aquifer is trapped and often under pressure, and its recharge area (the land from which water percolates to replace groundwater) may be hundreds of miles away. Most groundwater is considered a nonrenewable resource, because it has taken hundreds or even thousands of years to accumulate, and usually only a small portion of it is replaced each year by percolation of precipitation. The recharge of confined aquifers is particularly slow.

HOW WE USE WATER

Water consumption varies among countries. In some countries, acute water shortages limit water use per person to several gallons per day. In developed countries, per-capita water use may be as high as several *hundred* gallons per day, an amount that encompasses agricultural and industrial uses as well as direct individual consumption.

The greatest user of water worldwide is agriculture—for irrigation. Irrigation accounts for 68.3 percent of the world's total water consumption, industry for 23.1 percent, and domestic and municipal use for only 8.6 percent (Table 13–1).

Irrigation

Arid lands, or deserts, are fragile ecosystems in which plant growth is limited by lack of precipitation; **semiarid** lands receive more precipitation than deserts but are subject to frequent and prolonged droughts (Figure 5–17 shows where arid and semiarid lands occur worldwide). Farmers can increase the agricultural productivity of arid and semiarid lands with irrigation. Almost any crop can be grown in the desert if enough water is supplied to the soil.

Irrigation of arid and semiarid lands has become increasingly important worldwide in efforts to produce

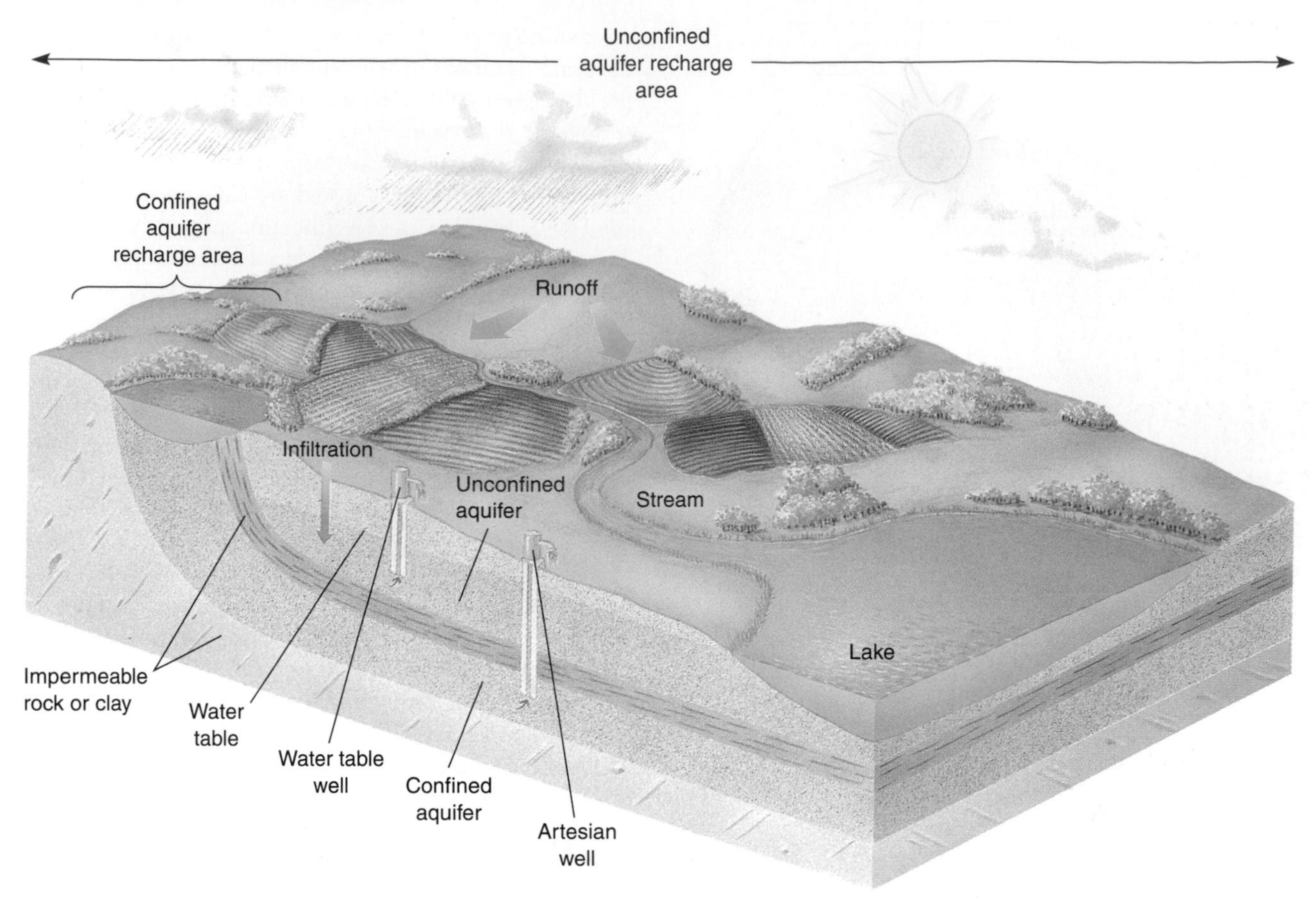

Figure 13–4

Groundwater. Excess surface water seeps downward through soil and porous rock layers until it reaches impermeable rock or clay. Groundwater that is recharged by surface water directly above is known as an unconfined aquifer. In a second type of aquifer, known as a confined aquifer, groundwater is stored between two impermeable layers and is often under pressure. Artesian wells, which produce water from confined aquifers, often do not require pumping because of this pressure.

Table 13–1
Water Usage (in cubic kilometers) in 1995

Region	Irrigation	Industry	Domestic/ Municipal
Africa	127.7	7.3	10.2
Asia*	1388.8	147.0	98.0
Australia-Oceania	5.7	0.3	10.7
Europe	141.1	250.4	63.7
North and Central America	298.1	255.5	54.8
South America	62.7	24.4	19.1
World total	2024.1	684.9	256.5
World total, as percent	68.3	23.1	8.6

*Latest data available are from 1987.

enough food for burgeoning populations (Figure 13–5). Since 1955 the amount of irrigated land has tripled to 248 million hectares (613 million acres) worldwide in 1993. Asia has more agricultural land under irrigation than do other regions, with China, India, and Pakistan accounting for most of it. It is projected that water use for irrigation will continue to increase in the 21st century, but at a slower rate than in the last half of this century.

WATER RESOURCE PROBLEMS

Water resource problems fall into three categories: too much, too little, and poor quality/contamination (Chapter 21 addresses the third category). Floods and droughts are part of natural climate variations and cannot be prevented. Human activities sometimes exacerbate them, however, and humans often court disaster when they make environmentally unsound decisions, such as building in an area that is prone to flooding.

Figure 13–5

Agricultural use of water. Center-pivot irrigation produces massive green circles. Each circular shape is caused by a long irrigation pipe that extends along the radius from the circle's center to its edge and slowly rotates around the circle, spraying the crop. (USDA/Natural Resources Conservation Service)

ENVIROBRIEF

WHERE THE WATER GOES

As one of the world's biggest water consumers, the United States in 1990 used a whopping 339 billion gallons of fresh water daily from lakes, streams, and underground sources. Still, U.S. water use that year was down almost 10 percent since 1980, due in part to better water conservation, new water-saving technologies, and higher water prices.

- Agricultural water use is mainly for irrigation, with a small amount attributed to farm animals and aquaculture. Recent water shortages and higher pumping costs have led to more efficient irrigation practices.
- Electric power production requires far more water each year than any other industrial purpose, but most of the water is available for reuse. Production of one kilowatt of electricity requires 4000 gallons of water.
- Although U.S. industrial output grew fourfold between 1950 and 1990, water use by U.S. industries declined by 19 percent. Industry in 1990 required more than 36 billion gallons of water a day, including 300 million gallons to produce a day's supply of newsprint, and 8.3 billion gallons to supply commercial users such as offices, restaurants, schools, and military bases.
- Domestic water use in 1990 reached a high of 25 billion gallons each day, but per-capita use had declined slightly since 1985, largely because of changes brought about by new building codes and water conservation efforts.

Too Much Water

Many ancient civilizations—ancient Egypt, for example—developed near rivers that periodically spilled over, inundating the surrounding land with water. When the water receded, a thin layer of sediment that was rich in organic matter remained and enriched the soil. These civilizations flourished partly because of their agricultural productivity, which in turn was the result of floods replenishing the soil's nutrients.

Modern floods can cause widespread destruction of property and sometimes loss of life (see Focus On: The Floods of 1993). Today's floods are more disastrous in terms of property loss than those of the past, but not because they involve more water. Human activities, such as the removal of water-absorbing plant cover from the soil and the construction of buildings on **flood plains** (areas bordering a river that are subject to flooding), increase the likelihood of both floods and flood damage.

Forests, particularly on hillsides and mountains, provide nearby lowlands with some protection from floods by trapping and absorbing precipitation. When woodlands are cut down, particularly when they are clearcut, the area cannot hold water nearly as well. Heavy rainfall then results in rapid runoff from the exposed, barren hillsides. This not only causes soil erosion, but also puts lowland areas at extreme risk of flooding.

When a natural area—that is, an area undisturbed by humans—is inundated with heavy precipitation, the plant-protected soil absorbs much of the excess water. What the soil cannot absorb runs off into the river, which may spill over its banks into the flood plain. However, because rivers meander, the flow is slowed, and the swollen waters rarely cause significant damage to the surrounding area. (See Figure 6–12 for a diagram of a typical river, including its flood plain.)

FOCUS ON

The Floods of 1993

The Mississippi River, which drains 31 states and 2 Canadian provinces, is one of the world's largest rivers. During the summer of 1993, the Mississippi and its tributaries flooded, spreading over 9.3 million hectares (23 million acres) of flood plains and engulfing farms and towns in nine Midwestern states (see figure). Fifty people were killed, and damage to property was estimated in excess of $12 billion. More than 70,000 people lost their homes, and 8.7 million acres of farmland were damaged. Large quantities of pesticides that washed off the fields were carried to the Gulf of Mexico where it was feared they would harm coastal ecosystems. The zebra mussel, an exotic species that has caused havoc wherever it has spread (see Chapter 1), was carried by floodwaters to new habitats.

The flood, considered by many experts to be the worst in U.S. history, was caused by above-average precipitation during the first six months of 1993, followed by prolonged rain in the summer. However, the damage was exacerbated by three practices: draining wetlands; building on flood plains; and constructing levees to hold back floodwaters.

For the past hundred years or so, people in the Midwest drained wetlands to produce farmland or land on which to build homes. The ability of wetlands, which act like a sponge, to moderate floods was simply unrecognized. It is probably no coincidence that Missouri, Illinois, and Iowa, the three states damaged the most by the 1993 floods, have less than 15 percent of their original wetlands.

Building on a flood plain increases the damage when floods occur. If a flood plain is covered by forest or other natural vegetation, the floodwaters spread over the land slowly, and the land absorbs much of the water. Because land in a developed flood plain is less able to absorb excess water, the water spreads more rapidly and extensively.

Hundreds of levees have been built along the Mississippi and its tributaries to hold floodwaters back from the flood plain. Although levees may save lives and property where they are built, they cause floodwaters upstream to surge, damaging farms and towns that are less protected.

FOLLOW–UP

The 1993 floods have rekindled an old debate: should the government rebuild damaged levees, or should it help relocate people away from the flood plain and restore the land to a more natural state so that it can better face the next flood?

Valmeyer and Hartsburg, two towns affected by the flooding, represent the two different strategies. The 900 residents of Valmeyer, Illinois, voted to use federal funds to relocate their town on a nearby hill. The town's old site on the flood plain will become a park and wetland. In contrast, the 131 residents of Hartsburg, Missouri, decided to stay in their original location on the flood plain. The levees that were built to protect the town and that failed during the 1993 floods have been rebuilt, and homes have been repaired.

The Flood Plain Management Task Force, assembled by the White House to reassess national flood policy in the aftermath of the floods, released its report in 1994. According to the report, people need to make wiser use of flood plains and rely less on levees to control flooding. (Two-thirds of the levees along the Mississippi River and its tributaries were damaged or destroyed by the floods of 1993.) Some flood plains need to be restored to their natural condition, which means that the towns built on them need to be moved to higher ground.

The authors of the report concluded that the presence of more wetlands would not have prevented flooding of this magnitude because the wetlands would have become saturated with water. The presence of more wetlands in the affected area would have indirectly reduced property damage from the floods of 1993, however, by keeping people and their property off the flood plain. This observation is significant because, according to a 1987 study, an additional 2 percent of the nation's flood plains are developed each year.

When an area is developed for human use, much of the water-absorbing plant cover is removed. Buildings and paved roads do not absorb water, so runoff, usually in the form of storm sewer runoff, is significantly greater (Figure 13–6). People who build homes or businesses on the flood plain of a river will most likely experience flooding at some point.

It is easier and more economical to ban or restrict development in a flood plain than to build on it and then try to prevent flooding by building such structures as retaining walls and levees (embankments). Increasingly, local governments, both in the United States and in the rest of the world, now zone flood plains to curtail development.

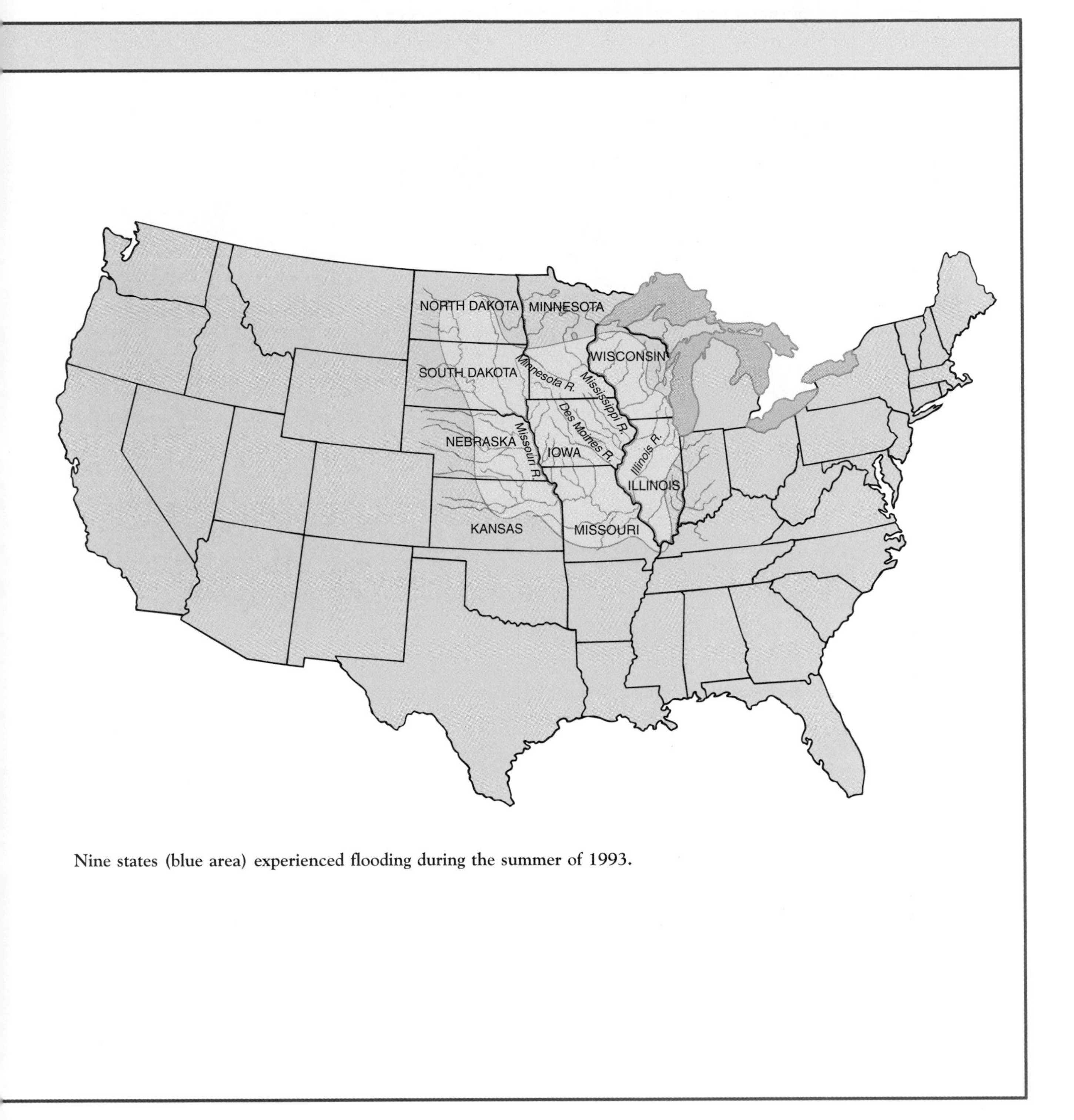

Nine states (blue area) experienced flooding during the summer of 1993.

Too Little Water

Forty percent of the world's population lives in arid or semiarid lands, primarily in Asia and Africa. These people spend substantial amounts of time and effort obtaining water. Each day they may have to walk many miles to a stream or river and carry back the heavy water.

Overpopulation in arid and semiarid regions intensifies the problem of water shortage. The immediate need for food prompts people to remove natural plant cover in order to grow crops on marginal land, that is, land subject to frequent drought and subsequent crop losses. Their livestock overgraze the small amount of plant cover in natural pastures. As a result of the lack of plants, the soil

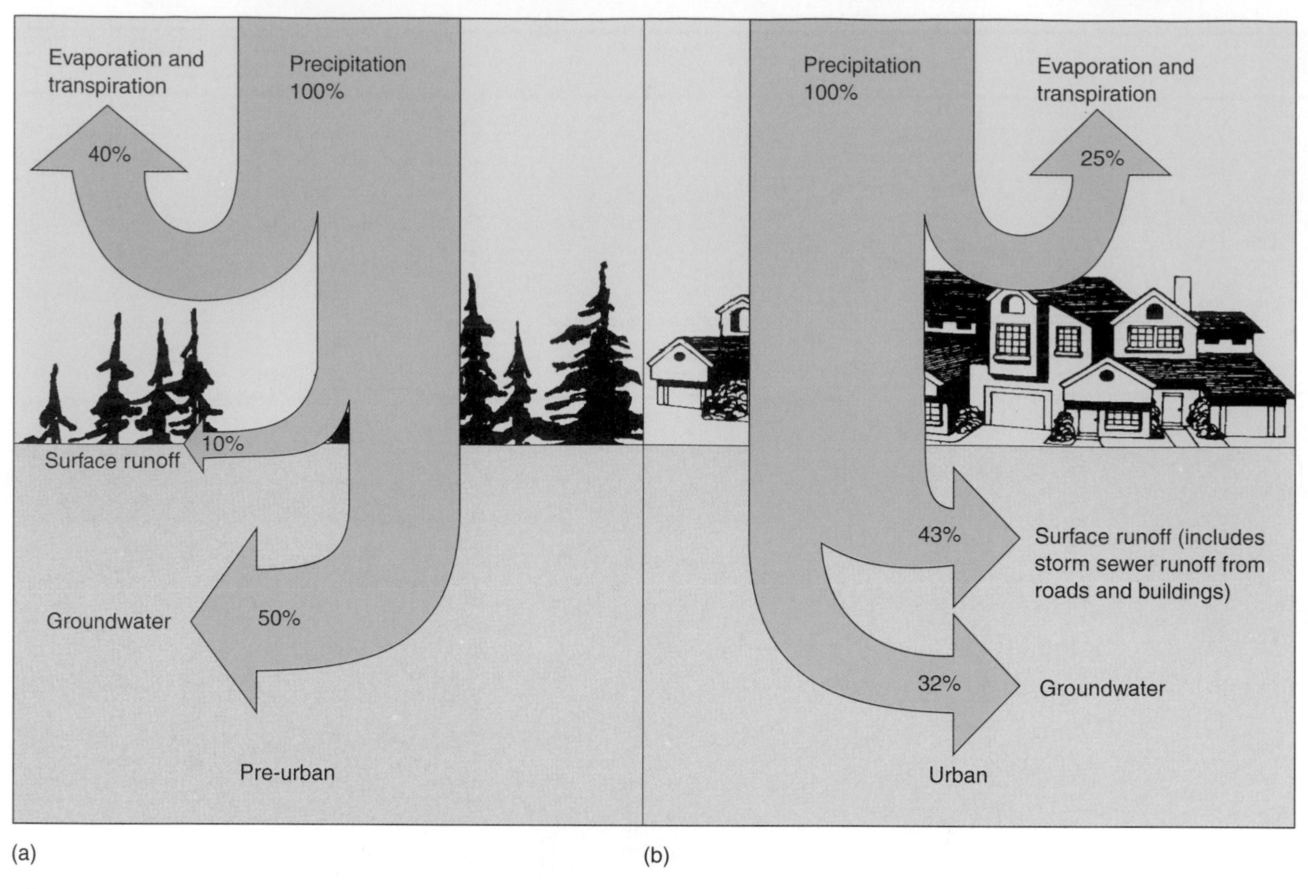

Figure 13–6

The development of an area changes the natural flow of water. The fate of precipitation in Ontario, Canada (a) before and (b) after urbanization. After Ontario was developed, runoff increased substantially, from 10 percent to 43 percent.

cannot absorb the water as well when the rains do come, and runoff is greater. Because the soil is not replenished by the precipitation that does fall, crop productivity is poor and the people are forced to cultivate food crops on additional marginal land.

Overdrawing surface waters

Removing too much fresh water from a river or lake can have disastrous consequences in local ecosystems. Humans can remove perhaps 30 percent of a river's flow without greatly affecting the natural ecosystem. In some places, however, considerably more than that amount is withdrawn for human use. In the arid American Southwest, for example, it is not unusual for 70 percent or more of surface water to be removed.

When surface water is overdrawn, wetlands dry up. Natural wetlands play many roles (see Chapter 17), not the least of which is serving as a breeding ground for many species of birds. Estuaries, where rivers empty into seawater, become saltier when surface waters are overdrawn (recall the chapter introduction), and this change in salinity greatly affects the productivity that is associated with estuaries.

Aquifer depletion

The removal by humans of more groundwater than can be recharged by precipitation or melting snow—called **aquifer depletion**—has several serious consequences. Prolonged aquifer depletion drains an aquifer dry, effectively eliminating it as a water resource. In addition, aquifer depletion from porous rock causes **subsidence**, or sinking, of the land on top. For example, some areas of the San Joaquin Valley in California have sunk almost 10 meters (33 feet) in the past 50 years due to aquifer depletion (Figure 13–7). **Salt water intrusion**, the movement of seawater into a freshwater aquifer, can occur along coastal areas when groundwater is depleted faster than it can be replenished (Figure 13–8).

Salinization of irrigated soil

Although irrigation improves the agricultural productivity of arid and semiarid lands, it sometimes causes salt to

Figure 13–7

Subsidence in San Joaquin, California. The markers on the utility pole indicate how the surface level has dropped over the years as groundwater was withdrawn for agriculture. (U.S. Geological Survey)

accumulate in the soil, a process called **salinization** (Figure 13–9). In a natural scenario, as a result of precipitation runoff, rivers carry salt away. Irrigation water, however, normally soaks into the soil and does not run off the land into rivers, so when it evaporates, the salt remains behind and accumulates in the soil. Salty soil results in a decline in productivity and, in extreme cases, renders the soil completely unfit for crop production. Chapter 21 discusses the problem of soil salinization in greater detail.

WATER PROBLEMS IN THE UNITED STATES

Compared to many countries, the United States has a plentiful supply of fresh water. According to the U.S. Geological Survey, approximately 15.9 trillion liters (4.2 trillion gallons) of water fall as precipitation per day in the continental United States. Approximately two-thirds of this water returns to the atmosphere through evaporation and transpiration, leaving roughly 5.3 trillion liters (1.4 trillion gallons) per day in the soil, surface waters, and groundwater (Figure 13–10). Most of this remaining water, approximately 4.9 trillion liters (1.3 trillion gallons) per day, makes its way into the ocean without ever being used. The relatively small amount of water that we borrow for our own purposes eventually returns to rivers and streams—for example, as treated or untreated sewage or as industrial wastes.

Despite the overall abundance of fresh water in the United States, many areas have severe water shortages because of geographical and seasonal variations (Figure 13–11). For example, arid and semiarid areas of the western United States normally receive little rainfall. Annual fluctuations in precipitation also occur. Droughts, higher-than-average precipitation rates, and other natural conditions cause problems in water availability throughout the country, and human activities sometimes exacerbate the difficulties.

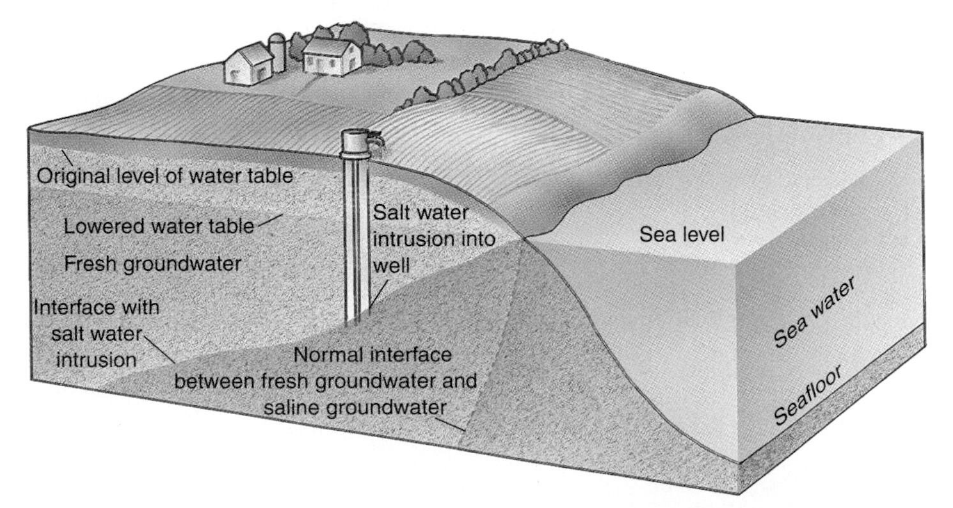

Figure 13–8

In coastal regions, aquifer depletion can cause salt water intrusion, which makes groundwater salty and unfit to drink. Normally the interface between fresh and salty groundwater is far from the site of the well. However, the removal of large amounts of fresh groundwater has caused the interface to migrate inland so that the well now draws up salty groundwater.

Figure 13–9

Salty soil that is unfit for agriculture. Irrigation water contains dissolved mineral salts. As the water evaporates from the soil's surface, the salts are left behind and gradually accumulate. (USDA/Natural Resources Conservation Service)

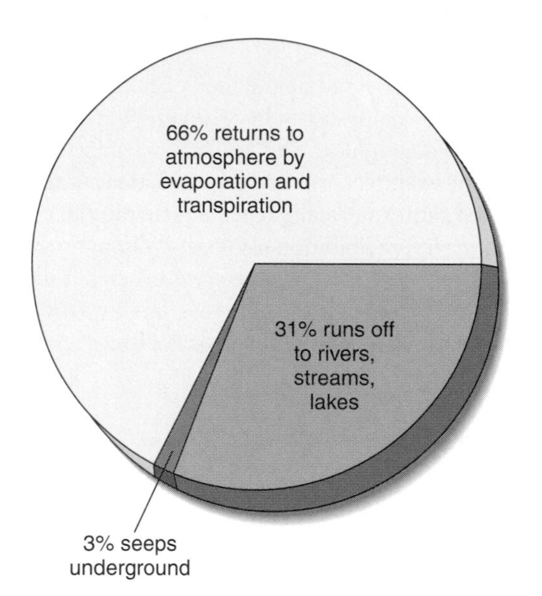

Figure 13–10

Water budget for the United States, excluding Alaska and Hawaii. Approximately 66 percent of the 15.9 trillion liters (4.2 trillion gallons) that fall on the continental United States as precipitation recycles into the atmosphere almost immediately by evaporation from wet surfaces and transpiration by plants. Most of the rest (about 31 percent) flows in rivers and streams to the ocean or to Canada and Mexico. Only a small percentage (about 3 percent) of the precipitation seeps into underground aquifers. Most water withdrawn for human use is cycled back to rivers and streams.

Surface Water

The increased use of U.S. surface water for agriculture, industry, and personal consumption during the past 40 years has caused many water supply and quality problems. Some U.S. regions that have grown in population during this period—for example, California, Nevada, Arizona, and Florida—have placed correspondingly greater burdens on their water supplies. If water consumption in these and other areas continues to increase, the availability of surface waters could become a regional problem in many places that have never before experienced water shortages.

Nowhere in the country are water problems as severe as they are in the West and Southwest. Much of this large region is arid or semiarid and receives less than 50 centimeters (about 20 inches) of precipitation annually. The West and Southwest consume an average of 44 percent of their renewable water, as compared to an average consumption of 4 percent of renewable water elsewhere in the United States (Figure 13–12). Historically, water in the West was used primarily for irrigation. However, with the rapid expansion of population in that region during the past 25 years, municipal, commercial, and industrial uses now compete heavily with irrigation for available water.

Until recently, the development of new sources of water met expanding water needs in the West and Southwest. Water was diverted from distant sources and transported via aqueducts—large conduits—to areas that needed it. As long ago as 1913, for example, Los Angeles started bringing in water from the Owens Valley, an area of California 400 kilometers (250 miles) north, along the east side of the Sierra Nevada. Dams were built and water-holding basins created to ensure a year-round supply. These solutions are no longer viable, because the closest, most practical water sources have already been used and because the public, which has come to expect inexpensive water, opposes paying for costly solutions such as dams and aqueducts.

Mono Lake

Mono Lake, a salty lake in eastern California, is a striking example of the effects of removing too much surface water (Figure 13–13). This lake is replenished by rivers and streams that are largely formed from snowmelt in the Sierra Nevada range. Evaporation provides the only natural outflow from the lake. Over time, Mono Lake is slowly becoming saltier as rivers deposit dissolved salts (recall that fresh water contains some salt) and as water, but not salt, is removed by evaporation.

Beginning in 1941, much of the surface water that would naturally feed Mono Lake was diverted to Los Angeles, 442 kilometers (275 miles) away. Over time, Mono Lake's water level subsided about 14 meters (46 feet), and its salinity increased dramatically, adversely affecting

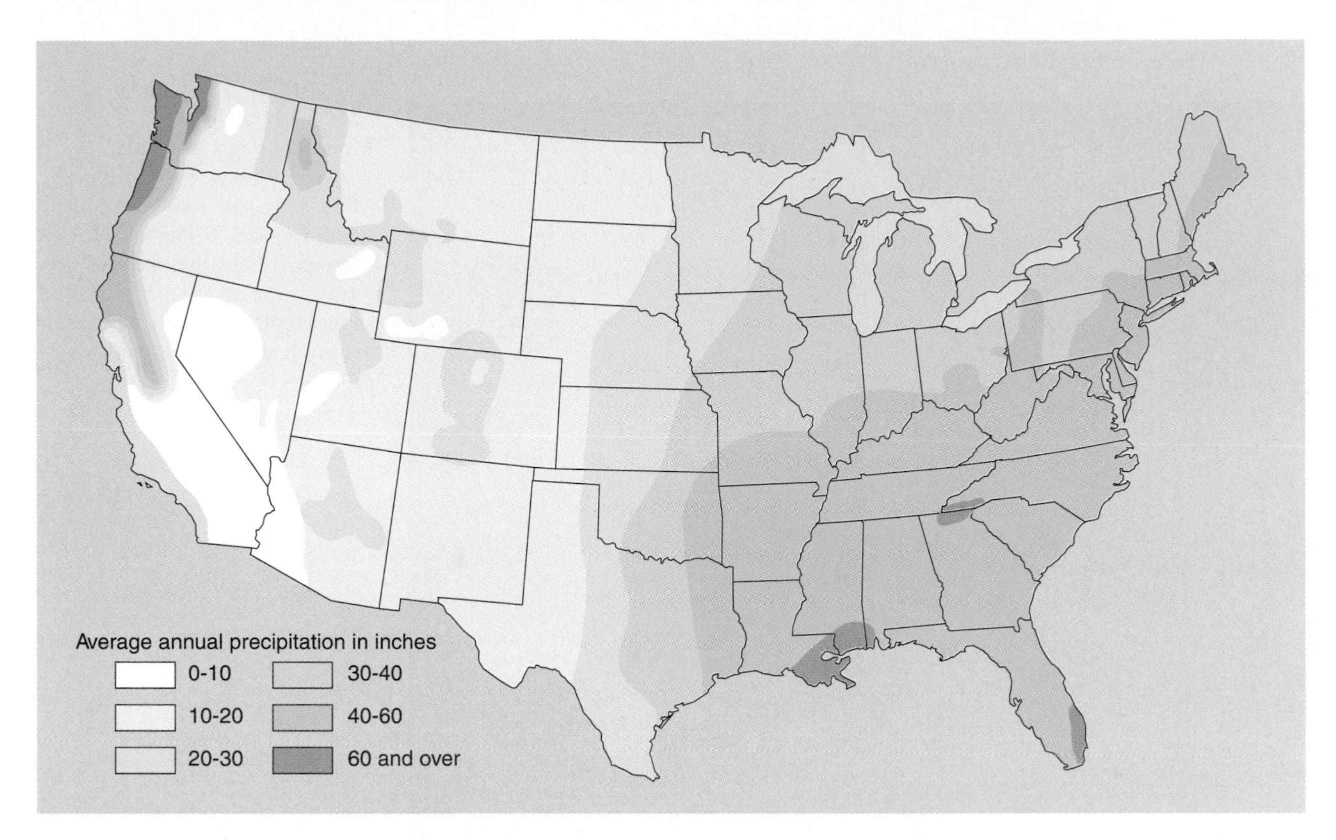

Figure 13–11

The average annual precipitation varies greatly across the United States.

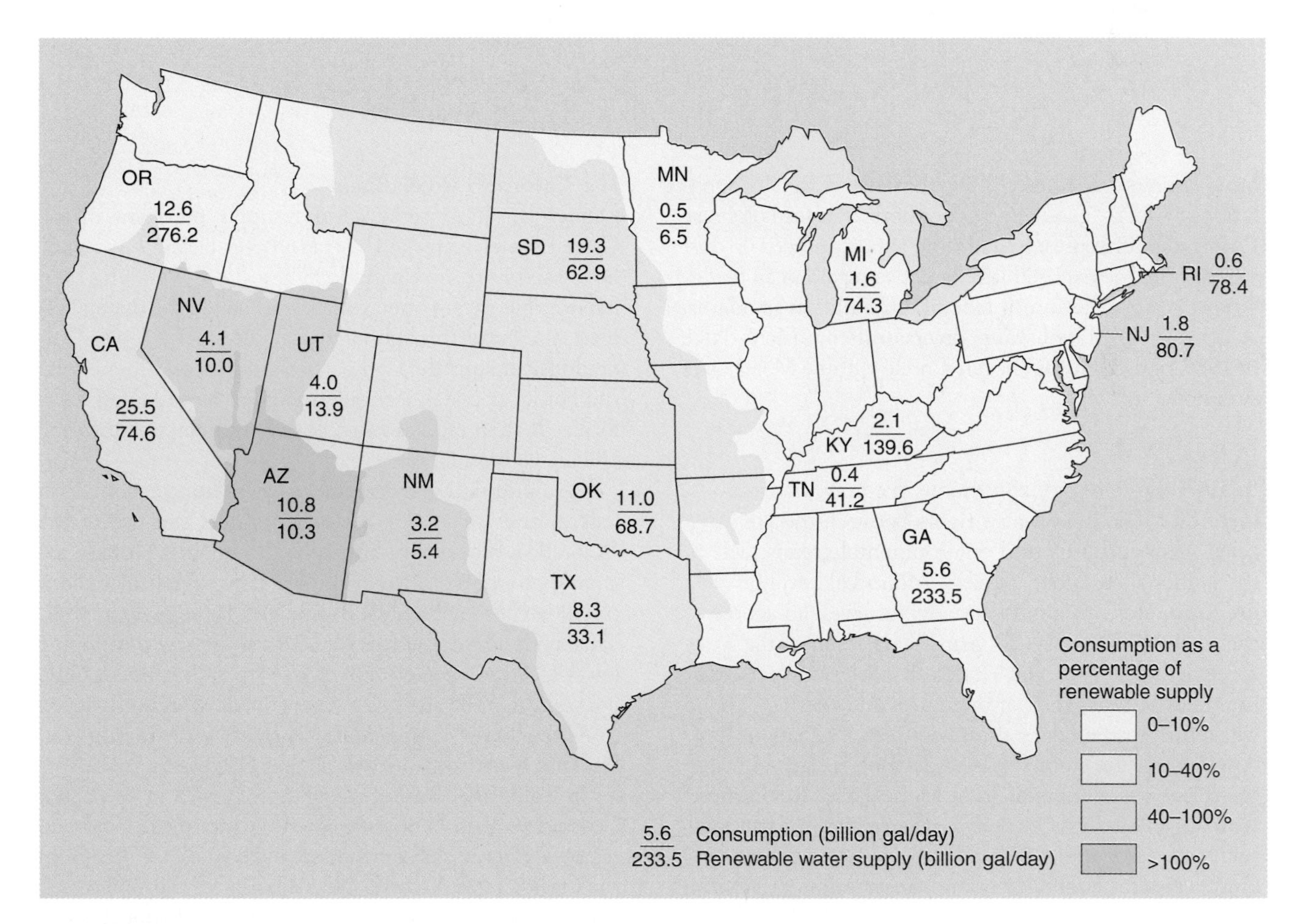

Figure 13–12

Water actually used (consumption) in the United States as a percentage of the renewable supply. The darker blue areas have severe water problems. Note that consumption actually exceeds renewable supply in the Southwest (the entire Colorado River basin).

Figure 13–13
Mono Lake in California. As Los Angeles diverted water from Mono Lake, the lake level subsided, exposing towers of calcium carbonate. Now that water is no longer being diverted, the lake should refill within the next twenty years. (W. Kleck/Terraphotographics)

brine shrimp and brine fly populations, as well as populations of ducks, geese, and other birds that feed on brine shrimp and brine flies. In addition, winds whipped up dust storms from the exposed lake bed, posing a human health hazard and a violation of federal air pollution standards. A court order halted water diversions from Mono Lake in 1989, and California began a review of the Mono Lake situation.

► *FOLLOW–UP*

In 1994 the state of California worked out an agreement on Mono Lake water rights between the Los Angeles water authority and environmental groups such as the National Audubon Society. Mono Lake will be allowed to return to almost its original level, an action that will probably take 20 years. As a result of the agreement, the National Audubon Society expects that hundreds of thousands of migratory and nesting birds will return to the lake's shores to nest. The city of Los Angeles will use state funds to develop reclaimed water to replace water supplies from Mono Lake. **Reclaimed water** is treated wastewater that is reused in some way, such as for irrigation, manufacturing processes that require water for cooling, wetlands restoration, or groundwater recharge.

The Colorado River Basin

One of the most serious water supply problems in the United States is in the Colorado River basin. The river's headwaters are in Colorado, Utah, and Wyoming, and major tributaries—often collectively called the upper Colorado—extend throughout these states. The lower Colorado River runs through part of Arizona and then along the border between Arizona and both Nevada and California. It then crosses into Mexico and empties into the Gulf of California.

An international agreement with Mexico, along with federal and state laws, severely restricts the use of the Colorado's waters. Traditionally, the upper Colorado region appropriated little of the water to which it was entitled, because it had few people and little development. This made more water available to the faster developing lower Colorado region, but it also gave that area a false impression of the size of its water supply. Water is diverted from the lower Colorado for the cities of Tucson and Phoenix in Arizona, as well as San Diego and Los Angeles in California. Recent population growth in the upper Colorado region is now threatening the lower Colorado region's water supply. Further, so much water is taken from the lower Colorado by people in the states through which it flows that the remainder is insufficient to meet Mex-

Figure 13–14

The Colorado River bed in San Luis Rio Colorado, Mexico. As a result of diversion for irrigation and other uses in the United States, the Colorado River often dries up before reaching the Gulf of California in Mexico. (Dan Lamont/Matrix)

ico's needs as set forth by international treaty (Figure 13–14). To compound the problem, as more and more water is used, the lower Colorado becomes increasingly saline as it flows toward Mexico.

Groundwater

Roughly half the population of the United States uses groundwater for drinking. Many large cities, including Tucson, Miami, San Antonio, and Memphis, have municipal wellfields and depend entirely or almost entirely on groundwater for their drinking water. In addition, many rural homes have private wells for their water supply. Groundwater is also used for industry and agriculture. Approximately 40 percent of the water used for irrigation in the United States comes from groundwater. Due to increased groundwater consumption during the past five decades, groundwater levels have dropped in many areas of heavy use across the United States. Aquifer depletion is particularly critical in three regions: southern Arizona, California, and the High Plains (a band of states extending from Montana and North Dakota south to Texas) because so much groundwater has been withdrawn for irrigation (Figure 13–15).

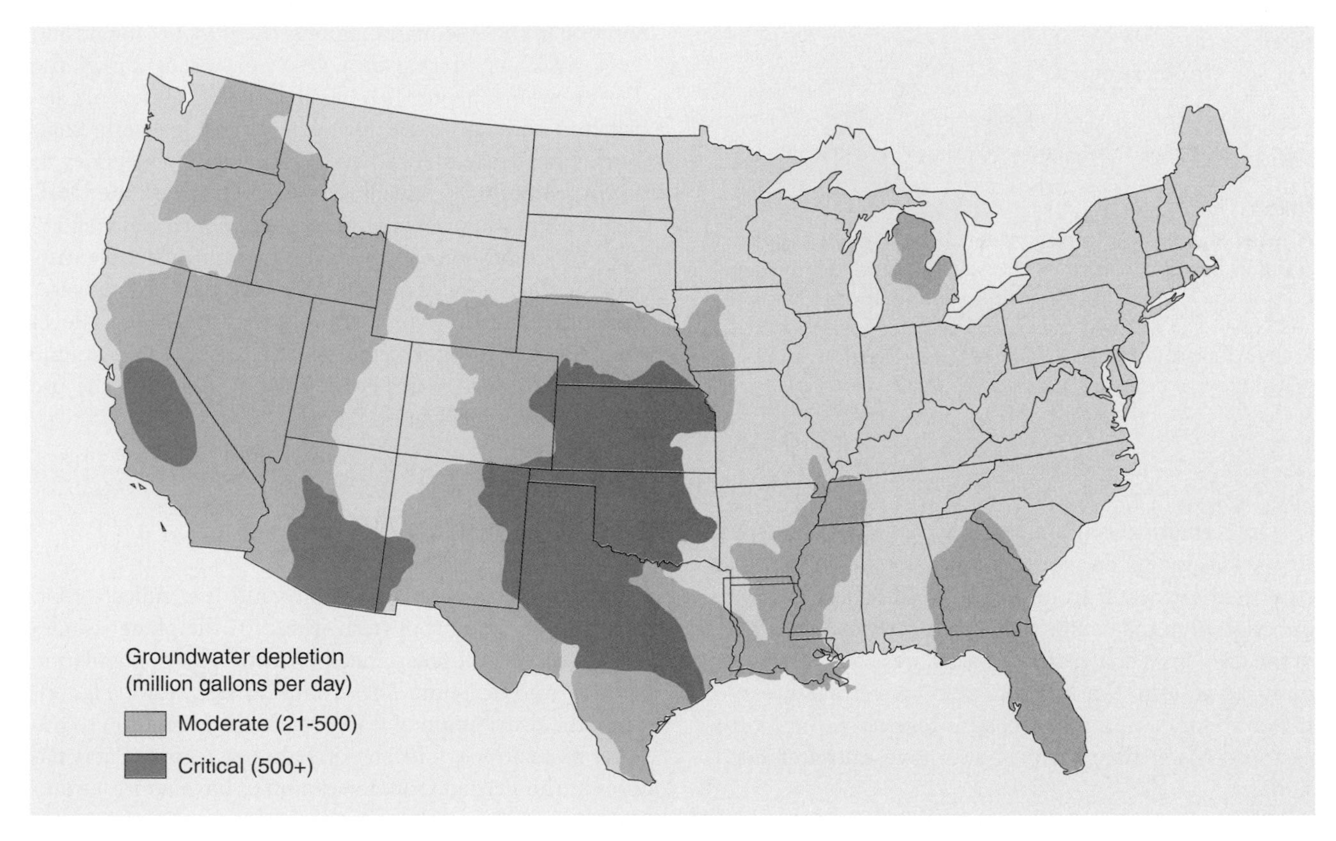

Figure 13–15

Aquifer depletion is a widespread problem in the United States, particularly in the High Plains, California, and southern Arizona.

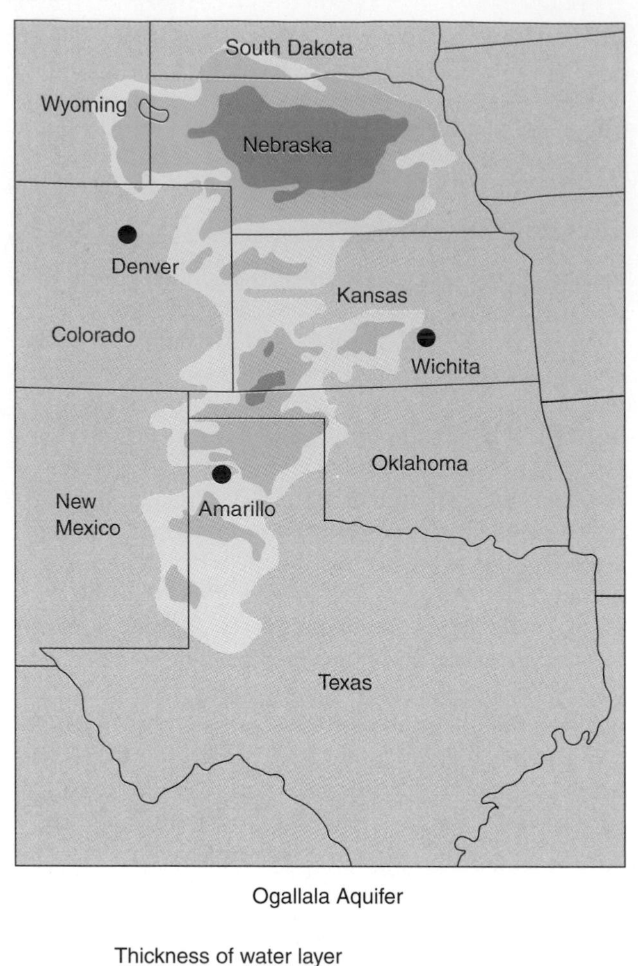

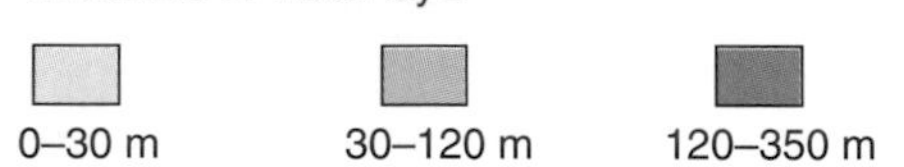

Figure 13–16

A massive deposit of groundwater, known as the Ogallala Aquifer, lies under eight Midwestern states. Extensive portions of the Ogallala Aquifer are located in Texas, Kansas, and Nebraska. Water in the Ogallala Aquifer takes hundreds or even thousands of years to be renewed after it is withdrawn to grow crops and raise cattle.

In certain coastal areas of Louisiana and Texas, the removal of too much groundwater has resulted in the intrusion of salt water from the Gulf of Mexico. Saltwater intrusion from the Pacific Ocean has occurred along parts of the California coast, along coastal areas of Puget Sound in northwestern Washington, and in certain areas of Hawaii. Florida and many coastal regions in the Northeast and Mid-Atlantic states also have saltwater intrusion.

The quality of an area's groundwater varies from fresh to saline, depending on the soil and rock characteristics of the area as well as the age of the water and its rate of recharge. Generally, groundwater is more saline in areas that have high evaporation rates and where saline water infiltrates, or penetrates, through rocks such as carbonate and sandstone, as it does in parts of North Dakota and Minnesota. Certain areas in the United States have poor groundwater quality due to naturally high localized concentrations of salts of such chemicals as fluoride and arsenic. Some areas of Nevada and Utah have unusually high levels of toxic minerals in their groundwater as a result of natural conditions, rather than pollution caused by humans. Chapter 21 discusses groundwater contamination caused by pollution.

The Ogallala Aquifer

The High Plains cover 6 percent of U.S. land, but produce more than 15 percent of its wheat, corn, sorghum, and cotton and almost 40 percent of its livestock. To achieve this productivity, it requires approximately 30 percent of the irrigation water used in the United States. Farmers on the High Plains rely on water from the **Ogallala Aquifer**, the largest groundwater deposit in the world (Figure 13–16).

In some areas farmers are drawing water from the Ogallala Aquifer as much as 40 times faster than nature replaces it. The depletion of the Ogallala has lowered the water table by more than 30 meters (100 feet) in some places, and where this has occurred, higher pumping costs have made it too expensive to irrigate. The amount of irrigated Texas farmland, for example, has declined by 11 percent in recent years. When farmers revert to dry-land farming in these semiarid regions, they risk economic and ecological ruin during droughts (see discussion of the Dust Bowl in Chapter 14). Ogallala water is unevenly distributed underground. Areas where the Ogallala is shallowest have experienced recent population declines as farms fail during dry spells. Those areas where the Ogallala is deepest, however, may not experience water shortages for another century. Most hydrologists (scientists who deal with water supplies) predict that groundwater will eventually drop in *all* areas of the Ogallala to a level that is uneconomical to pump. Their goal is to postpone that day through water conservation, including the use of water-saving irrigation systems.

GLOBAL WATER PROBLEMS

Data on global water availability and use indicate that, overall, the amount of fresh water on the planet is adequate to meet human needs, even taking population growth into account. These data do not, however, consider the distribution of water resources in relation to human populations. Citizens of Bahrain, a tiny island nation in the Persian Gulf, for example, have *no* freshwater supply and must rely completely on desalinization (removing salt) of salty ocean water for their fresh water.

Large variations in per-capita water use exist from country to country and from continent to continent, depending on the size of the human population and the available water supply. South America and Asia are the

Table 13–2
Freshwater Resources in 1995

Region	Total Renewable Water* (Cubic Kilometers)	Annual Per-Capita Water Availability (Thousand Cubic Meters)
Africa	3,996	5.5
Asia	13,207	3.8
Australia-Oceania	1,614	56.5
Europe	6,234	8.6
North and Central America	6,444	15.4
South America	9,526	29.8
World total	41,021	7.2

*Renewable water is generated by precipitation and includes runoff into rivers and groundwater recharge.

two continents with the greatest total water supply (Table 13–2). Together they receive more than one-half of the world's renewable fresh water (by precipitation). Although South America has more available water per person than Asia does, it does not have the potential to support as many people as its water supply would suggest. That is because most of the precipitation received by South America falls in the Amazon River basin, which has poor soil and is therefore unsuitable for large-scale agriculture. In contrast, because most of the precipitation in Asia falls on land suitable for agriculture, the water supply can support more people.

Humans need an adequate supply of water year round. Global water supply is complicated by the fact that **stable runoff**, the portion of runoff from precipitation that is available *throughout* the year, can be low despite the fact that total runoff is quite high. India, for example, has a wet season—June to September—during which 90 percent of its annual precipitation occurs. Most of the water that falls during India's wet season quickly drains away into rivers and is unavailable during the rest of the year; thus, India's stable runoff is low.

Variation in annual water supply is an important factor in certain areas of the world. The African Sahel region (just south of the Sahara Desert) has wet years and dry years, for example, and the lack of water during the dry years limits human endeavors during the wet years.

Drinking Water Problems

Many developing countries have insufficient water to meet the most basic drinking and household needs of their people. The World Health Organization (WHO) estimates that 1 billion people lack access to safe drinking water and almost 2 billion are without access to a satisfactory means of domestic wastewater and fecal waste disposal. They risk disease because the water they consume is contaminated by sewage or industrial wastes. Many of these people have to travel great distances to secure the water they need; this practice consumes large amounts of time, particularly for women and children, and tends to perpetuate poverty. WHO also estimates that 80 percent of human illness results from insufficient water supplies and poor water quality caused by lack of sanitation. Although many developing countries have installed or are installing public water systems, population increases tend to overwhelm efforts to improve the water supply.

The United States and other highly developed countries are involved in efforts to improve water quality and supply in countries with critical water problems. The U.S. Agency for International Development (AID) manages projects in areas vulnerable to prolonged drought, such as the Sahel region in Africa. AID has assisted in well digging and other measures that alleviate the effects of drought in the Sahel. In addition to contributions by individual governments, both the United Nations and the World Bank sponsor water management projects in developing countries.

Population Growth and Water Problems

As the world's population continues to increase, global water problems will become more serious. Population growth is already outstripping water supplies in countries such as India, where approximately 8000 villages have no local water. The water supply to some Indian cities—Madras, for example—has been so severely depleted that water is rationed from a public tap. Water supplies are also precarious in much of China, owing to population pressures. One-third of the wells in Beijing, for example,

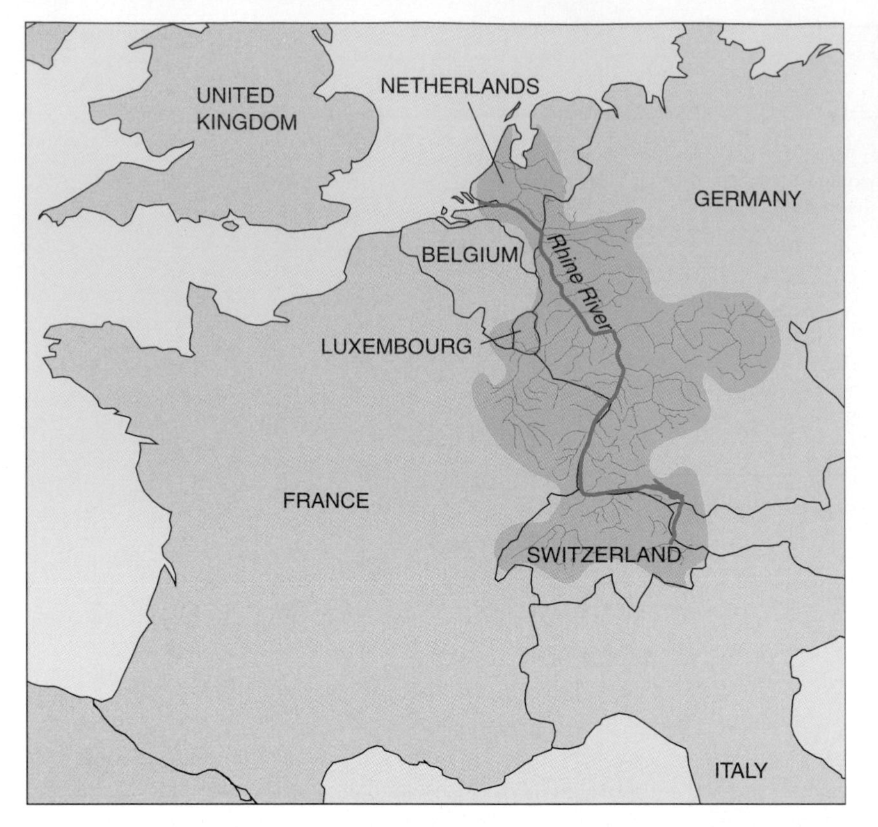

Figure 13–17

The Rhine River drains four European countries—Switzerland, France, Germany, and the Netherlands. Water management of such a river requires international cooperation.

have gone dry, and the water table continues to drop. Mexico is facing the most serious water shortages of any country in the Western Hemisphere. The main aquifer supplying Mexico City, for instance, is dropping by as much as 3.5 meters (11.2 feet) per year.

Shortages in global water supplies may also affect humans by limiting the amount of food they can grow. Recall that the main use of fresh water is for irrigation. As freshwater supplies are depleted by a growing human population, less water will be available for crops. Local or even widespread famines from water shortages are a very real danger.

Sharing Water Resources Among Countries

Global water supply is complicated by the fact that surface water is often an international resource. Three-fourths of the world's 200 or so major river basins are shared between at least two nations. The management of rivers that cross international boundaries requires international cooperation.

The river basin for the Rhine River in Europe, for example, is in Switzerland, Germany, France, and the Netherlands (Figure 13–17). Traditionally, Switzerland, Germany, and France used water from the Rhine for industrial purposes and then discharged polluted water back into the river. The Dutch then had to clean up the water so they could drink it. Today, these countries recognize that international cooperation is essential if the supply and quality of the Rhine River are to be conserved and protected.

The Aral Sea

The Aral Sea in Kazakstan, part of the former Soviet Union, is suffering from the same problem as Mono Lake. Like Mono Lake, it has no outflow other than evaporation. By the early 1980s, more than 95 percent of the inflow was diverted for irrigation of fertile farmland around the lake. Since 1960, the Aral Sea, which was once the world's fourth largest freshwater lake, has declined in area by 50 percent, and its total volume is down by 75 percent. Much of its biological diversity has disappeared—of the 24 species of fishes originally found there, only 4 remain—and it runs the risk of becoming a lifeless brine lake. Millions of people living near the Aral Sea have developed health problems ranging from allergies to throat cancer, presumably caused by winds that whip the salt on the receding shoreline into the air, causing blinding salt storms. Since the 1950s, such storms have increased sixtyfold. The salt is carried by the wind hundreds of miles from the Aral Sea, and where it is deposited it reduces the productivity of the land.

► *FOLLOW–UP*

Immediately following the breakup of the Soviet Union, plans to save the Aral Sea faltered as responsibility for

Figure 13–18

The Aral Sea in Kazakstan has shrunk from water diversion for irrigation. (a) Five Asian republics share the Aral basin, which includes two rivers—Syr Darya and Amu Darya—that empty into the Aral Sea. (b) As the water receded, these fishing boats became stranded far from the water's edge. (© 1989 David Turnley/Black Star)

its rescue shifted from Moscow to the five central Asian republics of the Aral basin. However, in 1994, the five republics—Uzbekistan, Kazakstan, Kyrgyzstan, Turkmenistan, and Tajikistan—established a joint fund to prevent the complete disappearance of the Aral Sea (Figure 13–18). In addition, the World Bank and the U.N. Environment Program approved a grant to the five republics to help address the ecological problems of the area.

Serious international water situations

The next century may well see countries facing one another in armed conflict over water rights. Tensions are high along the Mekong River basin, shared by Laos, Thailand, and Vietnam. Likewise, the Indus River basin is shared between Pakistan and India, two countries that have never been friendly, while India and Bangladesh quarrel over the Ganges River. Slovakia and Hungary both depend on the Danube River, whereas Mexico and the United States share water from the Colorado River.

One particularly troublesome international spot is the Jordan River between Israel and Jordan. Both of these Middle Eastern countries obtain fresh water from the Jordan River basin. Israel and Jordan are each experiencing large population increases during the 1990s, which could make their water situation even more critical. Neither country approves of the other increasing its allotment of water from the Jordan River, because an increase for one country would mean a smaller supply for the other.

North Africa also has a serious water-use situation, the Nile River. The nations of Ethiopia and Sudan require some of the Nile River's flow to increase their water supplies. Because almost all of Egypt's water supply comes from the Nile, the actions of Ethiopia and Sudan could imperil Egypt's freshwater supply at a time when its population is increasing. The United Nations engineered an international water-use agreement among the Nile River countries to diffuse this potentially dangerous water situation.

WATER MANAGEMENT

People have always considered water different from other resources. Resources such as coal or gold may be owned privately and sold as free-market goods, but we view water as public property.

Historically, both in the United States and in other countries, water rights were bound with land ownership. As more and more users compete for the same water, however, allocation decisions are increasingly being made by state or provincial governments. Some countries have separated land and water ownership so that water rights are sold separately.

Because rivers usually flow through more than one governmental jurisdiction, jurisdictions or states must develop agreements with each other about the management of a river's resources. Such interstate cooperation permits comprehensive rather than piecemeal management. In addition, these arrangements allow the water to be divided fairly between the jurisdictions, which then apportion their respective shares to individual users according to an established set of priorities.

Groundwater management is more complicated, in part because the extent of local groundwater supplies is not known. Some states manage groundwater, particularly where demand exceeds supply. Groundwater management includes issuing permits to drill wells, limiting the number of wells in a given area, and restricting the amount of water that may be pumped from each well.

The price of water varies, depending on how it is used. Historically, domestic use is most expensive and agricultural use is least expensive. In any case, the consumer rarely pays directly for the *entire* cost of water, which includes its transportation, storage, and treatment. State and federal governments heavily subsidize water costs, so we pay for some of the cost of water indirectly, through taxes.

Providing a Sustainable Water Supply

The main goal of water management is to provide a sustainable supply of high-quality water. Increasingly, state and local governments are considering the price of water as a mechanism to help ensure an adequate supply of water. Raising the price of water to users so that it reflects the actual cost generally promotes more efficient use of water.

Dams and reservoirs

Dams ensure a year-round supply of water in areas that have seasonal precipitation or snowmelt. Dams confine water in reservoirs, from which the flow is regulated (Figure 13–19). Dams have other benefits, including the generation of electricity. They provide flood control for areas downstream, because a reservoir can hold a large amount of excess water during periods of heavy precipitation and then release it gradually. Some of the reservoirs formed by dams also have recreational benefits: people swim, boat, and fish in them.

Many people, however, feel that the drawbacks of dams, including the cost of building them, far outweigh any benefits they provide. Consider the controversy surrounding the damming of the Missouri River, which flows from western Montana to St. Louis, Missouri, where it joins the Mississippi River and flows on to the Gulf of Mexico. The Missouri River, which is 3968 kilometers (2466 miles) long, is the longest river in the United States. It contains North America's three largest reservoirs (Lake Sakakawea in North Dakota, Lake Oahe in South Dakota, and Fort Peck Reservoir in Montana) and the world's largest compacted-earth dam (Fort Peck Dam). In fact, the Missouri River has a *series* of dams built by the Army Corps of Engineers that have provided both benefits and problems for people living along the river.

Figure 13–19
The Grand Coulee Dam on the Columbia River and its reservoir, the Franklin D. Roosevelt Lake. Dams help to regulate water supply, storing water that is produced in times when precipitation is plentiful to be used during dry periods. The many beneficial uses of dams include electricity generation and flood control, but they also destroy the natural river habitat and are expensive to build. (U.S. Department of Energy)

Since 1987 the Corps has opened up the northern dams in order to protect downstream navigation, including the shipping of $92 million of grain each year. In addition, people who live downstream count on the river for irrigation, electrical power, and individual water consumption. But the area along the northern Missouri River depends on the river for its multimillion-dollar fishing and tourism industry.

North Dakota, South Dakota, and Montana sued the Army Corps of Engineers for discharging water in an "arbitrary and capricious way." They won the decision, which was to have kept the Corps from releasing water from one of the large reservoirs in North Dakota, but the U.S. Court of Appeals reversed the ruling, and water flowed through the dams again.

The battle over water rights is becoming more heated and entangled as those who live upstream and those who live downstream fight to protect their interests. Although each side says it is willing to share, neither side seems willing to give up much. South Dakota, for example, wants the Corps to shorten the navigation season during years of drought so that upstream states will have enough. But residents downstream counter that their needs should be favored over those of the fishing and tourism industry up north.

In recent years scientists have come to understand many of the ways that dams alter river ecosystems downstream. For example, the Glen Canyon Dam, built in 1963, has profoundly affected the Colorado River in the Grand Canyon. Prior to the dam's construction, powerful spring floods used to deposit beaches and sandbars that provided nesting sites for birds and shallow waters for breeding fishes. The regulated flow of water since the Glen Canyon Dam opened changed the ecosystem to the detriment of some of the Grand Canyon's wildlife. The Bureau of Reclamation tried to rectify some of the changes that have occurred to the river by releasing an additional 117 billion gallons of water during a one-week period in the spring of 1996. The experimental flood, although small in comparison to some of the natural floods of the past, rebuilt some 50 beaches and sandbars that had disappeared since 1963 and enlarged most of the existing ones. It also killed exotic vegetation that had taken over the area and partly restored fish spawning habitats. Although it will take scientists years to analyze the full effects of the experimental flood, preliminary evidence indicates that rivers controlled by dams would benefit from periodic floods. The impact of dams on natural fish communities is discussed in Focus On: The Columbia River and the Pacific Northwest Salmon Debate. (Chapter 12 examines other issues associated with dams.)

Water diversion projects

One way to increase the natural supply of water to a particular area is to divert water from areas where it is in plentiful supply. This is done by pumping water through a system of aqueducts. Much of southern California receives its water supply via aqueducts from northern California (Figure 13–20). Water from the Colorado River is also diverted into southern California by aqueducts.

Large-scale water diversion projects are controversial and expensive. The Central Arizona Project, which pumps water 540 kilometers (336 miles) from the Colorado River to Phoenix and Tucson, was recently completed at a cost of almost $4 billion. As we saw earlier, a river or other body of water is damaged when a major portion of its water is diverted. Pollutants, which would have been diluted in the normal river flow, reach higher

(text continues on page 298)

FOCUS ON

The Columbia River and The Pacific Northwest Salmon Debate

The Columbia River is the fourth largest river in North America. Its river basin, which covers an area the size of France, includes seven states and two Canadian provinces. Such a large, complex river system has multiple uses. There are more than 100 dams within the Columbia River system, 19 of which are major generators of inexpensive hydroelectric power (see figure a). The Columbia River system supplies municipal and industrial water to several major urban areas, including Boise, Portland, Seattle, and Spokane. More than 1.2 million hectares (3 million acres) of agricultural land are irrigated with the Columbia's waters. Commercial ships navigate 805 kilometers (500 miles) of the river. Recreational uses of the river include boating, windsurfing, and swimming. The Columbia River system also offers sport and commercial fishing for salmon, steelhead trout, and other fishes.

As is often the case in natural resource management, a particular use of the Columbia River system may have a negative impact on other uses. For example, the dam impoundments along the Columbia River that generate electricity and control floods have adversely affected fish populations, particularly salmon. Salmon are migratory. They spawn in the upper reaches of freshwater rivers and streams. The young offspring, called smolts, migrate to the ocean, where they spend most of their adult lives. Salmon complete their life cycle by returning to their place of birth to reproduce and die.

The salmon population in the Columbia River system is only a fraction of what it was before the basin was developed. While several factors have contributed to their decline, the many dams that impede their migrations are widely considered the most significant. In addition, salmon have been overfished in the Pacific Ocean. Logging around salmon spawning streams contributes sediment pollution that degrades the salmon's habitat. Also, because streams in logged areas are no longer shaded, the water temperature becomes too hot for the developing salmon eggs.

Several projects to rebuild salmon populations were implemented during the 1980s and early 1990s, but none have been particularly effective. Many of the dams had

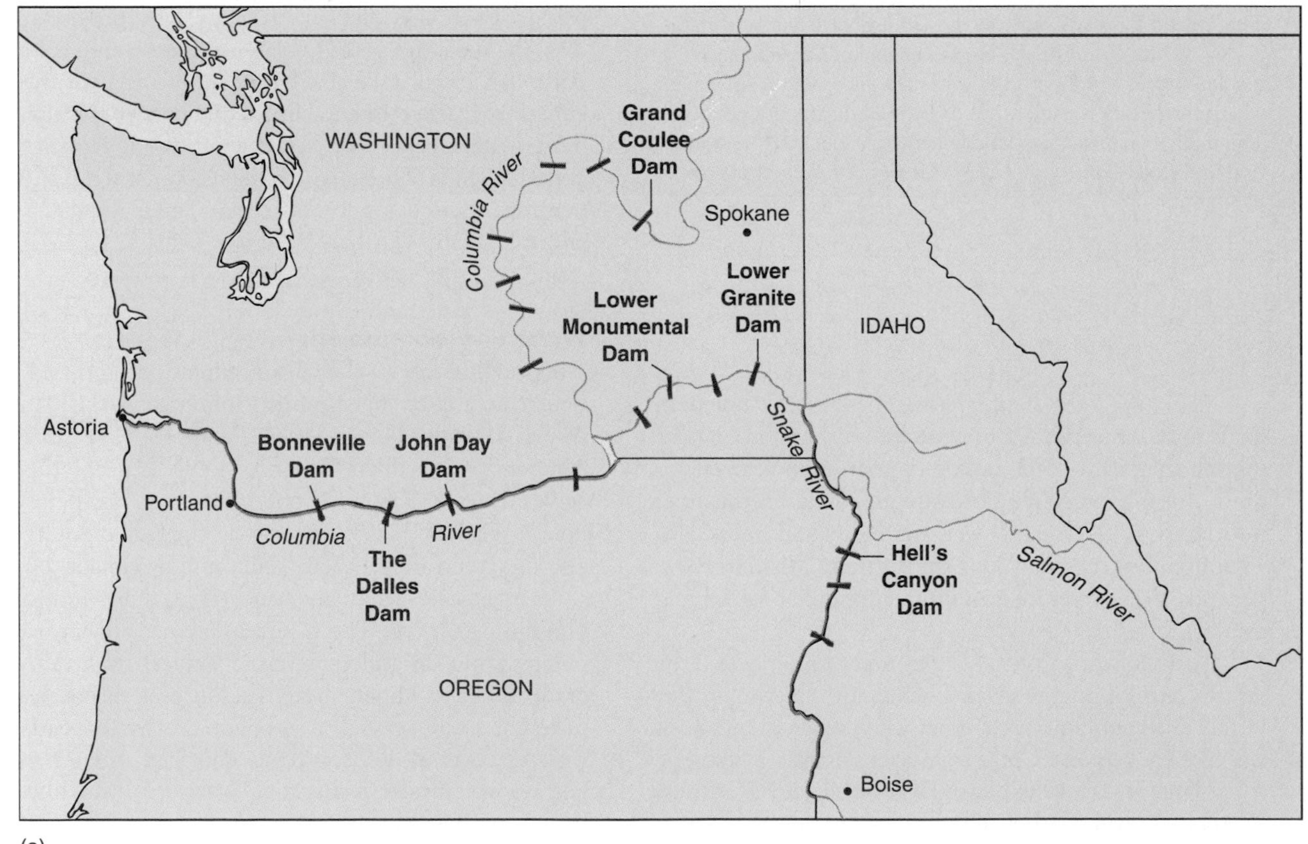

(a)
Major dams along the Columbia River and its tributaries.

(b)

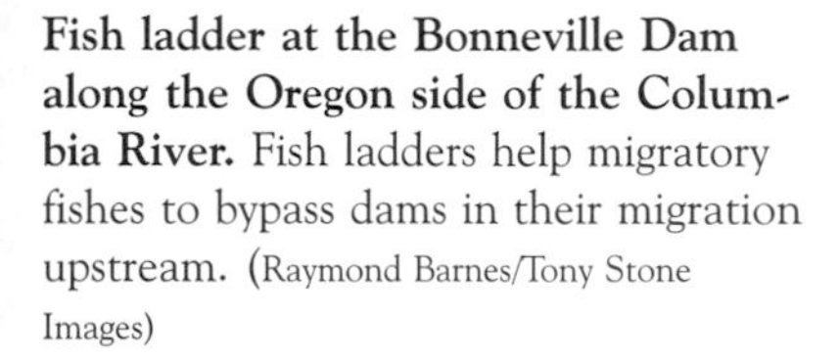

Fish ladder at the Bonneville Dam along the Oregon side of the Columbia River. Fish ladders help migratory fishes to bypass dams in their migration upstream. (Raymond Barnes/Tony Stone Images)

already installed fish ladders to enable some of the adult salmon to bypass the dams and continue their upstream migration (see figure b). The *downstream* migration of the young smolts is quite perilous, however: of every five that start the trip, only about one makes it to the Pacific Ocean. Prior to extensive dam building, smolts made the downstream trip to the Pacific in a few days, carried by the heavy spring flush of water from rains and melting snow. Now the same trip may take several months, with many of the fishes becoming stranded in the tranquil waters of reservoirs behind the dams. Many of the smolts are consumed by large predators in the reservoirs or destroyed by the turbine blades as they fall over the dams.

To increase the number of salmon, several hatcheries have been built upstream of the dams, in tributaries of the Columbia. Young fishes produced at these hatcheries are released to imprint the "smell" of the streams, enabling them to return there as adults to reproduce. Unfortunately, this effort has not reestablished natural spawning, in part because hatchery fishes appear to be genetically incompatible with wild populations.

To protect some of the remaining natural salmon habitats, several streams in the Columbia River system have been designated off limits for dam development. Special underwater screens and passages are being installed at dams to steer the smolts away from the turbines. Trucks and barges transport some of the young fishes around dams, while others swim safely over the dam because the electrical generators are periodically turned off to allow passage.

One of the more interesting approaches is the establishment of a "water budget" for the young fishes. Extra water, simulating spring snowmelt, is released from the dams to help wash the smolts downstream. Conservationists prefer the water-budget approach over trucking and barging, which have been done for almost 20 years without much success. However, other interest groups are opposed to increasing water flow for salmon. Farmers, for example, do not want the water "wasted" during spring months when water is plentiful; they want it saved for irrigation during summer months. The hydropower industry is also opposed because they want to save the water to generate electricity during the winter months, their time of peak demand.

► *FOLLOW-UP*

Because chinook and sockeye salmon are listed as endangered species, the Endangered Species Act requires the federal government to execute a comprehensive recovery plan for them. The recovery plan, released by the National Marine Fisheries Service in 1995, includes a variety of approaches. Not surprisingly, various elements of the plan were immediately criticized by the many groups with a vested interest: Native Americans who have treaties giving them the right to fish, commercial and sports fishermen, the hydropower industry, local industries that depend on a cheap supply of electrical power, shipping interests, farmers, loggers, and environmentalists.

The Endangered Species Act is up for renewal (see Chapter 16), and critics of this landmark legislation may use the Pacific Northwest as their rallying cry. They point out that salmon populations are stable in Alaska, so it does not matter if salmon are declining in the Pacific Northwest. They further argue that the salmon are not worth saving if the economic costs that bring about their recovery are too high.

(b)

Figure 13–20

Southern California, largely desert, relies on water diversion for the water needs of its millions of inhabitants. (a) The California Water Project includes 1042 kilometers (648 miles) of aqueducts to transfer large quantities of water to southern California. This map also shows some of the main reservoirs of the California Water Project. (b) An aqueduct near Fresno, California. (© 1997 Dembinsky Photo Associates)

concentrations when much of the flow has been removed. Fishes and other organisms may decline in number and diversity. Although no one denies that people must have water, opponents of water diversion projects contend that serious water conservation efforts would eliminate the need for additional large-scale water diversion.

Desalinization

Seawater and saline groundwater can be made fit to drink through the removal of salt, called **desalinization** (or **desalination**), by several methods. One of the most common is **distillation**—heating the salt water until the water evaporates, leaving behind a crust of salt; the water vapor is then condensed to produce fresh water. Another method, called **reverse osmosis**, involves forcing salt water through a membrane that is permeable to water but not to salt.

Desalinization is expensive because it requires a large input of energy, although recent advances in reverse osmosis technology have increased its efficiency so that it requires much less energy than distillation. Other expenses involved in desalinization projects include the cost of transporting the desalinized water from the site of production to where it will be used.

The disposal of salt produced by desalinization is also a concern. Simply dumping it back into the ocean, particularly near coastal areas that are highly productive, could cause a localized increase in salinity that would harm marine organisms.

Desalinization is a huge industry in the Middle East and North Africa, particularly in Saudi Arabia, because it is cost-competitive with alternative methods of obtaining fresh water in that arid region. The United States is also a major producer of desalted water—for example, at the Yuma Desalting Plant in Arizona. At present, however, desalinization is not a viable solution for the water supply problems of many developing nations, because it is too expensive.

Harvesting icebergs

For well over a decade, visionaries in arid areas such as Saudi Arabia, California, and Australia have toyed with the idea of towing icebergs from the Antarctic or Arctic so they could be melted down to supply fresh water. Skeptics voice concerns about the ownership of icebergs, particularly in Antarctica. The effects on marine organisms of introducing an iceberg into warm tropical waters are unknown.

The Department of Natural Resources in Alaska has begun issuing permits that allow icebergs to be collected. One entrepreneur plans to lift small icebergs from the sea, chop them into smaller pieces, and ship them to Japan, where they will be used as "gourmet" ice cubes. (Glacial ice makes better ice cubes than freezer ice because it is extremely pure and takes longer to melt.)

Australians are also serious about harvesting icebergs. They plan to wrap Antarctic icebergs in strong, lightweight fabrics to hold the water that melts during transport. Glacial ice will likely be cost-competitive with other freshwater sources in Australia, even taking into account the expense of harnessing, wrapping, and towing the icebergs.

WATER CONSERVATION

The right to an unlimited supply of water at a reasonable cost has always been assumed automatically by most Americans. However, population and economic growth have placed an increased demand on our water supply. Today there is more competition than ever before among water users whose priorities differ. Water conservation measures are necessary to guarantee sufficient water supplies. Most water users use more water than they really need, whether it is for agricultural, industrial, or direct personal consumption. With incentives, these users will lower their rates of water consumption. Many studies have shown that higher prices for water provide the motivation to conserve water. For example, farmers are more likely to invest in water-saving irrigation technologies if the money saved from decreased water consumption covers the expense of the initial installation.

Reducing Agricultural Water Waste

Irrigation generally makes inefficient use of water. Traditional irrigation methods, which have been practiced for more than 5000 years, involve flooding the land or diverting water to fields through open channels. Water flow must be increased in order to guarantee that the far end of the field or higher elevations of the field receive water. Less than 50 percent of the water applied to the soil by flood irrigation is absorbed by plants; the rest usually evaporates into the atmosphere.

One of the most important innovations in agricultural water conservation is **microirrigation**, also called **drip** or **trickle irrigation**, in which pipes with tiny holes bored in them convey water directly to individual plants (Figure 13–21). This reduces the water needed to irrigate crops by a substantial amount, usually 40 to 60 percent. Microirrigation also reduces the amount of salt left in the soil by irrigation water.

Another important water-saving measure in irrigation is the use of lasers to level fields. As a laser beam sweeps across a field, a field grader receives the beam and scrapes the soil, leveling it. Because farmers must use extra water to ensure that plants on higher elevations of a field receive enough, laser leveling of a field reduces the water required for irrigation.

The use of sound water management principles in agriculture reduces water consumption. Traditionally, farmers have been allotted specific amounts of water at specific times, with a "use it or lose it" philosophy. This approach encourages waste. If, instead, water needs are carefully monitored, often through computer controls, water can be applied in small, regulated quantities, thereby reducing overall consumption.

Although advances in irrigation technology are improving the efficiency of water use, many challenges remain. For one thing, sophisticated irrigation techniques are prohibitively expensive. Few farmers in developed countries, let alone subsistence farmers in developing nations, can afford to install them.

Reducing Water Waste in Industry

Electric power generators and many industries require water in order to function (recall that power plants heat wa-

Figure 13–21

Microirrigation. Cutaway view of soil shows a small tube at the root line. Tiny holes in the tube deliver a precise amount of water directly to the roots, eliminating much of the waste associated with traditional methods of irrigation. (USDA/Natural Resources Conservation Service)

ter to form steam, which turns the turbines). In the United States, five major industries consume almost 90 percent of industrial water (not including water being used for cooling purposes): chemical products, paper and pulp, petroleum and coal, primary metals, and food processing.

Stricter pollution control laws provide some incentive for industries to conserve water. Industries usually reduce their water use, and therefore their water treatment costs, by recycling water. The National Steel Corporation plant in Granite City, Illinois, for example, recycles approximately two-thirds of the 62 million gallons of water it uses daily; the used water is cleaned up before being discharged into a lake that spills into the Mississippi River.

It is likely that water scarcity, in addition to more stringent pollution control requirements, will encourage further industrial recycling. The potential for industries to conserve water by recycling is enormous.

Reducing Municipal Water Waste

Like industries, regions and cities can reduce their water consumption by recycling or reusing water before it is discharged. For example, individual homes and buildings can be modified to collect and store "gray water"—water that has already been used in sinks, showers, washing machines, and dishwashers (Figure 13–22). The gray water can then be reused to flush toilets, wash the car, or sprinkle the lawn.

Israel probably has the world's most highly developed system of treating and reusing municipal wastewater. Israel does this out of necessity, because all of its possible freshwater sources have already been tapped. The reclaimed water is used for irrigation, which allows higher quality fresh water to be channeled to cities. Used water contains pollutants, but most of these are nutrients from treated sewage and are therefore beneficial to crops.

Figure 13–22
Individual homes and buildings can be modified to collect and store "gray water," water that has already been used in sinks and showers. This "gray water" can be used when clean water is not required, for example, in flushing toilets, washing the car, and sprinkling the lawn.

Automated systems to purify and recycle wastewater have been developed and are cost-competitive with fresh water. In Tokyo, for example, the wastewater in the Mitsubishi office building is purified and recycled.

In addition to recycling and reuse, cities can decrease water consumption through other conservation measures. These include consumer education (both children and adults need to be taught the importance of using our limited supply of fresh water wisely), the use of water-saving household fixtures, and the development of economic incentives to save (see You Can Make a Difference: Conserving Water at Home). These measures have been used successfully to pull cities through dry spells; they are effective because individuals are willing to conserve for the common good during water crisis periods.

Increasingly, however, cities are examining ways to encourage individual water conservation methods all the time. The installation of water meters in residences in Boulder, Colorado, reduced water consumption by one-third. Before the installation, homeowners were charged a flat fee, regardless of their water use. For many apartment dwellers, water use is included in the rent; charging each apartment for its water use provides the incentive to use water more efficiently. In addition to installing water meters, a city might encourage water conservation by offering a rebate to any homeowner who installs a conserving device such as a water-saving toilet. Cities can help reduce municipal water consumption by passing building codes that specify such water conservation measures as installing low-flush toilets and water-saving faucets and shower heads. The water supply systems (pipes and water mains) in many urban areas are old and leaky; repairing them would improve the efficiency of water use.

Cities also promote water conservation by increasing the price of water to reflect its true cost; as water prices rise, people learn very quickly to conserve water. For example, charging more for water during dry periods encourages individuals to conserve water. Although the average cost of water to consumers is rising during the 1990s, many U.S. cities still do not charge consumers what the water actually costs them.

The city of Tucson, Arizona has broken its citizens' trend toward increased water consumption and now needs 25 percent less water per person than it did in the 1970s. In arid towns such as Tucson, simply replacing grass lawns with native desert plants substantially reduces water use by making lawn irrigation unnecessary.

SAVING WATER BY XERISCAPING

The landscaped yard shown in the photograph does not have the usual grass-covered lawn and water-hungry flowers, trees, and shrubs. Instead, this property in Berkeley, California, has been **xeriscaped**, that is, landscaped with rocks and plants that require little water. People in western states and in south Florida are increasingly xeriscaping their properties to conserve water. Xeriscaping has been shown to decrease household water consumption by up to 60 percent, reducing water bills and the energy-intensive maintenance that is required to mow, edge, and weed expansive lawns. Many homeowners do not wish to eliminate lawns entirely, but instead reduce the size of turfed areas and xeriscape the remaining land. The stone mulch spread around the base of the drought-resistant plants keeps the soil moist, reduces weed growth, and protects the soil from erosion when precipitation does occur. Dozens of attractive flower, tree, and shrub species are adapted to thrive with little water and so are perfect candidates for xeriscaping.

(Mark E. Gibson/Visuals Unlimited)

YOU CAN MAKE A DIFFERENCE

Conserving Water at Home

The average American uses 80 gallons of water per day at home (see figure). Many appliances, such as dishwashers, garbage disposals, and washing machines, need water. The growth of suburbs, with their expansive landscaping that requires watering, is also responsible for increased water use.

The cumulative effect of many people practicing personal water conservation measures has a significant impact on overall water consumption. You can practice these yourself. The bathroom is a good place to start because two-thirds of the water used in an average home is used for showers and baths and flushing toilets.

1. Install water-saving shower heads and faucets to cut down significantly on water flow. Low-flow shower heads, for example, reduce water flow from 5 to 9 gallons per minute to 2.5 gallons per minute. Replacing one old shower head saves a home $30 to $50 each year in water and energy savings.
2. Install a low-flush toilet or use a water displacement device in the tank of a conventional toilet. Low-flush toilets require only 2 gallons or less per flush, compared to 5 to 9 gallons for conventional toilets. To save water with a conventional toilet, fill an empty plastic laundry bottle with water and place it in the tank to displace some of the water. Do not put the bottle where it will interfere with the flushing mechanism; also, do not add bricks to the tank because they dissolve over time and can cause costly plumbing repairs.
3. If you are in the market for a washing machine, front-loading washing machines require less water than top-loading models.
4. Modify your personal habits to conserve water. Avoid leaving the faucet running. For example, allowing the faucet to run while shaving consumes an average of 20 gallons of water; you will use only 1 gallon if you simply fill the basin with water or run the water only to rinse your razor. You may save as much as 10 gallons of water a day by wetting your toothbrush and then turning off the tap while you brush your teeth, as opposed to running the water during the entire process. Also, most of us take longer showers than we need. Time yourself the next time you take a shower; if it is 10 minutes or longer, you can work on reducing your shower time.
5. Surprisingly, you will save water by using a dishwasher, which typically consumes about 12 gallons per run, instead of washing dishes by hand with the tap running—but only if you run the dishwasher with a full load of dishes. That 12 gallons of water is used regardless of whether the dishwasher is full or half empty.

Remember that wasting water costs you money. Conserving water at home reduces not only your water bill and your heating bill: if you are using less hot water, you are also using less energy to heat that water.

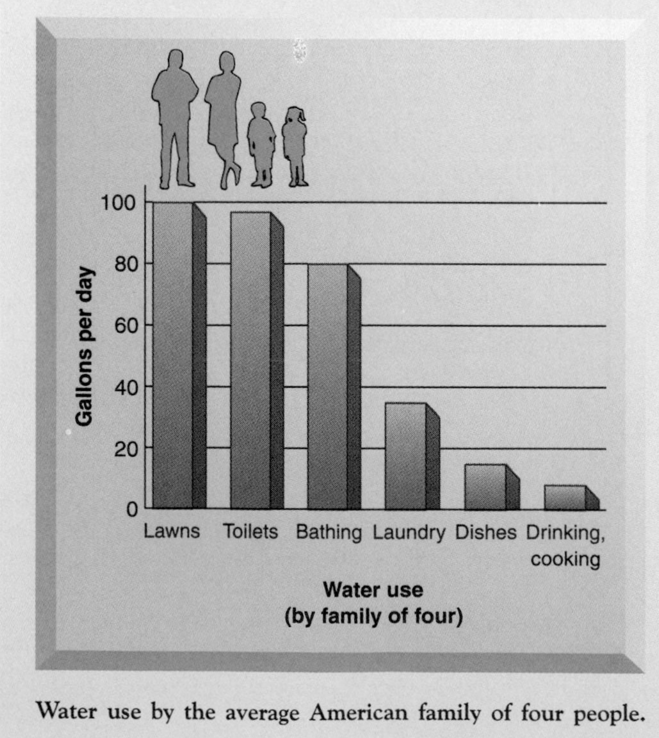

Water use by the average American family of four people.

SUMMARY

1. Water has many special properties, including a high heat capacity and a high dissolving ability. Many of the properties of water are the result of its polarity, which causes hydrogen bonding between water molecules.

2. Although a large portion of the Earth is covered by water, only a small percentage of it is available as fresh water. Fresh water occurs as groundwater, which is stored in aquifers, and as surface water.

3. Flooding occurs as a result of too much fresh water entering a particular area. Flood damage is exacerbated by the deforestation of hillsides and mountains and by the development of flood plains.

4. Although there is enough fresh water to support life on Earth, it is not evenly distributed. Some areas are barely able to support human life because of the shortage of water.

5. Water shortages can be dealt with in two ways. The amount of available water can be increased by damming rivers, tapping additional groundwater, diverting additional water from other areas, or desalinization. Alternatively, water conservation measures can make the present supply adequate.

6. Irrigation is the single largest use of water. Although irrigation increases agricultural productivity, it can contribute to water pollution, soil salinization, and depletion of water supplies. New agricultural techniques, including microirrigation and field leveling, can significantly cut agricultural water consumption.

7. After agriculture, the two largest users of water are industry and domestic/municipal water use. Water conservation, including recycling and reuse, can reduce both industrial and municipal water consumption.

8. Water management of both surface water and groundwater should have the long-term goal of developing a sustainable resource rather than the short-term goal of providing water in limitless supply.

THINKING ABOUT THE ENVIRONMENT

1. Why is the water accord discussed in the chapter introduction considered a model for water resource planning?

2. Discuss the dissolving ability of water as it relates to ocean salinity. To water pollution.

3. Are our water supply problems largely the result of too many people? Give reasons why you support or refute this idea.

4. Discuss the problems that result from withdrawing too much groundwater from a particular area.

5. Imagine you are a water manager for a Southwestern metropolitan district with a severe water shortage. What strategies would you use to develop a sustainable water supply?

6. How is water used in agriculture? Discuss two ways agricultural water can be conserved.

7. Which industries consume the most water? Discuss one way in which water can be conserved in industry.

8. Irrigation water is consumed, but water used by cities and industries is not. Explain the distinction and its significance.

9. Develop a detailed water conservation plan for your own personal daily use. How do you think you could save the most water?

10. Explain how water resource problems might contribute to economic or political instability.

11. Should housing be allowed on the flood plain of a river? Should taxpayers provide federal disaster assistance for those who choose to live on flood plains?

* 12. Water can be saved at home by repairing leaky faucets and toilets. Calculate how many gallons of water are wasted each year from a leak of one drop per second. (13,140 drops equals one gallon.)

RESEARCH PROJECTS

Investigate the water issues in your area. Contact your local water district and determine the perspectives of its water planners.

Read *Rivers of Eden: The Struggle for Water and the Quest for Peace in the Middle East*, by Daniel Hillel, Oxford University Press, 1994. Report to the class any insights you may have gained by reading this provocative book.

On-line materials relating to this chapter are on the ► World Wide Web at **http://www.saunderscollege.com/lifesci/environment2** (select Chapter 13 from the Table of Contents).

SUGGESTED READING

Abramovitz, J. N. "Freshwater Failures: The Crises on Five Continents," *WorldWatch* Vol. 8, No. 5, September–October 1995. Examines problems with five of the world's great freshwater systems.

Frederick, K. D. "Water as a Source of International Conflict," *Resources*, No. 123, Spring 1996. Competition for water is playing an increasingly important role in international relations.

Mairson, A. "The Great Flood of '93," *National Geographic*, January 1994. Dramatically portrays the devastation and suffering caused by the Midwest floods.

Maxwell, J. "Swimming with Salmon," *Natural History*, September 1995. Examines the status of wild salmon stocks in the Pacific Northwest.

Pearce, F. "Poisoned Waters," *New Scientist* Vol. 148, October 21, 1995. The Aral Sea has all but disappeared, leaving behind a host of suspected health problems.

Postel, S. "Where Have All the Rivers Gone?" *WorldWatch* Vol. 8, No. 3, May–June 1995. Highlights the overconsumption of water from such rivers as the Colorado and the Nile.

Platt, A. E. "Dying Seas," *WorldWatch* Vol. 8, No. 1, January–February 1995. Highlights the environmental ills of seven different seas: the Black, Mediterranean, Baltic, Caspian, Yellow, Bering, and South China Seas.

Water: A Portrait in Words and Pictures. Washington, D.C.: National Geographic, *special edition,* November 1993. Examines the excessive consumption of water by people in the United States.

Watson, B. "A Town Makes History by Rising to New Heights," *Smithsonian*, June 1996. The people of Valmeyer, Illinois, moved their town after the floods of 1993 destroyed it.

Zwingle, E. "Ogallala Aquifer: Wellspring of the High Plains," *National Geographic* Vol. 183, No. 3, March 1993. A picture essay about the ancient groundwater that irrigates the High Plains.

A section of China's Great Green Wall, in Xinjiang Province. (Wolfgang Kaehler)

Soils and Their Preservation

During its long history, China has had to contend with both invaders and devastating dust storms sweeping across its northern border. The Great Wall of China, a centuries-old earth and stone structure some 2400 kilometers (1500 miles) long, was built across northern China to defend it against Mongolian raids. Although China's Great Wall was only partially successful in keeping out invaders, a more recently constructed "wall" has been very effective in controlling the dust storms from the Gobi Desert. The new wall, often called the Great Green Wall, consists of some 300 million trees that were planted beginning in the 1950s. This wall of forest extends for some 4800 kilometers (3000 miles) and is 800 kilometers (500 miles) wide in places.

An average dust storm in northern China transports as much as 100 megatons of soil particles for hundreds or even thousands of kilometers. The dust damages crops, grounds air traffic, and causes many other inconveniences. The Great Green Wall, however, has reduced both the frequency and severity of dust storms because the trees slow the wind's velocity, and the moist forest floor discourages additional soil from being picked up and carried by the wind. During the 1950s Beijing, which is 300 miles downwind of the Gobi Desert, experienced 10 to 20 major dust storms per year. By the 1970s, the annual number had declined to fewer than five, and in the 1990s, even a single dust storm per year is unusual.

Soil is a valuable natural resource on which humans depend for food. Many human activities cause or accentuate soil problems, including erosion and mineral depletion. Removal of forests and overgrazing, for example, exacerbated the dust storms coming out of the Gobi Desert before the Great Green Wall was planted. The goals of soil conservation are to minimize soil erosion and maintain soil fertility so that this resource can be used in a sustainable fashion. Conserving our soil resources is critical to human survival because more than 99 percent of our food comes from the land. ◄

WHAT IS SOIL?

Soil is the ground underfoot, a relatively thin layer of the Earth's crust that has been modified by the natural actions of agents such as weather, wind, water, and organisms. It is easy to take soil for granted. We walk on and over it throughout our lives but rarely stop to think about how important it is to our survival.

Vast numbers and kinds of organisms inhabit soil and depend on it for shelter and food. Plants anchor themselves in soil, and from it they receive essential minerals and water. Thirteen of the 16 different elements essential for plant growth are obtained directly from the soil (Table 14–1). Terrestrial plants could not survive without soil, and because we depend on plants for our food, humans could not exist without soil, either.

How Soils Are Formed

Soil is formed from rock, called parent material, that is slowly broken down into smaller and smaller particles by chemical and physical **weathering processes** in nature. It takes a very long time, sometimes thousands of years, for rock to disintegrate into finer and finer mineral particles. To form 2.5 centimeters (one inch) of topsoil, for example, may require between 200 and 1000 years. Time is also required for organic material to accumulate in the soil. Soil formation is a continuous process that involves interactions between the Earth's solid crust and the biosphere. The weathering of parent material beneath soil that has already formed continues to add new soil. The thickness of soil varies from a thin film on very young lands, near the poles and on the tops of mountains, to

Table 14–1
Essential Elements for Plant Growth

Element	Source	Significant Function in Plants
Carbon	Air (as CO_2)	Part of important biological molecules
Hydrogen	Water	Part of important biological molecules
Oxygen	Water, air (as O_2)	Part of important biological molecules
Nitrogen	Soil	Part of important biological molecules
Phosphorus	Soil	In genetic molecules and energy molecules
Calcium	Soil	Part of cell walls
Magnesium	Soil	In chlorophyll molecules
Sulfur	Soil	In proteins and vitamins
Potassium	Soil	Involved in ionic balance of cells
Chlorine	Soil	Involved in ionic balance of cells
Iron	Soil	Involved in photosynthesis and respiration
Manganese	Soil	Involved in respiration and nitrogen metabolism
Copper	Soil	Involved in photosynthesis
Zinc	Soil	Involved in respiration and nitrogen metabolism
Molybdenum	Soil	Involved in nitrogen metabolism
Boron	Soil	Exact role unclear

more than 3 meters (10 feet) on very old lands, such as certain forests.

Organisms and climate both play essential roles in weathering, sometimes working together. For example, soil organisms produce acids that etch tiny cracks in the rock. In temperate climates, water from precipitation seeps into these cracks, which enlarge when the water freezes. Over many seasons, alternate freezing and thawing cause small pieces of the rocks to break off.

Topography, a region's surface features—such as the presence or absence of mountains and valleys—is also involved in soil formation. Steep slopes often have very little or no soil on them because soil and rock are continually transported down the slopes by gravity; runoff from precipitation tends to amplify erosion on steep slopes. Moderate slopes, on the other hand, may encourage the formation of deep soils.

SOIL STRUCTURE

Soil is composed of four distinct parts—mineral particles, organic matter, water, and air—and occurs in layers, each of which has a certain composition and special properties. The plants, animals, fungi, and microorganisms that inhabit soil interact with it, and minerals are continually cycled from the soil to organisms and back to the soil.

Components of Soil

The mineral portion, which comes from weathered rock, forms the basic soil material. It provides anchorage and essential minerals for plants, as well as pore space for water and air. Because the mineral compositions of rocks vary at different locations, the soils that develop from them vary in mineral composition and chemical properties. Rocks that are rich in aluminum form acidic soils, for example, whereas rocks that contain silicates of magnesium and iron form soils that may be deficient in calcium, nitrogen, and phosphorus.

The age of a soil also affects its mineral composition. In general, older soils are more weathered and lower in certain essential minerals. Large portions of Australia, South America, and India have old, infertile soils. In contrast, in geologically recent time, glaciers passed across much of the Northern Hemisphere, pulverizing bedrock and forming fertile soils.[1] Essential minerals are readily available in these geologically young soils and in young soils formed in areas of volcanic activity.

[1]The Pleistocene epoch, which began approximately 2 million years ago, was marked by four periods of glaciation. At their greatest extent, these ice sheets covered nearly 4 million square miles of North America, extending south as far as the Ohio and Missouri rivers.

Figure 14–1
Humus is partially decomposed organic material, primarily from plant and animal remains. Soil that is rich in humus has a loose, somewhat spongy structure with several properties, such as increased water-holding capacity, that are beneficial for plants and other organisms living in it. (USDA/Natural Resources Conservation Service)

The litter, droppings, and remains of plants, animals, fungi, and microorganisms in various stages of decomposition constitute the organic portion of soils. Microorganisms, particularly bacteria and fungi, gradually decompose this material, and the black or dark brown organic material that remains after much decomposition has occurred is called **humus** (Figure 14–1). Certain components of humus may persist in the soil for hundreds of years. Although humus is somewhat resistant to further decay, a succession of microorganisms gradually reduces it to carbon dioxide, water, and minerals. Detritus-feeding animals such as earthworms, termites, and ants also help break down humus.

As organic material is decomposed, essential minerals are released into the soil, where they may be absorbed by plant roots. Organic matter also increases the soil's water-holding capacity by acting as a sponge.

The pore space between soil particles occupies roughly 35 to 60 percent of a soil's volume and is filled with varying proportions of water (called **soil water**) and air (called **soil air**) (Figure 14–2); both are necessary to sustain all the organisms living in the soil.

Soil water contains low concentrations of dissolved mineral salts that enter the roots of plants as they absorb the water. Soil water not absorbed by plants moves down through the soil, carrying dissolved minerals with it. The removal of dissolved materials from the soil by water percolating downward is called **leaching**. The deposit of leached material in the lower layers of soil is known as **illuviation**. Iron and aluminum compounds, humus, and clay are some illuvial materials that can gather in the subsurface portion of the soil. Some substances completely

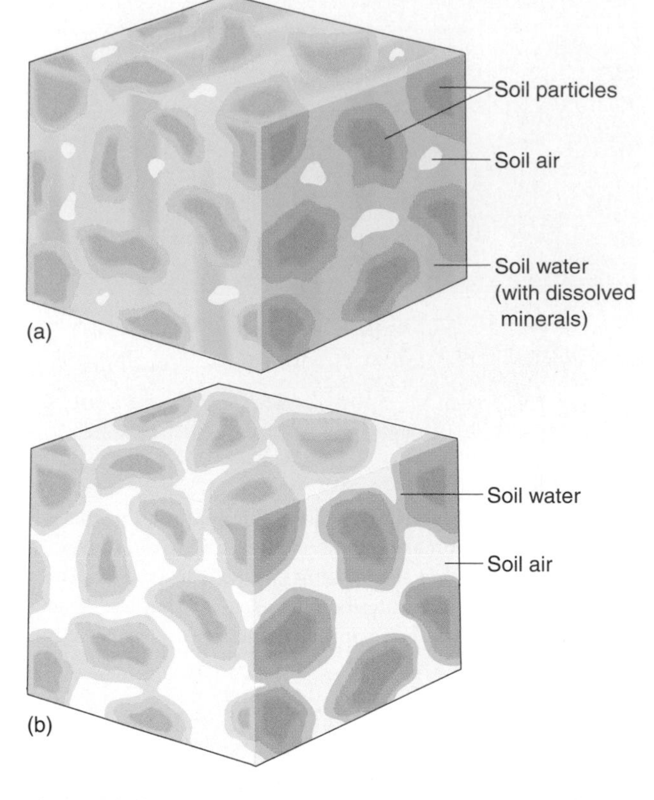

Figure 14–2

Pore space in soil is occupied by varying amounts of soil air and soil water. (a) In a wet soil, most of the pore space is filled with water. (b) In a dry soil, a thin film of water is tightly bound to soil particles, and most of the pore space is occupied by soil air.

THE FUTURE OF EVERGLADES SOIL

How are Everglades soils different from the vast majority of the world's soils? The mineral portion makes up the largest percentage by weight—at least 50 percent—of most of the world's soils. Most Everglades soils, however, only contain 15 percent or less of mineral particles. Instead, Everglades soils are formed primarily of organic material—heaps of partly decomposed sawgrass that has piled up for centuries over a hard layer of limestone bedrock. The organic material never decomposed completely because it was buried under water. When Everglades soil is exposed to oxygen, however, the organic material quickly decomposes to carbon dioxide and water.

In the early 20th century, most areas of the Everglades had a soil layer as thick as 3.7 meters (12 feet). In the Everglades Agricultural Area, where sugar and other crops have been grown for decades, the soil has subsided dramatically. Scientists estimate that by the year 2000, 45 percent of the soils in this area will be less than 0.3 meter (1 foot) thick, which is too thin to grow sugar and most other crops economically. This loss of Everglades soil is unavoidable. As the ground is drained and worked to raise crops, the soil is exposed to air, and decomposition is accelerated. (Other aspects of the Everglades are discussed in Chapter 6.)

leach out of the soil because they are so soluble that they migrate right down to the groundwater. It is also possible for water to move *upward* in the soil, transporting dissolved materials with it, as when the water table rises.

Soil air contains the same gases as atmospheric air, although they are usually present in different proportions. Generally, as a result of respiration by soil organisms, there is more carbon dioxide and less oxygen in soil air than in atmospheric air. Among the important gases in soil air are (1) oxygen, required by soil organisms for respiration; (2) nitrogen, used by nitrogen-fixing soil organisms (see Chapter 5); and (3) carbon dioxide, involved in soil weathering. Carbon dioxide dissolves in water to form carbonic acid, H_2CO_3, a weak acid that helps to weather rock.

Soil Horizons

A deep vertical slice, or section, through many soils reveals that they are organized into distinctive horizontal layers called **soil horizons**. A **soil profile** is a section from surface to parent material, showing the soil horizons (Figure 14–3).

The uppermost layer of soil, the **O-horizon**, is rich in organic material. Plant litter, including dead leaves and stems, accumulates in the O-horizon and gradually decays. In desert soils the O-horizon is often completely absent, but in certain organic-rich soils it may be the dominant layer.

Just beneath the O-horizon is the topsoil, or **A-horizon**, which is dark and rich in accumulated humus. The A-horizon has a granular texture and is somewhat nutrient-poor due to the gradual loss of many nutrients to deeper layers by leaching. In some soils, a heavily leached **E-horizon** develops between the A- and B-horizons.

The **B-horizon**, the light-colored subsoil beneath the A-horizon, is often a zone of illuviation in which minerals that leached out of the topsoil and litter accumulate. It is typically rich in iron and aluminum compounds and clay.

Beneath the B-horizon is the **C-horizon**, which contains weathered pieces of rock and borders the unweath-

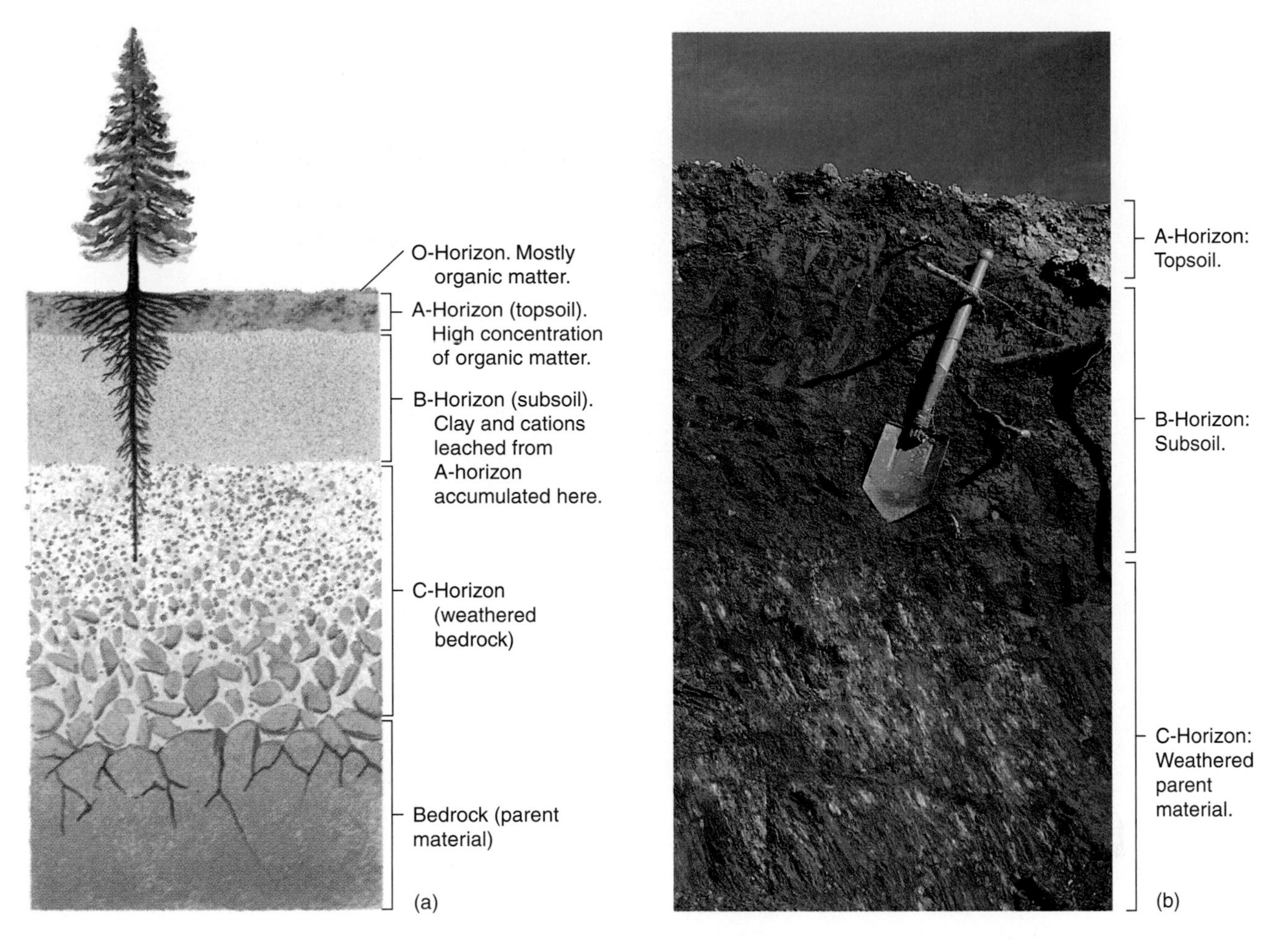

Figure 14–3

Many soils have multiple horizons or layers that are revealed in a soil profile. (a) A "typical" soil profile. Each horizon has its own chemical and physical properties. (b) This particular soil, located on a farm in Virginia, has no O-horizon because it is used for agriculture; the surface litter that would normally comprise the O-horizon was plowed into the A-horizon. The shovel gives an idea of the relative depths of each horizon. (Photo by Ray Weil, courtesy of Martin Rabenhorst)

ered solid parent material. The C-horizon is below the extent of most roots and is often saturated with groundwater.

Soil Organisms

Although soil organisms are usually hidden underground, their numbers are huge. Millions of soil organisms may inhabit just one teaspoon of fertile agricultural soil! In the soil ecosystem, bacteria, fungi, algae, worms, protozoa, insects, plant roots, and larger animals such as moles, snakes, and groundhogs all interact with each other and with the soil (Figure 14–4).

Earthworms, probably one of the most familiar soil inhabitants, eat soil and obtain energy and raw materials by digesting humus. **Castings**, bits of soil that have passed through the gut of an earthworm, are deposited on the soil surface. In this way, minerals from deeper layers in the soil are brought to upper layers. Earthworm tunnels serve to aerate the soil, and the worms' waste products and corpses add organic material to deeper layers of the soil.

Ants live in the soil in enormous numbers, constructing tunnels and chambers that help to aerate the soil. Members of soil-dwelling ant colonies forage on the surface for bits of food, which they carry back to their nests. Not all of this food is eaten, however, and its eventual decomposition helps increase the organic matter in the soil. Many ants are also indispensable in plant reproduction because they bury seeds in the soil.

Plants are greatly affected by the properties of soil, although most plants can tolerate a wide range of soil types. Soil, in turn, is affected by the types of plants that grow on it. As a result of the complex interactions among plants, climate, and soil, it is hard to specify cause and

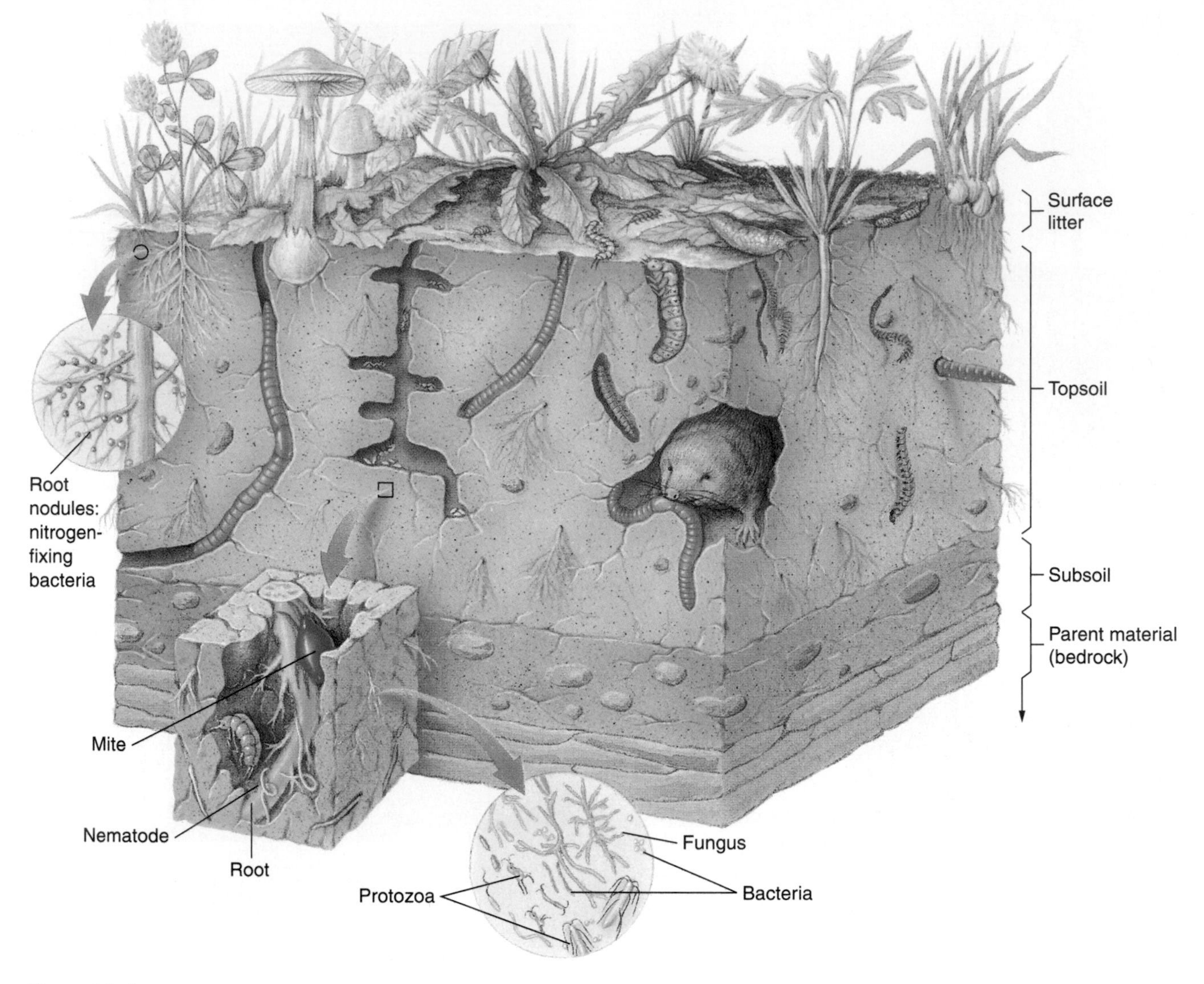

Figure 14–4

The diversity of life in fertile soil is remarkable. It includes plants, algae, fungi, earthworms, flatworms, roundworms, insects, spiders and mites, bacteria, and burrowing animals such as moles and groundhogs.

effect in their relationships. For example, are the plants growing in a certain locality because of the soil that is found there, or is the soil's type determined by the plants?

One very important symbiotic relationship in the soil occurs between fungi and the roots of vascular plants. These associations, called **mycorrhizae**, enable plants to absorb adequate amounts of essential minerals from the soil. The threadlike body of the fungal partner, called a **mycelium**, extends into the soil well beyond the roots. Minerals absorbed from the soil by the fungus are transferred to the plant, while food produced by photosynthesis in the plant is delivered to the fungus. Mycorrhizae have been demonstrated to enhance the growth of plants (see Figure 4–6).

Nutrient Recycling

In a balanced ecosystem, the relationships among soil and the organisms that live in and on it ensure soil fertility. As we saw in Chapter 5, essential minerals such as nitrogen and phosphorus are cycled from the soil to organisms and back again to the soil (Figure 14–5). Bacteria and fungi decompose plant and animal detritus and

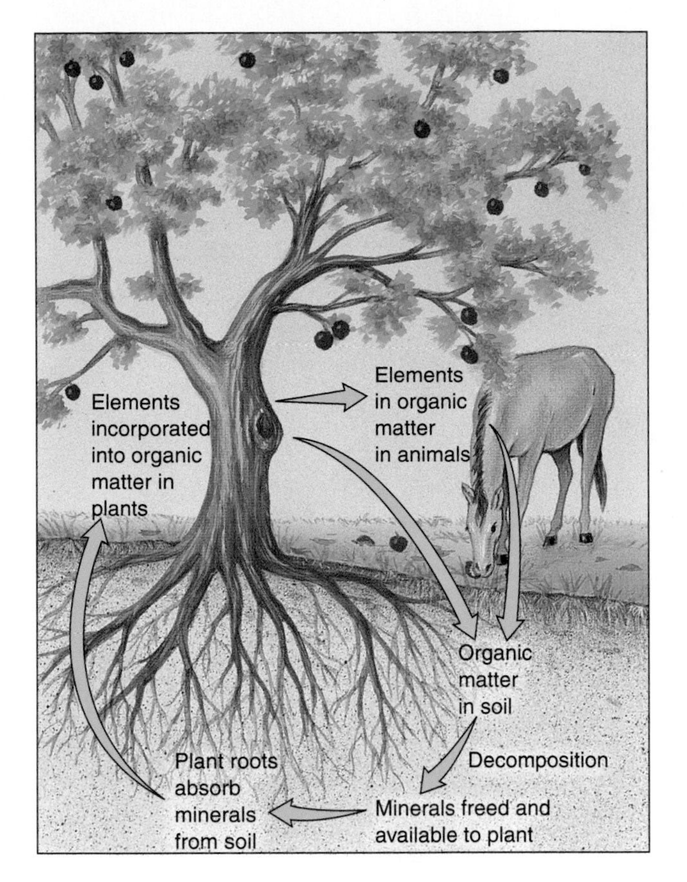

Figure 14–5
In a balanced ecosystem, mineral nutrients cycle from the soil to organisms and then back again to the soil.

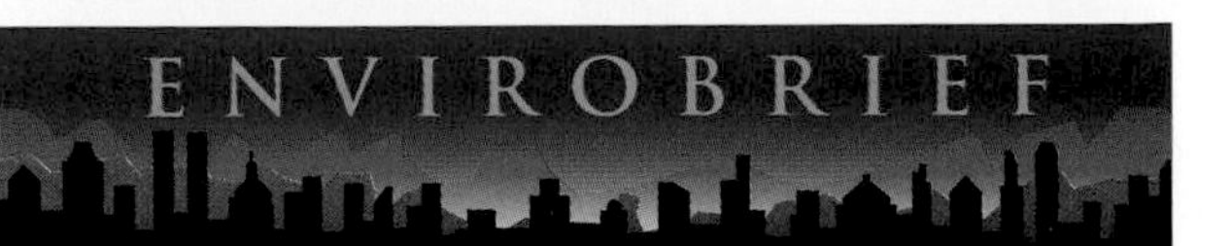

UNDERGROUND EVIDENCE OF CLIMATE CHANGE

Global carbon flows in a balanced cycle from plants to the soil and the atmosphere, from the soil to the atmosphere, and from the atmosphere back to plants. Soil's critical role in this balance is to receive about 60 billion metric tons of carbon each year, which is then transformed into carbon dioxide that is "exhaled" by soil bacteria, earthworms, fungi, and roots. The balance of the carbon cycle could be tilted, however, as levels of atmospheric carbon dioxide continue to increase dramatically with the burning of fossil fuels. If carbon dioxide's greenhouse effect raises atmospheric temperatures, soil's activity could accelerate, causing it to release even more carbon dioxide. Because carbon flux is so complex, it is extremely difficult for scientists to predict how higher concentrations of carbon dioxide will affect the balance between soil, air, and plants, especially when many other human influences could also alter the cycle. (We discuss global climate change in Chapter 20.)

wastes, releasing nutrients into the soil to be used again. Although leaching causes some minerals to be lost from the soil ecosystem to groundwater, the weathering of the parent material replaces much or all of them.

Mini-Glossary of Soil Terms

humus: Partly decomposed organic material in the soil; brown or black in color.

leaching: The movement of water and dissolved material downward through the soil.

illuviation: The deposit of a material into a lower layer of the soil from a higher layer. Illuviation is the result of leaching.

sand: Large (0.05 to 2-mm diameter) inorganic particles in the soil. Coarse enough to feel gritty.

silt: Medium-sized (0.002 to 0.05-mm diameter) inorganic particles in the soil. Too small to feel gritty.

clay: Small (<0.002-mm diameter) inorganic particles in the soil. Too small to settle out of suspension.

PHYSICAL AND CHEMICAL PROPERTIES OF SOIL

Texture and acidity are two parameters that help to characterize soils.

Soil Texture

The texture of a soil is determined by the amounts—that is, percentages by weight—of different-sized inorganic particles. Large particles (0.05 to 2 millimeters in diameter) are called **sand**, medium-sized particles (0.002 to 0.05 millimeters in diameter) are called **silt**, and small particles (<0.002 millimeters in diameter) are called **clay** (Figure 14–6). Sand particles are large enough to be seen easily with the naked eye; silt particles, which are about the size of flour particles, are barely visible with the naked eye; and clay particles are too small to be seen with an ordinary light microscope. These size assignments for sand, silt, and clay are arbitrary; they give soil scientists a way to classify soil texture. Obviously, soil particles form a continuum of sizes from very large to very small. Soil always contains a mixture of different-sized particles, but the proportions vary from soil to soil. **Loam**, which makes

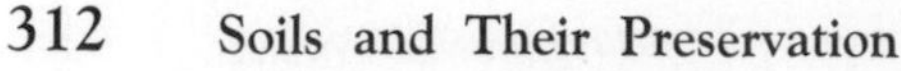

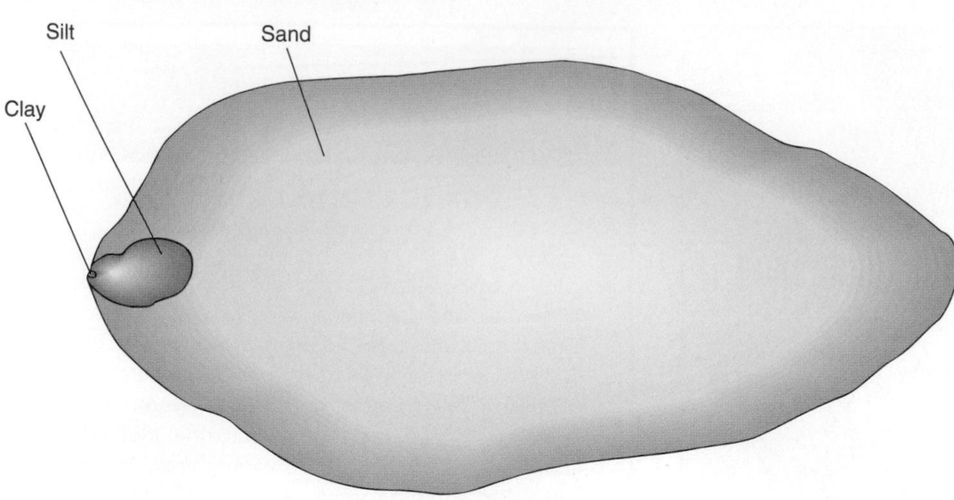

Figure 14–6
This diagram compares the relative sizes of sand, silt, and clay particles.

ideal agricultural soil for most climates, has approximately equal portions of sand, silt, and clay.

A soil's texture affects many of that soil's properties, which, in turn, influence plant growth (Table 14–2). Generally, the larger particles provide structural support, aeration, and permeability to the soil, whereas the smaller particles bind together into aggregates, or clumps, and hold nutrients and water.

Clay is particularly important in determining many soil characteristics because it has the greatest surface area for chemical reactions. If the surface areas of about 450 grams (1 pound) of clay particles were laid out side by side, they would occupy 1 hectare (2.5 acres). Each grain of clay has negative charges on its outer surface that attract and bind positively charged mineral ions (Figure 14–7).[2] Many of these mineral ions, such as potassium (K^+) and magnesium (Mg^{2+}), are essential for plant growth and are "held" in the soil for plant use by their interactions with clay particles.

Table 14–2
Soil Properties Affected by Soil Texture

	Soil Texture Type		
Soil Property	**Sandy soil**	**Loam**	**Clay soil**
Aeration	Excellent	Good	Poor
Drainage	Excellent	Good	Poor
Nutrient-holding capacity	Low	Medium	High
Water-holding capacity	Low	Medium	High
Workability (tillage)	Easy	Moderate	Difficult

Soil Acidity

Soil acidity is measured using the pH scale, which extends from 0 (extremely acidic) through 7 (neutral) to 14 (extremely alkaline). (See Appendix I for a discussion of pH.) The pH of most soils ranges from 4 to 8, but some soils are outside this range. The soil of the Pygmy Forest in Mendocino County, California, is extremely acidic, with a pH of 2.8 to 3.9. At the other extreme, certain soils in Death Valley, California, have a pH of 10.5.

Plants are affected by soil pH partly because the solubility of certain minerals varies with differences in pH, and soluble mineral elements can be absorbed by the plant, whereas insoluble forms cannot. At a low pH, for example, the aluminum and manganese in soil water are more soluble, sometimes becoming toxic to plants because they are absorbed by the roots in greater concentrations than are good for the plant. Other mineral salts that are essential for plant growth, such as calcium phosphate, become less soluble at higher pH's.

Soil pH greatly affects the availability of nutrients. An acidic soil has less ability to bind positively charged ions to it. As a consequence, certain mineral ions that are essential for plant growth, such as potassium (K^+), are leached more readily from acidic soil. The optimum soil pH for most plant growth is 6.5 to 7.5, because most nutrients needed by plants are available to the plants in that pH range.

Soil pH is affected by the types of plants growing on it. Soil litter composed of the needles of conifers, for ex-

[2]Soil minerals are often present in charged forms called **ions**. Mineral ions may be positively charged (K^+, for example) or negatively charged (NO_3^-, for example).

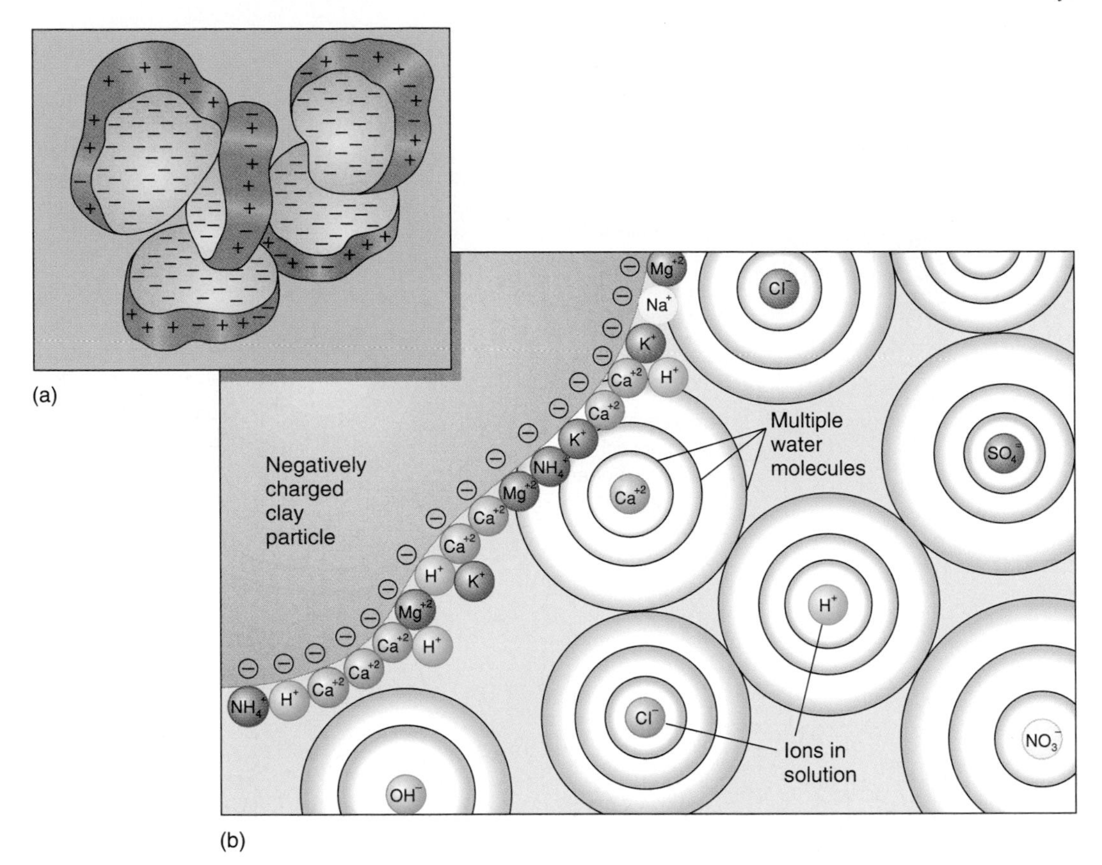

Figure 14–7

Clay particles help hold positively charged mineral ions in the soil where they can be absorbed by plant roots. (a) Positively charged ions are attracted to the broad, negatively charged surfaces of clay particles. (b) A closeup of part of a clay particle and the thin film of water around it. The numerous water molecules are represented by circles surrounding mineral ions that are dissolved in the soil water. Note the large number of positively charged ions attracted to the surface of the clay particle.

ample, contains acids that leach into the soil, lowering its pH.

MAJOR SOIL GROUPS

Variations in climate, vegetation, parent material, topography, and soil age throughout the world result in thousands of soil types that differ in color, depth, mineral content, acidity, pore space, and other properties. In the United States alone, as many as 17,000 different soil types have been identified. Here we focus on five soil groups that are very common: spodosols, alfisols, mollisols, aridisols, and oxisols (Figure 14–8).

Regions with colder climates, ample precipitation, and good drainage typically have soils called **spodosols**, with very distinct layers. A spodosol usually forms under a coniferous forest and has a layer of acidic litter composed primarily of needles; an ash-gray, acidic, leached E-horizon; and a dark brown, illuvial B-horizon. Spodosols do not make good farmland because they are too acidic and are nutrient-poor due to leaching.

Temperate deciduous forests grow on **alfisols**, soils with a brown to gray-brown A-horizon. Precipitation is great enough to wash much of the clay and soluble minerals out of the A- and E-horizons. When the deciduous forest is intact, soil fertility is maintained by a continual supply of plant litter such as leaves and twigs. When the soil is cleared for farmland, however, fertilizers (which contain nutrients such as nitrogen, potassium, and phosphorus) must be used to maintain fertility.

Mollisols, found primarily in temperate, semiarid grasslands, are very fertile soils. They possess a thick, dark brown to black A-horizon that is rich in humus. Some soluble minerals remain in the upper layers, because precipitation is not great enough to leach them into lower layers. Most of the world's grain crops are grown on mollisols.

Aridisols are found in arid regions of all continents. The lack of precipitation in these deserts precludes much

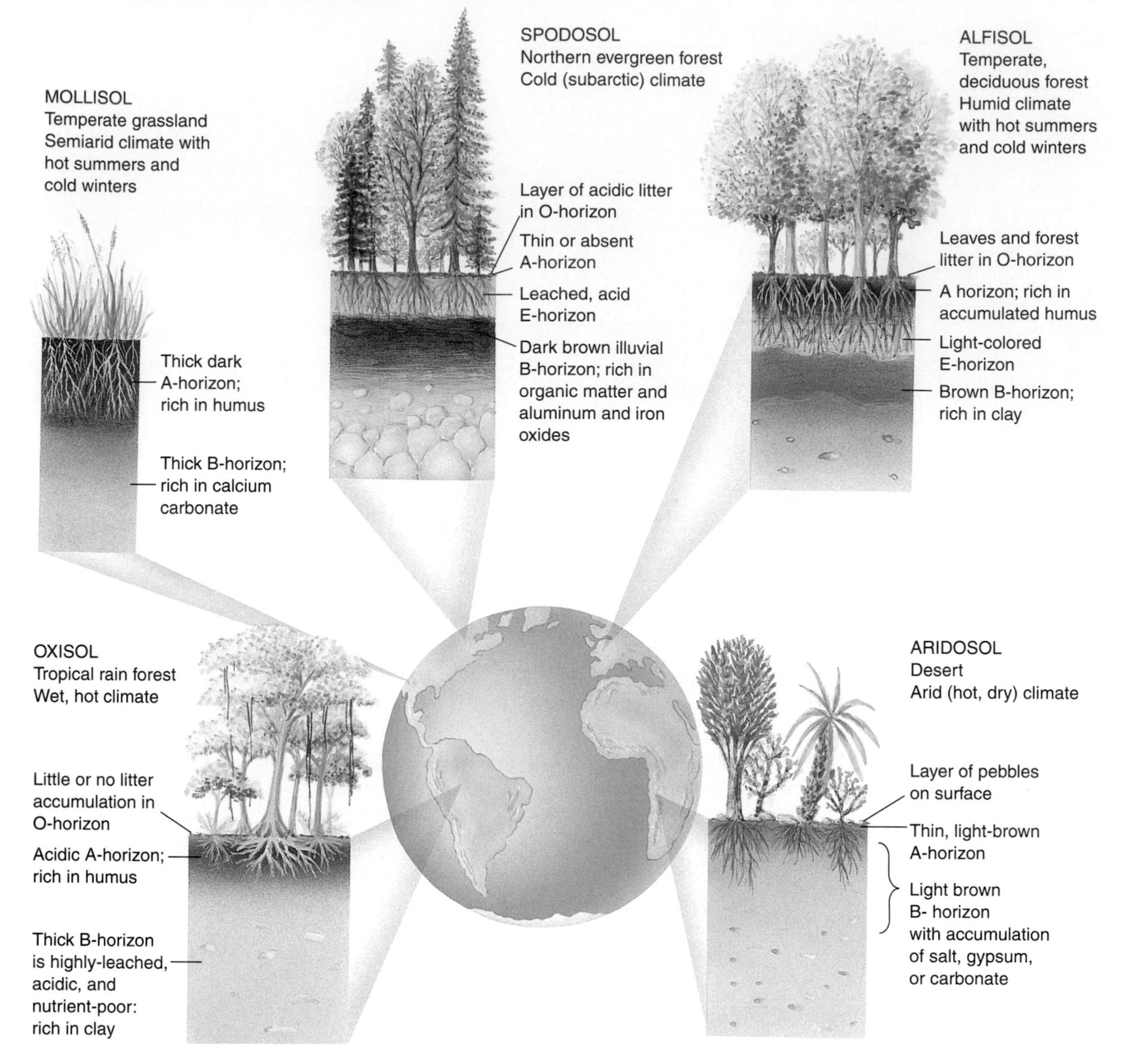

Figure 14–8

Five major soil groups are depicted, along with the vegetation and climate that are characteristic of each. Climate and vegetation help determine soil type.

leaching, and the lack of lush vegetation precludes the accumulation of much organic matter. As a result, aridisols do not usually have distinct layers of leaching and illuviation. Some aridisols provide rangeland for grazing animals, and crops can be grown on aridisols if water is supplied by irrigation.

Oxisols, which are low in nutrients, exist in tropical and subtropical areas with ample precipitation. Little organic material accumulates on the forest floor (O-horizon) because leaves and twigs are rapidly decomposed. The A-horizon is enriched with humus derived from the rapidly decaying plant parts. The B-horizon, which is quite thick, is highly leached, acidic, and nutrient-poor. Oddly enough, tropical rain forests, with their lush vegetation, grow on oxisols. Most of the minerals in tropical rain forests are locked up in the vegetation rather than in the soil. As soon as plant and animal remains touch the forest floor, they promptly decay, and the minerals are quickly reabsorbed by plant roots. Even wood, which may take years to be recycled in temperate soils, is decomposed rapidly (in a matter of months) in tropical rain forests by subterranean termites.

SOIL PROBLEMS

Human activities often cause or exacerbate soil problems, including erosion and mineral depletion of the soil, both of which occur worldwide.

Soil Erosion

Water, wind, ice, and other agents promote **soil erosion**, the wearing away or removal of soil from the land. Water and wind are particularly effective in removing soil. Rainfall loosens soil particles, which can then be transported away by moving water (Figure 14–9). Wind loosens soil and blows it away, particularly if the soil is barren and dry.

Because erosion reduces the amount of soil in an area, it limits the growth of plants. Erosion also causes a loss of soil fertility, because essential minerals and organic matter that are part of the soil are also removed. As a result of these losses, the productivity of eroded agricultural soils drops, and more fertilizer must be used to replace the nutrients lost to erosion.

Soil erosion is a national and international problem that rarely makes the headlines. To get a feeling for how serious the problem is, consider that approximately 4.0 billion metric tons (4.4 billion tons) of topsoil are lost *each year* from U.S. croplands and pasturelands as a result of soil erosion. The U.S. Department of Agriculture estimates that approximately one-fifth of U.S. cropland is vulnerable to soil erosion damage.

Figure 14–9
Soil erosion caused by water. Runoff from an open field carries soil sediments, fertilizers, and pesticides with it. Although the gullies in this field are quite conspicuous, most soil loss by erosion is unnoticeable. (USDA/Natural Resources Conservation Service)

THE PERILS OF COTTON

Most people think of cotton as a natural fiber that is not a threat to the environment. Not so. Cotton crops receive about 1.7 kilograms per hectare (1.5 pounds per acre) of insecticide–an amount equal to about one-fourth of the total insecticide applied to all major field crops. In addition to requiring large doses of pesticides and chemical fertilizers, cotton is commonly grown in hot, dry American states such as Arizona, California, and Texas, where it needs plenty of irrigation. The irrigating depletes groundwater reservoirs and contributes to soil erosion. It is estimated that 15 tons of topsoil are lost each year from Texas farmland that yields only a quarter of a ton of cotton fiber.

Humans often accelerate soil erosion with poor soil management practices. Here we consider soil erosion caused by agriculture in some detail, but it is important to realize that agriculture is not the only culprit. Removal of natural plant communities during the construction of roads and buildings also accelerates erosion. Unsound logging practices, such as clearcutting large forested areas, cause severe erosion.

Soil erosion has an impact on other resources as well. Sediment that gets into streams, rivers, and lakes affects water quality and fish habitats. If the sediment contains pesticide and fertilizer residues, they further pollute the water. Also, when forests are removed within the watershed of a hydroelectric power facility, accelerated soil erosion can cause the reservoir behind the dam to fill in with sediment much faster than usual. This process results in a reduction of electricity production at that facility.

Sufficient plant cover limits the amount of soil erosion: leaves and stems cushion the impact of rainfall, and roots help to hold the soil in place. Although soil erosion is a natural process, abundant plant cover makes it negligible in many natural ecosystems.

Wind erosion in grasslands

Semiarid lands, such as the Great Plains of North America, have low annual precipitation and are subject to periodic droughts. Prairie and steppe grasses, the plants that grow best in semiarid lands, are adapted to survive droughts.[3] Although the aboveground portions of the

[3]Prairies are tallgrass communities, whereas steppes, or plains, are shortgrass communities. Differences in annual precipitation account for the differences in species composition. Prairies receive more precipitation than steppes.

plant may die, the root systems can survive several years of drought. When the rains return, the root systems send up new leaves. Soil erosion is minimal because the dormant but living root systems hold the soil in place and resist the assault by water and wind.

The soils of semiarid lands are often of high quality, due largely to the accumulation over many centuries of a thick, rich humus. These lands are excellent for grazing and for growing crops on a small scale. Problems arise, however, when large areas of land are cleared for crops or when the land is overgrazed by animals (see Chapter 17). The removal of the natural plant cover opens the way for climatic conditions to "attack" the soil, and it gradually deteriorates from the onslaught of hot summer sun, occasional violent rainstorms, and wind. If a prolonged drought occurs under such conditions, disaster can strike.

The American Dust Bowl

The effects of wind on soil erosion were vividly experienced over a wide region of the central United States during the 1930s (Figure 14–10). Throughout the late 19th and early 20th centuries, much of the native grasses had been removed to plant wheat. Then, between 1930 and 1937, the semiarid lands stretching from Oklahoma and Texas into Canada received 65 percent less annual precipitation than was normal. The rugged prairie and steppe grasses that had been replaced by crops could have survived these conditions, but not the wheat. The prolonged drought caused crop failures, which left fields barren and particularly vulnerable to wind erosion.

Winds from the west swept across the barren, exposed soil, causing dust storms of incredible magnitude (Figure 14–11). Topsoil from Colorado and Oklahoma was blown eastward for hundreds of kilometers. Women hanging out clean laundry in Georgia went outside later to find it dust-covered. Bakers in New York City had to keep freshly baked bread away from open windows so it would not get dirty. The dust even discolored the Atlantic Ocean several hundred miles off the coast.

The Dust Bowl occurred during the Great Depression, and ranchers and farmers quickly went bankrupt. Many abandoned their dust-choked land and dead livestock and migrated west to the promise of California; their plight is movingly portrayed in the novel *The Grapes of Wrath* by John Steinbeck.

Although the United States no longer has a dust bowl, soil erosion is still a major problem. For example, a 1989 dust storm in western Kansas that occurred after months of drought was so severe that 150 miles of Interstate 70 were closed. We discuss the extent of soil erosion in the United States later in this chapter.

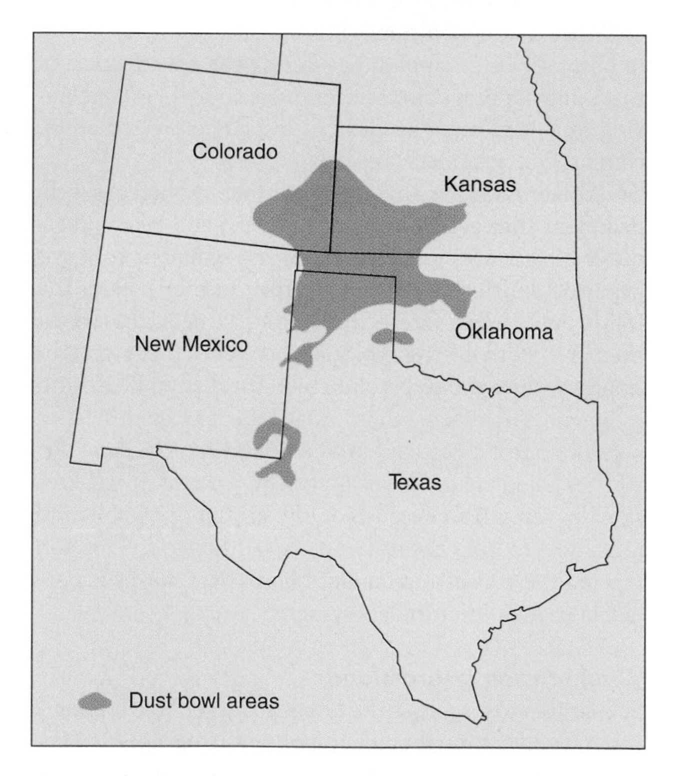

Figure 14–10

More than thirty million hectares (74 million acres) of land in the Great Plains were damaged during the Dust Bowl years. Shaded parts of Colorado, Kansas, Oklahoma, and Texas suffered the most extensive damage.

Mineral Depletion of the Soil

In a natural ecosystem, essential minerals cycle from the soil to organisms, and back again to the soil when those organisms die and decay. An agricultural system disrupts this pattern when the crops are harvested. Much of the plant material, containing minerals, is removed from the cycle, so it fails to decay and release its nutrients back to the soil. Thus, over time, soil that is farmed inevitably loses its fertility (Figure 14–12).

Mineral depletion in tropical rainforest soils

In tropical rain forests, the climate, the typical soil type, and the removal by humans of the natural forest community result in a particularly severe type of mineral depletion. Soils found in tropical rain forests are somewhat nutrient-poor because the nutrients are stored primarily in the vegetation. Any minerals that are released as dead organisms decay in the soil are promptly reabsorbed by plant roots and their mutualistic fungi. If this did not occur, the heavy rainfall would quickly leach the nutrients away. Nutrient reabsorption by vegetation is so effective that tropical rainforest soils can support luxuriant plant growth despite the relative infertility of the soil, as long as the forest remains intact.

(a)

(b)

Figure 14–11
The Dust Bowl years. (a) A family running for shelter during a dust storm. (b) An abandoned Oklahoma farm in 1937. Total devastation was often the aftermath of dust storms.
(a and b, U.S. Department of Agriculture)

When the forest is cleared, whether to sell the wood or to make way for crops or rangeland, its efficient nutrient recycling is disrupted. Removal of the vegetation that so effectively stores the forest's nutrients allows minerals to leach out of the system.

Crops can be grown on these soils for only a few years before the small mineral reserves in the soil are depleted. When cultivation is abandoned, the forest eventually returns to its original state, provided that there is a nearby forest to serve as a source of seeds. The regrowth of forest is a very slow process, however. (Chapter 17 discusses aspects of deforestation other than soil degradation.)

Soil Problems in the United States

Every five years the Natural Resources Conservation Service (NRCS), which was formerly called the Soil Conservation Service, measures the rate of soil erosion at thousands of sites across the United States. It also uses satellite data to estimate annual soil erosion.

These measurements and estimates indicate that erosion is a serious threat to cultivated soils in many regions throughout the United States, particularly in parts of southern Iowa, northern Missouri, western and southern Texas, and eastern Tennessee. The good news is that soil erosion in these and other U.S. croplands declined by about 25 percent between 1982 and 1992 (latest data available), although it is still at significant levels.

The Great Plains and deserts in the western states are particularly vulnerable to wind erosion. When this land is irrigated, crops can be grown without danger of failure. Without irrigation, however, the frequent and prolonged droughts increase the likelihood of crop failures, which result in bare, easily eroded soil. Because of persistent water shortages, particularly in the Southwest, many farmers are abandoning farming altogether. It may take centuries for the abandoned barren land to return to its natural state; until then, it will be susceptible to erosion, especially by wind.

Erosion of soil by water is particularly severe in the midwestern grain belt along the Mississippi and Missouri rivers, as well as in the central valley of California and in the hilly Palouse River region of the northwestern United States. The NRCS estimates that about 25 percent of agricultural land in the United States is losing topsoil faster than it can be regenerated by natural soil-forming processes. This loss is often so gradual that even farmers

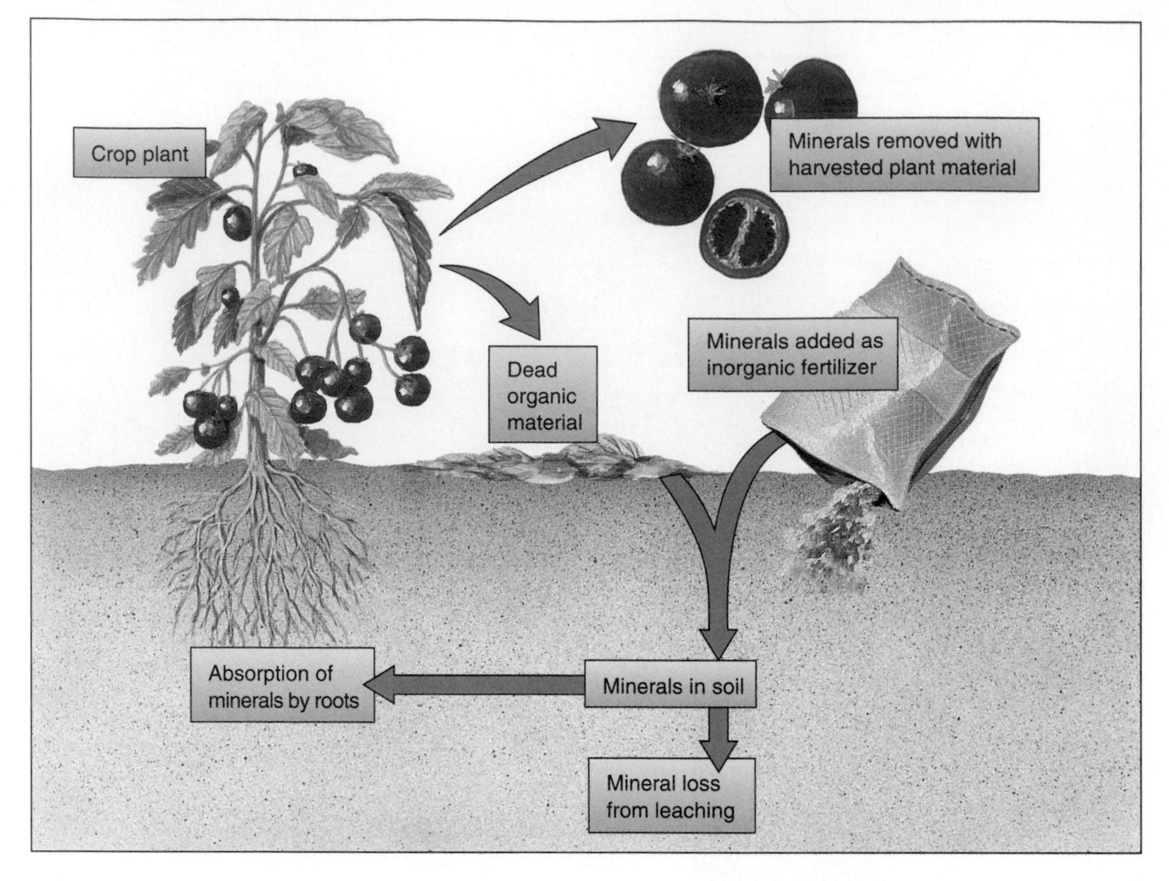

Figure 14–12

Why a soil that is farmed needs fertilizer. As plant and animal detritus decomposes in natural ecosystems, nutrients are cycled back to the soil for reuse. In agriculture, much of the plant material is harvested. Because the mineral nutrients in the harvested portions are unavailable to the soil, the nutrient cycle is broken. For this reason, fertilizer must be added to the soil periodically.

fail to notice it. For example, a severe rainstorm may wash away 1 millimeter (0.04 inch) of soil, which seems insignificant until the cumulative effects of many storms are taken into account. Twenty years of soil erosion amounts to the loss of about 2.5 centimeters (1 inch) of soil, an amount that could take hundreds of years to replace by natural soil-forming processes.

Worldwide Soil Problems

Soil erosion and mineral depletion are significant problems worldwide (Figure 14–13). Although estimates vary widely depending on what assumptions are made, soil erosion results in an annual loss of as much as 75 billion metric tons (83 billion tons) of topsoil around the world. Soil erosion is greatest in Asia, Africa, and Central and South America. It is estimated that more than one billion people depend on agricultural soils that are not productive enough to adequately support them. A combination of factors has caused this situation, including unsound farming methods, extensive soil erosion, and expanding deserts (see Chapter 17). Along with these factors, the needs of a rapidly expanding population exacerbate soil problems worldwide.

The first global assessment of soil conditions, released in 1992, summarized a three-year study of global soil degradation sponsored by the U.N. Environment Program. It reported that 1.96 billion hectares (4.84 billion acres) of soil—an area equal to 17 percent of the Earth's total vegetated surface area—have been degraded since World War II. Eleven percent of the Earth's vegetated surface—an area the size of China and India combined—has been degraded so badly that it will be very costly, or in some cases impossible, to reclaim it. Soil degradation

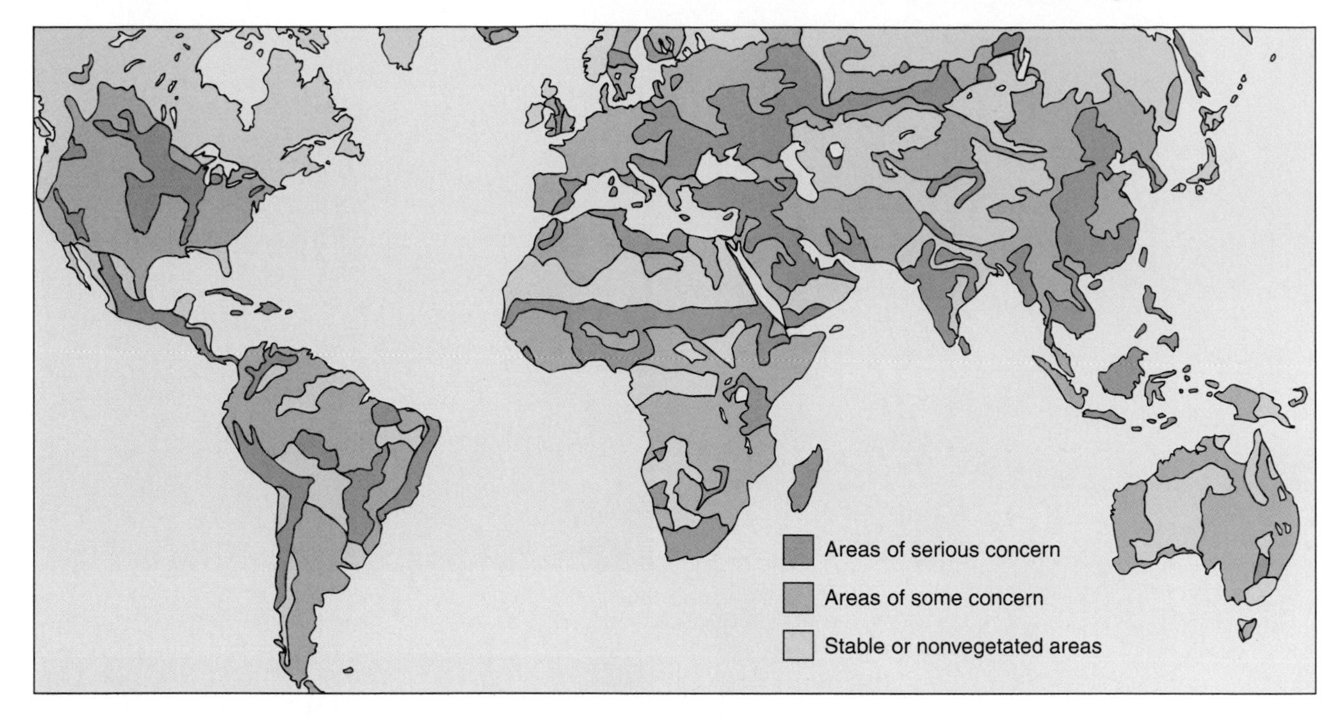

Figure 14–13
Worldwide soil degradation. Areas of serious concern, two-thirds of which are in Asia and Africa, indicate soils that range from moderately to extremely damaged.

is primarily attributed to poor agricultural practices, overgrazing, and deforestation.

Asia and Africa have the largest land areas with extensive soil damage, and in both places the problem is compounded by rapid population growth. The Sahelians in Africa, for example, must use their land to grow crops and animals for food or they will starve, but the soil is so overexploited that it is able to support fewer and fewer people. The day is approaching when the Sahel could become utterly unproductive desert (Figure 14–14). To reclaim the land would require restricting its use for many years so it could recover; but if these measures were taken, the Sahelians would have no means of obtaining food. Fortunately, organizations such as the International Council for Research in Agroforestry are developing techniques to lessen environmental degradation of the Sahel. (Agroforestry is discussed in Chapter 20.)

SOIL CONSERVATION AND REGENERATION

Conservation tillage, crop rotation, contour plowing, strip cropping, and terracing all help to minimize erosion and mineral depletion of the soil. Land that has been badly damaged by soil erosion can be successfully restored, but it is a costly, time-consuming process.

Conservation Tillage

Conventional methods of tillage, or working the land, include spring plowing, in which the soil is cut and turned in preparation for planting seeds; and harrowing, in which the plowed soil is leveled, seeds are covered, and weeds are removed. Conventional tillage prepares the land for crops, but in removing all plant cover it greatly increases the likelihood of soil erosion (Figure 14–15a). Fields that are conventionally tilled also contain less organic material and generally hold less water than does undisturbed soil.

Since the early 1980s, an increasing number of farmers have adopted a new approach called **conservation tillage**, in which residues from previous crops are left in the soil, partially covering it and helping to hold it in place. Several types of conservation tillage have been developed to fit different areas of the country and different crops. For example, one of these, called **no-tillage**, leaves

Figure 14–14

Cattle in Burkina Faso have eaten all the ground cover; the trees that remain will probably be stripped of branches to feed the hungry cattle. Overexploitation of the Sahel, a semiarid region south of the Sahara Desert, is increasing the amount of unproductive desert area. (Robert E. Ford/Terraphotographics)

the soil undisturbed prior to planting, when special machines cut a narrow furrow in the soil for seeds (Figure 14–15b).

Conservation tillage increases the organic material in the soil, which in turn improves water-holding capacity. Decomposing organic matter releases nutrients more gradually than when conventional tillage methods are employed. Although conservation tillage is an effective way of reducing soil erosion, it requires greater use of herbicides to control weeds. Research is needed to develop alternative methods of weed control for use with conservation tillage.

No-tillage is one of the fastest growing trends in American agriculture. It reduces soil erosion by 70 percent or more and increases crop yields. It is also less expensive than conventional tillage because equipment, fuel, and labor costs are less for no-tillage.

(a)

(b)

Figure 14–15

Conventional and conservation tillage. (a) Conventional tillage leaves almost no plant residues on the soil surface. If heavy spring rains come before the young plants become established, significant soil loss can occur. (b) Young soybeans are surrounded by decaying residues from the previous year's crop (rye). Conservation tillage reduces soil erosion because plant residues from the previous season's crops are left in the soil. (a, Grant Heilman; b, USDA/Natural Resources Conservation Service)

Crop Rotation

Farmers who practice effective soil conservation measures often use a combination of conservation tillage and **crop rotation**, the planting of a series of different crops in the same field over a period of years. When the same crop is grown continuously, pests for that crop tend to accumulate to destructive levels, so crop rotation lessens damage by insects and disease. Also, many scientific studies have shown that continuously growing the same crop over a period of years depletes the soil of certain essential nutrients faster and makes soils more prone to erosion. Crop rotation is therefore effective in maintaining soil fertility and in reducing erosion.

A typical crop rotation would be corn → soybeans → oats → alfalfa. Soybeans and alfalfa, both members of the legume family, increase soil fertility through their association with bacteria that fix atmospheric nitrogen into the soil. Thus, soybeans and alfalfa help produce higher yields of the grain crops with which they alternate in crop rotation.

Contour Plowing, Strip Cropping, and Terracing

Hilly terrain must be cultivated with care because it is more prone to soil erosion than flat land. Contour plowing, strip cropping, and terracing help control erosion of farmland with variable topography.

In **contour plowing**, fields are plowed and planted in curves that conform to the natural contours of the land, rather than in straight rows. Furrows run around, rather than up and down, hills (Figure 14–16).

Figure 14–16

Contour plowing and strip cropping are evident in this well-managed farm. Quite often crop rotations in such strips include a legume, which reduces the need for nitrogen fertilizers. (U.S. Department of Agriculture)

Figure 14–17

Terracing in the Luzon rice fields, Philippines. Terracing hilly or mountainous areas reduces the amount of soil erosion. However, some slopes are so steep they are totally unsuitable for agriculture. These areas should be left covered by natural vegetation to prevent extensive erosion. (David Cavagnaro)

Strip cropping, a special type of contour plowing, produces alternating strips of different crops along natural contours. For example, alternating a row crop such as corn with a closely sown crop such as wheat reduces soil erosion. Even more effective control of soil erosion is achieved when strip cropping is done in conjunction with conservation tillage.

In mountainous terrain, **terracing** produces level areas and thereby reduces soil erosion (Figure 14–17). Nutrients and soil are retained on the horizontal platforms instead of being washed away. Soils are preserved in a somewhat similar manner in low-lying areas that are diked to make rice paddies. The water forms a shallow pool, retaining sediments and nutrients.

Preserving Soil Fertility

The two main types of fertilizer are organic and inorganic. Organic fertilizers include such natural materials as animal manure, crop residues, bone meal, and compost. Organic fertilizers are complex, and their exact compositions vary. The nutrients in organic fertilizers become available to plants only as the organic material decomposes. For that reason, organic fertilizers are slow-acting and long-lasting. (For two very different discussions of compost, see You Can Make a Difference: Practicing Environmental Principles, and Meeting the Challenge: Municipal Solid Waste Composting.)

Inorganic fertilizers are manufactured from chemical compounds, and thus their exact compositions are

(text continues on page 323)

YOU CAN MAKE A DIFFERENCE

Practicing Environmental Principles

You can maintain and improve the soil in your own lawn and garden by using compost and mulch. Gardeners often dispose of grass clippings, leaves, and other plant refuse by either bagging it for garbage collection or burning it. But these materials do not have to be treated like wastes; they can be a valuable resource for making **compost**, a natural soil and humus mixture that improves not only soil fertility but also soil structure. Grass clippings, leaves, weeds, sawdust, coffee grounds, ashes from the fireplace or grill, shredded newspapers, potato peels, and eggshells are just some of the materials that can be **composted**, or transformed by microbial action to compost.

To make a compost heap, spread a 6- to 12-inch layer of grass clippings, leaves, or other plant material in a shady area, sprinkle it with an organic garden fertilizer or a thin layer of animal manure, and cover it with several inches of soil. Add layers as you collect more organic debris. Water the mixture thoroughly, and turn it over with a pitchfork each month to aerate it (see figure a). Although it is possible to make compost by just heaping it on the open ground in layers, it is more efficient to construct an enclosure. An enclosed compost heap is also less likely to attract animals.

When the compost is uniformly dark in color, is crumbly, and has a pleasant, "woodsy" odor, it is ready to use. The time it takes for decomposition will vary from one to six months depending on the climate, the materials you are using, and how often you turn it and water it.

Whereas compost is mixed into soil to improve the soil's fertility, **mulch** is placed on the surface of soil, around the bases of plants (see figure b). Mulch helps control weeds and increases the amount of water in the upper levels of the soil by reducing evaporation. It lowers the soil temperature in the summer and extends the growing season slightly by providing protection against cold in the fall. Mulch also decreases erosion by lessening the amount of precipitation runoff.

Although mulches can consist of inorganic materials such as plastic sheets or gravel, natural mulches of compost, grass clippings, straw, chopped corncobs, or shredded bark have the added benefit of increasing the organic content of the soil. Grass clippings are a very effective mulch when placed around the bases of garden plants because they mat together, making it difficult for weeds to become established. You must replace grass mulches often, however, because they decay rapidly. Some gardeners prefer mulches of more expensive materials such as shredded bark, because they take longer to decompose and are more attractive.

(a) (b)

Compost and mulch. (a) A compost heap. Composts form faster when the heap is located in a shady spot, kept damp, and aerated (turned over) frequently. (b) Mulches discourage the growth of weeds and help keep the soil damp. Organic mulches such as this shredded bark have added benefit of gradually decaying, thereby increasing soil fertility. (a and b, USDA/Natural Resources Conservation Service.)

known. Because they are soluble, they are immediately available to plants. However, inorganic fertilizers are available in the soil for only a short period of time because they quickly leach away.

It is environmentally sound to avoid or limit the use of manufactured fertilizers, for several reasons. First, because of their high solubility, inorganic fertilizers are very mobile and often leach into groundwater or surface runoff, polluting the water (see Chapter 21). Second, manufactured fertilizers do not improve the water-holding capacity of the soil as organic fertilizers do. Another advantage of organic fertilizers is that, in ways that are not yet completely understood, they change the types of organisms that live in the soil, sometimes suppressing the microorganisms that cause certain plant diseases.

Soil Reclamation

It is possible to reclaim land that is badly damaged from erosion. The United States has largely reversed the effects of the 1930s Dust Bowl, for example, and China has reclaimed badly eroded land in Inner Mongolia (northern China). Soil reclamation involves two steps: (1) stabilizing the land to prevent further erosion and (2) restoring the soil to its former fertility. In order to stabilize the land, the bare ground is seeded with plants; they eventually grow to cover the soil, holding it in place. For example, after the Dust Bowl, land in Oklahoma and Texas was seeded with drought-resistant native grasses. One of the best ways to reduce the effects of wind on soil erosion is by planting **shelterbelts**, rows of trees that lessen the impact of wind (recall the Great Green Wall of China; also see Figure 14–18).

The plants that have been established to stabilize the land start to improve the quality of the soil almost immediately, as dead portions are converted to humus. The humus holds mineral nutrients in place and releases them a little at a time; it also improves the water-holding capacity of the soil.

Restoration of soil fertility to its original level is a slow process, however. During the soil's recovery, use of the land must be restricted: it cannot be farmed or grazed. Disaster is likely if the land is put back to use before the soil has completely recovered. But restriction of land use for a period of several to many years is sometimes very difficult to accomplish. How can a government tell landowners that they may not use their own land? How can land use be restricted when people's livelihoods and maybe even their lives depend on it?

SOIL CONSERVATION POLICIES IN THE UNITED STATES

The disastrous effects of the Dust Bowl years on U.S. soils focused attention on the fact that soil is a valuable natural resource. The Soil Conservation Act of 1935 authorized the formation of the Soil Conservation Service (now called the NRCS); its mission is to work with the American people to conserve natural resources on private lands. To that end, the NRCS assesses soil damage and develops policies to improve and sustain our soil resource.

Historically, farmers have been more likely to practice soil conservation during hard financial times and periods of agricultural surpluses, both of which translate into lower prices for agricultural products. When prices are

Figure 14–18
Shelterbelts. Trees reduce the ability of the wind to pick up soil from farmland. This photograph also shows strip cropping. (U.S. Department of Agriculture)

MEETING THE CHALLENGE

Municipal Solid Waste Composting

The sanitary landfill, a modern replacement for the city dump, has been the recipient of most of the nation's solid waste for the past several decades, but it is not a long-term solution to waste disposal. For one thing, many landfills are filling up, and it is unlikely that enough replacement sites can be found. NIMBYism—the **n**ot **i**n **m**y **b**ackyard syndrome—frequently takes hold when government officials attempt to find new places for sanitary landfills.

Much of the bulky waste in sanitary landfills—paper, yard refuse, food wastes, and such—is organic and, given the opportunity, could decompose into compost. However, in sanitary landfills little of this material breaks down. Rapid and complete decomposition requires the presence of oxygen, and in a sanitary landfill, garbage is buried under a layer of soil so that little oxygen is available.

Of the several options for decreasing the quantity of trash and garbage in sanitary landfills (discussed in Chapter 23), one—**municipal solid waste composting**—has only received serious attention in this country beginning in the 1990s. Municipal solid waste composting is the large-scale composting of the entire organic portion of a community's garbage. Since approximately 75 percent by weight of household garbage is organic, municipal solid waste composting substantially reduces demand for sanitary landfills.

Numerous city and county governments are currently composting leaves and yard wastes in an effort to reduce the amount of solid waste sent to landfills (see figure). Although this endeavor alone is undeniably beneficial, municipal solid waste composting encompasses much more than yard wastes. It also involves composting food wastes, paper, and anything else in the solid waste stream that is organic.

Initial composting occurs quickly—in three to four days—because conditions such as moisture and the carbon-nitrogen ratio are continually monitored and adjusted (by adding water or fertilizer, for example) for maximum decomposition. The decay process is carried out by billions of bacteria and fungi, which convert the organic matter into carbon dioxide, water, and humus. So many decomposers eat, reproduce, and die in the compost heap that the drum heats up, killing off potentially dangerous organisms such as disease-causing bacteria. When the material emerges, it is placed outside for several months to cure, during which time additional decomposition occurs. Finally it is sold as compost.

The potential market for compost is largely unexploited. Compost can be used by professional nurseries, landscapers, greenhouses, and golf courses. The largest potential market, however, is the American farmer. Tons of compost could be used to reclaim the 167 million hectares (413 million acres) of badly eroded farmland in the United States; 59 metric tons (65 tons) of compost would be needed to apply 1 inch of compost to a single acre. Compost could also improve the fertility of badly eroded rangeland, forest land, and strip mines. There appears to be no shortage of markets for compost. Thus, it appears that, should certain technical problems be resolved, composting could become economically feasible.

Technical problems include concerns over the presence of pesticide residues and heavy metals in the compost. Pesticides sprayed on urban and suburban landscapes would naturally find their way into compost material on leaves, grass clippings, and other yard wastes. However, several scientific studies indicate that most pesticides are either decomposed by bacteria and fungi during composting or broken down by the high temperatures in the compost heap.

More troubling is the concern over heavy metals, such as lead and cadmium, that can enter compost from sewage sludge—which may contain industrial wastes—or consumer products such as batteries. (Sewage sludge is often added to compost because it is a rich source of nitrogen for the decomposing microorganisms.) One way to reduce heavy metal contamination in municipal compost is by sorting out heavy metal sources before everything is dumped into the composting drum.

Municipal yard wastes, such as branches and shrub clippings, are ground for composting at a composting facility in Lee's Summit, Missouri. (© Eric R. Berndt/Unicorn Stock Photos)

high, with a good market for agricultural products, farmers have more incentive to put every parcel of land into production, including marginal, highly erodible lands. During times when the farm economy has been strong, federal soil conservation programs have actually contributed to production on marginal lands by relying on voluntary rather than mandatory compliance. The federal government has traditionally used incentives, rather than penalties for noncompliance, to encourage soil conservation practices.

In a different approach, the Conservation Reserve Program (CRP), which was added to the Food Security Act (Farm Bill) of 1985, pays farmers an average of $50 per acre per year to stop producing crops on highly erodible farmland. It requires planting native grasses or trees on such land and then "retiring" it from further use for ten years. This land may not be grazed, nor may the grass be harvested for hay during that period.

The Conservation Reserve Program has benefitted the environment. Annual loss of soil on CRP lands that have been planted with grasses or trees has been reduced from an average of 7.7 metric tons of soil per hectare (8.5 tons per acre) to 0.6 metric tons per hectare (0.7 tons per acre). Because the vegetation is not disturbed once it is established, it provides biological habitat. Small and large mammals, birds of prey, and ground-nesting birds such as ducks have increased in number and kind on CRP lands. The reduction in soil erosion has also improved water quality and enhanced fish populations in surrounding rivers and streams.

► *FOLLOW-UP*

Congress was originally scheduled to vote on the Farm Bill in 1995 (this bill has traditionally been reauthorized every five years since it was first passed in the 1930s). However, the bill was delayed for a year while Congress and the Clinton Administration worked out certain details involving crop supports and conservation programs. In 1996 Congress passed the Farm Bill, officially called the Federal Agriculture Improvement and Reform Act, which continues the Conservation Reserve Program at its current level of 36.5 million acres until 2002.

SUMMARY

1. The formation of soil, the complex material in which plants root, involves interactions among parent material, climate, organisms, time, and topography. Soil, which is composed of inorganic minerals, organic materials, soil air, and soil water, is often organized into layers called horizons. The texture of a soil depends on the relative amounts of sand, silt, and clay in it. The soil properties of texture and acidity affect a soil's water-holding capacity and nutrient availability, which in turn determine how well plants grow.

2. Organisms are important not only in forming soil, but also in recycling nutrients. In a balanced ecosystem, the minerals removed from the soil by plants are returned when plants or animals that eat the plants die and are decomposed by soil microorganisms.

3. Soil erosion, the removal of soil from the land by the actions of water, wind, ice, and other agents, is a natural process that is often accelerated by human activities such as farming on arid land and deforestation. The Dust Bowl that occurred in the western United States during the 1930s is an example of accelerated wind erosion caused by human exploitation of marginal land for agriculture.

4. Mineral depletion occurs in all soils that are farmed. It is a particularly serious problem when tropical rain forests are removed, because the minerals in the soil are quickly leached out.

5. Conservation tillage, crop rotation, contour plowing, strip cropping, and terracing can be used to help control erosion and mineral depletion. Soil that has been badly damaged by erosion can be reclaimed by stabilizing the land (providing plant cover and shelterbelts) and by restoring soil fertility (restricting land use until the soil has time to recover).

THINKING ABOUT THE ENVIRONMENT

1. Charles Darwin once wrote that the land is plowed by earthworms. Explain.

2. Describe three ways in which mineral nutrients can be lost from the soil.

3. It could be said that, unlike other communities, tropical rain forests live *on* the soil rather than in it. What does this statement imply about tropical soils?

4. The American Dust Bowl is sometimes portrayed as a "natural" disaster brought on by drought and high winds. Present a case for the point of view that this disaster was not caused by nature as much as by humans.

5. Where does eroded soil go after it is transported by water, wind, or ice?

6. Some Everglades farmers are growing rice, which allows them to keep the soil flooded. Explain why keeping the soil flooded helps to preserve it.

7. How is human overpopulation related to worldwide soil problems?

8. Conservation tillage has many benefits, including reduction of soil erosion. However, certain pests that cause plant disease can reside in the plant residues left on the ground with conservation tillage. Knowing that disease-causing organisms are often quite specific for the plants they attack, recommend a way to control such disease organisms. Base your answer on the soil conservation methods discussed in this chapter.

9. President F. D. Roosevelt once sent a letter to the state governors in which he said, "A nation that destroys its soils, destroys itself." Explain.

* 10. In the chapter you learned that a severe rainstorm may wash away 1 millimeter of soil and that such losses add up to 2.5 centimeters over a 20-year period. If one millimeter of topsoil coming off a land area of one hectare weighs 13.0 metric tons, how much topsoil (in metric tons per hectare) is lost in 20 years? Convert your answer to English units (tons per acre) using the conversions in Appendix IV.

RESEARCH PROJECTS

The NRCS works with farmers to prevent the loss of topsoil. Assess the effectiveness of this or other soil conservation programs in your area.

On-line materials relating to this chapter are on the ► World Wide Web at **http://www.saunderscollege.com/lifesci/environment2** (select Chapter 14 from the Table of Contents).

SUGGESTED READING

"Composting: Nature's Recycling Program," *Consumer Reports*, February 1994. Discusses both backyard composting and municipal solid waste composting and provides detailed instructions for composting at home.

Gillis, A. M. "Shrinking the Trash Heap," *BioScience* Vol. 42, No. 2, February 1992. An examination of municipal solid waste composting.

Pimentel, D. et al. "Environmental and Economic Costs of Soil Erosion and Conservation Benefits," *Science* Vol. 267, February 24, 1995. A sobering review of the worldwide environmental threat of soil erosion.

Reganold, J. P., R. I. Papendick, and J. F. Parr. "Sustainable Agriculture," *Scientific American*, June 1990. Discusses alternative methods of agriculture that reduce a farmer's dependence on chemicals and improve and maintain the soil.

Richter, D. D. and D. Markewitz. "How Deep Is Soil?" *BioScience* Vol. 45, No. 9, October 1995. Soil—that part of Earth's crust that has been modified by weathering—is much deeper than was previously hypothesized.

Sachs, A. "Dust to Dust," *WorldWatch* Vol. 7, No. 1, January–February 1994. Could another dust bowl occur in the plains states? The author builds a case for such an event.

Simons, M. "Winds Toss Africa's Soil, Feeding Lands Far Away," *The New York Times*, 29 October 1992. Dust plumes from Africa cross the Atlantic Ocean and fertilize parts of North and South America.

Wilken, E. "Assault of the Earth," *WorldWatch* Vol. 8, No. 2, March–April 1995. Examines the serious problems of soil degradation.

World Resources Institute, U.N. Environment Program, U.N. Development Program, and World Bank. *World Resources 1996–97*. New York: Oxford University Press, 1996. Chapter 10 of this informative reference book discusses how to minimize soil degradation.

The Golden Sunlight Gold Mine in Montana is owned by a Canadian corporation.
(Calvin Larsen/Photo Researchers, Inc.)

Minerals: A Nonrenewable Resource

The General Mining Law of 1872 was established to encourage settlement in the sparsely populated western states. It allows companies or individuals—regardless of whether they are American or foreign—to stake mining claims on federal land. They can then purchase the land for $2.50 to $5 an acre, extract the valuable hardrock minerals such as gold, silver, copper, lead, or zinc, and keep all the profits. In contrast, 12.5 percent of profits on lumber, coal, oil, and natural gas obtained from federal land is paid to the government. In 1995, for example, the law permitted an American company (ASARCO) to obtain federal land in Arizona that contains copper and silver reserves worth an estimated $2.9 billion; the company paid $1745 for this land. The National Taxpayers Union has determined that the federal treasury loses hundreds of millions in tax dollars annually from the law's inequitable provisions. Although Congress put a hold on such sales, several hundred pending applications were filed before the hold went into effect and are therefore exempt.

The General Mining Law contains no provisions for environmental protection such as the replacement of topsoil and vegetation, or the reestablishment of biological habitat. As a result, hardrock mining has left a legacy of ravaged land, poisoned water, and lifeless ecosystems throughout the West that will cost between $32 billion and $72 billion to repair. For example, acid draining from tailings, which are loose rocks produced during the mining process, has made an estimated 19,300 kilometers (12,000 miles) of streams totally lifeless.

Fifty-two of the 550,000 abandoned mines in the West have been designated Superfund sites whose cleanup will be financed by the federal government, i.e., by American taxpayers. For example, after a Canadian firm (Galactic Resources Ltd.) extracted $105 million of gold from a mine in Summitville, Colorado, it declared bankruptcy in 1993, leaving behind an environmental disaster. Now a Superfund site, the Summitville mine will cost the federal government more than $140 million to clean up. Cleanup of all 52 Superfund sites will cost an estimated $12.5 billion to $17.5 billion.

In 1872, the same year that President Ulysses S. Grant signed the General Mining Law, Yellowstone was designated the first U.S. national park. Recently a proposed 200-acre mine site located in Montana less than three miles from the Yellowstone border and within the watershed that drains into Yellowstone threatened its pristine condition. A Canadian company (Noranda, Inc.) has the rights to this land, which is surrounded by lands designated wilderness, and they planned to establish the Crown Butte gold, silver, and copper mine there. Opponents of the mine, including the Superintendent of Yellowstone, worried that some of the pollution from the mine would contaminate Yellowstone, either accidentally or through negligence. Because Yellowstone is considered a national treasure, President Clinton stopped the Noranda mine by presidential order in 1996.

The General Mining Law may be revised by Congress sometime in the next several years. The mining industry, which is politically powerful in western states, opposes such reform (because it would decrease their profits) and has so far been successful in preventing it.

In this chapter we consider the distribution and abundance of minerals as well as the environmental damage caused by obtaining and processing them. We also examine our options for the future, when mineral reserves become depleted. ◄

USES OF MINERALS

Minerals, elements or compounds of elements that occur naturally in the Earth's crust, are such an integral part of our daily lives that we often take them for granted (Figure 15–1, Table 15–1). Steel, an essential building material, is a blend of iron and other metals. Beverage cans, aircraft, automobiles, and buildings all contain aluminum. Copper, which readily conducts electricity, is used for electrical and communications wiring. The concrete used in buildings and roads is made from sand, gravel, and cement, which contains crushed limestone. Sulfur, a com-

(a) (b) (c)

Figure 15–1

Minerals are an important part of our lives. (a) Concrete highways are made of sand, gravel, and crushed limestone. (b) Table salt is a nonmetallic mineral. (c) Copper, a metallic mineral, is often shaped into wire for electrical equipment or sheets for roofing, gutters, and downspouts. (a, Dennis Drenner; b, George Semple; c, Charlie Winters)

Table 15–1
Some Important Minerals and Their Uses

Mineral	Type	Some Uses
Aluminum (Al)	Metal element	Structural materials (airplanes, automobiles), packaging (beverage cans, toothpaste tubes), fireworks
Borax ($Na_2B_4O_7$)	Nonmetal	Diverse manufacturing uses—glass, enamel, artificial gems, soaps, antiseptics
Chromium (Cr)	Metal element	Chrome plate, pigments, steel alloys (tools, jet engines, bearings)
Cobalt (Co)	Metal element	Pigments, alloys (jet engines, tool bits), medicine, varnishes
Copper (Cu)	Metal element	Alloy ingredient in gold jewelry, silverware, brass, and bronze; electrical wiring, pipes, cooking utensils
Gold (Au)	Metal element	Jewelry, money, dentistry, alloys
Gravel	Nonmetal	Concrete (buildings, roads)
Gypsum ($CaSO_4$-$2H_2O$)	Nonmetal	Plaster of Paris, soil treatments
Iron (Fe)	Metal element	Basic ingredient of steel (buildings, machinery)
Lead (Pb)	Metal element	Pipes, solder, battery electrodes, pigments
Magnesium (Mg)	Metal element	Alloys (aircraft), firecrackers, bombs, flashbulbs
Manganese (Mn)	Metal element	Steel, alloys (steamship propellers, gears), batteries, chemicals
Mercury (Hg)	Liquid metal element	Thermometers, barometers, dental inlays, electric switches, streetlights, medicine
Molybdenum (Mo)	Metal element	High-temperature applications, lamp filaments, boiler plates, rifle barrels
Nickel (Ni)	Metal element	Money, alloys, metal plating
Phosphorus (P)	Nonmetal element	Medicine, fertilizers, detergents
Platinum (Pt)	Metal element	Jewelry, delicate instruments, electrical equipment, cancer chemotherapy, industrial catalyst
Potassium (K)*	Metal element	Salts used in fertilizers, soaps, glass, photography, medicine, explosives, matches, gunpowder
Common salt (NaCl)	Nonmetal	Food additive, raw material for synthetics
Sand (largely SiO_2)	Nonmetal	Glass, concrete (buildings, roads)
Silicon (Si)	Metal element	Electronics, solar batteries, ceramics, silicones
Silver (Ag)	Metal element	Jewelry, silverware, photography, alloys
Sulfur (S)	Nonmetal element	Insecticides, rubber tires, paint, matches, papermaking, photography, rayon, medicine, explosives
Tin (Sn)	Metal element	Cans and containers, alloys, solder, utensils
Titanium (Ti)	Metal element	Paints; manufacture of aircraft, satellites, and chemical equipment
Tungsten (W)	Metal element	High-temperature applications, light bulb filaments, dentistry
Zinc (Zn)	Metal element	Brass, metal coatings, electrodes in batteries, medicine (zinc salts)

*Potassium, which is very reactive chemically, is never found free in nature; it is always combined with other elements.

ponent of sulfuric acid, is an indispensable industrial mineral with many applications in the chemical industry. It is used to make plastics and fertilizers and to refine oil. Other important minerals include platinum, mercury, manganese, and titanium.

Human need and desire for minerals have influenced the course of history. Phoenicians and Romans explored Britain in a search for tin. One of the first metals to be used by humans, tin came into its own during the Bronze Age (3500 to 1000 B.C.), when tin and copper were combined to produce a tougher and more durable alloy known as bronze. The desire for gold and silver was directly responsible for the Spanish conquest of the New World. A gold rush in 1849 led to the settlement of California; more recently, the lure of gold in Amazonian and Indonesian rain forests has contributed to their destruction.

The Earth's minerals are elements or (usually) compounds of elements and have precise chemical compositions. For example, **sulfides** are mineral compounds in which certain elements are combined chemically with sulfur, and **oxides** are mineral compounds in which elements are combined chemically with oxygen.

Rocks are aggregates, or mixtures, of minerals and have varied chemical compositions. An **ore** is rock that contains a large enough concentration of a particular mineral that the mineral can be profitably mined and extracted. **High-grade ores** contain relatively large amounts of particular minerals, whereas **low-grade ores** contain lesser amounts.

Minerals can be metallic or nonmetallic. **Metals** are minerals such as iron, aluminum, and copper, which are malleable, lustrous, and good conductors of heat and electricity. **Nonmetallic minerals** lack these characteristics; they include sand, stone, salt, and phosphates.

Mini-Glossary of Mineral Terms

minerals: Elements and compounds that occur naturally in the Earth's crust. Minerals are either metals or nonmetals.

rock: A mixture of minerals that has varied chemical concentrations.

ore: Rock that contains a large enough concentration of a particular mineral for the mineral to be profitably mined and extracted. High-grade ores contain relatively large amounts of the desired minerals, and low-grade ores, relatively small amounts.

metals: Minerals that are malleable, lustrous, and good conductors of heat and electricity. Examples are gold, copper, and iron.

nonmetals: Minerals that are nonmalleable, nonlustrous, and poor conductors of heat and electricity. Examples are sand, salt, and phosphates.

NOT-SO-PRECIOUS GOLD

Annual global gold production in the mid-1990s was more than 2200 metric tons, up from 1105 metric tons in 1980. Demand for gold is booming throughout the world, and the environment is suffering a heavy blow from the increased mining. A new technology called cyanide heap leaching allows for profitable mining when minuscule amounts of gold are present, producing up to 3 million pounds of waste for every pound of gold produced. The highly toxic cyanide threatens waterfowl and fishes, as well as underground drinking water supplies. Small-scale miners use other extraction methods with destructive side effects: soil erosion, production of silt that clogs streams and threatens aquatic organisms, and contamination from mercury used to extract the gold. And the threats of gold mining do not end when the gold is carried away: if not disposed of properly, mining wastes cause such long-term nightmares as acid mine drainage and heavy-metal contamination.

MINERAL DISTRIBUTION AND ABUNDANCE

Certain minerals, such as aluminum and iron, are relatively abundant in the Earth's crust. Others, including copper, chromium, and molybdenum, are relatively scarce. Abundance does not necessarily mean that the mineral is easily accessible or profitable to extract, however. It is possible, for instance, that you have gold and other expensive minerals in your own backyard. However, unless the concentrations are large enough to make them profitable to mine, they will remain there.

Like other natural resources, mineral deposits in the Earth's crust are distributed unevenly. Some countries have extremely rich mineral deposits, whereas others have few or none. Although iron is widely distributed in the Earth's crust, for example, Africa has less than the other continents. Many copper deposits are concentrated in North and South America, particularly in Chile, whereas Asia has a relatively small amount of copper. The distribution of nickel is surprising in that a substantial portion of the world's known supply is found in the tiny island nation of Cuba. Much of the world's tin is in Malaysia and Indonesia, and most of the chromium reserves are in South Africa.

Formation of Mineral Deposits

Concentrations of minerals within the Earth's crust are apparently caused by several natural processes, including magmatic concentration, hydrothermal processes, sedimentation, and evaporation.

As molten rock cools and solidifies deep in the Earth's crust, it often separates into layers, with the heavier iron- and magnesium-containing rock[1] settling on the bottom and the lighter silicates (rocks containing silicon) rising to the top. Varying concentrations of minerals are often found in the different rock layers. This layering, which is thought to be responsible for some deposits of iron, copper, nickel, chromium, and other metals, is called **magmatic concentration**.

[1]Pure magnesium is a relatively lightweight element. Rock usually contains magnesium in the form of magnesium oxide, which is heavier.

Hydrothermal processes involve groundwater that has been heated in the Earth. This water seeps through cracks and fissures and dissolves certain minerals in the rocks, which are then carried along in the hot water solution. The dissolving ability of the water is greater if chlorine or fluorine is present, because these elements react with many metals (such as copper) to form salts (copper chloride, for example) that are soluble in water. When the hot solution encounters sulfur, a common element in the Earth's crust, a chemical reaction between the metal salts and the sulfur produces metal sulfides. Because metal sulfides are not soluble in water, they form deposits by settling out of the solution. Hydrothermal processes are responsible for deposits of minerals such as gold, silver, copper, lead, and zinc.

The chemical and physical weathering processes that break rock into finer and finer particles are important not only in soil formation (as we saw in Chapter 14) but in the production of mineral deposits. Weathered particles can be transported by water and deposited as sediment on riverbanks, deltas, and the sea floor in a process called **sedimentation**. During their transport, certain minerals in the weathered particles dissolve in the water. They later settle out of solution—for example, when the warm water of a river meets the cold water of the ocean—because less material dissolves in cold water than in warm water. Important deposits of iron, manganese, phosphorus, sulfur, copper, and other minerals have been formed by sedimentation.

Significant amounts of dissolved material can accumulate in inland lakes and in seas that have no outlet or only a small outlet to the ocean. If these bodies of water dry up by **evaporation**, a large amount of salt is left behind; over time, it may be covered with sediment and incorporated into rock layers. Significant worldwide deposits of common table salt (NaCl), borax ($Na_2B_4O_7$), potassium salts, and gypsum ($CaSO_4$-$2H_2O$) have been formed by evaporation.

How Large Is Our Mineral Supply?

Mineral reserves are mineral deposits that have been identified and are currently profitable to extract. In contrast, **mineral resources**, deposits of low-grade ores, are potential sources of minerals that are currently unprofitable to extract but may be profitable to extract in the future. Resources also include estimates of as-yet-unidentified deposits of minerals that may be confirmed in the future. The combination of a mineral's reserves and resources is called its **total resources** or its **world reserve base**. The World Resources Institute estimates the 1994 reserves of copper, for example, are 310 million metric tons, and its world reserve base is 590 million metric tons. Based on the 1994 rate of consumption, copper *reserves* will last until the year 2029. Table 15–2 gives the estimated reserves of selected minerals.

Table 15–2
1994 Reserves and Life Indices of Selected Minerals

Mineral	1994 Reserves (Thousand Metric Tons)	Life Index of World Reserves*
Aluminum	23,000,000	207
Cadmium	540	Not available
Copper	310,000	33
Iron ore	150,000,000	152
Lead	63,000	23
Mercury	130,000	45
Nickel	47,000	59
Tin	7,000	41
Zinc	140,000	20

*That is, an estimate of how many years reserves will last. It was calculated by dividing the 1994 estimate of reserves by production in 1994.

Estimates of mineral reserves and resources fluctuate with economic, technological, and political changes. If the price of a mineral on the world market falls, for example, certain borderline mineral reserves may slip into the resource category; increasing prices may restore them to the reserve category. When new technological methods decrease the cost of extracting mineral ores, mineral deposits that have been ranked in the resource category are reclassified as reserves. If the political situation in a country becomes so unstable that reserves cannot be mined, the reserves are reclassified as resources; another change in the political situation at a later time may cause the minerals to be placed in the reserve list again.

HOW MINERALS ARE FOUND, EXTRACTED, AND PROCESSED

The process of making mineral deposits available for human consumption occurs in several steps. First, a particular mineral deposit is located. Second, the mineral is extracted from the ground by mining. Third, the mineral is processed, or refined, by concentrating it and removing impurities. During the fourth and final step, the purified mineral is used to make a product.

Discovering Mineral Deposits

Geologists employ a variety of instruments and measurements to help locate valuable mineral deposits (Figure 15–2). Aerial or satellite photography sometimes discloses geological formations that are associated with certain

Figure 15–2

Vibrator trucks produce vibrations that bounce off underlying rock formations. Such geological surveys of inaccessible areas often provide information about mineral deposits. (George Steinmetz)

DIAMONDS UNDER THE TUNDRA

Diamonds had never been found in large quantities in the Western Hemisphere until 1989, when Canadian geologists pinpointed the site of a whole cluster of "pipes," the veins that carry diamonds to the surface from great depths. The deposits, likely worth billions of dollars, are in Canada's Northwest Territories, in a wild sub-Arctic region called the Barren Lands. Now the area bustles with the activity of more than 250 companies who had staked out 53 million acres for exploration by late 1994. The region became a prime target for diamond discovery when seismologists mapped the rocks as the world's oldest—diamonds are found only in ancient deposits. Geologists' new understanding of the characteristics of indicator minerals associated with diamonds led them to the pipes' specific locations. At least a few of the pipes are extremely rich in diamonds, producing 3 carats per metric ton of mined material, and estimates of the potential deposits run as high as 1000 pipes. Environmentalists are understandably concerned about the possible impact mining will have on the isolated area, which supports a caribou herd of 325,000. But jobs in the region are scarce and potential profits are huge, so drilling for diamonds is under way.

types of mineral deposits. Aircraft and satellite instruments that measure the Earth's magnetic field and gravity can reveal certain types of deposits. Geological knowledge of the Earth's crust and how minerals are formed is also used to estimate locations of possible mineral deposits. Once these sites are identified, geologists drill or tunnel for mineral samples and analyze their composition. Seismographs, which are used to detect earthquakes, also provide valuable clues about mineral deposits.

Deposits on the ocean floor cannot be estimated until detailed three-dimensional maps of the sea floor are produced, usually with the aid of depth-measuring devices. Sophisticated computer analysis is necessary to evaluate the complex data recorded by such devices.

Extracting Minerals

The depth of a particular deposit determines whether it will be extracted by **surface mining**, in which minerals are extracted near the surface, or **subsurface mining**, in which minerals that are too deep to be removed by surface mining are extracted. Surface mining is more common because it is less expensive than subsurface mining. However, because even surface mineral deposits occur in rock layers beneath the Earth's surface, the overlying layers of soil and rock, called **overburden**, must first be removed, along with the vegetation growing in the soil. Then giant power shovels scoop the minerals out.

There are two kinds of surface mining, open-pit and strip mining. Iron, copper, stone, and gravel are usually extracted by **open-pit surface mining**, in which a giant hole is dug (Figure 15–3). Large holes that are formed by open-pit surface mining are called quarries. In **strip mining**, a trench is dug to extract the minerals. Then a new trench is dug parallel to the old one; the overburden from the new trench is put into the old trench, creating a hill of loose rock known as a **spoil bank**.

Subsurface mining, which is done underground and is more complex, may be done with a shaft mine or a slope mine. A **shaft mine** is a direct vertical shaft to the vein of ore. The ore is broken apart underground and then hoisted through the shaft to the surface in buckets. A **slope mine** has a slanting passage that makes it possible to haul the broken ore out of the mine in cars rather than hoisting it up in buckets.

Subsurface mining disturbs the land less than surface mining, but it is more expensive and more hazardous for miners. There is always a risk of death or injury from ex-

Figure 15–3
Open-pit surface mining produces giant holes. This copper open-pit mine is near Tucson, Arizona. (Bruce F. Molnia/Terraphotographics)

plosions or collapsing walls, and prolonged breathing of dust in subsurface mines can result in lung disease.

Processing Minerals

Processing minerals often involves **smelting**, which is melting the ore at high temperatures to help separate impurities from molten metal. Purified copper, tin, lead, iron, manganese, cobalt, or nickel smelting is done in a blast furnace. Figure 15–4 shows a blast furnace used to smelt iron. Iron ore, limestone rock, and coke (modified coal used as an industrial fuel) are added at the top of the furnace, while heated air or oxygen is added at the bottom. The iron ore reacts with coke to form molten iron and carbon dioxide. The limestone reacts with impurities in the ore to form a molten mixture called slag. Both molten iron and slag collect at the bottom, but slag floats on molten iron because it is less dense than iron. Note the vent near the top of the iron smelter for exhaust gases. If air pollution control devices are not installed, many dangerous gases are emitted during smelting.

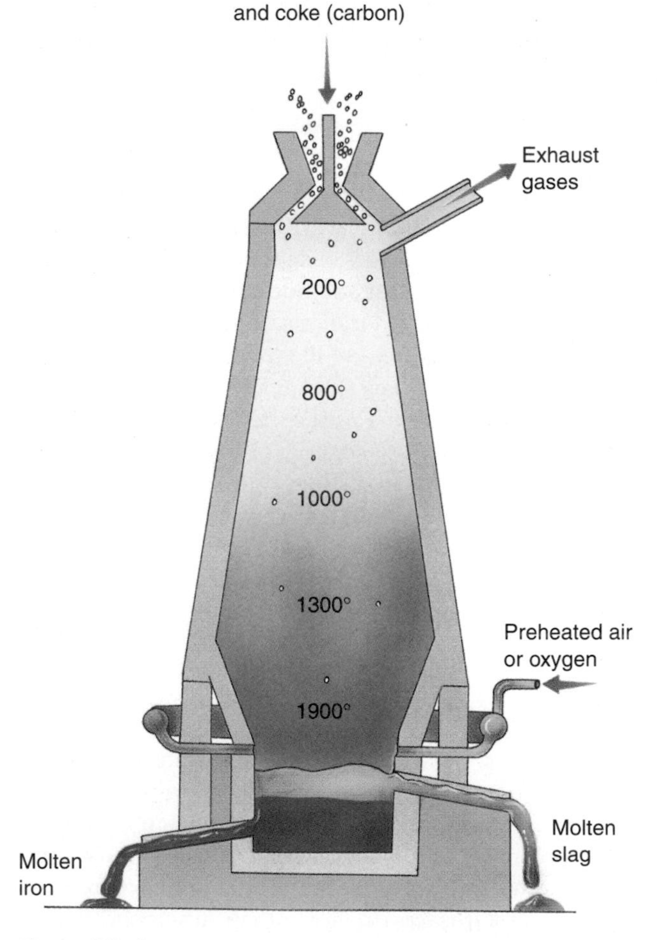

Figure 15–4
A blast furnace is used to smelt iron.

ENVIRONMENTAL IMPLICATIONS

There is no question that the extraction, processing, and disposal of minerals harm the environment. Mining disturbs and damages the land, and processing and disposal of minerals pollute the air, soil, and water. As noted in the discussion of coal in Chapter 10, pollution can be controlled and damaged lands can be fully or partially restored, but these remedies cost money. Historically, the environmental cost of extracting, processing, and disposal of minerals has not been incorporated into the actual price of mineral products to consumers (see Chapter 7).

Most developed countries have regulatory mechanisms in place to minimize environmental damage from mineral consumption, and many developing nations are in the process of putting them in place. These regulatory programs include policies to prevent or reduce pollution, restore mining sites, and exclude certain recreational and wilderness sites from mineral development.

Mining and the Environment

Mining, particularly surface mining, disturbs huge areas of land. In the United States, current and abandoned metal and coal mines occupy an estimated 9 million hectares (22 million acres). Because any vegetation that had grown there is destroyed by mining, this land is particularly prone to erosion, with wind erosion causing air pollution and water erosion polluting nearby waterways and damaging aquatic habitats.

Acids and other toxic substances in the spoil banks of mines are washed into soil and water by precipitation runoff. When such toxic compounds make their way into nearby lakes and streams, they adversely affect the numbers and kinds of aquatic life; groundwater can also be contaminated. This type of pollution is called **acid mine drainage**.

One effective way to correct acid mine drainage is to construct a series of marshes or ponds downstream from the mine. Plants such as cattails and bulrushes that are planted in the shallow waters of the wetlands help to trap acid in the sediments. Constructed wetlands typically take 50 to 100 years to neutralize the acid enough for aquatic life to return to rivers and streams downstream from acid mine drainage. This time estimate is based on observations of more than 800 wetland systems that have been constructed in Appalachia.

Environmental Impacts of Refining Minerals

Approximately 80 percent or more of mined ore consists of impurities that become wastes after processing. These wastes, called **tailings**, are usually left in giant piles on the ground or in ponds near the processing plants (Figure 15–5). Toxic substances and dust from tailings left exposed in this way can contaminate the air, soil, and water.

Smelting plants have the potential to emit large amounts of air pollutants during mineral processing. One impurity in many mineral ores is sulfur. Unless expensive pollution control devices have been added to smelters, the sulfur escapes into the atmosphere, where it forms sulfuric acid. The pollution control devices are the same as those used when sulfur-containing coal is burned—scrubbers and electrostatic precipitators (see Chapters 10 and 23). (See Focus On: Copper Basin, Tennessee on page 337, for a historical example of environmental degradation caused by smelting.)

Figure 15–5
Copper ore tailings from a mine in Anaconda, Montana. Toxic materials from mine tailings that are left in mountainous heaps can pollute the air, soil, and water. (R. Ashley/Visuals Unlimited)

Other contaminants found in many ores include the heavy metals lead, cadmium, arsenic, and zinc. These elements also have the potential to pollute the atmosphere during the smelting process. Cadmium, for example, is found in zinc ores, and emissions from zinc smelters are a major source of environmental cadmium contamination. In humans, cadmium is linked to high blood pressure; diseases of the liver, kidneys, and heart; and certain types of cancer. (The health effects of lead, another common emission from smelters, are discussed in Chapter 21.) In addition to airborne pollutants, smelters emit hazardous liquid and solid wastes that can cause soil and water pollution. Pollution control devices can prevent such hazardous emissions, however.

Restoration of Mining Lands

When a mine is no longer profitable to operate, the land can be reclaimed, or restored to a seminatural condition (Figure 15–6; also see Meeting the Challenge: Reclamation of Coal-Mined Land, in Chapter 10). Approximately two-thirds of the Copper Basin in Tennessee has been partially reclaimed, for example. The goals of reclamation include preventing further degradation and erosion of the land, eliminating or neutralizing local sources of toxic pollutants, and making the land productive for purposes other than mining. Restoration can also make such areas visually attractive.

A great deal of research is available on techniques of restoring lands that have been degraded by mining, called **derelict lands**. Restoration involves filling in and grading the land to its natural contours, then planting vege-

Figure 15–6

Part of a phosphate mine in Florida has been reclaimed and is currently used as a pasture (*background*). Restoration of mining lands makes them usable once again, or at the very least, stabilizes them so that further degradation does not occur. (William Felger/Grant Heilman)

tation to hold the soil in place. The establishment of plant cover is not as simple as throwing a few seeds on the ground. Often the topsoil is completely gone or contains toxic levels of metals, so special types of plants that can tolerate such a challenging environment must be used. According to experts, the main limitation on the restoration of derelict lands is not lack of knowledge but lack of funding.

Reclamation of areas that were surface mined for coal is required by the Surface Mining Control and Reclamation Act of 1977. No federal law is in place to require restoration of derelict lands produced by mines other than coal mines, however. Recall from the chapter introduction that the General Mining Law of 1872 makes no provision for reclamation.

MINERAL RESOURCES: AN INTERNATIONAL PERSPECTIVE

The economies of industrialized countries require the extraction and processing of large amounts of minerals to make products. Most of these developed countries rely on the mineral reserves in developing countries, having long since exhausted their own supplies. As developing countries become more industrialized, their own mineral requirements increase correspondingly, adding further pressure to a nonrenewable resource. In fact, more minerals have been consumed in the 50+ years since World War II than were consumed in the 5000+ years from the beginning of the Bronze Age to the middle of the 20th century.

U.S. and World Use

At one time, most of the developed nations had rich resource bases, including abundant mineral deposits, that enabled them to industrialize. In the process of industrialization, they have largely depleted their domestic reserves of minerals so that they must increasingly turn to developing countries. This is particularly true for Europe, Japan, and, to a lesser extent, the United States.

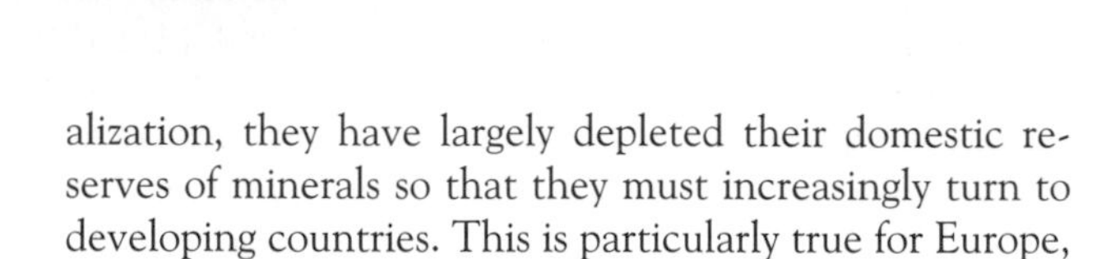

As with the consumption of other natural resources, there is a large difference in consumption of minerals between developed and developing countries. For example, the United States, which has about 4.6 percent of the world's population, consumes about 20 percent of many of the world's metals (Figure 15–7). It is too simplistic,

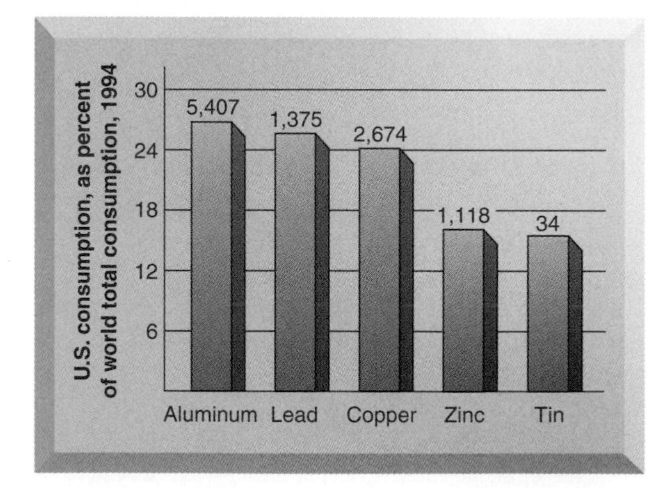

Figure 15–7

U.S. consumption of selected metals in 1994, as percent of world total consumption. The heavily industrialized United States, which has about 4.6 percent of the world's population, consumes a disproportionate share of many of the world's metals. Actual quantities consumed by the United States in 1994 are given at the top of each bar, in thousand metric tons.

FOCUS ON

Copper Basin, Tennessee

Travel to the southeast corner of Tennessee, near its borders with Georgia and North Carolina, and you will progress from lush forests to a panorama of red, barren hills baking in the sun (see figure). Few plant or animal species can be found—just 145 square kilometers (56 square miles) of hills with deep ruts gouged into them. The ruined land has a stark, otherworldly appearance. What is this place, and how did it come to be?

During the middle of the 19th century, copper ore was discovered near Ducktown in southeastern Tennessee. Copper mining companies extracted the ore from the ground and dug vast pits to serve as open-air smelters. They cut down the surrounding trees and burned them in the smelters to produce the high temperatures needed for the separation of copper metal from other contaminants in the ore. The ore contained great quantities of sulfur, which reacted with oxygen in the air to form sulfur dioxide. As sulfur dioxide from the open-air smelters billowed into the atmosphere, it reacted with water, forming sulfuric acid that fell as acid rain.

As a result of deforestation and acid rain, ecological ruin of the area occurred in a few short years. Any plants attempting a comeback after removal of the forests were quickly killed by the acid rain, which acidified the soil. Because plants no longer covered the soil and held it in place, soil erosion cut massive gullies in the gently rolling hills. Of course, the forest animals disappeared with the plants, which had provided their shelter and food. The damage did not stop here: soil eroding from the Copper Basin, along with acid rain, ended up in the Ocoee River, killing its entire aquatic community.

In the 1930s several government agencies, including the Tennessee Valley Authority and the U.S. Soil Conservation Service, tried to replant a portion of the area. They planted millions of loblolly pine and black locust trees as well as shorter ground-cover plants that tolerate acid conditions, but most of the plants died. The success of such efforts was marginal until the 1970s, when land reclamation specialists began using new techniques such as application of seed and time-released fertilizer by helicopter. These plants had a greater survival rate, and as they became established, their roots held the soil in place. Leaves dropping to the ground contributed organic material to the soil. The plants provided shade and food for animals such as birds and field mice, which slowly began to return.

Today approximately two-thirds of the Copper Basin has one sort of vegetation or another. If funding continues, it may be possible to have the entire area under plant cover in 10 to 20 years. Of course, the return of the complex forest ecosystem that originally covered the land before the 1850s will take a century or two—if it ever occurs.

Plant scientists and land reclamation specialists have learned a lot from the Copper Basin, and they will put this knowledge to use in future reclamation projects around the world.

Ducktown, Tennessee. Air pollution from a copper smelter in Tennessee killed the vegetation, and then water erosion carved gullies into the hillsides. (Pat Armstrong/Visuals Unlimited)

however, to divide the world into two groups, the mineral consumers (developed countries) and the mineral producers (developing countries). For one thing, four of the world's top five mineral producers are developed countries: the United States, Canada, Australia, and the Russian Federation. South Africa, a moderately developed, middle-income country, is the other mineral producer in the top five. Furthermore, many developing countries lack any significant mineral deposits.

Because industrialization increases the demand for minerals, countries that at one time met their mineral needs with domestic supplies become increasingly reliant on foreign supplies as development occurs. South Korea is one such nation. During the 1950s it exported iron, copper, and other minerals. South Korea experienced dramatic economic growth from the 1960s to the present and, as a result, must now import iron and copper to meet its needs.

Distribution Versus Consumption

Chromium, a metallic element, provides a useful example of worldwide versus national distribution and consumption. Chromium is used to make vivid red, orange, yellow, and green pigments for paints; chrome plate; and (combined with other metals) certain types of hard steel. There is no substitute for chromium in many of its important applications, including jet engine parts. Therefore, industrialized nations that lack significant chromium deposits, such as the United States, must import essentially all of their chromium. South Africa is one of only a few countries with significant deposits of chromium. Zimbabwe and Turkey also export chromium. Although worldwide reserves of chromium are adequate for the immediate future, the United States and several other industrialized countries are utterly dependent on a few, sometimes politically volatile, countries for their chromium supplies.

Many industrialized nations have stockpiled strategically important minerals to reduce their dependence on politically unstable suppliers. The United States and others have stockpiles of such metals as titanium, tin, manganese, chromium, platinum, and cobalt, mainly because these metals are critically important to industry and defense. These stockpiles are supposed to be large enough to provide strategic metals for a period of three years.

MINING WITH MICROBES

"Biomining" with microorganisms is becoming an economical and environmentally preferable way to recover metals. Traditional mining faces a problem of diminishing returns. As prime veins have been exhausted over the last century, ore grades have declined steadily, requiring mining companies to increase their scale of operation. But this has also increased the burden mining imposes on the environment, through leaching from waste dumps, massive open-pit excavations, smelter emissions, and the use of vast quantities of energy for processing. Microorganisms have proved very efficient for copper mining, allowing the U.S. copper industry to become internationally competitive. When mixed with sulfuric acid, a bacterium called *Thiobacillus ferrooxidans* promotes a chemical reaction that leaches copper into an acidic solution, releasing larger quantities of the metal more efficiently than traditional methods. Now, other important applications of biomining are emerging. Treating low-grade gold ores with bacteria such as *Thiobacillus* allows a 90 percent recovery of gold, compared to 75 percent recovery for the more expensive and energy-intensive conventional methods. Phosphates, used primarily for fertilizers and additives in some manufactured goods, have traditionally been extracted by inefficient burning at high temperatures or by wasteful acid treatment processes. New bioleaching processes can extract phosphates at room temperature.

Will We Run Out of Important Minerals?

A mineral's **life index of world reserves** is an estimate of the time it will take for the known reserves of that mineral to be expended (see Table 15–2). Life indices of world reserves are often meaningless because it is extremely difficult to forecast future mineral supplies. In the 1970s, projections of escalating demand and impending shortages of many important minerals were commonplace. There are three reasons that none of these shortages actually materialized. One, new discoveries of major deposits have occurred in recent decades—iron and aluminum deposits in Brazil and Australia, for example. Two, metals in many products were replaced by plastics, synthetic polymers, ceramics, and other materials. Three, a worldwide economic slump resulted in a lower consumption of minerals. Today on the world market there is even a glut of some minerals, which has caused their value to spiral downward. However, there is always the possibility that changes in the worldwide economic situation will contribute to mineral shortages.

One of the most significant economic factors in mineral production is the cost of energy. Mining and refining of minerals require a great deal of energy, particularly if the mineral is being refined from low-grade ore. For example, gold is currently being extracted from low-grade ores in Nevada. For every 0.9 metric ton (2000 pounds) of rock that is dug and crushed, as little as 0.7 gram (0.025 ounce) of gold is refined. Huge amounts of energy are required to dig and crush countless tons of rock. Higher energy prices result in higher production costs, which may cause the market price of minerals to increase. In turn, the higher prices decrease mineral consumption, extending supplies.

Economic factors aside, prediction of future mineral needs is also difficult because it is impossible to know when or if there will be new discoveries of mineral reserves or replacements for minerals (such as plastics). It is also impossible to know when or if new technological developments will make it economically feasible to extract minerals from low-grade ores.

With these reservations in mind, experts currently think mineral supplies—both metallic and nonmetallic—

FOCUS ON

Antarctica: Should We Mine It or Leave It Alone?

To date, no substantial mineral deposits have been found in Antarctica, although smaller amounts of valuable minerals have been discovered. Geologists think it likely that major deposits of valuable metals and oil are present and that they will be discovered in the future. Nobody owns Antarctica, and many nations have been involved in negotiations on the future of this continent and its possible mineral wealth.

The Antarctic Treaty, an international agreement that has been in effect since 1961, limits activity in Antarctica to peaceful uses such as scientific studies. Twenty-six nations are voting members to the Antarctic Treaty. During the 1980s, nearly a decade of delicate negotiations among 33 nations resulted in a pact that would have permitted exploitation of Antarctica's minerals. The pact, called the Convention on the Regulation of Antarctic Mineral Resource Activities, also required unanimous agreement in order to be ratified.

In 1989, several countries refused to support the pact because of concerns that *any* mineral exploitation would damage Antarctica's environment. As a result of these concerns, an international agreement, known as the Protocol to the Antarctic Treaty on Environmental Protection, or the Madrid Protocol, was established that includes a moratorium on mineral exploration and development for at least 50 years. The protocol will not become law until all 26 voting members of the Antarctic Treaty System have ratified it domestically. As of late 1997, 24 have done so, and 2 have not (Russia and Japan). Ratification by the United States was not approved by Congress and signed by President Clinton until 1996. U.S. ratification became official in 1997. It is thought that Russia and Japan will ratify the protocol in the near future.

Why is there concern about Antarctica's environment? For one thing, polar regions are extremely vulnerable to human activities. Even scientific investigations and tourists, with their trash, pollution, and noise, have negatively affected the wildlife along Antarctica's coastline. No one doubts that large-scale mining operations would wreak havoc on such a fragile environment.

Maintaining Antarctica in a pristine state is also important to us because this continent plays a pivotal role in regulating many aspects of the global environment, as for example, worldwide changes in sea level. By studying the natural environment of Antarctica, scientists gain valuable insights into such important environmental issues as global warming and ozone depletion (see Chapter 20).

will be adequate during the 21st century. During that period, however, several important minerals—silver, copper, mercury, tungsten, and tin, for example—may become increasingly scarce.

Another reasonable projection is that the prices of even relatively plentiful minerals, such as iron and aluminum, will increase during your lifetime. The depletion of large, rich, and easily accessible deposits of these metals will require mining and refining of low-grade ores, which will be more expensive.

INCREASING OUR MINERAL SUPPLIES

Some economists consider minerals to be an inexhaustible resource. Even though minerals are nonrenewable in the sense that there is only a limited supply of each kind in the Earth's crust, this economic view has some validity. As a resource becomes scarce, efforts intensify to discover new supplies, to conserve existing supplies of that resource, and to develop new substitutes for it. Although many reserves have been discovered and exploited, others that are as yet unknown may be found. In addition, the development of advanced mining technologies may make it possible to exploit known resources that are too expensive to develop using existing techniques.

Locating and Mining New Deposits

Many known mineral reserves have not yet been exploited. For example, although Indonesia is known to have many rich mineral deposits, its thick forests and malaria-carrying mosquitos have made accessibility to these deposits very difficult. Both northern and southern polar regions have had little mineral development. This is due in part to a lack of technology for mining in frigid environments. For example, normal offshore drilling rigs cannot be used in Antarctic waters, because the shifting ice formed during the harsh winter would tear the rigs apart. As new technologies become available, increasing pressure will be exerted to mine in northern Canada, Siberia, and Antarctica (see Focus On: Antarctica: Should We Mine It or Leave It Alone?).

Plans are afoot to exploit some of the rich mineral deposits in Siberia, although new technologies will have to be developed to make this feasible. Some of the ore deposits in Siberia have unusual combinations of minerals (for example, potassium combined with aluminum) that cannot be separated using existing technology.

Is there a possibility that currently unknown mineral deposits will be discovered at some future time? The U.S. Geological Survey thinks that undiscovered mineral deposits may exist, particularly in developing countries where detailed geological surveys have not been performed. It is likely that a detailed survey of the western portion of South America, along the Andes Mountains, will reveal significant mineral deposits. Geologists also presume that minerals will be found in the Amazon Basin, although in many ways the rain forest and thick overlying alluvial layers of the river basin make these deposits as inaccessible as those in Antarctica. Examination of certain areas deep in the rain forest to assess the likelihood of deposits being present is hampered by logistical problems; mining in the Amazon Basin would pose a grave environmental threat.

Geologists also consider it likely that deep deposits, those buried 1000 or more meters in the Earth's crust, will someday be discovered and exploited. The special technology required to mine deep deposits is not yet available.

Figure 15–8
Manganese nodules on the ocean floor. These potato-sized nodules have enticed miners, but the commercial feasibility of obtaining them is currently out of reach. (Science VU/Visuals Unlimited)

Minerals from the Ocean

The mineral reserves of the ocean may also provide us with future supplies. The sea floor may be mined, particularly where minerals have accumulated in the loose ocean sediments. Alternatively, minerals could be extracted from seawater.

Ocean floor

Although large deposits of minerals lie on the ocean floor, no seabed mining is currently done. The expense of obtaining these minerals is prohibitive, given our current technology. Many experts think that deep-sea mining will be feasible in a few decades, however, although environmental impacts of mining remain an obstacle to the exploration and development of the seabed (see Focus On: Safeguarding the Seabed).

Manganese nodules—small rocks the size of potatoes that contain manganese and other minerals, such as copper, cobalt, and nickel—are widespread on the ocean floor, particularly in the Pacific (Figure 15–8). According to the Marine Policy Center at the Woods Hole Oceanographic Institute, the estimates of these reserves are quite large: the Pacific Ocean, for example, may contain as much as 1.4 million metric tons of these minerals. However, dredging manganese nodules from the ocean floor would adversely affect sea life. Further, it is not clear which country, if any, has the legal right to these minerals, which are in international waters. Despite this concern, many industrial nations, including the United States, have staked out claims in a region of the Pacific known for its large number of nodules. Monitoring and policing their removal will almost certainly require international cooperation.

Seawater

Seawater, which covers approximately three-fourths of our planet, contains many different dissolved minerals. The total amount of minerals available in seawater is staggering, but their concentrations are very low. Currently, sodium chloride (common table salt), bromine, and magnesium can be profitably extracted from seawater. It may be possible in the future to profitably extract other minerals from seawater and concentrate them, but current mineral prices and technology make this impossible now.

Advanced Mining Technology

We have already mentioned that special technologies will be needed to mine minerals in inaccessible areas such as polar regions and deep deposits. Capitalizing on large, low-grade mineral deposits throughout the world will also require the development of special techniques. As minerals grow scarcer, economic and political pressure to exploit low-grade ores will increase. Obtaining high-grade

FOCUS ON

Safeguarding the Seabed

The ocean covers about two-thirds of our planet and constitutes a vast resource that is tapped by humans for food and scientific research. Its potential to supply us with minerals has been dreamed of for many years, but exploiting the ocean floor is controversial. Many people think it is inevitable that minerals will be mined from the floor of the deep sea. Others think the seabed should be declared off limits because of the potential ecological havoc mining could wreak on the diverse life forms inhabiting the ocean. Sea urchins, sea cucumbers, sea stars, acorn worms, sea squirts, sea lilies, and lamp shells are but a few of the animals known to inhabit the deep benthic environment. In addition to ecological problems, mining the loose sediments of the deep sea also poses ethical and legal questions. An international conference held in Berlin in 1991, the Dahlem Workshop, attempted to address some of these issues. The participants, mostly scientists, discussed such topics as how the mineral resources of the ocean floor would be harvested, to whom they belong, and how ocean mining should be regulated to protect the ocean's life forms.

These problems have been grappled with since the 1960s, when several industrialized countries, including the United States, expressed an interest in removing manganese nodules from the ocean floor. Their interest triggered the formation of an international treaty called the **United Nations Convention on the Law of the Sea** (UNCLOS), which became effective in November, 1994, after more than 60 countries ratified it. (As of late 1996, 109 countries had ratified this treaty.) This treaty initially viewed the minerals and organisms in the open seas as belonging to all of humankind. It required the establishment of an international group to oversee seabed mining and to sell mining rights to a particular country or investor. The money thus obtained would be used to help developing countries. Another requirement of the initial UNCLOS treaty was that any investor must bring in a developing country as a joint partner. The United States under President Reagan's leadership refused to ratify the treaty because it objected to the seabed mining provisions as being against free enterprise.

It seems likely that ocean mining will not become technologically feasible or profitable until sometime in the 21st century. Nevertheless, the fact that experts are trying to address the complex issues associated with undersea mining well in advance shows foresight. Their discussions may eventually result in international agreements on who can mine the ocean floor; when, where, and how they can extract minerals; and how the ocean environment will be protected for the sake of its many living things. Had such talks been instigated in advance of mining on land, many of the environmental problems associated with mining might have been avoided.

FOLLOW-UP

Because many of the provisions of the Law of the Sea Treaty benefit the United States, American military, commercial, and environmental interests were concerned that the United States had not ratified it. Accordingly, during the mid-1990s the Clinton Administration negotiated with the United Nations and many developing nations to modify the parts of the treaty to which it objected. The revised treaty, which provides incentives for seabed mining and does not seek to redistribute wealth from rich to poor nations, was submitted to the Senate for ratification in October, 1994, but has not been ratified as we go to press.

metals from low-grade ores is an expensive proposition, in part because a great deal of energy must be expended to obtain enough ore. Future technology may make such exploitation more energy-efficient, thereby reducing costs.

Even if advanced technology makes obtaining minerals from low-grade ores feasible, other factors may limit exploitation of this potential source. In arid regions, the vast amounts of water required during the extraction and processing of minerals may be the limiting factor. Also, the environmental costs may be too high, because obtaining minerals from low-grade ores causes greater land disruption and produces far more pollution than does the development of high-grade ores.

EXPANDING OUR SUPPLIES THROUGH SUBSTITUTION AND CONSERVATION

Because much of our civilization's technology depends on minerals, and because certain minerals may be unavailable or quite limited in the future, our society should ex-

NICKEL AS A CASH CROP

Although most plants do not tolerate soils rich in nickel, twist flower (*Streptanthus polygaloides*) thrives on it. This species is a hyperaccumulator, a rare plant that absorbs high quantities of a metal and stores it in its cells. Chemists at the U.S. Bureau of Mines may have found a way to capitalize on this unusual ability by cultivating twist flower to "mine" nickel. In early 1994 the Bureau of Mines planted 400,000 twist flower seeds in a one-acre California plot. After harvesting and burning the plants, chemists leached the ash to obtain 15 to 20 percent nickel by weight. The farming and extraction processes are harmless to the environment and may prove especially appropriate for areas where nickel ores are too low to merit traditional mining methods. Hyperaccumulating plants could even help improve environmental quality if used to remove heavy metals from contaminated soils.

tend existing mineral supplies as far as possible through substitution and conservation.

Finding Mineral Substitutes

The substitution of more abundant materials for scarce minerals is an important goal of manufacturing. The search for substitutes is driven in part by economics; one effective way to cut production costs is to substitute an inexpensive or abundant material for an expensive or scarce one. In recent years, plastics, ceramic composites, and high-strength glass fibers have been substituted for scarcer materials in many industries.

Earlier in this century, tin was a critical metal for can-making and packaging industries; since then, other materials have been substituted for tin, including plastic, glass, and aluminum. The amounts of lead and steel used in telecommunications cables have decreased dramatically during the past 35 years, while the amount of plastics has had a corresponding increase. In addition, glass fibers have replaced copper wiring in telephone cables.

Although substitution can extend our mineral supplies, it is not a cure-all for dwindling resources. Certain minerals have no known substitutes. Platinum, for example, catalyzes many chemical reactions that are important in industry. So far, no other substance has been found that possesses the catalyzing abilities of platinum.

Mineral Conservation

Our mineral supplies can be extended by conservation. In **recycling**, used items such as beverage cans and scrap iron are collected, remelted, and reprocessed into new products. The **reuse** of items such as beverage bottles, which can be collected, washed, and refilled, is another way to extend mineral resources. In addition to the introduction of specific conservation techniques such as recycling and reuse, public awareness and attitudes about resource conservation can be modified to encourage low waste.

Recycling

A large percentage of the products made from minerals—including cans, bottles, chemical products, electronic devices, and batteries—is typically discarded after use. The minerals in some of these products—batteries and electronic devices, for instance—are difficult to recycle. Minerals in other products, such as paints containing lead, zinc, or chromium, are lost through normal use.

However, we have the technology to recycle many other mineral products. Recycling of certain minerals is already a common practice throughout the industrialized world. Significant amounts of gold, lead, nickel, steel, copper, silver, zinc, and aluminum are now being recycled (Figure 15–9).

Figure 15-9
Recycling of scrap metal. The metal in these old, discarded automobiles will be fashioned into new products. (Courtesy of the Institute of Scrap Recycling Industries, Inc.)

Recycling has several advantages in addition to extending mineral resources. It saves unspoiled land from the disruption of mining, reduces the amount of solid waste that must be disposed (see Chapter 23), and reduces energy consumption and pollution. For example, recycling an aluminum beverage can saves the energy equivalent of about 180 milliliters (6 ounces) of gasoline. Recycling aluminum also reduces the emission of aluminum fluoride, a toxic air pollutant produced during aluminum processing.

More than two-thirds of the aluminum cans in the United States are currently being recycled. The aluminum industry, local governments, and private groups have established thousands of recycling centers across the country. It takes approximately six weeks for a can that has been returned to be melted, reformed, filled, and put back on a supermarket shelf. Clearly, even more recycling is possible. It may be that today's sanitary landfills will become tomorrow's mines, as valuable minerals and other materials are extracted from them.

Reuse

When the same product is used over and over again, as when beverage containers are collected, washed, and refilled, both mineral consumption and pollution are reduced. The benefits of reuse are even greater than those of recycling. For example, to recycle a glass bottle requires crushing it, melting the glass, and forming a new bottle. Reuse of a glass bottle simply requires washing it, which obviously expends less energy than recycling. Reuse is a national policy in Denmark, where nonreusable beverage containers are prohibited.

Several countries and states have adopted beverage container deposit laws, which require consumers to pay a deposit, usually a nickel, for each beverage bottle or can they purchase. The nickel is refunded when the container is returned to the retailer or to special redemption centers. In addition to encouraging recycling and reuse, thereby reducing mineral resource consumption, beverage container laws save tax money by reducing litter and solid waste. Countries that have adopted beverage container deposit laws include the Netherlands, Germany, Norway, Sweden, and Switzerland. Parts of Canada and the United States also have deposit laws.

Changing our mineral requirements

We can reduce our mineral consumption by becoming a low-waste society. Americans have developed a "throwaway" mentality in which damaged or unneeded articles are discarded (Figure 15–10). This attitude has been encouraged by industries looking for short-term economic profits, even though the long-term economic and environmental costs of such an attitude are high. Products that are durable and repairable enable us to consume fewer resources. Laws such as those requiring a deposit on beverage containers also reduce consumption by encouraging recycling and reuse.

The throwaway mentality has also been evident in manufacturing industries. Traditionally, industries consumed raw materials and produced not only goods but a large amount of waste that was simply discarded (Figure 15–11a). Increasingly, however, manufacturers are finding that the waste products from one manufacturing process can be used as raw materials in another industry.

(text continues on page 345)

Figure 15–10

The throw-away mentality of our industrial society is evident in this heap of discarded items. Much of this material can be recycled, and some of it could easily have been repaired. (Courtesy of the Institute of Scrap Recycling Industries, Inc.)

MEETING THE CHALLENGE

Industrial Ecosystems

Traditional industries operate in a one-way, linear fashion: natural resources from the environment → products → wastes dumped back into environment. However, natural resources such as minerals and fossil fuels are present in finite amounts, and the environment has a limited capacity to absorb waste. The field of **industrial ecology** has emerged to address these issues. An extension of the concept of sustainable manufacturing, industrial ecology seeks to use resources efficiently and regards "wastes" as potential products. Industrial ecology tries to create **industrial ecosystems** that compare in many ways to natural ecosystems.

Consider a pioneering industrial ecosystem in the Danish town of Kalundborg that consists of an electric power plant, an oil refinery, a pharmaceutical plant, a wallboard factory, a sulfuric acid producer, a cement manufacturer, fish farming, horticulture (greenhouses), and area homes and farms. At first glance, these entities appear to have little in common. However, they are linked to one another in complex ways that resemble a food web in a natural ecosystem (see figure). In this industrial ecosystem, the wastes produced by one company are sold to another company as raw materials for their processes, in a manner analogous to nutrient recycling in nature. Just as the interactions among producers, consumers, and decomposers in a natural ecosystem are complex, so too the interactions among the various components of Kalundborg's industrial ecosystem are intricate.

The coal-fired electric power plant originally cooled its waste steam and released it into the local fjord. The steam is now supplied to the oil refinery and the pharmaceutical plant, and additional surplus heat produced by the power plant warms greenhouses, the fish farm, and area homes; the need for 3500 oil-burning home heating systems has been eliminated as a result.

Surplus natural gas from the oil refinery is sold to the power plant and the wallboard factory. The power plant now saves 27,200 metric tons (30,000 tons of coal) each year by burning the less expensive natural gas. Before selling the natural gas, the oil refinery removes excess sulfur from it, which is required by air pollution control laws. This sulfur is sold to a company that uses it to manufacture sulfuric acid.

To meet environmental regulations, the power plant installed pollution control equipment to remove sulfur from its coal smoke. This sulfur, in the form of calcium sulfate, is sold to the wallboard plant—some 72,530 metric tons (80,000 tons) annually—and used as a substitute for gypsum, which is calcium sulfate that occurs naturally in the Earth's crust and is mined. The fly ash produced by the power plant goes to the cement manufacturer to be used for road-building.

Local farmers use the sludge from the fish farm as a fertilizer for their fields. The fermentation vats at the pharmaceutical plant also generate a high-nutrient sludge used by local farmers. Most pharmaceutical companies discard this sludge because it contains living microorganisms, but the Kalundborg plant heats the sludge to kill the microorganisms, thus converting a waste material into a commodity.

All of these interactions did not spring into existence at the same time; each represents a separately negotiated deal. It took a decade to develop the entire industrial ecosystem. Although these examples of industrial cooperation were initiated for economic reasons, each has distinct environmental benefits, from energy conservation to reduction of pollution.

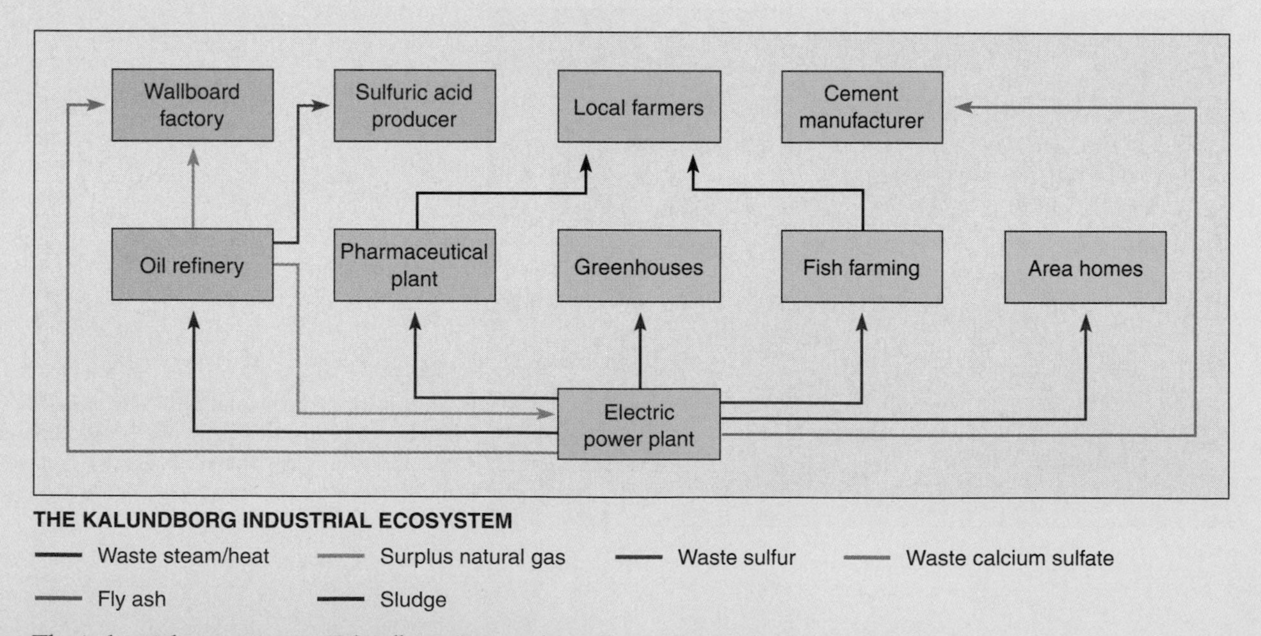

The industrial ecosystem in Kalundborg, Denmark, produces energy, food, and other products, thereby maximizing resource recovery and minimizing waste production.

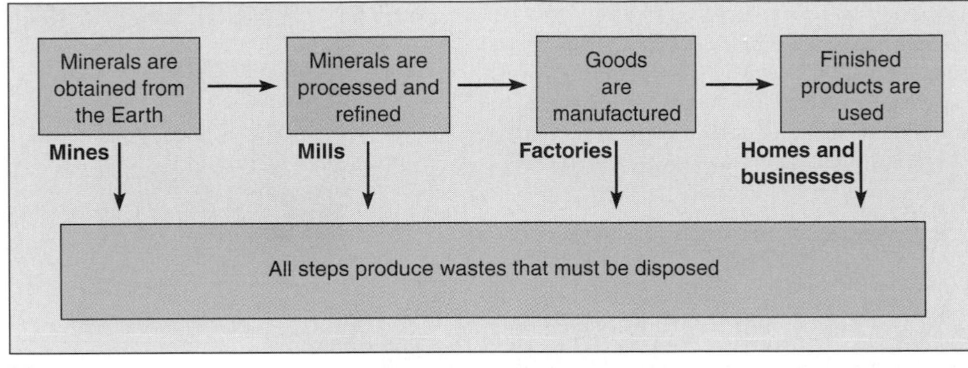

(a)

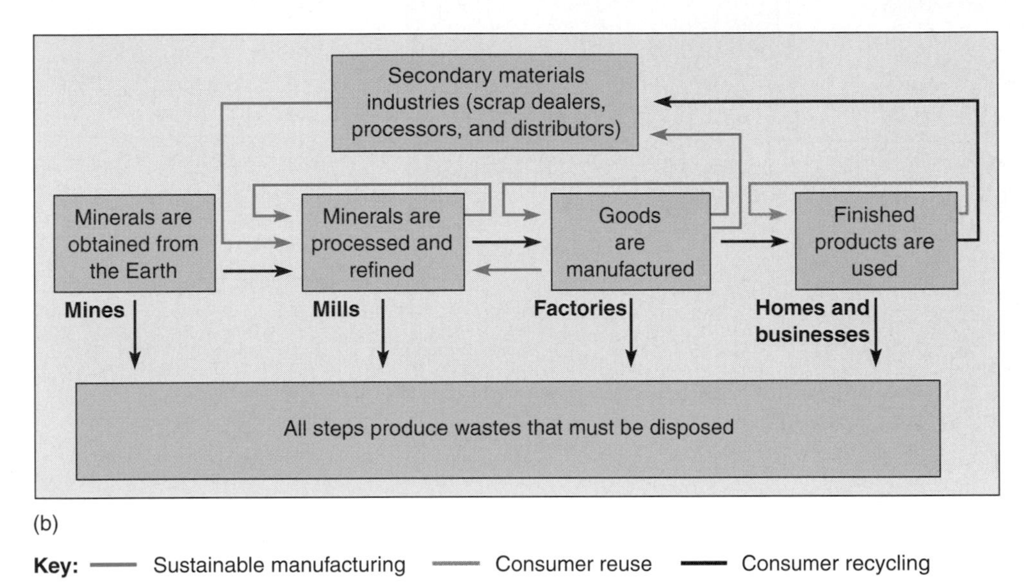

(b)

Key: —— Sustainable manufacturing —— Consumer reuse —— Consumer recycling

Figure 15–11

The flow of minerals in an industrial society. (a) The traditional flow of minerals is a one-way direction from the Earth to solid waste production. Massive amounts of solid waste are produced at all steps, from mining the mineral to discarding the used-up product. (b) The flow of minerals in a low-waste society is more complex, with recycling and reuse practiced at intermediate steps.

By selling these "wastes," industries gain additional profits and lessen the amounts of materials that must be thrown away. The chemical and petrochemical industries are among the first businesses to have pioneered the minimization of wastes by converting such wastes into useful products. For example, General Chemical buys a variety of used aluminum wastes from other companies and converts the aluminum in them to aluminum sulfate, a chemical used to treat municipal water supplies (see Chapter 21). Such minimization of waste by industry is known as **sustainable manufacturing** (see Figure 15–11b and Meeting the Challenge: Industrial Ecosystems). Sustainable manufacturing requires that companies provide information about their waste products to other industries so that any potential waste recovery can be implemented. However, many companies are reluctant to reveal the kinds of wastes they produce because their competitors may be able to deduce valuable trade secrets from the nature of their wastes. This difficulty will have to be overcome if sustainable manufacturing is to be fully implemented.

Dematerialization

As products evolve, they tend to become lighter in weight and often smaller. Washing machines manufactured in the 1960s were much heavier than comparable machines manufactured in the 1990s, for example. The same is true of other household appliances, automobiles, and elec-

Figure 15–12

Dematerialization. (a) The first general-purpose electronic calculator/computer was built at the University of Pennsylvania in 1946. It weighed 27 metric tons (30 tons), occupied 1800 square feet of floor space, and required 6 full-time technicians to operate. (b) A modern computer that has more applications than the historic computer is a fraction of its size. (a, UPI/Bettmann Archive; b, Courtesy of International Business Machines Corporation)

(a)

(b)

tronic items (Figure 15–12). This decrease in the weight of products over time is called **dematerialization**.

Although dematerialization gives the appearance of reducing consumption of minerals and other materials, it can sometimes have the opposite effect. Products that are smaller and lighter may also be of lower quality. Because repairing broken lightweight items is difficult and may cost more than the original products, consumers are encouraged by retailers and manufacturers to replace rather than repair the items. Thus, although the weight of materials being used to make each item has decreased, the number of such items being used in a given period of time may actually have increased.

SUMMARY

1. Minerals, which are essential to our industrial society, are naturally occurring elements and compounds. They may be metallic or nonmetallic. The Earth's crust has an uneven distribution of minerals, which are often found in concentrated, highly localized deposits due to magmatic concentration, hydrothermal processes, sedimentation, or evaporation.

2. Mineral reserves are deposits that have been identified and are profitable to extract, whereas mineral resources are deposits of low-grade ores that may or may not be profitable to extract in the future. A mineral's life index of world reserves is an estimate of the time remaining before the known reserves of a particular element will be depleted. It is extremely difficult to predict

life index times accurately, because mineral consumption changes over time. Economic factors influence all aspects of mineral consumption.

3. Steps in converting a mineral deposit into a usable product include locating the deposit, mining, and processing or refining. If the deposit is near the surface, it is extracted by open-pit surface mining or strip mining. Subsurface mines, either shaft mines or slope mines, are used to obtain minerals located deep in the Earth's crust.

4. Any form of mineral extraction and processing has negative effects on the environment. Surface mining disturbs the land more than subsurface mining, but subsurface mining is more expensive and dangerous. Mineral processing can cause air, soil, and water pollution. Derelict lands, which are extensively damaged due to mining, can be restored to prevent further degradation and to make the land productive for other purposes; land reclamation is extremely expensive, however.

5. Developed nations consume a disproportionate share of the world's minerals, but as developing countries industrialize, their need for minerals increases. The richest concentrations of minerals in highly industrialized countries have largely been exploited. As a result, these nations have increasingly turned to developing countries for the minerals they require. Sometimes developed nations must rely on politically volatile nations for strategically important minerals.

6. Mineral supplies can be extended in several ways. It is possible that new deposits will be identified. Advanced mining technology may make it possible to profitably extract minerals from inaccessible regions or from low-grade ores.

7. Substitution and conservation extend mineral supplies. Manufacturing industries continually try to substitute more common, less expensive minerals for those that are scarce and expensive. Mineral conservation includes recycling, in which discarded products are collected and reprocessed into new products, and reuse, in which a product is collected and used over again. In addition to conserving minerals, recycling and reuse cause less pollution and save energy when compared to the extraction and processing of virgin ores.

8. If our mineral supplies are to last and if our standard of living is to remain high, consumers must decrease their consumption. To accomplish this, manufacturers must make high-quality, durable products that can be repaired.

THINKING ABOUT THE ENVIRONMENT

1. Distinguish among the following ways in which mineral deposits may form: magmatic concentration, hydrothermal processes, sedimentation, and evaporation.

2. Explain why it is difficult to obtain an accurate appraisal of our total mineral resources.

3. Discuss several harmful environmental effects of mining and processing minerals.

4. Explain why obtaining minerals from low-grade ores is more environmentally damaging than extracting them from high-grade ores.

5. Historically, the cost of environmental damage arising from mining and processing minerals has not been included in the price of consumer products. Do you think it should be? Why or why not?

6. Consumer practices are a powerful way to make environmental statements. Explain the environmental rationale behind deciding not to purchase gold jewelry.

7. Some people in industry argue that the planned obsolescence of products, which means they must be replaced often, creates jobs. Others think that the production of smaller quantities of durable, repairable products would generate jobs and stimulate the economy. Explain each viewpoint.

8. Outline the benefits of beverage container deposit laws.

9. What is sustainable manufacturing, and whom does it benefit?

10. How does the industrial ecosystem at Kalundborg resemble a natural ecosystem? What are some of the environmental benefits of Kalundborg's industrial ecosystem?

* **11.** According to Figure 15–7, in 1994 the United States consumed 5,407,000 metric tons of aluminum, which represents 26.8 percent of the world total consumption. Calculate the world total consumption of aluminum in metric tons for 1994.

* **12.** The United States produced 374,000 metric tons of lead in 1994 but consumed 1,374,800 metric tons. Use this information to complete Table A. Note that figures are to be expressed in "thousand metric tons" in the table.

Table A
U.S. Production, Consumption, and Imports of Lead, 1994

Production (000 Metric Tons):	
Consumption (000 Metric Tons):	
Amount Imported (000 Metric Tons):	

RESEARCH PROJECTS

Japan is currently the world leader in deep-sea exploration, including the potential mining of manganese nodules. Investigate some of the recent Japanese discoveries and new equipment, including deep-sea robots and manned submersibles. Possible sources of information include the Japan Marine Science and Technology Center, the Woods Hole Oceanographic Institute, and the National Geographic Society.

Although the mining industry's image has been tarnished by its record of environmental damage in the past, many mining companies today accept responsibility for environmental damage caused by mining. Obtain a copy of *McLaughlin Mine: General Information Summary* (1990) from the Homestead McLaughlin Mine, San Francisco, California, and investigate their progressive environmental policies.

About 12 states currently have beverage container deposit laws. Using the Internet, try to determine why such laws are not in place throughout the United States and who opposes such laws.

Read the Eggert reference listed in the Suggested Reading as a starting point to an in-depth evaluation of the 1872 Mining Law. Then write a position paper on whether the proposed Noranda mine (discussed in the chapter introduction) and others should be allowed to move forward.

On-line materials relating to this chapter are on the World Wide Web at **http://www.saunderscollege.com/lifesci/environment2** (select Chapter 15 from the Table of Contents).

SUGGESTED READING

Baum, D. and Knox, M. L. "We Want People Who Have a Problem with Mine Wastes to Think of Butte," *Smithsonian*, November 1992. Butte, Montana, hopes that the techniques being developed to clean up their hazardous wastes from copper mining can be marketed to clean up other environmental disasters around the world.

Eggert, R. G. "Reforming the Rules for Mining on Federal Lands," *Resources* Vol. 117, Fall 1994. Examines the problems with the General Mining Law of 1872 and prospects for its reform.

Frosch, R. A. "Industrial Ecology: Adapting Technology for a Sustainable World," *Environment* Vol. 37, No. 10, December 1995. An in-depth evaluation of how industrial ecology recycles materials and minimizes wastes.

Frosch, R. A. "The Industrial Ecology of the 21st Century," *Scientific American*, September 1995. An overview of how industrial ecology recycles materials and minimizes waste.

Gillis, A. M. "Bringing Back the Land," *BioScience* Vol. 41, No. 2, February 1991. Scientists at Utah State University are pioneering rapid reclamation of arid land that has been surface mined.

Hodges, C. A. "Mineral Resources, Environmental Issues, and Land Use," *Science* Vol. 268, June 2, 1995. Provides an overview of current and projected mineral use, and of environmental issues associated with mineral extraction.

Young, J. E. "Aluminum's Real Tab," *WorldWatch* Vol. 5, No. 2, March–April 1992. Although cheap and invaluable to society, aluminum has hidden environmental costs that make it very expensive.

Young, J. E. "For the Love of Gold," *WorldWatch* Vol. 6, No. 3, May–June 1993. Highlights the environmental destruction caused by the world's increasing demand for gold.

Young, J. E. "Mining the Earth." In *State of the World: 1992, a WorldWatch Institute Report on Progress Toward a Sustainable Society*. New York: W.W. Norton & Company, 1992. Discusses the importance of minerals in the global economy and their environmental costs.

16

A mated pair of bald eagles. (©1997 Stan Osolinski/Dembinsky Photo Associates)

Preserving Earth's Biological Diversity

The American bald eagle—the symbol of the United States and an emblem of strength—was a common sight throughout colonial North America. More recently, however, the bald eagle fell on hard times. Its numbers dropped precipitously to only 417 nesting pairs in the lower 48 states in 1963, and it was in danger of extinction. Several factors contributed to its decline. As European settlers pushed across North America, they cleared many thousands of square kilometers of forest near lakes and rivers, thus destroying the bald eagle's habitat. Eagles were hunted for sport and because it was thought they preyed on livestock and commercially important fishes. In fact, bounties were offered for dead bald eagles as recently as 1952. In addition, eagles' numbers dwindled because they could not reproduce at high enough levels to ensure their population growth or their survival. This reproductive failure was the direct result of ingesting prey contaminated with the pesticide dichloro-diphenyl-trichloroethane (DDT). DDT caused the eagles' eggs to be so thin-shelled that they cracked open before the embryos could mature and hatch. Mercury, lead, and selenium were other environmental pollutants that harmed bald eagles.

Conservation efforts have helped the bald eagle make a remarkable comeback. In the mid-1970s, the first eagles to be bred in captivity were released in nature. In addition to raising birds in captive-breeding programs, biologists also remove eagle eggs from their nests in nature, raise the baby eagles in wildlife refuges, and return them to nature. Removal of eggs actually helps increase the number of eagles, because nesting eagles commonly lay more eggs to replace those that were removed. As a result of continuing efforts, the number of nesting pairs in the continental United States increased from 417 to 4452 pairs between 1963 and 1994. (Nesting pairs are counted because bald eagles mate for life.) In 1994 the bald eagle was removed from the "endangered" list and transferred to the less critical "threatened" list.

Today many states are reintroducing bald eagles to nature. Save the Eagle Project, a private, nonprofit conservation organization, works with the National Fish and Wildlife Foundation to restore bald eagle populations in the United States. Federal and state governments have supported such efforts, and private and corporate donors have also been generous in their support.

The world is a richer place because of the bald eagle. Today it symbolizes more than a country, for the bald eagle demonstrates that our biological heritage can be preserved if enough people care to do something about it. In this chapter we examine the importance of all forms of life and consider extinction, which has become an increasing threat to so many organisms. Finally, we explore what can be done to preserve our biological resources and save the many endangered species from disappearing forever.

HOW MANY SPECIES ARE THERE?

We do not know exactly how many species of organisms exist, but scientists estimate there may be something on the order of 13 to 14 million different species.[1] About 1.75 million species of organisms have been scientifically named and described. These include about 250,000 plant species, 42,000 vertebrate animals, and some 750,000 insects.

The variation among organisms is referred to as **biological diversity** or **biodiversity**, but the concept includes much more than simply the number of different species, called **species diversity** (Figure 16–1). It also takes into account **genetic diversity**, the genetic variety *within* a species—that is, the different populations that

[1]A species is a group of more or less distinct organisms that are capable of interbreeding with one another in the wild but do not interbreed with other organisms.

(a)

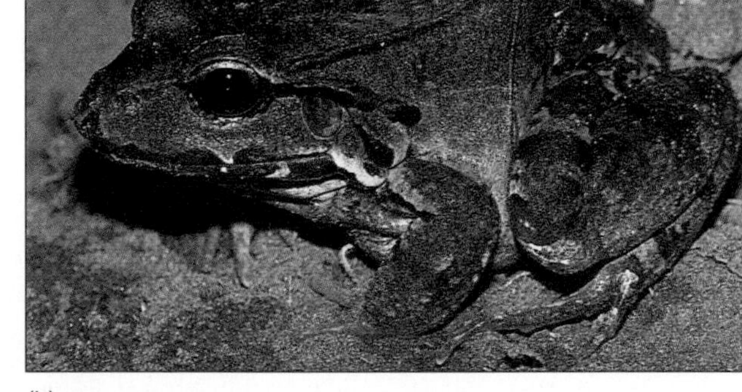

(b)

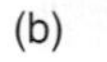

(c)

Figure 16–1

This sampling of three tropical rainforest amphibian species in Costa Rica demonstrates species diversity. (a) A red-eyed tree frog (*Agalychnis callidryas*) is active at night or when it rains. During the day, it curls up on the underside of a leaf, blending inconspicuously into the surrounding vegetation. (b) A smoky jungle frog (*Leptodactylus pentadactylus*) lives on the ground and blends into the litter on the floor of a tropical rain forest. (c) A dart-poison frog (*Dendrobates pumilio*) is brightly colored to advertise the fact that it is toxic. Also see Figure 4–4 for another poisonous rainforest frog species. (a, Dennis Paulson/Visuals Unlimited; b, © 1997 Dominique Braud/Dembinsky Photo Associates; c, James L. Castner)

Figure 16–2
The variation in corn kernels and ears is evidence of the genetic diversity in the species *Zea mays*. The genetic diversity of corn is primarily preserved in seed banks in Mexico. (David Cavagnaro)

make up a particular species (Figure 16–2). Biological diversity also includes **ecosystem diversity**, the variety of interactions among organisms in natural communities. For example, a forest community with its trees, shrubs, vines, herbs, insects, worms, vertebrate animals, fungi, bacteria, and other microorganisms has greater ecosystem diversity than does a cornfield. Ecosystem diversity also encompasses the variety of ecosystems found on Earth: the forests, prairies, deserts, coral reefs, lakes, coastal estuaries, and other ecosystems of our planet.

WHY WE NEED ORGANISMS

Humans depend on the contributions of thousands of different species for their survival. In primitive societies, these contributions are direct: plants, animals, and other organisms are the sources of food, clothing, and shelter. In industrialized societies most people do not hunt for their morning breakfasts or cut down trees for their shelter and firewood; nevertheless, we still depend on organisms.

Although all societies make use of many different kinds of plants, animals, fungi, and microorganisms, most species have never been evaluated for their potential usefulness. There are approximately 250,000 known plant species, but perhaps 225,000 of them have never been evaluated with respect to their industrial, medicinal, or agricultural potential. The same is true of most of the millions of microorganisms, fungi, and animals. Most people do not think of insects as an important biological resource, for example, but insects are instrumental in several important environmental and agricultural processes, including pollination of crops, weed control, and insect pest control. In addition, many insects produce unique chemicals that may have important applications for human society. Bacteria and fungi provide us with foods, antibiotics and other medicines, and important environmental services such as soil enrichment. Biological diversity represents a rich, untapped resource for future uses and benefits: many as-yet-unknown species may someday provide us with products, for example. A reduction in biological diversity decreases this treasure prematurely and permanently.

Ecosystem Stability and Species Diversity

You may recall from Chapters 3, 4, and 5 that the living world functions much like a complex machine. Each ecosystem is composed of many separate parts, the functions of which are organized and integrated to maintain the ecosystem's overall performance. The activities of all organisms are interrelated; we are bound together and dependent on one another, often in subtle ways (Figure 16–3).

Plants, animals, fungi, and microorganisms are instrumental in many environmental processes without which humans could not exist. Forests are not just a potential source of lumber; they provide watersheds from

Figure 16–3
The alligator plays an integral, but often subtle, role in its natural ecosystem. It helps maintain populations of smaller fishes by eating the gar, a fish that preys on them. Alligators dig underwater holes that other aquatic organisms use during periods of drought when the water level is low. The nest mounds they build are enlarged each year and eventually form small islands that are colonized by trees and other plants. In turn, the trees on these islands support heron and egret populations. The alligator habitat is maintained in part by underwater "gator trails," which help to clear out aquatic vegetation that might eventually form a marsh. (Connie Toops)

which we obtain fresh water, and they reduce the number and severity of local floods. Many species of flowering plants depend on insects to transfer pollen for reproduction. Animals, fungi, and microorganisms help to keep the populations of various species in check so that the numbers of one species do not increase enough to damage the stability of the entire ecosystem. Soil dwellers, from earthworms to bacteria, develop and maintain soil fertility for plants. Bacteria and fungi perform the crucial task of decomposition, which allows nutrients to recycle in the ecosystem.

You might think that the loss of some species from an ecosystem would not endanger the rest of the organisms, but this is far from true. Imagine trying to assemble an automobile if some of the parts were missing. You might be able to piece it all together so that it resembled a car, but it probably would not run as well. Similarly, the removal of organisms from a community makes an ecosystem run less smoothly. If enough organisms are removed, the entire ecosystem can collapse.

Genetic Reserves

The maintenance of a broad genetic base for economically important organisms is critical. During the 20th century, for example, plant scientists developed genetically uniform, high-yielding varieties of important food crops such as wheat. It quickly became apparent, however, that genetic uniformity resulted in increased susceptibility to pests and disease.

By crossing the "super strains" with more genetically diverse relatives, disease and pest resistance can be reintroduced into such plants. For example, a corn blight fungus that ruined the corn crop in the United States in 1970 was brought under control by crossing the cultivated, highly uniform U.S. corn varieties with genetically diverse ancestral varieties from Mexico. When some of the genes from Mexican corn were incorporated into the U.S. varieties, the latter became resistant to the corn blight fungus. (The global decline in domesticated plant and animal varieties is discussed in Chapter 18.)

Scientific Importance of Genetic Diversity

Genetic engineering, the incorporation of genes from one organism into an entirely different species (see Chapter 18), makes it possible to use the genetic resources of organisms on a much wider scale than had earlier been possible. The gene for human insulin, for example, has been engineered into bacteria, which subsequently become tiny chemical factories, manufacturing at a relatively low cost the insulin required in large amounts by diabetics. Genetic engineering, which has been available only since the mid-1970s, has already begun to provide us with new vaccines, more productive farm animals, and agricultural products with longer shelf life or other desirable characteristics.

Although we have the skills to transfer genes from one organism to another, we do not have the ability to *make* genes. Genetic engineering depends on a broad base of genetic diversity from which it can obtain genes. It has taken hundreds of millions of years for evolution to produce the genetic diversity found in organisms living on our planet today, a diversity that may hold solutions not only to problems we have today but to problems we have not even begun to conceive. It would be very unwise to allow such an important part of our heritage to disappear.

Medicinal, Agricultural, and Industrial Importance of Organisms

The genetic resources of organisms are vitally important to the pharmaceutical industry, which incorporates into its medicines many hundreds of chemicals derived from organisms. From extracts of cherry and horehound for cough medicines to certain ingredients of periwinkle and mayapple for cancer therapy, derivatives of plants play important roles in the treatment of illness and disease (Figure 16–4). Many of the natural products—that is, substances taken directly from organisms—that are promising anti-cancer drugs come from marine organisms such as tunicates, red algae, and sponges. All of the 20 best-selling prescription drugs in the United States are either natural products, natural products that have been

Figure 16–4

The rosy periwinkle produces chemicals that are effective against certain cancers. For example, drugs from the rosy periwinkle have increased the chance of surviving childhood leukemia from about 5 percent to more than 95 percent. (Doug Wechsler)

Figure 16–5
Many edible plants have not been used to any great extent, including the winged bean, a tropical legume that is nutritionally superior to many other foods. (Sigrid Salmela © 1993 Animals Animals/Earth Scenes)

slightly modified chemically, or manufactured drugs that were originally obtained from organisms.

The agricultural importance of plants and animals is indisputable, since we must eat to survive. However, the number of different kinds of foods we eat is limited when compared with the total number of edible species. There are probably many species that are nutritionally superior to our common foods. For example, quinoa, a plant long cultivated as food in the Andes Mountains in South America, looks and tastes somewhat like rice but has a much higher concentration of protein and is more nutritionally balanced. Winged beans are a tropical legume from Southeast Asia and Papua New Guinea (Figure 16–5). Because the seeds of the winged bean contain large quantities of protein and oil, they may be the tropical equivalent of soybeans. Almost all parts of the plant are edible, from the young, green fruits to the starchy storage roots.

Modern industrial technology depends on a broad range of genetic material from organisms, particularly plants, that are used in many products. Plants supply us with oils and lubricants, perfumes and fragrances, dyes, paper, lumber, waxes, rubber and other elastic latexes, resins, poisons, cork, and fibers. Animals provide wool, silk, fur, leather, lubricants, waxes, and transportation, and they are important in scientific research. The armadillo, for example, is used in research in Hansen's disease (leprosy) because it is one of only two species known to be susceptible to that disease (the other species is humans).

Insects secrete a large assortment of chemicals that represent a wealth of potential products. Certain beetles produce steroids with birth-control potential, and fireflies produce an antiviral compound that may be useful in treating viral infections. Centipedes secrete a fungicide over the eggs of their young that could help control the fungi that attack crops. Because biologists estimate that perhaps 90 percent of all insects have not yet been identified, insects represent a very important potential biological resource.

Aesthetic, Ethical, and Religious Value of Organisms

Organisms not only contribute to human survival and physical comfort; they also provide recreation, inspiration, and spiritual solace. Our natural world is a thing of beauty largely because of the diversity of living forms found in it. Artists have attempted to capture this beauty in drawings, paintings, sculpture, and photography, and it has inspired poets, writers, architects, and musicians to create works reflecting and celebrating the natural world.

The strongest ethical consideration involving the value of organisms is how humans perceive themselves in relation to other species. Traditionally, much of humankind have viewed themselves as the "masters" of the rest of the world, subduing and exploiting other forms of life for their benefit. An alternative view is that organisms have intrinsic value in and of themselves and that, as stewards of the life forms on Earth, we humans should watch over and protect their existence. As far as we know, the organisms on planet Earth are our only living companions in the Universe; therefore, we must treat them with reverence.

The conviction that all organisms have the right to exist and that humans should not cause the extinction of other organisms is known as **deep ecology**. The basic tenets of deep ecology are not new. The belief in the sacredness of life held by Eastern religions such as Buddhism and Taoism is similar to that of deep ecology. That biological diversity is valued by Judeo-Christian religions is evident in the story of Noah's Ark, which is considered to be the first "endangered species project."

ENDANGERED AND EXTINCT SPECIES

Extinction, the death of a species, occurs when the last individual member of a species dies. Extinction is an irreversible loss—once a species is extinct it can never reappear. Biological extinction is the eventual fate of all species, much as death is the eventual fate of all individuals. Biologists estimate that for every 2000 species that have ever lived, 1999 of them are extinct today.

During the span of time in which organisms have occupied Earth, there has been a continuous, low-level extinction of species, known as **background extinction**. At certain periods in the Earth's history, maybe five or six times, there has been a second kind of extinction, **mass extinction**, in which numerous species disappeared during a relatively short period of geological time. The course of a mass extinction episode may have taken millions of years, but that is a short time compared with the age of the Earth, estimated at 4.6 billion years.

The causes of background extinction and past mass extinctions are not well understood, but it appears that biological and environmental factors were involved. A major climate change could have triggered the mass extinction of species. Marine organisms are particularly vulnerable to temperature changes; if the Earth's temperature changed by just a few degrees, for example, it is likely that many marine species would have become extinct. It is also possible that mass extinctions of the past were triggered by catastrophes, such as the collision of the Earth and a large asteroid or comet. The impact could have forced massive quantities of dust into the atmosphere, blocking the sun's rays and cooling the planet.

Extinctions Today

Although extinction is a natural biological process, it can be greatly accelerated by human activities. The burgeoning human population has forced us to spread into almost all areas of the Earth, and whenever humans invade an area, the habitats of many organisms are disrupted or destroyed, which can contribute to their extinction. For example, the dusky seaside sparrow, a small bird that was found only in the marshes of St. Johns River in Florida, became extinct in 1987, largely due to human destruction of its habitat.

Currently, the Earth's biological diversity is disappearing at an alarming rate (Figure 16–6). Conservation biologists estimate that species are presently becoming extinct at a rate approximately 10,000 times the rate of background extinctions. The U.N. Global Biodiversity Assessment, released in late 1995 and based on the work of about 1500 scientists from around the world, estimates that more than 31,000 plant and animal species are currently threatened with extinction. As much as one-fifth of all species may become extinct within the next 30 years. Some biologists fear that we are entering the greatest period of mass extinction in the Earth's history. The Center for Plant Conservation, for example, estimates that about 4000 native plant species are of conservation concern in the United States; this number represents about 20 percent of U.S. plant species.

Endangered and Threatened Species

The legal definition of an **endangered species**, as stipulated in the Endangered Species Act, is a species in imminent danger of extinction throughout all or a significant portion of its range. The area in which a particular species is found is its **range**. A species is endangered when its numbers are so severely reduced that it is in danger of becoming extinct without human intervention.

When extinction is less imminent but the population of a particular species is quite low, the species is said to be **threatened**. The legal definition of a threatened species is one that is likely to become endangered in the foreseeable future, throughout all or a significant portion of its range.

Endangered and threatened species represent a decline in biological diversity, because as their numbers decrease, their genetic variability is severely diminished. Long-term survival and evolution depend on genetic diversity, so its loss adds to the risk of extinction for endangered and threatened species, as compared to species that have greater genetic variability.

Characteristics of endangered species

Many endangered species share certain characteristics that seem to have made them more vulnerable to extinction. These include (1) an extremely small (localized) range, (2) a large territory, (3) living on islands, and (4) low reproductive success, often the result of a small population size.

Many endangered species have a very limited natural range, which makes them particularly prone to extinction if their habitat is altered. The Tiburon mariposa lily, for example, is found nowhere in nature except on a single hilltop near San Francisco. Development of that area would almost certainly cause the extinction of this species.

Species that require extremely large territories in order to survive may be threatened with extinction when all or part of their territory is modified by human activity. For example, the California condor, a scavenger bird that lives off of carrion and requires a large, undisturbed territory—hundreds of square kilometers—in order to find adequate food, is on the brink of extinction. During

Figure 16–6

A small sample of Earth's species that are either classified as endangered or have become extinct during the last century or so. Declining biological diversity is a serious problem: officials at the U.S. Fish and Wildlife Service estimate that more than 500 U.S. species have gone extinct during the past 200 years. Of these, roughly 250 have gone extinct since 1980.

the five-year period from 1987 to 1992, it was no longer found in nature (Figure 16–7). A program to reintroduce zoo-bred California condors into the Los Padres National Forest, located about 100 miles northwest of Los Angeles, began in 1992. As of 1997, 17 released birds have survived their return to the wild in California. A second colony of six California condors was released north of the Grand Canyon in Arizona in January 1997.

Many island species that are *endemic* to certain islands (that is, they are not found anywhere else in the world) are endangered. These organisms often have small populations that cannot be replaced by immigration

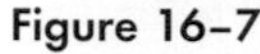

Figure 16–7

California condors, which are scavengers, require large territories in order to obtain enough food. They are critically endangered largely because development has reduced the size of their wilderness habitat. (Tom McHugh/Photo Researchers)

DISAPPEARING FROGS

Amphibians—frogs, toads, and salamanders—have bare and extremely permeable skin and lay gelatinous and unprotected eggs. Since the 1970s, many of the world's frog populations have dwindled. For example, at least 14 species of Australian rainforest frogs have become endangered or extinct since 1980. Likewise, populations of all seven native species of frogs and toads in Yosemite National Park have declined. Biologists are beginning to unravel some clues to these mysterious declines. One possible culprit appears to be increased UV radiation caused by ozone thinning. Dr. Andrew Blaustein and colleagues at Oregon State University reported in 1994 the results of their investigations on the effects of UV radiation on egg survival of some amphibian species. Amphibians possess photolyase, an enzyme that allows them to repair DNA damage caused by natural UV radiation. In the study, species suffering declines appear to be limited in their ability to repair such cellular damage. The researchers exposed the eggs of three species to natural radiation. Egg survival was high for the Pacific tree frog, which had the greatest photolyase activity and is not in decline. Egg survival was much lower (only 45 to 65 percent) for the Western toad and Cascades frog but increased dramatically when eggs were shielded from UV radiation. Experts caution that all amphibian declines are not caused by the same threat and are likely the result of multiple effects such as habitat loss, disease, and air and water pollution. However, scientists increasingly perceive amphibians to be **bellwether species**—that is, organisms that provide an early warning of environmental damage.

should their numbers be destroyed. Because they evolved in isolation from competitors, predators, and disease organisms, they have few defenses when such organisms are introduced, usually by humans, to their habitat. It is not surprising that of the 171 bird species that have become extinct in the past few centuries, 155 of them lived on islands. Endangered island species are discussed in several contexts later in the chapter.

Today, *island* refers not only to any land mass surrounded by water but also to any isolated habitat that is surrounded by an expanse of unsuitable territory. Accordingly, a small patch of forest surrounded by agricultural and suburban lands is considered an island. (National parks are discussed as islands in Chapter 17.)

In order for a species to survive, its members must be present within their range in large enough numbers for males and females to mate. The minimum population density and size that ensure reproductive success vary from one type of organism to another. However, for all organisms, if the population density and size fall below a critical minimum level, the population declines, becoming susceptible to extinction.

Endangered species often share other characteristics. Some have low reproductive rates; the female blue whale, for example, produces a single calf every other year. Some endangered species breed only in very specialized areas; the green sea turtle, for example, lays its eggs on just a few beaches. Highly specialized feeding habits can also endanger a species. In nature, the giant panda eats only bamboo, a plant all of whose members periodically flower and die together; when this occurs, panda populations face starvation. Like many other endangered species, giant pandas are also endangered because their habitat has been fragmented into small islands, and there are few intact habitats where they can survive.

HUMAN CAUSES OF SPECIES ENDANGERMENT AND EXTINCTION

Most species facing extinction today are endangered because of the destruction of habitats by human activities. We demolish habitats when we build roads, parking lots, and buildings; clear forests to grow crops or graze domestic animals; and log forests for timber. We drain marshes to build on aquatic habitats, thus converting them to terrestrial ones, and we flood terrestrial habitats when we build dams. Because most organisms are utterly dependent on a particular type of environment, habitat destruction reduces their biological range and ability to survive.

Even habitats that are left "totally" undisturbed and natural are degraded by human-produced acid rain, ozone depletion, and climate change. Acid rain is thought to have contributed to the decline of large stands of forest trees and the biological death of many freshwater lakes. Because ozone in the upper atmosphere shields the ground from a large proportion of the sun's harmful ultraviolet (UV) radiation, ozone depletion in the upper atmosphere represents a very real threat to all terrestrial life. Global climate change, which is caused in part by carbon dioxide released when fossil fuels are burned, is another threat. Such habitat modifications particularly reduce the biological diversity of species with extremely narrow and rigid environmental requirements. Organisms are also affected by other types of pollutants, including industrial and agricultural chemicals, organic pollutants from sewage, acid wastes seeping from mines, and thermal pollution from the heated wastewater of industrial plants. (See Chapters 19 through 23 for additional discussion of the adverse effects of pollutants on the environment.)

Figure 16–8
The introduction of exotic species often threatens native species. The brown tree snake, which reaches 5 to 8 feet in length as an adult, is an efficient predator. Since its accidental introduction on Guam, it has driven nine rainforest birds to extinction. Hawaii has mounted a campaign to prevent the snake's accidental introduction to its islands. Several brown tree snakes that hitched a ride in the wheel wells of planes have been discovered and killed. (Gordon H. Rodda/U.S. Fish and Wildlife Service)

The Problem of Exotic Species

Biotic pollution, the introduction of a foreign, or exotic, species into an area where it is not native, often upsets the balance among the organisms living in that area. The foreign species may compete with native species for food or habitat or may prey on them. Generally, an introduced competitor or predator has a greater negative effect on local organisms than do native competitors or predators (see Focus On: The Death of a Lake; also see Chapter 1). Although exotic species may be introduced into new areas by natural means, humans are usually responsible for such introductions, either knowingly or unknowingly. The blue water hyacinth, for example, was deliberately brought from South America to the United States because it has lovely flowers. Today it has become a nuisance in Florida waterways, clogging them so that boats cannot easily move and crowding out native species.

Islands are particularly susceptible to the introduction of exotic species. Less than 50 years ago the brown tree snake was accidentally introduced in Guam, an island in the West Pacific (Figure 16–8). Thought to have arrived from the Solomon Islands on a U.S. Navy ship shortly after the end of World War II, the brown tree snake thrived and is now estimated to number 3 million. It started consuming rainforest birds in large numbers, and as a result, 9 of the 11 native species of forest birds are extinct in nature. For example, the Guam rail, a flightless bird endemic to Guam, numbered about 80,000 in

FOCUS ON

The Death of a Lake

The world's second largest freshwater lake, Lake Victoria in East Africa, is larger than the state of West Virginia. Until relatively recently, it was home to about 400 different species of small, colorful fishes known as cichlids (pronounced *sik-lids*). The different species of cichlids in Lake Victoria had remarkably different eating habits. Some grazed on algae; some consumed dead organic material at the bottom of the lake; and others were predatory and ate insects, shrimp, and other cichlid species. These fishes, which thrived throughout the lake ecosystem, provided much-needed protein to the diets of 30 million humans living in the lake's vicinity.

Today, the aquatic community is very different in Lake Victoria. More than half of the cichlids and other native fish species have disappeared. As a result of their mass extinction, the algal population has increased explosively (the algae-eating cichlids are almost all extinct). When the dense algal blooms die, their decomposition uses up the dissolved oxygen in the water. The bottom zone of the lake, once filled with cichlids, is empty because the deep water contains too little dissolved oxygen, and any fishes venturing into the anaerobic zone suffocate. Local fishermen, who once caught and ate hundreds of different types of fishes, now catch only a few types (see figure).

Although overfishing and pollution are factors, the most important reason for the destruction of Lake Victoria's delicate ecological balance was the deliberate introduction of the Nile perch, a large and voracious predator, into the lake in 1960. It was thought by proponents of the introduction that the successful establishment of the Nile perch would stimulate the local economy and help the fishermen. For about 20 years, the Nile perch did not appear to have an appreciable effect on the lake. But in 1980, fishermen noticed that they were harvesting increasing quantities of Nile perch and decreasing amounts of native fishes. By 1985, most of the annual catch was Nile perch, which was experiencing a population explosion fueled by an abundant food supply—the cichlids.

The Lake Victoria story is far from finished. Once the Nile perch decimated the cichlids, it changed its feeding habits and now consumes freshwater shrimp and smaller Nile perch. A species that cannibalizes its own young for a major part of its food supply cannot persist indefinitely. Moreover, during the past decade or so, a large fishing industry, which processes Nile perch for export to countries such as Israel and the Netherlands, developed along the shores of Lake Victoria. Many people are concerned that overfishing could collapse the Nile perch population, ultimately causing protein malnutrition in the humans living around the lake basin.

A woman with a mixed catch of Nile perch (larger fishes) and cichlids (smaller fishes). (Mark Chandler, New England Aquarium)

Figure 16–9
After the slaughter of vast herds of bison, bone pickers scoured the Great Plains and sent the bones to plants where they were pulverized to make fertilizer. Shown is a mountain of bison skulls at a fertilizer plant in Detroit, Michigan. Today, the bison has made a comeback, and nationwide more than 250,000 of North America's heaviest land animal exist in parks, refuges, and ranches. (Burton Historical Collection, Detroit Public Library)

1968 but was extinct in nature by 1986; today it exists only in small captive populations safely isolated from the snakes. The snake has caused other problems as well. Although its venom does not harm adult humans, it is mildly poisonous to infants and children and sometimes crawls into beds and bites them while they are sleeping. The snake has also caused many power outages by climbing utility poles and shorting electrical lines.

Hunting

Sometimes species become endangered or extinct as a result of deliberate efforts to eradicate or control their numbers. Many of these species prey on game animals; some prey on livestock. Populations of large predators such as the wolf, mountain lion, and grizzly bear have been decimated by ranchers, hunters, and government agents. Predators of game animals and livestock are not the only animals vulnerable to human control efforts. Some animals are killed because their life styles cause problems for humans. The Carolina parakeet, a beautiful green, red, and yellow bird endemic to the southeastern United States, was extinct by 1920, exterminated by farmers because it ate fruit. Prairie dogs and pocket gophers, two more examples of animals killed by humans because of their life styles, have been poisoned and trapped because their burrows weaken the ground on which unwary cattle graze, and if the cattle step into the burrows, they may be crippled. As a result of sharply decreased numbers of prairie dogs and pocket gophers, the black-footed ferret, the natural predator of these animals, was not found in nature in the United States between 1986 and 1991. A successful captive-breeding program enabled scientists to release black-footed ferrets to the Wyoming prairie, beginning in 1991.

In addition to predator and pest control, hunting is done for three other reasons: (1) **subsistence hunters** kill animals for food; (2) **sport hunters** kill animals for recreation; and (3) **commercial hunters** kill animals for profit—for example, by selling their fur. Subsistence hunting has caused the extinction of certain species in the past but is not a major cause of extinction today, mainly because so few human groups still rely on hunting for their food supply. Sport hunting was one of several factors in the near extinction of the American bison (Figure 16–9),[2] but overall, has not been a major cause of extinction. Sport hunting is also strictly controlled in most countries.

Illegal commercial hunting, however, continues to endanger many larger animals, such as the tiger, cheetah, and snow leopard, whose beautiful furs are quite valuable. Rhinoceroses are slaughtered primarily for their horns, which are used for ceremonial dagger handles in the Middle East and for purported medicinal purposes in Asian medicine. Bears are killed for their gallbladders, which are used in Asian medicine to treat ailments from indigestion to heart ailments. Although these animals are

[2]Bison were also decimated by the U.S. Army, which killed bison to disrupt the food supply of the Plains Indians; by commercial hunters, who killed bison for their hides and tongues (considered a choice food); and by hunters who supplied meat to work crews for railroad companies.

Figure 16–10

These hyacinth macaws were seized in French Guiana in South America as part of the illegal animal trade there. (Jany Sauvanet/NHPA)

legally protected, the demand for their products on the black market has caused them to be hunted illegally. The American black bear's gallbladder, for example, can fetch $800 to $3000 on the black market.

In contrast to commercial hunting, in which the target organism is killed, **commercial harvest** is the removal of the live organism from nature. Commercially harvested organisms end up in zoos, aquaria, biomedical research laboratories, circuses, and pet stores. Several million birds are commercially harvested each year for the pet trade, but unfortunately many of them die in transit, and many more die from improper treatment after they are in their owners' homes. At least 40 parrot species are now threatened or endangered, in part because of excessive commercial harvest. Although it is illegal to capture endangered animals from nature, there is a thriving black market, mainly because collectors in the United States, Europe, and Japan are willing to pay extremely large amounts to obtain rare tropical birds (Figure 16–10). Imperial Amazon macaws, for example, fetch up to $20,000 each.

Animals are not the only organisms threatened by commercial harvest. Many unique and rare plants have been collected from nature to the point that they are endangered. These include carnivorous plants, wildflower bulbs, certain cacti, and orchids.

Where Is Declining Biological Diversity the Greatest Problem?

Although declining biological diversity is a concern throughout the United States, it is most serious in the states of California, Hawaii, Texas, and Florida, where many ecosystems are deteriorating at a rapid rate.

As serious as declining biological diversity is in the United States, it is even more serious abroad. Perhaps 40 percent of the world's species are concentrated in tropical rain forests, areas that are increasingly threatened by habitat destruction (see Focus On: Vanishing Tropical Rain Forests). This means that a few countries, primarily developing nations, hold most of the biological diversity that is so ecologically and economically important to the

SOLVING CRIMES AGAINST ORGANISMS

The U.S. Fish and Wildlife Service Forensics Laboratory in Ashland, Oregon, employs sophisticated technology to determine whether endangered species are being illegally killed or otherwise exploited. Investigators who specialize in such varied fields as forensics, criminal investigation, serology, morphology, and genetics solve crimes ranging from poaching of elk, to "head-hunting" of walruses for their tusk ivory, to the sale of bear gallbladders to Asian medical markets. The laboratory focuses on identifying animals, determining causes of death, and linking crimes to their perpetrators. Investigative techniques employed are those available to most modern crime labs: conventional tissue-typing, liquid chromatography, mass spectrometry, and DNA fingerprinting.

FOCUS ON

Vanishing Tropical Rain Forests

Tropical rain forests are found in South and Central America, central Africa, and Southeast Asia. Although only 7 percent of the Earth's surface is covered by tropical rain forests, as many as 40 percent of the Earth's species inhabit them. Habitat destruction is occurring throughout the Earth, but tropical rain forests are being destroyed faster than almost all other ecosystems. Using remote sensing surveys, scientists have determined that approximately one percent of tropical rain forests are being cleared or severely degraded each year. The forests are making way for human settlements, banana plantations, oil and mineral explorations, and other human activities.

Many species in tropical rain forests are endemic; the clearing of tropical rain forests therefore contributes to their extinction. These species are important in their own right, but the mass extinction that is currently taking place in tropical rain forests has indirect ramifications as far away as North America. Birds that migrate from North America to Central America and the Caribbean have been declining in numbers, but not all migratory birds are declining at the same rate; those birds that winter in tropical rain forests are declining at a much greater rate than birds that winter in tropical grasslands or tropical dry woodlands. Thus, tropical deforestation is also affecting organisms of the temperate region.

Tropical rain forests provide important ecological services that help to maintain their ecosystem. For example, much of the rainfall in tropical rain forests is generated by the forest itself. If half of the existing rain forest in the Amazon region of South America were to be destroyed, precipitation in the remaining forest would decrease. As the land became drier, organisms adapted to moister conditions would be replaced by organisms able to tolerate the drier conditions. Many of the original species, being endemic and unable to tolerate the drier conditions, would become extinct.

Perhaps the most unsettling outcome of tropical deforestation is its effect on the evolutionary process. In the Earth's past, mass extinctions were followed during the next several million years by the formation of many new species to replace those that died out. For example, after the dinosaurs became extinct, ancestral mammals evolved into the variety of running, swimming, flying, and burrowing mammals that exist today. The evolution of a large number of related species from an ancestral organism is called **adaptive radiation**. In the past, tropical rain forests have supplied the base of ancestral organisms from which adaptive radiations could occur. By destroying tropical rain forests, we may be reducing or eliminating nature's ability to replace its species through adaptive radiation. Tropical rain forests are also discussed in Chapters 6 and 17.

entire world. The situation is complicated by the fact that these countries are least able to afford the protective measures needed to maintain biological diversity. International cooperation will clearly be needed to preserve our biological heritage.

CONSERVATION BIOLOGY

Conservation biology is the study and protection of biological diversity. It includes two types of efforts that are being made to save organisms from extinction: in situ and ex situ. **In situ conservation**, which includes the establishment of parks and reserves, concentrates on preserving biological diversity *in nature*. A high priority of in situ conservation is the identification and protection of sites with a great deal of biological diversity. With increasing demands on land (see Chapter 17), however, in situ conservation cannot guarantee the preservation of all types of biological diversity. Sometimes only *ex situ* conservation can save a species. **Ex situ conservation** involves conserving biological diversity *in human-controlled settings*. The breeding of captive species in zoos and the seed storage of genetically diverse plant crops are examples of ex situ conservation.

Protecting Habitats

Many nations are beginning to appreciate the need to protect their biological heritage and have set aside areas for biological habitats. There are currently more than 3000 national parks, sanctuaries, refuges, forests, and other protected areas throughout the world. Some of these have been set aside to protect specific endangered species. The first such refuge was established in 1903 at Pelican Island, Florida, to protect the brown pelican. Today the National Wildlife Refuge System of the United States has land set aside in more than 500 refuges; although the bulk of the protected land is in Alaska, refuges exist in all 50 states.

Many protected areas have multiple uses. National parks may serve recreational needs, for example, and national forests may be open for logging, grazing, and farming operations. The mineral rights to many refuges are privately owned, and some refuges have had oil, gas, and other mineral development. Hunting is allowed in more than half of the wildlife refuges in the United States, and military exercises are conducted in several of them.

Certain parts of the world are critically short of protected areas. Protected areas are urgently needed in trop-

(a)

(b)

Figure 16–11

The University of Wisconsin-Madison Arboretum has pioneered restoration ecology. (a) The restoration of the prairie was at an early stage in November, 1935. (b) The prairie as it looks today. This picture was taken at approximately the same location as the 1935 photograph. (a, Courtesy of University of Wisconsin-Madison Arboretum; b, Courtesy of Virginia Kline, University of Wisconsin-Madison Arboretum)

ical rain forests, the tropical grasslands and savannas of Brazil and Australia, and dry forests that are widely scattered around the world. Desert organisms are underprotected in northern Africa and Argentina, and many islands and lake ecosystems need protection.

Restoring Damaged or Destroyed Habitats

Scientists can reclaim disturbed lands and convert them into areas with high biological diversity. The most famous example of ecological restoration has been carried out since 1934 by the University of Wisconsin-Madison Arboretum (Figure 16–11). During that time, several distinct natural communities of Wisconsin were carefully developed on damaged agricultural land. These communities include a tallgrass prairie, a dry prairie, and several types of pine and maple forests.

Restoration of disturbed lands not only creates biological habitats but also has additional benefits such as the regeneration of soil that has been damaged by agriculture or mining (see Chapters 10, 14, and 15). The disadvantages of restoration include the expense and the amount of time it requires to restore an area. Even so, restoration is an important aspect of conservation biology, as it is thought that restoration will deter many extinctions.

Zoos, Aquaria, Botanical Gardens, and Seed Banks

Zoos, aquaria, and botanical gardens often play a critical role in saving species that are on the brink of extinction. Eggs may be collected from nature, or the remaining few animals may be captured and bred in zoos and other research environments. Some of the endangered cichlid species from Lake Victoria are being maintained in aquaria in the United States, for example (see Focus On: The Death of a Lake).

Special techniques, such as artificial insemination and embryo transfer, are used to increase the number of animal offspring. In **artificial insemination**, sperm collected from a suitable male of a rare species is used to artificially impregnate a female, perhaps located in another zoo in a different city or even in another country. In **embryo transfer**, a female of a rare species is treated with fertility drugs, which cause her to produce multiple eggs. Some of these eggs are collected, fertilized with sperm, and surgically implanted into a female of a related but less rare species, who later gives birth to offspring of the rare species (Figure 16–12).

There have been a few spectacular successes in captive-breeding programs, in which large enough numbers of a species have been produced to re-establish small populations in nature (see Focus On: Reintroducing Endangered Animal Species to Nature). Whooping cranes had declined to the critically low population of 15 in 1941. As a result of captive breeding programs, the wild flock that winters in Texas numbered 158 in 1996, and most promising, included 28 chicks. Conservation biologists are hoping to have the whooping crane removed from the endangered species list and classified as only threatened by the year 2000.

Attempting to save a species on the brink of extinction is very expensive; therefore, only a small proportion

Figure 16–12

A newborn bongo calf and its surrogate mother, an eland. This young calf was transferred as an embryo to the uterus of the eland, where it completed development. The bongo, an elusive species inhabiting dense forests in Africa, is an endangered species. The larger and more common eland, a different but related species, inhabits open areas.

Figure 16–13

Small vials of seeds from the seed bank in Svalbard, Norway. (Courtesy of Nordiska Genbanken, Alnarp, Sverige)

of endangered species can be saved. Moreover, zoos, aquaria, and botanical gardens do not have the space to mount efforts to save *all* endangered species. This means that conservation biologists must prioritize which species to attempt to save. Clearly, it is more cost-effective to maintain natural habitats so that species will never become endangered in the first place.

Seed banks

More than 100 seed collections called **seed banks** exist around the world and collectively hold more than 3 million samples at very low temperatures (Figure 16–13). They offer the advantage of storing a large amount of plant genetic material in a very small space. Seeds stored in seed banks are safe from habitat destruction, climate change, and general neglect. There have even been some instances of seeds from seed banks being used to reintroduce to nature a plant species that had become extinct.

Some disadvantages to seed banks exist, however. First, many types of plants, such as avocados and coconuts, cannot be stored as seeds. The seeds of these plants do not tolerate being dried out, which is a necessary step before the seeds can be frozen. Second, seeds do not remain alive indefinitely and must be germinated periodically so that new seeds can be collected. Because accidents such as fires or power failures can result in the permanent loss of the genetic diversity represented by the seeds, biologists typically subdivide seed samples and store them in several different seed banks.

Perhaps the most important disadvantage of seed banks is that plants stored in this manner remain stagnant in an evolutionary sense. They do not evolve in response to changes in their natural environments. As a result, they may be less fit for survival when they are reintroduced into nature.

Despite their shortcomings, seed banks are increasingly viewed as an important method of safeguarding seeds for future generations. Other international efforts to preserve plant genetic diversity are also being planned and implemented. For example, some farmers may be paid to set aside some of their land for cultivating local varieties of crops, thereby preserving genetic diversity in agriculturally important plants (see Chapter 18).

Conservation Organizations

Conservation organizations are an essential part of the effort to maintain biological diversity (see Appendix II). They help to educate policy makers and the public about the importance of biological diversity. In certain instances they serve as catalysts by galvanizing public support for important biodiversity preservation efforts. They also provide financial support for conservation projects, from basic research to the purchase of land that is a critical habitat for a particular organism or group of organisms.

FOCUS ON

Reintroducing Endangered Animal Species to Nature

The ultimate goal of captive-breeding programs is to produce offspring in captivity and then release them into nature so that wild populations can be restored. However, only one out of every ten reintroductions using animals raised in captivity is successful. What guarantees that a reintroduced population will survive?

Whether such reintroductions actually succeed has been scientifically studied only in recent years. For example, the Hawaiian goose called the nene (pronounced *nay-nay*) was reintroduced to the islands of Hawaii and Maui beginning in the 1970s, but although more than 2100 birds have been released, a self-sustaining nene population has not developed on either island. Apparently, some of the same factors that originally caused the nene's extinction in nature, including habitat destruction and non-native predators, are responsible for the failure of the reintroduced birds.

Now before attempting a reintroduction, conservation biologists make a feasibility study. This includes determining (1) what factors originally caused the species to become extinct in nature, (2) whether these factors still exist, and (3) whether any suitable habitat still remains.

If the animal to be reintroduced is a social animal, a small herd is usually released together. This is accomplished by first placing the herd in a large, semi-wild enclosure that is somewhat protected from predators but requires the herd to obtain its own food. When the herd's behavior begins to resemble the behavior of wild herds, it is released.

Sometimes it is impossible to teach critical survival skills to animals raised in captivity. The effort to reintroduce captive-raised thick-billed parrots (see figure) to the Chiricahua Mountains of Arizona was canceled in 1993 because many of the 88 birds released from 1986 to 1993 were killed by hawks and other birds of prey. Wild thick-billed parrots are loud, sociable birds whose flocking instinct contributes to their survival because individual birds act as sentinels and loudly announce the presence of danger to the group. Those parrots raised in captivity lacked such social behavior, and despite efforts to teach them to stay together, they separated from the flock after they were released.

Once animals are released, they must continue to be monitored. If any animals die, their cause of death is determined in order to search for ways to prevent unnecessary deaths in future reintroductions.

A captive-raised, thick-billed parrot. (Noel Snyder)

Figure 16–14

When a protected area is set aside, it is important to know what the minimum size of that area must be so that it will not be affected by encroaching species from surrounding areas. Shown are 1 hectare and 10 hectare plots that are part of the Minimum Critical Size of Ecosystems Project, which is being conducted in Brazil by the World Wildlife Fund and Brazil's National Institute for Amazon Research. It is a long-term study of the effects of fragmentation on Amazonian rain forest. (© 1990 R. O. Bierregaard)

Table 16–1
Organisms Listed as Endangered or Threatened in the United States, 1996

Type of Organism	Number of Endangered Species	Number of Threatened Species
Mammals	55	9
Birds	74	16
Reptiles	14	19
Amphibians	7	5
Fishes	65	40
Snails	15	7
Clams	51	6
Crustaceans	14	3
Insects	20	9
Spiders	5	0
Flowering plants	405	90
Conifers	2	0
Ferns and other plants	26	2

The World Conservation Union (IUCN)[3] assists countries with hundreds of conservation biology projects. It and other conservation organizations are currently assessing how effective established wildlife refuges are in maintaining biological diversity (Figure 16–14). In addition, IUCN and the World Wildlife Fund[4] have identified major conservation priorities by determining which biomes and ecosystems are not represented by protected areas. IUCN maintains a data bank on the status of the world's species; its material is published in *The IUCN Red Data Books* about organisms and habitats.

Policies and Laws

In 1973 the **Endangered Species Act** was passed in the United States, authorizing the U.S. Fish and Wildlife Service (FWS) to protect from extinction endangered and threatened species in the United States and abroad. Many other countries now have similar legislation.

Since the passage of the Endangered Species Act, 955 species in the United States have been listed as endangered or threatened (Table 16–1). This act makes it illegal to sell or buy any product made from an endangered or threatened species. It further requires officials of the FWS to select critical habitats and design a recovery plan for each species listed. For example, the recent reintroduction of wolves to Yellowstone National Park, discussed in Chapter 1, is part of that species' recovery plan.

The Endangered Species Act, which was updated in 1982, 1985, and 1988, is scheduled to be reauthorized by Congress sometime in the late 1990s. It is considered one of the strongest pieces of environmental legislation in the United States, in part because species are designated as endangered or threatened entirely on biological grounds; currently, economic considerations cannot influence the designation of endangered or threatened species. The Endangered Species Act has also been one of the most controversial pieces of environmental legislation because it has interfered with some federally funded development projects.

Some critics—notably business interests—view the Endangered Species Act as an impediment to economic progress. To protect the habitat of the northern spotted owl, for example, the timber industry has been blocked

[3]Formerly called the International Union for Conservation of Nature and Natural Resources, the World Conservation Union still goes by the abbreviation IUCN.

[4]Outside of the United States, Canada, and Australia, the World Wildlife Fund is known as the World Wide Fund for Nature.

Table 16–2
Selected U.S. Animal Species That Are No Longer Classified as Endangered

Animal	Year of Reclassification or Delisting
Bald eagle	1994
Pacific gray whale	1994
Aleutian Canada goose	1990
American alligator	1987
Brown pelican	1985
Utah prairie dog	1984
Greenback cutthroat trout	1978

from logging old-growth forests in certain parts of the Pacific Northwest (see Chapter 7). Those who defend the Endangered Species Act point out that, of 34,000 past cases of endangered species versus "development," only 21 cases could not be resolved by some sort of compromise. When the black-footed ferret was reintroduced on the Wyoming prairie, for example, it was classified as an "experimental, nonessential species" so that its reintroduction would not block ranching and mining in the area. Thus, the ferret release program obtained the support of local landowners, support that was deemed crucial to its survival in nature.

Defenders of the Endangered Species Act agree that it is not perfect. Few endangered species have recovered enough to be reclassified as threatened or delisted (removed from protection; see Table 16–2), although the FWS, in a report to Congress in 1994, says that 41 percent of listed species are stable or improving. The law is geared more to saving a few popular or unique endangered species rather than the much larger number of less glamorous species that perform valuable ecosystem services. Yet it is the less glamorous organisms such as fungi and insects that play a central role in ecosystems and contribute most to their functioning. Conservationists would like to see the Endangered Species Act strengthened in such a way as to preserve whole ecosystems and maintain complete biological diversity rather than attempting to save individual endangered species. This approach offers protection to many declining species rather than to a single individual species.

The Biological Resources Division

The National Biological Service (NBS) was created in 1993[5] by combining science programs from seven different bureaus of the Department of the Interior. The mission of the BRD is to provide information and technologies to manage and conserve biological resources on federal lands, which represent 22 percent of all U.S. lands. To accomplish this mission, basic science is needed to describe the current status of the nation's organisms and to monitor any changes. The scientific knowledge gained will be used by the Department of the Interior in making land use decisions.

International policies and laws

The **World Conservation Strategy**, a plan designed to conserve biological diversity worldwide, was formulated in 1980 by the IUCN, the World Wildlife Fund, and the U.N. Environment Program. In addition to conserving biological diversity, the World Conservation Strategy seeks to preserve the vital ecosystem processes on which all life depends for survival and to develop sustainable uses of organisms and the ecosystems that they comprise. The most recent version of the World Conservation Strategy, published in 1991, also focuses on stabilizing the human population.

The biological diversity treaty produced by the 1992 Earth Summit was ratified by 153 nations as of late 1996 and is now considered binding (see Chapter 1). Under the conditions of the treaty, each signatory nation must inventory its own biodiversity and develop a **national conservation strategy**, a detailed plan for managing and preserving the biological diversity of that specific country.

The exploitation of endangered species can be somewhat controlled through legislation. At the international level, 128 countries participate in the Convention on International Trade in Endangered Species of Wild Flora and Fauna (CITES), which bans hunting, capturing, and selling of endangered or threatened species. Unfortunately, enforcement of this treaty varies from country to country, and even where enforcement exists, the penalties are not very severe. As a result, illegal trade in rare, commercially valuable species continues.

WILDLIFE MANAGEMENT

Efforts to handle wildlife populations and their habitats in order to ensure their sustained welfare are part of the science of **wildlife management**. Wildlife managers must know when and how to protect species that are endangered or of economic importance, and they must be able to set priorities, often in the face of conflicting goals. For example, a wildlife manager must sometimes decide whether it is better to manage an area to maintain maximum biological diversity or to protect a single "important" species within that area. Wildlife managers regulate an area by population control and habitat manipulation.

[5]The BRD was originally named the National Biological Survey and, later, the National Biological Service.

Population Control

The natural predators of many game animals have largely been eliminated in the United States. As a result of the disappearance of predators such as wolves and mountain lions, the populations of animals such as squirrels, ducks, and deer sometimes exceed the carrying capacity of their environment (see Chapter 8). When this occurs, the habitat deteriorates and many animals starve to death.

Sport hunting can effectively control overpopulation of game animals, provided restrictions are observed to prevent overhunting. Laws in the United States determine the time of year and length of hunting seasons for various species, as well as the number, sex, and size of each species that may be killed.

Habitat Management

Wildlife managers affect a particular species by manipulating the plant cover, food, and water supplies of its habitat. Because different animals predominate in different stages of ecological succession (see Chapter 4), controlling the stage of ecological succession of an area's vegetation encourages the presence of certain animals and discourages others. For example, quail and ring-necked pheasant are found in weedy, open areas that are characteristic of early-successional stages. Moose, deer, and elk predominate in partially open forest, such as an abandoned field or meadow adjacent to a forest; the field provides food, and the forest provides protective cover. Other animals, such as grizzly bears and bighorn sheep, require undisturbed climax vegetation. Wildlife managers control the stage of succession with techniques such as planting certain types of vegetation, burning the undergrowth with controlled fires, and building artificial ponds.

Management of Migratory Animals

International agreements must be established to protect migratory animals. Ducks, geese, and shorebirds, for example, spend their summers in Canada and their winters in the United States and Central America. During the course of their annual migrations, which usually follow established routes called **flyways**, they must have areas in which to rest and feed. Wetlands, the habitat of these animals, also must be protected in both their winter and summer homes.

Management of Aquatic Organisms

Fishes with commercial or sport value must be managed to ensure that they are not overexploited to the point of extinction. Freshwater fishes such as trout and salmon are managed in several ways. Fishing laws regulate the time of year, size of fish, and maximum allowable catch. Natural habitats are maintained to maximize population size; this includes pollution control. Ponds, lakes, and streams may be restocked with young hatchlings from hatcheries.

Traditionally, the ocean's resources have been considered common property, available to the first people to exploit them. As a result, many marine fishes have been severely reduced in numbers by commercial fishing. (Chapters 1 and 18 discuss this dwindling resource.)

Whales

During the 19th and 20th centuries, many whale species were harvested to the point of **commercial extinction**, meaning that so few remain that it is unprofitable to hunt them. Although commercially extinct species still have living representatives, their numbers are so reduced that they are endangered.

In 1946 the International Whaling Commission set an annual limit on killed whales for each whale species in an attempt to secure sustainable whale populations. Unfortunately, these limits were set too high, resulting in further population declines during the next 20 years. Conservationists began to call for a worldwide ban on commercial whaling; such a ban went into effect in 1986.

A WELCOME POPULATION BOOM

After exhibiting alarming declines in the 1980s and early 1990s, North American duck populations soared in 1994, increasing 24 percent over 1993 counts. The declines were attributed to a substantial loss of waterfowl habitat, made worse by a decade of drought conditions, while the rebound appears to be a direct result of successful conservation efforts. The Food Security Act (Farm Bill) of 1985 established the Conservation Reserve Program, which encourages farmers to retire land to prevent soil erosion (see Chapter 14). The more than 10 million acres set aside in the northern plains states under the Conservation Reserve Program provided prime breeding habitat for ducks. Increased grasslands also reduced the predator threat on ducks by providing them thick cover, and the wet conditions of 1993–94 further expanded available habitat. Breeding duck populations reached an estimated 32.5 million in 1994, well on the way to the North American Waterfowl Management's targeted recovery population of 36 million for the midcontinent breeding grounds.

UNUSUAL FISH TANKS

Obsolete army tanks are being dumped into the Gulf of Mexico off the coast of Alabama, much to the delight of the state's marine fisheries officials. The sunken tanks are part of a large artificial reef designed to produce more habitat for offshore fishes. With the cooperative efforts of state agencies, the military, the Environmental Protection Agency, and the National Marine Fisheries Service, the program may eventually sink several thousand tanks. Although preparing the tanks for being dumped—particularly ensuring that they are stripped of environmental hazards—is expensive, fisheries experts estimate that the reef could provide an economic return of $7 billion during the underwater life of the tanks. The larger local fish populations that result could improve catches up to 12-fold.

Scientists have since monitored whale populations and concluded that the ban is working. The populations of most whales, such as humpbacks and bowheads, appear to be growing. One species, the Pacific gray whale, has recovered sufficiently to be removed from the endangered species list and to be reclassified as only threatened. The North Atlantic right whale and southern blue whales, however, are still poised on the brink of extinction.

In 1994 the Southern Ocean Whale Sanctuary was established in Antarctic waters where most of the world's great whales feed and reproduce. This vast sanctuary, which bars commercial hunting, would remain should the current ban on whaling ever be lifted.

WHAT CAN WE DO ABOUT DECLINING BIOLOGICAL DIVERSITY?

Although our children and grandchildren are faced with inheriting a biologically impoverished world, we should view this problem as a challenge. People who are dedicated to preserving our biological heritage can reverse the trend toward extinction. It is important to realize that you do not have to be a biologist to make a contribution; some of the most important contributions come from outside the biological arena. Following is a partial list of actions that can be taken to help maintain the biological diversity that is our heritage.

Increase Public Awareness

The consciousness of both the public and legislators must be increased so that they understand the importance of biological diversity. A political commitment to protect organisms is necessary because no immediate or short-term economic benefit is obtained from conserving species. This commitment will have to take place at all political levels, from local to international. Law-making will not ensure the protection of organisms without strong public support. Thus, increasing public awareness of the benefits of biological diversity is critical.

Providing publicity on species conservation issues costs money. Funds raised by organizations such as the Sierra Club, The Nature Conservancy, and the World Wildlife Fund support such endeavors, but clearly more money is needed. As an individual, you can help preserve biological diversity by joining and actively supporting conservation organizations.

Support Research in Conservation Biology

Before an endangered species can be saved, its numbers, range, ecology, biological nature, and vulnerability to changes in its environment must be determined; basic research provides this information. We cannot preserve a given species effectively until we know how large a protected habitat must be established and what characteristics are essential in its design.

There are acute shortages of trained specialists in tropical forestry, conservation genetics, taxonomy, resource management, and similar disciplines. Many young people who are interested in these careers have selected others because of the dearth of funding for such research. (The funding covers training and salaries of skilled personnel, research equipment and supplies, and miscellaneous expenses such as transportation costs.)

As an individual, you can inform local and national politicians of your desire to have conservation research funded with tax dollars. When more funds are available, colleges, universities, and other research institutions will be able to justify adding faculty and research positions. As a result, more young people with interest in conservation will be able to undertake the necessary education.

Support the Establishment of an International System of Parks

A worldwide system of protected parks and reserves that includes every major ecosystem must be established. Conservationists estimate that a minimum of 10 percent of the world's land should be set aside for this purpose, and most countries have far less than the recommended minimum. The protected land would provide humans with

MEETING THE CHALLENGE

Wildlife Ranching as a Way to Preserve Biological Diversity in Africa

Africa's gazelles, giraffes, elephants, lions, and baboons are increasingly under pressure from farmers, who clear the wildlife from a section of land to make room for cattle and crops. As the expanding human population in Africa develops agriculture, wildlife are forced out. Farmers do not like wildlife because wildlife kill or spread disease among cattle and trample crops. Game animals are dangerous neighbors for humans, and dozens of villagers are killed or maimed by wildlife each year.

Zimbabwe has grappled with the cattle–people–wildlife issue and come up with an unorthodox solution—**wildlife ranching**, also known as **game farming**. Beginning in 1975 with the passage of Zimbabwe's Wildlife Act, private landowners may own wild animals. More and more farmers are converting from cattle and crops to wildlife ranching (see figure). African game earns more money than cattle in several ways. For one thing, tourists and photographers are willing to pay to take day-long hiking, canoeing, or horseback-riding safaris in which they observe and photograph wildlife. Sport hunters pay large trophy fees to stalk and kill non-endangered animals. The trophy fee to kill an elephant in Zimbabwe, for example, is $6000. (Elephants are not an endangered species in Zimbabwe.) Game yields beautiful hides and leather as well as lean, low-cholesterol meat.

Wildlife ranching, besides being financially attractive, is less harmful to the environment than traditional agriculture. Game animals, unlike cattle, eat a variety of plants and do not permanently damage the vegetation, so soil erosion is less of a problem. In addition, some wild animals require less water and are more resistant to disease than cattle.

As game farming takes hold in Zimbabwe and other African nations, African attitudes about wildlife are changing. People are more tolerant of wildlife since discovering its economic value: game animals earn up to three times more than cattle in dry areas.

Wildlife ranching represents a compromise between people who must use the land to earn a living and wildlife whose populations are declining. Africans engaged in wildlife ranching are earning more than they thought was possible, and wildlife herds are increasing in size. As long as wildlife is economically profitable, it will survive.

Wildlife ranching. Many species of hoofed animals, such as these impala, are increasingly being raised by wildlife ranchers. Some ranchers are trying crocodile farming as well. (© 1997 Stan Osolinski/Dembinsky Photo Associates)

other benefits in addition to the preservation of biological diversity. It would safeguard the watersheds that supply us with water, and it would serve as a renewable source of important biological products in areas with multiple uses. It would also provide people with unspoiled lands for aesthetic and recreational enjoyment. In addition to the establishment of new parks and reserves, particularly in developing nations, parks and reserves in developed nations must be expanded. As an individual, you can help establish parks by writing to national lawmakers.

Control Pollution

The establishment of parks and refuges will not be enough to prevent biological impoverishment if we continue to pollute the Earth, because it is impossible to protect parks and refuges from threats such as acid rain, ozone depletion, and climate change. Strong steps must be taken to curb the toxins we dump into the air, soil, and water—not only for human health and well-being but also for the well-being of the organisms that are so important to ecosystem stability. Specific recommendations on how you as an individual can help reduce pollution are discussed in Chapters 19 through 23.

Figure 16–15

An ethnobotanist consults with a Tirio Indian in Suriname about the uses of a rainforest plant. Ethnobotany, the study of the traditional uses of plants by indigenous people, helps pharmaceutical companies identify medicinal plants. Using this knowledge provides a shortcut in deciding which plants to test; studies show that plants identified by shamans and traditional plant users are up to 60 percent more likely to have medicinal value than those that are randomly collected. (Courtesy of Mark Plotkin, Conservation International)

Provide Economic Incentives to Tropical Nations

A few innovative economic incentives encourage the preservation of biological diversity (see Meeting the Challenge: Wildlife Ranching as a Way to Preserve Biological Diversity in Africa). Economic incentives are particularly critical because developing nations in the tropics, the repositories of most of the Earth's genetic diversity, do not have money to spend on conservation. Their governments are consumed with human problems such as overpopulation, disease, and crushing foreign debts.

One way to help such countries appreciate the importance of the genetic resources they possess is to allow them to charge fees for the use of that genetic material. Much of the money thus earned could be used to help alleviate human problems. And some of the money generated by genetic resources could be used to provide protection for organisms, thus preserving biological diversity for continued, sustained exploitation.

The Suriname Biodiversity Prospecting Initiative, announced in 1993 by Conservation International, is the first such agreement of its kind. This enterprise pays money to local people of Suriname, a small country in South America, for new drugs delivered from the plants that they identify in their rain forests (Figure 16–15). With support from the U.S. government, universities, and various nonprofit organizations, the initiative represents the first partnership between indigenous people and a private pharmaceutical company.

Other agreements have followed. The National Cancer Institute, for example, has guaranteed the government of Malaysia part of the profits of any drugs obtained from organisms in Malaysian rain forests. Malaysia has an economic incentive to preserve its remaining rain forests, because if a cure for cancer or AIDS exists there, Malaysia could earn millions of dollars.

A second way of providing economic incentives to developing nations is for developed countries to forgive or reduce debts owed by such nations. In exchange, the developing countries would agree to protect their biological diversity. Such forgiveness of debts provides a tangible reward for preserving a nation's species. The United States has arranged a debt-for-nature swap with Madagascar. It has agreed to purchase $1 million of Madagascar's national debt in exchange for government support of local conservation efforts, including the protection of endangered species.

Once again, you can help in the formulation of such policies. Let your lawmakers know where you stand. Join and support conservation groups. Campaign to preserve our biological heritage for future generations.

SUMMARY

1. Biological diversity, the number and variety of organisms, encompasses genetic diversity (the variety within a species), species diversity (the number of different species), and ecosystem diversity (variety within and among ecosystems). A reduction in biological diversity is occurring worldwide, but the problem is most critical in tropical areas.

2. Organisms are an important natural resource. They provide us with essential ecosystem services; bacteria and fungi, for example, perform the important task of decomposition. Organisms serve as sources of medicinal, agricultural, and industrial products. Genetic reserves are used for domesticated plant and animal breeding, both through traditional breeding methods and through genetic engineering. Organisms give us recreation, inspiration, and spiritual comfort.

3. When the last individual member of a species dies, it is said to be extinct. Extinction represents a permanent loss in biological diversity; once an organism is extinct, it can never exist again.

4. An organism whose numbers are severely reduced so that it is in danger of extinction is said to be endangered. When extinction is less imminent but numbers are quite low, a species is said to be threatened. Endangered and threatened species have limited natural ranges and low population densities. They may also have low reproductive rates or very specialized eating or reproducing requirements.

5. Human activities that contribute to a reduction in biological diversity include habitat destruction and disturbance, pollution, introduction of foreign species, pest and predator control, and illegal commercial hunting and harvest. Of these, habitat destruction is the most significant cause of declining biological diversity.

6. Efforts to preserve biological diversity in nature are known as in situ conservation. Ex situ conservation, which includes captive breeding and the establishment of seed banks, occurs in human-controlled settings.

7. Wildlife management includes the regulation of hunting and fishing and the management of food, water, and habitat. Wildlife management programs often have different priorities—for example, maintaining the population of a specific species versus managing a community to ensure maximum biological diversity.

8. There are several ways to reverse the trend of declining biological diversity. Both the public and lawmakers must become more aware of the importance of our biological heritage. Funding must be found for additional research in both basic and applied fields relating to conservation biology. A worldwide system of protected parks and reserves must be established, hopefully encompassing a minimum of 10 percent of the land area. Pollution, which is damaging to both humans and other organisms, must be brought under control. Economic incentives must be developed to help the developing nations that are the repositories of much of the world's biological diversity realize the value of their living resources.

THINKING ABOUT THE ENVIRONMENT

1. If you had the assets and authority to take any measure to protect and preserve biological diversity, but could take only one, what would it be?

2. Does being pro-nature mean that you are also anti-development?

3. Is biological diversity a renewable or nonrenewable resource? Why could it be seen both ways?

4. Why does the most recent version of the World Conservation Strategy include stabilizing the human population? How would stabilizing the human population affect biological diversity?

5. If we preserve species solely on the basis of their potential economic value—as a source of a novel drug, for example—does this mean that they lose their "value" once we have been able to capitalize on a newly discovered chemical? Why or why not?

6. According to this chapter, one of the ways in which you can help the world preserve its biological diversity is by talking about the problem, even to your friends. How will talking about biological impoverishment contribute to its reversal?

* 7. Annual global trade in animals is conservatively estimated at $10 billion, and one-third of that is illegal because it involves rare or endangered species. Calculate how much illegal trade, in millions of dollars, is done each day.

RESEARCH PROJECTS

Beginning in 1992, the African country Zimbabwe sawed off the horns of every rhinoceros in an effort to stop the illegal slaughter of these animals by poachers. However, the slaughter continues (because the horns grow back), and some in Zimbabwe now support the legalization of trade in rhino horns. Investigate the rhino problem, including why dehorning has not been effective in Zimbabwe. What are some of the ramifications of legalizing trade in rhino horns?

All trade of elephant products has been banned by CITES. Yet elephants are not threatened in South Africa and Zimbabwe, where elephant populations are sometimes too large and must be reduced by "culling," which is killing the older, less fit elephants. These countries would like to be able to sell the tusks of culled elephants on the international market, but other African countries fear that reopening the ivory trade would encourage poaching in countries where the elephant is endangered. Investigate the ecological and ethical dilemmas posed by African elephants.

On-line materials relating to this chapter are on the World Wide Web at **http://www.saunderscollege.com/lifesci/environment2** (select Chapter 16 from the Table of Contents).

SUGGESTED READING

Baskin, Y. "Curbing Undesirable Invaders," *BioScience* Vol. 46, No. 10, November 1996. The Norway/United Nations Conference on Alien Species calls for public education and new ways to assess the environmental and economic tradeoffs of exotic species.

Blaustein, A. R. and D. B. Wake. "The Puzzle of Declining Amphibian Populations," *Scientific American*, April 1995. Many frog, toad, and salamander species have experienced rapid population declines in recent years.

Chadwick, D. H. "Dead or Alive: The Endangered Species Act," *National Geographic* Vol. 187, No. 3, March 1995. The controversy surrounding the Endangered Species Act is explored.

Chester, C. C. "Controversy over Yellowstone's Biological Resources," *Environment* Vol. 38, No. 8, October 1996. Microorganisms that live in Yellowstone's hot springs have many commercially important applications. This article explores whether the park is entitled to some compensation for them.

Easter-Pilcher, A. "Implementing the Endangered Species Act," *BioScience* Vol. 46, No. 5, May 1996. An evaluation of how species are determined to be endangered or threatened.

Myers, N. "Environmental Services of Biodiversity," *Proceedings of the National Academy of Sciences* Vol. 93, April 1996. Examines the ecological and economic importance of biological diversity.

Posey, D. A. "Protecting Indigenous Peoples' Rights to Biodiversity," *Environment* Vol. 38, No. 8, October 1996. Examines the relationships among indigenous people, property rights, and bioprospecting.

Royte, E. "On the Brink: Hawaii's Vanishing Species," *National Geographic* Vol. 188, No. 3, September 1995. Hawaii has more endangered species than any other state in the United States.

Speart, J. "The Rhino Chainsaw Massacre," *Earth Journal*, January–February 1994. Excellent photographs highlight this article on conservation efforts to save the rhinoceros.

Vitousek, P. M., C. M. D'Antonio, L. L. Loope, and R. Westbrooks. "Biological Invasions as Global Environmental Change," *American Scientist* Vol. 84, September–October 1996. The introduction of exotic species harms ecosystems, threatens human health, and is economically disastrous.

Wheelwright, N. T. "Enduring Reasons to Preserve Threatened Species," *The Chronicle of Higher Education*, June 1, 1994. In this well written commentary, the author argues that we should try to preserve diverse species for the same reason we try to preserve the ceiling of the Sistine Chapel or the Taj Mahal—because organisms are "resplendently beautiful."

Mesquite Flat Dunes in Death Valley National Park. (Tony Stone)

Land Resources and Conservation

In late 1994, Congress established two new national parks, a national preserve, and several wilderness areas across 3.6 million hectares (9 million acres) of Southern California desert, now the largest protected region in the lower 48 states. Made official by President Clinton's signing of the California Desert Protection Act of 1994, the new Death Valley National Park, Joshua Tree National Park, and Mojave National Preserve were created. Once considered desert wasteland, the parks and preserve will protect at least 2000 plant and animal species, varied archeological sites, 90 mountain ranges, and sand dunes as high as 213 meters (700 feet; see figure).

In a decidedly modern twist on the role of national protected areas, the parks and the preserve will not function as simple nature sanctuaries but will instead incorporate private interests into management of the wilderness. Some existing mining and cattle ranching will be allowed to continue, recreational vehicles will be given some road access, and hunting will be permitted in the preserve, all as a result of Congressional amendments to support the private property rights of the region's residents.

Although the "marriage" of private users and the new National Park System management will very likely experience tensions, it represents a solution to about 20 years of wrangling over environmental issues versus the needs of desert dwellers. Residents' concerns have been considered, yet most of the area will be protected. Habitat for the threatened desert tortoise will be preserved, and motorcycle races will no longer destroy dunes. Future development, mining, and road construction will be prohibited so that generations of people can appreciate the beauty of the desert wilderness. ◀

IMPORTANCE OF NATURAL AREAS

Most of the Earth's land area has a low density of humans. These sparsely populated areas, known as **nonurban** or **rural lands**, include wilderness, forests, grasslands, and wetlands. Most people living in rural areas have jobs directly connected with natural resources—such as farming or logging. The many environmental services that are performed by rural lands enable the majority of humans to live in concentrated urban environments.

As you will see throughout this chapter, undisturbed land benefits us in many ways of which we are often not even aware. Consider the higher elevations of mountains, above the tree line, which have a distinctive ecosystem known as **alpine tundra** (Figure 17–1). The alpine tundra, which has strong winds, cold temperatures, and snow, is a harsh environment inhabited by few kinds of plants. Sparse populations of Tibetan herdsmen in the Himalayas are among the few groups of people living in alpine tundra. The western United States has approximately 1.2 million hectares (3 million acres) of mostly uninhabited alpine tundra. Although this ecosystem is worth preserving simply because it is unique and interesting, it also has an ecological value that is not readily apparent to most people. For example, melting snows in the highest portions of Utah's alpine tundra furnish 60 percent of the water in the state's streams, even during the hot days of summer.

Figure 17–1
Alpine tundra is mostly wilderness area in the United States. Alpine tundra performs a valuable service by regulating the flow of water from snowmelt in the mountains to lowland areas. (Lynn and Sharon Gerig/Tom Stack and Associates)

Unfortunately, the amount of U.S. alpine tundra that can be classified as undisturbed by human activities is decreasing. Sheep grazing, recreation, and mining all take their toll, particularly because the soil found in such regions is extremely thin and fragile.

Maintaining parcels of undisturbed land adjacent to agricultural and urban areas provides vital environmental services such as pest control, flood and erosion control, and groundwater recharge. Undisturbed land also breaks down pollutants and recycles wastes.

Natural environments provide homes for organisms. One of the best ways to maintain biological diversity and to protect endangered and threatened species is by preserving or restoring the natural areas to which these organisms are adapted.

Ecologists who conduct research on the complexities of ecosystems frequently use natural areas as outdoor laboratories. Geologists, zoologists, botanists, and soil scientists are some of the other scientists who use natural sites for scientific inquiry. Natural areas provide perfect settings for educational experiences not only in science, but also in history, because they can be used to demonstrate the way the land was when our ancestors settled here.

Certain unspoiled natural areas are also important for their recreational value, providing places for hiking, swimming, boating, rafting, sport hunting, and fishing. Wild areas are also important to the human spirit. Forest-covered mountains, rolling prairies, barren deserts, and other undeveloped areas are not only aesthetically pleasing, but help us to recover from the stresses of urban and suburban living. We can escape the tensions of the civilized world by retreating, even temporarily, to the solitude of natural areas.

CURRENT LAND USE IN THE UNITED STATES

About 55 percent of the land in the United States is privately owned by citizens, corporations, and nonprofit organizations, and about 3 percent by Native American tribes (Figure 17–2). The rest is owned by the federal government (about 35 percent of U.S. land) and by state and local governments (about 7 percent of U.S. land). Gov-

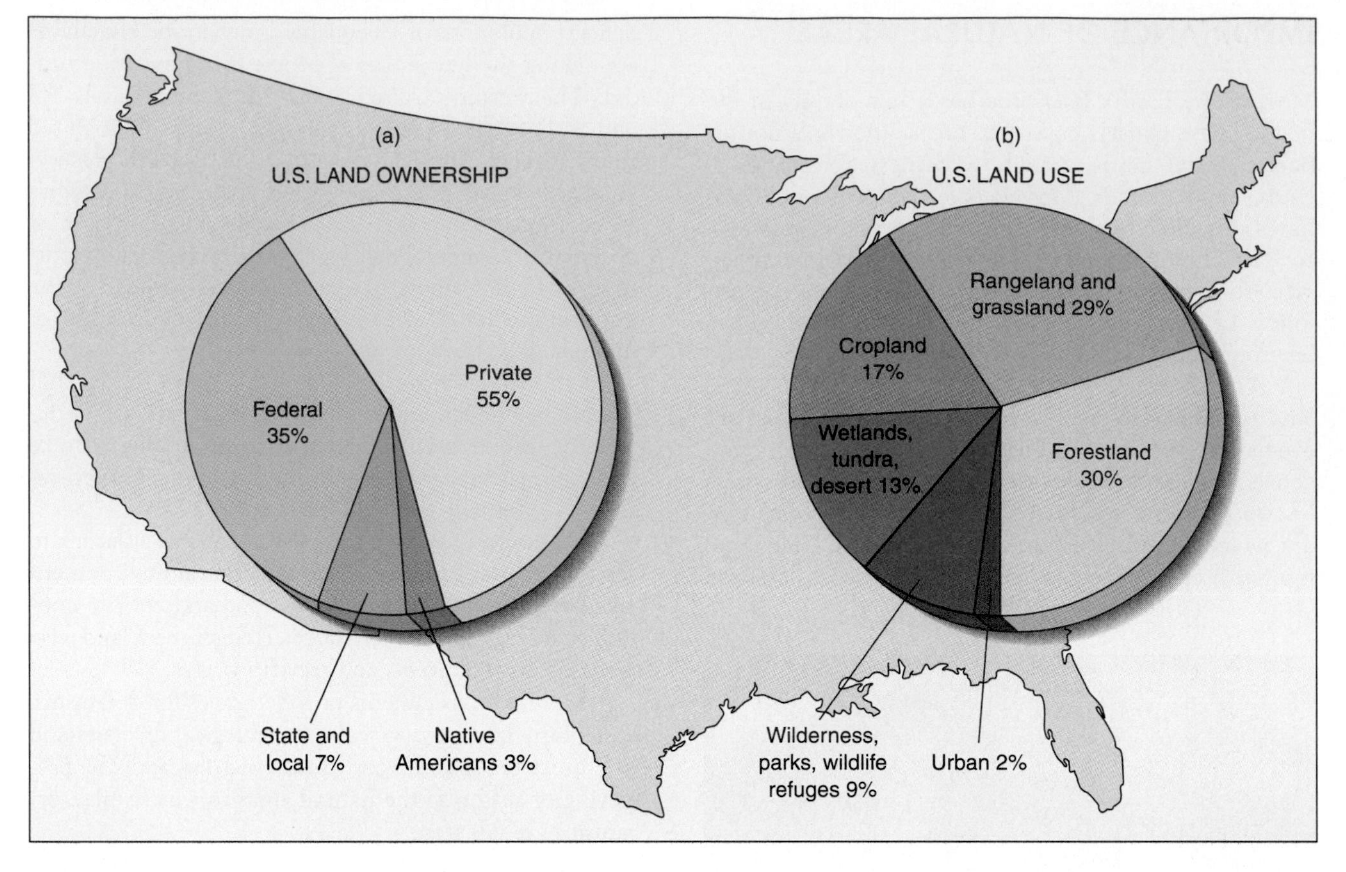

Figure 17–2
Land in the United States. (a) Land ownership. (b) How land is used.

ernment-owned land encompasses all types of ecosystems, from tundra to desert, and includes land that contains important resources such as minerals and fossil fuels, land that possesses historical or cultural significance, and land that provides critical biological habitat. Most of the federally owned land is in Alaska and 11 western states (Figure 17–3). It is managed primarily by four agencies, three in the U.S. Department of the Interior—the Bureau of Land Management (BLM), the Fish and Wildlife Service (FWS), and the National Park Service (NPS)—and one in the Department of Agriculture—the Forest Service (USFS) (Table 17–1).

Table 17–1
Administration of Federal Lands

Agency	Land Held	Area in Millions of Hectares (Acres)
Bureau of Land Management (Dept. of Interior)	National resource lands	109 (270)
U.S. Forest Service (Dept. of Agriculture)	National forests	77 (191)
U.S. Fish and Wildlife Service (Dept. of Interior)	National wildlife refuges	37 (92)
National Park Service (Dept. of Interior)	National Park System	34 (84)
Other—includes Department of Defense, Corps of Engineers (Dept. of the Army), and Bureau of Reclamation (Dept. of Interior)	Remaining federal lands	29 (72)

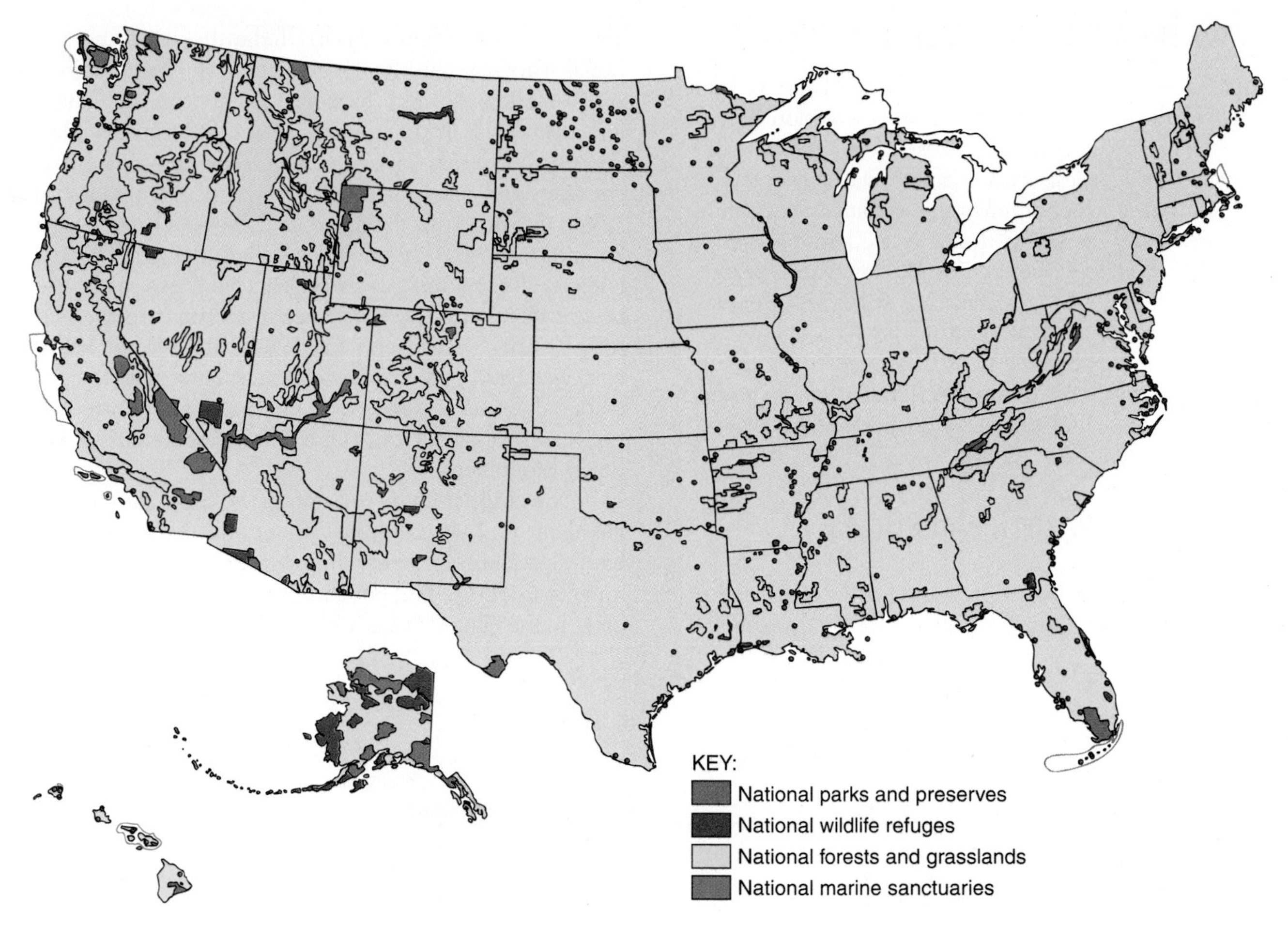

Figure 17–3

National parks and preserves, wildlife refuges, national forests and grasslands, and national marine sanctuaries in the United States. Other federal lands, such as military installations and research facilities, are not shown.

WILDERNESS

Regions of the Earth that have not been greatly disturbed by human activities and that humans visit but do not inhabit are known as **wilderness**. The Wilderness Act of 1964 authorized the U.S. government to set aside public wilderness areas, ranging from tiny islands to national parks that are several million hectares in size, as part of the National Wilderness Preservation System. Although mountains are the most common terrain to be safeguarded by this system, representative examples of other ecosystems have been set aside, including tundra, desert, and wetlands.

The Wild and Scenic Rivers Act was passed in 1968 to protect rivers with outstanding beauty, recreational value, or important wildlife. As of 1997, 154 river segments (less than 1 percent of the nation's total river systems) were protected by this act, with others being considered for inclusion. Rivers that have been given this designation have little or no development along their banks; most have no dams. Camping, swimming, boating, sport hunting, and fishing are permitted, but development of the shoreline is prohibited. Mining claims are permitted, however.

Millions of people visit U.S. wilderness areas each year, and some areas are overwhelmed by this use: soil and water pollution, litter and trash, and human congestion predominate in place of quiet, unspoiled land. Government agencies now restrict the number of people allowed into each wilderness area at one time, but it is likely that some of the most popular wilderness areas will require more intensive management. This would include the development of trails, outhouses, cabins, and campsites, amenities that are not encountered in true wilderness.

Do We Have Enough Wilderness?

Large tracts of wilderness, most of it in Alaska, have been added to the National Wilderness Preservation System since the passage of the Wilderness Act in 1964. The designation of wilderness areas is supported by people who view wilderness as a nonrenewable resource. They think it is particularly important to preserve additional land in the lower 48 states, where currently less than 2 percent of the total land area is specified as wilderness. Increasing the amount of federal land in the National Wilderness Preservation System is usually opposed by groups who operate businesses on public lands (such as timber, mining, ranching, and energy companies) and by their political representatives.

PARKS AND WILDLIFE REFUGES

The National Park System was originally composed of large, scenic areas in the West such as Grand Canyon and Yosemite Valley (Figure 17–4). Today, however, the National Park System has more cultural and historical sites—battlefields and historically important buildings and towns, for instance—than places of scenic wilderness. The National Park System currently has 369 different sites, and many of them have been purchased with money provided by the Land and Water Conservation Fund Act of 1965. Urban parks, such as the Golden Gate National Recreation Area near San Francisco, have also been established. The most recent additions to the National Park System are, in 1994, the Death Valley National Park, Joshua Tree National Park, and Mojave National Preserve (discussed in the chapter introduction), and, in 1996, Utah's Grand Staircase Escalante National Monument.

One of the primary roles of the NPS is to teach people about the natural environment and management of natural resources by providing nature walks and guided tours of its parks. The popularity and success of national parks in the United States (Table 17–2) have caused many

Figure 17–4
Yosemite National Park in California. This winter scene shows the Merced River flowing past the rock formation known as El Capitan. (Larry Ulrich/Tony Stone Images)

Table 17–2
The Ten Most Popular National Parks

National Park	Number of Recreation Visitors in 1995 (millions)
Great Smoky Mountains (North Carolina and Tennessee)	9.1
Grand Canyon (Arizona)	4.6
Yosemite (California)	3.9
Olympic (Washington)	3.7
Yellowstone (Wyoming, Montana, and Idaho)	3.1
Rocky Mountain (Colorado)	2.9
Acadia (Maine)	2.8
Grand Teton (Wyoming)	2.7
Zion (Utah)	2.4
Mammoth Cave (Kentucky)	1.9
Total visitors to all national parks	269.6

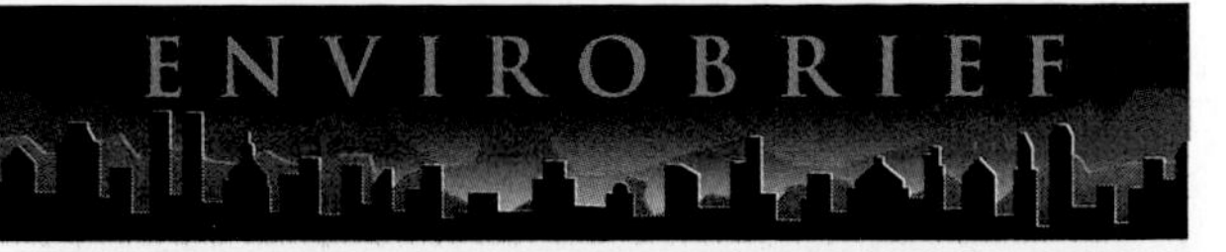

HOW PARKS PAY IN AFRICA

Tourism overtook coffee in 1988 as Kenya's largest source of foreign exchange. Currently, 7.5 percent of the country's total land area has been placed in 17 national parks and wildlife reserves. Along with generating employment and conserving wildlife, these areas preserve vegetation and aid soil conservation. The following statistics show the economic importance of tourists who come to see Kenya's wildlife:

- Number of tourists who visit Kenya each year, mainly to view wildlife: 735,000
- Amount of money they spend: (in U.S. dollars) $250,000,000
- Estimated annual tourist value of one lion in Kenya's Amboseli National Park: $27,000; of one of its elephant herds: $610,000
- Tourism is 50 times more valuable to Amboseli National Park than agriculture on the same land would be.

other nations to follow the example of the United States in establishing national parks (see Focus On: A National Park in Africa). As in the United States, these parks usually have multiple roles, from providing biological habitat to facilitating human recreation.

Threats to Parks

National parks are even more overcrowded than wilderness areas. All of the problems plaguing urban areas are found in popular national parks during peak seasonal use, including crime, vandalism, litter, traffic jams, and pollution of the soil, water, and air. Park managers have had to reduce visitor access to environmentally fragile park areas that have been overused.

Many people think that more funding is needed to maintain and repair existing parks. Facilities at some of the largest, most popular parks, such as Yosemite, the Grand Canyon, and Yellowstone, were last upgraded some 30 years ago. Although federal funding has not increased appreciably during the 1990s, more than 15 new parks and monuments have been added to the National Park System. Entrance fees currently account for about $100 million of the $900 million a year the Park Service spends. Food, lodging, and other concessions currently earn about $700 million annually, but pay less than 3 percent in franchise fees. The Park Service would like to charge higher entrance fees and higher concession fees to help meet operating costs.

Some national parks have imbalances in wildlife populations that involve declining populations of many species of mammals, including bears, white-tailed jackrabbits, and red foxes. For example, the number of grizzly bears in national parks in the western United States has greatly diminished. Grizzly bears require large areas of wilderness, and, clearly, human influences in national parks are a factor in their decline. More important, the parks may be too small to support grizzlies. Fortunately, grizzly bears have survived in sustainable numbers in Alaska and Canada.

Other mammal populations, notably elk, have been allowed to proliferate. Elk at Yellowstone National Park's northern range, for example, have increased from a population of 3100 in 1968 to 20,000 in 1996. Ecologists have documented that the elk have overgrazed the entire ecosystem, reducing the abundance of native vegetation, such as willow and aspen, and seriously eroding stream banks.

National parks are also affected by human activities beyond their borders. Pollution does not respect park

FOCUS ON

A National Park in Africa

Deforestation is occurring at an unprecedented rate throughout the tropics. In most of West Africa the forests have disappeared. Cameroon, a West African country, is fortunate in that its government is committed to forest preservation. Recently, three national parks were established in the southern rainforest area of the country. One of these, Korup National Park, has been extensively surveyed and has the richest biological diversity in Africa. It contains more than 400 tree species, 50 mammal species (including almost 25 percent of the primate species found in Africa), and more than 250 bird species.

Plans for Korup National Park during the next few decades include many aspects of conservation and sustainable utilization of forest resources. Villages now located in the park will be resettled outside its boundaries. Local people will be educated about environmental issues and the importance of conservation. Some of these people will become park staff members. A scientific research program is being established to help develop sustainable management practices for the park. Tourism, which depends upon preservation of the area's unique biological diversity, will be developed to help provide income for the local economy. Many international organizations are cooperating with the Cameroonian people in the establishment of Korup National Park. They include the British Overseas Development Administration, USAID, Parcs Canada, and the World Wide Fund for Nature. Korup National Park is a model of land conservation that other developing countries can emulate.

Korup National Park is a national treasure belonging to the Cameroonian people and to the world. (Jane Thomas/Visuals Unlimited)

boundaries (recall the Chapter 15 discussion of a proposed gold mine near Yellowstone National Park). Also, parks are increasingly becoming islands of natural habitat surrounded by human development (see Chapter 16). Development on the borders of national parks limits the areas in which wild animals may range, forcing them into isolated populations. Ecologists have found that, when environmental stresses occur, several small "island" populations are more likely to become threatened than a single large population occupying a sizable range.

Wildlife Refuges

The National Wildlife Refuge System, which was established in 1903 by President Theodore Roosevelt, is the most extensive network of lands and waters committed to wildlife in the world. As of 1996, the National Wildlife Refuge System contains 508 refuges in all 50 states and encompasses 37.4 million hectares (92.3 million acres) of land. The various refuges represent all major ecosystems found in the United States, from tundra to temperate rain forests to deserts, and are home to some of North America's most endangered species, such as the whooping crane. The mission of the National Wildlife Refuge System is to preserve lands and waters for the conservation of fishes, wildlife, and plants of the United States. Wildlife-dependent activities, such as hunting, fishing, wildlife observation, photography, and environmental education, are permitted on wildlife refuges as long as they are compatible with scientific principles of fish and wildlife management.

FORESTS

Forests, important ecosystems that provide many goods and services, occupy less than one-third of the Earth's total land area. Timber harvested from forests is used for fuel, construction materials, and paper products. Trees also supply nuts, fruits, and medicines. Forests provide employment for millions of people worldwide.

Forests influence local climate conditions. If you walk into a forest on a hot summer day, you will notice that the air is cooler and moister than it is outside the forest.

Figure 17–5

Forests play an important role in the hydrologic cycle by returning most of the water that falls as precipitation to the atmosphere by transpiration. In contrast, most precipitation is lost from deforested areas as runoff.

This is the result of a biological cooling process called **transpiration** in which water from the soil is absorbed by roots, transported through plants, and then evaporated from their leaves and stems. Transpiration also provides moisture for clouds, eventually resulting in precipitation (Figure 17–5).

Forests play an essential role in regulating global biogeochemical cycles, such as those for carbon and nitrogen. For example, photosynthesis by trees removes large quantities of carbon dioxide from the atmosphere and fixes it into carbon compounds. At the same time, oxygen is released into the atmosphere. Tree roots hold vast tracts of soil in place, reducing erosion. Forests are effective watersheds because they absorb, hold, and slowly release water; this provides a more regulated flow of water, even during dry periods, and helps to control floods and droughts. In addition, forests provide essential habitats for many organisms. (The importance of tropical forests, particularly tropical rain forests, as the repositories of most of the world's biological diversity was discussed in Chapter 16.)

Forest Management

When forests are managed, their species composition and other characteristics are altered. Specific varieties of commercially important trees are planted, and those trees that are not as commercially desirable are thinned out or removed. Traditional forest management often results in low-diversity forests. In the southeastern United States, for example, many forests of young pine that are grown for timber and paper production are all the same age and are planted in rows a fixed distance apart (Figure 17–6).

Figure 17–6

Tree plantations such as this pine plantation are monocultures, with trees of uniform size and age. (Kirtley-Perkins/Visuals Unlimited)

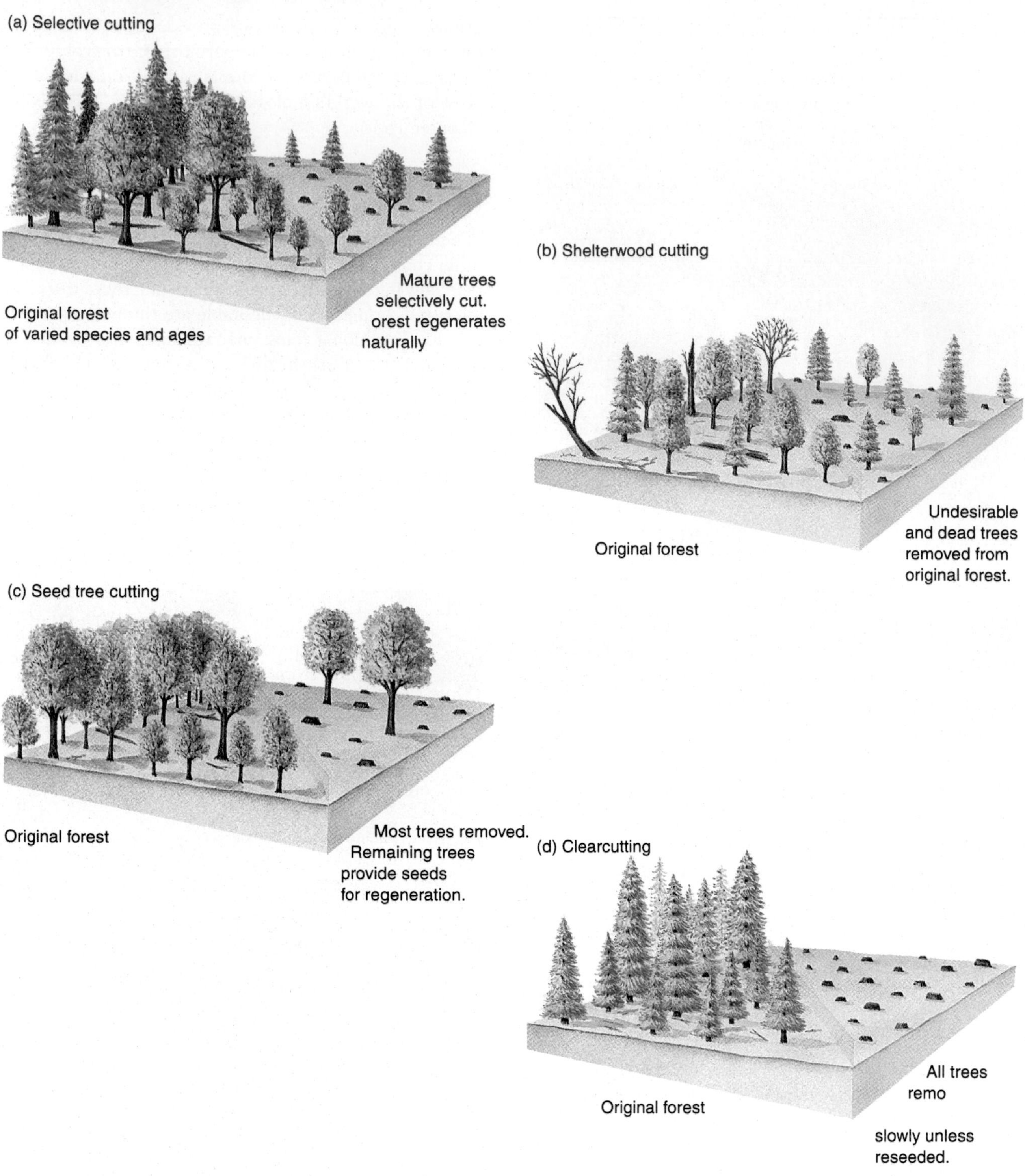

Figure 17–7

There are four major systems of tree harvesting. (a) In selective cutting, the older, mature trees are selectively harvested from time to time and the forest regenerates itself naturally. (b) In shelterwood cutting, less desirable and dead trees are harvested. As younger trees mature, they produce seedlings, which continue to grow as the now-mature trees are harvested. (c) Seed tree cutting involves the removal of all but a few trees, which are allowed to remain, providing seeds for natural regeneration. (d) In clearcutting, all trees are removed from a particular site. Clearcut areas may be reseeded or allowed to regenerate naturally.

These forests are essentially **monocultures**—areas covered by one crop, like a field of corn. One of the disadvantages of monocultures is that they are more prone to damage by insect pests and disease-causing microorganisms. Consequently, pests and diseases must be controlled in managed forests, usually by applying pesticides. Because managed forests contain few kinds of food, they cannot support the variety of organisms typically found in natural forests.

In recognition of the many environmental services performed by forests, a new method of forest management, known as **ecologically sustainable forest management**, is evolving. This broader approach seeks not only to conserve forests for the commercial harvest of timber and nontimber forest products, but also to sustain biological diversity, prevent soil erosion and protect the soil, and preserve watersheds that produce clean water. To achieve these goals, forestry operations are being designed that maintain a mix of trees in forests. When logging occurs, unlogged areas are set aside as sanctuaries for organisms, along with **wildlife corridors**, which are protected zones that connect unlogged areas. The purposes of wildlife corridors are to provide escape routes should they be needed and to allow animals to migrate so they can interbreed. (Small, isolated populations that are inbred may have a higher risk of extinction.) Wildlife corridors are also thought to allow large animals such as the Florida panther to maintain large territories. The effectiveness of wildlife corridors is currently being questioned by some scientists, and there is an urgent need for scientific evidence to resolve the issue.

The actual methods of ecologically sustainable forest management that distinguish it from traditional forest management are gradually being developed and vary from one forest ecosystem to another. Because trees have such long lifespans, scientists and forest managers of the future will judge the results of today's efforts.

Harvesting Trees

Trees are harvested in several ways—by selective cutting, shelterwood cutting, seed tree cutting, and clearcutting (Figure 17–7). **Selective cutting**, in which mature trees are cut individually or in small clusters while the rest of the forest remains intact, allows the forest to regenerate naturally. The trees left by selective cutting produce seeds that germinate to fill the void. Selective cutting has fewer negative effects on the forest environment than other methods of tree harvest, but it is not as profitable in the short term because timber is not removed in great enough quantities.

The removal of all mature trees in an area over a period of time is known as **shelterwood cutting**. In the first year of harvest, undesirable tree species and dead or diseased trees are removed. The forest is then left alone for perhaps a decade, during which the remaining trees continue to grow, and new seedlings become established. During the second harvest, many mature trees are removed. The forest is then allowed to regenerate on its own for perhaps another decade. A third harvest removes the remaining mature trees, but by this time a healthy stand of younger trees is replacing the mature ones. Little soil erosion occurs with this method of tree removal, even though more trees are removed than in selective cutting.

In **seed tree cutting**, almost all trees are harvested from an area; a scattering of desirable trees is left behind to provide seeds for the regeneration of the forest. **Clearcutting**, also called **even-age harvesting**, is the removal of all trees from an area (Figure 17–8). After the trees have been removed by clearcutting, the area is either allowed to reseed and regenerate itself naturally or is planted with specific varieties of trees. Timber companies prefer clearcutting because it is the most cost-effective way to harvest trees; also, relatively little road building has to be done to harvest a large number of trees. However, clearcutting over wide areas is ecologically unsound. It destroys biological habitats and increases soil erosion, particularly on sloping land. Obviously, the recreational benefits of forests are lost when clearcutting occurs.

Deforestation

The most serious problem facing the world's forests is **deforestation**, which is defined as the temporary or permanent clearance of forests for agriculture or other uses. When forests are destroyed, they no longer make valuable contributions to the environment or to the people who depend on them. Forest destruction, particularly in the tropics, threatens native people whose cultural and physical survival depends on the forests (Figure 17–9).

Deforestation results in decreased soil fertility and increased soil erosion. Uncontrolled soil erosion, particularly on steep deforested slopes, can affect the production of hydroelectric power as silt builds up behind dams. Increased sedimentation of waterways caused by soil erosion can also harm downstream fisheries. In drier areas, deforestation contributes to the formation of deserts (discussed shortly).

When a forest is removed, the total amount of surface water that flows into rivers and streams actually increases. However, because this water flow is no longer regulated by the forest, the affected region experiences alternating periods of flood and drought.

Deforestation contributes to the extinction of many species. Many tropical species, in particular, have very

Figure 17–8

Aerial view of a clearcut forest in Washington. (Gary Braasch/Tony Stone Images)

limited ranges within a forest, so they are especially vulnerable to habitat modification and destruction. Migratory species, including birds and butterflies, also suffer from deforestation.

Deforestation is thought to induce regional and global climate changes. Trees release substantial amounts of moisture into the air; about 97 percent of the water that roots absorb from the soil is evaporated directly into the atmosphere. This moisture falls back to the Earth in the hydrologic cycle (see Chapter 5). When a large forest is removed, rainfall may decline and droughts may become more common in that region. Tropical deforestation may contribute to an increase in global temperature by causing a release of stored carbon into the atmosphere as carbon dioxide, which in turn enables the air to retain heat (see Chapter 20).

Tropical Forests and Deforestation

There are two types of tropical forests: tropical rain forests and tropical dry forests. In places where the climate is very moist throughout the year—with about 200 or more centimeters (at least 79 inches) of precipitation annually—**tropical rain forests** prevail. Tropical rain forests

Figure 17–9

Native Indians from the Amazon River Basin watch television. The 136,000 Indians remaining in the rain forest now live on 82 million hectares (203 million acres) of land. Encroaching deforestation risks their traditional culture. Already an estimated 73,000 Amazonian Indians have left the rain forest, many settling in large cities. (Abril/Gamma Liaison)

PAPER WITHOUT TREES

Two agricultural products–kenaf, a fibrous plant cultivated since ancient Egypt, and straw–may reduce deforestation as they provide viable options to trees for making paper:

- Kenaf, the U.S. Department of Agriculture's chosen plant alternative for paper production, is 20 times more productive than southern pine, requires less bleaching because its fibers are whiter than wood's (so no chlorine is necessary), and produces a non-yellowing, high-quality paper.
- *Earth Island Journal*, the state of West Virginia, and Kinko's printing stores have already used kenaf in pilot projects.
- As much as two hundred million tons of straw burned or plowed under in the United States each year could be recycled as paper mill feedstock.
- Weyerhauser Company, which is testing straw, could eventually use as much as 400,000 tons a year for paper production by replacing as much as 20 percent of the wood chips it now uses.
- As of early 1997, six U.S. newspapers have started publishing on newsprint that contains some rice straw, rye grass, and fescue grass.

are found in Central and South America, Africa, and Southeast Asia, but almost half of them are in just three countries: Brazil in South America, Zaire in Africa, and Indonesia in Southeast Asia (Figure 17–10). In other tropical areas where annual precipitation is less but is still enough to support trees—including regions subjected to a wet season and a prolonged dry season—**tropical dry forests** occur. During the dry season, tropical trees shed their leaves and remain dormant, much as temperate trees do during the winter. India, Kenya, Zimbabwe, Egypt, and Brazil are a few of the countries that have tropical dry forests.

Most of the remaining undisturbed tropical forests, which lie in the Amazon and Congo river basins of South America and Africa, are being cleared and burned at a rate that is unprecedented in human history. Tropical forests are also being destroyed at an extremely rapid rate in southern Asia, Indonesia, Central America, and the Philippines.

Exact figures on rates of tropical forest destruction are unavailable. However, the Food and Agriculture Organization (FAO) of the United Nations released in late 1991 its second global assessment of tropical deforestation. A total of 87 tropical countries were evaluated in this study. The FAO estimated an average annual loss in forests of 0.9 percent per year from 1981 to 1990, and some areas, such as West Africa, experienced a loss of forests estimated at greater than 2 percent per year. If this rate of deforestation—which represents a worldwide annual loss of 16.9 million hectares (41.8 million acres)—

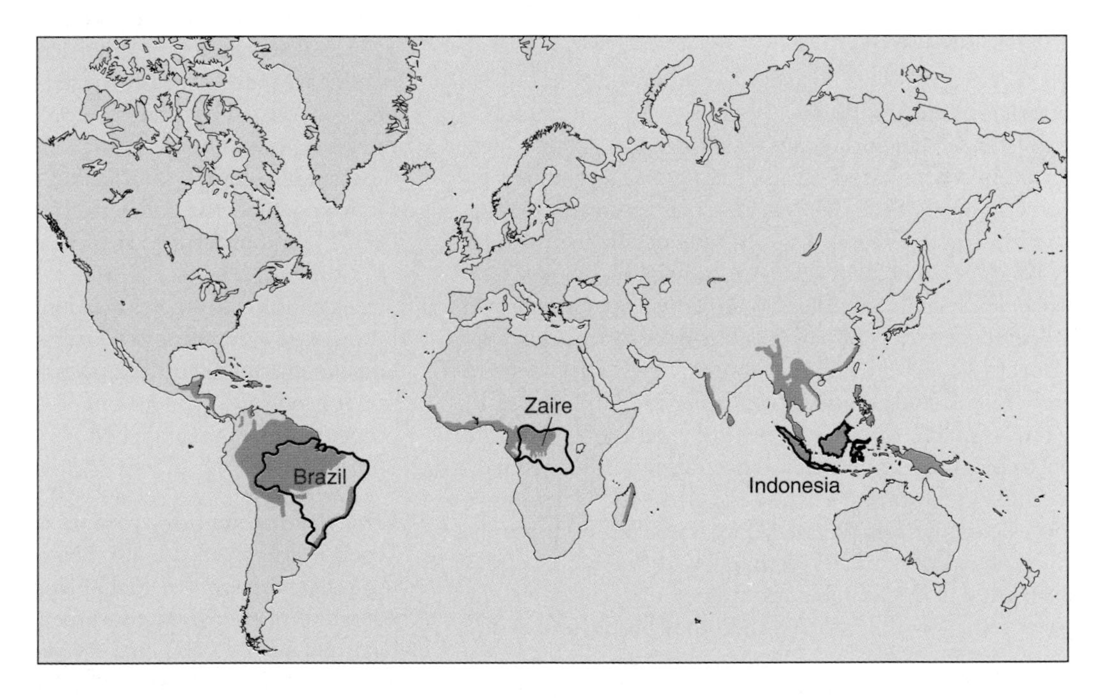

Figure 17–10

The distribution of tropical rain forests (green areas). Rain forests are located in Central and South America, Africa, and Southeast Asia.

continues, tropical forests will be all but gone by the second half of the next century.

Why Are Tropical Forests Disappearing?

Several studies show a strong statistical correlation between population growth and deforestation. More people need more food, and so the forests are cleared for agricultural expansion. However, tropical deforestation is a complex problem that cannot be attributed simply to population pressures. Moreover, the main causes of deforestation vary from place to place.

A variety of economic, social, and governmental factors interact to cause deforestation. Government policies sometimes provide incentives that favor the removal of forests. For example, the Brazilian government opened the Amazonian frontier for settlement, beginning in the late 1950s, by constructing the Belem-Brasilia Highway, which cut through the Amazon Basin. Sometimes economic conditions encourage deforestation. For example, the farmer who converts more forest to pasture can maintain a larger herd of cattle, which is a good hedge against inflation.

Keeping in mind that tropical deforestation is a complex problem, three agents—subsistence agriculture, commercial logging, and cattle ranching—are thought to be the most immediate causes of deforestation. Other reasons for the destruction of tropical forests include mining, particularly when ore smelters burn charcoal[1] produced from rainforest trees, and the development of hydroelectric power, which inundates large areas of forest.

Subsistence agriculture

Subsistence agriculture, in which enough food is produced by a family to feed itself, accounts for perhaps 60 percent of tropical deforestation. In many developing countries where tropical rain forests occur, the majority of people do not own the land on which they live and work. For example, in Brazil 5 percent of the farmers own 70 percent of the land. Most subsistence farmers have no place to go except into the forest, which they clear to grow food. Land reform would make the land owned by a few available to everyone, thereby easing the pressure on tropical forests by subsistence farmers. This scenario is unlikely, however, because wealthy landowners have more economic and political clout than landless peasants.

Subsistence farmers often follow loggers' access roads until they find a suitable spot. They first cut down the trees and allow them to dry, then they burn the area and plant crops immediately after burning; this is known as **slash-and-burn agriculture** (discussed further in Chapter 18). The yield from the first crop is often quite high because the nutrients that were in the burned trees are now available in the soil. However, soil productivity declines at a rapid rate, and subsequent crops are poor. In a very short time, the people farming the land must move to a new part of the forest and repeat the process. Cattle ranchers often claim the abandoned land for grazing, because land that is not rich enough to support crops can still support livestock.

Slash-and-burn agriculture done on a small scale with periods of 20 to 100 years between cycles is sustainable. The forest regrows rapidly after a few years of farming. But when *millions of people* try to obtain a living in this way, the land is not allowed to lie uncultivated long enough to recover.

Commercial logging

About twenty percent of tropical deforestation is the result of commercial logging, and vast tracts of tropical rain forests, particularly in Southeast Asia, are being harvested for export abroad. Most tropical countries allow commercial logging to proceed at a much faster rate than is sustainable. In the final analysis, tropical deforestation does not contribute to economic development; rather, it destroys a valuable natural resource.

Cattle ranching and agriculture for export

Approximately 12 percent of tropical deforestation is carried out to provide open rangelands for cattle (Figure 17–11). Cattle ranching is particularly important in Latin America. Much of the beef raised on these ranches, which are often owned by foreign companies, is exported to fast-food restaurant chains. After the forests are cleared, cattle can graze on the land for up to perhaps 20 years, after which time the soil fertility is depleted. When this occurs, shrubby plants, known as **scrub savanna**, take over the range.

A considerable portion of forest land cleared for plantation-style agriculture produces crops such as citrus fruits and bananas for export. Plantation-style agriculture is sustainable on forest soil as long as expensive fertilizers and other treatments are applied.

Why are tropical dry forests disappearing?

Tropical dry forests are also being destroyed at an alarming rate, primarily for fuel (Figure 17–12). Wood—perhaps half of the wood consumed worldwide—is used as heating and cooking fuel by much of the developing world. Often the wood cut for fuel is converted to charcoal, which is then used to power steel, brick, and ce-

[1]Partially burning wood in a large kiln from which air is excluded converts the wood into charcoal.

Figure 17–11

Cattle graze on Panamanian land that was formerly rain forest. Tropical rain forests are often burned to provide grazing land for cattle that are exported to countries such as the United States. (David Cavagnaro)

ment factories. Charcoal production is extremely wasteful: 3.6 metric tons (4 tons) of wood produce enough charcoal to fuel an average-sized iron smelter for only 5 minutes.

Boreal Forests and Deforestation

Although tropical forests have been depleted extensively during the second half of the 20th century, they are not the only forests being destroyed. Extensive deforestation by the logging of certain boreal forests began in the late 1980s and early 1990s. **Boreal forests**, also known as taiga, occur in Alaska, Canada, Scandinavia, and northern Russia. Boreal forests, which cover about 11 percent of the Earth's land area, comprise the world's largest biome.

Boreal forests, which are being harvested primarily by clearcut logging, are currently the primary source of the world's industrial wood and wood fiber. The annual loss of boreal forests has been estimated to encompass an area twice as large as the Amazonian rain forests of Brazil. About 1 million hectares (2.5 million acres) of Canadian forests are logged annually, and extensive tracts of Siberian forests in Russia are harvested, although exact estimates are unavailable. Alaska's boreal forests are also at risk because the U.S. government may increase logging on public lands in the future.

Figure 17–12

Women gather firewood in the Ranthambore National Park buffer zone. Note how the branches have been trimmed off the trees in the background. About 340,000 hectares (840,000 acres) of India's natural forests disappear per year, mostly for firewood consumption. (Martin Hervey/NHPA)

Temperate Forests

In recent years, temperate forests, particularly those in the eastern United States, have generally been holding steady or even expanding. In Vermont, for example, the amount of land covered by forests has increased from 35 percent in 1850 to 80 percent today. Expanding forests are the result of secondary succession on abandoned farms (see Chapter 4), commercial planting, and government protection. Although the returning forests do not have the biological diversity of virgin stands, many forest organisms have successfully become reestablished.

Forests in the United States

The USFS manages approximately 10 percent of the land in the United States. Forest Service lands have multiple uses, including timber harvest, livestock forage, water resources, recreation, and habitat for fishes and wildlife. The BLM also oversees some public forests, two-thirds of which lie in Alaska.

Recreation in the national forests ranges from camping at designated campsites to backpacking in the wilderness. Visitors to national forests also swim, boat, picnic, and observe nature. Effective maintenance and management of these areas for public use and enjoyment include trash removal, trail maintenance, and repair of damage caused by vandalism, but the funds allotted for these purposes do not keep pace with the increasing public use of national forests.

Slightly more than half of the forests in the United States are privately owned, and three-fourths of these private lands are in the eastern part of the country. Projected conversion of forests to agricultural, urban, and suburban lands over the next 40 years is expected to have the greatest impact in the South, although there will be considerable losses in other regions as well.

Issues Involving U.S. Forests

Because national forests are multiple-use lands, they are logged; used for grazing, farming, and mining; used for hunting, fishing, and other recreational pastimes; and preserved for their ecological benefits, including biological habitat. With so many different possible uses of U.S. forests, conflicts inevitably arise, particularly between timber interests and those who wish to preserve the trees for other purposes. Two issues involving national forests are representative of forest debates in the mid-1990s—clearcutting Alaska's Tongass National Forest and salvage logging of public forests.

Alaska's Tongass National Forest

Despite its northern location, the Tongass National Forest is a temperate rain forest (Figure 17–13; also see Chapter 6 for a description of temperate rain forests). This old-growth forest of giant Sitka spruce, yellow cedar, and hemlock provides habitat for a wealth of wildlife, such as grizzly bears and bald eagles. It is also a prime logging area because a single large Sitka spruce may yield as much as 10,000 board feet of high-quality timber. However, as in most national forests, it is expensive to log in the Tongass. To cover the high costs of operating, timber interests such as pulp mills have always relied on obtaining the timber from the federal government at below-market prices. This right was granted in 1954 by a 50-year contract that expired in the 1990s. In 1990 some members of Congress tried to pass the Tongass Timber Reform Act, which forced timber interests to pay market prices, but the legislation was bitterly opposed by other members of Congress, including Alaska Senator Murkowski. The compromise agreement, reached in 1996, provides timber to the mills at market prices. As a result of this legislation, it is thought that clearcut logging will continue in the Tongass, but at lower rates than in the past.

Salvage logging: a volatile U.S. issue

In 1995 a controversial bill was signed into law that permitted **salvage logging**, in which trees that are weakened by insects, disease, or fire are harvested in national forests. The law, which was supported by Western lawmakers who wanted to maintain employment at sawmills in the face of declining logging in their states, was based

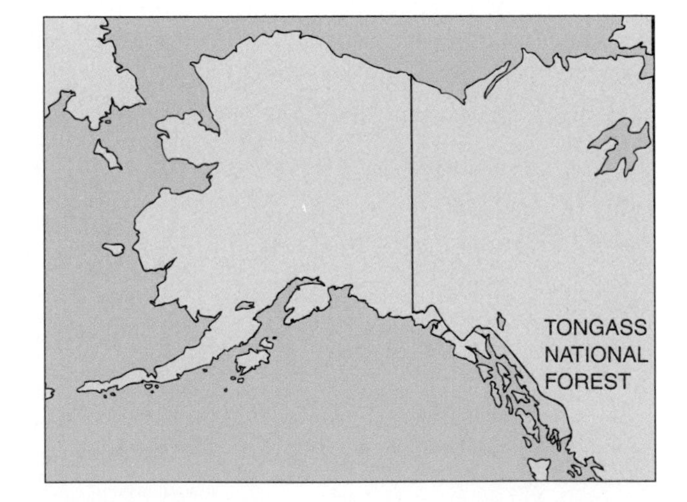

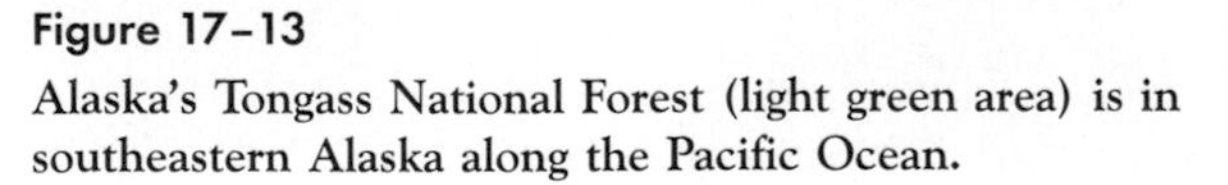

Figure 17–13

Alaska's Tongass National Forest (light green area) is in southeastern Alaska along the Pacific Ocean.

on the idea that publicly owned Western forests are overgrown and therefore more prone to fire and disease. It permitted timber companies to log in areas that are normally not open to logging, and it exempted these companies from complying with environmental laws such as the Clean Water Act and the Endangered Species Act. The law was a temporary measure that expired on December 31, 1996.

The law permitting salvage logging was opposed by forestry scientists, environmentalists, and taxpayer advocates. Forestry scientists point out that logging an area already damaged by fire actually delays or prevents the forest from recovering. Biologists emphasize that dead and diseased trees have important ecological roles in forests—they shelter and nurture organisms, help to maintain the physical structure of stream banks, and enrich the soil as they decay.

Environmentalists contend that healthy trees were harvested under the guise of salvage logging. They point out that the law resulted in dozens of tracts of forest, particularly in California and the Pacific Northwest, being clearcut in the interest of "protecting" them.

Taxpayer advocates, such as the National Taxpayers Union, opposed the law because it cost taxpayers money. The USFS annually spends millions of dollars more to harvest trees on public lands than it earns from timber sales. In 1994, for example, the federal government spent $100 million to build and maintain roads through national forests; these roads are primarily used by timber companies to haul lumber. Taxpayer advocates consider this a subsidy to the timber industry because they did not pay for any of the road construction or repair costs.

RANGELANDS

Rangelands are grasslands, in both temperate and tropical climates, that serve as important areas of food production for humans by providing fodder for domestic animals such as cattle, sheep, and goats (Figure 17–14). The predominant vegetation of rangelands includes grasses, forbs (small herbaceous plants other than grasses), and shrubs.

Intensive livestock practices in which animals are kept in small enclosures for some or all of their lives, rather than allowed to roam freely while grazing, are increasingly being adopted in many countries. Although these practices reduce the requirement for open spaces, rangelands are still important in the production of domestic animals. Rangelands are also mined for minerals and energy resources, used for recreation, and preserved for biological habitat and for maintaining soil and water resources.

Rangeland Degradation and Desertification

Grasses, the predominant vegetation of rangelands, have a **fibrous root system**, in which many roots form a diffuse network in the soil to anchor the plant. Plants with fibrous roots hold the soil in place quite well, thereby reducing soil erosion. If only the upper portion of the grass is eaten by animals, the fibrous roots can continue to develop, allowing the plants to recover and grow to their original size.

Figure 17–14
A rangeland in Montana. When the carrying capacity of rangeland is not exceeded, it is a renewable resource. (Grant Heilman, Grant Heilman Photography)

Figure 17–15
Overgrazing can contribute to nonproductive desert, as occurred in this arid region in western New Mexico and eastern Arizona. (J. Alcock/Visuals Unlimited)

The **carrying capacity** of a rangeland is the maximum number of animals the rangeland plants can sustain. When the carrying capacity of a rangeland is exceeded, grasses and other plants are **overgrazed**; that is, so much of the plant is consumed by the grazing animals that it cannot recover, and it dies. Overgrazing results in barren, exposed soil that is susceptible to erosion.

Most of the world's rangelands lie in semiarid areas that have natural extended periods of drought. Under normal conditions, native grasses in these drylands can survive a severe drought: the aerial portion of the plant dies back, but the extensive root system remains alive and holds the soil in place. When the rains return, the roots send forth new shoots.

When overgrazing occurs in combination with an extended period of drought, however, once-fertile rangeland can be converted to desert. The lack of plant cover due to overgrazing allows winds to erode the soil. Even when the rains return, the land is so degraded that it cannot recover. Water erosion removes the little bit of remaining topsoil, and the sand that is left behind forms dunes. This process, which induces unproductive desert-like conditions on formerly productive rangeland (or tropical dry forest), is called **desertification** (Figure 17–15). It reduces the agricultural productivity of economically valuable land, forces many organisms out, and threatens endangered species. There is still much that scientists do not understand about desertification and the processes that cause it. How, for example, do natural fluctuations in climate interact with population pressures and human activities to produce desert?

About 30 percent of the human population live in rangelands that border deserts. One of the consequences of desertification is a decline in agricultural productivity, and studies have shown that when an area cannot feed its people, they emigrate. For example, there are more Senegalese in certain regions of France than in their native villages in Africa. According to the United Nations Environment Programme, about 135 million people worldwide are in danger of displacement as a result of desertification.

Rangelands in the United States

Rangelands make up approximately 30 percent of the total land area in the United States and occur mostly in the western states and Alaska. Of this, approximately one-third is publicly owned and two-thirds is privately owned. Excluding Alaska, there are at least 89 million hectares (220 million acres) of public rangelands. Approximately 69 million hectares (170 million acres) are managed primarily by the BLM, which is guided by the Taylor Grazing Act of 1934, the Federal Land Policy and Management Act of 1976, and the Public Rangelands Improvement Act of 1978. The USFS manages an additional 20 million hectares (50 million acres).

Overall, the condition of public rangelands in the United States has slowly improved since the low point of the Dust Bowl in the 1930s (see Chapter 14). Much of this improvement can be attributed to fewer livestock being permitted to graze the rangelands after the passage of the Taylor Grazing Act in 1934. Better livestock management practices, such as controlling the distribution of animals on the range, as well as conservation measures, have contributed to rangeland recovery. But restoration has been slow and costly, and more is needed: a 1988 study by the General Accounting Office, the investigative arm of Congress, concluded that more than 50 percent of public rangelands were in either fair or poor condition.

Rangeland management includes seeding in places where plant cover is sparse or absent, constructing fences to prevent overgrazing (Figure 17–16), controlling weeds, and protecting biological habitats. Special incentives are sometimes offered to livestock operators to use public rangelands in a way that results in their overall improvement.

Figure 17–16

Fences play an important role in preventing overgrazing. The land to the left of the fence was ungrazed, whereas the grazed land on the right and in the distance has sparser vegetation. (USDA/Natural Resources Conservation Service)

Issues Involving U.S. Rangelands

The federal government distributes permits to private livestock operators that allow them to use public rangelands for grazing in exchange for a small fee. (In 1996 the monthly grazing fee was $1.35 per cow. This translates into $25 million paid to the federal government in 1996.) The permits are held for many years and are not open to free-market bidding—that is, only ranchers are allowed to obtain grazing permits.

Many environmental groups are concerned about the ecological damage caused by the overgrazing of public rangelands and want to reduce the number of livestock animals allowed to graze (see Meeting the Challenge: The Trout Creek Mountain Working Group). They want public rangelands to be managed primarily for other uses, such as biological habitat, recreation, and scenic value, rather than for privately owned livestock. To accomplish this goal, they would like to be able to purchase grazing permits and then set aside the land.

Conservative economists have joined environmentalists in criticizing management of federal rangelands. Taxpayers currently subsidize ranchers who graze their livestock on public rangelands by paying at least $52 million more than grazing fees (in 1996). This money is used to maintain the rangelands, including installing water tanks and fences, and to fix the damage done by the livestock. The National Taxpayers Union and other free-market groups want grazing fees to cover all costs of maintaining herds on publicly owned rangelands. These groups estimate that the federal government could save as much as $250 million over five years if this were done. Although critics of federal grazing policies have tried to reform range policy for decades, they have been largely blocked by ranchers and their supporters in Congress.

Another rangeland issue involves the 42,000 wild horses and burros that roam on western public rangelands. These animals are not native to North America but were introduced from Europe by the Spaniards. Because they have come to symbolize America's pioneer history, the wild horses and burros are protected. At the same time, they are very destructive to the natural ecosystem and must be managed so that they do not contribute to its deterioration. To prevent their populations from exceeding the carrying capacity of their rangeland home, the BLM started the Adopt-a-Horse program, in which between 5000 and 9000 wild horses and burros are annually removed from the range and given away. However, the Adopt-a-Horse program is expensive to operate. Animal scientists have recently developed a vaccine that provides one year of contraception for female horses, deer, and burros. The first vaccine trials on Western horses were conducted in 1992. Studies continue, and it may be that the vaccine will be used widely in a few years.

FRESHWATER WETLANDS

Wetlands are lands that are transitional between aquatic and terrestrial ecosystems (see Figure 6–15). As discussed

MEETING THE CHALLENGE

The Trout Creek Mountain Working Group

Confrontations among western ranchers trying to preserve their way of life, environmentalists trying to protect the environment, and officials trying to administer state and federal laws have been common news items during the mid-1990s. In southeastern Oregon, however, an unlikely partnership of 25 to 35 ranchers, environmentalists, and government officials make up the Trout Creek Mountain Working Group. The coalition was organized in 1988 to improve the habitat of a threatened species, the Lahontan cutthroat trout, in an area of federally owned rangeland in southeastern Oregon (see Figure). The Trout Creek Mountain Working Group's goal is to develop a long-term solution that will provide for both the ecological health of the land and the economic well-being of people living in the area. An important concept of the working group is that ranchers and other people are a part of the natural ecosystem.

The group adopted a broad approach, known as **ecosystem management**, to execute their task. Ecosystem management recognizes that organisms, including the Lahontan cutthroat trout, interact with one another and with their abiotic environment (soil, water, and air). Ecosystem management takes into account that ecosystems do not recognize political boundaries and may span federal, state, county, and private lands.

After a species has been declared threatened or endangered, the Endangered Species Act requires that federal officials design a plan to aid the species' recovery. For the Lahontan cutthroat trout, this requirement meant closing the streams to the anglers and limiting cattle from foraging and drinking along stream banks. Cattle trample and eat the smaller vegetation along a bank and increase the rate of soil erosion. Cattle feces pollute the water where the fish live. Cattle also like to eat aspen and willow shoots that grow along the edges of stream banks, and when these trees are grazed, the stream is shaded less, which in turn increases the water temperature. Because trout are adapted to cold temperatures, warmer temperatures decrease the trouts' reproductive success.

Although members of the Trout Creek Mountain Working Group were initially suspicious of one another, they continued to work together and gradually developed a compromise. The ranchers agreed to withdraw about 3000 cattle from environmentally sensitive parts of the stream for three years, from 1989 to 1992, a concession that was hard for them because it reduced their profits. As part of the compromise, environmentalists and government officials agreed to open parts of the stream to some cattle, beginning in 1992.

Today, the condition of the stream ecosystem, which is shared by both cattle and trout, has improved. Willow, aspen, alder, wild rose, and other plants are becoming reestablished along the stream banks. Also, the population of Lahontan cutthroat trout appears to be slowly increasing.

The work of the Trout Creek Mountain Working Group is far from over, and additional difficult decisions will have to be made. The group plans to continue to monitor the stream ecosystem and make adjustments as needed, taking into account both environmental and economic issues.

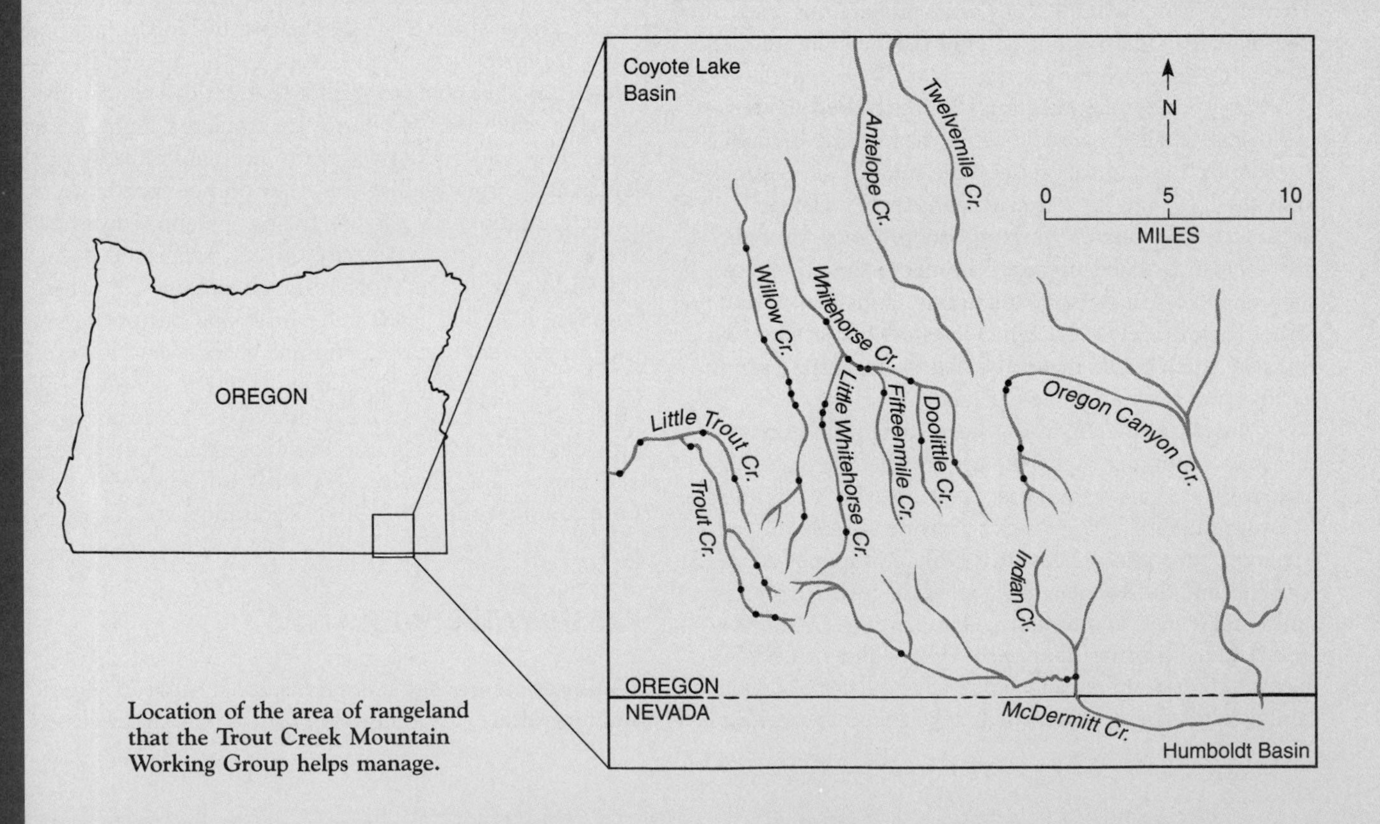

Location of the area of rangeland that the Trout Creek Mountain Working Group helps manage.

in Chapter 6, wetlands are usually covered by shallow water and have characteristic soils and water-tolerant vegetation. People used to think that the only benefit of wetlands was to provide food for migratory waterbirds. In recent years, however, the many ecosystem services that wetlands perform have come to be appreciated, and estimates of their economic value have increased accordingly. For example, wetlands help reduce damage from flooding because they hold excess water when rivers flood their banks. Wetlands also improve water quality by trapping and holding nitrates and phosphates from fertilizers, and they help to cleanse water that contains sewage. Freshwater wetlands produce many commercially important products including wild rice, blackberries, cranberries, blueberries, and peat moss. They are also sites for fishing, hunting, boating, bird-watching, photography, and nature study.

Wetlands are increasingly threatened by agriculture, pollution, engineering (such as dams), and urbanization. In the United States, wetlands have been steadily shrinking by an estimated 105,000 hectares (260,000 acres) per year since the mid-1980s. In the contiguous 48 states, of the more than 81 million hectares (200 million acres) of wetlands that originally existed, only 38 million hectares (95 million acres) remain. Most of the loss since the 1950s has been the result of farmers' converting wetlands to cropland. Urban and suburban development, dredging, and mining account for most of the remainder of the loss.

The loss of wetlands is legislatively controlled by a section of the 1972 Clean Water Act. This legislation, which is up for renewal in 1997, does a reasonably good job of protecting coastal wetlands but a poor job of protecting inland wetlands, which is what most wetlands are. The Emergency Wetlands Resources Act of 1986 authorizes the FWS to designate and acquire critically important wetlands. The Service is making an inventory and map of wetlands in the United States that is scheduled to be completed in 1998.

Currently, the United States is attempting to prevent any new net loss of wetlands by conserving existing wetlands and restoring those that have been lost. This means that development of wetlands is allowed only if a corresponding amount of previously converted wetlands is restored. The policy of no net loss of wetlands has only been partially successful, and wetlands loss has continued, although at a slower rate.

The policy is complicated by two factors: (1) confusion and dissent about the definition of wetlands (which was not spelled out in the original Clean Water Act) and (2) the question of who owns wetlands. In 1989 a team of government scientists developed a comprehensive, scientifically correct definition of wetlands. It provoked an outcry from farmers and real estate developers, who perceived it as a threat to their property values. Largely in response to their criticisms, politicians attempted to narrow the definition of wetlands in 1991 and again in 1995, removing marginal wetlands that are not as wet as swamps or marshes. This narrower definition, which ignores decades of wetlands research, would have excluded approximately one-half of existing wetlands in the United States from protection. In 1992 Congress asked the U.S. National Academy of Sciences to help settle the issue because it had become so controversial. The academy published its wetlands study in 1995, urging Congress to put the wetlands debate on a more scientific footing. Ironically, the study generated an even greater debate in Congress.

The federal government owns less than 25 percent of wetlands in the lower 48 states; the remaining 75 percent is privately owned. This means that private citizens control whether wetlands are protected and preserved or developed and destroyed. Because of the traditional rights of private land-ownership in the United States, landowners resent the federal government's telling them what they may or may not do with their properties. Property-rights advocates in Congress side with landowners in thinking the government has over-regulated wetlands.

It is therefore important that private landowners become informed of the environmental importance of wetlands and the critical need to maintain them in the broader public interest. Although some private owners do recognize the value of wetlands and voluntarily protect them, others do not. The federal government is examining proposals such as tax incentives and the outright purchase of wetlands to encourage their conservation.

In 1990 Congress passed the Food and Agriculture Conservation and Trade Act, which is a version of the Farm Bill that has been amended and renamed every 5 years or so since the 1930s. One of the provisions of this act is the establishment of the Wetlands Reserve Program (WRP), which seeks to restore 405,000 hectares (1 million acres) of privately owned freshwater wetlands that have previously been drained and converted to cropland. Farmers are offered financial incentives to restore the wetlands. However, the WRP is funded annually by Congress and is therefore subject to budget cuts.

COASTLINES

Coastal wetlands, also called saltwater wetlands, provide food and protective habitats for many aquatic animals (see Figures 6–16 and 6–17). They could be considered the ocean's nurseries because so many different marine fishes and shellfish spend the first parts of their lives there. In addition to being highly productive areas, coastal wetlands protect coastlines from erosion and reduce damage from hurricanes.

Historically, tidal marshes and other coastal wetlands have been regarded as wasteland, good only for breeding large populations of mosquitoes. Coastal wetlands throughout the world have been drained, filled in, or dredged out to turn them into "productive" structures such as industrial parks or marinas. Agriculture also contributes to the demise of coastal wetlands, which are drained for their rich soil. Other human endeavors that cause the destruction of coastal wetlands include fish farming and timber harvesting.

In the United States, people have belatedly recognized the importance of coastal wetlands and have passed legislation to slow their destruction. For example, intact coastal wetlands help tame the tides more inexpensively than engineering structures such as retaining walls.

The Coastal Barrier Resources Act of 1982 abolished most federal assistance programs, including federal flood insurance, for new development ventures on undeveloped coastal barriers (Figure 17–17). This law has helped to eliminate some of the contradictions in governmental policies regarding coastal wetlands.

Figure 17–17
Overdevelopment of coastal areas often contributes to their degradation. The past availability of federally funded insurance against damage has contributed to improper development of these environmentally sensitive areas, such as this part of Westhampton Beach, Long Island. (Mark S. Wexler)

Coastal Demographics

Many coastal areas are over-developed, highly polluted, and overfished. Although more than 50 countries have coastal management strategies, their focus is narrow: such plans usually deal only with the economic development of a thin strip of land that directly borders the ocean. Coastal management plans rarely integrate the management of both land and off-shore waters, nor do they take into account the main reason for coastal degradation—human numbers.

Some 3.8 billion people—about two-thirds of the world's population—live within 150 kilometers (93 miles) of a coastline. Demographers project that three-fourths of all humans—perhaps as many as 6.4 billion—will live in that area by 2025. Many of the world's largest cities are situated in coastal areas, and these cities are currently growing more rapidly than noncoastal cities.

If the world's natural coastal areas are not to become urban sprawl or continuous strips of tourist resorts during the next century, coastal management strategies must be developed that take into account projections of human population growth and distribution. Such comprehensive management plans will be difficult to formulate and execute because they must regulate coastal development and prevent resource degradation, both on land and in offshore waters. The key to successful planning is local community involvement. If the public understands the importance of natural coastal areas, they may become committed to the sustainable development of coastlines.

AGRICULTURAL LANDS

The United States has more than 121 million hectares (300 million acres) of **prime farmland**—land that has the soil type, growing conditions, and available water to

NATIONAL MARINE SANCTUARIES

The United States currently has 13 national marine sanctuaries to protect unique natural resources and historical sites along the Atlantic, Pacific, and Gulf of Mexico coasts. Home to many diverse organisms, these sanctuaries include coral reefs in the Florida keys, fishing grounds along the continental shelf, and deep submarine canyons, as well as shipwrecks and other sites of historical value. The sanctuaries are administered by the National Marine Sanctuary Program, which is part of the National Oceanic and Atmospheric Administration. Like many federal lands, they are managed for multiple purposes, including conservation, recreation, education, scientific research, and historical value. Commercial fishing is permitted in most of them.

Figure 17–18

Homes built on agricultural land. Loss of farmland to urban and suburban development is a problem in certain areas of the country. (American Farmland Trust)

produce food, forage, fiber, and oilseed crops. Certain areas of the country have large amounts of prime farmland. For example, 90 percent of the Corn Belt, a corn-growing region in the Midwest encompassing parts of six states, is considered prime farmland. Not all prime farmland is used to grow crops; approximately one-third contains roads, pastures, rangelands, forests, feedlots, and farm buildings.

Traditionally, farming was a family business. However, the family farm is rapidly being replaced by larger agribusiness conglomerates that can operate more efficiently. In 1994 the Census Bureau reported that there are currently 1.9 million U.S. farms, as compared to 6.8 million in 1940. At the same time, the average farm size has been increasing. Between 1987 and 1992, for example, the average farm size increased from 187 to 199 hectares (462 to 491 acres).

There was considerable concern a few years ago that much of our prime agricultural land was falling victim to urbanization and suburban sprawl by being converted to parking lots, housing developments, and shopping malls. However, most agricultural land is not in danger of urban intrusion. This does not mean that rural land conversion is of no consequence. Unquestionably, both natural ecosystems and agricultural lands adjacent to urban areas are being developed (Figure 17–18). In certain areas of the country, loss of rural land, including prime agricultural land, is a significant problem (Table 17–3).

Table 17–3
The Top 12 Farm Areas Threatened by Population Growth and Urban/Suburban Spread

Farm Areas (in order of priority)
California's Central Valley
South Florida
California's coastal region
Mid–Atlantic Chesapeake (Maryland to New Jersey)
North Carolina Piedmont
Puget Sound Basin (Washington)
Chicago–Milwaukee–Madison metro (Illinois/Wisconsin)
Willamette Valley (western Oregon)
Twin Cities metro (Minnesota)
Western Michigan (Lake shore)
Shenandoah/Cumberland Valleys (Virginia to Pennsylvania)
Hudson River/Champlain Valleys (New York to Vermont)

SUBURBAN SPRAWL AND URBANIZATION

In Chapter 9 we discussed **urbanization**, the concentration of humans in cities, relative to human population growth. Urbanization and its accompanying suburban sprawl particularly affect land use. Prior to World War II, jobs and homes were concentrated in cities, but during the 1940s and 1950s, jobs and homes began to move from urban centers to the suburbs. New housing and industrial and office parks were built on rural land surrounding the city, along with a suburban infrastructure that included new roads, schools, and the like. Further development extended the edges of the suburbs, cutting deeper and deeper into the surrounding rural land and causing environmental problems such as loss of wetlands, loss of biological habitat, air pollution, and water pollution. Meanwhile, people that remained in the city and older suburbs found themselves increasingly isolated from suburban jobs and decent housing. This pattern of land use, which has continued to the present, has increased the economic disparity between older neighborhoods and newer suburbs.

There is an urgent need for regional planning involving government and business leaders, environmentalists, inner-city advocates, suburbanites, and farmers to determine where new development should take place and where it should not. Also, most metropolitan areas need to make more efficient use of land that has already been developed, including central urban areas.

THE GRASS IS NO LONGER GREENER

In the United States, the lawn has long been a sacred symbol of suburbia, but now it is a target of criticism. Critics charge that not only do U.S. lawns turn the varied American landscape into virtually identical neighborhoods across the country, they require billions of dollars to maintain, pollute the environment, eliminate potential biological habitat, and consume natural resources:

- In the mid-1990s, lawn owners annually spent $750 million on grass seed alone.
- More land is taken up by lawns than any single crop.
- Lawns are treated with ten times more chemical pesticides per acre than are farms.
- Thirty-two of 34 common lawn pesticides have not been tested for long-term effects on humans or the environment.
- In the East, 30 percent of cities' water is used for lawns; in the West, it is as much as 60 percent.

LAND USE

Many environmental concerns converge in the matter of land use. Pollution, population issues, preservation of our biological resources, mineral and energy requirements, and production of food are all tied to land use.

Overriding all of these concerns are economic factors, such as the way privately owned land is taxed. For example, sometimes forest or agricultural land that is located near urban and suburban areas is taxed as potential urban land. Because of the higher taxes on this land, its owners fall under greater pressure to sell it, which ultimately hastens its development. However, if such land is taxed as forest or farmland, the lower taxes are an incentive for owners to hold onto the land and maintain it in its undeveloped condition. Thus, land use is largely controlled by economic factors.

Public Planning of Land Use

Examine the use of land where you live. You may be surrounded by high rises and factories or by tree-lined streets interspersed with open parkland. Regardless of your surroundings, it is likely that they got that way by accident. Most areas have a land-use plan that includes zoning, but rarely do land-use plans take into account all aspects of land as a resource both before and after development. The philosophy of most land-use plans is that development is good because it increases the tax base, even though the revenue from these taxes is usually consumed providing services to the developed area.

Land-use decisions are complex because they have multiple effects. For example, if a tract of land is to be developed for housing, then roads, sewage lines, and schools must be built nearby to accommodate the influx of people. This usually results in the opening of restaurants and shopping areas, which take up more land.

Public planning of land use must take into account all repercussions of the proposed land use, not just its immediate effects. It is helpful to begin with an inventory of the land, including its soil type, topography, types of organisms, endangered or threatened species, and historical or archaeological sites.

At this stage, the public planning commission attempts to understand the value of the land *as it currently exists*, as well as its potential value after any proposed change. In addition to providing people with open space for recreation and mental health, undeveloped land provides environmental services that must be recognized. All of these benefits should be compared with the possible economic benefits of development. In the long term, the best use of land may not be the use that provides immediate economic gain.

If the land will ultimately be developed, the development plan should be comprehensive. It should indicate which areas will remain open space, which will remain agricultural, and which will be zoned for high-, medium-, and low-density housing.

The Wise-Use Movement

How do we best manage the legacy of federal lands? Should federal lands be exploited so they benefit a handful of people today, or should they be conserved so they benefit the American people for generations to come? These questions have divided Americans into two groups, those who wish to exploit resources on federal lands now (a coalition of several hundred grassroots organizations known collectively as the wise-use movement) and those who wish to conserve the resources on federally owned lands (environmentalists).

People who support the **wise-use movement** think that the government has too many regulations protecting the environment and that property owners should be freed from the requirements of environmental laws. They believe the primary purpose of federal lands is to enhance economic growth. Some of their goals include the following:

1. Log all national forests, including old-growth forests.
2. Permit mining and commercial development of wilderness areas, wildlife refuges, and national parks.
3. Allow unrestricted development of wetlands.
4. Change the Endangered Species Act so that economic factors overrule scientific ones (recall from Chapter 16 that the definition of threatened and endangered species is currently based on scientific information only).
5. Sell resource-rich federal lands to private interests such as mining, oil, coal, ranching, and timber groups.

Many of the organizations that embrace the wise-use movement have environmentally friendly names. The National Wetlands Coalition, for example, consists primarily of real estate developers and oil and gas companies who wish to drain and develop wetlands. Similarly, logging companies support the American Forest Resource Alliance.

In contrast to the wise-use movement, the **environmental movement** views federal lands as a legacy of the American people. They think that:

1. The primary purpose of public lands is to protect biological diversity and ecosystem integrity.
2. Those who extract resources from public lands should pay the American people compensation equal to the fair market value of the resource and not be given any taxpayer-paid subsidies.
3. Those who use public lands should be held accountable for any environmental damage they cause.

CONSERVATION OF OUR LAND RESOURCES

Humans use an estimated 4.7 billion hectares (11.6 billion acres) for agriculture—that is, for raising crops and domestic animals. This area equals approximately 32 percent of the world's total land area (Figure 17–19). Another 4.4 billion hectares (11 billion acres)—about 30 percent of the land surface—are rock, ice, tundra, or desert and are considered unsuitable for long-term human use. This leaves 37 percent of the land surface as natural ecosystems—forests and grasslands—that could potentially be developed for human purposes. We have seen, however, that these natural ecosystems provide many valuable environmental services that are important to human survival.

Our ancestors considered natural areas as an unlimited resource to exploit. They appreciated prairies as valuable agricultural lands and forests as immediate sources of lumber and eventual farmlands. This outlook was practical as long as there was more land than people needed. But as the population increased and the amount of available land decreased, it was necessary to consider land as a limited resource. Increasingly, the emphasis has shifted from exploitation to preservation of the remaining natural areas.

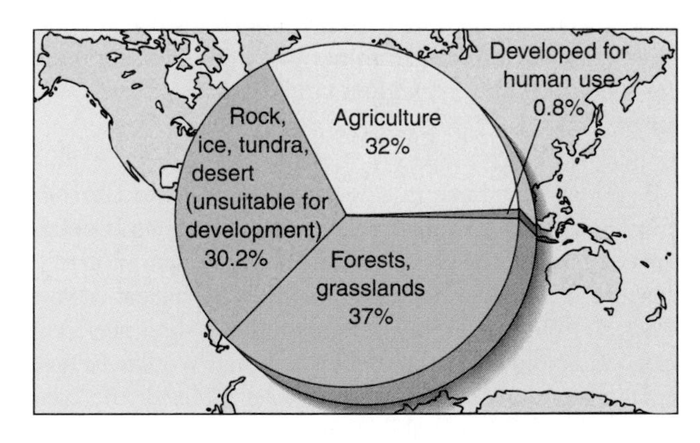

Figure 17–19
Worldwide land use.

Although all types of ecosystems must be conserved, several are in particular need of protection. Deforestation has become an international problem, along with desertification and erosion. The Defenders of Wildlife developed a numerical ranking of the most endangered ecosystems in the United States based on four criteria. These are: (1) the area lost or degraded since Europeans colonized North America, (2) the number of present examples of a particular ecosystem, or the total area, (3) an estimate of the likelihood that a given ecosystem will lose a significant area or be degraded during the next ten years, and (4) the number of threatened and endangered species living in that ecosystem. Table 17–4 lists the 21 most endangered ecosystems based on these criteria.

Government agencies, private conservation groups, and private citizens have begun to set aside natural areas for permanent preservation. Unfortunately, different federal and state policies on land use have often contradicted one another. For example, some programs are geared toward preserving wetlands, while other projects encourage drainage and development. Agricultural price supports boost the profits from food produced on converted wetlands, and farmers are encouraged to drain wetlands by federally supported, low-interest loans and technical assistance.

Another example of how government agencies sometimes work at cross purposes involves the national forests. One directive of the USFS is to oversee the harvest of timber from public forests. The FWS, however, has a legal mandate to maintain biological diversity, including endangered species such as the northern spotted owl, whose critical habitat is the same old-growth forest that may be harvested under the USFS's administration.

Table 17–4
The Top 21 Most Endangered Ecosystems in the United States

Ecosystems (in order of priority)
South Florida landscape
Southern Appalachian spruce-fir forest
Longleaf pine forest and savanna
Eastern grassland, savanna, and barrens
Northwestern grassland and savanna
California native grassland
Coastal communities in lower 48 states and Hawaii
Southwestern riparian forest
Southern California coastal sage scrub
Hawaiian dry forest
Large streams and rivers in lower 48 states and Hawaii
Cave and karst systems
Tallgrass prairie
California riparian forest and woodland
Florida scrub
Ancient eastern deciduous forest
Ancient forest of the Pacific Northwest
Ancient red and white pine forest, Great Lake states
Ancient ponderosa pine forest
Midwestern wetland
Southern forested wetland

SUMMARY

1. Natural areas provide us with many environmental services, including watershed management, soil erosion protection, climate regulation, and biological habitat. People benefit directly from natural areas by using them for scientific study, recreation, and renewal of the human spirit.

2. Areas that have not been greatly disturbed by human activities are called wilderness. Public lands designated as part of the National Wilderness Preservation System are protected from development. National parks have multiple roles, including recreation, ecosystem preservation, and biological habitat. The National Wildlife Refuge System contains lands and waters set aside for the conservation of fishes, wildlife, and plants of the United States.

3. Forests provide many environmental services, as well as commercially important timber. Forests that are intensively managed for commercial harvest have little species diversity. Trees may be harvested by selective cutting, shelterwood cutting, seed tree cutting, and clearcutting. Selective cutting and shelterwood cutting have less negative impact on forests but are not as economically profitable in the short term as the other methods.

4. The greatest problem facing world forests today is deforestation, which is the temporary or permanent removal of forests for agriculture or other uses. In the tropics, forests are destroyed to (1) provide colonizers with temporary agricultural land, (2) obtain timber, particularly for developed nations, (3) provide open rangelands for cattle, and (4) supply fuel. Extensive deforestation of cer-

tain boreal forests began in the late 1980s and early 1990s. Boreal forests are currently the world's primary source of industrial wood and wood fiber.

5. In rangelands, grasses, forbs, and shrubs predominate. Rangelands are often grazed by cattle, sheep, goats, and other domestic mammals. Provided the number of animals grazing on a particular area of rangeland is kept below the area's carrying capacity, the rangeland remains a renewable resource. However, overgrazing can result in barren, exposed soil that is susceptible to erosion. Overuse of rangelands or dry forests contributes to desertification, the development of unproductive desert-like conditions on formerly productive land.

6. Wetlands are transitional areas between aquatic and terrestrial ecosystems and may be either freshwater or coastal (saltwater). Wetlands provide habitat for many organisms, purify natural bodies of water, and recharge groundwater. Despite their environmental contributions, wetlands are often drained or dredged for other purposes. In the United States, wetlands are primarily converted to agricultural land.

7. Agricultural lands are former forests or grasslands that have been plowed for cultivation. In certain areas, agricultural lands are threatened by expanding urban and suburban areas. Increasingly, farming is done on large tracts of land by agribusiness conglomerates, which operate more efficiently than family farmers on small pieces of land.

8. The enlightened view of the interaction between humans and land is that humans are stewards of the land, managing it to achieve sustainable use. It is possible to become involved in local land-use planning by becoming familiar with land-use policies and laws.

9. Americans are divided over whether federal lands should be exploited now or conserved for future generations. The wise-use movement thinks the government has too many regulations that protect the environment, and the primary purpose of federal lands is to enhance economic growth. The environmental movement thinks that federal lands are a legacy of the American people that should be preserved to protect biological diversity and ecosystem integrity.

THINKING ABOUT THE ENVIRONMENT

1. Do you think additional lands should be added to the wilderness system? Why or why not?

2. Give at least five environmental benefits of nonurban lands. Why is it difficult to assign economic values to many of these benefits?

3. Suppose a valley contains a small city surrounded by agricultural land. The valley is encircled by mountain wilderness. Explain why the preservation of the mountain ecosystem would help support both urban and agricultural land in the valley.

4. Explain the relationship between eating a hamburger at a fast-food restaurant and tropical deforestation.

5. Some analysts think that an effective way to combat the influx of unwanted immigrants to such countries as the United States is to curb deforestation. Explain the connection.

6. How do the activities of the Trout Creek Mountain Working Group qualify as ecosystem management?

7. Explain how certain tax and zoning laws can increase the conversion of prime farmland to urban and suburban development.

8. Should private landowners have the right to do what they wish to their land, no matter how it affects the public? Present arguments for both sides of this issue.

* 9. Not counting agriculture, about 0.8 percent of the Earth's total surface area of 14.7 billion hectares is settled—that is, developed for homes, highways, industrial parks, shopping malls, and the like. Calculate the total land area, in hectares, that is settled. Convert your answer to acres, using Appendix IV to help you.

RESEARCH PROJECTS

Review the land-use plan for your community and assess how its policies will impact the environment for the next ten years.

In 1994, declining rates of deforestation in the Amazon Basin were attributed to the international attention bestowed on the problem by environmental groups. The victory was short-lived, however, because one year later, virgin rain forests were rapidly being cleared in parts of Brazil, Bolivia, Paraguay, and Uruguay for agriculture and development. Investigate why widespread deforestation has returned to the Amazonian Basin.

Investigate the tropical rain forests of Hawaii. What are the most serious problems facing them?

Some federal laws command the U.S. Department of the Interior to exploit the natural resources—forests, minerals, oil, and water—on federally owned lands. Other laws order the Department to protect and conserve these same resources for the long-term public good. Investigate how the current leadership of the Department of the Interior is handling this dichotomy.

On-line materials relating to this chapter are on the World Wide Web at **http://www.saunderscollege.com/lifesci/environment2** (select Chapter 17 from the Table of Contents).

SUGGESTED READING

Acharya, A. "Plundering the Boreal Forests," *World-Watch*, May–June 1995. Describes the current massive deforestation of forests in Canada and Siberia.

Chadwick, D. H. "Sanctuary: U.S. National Wildlife Refuges," *National Geographic* Vol. 190, No. 4, October 1996. Examines the nation's wildlife refuges and some of their problems, such as budget woes and conflicting secondary uses.

Doherty, J. "The Arctic National Wildlife Refuge: The Best of the Last Wild Places," *Smithsonian*, March 1996. Discusses the natural riches of the Arctic National Wildlife Refuge.

Hinrichsen, D. "Coasts in Crisis," *Issues in Science and Technology*, Summer 1996. This article summarizes the problems of coastal development and destruction.

Kusler, J. A., W. J. Mitsch, and J. S. Larson. "Wetlands," *Scientific American*, January 1994. Explains why wetlands are so difficult to recognize and preserve.

Mann, C. C. and M. L. Plummer. "Are Wildlife Corridors the Right Path?" *Science* Vol. 270, December 1, 1995. Although many ecologists support the establishment of wildlife corridors, some think that the concept has been overrated.

Mitchell, J. G. "Legacy at Risk," *National Geographic* Vol. 186, No. 4, October 1994. America's national parks are suffering from too much popularity.

Parfit, M. "California Desert Lands: A Tribute to Sublime Desolation," *National Geographic*, May 1996. Dramatic photographs accompany this article on the nation's newest national parks and national preserve.

Riebsame, W. E. "Ending the Range Wars?" *Environment* Vol. 38, No. 4, May 1996. Examines the often heated debates between environmental groups and ranchers and their supporters.

Sugal, C. "Labeling Wood: How Timber Certification May Reduce Deforestation," *WorldWatch* Vol. 9, No. 5, September–October 1996. Increasingly, the United States and Europe are demanding that products containing tropical woods come from sustainably harvested forests.

Zimmer, C. "How to Make a Desert," *Discover*, February 1995. Ecologists are studying the factors that contribute to desertification.

Walnut Acres Farm in Penn's Creek, Pennsylvania. (© Scott Goldsmith)

Food Resources: A Challenge for Agriculture

Since 1946 farmers Paul and Betty Keene have been producing organic foods at Walnut Acres in central Pennsylvania. At a time when other farmers were adopting highly mechanized operations that depended on commercial fertilizers and pesticides for crops and hormones and antibiotics for livestock, the Keenes kept their farming methods natural. They use animal manure and plant wastes to fertilize the soil and beneficial insects to control pests. Crop rotations and careful crop selections also help control insect pests and maintain soil fertility. Every fourth or fifth year they plant a cover crop, such as alfalfa or clover, and let the soil rest; the cover crop is tilled into the soil before the next growing season.

When the Keenes first started farming at Walnut Acres, they had 108 acres. Today Walnut Acres, one of the oldest, most respected organic farms in the country, consists of more than 100 people working 600 acres. The farm, which includes a cannery, bakery, mill, warehouse, and retail store, raises grains, such as wheat, oats, rye, soybeans, and corn; vegetables, such as sweet corn, tomatoes, peas, green beans, zucchini, cabbage, and beets; and livestock, such as cattle and chickens. Sales from its mail-order organic food company, which is the largest in the country, totaled $7 million in 1995. Organic farming, once on the agricultural fringe, has become a $2.3 billion a year industry, and more than 5300 farmers practice organic methods in the United States.

What are natural foods and organic foods? Typically, *natural foods* are any foods that are not highly processed and do not contain synthetic preservatives, artificial colors and flavors, refined sugars and synthetic sweeteners, or hydrogenated oils. Some natural foods are also organic. According to guidelines established by the Organic Food Production Act in 1990, *organic foods* are crops that are grown in soil that has been free of chemical fertilizers and pesticides for at least three years. If the land that the crops are grown on has been inspected, it can be labeled *certified organic*, which specifies the highest standards. Cattle and chickens that are labeled *free range* or *naturally raised* are raised in open pastures or fields rather than in cramped pens and are not treated with antibiotics (to control infections in such crowded conditions) or hormones (chemical messengers that regulate many of the body's functions, such as increasing body weight or milk output).

Because organic farming is more labor intensive and produces food on smaller quantities of scale than conventional agriculture, organic food is 5 percent to 30 percent more expensive. Although the appeal of organically grown food is undeniable, it must be remembered that the chemicals used in conventional agriculture allowed American agriculture to become the most efficient and productive food production system in the world. However, many environmental problems are associated with conventional agriculture, and there are concerns about whether conventional agriculture is sustainable.

The production of food in an environmentally sustainable way is one of the principal challenges facing humanity today. In this chapter we examine the magnitude and nature of the world's food problems. Most farmers in developing countries usually produce barely enough food to feed themselves and their families, with little left over as a reserve. Developed countries, for their part, have energy-intensive agricultural methods that produce high yields of food but cause serious environmental problems such as soil erosion and pollution. Overshadowing and exacerbating all food problems is the increase in the human population. The production of adequate food for the world's people will be an impossible goal until population growth is brought under control. ◄

HUMAN NUTRITIONAL REQUIREMENTS

Like other animals, humans must obtain their nutrients by consuming other organisms. The foods humans eat are composed of several major types of organic molecules that are necessary to maintain health—carbohydrates, proteins, and lipids. **Carbohydrates**, organic molecules such as sugars and starches, are important primarily because they are metabolized readily by the body in **cell respiration**, a process in which the energy of organic molecules is transferred to a molecule called adenosine triphosphate (ATP). The body uses the energy in ATP to contract muscles, produce heat, repair damaged tissues, grow, fight off infections, and reproduce. In other words, carbohydrates supply the body with the energy required to maintain life.

Proteins are large, complex molecules composed of repeating subunits called **amino acids**. Proteins perform several critical roles in the body. First, when the plant and animal proteins in food are digested, the body absorbs the amino acids, which are then reassembled in different orders to form human proteins. A substantial part of the human body, from hair and nails to muscles, is made up of protein. In addition, proteins are metabolized in cell respiration to release energy.

There are approximately 20 different amino acids required for human nutrition. The human body manufactures 12 of these for itself, using starting materials such as carbohydrates. However, human cells lack the ability to synthesize the other eight amino acids, called **essential amino acids**.[1] These must be obtained from food.

Lipids, a diverse group of organic molecules that includes fats and oils, are metabolized by cell respiration to provide the body with a high level of energy. Pound for pound, lipids deliver more energy when metabolized than either carbohydrates or proteins. In addition, lipids have several important roles in the body; some lipids are hormones, and others are essential components of cell membranes.

In addition to carbohydrates, proteins, and lipids, we humans require minerals, vitamins, and water in our diets. **Minerals** are inorganic elements, such as iron and calcium, and are essential for the normal functioning of the human body. Minerals are ingested in the form of salts dissolved in food and water. **Vitamins** are complex organic molecules that are required in very small quantities by living cells. Vitamins help to regulate metabolism and the normal functioning of the human body. Whereas plants synthesize most vitamins, humans and other animals must obtain vitamins from food.

WORLD FOOD PROBLEMS

In 1996 about 840 million people lacked access to the food needed to be healthy and lead productive lives. The two regions of the world with the greatest food insecurity are South Asia, with an estimated 270 million hungry people in 1995, and sub-Saharan Africa,[2] with an estimated 175 million.

[1]The eight essential amino acids are isoleucine, leucine, lysine, methionine, phenylalanine, threonine, tryptophan, and valine.

[2]Sub-Saharan Africa refers to all African countries located south of the Sahara Desert.

The average adult human must consume enough food to get approximately 2600 kilocalories per day.[3] If a person consumes less than this over an extended period of time, his or her health and stamina decline, even to the point of death. People who receive fewer calories than they need are said to be **undernourished**. Worldwide, an estimated 185 million children under the age of 6 are seriously underweight for their age.

The total number of calories consumed is not the only measure of good nutrition, however. People can receive enough calories in their diets but still be **malnourished** because they are not receiving enough of specific, essential nutrients such as proteins or vitamins. For example, a person whose primary food is rice can obtain enough calories, but a diet of rice lacks sufficient amounts of proteins, lipids, minerals, and vitamins to maintain normal body functions. Adults suffering from malnutrition are more susceptible to disease and have less strength to function productively than those who are well fed. Children who are malnourished do not grow or develop normally, and because malnutrition affects cognitive development, they do not do as well in school as children who are well fed.

The two most common diseases of malnutrition are marasmus and kwashiorkor. **Marasmus** (from the Greek word *marasmos*, meaning "a wasting away") is progressive emaciation caused by a diet low in both total calories and protein. Marasmus is most common among children in their first year of life—particularly children of very poor families in developing nations. Symptoms include a pronounced slowing of growth and extreme atrophy (wasting) of muscles. It is possible to reverse the effects of marasmus with an adequate diet.

Kwashiorkor (a native word in Ghana, meaning "displaced child") is malnutrition resulting from protein deficiency. It is common among children in all poor areas of the world. The main symptoms include edema (fluid retention); dry, brittle hair; apathy; stunted growth; and sometimes mental retardation. One of the most typical features of kwashiorkor is a pronounced swelling of the abdomen (Figure 18–1). Kwashiorkor can be treated by gradually restoring a balanced diet.

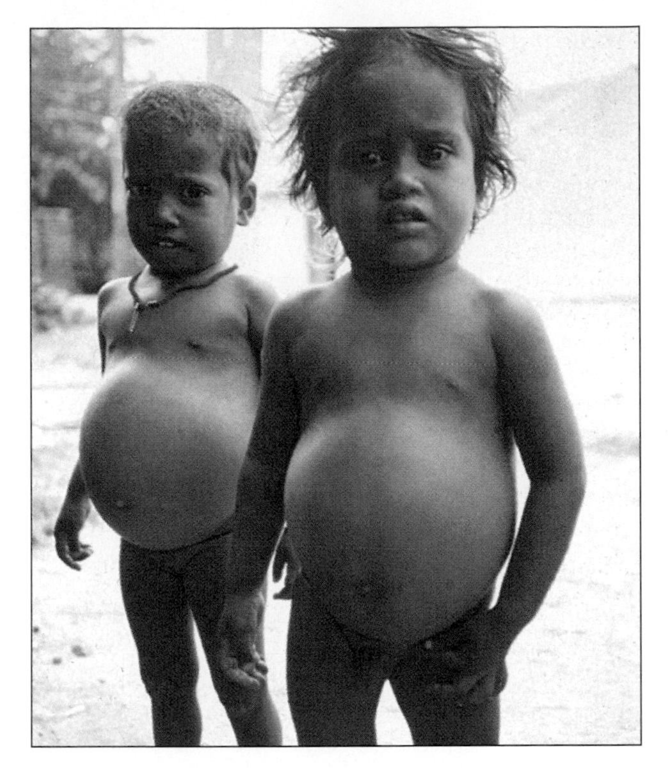

Figure 18–1
Millions of children suffer from kwashiorkor, a disease caused by severe protein deficiency. Note the characteristic swollen belly, which results from fluid imbalance. (U.N. Food and Agricultural Organization, photo by P. Pittet)

Crop failures caused by drought, war, flood, or some other catastrophic event may result in **famine**, a severe food shortage. Throughout human history, famine has struck one or more regions of the world every few years. The developing nations of Africa, Asia, and Latin America are most at risk. The worst African famine in history, which was caused in part by widespread drought, occurred from 1983 to 1985. Hardest hit were Ethiopia and Sudan, in which 1.5 million people died of starvation. The people living in this region lacked sufficient money to purchase food and did not have stored food reserves to protect them against several bad years of crop failures. More recently, drought and civil unrest in Somalia resulted in famine for an estimated 2 million Somalis. In 1993 the United Nations sent soldiers from 11 nations on a humanitarian mission to Somalia to stop warring factions from stealing food from relief missions. Famines get a great deal of media attention because of the huge and obvious amount of human suffering they cause. However, many more people die worldwide from undernutrition and malnutrition than from the starvation associated with famine.

Eating food in excess of that required is called **overnutrition**. Generally, a person suffering from overnutrition has a diet high in saturated (animal) fats, sugar, and salt. Overnutrition results in obesity, high blood pressure, and an increased likelihood of such disorders as diabetes and heart disease. In nutrition experiments, overnutrition in rodents resulted in a higher incidence of cancer compared to rodents fed a calorie-restricted diet, but evidence linking overnutrition to cancer in humans is sparse. Many human studies show a correlation between diets high in animal fat and red meat and certain kinds of cancer (colon and prostate). Overnutrition is most common among people in developed nations, although it is also emerging in some developing countries, particularly in urban areas. As people in developing countries earn more money, their diets shift from primarily cereal grains to more processed foods and livestock products.

[3]The average man requires 3000 kilocalories per day, whereas the average woman requires 2200 kilocalories per day.

Producing Enough Food

Producing enough food to feed the world's people is the largest challenge in agriculture today, and the challenge grows more difficult each year because the human population is continually expanding (Figure 18–2). Some demographers project an *annual* increase in the human population of 90 million *for the next 25 years*. Almost all of this growth will occur in developing countries.

Currently, 1.75 billion metric tons (1.93 billion tons) of grains such as wheat, corn, rice, and barley are required to feed the world's population for one year. Each year, an additional 28.5 million metric tons (31.4 million tons) of grain must be produced to account for that year's increase in population. During the 1980s and early 1990s, increases in food production did not keep pace with population growth in more than 50 developing countries. Although global food production can be increased in the short term, one fact remains: *the long-term solution to the food supply problem is control of the human population.*

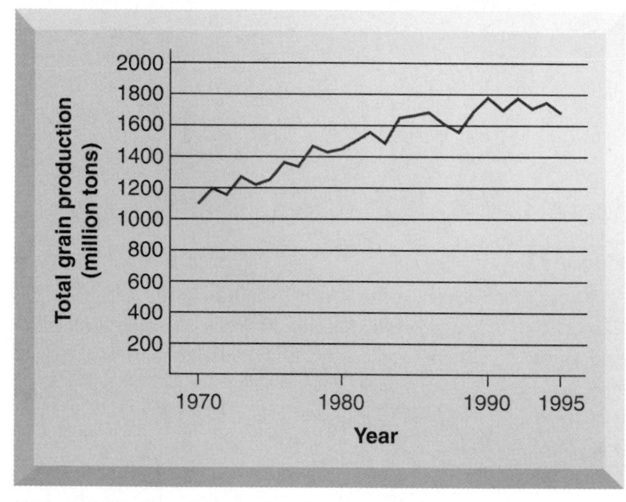

(a)

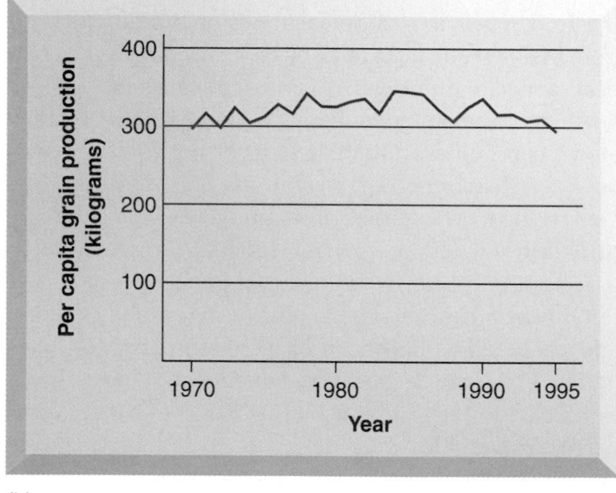

(b)

Figure 18–2

A comparison of (a) total world grain production and (b) grain production per person, 1970 to 1995. Food production has increased greatly over the past 25 years, but because the human population has also expanded, the amount of food produced per person has not increased.

Maintaining Grain Stockpiles

World grain carryover stocks are the amounts of rice, wheat, corn, and other grains remaining from previous harvests, as estimated at the start of a new harvest. World grain carryover stocks provide a measure of world food security. The amount of grain stockpiled in 1995 declined to 296 million tons, which is equivalent to feeding the world's people for 61 days (Figure 18–3). The United Nations issued a warning that these stockpiles, the lowest in two decades, are far below the minimum amounts needed to safeguard world security. When food is scarce and prices increase, the risk of political instability is a real concern in poor nations. Some experts think that it will take at least two to three bountiful harvests to rebuild stocks. Others are more pessimistic about the low grain stocks and think that they indicate the beginning of a period of agricultural scarcity and higher food prices.

There are two main reasons that world grain stocks have dropped in the past few years. One, bad weather has caused poor harvests in the United States, Russia, South Africa, and Latin America. Two, consumption of beef, pork, poultry, and eggs has increased, particularly in China, where people are becoming more affluent and can afford to diversify their diets (Figure 18–4). The increased consumption of meat and meat products in China, where one out of five of the world's people lives, has prompted

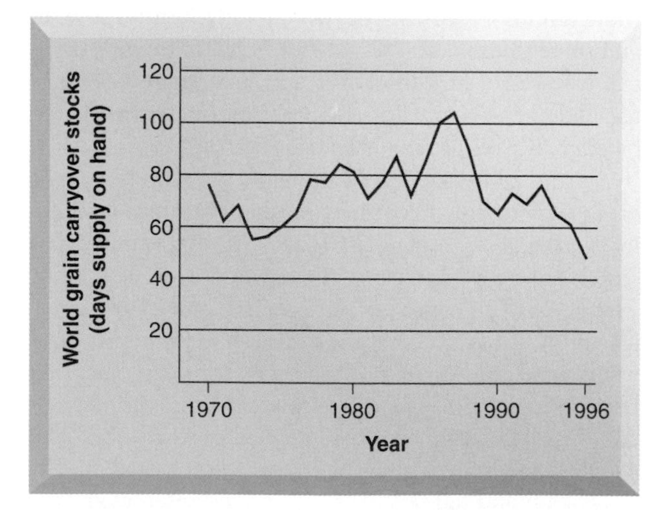

Figure 18–3

World grain carryover stocks, expressed as the number of days' supply on hand, 1970 to 1996.

Figure 18–4
A meat market in Guangzhou (Canton), China. As their incomes rise, people in China and many other Asian countries are eating more meat and meat products. Because livestock animals eat grains, the recent demand for meat has contributed to the decline in world grain carryover stocks.
(H. Donnezan/Photo Researchers, Inc.)

a surge in grain used to feed livestock. China was once one of the world's leading grain exporters, but since 1993 it has had to import grain. The decline in world grain stockpiles has sent prices soaring. U.S. corn prices increased, for example, from $2.11 a bushel at the end of 1994 to $3.69 in January 1996. Wheat prices in 1995 were 26 percent higher than in 1993.

Poverty and Food: Making Food Affordable for the Poor

The main cause of undernutrition and malnutrition is poverty. The world's poorest people—those living in developing countries in Asia, Africa, and Latin America—do not own land on which to grow food and do not have the money to purchase food. Over 1.1 billion people in developing countries have incomes equivalent to $1 dollar a day or less per person and are so poor that they cannot afford to eat enough food or enough of the right kinds of food.

Poverty and hunger are not restricted to developing nations, however. A task force of doctors in 1985 reported that 1 out of every 11 Americans was malnourished or undernourished and that at least half of these hungry Americans were children.

Cultural and Economic Effects on Human Nutrition

Even if the major food-producing countries could produce enough grain to feed the rest of the world, there would still be an enormous economic problem. It costs money to produce, transport, and distribute food. Asian, African, and Latin American countries, which have the greatest need for imported food, are least able to pay for it. On the other hand, the food-producing nations cannot afford to simply give food away and absorb the costs indefinitely. In addition, government inefficiency and bureaucratic red tape can add to food problems, sometimes making it difficult to distribute the food to the hungriest people and to ensure that it is not eaten by those who do not need it.

Thus, getting food to the people who need it is mainly an economic problem, and the solution is obvious, although not necessarily easy to implement: food must be produced in the area where it is to be consumed. If food production matches market demands in developing countries, not only are people fed, but economic growth occurs as agriculture generates incomes and employment.

Cultural acceptance of food is another important food problem. If, for example, a particular group of people have been eating rice for hundreds of generations, it is unlikely that they will eat corn even if they are very hungry. This may sound strange, but think how reluctant you might be to eat foods such as insect grubs or dog meat, which are exotic to you but are common fare in Africa and Asia (Figure 18–5). It is human nature to be suspicious of foods to which we are unaccustomed.

Thus, world food problems are many, as are their solutions. We need to increase the production of food, improve overall economic development, improve food distribution, and overcome cultural barriers to acceptance of foods. Developed nations can help developing countries become agriculturally self-sufficient by providing economic assistance and technical aid. But the ultimate solution to world hunger is tied to achieving a stable population in each nation at a level that it can support.

Figure 18–5
Mopani, a high-protein food in South Africa, are caterpillars of the emperor moth. (Anthony Bannister/NHPA)

Having a general grasp of world food problems, we now examine the history of agriculture from its beginnings to the present and consider some of the complex challenges facing agriculture today.

HISTORY OF AGRICULTURE

The earliest human societies were family and tribal units that relied on hunting and gathering to obtain food. This type of society was continually on the move, for they left an area as soon as it was depleted of food resources. The nomadic existence of hunter-gatherer groups prevented them from establishing large, materialistic societies.

The development of agriculture—the raising of plants and animals for food—approximately 10,000 years ago had significant effects on human societies. Agriculture provided a more reliable food supply, so a society's population could grow without depleting food in the area. In addition, cultivation of plants required that people give up their nomadic existence. This contributed to the establishment of permanent dwellings and eventually to the growth of villages and cities.

Although agriculture was more labor-intensive than hunting and gathering, not every member of the tribe had to be directly involved in obtaining food. In an agricultural society, the farmers produced enough to feed both themselves and some additional people, who could pursue other endeavors and barter with the farmers to obtain their food. This new freedom ultimately led to the development of the arts and sciences, trades, and other aspects of human culture. The success of agriculture in supporting people is evidenced by the fact that few societies reverted to hunting and gathering once they developed agriculture.

Where Did Agriculture Begin?

The first hunter-gatherer groups are thought to have formed settlements about 10,000 years ago. Evidence suggests that grain crops were the first foods to be grown, although a few early settlements may have raised goats, pigs, or other animals before they learned to grow plants.

Archaeological evidence supports the idea that plants were first domesticated in several different regions of the world. It seems likely that agriculture evolved independently in three main places: the Near East center, the Central/South America center, and the Far East center.

The **Near East center**, along the eastern end of the Mediterranean Sea (present-day Iran, Iraq, Syria, and Turkey), is sometimes called the *Fertile Crescent* or the *cradle of agriculture*. There is evidence that wheat was domesticated here from wild grasses as early as 10,000 years ago.

The maize (corn) culture developed in the **Central/South America center**. Cultivation of maize originated in western Mexico, where its wild grass ancestor teosinte is still found, and there is evidence of maize, squash, chili pepper, bean, and gourd cultivation in the central highlands of Mexico as early as 5000 years ago.[4] People living in the highlands of Peru, however, may have had agriculture even earlier.

Rice was domesticated in the **Far East center**, perhaps along the Yangtze River in central China. Rice grains and husks from several sites in the area have been dated, using radiocarbon methods, as 11,500 years old. Evidence also exists that plants other than rice—possibly peas or some type of beans—were domesticated in the Far East center as early as 9000 years ago.

Agriculture may have developed even earlier than the dates associated with wheat, maize, and rice cultivation. Wheat, maize, and rice are all propagated by seeds and require relatively sophisticated agricultural knowledge to cultivate. Growing other plants, such as bananas, yams, and potatoes, by vegetative propagation is easier and requires less agricultural knowledge. We have no way of knowing whether these plants were cultivated earlier than wheat, maize, or rice, because their propagation

[4]Radiocarbon dating of the oldest fossil corn kernels and cobs during the 1970s and 1980s established them as 7000 years old. The accelerator mass spectroscopy dating methods of the 1990s, however, place these samples at 5000 years old.

would have occurred in moist areas where archaeological evidence could not be preserved.

Other Early Advances in Agriculture

Agricultural societies in the Near East probably began to irrigate their fields more than 5000 years ago. There is also evidence of an extensive irrigation system in the Andes Mountains of Bolivia as early as 3500 years ago. Irrigation in arid and semiarid regions made it possible to produce larger crop yields on a given amount of land. This meant that fewer acres had to be cultivated to feed a population and, therefore, even fewer people had to be involved in producing food. As a result, more people were free to pursue the arts, religion, and trades. Another important advance in agriculture came with the use of domesticated animals to prepare fields for planting.

What Is a Domesticated Plant?

A plant that is **cultivated** is protected from natural competition with other plants and from plant-eating animals. The farmer selects desirable traits in a cultivated plant species and encourages their transmission by choosing the seeds of the plants having those traits to save for planting. The plants without the desired traits are not allowed to reproduce, so over time, such artificial selection brings about changes in the genetic makeup of a strain of plants. They may become so different from their indigenous ancestors that it is doubtful they could survive and compete successfully in the wild. Such plants are said to be **domesticated**.

The most extreme example of plant domestication is corn. The ears of maize found at the oldest archaeological sites were scarcely 1 inch long. Selective breeding over thousands of years is responsible for the development of large, compact ears covered by husks that prevent the seeds from being easily dispersed. Teosinte, the wild grass from which maize evolved, still exists in the highlands of Mexico but barely resembles modern-day corn. Its seeds, for example, are borne not compactly on a covered ear, but loosely on an uncovered stalk.

AGRICULTURE TODAY

If you were to travel around the world, you would find many different kinds of agriculture and types of food. Despite this diversity, there has been an overall trend—especially during the 20th century—toward greater uniformity in the plants and animals we eat; humans have come to rely on fewer and fewer types of plants and animals for the bulk of food production. There has also been an overall trend toward uniformity in agricultural practices, as farmers in developing nations have adopted techniques used in developed countries.

Plants and Animals That Stand Between People and Starvation

Although we do not know precisely how many different plant species exist, biologists estimate that there are about 250,000. Of these, just over 100 provide about 90 percent of the food that humans consume, either directly or indirectly. (Humans consume foods indirectly when cereal grains are used to feed livestock that humans eat as meat.) Just three species of plants—wheat, rice, and corn—provide the bulk of food for humans (Table 18–1).

Table 18–1
The 30 Most Important Food Crops

Food, in Descending Order of Production	Type of Crop
1. Wheat	Cereal grain
2. Rice	Cereal grain
3. Corn	Cereal grain
4. Potato	Ground crop*
5. Barley	Cereal grain
6. Sweet potato	Ground crop
7. Cassava (manioc)	Ground crop
8. Grape	Fruit, wine
9. Soybean	Legume
10. Oat	Cereal grain
11. Sorghum	Cereal grain
12. Sugarcane	Sugar plant
13. Millet	Cereal grain
14. Banana	Fruit
15. Tomato	Fruit
16. Sugar beet	Sugar plant
17. Rye	Cereal grain
18. Orange	Fruit
19. Coconut	Seed
20. Cottonseed	Oil plant
21. Apple	Fruit
22. Yam	Ground crop
23. Peanut	Legume
24. Watermelon	Fruit
25. Cabbage	Leaf crop
26. Onion	Ground crop
27. Bean	Legume
28. Pea	Legume
29. Sunflower seed	Oil plant
30. Mango	Fruit

*Ground crops include root, tuber, and bulb crops.

There are tens of thousands of kinds of plants that have been used as sources of food at one time or another. Many of these doubtless could be developed into important sources of food. We need to identify them, find out how to use them, and study their cultivation requirements.

Our dependence on so few species of plants for the bulk of our food puts us in an extremely vulnerable position. Should disease or some other factor wipe out one of the important food crops, humans might be threatened by severe famine.

Animals provide us with foods that are particularly rich in protein. These foods include fishes, shellfish, meat, eggs, milk, and cheese. Cows, sheep, pigs, chickens, turkeys, geese, ducks, goats, and water buffalo are the most important types of livestock. Although nutritious, livestock is an expensive source of food because animals are inefficient converters of plant food. For example, of every 100 calories of plant material a cow consumes, it burns off approximately 90 in its normal metabolic functioning. That means that only 10 calories out of 100 (10 percent) are stored in the cow to be consumed by humans. Meat consumption is high in affluent societies, so large portions of the crops grown and the fishes harvested in developed countries are used to produce livestock animals for human consumption. For example, almost half of the cereal grains grown in developed countries are used to feed livestock (see You Can Make a Difference: Vegetarian Diets, page 410).

The Principal Types of Agriculture

Agriculture can be roughly divided into two types: high-input agriculture and subsistence agriculture. Most farmers in developed countries and some in developing countries practice **high-input agriculture**, also called **industrialized agriculture**. It relies on large inputs of energy, in the form of fossil fuels, to produce and run machinery, irrigate crops, and produce chemicals such as fertilizers and pesticides (Figure 18–6a). High-input agriculture produces high yields of food per unit of farmland area, but not without costs. Several problems, such as soil degradation and increases in pesticide resistance in agricultural pests, are caused by high-input agriculture.

Figure 18–6
Three types of agricultural systems. (a) Industrialized agriculture requires high inputs of fossil fuels and money (to pay for expensive equipment, fertilizers, pesticides, and such). (b) Shifting agriculture and (c) nomadic herding are land-intensive forms of subsistence agriculture.

Figure 18–7
A livestock factory for chickens in Texas. The chickens remain indoors and are fed and watered by machine. Although livestock factories are very efficient and produce meat relatively inexpensively, they also cause environmental problems. (Paul S. Howell/Gamma Liaison)

High-input agriculture has favored the replacement of traditional family farms by large agribusiness conglomerates (see Chapter 17). In the United States, for example, most cattle, hogs, and poultry are now grown in feedlots and livestock factories (Figure 18–7). In livestock factories, the animals are confined to small pens in buildings the size of football fields. Such large concentrations of animals create many environmental problems, including air and water pollution. For example, the quantity of manure produced by several hundred thousand pigs causes a severe waste disposal problem. At hog factories, the manure is often stored in deep lagoons that have the potential to pollute the soil, surface water, and groundwater. In 1995, for example, major fish kills occurred when hog manure from a North Carolina livestock factory spilled into streams. People living near livestock factories are also concerned because the odor, which often exceeds federal and state guidelines for emissions, causes their property values to decline.

Most farmers in developing countries practice **subsistence agriculture**, the production of enough food to feed oneself and one's family, with little left over to sell or reserve for hard times. Subsistence agriculture, too, requires a large input of energy, but from humans and animals rather than from fossil fuels.

Some types of subsistence agriculture require large tracts of land. For example, **shifting agriculture**, also called slash-and-burn agriculture, involves clearing small patches of tropical forest to plant crops (Figure 18-6b; also see Chapter 17). Because tropical soils lose their productivity very quickly when they are cultivated (see Chapter 14), farmers using shifting agriculture must move from one area of forest to another every three years or so. **Nomadic herding**, in which livestock is supported by land that is too arid for successful crop growth, is another type of land-intensive subsistence agriculture (Figure 18–6c). Nomadic herders must continually move their livestock to find adequate food for them.

(text continues on page 412)

AMERICAN FARMERS: SCIENTISTS OF THE LAND

A new agricultural trend toward "precision farming" may quickly make farming both more profitable and less damaging to the environment:

- Computer technology allows farmers to map their land into dozens of plots, and to precisely determine location and characteristics of each plot from satellite signals.
- Soils can be thoroughly analyzed and yields tracked accurately.
- Data allow farmers to micromanage, permitting them to apply the right combination of nutrients needed to each parcel, in only the required amount.
- Farmers save money by reducing fertilizer costs and increasing yields, while leaching fewer nutrients into groundwater because excess fertilizer has never been applied.

YOU CAN MAKE A DIFFERENCE

Vegetarian Diets

A vegetarian is a person who does not eat the flesh of any animal, including that of fish and poultry. People adopt vegetarian diets for many reasons. Balanced vegetarian diets provide good nutrition without high levels of saturated fats or cholesterol, both of which cause health problems such as heart disease and obesity. Some studies in the United States indicate that Americans who are vegetarians live longer, healthier lives than non–vegetarians.

Some people become vegetarians because they are morally or philosophically opposed to killing animals, even for food. Certain religious groups, notably Hindus and Seventh Day Adventists, exclude animal products from their diets.

Other people convert to vegetarianism because of a sense of responsibility for land use and its wide repercussions. Fewer plants are required to support vegetarians than to support meat eaters. The amount of usable energy in the food chain is decreased by approximately 90 percent by adding an additional level—that is, the animals we eat—to the chain (see figure). Simply stated, if everyone were to become a vegetarian, much more food would be available for human consumption.

Vegetarians are divided into three groups—lactoovo-vegetarians, lacto-vegetarians, and vegans—based on whether they eat milk and/or egg products. *Lactoovo-vegetarians* eat milk, eggs, and foods made from milk and eggs. *Lacto-vegetarians* do not eat eggs but they do eat milk and milk products such as cheese, yogurt, and butter. *Vegans* exclude milk and eggs in all forms from their diets.

Some people are reluctant to switch to a vegetarian diet because they fear they will not get enough protein (plant foods have a lower percentage of protein than do animal foods). However, meat eaters in developed countries usually consume much more protein than they need. The problem with a vegetarian diet is usually not lack of protein, but obtaining the proper balance of essential amino acids. It is relatively easy for lactoovo-vegetarians and lacto-vegetarians to plan a healthy diet because milk and eggs contain all the essential amino acids. Milk is also rich in calcium, which is important for strong bones and teeth. Both milk and eggs contain vitamin B_{12}, which helps form red blood cells and helps nerves to function properly. Vegans, however, eat no animal products and rely exclusively on plant products to obtain their protein. Vegans must plan their diets carefully to obtain proper balance of amino acids, as well as enough calcium and vitamin B_{12}. Sesame seeds, broccoli, spinach, and other leafy green vegetables are rich in calcium, so these foods must be included in the vegetarian diet every day. Because vitamin B_{12} is found almost exclusively in animal products, most vegans take vitamin B_{12} supplements.

Although the human body can store excess lipids (as fat) and excess carbohydrates (as glycogen in the liver and as fat), it has no way to store excess amino acids. Therefore, all the amino acids essential for protein manufacture must be eaten together; it does no good to eat some of the essential amino acids for lunch and others for dinner.

A nutritious vegetarian diet includes a combination of foods that contains all the essential amino acids. A meal of rice and beans or corn and beans, for example, provides the proper complement of essential amino acids. Cookbooks and other references contain menus and recipes that give the vegetarian diet adequate amounts of high-quality protein. The following list provides an overview of combinations of foods that offer a proper balance of essential amino acids. At least one food from each column should be consumed at every meal.

Column I	Column II
Grains	Legumes
Barley, corn, oats, rice, rye, wheat	Fleshy beans, peas*
Nuts and Seeds	Dairy Products
Almonds, beechnuts, Brazil nuts, cashews, filberts, pecans, pumpkin seeds, sunflower seeds, walnuts	Cheese, cottage cheese, eggs, milk, yogurt

*There is a wide variety of beans and peas, including black beans, fava beans, kidney beans, lima beans, pinto beans, mung beans, navy beans, and soybeans (which are usually eaten as bean curd, or tofu).

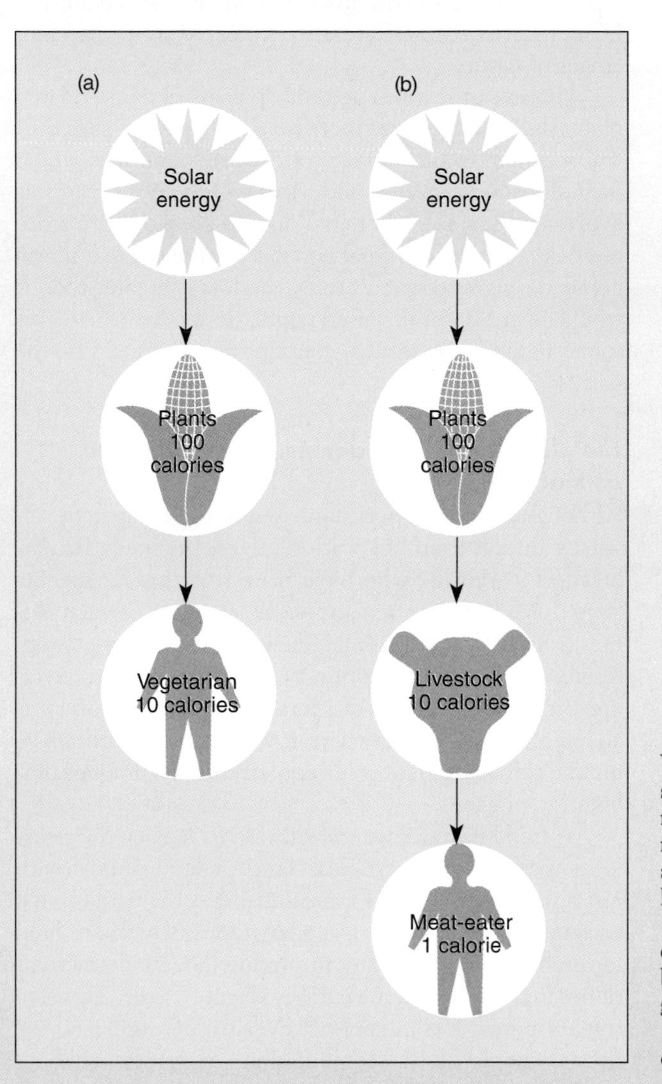

Vegetarians, because they are located on a shorter food chain, obtain more energy than meat-eaters do from the same amount of plant material. Reductions in energy as one moves along the food chain are the result of the second law of thermodynamics (discussed in Chapter 3). (a) A vegetarian will get no more than 10 calories of energy from every 100 calories that were fixed by plants in photosynthesis. (b) A meat-eater gets no more than 1 calorie of energy from the 100 calories fixed by the plants that the cattle consumed.

Polyculture is a traditional form of subsistence agriculture in which several different crops are grown at the same time. Native Americans practiced polyculture when they planted corn, bean, and squash seeds together in the same mound. Because the root systems of these plants grow to different depths, they do not compete with one another for water and essential minerals. In addition, the protein-rich bean crop fixes nitrogen that fertilizes the corn and squash plants naturally. In polyculture that is practiced in the tropics, fast- and slow-maturing crops are often planted together so that different crops are harvested throughout the year. For example, vegetable crops and cereal grains, which mature first, might be planted with papayas and bananas.

The Effect of Domestication on Genetic Diversity

Wild plant and animal populations usually have great genetic diversity (see Chapter 16), which contributes to species' long-term survival by providing the variation that enables each population to adapt to changing environmental conditions. When plants and animals are domesticated, much of this genetic diversity is lost, because the farmer selects for propagation only those plants and animals with the most desirable agricultural characteristics. At the same time, other traits that are not of obvious value to humans are selected against. Hence, many of the high-yielding crops produced by modern agriculture are genetically uniform. For example, most of the corn grown in the United States is of only a few different varieties (Table 18–2). Likewise, dairy cattle and poultry in the United States have low genetic diversity.

Table 18–2
Agricultural Diversity in the United States

Crop	Main Varieties in Production	Percentage of Total Crop Produced from These Varieties
Corn	6	71
Wheat	10	55
Soybeans	6	56
Rice	4	65
Potatoes	4	72
Peanuts	9	95
Peas	2	96

The loss of genetic diversity that accompanies modern agriculture usually does not prove disastrous to crop plants, because they do not have to survive in the wild under natural conditions. Under cultivation, they are watered, fertilized, and protected as much as possible from pests, including weeds, insects, and disease organisms. Domesticated animals are also protected from the challenges of nature.

However, the lower genetic diversity of domesticated plants and animals does increase the likelihood that they will succumb to new strains of disease organisms, which include bacteria, fungi, and viruses. (Disease organisms evolve quite rapidly.) When a disease breaks out in a domesticated plant or animal population, the entire uniform population is susceptible; thus, the loss is greater than it would be in a natural, varied population, in which at least some individuals would contain genes to resist the pathogen.

The global decline in domesticated plant and animal varieties

Although domestication contributes to less genetic diversity than is found in wild relatives, the many farmer-breeders worldwide who have been selecting for specific traits have developed many local varieties of each domesticated plant and animal. Each traditional variety represents the legacy of the hundreds of farmers who developed it over thousands of years. A traditional variety is adapted to the climate where it was bred and contains a unique combination of traits conferred by its unique combination of genes.

A trend is under way worldwide to replace the many local varieties of a particular crop or farm animal with just a few kinds. When farmers abandon their traditional varieties in favor of more modern ones, which are bred for uniformity and maximum production, the former varieties frequently become extinct (Figure 18–8). This represents a great loss in genetic diversity, because each variety's characteristic combination of genes gives it distinctive nutritional value, size, color, flavor, resistance to disease, and adaptability to different climates and soil types.

The gene combinations of local varieties are potentially valuable to agricultural breeders because they can be transferred to other varieties, either by traditional breeding methods or by genetic engineering. For example, U.S. wheat and barley crops were infested in the late 1980s by the Russian aphid. Over several years, aphid-resistant varieties were developed using genes of several varieties from the Middle East.

In order to preserve older, more diverse varieties of plants, many countries are collecting germplasm. **Germplasm** is any plant or animal material that may be used in breeding. It includes seeds, plants, and plant tis-

Figure 18–8

The cotswold, a rare sheep variety with long, curly wool. Cotswold sheep were originally introduced to England by the Roman Army. By 1954, only one flock existed worldwide, and the variety faced extinction. The Kelmscott Farm in Maine, home of this Cotswold sheep, is one of several farms in the United States that maintain rare breeds. (Courtesy Kelmscott Farms)

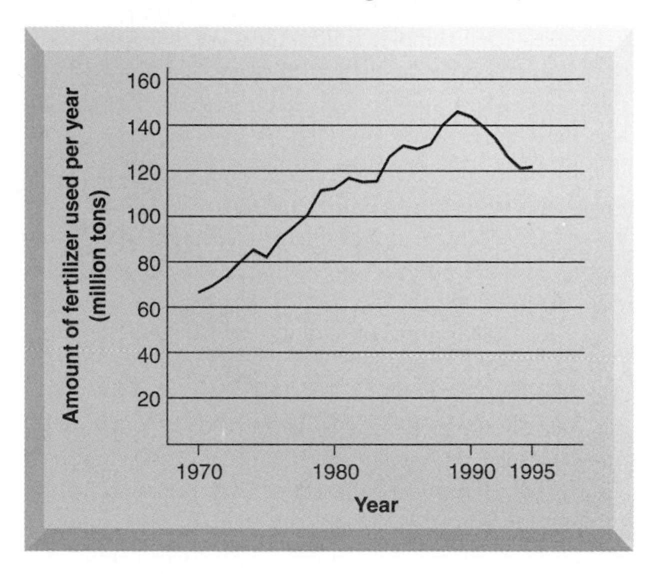

Figure 18–9

Worldwide fertilizer use, 1970 to 1995. The increasing use of fertilizers, which peaked in 1989, has resulted in higher crop yields. Worldwide fertilizer use declined from 1990 to 1994, but began to increase again in 1995.

sues of traditional crop varieties, and the sperm and eggs of traditional livestock breeds. The International Plant Genetics Resources Institute in Rome, Italy, is the scientific organization that oversees plant germplasm collections worldwide. National germplasm collections range in size from the U.S. National Plant Germplasm System in Colorado, which holds more than a quarter of a million different varieties, to the national gene bank in the African country Malawi, which holds about 8000 varieties of native crops and fruits. (See section on seed banks in Chapter 16 for more information about plant germplasm collections.) Several private organizations—such as Seeds of Change in New Mexico and Seed Savers Exchange in Iowa—save, propagate, and distribute traditional crop varieties.

Increasing Crop Yields

Until the 1940s, agricultural yields among various countries, both developed and developing, were generally equal. However, advances made by research scientists have caused a dramatic increase in food production in developed countries. Greater knowledge of plant nutrition has resulted in fertilization that promotes optimum yields (Figure 18–9). The use of pesticides to control insects, weeds, and disease-causing organisms has also improved crop yields. Animal production has been increased by the use of antibiotics to control disease. Selective breeding programs have resulted in agricultural plants and animals with more desirable features. For example, breeders developed wheat plants with larger, heavier grain heads (for higher yield). Because of the weight of the heads, other traits were gradually incorporated into wheat, such as shorter, thicker stalks, which prevent the plants from falling over during storms.

The green revolution

The introduction during the 1960s of high-yielding varieties of wheat and rice to Asian and Latin American countries gave these nations the chance to provide their people with adequate supplies of food. But the high-yielding varieties required intensive cultivation methods, including the use of fertilizers, pesticides, and mechanized machinery, in order to realize their potential. The production of more food per acre of cropland by using the new, high-yielding varieties and modern cultivation methods has been called the **green revolution**. Some of the success stories of the green revolution have been remarkable. For example, Indonesia used to import more rice than any other country in the world. Today Indonesia produces not only enough rice to feed its people, but also some for export.

Problems with the green revolution

The two most important problems associated with higher crop production are damage to the environment, which will be considered shortly, and the high energy costs that are built into this form of agriculture. Inorganic fertilizer requires a great deal of energy to manufacture, so its production costs are tied closely to the price of energy. Significant amounts of fossil fuels are required to provide power for farm equipment, such as tractors and combines. The installation and operation of irrigation systems, including construction of dams and canals, also require a substantial energy input. In general, it takes three times more energy per hectare to produce corn by irrigation than it does to produce the same amount of corn under rainfed conditions; the energy use in this example is for pumping groundwater from a depth of 30 meters (98 feet).

Rice and wheat are not the only crops to have been improved by the green revolution—high-yielding varieties of crops such as potatoes, barley, and corn have also been developed. Nonetheless, many important food crops remain to be improved by selective breeding and scientific research. People in Africa, for example, eat sorghum, millet, cassava, and sweet potatoes, none of which has been greatly improved by green revolution technology. Also, subsistence farmers, who represent a substantial segment of the agricultural community in most developing nations, have not benefitted from the green revolution. They need improved crops that respond to labor-intensive agriculture (human and animal labor) and that do not require large outlays of energy and capital.

Food Processing and Food Additives

Most of the food we eat is not harvested and used directly. Instead, after harvest it is processed, then offered for sale in grocery stores. There are two aspects to food processing. The first encompasses procedures such as drying, freezing, canning, pasteurizing, curing, irradiating, and refrigerating food to retard spoilage. The second involves adding **food additives**—chemicals that enhance the taste, color, or texture of the food; improve its nutrition; reduce spoilage and prolong shelf life; or maintain the food's consistency.

Sugar and salt are the two most common food additives. Although they are added to food primarily to make it taste better, in large amounts sugar and salt can be used to help preserve food—for example, fruit jelly and salted meat. Salt and sugar have been used as preservatives for centuries.

Essential amino acids and vitamins are sometimes added to foods to make them more nutritious than they would be naturally or to replace nutrients that are lost during processing. Both natural and synthetic **coloring agents** are used to make food visually appealing. Sodium propionate and potassium sorbate are two examples of **preservatives**, which are chemicals added to food to retard the growth of bacteria and fungi that cause food spoilage. Food can also spoil when lipids undergo oxidation. Food additives that prevent oxidation are called **antioxidants** and include **b**utylated **h**ydroxy**a**nisole (BHA) and **b**utylated **h**ydroxy**t**oluene (BHT).

Protection of the consumer

The Food and Drug Administration (FDA) is charged with the responsibility of monitoring food additives. In 1958 regulations were passed that require any new food additive to undergo extensive toxicity testing by the manufacturer. The results of such tests are then evaluated by the FDA to determine whether the additive is safe.

Additives that were in use prior to 1958, however, do not have to undergo such testing. In 1959 the FDA designated these chemicals "**g**enerally **r**ecognized **a**s **s**afe" and made up a list of them—usually called the GRAS (pronounced *grass*) list. All substances on the GRAS list have subsequently been reviewed. Some have been banned, including cyclamates (used for sweetening) and brominated vegetable oil. BHA, BHT, and several other substances on the GRAS list are undergoing further tests.

Are food additives bad?

The Center for Science in the Public Interest has compiled a list of food additives that may be harmful and should be avoided or eaten sparingly (Table 18–3). These include the preservatives BHA and BHT, several coloring agents (especially the red dyes), and nitrates and nitrites (see Focus On: Nitrates and Associated Compounds, page 416).

There are two opposing viewpoints about the safety of our food. Some people are very concerned about the large number and amounts of food additives in processed food. Their main fear is that some additives may be carcinogenic. Supporters of this view worry that, although the risk of developing cancer from exposure to each individual chemical may be quite small, the effects of hundreds of different food additives added together may be significant. Also, there is concern that these chemicals may interact synergistically (their total effect may be greater than the sum of their individual effects).

Others think that the health hazards from food additives are greatly exaggerated. They think it is important to put concerns about food additives and even pesticide residues in proper perspective. Everyone acknowledges that the chemicals in our foods do not pose anywhere

Table 18–3
Food Additives That May Be Harmful

Food Additive	Food	Possible Effect
BHA, BHT	Oils, potato chips, chewing gum	May be carcinogenic; allergic reactions in some
Citrus red dye #2	Skin of some oranges	May be carcinogenic
Nitrates, nitrites	Bacon, corned beef, hot dogs, ham, smoked fish, luncheon meats	Formation of carcinogenic *N*-nitroso compounds
Red dyes #3, 8, 9, 19, 37	Cherries (maraschino, fruit cocktail, candy)	May be carcinogenic
Saccharin	Diet foods	Known carcinogen in animal tests
Sulfites (sulfur dioxide, sodium sulfide, sodium bisulfite)	Dried fruit, canned and frozen vegetables, some beverages (e.g., wine), bread, salad dressings, and more	Severe allergic reactions

near the threat of cancer that smoking does, for example (see discussion of risk assessment in Chapter 2).

Moreover, much larger quantities of *natural* carcinogens are present in food. Nature is not benevolent; plants have evolved natural chemical defenses to discourage insects from consuming them. Some of these compounds, which can be present in large amounts, are toxins; some are carcinogenic. Thus, certain people contend that food additives pose a much smaller threat than do the natural, unavoidable chemicals already present in food.

THE ENVIRONMENTAL IMPACT OF AGRICULTURE TODAY

The practices of high-input farming have resulted in several environmental problems. The intensive use of fossil fuels produces air pollution, and according to the Environmental Protection Agency, agricultural chemicals such as fertilizers and pesticides are the single largest cause of water pollution in the United States. Some of these chemicals have been detected in water deep underground, as well as in surface waters; fishes and other aquatic organisms are killed by pesticide runoff into lakes, rivers, and estuaries.

To complicate matters, many insects, weeds, and disease-causing organisms have developed resistance to pesticides, forcing farmers to apply progressively larger quantities (Figure 18–10; also see Chapter 22). Residues of pesticides contaminate our food supply and reduce the number and diversity of beneficial microorganisms in the soil.

Soil erosion, which is exacerbated by large-scale mechanized operations, causes a decline in soil fertility, and the sediments lost due to erosion damage water quality. A 1989 study by the U.S. Department of Agriculture estimated that the annual cost of water pollution caused by eroded agricultural soils was $4 to $5 billion during

Figure 18–10
Colorado potato beetles eating potato leaves. As a result of being exposed to heavy applications of pesticides over the years, Colorado potato beetles are resistant to most insecticides that are registered for use on potatoes. (Grant Heilman, from Grant Heilman Photography)

FOCUS ON

Nitrates and Associated Compounds

Nitrates (compounds containing NO_3^-) and **nitrites** (compounds containing NO_2^-) occur in water and food as well as in the environment. Nitrates come both from natural sources (for example, they are produced by some soil bacteria) and from chemical fertilizers. Nitrates are present naturally in many vegetables, particularly the leafy green ones, such as spinach and lettuce, and the root crops, such as radishes and beets. Nitrates and nitrites are used as food additives and for curing meats. In addition, nitrates sometimes contaminate drinking water, with fertilizer being the most common source.

When nitrates get into the body, they are converted to nitrites, which reduce the blood's ability to transport oxygen. This condition is one of the causes of cyanosis (the "blue baby" syndrome), a serious disorder in very young children.

Some **N-nitroso compounds** (related nitrogen-containing compounds) are carcinogenic. *N*-nitrosodimethylamine, for example, is a potent carcinogen found in cured meats such as bacon, in smoked and salted fishes, and in tobacco smoke. Cheese and beer (brewed with water containing high levels of nitrates) are also sources of *N*-nitroso compounds.

Ingested nitrites are capable of reacting with other chemicals in food and in tobacco to form *N*-nitroso compounds in the stomach. We do not know how this occurs, nor do we know how they cause cancer. A great deal remains to be learned about the *N*-nitroso compounds found naturally in foods, as well. Interestingly, most fresh vegetables contain substances that inhibit the production of *N*-nitroso compounds.

What does all this mean to you? At this point, no one has an accurate picture of the risks involved. The presence of nitrates in fresh vegetables should probably not concern you since these plants also contain anti–*N*-nitroso compounds. You also need not worry about drinking water from municipal systems, because the level of nitrate in drinking water is monitored. If you drink well water, however, you should probably have its nitrate level checked periodically. In addition, it might be prudent to avoid excessively high amounts of cured meats such as bacon, ham, hot dogs, and luncheon meats.

the mid-1980s. Improper irrigation has also resulted in declining soil productivity as salts have accumulated in the soil.

Crop production requires enormous amounts of water. Agriculture consumes about 68 percent of the fresh water withdrawn from aquifers and surface waters for irrigation. Clearing grasslands and forests and draining wetlands to grow crops have resulted in habitat losses that reduce biological diversity. Many species have become endangered or threatened as a result of habitat loss caused by agriculture. A reduction of genetic diversity in agriculturally important crops and livestock has also occurred as a result of selective breeding.

Using More Land for Cultivation

In the temperate areas of the world, almost all the fertile land with an adequate supply of water is used for agriculture. Although in some regions the loss of prime farmland to urbanization is of concern, in temperate areas very little prime farmland not already under cultivation remains. Many tropical areas, on the other hand, have little prime agricultural land to start with. Those tropical areas with the greatest potential for cultivation and cropland expansion are in Latin America and sub-Saharan Africa.

One reason the United States had agricultural surpluses during the 1980s is that farmers brought large amounts of previously unused land into production. Unfortunately, much of this land was marginal as agricultural land, because it was prone to soil erosion caused by intermittent floods or frequent droughts (and therefore wind erosion when the ground cover was removed). Harvesting crops from highly erodible land is ecologically unsound and cannot be done indefinitely.

Other countries have paid a high price for cultivating land highly prone to erosion. The former Soviet Union, for example, began to cultivate a large area of marginal land during the 1950s. Although the initial production of cereal crops was high, by the 1980s much of this land had to be abandoned. The annual per-capita food production in Haiti, a very poor Caribbean nation, is half what it was in 1950 as a result of both population growth and loss of production on eroded soils.

From the 1940s to the 1980s, the amount of agricultural land was greatly increased through irrigation of dry land. During the last few years, however, there has been a worldwide decline in the rate of expansion of irrigation, and in the United States the amount of irrigated land has actually decreased. This change is due to the increasing cost of irrigation, the depletion of aquifers, the abandonment of salty soil, and the diversion of irrigation water to residential and industrial uses.

MOVING FOOD

We take for granted the presence of fresh fruits and vegetables in northern regions of the country during the winter. Some come from warm-climate states, many others from foreign countries. Food in general moves great distances before making its way to a supermarket shelf. The U.S. Department of Agriculture has estimated that the average processed food travels 1300 miles before being eaten. Trucks move 99 percent of all livestock, 88 percent of fresh fruits and vegetables, and 80 percent of fresh and frozen meats, dairy products, bakery goods, and beverages. The environmental cost is considerable. More than 4.25 million trucks are used primarily to transport food each year in the United States. They travel almost 50 million miles (a distance equal to more than 250 trips to the sun), burning almost $6 billion worth of fuel; they emit well over 4 million tons of pollutants into the air; and they cause hundreds of millions of dollars of damage to federal and state highways because of their extreme weight (one fully loaded 80,000-pound truck causes more damage to roads than 9600 automobiles). Yet our food-processing and distribution system compounds the problem. Large food processors often make deals with large growers in other parts of the country. So it is not uncommon, for example, to see tomatoes shipped all the way from California to a soup factory in New Jersey. Even if suitable tomatoes were grown outside the factory gate, corporate contracts preclude them from being purchased. This is one way in which economic arrangements clearly work against the environment.

ENVIROBRIEF

THE HIDDEN COST OF FOOD

The foods Americans eat travel an average of 1300 miles from field to plate.

One-fifth of U.S. municipal solid waste originates from food packaging; in the late 1980s, this was equivalent to 287 pounds per person each year.

Each pound of beef produced in the United States requires 360 gallons of water, mostly used for irrigation of grain crops, and about one-quarter of a gallon of gasoline for production of fertilizers and other farm inputs.

SOLUTIONS TO AGRICULTURAL PROBLEMS

Food production poses an environmental quandary: We must increase food production in order to eliminate world hunger, but growing more food damages the environment, which lessens our chances of increasing food production in the future. Fortunately, the dilemma is not as hopeless as it seems. Farming practices and techniques exist that can ensure a sustainable agricultural output at acceptable environmental costs. Farmers who have been practicing high-input agriculture can adopt these alternative agricultural methods, which cost less and are less damaging to the environment. Advances are also being made in sustainable shifting agriculture. In addition, new technologies, such as genetic engineering, and the widespread adoption of "new" crop plants promise greater productivity and variety in nutritious foods.

Alternative Agriculture, a Substitute for High-Input Agriculture

More and more farmers are trying forms of agriculture that cause fewer environmental problems than high-input agriculture (recall the chapter introduction). **Alternative agriculture**, also called **sustainable** or **low-input agriculture**, relies on beneficial biological processes and environmentally friendly chemicals (those that disintegrate quickly and do not persist as residues in the environment). In alternative agriculture, certain modern agricultural techniques are carefully combined with traditional farming methods. The sustainable farm consists of field crops, trees that bear fruits and nuts, small herds of livestock, and even tracts of forest. Such diversification protects the farmer against unexpected changes in the marketplace.

Instead of using large quantities of chemical pesticides, alternative agriculture controls pests by enhancing natural predator-prey relationships. For example, apple growers in Maryland monitor and encourage the presence of black ladybird beetles in their orchards because these insects feed voraciously on European red mites, a major pest of apples. As a general rule, alternative agriculture tries to maintain biological diversity on farms as a way to minimize pest problems. Providing hedgerows (rows of shrubs) between fields, for example, provides a habitat for birds and other insect predators.

Crop selection also helps control pests without heavy pesticide use. In parts of Oregon apples can be grown without major pest problems, but peaches are often infested by insects, whereas in western Colorado, apples have major pest problems but peaches do well. Therefore, apples would be the preferred crop for alternative agriculture in Oregon, as would peaches in Colorado.

The breeding of disease-resistant crop plants is an important part of alternative agriculture, as is the maintenance of animal health rather than the continual use of antibiotics to prevent disease.

An important goal of alternative agriculture is to preserve the quality of agricultural soil. For example, crop rotation, conservation tillage, and contour plowing help control erosion and maintain soil fertility (see Figures 14–15 and 14–16). Sloping hills that are converted to mixed-grass pastures erode less than do hills planted with field crops, thereby conserving the soil and supporting livestock. Water and energy conservation are also practiced in alternative agriculture.

Animal manure added to soil decreases the need for high levels of chemical fertilizers and cuts costs. Also, using biological nitrogen fixation to convert atmospheric nitrogen into a form that can be used by plants (see Chapter 5) lessens the need for nitrogen fertilizers.

Alternative agriculture is not a single program but rather a series of programs that are adapted for specific soils, climates, and farming requirements. For example, some alternative farmers—those who practice **organic agriculture**—use no pesticide chemicals; others use a system of **integrated pest management (IPM)**. In IPM a limited use of pesticides is incorporated with such practices as crop rotation, continual monitoring for potential pest problems, use of disease-resistant varieties, and biological pest controls (see Chapter 22).

Making Shifting Agriculture Sustainable

Traditional slash-and-burn agriculture is sustainable as long as there are few farmers and large areas of rain forest. Since relatively small patches of forest are cleared for raising crops, the trees quickly return when the land is abandoned and the farmer has moved on to clear another plot of forest. If the abandoned land lies fallow (idle) for a period of 20 to 100 years, the forest recovers to the point where subsistence farmers can again clear the forest for planting. Burning the trees releases nutrients into the soil so that crops can again be grown there. Today, however, too many people practice shifting agriculture, and as a result, more and more of tropical forests are being destroyed (see Chapter 17). Also, because so many people are trying to grow crops on rainforest land, the soil does not lie uncultivated between farming cycles long enough for it to recover.

Some researchers have been seeking ways to make former rainforest land retain its productivity for a longer period than is usual in shifting agriculture. Consider Papua New Guinea, a small island nation in which approximately 80 percent of the people are subsistence farmers. Research scientists in this country have developed methods to deal with some of the most troublesome problems associated with shifting agriculture: soil erosion, declining fertility, and attacks by insects and diseases. Their research, which is part of the Shifting Agriculture Improvement Program in Papua New Guinea, has helped forest plots remain productive for longer periods of time. Heavy mulching with organic material, such as weed and grass clippings, has lessened soil infertility and erosion. The composted mulch is then piled into rows that follow the contours of the land, further reducing erosion. Several crops are planted together, reducing insect damage. One of the crops is always a legume (such as beans), which helps restore nitrogen fertility to the soil. An extension program demonstrates these methods to farmers and distributes a book, *Subsistence Agriculture Improvement Manual*, to help educate rural farmers.

Genetic Engineering

The ability to take a specific gene from a cell of one kind of organism and place it into a cell of an unrelated organism, where it is expressed, is called **genetic engineering** or **biotechnology**. (Some people define biotechnology more broadly, to include any use of organisms to produce products.) Genetic engineering has begun to revolutionize medicine and has great potential to improve agriculture as well.

The goals of genetic engineering in agriculture are not new. Using traditional breeding methods, farmers and scientists have developed desirable characteristics in crop plants and agricultural animals for centuries. It takes time to develop such genetically improved organisms, however. For example, using traditional breeding methods, it might take 15 years or more to incorporate genes for disease resistance into a particular crop plant. Genetic engineering has the potential to accomplish the same goal in a fraction of that time.

Moreover, genetic engineering differs from traditional breeding methods in that desirable genes from *any* organism can be used, not just those from the species of the plant or animal that is being improved. For example, if a gene for disease resistance found in petunias would be beneficial in tomatoes, the genetic engineer can splice the petunia gene into the tomato plant (Figure 18–11). This could never be done by traditional breeding methods, because petunias and tomatoes belong to separate groups of plants and do not interbreed.

Genetic engineering could produce food plants that would be more nutritious because they would contain all the essential amino acids. (Currently, no single food crop has this trait.) Crop plants that were resistant to insect pests and viral diseases or that could tolerate drought, heat, cold, herbicides, or salty soil could be developed (Figure 18–12). For example, researchers at the U.S. De-

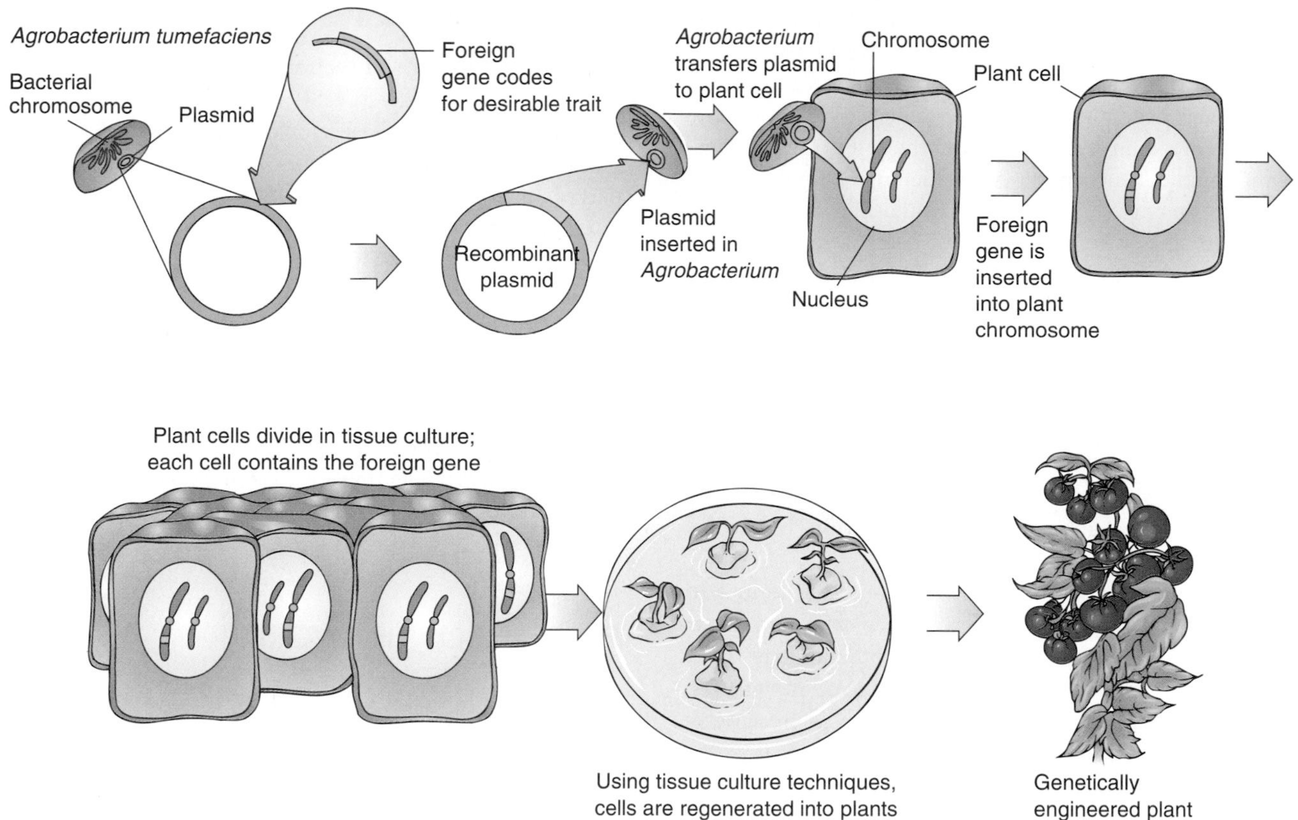

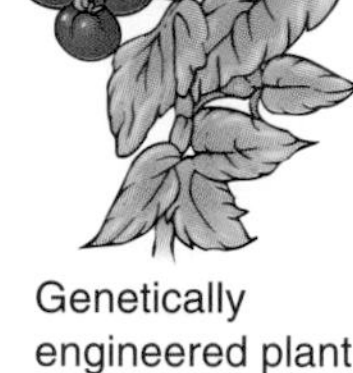

Figure 18–11

The plasmid of a bacterium called *Agrobacterium tumefaciens* can be used to introduce desirable genes from another organism into a plant. After the foreign DNA is spliced into the bacterial plasmid, the plasmid is inserted into the bacterium, which is then used to infect plant cells in culture. The foreign gene is inserted in the plant's chromosome, and genetically engineered plants are then produced from the cultured plant cells.

Figure 18–12

Genetic engineering has the potential to solve the potato/Colorado potato beetle problem discussed in Figure 18–10. A potato variety (Newleaf Russet Burbank, shown on the right) has been genetically engineered to resist Colorado potato beetles. Farmers who grow this variety do not have to apply chemical insecticides to control these insects. Note how unhealthy the control plants are (on left), which were not genetically engineered. Neither group received insecticide treatments. (Photo courtesy of Nature Mark)

partment of Agriculture have recently identified a gene in rye that codes for a protein that prevents plant roots from absorbing aluminum, a metal that often reaches toxic levels in acidic soils. Acidic soil is widespread in the tropics; for example, 51 percent of the soil in Latin America is acidified. Incorporation of the anti-aluminum gene into crop plants such as wheat would allow them to be grown in areas where they do not currently grow.

Genetic engineering has been used to develop more productive farm animals, including rapidly growing swine and fish. Perhaps the greatest potential contribution of genetic engineering in the animal arena, however, is in the production of vaccines against disease organisms that harm agricultural animals. For example, recombinant vaccines to protect cattle against a viral disease known as rinderpest have been developed. (Rinderpest is a deadly disease that has reached epidemic proportions in parts of Asia and Africa.)

Although genetic engineering has the potential to revolutionize agriculture, the changes will not occur overnight. Several hundred private biotechnology firms, as well as thousands of scientists in colleges, universities, and government research labs, are involved in agricultural genetic engineering. However, a great deal of research must be done before most of the envisioned benefits from genetic engineering are realized.

The safety of genetic engineering

While acknowledging that the potential uses of genetic engineering were important and beneficial, many scientists were initially concerned that someone might engineer an organism that would cause ecological or health problems. This possibility was recognized by those who developed genetic engineering and led them to propose stringent guidelines for making the new technology safe. Recent history has failed to bear out such concerns. Millions of experiments have demonstrated that genetic engineering experiments can be carried out safely.

Scientists recognize the practical importance of genetic engineering and generally agree that the potential threat to humans and the environment was overestimated. For example, the engineering of crops resistant to insects is widely perceived as posing a minimal threat to the environment. In fact, such genetically engineered crops *benefit* the environment because they reduce the use of chemical pesticides.

Although many regulations have been relaxed, guidelines still exist in areas of genetic engineering research in which there are unanswered questions about possible effects on the environment. Much research is currently being conducted to assess the effects of introducing genetically engineered organisms into natural environments. Although thousands of field tests of genetically engineered organisms have been performed to date, there have been no reports of genetically modified organisms causing damage to the environment.

Eating New Foods

A trip to the fresh food section of most supermarkets in North America today reveals several "new" fruits and vegetables. Many of these crops are gaining acceptance, despite our natural reluctance to try foods that are not part of our cultural background. Malanga, quinoa, carambola, chayote, and winged beans (see Figure 16–5) are just some of the exotic foods that are beginning to grace our tables.

Although the nutritional characteristics of such exotic foods are fairly well known, methods for profitably growing them in large quantities must be developed. Their adaptability to different climates, soil types, and water availability must be determined before farmers will be willing to risk growing them. Marketing research will have to be conducted to guarantee a wide market for these crops, should they be cultivated on a large scale.

FISHERIES OF THE WORLD

The ocean contains a valuable food resource (Figure 18–13). Just under 90 percent of the world's total marine catch is fishes, with clams, oysters, squid, octopus, and other mollusks representing an additional six percent of the total catch. Crustaceans, including lobsters, shrimp, and crabs, make up about three percent, and marine algae constitute the remaining one percent (Figure 18–14).

Figure 18–13

A full fish net is pulled on board a fishing vessel in the Bering Sea off the coast of Alaska. Fleets of fishing vessels are responsible for most of the world's fish harvest. Some of these ships are quite large and process the seafood on board. (Jack D. Swenson/Tom Stack and Associates)

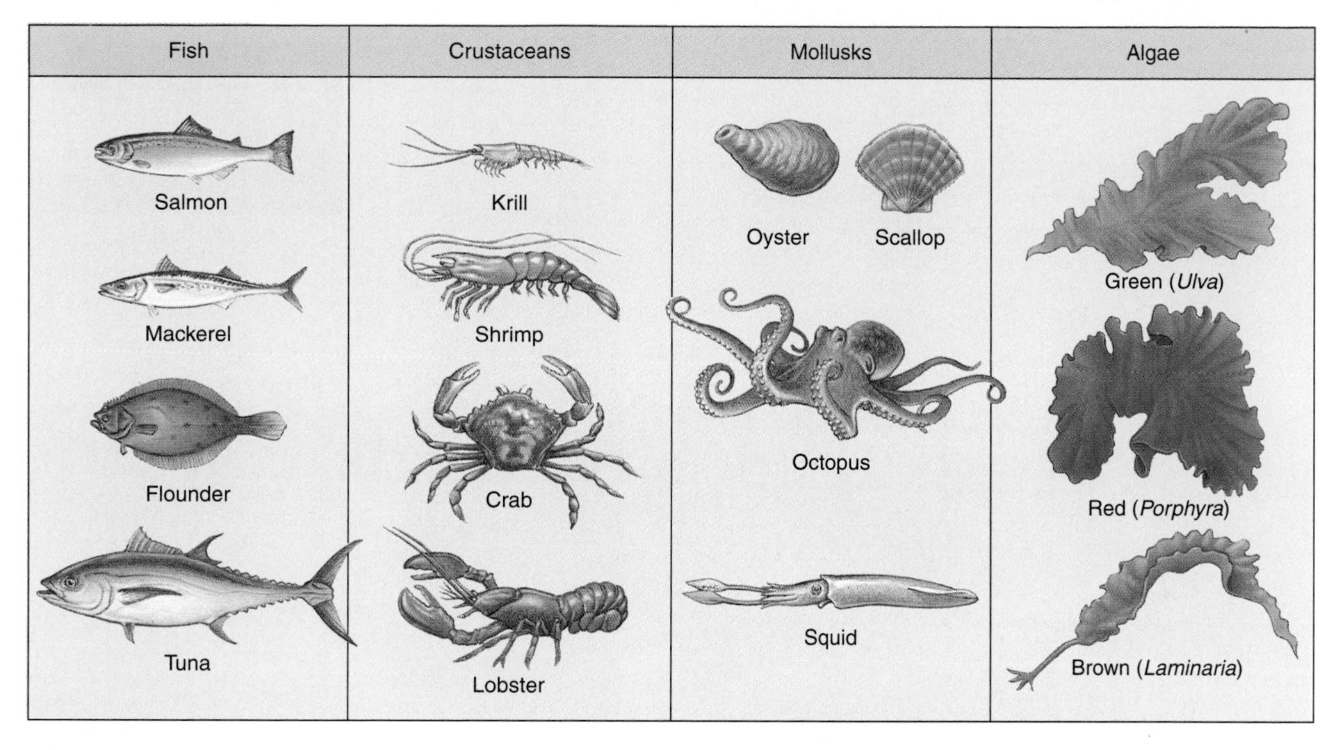

Figure 18–14
The major types of seafood that are commercially harvested include fishes, crustaceans, mollusks, and algae.

Fishes and other seafood are highly nutritious because they contain high-quality protein (protein with a good balance of essential amino acids) that is easily digestible. Worldwide, approximately 5 percent of the total protein in the human diet is obtained from fishes and other seafood; the rest is obtained from milk, eggs, meat, and plants. However, in certain countries, particularly in developing nations that border the ocean, seafood makes a much larger contribution to the total protein in the human diet.

Most of the world's marine catch is obtained by fleets of fishing vessels. In addition, numerous fishes are captured in shallow coastal waters and inland waters. The world's annual fish harvest increased substantially from 1950 (21 million tons) to 1994, when a record 109 million tons were caught. Certain fish stocks have declined severely, however, making it obvious that the ocean can yield only a limited number of fishes (recall the discussion of overfishing in Chapter 1). Each fish species has a maximum sustainable harvest level; if a particular species is overharvested, its numbers drop and harvest is no longer economically feasible. Table 18–4 shows some of the fish species that are at risk from overfishing.

Table 18–4
Selected Fish Species at Risk from Overfishing

Fish	Peak Year Catch in Tons (Year of Peak Catch)	1993 Catch in Tons
Haddock	914,300 (1970)	249,712
Atlantic cod	3,106,400 (1970)	1,134,147
Peruvian anchovy	13,059,900 (1970)	8,299,944
Cape hake	1,122,000 (1972)	199,252
Southern bluefin tuna	55,487 (1972)	14,355
Capelin	4,008,745 (1977)	1,742,149
Chub mackerel	3,412,602 (1978)	1,462,117
Japanese pilchard	5,428,922 (1988)	1,796,132

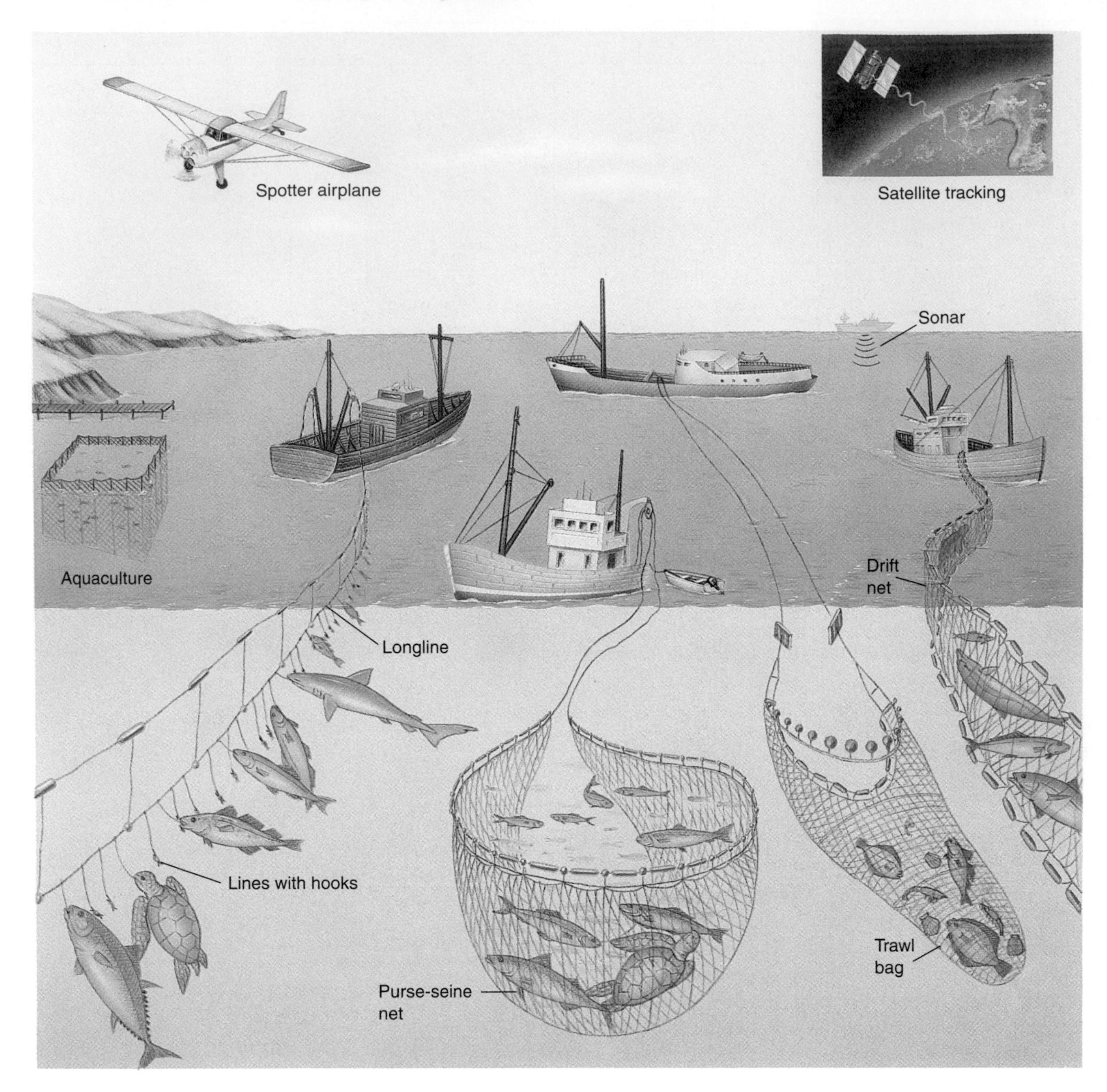

Figure 18–15
Modern fishing fleets use sophisticated equipment to capture fishes.

Problems and Challenges for the Fishing Industry

No nation lays legal claim to the open ocean. Consequently, resources in the ocean are more susceptible to overuse and degradation than are resources on the land, which individual nations own and for which they feel responsible (see Focus On: The Tragedy of the Commons in Chapter 2).

The most serious problem for marine fisheries is that many marine species have been overharvested to the point that their numbers are severely depleted. According to the U.N. Food and Agricultural Organization (FAO), 70 percent of the world's fish stocks are either fully exploited, overexploited, or depleted. There are two reasons why fisheries have experienced such pressure: one, the growing human population requires protein in their diets, leading to a greater demand; and two, technological advances in fishing gear have made it possible to fish so efficiently that every single fish is often removed from an area.

Sophisticated fishing equipment includes sonar, radar, computers, airplanes, and even satellites to locate fish schools (Figure 18–15). Some boats set out *longlines*, up to 128 kilometers (80 miles) of fishing lines with baited hooks. *Purse-seine nets* are huge nets set out by small powerboats that encircle large schools of tuna and other fishes; after the fishes are completely surrounded, the bottom of the net is closed to trap them. *Drift nets* are plastic nets up to 64 kilometers (40 miles) long that entangle thousands of fishes and other marine organisms; although drift nets have been banned by most countries, they continue to be used illegally. A *trawl bag* is a funnel-shaped net that is pulled along the bottom of the ocean to catch bottom-feeding fishes and shrimp; as much as 27 metric tons (30 tons) of fishes can be caught in a single net.

Fishermen tend to concentrate on a few fish species with high commercial value, such as menhaden, salmon, tuna, and flounder, while other fish species are caught and then discarded. The FAO reports that about 21 percent of all marine organisms caught are dumped back into the ocean; most of these unwanted fishes, dolphins, and sea turtles, collectively known as **bycatch**, are dead or soon die because they were out of the water too long. The United States and other countries are trying to reduce the amount of bycatch.

In response to overharvesting, many nations have adopted a policy of **ocean enclosure**, which puts the organisms within 320 kilometers (200 miles) of land under the jurisdiction of the country bordering the ocean (Figure 18–16). Open enclosure is supposed to prevent overharvesting by allowing nations to regulate the amounts of fishes and other seafood harvested from their waters. However, the United States and many other nations have a policy of **open management**, in which all fishing boats of that country are given unrestricted access to fishes in national waters.

One piece of good news for fisheries was reported in the journal *Science* in 1995. Fisheries biologists studied the population data of 128 depleted fish stocks and concluded that 125 of them can recover if fishing is carefully managed. Prior to this study, fisheries biologists had been concerned that many fish populations had declined so much that they would not be able to recover even if fishing were discontinued. Although this study provides good news about fish populations in the long term, it does not mean that recovery will occur rapidly. It may take years for some commercially important fish populations to rebound, which translates into severe economic hardship for the fishing industry.

Pollution and deteriorating habitat

One of the great paradoxes of human civilization is that the same ocean that is used to provide food to a hungry world is also used as a dumping ground. Pollution increasingly threatens the world's fisheries. Everything from accidental oil spills to the deliberate dumping of litter pollutes the water. Heavy metals such as lead, mercury, and cadmium are finding their way into aquatic food webs, where they are highly toxic to both fishes and the humans who eat fish.

Between 60 and 80 percent of all commercially important ocean fishes spend at least part of their lives in coastal areas. Tidal marshes, mangrove swamps, estuaries, and the like serve as spawning areas, nurseries, and feeding grounds. Coastal areas are also in high demand for recreational and residential development, however, and many of them are polluted. It is estimated that about 80 percent of global ocean pollution comes from human activities on land. As coastal ecosystems are degraded by development and pollution, the habitats of young marine animals are undermined, contributing to further depletion of fish stocks already suffering from overfishing.

► *FOLLOW–UP*

In the fall of 1995, the U.N. Environment Program sponsored a conference on protecting the marine environment from land-based activities. Representatives of more than 100 nations approved a "Global Program of Action" in which they pledged to reduce land-based water pollution, such as sewage, pesticides, organic pollutants, and sediments from erosion.

Aquaculture: Fish Farming

Aquaculture, the rearing of aquatic organisms, is more closely related to agriculture on land than it is to the fishing industry just described (Figure 18–17). Aquaculture is carried out both in fresh water and in marine water near the shore; the cultivation of marine organisms is sometimes called **mariculture**. To optimize the quality and productivity of their "crops," aquaculture farmers control the diets, breeding cycles, and environmental conditions of their ponds or enclosures. Aquaculturists try to reduce pollutants that might harm the organisms they

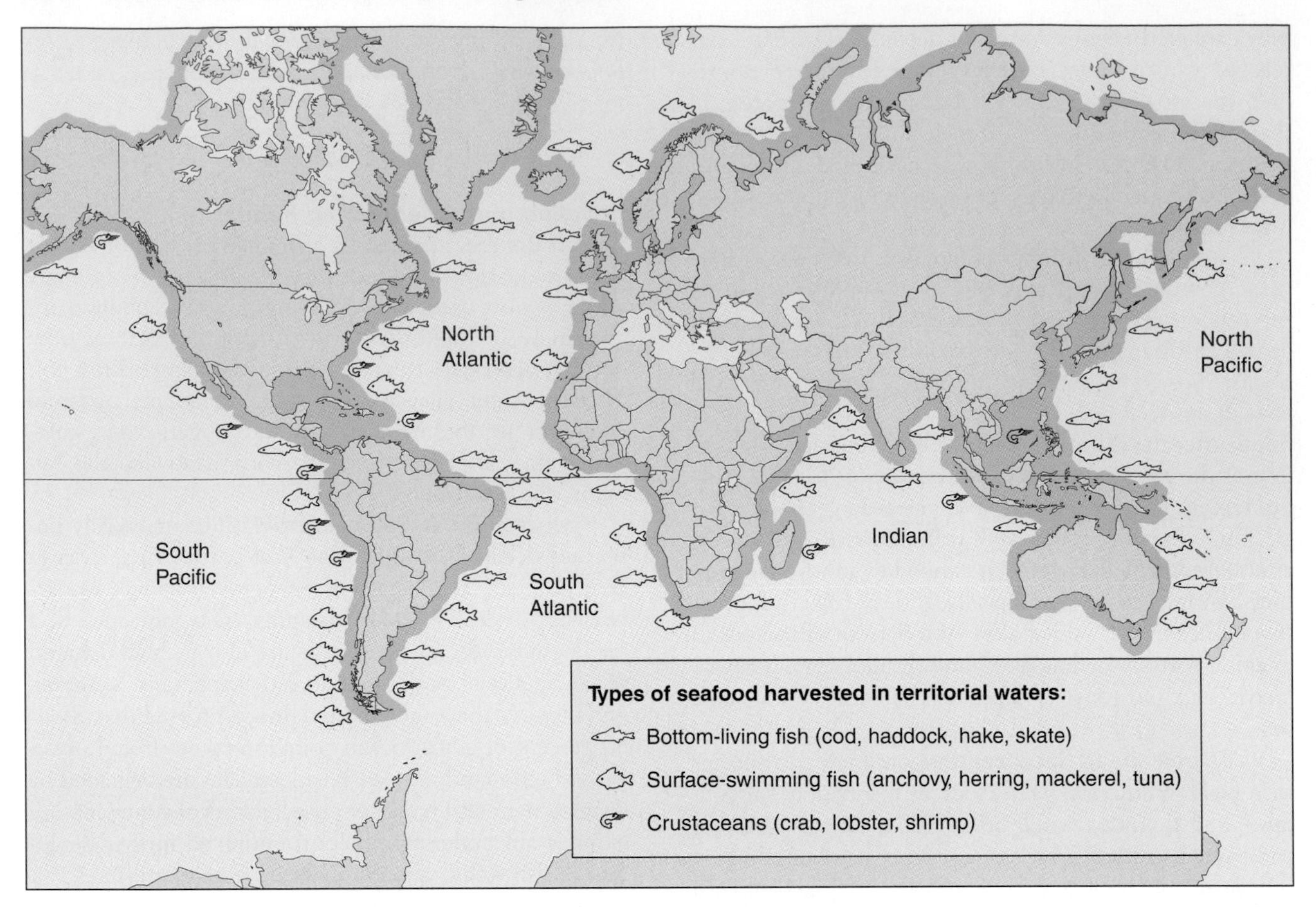

Figure 18–16

Many coastal nations have adopted a policy of open enclosure, in which they declare the 320 kilometers (200 miles) of ocean bordering their land (designated dark blue) as belonging to them.

are growing, and they keep them safe from potential predators.

Although aquaculture is an ancient practice that probably originated in China several thousand years ago, its enormous potential to provide food has only recently been appreciated. Aquaculture can contribute variety to the diets of people in developed countries. Inhabitants of developing nations can benefit even more from aquaculture: it may provide them with much-needed protein and even serve as a source of foreign exchange when they export such delicacies as aquaculture-grown shrimp.

According to the FAO, world aquaculture production more than doubled between 1984 and 1993, when 16.3 million tons of fishes and shrimp were harvested. Other important aquaculture crops include seaweeds, oysters, mussels, clams, lobsters, and crabs. Currently, aquaculture is growing at an annual rate of 10 to 15 percent, and one out of every five fish destined for human consumption comes from fish farms. The three nations with the largest aquaculture harvests are China, India, and Japan.

Aquaculture differs from fishing in several respects. For one thing, although the developed nations harvest more fishes from the ocean, the developing nations produce much more seafood by aquaculture. One reason for this is that developing nations have an abundant supply of cheap labor, which is a requirement of aquaculture because it, like land-based agriculture, is labor-intensive. Also, the limit on the size of a catch in fishing is the population of fishes found in nature, whereas the limit on aquacultural production of fishes and other seafood is largely the size of the area in which they are grown. In addition to being done inland, aquaculture is practiced in estuaries and in the ocean near the shore. Therefore, other uses of coastlines compete with aquaculture for available space. Developing countries that grow shrimp by aquaculture, for example, cut down the coastal mangroves that provide so many important environmental

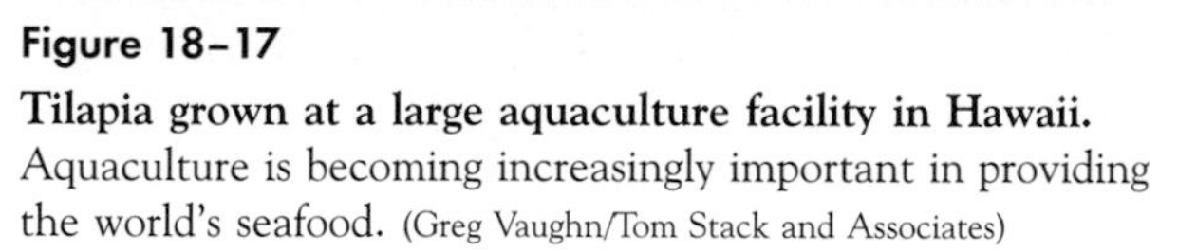

Figure 18–17

Tilapia grown at a large aquaculture facility in Hawaii. Aquaculture is becoming increasingly important in providing the world's seafood. (Greg Vaughn/Tom Stack and Associates)

ENVIROBRIEF

AQUACULTURE

Aquaculture is a $5 billion-a-year industry in the United States and accounts for 7 percent of fish and seafood consumed by Americans. All striped bass and rainbow trout available at American retail markets are produced by aquaculture, as is more than half the fresh salmon served in the United States. Catfish dominates American aquaculture; it currently accounts for 60 percent of all aquaculture-raised fishes produced in the United States. Tilapia (a mild fish similar in taste to sole) and oysters are also successfully grown by aquaculture in America.

benefits (see Chapter 6). Many marine fishes breed in the tangled roots of mangroves, for example, and there are concerns that an expansion of shrimp farming could contribute to a decline in marine fish populations. Because so many fishes are concentrated in a relatively small area, aquaculture produces wastes that can pollute the adjacent water and harm other organisms.

Although interest in aquaculture is increasing worldwide, several factors are slowing its expansion. Setting up and running an aquaculture facility is expensive. Also, scientific research is necessary in order to make aquaculture of certain organisms profitable. An organism's requirements for breeding must be established, for example, and ways to control excessive breeding must be available so that the population does not overbreed and produce many stunted individuals rather than fewer large ones. The population must be continually monitored for diseases, which have a tendency to spread rapidly in the crowded conditions that are characteristic of aquaculture.

One of the most important limits on aquaculture's potential is the receptivity of animals to the domestication process itself. Land animals such as cows, pigs, and sheep were domesticated over a period of thousands of years. During this time, there were undoubtedly failed attempts to domesticate many other animals, which for one reason or another could not be domesticated. The same is true of aquaculture: it is not simply a matter of observing a need for more tuna, for instance, and therefore opening an aquaculture facility that produces tuna. The organisms that are to be produced profitably by aquaculture must have certain traits that make their domestication possible. For example, aquatic organisms that are social by nature and do not exhibit territoriality or aggressive behavior are possible candidates for domestication.

SUMMARY

1. Humans require a balanced diet that includes carbohydrates, proteins, and lipids in addition to vitamins, minerals, and water. People who consume fewer calories than they need are said to be undernourished, whereas people who consume enough calories but whose diets are lacking in some specific nutrients are said to be malnourished.

2. In 1996 about 840 million people lacked access to the food needed to be healthy and to lead productive lives. The two regions of the world with the greatest food insecurity are South Asia and sub-Saharan Africa.

3. The greatest challenge in agriculture today is producing enough food to feed the world's population. Complicating factors include poverty, the problems of distributing food where it is needed, and cultural acceptance of nutritious but unfamiliar foods. The long-term solution to the problem of producing adequate food is the stabilization of the human population.

4. Agriculture developed independently at several centers beginning about 10,000 years ago. Early advances included domestication of plants and animals, irrigation, and use of animals for plowing. Although thousands of plant species are edible, today only about 100 plants provide the bulk of the world's food.

5. Developed countries and some developing countries rely on high-input agriculture, which uses large amounts of fossil fuels to power machinery, provide irrigation water, and produce fertilizers and pesticides. In high-input agriculture, yields of food per area unit of farmland are high, but so are environmental costs. Serious problems include soil erosion, which causes a decline in soil fertility as well as downstream sediment pollution, and water and soil pollution from pesticides and fertilizers.

6. Many of the world's farmers rely on subsistence agriculture, in which human and animal energy is used instead of fossil fuels. Subsistence farming produces less food per unit of farmland than high-input agriculture.

7. The challenges confronting agriculture are being met in a variety of ways. New methods are being developed to make agriculture sustainable. Genetic engineering is the high-technology answer to some of agriculture's problems. Our food base is also increasing through the introduction and general acceptance of "new" foods.

8. Seventy percent of the world's fish stocks are either fully exploited, overexploited, or depleted because of the growing human population and technological advances in fishing gear. Aquaculture is supplementing traditional fishing in the supply of high-quality protein from seafood.

THINKING ABOUT THE ENVIRONMENT

1. How does Walnut Acres, the farm described in the chapter introduction, mimic a natural ecosystem?

2. What age group in humans is usually most adversely affected by undernutrition, malnutrition, and famine? Why?

3. If modern-day domesticated corn were to be abandoned by agriculture, it is unlikely it would survive in the wild. Why? Give at least one trait of corn that is desirable from an agricultural viewpoint, but disastrous for survival in nature.

4. Why does decreased genetic diversity in farm plants and animals increase the likelihood of economic disaster from disease?

5. Describe the environmental problems associated with farming each of these areas: tropical rain forests; hillsides; arid regions.

6. Some scientists are genetically engineering herbicide resistance into crop plants so that when they apply herbicides, only the weeds die and not the crops. How might this specific example of genetic engineering have negative impacts on the environment?

7. Why is population control the most fundamental solution to world food problems? Explain your answer.

8. Explain why aquaculture is more like agriculture than it is like traditional fishing.

9. It has been suggested that toxicity testing of food additives should be performed by a research group independent of the food industry. Do you think this is a good idea? Why or why not?

* 10. It takes 7 kilograms of livestock grain such as corn to produce 1 kilogram of beef; 4 kilograms of livestock grain to produce 1 kilogram of pork; and 2.5 kilograms of livestock grain to produce 1 kilogram of poultry. If corn sells for $150 a ton, calculate how much it costs to produce 1 kilogram of each kind of meat. (One ton equals 907.2 kilograms.)

* 11. Complete Table A on world grain production in 1970 and 1995. What does this information tell you about how the average yield (amount of grain harvested per unit area of land) changed between 1970 and 1995?

Table A
World Grain Production, 1970 and 1995

	1970	1995	Percent Increase
Agricultural land (million hectares)	663	666	
Grain production (million tons)	1096	1680	
Tons of grain per hectare			

RESEARCH PROJECTS

Obtain a copy of *The Last Harvest: The Genetic Gamble that Threatens to Destroy American Agriculture*, by Paul Raeburn, Simon and Schuster, 1995. Read Raeburn's chapter entitled "Seed Banks and Seed Morgues," and assess the nature and seriousness of the problem.

Contact the Seed Savers Exchange (RR3, Box 239, Decorah, IA 52101, 319-382-5990) and ask for information about obtaining seeds and growing "heirloom" varieties of vegetable crops. If you are a gardener, put your knowledge to use by growing some of the older varieties in your own garden and by sharing the seeds with neighbors and friends.

On-line materials relating to this chapter are on the ► World Wide Web at **http://www.saunderscollege.com/lifesci/environment2** (select Chapter 18 from the Table of Contents).

SUGGESTED READING

Brown, L. "Facing Food Scarcity," *WorldWatch* Vol. 8, No. 6, December 1995. This article examines recent disturbing trends in world food production and offers several ways to counteract those trends.

Burns, G. et al. "The New Economics of Food," *Business Week*, May 20, 1996. Examines why the cost for grains and other farm products is projected to increase during the next few years.

Gardner, G. "Asia Is Losing Ground," *WorldWatch*, November–December 1996. Rapidly growing cities require space, and they are taking some of Asia's most productive cropland.

Klinkenborg, V. "A Farming Revolution: Sustainable Agriculture," *National Geographic*, December 1995. Sustainable agriculture uses a natural approach to get impressive yields with fewer chemicals.

Parfit, M. "Diminishing Returns: Exploiting the Ocean's Bounty," *National Geographic*, November 1995. As fish stocks decline, tensions increase over fishing grounds.

Plucknett, D. L. and D. L. Winkelmann. "Technology for Sustainable Agriculture," *Scientific American*, September 1995. Agriculture of the future must use environmentally benign technology to increase yields.

Powledge, F. "The Food Supply's Safety Net," *BioScience* Vol. 45, No. 4, April 1995. Examines efforts to obtain and maintain germplasm collections of traditional varieties of crop plants and livestock animals.

Prosterman, R. L., T. Hanstad, and L. Ping. "Can China Feed Itself?" *Scientific American*, November 1996. China, with its 1.2 billion people, may face serious food problems unless it changes certain policies.

Raloff, J. "Fishing for Answers: Deep Trawls Leave Destruction in Their Wake—But for How Long?" *Science News* Vol. 150, October 26, 1996. Deep trawls cause great ecological damage as they scour the ocean floor for seafood.

Rosegrant, M. W. and R. Livernash. "Growing More Food, Doing Less Damage," *Environment* Vol. 38, No. 7, September 1996. This article addresses the issue of how we can expand agricultural productivity without causing additional environmental damage.

Wolkomir, R. "Bringing Ancient Ways to Our Farmers' Fields," *Smithsonian*, November 1995. Certain traditional agricultural practices have great potential in providing a sustainable food supply for future generations.

► **Solid waste heaped on a barge on its way to Fresh Kills Landfill in Staten Island, New York.** *(Louie Psihoyos/Matrix International, Inc.)*

Environmental Concerns

Natural Capitalism

Paul Hawken is a businessman, environmentalist, and author of several books, including Growing a Business, The Ecology of Commerce, Seven Tomorrows, *and* The Next Economy. *Hawken also produced and hosted a 17-part PBS TV series, now shown in more than 115 countries, based on* Growing a Business. *He currently serves as chairman of The Natural Step, an educational foundation that assists world government and business leaders in achieving long-term competitive advantage through environmental sustainability.* ◀

PAUL HAWKEN

How has the modern industrial system not properly accounted for externalities like pollution, resource depletion, waste, and inefficiency?

Modern industrialism came into being in a world very different from today's: fewer people, less material well-being, and plentiful natural resources. Conventional economic theories will not guide our future for a simple reason: they have never placed "natural capital" on the balance sheet. Prices, costs, and other economic indicators change dramatically when natural capital is included as an integral part of the production process, not as a free amenity or as a putative infinite supply. Industries destroy natural capital because they have historically benefited from doing so.

What exactly is natural capital?

Everyone is familiar with the traditional definition of capital as accumulated wealth in the form of investments, factories, and equipment. Natural capital, on the other hand, comprises the resources we use, both nonrenewable (oil, coal, metal ore) and renewable (forests, fisheries, grasslands).

Why should big business recognize the value of "natural capital" as a limiting factor in the way they have traditionally recognized factors like labor, energy resources, machinery, and financial capital?

Although we usually think of renewable resources in terms of desired materials, such as wood, their most important value lies in the services they provide. They are not pulpwood but forest cover, not food but topsoil. Living systems feed us, protect us, heal us, clean the nest, and let us breathe. There are trillions of dollars of critical ecosystem services received annually by commerce.

Economic theory allows for endless substitutes, but there are no substitutes for many of the services provided by natural capital: production of oxygen, maintenance of biological diversity, purification of water and air, maintenance of wildlife, prevention of soil erosion, flood prevention, and more.

Some economists and others have noted the flaws of how gross domestic product is measured. What are some companies doing to incorporate factors like waste into their everyday operations?

Economic theory works only if you use financial efficiency as your benchmark and ignore physics, biology, and common sense. The key to resource efficiency is to understand products as a means to deliver a service to the customer, rather than thinking of them as things.

Ray Anderson, CEO of Interface, a $1 billion multinational carpet and flooring company, developed the "Evergreen Lease" to transform his product, carpet tiles, into a service. Old carpet tiles are replaced and recycled and made into new tiles as part of a lease fee, instead of being dumped in landfills. This is an example of how manufacturers can design and create products so that all components have value when they return—just as in nature—and not just when they leave the factory.

There are deep-rooted interests in maintaining the status quo. Big agriculture benefits from government programs including subsidies, tax breaks, price supports,

or use of public lands. How does the average citizen fit into this structure?

Citizens would benefit from a system that promotes use of labor over use of resources. To create a policy that supports resource productivity will require a shift away from taxing the social "good" of labor toward taxing the social "bads" of resource exploitation, pollution, fossil fuel consumption, and waste generation. The tax shift should be "revenue-neutral," meaning that for every dollar of taxation added to resources or waste, one dollar would be removed for labor taxes. The purpose is to change what is taxed, not who is taxed.

How long could this shift take to occur?

A shift toward taxing resources would require steady implementation, in order to give business a clear horizon to in which to make strategic investments. A time span of 15 to 20 years should be long enough to permit businesses to continue depreciating current capital investments over their useful lives.

The costs of natural capital will inevitably increase, so we should start to change the tax system now. Shifting taxes to resources won't—as some in industry will doubtless claim—mean diminishing standards of living. It will mean an explosion of innovation that will create products, techniques, and processes that are far more effective than those they replace.

Economic theory allows for endless substitutes, but there are no substitutes for many of the services provided by natural capital: production of oxygen, maintenance of biological diversity, purification of water and air, maintenance of wildlife, prevention of soil erosion, flood prevention, and more.
–Paul Hawken

Wouldn't this effort be just another case of citizens trying to change the tax system to achieve their particular goals?

Indeed, some economists will respond that we should let the markets dictate costs and that using taxation to promote particular outcomes is interventionist. But all tax systems are interventionist; the question is not whether to intervene but how to intervene.

A tax system should integrate cost with price, but we tend to disassociate the two. We know the price of everything but the cost of nothing. The buyers know the price but society bears the cost. For example, Americans pay about $1.50 per gallon at the gas pump, but gasoline actually costs up to $7 per gallon when you factor in all the costs.

Despite serious environmental problems to overcome you seem upbeat and optimistic. Why?

The benefits of resource productivity align almost perfectly with what American voters say they want: better schools, a better environment, safer communities, more economic security, stronger families and family support, freer markets, less regulation, fewer taxes, smaller governments, and more local control.

A resource productivity revolution in the next century is hard to fathom, but here's what it promises: progressively less material and energy use each year and where the quality of consumer services continues to improve; an economy where environmental deterioration stops and gets reversed as we invest in increasing our natural capital; and, finally, a society where we have more useful and worthy work available than people to do it.

Natural capitalism is not about social upheaval or sudden changes. It may not guarantee particular outcomes but it will ensure that economic systems more closely mimic biological systems, which have successfully adapted to dynamic changes over millennia.

▶ Web site: **www.hawken@well.com**
www.motherjones.com

Twilight along the Chattanooga waterfront. Chattanooga's air quality has improved dramatically during the past several decades. (Tennessee Photo Service)

Air Pollution

During the 1960s the federal government gave Chattanooga, Tennessee, the dubious distinction of having the worst air pollution in the United States. The air was so dirty in this manufacturing city that people driving downtown had to turn on their headlights in the middle of the day. The air soiled their white shirts so quickly that many businessmen brought extra ones to work. To compound the problem, mountains that surround the city kept the pollutants produced by its inhabitants from dispersing.

Today, less than 30 years later, the air in this scenic mid-sized city of 200,000 people is clean, and Chattanooga ranks high among U.S. cities in terms of air quality. Efforts by city and business leaders are credited with transforming Chattanooga's air. The city passed an Air Pollution Control Ordinance in 1969. It also established an air pollution control board and bureau to enforce regulations on emissions.

The new law allowed open burning by permit only, placed regulations on odors and dust, outlawed visible automotive emissions, set a 4 percent cap on sulfur content in fuel, and controlled the production of sulfur oxides. Limits were also set on industry's visible emissions, and businesses complied by installing expensive air pollution control devices.

By 1972, the measures taken had proven so effective that Chattanooga drew national attention for its clean-up effort. The National Air Pollution Control Association awarded Chattanooga First Place in their annual Cleaner Air Week ceremonies. The *Wall Street Journal*, the *New York Times*, ABC-TV, CBS-TV, and *U.S. News and World Report* did stories covering the success.

In 1984, Chattanooga was officially designated "in attainment" for particulates, one of the pollutants regulated by the Environmental Protection Agency (EPA); this designation meant that particulate levels were below the federal health standard for one year. The city reached attainment status for ozone in 1989. In fact, since the early 1980s in Chattanooga, the levels for all seven air pollutants regulated by the EPA have been better than the federal standards.

Chattanooga's air quality is an environmental success story that could be emulated by other American cities. The air we breathe is often dirty and contaminated with pollutants, particularly in urban areas. Air pollution also extends indoors, and the air we breathe at home, at the workplace, and in our automobiles may be more polluted than the air outdoors. Because air pollution causes a great many health and environmental problems, most developed nations and many developing nations have established air quality standards for numerous pollutants. ◀

THE ATMOSPHERE AS A RESOURCE

The atmosphere is an invisible layer of gases that envelops the Earth (see Chapter 5). The two atmospheric gases most important to humans and other organisms are carbon dioxide and oxygen. During photosynthesis, plants, algae, and certain bacteria use carbon dioxide to manufacture sugars and other organic molecules. During cell respiration, organisms use oxygen to break down food molecules and supply themselves with chemical energy.

We think of air as an unlimited resource, but perhaps we should reconsider. Ulf Merbold, a German space shuttle astronaut, felt very differently about the atmosphere after viewing it in space. "For the first time in my life, I saw the horizon as a curved line. It was accentuated by the thin seam of dark blue light—our atmosphere (Figure 19–1). Obviously, this was not the 'ocean' of air I had been told it was so many times in my life. I was terrified by its fragile appearance."

Figure 19–1

The "ocean of air" is actually an extremely thin layer compared to the size of the Earth. In this photo, shot over the ocean from a space shuttle, the atmosphere is the dark blue layer indicated by the bracket. (NASA)

TYPES, SOURCES, AND EFFECTS OF AIR POLLUTION

Air pollution consists of gases, liquids, or solids present in the atmosphere in high enough levels to harm humans, other organisms, or materials. Although air pollutants can come from natural sources—as, for example, when lightning causes a forest fire or a volcano erupts—human activities make a major contribution to global air pollution. From the standpoint of human health, probably more significant than the overall human contribution of air pollution is the fact that much of the air pollution released by humans is concentrated in densely populated urban areas.

Although many different air pollutants exist, we will focus our attention on the six most important types: particulates, nitrogen oxides, sulfur oxides, carbon oxides, hydrocarbons, and ozone (Table 19–1). Air pollutants are often divided into two categories, primary and secondary (Figure 19–2). **Primary air pollutants** are harmful chemicals that enter directly into the atmosphere. The major ones are carbon oxides, nitrogen oxides, sulfur dioxide, particulates, and hydrocarbons. **Secondary air pollutants** are harmful chemicals that form from other substances that have been released into the atmosphere. Ozone and sulfur trioxide are secondary air pollutants because both are formed by chemical reactions that take place in the atmosphere.

Sources of Outdoor Air Pollution

On a hot summer day in the Blue Ridge Mountains, which are part of the Appalachians, a blue haze hangs over the forested hills. This haze is caused by hydrocarbon emissions that come from the trees. Many plants produce a variety of hydrocarbons in response to heat. The hydrocarbon isoprene, for example, helps protect leaves from high temperatures. However, isoprene and other hydrocarbons are volatile and evaporate into the atmosphere, where they contribute to ozone formation, a key ingredi-

Table 19–1
Major Air Pollutants

Pollutant	Composition	Class	Primary or Secondary	Characteristics
Dust	Variable	Particulates	Primary	Solid particles
Lead	Pb	Particulates	Primary	Solid particles
Sulfuric acid	H_2SO_4	Particulates	Secondary	Liquid droplets
Nitrogen dioxide	NO_2	Nitrogen oxide	Secondary (mainly)	Reddish brown gas
Sulfur dioxide	SO_2	Sulfur oxide	Primary	Colorless gas with strong odor
Carbon monoxide	CO	Carbon oxide	Primary	Colorless, odorless gas
Methane	H–C(H)(H)–H or CH_4	Hydrocarbon	Primary	Colorless, odorless gas
Benzene	(ring structure) or C_6H_6	Hydrocarbon	Primary	Liquid with sweet smell
Ozone	O=O–O or O_3	Photochemical oxidant	Secondary	Pale blue gas with sweet smell

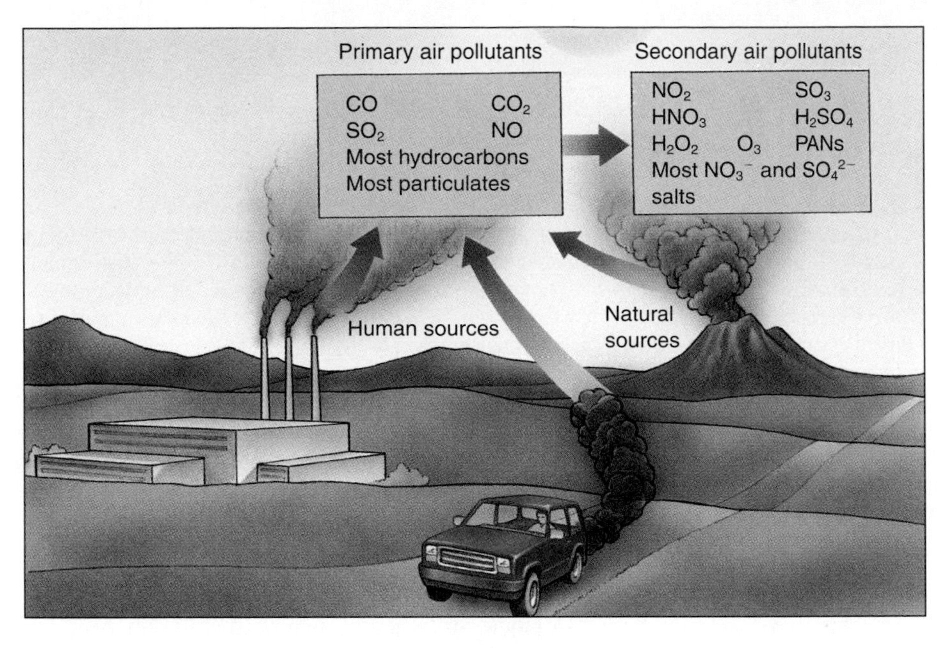

Figure 19–2

Primary and secondary air pollutants. Primary air pollutants are emitted, unchanged, from a source directly into the atmosphere, while secondary air pollutants form from chemical reactions involving primary air pollutants.

ent in photochemical smog (to be discussed shortly). The contribution of biologically generated hydrocarbon emissions in Atlanta, which is heavily wooded, is substantial (Figure 19–3).

On the island of Hawaii, the main cause of natural air pollution is a volcano named Kilauea, which has been emitting about 1000 tons of sulfur dioxide daily since 1986. Many residents blame their sore throats, headaches, bronchitis, and asthma conditions on the volcano, but studies conducted thus far have not been able to conclusively link local health problems to Kilauea.

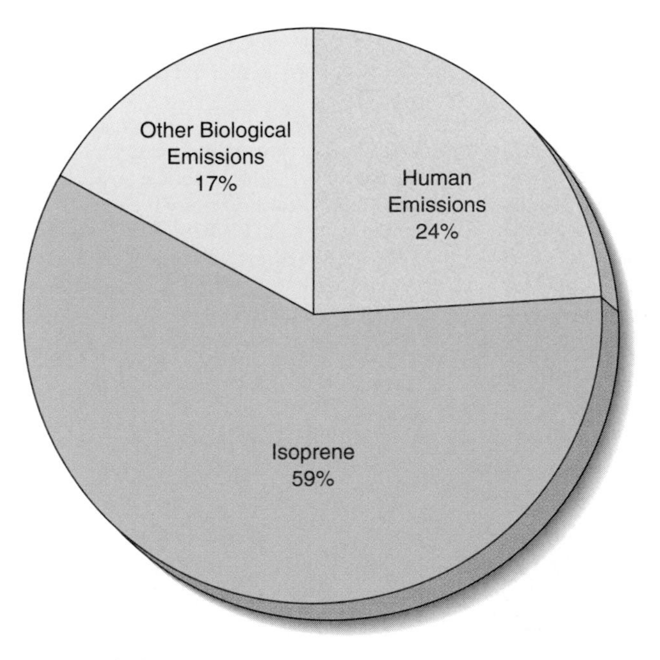

Figure 19–3

Hydrocarbon emissions in Atlanta. Isoprene, which is produced by plants, and other biological emissions contribute about three-fourths of all hydrocarbon emissions in the area.

The two main human sources of primary air pollutants are motor vehicles and industries (Figure 19–4). Automobiles and trucks release significant quantities of nitrogen oxides, carbon oxides, particulates, and hydrocarbons as a result of the combustion of gasoline (see Meeting the Challenge: Clean Cars, Clean Fuels). Electric power plants and other industrial facilities emit most of the particulate matter and sulfur oxides released in the United States; they also emit sizable amounts of nitrogen oxides, hydrocarbons, and carbon oxides. The combustion of fossil fuels, especially coal, is responsible for most of these emissions. The top three industrial sources of toxic air pollutants—that is, chemicals released into the air that are fatal to humans at specified concentrations—are the chemical industry, the metals industry, and the paper industry.

An Overview of the Effects of Air Pollution

Air pollution damages organisms, reduces visibility, and attacks and corrodes materials such as metals, plastics, rubber, and fabrics. The respiratory tracts of animals, including humans, are particularly harmed by air pollutants, which also worsen existing medical conditions such as chronic lung disease, pneumonia, and cardiovascular problems. The overall productivity of crop plants is reduced by most forms of air pollution, and when combined with other environmental stresses, such as low winter temperatures or prolonged droughts, air pollution causes plants to decline and die. Air pollution is involved in acid deposition, global temperature changes, and stratospheric ozone depletion (all discussed in Chapter 20).

Clean Cars, Clean Fuels

Performance and style have always been higher priorities for automobile manufacturers than reducing environmental pollutants. Concerned with an ever-increasing number of automobiles on California's congested highways—automobiles that burn too much gasoline and spew noxious emissions—California decided to legislate a clean, efficient car. A California law passed in 1990 mandates that at least 2 percent of the cars sold in that state by 1998 must have zero tailpipe emissions. By 2003, 10 percent of new cars must meet that criterion. Because more automobiles are bought in the state of California than in any other state, auto manufacturers cannot ignore the California mandate. In addition, many states along the Northeastern corridor have followed California's lead and toughened their emission standards, putting further pressure on carmakers.

Carmakers, however, have been unable to arrive at a consensus about what type of fuel should be used in low-emission cars. A company that invests heavily in a car designed for a fuel that does not become the industry standard could easily go bankrupt. A consensus among manufacturers is also needed in order to make the newly designed cars affordable. To that end, in 1993 the Clinton administration announced a 10-year Partnership for a New Generation of Vehicles program among the "Big Three" auto makers—Chrysler, General Motors, and Ford—and government agencies. It is hoped that this collaborative effort will result in the development by 2004 of a clean car with the same performance and cost of today's automobiles but three times the fuel efficiency. Improved fuel efficiency translates into lower emissions.

What are some of the fuels being considered by carmakers? An engine that runs on electricity is much cleaner and quieter than a gasoline engine (see figure). However, electric cars are cleaner than gasoline-powered cars only when the source of their electricity is natural gas or solar energy. If electricity comes from a coal-fired power plant, electric cars actually produce *more* emissions than gasoline-powered cars.

Methanol and ethanol are alcohol fuels that burn much cleaner than gasoline and can be made from renewable resources such as agricultural waste. Other possibilities are to reformulate gasoline so it burns with fewer emissions and to improve the design of the conventional, gasoline-powered internal-combustion engine so it is more efficient.

Liquid hydrogen is an extremely clean fuel, and some car designs have fuel cells that combine stored hydrogen with oxygen from the air to produce electricity. Mercedes-Benz unveiled a prototype fuel cell-powered minivan in 1995, and BMW displayed hydrogen-fueled cars at the 1996 World Hydrogen Energy Conference in Stuttgart. Other manufacturers are expected to follow the German car industry.

Futurists look ahead to the time in the not-so-distant future when solar hydrogen will power vehicles. Solar hydrogen fuel is produced when solar energy splits water molecules to produce hydrogen. Cars powered by solar hydrogen will require such extensive modification of existing designs, however, that they could not be a viable alternative until well into the 21st century.

Saturn's EV-1 is the nation's first modern electric passenger vehicle that is available to consumers. Estimated annual electricity costs for the EV-1 are between $126 (at 3¢ per kWh) and $420 (at 4¢ per kWh). A comparable gasoline-powered vehicle would have an estimated annual fuel cost of $696.

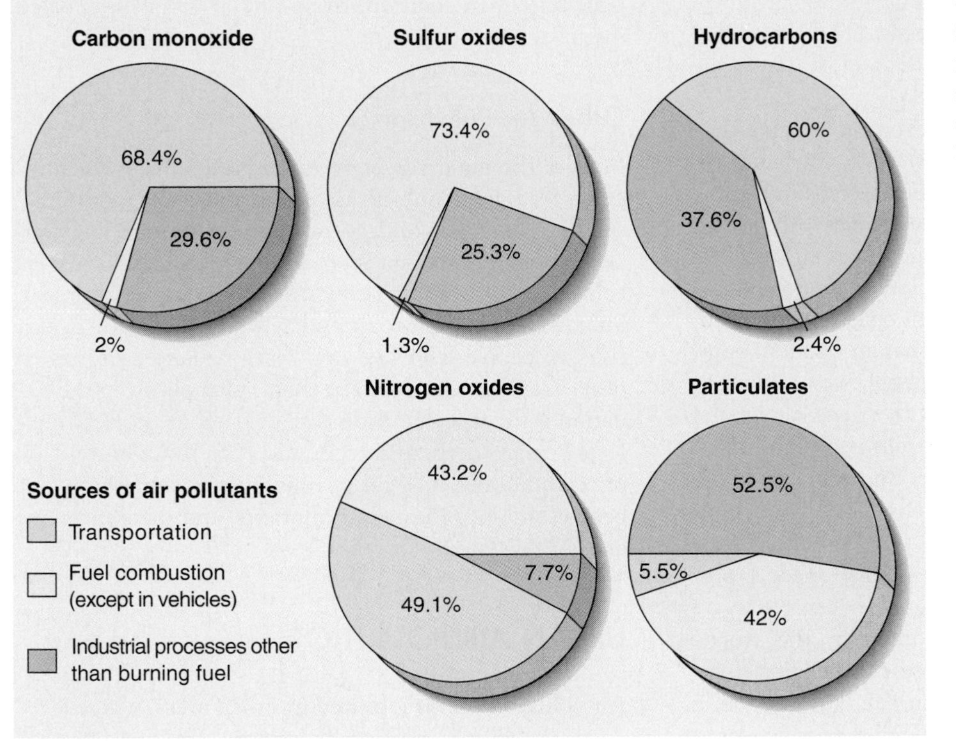

Figure 19–4

Sources of the five major primary air pollutants. Note that motor vehicles (transportation) and industrial fuel combustion are major contributors of pollutants.

Major Kinds of Air Pollutants

Particulate matter consists of thousands of different solid and liquid particles that are suspended in the atmosphere. **Solid particulate matter** is generally referred to as dust, whereas liquid suspensions are commonly called mists. Particulates include soil, soot, lead, asbestos, and sulfuric acid droplets. All particulate matter eventually settles out of the atmosphere, but it is possible for small particles, some of which are especially harmful to humans, to remain suspended in the atmosphere for weeks or even years.

Particulates reduce visibility by scattering and absorbing sunlight. Urban areas receive less sunlight than rural areas, partly as a result of greater quantities of particulate matter in the air. Particulate matter corrodes metals, erodes buildings and works of sculpture when the air is humid, and soils clothing and draperies. Smaller particles are inhaled into the respiratory system and can cause health problems; lead (see Chapter 21) and asbestos particles are especially harmful.

Nitrogen oxides are gases produced by the chemical interactions between nitrogen and oxygen. They consist mainly of nitric oxide (NO), nitrogen dioxide (NO_2), and nitrous oxide (N_2O). Nitrogen oxides inhibit plant growth and, when breathed, aggravate health problems such as asthma. They are involved in the production of photochemical smog, acid deposition, and global warming. Nitrogen oxides cause metals to corrode and textiles to fade and deteriorate.

Sulfur oxides are gases produced by the chemical interactions between sulfur and oxygen. Sulfur dioxide (SO_2), a colorless, nonflammable gas with a strong, irritating odor, is a major sulfur oxide emitted as a primary air pollutant. Another major sulfur oxide is sulfur trioxide (SO_3), a secondary air pollutant that forms when sulfur dioxide reacts with oxygen in the air. Sulfur trioxide, in turn, reacts with water to form another secondary air pollutant, sulfuric acid.

Sulfur oxides are very important in acid deposition, and they corrode metals and damage stone and other materials. Sulfuric acid and sulfate salts that are produced in the atmosphere from sulfur oxides damage plants and irritate the respiratory tracts of humans and other animals.

Carbon oxides are the gases carbon monoxide (CO) and carbon dioxide (CO_2). Carbon monoxide, a colorless, odorless, and tasteless gas produced in the largest quantities of any atmospheric pollutant except carbon dioxide, is poisonous and reduces the blood's ability to transport oxygen. Carbon dioxide, also colorless, odorless, and tasteless, traps heat in the atmosphere and is therefore involved in global climate change.

Hydrocarbons are a diverse group of organic compounds that contain only hydrogen and carbon. Some hydrocarbons are straight or branched chains, and some are cyclic (form rings); the simplest is methane (CH_4). The smaller hydrocarbons are gaseous at room temperature. Methane, for example, is a colorless, odorless gas that is the principal component of natural gas.[1] Medium-sized

[1]The odor of natural gas comes from sulfur compounds that are deliberately added so that humans can detect the presence of the gas.

hydrocarbons are liquids at room temperature, although many are volatile, or evaporate readily. The largest hydrocarbons are actually solids at room temperature—the waxy substance paraffin, for example.

There are many different hydrocarbons, and they have a variety of effects on human and animal health; some appear to cause no adverse effects, some injure the respiratory tract, and others cause cancer. All except methane are important in the production of photochemical smog. Methane is involved in global climate change.

Ozone (O_3) is a form of oxygen considered a pollutant in one part of the atmosphere but an essential component in another. In the stratosphere, which extends from 10 to 45 kilometers (6.2 to 28 miles) above the Earth's surface, oxygen reacts with ultraviolet radiation coming from the sun to form ozone. Stratospheric ozone prevents much of the solar ultraviolet radiation from penetrating to the Earth's surface. Unfortunately, stratospheric ozone is being attacked by human-made pollutants.

Unlike stratospheric ozone, ozone in the troposphere—the layer of atmosphere closest to the Earth's surface—is a human-made air pollutant. (Tropospheric ozone does not replenish the ozone that has been depleted from the stratosphere because it breaks down to form oxygen long before it drifts up to the stratosphere.) Tropospheric ozone is a secondary air pollutant that forms when sunlight catalyzes a reaction between nitrogen oxides and volatile hydrocarbons. The most harmful component of photochemical smog, ozone reduces air visibility and causes health problems. Ozone also stresses plants and reduces their vigor (Figure 19–5). Chronic exposure to ozone lowers crop yields and is a possible contributor to forest decline, which will be considered in more detail in Chapter 20. In addition, tropospheric ozone is involved in global climate change.

Figure 19–5

Plants exposed to ozone exhibit a variety of symptoms, including leaf damage and a lowered productivity. Compare the soybean leaf grown in clean air (right) with the one damaged by ozone (left). (Runk/Schoenberger from Grant Heilman)

Other Air Pollutants

Most of the hundreds of other air pollutants—which include lead, hydrochloric acid, formaldehyde, radioactive substances, and fluorides—are present in very low concentrations, although it is possible to have high local concentrations of specific pollutants. Some of these air pollutants are extremely toxic and may pose long-term health risks to people who live and work around factories or other facilities that produce them. Complicating the situation is the fact that little is known about the health effects of many of these compounds. As a result of this lack of scientific detail, legal air quality standards have not been established for many pollutants, and their emissions are not regulated.

URBAN AIR POLLUTION

Air pollution that is localized in urban areas, where it reduces visibility, is often called **smog**. The word "smog" was coined at the beginning of the 20th century for the smoky fog that was so prevalent in London because of coal combustion. Today there are several different types of smog. Traditional London-type smog—that is, smoke pollution—is sometimes called **industrial smog**. The principal pollutants in industrial smog are sulfur oxides and particulate matter. The worst episodes of industrial smog typically occur during winter months, when household fuel combustion is high.

Another important type of smog is **photochemical smog**, a brownish orange haze formed by chemical reactions involving sunlight (Figure 19–6). First noted in Los Angeles in the 1940s, photochemical smog is worst during the summer months. Both nitrogen oxides and hydrocarbons are involved in its formation. One of the photochemical reactions occurs among nitrogen oxides (largely from automobile exhaust), volatile hydrocarbons, and oxygen in the atmosphere to produce ozone; this reaction requires solar energy (Figure 19–7). The ozone formed in this way then reacts with other air pollutants, including hydrocarbons, to form more than 100 different secondary air pollutants (**p**eroxy**a**cyl **n**itrates [PANs], for example) which can injure plant tissues, irritate eyes, and aggravate respiratory illnesses in humans.

Although the main human source of photochemical smog is the automobile, you may be surprised to know that bakeries and dry cleaners are also significant contributors of the air pollutants that cause photochemical smog. When bread is baked, yeast byproducts are released that are converted to ozone by sunlight. The volatile fumes from dry cleaners also contribute to photochemical smog.

Figure 19–6

Photochemical smog in New York City when a thermal inversion trapped the polluted air near the ground. Metropolitan New York City is ranked in the top ten of U.S. cities with the worst air pollution. (William E. Ferguson)

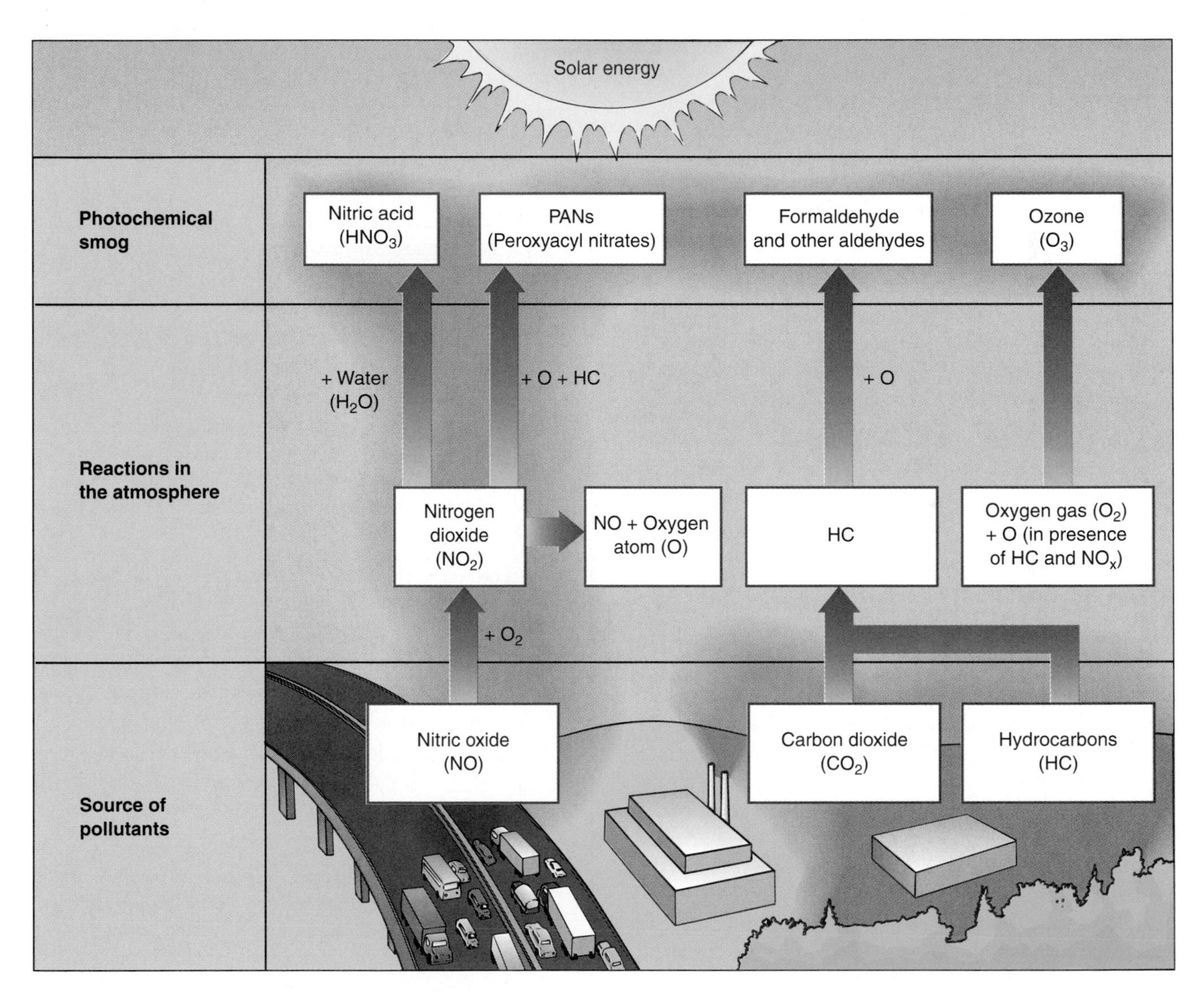

Figure 19–7

Photochemical smog consists of a complex mixture of pollutants. These include ozone, peroxyacyl nitrates (PANs), nitric acid, and organic compounds such as formaldehyde.

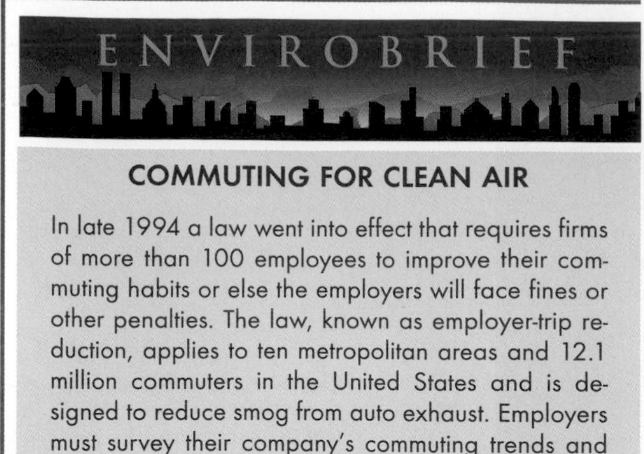

COMMUTING FOR CLEAN AIR

In late 1994 a law went into effect that requires firms of more than 100 employees to improve their commuting habits or else the employers will face fines or other penalties. The law, known as employer-trip reduction, applies to ten metropolitan areas and 12.1 million commuters in the United States and is designed to reduce smog from auto exhaust. Employers must survey their company's commuting trends and then provide incentives for increased use of carpools, bikes, and mass transit. The law requires that each city increase its people-per-car average* by 25 percent. For a city like Houston, where most workers drive to work solo, this will mean an average of 1.47 people per car. In Manhattan, where carpooling and mass transit are already well established, however, the goal is an average of 7.81 people per car. Unfortunately, the gains in air quality may be relatively small and the perceived inconveniences may be great, so employers will have to be creative to accomplish the program's goals. Parking fees, arranged carpools, company buses, shorter work weeks, and guaranteed rides home in emergencies are all potential incentives. Employer-trip reduction is only one of two dozen regulations associated with the 1990 Clean Air Act with which the ten cities must comply to meet a mandated 15 percent reduction in ozone-causing pollutants by 1996.

*People-per-car average includes all means of commuting.

How Climate and Topography Affect Air Pollution

Variation in temperature during the day usually results in air circulation patterns that help to dilute and blow away air pollutants. As the sun increases surface temperatures, the air near the ground is warmed. This heated air expands and rises to higher levels in the atmosphere, causing a low-pressure area near the ground; the surrounding air then moves into the low-pressure area. Thus, under normal conditions, air circulation patterns prevent toxic pollutants from increasing to dangerous levels near the ground.

However, during periods of **thermal inversion**, in which the air near the ground is colder than the air at higher levels (Figure 19–8), polluting gases and particulate matter remain trapped in high concentrations close to the ground, where people live and breathe. Thermal inversions usually persist for only a few hours before they are broken up by atmospheric turbulence. Sometimes, however, atmospheric stagnation caused by a stalled high-pressure air mass allows a thermal inversion to persist for several days.

Certain types of topography (surface features) increase the likelihood of thermal inversions. Cities located in valleys, near the coast, or on the leeward side of mountains (the side toward which the wind blows) are prime candidates for thermal inversions.

Los Angeles

Los Angeles, California, has some of the worst smog in the world. Mountains surround the Los Angeles basin on three sides, and the ocean is on its fourth side. This location, combined with a sunny climate, large population, and high automobile density, is conducive to the formation of stable thermal inversions that trap photochemical smog near the ground.

In 1969 California became the first state to enforce emission standards on motor vehicles, largely because of air pollution problems in Los Angeles. Today Los Angeles has stringent smog controls that regulate everything from alternative fuels for automobiles to lawn mower emissions to paint vapors.

TO MOW IS TO POLLUTE

Environmentalists and federal and state agencies are on the attack against the 89 million mowers and other small-engine garden tools in the United States. In addition to being just plain noisy, the machines threaten air and soil quality:

- According to the EPA, gas-powered mowers, blowers, and other small-engine machines produce 10 percent of total U.S. air pollution.
- On a per-horsepower basis, one hour of mowing with a gas-powered mower releases as many hydrocarbons as a car driven 50 miles.
- When refueling lawn equipment, users spill an estimated 17 million gallons of fuel each year, which is 6 million gallons more than that released in the *Exxon Valdez* spill.

In 1994 the EPA proposed national emission standards for mowers and other lawn machines that were expected to substantially reduce emissions of hydrocarbons and carbon monoxide. As of 1996, however, manufacturers have failed to agree to meet the EPA goals. Despite a lack of progress on regulating emissions, push mowers (also known as reel mowers), electric mowers, and ground covers that do not require mowing are rapidly gaining in popularity.

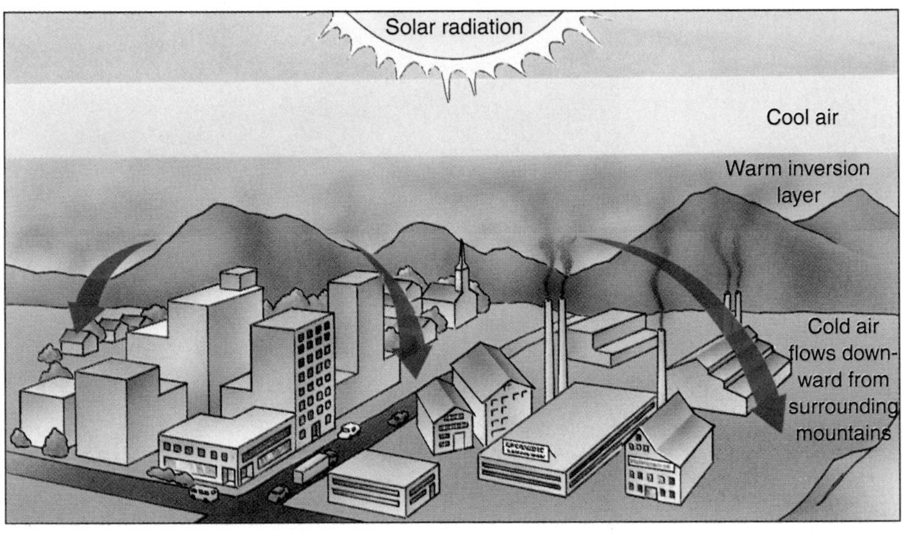

Figure 19–8

How a thermal inversion occurs. (a) Normally, warm, polluted air rises, diluting pollutants in the process. (b) When a thermal inversion occurs, a layer of warm air covers cooler air near the ground and the polluted air is trapped. Pollutants can increase to dangerous levels during thermal inversions.

► *FOLLOW–UP*

After several decades devoted to improving its air quality, Los Angeles has the cleanest skies since the 1950s. Smog levels in the mid-1990s are 25 to 30 percent lower than they were in the 1980s. Despite impressive progress, the city's air is still the dirtiest in the United States. At its current rate of improvement, however, Los Angeles should attain federal clean-air standards by 2010.

Dust Domes

Heat released by human activities such as fuel combustion is also highly concentrated in areas of high population density, known as **urban heat islands**. As a result, the air in these urban areas is warmer than the air in the surrounding suburban and rural countrysides. Urban heat islands affect local air currents and contribute to the buildup of pollutants, especially particulates, with which they form **dust domes** over cities.

CONTROLLING AIR POLLUTANTS

Many of the measures we have already discussed for energy efficiency and conservation (see Chapter 12) also help to reduce air pollution. Smaller, more fuel-efficient automobiles produce fewer emissions, for example. Appropriate technologies exist to control all the forms of air pollution discussed in this chapter except carbon dioxide.

Smokestacks that have been fitted with electrostatic precipitators, fabric filters, scrubbers, or other technologies remove particulate matter (Figure 19–9; see Chapters 10 and 23 for information on air pollution control devices). In addition, particulates are controlled by careful land-excavating activities, such as sprinkling water on dry soil that is being moved during road construction.

(a)

(b)

Figure 19–9

A comparison of emissions from a Delaware Valley steel company with the electrostatic precipitator turned off (a) and on (b). Electrostatic precipitators, in which particles are given a positive charge and then passed over a negatively charged collecting surface, are very effective at removing particulates from air. (a and b, John D. Cunningham/Visuals Unlimited)

Several methods exist for removing sulfur oxides from flue gases, but it is often less expensive to simply switch to a low-sulfur fuel such as natural gas or even to a non-fossil fuel energy source such as solar energy. Sulfur can also be removed from fuels before they are burned, as in coal gasification (see Chapter 10).

Reduction of combustion temperatures in automobiles lessens the formation of nitrogen oxides. Use of mass transit helps reduce automobile use, thereby decreasing nitrogen oxide emissions. Nitrogen oxides produced during high-temperature combustion processes in industry can be removed from smokestack exhausts. The release of nitrogen oxides from cultivated fields to which nitrogen fertilizers have been applied is reduced significantly when no-tillage is practiced (see Chapter 14).

Modification of furnaces and engines to provide more complete combustion helps control the production of both carbon monoxide and hydrocarbons. Catalytic afterburners, used immediately following combustion, oxidize any unburned gases; the use of catalytic converters to treat auto exhaust, for example, can reduce carbon monoxide and volatile hydrocarbon emissions by about 85 percent over the life of the car. Careful handling of

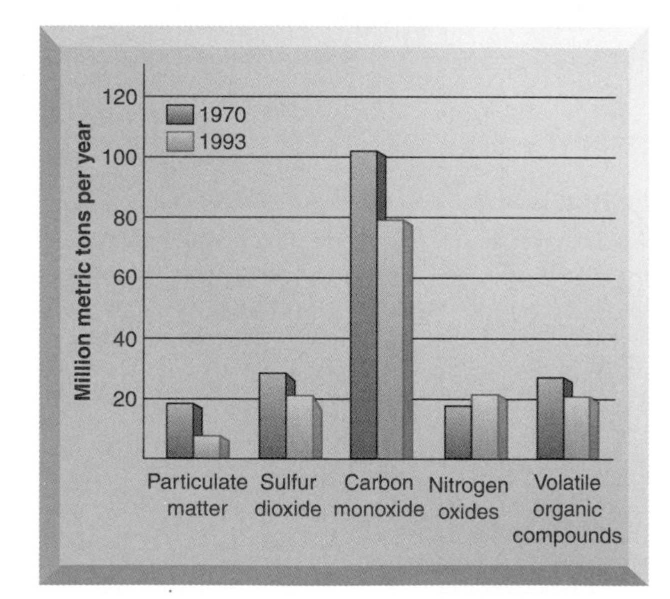

Figure 19–10

A comparison of 1970 and 1993 emissions in the United States. Particulate matter, sulfur dioxide, carbon monoxide, and volatile organic compounds showed decreases, and only nitrogen oxides did not decline.

hydrocarbons such as solvents and petroleum reduces air pollution from spills and evaporation.

AIR POLLUTION IN THE UNITED STATES

The Clean Air Act was first passed in 1970 and has been reauthorized (updated and amended) twice since then, in 1977 and 1990. Overall, air quality in the United States has slowly improved since 1970 (Figure 19–10). The most dramatic improvement has been in the amount of lead in the atmosphere, which showed a 97.5 percent decrease between 1970 and the 1990s, primarily because of the switch from leaded to unleaded gasoline. Levels of sulfur oxides, ozone, carbon monoxide, and particulates have also been reduced. For example, between 1970 and 1993, sulfur dioxide emissions declined by 27.5 percent. Although air quality has been gradually improving, the atmosphere in many urban areas still contains higher levels of pollutants than are recommended based on health standards. Photochemical smog continues to be a major problem in metropolitan areas (Table 19–2).

The Clean Air Acts of 1970, 1977, and 1990 have required progressively stricter controls of motor vehicle emissions. For example, the provisions of the 1990 Clean Air Act include the development of "superclean" cars, which emit lower amounts of nitrogen oxides and hydrocarbons, and the use of cleaner-burning gasolines in the country's most polluted cities; these changes are being phased in gradually by the year 2000. More recent automobile models do not produce as many pollutants as older models (unless the pollution control devices of the recent models have been deliberately tampered with). Yet despite the increasing percentage of newer automobile models on the road, air quality has not improved in some areas of the United States because of the large increase in the *number* of cars being driven. It is doubtful that the 1990 Clean Air Act will greatly alter the number of cars on the road. However, some consumers may be discouraged from purchasing new automobiles because of the cost increase that is expected to result from installing pollution control devices.

The 1990 Clean Air Act focuses on industrial airborne toxic chemicals in addition to motor vehicle emissions. Between 1970 and 1990, the airborne emissions of only seven toxic chemicals were regulated. In comparison, the 1990 Clean Air Act requires a 90 percent reduction in the atmospheric emissions of 189 toxic and cancer-causing chemicals by 2003. To comply with this

MEDICAL POLLUTION

The burning of medical waste is a serious source of air pollution, particularly in cities. American hospitals incinerate an astounding array of waste, including old bandages, soiled bedding, used syringes, contaminated plastics, and pathogenic remains like blood, body parts, and animal carcasses. From each occupied bed comes about 13 pounds of waste per day—an annual output estimated at 2.5 million tons nationwide. Medical waste volumes have increased with the advent of AIDS, mainly because medical personnel are protecting themselves with an elaborate assortment of disposable garments and shields, mostly made from plastic. In fact, disposability has become the standard for almost any conceivable medical item, from syringes and bed linens to bedpans and even telephones. The problem is the quality of incineration. Most of the 6000 incinerators at medical facilities throughout the United States are rudimentary by today's standards. They burn waste incompletely, causing emissions of acidic gases, heavy metals, volatile organic compounds, and dioxins that can be 10 to 100 times higher than those from municipal waste incinerators. The pollution is worst where population density is highest. New York City has almost 60 hospital incinerators in close proximity to each other.

Table 19–2
The Ten U.S. Urban Areas with the Worst Air Quality (Ozone Nonattainment Areas), 1992 to 1994

EXTREME (0.28 PPM OZONE AND GREATER)
Los Angeles South Coast Air Basin, California
VERY SEVERE (0.19 TO 0.28 PPM OZONE)
Houston, Galveston, and Brazoria, Texas
Southeast Desert, California
New York City and Long Island; Northern New Jersey; Connecticut
Milwaukee and Racine, Wisconsin
Chicago; Gary and Lake County, Indiana
SEVERE (0.18 TO 0.19 PPM OZONE)
Ventura County (between Santa Barbara and Los Angeles), California
Sacramento, California
Philadelphia; Wilmington, Delaware; Trenton, New Jersey
Baltimore, Maryland

requirement, both small businesses such as dry cleaners and large manufacturers such as chemical companies will have to install pollution control equipment. Sulfur dioxide and nitrogen oxide emissions from coal-fired power plants will be substantially reduced by the year 2000; these provisions of the Clean Air Act are the first U.S. legislation to target acid rain. The provisions for stratospheric ozone depletion initially required the United States to phase out the production of ozone-destroying chemicals by the year 2000; the phase-out date was subsequently moved up to 1996.

AIR POLLUTION IN DEVELOPING COUNTRIES

As developing nations become more industrialized, they also produce more air pollution. The leaders of most developing countries believe they must become industrialized rapidly in order to compete economically with developed countries. Environmental quality is usually a low priority in the race to develop. Thus, while air quality is slowly improving in the United States and other developed countries, it is deteriorating in developing nations. For example, Shenyang and neighboring cities in China have so many smokestacks belching coal smoke that residents can see the sun only a few weeks of the year. The rest of the time they are choked in a haze of orange-colored coal dust.

The growing number of automobiles in developing countries is also contributing to air pollution, particularly in urban areas. Many vehicles in these countries are old and have no pollution control devices. During the 1990s the most rapid proliferation of motor vehicles worldwide occurred in Latin America, Asia, and Eastern Europe. Brazil produced almost 40 percent more automobiles in 1994 than it did in 1992, for example, and China's auto production increased by almost 30 percent a year in the mid-1990s.

Lead pollution from heavily leaded gasoline is an especially serious problem in developing nations. In Cairo, for example, children's blood lead levels are more than two times higher than the level considered at-risk in the United States.

Mexico City

Mexico City, the world's fourth largest city, has the most polluted air of any major metropolitan area in the world (Figure 19–11). This is due in part to its great population growth in the past 40 years and in part to the city's location. Mexico City is in an elevated valley that is ringed on three sides by mountains; winds coming in from the open northern end are trapped in the valley. Air quality is at its worst from October to January, largely as a result of seasonal variations in atmospheric conditions.

The city has almost 3 million motor vehicles, 360 gasoline stations, and about 30,000 businesses, which the Mexican government says spew 3.94 million metric tons (4.35 million tons) of pollutants into the air each year. In addition, liquefied petroleum gas, which is the major source of energy for cooking and heating in Mexico City, leaks in unburned form into the atmosphere, releasing hydrocarbons that contribute significantly to the city's air pollution.

In 1990 Mexico embarked on an ambitious plan to improve Mexico City's air quality by gradually replacing old buses, taxis, delivery trucks, and cars with "cleaner" vehicles, such as those with catalytic converters, and by switching to unleaded gasoline. In addition, driving restrictions apply when the air quality is particularly poor, and exhaust emissions are periodically checked on autos.

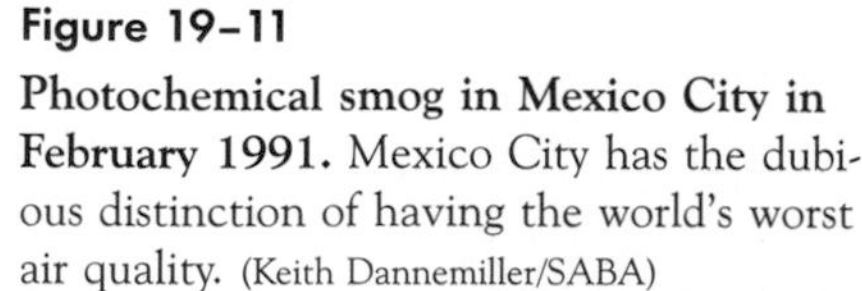

Figure 19–11

Photochemical smog in Mexico City in February 1991. Mexico City has the dubious distinction of having the world's worst air quality. (Keith Dannemiller/SABA)

► *FOLLOW-UP*

Although there have been reductions in lead, sulfur dioxide, and carbon monoxide, Mexico City's air pollution plan has not been totally effective. Critics say that the city government does not have the political will to aggressively enforce air pollution control measures in the face of widespread corruption. As a result, air quality in Mexico City was worse in 1996 than it was in 1986. In 1995, for example, ozone levels exceeded Mexico City's standards of acceptability on 324 days.

AIR POLLUTION IN REMOTE REGIONS

Certain hazardous air pollutants are distributed globally by atmospheric transport. Organochlorine compounds, such as the pesticides lindane and DDT, may be restricted from use or even banned by many industrialized countries. Yet because they are volatile, they move through the atmosphere from warmer to colder areas, where they are deposited on land and bodies of water. The process in which volatile chemicals evaporate from land as far away as the tropics and are transported by winds to higher latitudes, where they condense and fall to the ground, is known as the **global distillation effect**.

A study published in 1995 reported that many industrialized countries continue to be highly contaminated by persistent organochlorine compounds despite their restricted use. The effect is more pronounced at higher latitudes. Dangerous levels of certain pesticides have been measured in the Yukon (northwestern Canada) and in other arctic regions. These chemicals enter food webs and become concentrated in the body fat of animals at the top of the food chain (see discussion of biological magnification in Chapter 22). Fishes, seals, polar bears, and arctic people are particularly vulnerable.

Concern over protecting the Arctic led to the formation of the Arctic Council, consisting of Canada, the United States, Russia, Finland, Norway, Sweden, Denmark, and Iceland, in 1996. The group plans to meet in 1998 after additional environmental studies on arctic issues have been completed.

AIR POLLUTION AND HUMAN HEALTH

Generally speaking, exposure to low levels of pollutants such as ozone, sulfur oxides, nitrogen oxides, and particulates irritates the eyes and causes inflammation of the respiratory tract (Table 19–3). Evidence exists that many air pollutants also suppress the immune system, increasing susceptibility to infection. In addition, evidence continues to accumulate indicating that exposure to air pollution during respiratory illnesses may result in the development later in life of chronic respiratory diseases, such as emphysema and chronic bronchitis. Some other health problems that can result from long-term exposure to toxic air pollutants are cancer, chronic obstructive pulmonary disease, asthma, respiratory infections, and cardiovascular disease.

Health Effects of Specific Air Pollutants

Both sulfur dioxide and particulates irritate the respiratory tract and, because they cause the airways to constrict, actually impair the lungs' ability to exchange gases. People suffering from asthma and emphysema are very sensitive to sulfur dioxide and particulate pollution. Ni-

Table 19–3
Health Effects of Several Major Air Pollutants

Pollutant	Source	Effects
Particulate matter	Industries, motor vehicles	Aggravates respiratory illnesses; long-term exposure may cause increased incidence of chronic conditions such as bronchitis
Sulfur oxides	Electric power plants and other industries	Irritate respiratory tract; same effects as particulates
Nitrogen oxides	Motor vehicles, industries, heavily fertilized farmland	Irritate respiratory tract; aggravate respiratory conditions such as asthma and chronic bronchitis
Carbon monoxide	Motor vehicles, industries	Reduces blood's ability to transport O_2; headache and fatigue at lower levels; mental impairment or death at high levels
Ozone	Formed in atmosphere (secondary air pollutant)	Irritates eyes; irritates respiratory tract; produces chest discomfort; aggravates respiratory conditions such as asthma and chronic bronchitis

trogen dioxide also causes airway constriction and, in people suffering from asthma, an increased sensitivity to pollen and dust mites (microscopic animals found in household dust).

One of the largest studies ever conducted on the effects of air pollution on human health was published in 1995. This landmark Harvard study tracked the health records of more than one half million people in 151 cities from 1982 to 1989. The study compared mortality data with the amount of fine particulate matter (particles less than 2.5 micrometers in diameter) at each location. Because so many people were included in the study, scientists were able to cancel out the effects of tobacco, alcohol, poverty, and other factors that are related to death rates. The study found that people who live and work in the country's most polluted cities are 15 to 17 percent more likely to die prematurely than those living in cities with the cleanest air. Fine particles, which are emitted by motor vehicles, power plants, industries, and other human activities, are considered more dangerous than larger particles because they are breathed more deeply into the lungs.

Carbon monoxide combines with the blood's hemoglobin, reducing its ability to transport oxygen. At medium concentrations, carbon monoxide causes headaches and fatigue. As the concentration increases, reflexes slow down and drowsiness occurs; at a certain high level, carbon monoxide causes death. People at greatest risk from carbon monoxide include pregnant women, infants, and those with heart or respiratory diseases. For example, a four-year study in seven U.S. cities—Chicago, Detroit, Houston, Los Angeles, Milwaukee, New York, and Philadelphia—linked carbon monoxide concentrations in the air to increases in hospital admissions for congestive heart failure.

Ozone and the volatile compounds in smog cause a variety of health problems, including burning eyes, coughing, and chest discomfort. Ozone also brings on asthma attacks and suppresses the immune system.

Children and Air Pollution

Air pollution is a greater health threat to children than it is to adults. The lungs continue to develop throughout childhood, and air pollution can restrict lung development. In addition, a child has a higher metabolic rate than an adult and therefore needs more oxygen. To obtain this oxygen, a child breathes more air—about two times as much air per pound of body weight compared to an adult. This means that a child also breathes more air pollutants into the lungs. A 1990 study in which autopsies were performed on 100 Los Angeles children who died for unrelated reasons found that more than 80 percent had subclinical lung damage, which is lung disease in its early stages, before clinical symptoms appear. A long-term study is currently underway to determine the effects on children's developing lungs of chronic exposure to air pollution.

INDOOR AIR POLLUTION

The air in enclosed places such as automobiles, homes, schools, and offices may have significantly higher levels of air pollutants than the air outdoors. In congested traffic, for example, levels of harmful pollutants such as carbon monoxide, benzene, and airborne lead may be several times higher inside an automobile than in the air immediately outside. The concentrations of certain indoor air pollutants may be five times greater than those outdoors. Indoor pollution is of particular concern to urban residents because they typically spend 90 percent or more of their time indoors.

Because illnesses caused by indoor air pollution usually resemble common ailments such as colds, influenza, or upset stomachs, they are often not recognized. The most common contaminants of indoor air are radon, cigarette smoke, carbon monoxide, nitrogen dioxide (from gas stoves), formaldehyde (from carpeting, fabrics, and furniture), household pesticides, cleaning solvents, ozone (from photocopiers), and asbestos (Figure 19–12; also see Focus On: Smoking). In addition, viruses, bacteria, fungi (yeasts, molds, and mildews), dust mites, pollen, and other organisms or their toxic parts found in heating, air conditioning, and ventilation ducts are important forms of indoor air pollution.

Health officials are paying increasing attention to the **sick building syndrome**, the presence of air pollution inside office buildings that can cause eye irritations, nausea, headaches, respiratory infections, depression, and fatigue. The Labor Department estimates than more than 20 million employees are exposed to health risks from indoor air pollution. The EPA estimates that the annual medical costs for treating the health effects of indoor air pollution in the United States exceed $1 billion. When lost work time and diminished productivity are added to health care costs, the total annual cost to the economy may be as much as $50 billion. Fortunately, most building problems are relatively inexpensive to alleviate.

Radon

Not all environmental health hazards are the result of human activities. The most serious indoor air pollutant is likely to be **radon**, a colorless, tasteless radioactive gas produced naturally during the radioactive decay of uranium in the Earth's crust. Radon seeps through the

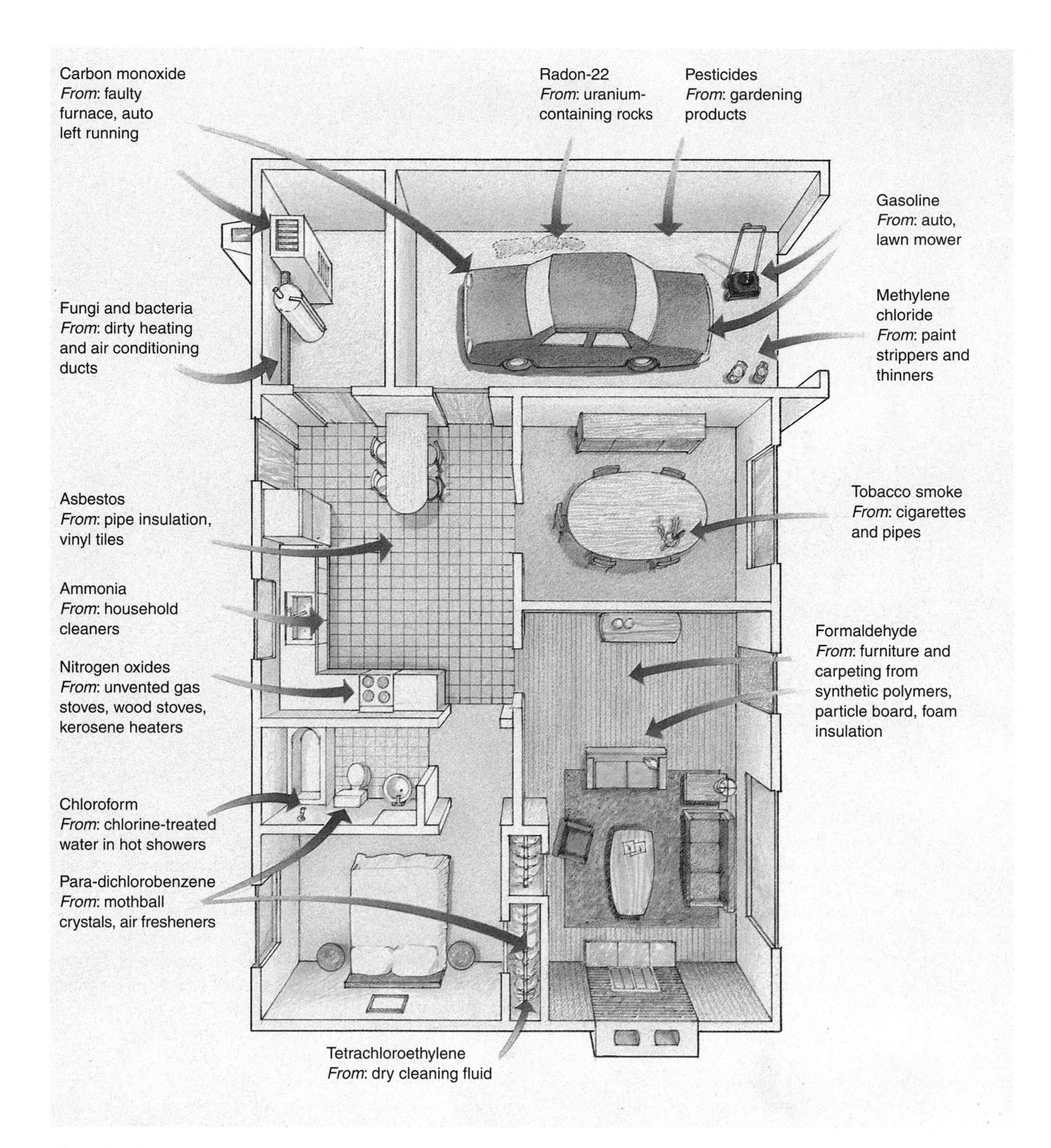

Figure 19–12

Indoor air pollution comes from a variety of sources. Homes may contain higher levels of toxic pollutants than outside air, even near polluted industrial sites.

Figure 19–13
How radon infiltrates a house. Cracks in basement walls or floors, openings around pipes, and pores in concrete blocks provide some of the entries for radon.

ground and enters buildings, where it sometimes accumulates to dangerous levels (Figure 19–13). Although radon is also emitted into the atmosphere, it gets diluted and dispersed and is of little consequence outdoors.

Radon and its decay products emit alpha particles, a form of ionizing radiation that is very damaging to tissue but cannot penetrate very far into the body. Consequently, radon can harm the body only when it is ingested or inhaled. The radioactive particles lodge in the tiny passages of the lungs and damage surrounding tissue. There is compelling evidence, based primarily on several studies of uranium miners, that inhaling large amounts of radon increases the risk of lung cancer. Several other studies suggest that people who are exposed to relatively low levels of radon over an extended time are also at risk for lung cancer. Of the 140,000 deaths from lung cancer in the United States each year, the EPA attributes about 13,000 to radon.[2]

[2]The actual number of deaths from lung cancer that can be attributed to radon is estimated to be between 5000 and 20,000. A reasonable middle-of-the-range estimate is 13,000.

Ironically, efforts to make our homes more energy-efficient have increased the radon hazard. Drafty homes waste energy but allow radon to escape outdoors so it does not build up inside. According to the EPA, the number of American homes with high enough levels of radon to warrant corrective action—that is, a radon level above 4 picocuries per liter of air—may be as great as 8 million, which represents almost 10 percent of all homes. Between 50,000 and 100,000 of these homes have radon levels in excess of the current occupational standard for underground miners. The highest radon levels in this country have been found in homes on a geological formation called the **Reading Prong**, which runs across Pennsylvania into northern New Jersey and New York. Iowa appears to have the most pervasive radon problem, where 71 percent of the homes tested in early 1989 had radon levels high enough to warrant corrective action.

Every home should be tested for radon because levels vary widely from home to home, even in the same neighborhood. Testing is inexpensive, and corrective actions are reasonably priced. Radon concentrations in homes can be minimized by sealing basement concrete floors and by ventilating crawl spaces and basements.

FOCUS ON

Smoking

Smoking, which causes serious diseases such as lung cancer, emphysema, and heart disease, is responsible for the premature deaths of nearly half a million people in the United States each year. Of the 140,000 deaths from lung cancer in the United States each year, about 120,000 of them are estimated to be caused by cigarette smoking. Smoking also contributes to cardiovascular disease, which contributes to heart attacks and strokes, and to cancer of the bladder, mouth, throat, pancreas, kidney, stomach, voice box, and esophagus.

Passive smoking—nonsmokers' chronic breathing of smoke from cigarette smokers——also increases the risk of cancer. In addition, passive smokers suffer more respiratory infections, allergies, and other chronic respiratory diseases than other nonsmokers. Passive smoking is particularly harmful to infants and young children, pregnant women, the elderly, and people with chronic lung disease. When parents of infants smoke, for example, the infant has double the chance of pneumonia or bronchitis in its first year of life.

There is good news and bad news about smoking. The good news is that fewer people in developed nations are smoking. A poll taken in 1995 found that about 25 percent of American adults said they were currently smoking, compared with a peak of 41 percent in the mid-1960s. Smoking has also declined in most European countries and in Japan.

The bad news is that more and more people are taking up the habit in China, Brazil, Pakistan, and other developing nations. In some countries, the smoking habit costs as much as 20 percent of a worker's annual income. Tobacco companies in the United States promote smoking abroad (see figure), and a substantial portion of our tobacco crop is exported.

Although fewer Americans are smoking, certain groups in our society still have high numbers of tobacco addicts, including certain minority groups and those with the least education. A need exists to continue educating these groups, as well as all young people (more than one million American children and teenagers take up smoking each year), about the dangers of smoking.

A cigarette ad in Myanmar (Burma). U.S. tobacco companies export the cigarette habit abroad to compensate for a lower consumption in America. (Andy Hernandez/Gamma Liaison)

Asbestos

Asbestos is a natural mineral that does not burn or conduct heat or electricity (Figure 19–14). These properties make asbestos valuable in construction and industry for fire retardant materials, electrical insulation, and roofing and pipe insulation. It is also used in automobile brake linings.

Asbestos has the ability to separate into long, thin fibers, which are so small and lightweight that they can remain suspended in air almost indefinitely. Asbestos fibers are easily inhaled into the lungs. The mucous membranes of the respiratory system remove some of the asbestos, but if more is inhaled than the body can handle, some of it lodges in the lungs.

Definite links exist between exposure to asbestos and several serious diseases, including lung cancer and **mesothelioma**, a rare and almost always fatal cancer of the body's internal linings. The lag time between exposure to asbestos and development of cancer is anywhere from 20 to 40 years. The danger from inhaling high levels of asbestos is well documented, but it is not known

Figure 19–14

Asbestos is a mineral that occurs naturally in the Earth's crust. Note its fibrous nature. When asbestos crumbles, microscopic fibers are released into the air. These fibers sometimes accumulate indoors, causing a health hazard when inhaled into the lungs. (Charlie Winters)

what, if any, danger exists from inhaling small amounts of asbestos.

The risk of developing cancer from exposure to asbestos fibers is extremely small for most people. The EPA estimates that between 3000 and 12,000 Americans die each year from asbestos-related cancer. Construction workers, fire fighters and building custodians are more likely than others to be exposed to asbestos fibers. Smoking also greatly increases the risk of developing disease from asbestos exposure.

Although health concerns about asbestos have been known since the 1920s, it was not until 1979 that the EPA began regulating its use. During the 1980s concern increased about the widespread presence of asbestos in schools, public and commercial buildings, and homes. Federal legislation (the Asbestos Hazard and Emergency Response Act of 1986) now requires the removal of asbestos from all school buildings. Under a mandate from the EPA, all uses of asbestos must be phased out by 1997.

By the early 1990s scientists, physicians, and other experts generally agreed that the threat of asbestos has been greatly exaggerated. The presence of asbestos in a building does not necessarily mean that people are at risk, unless the asbestos is exposed and crumbling. A much greater danger probably comes from the fibers that are released to the air when asbestos is removed improperly. As of the mid-1990s, most experts think that schools and other public buildings should be left alone. It is both safer and less expensive to seal the asbestos by painting it or covering it up instead of trying to remove it.

NOISE POLLUTION

Sound is caused by vibrations in the air (or some other medium) that reach the ears and stimulate a sensation of hearing. Sound is called **noise** when it becomes loud or disagreeable, particularly when it results in physiological or psychological harm. Of the 28 million Americans with hearing impairments, as many as 10 million could attribute their impairments at least in part to noise. Like other kinds of pollution in the environment, noise pollution can be reduced, although there is a cost associated with its reduction.

Most of the noise produced in the environment is of human origin. Vehicles of transportation, from trains to power boats to snowmobiles, produce a great deal of noise. Power lawn mowers, jets flying overhead, chain saws, jackhammers, and heavy traffic are just a few examples of the outside noise that assails our ears. Indoors, dishwashers, trash compactors, washing machines, televisions, and stereos add to the din.

Measuring Noise

The **intensity** (loudness) of sound is measured relative to a reference sound that is so low it is almost inaudible to the human ear. Relative loudness is expressed numerically using the **decibel (db)** scale or a modified decibel scale called the **decibel-A (dbA)** scale, which takes into account high-pitched sounds to which the human ear is more sensitive (Table 19–4). Sound that is barely audible, such as rustling leaves or breathing, is rated at 10 dbA. A quiet neighborhood during the day might have a background sound level equivalent to 50 dbA. Noise at 90 dbA (such as a motorcycle at close range) impairs hearing, and noise at 120 dbA (such as a chain saw) causes pain.

Effects of Noise

Prolonged exposure to noise damages hearing. The part of the ear that perceives sound is the **cochlea**, a spiral tube that resembles a snail's shell. Inside the cochlea are approximately 24,000 **hair cells** that detect differences in pressure caused by sound waves. When the hairlike projections of the hair cells move back and forth in response to sound, the auditory nerve sends a message to the brain. Loud, high-pitched noise injures the hair cells in the cochlea. Because injured hair cells are not replaced

Table 19-4
The Decibel-A Scale (dbA)

dbA	Example	Perception/General Effects
0		Hearing threshold
10	Rustling leaves, breathing	Very quiet
20	Whisper	Very quiet
30	Quiet rural area (night)	Very quiet–quiet
40	Library	Quiet
50	Quiet neighborhood (daytime)	Quiet–moderately loud
60	Average office conversation	Moderately loud
70	Vacuum cleaner, television	Moderately loud
80	Washing machine, typical factory	Very loud, intrusive
90	Motorcycle at 8 meters	Very loud; impaired hearing with prolonged exposure
100	Dishwasher (very close), jet flyover at 300 meters	Very loud–uncomfortably loud
110	Rock band, boom box held close to ear	Uncomfortably loud
120	Chain saw	Uncomfortably loud–painfully loud
130	Riveter	Painfully loud
140	Deck of aircraft carrier	Painfully loud
150	Jet at takeoff	Painfully loud–ruptured eardrum

by the body, prolonged exposure to loud noise results in permanent hearing impairment.

In addition to hearing loss, noise produces several physiological effects in the body. It increases the heart rate, dilates the pupils, and causes muscle contraction. Evidence exists that prolonged exposure to high levels of noise causes a permanent constriction of blood vessels, which can increase the blood pressure, thereby contributing to heart disease. Other physiological effects associated with noise pollution include migraine headaches, nausea, dizziness, and gastric ulcers. Noise pollution also causes psychological stress.

Controlling Noise

Obviously, noise pollution can be reduced by producing less noise. This can be accomplished in a variety of ways, from restricting the use of sirens and horns on busy city streets to engineering motorcycles, vacuum cleaners, jackhammers, and other noisy devices so that they produce less noise. The engineering approach is technologically feasible but is often avoided because consumers associate loud noise with greater power.

Putting up shields between the noise producer and the hearer can also help control noise pollution. Examples of sound shields include the noise barriers erected along heavily traveled highways and the noise-absorbing material installed around dishwashers. Earplugs are an effective way to protect oneself from unwanted noise; they differ from the preceding examples in that they shield the receiver rather than the noise producer.

ELECTROMAGNETISM AND HUMAN HEALTH

Scientists have been investigating a possible link between electric and magnetic fields associated with such common items as power lines, electric blankets, video displays, and microwave ovens and health hazards. We are subjected to electromagnetic fields outdoors (high-voltage transmission lines) as well as indoors (when electricity is used in household appliances, electrical wiring, and light fixtures).

Several studies have suggested that children who live near power lines may have an increased risk of leukemia.

In addition, a study of men who died from brain tumors found that a disproportionate number of them had careers (such as electricians, telephone workers, and electronics engineers) in which they would have received higher-than-normal exposures to electromagnetic fields. Some laboratory research has indicated that cells are altered by exposure to weak electric and magnetic fields, but no convincing biological data have yet been published that establish a definite link between electromagnetism and a higher risk of cancer.

Because of public concern about possible health hazards from electromagenetic fields, Congress requested the National Research Council (NRC)[3] to decide if protective regulations were needed. The NRC report, released in 1996 after more than 500 studies were reviewed, concluded that electromagnetic fields do not pose significant health risks.

[3]The National Research Council is a private, nonprofit society of distinguished scholars. It was organized by the National Academy of Sciences to advise the federal government on complex issues in science and technology.

SUMMARY

1. Air pollution comes from both natural sources and human activities, particularly motor vehicles and industries. Trees and other plants produce hydrocarbons in response to heat; the contribution of biologically generated hydrocarbon emissions in some areas is substantial. The six main types of air pollutants produced by human activities are particulates, nitrogen oxides, sulfur oxides, carbon oxides, ozone, and hydrocarbons.

2. Particulates are solid particles and liquid droplets suspended in the air. Nitrogen oxides, sulfur oxides, carbon oxides, and ozone are gaseous air pollutants. Hydrocarbons may be solids, liquids, or gases, depending on their molecular size. Hundreds of other chemicals, some of which are toxic, are emitted into the air.

3. Two kinds of smog occur, industrial smog and photochemical smog. Industrial smog is composed primarily of sulfur oxides and particulates. The formation of photochemical smog involves a complex series of chemical reactions involving nitrogen oxides, hydrocarbons, ozone, and sunlight.

4. Certain climates and topographies cause thermal inversions, in which the lower layers of air are cooler than higher layers. Thermal inversions that persist in congested urban areas cause air pollutants to accumulate to dangerous levels. An area of local heat production associated with high population density is known as an urban heat island. Heat islands affect air currents and can cause pollutants to accumulate in dust domes over cities.

5. Air quality in the United States has slowly improved since passage of the Clean Air Act. The most dramatic improvement has been in the amount of lead in the atmosphere, although levels of sulfur oxides, ozone, carbon monoxide, and particulates have also been reduced. While air quality is slowly improving in the United States and other developed countries, it is deteriorating in developing nations.

6. The process in which volatile chemicals evaporate from land as far away as the tropics and are transported by winds to higher latitudes, where they condense and fall to the ground, is known as the global distillation effect. As a result of the global distillation effect, some remote arctic regions are contaminated by volatile chemicals.

7. Air pollutants can have adverse effects on humans and other organisms. They can also damage materials and reduce visibility. All major forms of air pollution except carbon dioxide can be prevented or controlled with current technologies, although control may involve considerable expense.

8. The effect of air pollution on human health is a concern both outdoors—particularly in cities—and in enclosed places. In general, air pollutants irritate the eyes, inflame the respiratory tract, and suppress the immune system. Adults at greatest risk from air pollution include those with heart and respiratory diseases. Air pollution is a greater health threat to children than it is to adults because air pollution impedes lung development.

9. The most serious indoor air pollutant is radon, a radioactive gas that is produced naturally during the radioactive decay of uranium in the Earth's crust. Although

loose asbestos fibers in indoor air are an extremely dangerous threat to human health, asbestos is usually not in this form and thus is normally not a major health hazard.

10. Noise pollution not only has the potential to cause hearing impairment, but also alters many physiological processes in the body and causes psychological stress. Electric and magnetic fields produced by power lines, electrical wiring, and electrical appliances may be hazardous to human health and should be evaluated further.

THINKING ABOUT THE ENVIRONMENT

1. Which is a more stable atmospheric condition, cool air layered over warm air or warm air layered over cool air? Explain. Which condition is a thermal inversion?

2. Why might it be more effective to control smog in heavily wooded areas like Atlanta by reducing nitrogen oxides than by reducing volatile hydrocarbons?

3. List the six main kinds of air pollutants and briefly describe their sources and effects.

4. How have energy conservation efforts contributed to indoor air pollution and the sick building syndrome?

5. A recent study suggests that it might be safer for rock collectors to keep their home collections of minerals in display cases that are vented to the outdoors. Can you offer an explanation for this recommendation, based on what you have learned in this chapter?

6. One of the most effective ways to reduce the threat of radon-induced lung cancer is to quit smoking. Explain.

7. Why are some places considering restricting the loudness of music in public places, including nightclubs and concerts, to 85 dbA?

8. Some people have proposed that all smoking in public places be banned. Do you agree or disagree? Why?

* **9.** Graph the data in Table A on world sulfur emissions, 1950 to 1993. What long-term trend does the shape of the graph indicate? Does the graph show a recent (1990 to 1993) trend? If so, what?

Table A
World Sulfur Emissions, 1950 to 1993

Year	(Million Tons) World Sulfur Emissions
1950	30.1
1960	46.2
1970	57.0
1980	62.9
1990	68.7
1993	69.5

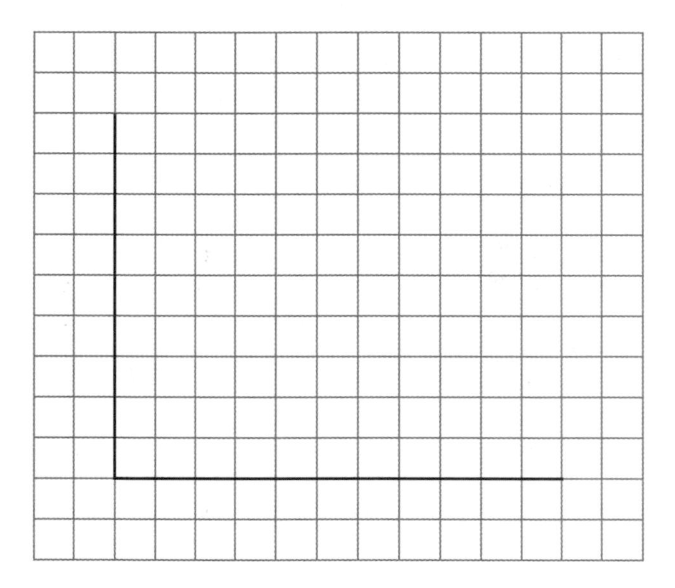

RESEARCH PROJECTS

Obtain annual data for the six major air pollutants in your county for the past five years, in terms of emissions and air quality measurements. Describe any overall trends, and construct a graph to support your assessment. Describe any associated air quality protection programs.

Investigate the current cigarette tax and price per pack in Canada. How has the Canadian tax affected teenage cigarette smoking? Spending on health care? Tax revenues?

On-line materials relating to this chapter are on the ► World Wide Web at **http://www.saunderscollege.com/lifesci/environment2** (select Chapter 19 from the Table of Contents).

SUGGESTED READING

"Air Pollution in the World's Megacities: A Report from the U.N. Environment Programme and the World Health Organization," *Environment* Vol. 36, No. 2, March 1994. An extensive assessment of urban air pollution in the world's largest cities.

Bartecchi, C. E., T. D. MacKenzie, and R. W. Schrier. "The Global Tobacco Epidemic," *Scientific American*, May 1995. The number of people who smoke is on the rise in developing nations.

Blum, D. "Nuclear Detectives," *Discover*, April 1993. Nuclear physics can be used to help determine the source of pollutants that have travelled many miles in the atmosphere.

Edwards, R. "A Tank of the Cold Stuff," *New Scientist*, November 23, 1996. The German car industry is committed to developing cars that are fueled by nonpolluting hydrogen.

"Electromagnetic Fields," *Consumer Reports*, May 1994. Examines a possible link between electromagnetic fields and cancer and suggests simple ways to reduce exposure to electromagnetic fields.

Hayes, R. D. "Ravaged Republics," *Discover*, March 1993. The air pollution in the Czech Republic and Slovakia is among the world's worst.

Lents, J. M., and W. J. Kelly. "Clearing the Air in Los Angeles," *Scientific American*, October 1993. Los Angeles has the worst air pollution of any city in the United States, but it is making progress in improving its air quality.

Monastersky, R. "Attack of the Vog: Natural Air Pollution Has Residents of Hawaii All Choked Up," *Science News* Vol. 147, May 6, 1995. Vog (volcanic smog) produced by a persistently erupting volcano on the island of Hawaii has been blamed for many of the residents' respiratory complaints.

Sperling, D. "The Case for Electric Vehicles," *Scientific American*, November 1996. The author thinks that electric vehicles will inevitably replace internal-combustion engines.

Wu, C. "Hybrid Cars: Renewed Pressure for Fuel-Efficient Vehicles," *Science News* Vol. 148, October 7, 1995. Explores hybrid electric vehicles, which have both electric motors and gasoline-powered engines.

Tree seedlings for a reforestation project in Guatemala. (CARE Photo by Rudolph von Bermuth)

Global Atmospheric Changes

During the 1990s, farmers in Guatemala are planting 52 million trees, paid for by Applied Energy Services (AES), an American company that recently built a coal-burning power plant in Connecticut. This project was undertaken to mitigate the adverse effects of carbon dioxide, an air pollutant produced in large quantities when coal and other fossil fuels are burned. (Carbon dioxide and certain other atmospheric pollutants trap solar heat in the atmosphere and may cause the Earth's climate to warm.) The trees will remove the same amount of carbon dioxide from the air that the new power plant will release into the air during its 40-year life span.

The tree-planting venture is actually a continuation of an established reforestation project by the Cooperative for American Relief Everywhere (CARE), an international development organization. CARE has been working in Guatemala since 1974 to help poor farmers improve their degraded lands through reforestation and soil conservation practices. However, CARE's funding was scheduled to expire in 1989. AES provided $2 million to continue the project, which is staffed by Peace Corps volunteers and involves about 40,000 Guatemalan landholders.

This reforestation project makes use of **agroforestry**, in which both forestry and agricultural techniques are used to improve degraded areas. In agroforestry, crops are often planted between the rows of tree seedlings. The trees grow for many years and provide several environmental benefits, such as reducing soil erosion, regulating the release of rainwater, and providing habitat for the natural enemies of crop pests. Over time, the degraded land slowly improves. When the trees are so tall that they shade out the crops, the forest provides the farmers with food (such as fruits and nuts), fuel wood, lumber, and other forestry products.

Although AES is to be commended for an innovative approach to the environmental problem of global warming, planting trees is only a partial solution. It has been calculated, for example, that an area larger than the United States would have to be reforested *each year* to offset both the world's annual emissions of CO_2 and the deforestation that is currently occurring (see Chapter 17). In order to reduce the possibility of disastrous consequences from accelerated global warming during the 21st century, other ways must be found to reduce our CO_2 emissions.

In this chapter we examine how air pollutants cause three serious regional and global changes: acid deposition (commonly called acid rain), global warming, and depletion of the stratospheric ozone shield. ◀

ACID DEPOSITION

What do lakes without fishes in the Adirondack Mountains, recently damaged Mayan ruins in southern Mexico, and dying red spruce trees in the Appalachian Mountains have in common? The answer is that their damage may be the result of acid precipitation or, more properly, **acid deposition**. Acid deposition is a type of air pollution. It includes sulfuric and nitric acids in precipitation (sometimes called **wet deposition**) as well as dry, sulfuric and nitric acid–containing particles that settle out of the air (sometimes called **dry deposition**).

Acid deposition is not a new phenomenon. It has been around since the Industrial Revolution began. The term "acid rain" was coined in the 19th century by a chemist who noticed that buildings in areas with heavy industrial activity were being worn away by rain.

Acid precipitation, including acid rain, sleet, snow, and fog, poses a serious threat to the environment. The Northern Hemisphere has been hurt the most—especially the Scandinavian countries, central Europe, Russia, China, and North America. In the United States alone, the damage from acid deposition has been estimated at $8 billion each year.

Measuring Acidity

The relative degree of acidity or alkalinity of a substance is expressed using the pH scale, which runs from zero to 14 (see Appendix I). A pH of 7 is neither acidic nor alkaline, whereas a pH less than 7 indicates an acidic solution. The pH scale is logarithmic, so a solution with a pH of 6 is ten times more acidic than a solution with a pH of 7, a solution with a pH of 5 is ten times more acidic than a solution with a pH of 6, and so on. A solution with a pH greater than 7 is alkaline, or basic.

For purposes of comparison, distilled water has a pH of 7, tomato juice has a pH of 4, vinegar has a pH of 3, and lemon juice has a pH of 2. Normally, rainfall is slightly acidic (with a pH from 5 to 6) because carbon dioxide and other naturally occurring compounds in the air dissolve in rainwater, forming dilute acids. However, the pH of precipitation in the northeastern United States averages 4 and is often 3 or even lower.

How Acid Deposition Develops

Acid deposition occurs when sulfur oxides and nitrogen oxides are released into the atmosphere (Figure 20–1). Motor vehicles are a major source of nitrogen oxides. Electrical power plants, large smelters, and industrial boilers are the main sources of sulfur dioxide emissions and produce substantial amounts of nitrogen oxides. Sulfur and nitrogen oxides, released into the air from tall smokestacks, can be carried long distances by atmospheric winds. For example, tall smokestacks allow England to "export" its acid deposition problem to the Scandinavian countries, and the midwestern United States to "export" its acid emissions to New England and Canada.

During their stay in the atmosphere, sulfur oxides and nitrogen oxides react with water to produce dilute solutions of sulfuric acid (H_2SO_4), nitric acid (HNO_3), and nitrous acid (HNO_2). Acid deposition returns these acids to the ground, causing the pH of surface waters and soil to decrease.

Effects of Acid Deposition

The link between acid deposition and declining aquatic animal populations is well established (Figure 20–2). A report published by the National Academy of Sciences concluded, for example, that acid deposition is responsible for the decline of fishes in lakes of the Adirondack Mountains and in Nova Scotia rivers. Studies in the 1980s of the Adirondack region indicate that 200 to 400 lakes may have lost fish populations due to acidification. Toxic metals such as aluminum become soluble in acidic lakes

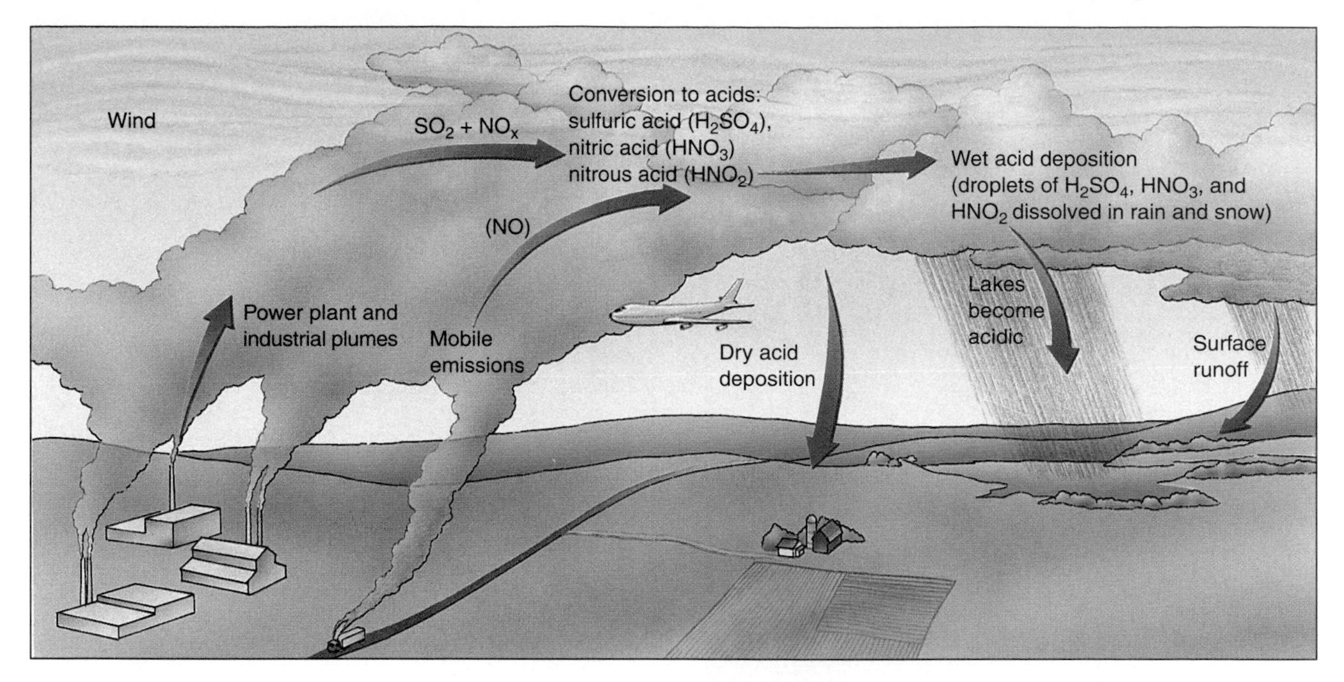

Figure 20–1
How acid deposition occurs.

and streams and enter food webs; this may explain the adverse effect that acidic water has on fishes.

Although fishes have received the lion's share of attention in regard to the acid deposition issue, other animals are also adversely affected. Several studies of the effects of acid deposition on birds found that birds living in areas with pronounced acid deposition were much more likely to lay eggs with thin, fragile shells that break or dry out before the chicks hatch. The inability to produce strong eggshells was attributed to reduced calcium in the birds' diets. Calcium is unavailable to them because in acidic soils it becomes soluble and is washed away, with

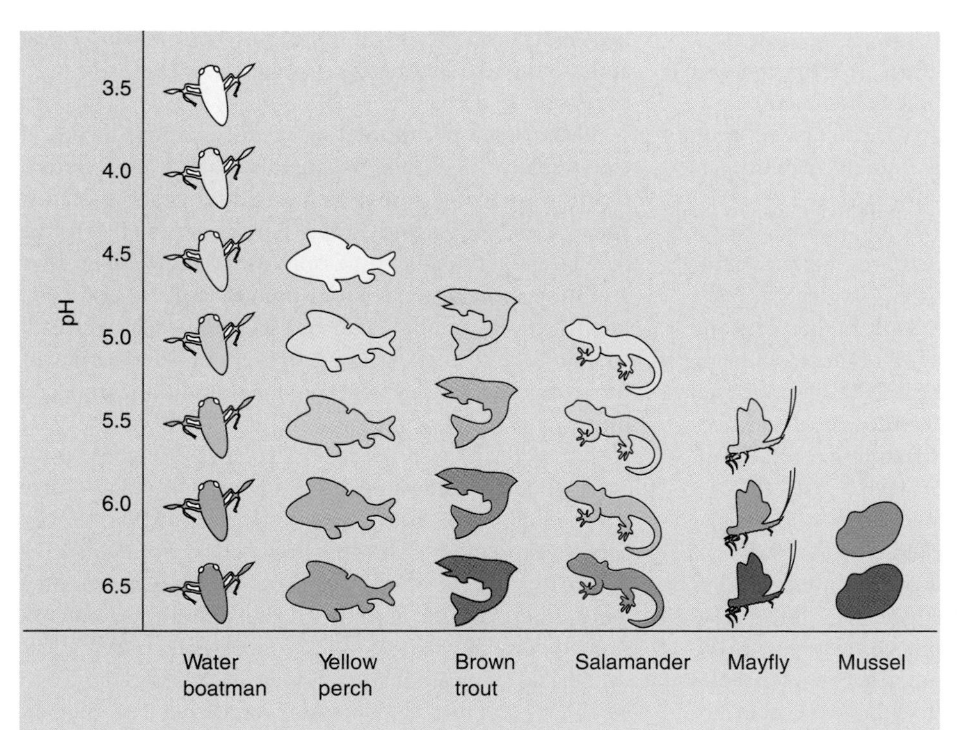

Figure 20–2
Different aquatic organisms have varying sensitivities to low pH (higher acidity) caused by acid deposition. Some organisms, such as the freshwater mussel, can tolerate very little acid, whereas others, such as the water boatman, have a wide pH tolerance.

Figure 20–3

Acid deposition eats away at stonework. Most of the writing on this English tombstone, carved in 1817, has been eaten away by acid deposition. (Bruce F. Molnia/Terraphotographics)

THE INCREASE IN ACID

- Number of strongly acidic lakes in the United States in the 1990s: 1130.
- Total number of lakes in the United States: approximately 28,000.
- Number of strongly acidic lakes in Canada in the 1990s: 31,000.
- Total number of lakes in Canada: approximately 1,140,000.
- Proportion of European forests showing damage attributable in part to acid rain in 1982: 8 percent.
- Proportion of European forests showing damage attributable in part to acid rain in 1988: 52 percent.

little left for absorption by plant roots. A lesser amount of calcium in plant tissues means a lesser amount of calcium in the insects and snails that eat the plants; moving along the food chain, the birds that eat these insects and snails have less calcium in their diets.

Acid deposition corrodes metals and building materials (Figure 20–3). It eats away at important monuments, such as the Washington Monument in Washington, D.C., and the Statue of Liberty in New York harbor. Historic sites in Venice and Rome and all across Europe are being worn away by acid deposition. Emissions from uncapped Mexican oil wells in the Gulf of Mexico cause acid deposition that has been tied to the destruction of ancient Mayan ruins in southern Mexico.

Acid deposition and forest decline

Forest surveys in the Black Forest of southwestern Germany indicate that up to 50 percent of trees in the areas surveyed are dead or severely damaged. (Damage is determined by monitoring the loss of leaves and needles from trees.) The same is true for trees in many other European forests. More than half of the red spruce trees in the mountains of the northeastern United States have died since the mid-1970s, and sugar maples in eastern Canada and the United States are also dying.

Many trees that are not dead are exhibiting symptoms of **forest decline**, characterized by gradual deterioration and often death of trees. The general symptoms of forest decline are reduced vigor and growth, but some plants exhibit specific symptoms, such as yellowing of needles in conifers. Forest decline is more pronounced at higher elevations, possibly because most trees growing at high elevations are at the border of their normal range and are therefore more susceptible to wind and low temperatures.

Many factors can interact to decrease the health of trees (Figure 20–4), and no single factor accounts for the recent instances of forest decline. Although acid deposition correlates well with areas that are experiencing tree damage, it is thought to be only partly responsible. Several other human-induced air pollutants have also been implicated, including toxic heavy metals such as lead, cadmium, and copper, and tropospheric (surface-level) ozone. Like acid deposition, these pollutants are produced by power plants, ore smelters, refineries, and motor vehicles. Insects and weather factors such as drought and severe winters—cold can injure susceptible plants—may also be important. To complicate matters further, the actual causes of forest decline may vary from one tree species to another and from one location to another. Thus, forest decline appears to result from the combination of multiple stresses—acid deposition, tropospheric ozone, ultraviolet radiation, insect attack, drought, and so on. When one or more stresses weaken a tree, then an additional stress may be decisive in causing its death.

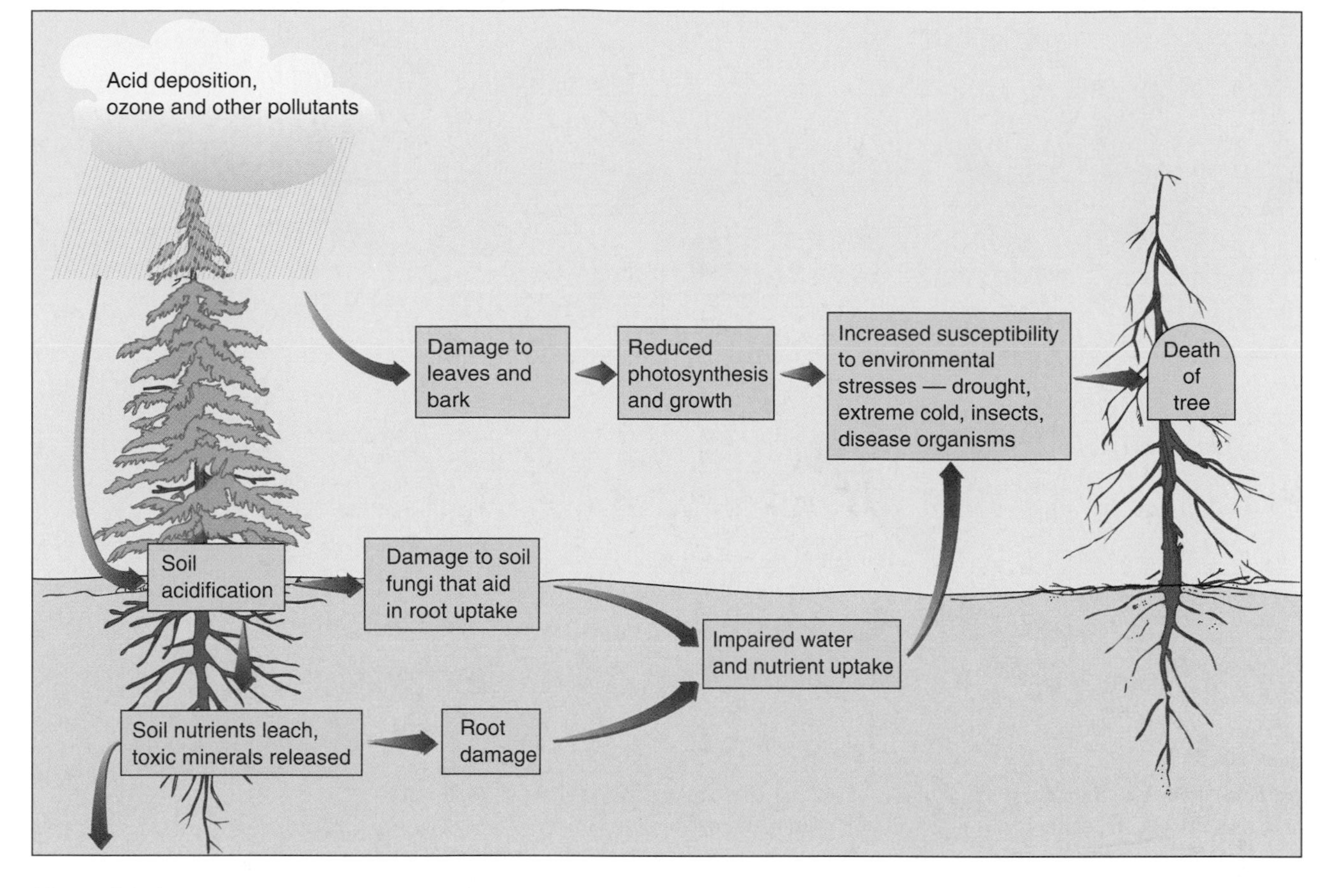

Figure 20–4
Air pollutants are one of several stresses that appear to interact, contributing to the decline and death of trees.

One way in which acid deposition harms plants is well established: acid deposition alters the chemistry of soils (Figure 20–5), which affects the development of plant roots as well as their uptake of dissolved minerals and water from soil. Essential plant minerals such as calcium and magnesium wash readily out of acidic soil, whereas others, such as nitrogen, become available in larger amounts. Also, heavy metals such as manganese and aluminum dissolve in acidic soil, becoming available for absorption in toxic amounts. A study completed in 1989 in Central Europe, which has experienced greater forest damage than North America, found a strong correlation between forest damage and soil chemistry altered by acid deposition.

► *FOLLOW-UP*

A 1996 report on the health of European forests has generated much controversy. The European Forest Institute (EFI), a private Norwegian research group, reported that European trees are not sickening and dying as numerous other studies have shown. EFI measured the heights and diameters of trees in certain forests of 12 European countries. Their data show that tree growth has increased during the past few decades, and they interpret this increase to mean that the trees are not in a state of decline. The growth trend appears to be occurring in parts of central and south Europe, where human-generated pollution is greatest.

Scientists often have differences of opinion when interpreting data, and the EFI measurements are no exception. Although EFI scientists conclude that their data dispute other reports of forest decline, many forestry scientists disagree. They think that tree height and diameter are not accurate indicators of tree health (as is the loss of leaves and needles). In fact, some researchers think that increased growth sometimes indicates that a tree is weak or in shock from other environmental stressors. As with other scientific uncertainties discussed throughout the text, additional research will be needed to resolve this issue.

The Politics of Acid Deposition

One of the factors that makes acid deposition so hard to combat is that it does not occur only in the locations where the gases that cause it are emitted. Acid deposi-

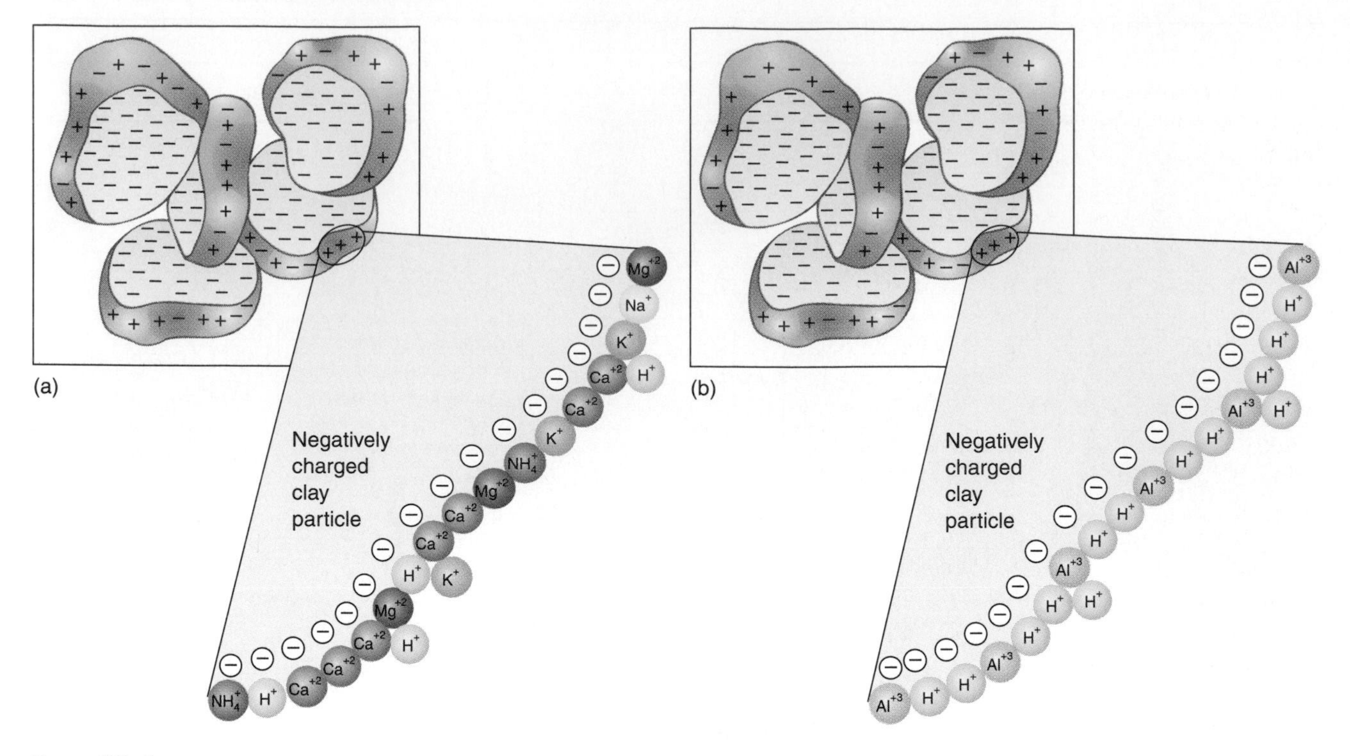

Figure 20–5

How acid alters soil chemistry. The surfaces of soil particles are negatively charged. (a) In normal soil, positively charged mineral ions such as calcium and magnesium are attracted to the negatively charged soil particles. (b) In acidified soil, hydrogen ions from acid deposition displace the positively charged mineral ions. In addition, aluminum ions released when the soil becomes acidified also adhere to the soil particles.

tion does not recognize borders between states or countries; it is entirely possible for sulfur and nitrogen oxides that were released in one spot to return to the surface hundreds of kilometers from their source.

The United States has wrestled with this issue. Several states in the Midwest and East—Illinois, Indiana, Missouri, Ohio, Pennsylvania, Tennessee, and West Virginia—produce between 50 and 75 percent of the acid deposition that contaminates New England and southeastern Canada. When legislation was formulated to deal with the problem, however, arguments ensued about who should pay for the installation of expensive air pollution devices to reduce emissions. Should the states emitting the gases be required to pay all the expenses to clean up the air, or should some of the cost be absorbed by the areas that stand to benefit most from a reduction in pollution?

Pollution abatement issues are quite complex within one country, but are magnified even more in international disputes. For example, England, which has large reserves of coal, uses coal to generate electricity. Gases from coal-burning power plants in England move eastward with prevailing winds and return to the surface as acid deposition in Sweden and Norway. England is not the only country responsible for acid deposition that crosses international borders. Japan receives acid deposition from China, whereas emissions from the United States produce acid deposition in Canada.

Controlling Acid Deposition

Although the science and the politics surrounding acid deposition are complex, the basic concept of control is straightforward: reducing emissions of sulfur and nitrogen oxides reduces acid deposition. Simply stated, if sulfur and nitrogen oxides are not released into the atmosphere, they cannot come down as acid deposition. Installation of scrubbers in the smokestacks of coal-fired power plants and use of clean-coal technologies to burn coal without excessive emissions effectively diminish acid deposition (see Chapter 10). In turn, a decrease in acid deposition prevents surface waters and soil from becoming more acidic than they already are.

► *FOLLOW–UP*

According to the Environmental Protection Agency's 1996 annual statement on air pollution, the level of sulfur dioxide in U.S. air dropped by 17 percent between 1994 and 1995, the result of cleaner-burning power plants and the use of reformulated gasoline. A report by the U.S. Geological Survey indicates that rainfall in the Midwest, Northeast, and Mid-Atlantic regions is less acidic than it was a few years ago. Acid deposition has also declined in parts of Canada and Europe.

Despite the fact that the United States, Canada, and many European countries have reduced emissions of sulfur, however, acidified forests and bodies of water have not responded as quickly as hoped. Trees in the U.S. Forest Service's Hubbard Brook Experimental Forest in New Hampshire, an area damaged by acid deposition, have grown very little since 1987, for example. Also, many northeastern streams and lakes, such as those in New York's Adirondack Mountains, remain acidic.

The main reason for the slow recovery is that the past 30 or more years of acid rain have profoundly altered soil chemistry in many areas (as discussed earlier). Essential plant minerals have been leached from forest soils and washed out into streams. Because soils take hundreds or even thousands of years to develop, it may be many decades before they recover from the effects of acid rain.

There are also concerns about nitrogen acidification. Many scientists are convinced that ecosystems will not recover from acid rain damage until substantial reductions in nitrogen oxide emissions occur. Nitrogen oxide emissions are harder to control than sulfur dioxide emissions because nitrogen oxides come mainly from automobiles. Engine improvements may help reduce nitrogen oxide emissions, but as the population continues to grow, the engineering gains will probably be offset by an increase in the number of motor vehicles. Thus, dramatic cuts in nitrogen oxide emissions will probably require expensive solutions such as increased mass transit, which would reduce the number of miles people drive, or switching from gasoline-powered cars to clean cars (see Chapter 19).

Data from the Hubbard Brook Forest also suggest that air pollution controls that are stricter than those mandated by the 1990 Clean Air Act will have to be initiated if forests are to recover in the near future.

GLOBAL WARMING

Earth's average surface temperature in 1995 edged to a record high and beat the previous record, which was set in 1990. The 1996 surface temperature, which was slightly cooler than in 1995, was the fifth warmest ever recorded. The warmest half-decade on record was 1991 through 1995, and the second warmest was 1986 through 1990. (Reliable records have been kept since the mid-nineteenth century.) The last two decades have been the warmest this century (Figure 20–6). Other evidence that suggests an increase in global temperature includes a slight increase in sea level, the retreat of glaciers worldwide, and the increased incidence of extreme weather events in certain regions.

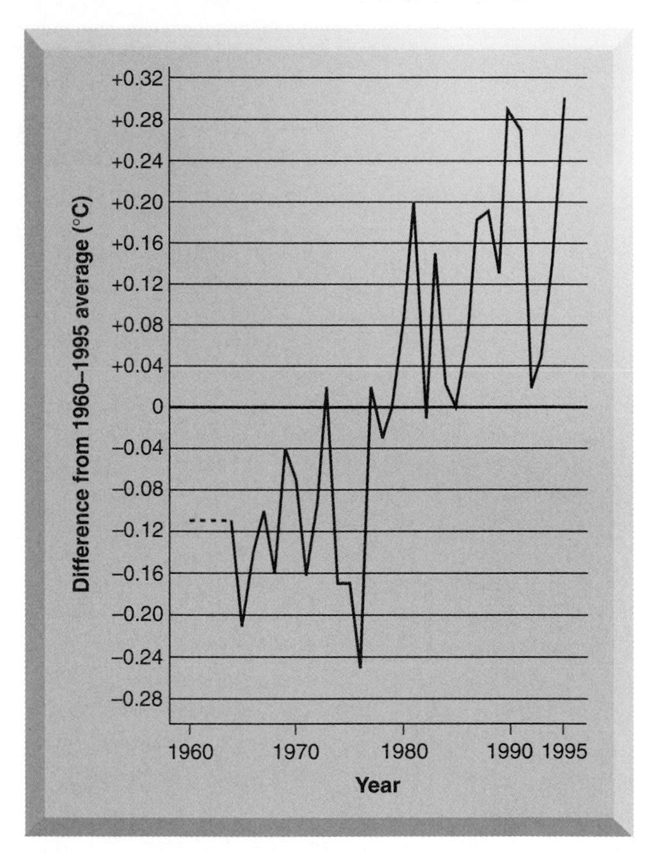

Figure 20–6

Global temperature at the Earth's surface, presented as the difference from the average 1960 to 1995 global temperature (°C). The horizontal line at 0 represents the 1960 to 1995 average, which was 15.09°C. Presented in this way, the data clearly show the warming trend of the last two decades.

In December 1995 the U.N. Intergovernmental Panel on Climate Change (IPCC) reported that human-produced air pollutants have played a role in recent climate change. This report, compiled by a group of climate experts, is the most definitive scientific statement about global warming released to date. Based on studies by numerous scientists, the IPCC projected a 1° to 3.5° C increase in global temperature by the year 2100, although the warming will not be uniform from region to region. Thus, Earth may become warmer during the 21st century than it has been for hundreds of thousands of years.

Most climate experts agree with the IPCC's assessment that the warming trend has already begun. Although most scientists agree that the world will continue to warm, they disagree over how rapidly the warming will proceed, how severe it will be, and where it will be most pronounced. (Recall from Chapter 2 that uncertainty and

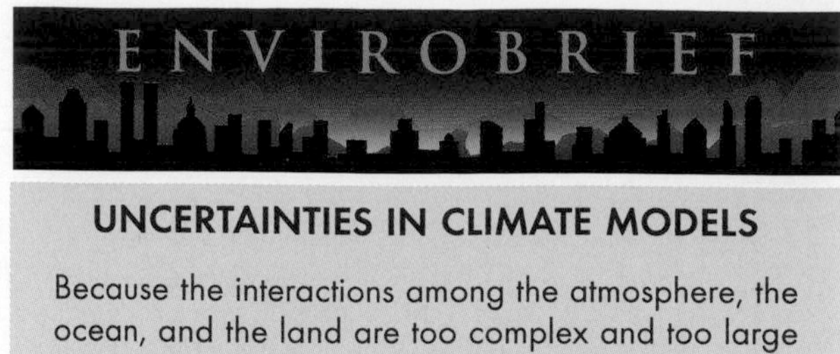

UNCERTAINTIES IN CLIMATE MODELS

Because the interactions among the atmosphere, the ocean, and the land are too complex and too large to study in a laboratory, climatologists develop models by using supercomputers. A model, however, is only as good as the data and assumptions on which it is based. Although global warming models have been increasingly refined over the past several years, some uncertainties remain. For example, global warming will probably cause greater evaporation from the Earth's surface, resulting in more clouds. If global warming causes more low-lying clouds to form, they may block some sunlight and decrease the warming trend. On the other hand, if global warming causes more high, thin cirrus clouds to form, they might intensify the warming. As new data about these and other uncertainties become available, they are used to make the models' predictions more precise.

Table 20–1
Increase in Selected Atmospheric Greenhouse Gases, Pre-Industrial Revolution to 1994

Gas	Concentration in Air: Estimated Preindustrial	1994
Carbon dioxide	280 ppm*	359 ppm
Methane	700 ppb**	1666 ppb
Chlorofluorocarbon-12	0 ppt***	509 ppt
Chlorofluorocarbon-11	0 ppt	261 ppt
Nitrous oxide	285 ppb	309 ppb

*ppm = parts per million
**ppb = parts per billion
***ppt = parts per trillion

debate are part of the scientific process and that scientists can never claim to know a "final answer.") As a result of these uncertainties about global warming, many people, including policy makers, are confused about what should be done. Yet the stakes are quite high because human-induced global warming has the potential to disrupt Earth's climate for a very long time.

The Causes of Global Warming

Carbon dioxide and certain other trace gases, including methane (CH_4), nitrous oxide (N_2O), chlorofluorocarbons (CFCs), and tropospheric ozone (O_3), are accumulating in the atmosphere as a result of human activities (Table 20–1 and Figure 20–7). The concentration of atmospheric carbon dioxide has increased from about 280 parts per million approximately 200 years ago (before the Industrial Revolution began) to 361 parts per million in 1995. And it is still rising, as are the levels of the other trace gases associated with global warming. For example, every time you drive your car, the combustion of gasoline in the car's engine releases carbon dioxide and nitrous oxide and triggers the production of tropospheric ozone; a single gallon of gasoline burns to produce almost 5.5 pounds of carbon dioxide. Every day, as tracts of rain forest are burned in the Amazon, carbon dioxide is released. CFCs, which are discussed later in the chapter, are released into the atmosphere from old, leaking refrigerators and air conditioners (Figure 20–8).

MISSING CARBON—HIDING IN THE FOREST

The world's northern woodlands, known as boreal forests, and the temperate forests farther south may provide the solution to one of the great unanswered questions of climate research. Carbon dioxide is released to the atmosphere when fossil fuels are burned. Much of the carbon dioxide remains in the atmosphere or is absorbed by the ocean, but scientists have never been able to account for a significant portion of the balance. Now it appears that northern forests may have acted as carbon "sinks" that use increased inputs of carbon dioxide for accelerated growth. This role can be reversed by human influences:

- Evidence suggests that deforestation can cause forests to become carbon "sources" that release carbon dioxide to the atmosphere because of accelerated decomposition of the exposed forest floor.
- Timber harvests are being shifted northward from the tropics, and the cutting of temperate and boreal forests could greatly affect the global storage of carbon.
- Countries with extensive boreal forests—particularly Canada and Russia—are reluctant to enter into any international conservation agreement, such as those established for tropical timber harvesting, to protect the northern forests.

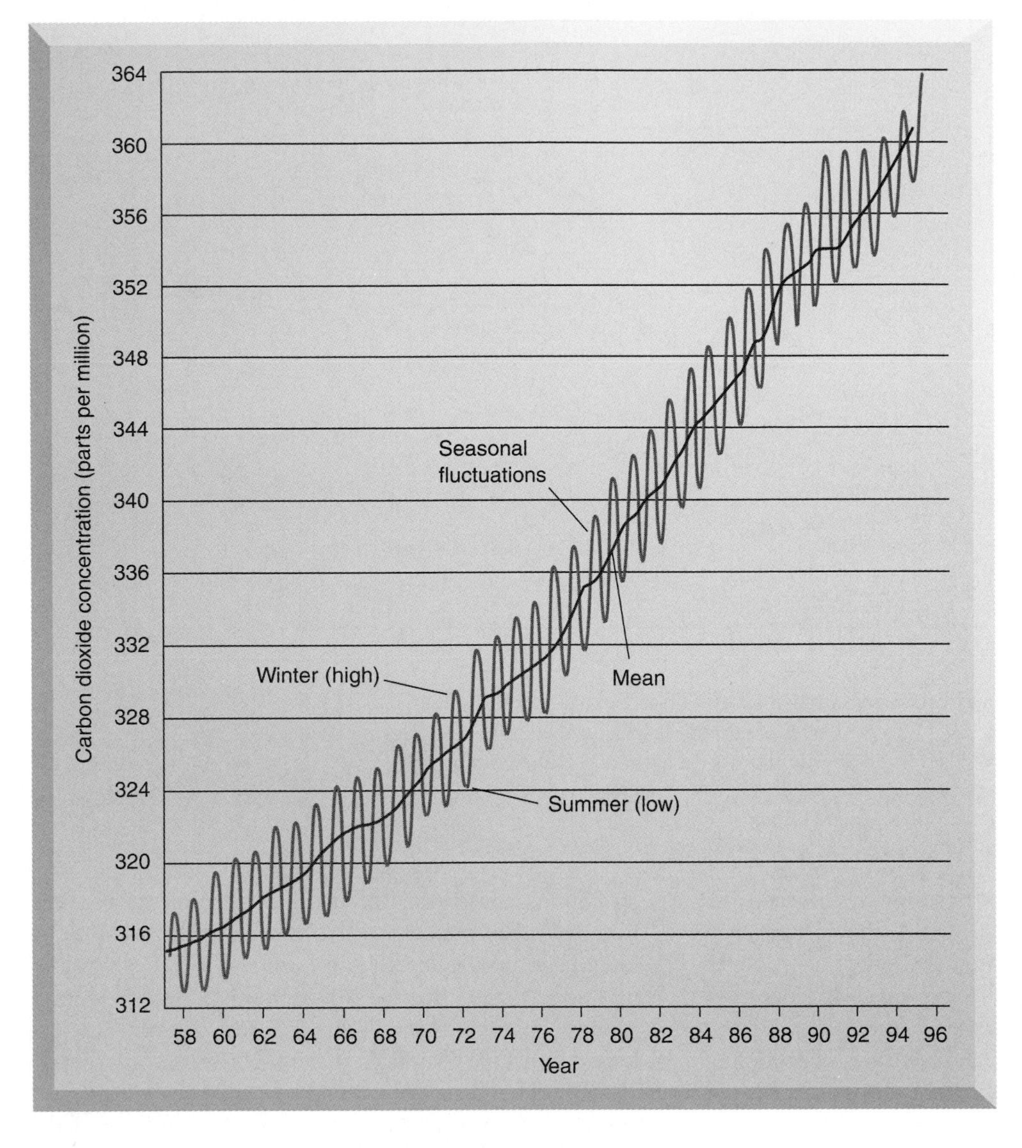

Figure 20–7
The concentration of carbon dioxide in the atmosphere has shown a slow but steady increase for many years. The measurements were taken at the Mauna Loa Observatory, Hawaii, far from urban areas where carbon dioxide is emitted by factories, power plants, and motor vehicles. The seasonal fluctuation corresponds to winter (a high level of CO_2) and summer (a low level of CO_2) and is caused by greater photosynthesis in summer. (Dave Keeling and Tim Whorf, Scripps Institution of Oceanography, LaJolla, California)

Figure 20–8
Old, improperly discarded refrigerators and air conditioners can leak CFCs into the atmosphere. (Jeff Greenberg/Visuals Unlimited)

Global warming occurs because these gases retain infrared radiation—that is, heat—in the atmosphere (Figure 20–9). Because this heat would normally dissipate into space, the atmosphere warms. Some of the heat from the atmosphere is transferred to the ocean and raises its temperature as well. As the atmosphere and ocean warm, the overall global temperature rises. Because carbon dioxide and other gases trap the sun's radiation in much the same way that glass does in a greenhouse, global warming produced in this manner is known as the **greenhouse effect**.

Other pollutants cool the atmosphere

One of the complications that makes global warming difficult to predict is that other air pollutants, known as atmospheric aerosols, tend to cool the atmosphere. **Aerosols**, which come from both natural and human sources, are tiny particles that are so small that they re-

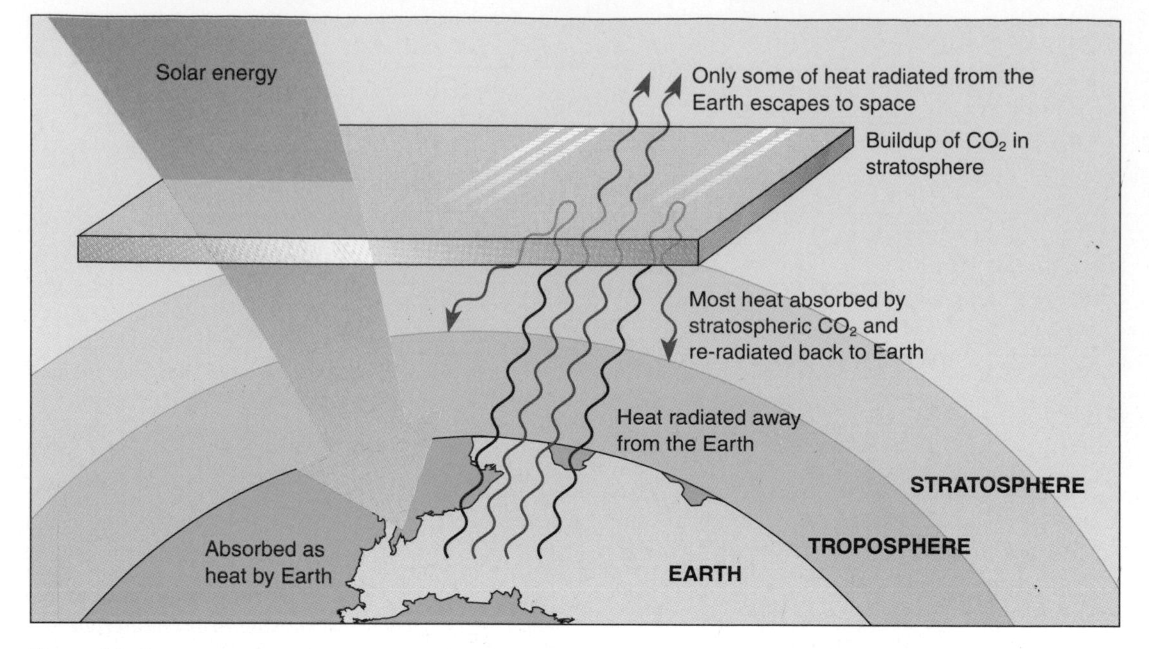

Figure 20–9

The buildup of carbon dioxide and other greenhouse gases causes the greenhouse effect, which results in global warming. The increased level of carbon dioxide in the stratosphere is depicted in this diagram as a sheet of glass to emphasize that in this situation CO_2 functions much like the glass walls and roof of a greenhouse.

main suspended in the atmosphere for days, weeks, or even months. Sulfur haze, the aerosol that causes acid deposition, also cools the planet by reflecting sunlight away from the Earth. Climatologists have evidence that sulfur haze significantly moderates greenhouse warming in industrialized parts of the world. Atmospheric cooling that occurs where aerosol pollution is greatest is known as the **aerosol effect.**

Sulfur emissions, which produce sulfur haze, come from the same smokestacks that pour forth carbon dioxide. In addition, volcanic eruptions can inject sulfur-containing particles into the atmosphere. The explosion of Mount Pinatubo in the Philippines in June 1991 was the largest volcanic eruption in the 20th century (Figure 20–10). The force of this eruption injected massive amounts of sulfur into the stratosphere (the layer of the

Figure 20–10

The eruption of Mount Pinatubo in June 1991, the largest volcanic eruption in the 20th century, spewed massive amounts of sulfur into the atmosphere. (Alberto Garcia/SABA)

atmosphere above the troposphere), where sulfur tends to remain longer than it does when emitted into the troposphere. Because sulfur in the stratosphere reduces the amount of sunlight that reaches the planet, this eruption caused the Earth to enter a short period of global cooling. Compared to the rest of the 1990s, 1992 and 1993 global temperatures were relatively cool.

Human-produced sulfur emissions should not be viewed as a panacea for the greenhouse effect, despite their cooling effect. For one thing, they are produced in heavily populated industrial areas, primarily in the Northern Hemisphere. Because they do not remain in the atmosphere for long, they do not disperse globally. Thus, sulfur pollution may cause regional, but not global, cooling.

In the grand scheme of things, the greenhouse gases will win out over sulfur haze. Greenhouse gases remain in the atmosphere for hundreds of years, whereas human-produced sulfur emissions remain for only days, weeks, or months. And carbon dioxide and other greenhouse gases help warm the planet 24 hours a day, whereas sulfur haze cools the planet only during the daytime. In addition, because sulfur emissions also cause acid deposition, most nations are trying to *reduce* their sulfur emissions, not maintain or increase them.

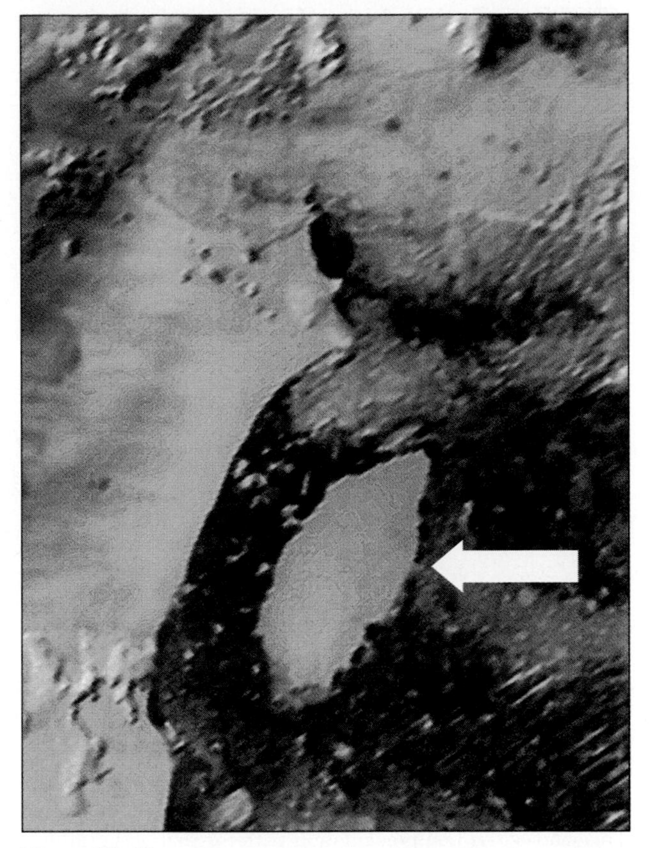

Figure 20–11
A satellite shot of part of Antarctica shows a massive iceberg (arrow) the size of Rhode Island that broke away from the Larsen Ice Shelf in January 1995. (Ted Scambos/NSIDC)

The Effects of Global Warming

Although current rates of fossil fuel combustion and deforestation are high, scientists think the warming trend will be slower than the increasing carbon dioxide might indicate. The reason is that the ocean takes longer than the atmosphere to absorb heat. Warming will probably be more pronounced in the second half of the 21st century than in the first half.

We now consider some of the potential effects of global warming, including changes in sea level, precipitation patterns, organisms, human health, and agriculture.

Rising sea level

If the overall temperature of the Earth increases by just a few degrees, there could be a major thawing of glaciers and the polar ice caps. In January 1995, for example, a huge chunk of the Larsen Ice Shelf broke away from the Antarctic Peninsula (Figure 20–11). (Antarctic ice shelves are thick sheets of floating ice that are fed mainly by glaciers that flow off the land.) This breakup coincided with a trend of atmospheric warming in the Antarctic during the past 50 years. A 16-year study from 1978 to 1994 showed that the area of ice-covered ocean in the Arctic has also retreated. In addition to sea-level rise caused by the retreat of glaciers and thawing of polar ice, the sea level will rise because thermal expansion of the ocean will probably occur. Water, like other substances, expands as it warms.

During the past century, the sea level has risen by about 18 centimeters (7 inches), most of it due to thermal expansion. The IPCC estimates that the sea level will rise by an additional 50 centimeters (20 inches) by 2100. Such a rise in sea level would flood low-lying coastal areas. Because of climatic changes, coastal areas that are not inundated will be more likely to suffer damage from weather events such as hurricanes and from coastal erosion. These likely effects are certainly a cause for concern, particularly since about two-thirds of the world's population lives within 150 kilometers (93 miles) of a coastline.

The countries that are most vulnerable to a rise in sea level—countries such as Bangladesh, Egypt, Vietnam, and Mozambique—have dense populations in low-lying river deltas. For example, a rising sea level could cause Bangladesh to lose as much as 18 percent of its land, displacing millions of people. Since 1970, the flooding and high waves accompanying tropical storms have resulted in more than 300,000 deaths in Bangladesh. A rising sea level caused by global warming would put even more people at risk in this densely populated nation.

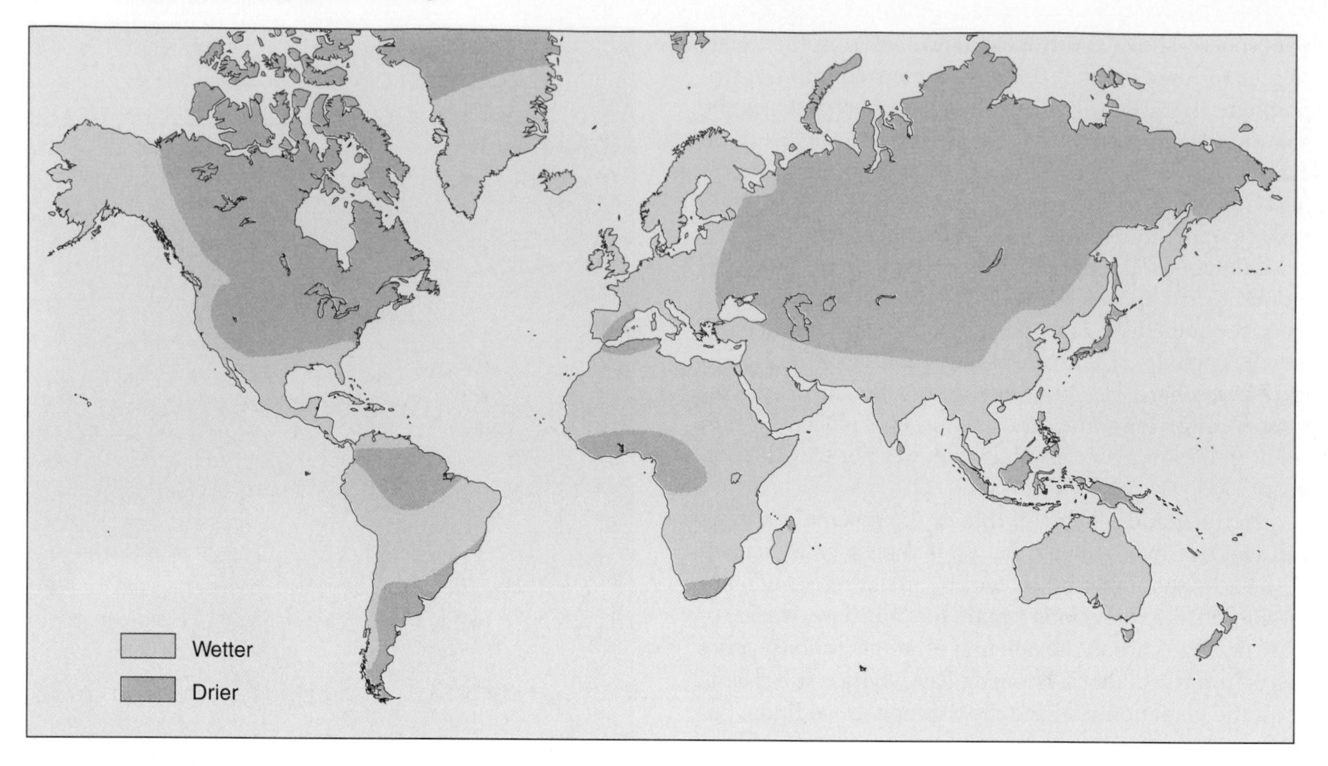

Figure 20–12

One possible scenario of how precipitation might be altered by warmer global temperatures. This map is based on precipitation patterns that occurred thousands of years ago when the Earth was warmer.

Changes in precipitation patterns

Precipitation patterns that occurred thousands of years ago when the Earth was warmer have been used to develop computer simulations of weather changes as global warming occurs. These simulations indicate that precipitation patterns will change, causing some areas to have more frequent droughts (Figure 20–12). At the same time, heavier snow and rainstorms may cause flooding in other areas. The frequency and intensity of storms may also increase. These changes are expected because as the atmosphere warms, more water evaporates, which in turn releases more energy into the atmosphere (recall the discussion of water's heat of vaporization in Chapter 13). This energy generates more powerful storms.

In 1995 climatologists at the National Climatic Data Center in Asheville, North Carolina, analyzed temperature and precipitation records for the past 100 years. They determined that hotter heat waves, more severe droughts, and heavier snow and rainstorms had occurred more frequently from 1980 to 1994. They then compared this trend to computer simulations of global warming and concluded that there is a 90 to 95 percent likelihood that recent precipitation extremes are the result of the increase in greenhouse gases in the atmosphere.

Changes in precipitation patterns could affect the availability and quality of fresh water in many locations. It is projected that areas that are currently arid or semiarid will have the most troublesome water shortages as the climate changes.

Effects on organisms

Biologists have started to examine some of the potential consequences of global warming for organisms. For example, one team of biologists placed infrared heaters above test plots in a Rocky Mountain meadow to mimic future warming. Changes in the kinds of plants growing in the plots were noted in a few months.

Other researchers determined that populations of zooplankton in the California Current have declined by 80 percent since 1951, apparently because the current has warmed slightly (Figure 20–13). (The California Current flows from Oregon southward along the California coast.) The decline in zooplankton has affected the entire food web. Populations of seabirds and plankton-eating fishes, for example, have also declined.

The first scientific evidence that a species has shifted its geographical range in response to climate change was published in 1996. A biologist at the University of Texas at Austin compared the current geographical range of a western butterfly (the Edith's checkerspot butterfly) to previously recorded observations. She found that the but-

Figure 20–13

A 40-year study of the California Current discovered that zooplankton such as this calenoid copepod (*Calanus* sp) respond dramatically to very small temperature changes. Zooplankton populations have declined by 80 percent in that time, the result of a slight warming of surface waters off the coast of California. (David Wrobel/Biological Photo Service)

terfly has shifted its range northward by about 160 kilometers (about 100 miles).

Each species reacts to changes in temperature differently (see Focus On: Two Species Affected by Global Warming). As warming accelerates in the 21st century, some species will undoubtedly become extinct, particularly those with narrow temperature requirements, those confined to small reserves and parks, and those living in fragile ecosystems. Other species may survive in greatly reduced numbers and ranges. Ecosystems considered to be at greatest risk of loss of species in the short term are polar seas, coral reefs, mountain ecosystems, coastal wetlands, tundra, and boreal and temperate forests. Water temperature increases of 1° to 2° C, for example, can cause coral bleaching that can contribute to the destruction of coral reefs.

Some species may be able to migrate to new environments or adapt to the changing conditions in their present habitats. Also, some species may be unaffected by global warming, whereas others may come out of global warming as winners, with greatly expanded numbers and range. Those organisms considered most likely to prosper include certain weeds, insect pests, and disease-carrying organisms that are already common in a wide range of environments.

Biologists generally agree that global warming will have an especially severe impact on plants because they cannot move about when environmental conditions change (Figure 20–14). Although seeds are dispersed, sometimes over long distances, seed dispersal has definite limitations in terms of the speed of migration. During past climate changes, tree species are thought to have migrated from 4 to 200 kilometers (2.5 to 124 miles) per century. If the Earth warms 1° to 3.5° C during the 21st century as projected, trees would have to migrate northward from 150 to 550 kilometers (93 to 342 miles). Moreover, soil characteristics, water availability, competition with other plant species, and human alterations of natural habitats all affect the rate at which plants can move into a new area.

Effects on human health

In terms of human health, the increase in carbon dioxide in the atmosphere and the resulting more frequent and more severe heat waves during summer months will cause an increase in the number of heat-related illnesses and deaths. Human health will also be indirectly affected, as for example, if mosquitos and other disease carriers expand their range into the newly warm areas and spread malaria, dengue, yellow fever, and viral encephalitis. For example, as many as 50 to 80 million additional cases of malaria could occur annually in tropical, subtropical, and temperate areas.

Effects on agriculture

Global warming will increase problems for agriculture, which is already beset with the challenge of providing enough food for a hungry world without doing irreparable damage to the environment. The rise in sea level will cause water to inundate river deltas, which are some of the world's best agricultural lands. Certain agricultural pests and disease-causing organisms will probably proliferate. As mentioned earlier in the chapter, scientists think that global warming will also increase the frequency and duration of droughts.

Current global warming models forecast that agricultural productivity will increase in some areas and decline in others. Some climate experts predict that tropical and subtropical regions—where many of the world's poorest people live—will be hardest hit by declining agricultural productivity.

International Implications of Global Warming

Global warming will be complicated by a variety of social and political factors. For example, how will we deal with the environmental refugees produced by global warming? Where will they go? Who will assist them to resettle? It will be difficult for all countries to develop a consensus on dealing with global warming, partly because global warming will clearly have greater impacts on some nations than on others. However, all nations must cooperate if we are to effectively deal with global warming.

Developed nations versus developing nations

Although greenhouse gases are produced primarily by the developed countries, their rate of production by certain

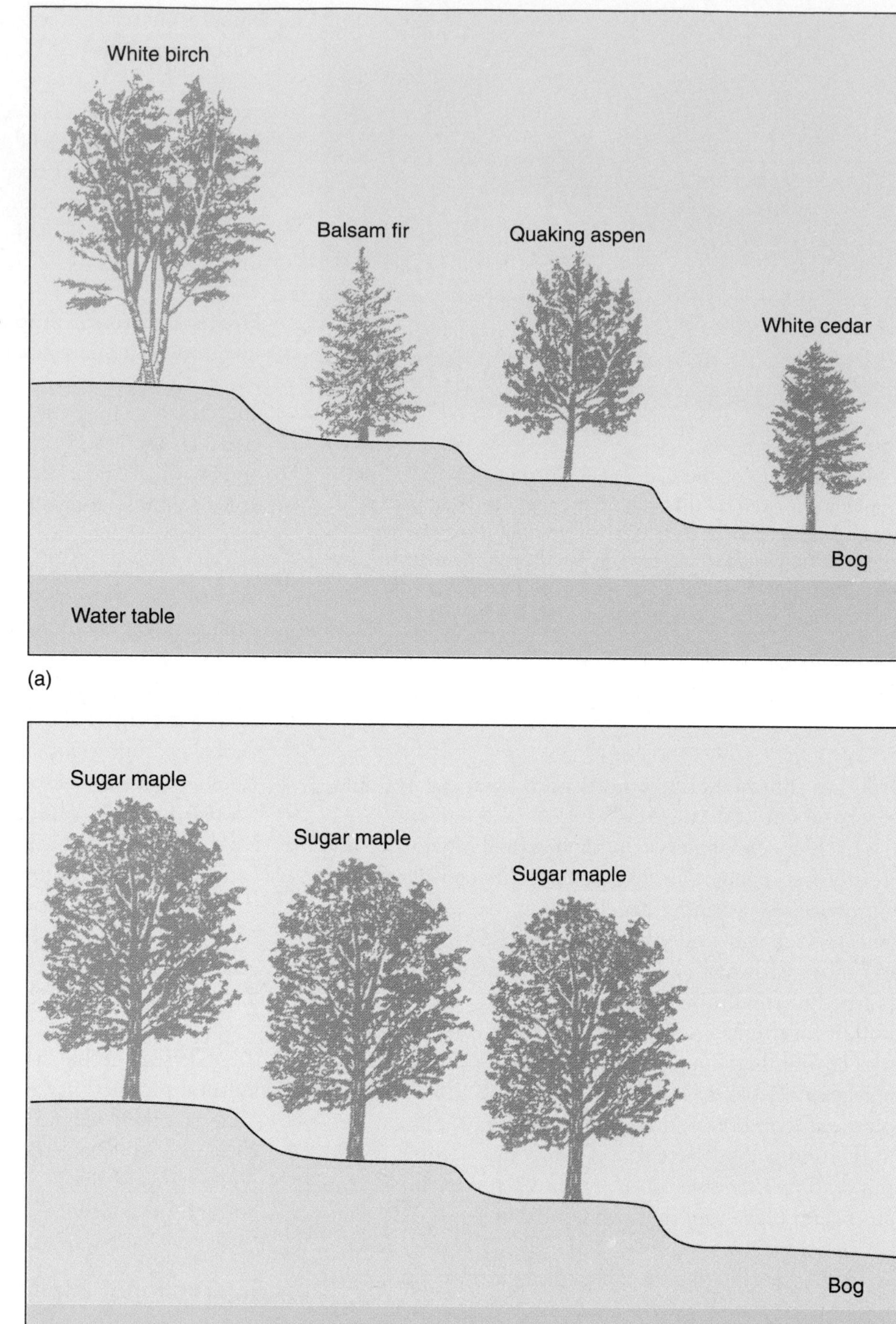

Figure 20–14
The possible species composition of a forest in northern Minnesota (a) now and (b) after global warming occurs.

developing countries is rapidly increasing (Figure 20–15). Furthermore, the developing nations may experience the greatest impacts of global warming. Because they have less technical expertise and fewer economic resources, these countries are the very ones that are least able to respond to the challenges of global warming.

Tensions are sure to mount among nations, especially between the developed and developing countries. The latter see increased use of fossil fuels as their route to industrial development. Developing countries will likely resist pressure from developed nations to decrease fossil fuel consumption.

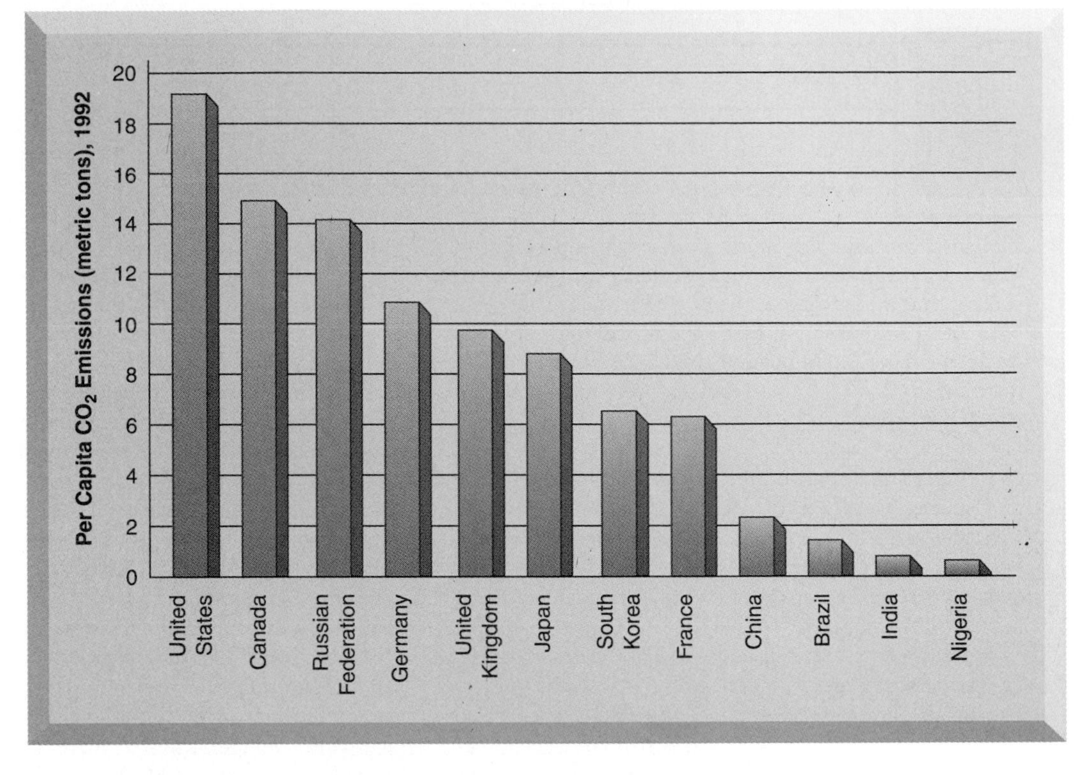

Figure 20–15

Per-capita CO_2 emissions in 1992 (latest data available) for twelve selected countries. Note that the developed nations currently produce a disproportionate share of carbon dioxide emissions. As developing nations such as South Korea industrialize, however, their per-capita CO_2 emissions increase.

How We Can Deal with Global Warming

Even if we were to immediately stop polluting the atmosphere with greenhouse gases, there would still be some climate change because of the greenhouse gases that have accumulated during the past 100 years. The amount and severity of global warming depend on how much additional greenhouse gas emissions we add to the atmosphere.

There are three basic ways to manage global warming—prevention, mitigation, and adaptation. *Prevention* of global warming can be accomplished by developing ways to prevent the buildup of greenhouse gases in the atmosphere. Prevention is the ultimate solution to global warming because it is permanent. *Mitigation* is the moderation or postponement of global warming, which buys us time to pursue other, more permanent solutions. It also gives us time to understand more fully how global warming operates so we can avoid some of its worst consequences. *Adaptation* is responding to changes brought about by global warming. Creating strategies to adapt to climate change implies an assumption that global warming is unavoidable.

Prevention of global warming

The development of alternatives to fossil fuel offers a permanent solution to the challenge of global warming caused by carbon dioxide emissions.[1] Alternatives are also necessary, given that fossil fuels are present in limited amounts. Some alternatives to fossil fuels—solar energy and nuclear energy—were discussed in Chapters 11 and 12.

The invention of technological innovations that trap the carbon dioxide being emitted from smokestacks would help prevent global warming and yet allow us to continue using fossil fuels (while they last) for energy. Incentives may be necessary to inspire such innovation. For example, several nations have imposed taxes on greenhouse gases. The taxes motivate emitters to improve efficiency and to develop carbon dioxide–free technologies.

The international community recognizes that it must stabilize CO_2 emissions. More than 165 nations, including the United States, have signed the climate change treaty developed by the 1992 U.N. Conference on Environment and Development (see Chapter 1 box on the Earth Summit). Its ultimate goal is to stabilize greenhouse gas concentrations in the atmosphere at levels low enough to prevent dangerous human influences on the

[1]Alternatives to the other greenhouse gases will also have to be developed, but carbon dioxide is the focus here because it is produced in the greatest quantities and has the largest total effect of all the greenhouse gases.

FOCUS ON

Two Species Affected by Global Warming

Biologists are only beginning to assess the possible impacts of global warming on the world's organisms, but it is clear that in many areas there will be major changes in both species diversity and species distribution. Two organisms—a mountain butterfly and an Asian vine widely introduced in the United States—provide us with a glimpse of what may be in store for many of the world's organisms.

A rare mountain butterfly may become one of the first species to face extinction solely because of global warming. This little insect (*Boloria acrocnema*), a member of a group of butterflies called fritillaries, lives in an extremely restricted area: a few high-altitude, northeast-facing, snow-moistened slopes in the rugged San Juan Mountains of western Colorado. Its habitat is surrounded by warmer, drier areas that the butterfly cannot tolerate. During the hot, dry summers of the late 1980s, one of the two known populations of this fritillary species disappeared, and less than half of the other population survived the decade. Researchers have concluded that the only reason for the butterfly's declining numbers was the unusually warm, dry weather. In the summer of 1991 the fritillary was listed as an endangered species, but this federal protection will not keep it from becoming extinct if the climate continues to warm.

Kudzu is a fast-growing, purple-flowered vine that has become a weed throughout much of the southern United States. It was imported to the United States from Japan and was planted widely throughout the South during the 1930s in an effort to halt soil erosion. Because it grows rapidly—as much as 30 centimeters (1 foot) per day under ideal growing conditions—kudzu has choked out many other plants, including trees. Kudzu has become a major nuisance from eastern Texas to Florida; its current range extends north through much of Pennsylvania. Botanists who have studied kudzu conclude that low winter temperatures are the only reason the plant has not spread farther north. Should global warming cause average winter temperatures to increase, the vine may well spread as far north as Michigan, New York, and Massachusetts.

climate. At the 1996 U.N. Climate Change Convention held in Geneva, Switzerland, developed countries agreed to establish legally binding timetables to cut emissions of greenhouse gases, beginning in the year 2000. As we go to press, however, the actual emission reduction commitments have not been decided.

Despite international resolve to reduce CO_2 emissions, which would mainly be accomplished by reducing our dependence on fossil fuels, there is much discussion about how fast we should respond. Should we postpone reducing CO_2 emissions until new technologies have been developed? Or will the environmental costs of postponement be too high? The scientific uncertainties associated with global warming make it difficult to know how we should proceed, but the potential seriousness of the problem cannot be ignored while we wait for more data.

Mitigation of global warming

One way to mitigate global warming involves forests. As you know, atmospheric carbon dioxide is removed from the air by growing forests, which incorporate the carbon into leaves, stems, and roots through the process of photosynthesis. In contrast, deforestation releases carbon dioxide into the atmosphere as trees are burned or decomposed. As discussed in the chapter introduction, we can mitigate global warming by planting and maintaining new forests. Some environmentalists have suggested that developed nations should pay developing nations to maintain their own tropical forests.

Increasing the energy efficiency of automobiles and appliances—and thus reducing the output of carbon dioxide—would also help mitigate global warming. This could be accomplished by establishing regulatory programs that increase minimum energy efficiency standards and by negotiating agreements with industry. Energy-pricing strategies, such as carbon taxes and the elimination of energy subsidies (see Chapter 10), are examples of other policies that could help mitigate global warming. Most experts think that by using existing technologies and developing such policies, greenhouse gas emissions can be significantly reduced with little cost to society.

Fertilizing the ocean with iron In 1987 an oceanographer proposed a novel way to remove some carbon dioxide from the atmosphere, by fertilizing parts of the ocean with iron. Certain areas of the ocean are rich in nutrients such as nitrogen but very low in iron; these areas also have very low numbers of phytoplankton (microscopic algae). The scientist reasoned that the addition of iron would, in effect, fertilize the ocean, stimulating large numbers of phytoplankton to grow. As they photosynthesized, the phytoplankton would remove carbon dioxide from the water, which would be replenished by carbon dioxide from the atmosphere. (It is hypothesized that when the phytoplankton die, they sink to the ocean floor. Thus, the carbon in their cells will be stored on the ocean floor.)

The iron hypothesis was first tested in 1995, when small amounts of dissolved iron were added to a 64-square-kilometer (26-square-mile) area of the equatorial Pacific Ocean. It stimulated a marked increase in the number of phytoplankton, so many that the ocean temporarily turned from blue to green. Although the first test of the iron hypothesis was successful, scientists caution that we should not start spreading iron into the ocean on a large scale until we understand all the ecological consequences of such an action.

Adaptation to global warming

Because the overwhelming majority of climate experts think that human-induced global warming is inevitable, government planners and social scientists are developing strategies to help us adapt to climate change. One of the most pressing issues is rising sea level. People living in coastal areas could be moved inland, away from the dangers of storm surges. This solution would have high societal and economic costs, however. An alternative, also extremely expensive, is the construction of dikes and levees to protect coastal land. Rivers and canals that spill into the ocean would have to be channeled to protect freshwater and agricultural land from salt water intrusion (see Chapter 13). The Dutch, who have been building dikes and canals for several hundred years, are offering their technical expertise to several developing nations that are particularly threatened by a rise in sea level.

We will also need to adapt to shifting agricultural zones. Many countries with temperate climates are evaluating semitropical crops to determine the best ones to substitute for traditional crops if and when the climate warms. Drought-resistant strains of trees are being developed by large lumber companies now, because the trees planted today will be harvested in the middle of the 21st century, when global warming may already be well advanced.

OZONE DEPLETION IN THE STRATOSPHERE

You may recall from Chapter 19 that ozone (O_3) is a form of oxygen that is a human-made pollutant in the troposphere but a naturally produced, essential component in the stratosphere, which encircles our planet some 10 to 45 kilometers (6 to 28 miles) above the surface. The stratosphere contains a layer of ozone that shields the surface from much of the ultraviolet radiation coming from the sun (Figure 20–16). **Ultraviolet (UV) radiation** is the name given to that part of the electromagnetic spectrum with wavelengths just shorter than visible light; it is a high-energy form of radiation that can be lethal to organisms in excessive amounts. Should ozone disappear from the stratosphere, the Earth would become uninhabitable for most forms of life.

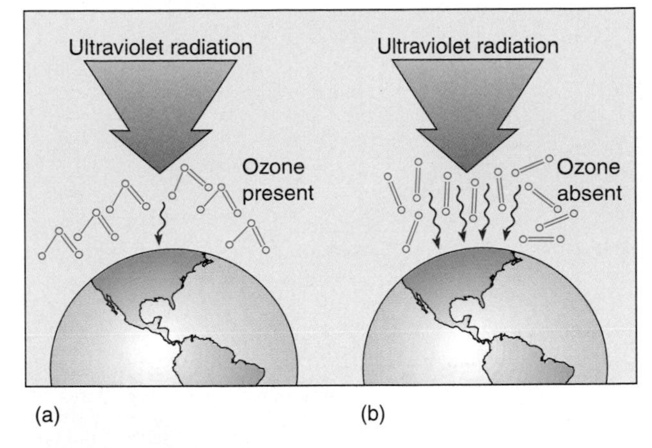

Figure 20–16
Ultraviolet radiation is largely blocked by the ozone layer. (a) Stratospheric ozone absorbs UV radiation, effectively shielding the surface. (b) When stratospheric ozone is absent, more high-energy UV radiation penetrates the atmosphere to the surface, where its presence harms organisms.

A thinning in the ozone layer over Antarctica was first observed and reported in 1985. This thin spot, which occurs each September and lasts for a couple of months, is popularly referred to as the "ozone hole" (Figure 20–17). There, ozone levels decrease by as much as 67 percent each year. During the mid-1990s, the hole encompassed about 23.3 million square kilometers (9 million square miles), an area about as large as North America. A smaller hole has been detected in the stratospheric ozone layer over the Arctic.

Probably the most disquieting news is that worldwide levels of stratospheric ozone have been decreasing for several decades. Ozone levels over Europe and North America, for example, have dropped by almost 10 percent since the 1950s. According to the World Meteorological Organization, ozone levels over the Northern Hemisphere reached an all-time low in 1996.

The Causes of Ozone Depletion

The primary culprits responsible for ozone loss in the stratosphere are a group of versatile compounds called **chlorofluorocarbons (CFCs)**. CFCs have been used as propellants for aerosol cans, as coolants in air condition-

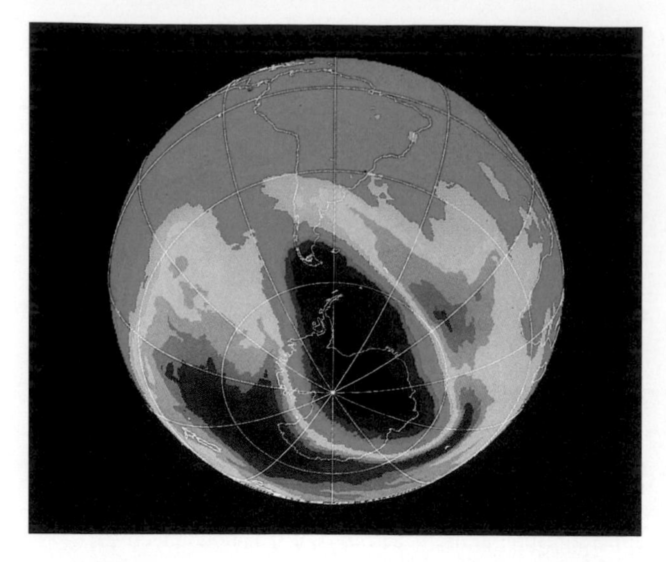

Figure 20–17

A computer-generated image of part of the Southern Hemisphere on October 17, 1994, reveals the ozone "hole" (black and purple areas) over Antarctica and the tip of South America. Relatively low ozone levels (blue and green areas) extend into much of South America as well as Central America. Normal ozone levels are shown in yellow, orange, and red. The ozone hole is not stationary but drifts about as a result of air currents. (NASA)

ers and refrigerators (for example, Freon), as foam-blowing agents for insulation and packaging (for example, Styrofoam), and as solvents. Additional compounds that also attack ozone include: halons (found in many fire extinguishers); methyl bromide (a pesticide used as a soil and crop fumigant); methyl chloroform (an industrial solvent); and carbon tetrachloride (used in many industrial processes, including the manufacture of pesticides and dyes).

How Ozone Depletion Takes Place

The scientific evidence linking CFCs and other human-made compounds to stratospheric ozone destruction includes thousands of laboratory measurements, atmospheric observations, and calculations by computer models. In 1995 the Nobel Prize in chemistry was awarded to Sherwood Rowland, Mario Molina, and Paul Crutzen, the scientists who first explained the connection between the thinning ozone layer and chemicals such as CFCs. This Nobel Prize was the first one ever given for work in environmental science.

CFCs and other chlorine-containing compounds slowly drift up to the stratosphere, where UV radiation breaks them down, releasing chlorine. Similarly, bromine is released by the breakdown of halons and methyl bromide. Under certain conditions found in the stratosphere, chlorine or bromine is capable of attacking ozone and converting it into oxygen. The chlorine or bromine is not altered by this process; as a result, a single chlorine or bromine atom can break down many thousands of ozone molecules. Human-produced pollution is not the only cause of ozone depletion. Volcanic eruptions such as that of Mount Pinatubo also accelerate ozone loss because the sulfur aerosols they spew into the atmosphere speed the breakdown of CFCs and halons.

The hole in the ozone layer that was discovered over Antarctica occurs annually between September and November, when the **circumpolar vortex**, a mass of cold air, circulates around the southern polar region, in effect isolating it from the warmer air in the rest of the world. The cold causes polar stratospheric clouds to form; these clouds contain ice crystals to which chlorine and other chemicals adhere, enabling them to attack ozone. When the circumpolar vortex breaks up, the ozone-depleted air spreads northward, diluting ozone levels in the stratosphere over South America, New Zealand, and Australia.

The Effects of Ozone Depletion

With depletion of the ozone layer, higher levels of UV radiation reach the surface (Figure 20–18). A study conducted in Toronto from 1989 to 1993, for example, showed that wintertime levels of UV-B (the damaging form of UV radiation) increased by more than 5 percent each year as a result of lower ozone levels. Excessive exposure to UV radiation is linked to several health problems in humans, including eye cataracts, skin cancer, and weakened immunity.

There is also substantial scientific documentation of crop damage from high levels of UV radiation. However, these experiments indicate what could happen if the ozone layer is significantly destroyed and UV levels increase dramatically. (The UV level in most studies is several times higher than it is expected to be in the year 2000 when stratospheric ozone is predicted to reach its lowest level.) No scientific evidence exists that plants are currently suffering from increased exposure to present levels of UV-B.

Scientists are also concerned because increased levels of UV radiation disrupt ecosystems. Recall from Chapter 3 that the productivity of Antarctic phytoplankton has declined as a result of increased exposure to UV-B. Also, the possible link between the widespread decline of amphibian populations and increased UV radiation is discussed in Chapter 16. Because organisms are interdependent, the negative effect on one species has ramifications throughout the ecosystem.

Protecting the Ozone Layer

In 1978 the United States, the world's largest user of CFCs, banned the use of CFC propellants in products

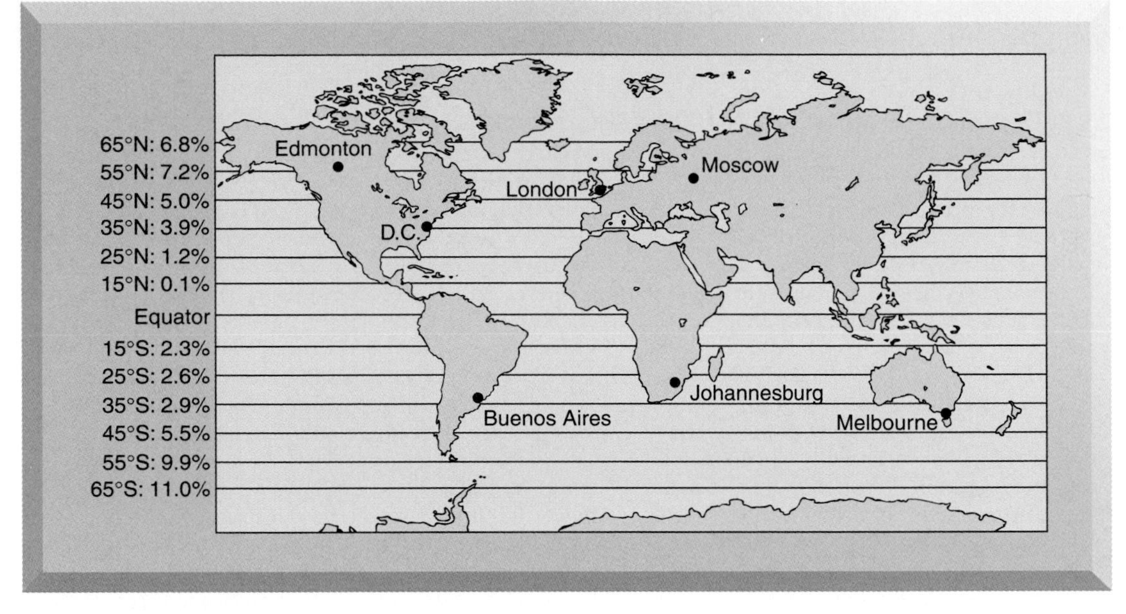

Figure 20–18

Average increases in exposure to harmful ultraviolet radiation per decade. This map is based on satellite data taken from 1978 to 1994. Note that higher latitudes are exposed to larger increases in DNA-damaging UV radiation (UV-B).

such as antiperspirants and hair sprays. Although this ban was a step in the right direction, it did not solve the problem. Most nations did not follow suit, and besides, propellants represent only the tip of the iceberg in terms of CFC use.

In 1987, representatives from many countries met in Montreal to sign the **Montreal Protocol**, an agreement to significantly reduce CFC production by 50 percent by 1998. Since 1987, more than 150 countries signed an agreement to phase out all use of CFCs by 2000. Despite these agreements, the environmental news about CFCs continually worsened in the early 1990s. In 1992, for example, scientists reported that decreases in stratospheric ozone occurred over the heavily populated mid-latitudes of the Northern Hemisphere in all seasons. As a result, some nations took even stricter measures to limit CFC production.

Industrial companies that manufacture CFCs quickly developed substitutes, such as hydrofluorocarbons (HFCs) and hydrochlorofluorocarbons (HCFCs) (see Meeting the Challenge: Business Leadership in the Phase-out of CFCs). HFCs do not attack ozone, although they are potent greenhouse gases. HCFCs attack ozone, although they are not as destructive as the chemicals they are replacing. Although production of HFCs and HCFCs is increasing rapidly, these chemicals are transitional substances that will only be used until industry develops nonfluorocarbon substitutes.

CFCs, carbon tetrachloride, and methyl chloroform production was completely phased out in the United States and other developed countries in 1996, except for a relatively small amount exported to developing countries (Table 20–2). Existing stockpiles can be used after the deadline, however. Developing countries are on a different timetable and will phase out CFC use by the year 2006. Currently, methyl bromide is scheduled to be phased out in 2010 (the United States will ban production in 2001), and HCFCs in 2030.

Unfortunately, CFCs are extremely stable, and those being used today will continue to deplete stratospheric ozone for 50 to 150 years. Scientists expect the Antarc-

Table 20–2
Estimated Annual World CFC Production for Selected Years, 1950 to 1995

Year	Total CFCs (Tons)
1950	42,000
1955	86,000
1960	150,000
1965	330,000
1970	640,000
1975	860,000
1980	880,000
1985	1,090,000
1990	820,000
1995	300,000

MEETING THE CHALLENGE

Business Leadership in the Phase-Out of CFCs

As scientific evidence began to accumulate linking CFCs to the thinning of the ozone layer, some business leaders predicted that the economy would suffer and our standard of living would decline when CFCs were banned. Other companies viewed the impending phase-out of CFCs as a challenge—a chance to achieve a competitive advantage by being among the first to adopt environmentally friendly products. These companies moved quickly and voluntarily to restrict their use of CFCs. Here we examine how three companies met the challenge: McDonald's, Nortel (formerly Northern Telecom), and York International. We then discuss how the chemical industry worked to develop safe, effective alternatives to CFCs.

Amid growing public concern over the effect of CFCs on the ozone layer, a group of schoolchildren petitioned McDonald's to stop using plastic-foam food containers made with CFCs as the foam-blowing agent. In August 1987, McDonald's told its suppliers of disposable foam packaging that they had 18 months to come up with alternatives. McDonald's announcement not only generated a great deal of positive publicity for itself, but galvanized the foam packaging industry, which voluntarily ended all use of CFCs by the end of 1988. Today all foam packaging made in the United States is completely ozone-friendly.

Nortel is a major electronics firm that used CFCs to clean circuit boards and other electronic components. In March 1988, the company's engineers met with government officials to develop a planned phase-out of CFCs. They decided to eliminate all use of CFCs in three years ("free in three"). Nortel tested possible CFC-substitutes. It also tried to reduce CFC consumption at its factories, which competed with one another to be the first to "get to zero." Meanwhile, Nortel worked within the electronics industry to modify the production process so that CFCs were not needed. Nortel's leadership not only made it one of the first electronics companies to stop using CFCs, but gave it unanticipated benefits because the new manufacturing processes turned out to be cheaper and more effective.

Manufacturers of refrigeration and air conditioning systems, including "chillers" for larger buildings, were initially skeptical that they could eliminate CFCs, which were used as refrigerants, without reducing energy efficiency. Nonetheless, the industry began an aggressive search for alternatives to CFCs. York International was the first company in the chiller industry to develop an alternative, HCFC-123. Other companies followed suit and stopped selling CFC-based chillers in the United States by January 1993. Many existing CFC chillers in buildings have been retrofitted or replaced by the new chillers, which are more energy efficient and therefore save owners money in electricity costs. The chiller story is not an unqualified success, however, because most new chillers still rely on HFCs or HCFCs. The chiller industry is still looking for suitable refrigerants that are more environmentally friendly than HFCs and HCFCs.

The international chemical industry worked hard to ensure that deadlines in the global phase-out of CFCs could be met. Seventeen major chemical companies collectively established two programs, the Alternative Fluorocarbons Environmental Acceptability Study (AFEAS) and the Programme for Alternative Fluorocarbon Toxicity Testing (PAFT). AFEAS was created to provide data on the potential environmental effects of CFC alternatives, and PAFT to provide information about potential effects on human health. Both programs relied on international cooperation among scientists from academic institutions, government research programs, and chemical companies. This collaborative effort enabled CFCs to be phased out more rapidly than would usually be the case when new chemicals have to be subjected to environmental and toxicity testing.

The CFC alternatives tested were hydrochlorofluorocarbons (HCFC-22 and HCFC-123) and hydrofluorocarbons (HFC-134a). Because these compounds contain hydrogen atoms, they tend to be removed in the lower atmosphere by natural processes. Hence, they do not have the potential to accumulate in the stratosphere that CFCs do. The environmental effects of these three CFC alternatives are summarized in Table A. HCFC-22 persists in the atmosphere for a long time (13.3 years) and has the greatest potential to deplete ozone and cause global warming, whereas HFC-134a persists in the atmosphere a long time (14 years) and also has a great global warming potential. Based on these and other data, HCFC-123 is widely considered the best CFC alternative that is currently available. HCFC-123 has a short lifespan in the atmosphere (1.4 years) and relatively low ozone-depletion and global-warming potentials.

Table A

A Comparison of the Environmental Effects of Three CFC Alternatives

	HCFC-22	HCFC-123	HFC-134a
Average time in atmosphere (years)	13.3	1.4	14
Relative potential to deplete stratospheric ozone	0.05	0.02	0.0
Relative potential to cause atmospheric warming	1700	93	1300

tic ozone hole, for example, to reappear each year until around 2050. However, initial gains have been made. Measurements of ten ozone-destroying chemicals in the lower atmosphere (at surface level) show they peaked in 1994 and began to decline in amount in 1996. It will take several years for this trend to become significant in the stratosphere. Ozone improvement that is directly attributed to the Montreal Protocol is not expected to be measurable until 2010. (The increase in stratospheric ozone must be several percent higher than natural fluctuations in global ozone levels before it can be attributed to the Montreal Protocol.)

Smuggling CFCS

The U.S. government imposed a tax on CFCs of $5.35 per pound, and the tax, along with a diminishing supply of CFCs, has caused wholesale prices to increase to as much as $10 a pound in 1996, up from $1 a pound in 1989. One of the purposes of the tax is to encourage consumers to switch to other CFC-substitutes. An automobile's air conditioning system can be modified to use HFCs for between $50 and $400, for example, but consumers might not consider doing this if their existing air conditioners remained inexpensive to maintain.

One of the problems with the increased cost of CFCs is that a thriving black market has developed. Beginning in the mid-1990s, thousands of tons of CFCs have been smuggled into the United States each year. Black marketeers buy CFCs in Russia and other countries and sell it in the United States for less than the retail price. According to the U.S. Customs Service, CFCs are the second-largest illegal import in Miami. (The largest is cocaine.) The use by consumers of smuggled CFCs discourages consumers from switching to CFC-alternatives and, more important, increases the damage to the ozone layer.

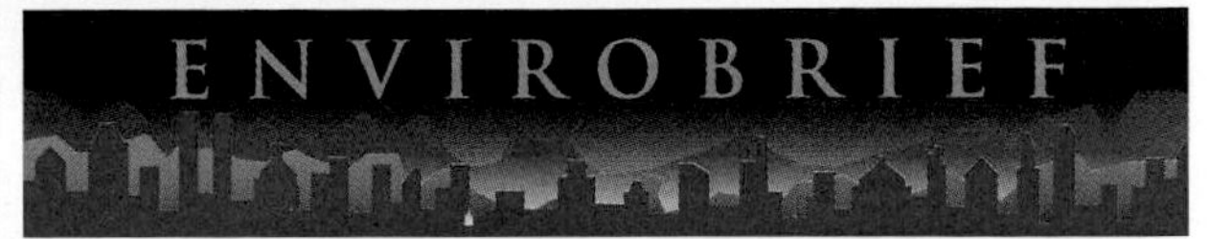

WHAT COULD BE BAD ABOUT REPAIRING THE OZONE LAYER?

Atmospheric chemists are finding that ozone repair may be somewhat of a double-edged sword, because an intact ozone layer might actually worsen planetary global warming. The problem is that ozone is a greenhouse gas that both turns UV radiation into heat and traps heat rising from the surface. Through computer simulations—the only way to model the complex relationship between ozone depletion and global warming—researchers have found that the depletion of stratospheric ozone has indeed cooled the planet, canceling out about 30 percent of the effect of greenhouse gases released in the 1980s. The cooling effect has resulted from both direct destruction of ozone and UV-triggered cloud formation that bounces sunlight back into space. Although the potential global warming associated with a healing ozone layer is unwelcome news, most atmospheric chemists agree that ozone layer depletion had to be stopped and that its repair will at least buy time for dealing with the other troubling effects on climate.

LINKS AMONG ACID DEPOSITION, GLOBAL WARMING, AND OZONE DEPLETION

Most environmental studies examine a single issue, such as acid deposition, global warming, or ozone depletion. Canadian researchers at the Experimental Lakes Area in Ontario, Canada, decided to take a different approach and explore the interactions of all three environmental problems simultaneously. In 1996, the scientists reported that North American lakes may be more susceptible to damage from UV radiation than the thinning of the ozone layer would indicate. It appears that the combined effects of acid deposition and climate warming increase how far UV radiation penetrates lake waters. Some of the possible effects of increased UV penetration are to disrupt photosynthesis in algae and aquatic plants and to cause sunburn damage in fishes.

How far UV radiation penetrates lake water is related to the presence of dissolved organic compounds. This organic material, which is present even in the clearest, least polluted lakes, comes from the decomposition of dead organisms. Dissolved organic material acts like a sunscreen by absorbing UV radiation so that it can only penetrate the water a few inches.

A warmer climate increases evaporation, which reduces the amount of water flowing into a lake from the surrounding watershed. Because most of a lake's dissolved organic compounds have been washed into the lake, even a slightly drier climate reduces how much organic material is present in the lake. Therefore, UV radiation penetrates deeper into lake waters as a result of climate warming.

Acid deposition also affects the amount of dissolved organic compounds in a lake. The presence of acid causes organic matter to clump together and settle on the lake floor. The removal of organic material from the water allows UV radiation to penetrate even further.

Scientists now know that environmental problems cannot be studied as separate issues because they often interact in surprising ways. As acid deposition, global warming, and ozone depletion are studied further, it is likely that other synergistic interactions among them will be discovered.

SUMMARY

1. Acid deposition, commonly called acid rain, is a serious regional problem caused by sulfur and nitrogen oxides. Acid deposition kills aquatic organisms and may contribute to forest decline by changing soil chemistry. Although some countries have reduced emissions of sulfur, it may be many decades before their acidified forests and bodies of water recover. Acid deposition also attacks materials such as metals and stone.

2. Carbon dioxide emissions from the combustion of fossil fuels cause the air to retain heat (infrared radiation), which warms the Earth. Other greenhouse gases are methane, nitrous oxide, CFCs, and tropospheric (surface-level) ozone. Air pollutants known as atmospheric aerosols tend to cool the atmosphere, but it is generally agreed that the effects of greenhouse gases will override those of aerosols.

3. The U.N. Intergovernmental Panel on Climate Change reported in 1995 that human-produced air pollutants have played a role in recent global warming, and most climate scientists think that the world will continue to warm during the next century. However, scientists disagree over the future rate and magnitude of climate change as well as what regional patterns may emerge.

4. During the 21st century, global warming could cause a rise in sea level, changes in precipitation patterns, death of forests, extinction of many species, and problems for agriculture. It could result in the displacement of millions of people, thereby increasing international tensions.

5. The challenge of global warming can be met by prevention (stop polluting the air with greenhouse gases), mitigation (slow down the rate of global warming), and adaptation (make adjustments to live with global warming). International negotiations are underway to develop the measures needed to reduce greenhouse gas emissions.

6. The ozone layer in the stratosphere helps to shield the Earth from harmful UV radiation. Plants and other organisms are damaged by exposure to increased UV radiation. In humans, excessive exposure to UV radiation causes cataracts, weakened immunity, and skin cancer.

7. The total amount of ozone in the stratosphere is slowly declining, and large ozone holes develop over Antarctica and the Arctic each year. The attack on the ozone layer is caused by chlorofluorocarbons (CFCs) and similar chlorine-containing compounds. International negotiations known as the Montreal Protocol resulted in an agreement to phase out CFC production.

THINKING ABOUT THE ENVIRONMENT

1. Some environmentalists contend that the wisest way to "use" fossil fuels is to leave them in the ground. How would this affect air pollution? Global warming? Energy supplies?

2. Discuss some of the possible causes of forest decline.

3. What effect will efforts to reduce acid deposition during the 21st century have on global warming?

4. Does the IPCC report unequivocally prove a link between global warming of the past century and rising greenhouse gases? Explain your answer.

5. Austrian biologists who study plant species living high in the Alps have found that the plants have migrated up the peaks at the rate of about 3 feet a decade in this century, apparently in response to climate warming. Assuming that warming continues into the 21st century, what will happen to the plants if they reach the tops of the mountains?

6. Why might developing countries be more reluctant than developed countries to curtail their use of fossil fuels in the interest of solving global warming?

7. Discuss and give examples of the three approaches to global warming: prevention, mitigation, and adaptation.

8. Why will adaptation to global warming be easier for developed nations than for developing nations?

9. Insurance companies are thinking of shifting hundreds of millions of dollars of their investments from fossil fuels to solar energy. Based on what you learned in this chapter, explain why insurance companies consider such an investment shift to be in their best interest.

10. The oil, coal, and utility industries actively oppose any binding international limits on greenhouse gas emissions. Based on what you have learned in this chapter, explain why these industries consider unlimited greenhouse gas emissions to be in their best self-interest.

11. The following statement was overheard in an elevator: "CFCs cannot be the cause of stratospheric ozone depletion over Antarctica because there are no refrigerators in Antarctica." Criticize the reasoning behind this statement.

12. Distinguish between the benefits of the ozone layer in the stratosphere and the harmful effects of ozone at ground level.

13. Explain how acid deposition, climate warming, and ozone depletion interact synergistically to increase the penetration of UV radiation in northern lakes.

* **14.** Using the data in Table B, calculate the mean global temperature in degrees centigrade for the 1960s, 1970s, 1980s, and 1990s (preliminary).

Mean global temperature, 1960s ___________
Mean global temperature, 1970s ___________
Mean global temperature, 1980s ___________
Mean global temperature, early 1990s ___________

Write a statement to summarize what the mean temperatures you have calculated demonstrate.

Table B
Mean Global Temperatures, 1960 to 1995

Year	Temp. °C	Year	Temp. °C	Year	Temp. °C	Year	Temp. °C
1960	14.98	1970	15.02	1980	15.18	1990	15.38
1961	15.08	1971	14.93	1981	15.29	1991	15.36
1962	15.02	1972	15.00	1982	15.08	1992	15.11
1963	15.02	1973	15.11	1983	15.24	1993	15.14
1964	14.74	1974	14.92	1984	15.11	1994	15.23
1965	14.88	1975	14.92	1985	15.09	1995	15.39
1966	14.95	1976	14.84	1986	15.16		
1967	14.99	1977	15.11	1987	15.27		
1968	14.93	1978	15.06	1988	15.28		
1969	15.05	1979	15.09	1989	15.22		

RESEARCH PROJECTS

Research has indicated that spreading dolomite lime on forest soils helps them recover more quickly from acid rain, and this option has been implemented in some European countries. Investigate the feasibility of spreading lime on northeastern forests in the United States.

On-line materials relating to this chapter are on the ► World Wide Web at **http://www.saunderscollege.com/lifesci/environment2** (select Chapter 20 from the Table of Contents).

SUGGESTED READING

Abate, T. "Swedish Scientists Take Acid-Rain Research to Developing Nations," *BioScience* Vol. 45, No. 11, December 1995. The Swedes, who have been studying acid rain longer than almost anyone else, are offering their expertise to developing nations threatened by acid rain caused by industrialization.

Charlson, R. J. and T. M. L. Wigley. "Sulfate Aerosol and Climatic Change," *Scientific American*, February 1994. Although sulfur emissions cool the planet, they are not a solution to global warming.

Dopyera, C. "The Iron Hypothesis," *Earth*, October 1996. Describes the recent test of a novel solution to global warming, in which the ocean was fertilized with iron to stimulate the growth of phytoplankton.

Flanagan, R. "Engineering a Cooler Planet," *Earth*, October 1996. Examines several possible ways to mitigate global warming, including the iron hypothesis.

Hedin, L. O. and G. E. Likens. "Atmospheric Dust and Acid Rain," *Scientific American*, December 1996. Acid rain is still a problem, despite the fact that acidic emissions have declined in recent years.

Monastersky, R. "Health in the Hot Zone," *Science News* Vol. 149, April 6, 1996. Scientists are beginning to assess how global warming may affect human health.

Muller, F. "Mitigating Climate Change: The Case for Energy Taxes," *Environment* Vol. 38, No. 2, March 1996. Developed nations may adopt carbon and energy taxes to help reduce their greenhouse gas emissions.

Parson, E. A. and O. Greene. "The Complex Chemistry of the International Ozone Agreements," *Environment* Vol. 37, No. 2, March 1995. Provides a chronology of ozone-protecting agreements and examines implementation by various countries.

Raloff, J. "When Nitrate Reigns: Air Pollution Can Damage Forests More Than Trees Reveal," *Science News* Vol. 147, February 11, 1995. How nitrogenous air pollutants are related to forest decline.

Wheelwright, J. "The Berry and the Poison," *Smithsonian*, December 1996. Examines the benefits and dangers of methyl bromide, a pesticide that attacks the ozone layer in the stratosphere.

White, R. "Climate Science and National Interests," *Issues in Science and Technology*, Fall 1996. This respected climate scientist examines the political complexities of responding to the threat of climate warming.

The wastewater treatment system for Arcata, California, is a popular area for hiking. (Ted Streshinsky/Photo 20–20).

Water and Soil Pollution

Fifteen thousand people inhabit the town of Arcata on the coast of northern California. Like many small towns, Arcata has an annual festival that attracts visitors to the city. Arcata's gala is very unusual, however. It is called the Flush with Pride Festival, and it celebrates the town's unique water treatment system.

Arcata's treatment of wastewater initially follows the steps used in most municipalities. The solid contaminants are allowed to settle out, and then the dissolved organic wastes are biologically degraded. However, conventional treatment does not remove other pollutants, such as nitrogen and phosphorus, because it is too expensive; such pollutants are usually left in the treated wastewater. Unfortunately, when treated wastewater is discharged into rivers, streams, or the ocean, these contaminants sometimes cause problems.

Arcata developed a way to remove such contaminants from treated water for a fraction of the cost of a normal advanced treatment plant and, at the same time, to increase the amount of ecologically important wetlands in the town's vicinity. Essentially, Arcata uses cattails, bulrushes, and other marsh plants to remove the contaminants by absorbing and assimilating them. The town hired biologists, who worked with city engineers to develop a wastewater refuge that occupies about 60.7 hectares (150 acres). After water spends some time being purified in the series of three marshes, it is pumped to a treatment center, where it is chlorinated to kill bacteria, dechlorinated to remove the chlorine, and finally released into nearby Humboldt Bay. Many organisms have moved into the wetlands, including fishes, muskrats, otters, ducks, seabirds, osprey, and falcons.

Arcata is not the only town that uses wetlands to filter municipal wastewater; an estimated 400 to 500 communities have built wetlands to treat wastewater. Orlando, Florida, for example, restored a

486-hectare (1200-acre) wetland that had been drained and used as a cow pasture since the late 1800s. This wetland now removes phosphorus and nitrogen contaminants from 49 million liters (13 million gallons) of treated city wastewater each day. It also serves as an excellent wildlife habitat.

As we saw in Chapter 13, water is required by all organisms for their survival. However, having water of good quality is just as important as having enough water. This chapter discusses some of the pollutants found in water and how we can improve water quality. Although the United States has made visible progress in cleaning up its water during the past 25 years, much remains to be done. In some areas, water quality has actually deteriorated, whereas in other areas, strong cleanup efforts have allowed us only to hold our ground, not make any gains. ◄

TYPES OF WATER POLLUTION

Water pollutants are divided into eight categories.

1. **Sediment pollution** is caused by soil particles that enter the water as a result of erosion.
2. **Sewage** is wastewater carried off by drains or sewers (from toilets, washing machines, and showers) and includes water that contains human wastes, soaps, and detergents.
3. **Disease-causing agents**, such as the organisms that cause typhoid and cholera, may be present in water; they come from the wastes of infected individuals.
4. **Inorganic plant and algal nutrients**, such as nitrogen and phosphorus, come from animal wastes and plant residues as well as fertilizer runoff.
5. **Organic compounds**, most of which are synthetic, are often toxic to aquatic organisms. Because many have unusual structures that are difficult for microorganisms to degrade, these compounds may persist in the environment for a long time.
6. **Inorganic chemicals** include such contaminants as the heavy metals mercury and lead.
7. **Radioactive substances** include the wastes from the mining and refinement of radioactive metals as well as the pollution caused by their use.
8. **Thermal pollution** occurs when heated water, produced during many industrial processes, is released into waterways.

Sediment Pollution

Sediments are particles suspended in a body of water that eventually settle out and accumulate on the bottom. Sediment pollution comes from agricultural lands, forest soils exposed by logging, degraded streambanks, overgrazed rangelands, strip mines, and construction. Control of soil erosion, which was discussed in detail in Chapter 14, reduces sediment pollution in waterways.

Sediment pollution causes problems by reducing light penetration, covering aquatic organisms, bringing insoluble toxic pollutants into the water, and filling in waterways. When sediment particles are suspended in the water, they make the water turbid (cloudy), which in turn decreases the distance that light can penetrate. Because the base of the food web in an aquatic ecosystem consists of photosynthetic algae and plants that require light for photosynthesis, turbid water lessens the ability of producers to photosynthesize. Extreme turbidity also reduces the number of photosynthesizing organisms, which in turn causes a decrease in the number of aquatic organisms that feed on the primary producers (Figure 21–1). When sediments build up to the point where they envelop coral reefs and shellfish beds, they can clog the gills and feeding structures of many aquatic animals.

Sediments adversely affect water quality by carrying toxic chemicals, both inorganic and organic, into the water. The sediment particles provide surface area to which some insoluble, toxic compounds adhere, so when sediments get into water, so do the toxic chemicals. Disease-causing agents can also be transported into water via sediments.

When sediments settle out of solution, they fill in waterways. This problem is particularly serious in reservoirs and in channels through which ships must pass. Thus, sediment pollution adversely affects the shipping industry.

Sewage

The release of sewage into water causes several pollution problems. First, because it carries disease-causing agents, water polluted with sewage poses a threat to public health (see the next section, on disease-causing agents). Sewage also generates two serious environmental problems in water, enrichment and oxygen demand. **Enrichment**, the fertilization of a body of water, is caused by the presence of high levels of plant and algal nutrients such as nitrogen and phosphorus. Because these nutrients get into waterways not only from sewage but also from other sources, we will consider enrichment later in the chapter and confine this discussion to oxygen demand.

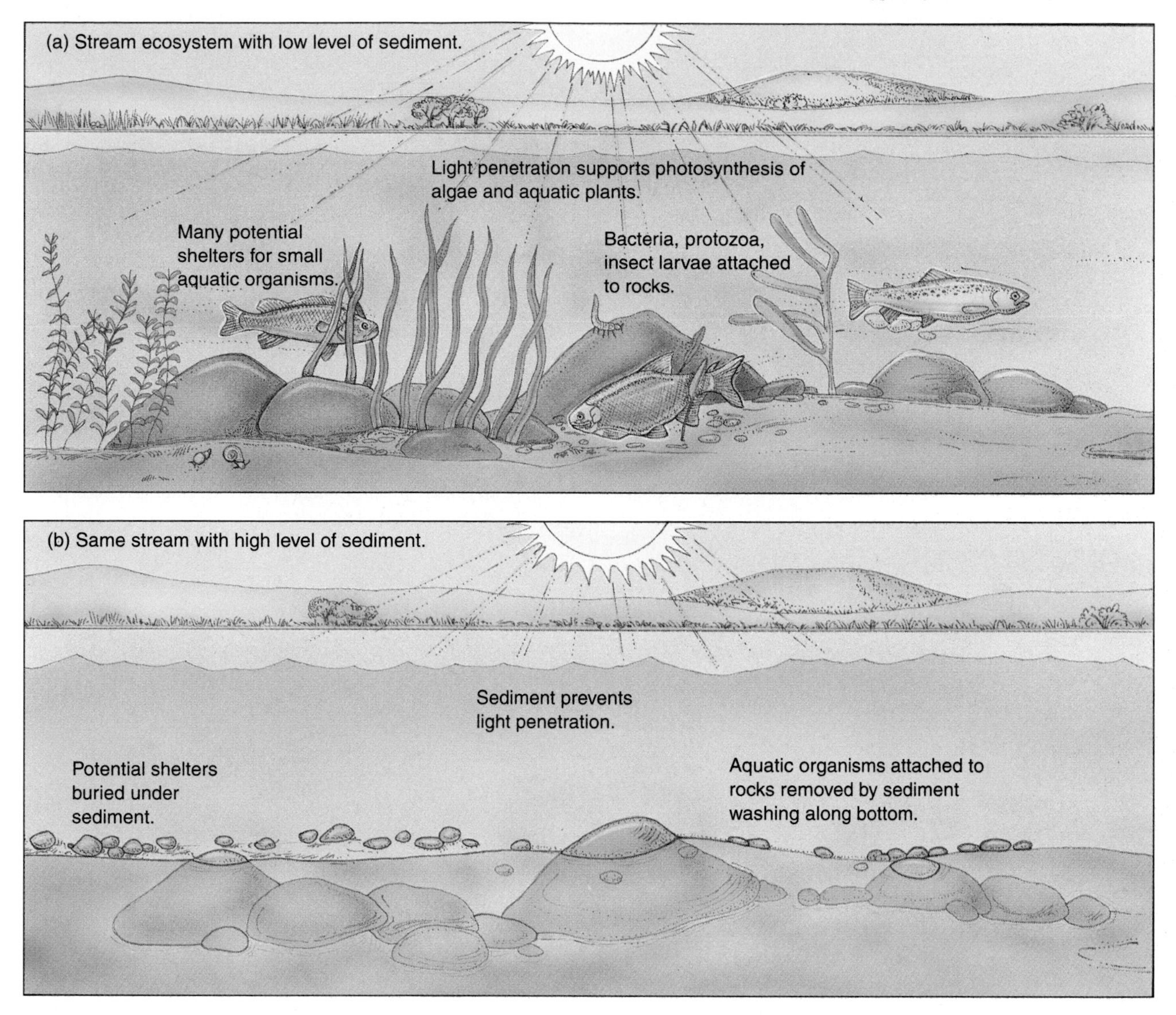

Figure 21–1

The effects of sediment pollution on a stream ecosystem. (a) A stream without sediment pollution can support diverse aquatic life. (b) How the same stream would appear after prolonged exposure to heavy sediment pollution.

Sewage and other organic materials are decomposed into carbon dioxide, water, and similar inoffensive materials by the action of microorganisms. This degradation process, known as **cell respiration**, requires the presence of oxygen. Dissolved oxygen is also used by most organisms living in healthy aquatic ecosystems, including fishes. But oxygen has a limited ability to dissolve in water, and when an aquatic ecosystem contains high levels of sewage or other organic material, the decomposing microorganisms use up most of the dissolved oxygen, leaving little for fishes or other aquatic animals. At extremely low oxygen levels, fishes and other animals leave or die.

Sewage and other organic wastes are measured in terms of their **biological oxygen demand (BOD)**, also known as biochemical oxygen demand, the amount of oxygen needed by microorganisms to decompose the wastes.[1] A large amount of sewage in water generates a high BOD, which robs the water of dissolved oxygen (Figure 21–2). When dissolved oxygen levels are low, anaerobic (without-oxygen) microorganisms also produce compounds that have very unpleasant odors, further deteriorating water quality.

Disease-Causing Agents

Municipal wastewater usually contains many bacteria, viruses, protozoa, parasitic worms, and other infectious

[1]BOD is usually expressed as milligrams of dissolved oxygen per liter of water for some number of days at a given temperature.

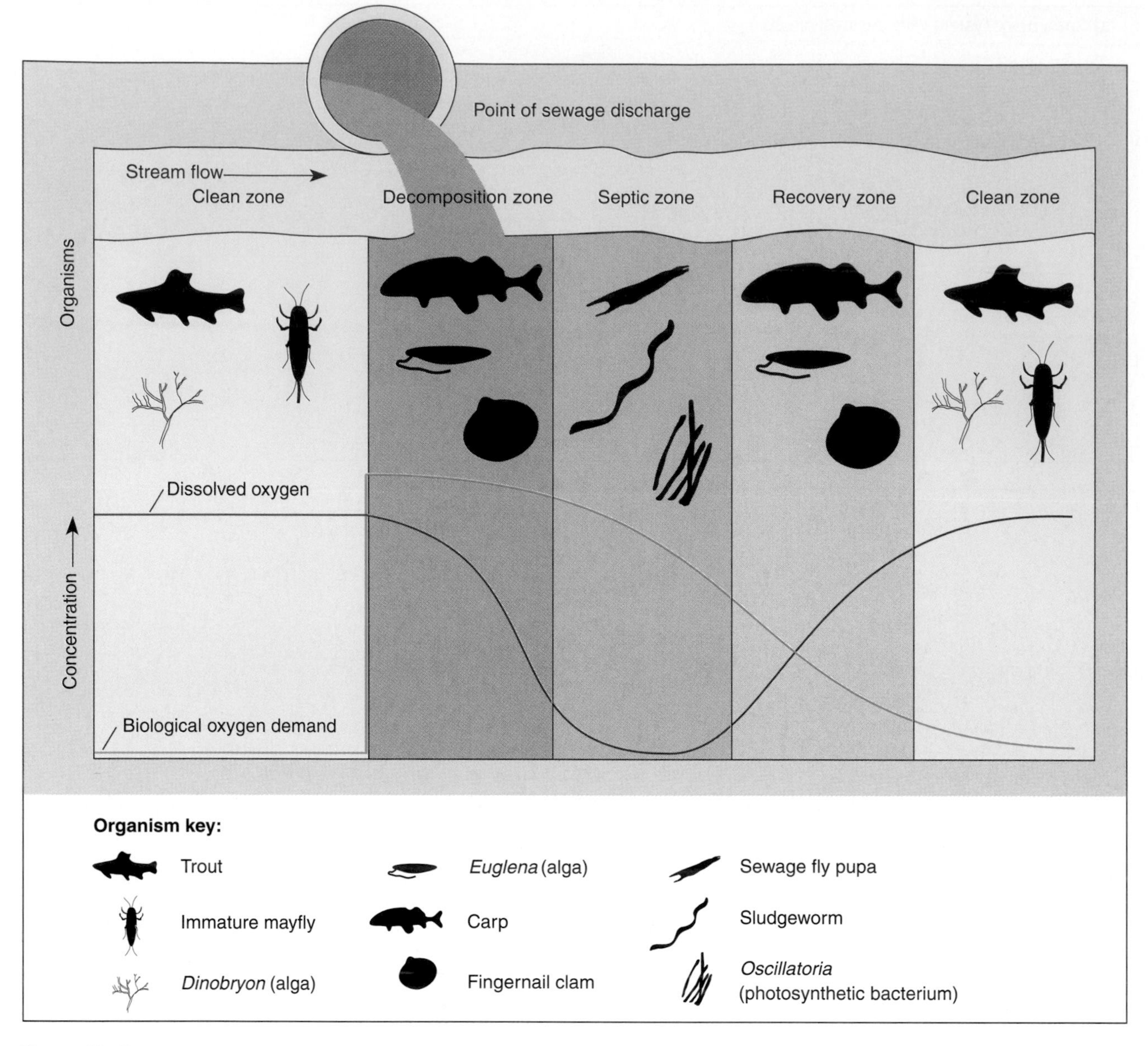

Figure 21–2

The effects of raw sewage on the dissolved oxygen, BOD, and organisms in a stream. Note how the stream gradually recovers as the sewage is diluted and degraded. (Organisms not drawn to scale.)

agents that cause human or animal diseases (Table 21–1). Typhoid, cholera, bacterial dysentery, polio, and infectious hepatitis are some of the more common diseases caused by bacteria or viruses that are transmissible through contaminated food and water. However, many human diseases are not transmissible through water (the AIDS virus, for example).

The vulnerability of our public water supplies to waterborne disease-causing agents was dramatically demonstrated in 1993 when a microorganism known as *Cryptosporidium* contaminated the water supply in the greater Milwaukee area. About 370,000 people developed diarrhea—making it the largest outbreak of a waterborne disease ever recorded in the United States—and several people with weakened immune systems died. Smaller outbreaks of contamination by salmonella and other bacteria have occurred in several cities and towns since the Milwaukee episode; these outbreaks have triggered additional concerns about the safety of our drinking water.

Monitoring for sewage

Because sewage-contaminated water is a threat to public health, periodic tests are made for the presence of sewage in our water supplies. Although many different microorganisms thrive in sewage, the common intestinal bacterium *Escherichia coli* is typically used as an indication of the amount of sewage present in water and as an indirect measure of the presence of disease-causing agents. *Es-*

Table 21-1
Some Human Diseases Transmitted by Polluted Water

Disease	Infectious Agent	Type of Organism	Symptoms
Cholera	*Vibrio cholerae*	Bacterium	Severe diarrhea, vomiting; fluid loss of as much as 20 quarts per day causes cramps and collapse
Dysentery	*Shigella dysenteriae*	Bacterium	Infection of the colon causes painful diarrhea with mucus and blood in the stools; abdominal pain
Enteritis	*Clostridium perfringens*, other bacteria	Bacterium	Inflammation of the small intestine causes general discomfort, loss of appetite, abdominal cramps, and diarrhea
Typhoid	*Salmonella typhi*	Bacterium	Early symptoms include headache, loss of energy, fever; later, a pink rash appears along with (sometimes) hemorrhaging in the intestines
Infectious hepatitis	Hepatitis virus A	Virus	Inflammation of liver causes jaundice, fever, headache, nausea, vomiting, severe loss of appetite; aching in the muscles occurs
Poliomyelitis	Poliovirus	Virus	Early symptoms include sore throat, fever, diarrhea, and aching in limbs and back; when infection spreads to spinal cord, paralysis and atrophy of muscles
Cryptosporidiosis	*Cryptosporidium* sp.	Protozoon	Diarrhea and cramps that last up to 22 days
Amoebic dysentery	*Entamoeba histolytica*	Protozoon	Infection of the colon causes painful diarrhea with mucus and blood in the stools; abdominal pain
Schistosomiasis	*Schistosoma* sp.	Fluke	Tropical disorder of the liver and bladder causes blood in urine, diarrhea, weakness, lack of energy, repeated attacks of abdominal pain
Ancylostomiasis	*Ancylostoma* sp.	Hookworm	Severe anemia, sometimes symptoms of bronchitis

cherichia coli is perfect for monitoring sewage because it is not present in the environment except from human and animal feces, where it is found in large numbers.

To test for the presence of *Escherichia coli* in water, the **fecal coliform test** is performed (Figure 21–3). A small sample of water is passed through a filter to trap all bacteria. The filter is then transferred to a petri dish that contains nutrients. After an incubation period, the number of greenish colonies present indicates the number of *E. coli*. Safe drinking water should contain no more than 1 coliform bacterium per 100 milliliters of water (about $\frac{1}{2}$ cup), safe swimming water should have no more than 200 per 100 milliliters of water, and general recreational water (for boating) should have no more than 2000 per

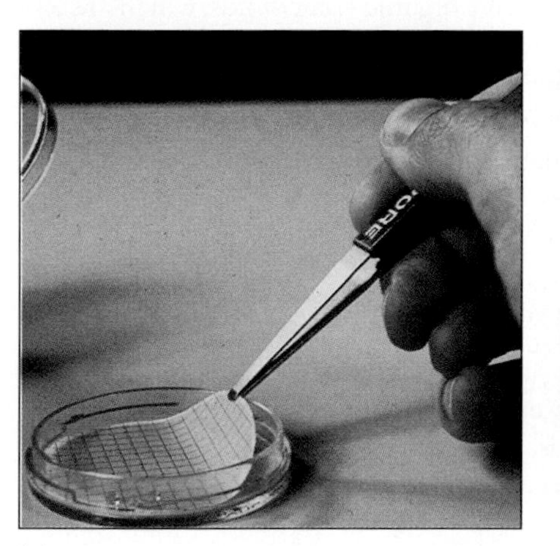

(a)

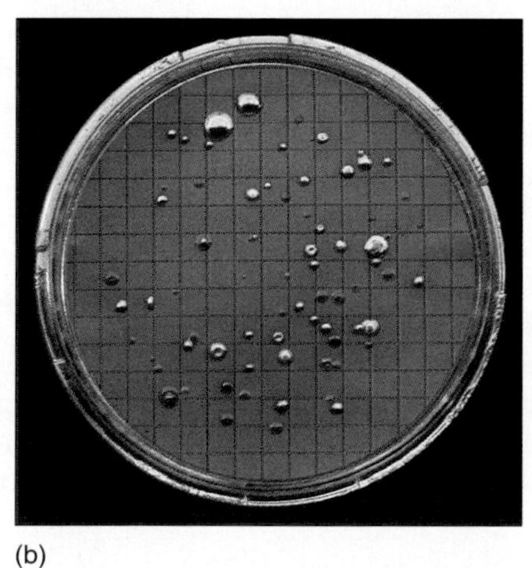
(b)

Figure 21–3
The fecal coliform test is used to indicate the likely presence of disease-causing agents in water. A water sample is first passed through a filtering apparatus. (a) The filter disk is then placed on a medium that supports coliform bacteria for a period of 24 hours. (b) After incubation, the number of bacterial colonies is counted. Each colony arose from a single coliform bacterium in the original water sample. (a,b, Courtesy of Millipore Corp.)

100 milliliters. In contrast, raw sewage may contain several million coliform bacteria per 100 milliliters of water. Although most strains of coliform bacteria do not cause disease, the fecal coliform test is a reliable way to indicate the likely presence of pathogens, or disease-causing agents, in water.

Inorganic Plant and Algal Nutrients

Fertilizer runoff from agricultural and residential land is a major contributor of inorganic plant and algal nutrients such as nitrogen and phosphorus to water, where they encourage excessive growth of algae and aquatic plants. Although algae and aquatic plants are the base of the food web in aquatic ecosystems, their excessive growth disrupts the natural balance between producers and consumers and causes other problems, including enrichment, bad odors, and a high BOD. The high BOD occurs when the excessive numbers of algae die and are decomposed by bacteria.

Fertilizer runoff from midwestern fields in such states as Iowa, Wisconsin, and Illinois eventually finds its way into the Mississippi River and, from there, into the Gulf of Mexico. These nutrients have created a huge "dead zone" in the Gulf of Mexico that is nearly as large as the state of New Jersey—some 17,500 square kilometers (7000 square miles) in area. No life exists in the dead zone because the water does not contain enough dissolved oxygen to support fishes, shrimp, or other aquatic organisms. This condition, known as **hypoxia**, occurs when algae grow rapidly due to the presence of nutrients in the water. When these algae die, they sink to the bottom and are decomposed by bacteria, which deplete the water of dissolved oxygen, leaving too little for other sea life. Although hypoxia occurs in many coastal areas around the world, the dead zone in the Gulf of Mexico is the largest one ever reported. The Environmental Protection Agency (EPA) began to study the problem in 1997 because of its threat to commercial fisheries and to the marine environment in general.

Organic Compounds

Most of the thousands of organic (carbon-containing) compounds found in water are synthetic chemicals that are produced by human activities; these include pesticides, solvents, industrial chemicals, and plastics. (Several examples of organic compounds sometimes found in polluted water are given in Table 21–2.) Some organic compounds find their way into surface water and groundwater by seeping from landfills, whereas others, such as pesticides, leach downward through the soil into groundwater or get into surface water by runoff from farms and residences. Many organic compounds are dumped directly into waterways by industries.

Table 21–2
Some Synthetic Organic Compounds Found in Polluted Water

Compound	Some Reported Health Effects
Aldicarb (pesticide)	Attacks nervous system
Benzene (solvent)	Blood disorders, leukemia
Carbon tetrachloride (solvent)	Cancer, liver damage; may also attack kidneys and vision
Chloroform (solvent)	Cancer
Dioxins (TCDD) (chemical contaminants)	May cause cancer; may harm reproductive, immune, and nervous systems
Ethylene dibromide (EDB) (fumigant)	Cancer; attacks liver and kidneys
Polychlorinated biphenyls (PCBs) (industrial chemicals)	Attack liver and kidneys; may cause cancer
Trichloroethylene (TCE) (solvent)	Induces liver cancer in mice
Vinyl chloride (plastics industry)	Cancer

Many synthetic organic compounds are toxic (see Chapter 23). Although very few have been tested extensively, some have been shown to cause cancer or birth defects, as well as a variety of other health disorders, in laboratory animals. The long-term health effects of drinking minute amounts of these substances are unknown.

There are several ways to control the presence of organic compounds in our water. Care should be taken by everyone, from individual homeowners to large factories, to prevent organic compounds from ever finding their way into water. Alternative organic compounds, which are less toxic and degrade more readily so that they are not as persistent in the environment, can be developed and used. Also, tertiary water treatment, considered later in this chapter, effectively eliminates many organic compounds in water.

Inorganic Chemicals

A large number of inorganic chemicals find their way into both surface water and groundwater from sources such as industries, mines, irrigation runoff, oil drilling, and urban runoff from storm sewers. Some of these inorganic pollutants are toxic to aquatic organisms. Their presence may make water unsuitable for drinking or other purposes. Here we consider lead and mercury, two inorganic chemicals that sometimes contaminate water and accumulate in the tissues of humans and other organisms (see discussion of bioaccumulation and biological magnification in Chapter 22).

Lead

People used to think of lead poisoning as affecting only inner-city children who ate paint chips that contained lead. Lead-based paint was banned in the United States in 1978, but the EPA estimates that more than three fourths of U.S. homes still contain some. Although the main source of lead poisoning in children is still lead-based paint, lead lurks in many other places in the environment as well. Low amounts of lead originate in natural sources such as volcanoes and wind-blown dust.

Most of the lead contaminating our world, however, can be traced to human activities. Anti-knock agents in gasoline are lead additives that are released into the atmosphere when the fuel is burned. Lead contaminates the soil, surface water, and groundwater when incinerator ash is dumped into ordinary sanitary landfills. It may be spewed into the atmosphere from old factories that lack air pollution control devices. We ingest additional amounts of lead from pesticide and fertilizer residues on produce, from food cans that are soldered with lead, and even from certain types of dinnerware on which our food is served.

According to the EPA, in the mid-1990s, more than 10 percent of all large and medium-sized municipal water supplies still contain lead levels that exceed the maximum permitted by the Safe Drinking Water Act (discussed later in this chapter). In addition, tap water often contains higher levels of lead than are in municipal water supplies; the extra lead comes from the corrosion of old lead water pipes or of lead solder in newer pipes.

Millions of Americans, many of them children, have damaging levels of lead in their bodies. According to the Agency for Toxic Substances and Disease Registry in Atlanta, 17 percent of American children—three to four million children—have blood lead levels that exceed 15 micrograms per deciliter of blood. A blood lead level above 10 micrograms per deciliter is considered dangerous.

The three groups of people at greatest risk from lead poisoning are middle-aged men, pregnant women, and young children. Middle-aged men with high levels of lead are more likely to develop **hypertension**, or high blood pressure. High lead levels in pregnant women increase the risk of miscarriages, premature deliveries, and stillbirths. Children with even low levels of lead in their blood may suffer from a variety of mental and physical impairments, including partial hearing loss, hyperactivity, attention deficit, lowered IQ, and learning disabilities. A 1996 study by the University of Pittsburgh School of Medicine reported a link between male juvenile delinquency and high bone lead concentrations. (Because lead accumulates in bones, bone lead levels more accurately reflect long-term exposure than do blood lead levels.)

POWERFUL POISONS

One of the more insidious forms of soil and water pollution comprises toxic elements and compounds that do not break down, but accumulate over time. Mercury and cadmium, two heavy metals that are toxic to humans, behave this way. Dry cell batteries, like those that power portable stereos and flashlights, are by far the greatest source of environmental mercury and cadmium. More than 2.5 billion dry cell batteries are purchased in the United States each year. Although they account for only 0.005 percent of the U.S. waste stream by weight, they contribute more than 50 percent of the mercury and cadmium found in trash. Approximately 14 million pounds of mercury enter U.S. landfills each year, mostly in batteries. Once released into the environment through incineration or as landfill leachate, mercury and cadmium accumulate in soil, water, and organisms.

Technically, batteries are recyclable. But the associated costs and other complications have discouraged a recycling market. Battery companies have reduced the amount of mercury in their products since 1990, but they have yet to remove the toxic metals altogether. If they do not do so voluntarily by the late 1990s, they may be forced to do so by law.

Mercury

Small amounts of mercury occur naturally in the environment, but because mercury is used in a variety of industrial processes, its presence is widespread. When industries release their wastewater, some metallic mercury may enter natural bodies of water along with the wastewater. The combustion of coal releases small amounts of mercury into the air; this mercury then moves from the atmosphere to the water via precipitation. Mercury sometimes enters water by precipitation after household trash containing batteries, paints, and plastics has been burned in incinerators. Once in a body of water, mercury settles into the sediments and is converted to methyl mercury

Figure 21–4

Mercury pollution is dangerously high in portions of the Florida Everglades. Because mercury readily enters the food web, fish advisories are posted where mercury pollution is a problem. (John Shaw/Tom Stack and Associates)

compounds, which readily enter the food web (Figure 21–4). Mercury accumulates in the muscles of tuna, swordfish, and shark—the top predators of the open ocean.

Methyl mercury compounds remain in the environment for a long time and are highly toxic to organisms, including humans. Prolonged exposure to methyl mercury compounds causes mental retardation in children and kidney failure. Mercury accumulation in the body also severely damages the nervous system.

Radioactive Substances

Radioactive materials get into water from several sources, including the mining and processing of radioactive minerals such as uranium and thorium. Many industries use radioactive substances; although nuclear power plants and the nuclear weapons industry use the largest amounts, medical and scientific research facilities also employ them. It is possible for radiation to inadvertently escape from any of these facilities, polluting the air, water, and soil. Accidents at nuclear power plants can release into the atmosphere large quantities of radiation, which eventually contaminate soil and water. Radiation from natural sources can also pollute groundwater.

Since the 1980s, low levels of radioactive substances have been discovered in the wastewater of about 12 sewage treatment plants in the United States. The EPA is currently conducting a study of radioactive materials in sewage sludge (a slimy solid mixture formed during the treatment of sewage) that should be completed by the year 2000. The health effects of exposure to radiation are discussed in Chapter 11.

Radon

Radon is a naturally occurring radioactive gas. It is produced in the Earth's crust and increases the risk of lung cancer when it is inhaled over long periods of time (see Chapter 19). It is also possible for radon to enter groundwater. When you shower, wash dishes, or wash clothes with radon-contaminated water, the radon gets into the air in your home. In addition, the EPA is concerned about possible health effects caused by a long-term exposure to radon in drinking water.

Thermal Pollution

Many industries, such as steam-generated electric power plants, use water to remove excess heat from their operations. Afterward, the heated water is allowed to cool a little before it is returned to waterways, but its temperature is still warmer than it was originally. The result is that the waterway is warmed slightly.

A rise in temperature of a body of water has several chemical, physical, and biological effects. Chemical reactions, including decomposition of wastes, occur faster, depleting the water of oxygen. Moreover, less oxygen dissolves in warm water than in cool water, and the amount of oxygen dissolved in water has important effects on aquatic life. There may be subtle changes in the activities and behavior of aquatic organisms in thermally polluted water, because temperature affects reproductive cycles, digestion rates, and respiration rates. For example, at warmer temperatures fishes require more food to maintain body weight; they also typically have shorter life spans and smaller populations. In cases of extreme thermal pollution, fishes and other aquatic organisms die.

EUTROPHICATION: AN ENRICHMENT PROBLEM

Normal lakes that have minimal levels of nutrients are said to be unenriched, or **oligotrophic**. An oligotrophic lake has clear water and supports small populations of aquatic organisms. **Eutrophication** is the enrichment of water by nutrients; a lake that is enriched is said to be **eutrophic**. The water in a eutrophic lake is cloudy and usually resembles pea soup because of the presence of vast numbers of algae and cyanobacteria that are supported by the nutrients (Figure 21–5).

Although eutrophic lakes contain large populations of aquatic animals, different kinds of organisms predominate there than in oligotrophic lakes. For example, an unenriched lake in the northeastern United States may contain pike, sturgeon, and whitefish; all three are found in the deeper, colder part of the lake, where there is a higher concentration of dissolved oxygen. In eutrophic lakes, on the other hand, the deeper, colder levels of wa-

(a)

(b)

Figure 21–5

The effect of enrichment on a lake or pond. (a) Crater Lake in Oregon, like other oligotrophic lakes, is low in nutrients. (b) Eutrophic lakes and ponds, such as this one in western New York, are often covered with slimy, smelly mats of algae and cyanobacteria. (a, Rich Buzzelli/Tom Stack and Associate; b, W. A. Banaszewski/Visuals Unlimited)

ter are depleted of dissolved oxygen because of the high BOD caused by decomposition on the lake floor. Therefore, fishes such as pike, sturgeon, and whitefish die out and are replaced by warm-water fishes, such as catfish and carp, that can tolerate lesser amounts of dissolved oxygen (Figure 21–6).

Over vast periods of time, oligotrophic lakes and slow-moving streams and rivers become eutrophic naturally. As natural eutrophication occurs, these bodies of water are slowly enriched and grow shallower from the immense number of dead organisms that have settled in the sediments over a long period. Gradually, plants such as water lilies and cattails take root in the nutrient-rich sediments and begin to fill the shallow waters, forming a marsh.

Eutrophication can be markedly accelerated by human activities. This fast, human-induced process is usually called **artificial eutrophication** to distinguish it from natural eutrophication, and it results from the enrichment of water by inorganic plant and algal nutrients—most commonly in sewage and fertilizer runoff.

Controlling Eutrophication

Water, sunlight, carbonate (dissolved carbon dioxide), nitrogen, phosphorus, and certain other inorganic nutrients are the main requirements for algal growth, which is limited by the essential material that is in shortest supply. Because phosphorus is often the limiting factor in fresh-

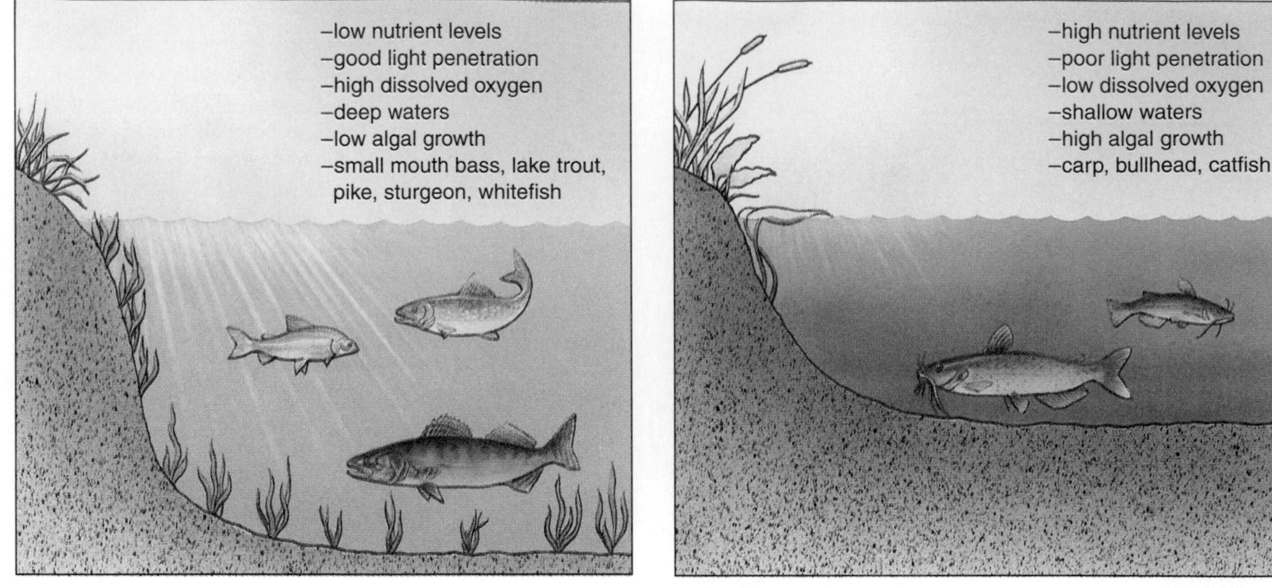

Figure 21–6
A comparison of the features of (a) an oligotrophic lake and (b) a eutrophic lake.

water lakes, the most effective way to slow artificial eutrophication is usually to limit the amount of phosphorus entering the aquatic system. For example, when treated sewage was no longer dumped in Lake Washington (see Chapter 2), the phosphorus level in the lake declined by about 75 percent and there was a corresponding drop in algal growth. Nitrogen (as nitrates) has also been implicated in eutrophication. In Sweden, for example, coastal water eutrophication of the Kattegat, the narrow waterway between Denmark and Sweden, has been linked to excessive nitrogen runoff from agriculture.

SOURCES OF WATER POLLUTION

Water pollutants come from both natural sources and human activities. For example, about 50 percent of the mercury that contaminates the biosphere comes from natural sources in the Earth's crust; the remainder is emitted by human activities. Nitrate pollution has both natural and human sources—the nitrate that occurs in soil and the inorganic fertilizers that are added to it, respectively. Although natural sources of pollution are sometimes of local concern, pollution caused by human activities is generally more widespread.

The sources of water pollution are classified into two types, point source pollution and nonpoint source pollution. **Point source pollution** is discharged into the environment through pipes, sewers, or ditches from specific sites such as factories or sewage treatment plants. **Nonpoint source pollution**, also called **polluted runoff**, is caused by land pollutants that enter bodies of water over large areas rather than at a single point. Nonpoint source pollution includes agricultural runoff, mining wastes, municipal wastes, and construction sediments. Soil erosion is a major cause of nonpoint source pollution.

Three major sources of human-induced water pollution are agriculture, municipalities (that is, domestic activities), and industries.

Water Pollution from Agriculture

According to the 1992 National Water Quality Inventory Report to Congress, which was prepared by the EPA, agriculture is the leading source of water quality impairment of surface waters nationwide. For example, 72 percent of the water pollution in rivers is attributed to agriculture. Agricultural practices produce several types of pollutants that contribute to nonpoint source pollution (Figure 21–7). Chemical pesticides that run off or leach into water are highly toxic and can adversely affect human health as well as the health of aquatic organisms. Fertilizer runoff causes water enrichment. Animal wastes and plant residues in waterways produce a high BOD and a high suspended solids level as well as water enrichment.

Soil erosion from fields and rangelands causes sediment pollution in waterways. In addition, some agricultural chemicals that are not very soluble in water (for example, certain pesticides) find their way into waterways by adhering to sediment particles. Thus, soil conservation methods not only conserve the soil but reduce water pollution.

(a)

(b)

Figure 21–7

Agriculture produces several water pollutants. (a) The runoff from feedlots and fields where cattle graze contributes animal wastes to water. (b) Liquid fertilizer is applied to newly seeded ground. Fertilizers and pesticides may contaminate both surface waters and groundwater. (a, US Department of Agriculture; b, Grant Heilman, from Grant Heilman Photography)

USING ORGANISMS TO DETECT WATER POLLUTION

Scientists use a variety of organisms, including shellfish, water birds, and fishes, to detect and monitor water pollution.

The U.S. government uses mussels to monitor toxic compounds, both organic and inorganic, in the sediments of coastal waters. The mussels are harvested from the water, and their tissues are analyzed for the presence of pesticides, heavy metals, and industrial chemicals.

The Canadian Wildlife Service monitors herring gull eggs, which accumulate toxic pollutants, around the Great Lakes. Comparisons of different areas in the Great Lakes region are possible because individual herring gulls tend to stay in one place, accumulating toxins from that area only.

A small tropical fish species monitors water pollution in the Stour River, in southern England, around the clock. Native to muddy waters in Nigeria, the elephant fish emits 300 to 500 pulses of electricity per minute to help it "see" in its environment. Interestingly, when water pollution is present, the fish emits more than 1000 pulses per minute. Scientists keep the little fish in individual tanks, through which warmed river water is pumped. If more than half of the fish suddenly increase their electrical emissions, which are monitored by computer, scientists are called in to test the water.

Municipal Waste Pollution

Although sewage is the main pollutant produced by cities and towns, municipal waste pollution also has a nonpoint source: urban runoff from storm sewers (Figure 21–8). The water quality of urban runoff from city streets is often worse than that of sewage. Urban runoff carries salt from roadways, untreated garbage, construction sediments, and traffic emissions (via rain that washes pollutants out of the air). It often contains such contaminants as asbestos, chlorides, copper, cyanides, hydrocarbons, lead, motor oil, organic wastes, phosphates, sulfuric acid, and zinc.

Municipal waste pollution from sewage is a greater problem in developing countries, many of which lack water treatment facilities, than in developed nations. Sewage from many densely populated cities in Asia, Latin America, and Africa is dumped directly into rivers or coastal harbors (see Focus On: Water Pollution Problems in Other Countries, page 492).

Industrial Wastes in Water

Different industries generate different types of pollutants. Food processing industries, for example, produce organic wastes that are readily decomposed but have a high BOD. Pulp and paper mills produce toxic compounds and sludge, although the industry has begun to adopt new manufacturing methods that produce significantly less toxic effluents.

Many industries in the United States treat their wastewater with advanced treatment methods. The electronics industry, for example, produces wastewater containing high levels of heavy metals such as copper, lead,

Figure 21–8

Everyday activities can result in polluted urban runoff. The largest single pollutant is organic waste, which, when it decays, removes dissolved oxygen from water. Fertilizers cause excessive algal growth, which further depletes the water of oxygen, harming aquatic organisms. Other everyday pollutants include used motor oil, which is often poured into storm drains, and heavy metals. These pollutants are carried from storm drains on streets to streams and rivers.

and manganese, but uses special techniques—ion exchange and electrolytic recovery—to reclaim those heavy metals. Plates with commercial value are produced from the recovered metals that would otherwise have become a component of hazardous sludge. Although U.S. industries do not usually dump very toxic wastes into water, disposal is still sometimes a problem.

GROUNDWATER POLLUTION

More than 50 percent of the people in the United States obtain their drinking water from groundwater, which is also withdrawn for irrigation and industry. In recent years attention has been drawn to the quality of the nation's groundwater, which can become contaminated in several ways. The most common pollutants—pesticides, fertilizers, and organic compounds—can seep into groundwater from municipal sanitary landfills, underground storage tanks, and intensively cultivated agricultural lands. For example, more than 250,000 underground tanks are thought to be leaking at service stations in the United States.

Contamination of groundwater is a relatively recent environmental concern. People used to think that the underlying soil and rock through which surface water must seep in order to become groundwater filtered out any contaminants, thereby ensuring the purity of groundwater. This assumption proved false when the quality of groundwater began to be monitored and contaminants were discovered at certain sites. It appears that the natural capacity of soil and rock to remove pollutants from groundwater varies widely from one area to another.

Currently, most of the groundwater supplies in the U.S. are of good quality, although there are some local problems that have led to well closures and raised public health concerns. Cleanup of polluted groundwater is very costly, takes years, and in some cases is not technically feasible. Compounding the cleanup problem is the challenge of safely disposing of the toxic materials removed from groundwater—which, if they are not handled properly, could contaminate groundwater once again.

IMPROVING WATER QUALITY

Water quality can be improved by removing contaminants from the water supply before and after it is used (Figure 21–9). Technology assists in both processes.

Purification of Drinking Water

The United States has nearly 60,000 municipal water facilities that serve 232 million people. Surface-water sources of municipal water supplies include streams, rivers, and lakes. Often artificial lakes called **reservoirs** are produced by building a dam across a river or stream. Reservoirs allow water to be accumulated and stored when there is an adequate supply for use during periods of drought.

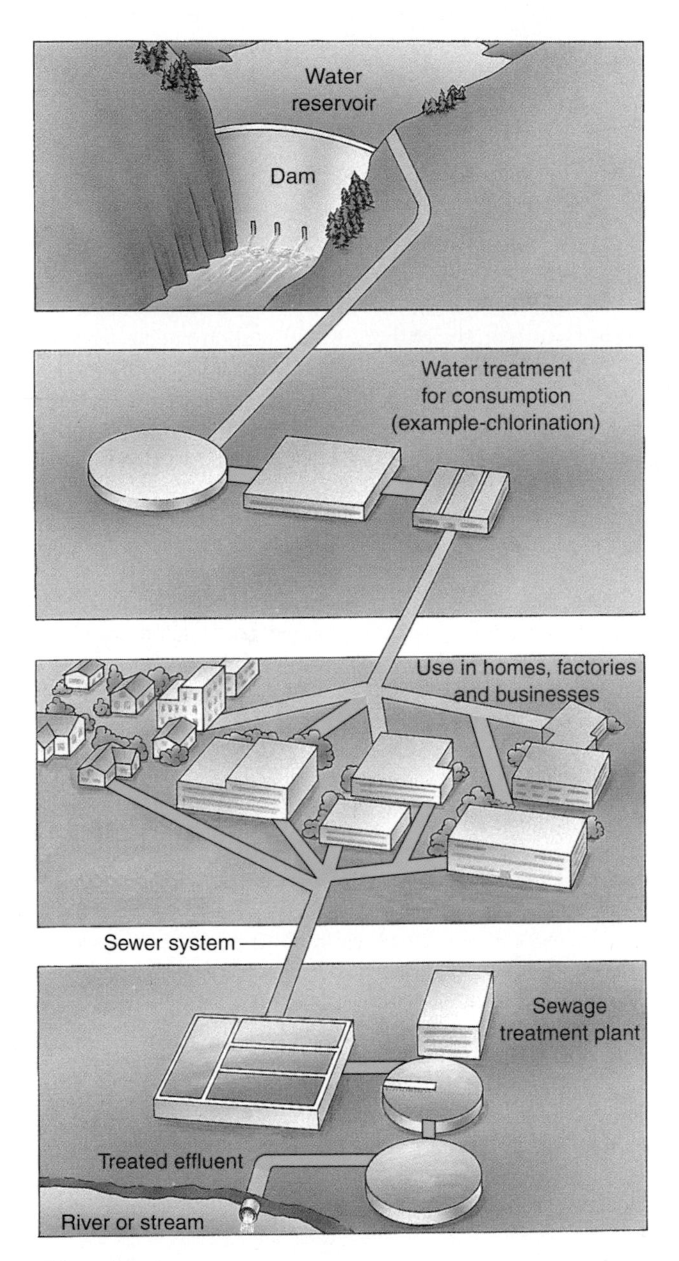

Figure 21–9

The water supply for a town is often stored in a reservoir. The water is treated before use so that it is safe to drink. After use, the quality of the water is fully or partially restored by sewage treatment.

In the United States, most municipal water supplies are treated before being used so that the water is safe to drink. Water that is turbid is treated with a chemical (aluminum sulfate) that causes the suspended particles to clump together and settle out. The water is then filtered through sand to remove any remaining suspended materials as well as many microorganisms. A few cities, such as Cincinnati, also pump the water through activated carbon granules to remove much of the organic compounds dissolved in the water.

In the final purification step, the water is disinfected to kill any remaining disease-causing agents. The most common way to disinfect water is by adding chlorine. A small amount of chlorine is left in the water to provide protection during its distribution through kilometers of pipes. Other disinfection systems use ozone or ultraviolet radiation in place of chlorine.

The chlorine dilemma

During the 19th century, drinking water supplies in the United States were often contaminated by waterborne, disease-causing organisms. The discovery that chlorine kills these organisms, however, has allowed 20th-century Americans to drink water with little fear of contracting typhoid, cholera, or dysentery. The addition of chlorine to our drinking water supply has undoubtedly saved millions of lives.

Now, however, chlorine has been tentatively linked to several kinds of cancer (rectal and bladder) and possibly to rare birth defects, triggering a national debate over the costs and benefits of chlorinating water. The concern is whether there is a long-term hazard from low levels of chlorine in drinking water.

Because there are few viable alternatives to chlorination, the EPA was initially reluctant to reduce the level of chlorine permissible in drinking water, despite the evidence of potential risks. The EPA did not want what happened in Peru to occur in the United States. In 1991 a terrible cholera epidemic swept much of Peru, infecting more than 300,000 people and killing at least 3500. This outbreak occurred when Peruvian officials decided to stop chlorinating much of the country's drinking water after they learned of the slightly increased cancer risk due to chlorination. Peru has since resumed chlorinating its drinking water.

After a detailed review of current scientific evidence linking chlorine to cancer, the EPA proposed in 1994 that water treatment facilities reduce the maximum permissible level of chlorine in drinking water. One alternative to chlorination is to filter water through activated carbon granules, as is done in Cincinnati; one-third less chlorine then needs to be used in the final step. Another alternative, the use of ozone, has been widely adopted in Europe, but preliminary research indicates that ozone may

(text continues on page 494)

FOCUS ON

Water Pollution Problems in Other Countries

In the late 1970s, only about one third of the world's people had access to safe drinking water, and most of these lived in developed countries. During the past 20 years, tremendous progress has been made in providing clean water. Yet in the late 1990s, 1.3 billion people still do not have access to safe drinking water, and about 2 billion people do not have access to adequate sanitation systems; most of these people live in rural areas of developing countries. Worldwide, at least 250 million cases of water-related illnesses occur each year, with 5 million or more of these resulting in death.

Almost every nation in the world faces problems of water pollution. To give you an international perspective on water pollution, we now examine some specific issues in South America, Africa, Europe, and Asia.

Lake Maracaibo, Venezuela Lake Maracaibo in Venezuela is the largest lake in South America. Larger than the state of Connecticut, Lake Maracaibo exemplifies most lakes and inland seas around the world. Lake Maracaibo is suffering from the effects of oil pollution and human wastes as well as contamination by farms and factories. About 10,000 oil wells have been drilled to tap the oil and natural gas under the lake (see figure a). An underwater network of old oil pipes about 15,400 kilometers (9600 miles) long leaks oil into the lake. Fertilizers and other agricultural chemicals drain into the lake from nearby farms, providing the nutrients for an overgrowth of algae. Until recently, raw sewage from Maracaibo city's 1.4 million people and many smaller communities was discharged directly into the water, contributing to the nutrient overload. Modern sewage-treatment facilities were installed during the 1990s to take care of human wastes, but Lake Maracaibo still needs to have its other pollution problems addressed.

Kwale, Kenya For many Africans, serious health problems are caused by surface water supplies that are contaminated with disease-causing organisms. A hand pump project sponsored by the World Bank and the U.N. Development Programme has solved the water safety problem for many Kenyans. For example, Kwale, Kenya, has been the site of cholera and diarrhea outbreaks in the past. Thanks to the installation of a village well with a hand pump, groundwater that is clean and healthful is now available to the inhabitants of Kwale.

Po River, Italy The Po River, which flows across northern Italy, empties into the Adriatic Sea. The Po is Italy's equivalent of the Mississippi River, and it is heavily polluted. Many cities, including Milan with 1.5 million residents, dump their untreated sewage into the Po, and Italian agriculture, which relies heavily on chemicals, is responsible for massive amounts of nonpoint source pollution.

Almost one-third of all Italians live in the Po River basin. The health of many Italians is threatened because the Po is the source of their drinking water. In addition, pollution from the Po has seriously jeopardized tourism and fishing in the Adriatic Sea. Although Italians recognize the problems of the Po and would like to do something about them, the improvement of water quality will be very difficult to implement because the river is under the jurisdiction of dozens of local and regional governments. The cleanup of the Po will require the implementation of a national plan over a period of several decades.

Ganges River, India The Ganges River symbolizes the spirituality and culture of the Indian people (see figure b). It is also highly polluted. Little of the sewage and industrial waste produced by the 350 million people who live in the Ganges River basin is treated. Another major source of contamination is the 35,000 human bodies per year that are cremated in the open air in Varanasi, the holy city of the Hindus. Bodies that are incompletely burned are dumped into the Ganges River, where their decomposition adds to the BOD of the river. Human corpses are also dumped into the river by people who cannot afford cremation costs.

The Indian government has initiated the Ganga Action Plan, an ambitious cleanup project that includes construction of water treatment plants in 29 large cities in the river basin. In addition, 32 electric crematoriums are being set up along the banks.

FOLLOW-UP

Although about $100 million has been spent constructing sewage treatment plants in major cities along the Ganges, most are not completed or are not working effectively. Raw sewage is still being discharged into the river by most of the 29 cities in the river basin. Costs have escalated, and many delays in the Ganga Action Plan have occurred.

One of the reasons the government plan has not worked is that it has not tried to get people involved at the community level. People along the Ganges have not been informed about what the government is trying to do, why it is important, and how they can help.

(a)

(b)

All countries face water pollution problems. (a) The late afternoon sun shines on some of the thousands of oil wells on Lake Maracaibo, Venezuela. This lake, the largest in South America, is dangerously polluted. (b) Bathing and washing clothes in the Ganges River are common practices in India. The river is contaminated by raw sewage discharged directly into the river at many different locations. (a, Stuart Franklin/Magnum Photos, Inc.; b, © 1997 Mike Barlow/Dembinsky Photo Associates)

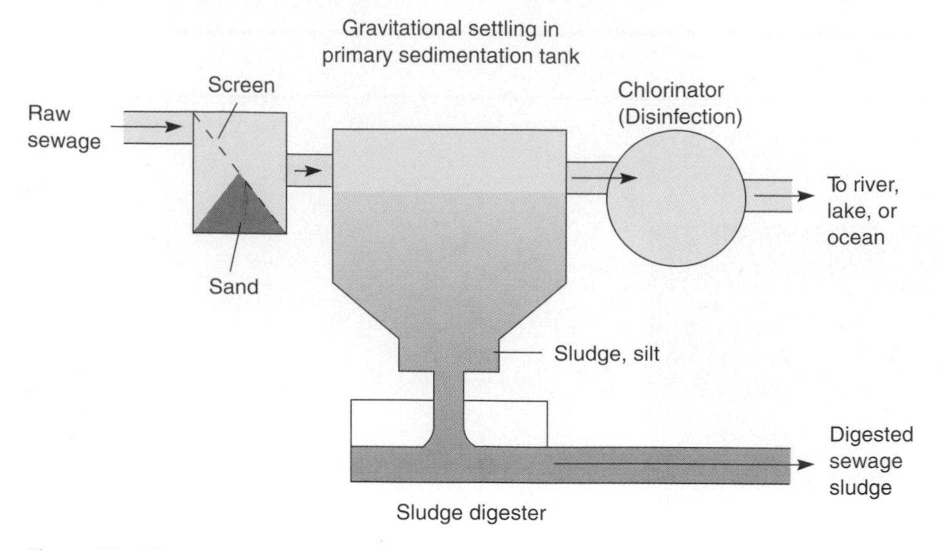

Figure 21–10
The steps involved in primary sewage treatment.

itself cause serious health effects. All of the alternatives are more expensive than chlorination.

Fluoridation

The addition of small amounts of fluoride to most municipal drinking water has been practiced since the mid-1940s to help prevent tooth decay. Fluoride has also been added to many toothpastes since that time, for the same reason.

This practice has been controversial, with opponents questioning the safety and effectiveness of fluoride and supporters saying it is completely safe and very effective in preventing decay. More than 40 years of research have failed to link fluoridation to cancer, kidney disease, birth defects, or any other serious medical condition. Most dental health officials think fluoride is the main reason for the 50 to 60 percent decrease in tooth decay observed in children during the past several decades. This observation is based on comparisons of cavity rates in schoolchildren between cities with fluoridation and without.

As of 1996, 62 percent of U.S. public water supplies are fluoridated. Currently, fluoridation is more common in the eastern half of the country than in the western half. California, however, recently (in 1995) mandated fluoridation.

Treating Wastewater

Wastewater, including sewage, usually undergoes several treatments at a sewage treatment plant to prevent environmental and public health problems. The treated wastewater is then discharged into rivers, lakes, or the ocean.

Primary treatment removes suspended and floating particles, such as sand and silt, by mechanical processes such as screening and gravitational settling. Primary treatment, however, does little to eliminate the inorganic and organic compounds that remain suspended in the wastewater. In the 1990s, wastewater treatment facilities for 10.8 percent of the U.S. population had primary treatment only (Figure 21–10).

Secondary treatment uses microorganisms to decompose the suspended organic material in wastewater. One of the several types of secondary treatment is trickling filters, in which wastewater trickles through aerated rock beds that contain bacteria and other microorganisms, which degrade (decompose) the organic material in the water. In another type of secondary treatment, water is aerated and circulated through bacteria-rich particles; the bacteria degrade suspended organic material. After several hours, the particles and microorganisms are allowed to settle out, forming **sewage sludge**, a slimy mixture of bacteria-laden solids. Water that has undergone primary and secondary treatment is clear and free of organic wastes such as sewage. In the 1990s, wastewater treatment facilities for 62 percent of the U.S. population had both primary and secondary treatments (Figure 21–11).

Even after primary and secondary treatments, wastewater still contains pollutants, such as dissolved minerals, heavy metals, viruses, and organic compounds. Advanced wastewater treatment methods, also known as **tertiary treatment**, include a variety of biological, chemical, and physical processes. Tertiary treatment must be employed to remove phosphorus and nitrogen, the nutrients most commonly associated with enrichment. Tertiary treatment can also be used to purify wastewater so that

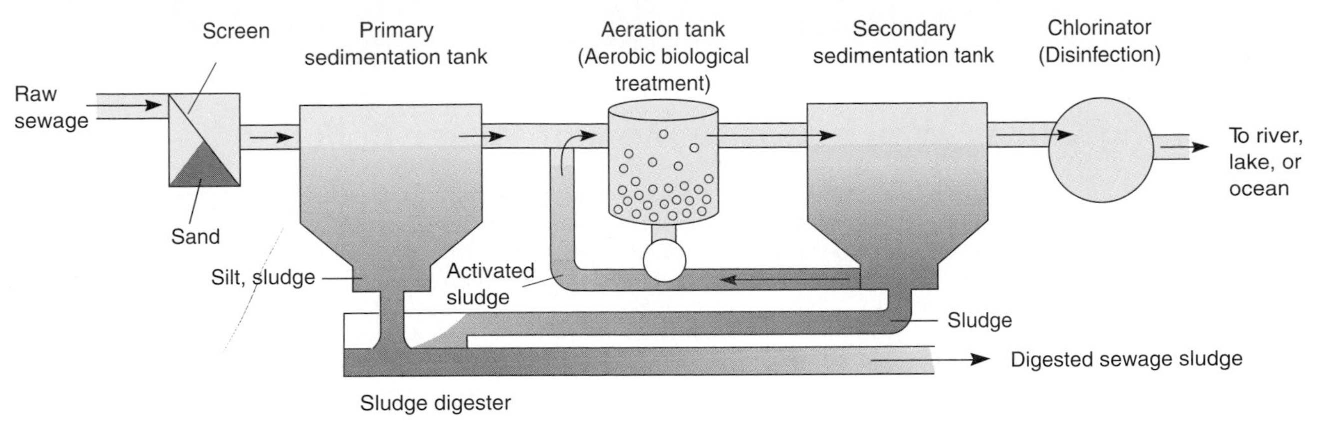

Figure 21–11
The steps involved in primary and secondary sewage treatments.

it can be reused in communities where water is scarce. In the 1990s, wastewater treatment facilities for 26.7 percent of the U.S. population had primary, secondary, and tertiary treatments (Figure 21–12).

Disposal of sewage sludge

A major problem associated with wastewater treatment is disposal of the sewage sludge that is formed during primary and secondary treatments. Five possible ways to handle sewage sludge are anaerobic digestion, application to soil as a fertilizer, incineration, ocean dumping, and disposal in a sanitary landfill. In anaerobic digestion, the sewage sludge is placed in large circular digesters and kept warm (about 35° C, or 95° F), which allows anaerobic bacteria to break down the organic material into gases such as methane and carbon dioxide; the methane can be trapped and burned to heat the digesters. After a few weeks of digestion, the sewage sludge resembles humus and can be used as a soil conditioner.

Sewage sludge can also be used as a fertilizer. It has the advantage of being rich in plant nutrients, although sometimes it contains too many heavy metals to be used commercially. This happens when sewer systems mix industrial waste, which may contain toxic substances, with household waste. Farmers have long used sewage sludge to fertilize hay and feedgrain crops. However, many farm-

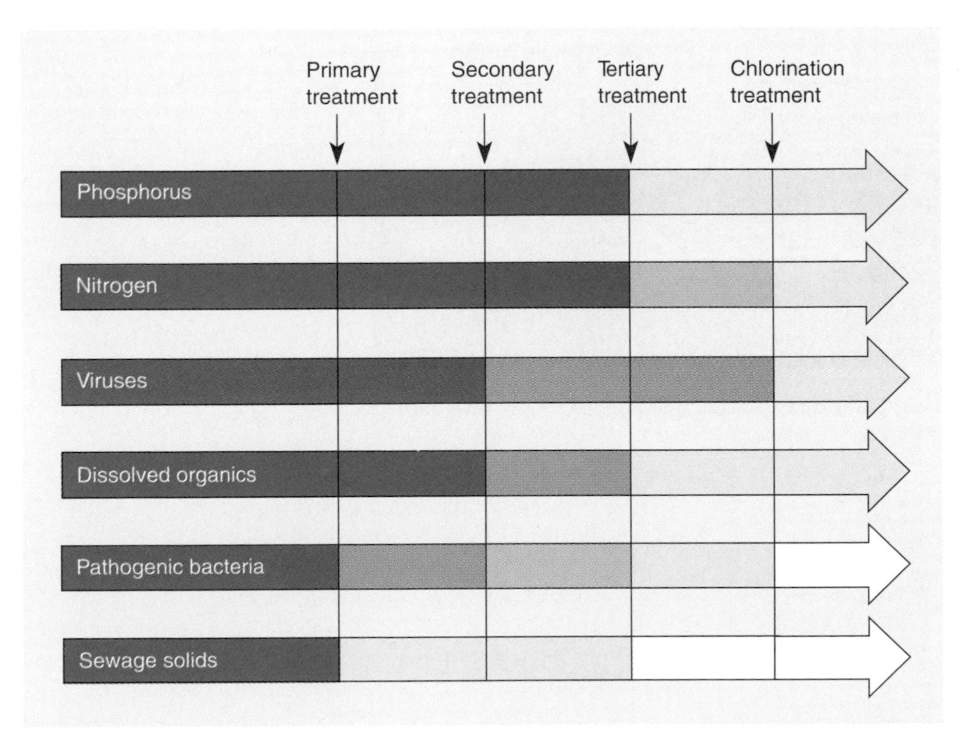

Figure 21–12
The effectiveness of primary, secondary, and tertiary sewage treatments in removing various water pollutants. Note how ineffective secondary treatment is in removing certain contaminants, such as phosphorus and nitrogen. Also note that even after tertiary treatment, some pollutants are still detectable.

MEETING THE CHALLENGE

Using Citizen Watchdogs to Monitor Water Pollution

Lack of funds and insufficient staff prevent well intentioned government agencies from effectively monitoring and enforcing environmental laws such as the Safe Drinking Water Act. For example, the San Francisco Bay Conservation and Development Commission is the regulatory agency charged with patrolling 1600 kilometers (1000 miles) of shoreline and 1500 square kilometers (600 square miles) of water to ascertain that no one is illegally polluting San Francisco Bay. In addition, it handles hundreds of cases arising from its monitoring activities. The agency is severely understaffed and unable to adequately protect the bay.

San Francisco Bay, like most other aquatic ecosystems, is endangered by the combined impact of many different pollution sources rather than from a single disaster, such as occurred when the *Exxon Valdez* spilled oil in Alaska (see Chapter 10). Thus, continual monitoring is necessary to ensure that many small polluters do not collectively do irreparable harm to the bay.

A growing number of private citizens have become actively involved in monitoring and enforcing environmental laws. Provisions in the Safe Drinking Water Act and other key environmental laws allow citizens to file suit when the government does not enforce the laws. Citizen action groups also pressure firms to clean up. For example, as a direct result of citizen action groups, Monsanto launched a comprehensive effort that has resulted in reducing its toxic emissions by more than 90 percent.

In San Francisco Bay, some 300 citizen watchdogs called "Bay-Keepers" monitor the bay from boats (see figure), airplanes, and helicopters. Law students at local universities advise the keepers on issues of litigation. The Bay-Keepers are modeled after the River-Keepers of the Hudson River, who organized in 1983 and were probably the first such group to monitor polluters. Similar keeper groups have sprung up across the country. For example, keepers monitor Long Island Sound, the Delaware River, and Puget Sound. A Swamp Squad in the Chicago area monitors construction in wetlands and contacts developers who appear to be in violation of federal laws.

A Bay-Keeper monitors a steel plant that had illegally dumped debris into San Francisco Bay in the past. (Felix Rigau Photography)

PLANTING LIFE BACK INTO A DEAD MARSH

The Arthur Kill, which is the body of water separating Staten Island and New Jersey, suffered a major heating oil spill in 1990 that destroyed life in a large area of marsh on the New York side. Although even many environmentalists thought the damage was too great for any sort of quick recovery, the New York Parks Department launched a labor-intensive effort that appears to be producing results:

- Bundles of native salt marsh cordgrass, *Spartina alterniflora,* have been planted across several acres of damaged marsh.
- Increased oxygen drawn to soil by the cordgrass appears to be supporting bacteria that are consuming the fuel oil.
- The cordgrass also provides food for zooplankton, which in turn support a food web that includes mussels and crabs, small fishes, herons, falcons, and muskrats.
- Recovery seems well under way in the Arthur Kill; as of late 1994, oil saturation had been reduced by 70 percent, and many former animal residents had returned.

ers have been reluctant to use it on crops for human consumption because consumers might not purchase the food grown in sewage sludge out of concern that it may pose a threat to human health. In 1996, however, the National Research Council[2] announced that properly treated sludge could be used safely to fertilize food crops.

Although sewage sludge can be used to condition soil, it is generally treated as a solid waste. Dried sewage sludge is often incinerated, which may contribute to air pollution, although sometimes the heat produced by this process is used constructively.

Coastal cities such as New York traditionally dumped their sewage sludge into the ocean. However, in 1988 the U.S. Congress passed legislation—the Ocean Dumping Ban Act—that barred ocean dumping of sewage sludge and industrial waste beginning in 1991. All cities stopped disposing of municipal sludge in the ocean on schedule except New York City, which continued to dump its sludge until June, 1992.

Alternatively, sewage sludge can be disposed of in sanitary landfills. As landfill space becomes more costly (see Chapter 23), many cities are looking for alternative ways to handle sewage sludge. Houston plans to use a special process to reduce its sewage sludge to sterile ash with approximately 5 percent the original volume, thus cutting the cost of disposal. Meanwhile, Texas is testing sludge ash as a paving material.

The problem with combined sewer overflow

Some 1100 cities across the United States—such as New York, San Francisco, and Boston—have a **combined sewer system** in which human and industrial wastes are mixed with urban runoff from storm sewers before flowing into the sewage treatment plant. A problem arises when there is heavy rainfall or a large snowmelt because even the largest sewage treatment plant can only process a given amount of wastewater each day. When too much water enters the system, the excess, known as **combined sewer overflow**, flows into nearby waterways without being treated. Combined sewer overflow, which contains raw sewage, has been illegal since passage of the Clean Water Act of 1972 (discussed shortly), but cities have only recently begun to address the problem.

Some cities, such as St. Paul, Minnesota, have installed two separate sewers, one for sewage and industrial wastes and one for urban runoff. However, such an installation is expensive and requires that every street be dug up. Other cities, such as Birmingham, Michigan, are keeping their combined sewer systems but installing huge storage tanks to hold the overflow until it can be treated. The one in Birmingham is larger than a football field and more than 9.1 meters (30 feet deep). Such tanks may be less expensive to install than separate sewer systems, but there are concerns that they may produce offensive odors on hot summer days; moreover, after several days of heavy rain or snow, it is possible that the tank itself could overflow.

LAWS CONTROLLING WATER POLLUTION

Many governments have passed legislation aimed at controlling water pollution. Point source pollutants lend themselves to effective control more readily than nonpoint source pollutants. Governments generally control point source pollution in one of two ways, either by imposing penalties on polluters (a common approach in the United States) or by taxing polluters to pay for the cleanup (common in Japan).

Although most countries have passed laws to control water pollution, monitoring and enforcement are difficult, even in developed countries. Typically, too few resources are allotted for enforcement. For example, in 1996 the 70 enforcement agents in the Office of Drinking Water of the EPA were expected to handle more than 80,000 complaints about drinking water safety. (Also, see Meeting the Challenge: Using Citizen Watchdogs to Monitor Water Pollution.)

[2]The National Research Council is a private, nonprofit society of distinguished scholars. It was organized by the National Academy of Sciences to advise the federal government on complex issues in science and technology.

The United States has attempted to control water pollution through legislation since the passage of the Refuse Act of 1899, which was intended to reduce the release of pollutants into navigable rivers. The three federal laws that have the most impact on water quality today are the Safe Drinking Water Act, the Clean Water Act (originally called the Water Pollution Control Act), and the Water Quality Act. The most important environmental law for regional water quality is the Great Lakes Toxic Substance Control Agreement.

Safe Drinking Water Act

Prior to 1974, individual states set their own standards for drinking water, which of course varied a great deal from state to state. In 1974 the Safe Drinking Water Act was passed, which set uniform federal standards for drinking water in order to guarantee safe public water supplies throughout the United States. This law required the EPA to determine the **maximum contaminant level**, which is the maximum permissible amount of any water pollutant that might adversely affect human health.

During the mid-1990s the Natural Resources Defense Council analyzed the nation's largest public water supplies. It determined that the Safe Drinking Water Act needs to be strengthened to protect our supply of drinking water. EPA administrator Carol Browner stated in 1996 that one out of five public water systems reports violations of public health standards. Currently, most water suppliers take few or no steps to prevent the contamination of the watershed or groundwater that they draw from. The vast majority of water utilities do not use modern water-treatment technologies such as activated carbon granules or ozone to reduce chemical contamination by pesticides, arsenic, and chlorine disinfection byproducts. Also, the average water pipe in the United States is 100 or more years old before it is replaced. Many aging pipes are cracked, which permits contaminated water to seep into them and increases the risk of waterborne diseases.

The Safe Drinking Water Act was amended in 1986 and in 1996. The 1996 version requires municipal water suppliers to tell consumers what contaminants are present in their city's water and if these contaminants pose a health risk. The law also requires the EPA to review risks posed by radon and arsenic in drinking water and to revise its drinking water standards for each contaminant accordingly.

Clean Water and Water Quality Acts

The quality of rivers and lakes in the United States is most affected by the Clean Water Act and the Water Quality Act of 1987. The Clean Water Act was originally passed as the Water Pollution Control Act of 1972; it was amended and renamed the Clean Water Act of 1977, and additional amendments were made in 1981 and 1987. The Clean Water Act should be reauthorized by Congress sometime in the late 1990s. Under the provisions of these acts, the EPA is required to set up and monitor **national emission limitations**—the maximum permissible amounts of water pollutants that can be discharged into rivers, lakes, and the ocean from sewage treatment plants, factories, and other point sources.

Overall, the Clean Water Act has been effective at improving the quality of water from point sources, despite the relatively low fines it imposes on polluters. It is not hard to identify point sources, which must obtain permits from the National Pollutant Discharge Elimination System to discharge untreated wastewater. The United States has improved its water quality in the past several decades, thereby demonstrating that the environment can recover rapidly once pollutants are eliminated.

In 1986, the EPA conducted a national water quality inventory, which showed that nonpoint source pollution was a major cause of water pollution. Although the Water Quality Act of 1987 had provisions to reduce nonpoint source pollution, it is much more difficult and expensive to control than point source pollution. To date, U.S. environmental policies have failed to effectively address nonpoint source pollution, which could be reduced by regulating land use, agricultural practices, and many other activities. The problem is that such regulation would require the interaction and cooperation of many government agencies, environmental organizations, and private citizens. Such coordination is enormously challenging.

Great Lakes Toxic Substance Control Agreement

Millions of Americans and Canadians depend on the Great Lakes for drinking water, fishing, industry, irrigation, swimming, boating, and waste disposal. Yet human use has impaired the water quality that is essential for many of the services provided by the Great Lakes (see Focus On: The Great Lakes, page 500).

Canada and the United States have cooperated to improve the condition of the Great Lakes. In 1972 a joint pollution control program was enacted. Since then, $20 billion has been spent to clean up the lakes. The Great Lakes Toxic Substance Control Agreement was signed by the eight Great Lakes states in 1986 and by the two Great Lakes Canadian provinces in 1988. This legislation is designed to reduce pollution in the lakes by developing coordinated programs among the eight states and two

POLLUTION IN THE WAKE OF MOTORBOATS

The EPA has finalized regulations on exhaust emissions from motorboats. The recreational boating industry now faces the task of complying with the new rules, to go into effect in 1998, that limit sources of water and air pollution:

- As much as one-third of the fuel used by the two-stroke engine commonly used for outboard motors is expelled into the water.
- The engine exhaust mixture releases compounds known to be toxic to aquatic organisms, as well as hydrocarbons that contribute to photochemical smog.
- Total hydrocarbons emitted by boat engines range from 149 million gallons a year—a boating industry estimate—to 272 to 414 million gallons a year, as estimated by the EPA.
- The new EPA regulations will cut hydrocarbon emissions by 75 percent.
- Manufacturers of outboard engines are phasing out two-stroke engines and developing new technologies to increase fuel efficiency and limit exhaust.

provinces. Only the early steps of this agreement have been implemented, and much remains to be done.

Laws That Protect Groundwater

Several federal laws attempt to control groundwater pollution. The Safe Drinking Water Act contains provisions to protect underground aquifers that are important sources of drinking water. In addition, underground injection of wastes is regulated by the Safe Drinking Water Act in an effort to prevent groundwater contamination. The Resource, Conservation, and Recovery Act of 1976 (see Chapter 23) deals with the storage and disposal of hazardous wastes and helps prevent groundwater contamination. Several miscellaneous laws related to pesticides, strip mining, and cleanup of abandoned hazardous waste sites also indirectly protect groundwater.

The many laws that directly or indirectly affect groundwater quality, which were passed at different times and for different reasons, provide a disjointed and, at times, inconsistent protection of groundwater. The EPA makes an effort to coordinate all these laws, but groundwater contamination still occurs.

SOIL POLLUTION

Soil pollution is important not only in its own right but because so many soil pollutants tend to get into surface water, groundwater, and air. For example, selenium, an extremely toxic natural element that is found in many western soils, leaches off irrigated farmlands and poisons nearby lakes, ponds, and rivers, causing death and deformity in thousands of migratory birds and other organisms annually. Most of this chapter's information on water pollution applies to soil pollution as well.

Except for salts, petroleum products, and heavy metals, the chief soil pollutants originate as agricultural chemicals, including fertilizers and pesticides. Until recently, it was assumed that most pesticides in current use had few long-term effects because tests had indicated that they evaporated, decomposed, or leached out of the soil in a relatively short time. However, in the late 1980s, soil scientists developed a new method of testing for contaminants and determined that many agricultural chemicals persist in the soil by seeping into tiny cracks, called micropores, and adhering to soil particles. It appears that the soil may act like a reservoir that stores contaminants and continuously releases them to surface water and groundwater, as well as topsoil, over a long period.

Some heavy metals have been slowly accumulating on farmland soils. One of the most toxic is cadmium, which is present in trace amounts in certain fertilizers. A buildup of cadmium in soil would probably not be occurring if we were not applying so much fertilizer to increase food production.

Fortunately, there is a growing interest in alternative farming practices that reduce the need for large chemical applications to the soil (see Chapter 18). Alternative agriculture may help solve the dual problems of producing enough food and preventing environmental contamination.

Irrigation and Salinization of the Soil

Soils found in arid and semiarid regions often contain high natural concentrations of inorganic compounds as mineral salts. In these areas, the amount of water that drains into lower soil layers is minimal because the little precipitation that falls quickly evaporates, leaving behind the salt. In contrast, humid climates have enough precipitation to leach salts out of the soils and into waterways and groundwater.

FOCUS ON

The Great Lakes

The five Great Lakes of North America—Lakes Superior, Michigan, Huron, Erie and Ontario—formed about 10,500 years ago when the melting waters of retreating glaciers drained into the lake basins that were carved from river valleys by the glaciers. The Great Lakes, which are connected to one another, collectively hold about one-fifth of the world's fresh surface water (see figure). The combined area of the Great Lakes is 244,000 square kilometers (94,200 square miles).

Industrial wastes, sewage, fertilizers, and other pollutants have contaminated the Great Lakes since the mid-1800s. More than 33 million people live in the Great Lakes watershed, an area that is home to agriculture, trade, industry, and tourism. Important industrial cities—such as Duluth, Milwaukee, Chicago, Cleveland, Erie, Buffalo, and Toronto—are located along the lake shorelines. At least 38 million people obtain their drinking water from the Great Lakes.

During the 1960s, pollution in the Great Lakes became a highly visible problem, particularly in Lake Erie,* Lake Ontario, and shallow parts of Lake Huron and Lake Michigan. Thousands of toxic chemicals polluted the lakes. Eutrophication was pronounced, bacterial counts were high enough to be a health hazard, and fish kills were common. Birth defects—such as missing brains, internal organs located outside the body, and deformed feet and wings—were observed in almost 50 percent of the animal species studied.

Since 1972, Canada and the United States have made a concerted effort to reduce pollution in the Great Lakes. Sewage treatment plants in the Great Lakes basin have been improved, and fertilizer runoff from farms, phosphate detergents, and industrial pollutants have been reduced.

As of the late 1990s, the Great Lakes are in better shape than they have been in recent memory. The U.S. EPA and Environment Canada concluded in the *State of the Great Lakes 1995* that water quality and human health are slowly improving in the Great Lakes region. Levels of DDT in women's breast milk have declined since 1967, for example, as have levels of PCBs in trout, Coho salmon, and herring gulls. Some animal populations, such as double-crested cormorants and bald eagles, have rebounded.

Many problems remain, however. Forty-three shoreline areas are so polluted that they are designated "areas of concern" by the U.S. EPA and Environment Canada. Zebra mussels, sea lampreys, and other exotic species have proliferated and threaten native species (recall the discussion of exotic species in Chapter 1). Zebra mussels, which are freshwater shellfish native to the Caspian Sea and first found in North America in 1985 or 1986, clog water-intake pipes for drinking, irrigation, utilities, and industries. Zebra mussels also encrust the bottoms of ships, piers, fish traps and nets, and navigation buoys. They have also led to local extinctions of native mussels in some regions. Sea lampreys, fish with long, eel-like bodies and round, sucking mouths, prey on lake trout, whitefish, and other large fishes.† Lampreys attach their mouths by suction to the bodies of other fishes and then feed on their blood and body fluids. Shoreline development continues to encroach on natural areas, and contributes to flooding and shoreline erosion during storms.

Persistent toxic compounds remain in the lakes, much of them coming from air pollution, such as the spraying of insecticides and the incineration of wastes, or from contaminated lake sediments. Mercury contamination, for example, is on the increase. The presence of certain toxic chemicals causes hormonal changes that are linked to reproductive failures, abnormal development, and abnormal behaviors in some fishes, birds, and mammals (recall the discussion of endocrine disrupters in Chapter 1). Because persistent toxic chemicals have accumulated in food webs (see Table A), fish consumption advisories are issued warning people who eat Great Lakes fishes of potential health problems.

*Lake Erie is the most polluted of the Great Lakes because it has the largest human population on its shores, and it is the shallowest of the Great Lakes and therefore contains the smallest volume of water.

†The possibility of sea lamprey control is more promising since the recent development of a chemical that sterilizes male sea lampreys. Male sea lampreys are trapped, sterilized and released, after which they mate with females, who subsequently do not bear offspring.

Table A
Major Contaminants That Persist in the Great Lakes

Chemicals Usually Found*
Mercury (heavy metal)
Dieldrin (pesticide)
PCBs (industrial chemicals)
Chlordane (pesticide)
DDT (pesticide)
Dioxins (chemical contaminants)
Furans (industrial chemicals)

*Based on contaminant monitoring of fish tissues at multiple sites in the Great Lakes and tributary river mouths.

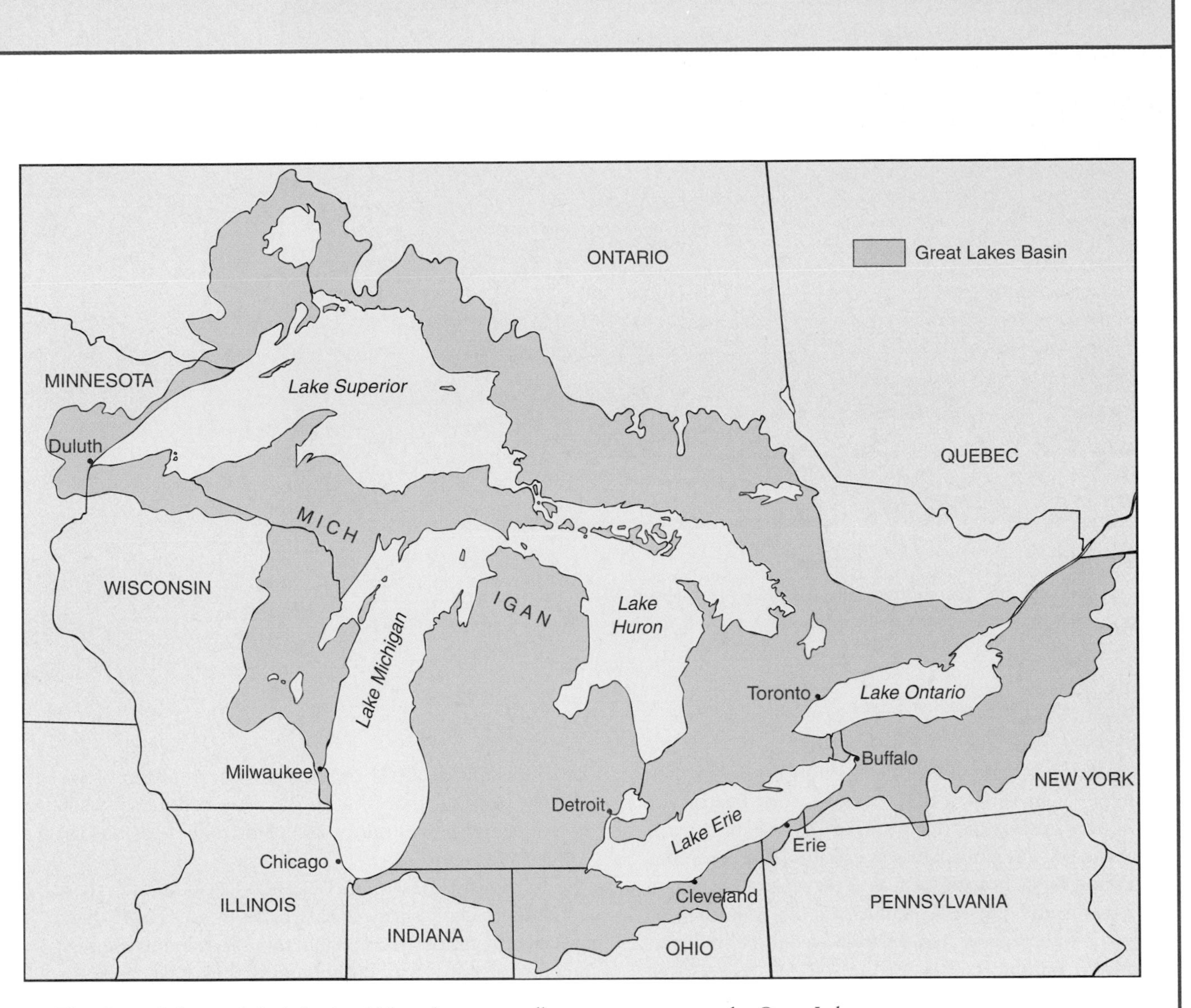

The Great Lakes and their basin. Although many small streams empty onto the Great Lakes, they drain a relatively small area. Lake Superior is the largest body of fresh water in the world; it is also the least polluted of the Great Lakes. Lake Michigan is the only Great Lake located entirely within the United States; all others are located in both the United States and Canada. Lake Huron is the second least polluted of the Great Lakes. Water in the Great Lakes flows eastward and eventually drains into the Atlantic Ocean via the St. Lawrence River. The two easternmost lakes, Lake Erie and Lake Ontario, therefore contain their own pollution plus pollution from the other three lakes.

Figure 21–13
This irrigated soil has become too salty for plants to tolerate. When irrigated land becomes salinized, its agricultural value is reduced or eliminated. (USDA/Natural Resources Conservation Service)

Irrigation of agricultural fields often results in their becoming increasingly saline (salty), an occurrence known as **salinization** (Figure 21–13). Irrigation water contains small amounts of dissolved salts. The continued application of such water, season after season, year after year, leads to the gradual accumulation of salt in the soil. When the water evaporates, the salts are left behind, particularly in the upper layers of the soil—the layers that are most important for agriculture. Given enough time, the salt concentration can rise to such a high level that plants are poisoned or their roots dehydrated. Also, when irrigated soil becomes waterlogged, salts may be carried by capillary movement from groundwater to the soil surface, where they are deposited as a crust of salt.

How soil salinization affects plants

Most plants cannot obtain all the water they need from saline soil. The cause is a water balance problem that exists because water always moves from an area of higher concentration (of water) to an area of lower concentration.[3] Under normal conditions, the dissolved materials in plant cells give them a lower concentration of water than that in soil. As a result, water moves from the soil into plant roots. When soil water contains a large quantity of dissolved salts, however, its concentration of water may be lower than that in the plant cells; if so, water consequently moves *out* of the plant roots and into the saline soil (Figure 21–14).

[3]The concentration of water is determined by the amount of dissolved materials in it. For example, a solution containing 10 percent dissolved salt has 90 percent water.

Obviously, most plants cannot survive under these conditions. Plant species that thrive in saline soils have special adaptations that enable them to tolerate the high amount of salt. Most crops, unless they have been genetically selected to tolerate high salt, are not productive in saline soil.

Soil Remediation

Until relatively recently, the only sure way to remove contaminants from soil was to excavate (dig up) the soil from an entire field and incinerate it. This solution, besides being impractical, killed all beneficial soil organisms. As a result of research in soil remediation, other techniques are available to clean up contaminated soil. Four of these techniques are dilution, vapor extraction, bioremediation, and phytoremediation.

Dilution involves running large quantities of water through contaminated soil in order to leach out pollutants such as excess salt. Although this sounds like a straightforward process, it is extremely difficult, and in many cases impossible, to accomplish. Many soils do not have good drainage properties, so adding lots of water

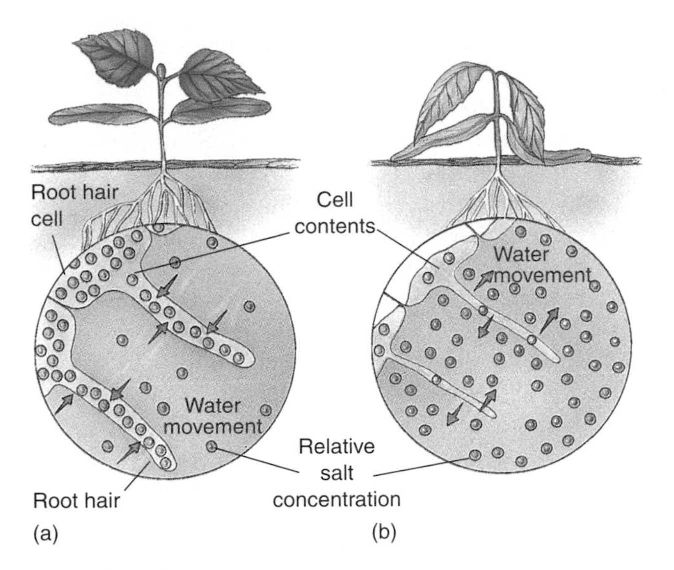

Figure 21–14

The effect of salinized soil on water absorption by the roots of plants. (a) Normally, the water concentration inside plant cells is lower than that in the soil, resulting in a net movement of water (blue arrows) into root cells. (b) When soil contains a high amount of salt, its relative water concentration can be lower than the water concentration inside cells. This causes water to move out of roots into the soil, even when the soil is wet.

simply causes them to become waterlogged. Dilution also results in polluted water, which poses a disposal problem, and is expensive when large volumes of soil have been contaminated. Moreover, it is not feasible in arid and semiarid regions because water is in short supply.

In **vapor extraction**, air is injected into or pumped through soil to remove organic compounds that are volatile (evaporate quickly). Vapor extraction, which is usually done for six months to a year, does not require soil excavation and has been used successfully thousands of times. The oxygen in the injected or pumped air also helps to clean up the soil by stimulating certain microorganisms to degrade non-volatile organic compounds.

Soil remediation specialists increasingly use microorganisms to degrade organic contaminants such as oil and sewage sludge. The use of bacteria and other microorganisms to clean up soil and water pollution, known as **bioremediation**, is discussed further in Chapter 23.

Plants can clean up polluted soils, particularly those contaminated by salts or heavy metals such as cobalt and zinc. The roots of certain plants absorb these contaminants, which subsequently accumulate in the stems and leaves. The use of plants to clean up polluted soil, known as **phytoremediation**, is a relatively inexpensive soil-remediation technique, although it requires several to many years before the soil is clean.

SUMMARY

1. The eight main types of water pollution are sediments, sewage, disease-causing agents, inorganic plant and algal nutrients, organic compounds, inorganic chemicals, radioactive substances, and thermal pollution. Sediment pollution, which is caused primarily by soil erosion, increases water turbidity, thereby reducing the photosynthetic productivity of the water. Sewage contributes to enrichment (fertilization of water) and produces an oxygen demand as it is decomposed. Disease-causing agents are transmitted in sewage.

2. Inorganic plant and algal nutrients, such as nitrogen and phosphorus, contribute to enrichment. Many organic compounds are synthetic and do not decompose readily; some of these are quite toxic to organisms. Inorganic chemicals include toxins such as lead and mercury. Radioactive substances and heat can also pollute water.

3. Eutrophication, the enrichment of oligotrophic bodies of water by nutrients, results in high photosynthetic productivity, which supports an overpopulation of algae. Eutrophic bodies of water tend to fill in rapidly as dead organisms settle to the bottom. Eutrophication also kills fishes and causes a decline in water quality as large numbers of algae die and decompose rapidly.

4. Water pollutants come from both natural sources and human activities. Pollution that enters the water at specific sites, such as pipes from industrial or sewage treatment plants, is called point source pollution. Nonpoint source pollution, also called polluted runoff, comes from the land rather than from a single point of entry. Three major sources of human-induced water pollution are agriculture, municipalities (sewage and urban runoff), and industries.

5. Groundwater can become contaminated by pollutants that seep from sanitary landfills, underground storage tanks, and agricultural operations. Currently, most of the groundwater supplies in the United States are of good quality, although there are some local problems. Because cleanup of polluted groundwater is very costly, takes years, and in some cases is not technically feasible, it is impor-

tant to prevent groundwater contamination from occurring in the first place.

6. In the United States, most municipal water supplies are treated before being used so that the water is safe to drink. Water is usually treated with aluminum sulfate to cause suspended particles to clump and settle out, filtered through sand, and disinfected by adding chlorine. Because there is concern over whether low levels of chlorine in drinking water pose a health hazard, the EPA has proposed that water treatment facilities reduce the maximum permissible level of chlorine in drinking water.

7. Wastewater treatment includes primary treatment (the physical settling of solid matter), secondary treatment (the biological degradation of organic wastes), and tertiary treatment (the removal of special contaminants such as organic chemicals, nitrogen, and phosphorus). One of the most pressing problems of wastewater treatment is disposal of the sewage sludge that results from primary and secondary treatments.

8. Laws attempt to control water pollution. The Safe Drinking Water Act requires the EPA to establish maximum levels for water pollutants that might affect human health. The Clean Water Act and the Water Quality Act require the EPA to establish national emission limitations for wastewater that is discharged into U.S. surface waters. Legislation has been more effective in controlling point source pollution than in controlling nonpoint source pollution. The many laws that address groundwater pollution operate in isolation from one another and at times at cross purposes. The Great Lakes Toxic Substance Control Agreement is designed to reduce pollution in the Great Lakes by developing coordinated programs among the eight states and two Canadian provinces that border the lakes' ecosystem.

9. Soil pollution is important not only in its own right but because so many soil pollutants pollute surface water, groundwater, and air. The chief soil pollutants are salts, petroleum products, heavy metals, and agricultural chemicals.

10. Soil salinization, a common problem in irrigated arid and semiarid regions, makes soil unfit for growing most crops. It is extremely difficult to remove excess salts from salinized soils.

THINKING ABOUT THE ENVIRONMENT

1. Explain why untreated sewage may kill fishes when it is added directly to a body of water.

2. How do midwestern farmers threaten the livelihood of fishermen in the Gulf of Mexico?

3. Tell whether each of the following represents point source pollution or nonpoint source pollution: fertilizer runoff from farms, thermal pollution from a power plant, urban runoff from storm sewers, sewage from a ship, erosion sediments from deforestation.

4. Compare the potential pollution problems of groundwater and surface water used as sources of drinking water.

5. Is the Clean Water Act related in any way to the quality of public drinking water in the United States? Explain your answer. (Recall that the Safe Drinking Water Act regulates the quality of drinking water.)

6. The United States has a Clean Air Act and a Clean Water Act. Should we also have a Clean Soil Act? Present at least one argument for and one argument against such legislation.

7. Explain why saline soils are physiologically dry for plants even though they may be physically wet.

* 8. An asphalt parking lot measures 100 by 500 meters. Calculate its area. If 2 centimeters of rain fall on the parking lot during a heavy rainstorm, how much water (in cubic meters) will run off? Where does the runoff go?

RESEARCH PROJECTS

Contact your local water company and determine the cost per gallon of tap water. What does it cost to treat each gallon of wastewater? Does your water company have any long-term plans to increase the water supply? To improve water quality?

Prepare a map of the watershed in which your school is located. How large is the watershed? What major river is in it, and is it part of a larger drainage basin?

On-line materials relating to this chapter are on the ► World Wide Web at **http://www.saunderscollege.com/lifesci/environment2** (select Chapter 21 from the Table of Contents).

SUGGESTED READING

Adler, T. "The Expiration of Respiration," *Science News* Vol. 149, February 10, 1996. Hypoxia, which is the presence of low oxygen levels in coastal waters, has become a major environmental issue in recent years.

Boyle, R. H. "Life—or Death—for the Salton Sea," *Smithsonian*, June 1996. The Salton Sea in southeastern California, which formed 90 years ago when the Colorado River changed course, began as a freshwater lake but is now saltier than the Pacific Ocean. It is also heavily polluted by raw sewage from Mexico and agricultural runoff from nearby farms.

Briscoe, J. "When the Cup Is Half Full," *Environment* Vol. 35, No. 4, May 1993. Examines water and sanitation services in developing countries.

Foran, J. A., and R. W. Adler. "Cleaner Water, But Not Clean Enough," *Issues in Science and Technology*, Winter 1993–94. Although the Clean Water Act has helped improve the quality of the nation's lakes, rivers, and coasts, much work still remains.

Harris, T. "The Scene of the Crime," *Amicus*, Summer 1993. Examines selenium, one of nature's most toxic elements, and its devastating effects on water birds. Selenium leaches into waterways from irrigated soils in California and other western states.

Malle, K. "Cleaning Up the River Rhine," *Scientific American*, January 1996. Switzerland, France, Germany, and the Netherlands have worked together to clean up the once badly polluted Rhine River, and many fish species are making a comeback.

Mitchell, J. G. "Our Polluted Runoff," *National Geographic* Vol. 189, No. 2, February 1996. Polluted runoff is the main reason why many of America's waters do not meet the Clean Water Act's goals.

Platt, A. "Water-borne Killers," *WorldWatch*, March/April 1996. Infectious diseases are expanding worldwide, largely because many people do not have water clean enough to drink or even bathe in.

Raloff, J. "Something's Fishy," *Science News* Vol. 146, July 2, 1994. The author hypothesizes that the many recent epidemics involving marine mammals have occurred and been so deadly because the animals' immune systems were weakened. The reason: these animals lived in polluted water and accumulated toxic organic chemicals in their bodies.

Regier, H. and H. Shear. "Restoring the Great Lakes Cornucopia," *People & the Planet* Vol. 4, No. 2, 1995. Although the Great Lakes have been badly polluted by human activities, they are slowly being rehabilitated.

Sampat, P. "The River Ganges Long Decline," *WorldWatch* Vol. 9, No. 4, July/August 1996. Examines the contamination of the Ganges River.

22

The salad vac is an environmentally safe method of controlling certain insect pests. (Richard Steven Street)

The Pesticide Dilemma

Picture a giant vacuum cleaner slowly moving over rows of strawberry or vegetable crops and sucking insects off the plants. Such a machine, given names such as "salad vac" and "bug vac," was invented by an entomologist (a biologist who studies insects) as a substitute for the chemical poisons we call pesticides. Each vacuuming eliminates the need for one application of pesticide.

More than 40 California growers use the farm-sized vacuum cleaner to remove lygus bugs, leafhoppers, Colorado potato beetles, and other insect pests from their strawberries and other crops. Some beneficial insects are also killed by the bug vac, but fewer than would be killed if chemical insecticides were sprayed.

Increasing public concern about pesticide residues on food has caused farmers to look seriously at other ways to control pests—even by vacuuming insects. In this chapter we examine the types and uses of pesticides, their benefits and disadvantages. Pesticides have saved millions of lives by killing insects that carry disease and by increasing the amount of food we grow. Modern agriculture depends on pesticides to produce blemish-free fruits and vegetables at a reasonable cost to farmers (and therefore to consumers).

However, pesticides also cause environmental and health problems, and it appears that in many cases their harmful effects outweigh their benefits. Pesticides rarely affect the pest species alone, and the balance of nature, such as predator-prey relationships, is upset. Certain pesticides concentrate at higher levels of the food chain. Humans who apply and work with pesticides may be at risk from pesticide poisoning (short-term) and cancer (long-term), and people who eat traces of pesticide on food are concerned about their long-term effects. In this chapter we also consider some alternatives to pesticides and the pesticide laws that are supposed to protect our health and the environment. ◀

WHAT IS A PESTICIDE?

Any organism that interferes in some way with human welfare or activities is called a **pest**. Some weeds, insects, rodents, bacteria, fungi, nematodes (microscopic worms), and other pest organisms compete with humans for food; other pests cause or spread disease. The definition of "pest" is subjective; a mosquito may be a pest to you, but it is not a pest to the bat or bird that eats it. People try to control pests, usually by reducing the size of the pest population. Toxic chemicals called **pesticides** are the most common way of doing this, particularly in agriculture. Pesticides can be grouped by their target organisms, that is, by the pests they are supposed to eliminate. Thus, **insecticides** kill insects, **herbicides** kill plants, **fungicides** kill fungi, and **rodenticides** kill rodents such as rats and mice.

Agriculture is the sector that uses the most pesticides in the United States—approximately 74 percent of the estimated 0.43 million metric tons (0.48 million tons) used each year. U.S. farmers spend $4.1 billion each year on pesticide treatments for crops. Other major pesticide users are government and industries (almost 13 percent of total pesticide use) and households (almost 13 percent).

The "Perfect" Pesticide

The ideal pesticide would be a **narrow-spectrum pesticide**, which would kill only the organism for which it was intended and not harm any other species. The perfect pesticide would also readily be degraded, or broken down, either by natural chemical decomposition or by biological organisms, into safe materials such as water, carbon dioxide, and oxygen. Finally, the ideal pesticide would stay exactly where it was put and would not move around in the environment.

Unfortunately, there is no such thing as an ideal pesticide. Most pesticides are **broad-spectrum pesticides**, which kill a variety of organisms in addition to the target pest. Some pesticides either do not degrade readily or else break down into compounds that are as dangerous as, if not more dangerous than, the original pesticide. And pesticides move around a great deal throughout the environment.

First-Generation and Second-Generation Pesticides

Before the 1940s, pesticides were of two main types, inorganic compounds (also called minerals) and organic compounds. Inorganic compounds that contain lead, mercury, and arsenic are extremely toxic to pests but are not used much today, in part because their chemical stability—they are not broken down by natural processes—allows them to persist and accumulate in soil and water. This accumulation poses a threat to humans and other organisms, which, like the target pests, are susceptible to inorganic compounds.

Plants provide humans with several natural organic compounds that are poisonous, particularly to insects. Such plant-derived pesticides are called **botanicals**. Examples of botanicals include nicotine from tobacco, pyrethrum from chrysanthemum flowers (Figure 22–1), and rotenone from roots of the derris plant, all of which are used to kill insects. Botanicals are easily degraded by microorganisms and, therefore, do not persist in the environment. **Synthetic botanicals** are human-made insecticides produced by chemically modifying the structure of natural botanicals. An important group of synthetic botanicals are the pyrethroids, which are chemically similar to pyrethrum. Pyrethroids do not persist in the environment; they are relatively nontoxic to mammals, but extremely toxic to fishes and aquatic invertebrates.

In the 1940s a large number of synthetic organic pesticides began to be produced. Earlier pesticides—both inorganic compounds and botanicals—are called **first-generation pesticides** to distinguish them from the vast array of synthetic poisons in use today, called **second-generation pesticides**. The insect-killing ability of **d**ichloro-**d**iphenyl-**t**richloroethane (DDT), the first of the second-generation pesticides, was recognized in 1939 (Figure 22–2). Today there are thousands of pesticide products, made up of combinations of more than 1000 different chemicals.

The Major Kinds of Insecticides

The largest category of pesticides, the insecticides, are usually classified into groups based on chemical structure. Three of the most important kinds of second-generation insecticides are the chlorinated hydrocarbons, organophosphates, and carbamates (Table 22–1).

DDT is an example of a **chlorinated hydrocarbon**, an organic compound containing chlorine. After DDT's insecticidal properties were recognized, many more chlorinated hydrocarbons were synthesized as pesticides. Generally speaking, chlorinated hydrocarbons are broad-spectrum insecticides. They are slow to degrade and therefore persist in the environment (even inside organisms) for many months or even years. They were widely used from the 1940s to the 1960s, but since then they have been banned or their use has largely been restricted, mainly because of problems associated with their persistence in the environment. Many people first became aware of the problems with pesticides in 1963, when Rachel Carson published her book *Silent Spring*.

Organophosphates, organic compounds that contain phosphorus, were developed during World War II as an outgrowth of German research on nerve gas. Organophosphates are more poisonous than other types of insecticides, and they are also toxic to animals other than insects. The toxicity of organophosphates in humans

Figure 22–1

Chrysanthemum flowers, shown here as they are harvested in Rwanda, are the source of the insecticide pyrethrum. Botanicals are chemicals from plants that can be used as pesticides. (Robert E. Ford/Terraphotographics)

is comparable to that of some of our most dangerous poisons—arsenic, strychnine, and cyanide. Organophosphates do not persist in the environment as long as chlorinated hydrocarbons because organophosphates are usually degraded more easily by microorganisms. As a result, they have generally replaced the chlorinated hydrocarbons in large-scale uses such as agriculture, although they are not widely available to consumers because of their high level of mammalian toxicity. Malathion is an example of an organophosphate.

The third group of insecticides, the **carbamates**, are broad-spectrum insecticides, derived from carbamic acid, that are generally not as toxic to mammals as the organophosphates, although they still show broad, nontarget toxicity. Two common carbamates are carbaryl (trade name Sevin) and propoxur (trade name Baygon).

Table 22–1
Selected Pesticides

Kind of Pesticide	Examples	Persistence in the Environment
Insecticides		
Botanicals	Nicotine, pyrethrum, rotenone	Days to weeks
Chlorinated hydrocarbons	Aldrin, benzene hexachloride, chlordane, DDT, dieldrin, endrin, heptachlor, kepone	Years
Organophosphates	Malathion, parathion, diazinon	Generally weeks
Carbamates	Carbaryl (Sevin), carbofuran, propoxur (Baygon), methylcarbamate (Zectran)	Days to weeks
Herbicides		
Selective herbicides	Atrazine, 2,4-D, 2,4,5-T, picloram, silvex	Days to weeks
Nonselective herbicides	Paraquat	Days to weeks
Fungicides	Captan, methyl bromide, zeneb	Days

Figure 22–2

DDT is sprayed to control mosquitoes at New York's Jones Beach State Park in 1945. In the early years of its use, DDT and other pesticides were used in ways that would be unacceptable now. Note the sign on the truck. The harmful effects on the environment of DDT were not known until many years later. (UPI/Corbis-Bettmann)

The Major Kinds of Herbicides

Chemicals that kill or inhibit the growth of unwanted vegetation such as weeds in crops or lawns are called herbicides (Table 22–1). Like insecticides, herbicides can be classified into groups based on chemical structure, but this method is cumbersome because there are at least 12 different chemical groups that are used as herbicides. It is more common to group herbicides according to how they act and what they kill. **Selective herbicides** kill only certain types of plants, whereas **nonselective herbicides** kill all vegetation. Selective herbicides can be further classified according to the types of plants they affect. **Broad-leaf herbicides** kill plants with broad leaves but do not kill grasses; **grass herbicides** kill grasses but are safe for most other plants.

Two common herbicides with similar structures are **2,4-d**ichlorophenoxyacetic acid (2,4-D) and **2,4,5-t**richlorophenoxyacetic acid (2,4,5-T), both developed in the United States in the 1940s. These broad-leaf herbicides are similar in structure to a natural growth hormone in certain plants and disrupt the plants' natural growth processes; they kill plants such as dandelions but do not harm grasses. You may recall from Chapter 18 that many of the world's important crops, such as wheat, corn, and rice, are cereal grains, which are grasses. Both 2,4-D and 2,4,5-T can be used to kill weeds that compete with these crops, although 2,4,5-T is no longer used in this country. (See Focus On: The Use of Herbicides in the Vietnam War for a discussion of the long-term effects of using these herbicides.)

FOCUS ON

The Use of Herbicides in the Vietnam War

One of the controversial aspects of the Vietnam War was the defoliation program carried on by the United States in South Vietnam. From 1962 to 1971, the United States sprayed more than 12 million gallons of herbicides over about 4.5 million acres of South Vietnam to expose hiding places and destroy crops planted by the Vietcong and North Vietnamese troops (see figure). The three mixtures of herbicides used were designated Agent White, Agent Blue, and Agent Orange.

The negative impact of these herbicides on the environment is still being felt today. It is estimated that 20 to 50 percent of the ecologically important mangrove forests of South Vietnam were destroyed. The forests have been replaced by shrubs, and it may take decades for the forests to return. Approximately 30 percent of the nation's commercially valuable hardwood forests were killed, and bamboo and weedy grasses have replaced them.

In addition to the ecological damage, the herbicide sprays caused health problems in the native people and members of the U.S. military who were exposed to them in the Vietnamese jungles. Agent Orange, a mixture of two herbicides (2,4-D and 2,4,5-T), also contained minute amounts of **dioxins**, a group of dangerous poisons formed during the manufacture of these herbicides. (The contamination of Agent Orange by dioxins was discovered after the war.)

High doses of dioxins have been shown to cause birth defects in animals. Reportedly, the number of birth defects, stillbirths, female reproductive disorders, and several cancers have increased in Vietnam since the war, particularly in areas where the herbicide was sprayed. Although medical researchers caution that it is very difficult to prove a direct cause and effect between Agent Orange and these medical problems, statistical evidence indicates that Agent Orange is at least partly responsible.

It also appears that American veterans who were exposed to high levels of Agent Orange have more health problems than do other veterans. The U.S. Department of Veterans Affairs currently recognizes ten medical conditions linked to dioxin exposure. These include a variety of soft-tissue cancers, skin diseases, urological disorders, and birth defects.

The spraying of herbicides over a forested area beside a South Vietnamese highway during the Vietnam War. This photo was taken in 1966. (Archive Photos)

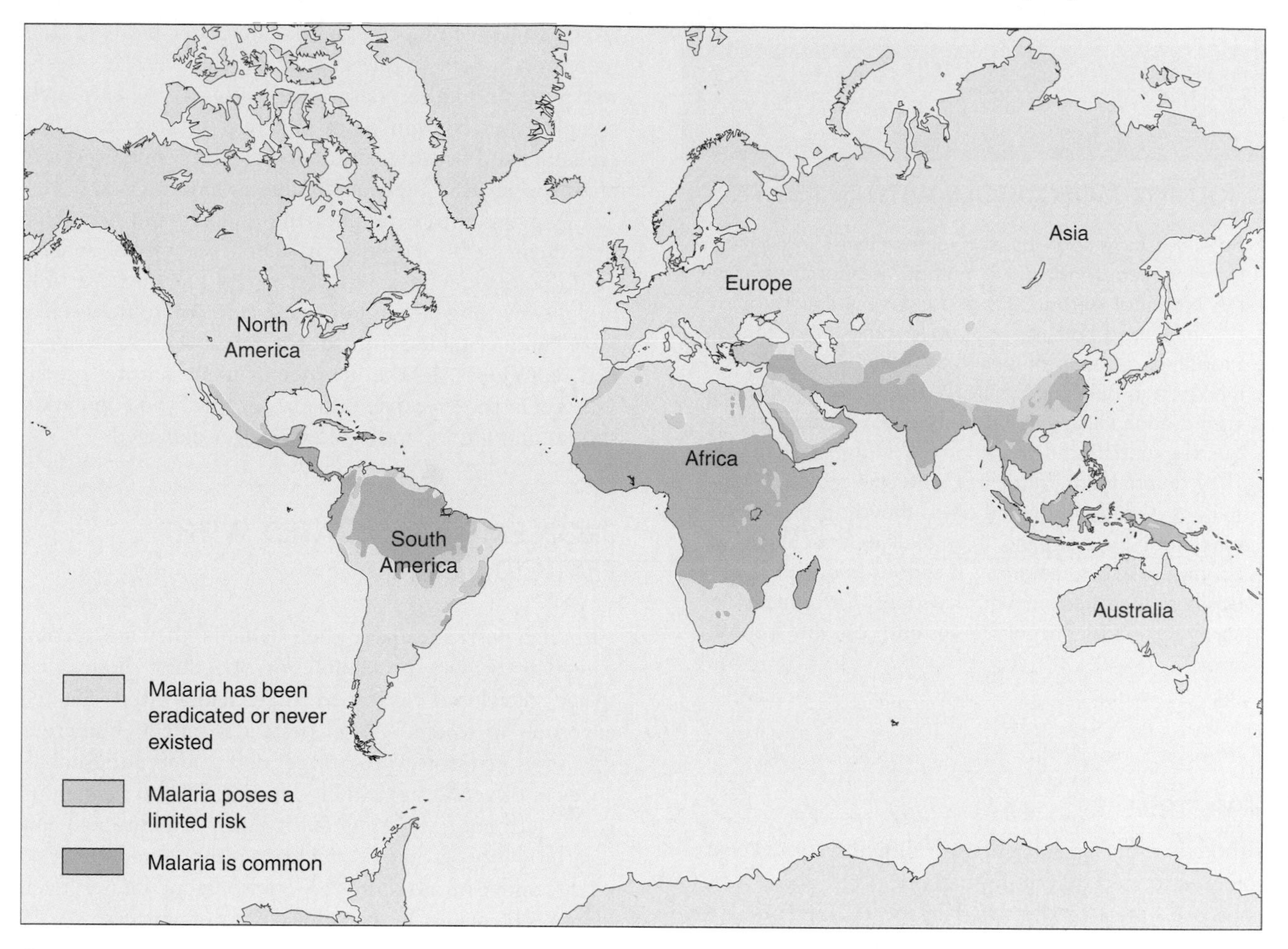

Figure 22–3
Where malaria occurs. Worldwide, millions of lives have been saved by insecticides sprayed to control mosquitoes.

BENEFITS OF PESTICIDES

Each day a war is waged as farmers, struggling to produce bountiful crops, battle insects and weeds. Similarly, health officials fight their own war against the ravages of human diseases transmitted by insects. One of the most effective weapons in the arsenals of farmers and health officials is the pesticide.

Disease Control

Several devastating human diseases are transmitted by insects. Fleas and lice, for example, carry the microorganism that causes typhus in humans. Malaria, which is also caused by a microorganism, is transmitted to millions of humans each year by female *Anopheles* mosquitoes (Figure 22–3). Worldwide, approximately 100 million people currently suffer from malaria.

Pesticides, particularly DDT, have helped control the population of mosquitoes, thereby reducing the incidence of malaria. Consider Sri Lanka. In the early 1950s, more than 2 million cases of malaria were reported in Sri Lanka each year. When spraying of DDT was initiated to control mosquitoes, malaria cases dropped to almost zero. When DDT spraying was discontinued in 1964, malaria reappeared almost immediately; by 1968, its annual incidence had increased to greater than 1 million cases per year. Despite the negative effects of DDT on the environment and organisms, the Sri Lankan government decided to begin spraying DDT once again in 1968. Today, insecticides are still used in Sri Lanka to help control malaria, although DDT has been replaced by other, less persistent pesticides such as malathion.

KILLING MOSQUITOES WITH SATELLITES

Satellites may soon assist health officials in fighting the mosquitoes that cause malaria. Researchers from the National Aeronautics and Space Administration (NASA) used satellite images to pinpoint potential breeding grounds of one type of malaria-carrying mosquito in Belize in Central America. Scientists then visited each village in the study area, collected mosquitoes, and tested them for the malarial parasite. They determined that almost all of the malarial mosquitoes were found in the areas they had previously identified. By adopting this satellite technology, a country in which malaria is a serious health problem could reduce the amount of widespread insecticide spraying and concentrate instead on the trouble spots.

Crop Protection

Although exact assessments are difficult to make, it is widely estimated that more than one-third of the world's crops are eaten or destroyed by pests. Given our expanding population and world hunger, it is easy to see why control of agricultural pests is desirable.

Pesticides reduce the amount of a crop that is lost through competition with weeds, consumption by insects, and diseases caused by plant **pathogens** (microorganisms, such as fungi and bacteria, that cause disease). Although many insect species are beneficial from a human viewpoint (two examples are honeybees, which pollinate crops, and ladybugs, which prey on crop-eating insects), a large number are considered pests. Of these, about 200 species have the potential to cause large economic losses in agriculture. For example, the Colorado potato beetle is one of many insects that voraciously consume the leaves of the potato plant, reducing the plant's ability to produce large tubers for harvest.

Serious agricultural losses are minimized in the United States and other developed nations primarily by the heavy application of pesticides. Pesticide use is usually justified economically, in that farmers save an estimated \$3 to \$5 in crops for every \$1 that they invest in pesticides. In developing countries where pesticides are not used in appreciable amounts, the losses due to agricultural pests can be considerable.

Why are agricultural pests found in such great numbers in our fields? Part of the reason is that agriculture is usually a **monoculture**; that is, only one type of plant is grown on a given piece of land. The cultivated field thus represents a very simple ecosystem. In contrast, forests, wetlands, and other natural ecosystems are extremely complex and contain many different species, including predators and parasites that control pest populations and plant species that are not suitable as food for pests.

A monoculture reduces the dangers and accidents that might befall a pest as it searches for food. A Colorado potato beetle in a forest would have a hard time finding anything to eat, but a 500-acre potato field is like a big banquet table set just for the pest. It eats, prospers, and reproduces. In the absence of many natural predators and in the presence of plenty of food, the population thrives and grows, and more crops get damaged.

PROBLEMS ASSOCIATED WITH PESTICIDE USE

Although pesticides have their benefits, they are accompanied by several problems. For one thing, many pest species develop a resistance to pesticides after repeated exposure to them. Also, pesticides affect numerous species in addition to the target pests, generating imbalances in the ecosystem (including agricultural fields) and posing a threat to human health. And, as mentioned earlier, the ability of some pesticides to resist degradation and readily move around in the environment causes even more problems for humans and other organisms.

Development of Genetic Resistance

The prolonged use of a particular pesticide can cause a pest species to develop genetic resistance to the pesticide. **Genetic resistance** is any inherited characteristic that decreases the effect of a pesticide on a pest.

In the 50 years during which pesticides have been widely used, more than 500 species of insects have developed genetic resistance to certain pesticides, and at least 17 species are resistant to *all* major classes of insecticides (Figure 22–4). For example, some insects that attack cotton have become so resistant to insecticide applications that chemical control is no longer effective (Table 22–2). Insects are not the only pests to develop genetic resistance; at least 84 weed species are currently resistant to certain herbicides.

How does genetic resistance to pesticides occur? Each time a pesticide is used to control a pest, some survive. The survivors, because of certain genes they already possess, are genetically resistant to the pesticide, and they pass on this trait to future generations. Thus, evolution—any cumulative genetic change in a population of organisms—occurs, and the pest population contains a larger percentage of pesticide-resistant pests than before.

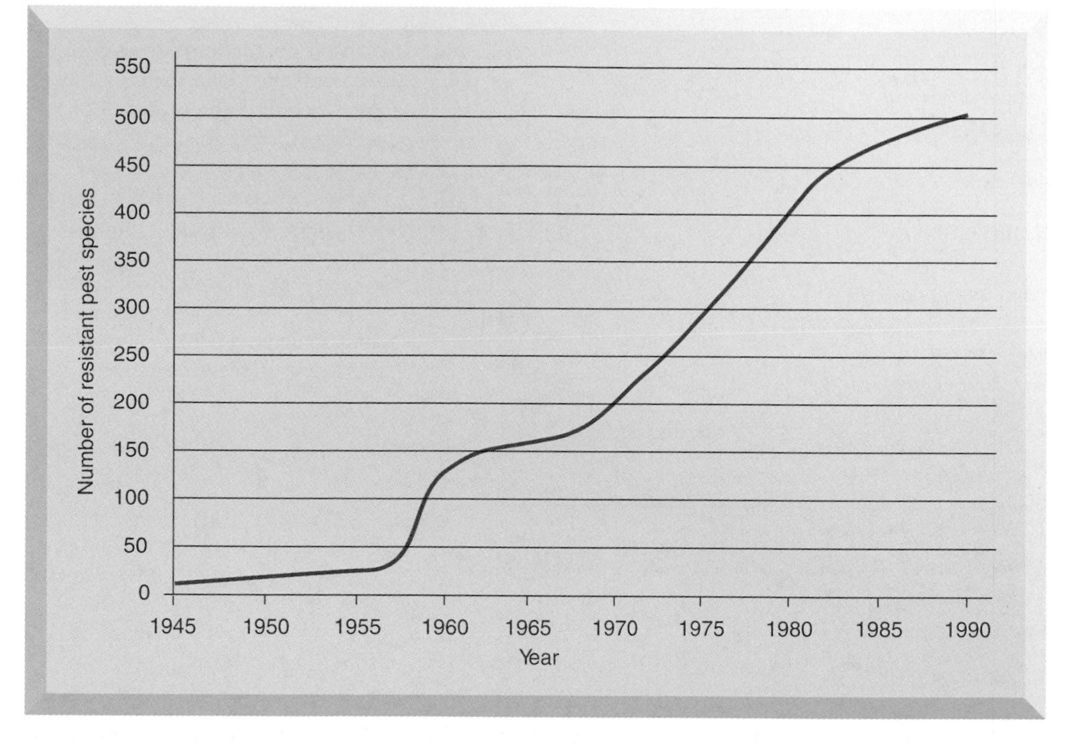

Figure 22-4
There has been a dramatic increase in the number of insect species exhibiting genetic resistance to insecticides.

Insects and other pests are constantly evolving. The short generation times (the period between the birth of one generation and that of another) and large populations that are characteristic of most pests favor rapid evolution, which allows the pest to quickly adapt to the pesticides used against it. As a result, an insecticide that kills a large portion of an insect population becomes less effective after prolonged use because the survivors and their offspring are genetically resistant.

Manufacturers of chemical pesticides often respond to genetic resistance by recommending that the pesticide be applied more frequently or in large doses. Alternatively, they recommend switching to a new, often more expensive, pesticide. These responses result in a predicament

Table 22-2
Change in Pesticide Effectiveness Caused by Genetic Resistance

	Average Amounts of Pesticide Needed to Kill Two Insect Pests on Cotton*			
	Bollworm		Tobacco Budworm	
Compound	1960	1965	1960	1965
DDT	0.03	1,000 +	0.13	16.51
Endrin	0.01	0.13	0.06	12.94
Carbaryl	0.12	0.54	0.30	54.57

*In milligrams of pesticide per gram of caterpillars (larvae).

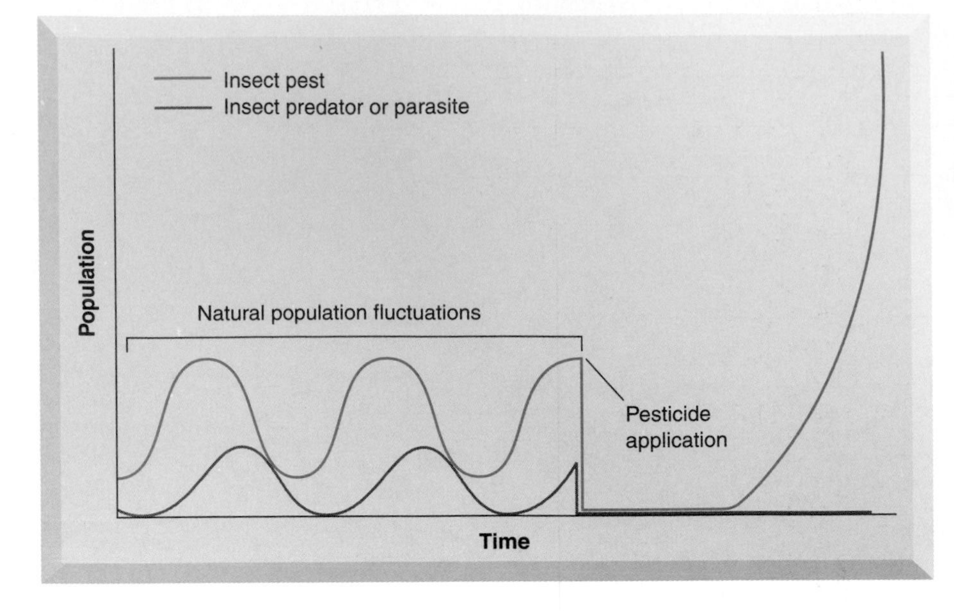

Figure 22-5
Natural population fluctuations are controlled by a variety of factors, including the presence of predators and parasites. When pesticides are applied, the predators and parasites of the pest species are also affected. The effect on predator/parasite populations disrupts the normal interactions between species and can cause a huge increase in the population of the pest species a short time after the pesticide is applied.

that has come to be known as the **pesticide treadmill**, in which the cost of applying pesticides increases while their effectiveness decreases.

Resistance management

Resistance management is a relatively new approach to dealing with genetic resistance. It encompasses efforts to delay the development of genetic resistance in insect pests so that the period of time in which an insecticide is useful is maximized. Strategies of resistance management vary depending on the insect involved. One strategy of resistance management is to maintain a nearby "refuge" of untreated plants where the insect pest can avoid being exposed to the insecticide. Those insects that live and grow in the refuge remain susceptible to the insecticide. When susceptible insects migrate into the area being treated with insecticide, they dilute the genetically resistant population. The millions of susceptible individuals interbreed with the small number of resistant insects, delaying the development of genetic resistance in the population as a whole.

Imbalances in the Ecosystem

One of the worst problems associated with pesticide use is that pesticides affect species other than the pests for which they are intended. Beneficial insects such as honeybees, which pollinate plants, and ladybugs, which consume many insect pests, are killed as effectively as insect pests. In a study of the effects of spraying the insecticide dieldrin to kill Japanese beetles, scientists found a large number of dead animals in the treated area, such as various birds, rabbits, ground squirrels, cats, and beneficial insects. (Use of dieldrin in the United States has since been banned.)

Organisms do not have to be killed to be negatively affected by pesticides. Quite often the stress of carrying pesticides in its body makes an organism more vulnerable to other diseases or stresses in its environment.

Because the natural enemies of pests—organisms that prey on the pests to survive—often starve or migrate in search of food after pesticide has been sprayed in an area, pesticides are indirectly responsible for a large reduction in the populations of natural enemies of pests. Pesticides also kill natural enemies directly, because predators consume a lot of the pesticide by consuming the pests. After a brief period, the pest population rebounds and gets larger than ever, partly because no natural predators are left to keep its numbers in check (Figure 22–5).

Despite a 33-fold increase in pesticide use in the United States since 1945, crop losses due to pests have

Table 22-3
Percentage of Crops Lost Annually to Pests in the United States

Period	Insects	Diseases	Weeds
1989	13.0	12.0	12.0
1974	13.0	12.0	8.0
1951 to 1960	12.9	12.2	8.5
1942 to 1951	7.1	10.5	13.8

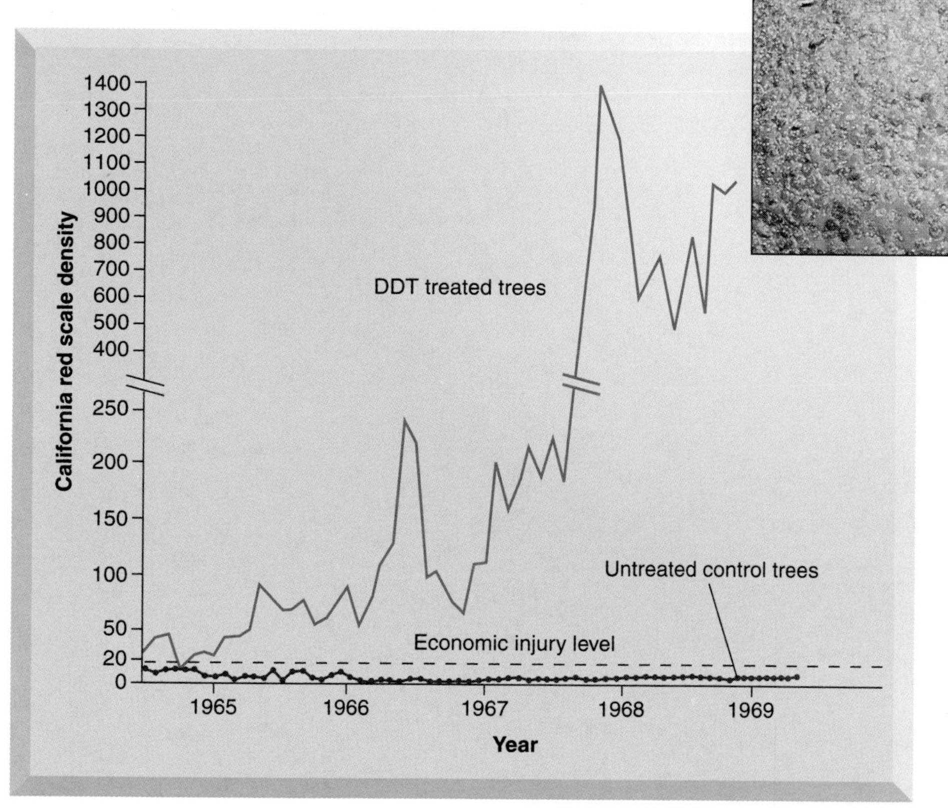

Figure 22–6

Pesticide use sometimes makes new pest species. (a) Red scale on green (unripened) citrus fruit. An infestation of red scale insects on lemons occurred after DDT was sprayed to control a different pest. Prior to DDT treatment, red scale was not a problem on citrus crops. (b) A comparison of red scale populations on DDT-treated and untreated trees. (a, William E. Ferguson)

not declined significantly (Table 22–3). Increasing resistance to pesticides in pests and the destruction of the natural enemies of pests provide a partial explanation. Changes in agricultural practices are also to blame; for example, crop rotation, a proven way of controlling certain pests, is not practiced as much today as it was several decades ago (see Chapter 14).

Creation of new pests

In some instances, the use of a pesticide has resulted in a pest problem that did not exist before. Creation of new pests—that is, turning nonpest organisms into pests—is possible because the natural predators, parasites, and competitors of a certain organism may be largely killed by a pesticide, allowing the organism's population to multiply. The use of DDT to control certain insect pests on lemon trees, for example, was documented as causing an outbreak of a scale insect (a sucking insect that attacks plants) that had not been a problem before spraying (Figure 22–6). In a similar manner, the European red mite became an important pest on apple trees in the northeastern United States after the introduction of pesticides.

Persistence, Bioaccumulation, and Biological Magnification

Certain problems of chlorinated hydrocarbon pesticide use were first demonstrated by the effects of DDT on many bird species. Falcons, pelicans, bald eagles, ospreys, and many other birds are very sensitive to traces of DDT in their tissues. A substantial body of scientific evidence indicates that one of the effects of DDT on these birds is that they lay eggs with extremely thin, fragile shells that usually break during incubation, causing the chicks' deaths. After 1972, the year DDT was banned in the

(a)

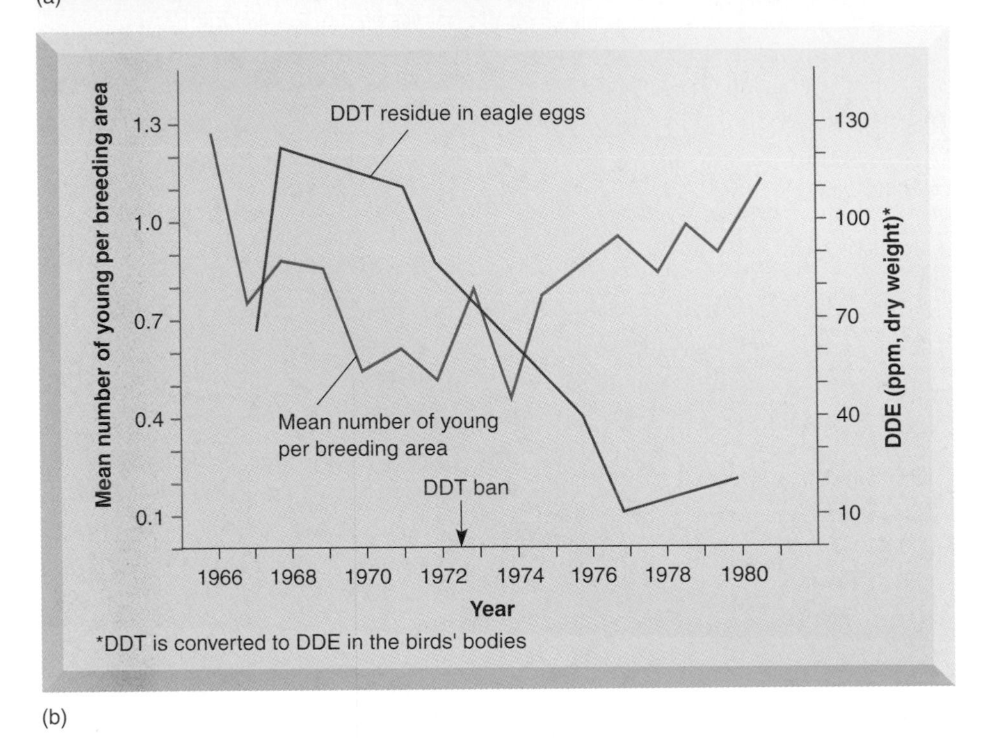

(b)

Figure 22–7
The effect of DDT on birds. (a) Many birds, such as the bald eagle, suffered reproductive failure after DDT accumulated in their tissues and interfered with their ability to produce strong eggshells. (b) A comparison of the number of successful bald eagle offspring with the level of DDT residues in their eggs. Note that reproductive success improved after DDT levels decreased. (a, A. Carey/VIREO)

United States, the reproductive success of many birds improved (Figure 22–7).

The impact of DDT on birds is the result of three characteristics of DDT—its persistence, bioaccumulation, and biological magnification. Some pesticides, particularly chlorinated hydrocarbons, are extremely stable and may take many years to be broken down into less toxic forms. The **persistence** of synthetic pesticides is a result of their novel chemical structures. Natural decomposers such as bacteria have not evolved ways to degrade synthetic pesticides, so they accumulate in the environment and in the food web.

When a pesticide is not metabolized (broken down) or excreted by an organism, it simply gets stored, usually in fatty tissues. Over time, the organism may accumulate high concentrations of the pesticide. The buildup of such a pesticide in an organism's body is known as **bioaccumulation** or **bioconcentration**.

Organisms at higher levels on food webs tend to have greater concentrations of bioaccumulated pesticide stored in their bodies than those lower on food webs. The increase in pesticide concentrations as the pesticide passes through successive levels of the food web is known as **biological magnification** or **biological amplification**.[1]

As an example of the concentrating characteristic of persistent pesticides, consider a hypothetical food chain: plant → small fish → large fish → heron (Figure 22–8). When a pesticide such as DDT is sprayed on plants, some of it gets into aquatic waterways; its concentration in water is extremely dilute, perhaps on the order of 0.00005 parts per million (ppm). The algae and aquatic plants contain a greater concentration of DDT—0.04 ppm. Each

[1]Other toxic substances besides pesticides may exhibit bioaccumulation and biological magnification, including radioactive isotopes, heavy metals such as mercury, and industrial chemicals such as PCBs (see Chapters 11, 21, and 23).

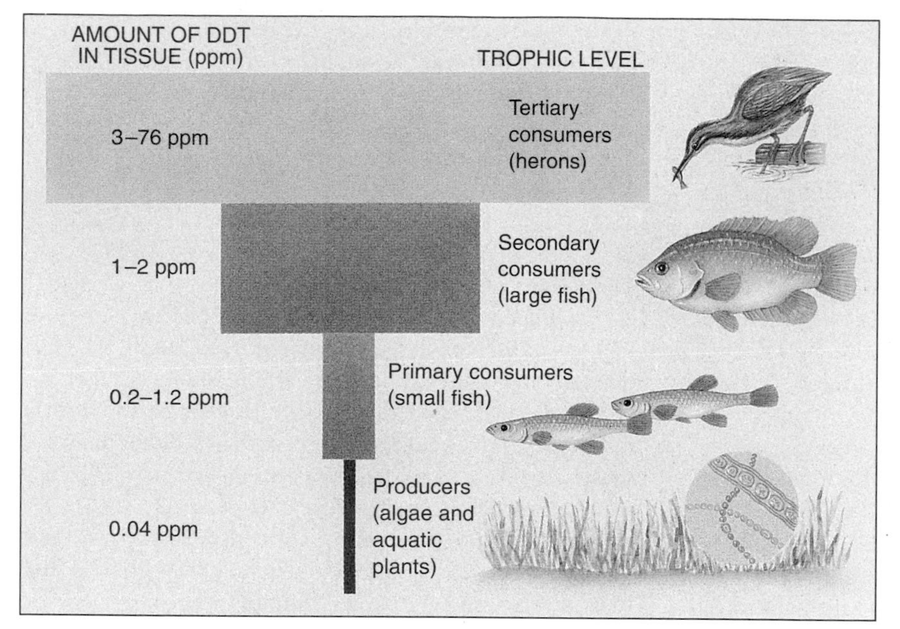

Figure 22–8

The biological magnification of DDT in an aquatic ecosystem. Note how the level of DDT concentrates in the tissues of various organisms as you move along the food chain from producers to consumers. The heron at the end of the food chain has approximately one million times more DDT in its tissues than the concentration of DDT in the water.

small fish grazing on the algae and plants concentrates the pesticide in its tissues to 0.2 to 1.2 ppm. A large fish that eats many small fishes laced with pesticide ends up with a pesticide level of 1 to 2 ppm. The top carnivore in this example, a heron, will have a pesticide value of 3 to 76 ppm from eating contaminated fishes. Although this example involves a bird at the top of the food chain, it is important to recognize that *all* top carnivores, from fishes to humans, are at risk from biological magnification. Because of this risk, currently approved pesticides have been tested to assure they do not persist and accumulate in the environment.

Mobility in the Environment

Another problem associated with pesticides is that they do not stay where they are applied, but tend to move through the soil, water, and air, sometimes long distances (Figure 22–9). For example, fishes can be affected by pesticides that were applied to agricultural lands and then

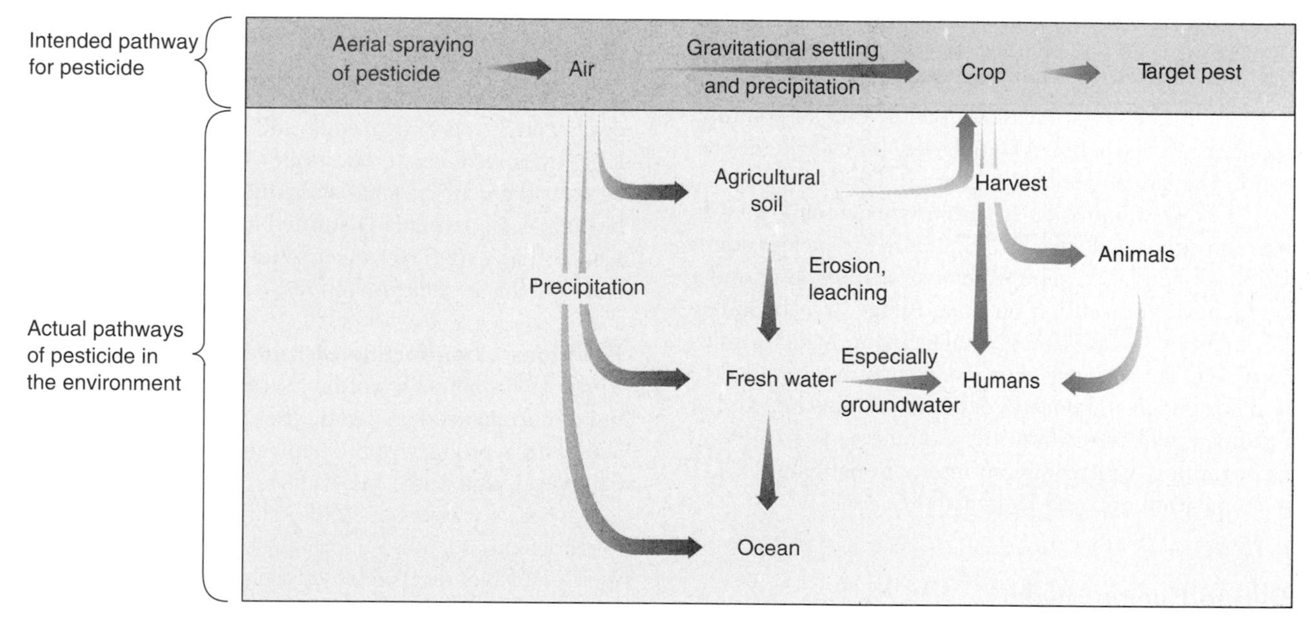

Figure 22–9

The actual pathways of pesticides in the environment. (The *intended* pathway of pesticides is marked in orange at the top of the figure.)

washed into rivers and streams when it rained. If the pesticide level in its aquatic ecosystem is high enough, the fishes may be killed. If the level is sublethal—that is, not enough to kill the fishes—the fishes may still suffer from undesirable effects such as bone degeneration.

Pesticide mobility is also a problem for humans. In 1994 the Environmental Working Group, a private environmental organization, analyzed herbicides in drinking water by evaluating 20,000 water tests performed by state and federal government inspectors. Their study revealed that 14.1 million Americans drink water containing traces of five widely used herbicides. Because these herbicides are often used on corn and soybeans, the study focused on the midwestern states where these crops are commonly grown. The study concluded that 3.5 million people living in the Midwest face a slightly elevated cancer risk because of their exposure to the herbicides. The Environmental Protection Agency (EPA) recently reduced the use of two of the herbicides (alachlor and metachlor) by approving a safer substitute, and is reviewing the use of the other three (atrazine, cyanizine, and simazine).

Mobility in the atmosphere

In the mid-1990s, biologists announced that trace amounts of DDT were measured in conifer needles and leaves of other trees growing in two remote forests in New Hampshire that had never been sprayed with the insecticide. Because greater levels of DDT occurred in leaves of trees growing at higher elevations and on slopes facing the prevailing west winds, biologists think that the DDT is deposited from the atmosphere. Although DDT was banned in the United States in 1972, it is still widely used in many other countries, such as Mexico, India, Eastern Europe, and Southeast Asia. The measured DDT levels are too low to pose a direct threat to the forest ecosystem, but they are troubling because, like acid rain and pollution in the Great Lakes, they show that air pollution does not respect political boundaries. (Recall the discussion of the global distillation effect in Chapter 19.)

A U.N. Commission is currently examining ways to limit the airborne flow of DDT and other pesticides across national boundaries. The Commission is working under the Geneva Convention on Long-Range Transboundary Air Pollution, a 1979 treaty established to regulate emissions from smokestacks, but more recently expanded to include agricultural releases of pesticides. It is hoped that a protocol will result from the Commission's work that, when ratified, will require an international ban of DDT and similar chlorinated hydrocarbons.

Risks to Human Health

Exposure to pesticides can also damage human health. Pesticide poisoning caused by short-term exposure to high levels of pesticides can result in harm to organs and even death (see Focus On: The Bhopal Disaster), whereas long-term exposure to lower levels of pesticides can cause cancer. There is also concern that exposure to trace amounts of certain pesticides can disrupt the human endocrine (hormone) system. Children are considered to be at greater risk from exposure to household pesticides than adults.

Short-term effects of pesticides

Approximately 67,000 people are poisoned by pesticides in the United States each year. Most of these are farm workers or others whose occupations involve daily contact with large quantities of pesticides. A person with a mild case of pesticide poisoning may exhibit symptoms such as nausea, vomiting, and headaches. More serious cases, particularly organophosphate poisonings, may result in permanent damage to the nervous system and other body organs. Although the number is low in the United States, people do die from overexposure to pesticides. Almost any pesticide can kill a human if the dose is large enough.

The World Health Organization estimates that globally, more than 3 million people are poisoned by pesticides each year; of these, about 220,000 die. The incidence of pesticide poisoning is highest in developing countries, in part because they often use dangerous pesticides that have been banned by developed nations. Also, pesticide users in developing nations often are not trained in the safe handling and storage of pesticides.

Long-term effects of pesticides

Many studies of farm workers and others who are exposed to low levels of pesticides over many years show an association between cancer and long-term exposure to pesticides. A type of lymphoma (a cancer of the lymph system) has been associated with the herbicide 2,4-D, for example. Other pesticides have been linked to a variety of cancers, such as leukemia and cancers of the brain, lungs, and testicles. In addition, the children of agricultural workers are at greater risk than other children for birth defects, particularly stunted limbs. There is also evidence that exposure to pesticides may compromise the body's ability to fight infections.

Pesticides as endocrine disrupters

In the 1990s, many scientific articles were published that linked certain pesticides and other persistent toxic chemicals with reproductive problems in animals. Recall from Chapter 1 that some male alligators living in a Florida lake where a massive pesticide spill occurred in 1980 had birth defects, such as abnormally small penises (see Figure 1–4). River otters exposed to synthetic chemical pollutants were also found to have abnormally small penises. Female sea gulls in Southern California exhibited behavioral aberrations by pairing with one another rather than with males during the mating season. In many cases sci-

FOCUS ON

The Bhopal Disaster

In December 1984, the world's worst industrial accident occurred at a Union Carbide pesticide plant in the Indian city of Bhopal. As much as 36 metric tons (40 tons) of methyl isocyanate (MIC) gas, which is used to produce carbamate pesticides, erupted from an underground storage tank after water leaked in and caused an explosive chemical reaction. Some of the MIC, which is itself highly toxic, converted in the atmosphere to hydrogen cyanide, which is even more deadly. The toxic cloud settled over 78 square kilometers (30 square miles), exposing up to 600,000 people.

According to official counts, about 2500 people were killed outright from exposure to the deadly gas (see figure). Another 2500 have since died. An international group of medical specialists (the International Medical Commission on Bhopal) estimated in 1996 that between 50,000 and 60,000 people have serious respiratory, ophthalmic, intestinal, reproductive, and neurological problems. Young women who were exposed to the gas have been unable to marry because it is widely assumed they are sterile. Many also suffer from psychological disorders, such as posttraumatic stress and pathological guilt over being unable to save loved ones.

Union Carbide agreed in 1989 to pay $470 million in compensation. It is also spending more than $100 million to build a hospital for victims. The Indian government is demanding a larger settlement because they say the accident was caused by negligence, but Union Carbide claims the accident was due to sabotage by a disgruntled Indian employee. As of 1996, 12 years after the disaster, most of the victims have not received compensation. The Indian government, which is disbursing the $470 million paid by Union Carbide, is still considering the thousands of claims and will probably not allocate most of the money until the year 2000.

An Indian child who was a victim of the Bhopal disaster in 1984 is buried. (Ragu Rai/Magnum Photos, Inc.)

entists have been able to reproduce the same abnormal symptoms in the laboratory, thereby verifying that the defects are caused by certain persistent chemicals (Table 22–4).

But it was the suggestion in the 1996 book, *Our Stolen Future*, by Theo Colburn, Dianne Dumanoski, and John Peterson Myers, that persistent toxic chemicals in the environment are disrupting *human* hormone systems that ignited a barrage of media attention. Dr. Theo Colburn, a senior scientist at the World Wildlife Fund, hypothesized that ubiquitous chemicals in the environment are linked to disturbing trends in human health. These include increases in breast and testicular cancer, increases in male birth defects, and decreases in sperm counts (see Chapter 1).

Although the hypothesis is supported by many scientific studies, a direct cause-and-effect relationship between chemicals in the environment and adverse effects on the human population remains to be established. Some scientists think the potential danger is so great, however, that persistent chemicals such as DDT should be internationally banned immediately. Other scientists are more cautious in their assessment of the danger but still think the problem should not be disregarded.

Our Stolen Future ignited public concern and triggered scientific investigations by universities, governments, and industries on how synthetic chemicals interfere with the actions of human hormones. It will take several years before these studies can tell us if these chemicals are acting as endocrine disrupters in humans.

Table 22–4
*Some Pesticides That Are Known Endocrine Disrupters**

Pesticide	General Information
Atrazine	Herbicide; still used
Chlordane	Insecticide; banned in United States in 1988
DDT	Insecticide; banned in United States in 1972
Endosulfan	Insecticide; still used
Kepone	Insecticide; banned in United States in 1977
Methoxychlor	Insecticide; still used

*Based on experimental research with laboratory animals.

Meanwhile, an international effort is currently underway to ban all production and use of nine pesticides suspected of being endocrine disrupters. More than 100 countries participating in the 1995 U.N. Conference for the Protection of the Marine Environment agreed to develop a legally binding ban of these pesticides and three industrial chemicals with similar effects.

Pesticides and children

In recent years, there has been increased attention to the health effects of household pesticides on children because it appears that household pesticides are a greater threat to children than to adults. For one thing, children tend to play on floors and lawns where they are exposed to greater concentrations of pesticide residues. Also, children may be more sensitive to pesticides because their bodies are still developing. Several preliminary studies suggest that exposure to household pesticides may cause brain cancer and leukemia in children, but more research must be done before any firm conclusions can be made.

The EPA estimates that 84 percent of American homes use pesticide products, such as pest strips, bait boxes, bug bombs, flea collars, pesticide pet shampoos, aerosols, liquids, and dusts. More than 20,000 different household pesticides are manufactured, and these contain over 300 active ingredients and 1700 inert ingredients.[2] In 1993, poison control centers nationwide received 130,200 reports of exposure and possible poisoning from household pesticides. More than half of these incidents involved children.

There is also concern about the ingestion of pesticide residues by children. The National Research Council[3] published a three-year study in 1993 called *Pesticides in the Diets of Infants and Children*. It called for additional research on how pesticide residues on food affect the young because it is not known if infants and children are more, or less, susceptible than adults. Also, because current pesticide regulations are intended to protect the health of the general population, the report stated that infants and children may not be adequately protected.

ALTERNATIVES TO PESTICIDES

Given the many problems associated with pesticides, it is clear that they are not the final solution to pest control. Fortunately, pesticides are not the only weapon in our arsenal. Alternative ways to control pests include cultivation methods, biological controls, genetic controls, pheromones and hormones, quarantine, and irradiation. A combination of these methods in agriculture, often including a limited use of pesticides as a last resort, is known as integrated pest management (IPM). IPM is the most effective way to control pests.

Using Cultivation Methods to Control Pests

Sometimes agricultural practices can be altered in such a way that a pest is adversely affected or discouraged from causing damage. Although some practices, such as the insect vacuum mentioned at the beginning of the chapter, are quite new, other cultivation methods that discourage pests have been practiced for many years. The proper timing of planting, fertilizing, and irrigating promotes healthy, vigorous plants that are more able to resist pests because they are not being stressed by other environmental factors. A technique that has been used with success in alfalfa crops is strip cutting, in which only one segment of the crop is harvested at a time. The unharvested portion of the crop provides an undisturbed habitat for natural predators and parasites of the pest species. The rotation of crops also helps control pests. When corn is not planted in the same field for two years in a row, for example, the corn rootworm is effectively controlled.

Biological Controls

Biological controls involve the use of naturally occurring disease organisms, parasites, or predators to control

[2]Many pesticide products contain as much as 99 percent inert ingredients. Because pesticide manufacturers decide whether an ingredient is inert or active, some pesticide products contain active ingredients that are considered inert in other products. Many inert ingredients are generally recognized as safe (examples include pine oil, ethanol, silicone, and water). Others are known to be toxic (examples include asbestos, benzene, formaldehyde, lead, and cadmium).

[3]The National Research Council is a private, nonprofit society of distinguished scholars. It was organized by the National Academy of Sciences to advise the federal government on complex issues in science and technology.

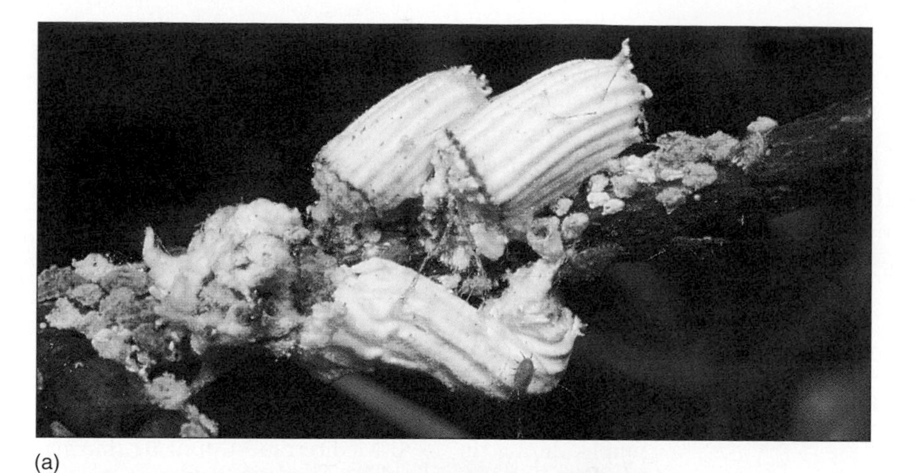

(a)

(b)

Figure 22–10

The biological control of cottony-cushion scale. (a) The cottony-cushion scale is an insect that attacks the stems and bark of several important crops. The white, ridged mass on top of each female contains up to 800 eggs. The males are reddish-brown. (b) A vedalia beetle larva feeds on cottony-cushion scale. Both adults and larvae of the vedalia beetle control the cottony-cushion scale. (a and b, Peter J. Bryant/Biological Photo Service)

pests. As an example, suppose that an insect species is accidentally introduced into a country where it was not found previously, and becomes a pest. It might be possible to control this pest by going to its native country and identifying an organism there that is an exclusive predator or parasite of the pest species. That predator or parasite, if successfully introduced, may be able to lower the population of the pest species so that it is no longer a problem.

Cottony-cushion scale provides an example of successfully using one organism to control another. The cottony-cushion scale is a small insect that sucks the sap from the branches and bark of many fruit trees, including citrus trees (Figure 22–10). It is native to Australia but was accidentally introduced to the United States in the 1880s. An American entomologist went to Australia and returned with several possible biological control agents. One, the vedalia beetle, was found to be very effective in controlling scale, which it eats voraciously and exclusively. Within two years of its introduction, the vedalia beetle had eliminated the cottony-cushion scale from citrus orchards. Today both the cottony-cushion scale and the vedalia beetle are present in very low numbers, and the scale is not considered an economically important pest.

Since the successful introduction of a biological control for cottony-cushion scale, the U.S. Department of Agriculture's Agricultural Research Service (ARS) has successfully introduced 11 biological controls for specific pests. The ARS is currently investigating possible biological controls for about a dozen other insect and weed pests.

Although some examples of biological control are quite spectacular, finding an effective parasite or predator is very difficult. And just because a parasite or predator has been identified does not mean it can become successfully established in a new environment. Slight variations in environmental factors such as temperature and moisture can alter the effectiveness of the biological control organism in its new habitat. Care must also be taken to ensure that the introduced control agent does not attack unintended hosts and become a pest itself. To guard against this when the control agent is an insect, scientists put the insects in cages with samples of important crops, ornamentals, and native plants to determine if the insects will eat the plants when they are starving.

Insects are not the only biological control agents. Bacteria that harm insect pests have also been used successfully as biological controls. *Bacillus popilliae*, which causes milky spore disease in insects, can be applied as a dust on the ground to control the grub (larval) stage of Japanese beetles. *Bacillus thuringiensis* (*Bt*), which produces a toxin that is poisonous to some insects when they eat it, is used against insects such as the cabbage looper, a green caterpillar that damages many vegetable crops, and the corn earworm (see Focus On: *Bt*, Its Potential and Problems). Viruses can also be introduced as biological control agents.

Genetic Controls

Like biological control, genetic control of pests involves the use of organisms. Instead of using another species to reduce the pest population, however, genetic control strategies suppress pests in other ways. One is to reduce the pest population by sterilizing some of its members; another is to breed crop plants and domesticated animals so that they can resist pests.

The sterile male technique

Altering the genetic makeup of an insect pest is one type of genetic control. One of the most common methods of alteration is the **sterile male technique**, in which large numbers of males are sterilized, usually with radiation or chemicals.[4] They are then released into the wild, where they reduce the reproductive potential of the pest population by mating with normal females, who then lay eggs that never hatch. As a result, of course, the population of the next generation is much smaller.

One disadvantage of the sterile male technique is that it must be carried out continually to be effective. If sterilization is discontinued, the pest population rebounds to a high level in a few generations (which, you will recall, are very short). The procedure is also expensive, as it requires the rearing and sterilization of large numbers of insects in a laboratory or production facility. For example, during the 1990 Mediterranean fruit fly (medfly) outbreak in California, as many as 400 million sterile male medflies were released each week.

Developing resistant crops

Selective breeding has been used to develop many varieties of crops that are resistant to disease organisms or insects. Traditional breeding of crop plants typically involves identifying individual plants that are in an area where the pest is common but do not appear to be damaged by the pest. These individuals are then crossed with standard crop varieties in an effort to produce a pest-resistant version. It may take 10 to 20 years to develop a resistant crop variety, but the benefits are usually worth the time and expense.

Genetic engineering offers great promise in the field of breeding pest-resistant plants (see Chapter 18). For example, a gene from the soil bacterium *Bacillus thuringiensis* (already discussed) has been introduced into cotton plants. Caterpillars that eat cotton leaves from these genetically altered plants die or exhibit stunted growth.

In 1994 scientists from three different research labs announced the discovery of a "family" of plant genes, one of which defends against a bacterium, another against a virus, and a third against a fungus. The genes were isolated from three unrelated plant species, but they all share similarities in their DNA. These similarities mean it may be relatively easy to transfer the genes to other plants, offering the likelihood of future protection against bacterial, viral, and fungal infections.

Although selective breeding has resulted in many disease-resistant crops, it has not been an unqualified success. Fungi, bacteria, and other plant pathogens evolve rapidly. As a result, they can quickly adapt to the disease-resistant host plant, meaning that the new pathogen strains can cause disease in the formerly disease-resistant plant variety. Plant breeders, then, are in a continual race to keep one step ahead of plant pathogens.

[4]Males are sterilized rather than females because the male insects mate several to many times, but the females mate only once. Thus, releasing a single sterile male may prevent successful reproduction by several females, whereas releasing a single sterile female would prevent successful reproduction by only that female.

A LETHAL "BUG JUICE"

The cassava hornworm is a serious pest of cassava, a root crop that supports 500 million of the world's poorest people. An innovative, low-tech pesticide for use against the hornworm was developed through studies conducted by plant and agriculture scientists at Cornell University and in Colombia and Brazil:

- When 12 hornworms infected with a particular virus lethal to that species are mixed with water in a blender, the resulting "milkshake" is a potent insecticide against other hornworms.
- This size batch of homemade pesticide effectively treats 2.5 acres.
- When sprayed on cassava plant leaves, the blended pesticide is 60 to 100 percent effective at killing hornworms, who eat the virus and die a few days later.
- The unusual milkshake is harmless to other insects, other animals, and humans.

FOCUS ON

Bt, Its Potential and Problems

The common soil bacterium *Bacillus thuringiensis*, or *Bt* (see figure), serves as a natural pesticide that is toxic to insects and yet environmentally "friendly." When eaten by insects, *Bt* produces spores that release a poison. The poison does not persist in the environment and is not known to harm mammals, birds, or other non-insect species. It has been marketed since the 1950s but was not sold on a large scale until recently, mainly because there are many different varieties of the *Bt* bacterium, and each variety is toxic to only a small group of insects. For example, the *Bt* variety that works against corn borers would not be effective against Colorado potato beetles. As a result, *Bt* was not economically competitive against chemical pesticides, each of which could kill many different kinds of pests on many different crops.

The potential of *Bt*'s toxin as a natural pesticide has greatly increased because of improvements made by genetic engineers. For example, genetic engineers modified the gene coding for the toxin so that it affects a wider range of insect pests. Also, genetic engineers have inserted the *Bt* gene that codes for the toxin into more than 30 crop species, including corn, tomato, and cotton. *Bt* corn, for example, produces a continuous supply of toxin, which provides a natural defense against insects that eat corn plants. Likewise, the genetically altered tomato and cotton are more resistant to pests such as the tomato pinworm and the cotton bollworm.

The future of the *Bt* toxin as an effective substitute for chemical pesticides is not completely secure, however. Beginning in the late 1980s, several farmers began to notice that *Bt* was not working as well against the diamondback moth as it had in the past. All of the farmers who reported this reduction in effectiveness had used *Bt* frequently and in large amounts on their fields. Also, in 1996, many farmers growing *Bt* cotton reported that their crop succumbed to the cotton bollworm, which *Bt* is supposed to kill. It appears that certain insects, such as the diamondback moth and possibly the cotton bollworm, have developed resistance to this natural toxin in much the same way that they develop resistance to chemical pesticides. If *Bt* continues to be used in greater and greater amounts, it is likely that more insect pests will develop genetic resistance to it, greatly reducing *Bt*'s potential as a natural pesticide.

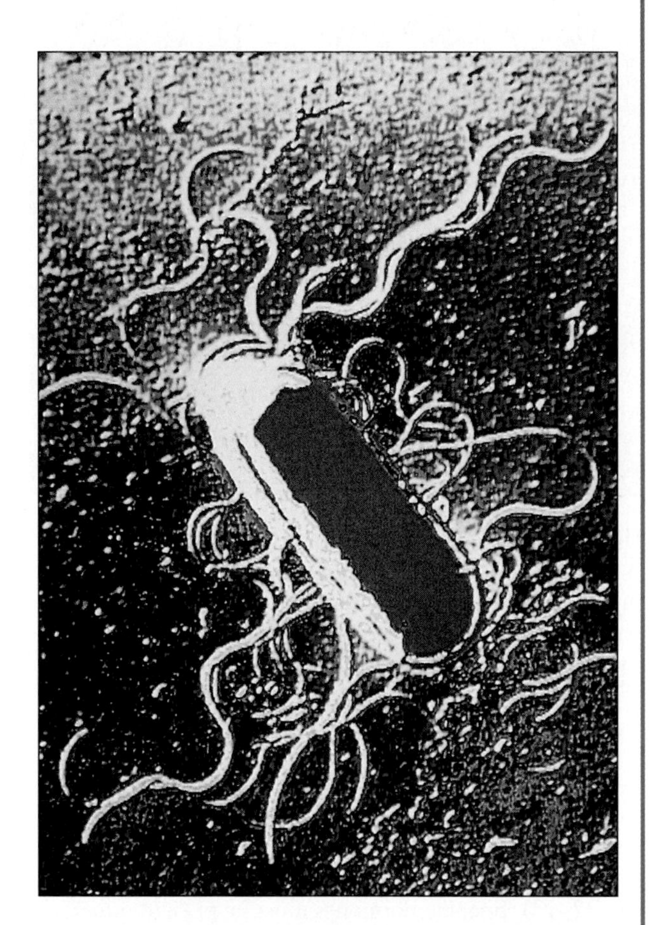

Electron micrograph of the flagellated bacterium, *Bacillus thuringiensis*. This bacterium contains genes that produce natural toxins against insects. These genes have been genetically engineered into several crop plants.
(Photo Researchers, Inc.)

Pheromones and Hormones

Pheromones are natural substances produced by animals to stimulate a response in other members of the same species. Pheromones are commonly called sexual attractants because they are often produced by an individual to attract members of the opposite sex for mating. Each insect species produces its own specific pheromone, so once the chemical structure is known, it is possible to make use of pheromones to control individual pest species. Pheromones have been successfully used to lure insects such as Japanese beetles to traps, where they are killed (Figure 22–11). Alternatively, pheromones can be released into the atmosphere to confuse insects so that they cannot locate mates.

Figure 22–11

A Japanese beetle trap uses a sexual attractant to lure Japanese beetles, which fall into the bag and die. Such insect traps draw large numbers of Japanese beetles. (John D. Cunningham/Visuals Unlimited)

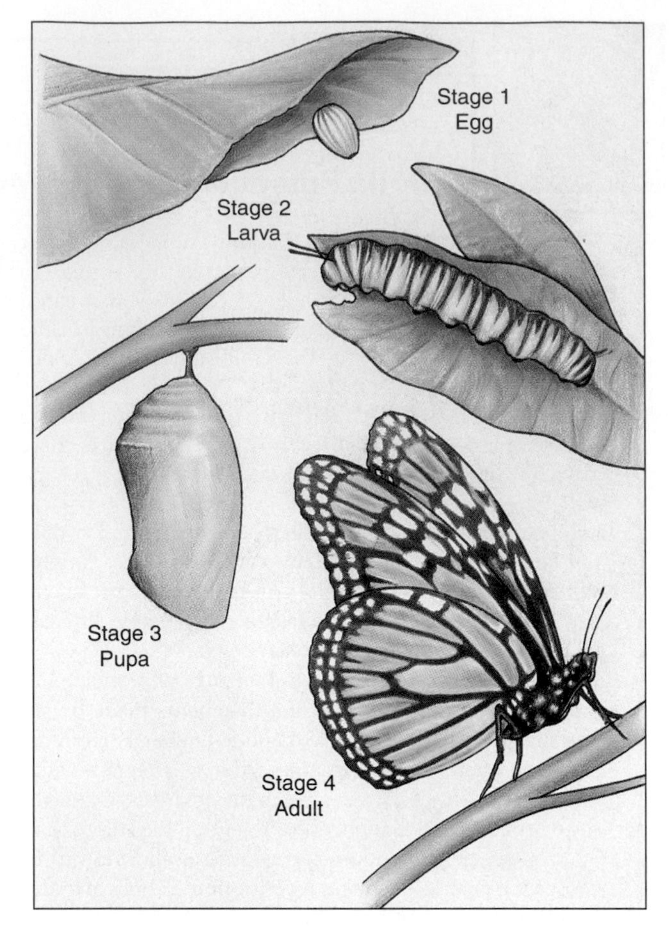

Figure 22–12

Different stages in the life cycle of an insect with complete metamorphosis. The egg hatches into a larva (caterpillar), which later becomes a pupa. Metamorphosis is complete when the pupa becomes an adult insect. Understanding the life cycle of an insect pest is necessary to effectively control its numbers.

Insect **hormones** are natural chemicals produced by insects to regulate their own growth and metamorphosis—the process by which an insect's body changes in form from a larva to a pupa and from a pupa to an adult (Figure 22–12). Specific hormones must be present at certain times in the life cycle of the insect; if they are present at the wrong time, the insect develops abnormally and dies. Many such insect hormones have been identified, and synthetic hormones with similar structures have been made. The possibility of using these substances to control insect pests is being actively pursued by entomologists.

The first insect hormone to be approved for use is a synthetic version of the insect hormone ecdysone, which causes molting. Known as MIMIC, the hormone triggers abnormal molting in insect larvae (caterpillars) of moths and butterflies. Since some beneficial insects can be affected by MIMIC as well, its use has limitations.

Quarantine

Governments attempt to prevent the importation of foreign pests and diseases by practicing **quarantine**, or restriction of the importation of exotic plant and animal material that might harbor pests. If a foreign pest is accidentally introduced, quarantine of the area where it is detected helps prevent its spread. For example, if a foreign pest is detected on a farm, the farmer may be required to destroy the entire crop.

Quarantine is an effective, although not foolproof, means of control. The U.S. Department of Agriculture has blocked the accidental importation of medflies on more than 100 separate occasions. The few occasions when quarantine failed and these insects successfully passed into the United States have done millions of dollars' worth of crop damage and required the expenditure of additional millions to eradicate the pests.

The adult medfly lays its eggs on 250 different fruits and vegetables; when the eggs hatch, the maggots feed on the fruits and vegetables and turn them into a disgusting mush. Eradication efforts, which include the use of helicopters to spray the insecticide malathion over hundreds of square kilometers and the rearing and releasing of millions of sterile males to breed the medfly out of existence, are extremely costly.

► *FOLLOW-UP*

Many experts think that the repeated finds of medflies in California indicate that, rather than being accidentally introduced each year, the medfly has become established in the state. If so, there are potentially disastrous consequences for California's $18 billion agricultural economy. Other countries could stop importing California produce or require expensive inspections and treatments of every shipment in order to prevent the importation of the medfly into their countries.

Integrated Pest Management

Many pests cannot be controlled effectively with a single technique; a combination of control methods is often more effective. **Integrated pest management (IPM)** combines the use of biological, cultural, and chemical controls that are tailored to the conditions and crops of an individual farm. Nonchemical controls, including crops designed to resist pests, are used as much as possible, and pesticides are used sparingly and only when other methods fail (see Meeting the Challenge: Reducing Agricultural Pesticide Use by 50 Percent in the United States). Thus, IPM allows us to control pests with a minimum of environmental disturbance.

In order to be effective, IPM requires thorough knowledge of the life cycles of the pests and their hosts as well as all of their interactions. The timing of treatments is critical and is determined by carefully monitoring the concentration of pests. IPM also optimizes natural controls by using agricultural techniques that discourage pests. IPM is an important part of the "new" agricultural methods known as alternative agriculture (see Chapter 18).

There are two fundamental premises associated with IPM. First, IPM is the *management* rather than the eradication of pests. Farmers who have adopted the principles of IPM allow a low level of pests in their fields and accept a certain amount of economic damage caused by the pests. These farmers do not spray pesticides at the first sign of a pest. Second, IPM requires that farmers be educated so they can adopt the strategies that will work best in their particular situations. Managing pests is much more complex than trying to eradicate them.

Cotton, which is attacked by many insect pests, has responded well to IPM. Cotton has the heaviest insecticide application of any crop: although only about 1 percent of agricultural land in the United States is used for this crop, cotton accounts for almost 50 percent of all the insecticides used in agriculture! By simple techniques such as planting a strip of alfalfa adjacent to the cotton field, the need for chemical pesticides is lessened. Lygus bugs, a significant pest of cotton, move from the cotton field to the strip of alfalfa, which they prefer as a food.

IPM has been most successful in controlling insect pests. Some scientists are now trying to develop IPM techniques to effectively control weeds with a minimal use of herbicides. There is also a widely recognized need to develop IPM techniques for urban and suburban environments.

Integrated pest management in Asia

Since the 1950s, rice farmers in many Asian countries applied pesticides on their paddies several times during each growing season. They were encouraged to do so by both scientists and chemical pesticide companies. In the 1980s and 1990s, however, a quiet revolution against the indiscriminate use of pesticides has occurred. It is promoted by the International Rice Research Institute (IRRI), the world's largest scientific establishment devoted to rice cultivation.

The IRRI has gone into different rice-growing areas in Asia and asked local farmers to conduct a group experiment. Some farmers are asked to not spray *any* pesticide during the first 40 days of the growing season. These farmers, who have received some training in IPM, are also told to use the more traditional cultivation practices that were abandoned when pesticides were introduced. Other farmers are asked to spray pesticides on their rice crops as usual, including the two, three, or more "preventative" treatments they usually apply during the early part of the growing season; this group serves as the experiment's control. When the crops are harvested, the rice yields on treated and untreated fields are compared. Many farmers are astonished to learn that the untreated fields have yields that are as large as or larger than those of the treated fields (Figure 22–13).

The prevailing agricultural philosophy—to wipe out pests while their populations are low by repeatedly applying pesticides throughout the growing season—is expensive, damaging to the environment, and, as the rice farmers have discovered, ineffective. Farmers report seeing many more natural enemies of rice pests, such as spiders, frogs, and carnivorous beetles, in the untreated fields, than in fields that are regularly sprayed with pesticides. Predator populations, which are normally held in check by pesticides, have a chance to grow when paddies are untreated.

Irradiating Foods

It is possible to prevent insects and other pests from damaging harvested food without using pesticides. The food can be exposed to ionizing radiation (usually gamma rays from cobalt 60), which kills many microorganisms, such as *Salmonella*, a bacterium that causes food poisoning. Numerous countries—for example, Canada, much of western Europe, Japan, Russia, and Israel—extend the shelf lives of foods with irradiation. The U.S. Food and Drug Administration (FDA) approved this process for fruits and vegetables as well as certain meats in 1986, and the

MEETING THE CHALLENGE

Reducing Agricultural Pesticide Use By 50 Percent in The United States

Pesticides have benefitted farmers (by increasing agricultural productivity) and consumers (by lowering food prices). But pesticide use has had its price—not necessarily in economic terms, for it is difficult to assign monetary values to many of its effects—but in terms of health problems and damage to agricultural and natural ecosystems. Society is increasingly concerned about pesticide use. For example, Proposition 128 on the 1990 California ballot asserted that no pesticide known to cause cancer could be applied to foods. Although this proposition was defeated, the fact that it was even proposed and considered by voters indicates a high level of public concern over pesticides.

Is it feasible to ban all pesticides known to cause cancer in laboratory animals? Probably not—at least not now. In many cases, substitute pesticides either do not exist, are less effective, or are considerably more expensive. A pesticide ban would increase food prices, although estimates on the magnitude of that increase vary considerably from one study to another. A pesticide ban would also cause considerable economic hardship for certain growers, although other farmers might benefit. For example, some insects are more troublesome in certain areas than in others. A farmer growing a crop in an area where an insect was very harmful might not be able to afford the crop losses that would occur without the use of a banned pesticide. Growers in areas where the insect was less of a problem could then increase their production of that crop, benefiting financially from the first farmer's loss.

Since it is impractical to ban large numbers of pesticides right now, is there another way to provide greater protection to the environment and human health without reducing crop yields? Governments in Sweden, Denmark, the Netherlands, and the Province of Ontario think so. Sweden achieved a 50 percent reduction in pesticide use in 1992 and is now on a second program to reduce pesticide use by another 50 percent. Denmark, the Netherlands, and the Province of Ontario are also implementing similar programs to reduce pesticide use by 50 percent during the 1990s. Strategies to reduce pesticide use include applying pesticide only when needed, using improved application equipment, and adopting IPM practices.

Too often pesticides are applied unnecessarily to prevent a possible buildup of pests. The regular use of pesticides regardless of whether pests are a problem or not is known as **calendar spraying**. Pesticide use can be decreased by continually monitoring pests so that pesticides are applied only when pests become a problem; this technique is known as **scout-and-spray**. For example, a 1991 study at Cornell University determined that a monitoring program might reduce the use of insecticides on cotton by 20 percent.

The use of aircraft to apply pesticide is extremely wasteful because 50 to 75 percent of the pesticide does not reach the target area, but instead drifts in air currents until it settles on soil or water (see figure). Pesticides applied on land with traditional methods also drift in air currents. Advances in the design of equipment for applying pesticides could reduce pesticide use considerably. For example, a recently developed rope-wick applicator reduces herbicide use on soybean fields by approximately 90 percent.

Pesticide use can also be reduced considerably through alternative pest control strategies. The widespread adoption of IPM makes it feasible for the United States to reduce pesticide use by 50 percent within a five- to ten-year period.

Aerial spraying of insecticide on corn. Most of the pesticide applied by aircraft never reaches the target area. (Grant Heilman, Grant Heilman Photography)

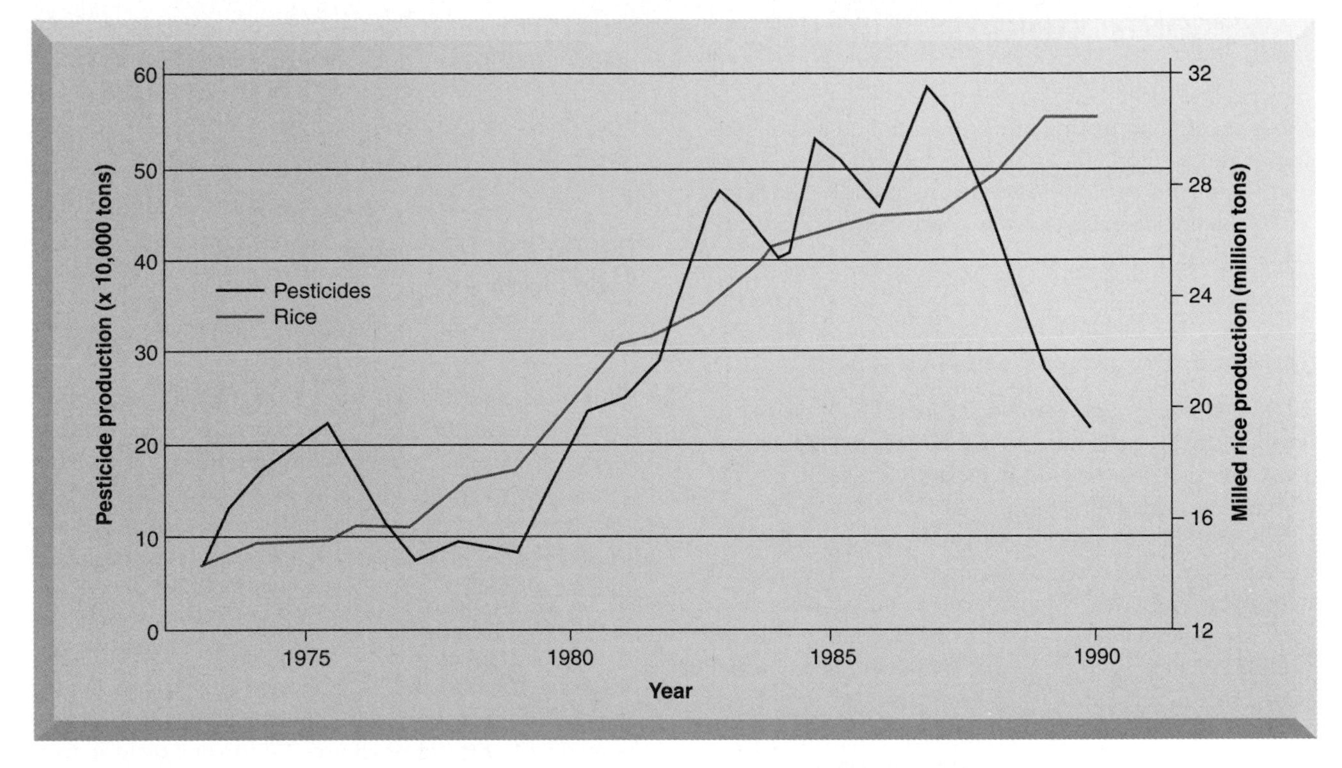

Figure 22–13

Rice production and pesticide use in Indonesia, 1972 to 1990. The decline in pesticide use in the late 1980s and early 1990s did not cause a decrease in rice yields. Instead, rice production increased by 12 percent during the four years following the new policy. Indonesia was the first Asian country to widely embrace integrated pest management. It banned the use of many pesticides on rice and trained more than 200,000 farmers in IPM techniques.

first irradiated food in the United States (strawberries) was sold in January 1992.

The irradiation of foods is somewhat controversial. Some consumers are concerned because they fear that irradiated food is radioactive. This is not true. In the same way that you do not become radioactive when your teeth are x-rayed, irradiated food is not radioactive. Nevertheless, many U.S. consumers refuse to purchase food that is labeled as having been irradiated.

Critics of irradiation are concerned because irradiation forms traces of certain chemicals called free radicals, some of which have been demonstrated to be carcinogenic in laboratory animals. They also point out that we do not know the long-term effects of eating irradiated foods. Proponents of irradiation argue that free radicals normally occur in food and are also produced by cooking processes such as frying and broiling. They assert that more than 1000 investigations of irradiated foods, conducted worldwide for more than three decades, have demonstrated that it is safe. Furthermore, irradiation lessens the need for pesticides and food additives.

ENVIROBRIEF

USING HEAT INSTEAD OF POISON

An entomologist who retired from the University of California at Los Angeles has invented an environmentally benign form of home pest control that kills bugs by heating them past their level of tolerance. Thermal pest eradication, a process now patented by Isothermics, Inc., of Anaheim, California, gets rid of drywood termites, cockroaches, carpenter ants, and other damaging insects in homes and other buildings.

The infested structure is draped with tarpaulins; special heaters boost the indoor temperature to 66° C (150° F) for about 6 hours; and because the insects cannot sweat, they die, often bursting in the process. Inspired by an ancient method of using hot rocks to drive vermin from clothing, the heat treatment is a nonchemical alternative to insecticidal fumigation. It is being tried by several pest control companies in California.

LAWS CONTROLLING PESTICIDE USE

Several laws have been passed to regulate pesticides in the interest of protecting human health and the well-being of the environment. These include the Food, Drug, and Cosmetics Act; the Federal Insecticide, Fungicide, and Rodenticide Act; and the Food Quality Protection Act of 1996.]

The Food, Drug, and Cosmetics Act

The Food, Drug, and Cosmetics Act (FDCA), passed in 1938, recognized the need to regulate pesticides found in food but did not provide a means of regulation. The FDCA was made more effective in 1954 with the passage of the Pesticide Chemicals Amendment (also called the Miller Amendment), which required the establishment of acceptable and unacceptable levels of pesticides in food.

An updated FDCA, passed in 1958, contained an important section known as the Delaney Clause, which stated that no substance capable of causing cancer in test animals or in humans will be permitted in processed food. Processed foods are foods that are prepared in some way—such as frozen, canned, dehydrated, or preserved—before being sold. The Delaney Clause recognized that pesticides tend to concentrate in condensed processed foods, such as tomato paste and applesauce.

The Delaney Clause, although commendable for its intent, had two inconsistencies. First, it did not cover pesticides on raw foods such as fresh fruits and vegetables, milk, meats, fish, and poultry. As an example of this double standard, residues of a particular pesticide might be permitted on fresh tomatoes but not in tomato ketchup. Second, because the EPA lacks sufficient data on the cancer-causing risks of pesticides that have been used for a long time, the Delaney Clause applied only to pesticides that were registered after strict tests were put into effect in 1978. This situation gave rise to one of the paradoxes of the Delaney Clause: there have been cases in which a newer pesticide that posed minimal risk was banned because of the Delaney Clause, but an older pesticide, which the newer one was to have replaced, was still used despite the fact that it is many times more dangerous.

When the Delaney Clause was passed, the technologies for detecting pesticide residues could only reveal high levels of contamination. Modern scientific techniques are so sensitive, however, that it is almost impossible for any processed food to meet the Delaney standard. As a result, the EPA found it difficult to enforce the strict standard required by the Delaney Clause. In 1988 the EPA began granting exceptions that permitted a "negligible risk" of one case of cancer in 70 years for every one million people. However, the EPA was taken to court because of its failure to follow the Delaney Clause as currently written, and the U.S. courts decided in 1994 that no exceptions can be granted unless Congress modifies the Delaney Clause. (One of the key provisions of the 1996 Food Quality Protection Act, discussed shortly, revises the Delaney Clause.)

The Federal Insecticide, Fungicide, and Rodenticide Act

The Federal Insecticide, Fungicide, and Rodenticide Act (FIFRA) was originally passed in 1947 to regulate the effectiveness of pesticides—that is, to prevent people from buying pesticides that did not work. FIFRA has been amended over the years to require testing and registration of the active ingredients of pesticides. Also, any pesticide that does not meet the tolerance standards established by the FDCA must be denied registration by FIFRA.

In 1972 the EPA was given the authority to regulate pesticide use under the terms of the FDCA and FIFRA. Since that time, the EPA has banned or restricted the use of many chlorinated hydrocarbons. In 1972 the EPA banned DDT for almost all uses. Aldrin and dieldrin were outlawed in 1974 after more than 80 percent of all dairy products, fish, meat, poultry, and fruits were found to contain residues of these insecticides. The banning of kepone occurred in 1977 and of chlordane and heptachlor in 1988.

A two-year study by the National Research Council concluded in 1987 that U.S. laws regarding pesticide residues in food are not adequate to protect the public from cancer-causing pesticides (Table 22–5). It included several recommendations that were made into law—an amended FIFRA—in 1988. The 1988 law requires reregistration of older pesticides before the end of the 1990s, which will subject them to the same toxicity tests that new pesticides face.

Not everyone is happy with the 1988 law. Although it is stricter than previous legislation, it represents a compromise between agricultural interests, including pesticide manufacturers, and those opposed to all uses of pesticides. The new law does not address a very important issue, the contamination of groundwater by pesticides. It also fails to address the establishment of standards for pesticide residues on foods and the safety of farm workers who are exposed to high levels of pesticides.

The Food Quality Protection Act of 1996

The Food Quality Protection Act of 1996 amends both the FDCA and FIFRA. It revises the Delaney Clause by establishing identical pesticide residue limits—those that pose a negligible risk—for both raw produce and processed foods. The new law requires that the increased

Table 22–5
Worst-Case Estimates of Risk of Cancer from Pesticide Residues on Food

Food	Cancer Risk*
Tomatoes	8.75×10^{-4}†
Beef	6.49×10^{-4}
Potatoes	5.21×10^{-4}
Oranges	3.76×10^{-4}
Lettuce	3.44×10^{-4}
Apples	3.23×10^{-4}
Peaches	3.23×10^{-4}
Pork	2.67×10^{-4}
Wheat	1.92×10^{-4}
Soybeans	1.28×10^{-4}

*Please note that these figures are worst-case estimates. Four assumptions were made in arriving at these values: (1) the entire U.S. crop (of tomatoes, for example) is treated (2) with *all* pesticides that are registered for use on that crop; (3) the pesticides are applied the maximum number of times (4) at the maximum rate, or amount, each time.
†As an example of how to interpret these figures, tomatoes are estimated to cause an average of 8.75 deaths from cancer for every 10,000 people.

susceptibility of infants and children to pesticides be considered when establishing pesticide residue limits. The pesticide limits are established for all health risks, not just cancer. For example, the EPA must develop a program to test pesticides for endocrine-disrupting properties. Another key provision of the Food Quality Protection Act is that it reduces the time it takes to ban a pesticide considered dangerous—from 10 years to 14 months.

THE MANUFACTURE AND USE OF BANNED PESTICIDES

As mentioned earlier in the chapter, some American companies manufacture pesticides that have been banned or heavily restricted in the United States and export them to developing countries, particularly in Asia, Africa, and Latin America. According to the Foundation for Advancements in Science and Education, between 1992 and 1994, the United States exported at least 7700 metric tons (7000 tons) of banned pesticides. Other nations also export banned pesticides.

The U.N. Food and Agriculture Organization (FAO) is attempting to help developing nations become more aware of dangerous pesticides. It has established a "red alert" list of more than 50 pesticides that have been banned in 5 or more countries. The FAO further requires that the manufacturers of these pesticides inform importing countries about why such pesticides were banned. The United States supports these international guidelines and exports banned pesticides only with the informed consent of the importing country. However, many foreign farmers never receive these guidelines.

The Importation of Food Tainted with Banned Pesticides

The fact that many dangerous pesticides are no longer being used in the United States is no guarantee that traces of those pesticides are not in our food. Although many pesticides have been restricted or banned in the United States, they are widely used in other parts of the world. Much of our food—some 1.2 million shipments annually—is imported from other countries, particularly those in Latin America. Some produce contains traces of banned pesticides such as DDT, dieldrin, chlordane, and heptachlor.

It is not known how much of the food coming into the United States is tainted with pesticides. The FDA monitors toxic residues on incoming fruits and vegetables, but it is only able to inspect about one percent of the food shipments that enter the country each year. In addition, the General Accounting Office reports that some food importers illegally sell food after the FDA has found it to be tainted with pesticides. When caught, these companies face fines that are not severe enough to discourage such practice.

CHANGING ATTITUDES

Heavy pesticide use can be attributed partly to the consumer, who has come to expect perfect, unblemished produce. There is no question that pesticides help farmers grow crops that are more visually appealing, but it comes to this: would you rather buy apples that are smaller and have an occasional blemish or worm but are pesticide-free, or apples that are free from all imperfections but contain traces of pesticides? Until consumers change their attitudes and demand produce that is grown without pesticides, farmers will continue to use pesticides (Figure 22–14).

Many farmers are exploring alternatives to pesticides on their own because they come into contact with pesticides on a regular basis. These farmers are aware of the dangers and problems associated with pesticide use and are concerned for their own safety and that of their families. They do not want to inhale pesticide or let it settle on their skin when they are applying it. They do not want to drink well water or eat food with traces of pesticide any more than the typical consumer does.

Figure 22–14
Pesticide-free produce in a supermarket. If consumers demand food that is grown in the absence of pesticides, then farmers will grow more pesticide-free crops. (Steve Feld)

Pesticide Risk Assessment

Although the effects of long-term exposure to low levels of carcinogenic pesticides should be of concern to all informed consumers, panic, which is often fueled by sensational news reports, is not justified. It is important to keep a balanced perspective when considering pesticides. The threat of cancer from consuming pesticide residues on our food is quite small compared to the threat of cancer from smoking cigarettes or from overexposure to ultraviolet radiation from the sun.

However, it is difficult for consumers to make informed decisions about the risks of pesticide residues on food, because we have no way of knowing what kinds of pesticides have been used on the foods we eat. A requirement that all foods list such chemicals would go a long way toward helping us assess risks, as well as discouraging heavy pesticide use.

SUMMARY

1. Pesticides are toxic chemicals that are used to kill pests, such as insects, weeds, fungi, nematodes, and rodents. Most are broad-spectrum pesticides, which kill a variety of organisms besides the target organism.

2. Insecticides are classified into groups based on their chemical structure—such as chlorinated hydrocarbons, organophosphates, and carbamates. Herbicides may be classified as either nonselective or selective, with selective herbicides including broad-leaf herbicides and grass herbicides.

3. Pesticides provide important benefits to humans, such as the prevention of diseases that are transmitted by insects, such as malaria. In addition, pesticides reduce crop losses from pests, thereby increasing agricultural productivity.

4. Some serious health problems are associated with pesticide use. Humans may be poisoned by exposure to large amounts, whereas lower levels of many pesticides pose a long-term threat of cancer. Concern is increasing that certain persistent pesticides that interfere with the actions of natural hormones may be responsible for increases in breast and testicular cancers, increases in male birth defects, and decreases in sperm counts. It appears that household pesticides are a greater threat to children than to adults.

5. In addition to its negative effect on human health, pesticide use has caused environmental problems. These include the development of genetic resistance in the pest, adverse effects on nontarget species, the creation of new pests, and mobility in the environment. Persistence, bioaccumulation, and biological magnification are other problems associated with some types of pesticides.

6. The need for pesticides in agriculture can be lessened or even eliminated by using quarantine, biological controls, genetic controls, and pheromones and hormones. Cultivation techniques such as crop rotation are also effective in controlling pests. Integrated pest management (IPM) in agriculture stresses biological controls and cultivation methods along with the judicious use of pesticides. Irradiation of food can be used to control pests after food has been harvested.

7. Pesticide registration and use are controlled by the Food, Drug, and Cosmetics Act (FDCA); the Federal Insecticide, Fungicide, and Rodenticide Act (FIFRA); and the Food Quality Protection Act.

THINKING ABOUT THE ENVIRONMENT

1. What is the dilemma referred to in the title of this chapter?

2. Overall, do you think the benefits of pesticide use outweigh its disadvantages? Give at least two reasons for your answer.

3. Sometimes pesticide use can increase the damage done by pests. Explain.

4. How is the buildup of insect resistance to pesticides similar to the increase in bacterial resistance to antibiotics?

5. Biological control is often much more successful on a small island than on a continent. Offer at least one reason why this might be the case.

6. It is more effective to use the sterile male technique when an insect population is small than when it is large. Explain.

7. Which of the following uses of pesticides do you think are most important? Which are least important? Explain your views.

(a) Keeping roadsides free of weeds
(b) Controlling malaria
(c) Controlling crop damage
(d) Producing blemish-free fruits and vegetables

8. Why is pesticide misuse increasingly being viewed as a *global* environmental problem?

* **9.** A water sample was measured and found to have 0.00005 ppm of DDT. The plankton living in the water were then measured and found to have 800 times that amount of DDT. What was the concentration of DDT, in ppm, in the plankton?

* **10.** Worldwide, the pesticide industry sold $27.8 billion in pesticides in 1994. Sales in 1998 are projected to be $34 billion. Calculate the average percent *annual* increase in pesticide sales from 1994 to 1998.

* **11.** Table A shows the five most common active ingredients in household pesticides, along with the number of U.S. households that use them. Use the data to draw a bar graph in which the number of households is represented as "Millions of Households."

Table A
The Five Most Common Household Pesticide Ingredients

Active Ingredient	Number of Households
Piperonyl butoxide	27,335,000
Pyrethrins	22,739,000
MGK-264	19,532,000
Propoxur	18,749,000
DEET	17,227,000

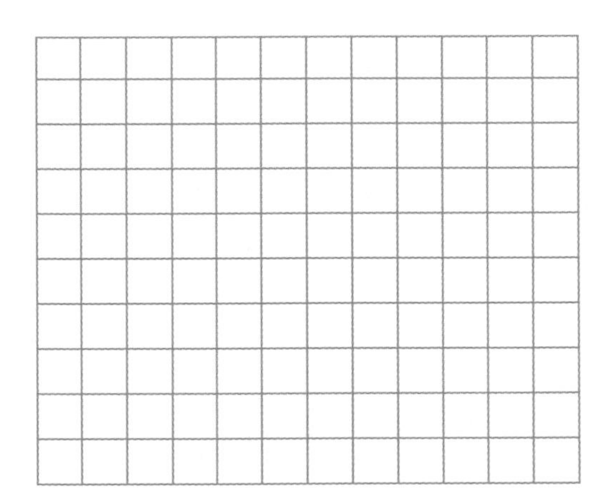

RESEARCH PROJECTS

Conduct an investigation of annual pesticide use in a major agricultural area near your school. Identify the pesticides used, their intended effects, and their persistence in the environment. Ask an agribusiness or small farmer if there are any practical alternatives to the pesticides currently used.

Obtain from your local supermarket a list of exporters of fresh produce sold by the store. Identify the nations of origin and compare those nations' environmental protection laws to those of the United States. Describe efforts to maintain safety standards in food imports.

Find out what pesticides are used on your campus and how often and how extensively they are applied. What pests are they supposed to control? Are they effective?

On-line materials relating to this chapter are on the World Wide Web at **http://www.saunderscollege.com/lifesci/environment2** (select Chapter 22 from the Table of Contents).

SUGGESTED READING

Carson, R. L. *Silent Spring*. Boston: Houghton Mifflin, 1962. The classic book that first alerted the public to the environmental dangers of pesticide use.

Colburn, T., D. Dumanoski, and J. P. Myers. *Our Stolen Future*. New York: Dutton, 1996. Drawing on work from many fields, this controversial book records the rapidly unfolding story of commonly used chemicals that disrupt the activities of hormones.

Ffrench-Constant, R. H. "The Zap Trap," *The Sciences*, March/April 1995. Explains why genetic resistance will always be a problem and argues for using insecticides only when they are absolutely necessary.

Gardner, G. "IPM and the War on Pests," *WorldWatch*, March/April 1996. A fascinating history of integrated pest management.

Holmes, B. "The Perils of Planting Pesticides," *New Scientist* Vol. 139, No. 1888, 28 August 1993. Increased use of *Bt* toxins has resulted in greater insect resistance to these natural pesticides.

Leary, W. E. "Natural Weapons Against an Ancient Plague," *The New York Times*, 4 January 1994. Biological control agents can be used to help control locusts without harming the environment.

Madeley, J. "Beyond the Pestkillers," *New Scientist* Vol. 142, No. 1924, May 7, 1994. Asian rice farmers are reducing their use of pesticides.

Pimentel, D., et al. "Environmental and Economic Costs of Pesticide Use," *BioScience* Vol. 42, No. 10, November 1992. The authors argue that responsible pesticide policies must take into account the hidden, indirect costs associated with pesticide use.

Raloff, J. "The Pesticide Shuffle," *Science News* Vol. 149, No. 11, March 16, 1996. Examines the airborne spread of pesticides across national boundaries.

Walker, S. "Back to Nature, Down on the Farm," *The Christian Science Monitor*, June 29, 1993. U.S. farmers are increasingly adopting low-input, sustainable agriculture and reducing their use of chemicals.

Wilson, J. D. "Resolving the 'Delaney Paradox,'" *Resources* Issue 125, Fall 1996. Discusses the 1996 Food Quality Protection Act.

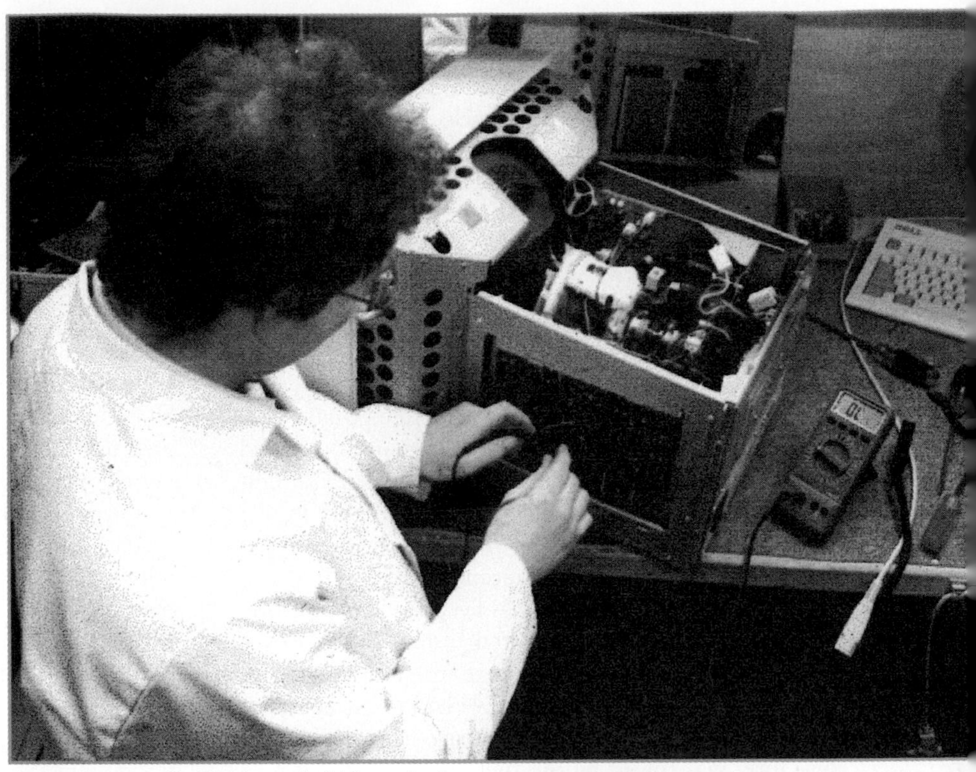

An engineer at R. Frazier refurbishes obsolete computer equipment. (Courtesy R. Frazier, Ltd.)

Solid and Hazardous Wastes

In the United States and other developed countries, the average computer is replaced every three years, not because it is broken but because rapid technological developments and new generations of software make it obsolete. Old computers may still be in working order, but they have no resale value and are even difficult to give away. As a result, they frequently are thrown away with the trash. According to one estimate, there will be more than 150 million computers disposed of in sanitary landfills around the world by 2005. This disposal represents a huge waste of the high-quality plastics and metals (aluminum, copper, tin, nickel, palladium, silver, and gold) that make up computers.

In addition, there *is* a market for old computers because many businesses cannot afford the current generation of computers but could benefit from inexpensive, entry-level computers. This situation has provided a business opportunity for certain companies. For example, R. Frazier, Ltd., in Scotland reclaims and resells old computers and electronic equipment they obtain from waste collection agencies in Europe. They also receive obsolete equipment from computer and telecommunications equipment manufacturers. Computer equipment is sorted, tested, and, where possible, repaired. These products are sold worldwide.

Computer equipment beyond economic repair is broken down for individual components and only as a last resort are they stripped for recycling. The plastics in broken components are recycled to make such items as park benches and shelving. Metals are also recycled. R. Frazier is so efficient in reusing and recycling old computers that less than 5 percent of all equipment they receive ends up in landfills.

R. Frazier has shown that it is possible to practice environmentally friendly business and profit economically. R. Frazier has grown in three years from a staff of 2 to

160 at its European headquarters in Dumfries. It now has subsidiaries in Malaysia and New Zealand; joint ventures in the Netherlands and Germany; and a sister company, R. Frazier, Inc., in the United States.

There is no single solution to the challenge of solid waste disposal today. A combination of source reduction, reuse, recycling, composting, burning, and burying in sanitary landfills is currently the optimal way to manage solid waste. In this chapter we examine the problems and the opportunities associated with the management of both solid and hazardous waste. ◄

THE SOLID WASTE PROBLEM

Every man, woman, and child in the United States produces an average of 2.0 kilograms (4.3 pounds) of solid waste per day, which corresponds to a total of 177.5 million metric tons (195.7 million tons) per year in the United States (Figure 23–1). And the problem worsens each year as the U.S. population increases.

The solid waste problem has been made abundantly clear by several highly publicized instances of garbage barges wandering from port to port and from country to country, trying to find someone willing to accept their cargo. In 1987, for example, a garbage barge from Islip, New York, was towed by the tugboat *Break of Dawn* to North Carolina. When North Carolina refused to accept the solid waste, the *Break of Dawn* set off on a journey of many months. In total, six states and three countries rejected the waste, which was eventually returned to New York to be incinerated.

A more ominous example of our solid waste problem is the story of the *Khian Sea*, a Bahamian ship that was hired by the city of Philadelphia in 1986. It was to transport thousands of tons of incinerated ash to Panama to be buried under a road for a new tourist resort. Incinerated ash contains toxic chemicals that could have adversely affected the fragile wetlands through which the road was being built, so Panama rejected the waste. The *Khian Sea* spent the next two years wandering, attempting to give its cargo to countries on five different continents. The ship reappeared in 1988, empty of cargo and with no explanations. Whether the ash was illegally dumped in some foreign nation or at sea is unknown.

Waste is an unavoidable consequence of prosperous, high-technology, disposable economies. It is a problem not only in the United States but in other industrialized nations. Many products that have the potential to be repaired, reused, or recycled are simply thrown away. Others, including paper napkins and disposable diapers, are designed to be used once and then discarded. Packaging, which not only makes a product more attractive and therefore more likely to sell, but protects it and keeps it sanitary, also contributes to waste. Nobody likes to think

Figure 23–1

The solid waste produced by an average American family of four in one year. The cans, bottles, and newspapers on the left are what the average American household now recycles. The trash cans and bags are filled with the solid waste they discard after recycling. (Jose Azel/Aurora)

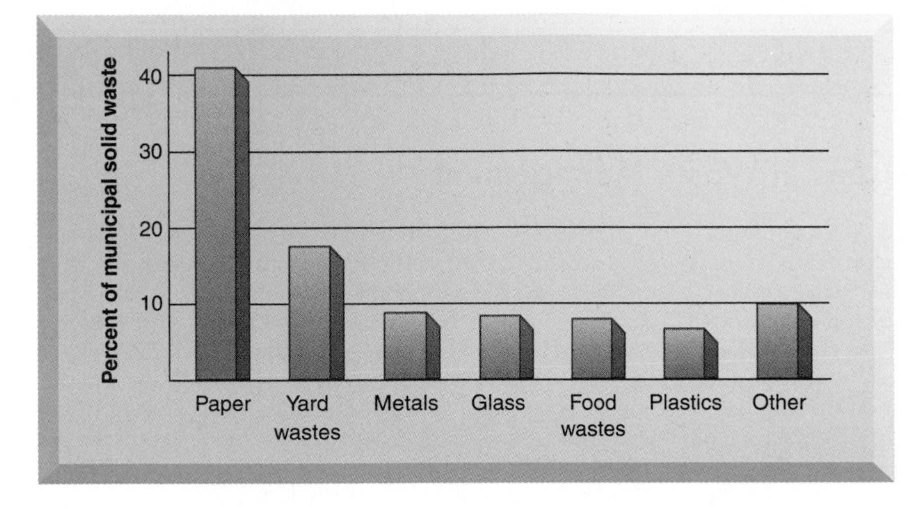

Figure 23–2
What is in our solid waste? Municipal solid waste is composed of paper, yard waste, metal, glass, plastic, food scraps, and other materials such as rubber and leather.

about solid waste, but the fact is that it is a pressing concern of modern society—we keep producing it, and places to dispose of it safely are dwindling in number.

TYPES OF SOLID WASTE

Municipal solid waste consists of solid materials discarded by homes, office buildings, retail stores, restaurants, schools, hospitals, prisons, libraries, and other commercial and institutional facilities. Municipal solid waste is a heterogeneous mixture composed primarily of paper and paperboard, yard waste, glass, metals, plastics, food waste, and other materials such as rubber, leather, and textiles (Figure 23–2). The proportions of the major types of solid waste in this mixture change over time. Today's solid waste contains more paper and plastics than in the past, whereas the amounts of glass and steel have declined.

Municipal solid waste is a relatively small portion of all the solid waste produced. **Nonmunicipal solid waste**, which includes wastes from industry, agriculture, and mining, is produced in substantially larger amounts than municipal solid waste. Approximately 98.5 percent of solid waste in the United States is from nonmunicipal sources (Figure 23–3).

DISPOSAL OF SOLID WASTE

Solid waste has traditionally been regarded as material that is no longer useful and that should be disposed of. There are four ways to get rid of solid waste: dump it, bury it, burn it, or compost it. Alternatively, we can dispose of solid waste by exporting it to another country (see Focus On: International Issues in Waste Management).

Open Dumps

The old method of solid waste disposal was dumping. Open dumps were unsanitary, malodorous places in which disease-carrying vermin such as rats and flies proliferated. Methane gas was released into the surrounding air as microorganisms decomposed the solid waste, and fires pol-

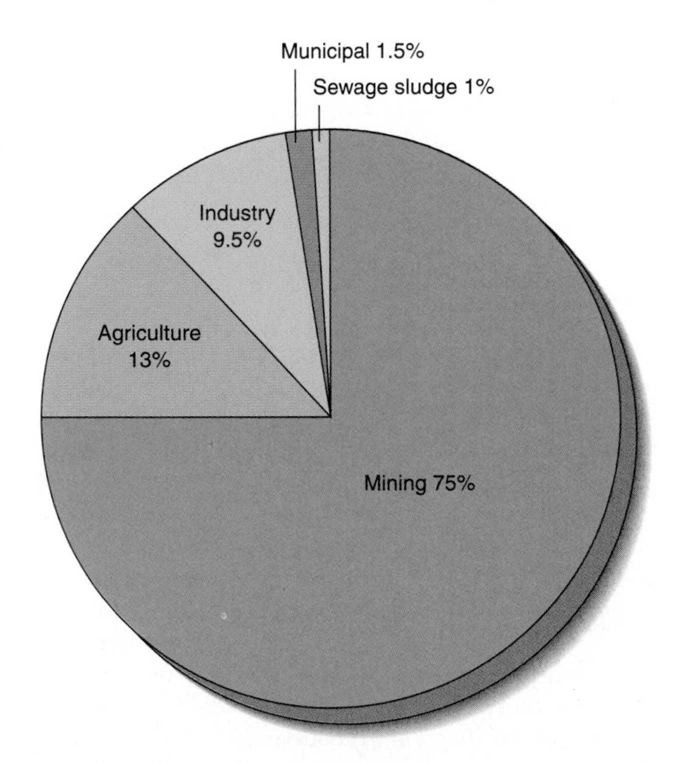

Figure 23–3
Sources of solid waste in the United States. The three largest producers of solid waste are mining, agriculture, and industry. Municipal solid waste contributes only 1.5 percent to the solid waste stream.

FOCUS ON

International Issues in Waste Management

Although there are ways to reduce and dispose of waste in an environmentally sound manner, industrialized countries have sometimes chosen to send their waste to other countries. (As industrialized nations develop more stringent environmental standards, disposing of hazardous waste at home becomes much more expensive than the cost of sending it to a developing nation for disposal.) Some waste is exported for legitimate recycling, but some has been exported strictly for disposal.

The export of both solid and hazardous wastes by the United States, Japan, and the European Economic Community is one of the most controversial aspects of waste management today. Usually the recipients of this waste are developing nations in Africa, Central and South America, and the Pacific Rim of Asia, although Eastern and Central Europe and countries of the former Soviet Union are also important sites. The governments, industries, and citizens of these countries are often inexperienced and ill-equipped to handle such materials. As a result, the waste often causes the same types of environmentally hazardous sites in developing nations that industrialized nations are trying to clean up at home.

In 1989 the U.N. Environment Programme developed a treaty known as the Basel Convention to restrict the international transport of hazardous waste. It allows countries to export hazardous waste only with prior consent of the importing country as well as any countries that the waste passes through in transit. The Basel Convention became official in 1992, and 100 countries have ratified it as of late 1996. (The United States signed the treaty in 1989 but has not yet ratified it because Congress has not passed the necessary legislation.)

In 1995 the Basel Convention was amended to ban the export of *any* hazardous waste between industrial and developing countries; this amendment goes into effect after 1997. Many industrialized nations oppose such a ban. In 1995, for example, the European Union announced plans to circumvent the international treaty by redefining many kinds of waste that were formerly classified hazardous as nonhazardous. Redefining waste is legal because the Basel Convention currently does not provide sufficient detail about what waste is hazardous and what is not.

luted the air with acrid smoke. Liquid that oozed and seeped through the solid waste heap ultimately found its way into the soil, surface water, and groundwater. Hazardous materials that were dissolved in this liquid often contaminated soil and water.

Sanitary Landfills

Open dumps have largely been replaced by sanitary landfills, which receive 55.7 percent of the solid waste generated in the United States today (Figure 23–4). Sanitary landfills differ from open dumps in that the solid waste is placed in a hole, compacted, and covered with a thin layer of soil every day (Figure 23–5). This process reduces the number of rats and other vermin usually associated with solid waste, lessens the danger of fires, and decreases the amount of odor. If a sanitary landfill is operated in accordance with solid waste management-approved guidelines, it does not pollute local surface and groundwater. Safety is ensured by the presence of layers of compacted clay and plastic sheets at the bottom of the landfill, which prevent liquid waste from seeping into groundwater (Figure 23–6). Newer landfills possess a double liner system (plastic, clay, plastic, clay) and use sophisticated systems to collect leachate (liquid that seeps through the solid waste) and gases that form during decomposition.

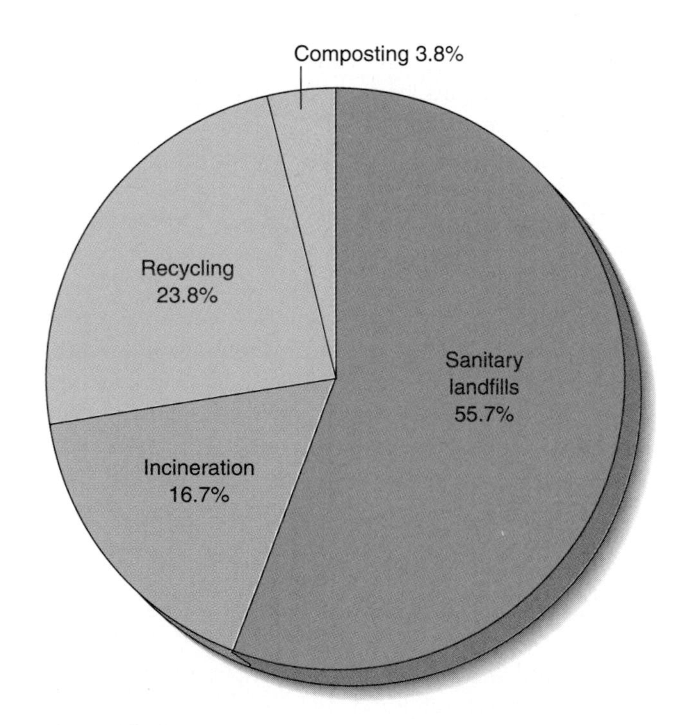

Figure 23–4

How our municipal solid waste was disposed of in 1996. The recovery of waste from recycling and composting is expected to increase during the next several years.

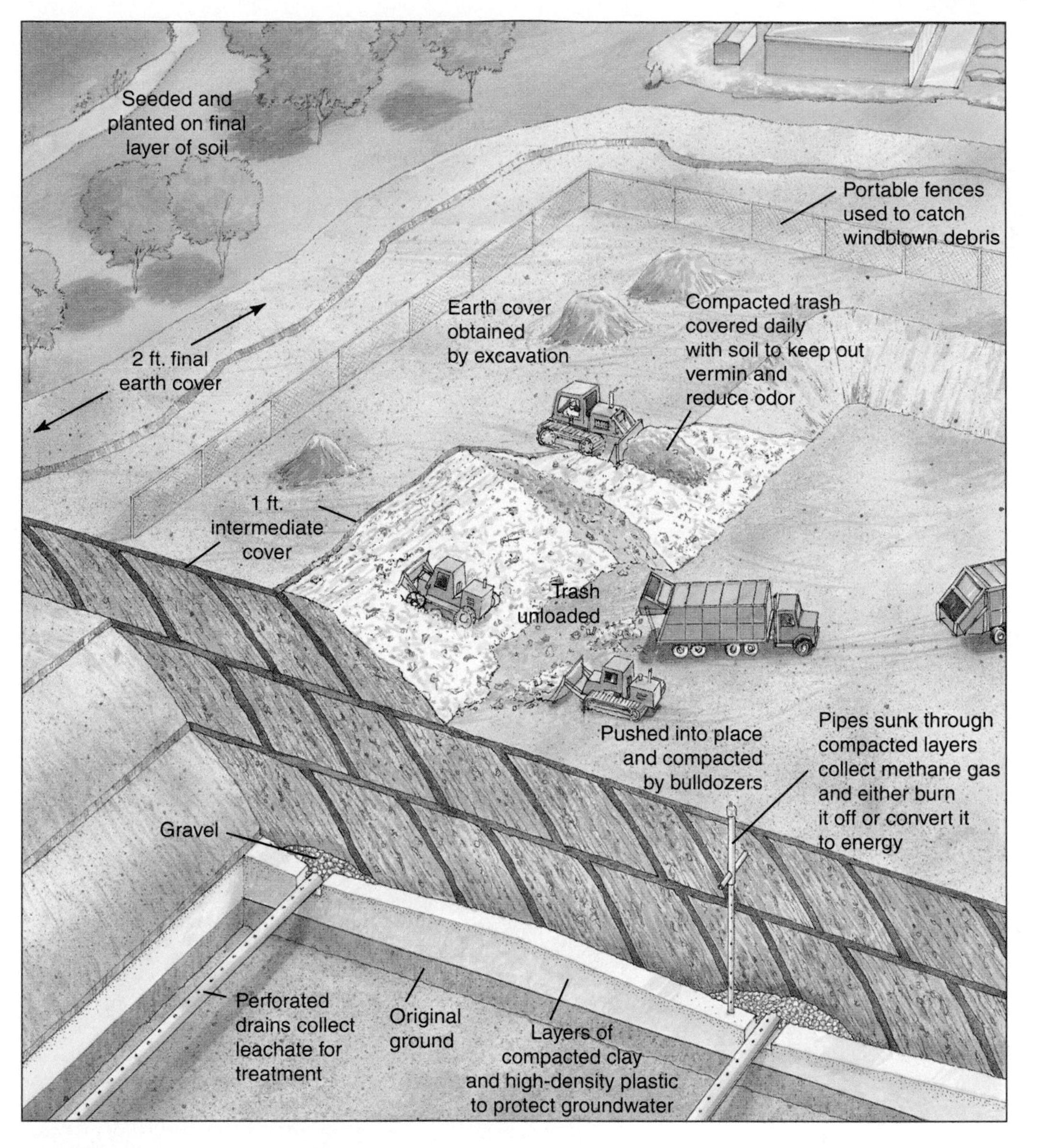

Figure 23–5

A sanitary landfill is much more than a hole in the ground. Solid waste is spread in a thin layer, compacted into small sections, called cells, and covered with soil. Sanitary landfills constructed today have protective liners of compacted clay and plastic and sophisticated leachate collection systems that minimize environmental problems such as contamination of groundwater.

The location of an "ideal" sanitary landfill is based on a variety of factors, including the geology of the area and the proximity of nearby bodies of water and wetlands. The landfill should be far enough away from centers of dense population to be inoffensive but close enough so as not to require high transportation costs.

Problems associated with sanitary landfills

Although the operation of sanitary landfills has improved over the years with the passage of stricter and stricter guidelines, very few landfills are ideal. Most sanitary landfills in operation today do not meet current legal standards.

One of the problems associated with sanitary landfills is the production of methane gas by microorganisms that decompose organic material anaerobically (in the absence of oxygen). This methane may seep through the solid waste and accumulate in underground pockets, creating the possibility of an explosion. It is even possible for methane to seep into basements of nearby homes, which is an extremely dangerous situation. Conventionally, landfills collected the methane and burned it off in flare

Figure 23–6
A modern sanitary landfill contains plastic liners (shown) to trap leachate and prevent it from draining any deeper into the soil and contaminating groundwater. Beginning in 1996, landfills in the United States must have a double lining of plastic, clay, plastic, clay. (Terry Wild Studio)

systems. A growing number of landfills, however, have begun to use the methane for gas-to-energy projects. About 120 landfills in North America currently use methane to generate electricity.

Another problem associated with sanitary landfills is the potential contamination of surface water and groundwater by leachate that seeps from unlined landfills or through cracks in the lining of lined landfills. Even household trash contains toxic chemicals that can be carried into groundwater. Pollutants in the leachate may include heavy metals, pesticides, and organic compounds.

Although we currently dispose of about 55.7 percent of our solid waste in landfills, they are not a long-term remedy for waste disposal because they are filling up. During the period from 1978 to 1993, the number of landfills in operation decreased from 20,000 to 5300. Many of the landfills closed because they had reached their capacity. Other landfills closed because they did not meet state or federal operating standards.

Fewer new sanitary landfills are being opened to replace the closed ones. The reasons for this are many and complex. Many desirable sites have already been taken. Also, people living near a proposed site are usually adamantly opposed to a landfill near their homes. Recall from Chapter 11 that the opposition of people to the location of hazardous facilities near their homes is known as the **n**ot **i**n **my** **b**ack**y**ard, or NIMBY, response. This attitude is partly the result of past problems with landfills—everything from offensive odors to dangerous contamination of drinking water. It is also caused by the fear that property values will be lowered by a nearby facility.

Closing a sanitary landfill

Considerable expense is involved in closing a sanitary landfill once it is full. The possibility of groundwater pollution and gas explosions remains for a long time, so the Environmental Protection Agency (EPA) currently requires owners to continuously monitor a landfill for 30 years after the landfill is closed. In addition, no homes or

SOLID WASTE BREAKDOWN

Estimated length of time it will take for the following items to decompose, if left untreated and exposed to the elements:

Litter paper: 2 to 4 weeks
Cotton rags: 1 to 5 months
Orange peels: 6 months
Woolen socks: 12 months
Filter-tip cigarette butts: 10 to 12 years
Plastic bags: 10 to 20 years
Leather shoes: 25 to 40 years
Aluminum cans: 200 to 500 years

If buried in a sanitary landfill, decomposition may take significantly longer.

HOUSEHOLD HAZARDOUS WASTE

Each year, the typical American home throws away about 20 pounds of hazardous household waste—automotive products, household cleaners, paints and solvents, and pesticides:

- When improperly disposed, household hazardous waste can cause fires and explosions, caustic burns when people are exposed, and environmental contamination.
- Users of hazardous materials should buy only the quantity they need, take advantage of local hazardous waste disposal days, store any leftover waste carefully, and try alternatives to hazardous substances, such as using 1/2 cup vinegar in a bucket of hot water instead of floor cleaners.
- Hazardous household waste should *never* be flushed down a toilet, put in with other solid waste for collection, or poured into a ditch, yard, or storm drain.

other buildings can be built on a closed sanitary landfill for many years.

The special problem of plastic

The amount of plastic in our solid waste is growing faster than any other component of municipal solid waste. More than half of this plastic is from packaging. Plastics are chemical **polymers** that are composed of chains of repeating carbon compounds. The properties of the many types of plastics—polypropylene, polyethylene, and polystyrene, to name a few—differ based on their chemical compositions.

One of the characteristics of most plastics is that they are chemically stable and do not readily break down, or decompose. This characteristic, which is essential in the packaging of certain products, such as food, causes long-term problems: most plastic debris will probably last for centuries.

In response to concerns about the volume of plastic waste, some places have actually banned the use of certain types of plastic, such as the polyvinyl chloride employed in packaging. Special plastics that have the ability to degrade or disintegrate have been developed. Some of these, however, are **photodegradable**; they break down only after being exposed to sunlight, which means they will not break down in a sanitary landfill. Other plastics are **biodegradable**—that is, decomposed by microorganisms such as bacteria. Whether biodegradable plastics actually break down under the conditions found in a sanitary landfill is not yet clear, although preliminary studies indicate that they probably do not.

Figure 23–7
Piles of discarded, worn out tires in a tire dump in Ohio.
(Mark C. Burnett/Photo Researchers, Inc.)

The special problem of tires

One of the most difficult materials to manage is rubber. Discarded tires—some 250 million each year in the United States—are made of vulcanized rubber, which cannot be melted and reused. An estimated 2 to 3 billion tires have been discarded in sanitary landfills and tire dumps (Figure 23–7), as well as along roadsides and in vacant lots. Disposal of tires in sanitary landfills is a real problem because tires, being relatively large and light, have a tendency to move upward through the accumulated solid waste. After a period of time, they work their way to the surface of the landfill. These tires are a fire hazard and collect rainwater, providing a good breeding place for mosquitoes. Accordingly, most states either ban tires from sanitary landfills or require that they be shredded to save space and prevent water from pooling in them.

Incineration

When solid waste is incinerated, two positive things are accomplished. First, the volume of solid waste is reduced by up to 90 percent. The ash that remains is, of course, much more compact than solid waste that has not been burned. Second, incineration produces heat that, if properly channeled, can produce steam to warm buildings or generate electricity. In 1996 the United States had about 150 waste-to-energy incinerators, which burned 16.7 percent of the nation's solid waste. By comparison, only one percent of U.S. solid waste was incinerated in 1970.

Some materials are best removed from solid waste before incineration occurs. For example, glass does not burn, and when it melts, it is difficult to remove from the incinerator. Although food waste burns, its high moisture content often decreases the efficiency of incineration, so it is better to remove it before incineration. The best materials for incineration are paper, plastics, and rubber.

Paper is a good candidate for incineration because it burns readily and produces a great amount of heat. One potential complication is the presence in the ink and paper of toxic compounds, which might be emitted during incineration. Some types of paper release dioxins into the atmosphere when burned; dioxins are toxic compounds that are discussed later in the chapter.

Plastic produces a lot of heat when it is incinerated: a kilogram of plastic waste yields almost as much heat as a kilogram of fuel oil. As with paper, there is concern about some of the pollutants that might be emitted during the incineration of plastic. Polyvinyl chloride, a common component of many plastics, may release dioxins and other toxic compounds when incinerated. More research is needed to determine whether these or other hazardous substances are actually produced during the incineration of plastics.

One of the best uses for old tires is incineration, because burning rubber produces much heat. Some electric utilities in the United States and Canada are burning tires instead of or in addition to coal; tires produce as much heat as coal and often generate less pollution.

Types of incinerators

The three types of incinerators are mass burn, modular, and refuse-derived fuel. **Mass burn incinerators** are large furnaces that burn all solid waste except for unburnable items such as refrigerators. Most mass burn incinerators are large and are designed to recover the energy produced from combustion (Figure 23–8). **Modular incinerators** are smaller incinerators that burn all solid waste. They are assembled at factories and so are less expensive to build. In **refuse-derived fuel incinerators**, only the combustible portion of solid waste is burned. First, noncombustible materials such as glass and metals are removed by machine or by hand. The remaining solid waste, including plastic and paper, is shredded or shaped into pellets and burned.

Problems associated with incineration

The possible production of toxic air pollutants is the main reason people oppose incineration. Incinerators can pollute the air with carbon monoxide, particulates, heavy metals, and other toxic materials unless expensive air pollution control devices are used. Such devices include **lime**

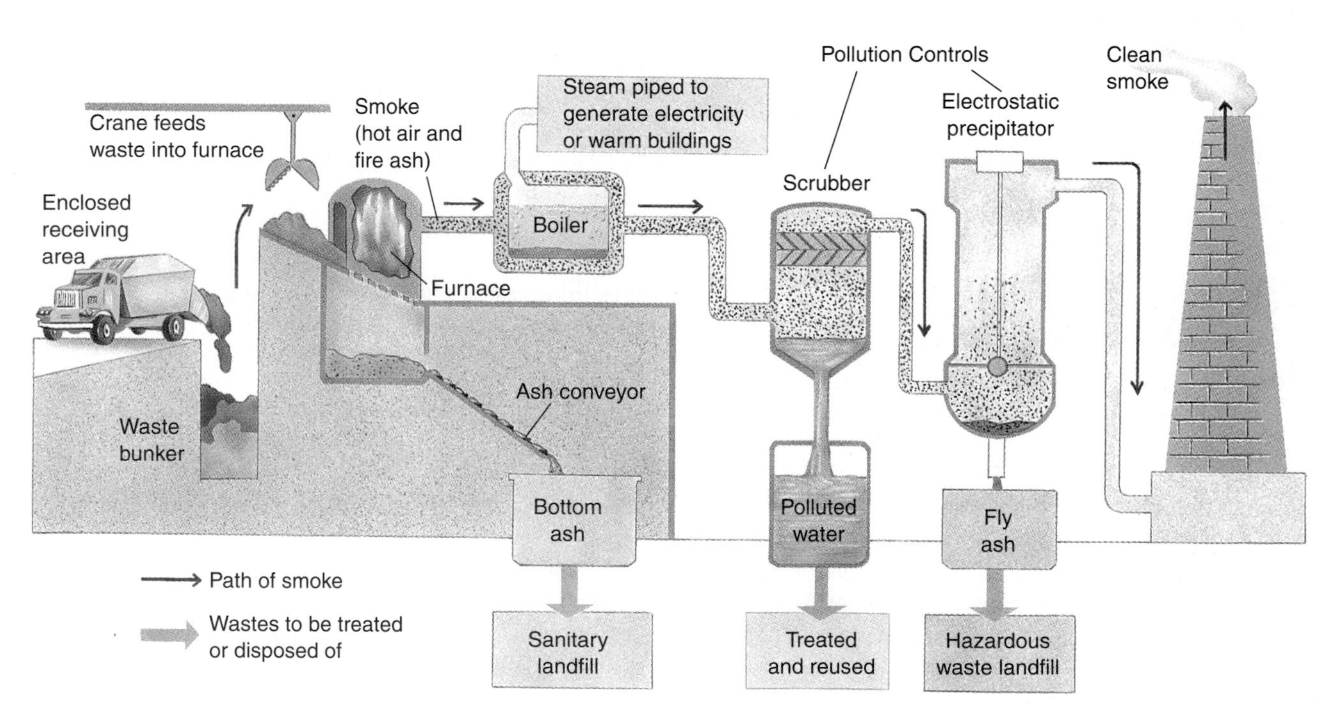

Figure 23–8

A mass burn, waste-to-energy incinerator. Modern incinerators have pollution control devices such as lime scrubbers and electrostatic precipitators to trap dangerous and dirty emissions.

scrubbers, towers in which a chemical spray neutralizes acidic gases, and **electrostatic precipitators**, which give ash a positive electrical charge so that it adheres to negatively charged plates rather than going out the chimney.

Incinerators produce large quantities of ash that must be disposed of properly. Two kinds of ash are produced, bottom ash and fly ash. **Bottom ash**, also known as **slag**, is the residual ash left at the bottom of the incinerator when combustion is completed. **Fly ash** is the ash from the flue (chimney) that is trapped by electrostatic precipitators. Fly ash usually contains more toxic materials, including heavy metals and possibly dioxins, than bottom ash.

Currently, both types of incinerator ash are best disposed of in hazardous waste landfills. What happens to the toxic materials in incinerator ash when it is placed in an average sanitary landfill is unknown, but there are concerns that the toxic materials could contaminate groundwater. In 1994 the Supreme Court considered a case involving a Chicago waste-to-energy incinerator that was disposing its ash, which contained very high levels of lead and cadmium, in a regular sanitary landfill. The Court ruled that incinerator ash must be tested regularly for toxic materials. When the hazardous materials in incinerator ash exceed acceptable levels, the ash must be disposed of in licensed hazardous waste sites.

Site selection for incinerators, like that for sanitary landfills, is controversial. People may recognize the need for an incinerator, but they do not want it near their homes. Another drawback of incinerators is their high cost. Prices have been escalating because costly pollution control devices are now required.

Composting

Yard waste—grass clippings, branches, and leaves—is a large component of municipal solid waste (see Figure 23–2). As space in sanitary landfills becomes more limited, other ways to dispose of yard waste are being developed and implemented. One of the best ways is to convert organic waste into soil conditioners such as compost or mulch (see You Can Make a Difference: Practicing Environmental Principles, in Chapter 14). Food scraps, sewage sludge, and agricultural manure are other forms of solid waste that can be used to make compost. Compost and mulch can be used in public parks and playgrounds, for landscaping, or as part of the daily soil cover at sanitary landfills; or they can be sold to gardeners.

Composting as a way to manage solid waste first became popular in Europe. Several municipalities in the United States have composting facilities as part of their comprehensive solid waste management plans (Figure 23–9), and about ten states have banned yard waste from sanitary landfills. This trend is likely to continue, making composting even more desirable. The United States currently composts 3.8 percent of municipal solid waste.

WASTE PREVENTION

The three goals of waste prevention, in order of priority, are: (1) *reduce* the amount of waste as much as possible; (2) *reuse* products as much as possible; and (3) *recycle* materials as much as possible. Reducing the amount of waste includes purchasing products that have less packaging and that last longer or are repairable. Consumers can also reduce waste by decreasing their consumption of products. Before deciding to purchase a product, a consumer should ask, "Do I really *need* this product, or do I merely *want* it?" Reuse is a lower priority than reducing waste because reuse of products requires more energy and produces more pollution than not having the product in the first place. Reuse saves energy and reduces pollution significantly as compared to recycling, however.

Figure 23–9

Leaves, twigs, and other yard wastes are collected and delivered to the Ann Arbor Municipal Recycling Center in Michigan. Composting is a great way to dispose of yard waste without filling up valuable landfill space. (The plastic trash bags will be screened out later.)

MEETING THE CHALLENGE

Reusing and Recycling Old Automobiles

Americans discard more than 11 million cars and trucks each year. Although about 75 percent by weight of a retired car is easily reused (as secondhand parts) or recycled (as scrap metal), the remaining 25 percent is not easy to recycle and usually ends up in sanitary landfills. Approximately 2000 companies are involved in automotive recycling in the United States.

Because automobiles typically contain about 600 different materials—glass, metals, plastics, fabrics, rubber, foam, leather, and so on—ways to reuse or recycle old parts is a complex problem. Economics is an important aspect of the problem, as reuse and recycling companies must make money in their recycling endeavors.

Workers typically begin disassembling a used car by draining all fluids—such as antifreeze, gasoline, transmission fluid, Freon, oil, and brake fluid—and recycling the fluids or processing them for disposal (see figure). Reusable parts and components, such as the engine, tires, and battery, are then removed, cleaned, tested, and inventoried before being sold as used parts. Body shops, new and used car dealers, repair shops, and auto and truck fleets are the main buyers of used parts. In addition to being reused, some parts are disassembled for their materials. Catalytic converters, for example, are disassembled because they contain valuable amounts of platinum and rhodium.

An automotive recycling facility then sends the remaining vehicle "shell" to a scrap processor for "scrapping." At the scrap processing facility, a giant machine shreds the entire automobile into pieces the size of a jar lid. Magnets and other machines sort the pieces into piles of steel, iron, copper, aluminum, and "fluff," which consists of the remaining materials, such as plastic, rubber, and glass.

About 37 percent of the iron and steel scrap reprocessed in the United States comes from old automobiles. Recycling iron and steel saves energy and reduces pollution. According to the EPA, using scrap iron and steel produces 86 percent less air pollution and 76 percent less water pollution than mining and refining iron ore.

Recycling plastic is the biggest challenge in auto recycling. Plastic is lightweight, and, as a result, automakers are using more and more plastic to improve fuel efficiency. Also, no industry standards for plastic parts currently exist, so the kinds and amounts of plastic from which cars are made vary a great deal. As many as 15 different plastics comprise today's dashboards, for example, and because many of these plastics are chemically incompatible, they cannot be melted together for recycling.

Auto manufacturers in the United States have begun to study the challenge of reusing and recycling old cars. The Vehicle Recycling Partnership was formed in 1991 between General Motors, Ford, and Chrysler Corporations to study how auto recycling can be simplified. Legislators are watching U.S. automakers' efforts as well, perhaps with the idea of someday devising legislation that would require U.S. manufacturers to accept scrapped autos from consumers after they have outlived their usefulness.

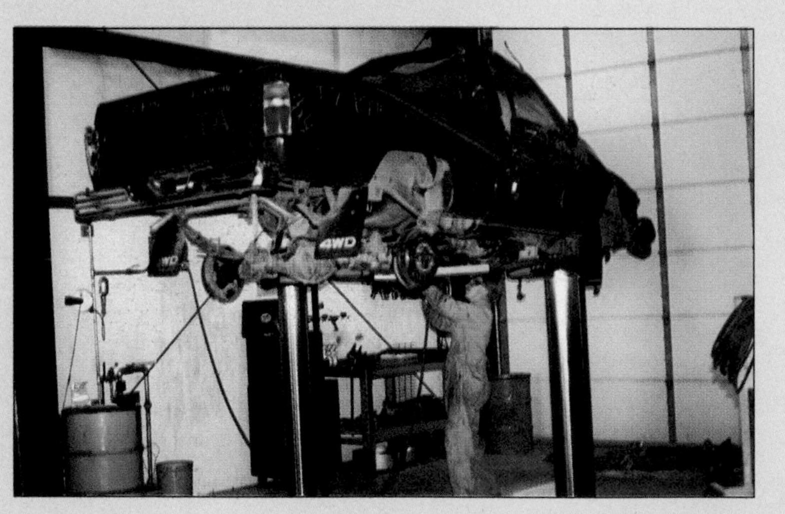

A worker begins disassembling a retired car by draining all fluids and recycling them or processing them for disposal. (Automotive Recyclers Association)

We already discussed dematerialization, reuse, and recycling in Chapter 15, in the context of resource conservation. Now we examine the impact of these practices on solid waste (see Meeting the Challenge: Reusing and Recycling Old Automobiles). We then consider the fee-per-bag approach that many communities have adopted to encourage waste prevention.

Reducing the Amount of Waste: Source Reduction

Industries can often design and manufacture their products in ways that decrease not only the volume of solid waste but the amount of hazardous materials in the solid waste that remains. Such a strategy is known as **source reduction**, and it is the most under-utilized aspect of waste management. Source reduction is accomplished in a variety of ways. Innovations and product modifications can reduce the waste produced after a product has been used by the consumer. Dry-cell batteries, for example, contain much less mercury today than they did in the early 1980s. The 35 percent reduction in the weight of aluminum cans since the 1970s provides another example of source reduction.

AN INDUSTRIAL ECOSYSTEM

The J. R. Simplot Company supplies more than half of the french fries sold by McDonald's each year. Jack Richard Simplot, the founder of the Idaho-based firm, has found a profitable use for the company's waste with an extraordinary waste reduction system. Because french fries are a quality-controlled product, approximately 50 percent of the potatoes processed by Simplot's company ended up as waste, until he learned that the peelings could be mixed with grain to feed cattle. As of 1994, company potato residues feed more than 250,000 animals each year. But that is only the beginning. Potato particles contained in processing water are anaerobically digested to produce methane for power plants, and the nutrient-bearing process water is used to irrigate and fertilize agricultural crops. Energy in the form of fuel-grade ethanol is also derived from processing waste and cull potatoes. The sludge from the ethanol process is converted to low-cost, high-protein livestock feed. Manure from the cattle feedlots is applied to farmland to help nourish the hay and grain crops used to feed the cattle.

Dematerialization, the progressive decrease in the size and weight of a product as a result of technological improvements, was discussed in Chapter 15. *Dematerialization results in source reduction only if the new product is as durable as the one it replaced.* If smaller, lighter products have shorter life spans so that they have to be thrown out and replaced more often, source reduction is not accomplished.

Reusing Products

One example of reuse is refillable glass beverage bottles. Years ago, refillable beverage bottles were used a great deal in the United States (see Chapter 15). Today their use is rare, although about ten states still have them. In order for a glass bottle to be reused, it must be considerably thicker (and therefore heavier) than one-use bottles. Because of the increased weight, transportation costs are higher. In the past, reuse of glass bottles worked because there were many small bottlers scattered across the United States, helping to minimize transportation costs. Today there are approximately one-tenth as many bottlers. Because of the centralization of bottling beverages, it is economically difficult to go back to the days of refillable bottles. The price of beverages might have to increase to absorb increased transportation costs.

Although the quantity of reusable glass bottles in the United States has declined, certain countries still reuse glass to a large extent. In Japan, almost all beer and sake bottles are reused as many as 20 times; bottles in Ecuador may remain in use for ten years or longer. European countries such as Denmark, Finland, Germany, the Netherlands, Norway, Sweden, and Switzerland have passed legislation that promotes the refilling of beverage containers. Parts of Canada and the United States also have deposit laws.

Can refillable bottles make a comeback in the United States?

Some beverage producers have recently reversed the trend toward disposable bottles. Some dairies have started delivering milk directly to consumers in bottles made of acrylic plastic that can be reused 50 to 100 times before being disposed of. A few breweries have also switched from disposable back to refillable bottles.

Despite these examples of a shift toward refillables, however, it is unlikely that the United States will switch to refillable bottles on a large scale unless legislators and consumers take action. If a federal deposit law—a national "bottle bill"—were passed, for example, consumers would return bottles to get the small deposit, such as a nickel or a dime per bottle. Other legislation, such as a requirement that bottlers sell a certain percentage of their product in refillable bottles, would also encourage reuse.

Recycling Materials

It is possible to collect and reprocess many materials found in solid waste into new products of the same or different type. Recycling is preferred over incineration and landfill disposal because it conserves our natural resources and is more environmentally benign. For example, every ton of recycled paper saves 17 trees, 7000 gallons of water, 4100 kilowatt hours of energy, and 3 cubic yards of landfill space. Recycling also has a positive effect on the economy by generating jobs and revenues (from selling the recycled materials).

Solid waste is a mixture of many different materials that must be separated from one another before recycling can be accomplished. It is easy to separate materials such as glass bottles and newspapers, but the separation of materials in items with complex compositions is difficult. Some food containers are composed of thin layers of metal foil, plastic, and paper, for example. Trying to separate these layers is a daunting prospect, to say the least.

The number of communities with recycling programs has increased remarkably during the 1990s, and the average American family of four recycles over 1000 pounds of aluminum and steel cans, plastic bottles, glass containers, newspapers, and cardboard each year (see Figure 23–1). Some 65 million people now live in areas with curbside-collection recycling programs. Other communities collect solid waste as a mixture and take it to special resource recovery facilities (Figure 23–10), where it is either hand-sorted or separated using a variety of technologies, including magnets, screens, and conveyor belts. The United States currently recycles 23.8 percent of its municipal solid waste.

The most successful large-scale recycling program in the United States is in Seattle, Washington. By 1995 Seattle had a residential recycling rate of 48 percent, as compared to a 24 percent recycling rate in 1988. Seattle's ultimate goal is to recycle 60 percent of its solid waste.

Most people think that recycling involves merely separating certain materials from the solid waste stream, but that is only the first step in recycling. In order for recycling to work, there must be a market for the recycled goods, and the recycled products must be used in preference to virgin products. Prices paid by processors for old newspaper, used aluminum cans, used glass bottles, and the like vary significantly from one year to the next, largely depending on the demand for recycled products.

(a)

(b)

Figure 23–10

Resource recovery. (a) Plastic containers can be separated from other solid waste and recycled. (b) A technician at a recycling facility in Pennsylvania holds a sample of shredded waste plastic, which can be used to make polyester carpet and other products. (a, © 1997 Larime Photographic/Dembinsky Photo Associates; b, Hank Morgan/Photo Researchers, Inc.)

Paper

Americans recycle about 33 percent of their paper; the rest ends up in landfills or incinerators. This is a dismal record compared to most other developed countries. Denmark, for example, recycles 97 percent of its paper. Part of the reason paper is not recycled more in the United States is that many of our paper mills are old and are not equipped to process waste paper. The number of mills that can handle waste paper is increasing, however, in part because of consumer demand. Between 1990 and 1994, for example, 85 new paper mills with recycling capabilities were built in the United States.

In addition to a slow increase in paper recycling in the United States, there is a growing demand for U.S. waste paper in other countries. Mexico, for example, imports a large quantity of waste paper and cardboard from the United States. Used paper is also in great demand in China, Taiwan, and Korea.

As mentioned earlier, paper recycling does not work unless there is a market for recycled paper products. Sometimes the demand can be created by legislation. In Toronto, Canada, for example, the city council passed a law, effective in 1991, that daily newspapers must contain at least 50 percent recycled fiber, or the publishers will not be allowed to have vending boxes on city streets. President Clinton issued an executive order, effective in 1994, that all paper purchased by the federal government must contain 20 percent post-consumer content; by 1998, recycled paper purchased by the federal government must contain 30 percent post-consumer content.

Figure 23–11

Recycling of glass as glassphalt. Shown is a closeup of a Baltimore street paved with glassphalt, which contains a mixture of broken glass from different colored containers. At night the road sparkles, as light from automobile headlights reflects off the pieces of glass. (Courtesy of Baltimore Department of Public Works)

Glass

Glass is another component of solid waste that is appropriate for recycling. Recycled glass costs less than glass made from virgin materials. Glass food and beverage containers are crushed to form **cullet**, which can be melted and used by glass manufacturers to make new products without any special adaptations in their factories. Although cullet is much more valuable when glass containers of different colors are separated before being crushed, cullet made from a mixture of colors can be used to make glassphalt, a composite of glass and asphalt that makes an attractive roadway (Figure 23–11).

Aluminum

The recycling of aluminum (see Chapter 15) is one of the best success stories in U.S. recycling, largely because of economic factors. Making a new aluminum can from a recycled one requires a fraction of the energy it would take to make a new can from raw metal. Because energy costs are high, there is strong economic incentive to recycle aluminum. According to the Can Manufacturers Institute, approximately two out of three aluminum cans—some 62.7 billion—were recycled in 1995, saving 19.3 million barrels of oil.

Metals other than aluminum

Other recyclable metals include lead, gold, iron and steel, silver, and zinc. One of the obstacles to recycling metal products discarded in municipal solid waste is that their metallic compositions are often unknown. It is also difficult to extract metal from products, such as stoves, that contain other materials besides metal (plastic, rubber, or glass, for instance). In contrast, any waste metal produced at factories is recycled easily because its composition is known.

The economy has a large influence on whether metal is recycled or discarded. Greater recycling occurs when the economy is strong than when there is a recession. Thus, although the supply of metal waste is fairly constant, the amount of recycling varies from year to year.

One exception to this generalization is steel. Before the 1970s, almost all steel was produced in mills that processed raw ores. Starting in the 1970s and continuing to the present, however, "mini-mills" that produce steel products from up to 100 percent scrap became increasingly important. According to the Institute of Scrap Recycling Industries, new steel products in the mid-1990s contained, on average, 58 percent recycled scrap steel.

Plastic

About 2.5 percent of plastic is recycled. One reason plastic is not recycled much is that, depending on the economic situation, it is sometimes less expensive to make it from raw materials (petroleum and natural gas) than to recycle it. In other words, plastic recycling—indeed all recycling—is driven by economics. However, public pressure to recycle plastic has been so strong that the demand for recycled plastic is currently greater than the supply. As a result, more plastic is being recycled each year. Increasingly, local and state governments are supporting or requiring the recycling of plastic.

Polyethylene terphthalate (PET), the plastic used in soda bottles, is recycled more than any other plastic. Industry sources declare a 35 to 40 percent recycling rate for the 9.5 billion PET bottles sold annually, which is incorporated into such diverse products as carpeting, automobile parts, tennis ball felt, and polyester cloth; it takes about 25 plastic bottles to make one polyester sweater.

Polystyrene (one form of which is Styrofoam) is an example of a plastic with great recycling potential. Cups, silverware, and packaging material that are made of polystyrene can be recycled into a variety of products, such as coat hangers, flower pots, foam insulation, and toys. Because approximately 2.3 billion kilograms (5 billion pounds) of polystyrene are produced each year, large-scale recycling would make a major dent in the amount of polystyrene that ends up in landfills. (By comparison, 0.7 billion kilograms, or 1.5 billion pounds, of PET is produced each year.)

One of the challenges associated with recycling plastic is that there are many different kinds. Forty-six different plastics are common in consumer products, and many products contain multiple kinds of plastic. A plastic ketchup bottle, for example, may have up to six layers of different plastics bonded together. In order to effectively recycle high-quality plastic, the different types must be sorted or separated. If two or more resins are recycled together, the resultant plastic is of lower quality.

Low-quality plastic mixtures are used to make a construction material similar to wood. This "plastic lumber" is particularly useful for outside products, such as fence posts, planters, highway retaining walls, and park benches, because of its durability.

Tires

The number of products that can be made from old tires is limited, although some research in product development is being carried out, and almost all states have established tire recycling programs. Uses for old tires include playground equipment; trash cans, garden hoses, and other consumer products; reef building on coastlines; and rubberized asphalt for pavement.

The Fee-Per-Bag Approach

Consumers have traditionally paid the same amount in taxes or collecting fees whether they put out one bag or twenty bags of solid waste. Many communities have reduced the volume of solid waste and encouraged reuse and recycling by charging households for each container of solid waste—the **fee-per-bag approach**.

Two environmental economists examined the weight and volume of weekly solid waste and recycling in Charlottesville, Virginia, before and after a charge of $.80 per 32 gallon bag of solid waste was implemented in 1992. They determined that charging consumers for each bag of solid waste that they discard is an effective way to reduce the volume of solid waste. The average Charlottesville household reduced the weight of its solid waste by 14 percent and the volume of its solid waste by 37 percent. Increased recycling accounted for much of the decrease in solid waste.

Similar results were obtained in a 1996 study of nine communities in California, Illinois, and Michigan. However, the fee-per-bag approach has raised concerns about people illegally disposing of solid waste by burning or dumping along back roads.

HAZARDOUS WASTE

Beginning in 1977 with the discovery that toxic waste from an abandoned chemical dump had contaminated homes and possibly people in Love Canal, New York, toxic waste has held national attention. Lois Gibbs, a housewife in Love Canal, led a successful crusade to evacuate the area after she discovered what seemed to be a high number of serious illnesses, particularly among children, in the neighborhood.[1] In 1978, Love Canal had the dubious distinction of being the first area ever declared a national emergency disaster area because of toxic waste; some 700 families were evacuated.

From 1942 to 1953, a local industry, Hooker Chemical Company, disposed of 19,945 metric tons (22,000 tons) of toxic chemical waste in the 914-meter-long (3000 feet) Love Canal. When the site was filled, Hooker added topsoil and donated the land to the local Board of Education. A school and houses were built on the site, which

[1]Today, Ms. Gibbs is executive director for the Citizens' Clearinghouse for Hazardous Waste.

began oozing toxic waste several years later. Over 300 chemicals, many of them carcinogenic, have been identified in Love Canal's toxic waste.

After almost ten years of cleanup, the EPA and the New York Department of Health declared the area safe to be resettled, although some chemicals remain in the canal. The area has been renamed Black Creek Village.

The Love Canal disaster generated an immediate concern about the hazardous waste in thousands of old landfills, dumps, and junkyards across the United States, and that concern has been with us in one form or another ever since. In the summer of 1988, for example, hazardous waste made headlines again when medical waste that had been illegally dumped in the ocean washed onto beaches from Maine to Florida. Some of this refuse, including used syringes and vials of blood contaminated with the AIDS virus, was extremely hazardous.

Other industrialized countries have the same problems with toxic waste management that Americans do. How shall we deal with the bewildering array of hazardous waste that is produced by mining, industrial processes, incinerators, military activities, and thousands of small businesses in ever-increasing amounts? How do we clean up the hazardous materials that are already contaminating our world?

Types of Hazardous Waste

Any discarded solid, liquid, or gas that threatens human health or the environment is known as **hazardous** or **toxic waste**. Hazardous waste, which accounts for about one percent of the solid waste stream in the United States, includes chemicals that are dangerously reactive, corrosive, explosive, or toxic (Table 23–1).

More than 700,000 different chemicals are known to exist. How many are hazardous is unknown, because most have never been tested for toxicity, but without a doubt, hazardous substances number in the thousands. They include a variety of acids, dioxins, abandoned explosives, heavy metals, infectious waste, nerve gas, organic solvents, PCBs, pesticides, and radioactive substances. Many of these chemicals have already been discussed, particularly in Chapters 1, 11, 19, 21, and 22, which examine endocrine disrupters, radioactive waste, air pollution, water and soil pollution, and pesticides. Here we discuss dioxins, PCBs, and radioactive and toxic wastes produced during the Cold War.

Dioxins

Dioxins are a group of 75 similar chemical compounds that are formed as unwanted by-products during the com-

Table 23–1
Examples of Hazardous Waste

Hazardous Material	Possible Sources
Acids	Ash from power plants and incinerators; petroleum products
CFCs	Coolant in air conditioners and refrigerators
Cyanides	Metal refining; fumigants in ships, railway cars, and warehouses
Dioxins	Emissions from incinerators and pulp and paper plants
Explosives	Old military installations
Heavy metals	Paints, pigments, batteries, ash from incinerators, sewage sludge with industrial waste, improper disposal in landfills
Arsenic	Industrial processes, pesticides, additives to glass, paints
Cadmium	Rechargeable batteries, incineration, paints, plastics (?)
Lead	Lead-acid storage batteries, stains and paints; TV picture tubes and electronics discarded in landfills
Mercury	Paints, household cleaners (disinfectants), industrial processes, medicines, seed fungicides
Infectious wastes	Hospitals, research labs
Nerve gas	Old military installations
Organic solvents	Industrial processes, household cleaners, leather, plastics, pet maintenance (soaps), adhesives, cosmetics
PCBs	Older appliances (built before 1980); electrical transformers and capacitors
Pesticides	Household products
Radioactive wastes	Nuclear power plants, nuclear medicine facilities, weapons factories

bustion of chlorine compounds. Some of the known sources of dioxins are medical waste and municipal waste incinerators, iron ore mills, copper smelters, coal combustion, pulp and paper plants that use chlorine for bleaching, and chemical accidents. Incineration of medical and municipal wastes accounts for 70 to 95 percent of known human emissions. Hospital-waste incinerators are probably the largest polluters because they are so numerous (there are more than 6000 of them in the United States), and they generally have unsophisticated pollution controls. Motor vehicles, barbecues, and cigarette smoke also emit dioxins, although their contribution is minor. Forest fires and volcanic eruptions are natural sources of dioxins.

Dioxins are emitted in smoke and then settle on plants, the soil, and bodies of water; from there they get into the food web. When dioxins are ingested by animals, including humans, they are stored in their fatty tissues. Humans are primarily exposed when they eat contaminated meat, dairy products, and fish. Because dioxins are so widely distributed in the environment, virtually everyone contains dioxins in their body fat.

Just how dangerous dioxins are to humans is controversial. Dioxins are known to cause several kinds of cancer in laboratory animals, but the data are conflicting on their carcinogenicity in humans. During the 1990s, the EPA conducted a three-year reassessment of dioxins in an attempt to develop a scientific consensus on the dangers of dioxins. More than 100 scientists worked on the report, which was published in 1994. The report confirmed that dioxins probably cause cancer in humans and raised other concerns about the effects of dioxins on the human reproductive, immune, and nervous systems. For example, dioxins may delay fetal development, cause endometriosis (the growth of uterine tissue in abnormal locations in the body) in women, and decrease sperm production in men. Dioxins have also been linked to an increased risk of heart disease. Because human milk contains dioxins, nursing infants, who feed almost exclusively on milk, are considered particularly at risk.

PCBs

Polychlorinated biphenyls (PCBs) are a group of 209 industrial chemicals composed of carbon, hydrogen, and chlorine. They are clear or light yellow, oily liquids or waxy solids. PCBs were manufactured in the United States between 1929 and 1979 for a wide variety of uses. They have been employed as cooling fluids in electrical transformers, electrical capacitors, vacuum pumps, and gas-transmission turbines. PCBs were also used in hydraulic fluids, fire retardants, adhesives, lubricants, pesticide extenders, inks, and other materials.

The first indication that PCBs were dangerous occurred in 1968, when hundreds of Japanese who ate rice bran oil that had accidentally been contaminated with PCBs experienced serious effects, including liver and kidney damage. A similar mass poisoning, also attributed to PCB-contaminated rice oil, occurred in Taiwan in 1979. Since then, toxicity tests conducted on animals indicate that PCBs harm the skin, eyes, reproductive capacity, and gastrointestinal system. PCBs are also endocrine disrupters because they interfere with hormones released by the thyroid gland. Children exposed to PCBs before birth have been demonstrated to have certain intellectual impairments, such as poor reading comprehension, memory problems, and difficulty paying attention. PCBs may also be carcinogenic; they have been shown to cause liver cancer in rats.

Some of the properties of PCBs that make them so useful in industry also make them dangerous to organisms and difficult to eliminate from the environment. PCBs are chemically stable and resist chemical and biological degradation. Like DDT and dioxins, PCBs accumulate in fatty tissues and are subject to biological magnification in food webs (see discussion of bioaccumulation and biological magnification in Chapter 22).

Prior to their banning by the EPA in the 1970s, large quantities of PCBs were dumped into landfills, sewers, and fields. Such improper disposal is one of the reasons PCBs are still a threat today. Also, when sealed electrical transformers and capacitors leak or catch fire, PCB contamination of the environment occurs.

One of the most effective ways to destroy PCBs is by high-temperature incineration. However, incineration is not practical for the removal of PCBs that have leached into the soil and water because, among other difficulties, the cost of incinerating large quantities of soil is prohibitively high. Another way to remove PCBs from soil and water is to extract them by using solvents. This method is undesirable for two reasons: first, the solvents themselves are hazardous chemicals, and second, extraction is also costly because wells must be dug to collect the chemically contaminated leachate in order to decontaminate it.

More recently, researchers have discovered several bacteria that can degrade PCBs at a fraction of the cost of incineration. One of the limitations of this method is that if PCB-eating bacteria are sprayed on the surface of the soil, they cannot decompose the PCBs that have already leached deep into the soil or groundwater systems. Although these microorganisms show promise as a help in removing PCBs from the environment, additional research will have to be conducted to make the biological degradation of PCBs practical. (Additional discussion about using bacteria to break down hazardous waste occurs later in this chapter.)

Radioactive and toxic wastes produced during the Cold War: Hanford Nuclear Reservation

America's nuclear weapons facilities may not be actively manufacturing nuclear weapons anymore, but they present us with a greater challenge: cleaning up and disposing of radioactive and toxic wastes that have accumu-

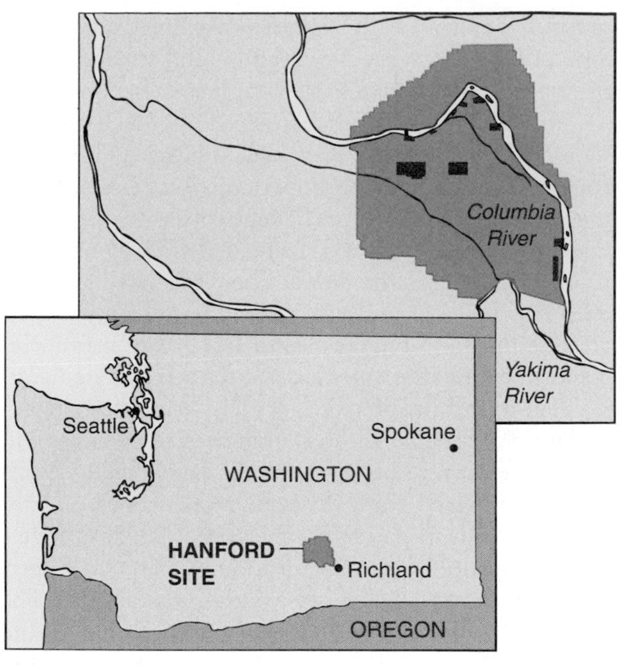

Figure 23–12

Location of Hanford Nuclear Reservation in Washington state. Hanford has nine nuclear reactors located along the Columbia River, along with tons of radioactive waste.

lated at numerous sites around the country since the 1940s. We focus our discussion on the Hanford Nuclear Reservation, a 1400-square-kilometer (560-square-mile) area on the Columbia River in south central Washington state (Figure 23–12). Hanford is the largest, most seriously contaminated site in the U.S. nuclear weapons infrastructure.

The sheer immensity of the cleanup task at Hanford is daunting. Tons of highly radioactive solid and liquid wastes were stored or dumped into trenches, pits, tanks, ponds, and underground cribs—a total of 1377 waste sites. (These methods of disposal were considered acceptable at the time.) For example, two concrete pools of water store more than 100,000 spent fuel rods. As they corrode, the rods release highly radioactive uranium, plutonium, cesium, and strontium into the water. Because these pools are leaking, soil and groundwater have been contaminated, and the Columbia River is in danger.

The Columbia River is also threatened by the sixty million gallons of toxic chemical and radioactive liquid wastes that are stored in 177 tanks. Liquids in some of these tanks are so reactive that they boiled for years from the heat of their own radioactivity or chemical activity. Most are now covered by semi-solid crusts that formed from chemical reactions within the mixtures. Some of the tanks are potentially explosive; for example, chemical reactions in 18 of the 177 tanks produce hydrogen and other toxic gases, which accumulate under the crust, increasing in pressure until they burst through it in gigantic "belches." Many of these tanks are also leaking their poisons into the ground.

Cleanup of this toxic waste nightmare is complicated by the fact that the *extent* of radioactive pollution and *kinds* of toxic mixtures present are not well known. As a result, scientists and engineers are conducting studies to assess the damage, prioritize the cleanup process, and determine how best to proceed for each type of contamination. Assuming that assessment is conducted properly, the extra years and millions of dollars that it will require will, in the end, save taxpayers billions of dollars and ensure that the most dangerous sites are attended to first. The cleanup, which is being monitored by the Military Production Network and the Military Toxics Project, will take decades to complete.

THE MILITARY AND THE ENVIRONMENT

The U.S. military has undergone a radical change in its view on the environment:

"We're in the business of protecting the nation, not the environment."—Comment in 1984 by a Virginia military base commander.

"I want the Department of Defense (DOD) to be the federal leader in agency environmental compliance and protection."—Memorandum in 1989 by Dick Cheney, Secretary of Defense from 1989 to 1993.

- The DOD has identified more than 10,000 actual or suspected hazardous waste sites on active military installations. Sixty percent of these sites are contaminated by fuel or solvents, much of which have leaked into the ground from underground storage tanks. Thirty percent of these sites contain explosive compounds, heavy metals, and other hazardous waste.
- More than 100 DOD facilities have been placed on the Superfund National Priorities List.
- The DOD estimates \$25 to \$30 billion will have to be spent to clean up contaminated sites on military installations.
- DOD's largest, most expensive cleanup project is the U.S. Army's Rocky Mountain Arsenal in Colorado, which was formerly a chemical weapons plant. Contaminated by pesticides, mustard gas, nerve agents, heavy metals, and incendiaries, it will remain a wildlife refuge after the cleanup is completed.

MANAGEMENT OF HAZARDOUS WASTE

Humans have the technology to manage toxic waste in an environmentally responsible way, but it is extremely expensive. Although great strides have been made in educating the public about the problems posed by hazardous waste, we have only begun to address most of the issues of hazardous waste disposal. No country currently has an effective hazardous waste management program, but several European countries have led the way by producing smaller amounts of hazardous waste and by using fewer hazardous substances.

Chemical Accidents

When a chemical accident occurs in the United States, whether at a factory or during the transportation of hazardous chemicals, the EPA's Emergency Response Notification System (ERNS) is notified. More than 50 percent of all chemical accidents reported to ERNS involve oil, gasoline, or other petroleum spills. The remaining accidents involve some 800 other hazardous chemicals, such as PCBs, ammonia, sulfuric acid, and chlorine. A study of the ERNS database from 1988 to 1992 indicated that more than 308,250 metric tons (340,000 tons) of toxic chemicals were released into the environment during reported accidents. The three states with the greatest number of toxic chemical accidents are California (4820 accidents reported during the 1988 to 1992 study period), Texas (4532 accidents), and Louisiana (2505 accidents).

Chemical safety programs have traditionally stressed accident mitigation and adding safety systems to existing procedures. More recently, however, industry and government agencies have stressed accident prevention through the **principle of inherent safety**, in which industrial processes are redesigned to involve less toxic materials so that dangerous accidents are prevented. The principle of inherent safety is an important aspect of source reduction.

Current Management Policies

Currently, two federal laws dictate how hazardous waste should be managed: (1) the Resource Conservation and Recovery Act, which is concerned with managing hazardous waste that is being produced now; and (2) the Superfund Act, which provides for the cleanup of abandoned and inactive hazardous waste sites.

The Resource Conservation and Recovery Act (RCRA) was passed in 1976 and amended in 1984. Among other things, RCRA instructs the EPA to identify which waste is hazardous and to provide guidelines and standards to states for hazardous waste management programs. RCRA bans hazardous waste from disposal on land unless it has been treated to meet EPA's standards of reduced toxicity. In 1992 the EPA initiated a major reform of RCRA to expedite cleanups and streamline the permit system to encourage hazardous waste recycling.

In 1980 the Comprehensive Environmental Response, Compensation, and Liability Act (CERCLA), commonly known as the Superfund Act, established a program to tackle the huge challenge of cleaning up abandoned and illegal toxic waste sites across the country. At many of these sites, hazardous chemicals have migrated deep into the soil and have polluted groundwater. The greatest threat to human health from toxic waste sites comes from drinking water laced with such contaminants.

The Superfund program got off to a slow start, in part because the EPA, which was authorized to administer it, experienced severe budget cuts. An amendment passed in 1986 provided the EPA with more money and instructions to work faster. By the end of 1995, Congress had appropriated $14.9 billion to implement the Superfund Act, and $1.4 billion were authorized for 1996. The Superfund Act is under review by Congress and should be amended and reauthorized sometime in the late 1990s. Meanwhile, as Congress struggles with the legislation, the authority to collect the tax on oil and chemical companies that helps to fund the Superfund program expired at the end of 1995, halting many Superfund projects.

Cleaning Up Existing Toxic Waste

The federal government estimates that the United States has more than 400,000 hazardous waste sites with leaking chemical storage tanks and drums (both above and below ground), pesticide dumps, and piles of mining waste. This figure does not include the hundreds or thousands of toxic waste sites at military bases and nuclear weapons facilities.

By the mid-1990s, almost 40,000 sites were in the CERCLA inventory, which means that they have been identified by the EPA as possibly qualifying for cleanup (Figure 23–13). These sites are not identified according to any particular criteria; some are dumps that local or state officials have known about for years, and others are identified by concerned citizens. The sites in the inventory are evaluated and ranked to identify those with extremely serious hazards. The ranking system uses data from preliminary assessments, site inspections, and expanded site inspections, which include contamination tests of soil and groundwater and sampling of toxic waste.

The sites that pose the greatest threat to public health and the environment are placed on the **Superfund National Priorities List**, which means that the federal government will assist in their cleanup. As of 1996, 1387 sites were on the National Priorities List, and the number continues to grow as other sites are evaluated. One-fifth of these sites are open dumps or sanitary landfills that operated before state and federal regulations

(a)

(b)

Figure 23–13
Cleaning up hazardous waste. (a) Toxic waste in deteriorating drums at a site near Washington, D.C. The metal drums in which much of the waste is stored have corroded and started to leak. Old toxic waste dumps are commonplace around the United States. (b) Cleanup of a hazardous waste site near Minneapolis, Minnesota. Removal and destruction of the waste are complicated by the fact that usually nobody knows what chemicals are present. (a, USDA/Natural Resources Conservation Service; b, Gary Milburn/Tom Stack and Associates)

were imposed and that received solid waste that would be considered hazardous today. The five states with the greatest number of sites on the priorities list are New Jersey (101 sites as of January 1997), Pennsylvania (95 sites), New York (74 sites), Michigan (73 sites), and California (67 sites).

One reason for the urgency about cleaning up the sites on the National Priorities List is their locations. Most were originally in rural areas on the outskirts of cities. With the growth of cities and their suburbs, however, many of the dumps are now surrounded by residential developments. About 27 million Americans, including over 4 million children, live within 6.4 kilometers (4 miles) of one or more Superfund sites.

Because the federal government cannot assume major responsibility for cleaning up every old dumping ground in the United States, cleanup costs for each site are shared by the current landowner, prior owners, anyone who has dumped waste on the site, and anyone who has transported waste to the site. For some sites, as many as several hundred different parties are considered liable for cleanup costs. Despite the urgency of cleaning up sites on the National Priorities List, it will take many years to complete the job.

As of January 1997, 137 hazardous waste sites had been cleaned up enough to be deleted from the National Priorities List, and 416 other sites had been partially corrected. The typical hazardous waste site on the National Priorities List takes 12 years and $30 million to be cleaned.

The biological treatment of hazardous contaminants

A variety of methods are employed to clean up soil contaminated with hazardous waste (recall the section on soil remediation in Chapter 21). Because most of these processes are prohibitively expensive, innovative approaches such as bioremediation and phytoremediation are being developed to deal with hazardous waste. **Bioremediation** is the use of bacteria and other microorganisms to break down hazardous waste, whereas **phytoremediation** is the use of plants to absorb and accumulate toxic materials from the soil. (The prefix *phyto* comes from the Greek word for "plant.")

To date, more than 1000 different species of bacteria and fungi are being used to clean up various forms of pollution. Bioremediation takes a little longer to work than traditional hazardous waste disposal methods, but it accomplishes the cleanup at a fraction of the cost. In bioremediation, the contaminated site is exposed to an army of microorganisms, which gobble up the poisons such as petroleum and other hydrocarbons and leave behind harmless substances such as carbon dioxide, water, and chlorides. Bioremediation encourages the natural processes by which bacteria consume organic molecules such as hydrocarbons. During bioremediation, conditions at the hazardous waste site are modified so that the desired bacteria will thrive in large enough numbers to be effective. Environmental engineers, for example, might pump air through the soil (to increase its oxygen level) and add a few soil nutrients such as phosphorus or nitrogen. They might also install a drainage system to pipe any contaminated water that leached through the soil back to the surface for another exposure to the bacteria.

In phytoremediation, plant species that are known to remove specific toxic materials from the soil are grown at a contaminated site. As the roots penetrate the soil, they selectively absorb the toxins, which accumulate in both root and shoot tissues. Later, the plants are harvested and disposed of in a hazardous waste landfill.

The field of phytoremediation is in its infancy, but already, specific plants have been identified that remove such hazardous materials as trinitrotoluene (TNT), radioactive strontium and uranium, selenium, lead, and other heavy metals (Table 23–2). Researchers are also working to genetically engineer certain plants to do an even better job of accumulating toxins.

Table 23–2
Selected Examples of Phytoremediation

Plant	Materials That Plant Absorbs and Accumulates	Location Under Study
Indian mustard	Strontium-90	Near Chernobyl nuclear power plant in Ukraine
	Cadmium	Abandoned lead mine in England
	Chromium	Contaminated soil from an old chromium smelter in New Jersey
Sunflower	Strontium-90, cesium-137	Former Department of Energy facility in Ashtabula, Ohio; near Chernobyl nuclear power plant
Cattail	Selenium	Oil refinery in Point Richmond, California
Parrot feather, stonewort	TNT	EPA project in Athens, Georgia

Phytoremediation is much cheaper than conventional methods to clean up hazardous waste sites, but it does have limitations. For example, plants cannot remove contaminants that are present in the soil deeper than their roots normally grow. There is also concern that insects and other animals might eat the plants, thereby introducing the toxins into the food web.

Managing the Toxic Waste We Are Producing Now

Many people think, incorrectly, that the establishment of the Superfund has eliminated the problem of toxic waste. The Superfund deals only with hazardous waste produced in the past; it does nothing to eliminate the large amount of toxic waste that continues to be produced today. There are three ways to manage hazardous waste: (1) source reduction, (2) conversion to less hazardous materials, and (3) long-term storage.

As with municipal solid waste, the most effective of the three is source reduction—that is, reducing the amount of hazardous materials used in industrial processes and substituting less hazardous or nonhazardous materials for hazardous ones. **Environmental chemistry**, also known as **green chemistry**, is an increasingly important subdiscipline of chemistry in which commercially important chemical processes are redesigned to significantly reduce environmental harm. For example, chlorinated solvents are widely used in electronics, dry cleaning, foam insulation, and industrial cleaning. It is sometimes possible to accomplish source reduction by substituting a less hazardous water-based solvent for a chlorinated solvent. Substantial source reduction of chlorinated solvents can also be realized by reducing solvent emissions. Most chlorinated solvent pollution gets into the environment by evaporation during industrial processes. Installing solvent-saving devices not only benefits the environment but provides economic gains, because smaller amounts of chlorinated solvents must be purchased. No matter how efficient source reduction becomes, however, it will never entirely eliminate hazardous waste.

The second best way to deal with hazardous waste is to reduce its toxicity. This can be accomplished by chemical, physical, or biological means, depending on the nature of the hazardous waste. One way to detoxify organic compounds is by high-temperature incineration. The high heat of combustion reduces these dangerous compounds, such as pesticides, PCBs, and organic solvents, into safe products such as water and carbon dioxide. The incineration ash is hazardous, however, and must be disposed of in a landfill designed specifically for hazardous materials. New methods to reduce the toxicity of hazardous waste are also being developed—such as the plasma torch, which produces such high temperatures (up to 10,000° C) that hazardous waste is almost completely converted to harmless gases (Figure 23–14). In comparison, conventional incinerators produce temperatures no higher than 2000° C.

Figure 23–14
The plasma torch produces a flame temperature up to 10,000 °C. At this temperature, toxic molecules such as PCBs and dioxins are broken down into nonhazardous or less hazardous gases such as hydrogen and nitrogen. The main drawback to the plasma torch is that is it expensive to operate. (Courtesy of Westinghouse Electric Corporation)

Hazardous waste that is produced in spite of source reduction and that is not completely detoxified must be placed in long-term storage. Landfills that are equipped to store hazardous substances have many special features, including several layers of compacted clay and plastic liners at the bottom of the landfill to prevent leaching of hazardous substances into surface water and groundwater. Liquid that percolates through a hazardous waste landfill is collected and treated to remove contaminants. The entire facility and nearby groundwater deposits are carefully monitored to make sure there is no leaking. Only solid chemicals (not liquids) that have been treated to detoxify them as much as possible are accepted at hazardous waste landfills. These chemicals are placed in sealed barrels before being stored in the hazardous waste landfill.

Very few facilities are certified to handle toxic waste (Figure 23–15). In 1997 there were only 15 commercial hazardous waste landfills in the United States. As a result, most of our hazardous waste is still placed in sanitary landfills, burned in incinerators that lack the required pollution control devices, or discharged into sewers.

ENVIRONMENTAL JUSTICE

Since well publicized studies in 1987 and 1992, opposition has swelled against the environmental inequities faced by low-income communities in both rural and urban areas. Poor minority neighborhoods are more likely to have hazardous waste facilities, sanitary landfills, sewage treatment plants, and incinerators in their neighborhoods.

Because people in low-income communities frequently lack access to sufficient health care, they may not be treated adequately for exposure to environmental haz-

Figure 23–15

A hazardous waste landfill near Detroit, Michigan, is designed to accommodate hazardous waste. The bottom of the landfill has several feet of compacted clay covered by three plastic liners (shown). Barrels of hazardous waste are then placed on the liners and covered with soil. (Dennis Barnes)

ards. The high incidence of asthma in many minority communities is an example of a health condition that may be caused or exacerbated by exposure to environmental pollutants. In addition, low-income communities may not receive equal benefits from federal cleanup programs.

Grassroots and community action efforts within individual communities brought the cause for environmental justice to the attention of public officials. **Environmental justice** means that every citizen, regardless of age, race, gender, social class, or other factor, is entitled to adequate protection from environmental hazards. The environmental justice movement is calling for special efforts to clean up hazardous sites in low-income neighborhoods, from inner-city streets to Indian reservations.

In 1994, President Clinton signed an executive order requiring all federal agencies to develop strategies and policies to ensure that their programs do not discriminate against poor and minority communities when decisions are made about where such facilities are to be located. The National Environmental Justice Advisory Council, also established in 1994, provides grants to help low-income communities around the country identify and address local environmental problems.

FUTURE MANAGEMENT OF SOLID AND HAZARDOUS WASTE

The most effective way to deal with solid waste is with a combination of techniques. In **integrated waste management**, a variety of options that minimize waste, including the 3 R's of waste prevention (*reduce*, *reuse*, and *recycle*), are incorporated into an overall waste management plan (Figure 23–16). Even on a large scale, recycling and source reduction will not entirely eliminate the need for disposal facilities such as incinerators and landfills. They will, however, substantially reduce the amount of solid waste requiring disposal in incinerators and landfills.

Changing Attitudes

In the United States we have become accustomed to the convenience of a throwaway society. The products we purchase rarely, if ever, have disposal expenses built into their prices, so we are not aware of their true cost to our communities. And many municipalities charge all citizens equally for trash collection, regardless of whether they generate a lot or a little solid waste.

If all other factors are equal, most Americans do not select a particular product because it produces less waste than another. Most of us like the neatness, squeezability, and unbreakability of plastic ketchup bottles, and we prefer the convenience of disposable diapers over cloth diapers. (See Focus On: The Environmental Costs of Diapers for an environmental comparison of cloth and disposable diapers.) We value attractive packaging. In contrast, Europeans are generally more willing than Americans to accept new products or changes in packaging that produce less solid waste. Europeans' recycling technologies are also much more advanced.

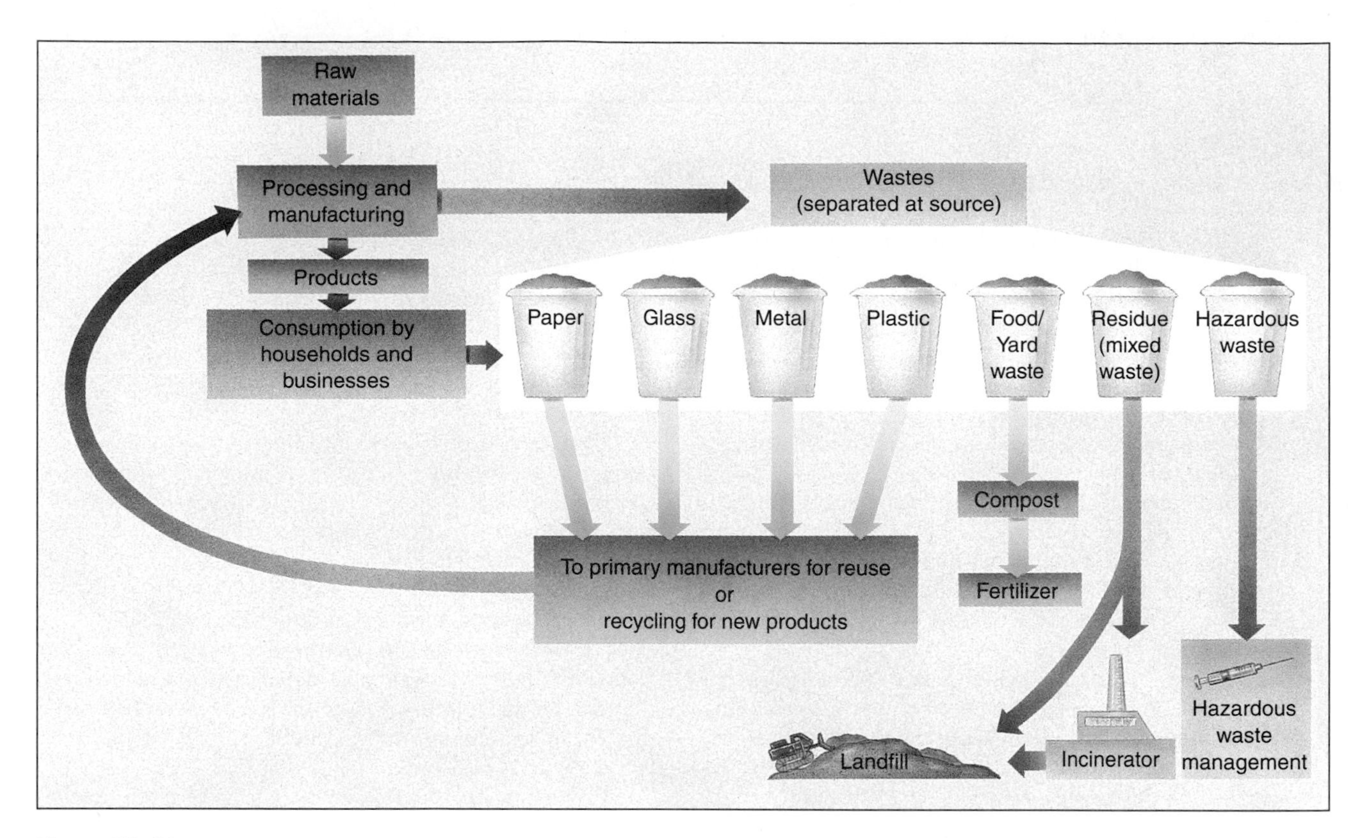

Figure 23–16

Integrated waste management makes use of source reduction, reuse, recycling, and composting, in addition to incineration and disposal in landfills.

Locating Future Waste-Handling Facilities

Even if we reduce municipal solid waste to near zero through source reduction, reuse, and recycling, there will still be a need for sanitary landfills, if only to contain the residues produced during recycling and incineration. In addition, facilities that handle hazardous waste will always be needed, even though source reduction decreases the amounts of such materials being produced.

The selection of a site does not have to be a political bombshell if communities are fully informed and are allowed to participate throughout the selection process. In the past, communities typically found out that they were being considered for such a site only when the waste disposal company applied for a building permit in their area. At such a late date, public opposition is usually strong and successful.

More states and communities should follow the example of the state of Washington. During the late 1980s the need for a hazardous waste disposal facility in Washington became apparent. From the start of its search for a site, the private waste disposal company ECOS let everyone know what it was planning. First the entire state area was evaluated, and environmentally sensitive areas (such as wetlands), areas with historical or archaeological importance, and inaccessible areas were eliminated from consideration. ECOS then requested community participation through a series of statewide press releases. Meetings were organized in nine of the communities that expressed interest in hosting such a facility. Citizen advisory committees then formed to negotiate for strict environmental precautions and economic benefits for the community. Two towns voted in favor of locating such a site in their vicinity, and in 1989 the community of Lind was selected. Only then did the waste disposal company apply for a permit.

FOCUS ON

The Environmental Costs of Diapers

Should babies wear disposable diapers or cloth diapers? New parents must choose whether to use about 6000 disposables or 50 cloth diapers between a child's birth and his or her toilet training. Until the late 1980s, convenience and cost were the only issues parents considered in choosing diapers: disposables are more convenient, and cloth diapers are less expensive. More recently, the environmental effects of diapers have become a consideration for many parents.

Approximately 16 billion disposable diapers are produced in the United States each year, making up more than 85 percent of the diaper market. Disposables ultimately end up in sanitary landfills, where they account for about 1.2 percent of the country's total municipal solid waste. In the early 1990s, as the nation's solid waste problem became more acute, parents were urged by certain environmental groups to convert to cloth diapers. Several states with particularly acute sanitary landfill shortages contemplated an outright ban on disposables in favor of cloth diapers, which contribute only about one-hundredth as much solid waste as disposables do.

It turns out that the diaper issue is not as clear-cut as originally thought, however. Analysis of the entire life cycles of both disposable and cloth diapers reveals that both types have environmental costs. Cloth diapers begin their lives as cotton growing in a field, and during this stage the cotton requires pesticides, which contribute to air, soil, and water pollution. Cotton plants also have the potential to erode the soil, and some cotton fields are irrigated, which can contribute to groundwater depletion and soil salinization. Energy is required to manufacture cotton diapers and to transport them to stores or homes, and the trucks and cars that transport them produce air pollution. After being used, cloth diapers are washed and reused. Washing cotton diapers requires lots of water, and detergents, bleaches, and babies' wastes produce water pollution. Energy is required to run the washing machine and dryer.

Disposable diapers begin their lives as trees. The harvesting of the trees has potential to cause soil erosion and contribute to deforestation. The production of paper from trees requires many chemicals that can pollute both air and water. Disposable diapers also contain plastic, which is made from petroleum that must be extracted from the ground and refined, producing both air and water pollution. As with cloth diapers, energy is required to manufacture and transport disposable diapers, and the vehicles that do the transporting pollute the air. Disposable diapers usually end up in sanitary landfills, which are filling up rapidly in some states.

Comparing the environmental effects of cloth diapers and disposable diapers is a bit like comparing apples and oranges, but it is clear that both types have environmental costs. Neither cloth nor disposable diapers have a clear-cut environmental advantage. Since neither diaper is environmentally superior, perhaps the best suggestion to parents is: if you live in an area where landfill space is becoming a problem (such as the Northeast and Midwest), try using cloth diapers; if you live in an area where water is in short supply (such as in the Southwest), try using disposable diapers.

SUMMARY

1. One of the most urgent problems of industrialized nations is the disposal of solid and hazardous wastes, which increase in quantity each year. Traditionally, solid waste has been disposed of in open dumps and, more recently, in sanitary landfills.

2. Sanitary landfills are less likely to harbor disease-carrying vermin than open dumps. However, like dumps, most landfills have the potential to contaminate soil, surface water, and groundwater. Most people oppose the establishment of sanitary landfills near their homes.

3. Incineration reduces solid waste to a fraction of its original volume, and the heat produced during incineration can be used to warm buildings or generate electricity. One of the drawbacks of incineration is the great expense of installing pollution control devices on the incinerators. These controls make the gaseous emissions from incinerators safe (as far as we know) but make the ash that remains behind more toxic. Another problem associated with incineration is finding appropriate sites, as most people oppose putting incinerators near residential areas.

4. Composting is increasingly being used to reduce the amount of organic waste, particularly yard waste, in the solid waste stream.

5. The three goals of waste prevention are to reduce the amount of waste as much as possible, reuse products as much as possible, and recycle materials as much as possible. Reducing the amount of waste includes purchasing fewer products and purchasing products that have less packaging and that last longer or are repairable.

6. One example of reuse is refillable glass beverage bottles. Reuse of refillable bottles is not as widespread in the United States as it was a few years ago. Other countries reuse containers to a greater extent than the United States in order to conserve resources as well as reduce solid waste.

7. Recycling involves collecting and reprocessing materials into new products. Many communities are recycling paper, glass, metals, and plastic.

8. Source reduction, which has great potential to reduce the volume of solid waste, can be accomplished in a variety of ways; these include substituting raw materials that introduce less hazardous waste or solid waste during the manufacturing process and reusing and recycling hazardous and solid wastes at the plants where they are generated. Consumers can practice source reduction by decreasing their consumption of products and by purchasing durable products that are designed to generate less solid waste.

9. Hazardous wastes are solids, liquids, and gases that pose a real or potential threat to the environment or to human health. In the past, hazardous waste was dumped, along with other solid waste, in open dumps and sanitary landfills. As a result, we are faced with the daunting prospect of cleaning up old toxic waste dumps and the soil and water they have polluted.

10. Hazardous waste being produced today is supposed to be disposed of (1) in special landfills designed to minimize the risk of environmental contamination or (2) by high-temperature incineration. Unfortunately, very few sites are approved for the disposal of hazardous waste. Source reduction is the best way to reduce hazardous waste.

11. There is no single solution to the problem of solid and hazardous wastes. The best approach is to use integrated waste management, which is a combination of techniques including source reduction, reuse, recycling, and composting, to reduce the amount of waste to be discarded. Combustible materials that remain can then be incinerated to further reduce the volume of solid waste. Incinerators must be equipped with pollution control devices to ensure that the pollutants are not simply being transferred from one medium to another. The remaining solid waste, a small fraction of the original solid waste, should be disposed of in an environmentally safe sanitary landfill.

THINKING ABOUT THE ENVIRONMENT

1. List what you think are the best ways to treat each of the following types of solid waste, and explain the benefits of the processes you recommend: paper, glass, metals, food waste, yard waste.

2. Keep an inventory of the plastic materials you discard during one week. Include all forms of packaging (for instance, meat trays and milk cartons) as well as disposable items such as Styrofoam cups.

3. How do industries such as Goodwill, which accepts donations of clothing, appliances, and furniture for resale, affect the volume of solid waste?

4. It could be argued that a business that collects and sells its waste paper is not really recycling unless it also buys products made from recycled paper. Explain.

5. Why is creating a demand for recycled materials sometimes referred to as "closing the loop"?

6. Suppose hazardous chemicals were suspected to be leaking from an old dump near your home. Outline the steps you would take to (1) have the site evaluated to determine if there is a danger and (2) mobilize the local community to get the site cleaned up.

7. Why must a sanitary landfill always be included in any integrated waste management plan?

8. The Organization for African Unity has vigorously opposed the export of hazardous waste from industrialized countries to developing nations. They call this practice "toxic terrorism." Explain.

* 9. In the mid-1990s, new steel products contained, on average, 58 percent recycled steel. If a total of 98.1 million tons of steel was produced in 1994, how many million tons of that amount were scrap (recycled)?

RESEARCH PROJECTS

Investigate how Walt Disney World in Orlando, Florida recovers food waste from its restaurants. How much food waste is collected daily? What is it used for?

Obtain a map of Superfund sites in your state. Document the number of people living within a short distance (such as 4 miles) of the sites and assess the progress of cleaning up each site.

Obtain copies of "Lessons from Superfund," by Thomas P. Grumbly (*Environment* Vol. 37, No. 2, March 1995) and "In Praise of Superfund," by Charles de Saillan (*Environment* Vol. 35, No. 8, October 1993). These two short articles reach very different conclusions about the effectiveness of the Superfund Program. Write a 2- to 4-page paper that evaluates each view.

On-line materials relating to this chapter are on the World Wide Web at **http://www.saunderscollege.com/lifesci/environment2** (select Chapter 23 from the Table of Contents).

SUGGESTED READING

Adler, T. "Botanical Cleanup Crews," *Science News* Vol. 150, July 20, 1996. How plants can help clean up contaminated soil, from sunflowers that remove radioactive materials to poplars that remove a once-commonly-used dry cleaning fluid.

Fishlock, D. "The Dirtiest Place on Earth," *New Scientist*, February 19, 1994. Cleaning up Hanford Nuclear Reservation is a major environmental challenge.

Grove, N. "Recycling," *National Geographic* Vol. 186, No. 1, July 1994. A well researched article on the current trend to recycle materials from paint to cans and bottles.

Hearne, S. A. "Tracking Toxics: Chemical Use and the Public's 'Right-to-Know'," *Environment* Vol. 38, No. 6, July/August 1996. An in-depth discussion of the Toxics Release Inventory, which is part of the 1986 Emergency Planning and Community Right-to-Know Act.

Hoffman, A. J. "An Uneasy Rebirth at Love Canal," *Environment* Vol. 37, No. 2, March 1995. A balanced history of Love Canal's contamination and cleanup.

Rathje, W. L. "Once and Future Landfills," *National Geographic*, May 1991. All you ever wanted to know about sanitary landfills.

Rogan, W. J. "Environmental Poisonings of Children—Lessons from the Past," *Environmental Health Perspectives* Vol. 103, Supplement 6, September 1995. This article examines why children are more vulnerable to environmental chemicals than are adults.

Ryan, M. "Los Angeles 21, New York, 5. . . .," *World-Watch* Vol. 6, No. 2, March/April 1993. Together, the cities of Los Angeles and New York generate 8 percent of the nation's municipal solid waste. Los Angeles is doing much more than New York to reduce its share.

Snyder, J. D. "Off-the-Shelf Bugs Hungrily Gobble our Nastiest Pollutants," *Smithsonian*, April 1993. Microorganisms can detoxify a variety of hazardous waste.

Zorpette, G. "Hanford's Nuclear Wasteland," *Scientific American*, May 1996. Considers both the extent of contamination at Hanford Nuclear Reservation and the shortcomings of the agreement among the DOE, EPA, and state of Washington to clean it up.

► Walkers hike near Bossons Glacier in France. *(Jess Stock/Tony Stone Images)*

PART 7

Tomorrow's World

The Green University

Dr. Irwin Price is Executive Dean of the George Washington University Virginia Campus in the Northern Virginia Technology Corridor. He develops and maintains close contact with industry and government leaders, and he initiated the University's effort to become the first Green University in the country. Dr. Price worked as an engineer for several years before acquiring an M.B.A. and eventually his Ph.D. in Economics. He taught at Boston University for more than 15 years and was Dean of Metropolitan College at B.U. before coming to George Washington in 1990. ◄

IRWIN PRICE

How did the Green University Initiative begin at George Washington University?

In 1994 I formed a group of industry, government, and university people who wanted to work collectively on developing environmental technologies and practices. At one of our meetings, a representative from the U.S. Environmental Protection Agency suggested that GW might consider the idea of becoming the first green university. That weekend, two of my neighbors—who work for EPA—and I brainstormed the idea. I conceptualized it and wrote a summary that outlined the concept of a "Green University." Stephen Trachtenberg, President of GW, thought it was a great idea, and we started the process of building widespread support within the university for this effort.

How did this idea crystallize into a mission with objectives?

President Trachtenberg set up the Task Force, out of which came six subcommittees: Academic Programs, Research, Infrastructure and Facilities, Environmental Health, International Issues, and Outreach. An organizational structure was established that identified the Institute for the Environment as the coordination body that tries to ensure that we do not duplicate our activities, and that we constantly see how one activity affects the others. We have students, faculty, staff, and administrators in each of our committees, which are usually co-chaired by a faculty member or staff person and a student.

One of the more visible outcomes was the establishment of new courses where the students combine their environmental studies with actual work in the community. The courses have been very successful, and increasingly, administrative units and academic departments are looking for these students to come and work with them.

What are some of the major successes?

We have had many successes, but in my view, the biggest is conceptual. We have agreement across all areas of the university that we want to continue to make GW better environmentally. So in our coordinating groups, we have active members from facilities, students, and the faculty. No one group dominates, and each takes pride in the accomplishments of its sector. We have participation from various levels, from the VP level to the people who accomplish specific tasks. It has truly become a grass-roots effort.

For instance, we have made paper reduction a high priority because we were amazed at how much paper a university like ours consumed. We looked for units, both administrative and academic, to model how we could use less paper. Each unit has a volunteer who is responsible for continuing the effort, and to see how we can achieve even greater reductions. The units get to keep the funds not expended on paper, so there is an incentive, and something to show other units so they will follow suit.

One of our other major goals was for GW to model for other universities, to define what it means to green itself. And we are gratified at how interested other schools and colleges are, and how we have been able to work with them. A university has a unique opportunity to impact the future—because not only is it a large institution—but when it changes its culture regarding the environment, it influences countless students who will become the leaders of the next generation.

What is the nature of your arrangement with the Environmental Protection Agency? How does each organization benefit?

On December 12, 1994, Carol Browner, Administrator of the EPA, and Stephen Joel Trachtenberg, President of GW, signed an agreement that sets up GW as a model university. In the preamble of the agreement, it states that together we can enhance leadership and stewardship in environmental management and sustainable development. That is the mutual goal. We are indeed fortunate that EPA is in the same town as us. We have let our campus be its laboratory.

For instance, as we wanted to develop a new recycling program, EPA assigned its expert in recycling to work directly with us. When we design a new university building, we involve EPA experts who have developed lists of equipment providers to make the building as environmentally sound as possible. In our outreach to the local community we focus on issues like environmental justice that EPA has identified as a high priority. Thus, both organizations benefit by learning from each other, and by reinforcing the high priorities each has established.

> **We have participation from various levels, from the VP level to the people who accomplish specific tasks. It has truly become a grass-roots effort.**
>
> *Irwin Price*

How has the Green University Initiative blended with academic programs and research?

As an institution of higher education, GW creates a learning environment that prepares students to become sensitive to environmental issues and sustainability. Environmental issues cannot be addressed solely with traditional academic disciplines, So we are committed to developing and promoting multidisciplinary courses and programs. One is called "Sustainability/Green University Practicum," available to all students, and taught by faculty from over 10 disciplines.

Another important component of the Green University framework is "learning by doing," which encourages faculty to plan more course projects involving students. Examples include: a business school student's work on designing and maintaining the Green University Web site; a pilot project to test lighting sensors as a way of conserving energy; an undergraduate engineering class trained by the EPA Green Lights consulting group; and an internship program that places students at federal agencies and private and non-governmental environmental institutions.

How do you assess your progress to date?

It actually took us a long time to: (1) decide what was important to measure, and (2) to figure out how to measure it. And then we had to do benchmarking. It has not been as easy as we had hoped, in part, because we are not in a static environment, but in a very dynamic one. When we are examining energy consumption, new facilities are coming on line, equipment is being replaced, and so on. When we are reviewing our academic programs, new courses and programs are always being created. But by the measures we have established, we have accomplished a great deal already.

What are your future objectives?

We are now trying to formalize our efforts to expand our outreach and network with other academic institutions. The EPA/GW partnership should reach out and create a mechanism where we can all learn from each other. Some schools have made great strides in recycling, others have developed great academic programs in environmental sciences, others have built wonderful outreach programs with their communities. We need to join these efforts so that each school has working models to emulate, and doesn't have to invent the process on its own. We are creating university-wide mechanisms in pursuing common environmental goals. And we are eager to share our experiences, but we want to learn from others. Our next vision is to create a web of universities, world-wide, with this common interest in making the global environment better.

▶ Web site: **http://www.gwu.edu/~greenu**

The choices we make today will determine whether future generations will inherit a world of prosperity or decline. (© 1993 Lawrence Migdale)

Tomorrow's World

You have been presented with a broad overview of the environment, including both the principles necessary to understand how ecosystems operate and a detailed review of the individual issues that are having an important influence on today's world. Now we wish to speak directly to you, not only as students but also as citizens. Our world is in very serious trouble today, presenting us with an enormous challenge. What follows represents where we stand, our opinions on what must be done to meet today's critical environmental challenge. There are reasonable people who hold other views, but we believe that we would not be doing our job as responsible teachers if we did not lay out our views for you. Read, think, discuss, and come to your own conclusions. And then, if our world is to have any future—act. ◄

THE PROBLEMS WE FACE

We who live in the United States at the end of the 20th century enjoy an abundance of what the world has to offer; in fact, we are among the wealthiest people who have ever existed on the face of the Earth, with the highest standard of living (shared with a few other rich countries). Since we control about 25 percent of the world's economy, it is obvious that we depend on many other nations for our prosperity; but in our actions, we often seem to miss this relationship and to underestimate our effects on the environment that supports us. It is as if it would all be there forever, and our actions did not matter. In fact, nothing could be farther from the truth.

When the century began, neither human numbers nor technology had the power to radically alter planetary systems. As the century closes, not only do vastly increased human numbers and their activities have that power, but major unintended changes are occurring in the atmosphere, in soils, in waters, among plants and animals, and in the relationships among all of these. The rate of change is outstripping the ability of scientific disciplines and our current capabilities to assess and advise. It is frustrating the attempts of political and economic institutions, which evolved in a different, more fragmented world, to adapt and cope. [Bruntland report, World Commission on Environment and Development, 1987, p. 22, Oxford University Press]

During the second half of the 20th century, it has become obvious that we are not managing the world's productive systems in such a way that they will continue to sustain us. For example, since the end of World War II, we have lost to erosion and general mismanagement about one-fourth of the world's topsoil, changed the characteristics of the atmosphere profoundly, and cut about a third of the forests that existed then without replacing them. We are feeding more than twice as many people with about 80 percent of the agricultural lands that were under cultivation in 1950, having lost the remainder to erosion, salinization associated with overirrigation, urban sprawl, and other causes. Global warming, in the opinion of the great majority of scientists who have attempted to analyze the trends in this area, is underway, and the stratospheric ozone layer that protects us from cancer-causing ultraviolet radiation has been seriously depleted. As the productivity of the Earth declines further, our general prospects for healthy, productive lives will decrease accordingly. Taken together, these trends constitute a clear call to action.

In this textbook we have explored these problems individually. Now we shall examine where our world stands today, and discuss some ways in which we might appropriately address the future.

ISSUE 1: POPULATION PRESSURES

In Chapters 8 and 9, we reviewed the growth of the world population from 2.5 billion in 1950 to nearly 5.8 billion in 1996; it is continuing to increase at a rate of nearly 90 million additional people each year. While this astonishingly rapid growth has been taking place, the proportion of people living in the industrialized countries of the world has dropped from an estimated 33 percent, or one out of every three people in the world, in 1950 to an estimated 20 percent, or one out of five, by the end of the 1990s. This proportion is projected to fall to about one sixth by the year 2020. In other words, the proportion of the world's people living in industrialized countries will have dropped in a period of 70 years—the space of an average human lifetime—from one out of three to one out of six.

Achieving Stability

The rate of world population growth peaked at about 2 percent in 1965 and fell to about 1.5 percent by the late 1990s; and the absolute number of people added has also been decreasing year by year from the mid-1990s onward. The World Bank has recently calculated that world population could stabilize at a level of approximately 10.4 billion people by the end of the next century; United Nations projections for a stable population range from 7.8 to 12.5 billion people. In order to be able to restrict the global population even to one of these enormous totals, however, there needs to be sustained worldwide attention to family planning. Our population will not automatically reach a level of 7.8 billion or 12.5 billion people and then stop growing. If we continue to pay consistent attention to this problem and devote the resources that are necessary to bring about success, we can make it happen. If we do not, we simply will not achieve population stability.

In addition, it is important to note that the difference between 7.8 and 12.5 billion—the range of the estimates—is nearly equal to the *entire world population* in the late 1990s, a population that, as we have seen in this text, is already high enough to be consuming the world's resources at an unsustainable rate! This realization reinforces the need for sustained attention by all governments to the need for family planning throughout the world; it is important both in industrialized countries, which consume most of the world's resources even though they constitute a rapidly shrinking minority of the world's population, and in developing countries.

Urbanization

At the beginning of the Industrial Revolution, approximately in 1800, only 3 percent of the world's people were living in cities; fully 97 percent were rural, living on farms or in small towns. In the two centuries since then, pop-

(a)

(b) (Peter Ginter/Material World)

ulation distribution has changed radically—toward the cities. More people will live in Mexico City by the end of the 1990s than were living in all the cities of the world combined 200 years earlier. This is a staggering difference in the way people live. About 43 percent of the world's population now lives in cities, with a very high proportion of the people who are added to the population also being city dwellers. For example, although the population of Brazil has grown rapidly to its present 160.5 million people, the population of rural areas in Brazil is actually falling. In industrialized countries like the United States, where approximately three-quarters of the people live in cities, and in other countries, it is relatively difficult for urban dwellers to appreciate and take steps to preserve the biological productivity on which our common future ultimately depends.

The Challenge Is Now

The next 25 years will be crucial to the pattern of growth of human populations, and thus to their impact on global ecology. During this period, more than two billion people are likely to be added to a world population that is already far beyond any level attained earlier. About 98 percent of the increase will occur in developing countries. From every point of view, this quarter century is likely to be the most stressful that the world has ever experienced.

(Louis Psihoyos/Matrix)

We need to take action now in order to affect the future positively. With the addition of more than two billion people to the 4.6 billion who now live in developing countries, we clearly have in place the recipe for a global disaster of unprecedented dimensions.

SPATIAL SCALE AND SUSTAINABLE SOLUTIONS

Identifying the spatial scale of an environmental problem—pinpointing its location and extent—is critical to finding effective remedies. As the population of Hyderabad, India, grew from 3.2 million in 1985 to 5.2 million in 1995, the commissioner of the city's Municipal Corporation considered the spatial scale of each urban problem: Localized problems are best solved at the local level, and widespread problems require more complex, comprehensive actions.

- Garbage collection can be managed effectively at the household and community level. In one pilot neighborhood, the city paid someone a small amount per household to collect garbage and deposit it in central bins. The residents' association matched the payment. Total costs are much less than government collection, and 170 neighborhoods have since created residents' associations to control garbage removal.
- Traffic congestion is a broad activity and cannot be solved at the local level. Wasted fuel, pollution, accidents, and time delays transcend city boundaries. Higher level policymakers in Hyderabad are working with local officials to modify road networks and land-use patterns, increase the diversity of transportation choices, and increase the costs of owning and driving a motor vehicle.

ISSUE 2: RICH VERSUS POOR

In today's world, the distribution of resources is very unequal. Those who live in industrialized countries, a rapidly shrinking 20 percent of the global population, control about 85 percent of the world's finances, as measured by summing gross domestic products. The per-capital GNP in industrialized countries in 1996 was about $18,130; the per-capita GNP of the other 80 percent of the people in the world was about $1,090. In general, developing countries are using up their own resources, which otherwise might have been renewable, much more rapidly than are the citizens of industrialized countries, who often consume other people's resources. As a result, the disparity in wealth and access to resources between industrialized and developing nations only grows greater with time.

The difference between rich and poor is evident at every level of consumption. For example, the 80 percent of the world's people who live in developing countries utilize only about 32 percent of the world's commercial energy. In terms of pollution, the picture is similar. For example, the United States, with less than 5 percent of the world's population, produces almost 22 percent of the world's carbon dioxide emissions. As nations such as China continue to industrialize, the world totals are likely to climb rapidly unless effective international agreements are in place to control the emission of these gases. Of particular concern are the estimated 1.5 billion people who depend on firewood as their principal source of fuel, and consume the trees and shrubs in their vicinity at a continually increasing rate. About one-third of these people lack access to dependable supplies of fresh water, a condition that condemns mainly women and children to a lifetime of carrying wood and water to smoky, unhealthful dwellings, where their days are spent tending cooking fires, without any hope of improving their lot. For all the raw materials that contribute to living standards, such as

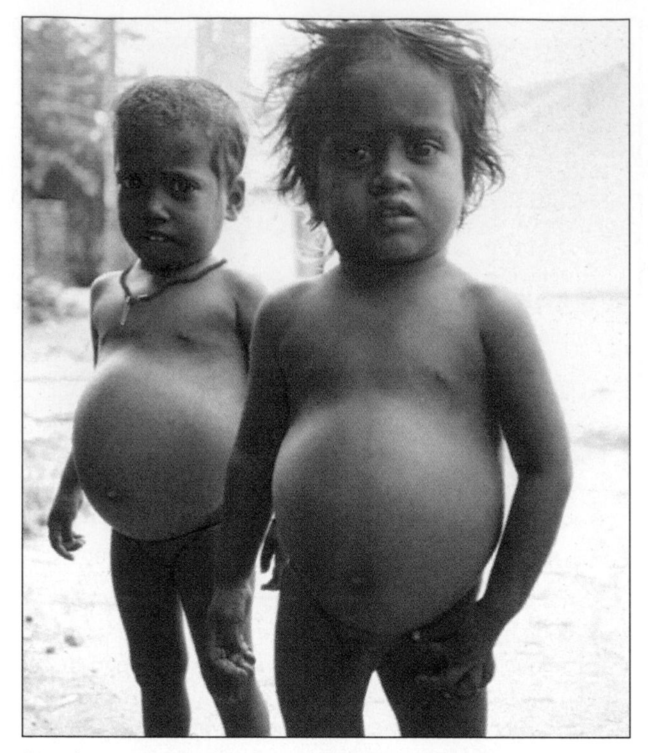

(United Nations Food and Agricultural Organization (Photo by P. Pillet))

iron, nickel, and aluminum, the 80 percent of the world's population who live in developing countries typically consume less than a quarter of these materials.

Poverty

The developing world experiences far greater poverty than most people living in industrialized countries can appreciate. According to the World Bank, about 1.5 billion of the citizens of developing countries live in absolute poverty, with incomes of less than $1 per day. These people are frequently unable to obtain the necessities of life—food, shelter, and clothing. About 840 million people do not get enough food to be healthy and lead productive lives. In addition, UNICEF has estimated that about 13 million infants per year, or more than 35,000 each day, starve to death or die of diseases related to starvation. Against such a background, it is incredible that many authorities assert that the world is functioning well and that we are fortunate to have avoided the widespread starvation that Paul Ehrlich[1] and other "ecological extremists" in the 1960s predicted would occur by now. How could "worldwide disaster" be defined if we are not experiencing one now?

One of the greatest barriers to equalizing the gap between industrialized and developing nations is the lack of trained professionals in the latter group. Only about 6 percent of the world's scientists and engineers live in these countries. Considering that a majority of those scientists and engineers live in only four nations (China, Brazil, Mexico, and India), you can see that for most developing countries, trained technical personnel are virtually lacking. Therefore, how are they to decide, based on their own knowledge, whether to join international agreements concerning the environment or how best to manage their own natural resources?

Developing countries are home to at least 80 percent of the world's biodiversity—the plants, animals, fungi, and microorganisms on which we all depend. How will developing countries use, manage, and conserve these precious resources? Addressing the inequality in scientific and technical personnel between industrialized and developing countries must become one of the most pressing items on the agenda of tomorrow's world.

(Les Stone/SYGMA)

[1]Paul Ehrlich is Bing Professor of Population Studies at Stanford University and an internationally recognized expert and author on population biology.

Immigration

Another important way in which we are linked to the developing world is revealed by the massive immigration of poor people from the tropics and subtropics into the industrialized nations of the temperate zone. The U.S. Immigration and Naturalization Service estimates that approximately 1.6 million illegal aliens were apprehended at the Mexican border alone in 1996, suggesting that perhaps twice that many may have entered successfully. The U.S. Census Bureau estimates that immigration now accounts for 30 percent of population growth in the United States and will account for *all* growth by the 2030s if present trends continue.

The same pattern of migrating to flee poverty can be seen worldwide. The United Nations has estimated that more than 10 million Africans have left their homes in search of food, often crossing national borders. The number of people seeking to enter the United States and other industrialized countries is likely to increase greatly. This pattern is the direct result of mounting populations and economic pressures in the developing countries. If we continue to ignore the driving forces that underlie immigration, we will never succeed in decreasing the numbers.

Political Instability

Finally, the most immediate danger of problems in the developing world concerns political instability. Industrialized nations can no longer solve the problems of these countries by direct intervention or, at the other extreme, by pretending that there is no problem. On the other hand, we have a direct interest in promoting stability throughout the world, because the global economy is interconnected in a very complex way. The United States, the richest nation on Earth and therefore the one with the most to lose from global instability, should logically invest more in international development than any other nation, but it invests the *least* per capita of industrialized nations.

Here ecology and politics meet head on. True stability in developing countries will involve the incorporation of poor people into the economies of their regions, as well as the management of natural resources so that they will continue to be productive in the long term. Political stability can be achieved in no other way, and it cannot be achieved without the active cooperation of industrialized nations.

ISSUE 3: OUR MANAGEMENT OF THE BIOSPHERE

The rapidly growing global human population is a dominant ecological force without precedent in the world's history. The bright lights of densely populated urban centers visible almost everywhere on Earth in nighttime satellite pictures (see Figure 1–1), the large-scale changes we are causing in the composition of the atmosphere, the very fact that there is no longer a square inch anywhere on Earth where chemical pollutants do not fall—all of these relationships signal our huge and growing impact on the productive capacity of the Earth.

(Donna DeCesare/Impact Visuals)

Productivity

Peter Vitousek and his colleagues at Stanford University calculated that, in the middle 1980s, the proportion of total net photosynthetic productivity on land that humans used directly, wasted, or diverted was about 40 percent of the total amount available! This is an alarming figure in view of the extremely rapid growth of the human population and the universal desire for improved living standards. No change in law or behavior can increase the amount of light coming from the sun. We can address the problem of consumption and resource management now, or be forced to address an even more desperate problem later.

Global Warming

What are some of the consequences of the pressure we are exerting on the global environment? You have encountered many of them in this text. One of the most widely discussed is the rapid enhancement of the greenhouse effect. Both industrial and developing countries contribute to major increases in carbon dioxide in the atmosphere, as well as to the increasing amounts of nitrous oxide, ozone, methane, and CFCs, which also trap radiant energy from the sun. Although the Earth's climate has been relatively stable during the present interglacial period (the past 10,000 years), human activities have begun to cause it to change. Surface temperature has in-

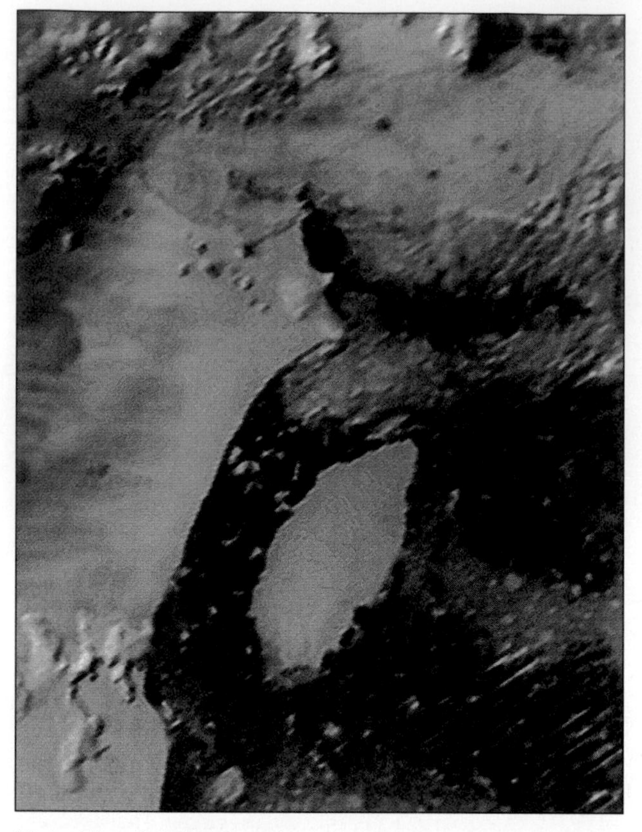

(National Snow & Ice Data Center)

creased by about 0.5°C over the past century and could increase by 1° to 3.5°C by 2100.

The consequences of these potential changes are very serious, because modern society has evolved and successfully adapted to the Earth's conditions as they are. Although news reports sometimes trivialize the effects of rising temperatures, presenting them as though we could cope with them by building dikes, turning up our air conditioners, changing our farming systems, and the like, these are simplistic views that do not take into account all the effects of global warming. Keeping in mind the fact that the change from the last ice age to the present was accompanied by an increase in global temperatures of 5°C puts the consequences of the present changes, the most rapid of the last 10,000 years, into clear perspective. A melting of the Antarctic and Greenland ice caps would cause a rise in sea level worldwide of about 79.2 meters (260 feet), which would inundate the places where a majority of the world's population now lives. In addition, climatic changes of that magnitude will significantly alter patterns of rainfall and temperature, which will have a major, unpredictable impact worldwide. In the face of these very serious problems, however, it is significant that the majority of energy experts think that significant reductions in greenhouse gas emissions are technically feasible. Such reductions must be addressed urgently.

Loss of Agricultural Land

Overuse of agricultural lands has caused their rapid loss worldwide. It is calculated that more than 15 percent of the world's vegetated surface has been degraded since World War II, its vegetation cover destroyed and its topsoil often lost as well; an additional 25 percent is at risk. Worldwide, we are growing our food and other agricultural products (such as cotton and biomass) on an area about the size of South America; since 1950, we have lost about a quarter of the topsoil on these lands, some 20 percent of which no longer are able to produce crops. Very little additional land—land that is not now cultivated—is suitable for agriculture, so that there is far less per person now, with 5.8 billion people in the world, than there was in 1950, when there was 25 percent more productive farmland to supply the needs of 3.3 billion fewer people. Obviously, improving agriculture is one of the highest priorities involved in achieving future global sustainability, although in a rich country like the United States, where food is very inexpensive, it is difficult to understand this relationship.

A Hungry World

Grain production per capita has largely kept pace with human population growth over the past 40 years (see Figure 18–2b, for example). In Europe and the United States, we tend to overproduce, and as a result, we largely ignore the fact that much of our agricultural productivity is taking place at high environmental cost. These include soil and water pollution, air pollution (from fossil fuel use), soil erosion and loss of soil fertility, and aquifer depletion. Our overproduction, in turn, tends to cause us to fail to acknowledge the inadequate productivity in much of the world and the serious need for additional food associated with rapidly growing populations in developing countries.

Many improvements in agriculture will be needed in order to feed the world's people adequately in the future. For example, no-till agriculture, in which the topsoil is protected, is one important element of the strategy; this practice is growing rapidly throughout the world. Precision farming, in which satellite data are used to determine the most appropriate levels of fertilizer application for individual small areas, provides an important key to the agriculture of the future. Genetic engineering should be used fully to produce the best possible crops for each situation; it is an important way to achieve the traditional goals of genetic modification of crops, which have been pursued for more than 10,000 years, more rapidly than would be possible otherwise. The need for sustainable agricultural productivity must be addressed vigorously throughout the world if we are to be able to feed people adequately and to allow for rising dietary standards, such as the inclusion of more meat and other high-quality protein in diets, particularly in developing countries.

ISSUE 4: CUTTING THE WORLD'S FORESTS

The world's forests are being cut, burned, or seriously disturbed at a frightening rate. Tropical lowland moist forests, or rain forests—biologically the richest areas on Earth—have already been reduced to half their original area. What remains in Asia, Africa, and Latin America collectively amounts to less than two-thirds of the area of the United States and is being destroyed so quickly that most will be gone within 30 years.

We know very little about replacing most tropical forests with productive agriculture and forestry. For many tropical soils, the combination of inefficient or short-term exploitation with disorganized logging and clearing (often by burning) results in irreversible destruction of their potential productivity. Methods of forest clearing that were suitable when population levels were lower and forests had time to recover from temporary disturbances simply do not work any longer: They convert a potentially sustainable resource (forests growing in infertile soil) into an unsustainable one. Clearing the woods and prairies of Eurasia and North America traditionally led to the establishment of productive farms, for the soil was rich; in contrast, clearing the forests of tropical Africa or Latin America often creates wastelands. The relative infertility of many tropical soils, their thin and easily disturbed surface layer of organic matter, and the high temperature and precipitation levels of tropical regions often combine to make the attainment of sustainable agriculture or forestry systems extremely difficult.

The world's forests are being lost for two principal reasons. First, they are being converted to cash. Like natural resources all over the world, they are being exploited and sold, and the resulting funds invested. Old-growth forests in Oregon, Washington, Alaska, Canada, and Siberia are being lost for the same reasons, with the cutting actually being subsidized by the central government in each case. Until concepts of natural productivity and the central role of sustainability are built more securely into economic calculations, susceptible natural resources will continue to be consumed, driven by short-term economics, and often lost permanently.

The second reason that the world's forests are being lost is the pressures created by rapid population growth and widespread poverty, often leading to unwise or unplanned use of resources. In many developing countries, the forests have traditionally served as a "safety valve" for the poor, who, by consuming small tracts of forest on a one-time basis and moving on, find a source of food, shelter, and clothing for themselves and their families. But now the numbers of people in developing countries are too great for their forests to support—and over the next 25 years, more than two billion people are likely to be added to the populations living there.

To give but one example of the impact of relentless population pressure on the forests, consider firewood. Roughly 1.5 billion people in developing countries depend on firewood as their major source of fuel. The demand greatly exceeds the supply. In India, for example,

(Gary Braasch/Tony Stone)

(Martin Hervey/NHPA)

where forests can sustain an annual harvest of only 39 million metric tons of wood, the annual demand is for 133 million metric tons. As the price of fuel wood rises sharply, its use is increasingly being restricted to relatively affluent people, with the poor in many regions losing the ability to cook their food.

Tropical forests are rapidly being cleared and destroyed, not only because of the needs of the people who live in or near them, but also because of the demands of the global economy. Many products—foods, such as beef, bananas, coffee, and tea; medicines; and hardwoods, for example—come to the industrialized world from the tropics. As timber is harvested or trees are cleared for other reasons, however, only a very limited amount of replanting is taking place. It is estimated that only one tree is planted for every ten cut in the Latin American tropics, and in Africa the proportion is much lower. The industrial-world consumption of tropical hardwoods is 15 times higher than it was in 1950. Few nations have developed forestry plans, and there is almost no international coordination of forestry policies.

ISSUE 5: LOSS OF BIODIVERSITY

One of the most serious of all global environmental problems in terms of its consequence for humans is the loss of biological diversity. We depend directly on the Earth's plants, animals, fungi, and microorganisms for food, medicine, shelter, and clothing, and their loss will progressively hamper our ability to construct a sustainable world for the future.

Although the great majority of species have not been catalogued, we can estimate the rate of extinction by considering what we know about extinctions of such well known groups as birds and mammals over the past three centuries. By doing so, and comparing these losses with the average life span for vertebrate species of about four million years (as measured in the fossil record), we can calculate that the rate of extinction over the past 300 years has reached 100 to 1000 times the background rate—up to 1000 species are being lost for each additional one that evolves. As tropical forests and other habitats are cut and otherwise destroyed or disturbed, we can expect this rate to climb to 10,000 times the background rate early in the next century. Such a rate would mean the loss of perhaps a fifth of all the species in the world—some two million species—over the first few decades of the 21st century, and the loss of perhaps six million species—two-thirds of the estimated world total—by the end of that century. Such a rate of extinction would be equivalent to that which occurred at the end of the Cretaceous Period some 66 million years ago, and would amount to a major structural change in the nature of life on Earth—entirely driven by the activities of a single species of the estimated 13 to 14 million that exist today.

Over the next 25 years, we can expect the rate of extinction to climb from dozens of species a day to hundreds of species a day, the great majority of them unknown to science. How big a loss is this? Unfortunately, we still have very limited knowledge about the world's biological diversity. The world has but one living library; people have read very few of its books and don't even have a complete catalogue of the volumes it contains—but the

(Burton Historical Collection, Detroit Public Library)

library is being burned unread. We have given scientific names to only about 1.75 million kinds of organisms. Although we know flowering plants, vertebrate animals, butterflies, and a few other groups reasonably well, we have only a small amount of information about most of the other living inhabitants of this planet, so we are ignorant of much of what we are losing.

The loss is driven by the operation, throughout the world, of an outmoded economic system that fails to place value on what cannot immediately be turned into cash. National economic planners have not taken into account the very real cost of irreversible loss of productivity, and as a result, the world continues its routine business, dealing with natural resources as if somehow they will be renewed. To act in this way is a tragic error for which our descendants should not forgive us.

Why the Loss is Important

Organisms are solely responsible for capturing a small fraction of the Sun's energy, and transforming it into chemical bonds, in which form it supports all life activities on Earth. Only about 300,000 species (250,000 plants; about 50,000 algae; and a few species of bacteria) have the ability to photosynthesize and thus to perform this critical transformation. In addition, individual organisms produce our food, and there are many thousands that might be used to feed a hungry world if we understood their properties better. For example, the gene pool of commercial corn (an annual) has been enriched by the discovery, in a small forest clearing in the mountains of Jalisco, Mexico, of a wild, perennial relative. Organisms are also the major sources of our medicines, with plants and other groups serving as natural biochemical factories. For example, oral contraceptives for many years were produced from a species of Mexican yam; muscle relaxants used in surgery worldwide come from an Amazonian vine; and a drug used to treat Hodgkin's disease comes from the rosy periwinkle, a native of Madagascar. Much of plant diversity is linked to the variety of chemical defenses they have evolved to deter herbivores, and many of these chemical compounds are of great value to us. In addition to their use as foods and medicines, many plants have extraordinary possibilities as sources of oils, waxes, fibers, and other commodities of interest to modern industrial societies.

We might also consider the use of organisms as sources of genes for genetic engineering, the process by which the genetic modification of organisms can be made more precise than has been possible before. Every species has tens of thousands of genes, and these genes, still poorly understood, will be the basis on which biology contributes to the formation of a sustainable world in the future. Both because of the direct use of individual species, and their potential contribution to biotechnology, the conservation of species, as stated by the International Convention on Biodiversity adopted following the 1992 Earth Summit in Brazil, is of fundamental importance.

In addition to these individual economic values, organisms in communities and ecosystems provide what are known as ecosystem services. Organisms protect topsoil, help to regulate local water cycles and regional climates, absorb pollution, and in general make the world function, even at times under severe stress. These ecosystem services throughout the world make a major economic contribution, and yet they tend to be taken for granted as we often carelessly rip apart the very fabric of the systems that support us. We know little about the functioning of individual species in ecosystems; yet such knowledge is the basis of global sustainability and the best hope for our future on Earth.

Finally, moral and ethical considerations call into question our thoughtless extermination of such a major proportion of the organisms that share our planet with us. The imminent loss of as many as 50,000 of the world's 250,000 plant species, for example, not only will substantially reduce the possible organisms that we might use for various purposes; it will make it much more difficult for us to repair and restore damaged ecosystems. In addition, it will simply greatly impoverish the Earth and make it a much less attractive and interesting place for our children and grandchildren than it is for us. To a religious person, this amounts to the desecration of Creation; to any thoughtful person, allowing a major proportion of the biodiversity on Earth to disappear forever without caring is one of the most ignorant things that we could conceivably do.

(Courtesy of Nordiska Genbanken, Alnarp, Sverige)

TRANSITION ECONOMIES

Many countries in Eastern Europe, formerly part of the Soviet Union, have promoted heavy industry since World War II. The resultant high levels of pollution have spawned widespread health problems and have exacerbated environmental deterioration.

- Diseases associated with poverty, such as diphtheria, tuberculosis, and hepatitis, are on the rise, demonstrating the decline in sanitary conditions, hygiene, and nutrition.
- Ambient lead levels are high, especially in mining towns. Children exposed to even low doses of lead can incur learning problems and brain damage.
- Life expectancies in urban areas of Poland and the Czech Republic are lower than in their rural areas, in contrast to usual demographic trends.

This region is making the slow transition from unchecked industrial pollution to more modern means of economic growth. Many municipalities are implementing plans to combat industrial pollution. Poland's Minister of Environmental Protection ordered the most polluting industries to reduce emissions and has expanded the air quality monitoring system. The town of Novokuznetsk, Russia, has formalized a partnership with the city of Pittsburgh, Pennsylvania, to exchange information about the links between pollution and health.

ISSUE 6: WORKING TOGETHER

One of the key steps in any rational approach to world problems is to promote *internationalism*—a widespread understanding that all of our human problems are interconnected. From the vantage point of a country such as the United States, attaining this objective is more difficult than it sounds; like most people everywhere, we tend to operate as if we were alone.

Narrow Perspectives

Americans are growing more isolated from the rest of the world just when their dependency is becoming critical. We worry a great deal about Japan and Korea as economic competitors, but how many Americans study Japanese or Korean language, history, or culture? It is ironic that the wealthiest nation on Earth would be turning its back on the rest of the world just as most nations are opening up. We have by far the most to lose; anything we can do to foster an international outlook will be of fundamental importance to our common future.

As an example of what might be done, consider the possibilities for improving relationships between North America and Latin America. As industrialized countries, the United States and Canada comprised 295 million people in 1996, their populations growing at 0.6 percent per year, with average per-capita GNP of about $25,000. At the same time, the developing countries of Latin America and the Caribbean comprised 486 million people, their populations growing at 1.9 percent per year, with a per-capita GNP of about $3300. By early in the 21st century, there will be twice as many people in Latin America as in the United States and Canada. What kinds of understandings and relationships shall we forge to expedite our common task of managing the lands and waters of the Western Hemisphere for our common benefit?

It is very much in the immediate and long-term interest of the United States and Canada to encourage the cooperation that is growing among the nations of Latin America as well as their efforts to help themselves. Positive political and economic actions should be taken on a cooperative basis, with development funds, education and other assistance, and low-cost loans regarded as a matter of enlightened self-interest, not charity. When the United States deals with its southern neighbors, the issues emphasized tend to be confrontational ones, including drug trafficking, illegal immigration, debt repayment, protective trade practices, and national policies toward repression of political conflict. In place of these issues, we should begin to move the agenda toward cooperative initiatives in science, technology, and the humanities; toward sustainable development and conservation; and toward mutual cultural understanding. Deep and mutually considerate exchanges of this sort would prepare us much better for the increasingly intense interactions that are certain to characterize our relationships in the next century.

The Status of Women

The status of women worldwide is a matter for the most serious consideration. Although substantial numbers of women have entered the work forces of industrialized countries, their opportunities are still much more limited than those available to men. Worldwide, no single factor will influence the rate of population growth more than assigning a higher status to women. Family planning is very much in the interest of individual women's health and welfare, but access to education and contraceptives is becoming increasingly difficult. As the populations of developing countries have increased dramatically, assistance for dealing with the problem has decreased. An appropriate response to these issues was developed at the U.N. International Conference on Population and Development, held in Cairo, Egypt, in September 1994, and should be adopted widely.

(Larry C. Price)

As Nafis Sadik points out (see Interview p. 154), women hold a paradoxical place in many societies. As part of their traditional duties as mothers and wives, they are expected to bear the whole responsibility for child care; at the same time, they are often expected to contribute significantly to the family income by direct labor. In many developing countries women have few rights and little legal ability to protect their property, their rights to their children, their income, or anything else. Improving the status of women is a crucial aspect of development. As Lester Brown[2] puts it, "Until female education is widespread—until women gain at least partial control over the resources that shape their economic lives—high fertility, poverty, and environmental degradation will persist in many regions."

Of even more fundamental importance than the limited rights of women is the simple fact that the human race will never be able to accomplish its best and deal most effectively with the planet that sustains us all, as

[2]Lester Brown is founder of the Worldwatch Institute, a research organization providing information on global environmental issues to concerned citizens and policy-makers.

(Marc Sherman)

long as we discriminate based on gender, race, or any other arbitrary characteristic. We need all of the abilities that each of us can bring to bear for the common benefit, and this relationship alone, not even considering simply social justice or morality, ought to be sufficient to make each one of us avoid discrimination in all of its forms.

The Poor

One out of four people in the world lives in abject poverty. From day to day these people have no idea where their next meal is coming from, whether their children will be properly clothed, or whether they will find adequate shelter. Rich nations tend to view poor ones as if they should somehow pull themselves together and do a better job of feeding their people, while the rich nations are willing to provide only a limited amount of help to them. Nor do people typically work together within developing countries. The more affluent classes, and those centered in the cities, tend to dominate the rural poor; in general, laws and customs coincide to ensure the continuation of such domination. The outcome of this negligence of the poor is that their ranks are swelling rapidly.

What is wrong with such attitudes? First of all, they are morally indefensible. Faced with the facts, most people would find unacceptable the deaths of 35,000 babies each day, most of which could have been avoided by access to adequate supplies of food or simple medical techniques and supplies. There is no justification for the bigoted view that these deaths prevent the world from becoming overpopulated; the world is already overpopulated, and for us to allow so many to starve, to go hungry, and to live in absolute poverty (and we do allow it, whether we choose to look at the problem squarely or not) is to threaten the future of the global ecosystem that sustains us all. Everyone must have a reasonable share of the Earth's productivity, in the face of our current numbers and their projections, or our civilization will simply come unraveled.

WHAT SHOULD BE DONE: AN AGENDA FOR TODAY

The six problems just discussed carry an urgent message to us all, that the explosive and unevenly distributed growth of an unprecedented human population is putting unsupportable strains on the global ecosystem. The long period of transition from a hunter-gatherer society, in which widely dispersed humans were one of millions of species of organisms, to a modern industrial society, in which humans consume, co-opt, or waste 40 percent of the total global productivity, has now reached a point where our traditional modes of operation seem increasingly unlikely to lead us to stability and prosperity. In the light of this grave threat

(Jose Azel/Aurora)

to our common future, we must weigh our options carefully as we seek ways to avoid poisoning the air we breathe and the water we drink, to avoid exterminating the organisms that we will need to build a stable life for future generations, and to collectively and sustainably use the biosphere. There are a number of critical areas where significant progress can be made over the next few years, and we shall now discuss some of the concerns that we consider to be top priorities for global attention.

1. **Population stability** must be attained throughout the world, and the industrialized countries must assist others in carrying out their plans in this respect. It is especially important for industrialized countries to attain stable population levels, since their people consume such a disproportionately large share of what the world is capable of producing. There is no hope for a peaceful world without overall population stability, and no hope for regional economic sustainability without regional population stability. International assistance for family planning should be increased, with the United States continuing to play a leading role in the overall effort, as urged by the participants in the U.N. International Conference on Population and Development in 1994.
2. **Women's rights** need to find fuller expression everywhere. Ensuring that all women have access to the full range of human opportunities would address environmental problems in two significant ways. First, it would greatly accelerate our progress toward global population stability. Second, because we will need all of the human talent available in order to build a sustainable society, we cannot afford to limit any individual's potential for contributing to this effort.
3. **New energy sources** must be sought. The greenhouse effect associated with increasing concentration of carbon dioxide and other gases in the atmosphere

(Courtesy of Keneteck Windpower, Inc.)

(Solar Electric Light Fund)

ought to stimulate research into alternative energy sources. Even if we choose to act as if the supplies of fossil fuels such as coal and oil are infinite, burning them in the quantities we do now is certainly not an environmentally sound practice. To make major changes in our sources of energy will require so high a level of readjustment that the effort must be begun well before it is to have a significant impact. Renewable energy sources such as solar/hydrogen power, and biomass production should be explored while there is still time to accomplish changes.

4. **A comprehensive energy plan** must be implemented for the world. At present, the four-fifths of the world's people who live in developing countries use about 32 percent of the world's commercial energy. Their numbers are growing rapidly, however, and even without any increase in their standard of living, these countries will use much larger quantities of energy in the future. Of special concern in this connection is China, a rapidly industrializing country of more than 1.2 billion people, with the world's second-largest economy and the largest deposits of coal anywhere. If we reach a global consensus on limiting carbon dioxide in the atmosphere, those of us who live in the industrial nations of the world will find ourselves insisting that developing countries not burn coal as they industrialize, or that they take steps to remove gases from coal smoke that are far more rigorous and expensive than any we have used in our own countries. Such a strategy seems unlikely to prevail—unless we all share in paying for it. The key is for those of us who live in the industrialized nations to realize that the implementation of a comprehensive energy plan for the developing world is a necessary ingredient for our own future security, as well as for global stability. This energy plan must involve reducing our own profligate use of energy, which certainly cannot be sustained—much less used as a model for the rest of the world.

5. **Regional cooperation** will be necessary to solve many pollution and conservation problems. Acid precipitation, for example, is a problem that must be solved regionally. A country, such as the United Kingdom or the United States, that "exports" acid precipitation to other countries by erecting tall smokestacks with inadequate scrubbers to remove sulfates saves money, but it causes other countries to incur substantial costs. When the problem is viewed regionally, there is an overall economic gain to ending the pollution. The Earth Summit that took place in Brazil in 1992 was a significant step in the process of considering common problems in an appropriate, governmental, collective context. Such efforts, which should be supported strongly, have begun to form a framework for the actions that will be necessary to achieve a secure future.

6. **Soil and water** must be better conserved. Plans for the sustainable use of the soil and water of all regions of the world must be developed, with provisions for sustainable forms of agriculture and forestry. Developing countries can achieve stability only if their best lands—those capable of sustainable productivity—are developed properly and if appropriate land-use schemes are implemented. Many experts agree that all of the agricultural and forestry needs of the poor people who live in the tropics could certainly be met by proper development of lands that have already been deforested. It is a profound human tragedy that we are not accomplishing this development. Instead, the rural poor are ignored, left to "mine" undisturbed forests on a one-time basis, and so convert potentially renewable resources into nonrenewable wasteland.

7. **Biodiversity** must be truly protected throughout the world. Our entire sustainable use of the world's resources depends on the wise management of biodiversity, because we obtain many different products from unique kinds of organisms, and we depend on the management of communities of organisms, regardless of how poorly we understand them, for the global preservation of soil, water, and air.
8. **Improving agriculture** is an important part of the global strategy that we must undertake in order to achieve sustainability. Retarding the loss of topsoil and the degradation of agricultural lands; conserving water; growing crops nearer where they are consumed to conserve energy; and reducing the use of chemicals through alternative agriculture, integrated pest management, and genetic engineering will all have major impacts on global sustainability. We depend entirely on agriculture, and have under cultivation a decreasing area of land the size of the continent of South America—so that the extreme necessity of improving the sustainability of agriculture in relation to meeting human needs and improving global sustainability should be obvious to everyone.
9. **Individual values** could use improvement, particularly among those making economic decisions. Often, business people acquire assets and manage them guided strictly by economic considerations, without thinking about the serious environmental consequences of their actions. Many of these consequences have been discussed in this book. New methods of conducting business that internalize environmental costs are emerging, as are new work patterns such as telecommuting, in which travel is minimized.

TELECOMMUTING: GOOD FOR THE PLANET

Telecommuting is the term that describes working at home by using computers, modems, and fax machines linked through telephone lines to an office in another location. It is a growing trend. In the future, it will be common for people to work at home most of the time and travel to the office only when they need to meet with fellow workers.

In 1990, the environmental consulting firm Arthur D. Little studied the potential impact in the United States if telecommunications (including videoconferencing) were substituted for transportation for just 10 to 20 percent of the time spent in working, shopping, business meetings, and the movement of electronic data. The benefits: $23 billion saved (including $610 million through reduced airborne pollution, $1.5 billion through reduced energy expenditures, and $15.38 billion through increased productivity), 1.8 million tons of pollution avoided, 3.5 million gallons of gasoline not burned, and 3.1 billion hours of personal time gained.

(Courtesy of Kelmscott Farm)

WHAT KIND OF WORLD DO WE WANT?

Perhaps the most important single lesson to have learned from this textbook is that those of us who live in industrialized countries are the core of the problems facing the global ecosystem today. We number about 20 percent of the world's people, and our activities alone are more than sufficient to create global instability. Industrialized countries consume a disproportionate share of resources, indicating clearly that the industrialized countries of the world must act forcefully to reduce their levels of consumption if we are all going to be able to attain stability.

As we continue to strive aggressively to increase our high standard of living in industrialized countries, from a level that would be considered utopian by most people on Earth, we drain resources from the entire globe and thus contribute to a future in which neither we nor our children nor our grandchildren will be able to live in anything like the affluence that we experience now.

(NASA)

The citizens of industrialized nations often seem to assume implicitly that overall prosperity can be created as a result of science and technology and that primary productivity will take care of itself. If we are to improve the world situation, however, we must find ways to support the farmers and others who produce the crops on which we ultimately depend, and to involve these citizens in the affairs of the countries where they live.

To put it concisely, we must radically change our view of the world and adopt new ways of thinking, or we will perish together. We must learn to understand, respect, and work with one another, regardless of the differences that exist among us. The most heartening aspect of the situation that we confront is that people, given the motivation, do have the ability to make substantial changes.

At the deepest level, the most critical environmental problems, from which all others arise, are our own attitudes and values. We are totally out of touch and out of balance with the natural world, and until we reconnect and readjust in some significant way, all solutions will be stopgap ones. As a society we don't feel a part of the global ecosystem; we feel separate, above it, and therefore in a position to consume and abuse without thought of consequences. This book has been about consequences. In the last 30 years and especially in the last decade, we have come to recognize the nature of the impact of human activity on the biosphere. We now understand this impact enough to know that we *cannot* continue to act as we have been acting and expect any sort of viable future for our species. If all

(Larry Ulrich/Tony Stone Images)

(Carl Ho/Reuters)

we do as a result of this new knowledge is to make some shifts in consumer choices and write a few letters, it won't be nearly enough.

Your generation must become the next pioneers. A pioneer is one who ventures into unexplored territory, a process that is simultaneously terrifying and profoundly exciting. The unexplored territory in this case is the development of a truly different way for humans to exist in the world. No models exist for this kind of change. You must forge a new revolution, akin in scope and effect to the Agricultural Revolution or the Industrial Revolution, yet totally different because it must be deliberate. You must help create the political will for it with your numbers and your commitment. You must create the economic power with your thoughtful decisions as both consumers and leaders. You must create social change with your acceptance and respect of the differences among peoples.

This change will require reconnecting to the natural world. That means, at a personal level, taking opportunities to be in the wilderness (even if it is a city park or a backyard garden), to listen to the wind, and to look at the exquisite variety of plants and rocks and insects with which we coexist. Humans evolved in nature. Our immensely complex and multidimensional brains developed precisely because we were able to interact with growing things, weather patterns, and other animals. The world we have created now screens us from all that. The sophisticated devices we imagined and manufactured—things such as televisions, computers, and automobiles—have come to define our world. One of your challenges will be to use technology as a tool but not to let it define your interaction with the world.

The new revolution will involve revaluing ourselves and our lives according to a different set of ideals. Wealth and material possessions have come to mean success, at tremendous cost to the planet. Such ideas are deeply imbedded and extremely compelling and will be difficult to change. You may need to throw off some of the myths of Western culture, such as the belief that "faster" and "more" inevitably mean "better." You will at least be called on to examine those myths very deeply and very thoughtfully and to decide what they mean to you.

It will require reinventing economic constructs, such as building the cost of environmental impact and damage into our accounting systems, which will cause market forces to work in favor of environmental protection. Business activities must involve developing cooperative partnerships with people all over the world and making decisions based on long-term benefits to the environment.

This is an overwhelming responsibility. The choices we make now will have a greater impact on the future than those that any generation has had before, whether these choices lead us further into or out of ecological collapse. Even choosing to do nothing will have profound consequences for the future. At the same time, it is an incredible opportunity. This is a time in history when the best of human qualities—vision, courage, imagination, and concern—will play a critical role in establishing the contours of tomorrow's world.

APPENDIX I

REVIEW OF BASIC CHEMISTRY

ELEMENTS

All matter, living and nonliving, is composed of chemical **elements,** substances that cannot be broken down into simpler substances by chemical reactions. There are 92 naturally occurring elements, ranging from hydrogen (the lightest) to uranium (the heaviest). In addition to the naturally occurring elements, about 17 elements heavier than uranium have been made in laboratories by bombarding elements with subatomic particles.

Instead of writing out the name of each element, chemists use a system of abbreviations called **chemical symbols**—usually the first one or two letters of the English or Latin name of the element. For example, O is the symbol for oxygen, C for carbon, Cl for chlorine, and Na for sodium (its Latin name is *natrium*).

Atoms

The atom is the smallest subdivision of an element that retains the characteristic chemical properties of that element. Atoms are almost unimaginably small, much smaller than the tiniest particle visible under a light microscope.

An atom is composed of smaller components called subatomic particles—protons, neutrons, and electrons. **Protons** have a positive electrical charge; **neutrons** are uncharged particles with about the same mass as protons. Protons and neutrons make up almost all the mass of an atom and are concentrated in the atomic nucleus. **Electrons** have a negative electrical charge and an extremely small mass (only about 1/1800 of the mass of a proton). The electrons spin in the space surrounding the atomic nucleus.

Each kind of element has a fixed number of protons in the atomic nucleus. This number, called the **atomic number,** determines the chemical identity of the atom. The total number of protons plus neutrons in the atomic nucleus is termed the **atomic mass.** For example, the element oxygen has eight protons and eight neutrons in the nucleus; it therefore has an atomic number of 8 and an atomic mass of 16.

Where an atom is uncombined, it contains the same number of electrons as protons. Some kinds of chemical combinations and certain other circumstances change the number of electrons, but chemical reactions do not affect anything in the atomic nucleus. Because electrons and protons have equal but opposite charges, an uncombined atom is electrically neutral.

The way electrons are arranged around an atomic nucleus is referred to as the atom's **electronic configuration.** Knowing the locations of electrons enables chemists to predict how atoms can combine to form different types of chemical compounds.

An atom may have several **energy levels,** or **electron shells,** where electrons are located. The lowest energy level is the one closest to the nucleus. Only two electrons can occupy this energy level. The second energy level can accommodate a maximum of eight electrons. Although the third and outer shells can each contain more than eight electrons, they are most stable when only eight are present. We may consider the first shell complete when it contains two electrons and every other shell complete when it contains eight electrons.

The energy levels correspond roughly to physical locations of electrons, called **orbitals.** There may be several orbitals within a given energy level. Electrons are thought to whirl around the nucleus in an unpredictable manner, now close to it, now farther away. Orbitals represent the places where electrons are most probably found.

Isotopes

Isotopes are atoms of the same element that contain the same number of protons but different numbers of neutrons. Isotopes, therefore, have different atomic mass numbers. The three isotopes of hydrogen contain zero, one, and two neutrons, respectively. Elements usually occur in nature as mixtures of isotopes.

All isotopes of a given element have essentially the same chemical characteristics. Some isotopes with excess neutrons are unstable and tend to break down, or decay, to a more stable isotope (usually of a different element). Such isotopes are termed **radioisotopes,** since they emit high-energy radiation when they decay.

Molecules

Two or more atoms may combine chemically to form a **molecule.** When two atoms of oxygen combine, for example, a molecule of oxygen is formed. Different kinds of atoms can combine to form **chemical compounds.** A chemical compound is a substance that consists of two or more different elements combined in a fixed ratio. Water is a chemical compound in which each molecule consists of two atoms of hydrogen combined with one atom of oxygen.

Chemical Bonds

The chemical properties of an element are determined primarily by the number and arrangement of electrons in the *outermost* energy level (electron shell). In a few elements, called the **noble gases,** the outermost shell is filled. These elements are chemically inert, meaning that they will not readily combine with other elements. The electrons in the outermost energy level of an atom are referred to as **valence electrons.** The valence electrons are chiefly responsible for the chemical activity of an atom. When the outer shell of an atom contains fewer than eight electrons, the atom tends to lose, gain, or share electrons to achieve an outer shell of eight. (The exceptions are zero or two electrons in the case of the lightest elements, hydrogen and helium.)

The elements in a compound are always present in a certain proportion. This reflects the fact that atoms are attached to each other by chemical bonds in a precise way to form a compound. A **chemical bond** is the attractive force that holds two atoms together. Each bond represents a certain amount of potential chemical energy. The atoms of each element form a specific number of bonds with the atoms of other elements—a number dictated by the number of valence electrons.

Ions

Some atoms have the ability to gain or lose electrons. Because the number of protons in the nucleus remains unchanged, the loss or gain of electrons produces an atom with a net positive or negative charge. Such electrically charged atoms are termed **ions.**

CHEMICAL FORMULAS

A **chemical formula** is a shorthand method for describing the chemical composition of a molecule. Chemical symbols are used to indicate the types of atoms in the molecule, and subscript numbers are used to indicate the number of each atom present. The chemical formula for molecular oxygen, O_2, tells us that each molecule consists of two atoms of oxygen. This formula distinguishes it from another form of oxygen, ozone, which has three oxygen atoms and is written O_3. The chemical formula for water, H_2O, indicates that each molecule consists of two atoms of hydrogen and one atom of oxygen. Note that when a single atom of one type is present it is not necessary to write 1; it is not necessary to write H_2O_1.

CHEMICAL EQUATIONS

The chemical reactions that occur between atoms and molecules—for example, between methane and oxygen—can be described on paper by means of **chemical equations.**

$$\underset{\text{Methane}}{CH_4} + \underset{\text{Oxygen}}{2\,O_2} \longrightarrow \underset{\text{Carbon dioxide}}{CO_2} + \underset{\text{Water}}{2\,H_2O}$$

Methane is broken down in this reaction.

In a chemical reaction, the **reactants** (the substances that participate in the reaction) are written on the left side of the equation and the **products** (the substances formed by the reaction) are written on the right side. The arrow means *yields* and indicates the direction in which the reaction tends to proceed. The number preceding a chemical symbol or formula indicates the number of atoms or molecules reacting. Thus, 2 O_2 means two molecules of oxygen. The absence of a number indicates that only one atom or molecule is present. Thus, the equation can be translated into ordinary language as, "One molecule of methane reacts with two molecules of oxygen to yield one molecule of carbon dioxide and two molecules of water."

ACIDS AND BASES

An **acid** is a compound that ionizes in solution to yield hydrogen ions (H^+)—that is, protons—and a negatively charged ion. Acids turn blue litmus paper red and have a sour taste. Hydrochloric acid (HCl) and sulfuric acid (H_2SO_4) are examples of acids.

$$\underset{\text{Hydrochloric acid}}{HCl} \xrightarrow{\text{in water}} \underset{\text{Hydrogen ion}}{H^+} + \underset{\text{Chloride ion}}{Cl^-}$$

The strength of an acid depends on the degree to which it ionizes in water, releasing hydrogen ions. Thus, HCl is a very strong acid because most of its molecules dissociate, producing hydrogen and chloride ions.

Most bases are substances that yield a hydroxide ion (OH^-) and a positively charged ion when dissolved in water. Bases turn red litmus paper blue. Sodium hydroxide (NaOH) and aqueous ammonia (NH_4OH) are examples of bases.

$$\underset{\text{Sodium hydroxide}}{NaOH} \xrightarrow{\text{in water}} \underset{\text{Sodium ion}}{Na^+} + \underset{\text{Hydroxide ion}}{OH^-}$$

Bases react with hydrogen ions and remove them from solution.

pH

Since the concentration of hydrogen or hydroxide ions is usually small, it is convenient to express the degree of acidity or alkalinity in a solution in terms of **pH,** formally defined as the negative logarithm of the hydrogen ion concentration. The pH scale is logarithmic, extending from 0, the pH of a very strong acid, to 14, the pH of a very strong base. The pH of pure water is 7, neither acidic nor alkaline (basic), but neutral. Even though water does ionize slightly, the concentrations of H^+ ions and OH^- ions are exactly equal; each of them has a concentration of 10^{-7}, which is why we say that water has a pH of 7. Solutions with a pH of *less* than 7 are acidic and contain more H^+ ions than OH^- ions. Solutions with a pH *greater* than 7 are alkaline and contain more OH^- ions than H^+ ions.

Because the scale is logarithmic (to base 10), a solution with a pH of 6 has a hydrogen ion concentration that is ten times greater than a solution with a pH of 7, and is much more acidic. A pH of 5 represents another tenfold increase. Therefore, a solution with a pH of 4 is 10×10 or 100 times more acidic than a solution with a pH of 6.

The contents of most animal and plant cells are neither strongly acidic nor alkaline but are an essentially neutral mixture of acidic and basic substances. Most life cannot exist if the pH of the cell changes very much.

APPENDIX II

HOW TO MAKE A DIFFERENCE

Individuals and groups all around the world are beginning to attack complex environmental problems. Youths from different countries and cultures are working together to clean up the Mediterranean Sea. Women in India are protecting trees from lumberjacks and planting more trees. Political parties that address environmental issues are gaining increasing political clout throughout Europe. Grassroots environmental groups are organizing in Russia to stop development projects that are environmentally unsound.

In this book we have explored some of the complexities of our environment, and we have examined many problems that are the direct result of human action. Some environmental problems seem almost impossible to address, yet the quality of life and the kind of world we want this place to be, not only for our children and grandchildren but for ourselves, depend largely on the actions we take today.

WHAT YOU CAN DO AS AN INDIVIDUAL

Throughout this text we have highlighted things you can do as an individual to conserve energy and other natural resources and reduce waste. Although individual savings are small, they represent substantial conservation when combined with other individual contributions. Knowing what you now know about the environment, you should try to break personal habits that harm the environment.

Because you are taking a course in environmental science, you should be well informed on most environmental issues. Make an effort to stay informed and share your knowledge with others. You can help shape environmental policies of the future, but you cannot be effective if you don't know what you're talking about.

Most important, should you be particularly concerned about one or more environmental issues, don't just fret over them . . . **get committed.** Join appropriate environmental groups whose collective power carries more weight than single voices. Become politically involved on local, state, and national levels. Help to elect leaders who support a sustainable Earth society and to influence those leaders in office to support the environmental causes you advocate. Be persistent in your efforts.

In summary, there are three things you can do to help preserve and improve the environment: (1) Lower your consumption and reduce waste so that your life style reflects the conservation of natural resources; (2) stay informed on environmental topics and share your knowledge; (3) get involved in environmental issues.

ENVIRONMENTAL ORGANIZATIONS YOU CAN JOIN

The following is an incomplete listing of environmental organizations. Some are grassroots groups that depend on their constituents—people like yourself—for funding and momentum. Others are private, nonprofit organizations that are funded by foundations or other groups. A few are trade associations or professional societies.

Something most of these groups have in common is that they offer internships for college students. Some groups have organized internship programs, while others have internships available only occasionally. Most offer paid internships, but a few are too small to be able to afford any stipend. You can generally arrange to receive college credit for paid or unpaid internships. When this list was being compiled, a number of the organizations were actively seeking interns, but most environmental organizations do not advertise such positions. Therefore, you must contact the organizations you are most interested in to see if they are currently offering anything of interest to you.

Advocates for Youth, 1025 Vermont Ave., NW, Suite 200, Washington, DC 20005 (202)347-5700
A nonprofit organization that directs its efforts at the prevention of early childbirth both in the United States and other countries. Internships available.

Air Pollution Control Association, 441 Smithfield St., Pittsburgh, PA 15222 (412)562-7300
An international professional organization for people involved in air pollution control. No internships available.

American Association for World Health, 1825 K Street, Suite 1208, Washington, DC 20006 (202)466-5883
Educates the public about health problems, both national and global. Internships available.

American Council for an Energy-Efficient Economy, 1001 Connecticut Avenue, NW, Suite 801, Washington, DC 20036 (202)429-8873
Nonprofit organization dedicated to advancing energy efficiency as a means of promoting both economic prosperity and environmental protection.

American Farmland Trust, 1920 N Street, NW, Suite 400, Washington, DC 20036 (202)659-5170
http://www.farmland.org
National nonprofit membership organization working to stop the loss of productive farmland and promote farming practices that lead to a healthy environment. Offers unpaid internships.

American Fisheries Society, 5410 Grosvenor Lane, Bethesda, MD 20014 (301)897-8616
Oldest and largest scientific body dedicated to the advancement of fisheries, science, and the conservation of renewable aquatic resources. Nonprofit. Offers summer internships.

American Forestry Association, 1516 P St., NW, Washington, DC 20005 (202)667-3300
http:// www.amfor.org
Emphasizes conservation of soil and forests but also involved in air and water pollution issues. Their Global ReLeaf Program addresses the issue of global climate change. Internships available.

American Lung Association, 1740 Broadway, New York, NY 10019 (800)LUNG USA info@lungusa.org
http://www.lungusa.org
Educates schools and the general community on the health effects of indoor and outdoor air pollution. Local internships available.

American Rivers, Inc., 1025 Vermont Ave, NW, Suite 720, Washington, DC 20005 (202)347-9224
amrivers@amrivers.org
Focuses on the preservation of the nation's rivers and landscapes.

Center for Acid Rain and Clean Air Policy Analysis, 444 N. Capitol Street, Suite 602, Washington, DC 20001 (202)624-7709
An energy and environmental research and advocacy organization founded in 1985 by bipartisan state governors. Internships available.

Center for Environmental Education, 881 Alma Real Drive, Suite 300, Pacific Palisades, CA 90272 (310)454-4585
Nonprofit organization housing the nation's most comprehensive K–12th grade environmental resource and curricula library. Seeks to ensure that every U.S. student receives an environmental education by gathering and providing a comprehensive collection of the best materials available. Offers internships.

Center for Environmental Information, 50 Main Street West, Rochester, NY 14614-1218 (716)262-2870
Provides information, communications services, publications, and educational programs. Offers internships.

Center for Marine Conservation, 1725 DeSales NW, Suite 600, Washington, DC 20009 (202)429-5609
http://www.cmc-ocean.org
Educates the public on topics such as marine pollution issues and marine wildlife sanctuaries. Internships available.

Center for Plant Conservation (c/o Missouri Botanical Gardens), P.O. Box 299, St. Louis, MO 63166 (314)577-9450
Conserves plant species native to the United States.

Center for Science in the Public Interest, 1875 Connecticut Ave, NW, Suite 300, Washington, DC 20009 (202)332-9110
Concerned with sustainable agriculture, alcohol, food, and nutrition.

Citizens' Clearinghouse for Hazardous Waste, P.O. Box 6806, Falls Church, VA 22040 (703)237-2249
cchw@essential.org
http://www.essential.org/cchw
Assists citizens' groups in the fight against toxic polluters and for environmental justice.

Clean Water Action Project, 1320 18th Street NW, Washington, DC 20036 (202)457-1286
Lobbies for stricter water pollution controls and safe drinking water.

Climate Institute, 120 Maryland Ave NE, Washington, DC 20002 (202)547-0104 climateinst@igc.apc.org
http://www.climate.org
Devoted to helping maintain the balance between climate and life on earth. Nonprofit. Offers internships.

Concern, Inc., 1794 Columbia Road NW, Washington, DC 20009 (202)328-8160 concern@igc.org
http://www.sustainable.org
Focuses on environmental education and sends materials to schools and public and private organizations.

Conservation International Foundation, 1015 18th Street NW, Washington, DC 20036 (202)429-5660
http://www.conservation.org
This nonprofit group provides funds and technical support for a variety of projects related to animal and plant conservation.

Conservation Law Foundation, 62 Summer Street, Boston, MA 02110 (617)350-0990
Nonprofit, public interest, member-supported, environmental advocacy organization founded in 1966 that

uses the law to safeguard the health of New England communities and to protect its natural resources from pollution, mismanagement, and exploitation. Offers internships.

Consumer Federation of America, 1424 16th Street NW, Washington, DC 20036 (202)387-6121
An advocacy group composed of several hundred consumer organizations that lobbies on a broad range of energy and indoor air pollution issues. Internships available.

Cousteau Society, 870 Greenbriar Center, Suite 402, Chesapeake, VA 23320 (757)523-9335
tcsza@igc.apc.org
http://cousteau.edi.fr/cousteau
Works to protect and improve the quality of life for present and future generations.

Defenders of Wildlife, 1101 14th Street, NW, Suite 1400, Washington, DC 20005 (202)682-9400
info@defenders.org
http:// www.defenders.org
Works to preserve, enhance, and protect the diversity of wildlife and the habitats critical for their survival. Internships available.

Earth Island Institute, 300 Broadway, Suite 28, San Francisco, CA 94133 (415)788-3666
Develops and supports projects that counteract threats to the biological and cultural diversity that sustains the environment. Through education and activism these projects promote the conservation, preservation, and restoration of the earth.

Environmental Action, Inc., 6930 Carroll Ave, Suite 600, Takoma Park, MD 20912 (301)891-1100
eaf@igc.apg.org
http://econet.apc.org/eaf
Environmental advocacy group that was founded by the originators of Earth Day 1970. Recent concerns include solid waste and energy issues. Internships available.

Environmental Defense Fund, Inc., 257 Park Avenue S., 16th Floor, New York, NY 10010 (212)505-2100
http://www.edf.org
Other offices are in Washington DC; Oakland, CA; Boulder, CO; Raleigh, NC; Austin, TX; and Richmond, VA. A national nonprofit organization that uses science, economics, and law to develop economically viable solutions to environmental problems. Student internships are offered at all offices.

Environmental Industry Associations, 4301 Connecticut Ave., NW, Suite 300, Washington, DC 20008 (202)244-4700 envasns@envasns.org
http://www.envasns.org
A trade association that works with federal and state governments and provides educational services on solid-waste management and resource recovery.

Environmental Law Institute, 1616 P Street, NW, 2nd Floor, Washington, DC 20036 (202)939-3800
eli@igc.apc.org.
http:// www.eli.org
A nonprofit grassroots organization involved in finding legal solutions to environmental problems. Internships available.

Friends of the Earth, 1025 Vermont Ave NW, Suite 300, Washington, DC 20005 (202)783-7400
foe@foe.org http:// www.foe.org
An environmental activist group distinguished by its international approach to global problems such as ozone depletion, drinking water contamination, and hazardous waste disposal, as well as its advocacy work on Capitol Hill. Internships available.

Greenpeace USA, 1436 U Street, NW, Washington, DC 20009 (202)462-1177 http://www.greenpeace.org
An international organization that currently focuses on ocean ecology, hazardous wastes, disarmament, and atmospheric pollution issues. Internships available

INHALE, 1360 W. 9th Street, Suite 400, Cleveland, OH 44113 (216)523-1111
Prevents air pollution by purchasing and retiring marketable, government-issued, air pollution rights. INHALE improves the air we breathe and reduces acid rain. Offers internships.

International Planned Parenthood Federation, 902 Broadway, New York, NY 10010 (212)995-8800
Headquartered in London, this organization helps governments in over 120 different countries establish family planning services. Internships available.

Izaak Walton League of America, 707 Conservation Lane, Gaithersburg, MD 20878 (301)548-0150
general@iwla.org http://www.iwla.org
The conservation of wildlife, renewable natural resources, and water quality are the concerns of this group.

Kids for Saving Earth Worldwide (KSE), P.O. Box 421118, Plymouth, MN 55442 (612)559-0602
kseww@aol.com
To educate and empower children to help the earth's environment through action-oriented programs. Elementary through junior high environmental educational materials available. Nonprofit. Offers internships.

The Land Institute, 2440 E. Water Well Road, Salina, KS 67401 (913)823-5376
theland@igc.apc.org

http://funnelweb.utcc.utk.edu/~samuels/theland/tli.htm
Seeks to develop an agriculture that will save soil from being lost or poisoned, while promoting a community life that is both prosperous and enduring.

League of Conservation Voters, 1707 L Street, NW, Suite 750, Washington, DC 20036 lcv@lcv.org
http://www.lcv.org
A nonpartisan political action group that works to elect pro-environmental candidates to Congress. They also publish a scorecard that rates members of Congress on their environmental votes. Internships available.

National Audubon Society, 700 Broadway, New York, NY 10003 (212)979-3000
http://www.igc.apc.org/audubon
Also 9 regional offices and over 500 local chapters. Works to conserve and restore natural ecosystems, focusing on birds and other wildlife, for the benefit of humanity and the earth's biological diversity. Currently runs over 80 wildlife sanctuaries. Internships available.

National Fish and Wildlife Foundation, 1120 Connecticut Ave, NW, Suite 900, Washington, DC 20036 (202)857-0166 http://www.nfwf.org
Nonprofit organization dedicated to the conservation of natural resources: fish, wildlife, and plants. The foundation invests in the best possible solutions to natural resource management by awarding challenge grants using its federally appropriated funds to match sector funds. Offers internships.

National Parks and Conservation Association, 1776 Massachusetts Ave, NW, Suite 200, Washington, DC 20036 (202)223-6722
Acquisition and protection of national parks. Active in environmental issues as they relate to the national parks.

National Resources Defense Council, 40 West 20th Street, New York, NY 10011 (212)727-4400
http://www.nrdc.org
Offices in Washington, DC; Los Angeles; San Francisco; and Hawaii. Organization and litigation on a number of environmental issues, including public health, air, and energy issues; nuclear weapons; and land, water, and coastal issues. Internships available.

National Wildflower Research Center, 4801 La Crosse Ave, Austin, TX 78739 (512)292-4200
Only national, nonprofit organization that works toward the preservation and re-establishment of native plants and planned landscapes. Emphasis on environmental education. Offers internships.

National Wildlife Federation, 8925 Leesburg Pike, Vienna, VA 22180 (703)790-4000
http://www.nwf.org/nwf
The largest private conservation organization in the world. Conservation education is its primary mission. Its conservation internship program is primarily intended for recent college graduates, but college seniors sometimes qualify.

Natural Resources Council of America, 1025 Thomas Jefferson Street, NW, Suite 109, Washington, DC 20007 nrca@msn.com
A consortium that serves over 75 different environmental groups. Paid and unpaid internships available.

The Nature Conservancy, 1815 North Lynn Street, Arlington, VA 22209 (703)841-5300 http://www.tnc.org
Global preservation of natural diversity. This group finds, protects, and maintains the best examples of communities, ecosystems, and endangered species in the natural world. Internships available.

North American Association for Environmental Education, 1255 23rd Street, NW, Suite 400, Washington, DC 20037 (202)884-8912
Integrated network of professionals in the field of environmental education dedicated to promoting environmental education and supporting the work of environmental educators around the world.

Physicians for Social Responsibility, 1101 14th Street, NW, Suite 700, Washington, DC 20005 (202)898-0150
Focuses on the effects of nuclear war and nuclear weapons on human health.

Planet/Drum Foundation, P.O. Box 31251, San Francisco, CA 94131 (415)285-6556
planetdrum@igc.apc.org
Works to promote and encourage ecologically sustainable living in urban and rural areas. Internships available.

Planned Parenthood Federation of America, 810 Seventh Avenue, New York, NY 10019 (212)541-7800
http://www.ppfa.org/ppfa
Addresses the family planning needs of nearly eight million men and women around the world each year. Internships available.

Population Action International, 1120 19th Street NW, Washington, DC 20036 (202)659-1833 pai@popact.org
A private, nongovernmental organization that seeks to increase political and financial support for effective population policies and programs grounded in individual rights.

Population-Environment Balance, Inc., 1325 G Street, NW, Suite 1003, Washington, DC 20005 (202)955-5700
Aims to stabilize the population of the United States and to safeguard the land's carrying capacity. Also supports a responsible immigration policy, increased funding for contraceptive research and availability, and a woman's right to reproductive choice.

The Population Institute, 107 2nd Street NE, Washington, DC 20002 (202)544-3300 popline@primanet.com http://www.populationinstitute.org
Education, lobbying, and public policy on population growth, particularly in developing nations. Internships available.

Population Reference Bureau, 1875 Connecticut Ave, NW, Suite 520, Washington, DC 20009 (202)483-1100 http://prb.org/prb
A private, nonprofit educational organization that disseminates demographic and population information. PRB is a clearinghouse on U.S. and international population matters. Internships are only available for recent college graduates.

Population Resource Center, 1725 K Street NW, Suite 1102, Washington, DC 20006 (202)467-5030 prcdc@aol.com
Deals with population and demographic issues, both domestically and internationally. Internships available.

The Public Citizen, 1600 20th Street NW, Washington, DC 20009 (202)588-1000 public_citizen@citizen.org http://www.essential.org/public_citizen/
Political action group founded by Ralph Nader. Composed of several sister organizations with different missions, including Congress Watch, Critical Mass (energy project), and the Health Research Group. Internships available.

Public Forestry Foundation, P.O. Box 371, Eugene, OR (503)687-1993
Promotes sustainable forestry through direct interaction with forest managers and concerned citizens. Offers internships.

Rainforest Action Network, 450 Sansome Street, Suite 700, San Francisco, CA 94111 (415)398-4404
Works to protect the earth's rain forests and support the rights of their inhabitants through education, grassroots organizing, and nonviolent direct action. Offers internships.

Renew America, 1400 16th Street, NW, Washington, DC 20036 (202)232-2252 renewamerica@igc.apc.org http://solstice.crest.org/renew_america
A nonprofit organization that focuses on educational and information networking on national and global environmental issues. Internships available.

Resources for the Future, 1616 P Street NW, Washington, DC 20036 (202)328-5000 info@rff.org http://www.rff.org
Concerned with the quality of the environment and the conservation of natural resources. Summer internships available.

Sea Shepherd Conservation Society, 3107A Washington Blvd., Marina Del Rey, CA 90401 (310)394-3198
Offers internships.

Sierra Club, 85 2nd Street, 2nd Floor, San Francisco, CA 94105 (415)977-5500 http://www.sierraclub.org
Interested in conserving the natural environment by influencing public policy decisions. Sierra Club members explore, enjoy, and protect the wild places of the Earth. Internships are offered in the San Francisco and Washington, DC, offices.

Smithsonian Institution, 1000 Jefferson Drive, SW, Washington, DC 20560 (202)357-1300 http://www.si.edu
Sponsors a wide variety of environmental education and research programs.

Student Conservation Association, Inc., P.O. Box 550, Charlestown, NH 03603 (603)543-1700
Resource assistant program places college students in internship positions at such places as national parks. Opportunities are available throughout the year.

Union of Concerned Scientists, 1616 P Street, NW, Washington, DC 20036 (202)332-0900 ucs@ucsusa.org
Nuclear power safety and arms control are the main focuses of this group.

Water Environment Federation, 601 Wythe Street, Alexandria, VA 22314 (703)684-2400 dtrouba@wef.org http://www. wef.org
Focuses primarily on wastewater treatment and water quality. A few internships are offered, and scholarships are sometimes offered to both undergraduate and graduate students.

The Wilderness Society, 900 17th Street NW, Washington, DC 20006 (202)833-2300 tws@tws.org
Wilderness, parks, and public lands are the focuses of this group. Internships available.

Wildlife Habitat Enhancement Council, 1010 Wayne Ave, Suite 920, Silver Spring, MD 20910 (301)588-8994 whc@cais.com http://www.igc.org/igc/econet/index.html
A nonprofit consortium of corporations and conservation groups that makes recommendations on how to enhance corporate-held lands (such as rights-of-way under power transmission lines) for wildlife. Research assistant programs are sometimes available.

Wildlife Management Institute, 1101 14th Street, NW, Suite 801, Washington, DC 20005 (202)371-1808 wmihq@aol.com

A scientific and educational organization that works to improve the professional management and wise use of wildlife and other natural resources.

Wildlife Society, 5410 Grosvenor Lane, Bethesda, MD 20814 (301)897-9770 tws@wildlife.org
A professional society for wildlife biologists.

World Environment Center, 419 Park Avenue S., 18th Floor, New York, NY 10016 (212)683-4700
A liaison between government and industry leaders to resolve environmental problems caused by industry.

World Resources Institute, 1709 New York Ave NW, Suite 700, Washington, DC 20006 (202)638-6300
A policy research center that focuses on the environment and development.

World Society for the Protection of Animals, P.O. Box 190, Boston, MA 02130 (617)522-7000 wspa@igc.apc.org
Promotes effective means for the protection of animals, for the prevention of cruelty, and the relief of suffering to animals. Offers internships.

World Wildlife Fund, 1250 24th Street, NW, Washington, DC 20037 (202)293-4800 http://www.wwf.org
Finds ways to save endangered species, including the acquisition of wildlife habitat. Internships are offered from time to time, primarily for college graduates.

Worldwatch Institute, 1776 Massachusetts Avenue, NW, Washington, DC 20036 (202)452-1999
Nonprofit, public interest research institute concentrating on global and environmentally related issues.

WorldWide Network, 1627 K Street, NW, Suite 300, Washington, DC 20006 (202)347-1514
An international network for women involved in environmental issues, from women in rural villages to national leaders. Closely affiliated with the United Nations Environment Program (UNEP). Internships available.

Zero Population Growth, 1400 16th Street, NW, Washington, DC 20036 (202)332-2200 zpg@igc.apc.org http://www.zpg.org
The largest nonprofit membership group concerned with the human population's impact on resources and the environment. Internships available.

GET POLITICALLY INVOLVED

It is important to let your elected officials know about your concerns. Writing to your senator or congressional representative is a very effective way to help influence the formation of policies and laws affecting environmental issues. It also helps to stay in touch when laws that have been passed are being implemented.

How to Write to Elected Officials

When addressing a letter, you should use the appropriate heading. If you are writing to the President of the United States, address your letter to: The President, The White House, 1600 Pennsylvania Avenue NW, Washington, DC 20500, Dear Mr. President. A letter to a senator should be addressed to The Honorable _______, Senate Office Building, Washington, DC 20510, Dear Senator _______. When writing to your representative, address the letter to The Honorable _______, House Office Building, Washington, DC 20515, Dear Representative _______.

You can also contact them via the World Wide Web:
www.whitehouse.gov
www.house.gov
www.senate.gov

The Do's. Letters are more effective if you keep them short (1 page maximum) and to the point. Talk about the item of concern as specifically as possible: how it impacts you and others, what your position is on the matter, and why. If possible, make a specific request of the official; that is, let them know what you would like them to do. If you are writing about a particular bill, it would be helpful to know the number or name of the bill. Also, make sure you include your name and return address in case they wish to respond.

The Don'ts. Letters are more effective if you avoid being contentious or rude. Also, don't come across as high and mighty ("as a taxpayer who put you in office . . ."). Don't introduce more than one issue or allow the letter to ramble on and on.

Federal and International Agencies You Can Contact

The following agencies can be contacted for information about specific environmental bills, laws, and issues.

Agency for International Development, 320 21st Street, NW, Washington, DC 20523 (202)647-4000

Biological Resources Division, 12201 Sunrise Valley Drive, Reston, VA 22092 (703)648-4050
http://www.nbs.gov

Bureau of the Census, 4700 Silverhill Road, Suitland, MD 20746 (301)452-5120 http://www.census.gov

Bureau of Land Management, U.S. Department of Interior, Interior Building, Room 5600, Washington, DC 20240 (202)452-5120

Bureau of Mines, U.S. Department of Interior, 2401 E Street, NW, Washington, DC 20241

Bureau of Reclamation, U.S. Department of Interior, Interior Building, Room 7654, Washington, DC 20240 (202)208-2553

Congressional Research Service, Library of Congress, 101 Independence Avenue, SE, Washington, DC 20540 (202)707-5700

Council on Environmental Quality, 722 Jackson Place, NW, Washington, DC 20006 (202)456-6224 http://www.whitehouse.gov/CEQ

Department of Agriculture, 14th and Independence Avenue, SW, Washington, DC 20250 (202)720-2791 http://www.usda.gov

Department of Energy, Forrestal Building, 1000 Independence Avenue, Washington, DC 20585 (202)586-5000 http://www.doe.gov

Department of the Interior, 18th and C Street, NW, Washington, DC 20240 (202)208-3100 http://www.doi.gov

Environmental Protection Agency, 401 M Street, SW, Washington, DC 20240 (202)260-2090 http://www.epa.gov

Federal Energy Regulatory Commission, 888 1st Street, NE, Washington, DC 20426 (202)208-0200 http://www.fedworld.gov/ferc/ferc.html

Fish and Wildlife Service, 130 Webb Building, 4401 North Fairfax Drive, Arlington, VA 22203 (703)358-1711 http://www.fws.gov

Food and Drug Administration, U.S. Department of Health and Human Services, 5600 Fishers Lane, Rockville, MD 20852 (800)532-4440 http://www.fda.gov

Forest Service, U.S. Department of Agriculture, P.O. Box 96090, Washington, DC 20090 (202)205-8333

Geological Survey, U.S. Department of Interior, 12201 Sunrise Valley Drive, Reston, VA 22092 (703)648-4000 http://www.usgs.gov

Government Printing Office, 732 North H Street, NW, Washington, DC 20401 (202)512-0000

International Union for the Conservation of Nature and Natural Resources, Av. Du Mont Blanc, CH-1196 Gland, Switzerland 022-64-71-81

National Academy of Sciences, 2101 Constitution Avenue, NW, Washington, DC 20418 (202)334-2000 http://www.nas.edu

National Oceanic and Atmospheric Administration, 14th and Constitution Avenue, NW, Washington, DC 20230 (301)713-4000 http://www.rdc.noaa.gov

National Park Service, U.S. Department of Interior, P.O. Box 37127, Washington, DC 20013-7127 (202)208-3100 http://www.nps.gov

National Response Center, 2100 2nd Street SW, Washington, DC 20593 (202)426-2675 http://www.dot.gov/dotinfo/uscg/hq/nrc

National Science Foundation, 4201 Wilson Blvd., Arlington, VA 22230 (703)306-1234 http://www.nsf.gov

Natural Resources Conservation Service, U.S. Department of Agriculture, P.O. Box 2890, Washington, DC 20013 (202)205-0026 http://www.ncg.nrcs.usda.gov

Nuclear Regulatory Commission, 11545 Rockville Pike, Rockville, MD 20852 (301)415-7000

Occupational Safety and Health Administration, U.S. Department of Labor, 200 Constitution Avenue, NW, Washington, DC 20210 (202)576-6339 http://www.osha.gov

United Nations Environmental Program, P.O. Box 30552, Nairobi, Kenya, or 2 United Nations Plaza, Room 803, New York, NY 10017 (212)963-8139 http://www.unep.org

APPENDIX III

GREEN COLLAR PROFESSIONS

Careers relating to the environment are varied and their number is ever increasing. The following list of "green collar" jobs, organized into three general categories, is intended only as a broad sample. It should be noted that many jobs relating to the environment are by nature cross-disciplinary. With a creative approach, you can contribute to the welfare of our environment through almost any career path.

ENVIRONMENTAL PROTECTION (SOLID AND HAZARDOUS WASTE MANAGEMENT, POLLUTION CONTROL)

air quality engineer (i.e., analyzing and controlling air pollution)
atmospheric scientist (i.e., measuring the chemical composition of the atmosphere)
biostatistician (i.e., measuring the statistical relation between injury or disease and exposure to pollution or toxins)
chemical engineer (i.e., designing systems for chemical waste disposal)
chemist (i.e., testing toxins and their interactions for possible environmental effects)
electrical engineer (i.e., designing energy-efficient power sources)
emergency response specialist (i.e., managing the emergency cleanup of chemical fires or spills)
environmental engineer (i.e., maintaining public water resources)
environmental health scientist (i.e., monitoring the health of ecosystems)
environmental protection specialist (i.e., reviewing government contracts for compliance with EPA regulations)
geological engineer (i.e., designing environmentally safe sanitary landfills; selecting hazardous waste disposal sites)
hazardous waste manager (i.e., transporting and disposing of hazardous materials)
health physicist (i.e., developing protective measures for radiation exposure)
industrial hygienist (i.e., establishing procedures to minimize worker health risks associated with hazardous materials)
mechanical engineer (i.e., designing machinery to comply with environmental regulations)
meteorologist (i.e., monitoring the effects of atmospheric pollutants on weather patterns)
microbiologist (i.e., studying the effects of toxins on microorganisms as part of natural food chains; discovering microorganisms to be used in bioremediation)
mining engineer (i.e., designing mining tunnels to minimize dangerous methane gas leaks and ground water contamination)
noise-control specialist (i.e., developing methods of neutralizing noise at heavy-industry manufacturing sites)
nuclear engineer (i.e., improving design safety of nuclear power plants; disposing of nuclear waste)
oceanographer (i.e., testing ocean water for the presence of sludge and industrial wastes)
risk manager (i.e., assessing potential environmental risks of proposed industrial procedures, chiefly for insurance purposes)
safety engineer (i.e., designing environmentally safe manufactured products)
soil scientist (i.e., testing soil for toxins leaked from landfill sites)
solid waste manager (i.e., managing the distribution, transportation, and disposal of solid wastes)
toxicologist (i.e., determining the link between human illness or disease and exposure to toxic substances)
water quality technologist (i.e., testing for contaminants at water treatment plants)

NATURAL RESOURCE MANAGEMENT (LAND AND WATER CONSERVATION, FISHERY AND WILDLIFE MANAGEMENT, FORESTRY, PARKS, AND OUTDOOR RECREATION)

agricultural engineer (i.e., designing agricultural systems for soil and water conservation)
agronomist (i.e., developing alternatives to chemical fertilizers and pesticides for use in food crop cultivation)
botanist (i.e., identifying new plant species; using genetic engineering to develop new, commercially valuable plant breeds)
ecologist (i.e., studying how natural and human-altered ecosystems work)
fishery biologist (i.e., studying the effects of changing environmental conditions on the survival and growth of fish)
fishery manager (i.e., managing fish hatcheries)

forester/park ranger (i.e., planting and maintaining populations of tree species)
hydrologist/geologist (i.e., managing water resources for agricultural use)
landscape architect (i.e., designing parks, recreational, and other public facilities that preserve or restore natural ecosystems)
petroleum engineer (i.e., ensuring that oil well sites are returned to their original condition after drilling is completed)
wildlife biologist (i.e., researching wildlife populations)
wildlife manager (i.e., restoring or preserving wildlife populations and habitats)
zoologist (i.e., studying the habitat requirements of animal species to prevent their extinction)

COMMUNICATIONS AND PUBLIC AFFAIRS

architect (i.e., designing energy-efficient buildings)
civil engineer (i.e., planning public works projects, including highway and sewage restoration)
computer specialist (i.e., developing and operating computer systems that monitor or simulate environmental problems)
demographer (i.e., researching statistics on human population growth rates to assist efforts to stabilize population growth)
educator (i.e., teaching environmental studies at the undergraduate or graduate level or conducting employee training programs)
environmental lawyer (i.e., specializing in the interpretation of new environmental legislation, regulations, and enforcement procedures)
journalist (i.e., reporting on environmental issues)
interpretive naturalist (i.e., conducting educational tours in natural settings)
occupational/environmental physician (i.e., treating patients exposed to radioactive or toxic materials)
technical writer (i.e., preparing environmental impact statements)
urban/community planner (i.e., designing residential areas to preserve natural ecosystems)

APPENDIX IV

UNITS OF MEASURE: SOME USEFUL CONVERSIONS

Some Common Prefixes

Prefix	*Meaning*	*Example*
giga	1,000,000,000	1 gigaton = 1,000,000,000 tons
mega	1,000,000	1 megawatt = 1,000,000 watts
kilo	1000	1 kilojoule = 1000 joules
centi	0.01	1 centimeter = 0.01 meter
milli	0.001	1 milliliter = 0.001 liter
pico	0.000000000001	1 picocurie = 0.000000000001 curie

Length: Standard Unit = Meter

1 meter = 39.37 in
1 inch = 2.54 cm
1 mile = 1.609 km

Area: Standard Unit = Square meter (m^2)

1 hectare = 10,000 square meters
= 2.471 acres
1 acre = 0.405 hectare
1 square kilometer = 0.4 square mile

Volume: Standard Unit = Cubic meter (m^3)

1 liter = 1000 cm^3 = 1.057 qt (U.S.)
1 gallon (U.S.) = 3.785 L

Mass: Standard Unit = Kilogram

1 kilogram = 2.205 lb
1 metric ton = 1.103 ton
1 pound = 453.6 g

Energy: Standard Unit = Joule

1 joule = 0.24 cal
1 calorie = 4.184 J
1 British thermal unit = 252 cal

Electrical Power: Standard Unit = Watt

1 watt = 1 J/second

Pressure: Standard Unit = Pascal

1 bar = 10^5 Pa
1 atm = 1.01 bar = 1.01×10^5 Pa
1 millibar = 1.45×10^{-2} lb/in^2

Temperature: Standard Unit = Centigrade

$°C = (°F - 32) \times 5/9$
$°F = °C \times 9/5 + 32$
$1°C = 1.8°F$

APPENDIX V

GRAPHING

Much of the study of environmental science involves learning about relationships between variables. For instance, there is a definite relationship between the amount of pollution discharged and the cost of the environmental damage caused by the pollution. Often such relationships can be expressed and understood through graphs.

A **graph** is a diagram that expresses a relationship between two or more quantities. In some cases there is a definite cause-and-effect relationship, whereas in others the association is not as direct. Graphic presentation of data may not explain the *reason* for the relationship (as, for example, the worldwide increase in fertilizer use since 1959), but the shape of it can provide clues. A graph puts into visual form abstract ideas or experimental data, so that their relationships become more apparent.

VARIABLES

The related quantities displayed on a graph are called **variables.** The simplest sort of graph uses a system of coordinates or axes to represent the values of the variables. Usually the relative size of a variable is represented by its position along the axis. Numbers along the axis allow the reader to estimate the values.

If the relationship being plotted is one of cause and effect, the variable that expresses the cause is called the **independent variable.** Usually this is represented by the horizontal axis, which is called the ***x*-axis.** The variable that changes as a result of changes in the independent variables is the **dependent variable.** It is usually represented on the vertical axis, which is called the ***y*-axis.** The two axes are arranged at right angles to each other and cross at a point called the origin (see Figure a).

To show the relationship between two variables that are directly related at some specific value, such as point A in Figure a, the value on the *x*-axis (x_1) is extended vertically, and the corresponding value on the *y*-axis (y_1) is extended horizontally. The point A at which these lines cross is determined by their relationship.

If another pair of points (x_2 and y_2) is chosen, their point of intersection on the graph can also be plotted; this is point *B*. A line drawn between points A and *B* can then give information about how all other *x* and *y* values on this graph should relate to each other.

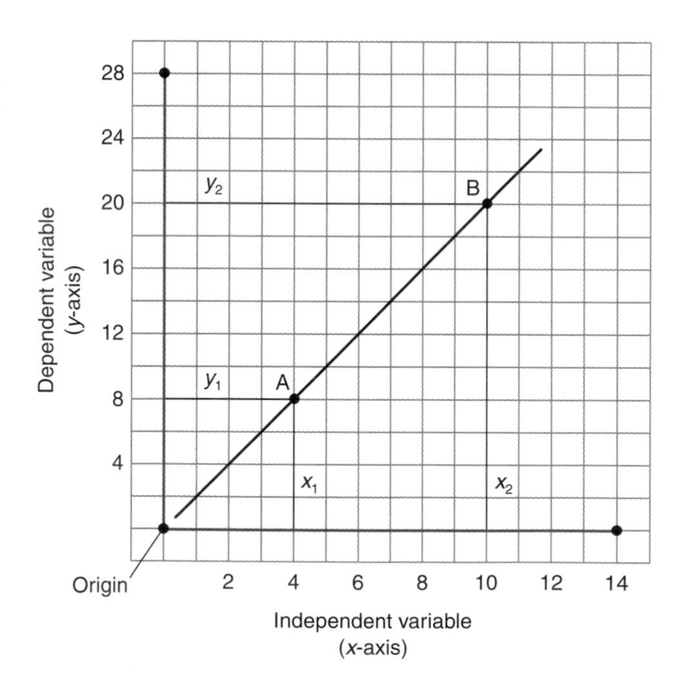

Figure a

A typical line graph. Point A is obtained by extending the corresponding values on the *x*- and *y*-axes until they intersect.

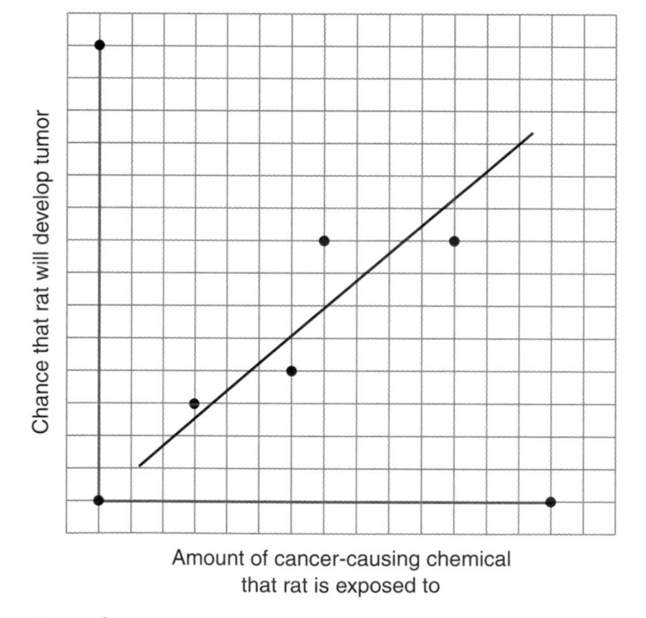

Figure b

In a direct relationship, *y*-values increase as *x*-values increase, producing an upward sloping line or curve. Data points often do not fall exactly on a straight line, but are scattered about an ideal line, which is drawn to show the general relationship..

TYPES OF RELATIONSHIPS

As you may have guessed, this explanation represents a very simple case in which some important assumptions were made. We first assumed that for every *x*-value there was only one *y*-value. We further assumed that all of the *y*-variables were directly related to all of the *x* variables. This is the simplest kind of relationship that a graph can represent. It is called a **direct relationship:** the *y*-values get larger as the *x*-values get larger. An example of this type of graph is shown in Figure b. **Inverse relationships** are also common. In inverse relationships, the *y*-values get *smaller* as the *x*-values get larger (see Figure c). Most relationships found in environmental science are not simple. Over some ranges, a relationship may be direct or inverse, and then it may change as a wider range of variables is considered.

Take a few moments to flip through the pages of this book. You will see many graphs. Some express simple relationships over their entire range of data, whereas others are more complicated, expressing several relationships at once. In some cases there are several lines on the graph, each describing some aspect of the idea being presented. Some data are presented as bar graphs or pie charts instead of lines to illustrate relationships. Whatever their form, all these graphs are designed to present important relationships in the clearest possible way. When you learn to interpret information presented graphically, you are well on your way to understanding environmental science.

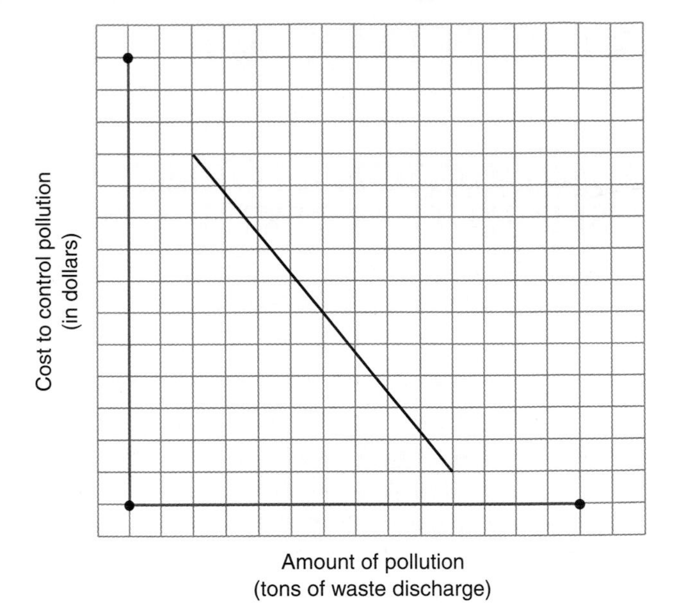

Figure c

In an inverse relationship, *y*-values decrease as *x*-values increase, producing a downward sloping line or curve.

APPENDIX VI

ABBREVIATIONS, FORMULAS, AND ACRONYMS USED IN THIS TEXT

AES — Applied Energy Systems
AID — U.S. Agency for International Development
ARS — USDA's Agricultural Research Service
ATP — adenosine triphosphate
BHA — butylated hydroxyanisole
BHT — butylated hydroxytoluene
BLM — Bureau of Land Management
BOD — biological/biochemical oxygen demand
BRD — Biological Resources Division
Bt — *Bacillus thuringiensis*
BTU — British thermal unit
CARE — Cooperative for American Relief Everywhere
CERCLA — Comprehensive Environmental Response, Compensation, and Liability Act
CFCs — chlorofluorocarbons
CH_4 — methane
CITES — The Convention on International Trade in Endangered Species of Wild Flora and Fauna
CO — carbon monoxide
CO_2 — carbon dioxide
CO_3^{2-} — carbonate
CRP — Conservation Reserve Program
db — decibel
dbA — decibel-A
DDT — dichloro-diphenyl-trichloroethane
DNA — deoxyribonucleic acid
DOD — U.S. Department of Defense
DOE — U.S. Department of Energy
EDB — ethylene dibromide
EFI — European Forest Institute
EIS — environmental impact statement
ENSO — El Niño–Southern Oscillation
EPA — U.S. Environmental Protection Agency
ERC — emission reduction credit
ERNS — EPA's Emergency Response Notification System
ESA — Endangered Species Act
FAO — U.N. Food and Agriculture Organization
FDA — U.S. Food and Drug Administration
FDCA — Food, Drug, and Cosmetics Act
FIFRA — Federal Insecticide, Fungicide, and Rodenticide Act
FWS — U.S. Fish and Wildlife Service
GDP — gross domestic product
GNP — gross national product
GPP — gross primary productivity
GRAS — generally recognized as safe
HCFCs — hydrochlorofluorocarbons
HCO_3^- — bicarbonate
HFCs — hydrofluorocarbons
HNO_2 — nitrous acid
HNO_3 — nitric acid
H_2SO_4 — sulfuric acid
IAEA — International Atomic Energy Agency
IIASA — International Institute for Applied Systems Analysis
IPCC — U.N. Intergovernmental Panel on Climate Change
IPM — integrated pest management
IRCA — Immigration Reform & Control Act
IRRI — International Rice Research Institute
IUCN — World Conservation Union (formerly International Union for the Conservation of Nature and Natural Resources)
K — carrying capacity
kWh — kilowatt hour
medfly — Mediterranean fruit fly
MIC — methyl isocyanate
mpg — miles per gallon
mph — miles per hour
NASA — National Aeronautics and Space Administration
NDP — net domestic product
NEPA — National Environmental Policy Act
NH_3 — ammonia
NIMBY — not in my backyard
NIMTOO — not in my term of office
NNP — net national product
NO — nitric oxide
N_2O — nitrous oxide
NO_2 — nitrogen dioxide
NO_2^- — nitrite
NO_3^- — nitrate
NPP — net primary productivity

NPS	National Park System
NRC	National Research Council
NRCS	National Resources Conservation Service (formerly Soil Conservation Service)
NWS	National Weather Service
O_3	ozone
OH^-	hydroxyl ion
OPEC	Organization of Petroleum Exporting Countries
OTEC	ocean thermal energy conversion
PANs	peroxyacyl nitrates
PCBs	polychlorinated biphenyls
PET	polyethylene terphthalate
PO_4^{3-}	phosphate
ppm	parts per million
PV	photovoltaic
r	growth rate
RCRA	Resource Conservation and Recovery Act
SMCRA	Surface Mining Control & Reclamation Act
SO_2	sulfur dioxide
SO_3	sulfur trioxide
TCDD	tetrachlorodibenzo-p-dioxin (a dioxin)
TCE	trichloroethylene
TNT	trinitrotoluene
2,4-D	2,4-dichlorophenoxyacetic acid
2,4,5-T	2,4,5-trichlorophenoxyacetic acid
U.N.	United Nations
UNCLOS	U.N. Convention on the Law of the Sea
USDA	U.S. Department of Agriculture
USFS	U.S. Forest Service
UV	ultraviolet
VOC	volatile organic compound (or chemical)
WHO	World Health Organization
WRP	Wetlands Reserve Program

APPENDIX VII

SOLUTIONS TO QUANTITATIVE "THINKING ABOUT THE ENVIRONMENT" QUESTIONS

Chapter 1, Question 7.

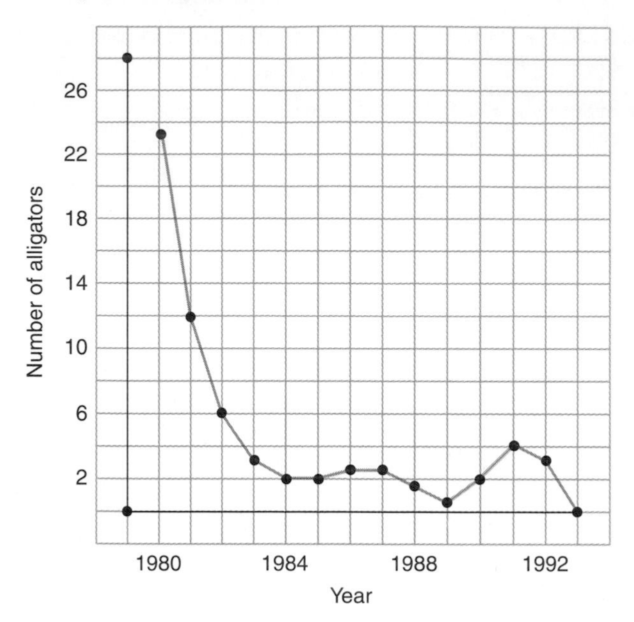

Chapter 2, Question 9.

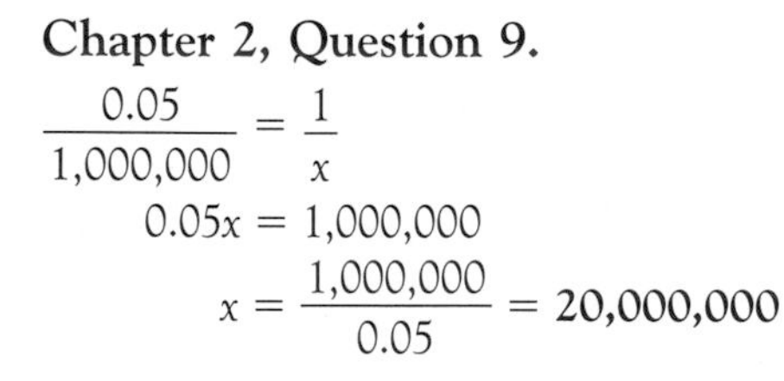

$$\frac{0.05}{1{,}000{,}000} = \frac{1}{x}$$

$$0.05x = 1{,}000{,}000$$

$$x = \frac{1{,}000{,}000}{0.05} = \mathbf{20{,}000{,}000}$$

Chapter 3, Question 11.

A **bar graph** would be more appropriate because the data compare different terrestrial ecosystems. Connecting the values in a line graph would not provide additional insight into their relationships.

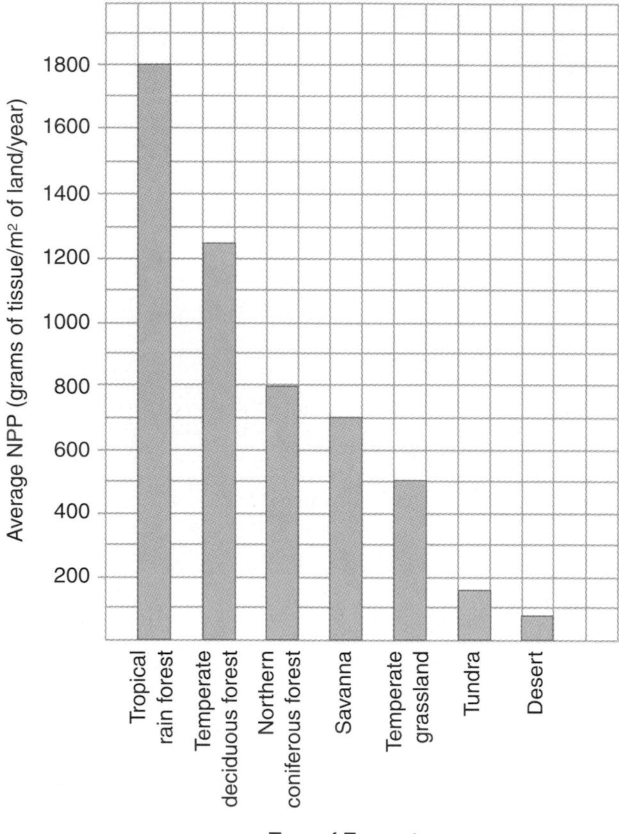

Chapter 4, Question 9.

a. **Day 2**
b. **Days 3 to 5**
c. **Day 18**

Chapter 5, Question 11.

$°F = °C \times 9/5 + 32$

$°F = -45 \times 9/5 + 32 = -81 + 32 = \mathbf{-49°F}$

$°F = -75 \times 9/5 + 32 = -135 + 32 = \mathbf{-103°F}$

Chapter 5, Question 12.

$$\frac{4{,}100{,}000}{79{,}107} = \textbf{51.8 acres per fire}$$

Chapter 6, Question 11.

1 meter = 39.37 inches = 3.28 feet
200 m × 3.28 = **656 feet**
2000 m × 3.28 = **6560 feet**
4000 m × 3.28 = **13,120 feet**

Chapter 7, Question 7.

a. The amount of pollution indicated in red is **more** than the economically optimum amount of pollution. The marginal cost of pollution is **higher** than if the pollution level was at its optimum. The cost of pollution abatement is **lower** than if the pollution level was at its optimum.
b. The amount of pollution indicated in red is **less** than the economically optimum amount of pollution. The marginal cost of pollution is **lower** than if the pollution level was at its optimum. The cost of pollution abatement is **higher** than if the pollution level was at its optimum.

Chapter 8, Question 13.

$$\frac{15{,}500{,}000}{990} = \textbf{15,657 people per square mile in the Netherlands}$$

$$\frac{265{,}200{,}000}{3{,}615{,}200} = \textbf{73 people per square mile in the United States}$$

The Netherlands has the greater population density.

Chapter 8, Question 14.

$r = b - d$

$$r = \frac{2700}{100{,}000} - \frac{900}{100{,}000}$$

$r = 0.027 - 0.009 = 0.018 =$ **1.8% per year**

$$t_d = \frac{70}{r} = \frac{70}{1.8} = \textbf{38.9 years}$$

Chapter 8, Question 15.

949,600,000 × 0.019 = 18,042,400

  949,600,000
+ 18,042,400
967,642,400 people in India in 1997

The easiest way to determine the population of India in the year 2000 is to do similar calculations for each year:
967.6 million × 0.019 = 18.4 million

  967.6 million
+ 18.4 million
986.0 million in 1998

986 million × 0.019 = 18.7 million

  986.0
+ 18.7
1004.7 million in 1999

1004.7 million × 0.019 = 19.1 million

  1004.7
+ 19.1
1023.8 million = **1.02 billion people in India in 2000**

$$t_d = \frac{70}{r} = \frac{70}{1.9} = 36.8 \text{ years}$$

1996 + 36.8 = 2032.8 = **the year 2032**

Chapter 8, Question 16.

$r = b - d$
$r + d = b$
$d = b - r$

$$d = \frac{24}{1000} - \frac{15}{1000} = \mathbf{\frac{9}{1000}}$$

Chapter 9, Question 9.

Let x = amount of CO_2 per capita in India
Then $30x$ = amount of CO_2 in the United States

For India, $x \times 949{,}600{,}000 = 949{,}600{,}000x$

For the United States,
$30x \times 265{,}200{,}000 = 7{,}956{,}000{,}000x$

The United States produces more CO_2 than India.

$$\frac{949{,}600{,}000x}{7{,}956{,}000x} = 8.4$$

The United States produces 8.4 times more CO_2 pollution than India.

Chapter 10, Question 8.

Table A–1
Per Capita Gasoline Consumption in the United States, 1950 to 1993

Year	Gasoline (Million Barrels/Day)	Gasoline (Million Gallons/Day)	Population (Millions)	Gasoline Consumption (Gallons/Day/Person)
1950	2.72	114.24	151.3	0.8
1955	3.66	153.72	165.1	0.9
1960	4.13	173.46	179.3	1.0
1965	4.59	192.78	193.5	1.0
1970	5.78	242.76	203.2	1.2
1975	6.67	280.14	215.5	1.3
1980	6.58	276.36	226.5	1.2
1985	6.83	286.86	238.7	1.2
1990	7.23	303.66	248.7	1.2
1993	7.48	314.16	257.9	1.2

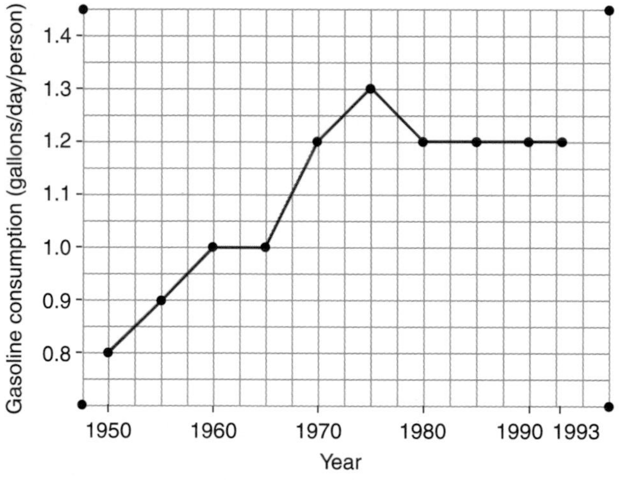

Chapter 10, Question 9.

$$\begin{array}{r} 314.16 \times 10^6 \text{ gallons/day} \\ \times \quad 365 \text{ days a year} \\ \hline 114{,}668.4 \times 10^6 \text{ gallons/year} = 114.67 \times 10^9 \text{ gallons/year} \end{array}$$

$$\frac{\$40 \times 10^9}{114.67 \times 10^9} = \mathbf{\$0.35 \text{ per gallon}}$$

Chapter 11, Question 7.

atomic number = 92 = **number of protons**

$$\begin{array}{r} 235 \\ -\ 92 \\ \hline 143 \end{array}$$ **neutrons**

Chapter 11, Question 8.

a. 1/2 × 1000 grams = **500 grams**

b. 1 half life = 500 grams decayed
 2 half lives = 750 grams decayed

2 × 250,000 = **500,000 years**

c. $\frac{1{,}000{,}000}{250{,}000} = 4$ half lives

1st half life: 1000 g → 500 g
2nd half life: 500 g → 250 g
3rd half life: 250 g → 125 g
4th half life: 125 g → **62.5 g**

Chapter 12, Question 12.

$$\begin{array}{r} 2170 \\ -\ 1930 \\ \hline 240 \end{array}$$

$$\frac{240}{1930} = 0.124 = \mathbf{12.4\%}$$

$$\begin{array}{r} 2510 \\ -\ 2170 \\ \hline 340 \end{array}$$

$$\frac{340}{2170} = 0.157 = \mathbf{15.7\%}$$

$$\begin{array}{r} 2990 \\ -\ 2510 \\ \hline 480 \end{array}$$

$$\frac{480}{2510} = 0.191 = \mathbf{19.1\%}$$

$$\begin{array}{r} 3680 \\ -\ 2990 \\ \hline 690 \end{array}$$

$$\frac{690}{2990} = 0.231 = \mathbf{23.1\%}$$

$$\begin{array}{r} 4880 \\ -\ 3680 \\ \hline 1200 \end{array}$$

$$\frac{1200}{3680} = 0.326 = \mathbf{32.6\%}$$

Table A-2
Global Wind Power Generating Capacity, 1990 to 1995

Year	Capacity (Megawatts)	Percent Increase Over Preceding Year
1990	1930	11.6
1991	2170	12.4
1992	2510	15.7
1993	2990	19.1
1994	3680	23.1
1995 (preliminary)	4880	32.6

Chapter 13, Question 12.

1 drop/second × 60 seconds = 60 drops/minute
60 drops/minute × 60 minutes = 3600 drops/hour
3600 drops/hour × 24 hours = 86,400 drops/day
86,400 drops/day × 365 = 31,536,000 drops/year

$$\frac{31{,}536{,}000}{13{,}140} = \mathbf{2400\ gallons/year}$$

Chapter 14, Question 10.

$$\frac{1\text{ mm}}{13\text{ metric tons}} = \frac{250\text{ mm}}{x}$$

x = **3250 metric tons**
3250 × 1.103 = **3584.8 tons (English units)**

Chapter 15, Question 11.

$$\frac{5.407 \times 10^6}{26.8} = \frac{x}{100}$$

$$26.8x = 540.7 \times 10^6$$

$$x = \frac{540.7 \times 10^6}{26.8} = 20.175 \times 10^6 = \mathbf{20{,}175{,}000}$$

Chapter 15, Question 12.

Table A-3
U.S. Production, Consumption, and Imports of Lead, 1994

Production(000 metric tons)	374
Consumption (000 metric tons)	1374.8
Amount imported (000 metric tons)	1000.8

Chapter 16, Question 7.

1/3 × \$10 billion = \$3.333 billion = **\$3333** million
\$3333 billion/365 = **\$9.13 million**

Chapter 17, Question 9.

0.8% = 0.008
14.7 billion hectares × 0.008 = 0.1176 billion hectares = **117.6 million hectares**
117.6 million hectares × 2.471 = **290.6 million acres**

Chapter 18, Question 10.

$$\frac{\$150}{907.2\text{ kg}} = \frac{x}{7\text{ kg}}$$

$907.2x = 1050$
x = **\$1.16 (beef)**

$$\frac{\$150}{907.2\text{ kg}} = \frac{x}{4\text{ kg}}$$

$907.2x = 600$
x = **\$0.66 (pork)**

$$\frac{\$150}{907.2\text{ kg}} = \frac{x}{2.5\text{ kg}}$$

$907.2x = 375$
x = **\$0.41 (poultry)**

Chapter 18, Question 11.

Table A-4
World Grain Production, 1970 and 1995

	1970	1995	Percent Increase
Agricultural Land (million hectares)	663	666	0.45%
Grain Production (million tons)	1096	1680	53.3%
Tons of grain per hectare	1.65	2.52	52.7%

The average grain yield increased by 52.7 percent between 1970 and 1995.

Chapter 19, Question 9.

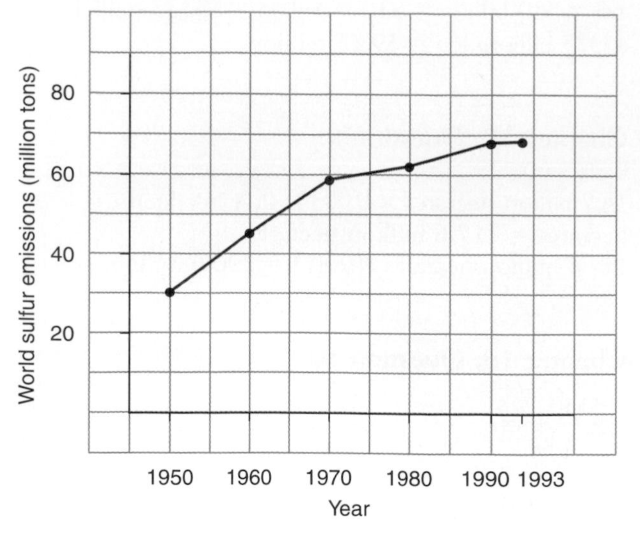

Long term, the graph indicates that world sulfur emissions are increasing. Recently, however, the increase in sulfur emissions has slowed.

Chapter 20, Question 13.

Sum of the ten yearly temperatures in the 1960s = 149.64

$$\frac{149.64}{10} = \textbf{14.96°C (mean global temperature, 1960s)}$$

Perform similar calculations for the 1970s, 1980s, and 1990s:

Mean global temperature, 1970s = **15.00°C**
Mean global temperature, 1980s = **15.19°C**
Mean global temperature, 1990s (preliminary) = **15.32°C**
Since 1960, the mean global temperature for each decade has been higher than the decade preceding it.

Chapter 21, Question 8.

100 m × 500 m = **50,000 square meters**
2 centimeters = 0.02 meter
50,000 square meters × 0.02 meter = **1000 cubic meters of water**
The runoff goes into storm sewers (primarily) and surrounding land (a little).

Chapter 22, Question 9.

0.00005 ppm × 800 = **0.04 ppm in plankton**

Chapter 22, Question 10.

$34.0 billion
−27.8 billion
$ 6.2 billion increase, 1994 to 1998

$$\frac{6.2}{27.8} = 22.3\%, \text{ 1994 to 1998}$$

$$\frac{22.3\%}{4 \text{ yrs}} = \mathbf{5.58\%}$$

Chapter 22, Question 11.

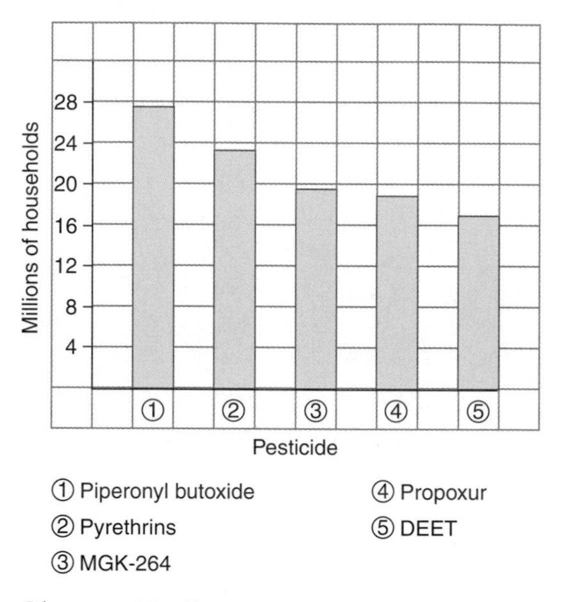

Chapter 23, Question 9.

0.58 × 98.1 × 10^6 tons = **56.9 × 10^6 tons, or 56.9 million tons**

APPENDIX VIII

WORLD POPULATION DATA SHEET ASSIGNMENT

Use the Population Reference Bureau's World Population Data Sheet from inside the back cover and use it to answer the following questions. Then place numbers corresponding to the answers on the map provided on the next page. Use the world map located inside the front cover to help locate the country or political entity.

1. What is the estimated world population in *billions*?
2. What is the current birth rate for the world?
3. Which continent has the highest birth rate, and what is it?
4. Which country or countries has/have the highest birth rate, and what is it?
5. Which continent has the lowest birth rate, and what is it?
6. Which country or countries has/have the lowest birth rate, and what is it?
7. What is the current death rate for the world?
8. Which country or countries has/have the highest death rate, and what is it?
9. How is natural increase (annual, %) calculated?
10. Which country or countries does/do not have a positive natural increase (i.e., either stable or decreasing in population)?
11. At the current rate of natural increase, which country or countries has/have the shortest doubling time, and what is it?
12. At the current rate of natural increase, which country or countries has/have the longest doubling time, and what is it?
13. What is infant mortality rate?
14. Which country or countries has/have the world's highest infant mortality rate, and what is it?
15. Which country or countries has/have the world's lowest infant mortality rate, and what is it?
16. What is total fertility rate?
17. Which country or countries has/have the world's highest total fertility rate, and what is it?
18. Which country or countries has/have the world's lowest total fertility rate, and what is it?
19. Which country or countries has/have the highest percentage of the population under age 15, and what is it?
20. Which country or countries has/have the highest percentage of the population over age 65, and what is it?
21. Which country or countries has/have the longest life expectancy, and what is it?
22. Which country or countries has/have the shortest life expectancy, and what is it?
23. Which country or countries has/have the highest percentage of the population living in urban areas, and what is the percentage?
24. Which country or countries has/have the highest per-capita GNP, and what is it?
25. Which country or countries has/have the lowest per-capita GNP, and what is it?

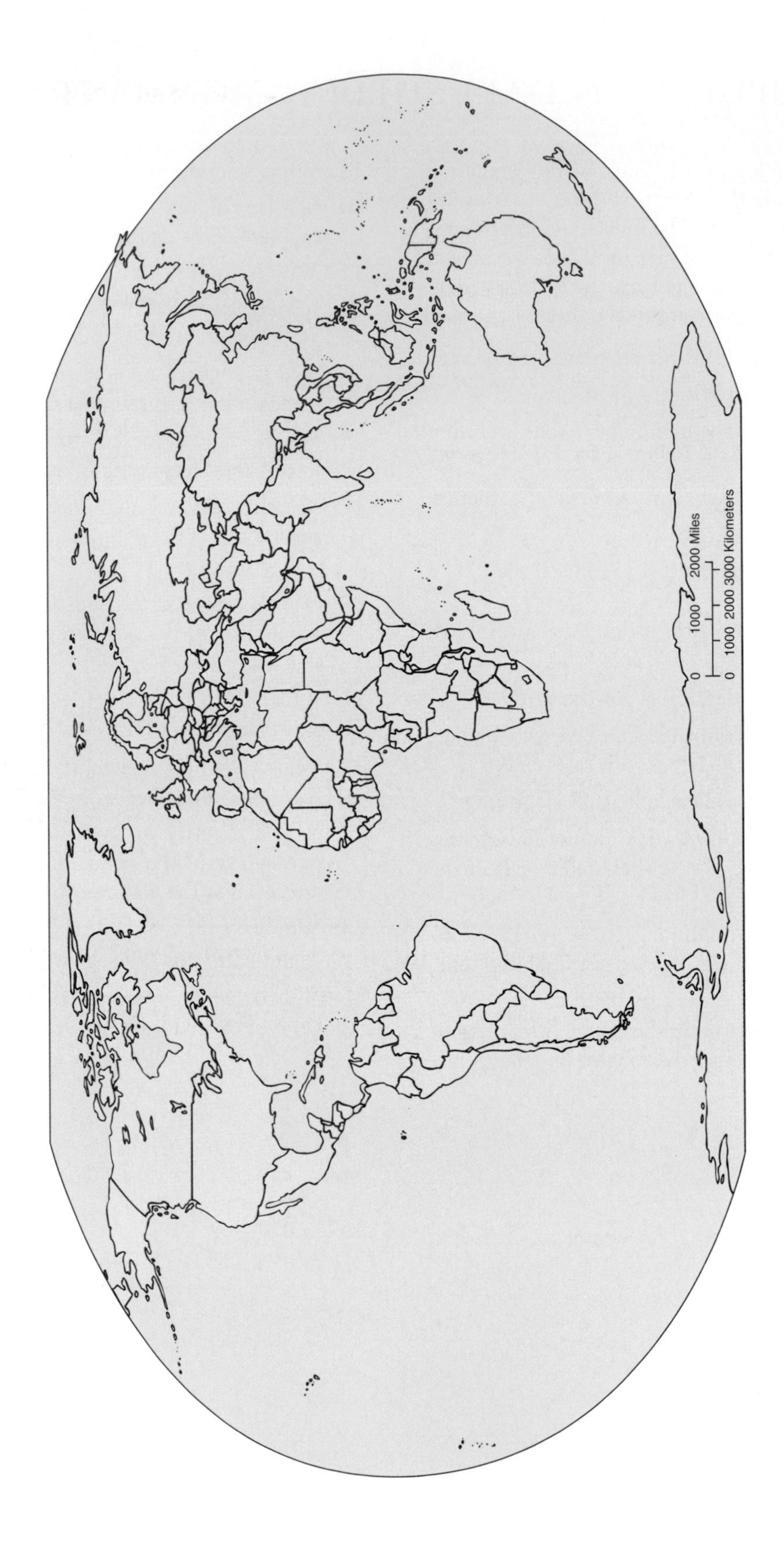
0 1000 2000 Miles
0 1000 2000 3000 Kilometers

REFERENCES

Table 1-1
Data from Dr. Alan Woodward, Florida Game and Fresh Water Fish Commission. Reprinted by permission.

Page 6, "Population of Tokyo and of world's 10 largest cities, number of megacities projected by 2010"
From *World Urbanization Prospects,* United Nations.

Page 7, "Number of people who consume less than 80% of the recommended levels of food"
From FOOD, AGRICULTURE AND NUTRITION: MEETING THE SUMMIT GOALS. Background Paper #14 for the World Food Summit 1996.

Page 7, "Number of People Living in Extreme Poverty"
From The International Bank for Reconstruction and Development/World Bank, 1996, Washington, D.C.

Figure 1-11
Data by Dave Keeling and Tim Whorf, Scripps Institution of Oceanography, La Jolla, California. Reprinted by permission of the Carbon Dioxide Information Analysis Center.

Table 2-2
Excerpted in part from "Risk Assessment and Comparisons: An Introduction," by R. Wilson and E.A. Crouch and from "Ranking Possible Carcinogenic Hazards," by B.N. Ames, R. Magaw, and L.S. Gold in *Science,* Vol. 236, April 17, 1987. Copyright © 1987 by the American Association for the Advancement of Science. Reprinted by permission.

Figures 2-6 and 2-7
From *The Uses of Ecology: Lake Washington and Beyond* by W.T. Edmondson. Copyright 1991 by the University of Washington Press. Adapted and redrawn by permission.

Table 3-1
Data adapted from R.H. Whittaker and G.E. Likens, *Human Ecology* 1:357–369, 1973.

Figure 4-9 b, c, d
Data for graphs adapted from *The Struggle for Existence,* study by G.F. Gause, 1934.

Figure 4-10
Adapted after MacArthur and MacArthur, 1961.

Table 5-1
Figures based on data from World Almanac, Munich RE, Lloyd's of London, and Geological Survey of United States Department of Interior.

Table 5-15
Data adapted from the Climate Analysis Center (CAC), National Weather Service and NOAA, Camp Springs, MD.

Page 105, "Number of Wildfires, Number of Acres Burned in 1994"
From National Interagency Fire Center, Boise, Idaho.

Figure on page 125
From *Everglades: The Ecosystem and Its Restoration* by Steven Davis and John C. Ogden. Copyright © 1994 by St. Lucie Press, an imprint of CRC Press, Boca Raton, Florida. Adapted and redrawn by permission of CRC Press LLC.

Page 134, "Definition of Economics"
Michael Parkin, ECONOMICS, © 1990 by Addison-Wesley Publishing Company.

Table 7-2
Updated figures for "Sulfur Dioxide Emissions in Selected Countries" from *World Resources* 1996–1997. Reprinted by permission of The World Resources Institute.

Box on page 138, "Focus On: Natural Resources, the Environment, and the National Income Accounts"
Excerpted from "The Economy and the Environment: Revising the National Accounts," by Jonathan Levin, *IMF Survey,* June 4, 1990. Reprinted by permission of the International Monetary Fund: Washington.

Figure on page 161
Adapted and redrawn from data by Dr. Rolf O. Peterson of Michigan Technological University.

Table 9-1
"World's Ten Largest Cities" from *World Urbanization Prospects: 1994.* Reprinted and updated by permission.

Figures 10-1, 10-4, 10-9, 10-12, 10-13
Data for graphs adapted from *World Resources* 1996–1997 and based on multiple sources (World Resources Institute, United Nations, and World Bank).

Table 10-1
Adapted from *Modern Physical Geology* by Graham R. Thompson and Jonathan Turk. Copyright © 1997 by Saunders College Publishing, reproduced by permission of the publisher.

Table 10-3
U.S. Department of Energy.

Figure 10-16
Adapted from *Energy* by Roger A. Hinrichs. Copyright © 1992 by Saunders College Publishing, reproduced by permission of the publisher.

Figure on page 236
From NCRP Report No. 93, 1987. Reprinted by permission of the National Council on Radiation Protection and Measurements.

Table 11-2
Adapted from *Energy: Its Use and the Environment,* Second Edition, by Roger A. Hinrichs. Copyright © 1996 by Saunders College Publishing, reproduced by permission of the publisher.

Table 11-3
Number of commercial reactors from International Atomic Energy Agency. Spent fuel inventories adapted in part from data in *WorldWatch Paper 106,* 1990–1991, and other sources.

Figure 11-11
Data by Arjun Makhijani of the Institute for Energy and Environmental Research (IEER). Used by permission.

Table 12-2
Updated data to 1995 "Generating Costs of Electric Power Plants." Reprinted by permission of Investor Responsibility Research Center.

Table 12-3
Data from the Rocky Mountain Institute, 1739 Snowmass Creek, Snowmass, CO 81654-9199 (970)927–3851. Reprinted by permission.

Table 13-1
World Resources Institute, United Nations Environment Programme, United Nations Development Programme, and World Bank, 1996–97.

Table 13-2
World Resources Institute, United Nations Environment Programme, United Nations Development Programme, and World Bank 1996–97.

Figure 13-6
Data adapted from World Resources Institute (in collaboration with United Nations Environment Programme), *World Resources,* 1988–89.

Figure 13-15
From *Water Scarcity: Impacts on Western Agriculture* by Ernest Engelbert and Ann Foley Scheuring. Copyright © 1984 by The Regents of the University of California. Reprinted by permission.

Figure 14-13
Modified map illustration by Rob Wood from *Discover,* January 1993. Copyright © 1993 by The Walt Disney Company. Redrawn by permission of *Discover Magazine.*

Table 15-2
Data adapted from *World Resources* 1996–1997 and based on multiple sources (World Resources Institute, United Nations Environment Programme, United Nations Development Programme, and World Bank).

Figure 15-4
Modified from *Chemistry* by Radel and Navidi. Copyright © 1994, 1990 West Publishing Company. Reprinted by permission of Brooks/Cole Publishing Company, Pacific Grove, CA, a division of International Thompson Publishing, Inc.

Figure 15-7
Data adapted from *World Resources* 1996–1997 and based on multiple sources (World Resources Institute, United Nations Environment Programme, United Nations Development Programme, and World Bank).

Table 17-2
"Fact-Bytes: The Ten Most Popular National Parks," from National Park Service Statistical Abstract: 1995. *Socio-Economic Studies,* National Park Service, Department of Interior.

Table 17-3
"Top 12 Farm Areas Threatened by Population Growth," from American Farmland Trust. Reprinted by permission.

Table 17-4
Adapted by permission from "Endangered Ecosystems: Status Report on America's Vanishing Habitat and Wildlife," by Reed F. Noss and Robert L. Peters. Copyright © 1995 by Defenders of Wildlife.

Table 18-2
From "Agriculture's Diminishing Diversity," by W. Reichart in *Environment* 24:9, 1982. Published by Heldref Publications, 1319 Eighteenth Street, N.W., Washington, D.C. 20036-1802. Copyright © 1982. Reprinted by permission of the Helen Dwight Reid Educational Foundation.

Table 18-4
U.N. Population Division and Food and Agriculture Organization.

Chapter 19 quote on page 433
From Ulf Merbod of The German Aerospace Liaison. Reprinted by permission.

Figure 19-3
Data from the Environmental Protection Agency in *Science,* Vol. 268, May 5, 1995.

Figure 19-10
Based on 1970 data from the Environmental Protection Agency.

Table 20-1
Adapted from *World Resources* 1996–1997 and based on multiple sources (World Resources Institute, United Nations Environment Programme, United Nations Development Programme, and World Bank).

Figure 20–12
From "Climate Change and Society," by W.W. Kellogg and R. Schware. Copyright © 1981 by Westview Press. Redrawn by permission of W.W. Kellogg.

Figure 20–15
Adapted from *World Resources* 1996–1997 and based on multiple sources (World Resources Institute, United Nations Environment Programme, United Nations Development Programme, and World Bank).

Figure 20–18
From "Sun-Ozone Shield Effect More Accurately Analyzed," by Kathy Sawyer in *The Washington Post,* August 2, 1996. Copyright © 1996 by *The Washington Post.* Redrawn by permission.

Table 22–2
From "Controlling Cotton's Insect Pests: A New System," by P.L. Adkisson in *Science,* Vol. 82, April 2, 1982. Copyright © 1982 by the American Association for the Advancement of Science. Excerpted by permission.

Figure 22–3
Adapted from "Environmental and Economic Effects of Reducing Pesticide Use," by David Pimental from *Bioscience* 41:6, June 1991. Copyright © 1991 by the American Institute of Biological Sciences. Reprinted by permission.

Figure 22–3b
From "International Travel and Health. Vaccination Requirements and Health Advice: Situation as of January 1, 1992." Reproduced by permission of the World Health Organization: Geneva.

Table 22–5
Excerpted from "Worst Case Estimates of Risk of Cancer from Pesticide Residues on Food," from *Regulating Pesticides in Food: The Delaney Paradox.* Copyright © 1987 by the National Academy of Sciences. Courtesy of the National Academy Press, Washington, D.C.

Figure 22–6b
From *Biological Control by Natural Enemies,* 1974 by Paul Debach. Copyright © 1974 by Cambridge University Press. Used by permission of the publisher.

Figure 22–7b
From "Ban of DDT and Subsequent Recovery of Reproduction in Bald Eagles," by J.W. Grier from *Science,* Vol. 218, December 17, 1982. Copyright © 1982 by the American Association for the Advancement of Science. Redrawn by permission.

Figure 22–13
From "IPM and the War on Pests," by Gary Gardner in *WorldWatch,* Vol 9, No. 2, March/April 1996. Permission to adapt and redraw by permission of *WorldWatch.*

Table page 531
U.S. Department of Commerce.

Box on page 556
"Disposable Diapers: Environmental Cost of Diapers"
Copyright 1991 by Consumers Union of U.S., Inc., N.Y. 10703-1057. Reprinted by permission from *Consumer Reports,* August 1991.

Appendix V
Excerpts from *Human Physiology,* Third Edition by Rodney Rhoades and Richard Pflanzer. Copyright © 1996 by Saunders College Publishing. Reproduced by permission of the publisher.

Population figures quoted throughout text, and tables and figures in Chapters 8 and 9
Updated from the *1996 Population Data Sheet.* Reprinted by special arrangement with the Population Reference Bureau, Inc., Washington, D.C.

Data for graphs and tables in Chapters 9, 12, 18, 19, and 20
Adapted and redrawn from *Vital Signs 1996,* WorldWatch Institute, and based on multiple sources.

The following organizations provided information used throughout the text:
Geological Survey of U.S. Department of Interior
International Atomic Energy Agency
National Park Service
National Weather Service
Population Reference Bureau
United Nations
U.N. Environment Programme
U.S. Department of Commerce
U.S. Department of Energy
U.S. Department of Food and Agriculture
U.S. Department of Interior
U.S. Environmental Protection Agency
The World Bank Group
World Health Organization, Geneva
World Resources Institute
WorldWatch Institute

GLOSSARY

abiotic Nonliving. Compare *biotic.*

acid A substance that releases hydrogen ions (protons) in water. Acids have a sour taste and turn blue litmus paper red. Compare *base.*

acid deposition A type of air pollution that includes acid that falls from the atmosphere to the Earth as precipitation or as dry acidic particles. See *acid precipitation.*

acid mine drainage Pollution caused when acids and other toxic compounds wash from mines into nearby lakes and streams.

acid precipitation Precipitation that is acidic as a result of both sulfur and nitrogen oxides forming acids when they react with water in the Earth's atmosphere; partially due to the combustion of coal; includes acid rain, acid snow, and acid fog.

active solar heating A heating method in which a series of collection devices mounted on a roof or in a field are used to absorb solar energy for space heating or heating water. Pumps or fans distribute the collected heat. Compare *passive solar heating.*

adaptive radiation The evolution of a large number of related species from an unspecialized ancestral organism.

aerobic respiration The process by which cells utilize oxygen to break down organic molecules into waste products (such as carbon dioxide and water) with the release of energy that can be used for biological work.

aerosol Tiny particles of natural and human-produced air pollution that are so small they remain suspended in the atmosphere for days or even weeks.

aerosol effect Atmospheric cooling that occurs as a result of aerosol pollution.

age structure The number and proportion of people at each age in a population.

agroforestry Forestry and agricultural techniques that are used to improve degraded areas; in agroforestry trees and crops are often planted together.

A-horizon The topsoil; located just beneath the O-horizon of the soil. The A-horizon is rich in various kinds of decomposing organic matter.

air pollution Various chemicals (gases, liquids, or solids) present in high enough levels in the atmosphere to harm humans, other animals, plants, or materials. Excess noise and heat are also considered air pollution.

alcohol fuel A liquid fuel substitute such as methanol or ethanol that may eventually replace gasoline.

algae Single-celled or simple multicellular photosynthetic organisms; important producers in aquatic ecosystems.

allelopathy An adaptation in which toxic substances secreted by roots or shed leaves inhibit the establishment of competing plants nearby.

alpine tundra A distinctive ecosystem located in the higher elevations of mountains, above the tree line. Compare *tundra.*

alternative agriculture Agricultural methods that rely on beneficial biological processes and environmentally friendly chemicals rather than conventional agricultural techniques. Also called *sustainable* or *low-input agriculture.* Compare *high-input agriculture.*

altitude The height of a thing above sea level. Compare *latitude.*

amino acids Organic compounds that are linked together to form proteins. See *essential amino acid.*

ammonification The conversion of nitrogen-containing organic compounds to ammonia (NH_3) by certain bacteria (ammonifying bacteria) in the soil; part of the nitrogen cycle.

annual A plant that grows, reproduces, and dies in one growing season. Compare *biennial* and *perennial.*

antagonism A phenomenon in which two or more pollutants interact in such a way that their combined effects are less severe than the sum of their individual effects. Compare *synergism.*

anthracite The highest grade of coal; has the highest heat content and burns the cleanest of any grade of coal. Also known as *hard coal.* Compare *bituminous coal* and *lignite.*

anticline An upward folding of rock layers, or strata.

antioxidant Food additive that prevents oxidation (breakdown) of food molecules.

aquaculture The rearing of aquatic organisms (fish and shellfish), either freshwater or marine, for human consumption. See *mariculture.*

aquatic Pertaining to the water. Compare *terrestrial.*

aquifer The underground caverns and porous layers of underground rock in which groundwater is stored. See *confined aquifer* and *unconfined aquifer.*

aquifer depletion The removal by humans of more groundwater than can be recharged by precipitation or melting snow.

arctic tundra See *tundra.*

arid land A fragile ecosystem in which lack of precipitation limits plant growth. Arid lands are found in both temperate and tropical regions. Also called *desert.*

artesian aquifer See *confined aquifer.*

artificial eutrophication Overnourishment of an aquatic ecosystem by nutrients such as nitrates and phosphates. In artificial eutrophication, the pace of eutrophication is rapidly accelerated due to human activities such as agriculture and discharge from sewage treatment plants. Also called *cultural eutrophication.* See *eutrophication.*

artificial insemination A technique in which sperm collected from a suitable male of a rare species is used to artificially impregnate a female (perhaps located in another zoo in a different city or even in another country).

assimilation The conversion of inorganic nitrogen (nitrate, NO_3^-, or ammonia, NH_3) to the organic molecules of organisms; part of the nitrogen cycle.

atmosphere The gaseous envelope surrounding Earth.

atom The smallest quantity of an element that can retain the chemical properties of that element; composed of protons, neutrons, and electrons.

atomic mass A number that represents the sum of the number of protons and neutrons in the nucleus of an atom. The atomic mass represents the relative mass of an atom. Compare *atomic number.*

atomic number A number that represents the number of protons in the nucleus of an atom. Each element has its own characteristic atomic number. Compare *atomic mass.*

autotroph See *producer.*

background extinction The continuous, low-level extinction of species that has occurred throughout much of the history of life. Compare *mass extinction.*

bacteria Unicellular, prokaryotic microorganisms. Most bacteria are decomposers, but some are autotrophs, and some are parasites.

base A compound that releases hydroxyl ions (OH^-) when dissolved in water. A base turns red litmus paper blue. Compare *acid.*

bellwether species An organism that provides an early warning of environmental damage. Examples include lichens, which are very sensitive to air pollution, and amphibians.

benthic environment The ocean floor.

benthos Bottom-dwelling marine organisms that fix themselves to one spot, burrow into the sand, or simply walk about on the ocean floor.

B-horizon The light-colored, partially weathered soil layer underneath the A-horizon; subsoil. The B-horizon contains much less organic material than the A-horizon.

biennial A plant that requires two years to complete its life cycle. Compare *annual* and *perennial.*

bioaccumulation The buildup of a persistent pesticide in an organism's body. Also called *bioconcentration.*

biochemical oxygen demand See *biological oxygen demand.*

bioconcentration See *bioaccumulation.*

biodegradable A chemical pollutant decomposed (broken down) by organisms or by other natural processes. Compare *nondegradable.*

biodiversity See *biological diversity.*

biogas A clean fuel, usually composed of a mixture of gases, whose combustion produces fewer pollutants than either coal or biomass. Biogas is produced from plant matter of one form or another.

biogas digester A vat that uses bacteria to break down household wastes, including sewage. The gas produced by the bacteria is burned as a fuel.

biogeochemical cycle Process by which matter cycles from the living world to the nonliving physical environment and back again. Examples of biogeochemical cycles include the carbon cycle, the nitrogen cycle, and the phosphorus cycle.

biological amplification See *biological magnification.*

biological control A method of pest control that involves the use of naturally occurring disease organisms, parasites, or predators to control pests. Also called *biological pest control.*

biological diversity Number and variety of organisms; includes genetic diversity, species diversity, and ecological diversity. Also called *biodiversity.*

biological magnification The increased concentration of toxic chemicals such as PCBs, heavy metals, and pesticides in the tissues of organisms at higher trophic levels in food webs. Also called *biological amplification.*

biological oxygen demand (BOD) The amount of oxygen needed by microorganisms to decompose (by aerobic respiration) the organic material in a given volume of water. Also called *biochemical oxygen demand.*

biological pest control See *biological control.*

biomass (1) A quantitative estimate of the total mass, or amount, of living material. Often expressed as the dry weight of all the organic material that comprises organisms in a particular ecosystem. (2) Plant and animal materials used as fuel.

biome A large, relatively distinct terrestrial region characterized by a similar climate, soil, plants, and animals, regardless of where it occurs on Earth; because it is so large in area, a biome encompasses many interacting ecosystems.

bioremediation A method employed to clean up a hazardous waste site that uses microorganisms to break down the toxic pollutants.

biosphere All of the Earth's organisms. Includes all the communities on Earth. Compare *ecosphere.*

biotechnology See *genetic engineering.*

biotic Living. Compare *abiotic.*

biotic pollution The accidental or intentional introduction of a foreign, or exotic, species into an area where it is not native.

biotic potential The maximum rate at which a population could increase under ideal conditions.

birth rate The number of births per 1000 people per year. Also called *crude birth rate* and *natality.*

bituminous coal The most common form of coal; produces a high amount of heat and is used extensively by electric power plants. Also called *soft coal.* Compare *anthracite* and *lignite.*

bloom Large algal populations caused by the sudden presence of large amounts of essential nutrients (such as nitrates and phosphates) in surface waters.

boreal forest See *taiga.*

botanical A plant-derived chemical used as a pesticide.

bottom ash The residual ash left at the bottom of an incinerator when combustion is completed. Also called *slag.* Compare *fly ash.*

breeder nuclear fission A type of nuclear fission in which nonfissionable U-238 is converted to fissionable P-239. Thus, a breeder nuclear reactor produces more fuel than it consumes.

broad leaf herbicide A herbicide that kills plants with broad leaves but does not kill grasses (such as corn, wheat, and rice).

bycatch Unwanted fish, dolphins, and sea turtles that are caught along with commercially valuable fish and then dumped, dead or dying, back into the ocean.

calendar spraying The regular use of pesticides regardless of whether pests are a problem or not. Compare *scout-and-spray.*

calorie A unit of heat energy; the amount of heat required to raise 1 g of water 1°C.

cancer A malignant tumor anywhere in the body. Cancers tend to spread throughout the body.

carbamates A class of broad-spectrum pesticides that are derived from carbamic acid.

carbohydrate An organic compound containing carbon, hydrogen, and oxygen in the ratio of 1C:2H:1O. Carbohydrates include sugars and starches, molecules metabolized readily by the body as a source of energy.

carbon cycle The worldwide circulation of carbon from the abiotic environment into organisms and back into the abiotic environment.

carcinogen Any substance that causes cancer or accelerates its development.

carnivore An animal that feeds on other animals; flesh-eater. Compare *herbivore* and *omnivore.* See *secondary consumer.*

carrying capacity (*K*) The maximum number of individuals of a given species that a particular environment can support sustainably (long-term).

cell The basic structural and functional unit of life. Simple organisms are composed of single cells, whereas complex organisms are composed of many cells.

cell respiration A process in which the energy of organic molecules is released within cells. Aerobic respiration is a type of cell respiration.

CFCs See *chlorofluorocarbons.*

chain reaction A reaction maintained because it forms as products the very materials that are used as reactants in the reaction. For example, a chain reaction occurs during nuclear fission when neutrons collide with other U-235 atoms and those atoms are split, releasing more neutrons, which then collide with additional U-235 atoms.

chaparral A biome with a Mediterranean climate (mild, moist winters and hot, dry summers). Chaparral vegetation is characterized by small-leaved evergreen shrubs and small trees.

chlorinated hydrocarbon A synthetic organic compound that contains chlorine and is used as a pesticide (for example, DDT) or an industrial compound (for example, PCBs).

chlorofluorocarbons Human-made organic compounds composed of carbon, chlorine, and fluorine that have a number of industrial and commercial applications but are being banned because they attack the stratospheric ozone layer. Also called *CFCs.*

chlorophyll A green pigment that absorbs radiant energy for photosynthesis.

C-horizon The largely unweathered layer in the soil located beneath the B-horizon. The C-horizon borders solid parent material.

circumpolar vortex A mass of cold air that circulates around the southern polar region, in effect isolating it from the warmer air in the rest of the world.

clay The smallest inorganic soil particles. Compare *silt* and *sand.*

clean-coal technologies New methods for burning coal that do not contaminate the atmosphere with sulfur oxides and that produce significantly fewer nitrogen oxides.

clearcutting A forest management technique that involves the removal of all trees from an area at a single time. Also called *even-age harvesting.* Compare *seed-tree cutting, selective cutting,* and *shelterwood cutting.*

climate The average weather conditions that occur in a place over a period of years. Includes temperature and precipitation. Compare *weather.*

coal A black combustible solid found in the Earth's crust. Formed from the remains of ancient plants that lived millions of years ago. Used as a fuel. See *fossil fuel.*

coal gasification The technique of producing combustible gases (such as carbon monoxide and hydrogen) from coal.

coal liquefaction The process by which coal is used to produce a nonalcohol liquid fuel.

coastal wetlands Marshes, bays, tidal flats, and swamps that are found along a coastline. See *mangrove forest, salt marsh* and *wetlands.*

cochlea The part of the ear that perceives sound.

coevolution The interdependent evolution of two or more species that occurs as a result of their interactions over a long period of time. Flowering plants and their animal pollinators are an example of coevolution because each has profoundly affected the other's characteristics.

cogeneration A nontraditional energy technology that involves recycling "waste" heat so that two useful forms of energy (electricity and either steam or hot water) are produced from the same fuel.

coloring agent Natural or synthetic food additives used to make food visually appealing.

combined sewer overflow A problem that arises in a combined sewer system when too much water (from a heavy rainfall or snowmelt) enters the system. The excess flows into nearby waterways without being treated.

combined sewer system A municipal sewage system in which human and industrial wastes are mixed with urban runoff from storm sewers before flowing into the sewage treatment plant.

combustion The process of burning by which organic molecules are rapidly oxidized, converting them into carbon dioxide and water with an accompanying release of heat and light.

command and control Pollution control laws that work by setting pollution ceilings. Examples include the Clean Water and Clean Air Acts.

commensalism A type of symbiosis in which one organism benefits, and the other one is neither harmed nor helped. See *symbiosis.* Compare *mutualism* and *parasitism.*

commercial extinction Depletion of the population of a commercially important species to the point that it is unprofitable to harvest.

commercial harvest The harvest of commercially important organisms from the wild. Examples include the commercial harvest of parrots (for the pet trade) and cacti (for houseplants).

commercial hunting Killing commercially important animals for profit (from the sale of their furs, meat, and so on). Compare *sport hunting* and *subsistence hunting.*

community An association of different species living together in a defined habitat with some degree of mutual interdependence. Compare *ecosystem.*

competition The interaction among organisms that vie for the same resources in an ecosystem (such as food, living space, or other resources).

competitive exclusion The concept that no two species with identical living requirements can occupy the same ecological niche indefinitely. Eventually, one species will be excluded by the other as a result of interspecific (between-species) competition for a resource in limited supply.

compost A natural soil and humus mixture that improves soil fertility and soil structure.

confined aquifer A groundwater storage area trapped between two impermeable layers of rock. Also called *artesian aquifer*. Compare *unconfined aquifer*.

conifer Any of a group of woody trees or shrubs (gymnosperms) that bear needle-like leaves and seeds in cones.

conservation biology The study and protection of biological diversity; includes *in situ* and *ex situ* conservation.

conservation tillage A method of cultivation in which residues from previous crops are left in the soil, partially covering it and helping to hold it in place until the newly planted seeds are established. See *no-tillage*. Compare *conventional tillage*.

conservationist A person who supports the conservation of natural resources.

consumer Organism that cannot synthesize its own food from inorganic materials and therefore must use the bodies of other organisms as sources of energy and body-building materials. Also called *heterotroph*. Compare *producer*.

consumption overpopulation Circumstance in which each individual in a population consumes too large a share of resources—that is, much more than is needed to survive. Consumption overpopulation results in pollution, environmental degradation, and resource depletion. See *overpopulation*. Compare *people overpopulation*.

containment building A safety feature of nuclear power plants that provides an additional line of defense against any accidental leak of radiation.

continental shelf The submerged, relatively flat ocean bottom that surrounds continents. The continental shelf extends out into the ocean to the point where the ocean floor begins a steep descent.

contour plowing Plowing that matches the natural contour of the land—that is, the furrows run around rather than up and down a hill.

contraceptive A device or drug used to intentionally prevent pregnancy.

control An essential part of every scientific experiment in which the experimental variable remains constant. The control provides a standard of comparison in order to verify the results of an experiment. See *variable*.

conventional tillage The traditional method of cultivation in which the soil is broken up by plowing before seeds are planted. Compare *conservation tillage*.

cooling tower Part of an electric power generating plant within which heated water is cooled.

Coriolis effect The tendency of moving air or water to be deflected from its path to the right in the Northern Hemisphere and to the left in the Southern Hemisphere. Caused by the direction of the Earth's rotation.

corridor See *wildlife corridor*.

cost-benefit analysis A mechanism that helps policy makers make decisions about environmental issues. Compares estimated costs of a particular action with potential benefits that would occur if that action were implemented.

crop rotation The planting of different crops in the same field over a period of years. Crop rotation reduces mineral depletion of the soil because the mineral requirements of each crop vary.

crude birth rate See *birth rate*.

crude death rate See *death rate*.

crude oil See *petroleum*.

cullet Crushed glass food and beverage containers that are melted to make new products.

cultural eutrophication See *artificial eutrophication*.

datum (pl. *data*) The information, or facts, with which science works and from which conclusions are inferred.

death rate The number of deaths per 1000 people per year. Also called *crude death rate* and *mortality*.

debt-for-nature swap The cancellation of part of a country's foreign debt in exchange for their agreement to protect certain land (or other resources) from detrimental development.

decibel (dB) A numerical scale that expresses the relative loudness of sound.

decibel-A (dBA) A modified decibel scale that takes into account high-pitched sounds to which the human ear is more sensitive.

decommission The dismantling of an old nuclear power plant after it closes. Compare *entombment*.

decomposer A heterotroph that breaks down organic material and uses the decomposition products to supply it with energy. Decomposers are microorganisms of decay. Also called *saprotroph*. Compare *detritivore*.

deductive reasoning Reasoning that operates from generalities to specifics and can make relationships among data more apparent. Compare *inductive reasoning*.

deep ecology The conviction that all creatures have the right to exist and that humans should not cause the extinction of other organisms.

deforestation The temporary or permanent clearance of forests for agriculture or other uses.

delta A deposit of sand or soil at the mouth of a river.

demand-side management A way that electric utilities can meet future power needs—by helping consumers conserve energy and increase energy efficiency. In demand-side management, consumers save money because they use less energy, and utilities save money because they do not have to build new power plants or purchase additional power.

dematerialization The decrease in the size and weight of a product as a result of technological improvements that occur over time.

demographics The applied science that deals with population statistics (such as density and distribution).

denitrification The conversion of nitrate (NO_3^-) to nitrogen gas (N_2) by certain bacteria (denitrifying bacteria) in the soil; part of the nitrogen cycle.

density-dependent factor An environmental factor whose effects on a population change as population density changes; density-dependent factors tend to retard population growth as population density increases and enhance population growth as population density decreases.

density-independent factor An environmental factor that affects the size of a population but is not influenced by changes in population density.

derelict land Land area degraded by mining.

desalination See *desalinization.*
desalinization The removal of salt from ocean or brackish (somewhat salty) water. Also called *desalination.*
desert See *arid land.*
desertification Degradation of once-fertile rangeland (or tropical dry forest) into nonproductive desert. Caused partly by soil erosion, deforestation, and overgrazing.
detritivore An organism (such as an earthworm or crab) that consumes fragments of dead organisms. Also called *detritus feeder.* Compare *decomposer.*
detritus Organic matter that includes dead organisms (such as animal carcasses and leaf litter) and wastes (such as feces).
detritus feeder See *detritivore.*
deuterium An isotope of hydrogen that contains one proton and one neutron per atom. Compare *tritium.*
developed country A country industrialized and characterized by a low fertility rate, low infant mortality rate, and high per-capita income. Developed countries include the United States, Canada, Japan, and European countries. Also called *highly developed country*. Compare *developing country.*
developing country A country not highly industrialized and characterized by a high fertility rate, high infant mortality rate, and low per-capita income. Most developing countries are located in Africa, Asia, and Latin America, and fall into two subcategories: *moderately developed* and *less developed.* Compare *developed country.*
dilution A technique of soil remediation that involves running large quantities of water through contaminated soil in order to leach out pollutants such as excess salt.
dioxin Any of a family of extremely toxic chlorinated hydrocarbon compounds that are formed as by-products in certain industrial processes.
disease A departure from the body's normal healthy state as a result of infectious organisms, environmental stresses, or some inherent weakness.
distillation A heat-dependent process used to purify or separate complex mixtures. Saltwater or brackish water may be distilled to remove the salt from the water. Compare *reverse osmosis.*
DNA Deoxyribonucleic acid. Present in a cell's chromosomes, DNA contains the genetic information for all organisms.
domesticated Adapted to humans. Describes plants and animals that, during their association with humans, have become so altered from their original ancestors that it is doubtful they could survive and compete successfully in the wild.
doubling time The amount of time it takes for a population to double in size, assuming that its current rate of increase doesn't change.
drainage basin The area of land drained by a river system. Also called *watershed.*
drip irrigation See *microirrigation.*
dry deposition A form of acid deposition in which dry, sulfate-containing particles settle out of the air. Compare *wet deposition.*
dust bowl A semiarid region that has become desert-like as a result of extended drought and severe dust storms.
dust dome A dome of heated air that surrounds an urban area and contains a lot of air pollution. See *urban heat island.*

ecological diversity Biological diversity that encompasses the variety among ecosystems—forests, grasslands, deserts, lakes, estuaries, and oceans, for example. Compare *genetic diversity* and *species diversity.*
ecological niche See *niche.*
ecological pyramid A graphic representation of the relative energy value at each trophic level. See *pyramid of biomass, pyramid of energy,* and *pyramid of numbers.*
ecological risk assessment The process by which the ecological consequences of human activities are estimated. See *risk assessment.*
ecological succession See *succession.*
ecologically sustainable forest management A new method of forest management that seeks not only to conserve forests for the commercial harvest of timber and nontimber forest products, but to sustain biological diversity, prevent soil erosion and protect the soil, and preserve watersheds that produce clean water.
ecology A discipline of biology that studies the interrelationships between organisms and their environment.
economics The study of how people (individuals, businesses, or countries) use their limited economic resources to fulfill their needs and wants. Economics encompasses the production, consumption, and distribution of goods.
ecosphere The interactions among and between all the Earth's organisms and the air (atmosphere), land (lithosphere), and water (hydrosphere) that they occupy. Compare *biosphere.*
ecosystem The interacting system that encompasses a community and its nonliving, physical environment. Compare *community.*
ecosystem management A management focus that emphasizes restoring and maintaining ecosystem quality rather than emphasizing individual species. Ecosystem management recognizes that organisms interact with one another and with their abiotic environment (soil, water, and air).
ecotone The transitional zone where two ecosystems or biomes intergrade.
edge effect The ecological phenomenon in which ecotones between adjacent communities often have more kinds of species or greater population densities of certain species than either adjoining community.
E-horizon A heavily leached soil area that sometimes develops between the A- and B-horizons.
electrostatic precipitator An air pollution control device that gives ash a positive electrical charge so that it adheres to negatively charged plates.
embryo transfer A technique in which a female of a rare species is treated with fertility drugs, which cause her to produce multiple eggs; some of these eggs are collected, fertilized with sperm, and surgically implanted into a female of a related but less rare species, which later gives birth to offspring of the rare species.
emigration A type of migration in which individuals leave a population and thus decrease its size. Compare *immigration.*

emission charge A government policy that controls pollution by charging the polluter for each given unit of emissions; that is, by establishing a tax on pollution.

emission reduction credit (ERC) A waste-discharge permit that can be bought and sold by companies that produce emissions.

endangered species A species whose numbers are so severely reduced that it is in imminent danger of becoming extinct in all or a significant part of its range. Compare *threatened species.*

endocrine disrupter A chemical that interferes with the actions of the endocrine system (the body's hormones). Includes certain plastics such as polycarbonate; chlorine compounds such as PCBs and dioxin; the heavy metals lead and mercury; and some pesticides such as DDT, kepone, chlordane, and endosulfan.

energy The capacity or ability to do work.

energy conservation Saving energy by reducing energy use and waste. Carpooling to work or school is an example of energy conservation. Compare *energy efficiency.*

energy efficiency Using less energy to accomplish the same task. Purchasing an appliance such as a refrigerator that uses less energy to keep food cold is an example of energy efficiency. Compare *energy conservation.*

energy flow The passage of energy in a one-way direction through an ecosystem.

energy intensity A country's or region's total energy consumption divided by its gross natural product.

enrichment The process by which uranium ore is refined after mining to increase the concentration of fissionable U-235.

entombment An option after the closing of an old nuclear power plant in which the entire power plant is permanently encased in concrete. Compare *decommission.*

entropy A measure of the randomness or disorder of a system.

environment All the external conditions, both abiotic and biotic, that affect an organism or group of organisms.

environmental chemistry The subdiscipline of chemistry that deals with redesigning commercially important chemical processes to significantly reduce environmental harm. Also called *green chemistry.*

environmental impact statement (EIS) A statement that accompanies federal recommendations or proposals and is supposed to help federal officials and the public make informed decisions. Required by the National Environmental Policy Act of 1970.

environmentalist A person who works to solve environmental problems such as overpopulation; pollution of the Earth's air, water, and soil; and depletion of natural resources. Environmentalists are collectively known as the environmental movement.

environmental justice The right of every citizen, regardless of age, race, gender, social class, or other factors, to adequate protection from environmental hazards.

environmental movement See *environmentalist.*

environmental resistance Limits set by the environment that prevent organisms from reproducing indefinitely at their biotic potential; includes the limited availability of food, water, shelter, and other essential resources, as well as limits imposed by disease and predation.

environmental science The interdisciplinary study of how humanity affects other organisms and the nonliving physical environment.

environmental sustainability See *sustainability.*

epicenter The site at the Earth's surface that is located directly above an earthquake's focus.

epiphyte A small organism that grows on another organism but is not parasitic on it. Small plants that live attached to the bark of a tree's branches are epiphytes.

essential amino acid Any of the eight amino acids that must be obtained in the diet because humans cannot synthesize them from simpler materials.

estuary A coastal body of water that connects to oceans, in which fresh water from the land mixes with saltwater from the oceans.

ethanol A colorless, flammable liquid, C_2H_5OH. Also called *ethyl alcohol.*

ethyl alcohol See *ethanol.*

eutrophic lake A lake enriched with nutrients such as nitrate and phosphate and consequently choked with plant or algal organisms. Water in a eutrophic lake contains very little dissolved oxygen. Compare *oligotrophic lake.*

eutrophication The enrichment of a lake or pond by nutrients. Eutrophication that occurs naturally is a very slow process in which the lake gradually fills in and converts to a marsh, eventually disappearing. See *artificial eutrophication.*

evaporation The conversion of water from a liquid to a vapor. Also called *vaporization.*

evaporite deposit A massive salt deposit that forms when a body of water dries up.

even-age harvesting See *clearcutting.*

evolution The cumulative genetic changes in populations that occur during successive generations. Evolution explains the origin of all the organisms that exist today or have ever existed.

ex situ conservation Conservation efforts that involve conserving biological diversity in human-controlled settings. Compare in *situ conservation.*

exosphere The outermost layer of the atmosphere, bordered by the thermosphere and interplanetary space.

exponential growth Growth that occurs at a constant rate of increase over a period of time. When the increase in number versus time is plotted on a graph, exponential growth produces a characteristic J-shaped curve.

extinction The elimination of a species from Earth; occurs when the last individual member of a species dies.

facultative parasite An organism that is normally saprotrophic but, given the opportunity, becomes parasitic. Compare *obligate parasite.*

fall turnover A mixing of the lake waters in temperate lakes, caused by falling temperatures in autumn. Compare *spring turnover.*

family planning Providing the services, including information about birth control methods, to help people have the number of children they want.

famine Widespread starvation caused by a drastic shortage of food. Famine is caused by crop failures that are brought on by drought, war, flood, or some other catastrophic event.

fault Fractures in the crust along which rock moves forward and backward, up and down, or from side to side. Fault zones are often found at plate boundaries.

fecal coliform test A water quality test for the presence of *E. coli* (fecal bacteria common in the intestinal tracts of people and animals). The presence of fecal bacteria in a water supply indicates a chance that pathogenic organisms may be present as well.

fee-per-bag approach An effective way to reduce the weight and volume of municipal solid waste by charging households for each container of garbage.

first law of thermodynamics Energy cannot be created or destroyed, although it can be transformed from one form to another. Compare *second law of thermodynamics.*

fission A nuclear reaction in which large atoms of certain elements are each split into two smaller atoms with the release of a large amount of energy. Compare *fusion.*

flood plain The area bordering a river that is subject to flooding.

flowing-water ecosystem A freshwater ecosystem such as a river or stream in which the water flows.

fluidized-bed combustion A clean-coal technology in which crushed coal is mixed with particles of limestone in a strong air current during combustion. The limestone neutralizes the acidic sulfur compounds produced during combustion.

fly ash The ash from the flue (chimney) that is trapped by electrostatic precipitators. Compare *bottom ash.*

flyway An established route that ducks, geese, and shorebirds follow during their annual migrations.

focus The site, often far below the Earth's surface, where an earthquake begins.

food additive A chemical added to food because it enhances the taste, color, or texture of the food, improves its nutrient value, reduces spoilage, prolongs shelf life, or helps maintain its consistency.

food chain The successive series of organisms through which energy flows in an ecosystem. Each organism in the series eats or decomposes the preceding organism in the chain.

food web A complex interconnection of all the food chains in an ecosystem.

forest decline A gradual deterioration (and often death) of many trees in a forest. The cause of forest decline is unclear at the present time, and it may involve a combination of factors.

forest edge The often sharp boundary between the forest and surrounding farmlands or residential neighborhoods. Songbird nests near a forest edge are more vulnerable to predation by small mammalian predators and egg-eating birds as well as nest parasitism by brown-headed cowbirds.

forest management See *ecologically sustainable forest management.*

fossil fuel Combustible deposits in the Earth's crust. Fossil fuels are composed of the remnants of prehistoric organisms that existed millions of years ago. Examples include oil, natural gas, and coal.

freshwater wetlands Lands that are usually covered by shallow fresh water for at least part of the year and that have a characteristic soil and water-tolerant vegetation. Include marshes and swamps.

fundamental niche The potential ecological niche that an organism could have if there were no competition from other species. See *niche.* Compare *realized niche.*

fungicide A toxic chemical that kills fungi.

fusion A nuclear reaction in which two smaller atoms are combined to make one larger atom with the release of a large amount of energy. Compare *fission.*

Gaia hypothesis The collective name for a series of hypotheses that Earth's organisms adjust the environment to improve conditions for life.

game farming See *wildlife ranching.*

gas hydrates Reserves of ice-encrusted natural gas that are located in porous rock deep in the Earth's crust.

genetic diversity Biological diversity that encompasses the genetic variety among individuals within a single species. Compare *ecological diversity* and *species diversity.*

genetic engineering The ability to take a specific gene from one cell and place it into another cell where it is expressed. Also called *biotechnology.*

genetic resistance An inherited characteristic that decreases the effect of a pesticide on a pest. Over time, the repeated exposure of a pest population to a pesticide causes an increase in the number of individuals that can tolerate the pesticide.

geothermal energy Heat produced deep in the Earth from the natural decay of radioactive elements.

germplasm Any plant or animal material that may be used in breeding; includes seeds, plants, and plant tissues of traditional crop varieties and the sperm and eggs of traditional livestock breeds.

global commons Those resources of our environment that are available to everyone but for which no single individual has responsibility.

global distillation effect The process whereby volatile chemicals evaporate from land as far away as the tropics and are carried by air currents to higher latitudes, where they condense and fall to the ground.

grain stockpiles See *world grain carryover stocks.*

green architecture The practice of designing and building homes with environmental considerations, such as energy efficiency, recycling, and conservation of natural resources, in mind.

green chemistry See *environmental chemistry.*

green revolution The period of time during the 20th century when plant scientists developed genetically uniform, high-yielding varieties of important food crops such as rice and wheat.

greenhouse effect The global warming of our atmosphere produced by the buildup of carbon dioxide and other greenhouse gases, which trap the sun's radiation in much the same way that glass does in a greenhouse. Greenhouse gases allow the sun's energy to penetrate to the Earth's surface but do not allow as much of it to escape as heat.

gross primary productivity The rate at which energy accumulates in an ecosystem (as biomass) during photosynthesis. Compare *net primary productivity.*

groundwater The supply of fresh water under the Earth's surface. Groundwater is stored in underground caverns and porous layers of underground rock called aquifers. Compare *surface water.*

growth rate (r) The rate of change of a population's size, expressed in percent per year. In populations with little or no migration, it is calculated by subtracting the death rate from the birth rate. Also called *natural increase* in human populations.

gyre A circular, prevailing wind that generates circular ocean currents.

habitat The local environment in which an organism, population, or species lives.

hair cell One of a group of cells inside the cochlea of the ear that detects differences in pressure caused by sound waves.

half-life See *radioactive half-life.*

hard coal See *anthracite.*

hazardous waste Any discarded chemical (solid, liquid, or gas) that threatens human health or the environment. Also called *toxic waste.*

herbicide A toxic chemical that kills plants.

herbivore An animal that feeds on plants or algae. Compare *carnivore* and *omnivore.* See *primary consumer.*

heterocysts Oxygen-excluding cells of cyanobacteria that fix nitrogen.

heterotroph See *consumer.*

high-grade ore An ore that contains relatively large amounts of a particular mineral. Compare *low-grade ore.*

high-input agriculture Agriculture that relies on large inputs of energy in the form of fossil fuels. Also called *industrialized agriculture.* Compare *alternative agriculture.*

high-level radioactive waste Any radioactive solid, liquid, or gas that initially gives off large amounts of ionizing radiation. Compare *low-level radioactive waste.*

highly developed country See *developed country.*

horizons See *soil horizons.*

host The organism in a parasitic relationship that nourishes a parasite. See *parasitism* and *parasite.*

hot spot A rising plume of magma that flows out through an opening in the crust. Volcanic islands such as the Hawaiian Islands formed as the Pacific plate moved over a hot spot.

humus Black or dark brown decomposed organic material.

hydrocarbons A diverse group of organic compounds that contain only hydrogen and carbon.

hydrogen bond A bond between water molecules, formed when the negative (oxygen) end of one water molecule is attracted to the positive (hydrogen) end of another water molecule. Hydrogen bonding is the basis for a number of water's physical properties.

hydrologic cycle The water cycle, which includes evaporation, precipitation, and flow to the seas. The hydrologic cycle supplies terrestrial organisms with a continual supply of fresh water.

hydrology The science that deals with Earth's waters, including the availability and distribution of fresh water.

hydropower The energy of flowing or falling water used to generate electricity.

hydrosphere The Earth's supply of water (both liquid and frozen, fresh and salty).

hypertension High blood pressure.

hypothesis An educated guess that might be true and is testable by observation and experimentation. Compare *theory.*

hypoxia Low dissolved oxygen concentrations that occur in many bodies of water when nutrients stimulate the growth of algae that subsequently die and are decomposed by oxygen-using bacteria. Hypoxia often causes too little oxygen for other sea life.

illuviation The deposition of leached material in the lower layers of soil from the upper layers; caused by leaching.

immigration A type of migration in which individuals enter a population and thus increase the population size. Compare *emigration.*

indicator species A species that indicates particular conditions. For example, a large population of *Euglena* indicates polluted water.

inductive reasoning Reasoning that uses specific examples to draw a general conclusion or discover a general principle. Compare *deductive reasoning.*

industrial ecosystem A complex web of interactions among various industries, in which "wastes" produced by one industrial process are sold to other companies as raw materials. Finding profitable uses for a company's wastes is an extension of sustainable manufacturing.

industrial smog The traditional, London-type smoke pollution, which consists principally of sulfur oxides and particulate matter. Compare *photochemical smog.*

industrialized agriculture See *high-input agriculture.*

infant mortality rate The number of infant deaths per 1000 live births. (An infant is a child that is in its first year of life.)

infectious disease A disease caused by a microorganism (such as a bacterium or fungus) or infectious agent (such as a virus). Infectious diseases can be transmitted from one individual to another.

infrared radiation Electromagnetic radiation with wavelengths longer than visible light but shorter than microwaves. Humans perceive infrared radiation as invisible waves of heat energy.

inherent safety See *principle of inherent safety.*

inorganic chemical A chemical that does not contain carbon and is not associated with life. Inorganic chemicals that are pollutants include mercury compounds, road salt, and acid drainage from mines.

inorganic fertilizer Plant nutrients (especially nitrates, phosphates, and potassium) that are manufactured commercially.

inorganic plant nutrient A nutrient such as phosphate or nitrate that stimulates plant or algal growth. Excessive amounts of inorganic plant nutrients, which may come from animal wastes and plant residues as well as fertilizer runoff, can cause both soil and water pollution.

insecticide A toxic chemical that kills insects.

in situ conservation Conservation efforts that concentrate on preserving biological diversity in the wild. Compare *ex situ conservation.*

integrated pest management (IPM) A combination of pest control methods (biological, chemical, and cultivation) that, if used in the proper order and at the proper times, keep the size of a pest population low enough that it does not cause substantial economic loss.

integrated waste management A combination of the best waste management techniques into a consolidated program to deal effectively with solid waste.

intertidal zone The shoreline area between the low tide mark and the high tide mark.

inversion See *thermal inversion.*

ionizing radiation Radiation that contains enough energy to eject electrons from atoms, forming positively charged ions. Ionizing radiation can damage living tissue.

isotope An alternate form of the same element that has a different atomic mass. That is, an isotope has a different number of neutrons but the same number of protons and electrons.

K-selected species A species whose reproductive strategy is that of *K* selection. Also called K *strategist.*

K selection A reproductive strategy in which a species typically has a large body size, slow development, long life-span, and does not devote a large proportion of its metabolic energy to the production of offspring.

K strategist See K-*selected species.*

keystone species A species that is crucial in determining the nature and structure of the entire ecosystem in which it lives; other species of a community depend on or are greatly affected by the keystone species, whose influence is much greater than would be expected by its relative abundance.

kinetic energy The energy of a body that results from its motion. Compare *potential energy.*

krill Tiny shrimplike animals that are important in the Antarctic food web.

kwashiorkor Malnutrition, most common in infants and in young children, that results from protein deficiency. Compare *marasmus.*

latitude The distance, measured in degrees north or south, from the equator. Compare *altitude.*

lava Magma (molten rock) that reaches the Earth's surface.

leaching The process by which dissolved materials (nutrients or contaminants) are washed away or carried with water down through the various layers of the soil.

less developed country A developing country with a low level of industrialization, a very high fertility rate, a very high infant mortality rate, and a very low per-capita income (relative to highly developed countries). Compare *moderately developed country* and *highly developed country.*

life index of world reserves An estimate of the time remaining before the known reserves of a particular nonrenewable resource are expended (at current rates of extraction and use).

lignite A grade of coal that is brown or brown-black and has a soft, woody texture (softer than bituminous coal). Compare *anthracite* and *bituminous coal.*

lime scrubber An air pollution control device in which a chemical spray neutralizes acidic gases. See *scrubbers.*

limiting factor An environmental factor that restricts the growth, distribution, or abundance of a particular population; limiting factors restrict an organism's ecological niche.

limnetic zone The open-water area away from the shore of a lake or pond that extends down as far as sunlight penetrates.

lipid A diverse group of organic molecules that are metabolized by cell respiration to provide the body with a high level of energy. Lipids are commonly called fats and oils.

liquefied petroleum gas A mixture of liquefied propane and butane. Liquefied petroleum gas is stored in pressurized tanks.

lithosphere The soil and rock of Earth's crust.

littoral zone The shallow-water area along the shore of a lake or pond.

loam A soil that has approximately equal portions of sand, silt, and clay. A loamy soil that also contains organic material makes an excellent agricultural soil.

low-grade ore An ore that contains relatively small amounts of a particular mineral. Compare *high-grade ore.*

low-input agriculture See *alternative agriculture.*

low-level radioactive waste Any radioactive solid, liquid, or gas that gives off small amounts of ionizing radiation. Compare *high-level radioactive waste.*

magma Molten rock formed within the Earth.

malnutrition A condition caused when a person does not receive enough specific essential nutrients in the diet. See *marasmus* and *kwashiorkor.* Compare *overnutrition* and *undernutrition.*

manganese nodule A small (potato-sized) rock that contains manganese and other minerals. Manganese nodules are common in the ocean floor.

mangrove forest Swamps of mangrove trees that grow along many tropical coasts.

marasmus A condition of progressive emaciation that is especially common in children and is caused by a diet low in both total calories and protein. Compare *kwashiorkor.*

marginal cost of pollution The cost in environmental quality of a unit of pollution emitted into the environment.

marginal cost of pollution abatement The cost to dispose of a unit of pollution in a nonpolluting way.

mariculture The rearing of marine organisms (fish and shellfish) for human consumption. See *aquaculture.*

marine snow Organic debris that drifts into the darkened regions of the oceanic province from the upper, lighted regions.

marsh A wetland that is treeless and dominated by grasses; freshwater marshes are found inland along lakes and rivers, and salt marshes are found in bays and rivers nears the ocean or in protected coastal areas. See *salt marsh.*

mass extinction The extinction of numerous species during a relatively short period of geological time. Compare *background extinction.*

maximum contaminant level The maximum permissible amount (by law) of a water pollutant that might adversely affect human health.

meltdown The melting of a nuclear reactor's vessel. A meltdown would cause the release of a substantial amount of radiation into the environment. See *reactor vessel.*

mesosphere The layer of the atmosphere between the stratosphere and the thermosphere. It is characterized by the lowest atmospheric temperatures.

metal An element that is malleable, lustrous, and a good conductor of heat and electricity. See *mineral.* Compare *nonmetal.*

methane The simplest hydrocarbon, CH_4, which is an odorless, colorless, flammable gas.

methanol A colorless, flammable liquid, CH_3OH. Also called *methyl alcohol.*

methyl alcohol See *methanol.*

microclimate Local variations in climate produced by differences in elevation, in the steepness and direction of slopes, and in exposure to prevailing winds.

microirrigation A type of irrigation that conserves water. In microirrigation pipes with tiny holes bored into them convey water directly to individual plants. Also called *drip* or *trickle irrigation.*

mineral An element, inorganic compound, or mixture that occurs naturally in the Earth's crust. See *metal* and *nonmetal.*

mineral reserve A mineral deposit that has been identified and is currently profitable to extract. Compare *mineral resource.*

mineral resource Any undiscovered mineral deposits or a known deposit of low-grade ore that is currently unprofitable to extract. Mineral resources are potential resources that may be profitable to extract in the future. Compare *mineral reserve.*

moderately developed country A developing country with a medium level of industrialization, a high fertility rate, a high infant mortality rate, and a low per-capita income (relative to highly developed countries). Compare *less developed country* and *highly developed country.*

monoculture The cultivation of only one type of plant over a large area.

Montreal Protocol International negotiations that resulted in a timetable to phase out CFC production.

mortality See *death rate.*

mulch Material placed on the surface of soil around the bases of plants. A mulch helps to maintain soil moisture and reduce soil erosion. Organic mulches have the additional advantage of decomposing over time, thereby enriching the soil.

municipal solid waste Solid waste generated in homes, office buildings, retail stores, restaurants, schools, hospitals, prisons, libraries, and other commercial and institutional facilities. Compare *nonmunicipal solid waste.*

municipal solid waste composting The large-scale composting of the entire organic portion of a community's solid waste.

mutation A change in the DNA (a gene) of an organism. A mutation in reproductive cells may be passed on to the next generation, where it may result in birth defects or genetic disease.

mutualism A symbiotic relationship in which both partners benefit from the association. See *symbiosis.* Compare *parasitism* and *commensalism.*

mycorrhizae A symbiotic association between a fungus and the roots of a plant. Most plants form mycorrhizal associations with fungi. This mutualistic relationship enables plants to absorb adequate amounts of essential minerals from the soil.

narrow-spectrum pesticide An "ideal" pesticide that only kills the organism for which it is intended and does not harm any other species. Compare *wide-spectrum pesticide.*

natality See *birth rate.*

national emission limitations The maximum permissible amount (by law) of a particular pollutant that can be discharged into the nation's rivers, lakes, and oceans from point sources.

National Environmental Policy Act (NEPA) A U.S. federal law, passed in 1969, that is the cornerstone of U.S. environmental policy. It requires that the federal government consider the environmental impact of any construction project funded by the federal government.

natural gas A mixture of gaseous hydrocarbons (primarily methane) that occurs, often with oil deposits, in the Earth.

natural increase See *growth rate.*

natural resources Goods and services—coal, fresh water, clean air, arable land, and organisms, for example—that are supplied by the environment.

natural selection The tendency of organisms that possess favorable adaptations to their environment to survive and become the parents of the next generation; evolution occurs when natural selection results in genetic changes in a population. The mechanism of evolution first proposed by Charles Darwin.

nekton Relatively strong-swimming aquatic organisms such as fish and turtles. Compare *plankton.*

neo-Malthusians Economists who hold that developmental efforts are hampered by a rapidly expanding population.

neotropical birds Birds that spend the winter in Central and South America and the Caribbean (*neotropical* means the tropics in the New World) and then migrate north to the United States and Canada to breed during the summer.

neritic province Open ocean from the shoreline to a depth of 200 meters.

nest parasitism A behavior practiced by brown-headed cowbirds in which females lay their eggs in the nests of other bird species and leave all parenting jobs to the hosts.

net primary productivity Energy that remains in an ecosystem (as biomass) after cell respiration has occurred. Compare *gross primary productivity.*

niche The totality of an organism's adaptations, its use of resources, and the life style to which it is fitted. The niche describes how an organism utilizes materials in its environment as well as how it interacts with other organisms. Also called *ecological niche.*

nitrification The conversion of ammonia (NH_3) to nitrate (NO_3^-) by certain bacteria (nitrifying bacteria) in the soil; part of the nitrogen cycle.

nitrogen cycle The worldwide circulation of nitrogen from the abiotic environment into organisms and back into the abiotic environment.

nitrogen fixation The conversion of atmospheric nitrogen (N_2) to ammonia (NH_3) by nitrogen-fixing bacteria and cyanobacteria; part of the nitrogen cycle.

noise A loud or disagreeable sound.

nomadic herding A type of subsistence agriculture in which livestock is supported on land too arid for successful crop growth. Nomadic herding is land intensive because the herders must continually move the animals in order to provide them with enough forage.

nondegradable A chemical pollutant (such as the toxic elements mercury and lead) that cannot be decomposed (broken down) by organisms or by other natural processes. Compare *biodegradable.*

nonmetal A mineral that is nonmalleable, nonlustrous, and a poor conductor of heat and electricity. See *mineral.* Compare *metal.*

nonmunicipal solid waste Waste generated by industry, agriculture, and mining. Compare *municipal solid waste.*

nonpoint source pollution Pollutants that enter bodies of water over large areas rather than being concentrated at a single point of entry. Examples include agricultural fertilizer runoff and sediments from construction. Also called *polluted runoff*. Compare *point source pollution.*

nonrenewable resources Natural resources that are present in limited supplies and are depleted by use; include minerals such as copper and tin and fossil fuels such as oil and natural gas. Compare *renewable resources.*

nonurban land See *rural land.*

no-tillage A method of conservation tillage that leaves both the surface and subsurface soil undisturbed. Special machines punch holes in the soil for seeds. See *conservation tillage.* Compare *conventional tillage.*

nuclear energy The energy released from the nucleus of an atom in a nuclear reaction (fission or fusion) or during radioactive decay.

nuclear reactor A device that initiates and maintains a controlled nuclear fission chain reaction in order to produce energy (electricity).

obligate parasite An organism that can only exist as a parasite. Compare *facultative parasite.*

oceanic province That part of the open ocean that is deeper than 200 meters and comprises most of the ocean.

ocean temperature gradient The differences in temperature at various ocean depths.

Ogallala Aquifer A massive groundwater deposit under eight midwestern states, including Nebraska, Kansas, Oklahoma, and Texas.

O-horizon The uppermost layer of certain soils, composed of dead leaves and other organic matter.

oil See *petroleum.*

oil shales Sedimentary "oily rocks" that must be crushed, heated to high temperatures, and refined after they are mined in order to yield oil.

oligotrophic lake A deep, clear lake that has minimal nutrients. Water in an oligotrophic lake contains a high level of dissolved oxygen. Compare *eutrophic lake.*

omnivore An animal that eats a variety of plant and animal material. Compare *herbivore* and *carnivore.*

oncogene Any of a number of genes that usually are involved in cell growth or division and that cause the formation of a cancer cell when mutated.

open management A policy in which all fishing boats of a particular country are given unrestricted access to fish in their national waters.

open-pit mining See *surface mining.*

optimum amount of pollution The amount of pollution that is economically most desirable. It is determined by plotting two curves, the marginal cost of pollution and the marginal cost of pollution abatement. The point where the two curves meet is the optimum amount of pollution from an economic standpoint.

ore Rock that contains a large enough concentration of a particular mineral that it can be profitably mined and extracted.

organic agriculture Growing crops and livestock without the use of synthetic pesticides or inorganic fertilizers. Organic agriculture makes use of natural organic fertilizers (such as manure and compost) and chemical-free methods of pest control.

organic compound A compound that contains the element carbon and is either naturally occurring (in organisms) or synthetic (manufactured by humans). Many synthetic organic compounds persist in the environment for an extended period of time, and some are toxic to organisms.

organophosphate A synthetic organic compound that contains phosphorus and is very toxic; used as an insecticide.

overburden Overlying layers of soil and rock over mineral deposits. The overburden is removed during surface mining.

overgrazing The destruction of an area's vegetation that occurs when too many animals graze on the vegetation, consuming so much of it that it does not recover.

overnutrition A condition caused by eating food in excess of that required to maintain a healthy body. Compare *malnutrition* and *undernutrition.*

overpopulation A situation in which a country or geographical area has more people than its resource base can support without damaging the environment. See *people overpopulation* and *consumption overpopulation.*

oxide A compound in which oxygen is chemically combined with some other element.

ozone A blue gas, O_3, that has a distinctive odor. Ozone is a human-made pollutant in one part of the atmosphere (the troposphere) but a natural and essential component in another (the stratosphere).

parasite Any organism that obtains nourishment from the living tissue of another organism (the host). See *parasitism* and *host.*

parasitism A symbiotic relationship in which one member (the parasite) benefits and the other (the host) is adversely affected. See *symbiosis.* Compare *mutualism* and *commensalism.*

particulate matter Solid particles and liquid droplets suspended in the atmosphere.

parts per billion The number of parts of a particular substance found in one billion parts of air, water, or some other material. Abbreviated ppb.

parts per million The number of parts of a particular substance found in one million parts of air, water, or some other material. Abbreviated ppm.

passive solar heating A heating method that uses the sun's energy to heat buildings or water without requiring mechanical devices (pumps or fans) to distribute the collected heat. Compare *active solar heating.*

pathogen An agent (usually a microorganism) that causes disease.

people overpopulation The situation in which there are too many people in a given geographical area. Even if those people use few resources per person (the minimum amount they need to survive), people overpopulation results in pollution, environmental degradation, and resource depletion. See *overpopulation.* Compare *consumption overpopulation.*

perennial A plant that lives for more than two years. Compare *annual* and *biennial.*

permafrost Permanently frozen subsoil characteristic of frigid areas such as the tundra.

persistence A characteristic of certain chemicals that are extremely stable and may take many years to be broken down into simpler forms by natural processes. Certain pesticides, for example, exhibit persistence and remain unaltered in the environment for years.

pest Any organism that interferes in some way with human welfare or activities.

pesticide Any toxic chemical used to kill pests. See *fungicide, herbicide, insecticide,* and *rodenticide.*

pesticide treadmill A predicament faced by pesticide users, in which the cost of applying pesticides increases because they have to be applied more frequently or in larger doses, while their effectiveness decreases as a result of genetic resistance in the target pest.

petrochemicals Chemicals obtained from crude oil that are used in the production of such diverse products as fertilizers, plastics, paints, pesticides, medicines, and synthetic fibers.

petroleum A thick, yellow to black, flammable liquid hydrocarbon mixture found in the Earth's crust. When petroleum is refined, the mixture is separated into a number of hydrocarbon compounds, including gasoline, kerosene, fuel oil, lubricating oils, paraffin, and asphalt. Also called *crude oil.*

pH A number from 0 to 14 that indicates the degree of acidity or alkalinity of a substance.

pheromone A substance secreted by one organism into the environment that influences the development or behavior of other members of the same species.

phosphorus cycle The process by which phosphorus cycles from the land to sediments in the ocean and back to the land.

photochemical smog A brownish orange haze formed by complex chemical reactions involving sunlight, nitrogen oxides, and hydrocarbons. Some of the pollutants in photochemical smog include peroxyacyl nitrates (PANs), ozone, and aldehydes. Compare *industrial smog.*

photodegradable Breaking down upon exposure to sunlight.

photosynthesis The biological process that captures light energy and transforms it into the chemical energy of organic molecules (such as glucose), which are manufactured from carbon dioxide and water. Photosynthesis is performed by plants, algae, and several kinds of bacteria.

photovoltaic (PV) solar cell A wafer or thin-film device that generates electricity when solar energy is absorbed.

phytoplankton Microscopic floating algae that are the base of most aquatic food chains. See *plankton.* Compare *zooplankton.*

phytoremediation A method employed to clean up a hazardous waste site that uses plants to absorb and accumulate toxic materials.

pioneer community The first organisms (such as lichens or mosses) to colonize an area and begin the first stage of ecological succession. See *succession.*

plankton Small or microscopic aquatic organisms that are relatively feeble swimmers and thus, for the most part, are carried about by currents and waves. Composed of phytoplankton and zooplankton. Compare *nekton.*

plate boundary Any area where two tectonic plates meet. Plate boundaries are often sites of intense geological activity.

plate tectonics The theory that explains how the Earth's crustal plates move and interact at their boundaries.

point source pollution Water pollution that can be traced to a specific spot (such as a factory or sewage treatment plant) because it is discharged into the environment through pipes, sewers, or ditches. Compare *nonpoint source pollution.*

polar easterly A prevailing wind that blows from the northeast near the North Pole or from the southeast near the South Pole.

polluted runoff See *nonpoint source pollution.*

pollution An unwanted change in the atmosphere, water, or soil that can harm humans or other organisms.

polychlorinated biphenyls (PCBs) A group of toxic, oily, synthetic industrial chemicals. PCBs are chlorinated hydrocarbons (composed of carbon, hydrogen, and chlorine) that persist in the environment and exhibit biological magnification in food chains.

polyculture A traditional form of subsistence agriculture in which several different crops are grown at the same time. In the tropics, fast- and slow-maturing crops are often planted together so that different crops can be harvested throughout the year.

polymer A compound composed of repeating subunits.

population A group of organisms of the same species that live in the same geographical area at the same time.

population density The number of individuals of a species per unit of area or volume at a given time.

population ecology That branch of biology that deals with the numbers of a particular species that are found in an area and how and why those numbers change (or remain fixed) over time.

potential energy Stored energy that is the result of the relative position of matter instead of its motion. Compare *kinetic energy.*

preservative A chemical added to food to retard the growth of bacteria and fungi that cause food spoilage.

prevailing wind A major surface wind that blows more or less continually.

primary air pollutant A harmful chemical that enters directly into the atmosphere either from human activities or natural processes (such as volcanic eruptions). Compare *secondary air pollutant.*

primary consumer A consumer that eats producers. Also called *herbivore.* Compare *secondary consumer.*

primary succession An ecological succession that occurs on land that has not previously been inhabited by plants; no soil is present initially. Compare *secondary succession.*

primary treatment Treating wastewater by removing suspended and floating particles (such as sand and silt) by mechanical processes (such as screens and physical settling). Compare *secondary treatment* and *tertiary treatment.*

principle A scientific theory that has withstood repeated testing and has our highest level of confidence.

principle of inherent safety Chemical safety programs that stress accident prevention by redesigning industrial processes to involve less toxic materials (so that dangerous accidents are prevented).

producer An organism (such as a chlorophyll-containing plant) that manufactures complex organic molecules from simple inorganic substances. In most ecosystems, producers are photosynthetic organisms. Also called *autotroph.* Compare *consumer.*

profundal zone The deepest zone of a large lake.

pronatalists Those who are in favor of population growth.

protein A large, complex organic molecule composed of amino acid subunits; proteins are the principal structural components of cells.

pyramid of biomass An ecological pyramid that illustrates the total biomass (for example, the total dry weight of all organisms in a community) at each successive trophic level. See *ecological pyramid.* Compare *pyramid of energy* and *pyramid of numbers.*

pyramid of energy An ecological pyramid that shows the energy flow through each trophic level of an ecosystem. See *ecological pyramid.* Compare *pyramid of biomass* and *pyramid of numbers.*

pyramid of numbers An ecological pyramid that shows the number of organisms at each successive trophic level in a given ecosystem. See *ecological pyramid.* Compare *pyramid of biomass* and *pyramid of energy.*

quarantine Practice in which the importation of exotic plant and animal material that might be harboring pests is restricted.

r-selected species A species whose reproductive strategy is that of r selection. Also called r *strategist.*

r selection A reproductive strategy in which a species typically has a small body size, rapid development, short life-span, and devotes a large proportion of its metabolic energy to the production of offspring.

r strategist See r-*selected species.*

radiation The emission of fast-moving particles or rays of energy from the nuclei of radioactive atoms.

radioactive Atoms of unstable isotopes that spontaneously emit radiation.

radioactive decay The process in which a radioactive element emits radiation and, as a result, its nucleus changes into the nucleus of a different element.

radioactive half-life The period of time required for one-half of a radioactive substance to change into a different material.

radioisotope An unstable isotope that spontaneously emits radiation.

radon A colorless, tasteless, radioactive gas produced during the radioactive decay of uranium in the Earth's crust.

rain shadow An area on the downwind side of a mountain range with very little precipitation. Deserts often occur in rain shadows.

range The area of the Earth in which a particular species occurs.

reactor vessel A huge steel potlike structure encasing the uranium fuel in a nuclear reactor. The reactor vessel is a safety feature designed to prevent the accidental release of radiation into the environment.

realized niche The life style that an organism actually pursues, including the resources that it actually utilizes. An organism's realized niche is narrower than its fundamental niche because of competition from other species. See *niche.* Compare *fundamental niche.*

reclaimed water Treated wastewater that is reused in some way, such as for irrigation, manufacturing processes that require water for cooling, wetlands restoration, or groundwater recharge.

recycling Conservation of the resources in used items by converting them into new products. For example, used aluminum cans are recycled by collecting, remelting, and reprocessing them into new cans. Compare *reuse.*

renewable resources Resources that are replaced by natural processes and can be used forever, provided they are not overexploited in the short term. Examples include fresh water in lakes and rivers, fertile soil, and trees in forests. Compare *nonrenewable resources.*

replacement-level fertility The number of children a couple must produce in order to "replace" themselves. The average number is greater than two because some children die before reaching reproductive age.

reservoir An artificial lake produced by building a dam across a river or stream; allows water to be stored for use.

resistance management A relatively new approach to dealing with genetic resistance so that the period of time in which a pesticide is useful is maximized; involves efforts to delay the development of genetic resistance in pests.

resource recovery The process of removing any material—sulfur or metals, for example—from polluted emissions or solid waste and selling it as a marketable product.

reuse Conservation of the resources in used items by using them over and over again. For example, glass bottles can be collected, washed, and refilled again. Compare *recycling.*

reverse osmosis A desalinization process that involves forcing saltwater through a membrane permeable to water but not to salt. Compare *distillation.*

riparian buffer The area adjacent to the bank of a river or stream that is left in its vegetated state to protect the aquatic ecosystem from sedimentation caused by soil erosion; riparian buffers protect the habitat of salmon, trout, and other aquatic species.

risk assessment The process of estimating the harmful effects on human health or the environment of exposure to a particular danger. Risk estimates are most useful when they are compared with one another. See *ecological risk assessment.*

risk management Determining whether there is a need to reduce or eliminate a particular risk and, if so, what should be done. Based on data from risk assessment as well as political, economic, and social considerations.

rodenticide A toxic chemical that kills rodents.

runoff The movement of fresh water from precipitation and snowmelt to rivers, lakes, wetlands, and ultimately, the ocean.

rural land Sparsely populated areas, such as wilderness, forests, grasslands, and wetlands. Also called *nonurban land.*

salinity The concentration of dissolved salts (such as sodium chloride) in a body of water.

salinity gradient The difference in salt concentrations that occurs at different depths in the ocean and at different locations in estuaries.

salinization The gradual accumulation of salt in a soil, often as a result of improper irrigation methods. Most plants cannot grow in salinized soil.

salt marsh A wetland dominated by grasses in which the salinity fluctuates between that of seawater and fresh water. Salt marshes are usually located in estuaries.

saltwater intrusion The movement of seawater into a freshwater aquifer located near the coast; caused by groundwater depletion.

salvage logging A controversial logging method in which trees that are weakened by insects, disease, or fire are harvested. Forestry scientists disprove of salvage logging, particularly in fire-damaged areas, and say it delays or prevents the forest from recovering.

sand Inorganic soil particles that are larger than clay or silt. Compare *clay* and *silt.*

sanitary landfill The disposal of solid waste by burying it under a shallow layer of soil.

saprotroph See *decomposer.*

savanna A tropical grassland with widely scattered trees or clumps of trees; found in areas of low rainfall or seasonal rainfall with prolonged dry periods.

scientific method The way a scientist approaches a problem (by formulating a hypothesis and then testing it by means of an experiment).

sclerophyllous leaf A hard, small, leathery leaf that resists water loss; characteristic of perennial plants adapted to extremely dry environments.

scout-and-spray The use of pesticides that are applied only when pests become a problem; requires continual monitoring. Compare *calendar spraying.*

scrubbers Desulfurization systems that are used in smokestacks to decrease the amount of sulfur released in the air by 90 percent or more. One type is a *lime scrubber.*

second law of thermodynamics When energy is converted from one form to another, some of it is degraded into a lower quality, less useful form. Thus, with each successive energy transformation, less energy is available to do work. Compare *first law of thermodynamics.*

secondary air pollutant A harmful chemical that forms in the atmosphere when a primary air pollutant reacts chemically with other air pollutants or natural components of the atmosphere. Compare *primary air pollutant.*

secondary consumer An organism that consumes primary consumers. Also called *carnivore.* Compare *primary consumer.*

secondary succession An ecological succession that takes place after some disturbance destroys the existing vegetation; soil is already present. Compare *primary succession.*

secondary treatment Treating wastewater biologically, by using microorganisms to decompose the suspended organic material; occurs after primary treatment. Compare *primary treatment* and *tertiary treatment.*

sediment pollution Soil particles that enter the water as a result of erosion.

sedimentation (1) Letting solids settle out of wastewater by gravity during primary treatment. (2) The process in which eroded particles are transported by water and deposited as sediment on river deltas and the sea floor. If exposed to sufficient heat and pressure, sediments can solidify into sedimentary rock.

seed-tree cutting A forest management technique in which almost all trees are harvested from an area in a single cutting, but a few desirable trees are left behind to provide seeds for the regeneration of the forest. Compare *clearcutting, selective cutting,* and *shelterwood cutting.*

seismic waves Vibrations that travel through rock as a result of an earthquake.

selective cutting A forest management technique in which mature trees are cut individually or in small clusters while the rest of the forest remains intact so that the forest can regenerate quickly (and naturally). Compare *clearcutting, seed-tree cutting,* and *shelterwood cutting.*

semi-arid land Land that receives more precipitation than a desert but is subject to frequent and prolonged droughts.

sewage Wastewater carried off by drains or sewers.

sewage sludge A slimy mixture of bacteria-laden solids that settles out from sewage wastewater during primary and secondary treatments.

shelter belt A row of trees planted as a windbreak to reduce soil erosion of agricultural land.

shelterwood cutting A forest management technique in which all mature trees in an area are harvested in a series of partial cuttings over a period of time. (Typically two or three harvests are made over a decade.) Compare *clearcutting, seed-tree cutting,* and *selective cutting.*

shifting agriculture Agriculture that involves clearing a small patch of tropical land to plant crops. Typically, the soil is depleted of nutrients within a few years and the plot must be abandoned. See *slash-and-burn agriculture.*

sick building syndrome Eye irritations, nausea, headaches, respiratory infections, depression, and fatigue caused by the presence of air pollution inside office buildings.

silt Medium-sized inorganic soil particles. Compare *clay* and *sand.*

slag See *bottom ash.*

slash-and-burn agriculture A type of shifting agriculture in which the forest is cut down, allowed to dry, and burned; the crops that are planted immediately afterwards thrive because the ashes provide nutrients. In a few years, however, the soil is depleted and the land must be abandoned. See *shifting agriculture.*

smelting Process in which ore is melted at high temperatures to help separate impurities from the molten metal.

smog Air pollution caused by a variety of pollutants. See *industrial smog* and *photochemical smog.*

soft coal See *bituminous coal.*

soil The uppermost layer of the Earth's crust, which supports terrestrial plants, animals, and microorganisms. Soil is a complex mixture of inorganic minerals (from the parent material), organic material, water, air, and organisms.

soil erosion The wearing away or removal of soil from the land; caused by wind and flowing water. Although soil erosion occurs naturally from precipitation and runoff, human activities (such as clearing land) accelerate it.

soil horizons The horizontal layers into which many soils are organized. May include the O-horizon (surface litter), A-horizon (topsoil), E-horizon, B-horizon (subsoil), and C-horizon (weathered parent material).

soil profile A section through the soil from the surface to the parent material that reveals the horizons.

soil salinization A process in which salt accumulates in the soil.

solar energy Energy from the sun. Solar energy includes both direct solar radiation and indirect solar energy (such as wind, hydropower, and biomass).

solar pond A technique to harvest the sun's energy by using a pond of water to collect solar energy.

solar thermal electric generation A means of producing electricity in which the sun's energy is directed by mirrors onto a fluid-filled pipe; the heated fluid is used to generate electricity.

sound Vibrations in the air (or some other medium) that reach the ears and stimulate a sensation of hearing.

source reduction An aspect of waste management in which products are designed and manufactured in ways that decrease not only the volume of solid waste but the amount of hazardous materials in the solid waste that remains.

species A group of similar organisms that are able to interbreed with one another but unable to interbreed with other sorts of organisms.

species diversity Biological diversity that encompasses the number of different species in an area (or community). Compare *genetic diversity* and *ecological diversity.*

spoil bank A hill of loose rock created when the overburden from a new trench is put into the old (already excavated) trench during strip mining.

sport hunting Killing animals for recreation. Compare *commercial hunting* and *subsistence hunting.*

spring turnover A mixing of the lake waters in temperate lakes that occurs in spring as ice melts and the surface water reaches 4°C, its temperature of greatest density. Compare *fall turnover.*

stable runoff The share of runoff from precipitation available throughout the year. Most geographical areas have a heavy runoff during a few months (the spring months, for example) when precipitation and snowmelt are highest. Stable runoff is the amount that can be depended on every month.

standing-water ecosystem A body of fresh water that is surrounded by land and whose water does not flow; a lake or pond.

sterile male technique A method of insect control that involves rearing, sterilizing, and releasing large numbers of males of the pest species.

strata Layers of rock.

stratosphere The layer of the atmosphere between the troposphere and the mesosphere. It contains a thin ozone layer that protects life by filtering out much of the sun's ultraviolet radiation.

strip cropping A type of contour plowing that produces alternating strips of different crops that are planted along the natural contours of the land.

strip cutting An agricultural harvesting technique that involves harvesting only one segment of the crop at a time.

strip mining See *surface mining.*

structural traps Underground geological structures that tend to trap oil or natural gas if it is present.

subduction The process in which one tectonic plate descends under an adjacent plate.

subsidence The sinking or settling of land caused by aquifer depletion (as groundwater supplies are removed).

subsistence agriculture The production of enough food to feed oneself and one's family with little left over to sell or reserve for bad times.

subsistence hunting Killing wild animals for food and furs and other products needed for survival. Compare *commercial hunting* and *sport hunting.*

subsurface mining The extraction of mineral and energy resources from deep underground deposits. Compare *surface mining.*

succession The sequence of changes in a plant community over time. Also called *ecological succession.*

sulfide A compound in which an element is combined chemically with sulfur.

surface mining The extraction of mineral and energy resources near the Earth's surface by first removing the soil, subsoil, and overlying rock strata. Also called *strip mining* and *open-pit mining.* Compare *subsurface mining.*

surface water Fresh water found on the Earth's surface in streams and rivers, lakes, ponds, reservoirs, and wetlands. Compare *groundwater.*

survivorship The proportion of individuals in a population that survive to a particular age; usually presented as a survivorship curve.

sustainability The ability of the environment to function indefinitely without going into a decline from the overuse of natural systems (such as soil, water, air, biological diversity) that maintain life. Also called *environmental sustainability.*

sustainable agriculture See *alternative agriculture.*

sustainable development See *sustainable economic development.*

sustainable economic development Economic growth in which natural resources are not overused and excessive pollution is not generated so that our future economic growth is not threatened. See *sustainability.*

sustainable manufacturing A manufacturing system based on minimizing waste by industry. Sustainable manufacturing involves such practices as recycling, reuse, and source reduction.

swamp A wetland that is dominated by trees; freshwater swamps are found inland, and saltwater swamps occur along protected coastal areas. See *mangrove forest.*

symbionts The partners of a symbiotic relationship.

symbiosis An intimate relationship between two or more organisms of different species. See *commensalism, mutualism,* and *parasitism.*

synergism A phenomenon in which two or more pollutants interact in such a way that their combined effects are more severe than the sum of their individual effects. Compare *antagonism.*

synfuel A liquid or gaseous fuel synthesized from coal and other sources and used in place of oil or natural gas. Also called *synthetic fuel.*

synthetic botanical Any of a group of human-made insecticides that are produced by chemically modifying the structure of natural botanicals. See *botanical.*

synthetic fuel See *synfuel.*

taiga A region of coniferous forests (such as pine, spruce, and fir) in the Northern Hemisphere. The taiga is located just south of the tundra and stretches across both North America and Eurasia. Also called *boreal forest.*

tailings Piles of loose rock produced when a mineral such as uranium is mined and processed (extracted and purified from the ore).

tar sand An underground sand deposit so heavily permeated with tar or heavy oil that it doesn't move. The oil can be separated from the sand by heating.

temperate deciduous forest A forest biome that occurs in temperate areas where annual precipitation ranges from about 75 cm to 125 cm.

temperate grassland A grassland characterized by hot summers, cold winters, and less rainfall than is found in a temperate deciduous forest biome.

temperate rain forest A coniferous biome characterized by cool weather, dense fog, and high precipitation. Found on the north Pacific coast of North America.

terracing A soil conservation method that involves building dikes on hilly terrain in order to produce level, terraced areas for agriculture.

terrestrial Pertaining to the land. Compare *aquatic.*

tertiary treatment Advanced wastewater treatment methods that occur after primary and secondary treatments and include a variety of biological, chemical, and physical processes. Compare *primary treatment* and *secondary treatment.*

theory A widely accepted, fundamental explanation that is supported by a large body of observations and experiments. Compare *hypothesis.*

thermal inversion A layer of cold air temporarily trapped near the ground by a warmer, upper layer. If a thermal inversion persists, air pollutants may build up to harmful or even dangerous levels. Also called *inversion.*

thermal pollution Water pollution that occurs when heated water produced during many industrial processes is released into waterways.

thermal stratification The marked layering (separation into warm and cold layers) of temperate lakes during the summer. See *thermocline.*

thermocline A marked and abrupt temperate transition in temperate lakes between warm surface water and cold deeper water. See *thermal stratification.*

thermodynamics The branch of physics that deals with energy and its various forms and transformations.

thermosphere The layer of the atmosphere between the mesosphere and the exosphere. Temperatures are very high due to the absorption of x-rays and short-wave ultraviolet radiation.

threatened species A species in which the population is low enough for it to be at risk of becoming endangered in the foreseeable future throughout all or a significant portion of its range. Compare *endangered species.*

topography A region's surface features (such as the presence or absence of mountains and valleys).

total fertility rate The average number of children born to a woman during her lifetime.

total resources The combination of a mineral's reserves and resources. See *mineral reserve* and *mineral resource.* Also called *world reserve base.*

toxic waste See *hazardous waste.*

toxicology The science of poisons, including their effects and any antidotes.

trade wind A prevailing tropical wind that blows from the northeast (in the Northern Hemisphere) or from the southeast (in the Southern Hemisphere).

transpiration The evaporation of water vapor from plants.

trickle irrigation See *microirrigation.*

tritium An isotope of hydrogen that contains one proton and two neutrons per atom. Compare *deuterium.*

trophic level Each level in a food chain. All producers belong to the first trophic level, all herbivores belong to the second trophic level, and so on.

tropical cyclone Giant, rotating tropical storms with winds of at least 119 kilometers per hour (74 mph); the most powerful tropical cyclones have wind velocities greater than 250 kilometers per hour (155 mph). Called hurricanes in the Atlantic, typhoons in the Pacific, and cyclones in the Indian Ocean.

tropical dry forest A tropical forest where enough precipitation falls to support trees, but not enough to support the lush vegetation of a tropical rain forest. Many tropical dry forests occur in areas with pronounced rainy and dry seasons.

tropical rain forest A lush, species-rich forest biome that occurs in tropical areas where the climate is very moist throughout the year. Tropical rain forests tend to be characterized by old, infertile soils.

troposphere The atmosphere from the Earth's surface to the stratosphere. It is characterized by the presence of clouds, turbulent winds, and decreasing temperature with increasing altitude.

tundra The treeless biome in the far north that consists of boggy plains covered by lichens and small plants such as mosses. The tundra is characterized by harsh, very cold winters and extremely short summers. Also called arctic tundra. Compare *alpine tundra.*

ultraviolet radiation That part of the electromagnetic spectrum with wavelengths just shorter than visible light; a high-energy form of radiation that can be lethal to organisms in excessive amounts. Also called *UV radiation.*

unconfined aquifer A groundwater storage area located above a layer of impermeable rock. Water in an unconfined aquifer is replaced by surface water that drains down from directly above it. Compare *confined aquifer.*

undernutrition A condition caused when a person receives fewer calories in the diet than are needed; an undernourished person eats insufficient food to maintain a healthy body. Compare *overnutrition* and *malnutrition.*

upwelling A rising ocean current that transports colder, nutrient-laden water to the surface.

urban growth The rate at which a city's population grows.

urban heat island Local heat buildup in an area of high population density. See *dust dome.*

urbanization The increasing convergence of people from rural areas into cities.

UV radiation See *ultraviolet radiation.*

vapor extraction A technique of soil remediation that involves injecting or pumping air into soil to remove organic compounds that are volatile (evaporate quickly).

vaporization See *evaporation.*

variable A factor that influences a process. In scientific experiments, all variables are kept constant except for one. See *control.*

vector An organism that transmits a pathogen from one organism to another.

vitamin A complex organic molecule required in very small quantities for the normal metabolic functioning of living cells.

vitrification A method of safely storing high-level radioactive liquid wastes in solid form—as enormous glass logs.

waste-discharge permit A government policy that controls pollution by issuing permits allowing the holder to pollute a given amount. Holders are not allowed to produce more emission than the permit allows.

water table The uppermost level of an unconfined aquifer, below which the ground is saturated with water.

watershed See *drainage basin.*

weather The general condition of the atmosphere (temperature, moisture, cloudiness) at a particular time and place. Compare *climate.*

weathering process A chemical or physical process that helps form soil from rock; during weathering, the rock is gradually broken down into smaller and smaller particles.

westerly A prevailing wind that blows in the mid-latitudes from the southwest (in the Northern Hemisphere) or from the northwest (in the Southern Hemisphere).

Western diseases A group of noninfectious diseases that are generally more commonplace in industrialized countries, including obesity and heart disease.

wet deposition A form of acid deposition in which acid falls to the Earth as precipitation. Compare *dry deposition.*

wetlands Lands that are transitional between aquatic and terrestrial ecosystems and are covered with water for at least part of the year.

wide-spectrum pesticide A pesticide that kills a variety of organisms in addition to the pest against which it is used. Most pesticides are wide-spectrum pesticides. Compare *narrow-spectrum pesticide.*

wilderness Any area that has not been greatly disturbed by human activities and that humans may visit but do not inhabit.

wildlife corridor Protected zones that connect unlogged or undeveloped areas; wildlife corridors are thought to provide escape routes and allow animals to migrate so they can interbreed.

wildlife management Efforts to handle wildlife populations and their habitats in order to assure their sustained welfare.

wildlife ranching An alternative use of African land in which private landowners maintain herds of wild animals and earn money from tourists, photographers, and sport hunters as well as from animal hides, leather, and meat. Also called *game farming.*

wind Surface air currents that are caused by the solar warming of air.

wind farm An array of wind turbines for utilizing wind energy by capturing it and converting it to electricity.

wise-use movement A coalition of several hundred grassroots organizations who think that the government has too many environmental regulations and that federal lands should be used primarily to enhance economic growth.

world grain carryover stocks The amounts of rice, wheat, corn, and other grains remaining from previous harvests, as estimated at the start of a new harvest; these provide a measure of world food security.

world reserve base See *total resources.*

zero population growth When the birth rate equals the death rate. A population with zero population growth remains the same size.

zooplankton The nonphotosynthetic organisms—tiny shrimp, larvae, and other drifting animals—that are part of the plankton. See *plankton.* Compare *phytoplankton.*

zooxanthellae Algae that live inside coral animals and have a mutualistic relationship with them.

INDEX

Note: Page numbers in **boldface** indicate where a key term is introduced. Page numbers in *italics* indicate a figure or illustration; *t* denotes a table; *n* refers to a footnote.